Applying Maths in the Chemical and Biomolecular Sciences

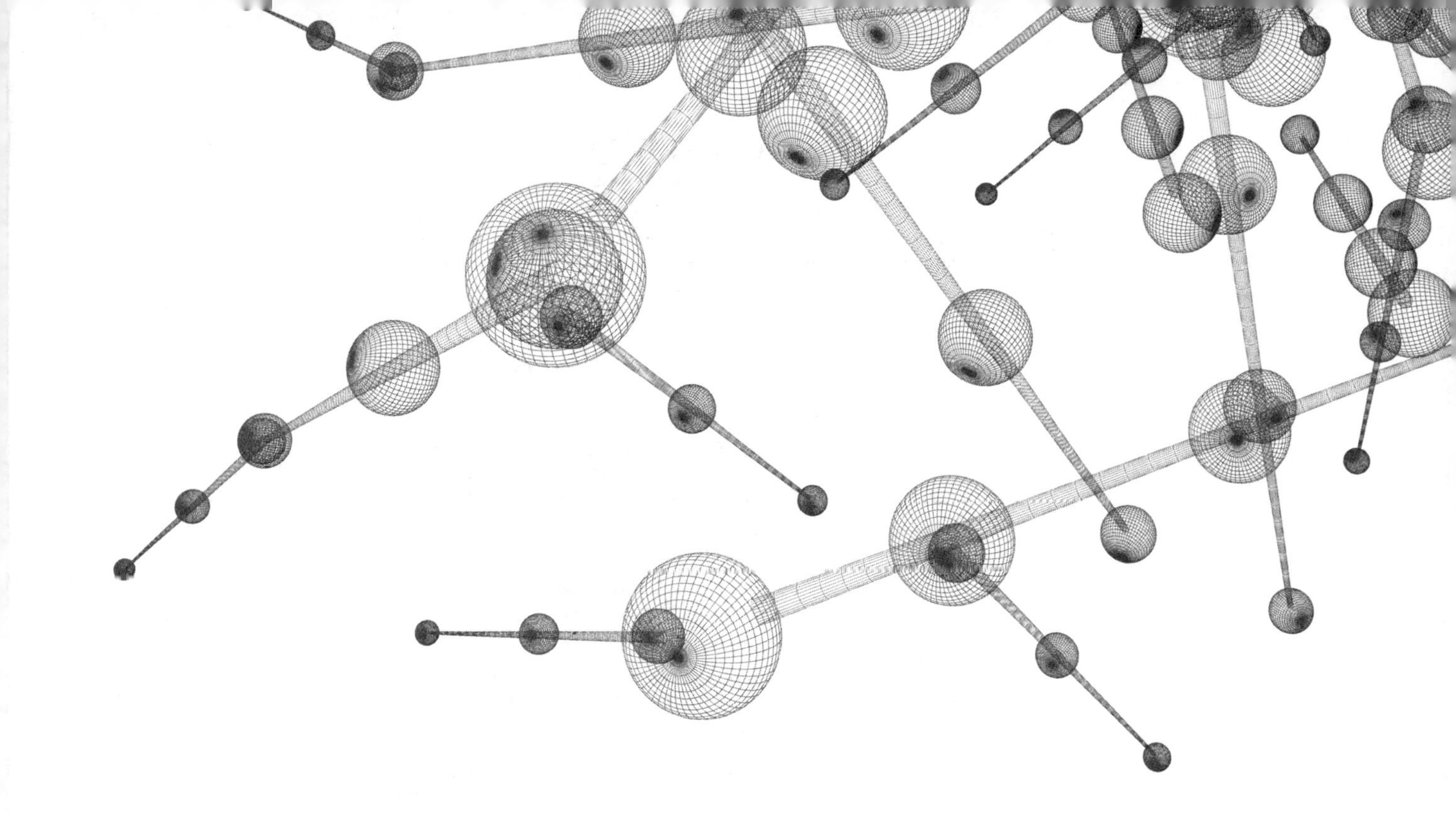

Applying Maths in the Chemical and Biomolecular Sciences

An example-based approach

Godfrey Beddard

OXFORD
UNIVERSITY PRESS

Great Clarendon Street, Oxford OX2 6DP

Oxford University Press is a department of the University of Oxford.
It furthers the University's objective of excellence in research, scholarship,
and education by publishing worldwide in

Oxford New York

Auckland Cape Town Dar es Salaam Hong Kong Karachi
Kuala Lumpur Madrid Melbourne Mexico City Nairobi
New Delhi Shanghai Taipei Toronto

With offices in

Argentina Austria Brazil Chile Czech Republic France Greece
Guatemala Hungary Italy Japan Poland Portugal Singapore
South Korea Switzerland Thailand Turkey Ukraine Vietnam

Published in the United States
by Oxford University Press Inc., New York

First published by Oxford University Press 2009

British Library Cataloguing in Publication Data
Data available

Library of Congress Cataloging in Publication Data

Typeset by Graphicraft Limited, Hong Kong
Printed in China
on acid-free paper by
C&C Offset Printing Co. Ltd.

ISBN 978–0–19–923091–4

1 3 5 7 9 10 8 6 4 2

Preface

This textbook is primarily intended for final year undergraduate and postgraduate students, although the more elementary parts of the subject are included so as to make the book complete in itself. It is not written with any particular science degree in mind although the emphasis is towards the chemistry and physics of molecules with examples ranging from hydrogen to proteins and DNA.

The mathematics described is extensively supported by using the computer algebra program Maple either as single line calculations or as algorithms for more complex problems such as Monte Carlo simulation or algebraic and numerical solutions of differential equations.

There are over 580 problems included in the text, some of which are purely mathematical but most are of a more scientific nature. The solutions are all fully worked out and are to be found on the web at **www.oxfordtextbooks.co.uk/orc/beddard**

Acknowledgements

I would like to thank David Salt for reading and critically commenting on early drafts of several chapters, David Fogarty for the experimental data used in Chapter 13, Marcelo de Miranda and Gavin Reid for their constructive comments on numerous topics, and to Tom and Dawn Beddard for help with compiling the index and Mathematica syntax. I also thank the authors of the books and papers which have formed the basis of several questions and diagrams all of whom I have tried to acknowledge in the text. I also extend my thanks to Jonathan Crowe at Oxford University Press who has kept faith in this project and whose help and advice has been greatly appreciated. Finally I thank my family who have received much less of my time and attention than is their due and to whom this book is dedicated.

Godfrey Beddard
School of Chemistry
University of Leeds

Contents

Appendix 1 A Maple™ Language Crib

Glossary of Selected Mathematical Symbols

Symbols	Meaning
$a = b$	Equality, with numbers $\pi = 3.14159 \cdots$ dots are added as necessary.
$a \neq b$	a is not equal to b
$a \equiv b$	Identity; a is identical to b $(a+b)^2 \equiv a^2 + 2ab + b^2$. Rarely used.
$a < b$	a is less than b,
$a > b$	a is greater than b.
$\leq$	Less than or equal to
$\geq$	Greater than or equal to
$\ll$	Much less than
$\gg$	Much greater than
$a \approx b$	a is approximately equal to b.
$a \sim b$	a is of the order of b, or a changes at the same rate as b.
373 K $\triangleq$ 100 °C	Indicates change of units to equivalent value
$\pm a$	Values are plus and minus a
$\mp a$	Values are minus and plus a
$a \wedge b$	a raised to power b. Used only in computer languages.
$\frac{a}{b}$, a/b, $a \div b$	a is divided by b.
$\angle$	Angle
$a \perp b$, $a \parallel b$	a is perpendicular and parallel to b respectively
∞	Infinity
$\rightarrow$	Tends to or approaches, and used as in $a \rightarrow \infty$ or $a \rightarrow 0$
Σ	Summation, $\sum_{i=0}^{n} x_i = x_0 + x_1 + x_2 + \cdots + x_n$
$O(x^n)$	Big 'O' notation. Used in series expansion to indicate that the next, un-written terms do not grow faster than x^n.
Π	Product, $\prod_{i=0}^{n} x_i = x_0 x_1 x_2 \cdots x_n$
!	Factorial, $n! = 1.2.3.4.5 \cdots n$
$\delta(x)$	Delta function $\delta(x) = 1$ if $x = 0$ else is zero
δ_{nm}	Kronecker delta, $\delta_{nm} = 1$ if $m = n$ else is zero. m and n are integers
$f(x)$	Function of x
$f'(x), f''(x)$	First and second derivatives with respect to x
$\dot{f}, \ddot{f}$	First and second derivatives, usually with respect to time

Other more complex symbols such as for differentiation are given when their usage is described in the text.

Table of Scientific Constants

Quantity	symbol	value	units
Avogadro constant	N_A	6.02214×10^{23}	mol^{-1}
Faraday constant	F	96485.4	$C\ mol^{-1}$ ($A\ s\ mol^{-1}$)
Molar gas constant	R	8.31447	$J\ mole^{-1}\ K^{-1}$
Boltzmann constant	k_B	1.38065×10^{-23}	$J\ K^{-1}$
Planck constant	h	6.62607×10^{-34}	J s
Planck constant/2π	$\hbar$	1.05457×10^{-34}	J s
Rydberg constant	R_∞	1.09737×10^{7}	m^{-1}
Speed of light in vacuum[1]	c	2.99792458×10^{8}	$m\ s^{-1}$
Elementary charge	e	1.60218×10^{-19}	C
Electron mass	m_e	9.10938×10^{-31}	kg
Atomic mass constant	u	1.66054×10^{-27}	kg
Permittivity of free space[2]	ε_0	8.85419×10^{-12}	$C^2\ J^{-1}\ m^{-1}$ ($A^2\ s^4\ kg^{-1}\ m^{-3}$)
Permeability of free space[2,3]	μ_0	$4\pi \times 10^{-7}$	$J\ s^2\ C^{-2}\ m^{-1}$ ($kg\ m\ A^{-2}\ s^{-2}$)
Bohr magneton	μ_B	9.27401×10^{-24}	$J\ T^{-1}$ ($A\ m^2$)
Nuclear magneton	μ_n	5.05079×10^{-27}	$J\ T^{-1}$ ($A\ m^2$)
Bohr radius	a_0	5.29178×10^{-11}	m
Hartree energy	E_h	4.35974×10^{-18}	J

[1] The speed of light is *defined* at exactly this value, [2] 'free space' is also called 'vacuum', [3] defined as this value.

The numerical value of a quantity and its units can be obtained from Maple. Some examples are;

```
> with(ScientificConstants);
  evalf(Constant( h )); 
```
$6.626068760\ 10^{-34}$

```
  GetUnit(Constant( h )) ;
```
$$\left[\frac{m^2 kg}{s}\right]$$

```
> GetConstant( h ) ;
```
Planck_constant, symbol = *h*, *value* = $6.626068760\ 10^{-34}$, *uncertainty* = $5.200000000\ 10^{-41}$, *units* = *J s*

```
> convert(joule, dimensions, base=true);
```
$$\frac{length^2\ mass}{time^2}$$

Conversion Table: Energy Units and Related Quantities

	J	kJ mol^{-1}	eV	Hz	cm^{-1}
J	1	6.02214×10^{20}	6.24151×10^{18}	1.50909×10^{33}	5.03411×10^{22}
kJ mol^{-1}	1.66054×10^{-21}	1	1.03643×10^{-2}	2.50607×10^{12}	83.5935
eV	1.60218×10^{-19}	96.4853	1	2.41799×10^{14}	8065.54
Hz	6.62608×10^{-34}	3.99031×10^{-13}	4.13567×10^{-15}	1	3.33564×10^{-11}
cm^{-1}	1.98645×10^{-23}	1.19627×10^{-2}	1.23984×10^{-4}	2.99792×10^{10}	1

To convert 6 eV into cm^{-1}, read along from eV in the left column and multiply by the number under cm^{-1} in the top row; e.g. $6\ eV \times 8065.54\ cm^{-1}/1\ eV = 48393.34\ cm^{-1}$.

Table of Derived Units and Quantities

Quantity or Unit	Symbol	SI base units
Force (newton)	N	kg m s^{-2}
Energy, (force × distance), work, heat, (joule)	J	kg m^2 s^{-2}
Pressure (force/area) (pascal)	Pa	N m^{-2}
Electrical charge (coulomb)	C	A s
Electrical potential (volt)	V	J C^{-1}
Electrical resistance (ohm)	Ω	V A^{-1}
Electrical capacitance (farad)	F	C V^{-1}
Inductance (henry)	H	V A^{-1} s = m kg s^{-2} A^{-2}
Angular velocity	ω	rad s^{-1} (or s^{-1})
Frequency (hertz)	Hz	s^{-1}
Wavenumber	cm^{-1}	m^{-1} but conventionally cm^{-1}
Heat capacity, entropy	C_p, C_V, S	J K^{-1}
Molar heat capacity, molar entropy	C_p, C_V, S	J K^{-1} mol^{-1}
Concentration	[c]	mol m^{-3} (Conventionally mol dm^{-3})
Viscosity*	η	Pa s = kg m^{-1} s^{-1}

*Viscosity is still commonly quoted in centipoise (cP) where 1 cP = 10^{-3} Pa s and this unit is used because water has a viscosity of ≈1 cP at room temperature, and ethylene glycol about 18 cP.

Table of Prefixes

10^{-1}	10^{-2}	10^{-3}	10^{-6}	10^{-9}	10^{-12}	10^{-15}	10^{-18}	10^{-21}	10^{-24}
deci	centi	milli	micro	nano	pico	femto	atto	zepto	yocto
d	c	m	μ	n	p	f	a	z	y
10	10^2	10^3	10^6	10^9	10^{12}	10^{15}	10^{18}	10^{21}	10^{24}
deca	hecto	kilo	mega	giga	tera	peta	exa	zeta	yotta
D	H	K	M	G	T	P	E	Z	Y

Numbers, Equations, Operators, and Algorithms

1

In this chapter, some of the language of mathematics is introduced. Many of the calculations in this book are done using a computer mathematics program called Maple. The syntax used and many of the instructions are described in Appendix 1 but some of the basic operations are also described here and the different usage compared to the way you would normally write mathematics. Mathematical symbols are listed in the Glossary.

1.1 Symbols and basics

When you learn mathematics, it is normal to use x and y to do most of the algebraic manipulations. In science, x and y are almost never used and this simple change can make a simple equation look complicated. This is something that you just have to get used to, but is made easier if you think what the variables mean; then perhaps you will be confident, say, to differentiate with respect to temperature T rather than feeling that this can only be done with variable x. How ingrained using x and y are, is not, however, to be underestimated, for even experienced practitioners will on occasion revert to these when faced with a difficult calculation.

1.1.1 Equations

There is one fundamental rule for manipulating an equation and that is to make sure that whatever is done to one side is done to the other. The equality sign =,[1] means that the left-hand side of any equation is always equal to the right; for example the ideal gas law equation

$$pV = nRT \tag{1.1}$$

means that the product of pressure p and volume V is always equal to the number of moles multiplied by the gas constant R and the temperature T. To isolate the pressure on the left-hand side, both sides of the equation must be divided by the volume,

$$\frac{pV}{V} = \frac{nRT}{V}$$

which is then cancelled through on the left giving $p = nRT/V$. To avoid making mistakes, the best approach is to do any manipulations step by step even if they seem trivial. Repeating a calculation because a slip has been made in an earlier step takes far longer than it does being thorough at each stage.

1.1.2 Multiplication and division

Slightly different notations are used when mathematics is written by hand, printed in a book, or interpreted by a computer language. One instance is the different symbols used to indicate multiplication; for example, multiplying together three quantities can be written as pqr, $p \times q \times r$, $p \cdot q \cdot r$ or $p.q.r$. In the last two examples, the point could be

[1] The = sign was probably first used by Robert Recorde in an algebra textbook of 1557.

confused with a decimal point if numbers were involved. In many computer languages, including Maple used in this book, multiplication is indicated by the symbol, *, for example, $a{*}b$. Raising a number to a power is written as p^3 or p^(1/3) for the cube root.

In mathematics, the brackets (), {}, [] are each used to indicate multiplication, and if an equation dictates that brackets must be nested then different ones may be used for clarity. In a computer language, different brackets are usually used to indicate different types of arrays or functions, but exactly which is used depends on the language.

Each term inside a bracket, is multiplied by any terms outside; for example, $a(b - c) = ab - ac$ and $(a - d)(b - c) = ab - ac - db + dc$. If a bracketed expression is multiplied by itself then this is written as, for example,

$$(a-d)^2 = (a-d)(a-d) = a^2 - 2ad + d^2$$

In Maple and other languages, $a(b-c)$ would be written as $a{*}(b-c)$. The reason for the multiplication sign is that one or more letters, such as a, followed by a bracket could indicate a function, array, vector, or matrix. Which one depends on how the name, in this case, a, was defined.

In printed equations, division is sometimes ambiguous because it is not clear what is to be divided; for example, a/bc can mean $\dfrac{a}{bc}$ or $\dfrac{a}{b}c$, which is $(a/b)c$; it depends on the convention being used as to which operation takes precedence. The safest way, although a little clumsy, is to write $a/(bc)$, making it clear that a is divided by bc. In Maple and other computer languages, $a/b{*}c$ means $(a/b){*}c$, so adding brackets is essential to get the calculation you expect.

1.1.3 Approximation: never do this in the middle of a calculation

Sometimes when one variable is small compared to another it can be ignored. The expansion of the exponential is $e^{-x} = 1 - x + \dfrac{x^2}{2!} - \dfrac{x^3}{3!} - \cdots$. Suppose that x is small compared to 1 then x^2 is very small and x^3 tiny, then it is possible to approximate $e^{-x} \approx 1 - x$. However, to do this in the middle of a calculation is invariably fatal. For example, in the expression $y = x + \dfrac{(N - x^2)}{2x}$ suppose that $x < N$. It would be wrong to make this $y = x + \dfrac{N}{2x}$ by arguing that as x^2 is much less than N, it can be ignored, because when it is expanded out the expression becomes $y = \dfrac{1}{2}\left(x + \dfrac{N}{x}\right)$ and no comparison of the size of x and N is possible.

1.1.4 Assignments in Maple

In Maple when something has to be calculated, it can be typed directly and the line ended with a semicolon; for example, `2.5^6;` or `sin(Pi/6);` and the result is produced on the worksheet when the 'enter' or 'return' key is pushed. However, if the result is needed later on, then the calculation must be assigned a name. This name could be almost anything but is usually a letter and numbers, `f01`, `my_result` and so forth. The assignment operation is the symbol `:=`, so, `the_sine:= sin(Pi/6);` would produce sin(π/6) every time `the_sine;` is typed; thus, `6*the_sin;` is the same as 6 sin(π/6). The assignment does not produce an equation; it makes the *name* on the left of the `:=` sign produce the calculation shown on the right. Thus a statement such as

```
> eqn1:= x^2 + 3*x = 2;
```

is acceptable. The equation is wholly on the right of `:=` and this equation has the name `eqn1`, therefore whenever `eqn1;` is typed the equation is produced in response. In Maple, the semicolon at the end of the line completes the line and prints an answer. If a colon is used instead of a semicolon, the result is still calculated but not printed. This is useful for long calculations when intermediate results may not be needed.

Details of Maple syntax is shown in Appendix 1. Because many different symbols are used in maths, and some only infrequently, these are listed in the Glossary with their meanings.

1.1.5 **Functions**

The equation $y = 1/x$ may be written in functional form as $f(x) = 1/x$, thus, if $x = 2$ the function has the value of 1/2, or $f(2) = 1/2$. The function *operates* on its argument x, which here is 2, to produce its reciprocal. The x in the function, either as its argument or in the body of the function is, however, simply a symbol, and is arbitrary; it is the algebraic expression or operation itself that determines what happens. Therefore, writing $f(w) = 1/w$ is entirely the same as $f(x) = 1/x$; if a function is evaluated with a series of numbers then it does not matter whether w or x was used to represent the mathematical operation; the result is the same. Similarly, as x is only symbolic, it is possible to write a composite function where the argument is itself a function; e.g. $f(x^2) = 1/x^2$ or $f(e^x) = 1/e^x$, which means that the function f *operates* on x^2 or e^x to produce its reciprocal. Formally, this would be written as $f[g(x)] = 1/g(x)$ where $g(x)$ is the 'inside' function.

You can consider a function to be a *rule* that converts x or w, or whatever symbol is convenient to use, into some other mathematical form. You can also consider it as a *mapping* that transforms x into a new form. The function log follows the rules to make the log of a number, and is written as $f(x) = \ln(x)$, and formally the range $x \ge 0$ should be added, but usually this is assumed to be known. A different example is illustrated by the functions used to find the real and imaginary part of a complex number. Complex numbers z have two parts; the 'real' and the 'imaginary', which is multiplied by $i = \sqrt{-1}$; for example, $z = 2 + 3i$, see Chapter 2. The real function is written $f(z) = \text{Re}(z)$ and the imaginary function $f(z) = \text{Im}(z)$ and if $z = 2 + 3i$ then $\text{Re}(z) = 2$ and $\text{Im}(z) = 3$.

While some functions are 'even' and others 'odd', many are neither. Even functions have the property that they do not change sign when their argument changes sign, $f(x) = f(-x)$, but odd functions do change sign $f(x) = -f(x)$; x^2 is an even, x^3 an odd function. A function's odd or even properties are very useful in evaluating integrals.

Inverse functions also exist, which means that if a function $f(x)$ changes x to something else then the inverse function $f^{-1}(x)$ changes it back. Suppose that the function *Fahr* changes temperature in degrees centigrade (°C) to Fahrenheit (°F), $Fahr(C) = 9C/5 + 32$; the inverse, *Cent*, takes °F and converts it into °C as $Cent(F) = 5(F - 32)/9$. This last function could be written as $Fahr^{-1}(F)$ because it is the inverse function. Finding the inverse function can be quite involved and is described in Section 1.6 at greater length.

To summarize, a function transforms its argument into some other mathematical object, which can be of almost every possible type. The principle is the same whether the function produces the reciprocal, or square or an integral or a series and so forth. The function may also be called a *transform*, an *operator*, or a *mapping* of the argument (x) onto some other object.

1.1.6 **Functions in Maple**

In Maple and other languages, a function has a particular syntax that is related to that used in mathematics, but with some important differences. A function is first defined using an assignment, then, when used it has a pair of brackets added to it. The function $f(x) = 2x + 3x^2$ is defined as

```
> f:= x -> 2*x + 3*x^2;
```

$$f := x \to 2\,x + 3\,x^2$$

and used as

```
> f(x);                         2 x + 3 x^2
> f(2);                         16
> f( sin(x) );                  2 sin(x) + 3 sin(x)^2
```

If the function has two arguments then this is written as

```
> f:=(x, y)-> 2*x + 3*y^2;
```

$$f := (x,y) \mapsto 2\,x + 3\,y^2$$

```
> f(x, y);                      2 x + 3 y^2
```

Be aware that if you write `g(x):=3;` without first defining `g` as a function Maple will not complain, it will assume that g has been defined somewhere and replace `g(x)` with 3 in any subsequent expressions. However, if instead you use `g(s)` etc., in an expression it does not replace this with 3 but uses `g(s)`, which usually has no meaning.

Some functions, particularly the named polynomials, Hermite, Legendre, and so forth, use integers as well as x as arguments. The x is always in brackets, but the integers can be written as subscripts or superscripts; a Hermite is written as $H_n(x)$, with integer n, an associated Legendre polynomial as $P_\ell^m(x)$ with m and ℓ as integers. In Maple all the arguments are placed in brackets;

```
> H(n, x); P(L, m, x);
```

However, with these complex functions, Maple uses its own algorithms to calculate them and the exact syntax has to be looked up in the 'help' menu, and the order of the arguments checked against the definition in the equation you are calculating.

1.1.7 Algorithms

There is no one definition of an algorithm. We can consider this as a logical, fixed set of rules for solving a problem that may or may not be mathematical. Starting with a known set of initial conditions, an algorithm proceeds to some other known set of conditions that ends the process and does so in a finite number of steps. A game could be considered to follow an algorithm in that it follows a fixed set of rules and starts and ends in a definite way. An algorithm could be the sequence of operations undergone to determine the rotation, reflection, and other symmetry properties of a molecule, or it could simply be the steps producing an algebraic or numerical calculation. This leads to a more restrictive and more common definition of an algorithm as a series of instruction with which to perform a calculation on a computer.

1.1.8 Numbers

So familiar are numbers and counting that we hardly give them much thought and automatically use different types of numbers and ways of counting as the situation dictates. Integer numbers are either prime, when the number is only divisible by itself and 1, or composite and the product of two other integers. Besides the positive and negative integers, there are the rational numbers or fractions, generated from their ratios, n/m, such as 1/2, 2/3, etc. A fraction is called *proper* when the numerator is less than the denominator, 2/3 for instance, and a fraction *improper* if the numerator is greater.

It is usual to express fractions as decimal expansions such as 0.500, or 1.36348 ... etc., which are called *real* numbers. In some countries, real numbers are expressed with a comma 0,500 or 1,36348 instead of a decimal point, and the comma is the international system of units (SI) recommended symbol, although it is never used in the English-speaking world.

In fractions, such as 98/77, there are repeats of the sequence of digits; this fraction has a 2 digit repeat and is 1.272727 ... while $98/78 = 1.256410256410\ldots$ has a 6 digit repeat. The *irrational fractions* can also be expanded in decimal notation, $\sqrt{2} \approx 1.4142135\ldots$, however, the sequence of digits does not repeat itself.

The decimal numbers, as the name implies, use a base of 10 for counting, however, we are also quite happy using other counting systems; base 60 to count time, with 60 seconds in a minute and 60 minutes in an hour, as well as base 12 or 24 for counting the hours. Angles such as latitude and longitude, essential for navigation, are based on the 360° of a circle with the degrees each split into 60 minutes and then into 60 seconds of arc. Engineers used to use grads, which divide the circle into 400 parts, and perhaps some still do. In measuring weights, the old Imperial units, pounds (lb) and ounces (oz), where 16 oz = 1 lb and 14 lb = 1 stone are still used, although this is being replaced by the decimal units based kg and g. Distances in the UK and in North America are measured in miles, an arbitrary measure of 5280 ft, based on a yard of 3 ft with each foot containing 12 inches, although the km, m, and cm are replacing these older measurements. Eight km is approximately 5 miles and 1 inch $\triangleq$ 25.4 mm. A distance of 3 yards, 1 ft and 3 inches is often written as 3, 1′ 3″ and a similar notation used for degrees, minutes, and seconds of arc, 25° 12′ 3″.

Because computers are now so fast in performing calculations, it is almost never necessary to write code at the binary level, which is base 2. Binary means calculating at the bit and byte level; the numbers used are 0 and 1 only. In octal, numbers based on 0 to 7 only are used and in hexadecimal numbers are on base 16. These are chosen by convention to have the numbers and letters, 0, 1, 2, 3, 4, 5, 6, 7, 8, 9, A, B, C, D, E, F. Thus 10 in decimal is A in hexadecimal, 15 in decimal is F in hex and 16 in decimal is 10, pronounced as '1 zero', in hex. It is not difficult to do arithmetic in hexadecimal, just awkward because we are so used to decimals and often change to decimal to do the calculation and convert back, this being easier than learning the hexadecimal multiplication table. Hexadecimal arithmetic is used only rarely now as programming a computer is usually done at a high level. Once hex' was necessary not only to get speed in a calculation on a computer using words 16 bits wide, but also to 'talk' to an external instrument *via* pulses generated by repeatedly sending a 1 or 0 to a certain address that is then mapped onto the pins of an output socket on the computer. The table shows a comparison of different number systems in the order decimal, octal, hexadecimal, and binary.

0	1	2	3	4	5	6	7	8	9	10	11	12	13	14	15	16	17		20	30	40	50
0	1	2	3	4	5	6	7	10	11	12	13	14	15	16	17	20	21		24	36	50	62
0	1	2	3	4	5	6	7	8	9	A	B	C	D	E	F	10	11		14	1A	28	32
0	1	10	101	100	101	110	111	1000	1001	1010	1011	1100	1101	1110	1111	10000	10001		10100	11110	101000	111010

1.2 Integers, real, and irrational numbers

Integers are the numbers that are colloquially called whole numbers, −2, 0, 3, etc. and they extend from minus infinity to plus infinity; $-\infty$ to $+\infty$ and although there is an infinity of them they are what mathematicians call *countable*. Integers are clearly not the only numbers as we are familiar with the real numbers such as 1.2, 3.56, and so on, of which there is also an infinite number, but many, many more than there are integers. If two different real numbers are chosen, an infinite number of others can be squeezed in between them simply by adding more decimal places; this and other interesting discussions on numbers and algorithms are explained clearly in *The Emperor's New Mind* (Penrose 1990). Surprisingly, real numbers can usually be reduced to a rational fraction which is a ratio of two integers, e.g. 0.751 = 326/533, but, for some numbers, π, e, sqrt(2) there is no ratio of integers that will accurately form the number and they are called *irrational*, i.e. not being logical! You may recall that $\pi \approx 22/7$, which would seem to contradict this statement, but 22/7 is a very poor approximation to π.

When you use a computer to perform calculations, a distinction is made whether a real or integer number is used. This is a common feature of programming languages and also in computer algebra, for example, the integer division 2/3 is exact and in Maple, it remains in the calculation as such. In other languages integer division can result in the answer 2/3 = 0 rather than $0.66\dot{6}$ or remaining as an exact ratio so this has to be checked carefully.

We look next at some of the irrational numbers as their calculation has fascinated mathematicians over the centuries and they allow us to use some interesting geometry and to write some algorithms.

1.2.1 Pi (π) and an algorithmic way of calculating its value

Whole books have been devoted to the number π = 3.1415926, such as *The Joy of Pi* (Blatner 1999), and calculations performed to study the pattern of its digits, which now extend to many thousands of millions of decimal places. As far as is known, the pattern of digits does not repeat itself; if they did it would mean that π could be written as a fraction of two whole numbers and would no longer be an irrational number. π is also a *transcendental* number, which is one that is not the solution (we say *root*) of any polynomial equation, and it is not considered to be algebraic; it transcends algebra and this guarantees that it is irrational.

π appears naturally in many algebraic formulae that have no obvious geometric interpretation. Some estimations involve summing series, others dropping needles across a square grid (Buffon, 1707–1788), and yet others, going back at least to Archimedes, measure either the length of one polygon that is just larger, or one that is just smaller, than a circle. The proof of this method is very elegant and leads us in a direction that is not obvious before coming to its conclusion. We shall follow it through and then write an

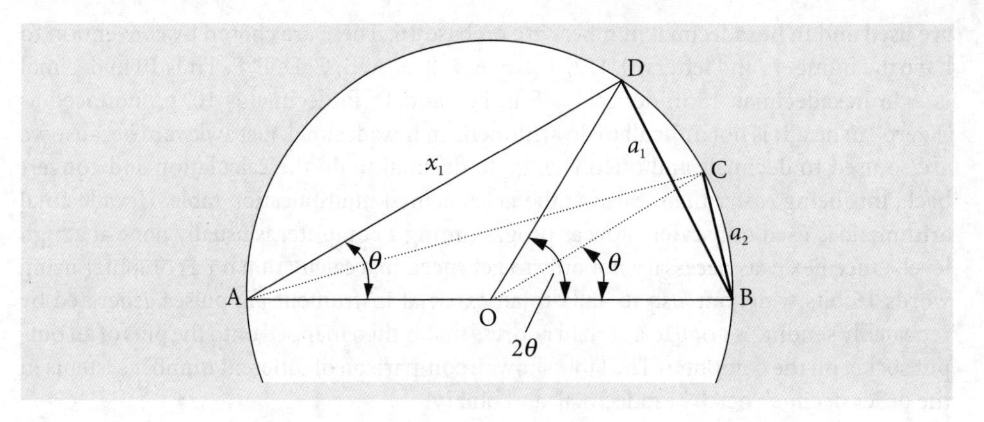

Fig. 1.1 The hexagon has a side of length DB = a_1 the dodecagon CB = a_2.

algorithm with which to calculate π. We consider ourselves to be advanced and modern, but looking at the subtlety in this and other proofs, we soon realize how brilliant were those who, in the past, worked them out.

To calculate π, the idea is first to draw a circle and then place a square inside this touching its circumference. Next, a pentagon, hexagon, and "n-agons" are drawn in the same way, and as the number of sides gets larger, the closer the straight-sided polygon will get to the circumference of the circle. And, if another polygon is drawn just outside the circle, then the two values of the polygons' circumferences define upper and lower bounds to the value of π. However, actually drawing these polygons on a piece of paper, or the computer, is not necessary as geometry and algebra can be used. This was how the first calculation of an accurate value of π was achieved, and furthermore, to 15 decimal places, and was described by the Iranian astronomer Jamshīd al-Kāshī in 1424 in his book *A Treatise on the Circumference.*

The calculation starts by sketching a hexagon in the circle, then a 12-sided figure (a dodecagon). Just one segment of each is shown in Fig. 1.1. The drawing need not be very accurate and only represents the shapes. Geometry is used to assert their true shapes.

Figure 1.1 shows the straight edge of one piece of the hexagon, DB = a_1 and a dodecagon CB = a_2. The plan al-Kāshī had was to find a_1 and then to determine a_2 in terms of a_1 and then continue to a '24-gon' and find the length of one of its segments, calling this a_3, which is known in terms of a_2 and so on. By adding up the lengths of polygons with more and more sides, better and better estimates of π are obtained.

It has been known since the time of Euclid, that for a triangle drawn in a circle the angle at the centre of a circle is twice that at the circumference; therefore, if ∠DAB is θ, then ∠DOB is 2θ. An angle at the centre and circumference is shown in Fig. 1.2:

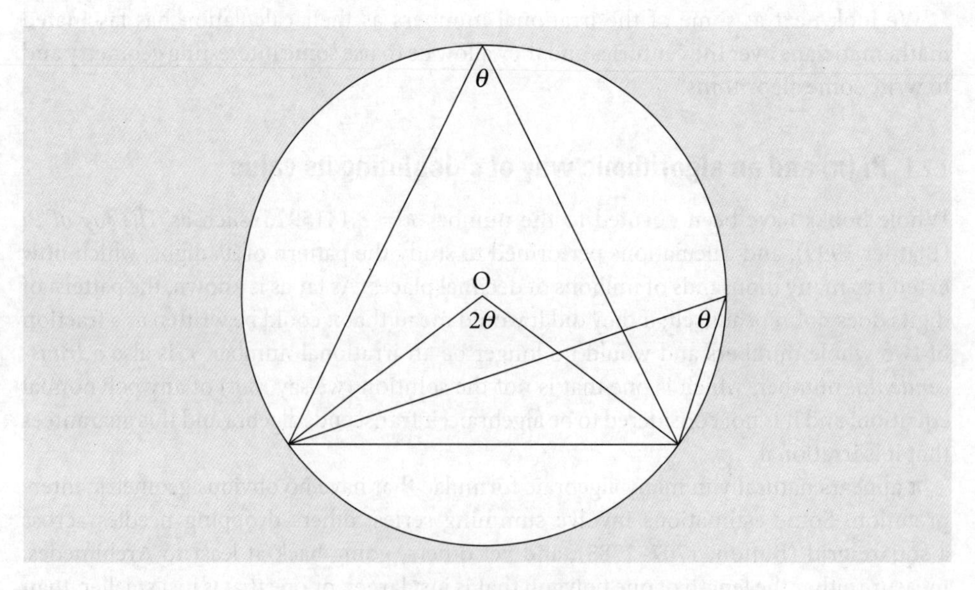

Fig. 1.2 The angle at the centre is twice that at the circumference.

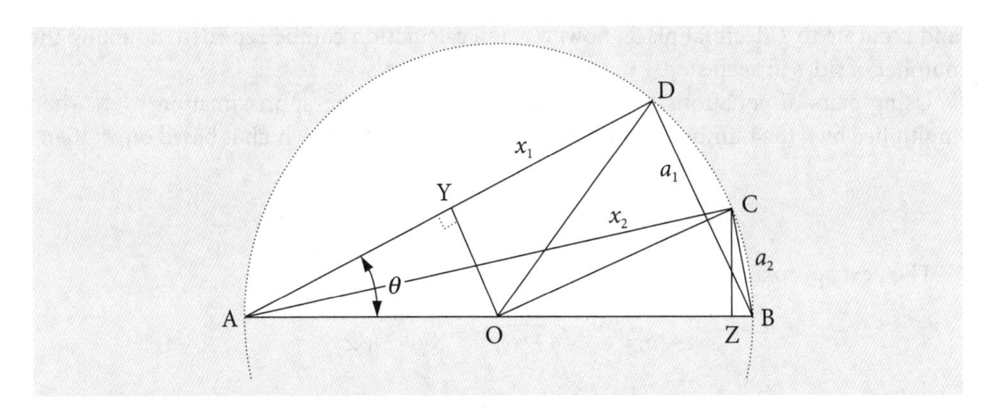

Fig. 1.3 Geometric construction to calculate π but not drawn to scale.

Using this theorem also means that ∠ADB is a right angle, since if $2\theta = 180°$, then $\theta = 90°$ and ∠ACB is therefore also a right angle. You may also notice in Fig. 1.1 that AD is parallel to OC, that ∠ADO is also θ and that OC and DB are at right angles since the arc DC = CB and therefore OC bisects DB.

Now we can start the calculation of π. Remember that we have to find a_2 in terms of a_1. To find a_1 we have to find the length of one side of a hexagon, BD. But in a hexagon, the total length of the six sides is easily found since each segment, such as ODB is an equilateral triangle, $2\theta = 60°$, and if the circle has a radius of 1, then OA and OB = 1, since OC, OD and BD ≡ a_1 thus, $a_1 = 1$.

The total length of the hexagon, which approximates 2π the circumference of the circle, is 6 which produces the rather unimpressive value π = 3; this can be vastly improved on. Al-Kāshī's brilliant reasoning is now outlined. His method was to find a relationship between x_1 (AD) and x_2 (AC), and use this to find a_2 (CB) knowing that $a_1 = 1$ and to repeat this process for polygons with progressively larger numbers of sides. To find a_2, two lines CZ and OY, are drawn that are, respectively, perpendicular to AB and AD as shown in Fig. 1.3.

Next, convince yourself, for example by drawing them out, that the triangles ABC and ACZ are similar because their angles are similar. The angle ACB is 90° because an angle at the circumference is half that at the centre; see Fig. 1.2, and AZC is also a right angle. Therefore the ratio of the sides AZ/AC is the same as AC/AB or as an equation $AZ/x_2 = x_2/AB$ from which

$$x_2^2 = AZ \times AB = 2(1 + OZ), \tag{1.2}$$

the length AB being 2 because the radius of the circle is 1. Our task now is to find OZ.

The angles in triangles AOY and OZC are the same, see Fig. 1.1, and the triangles are also the same size as both have the radius as their hypotenuse, AO and OC. Notice also that ∠AOY = 90 − θ and therefore bisects the angle ∠AOD making AY = YD and $AY = x_1/2 = OZ$.

Substituting for OZ produces

$$x_2^2 = 2 + x_1. \tag{1.3}$$

A relationship between the length x_1 or x_2 and a_2 has to be found. Triangle ADB is right angled and as the hypotenuse is 2 (the diameter), using Pythagoras's theorem is

$$a_1^2 = 4 - x_1^2$$

and because $a_1 = 1$ then $x_1^2 = 3$. Using equation (1.3) produces x_2 as

$$x_2^2 = 2 + \sqrt{3} = 3.7320.$$

The dodecahedron value for π is calculated by using the right-angled triangle ACB to obtain $a_2^2 = 4 - x_2^2$. Hence $a^2 = 0.5177$, and as there are 12 sides to the dodecahedron and the circumference is 2π, we obtain $2\pi = na_2$ or $\pi = 12/2 \times 0.5177 = 3.106$. Not too bad a result and accurate to 1 decimal place, however, the calculation can be repeated, doubling the number of sides in each step.

Using pairs of equations, the parameter a_n becomes the approximation to π, when multiplied by n the number of sides, e.g. using the value for x_2 which is based on x_1, then

$$a_3 = \frac{24}{2}\sqrt{4 - x_3^2} \qquad \text{with} \qquad x_3^2 = 2 + x_2,$$

The next approximation is

$$a_4 = \frac{48}{2}\sqrt{4 - x_4^2}, \qquad x_4^2 = 2 + x_3,$$

and so on making the ith step, with an n_i-sided polygon,

$$a_{n+1} = \frac{n_i}{2}\sqrt{4 - x_{i+1}^2}, \qquad x_{i+1}^2 = 2 + x_i.$$

This calculation of π also means that Al-Kāshī knew how to evaluate square roots. His method is written in Maple as Algorithm 1.1, if you are not familiar with the syntax then look at the Maple appendix now.

Any algorithm is a process or 'machine' used to arrive at an answer and starts by defining the parameters needed, then setting their initial values. A termination criterion has also to be defined to stop the calculation. The calculation is then performed step by step, repeating steps as necessary and the results printed as the calculation proceeds or just at the end as necessary. If the calculation involves repeated evaluations, as in this case, then a loop is needed to do this.

In the Maple code, note that the symbols `:=` while looking like an equality sign *do not* indicate an equation but rather that an *assignment* is being made; for example, the statement `m:= 20:` means 'set aside some space in the computer's memory, give it the *name* m, and when that name is called it will produce the number 20'. The instruction `for i from 1 to m do .. end do` loop repeats the calculation m times starting at 1, automatically incrementing the value of `i`. In this example, this is done 20 times, which corresponds to using a 6×2^{26}-gon which has 402,653,184 sides! The arrays x, n, and a hold the intermediate values; n contains the number of sides of each polygon starting with a hexagon `n[1]:= 6;`. The line `Digits:= 30:` makes the calculation accurate by calculating to more digits (30) than we are likely to need and the `evalf` command produces an arithmetic answer instead of an algebraic one and to 12 decimal places. Only some of the answers are reproduced here, but it is clear that the true value of π is being approached and the last approximate value listed is accurate to 12 decimal places.

Algorithm 1.1 Al-Kāshī's method of calculating π

```
> restart:
> Digits:= 30:                                    # set precision
  m:= 20:                                         # number repeat calc's
  x:= Array(1..30):                               # define arrays
  n:= Array(1..30):
  a:= Array(1..30):
  x[1]:= sqrt(3.0):                               # initial x
  n[1]:= 6:                                       # sides of polygon
  print( 1,`first guess`,3*sqrt(4.0 - x[1]^2 ) );
  for i from 1 to m do
    x[i+1]:= sqrt( 2 + x[i] ):
    n[i+1]:= 2*n[i];
    a[i+1]:= n[i+1]/2*sqrt( 4 - x[i+1]^2 ):
    print( i+1,`approx pi`, evalf(a[i+1],12) );
  end do:                                         # end looping when i=m
```

1, *first guess*, 3.00000000000
4, *approx pi*, 3.13935020305
7, *approx pi*, 3.14155760791
10, *approx pi*, 3.14159210600
13, *approx pi*, 3.14159264503

16, *approx pi*, 3.14159265346
19, *approx pi*, 3.14159265359

Al-Kāshī repeated his calculation of π until its error was the same ratio as the width of a hair is to the size of the universe. From historians' estimates of what was considered to be the size of the universe in the thirteenth century, this means that he calculated π to 15 decimal places, which is a genuinely extraordinary calculation and similar in accuracy to our answer.

Such has been the fascination of π to mathematicians that there are many algebraic formulae which contain π and others with which to calculate it. Some examples are shown below.

$$e^{i\pi} = -1, \qquad \int_0^\infty \frac{\ln(x)^2}{1+x^2}dx = \left(\frac{\pi}{2}\right)^3, \qquad \frac{\pi^2}{6} = \frac{1}{1^2} + \frac{1}{2^2} + \frac{1}{3^2} \cdots = \sum_{n=1}^{\infty} \frac{1}{n^2}.$$

In these and many other formulae producing π, it is difficult to appreciate why π appears; some are far more efficient at calculating π than the method described, but do not have the appeal of the geometric approach.

1.2.2 Square Roots √

The square root is familiar as the solution to an equation such as $n^2 = 3$. The square root is also the length of the diagonal of a square or rectangle and therefore has a geometric interpretation. The Pythagoreans discovered that the length of the diagonal of a square of unit side, i.e. length 1, is $\sqrt{2}$; see Fig. 1.4. Although it is hard to imagine this today, this calculation caused a scandal since this number was not 'pure' in the sense that it was not the ratio of two whole numbers. And the scandal still echoes down the centuries. Numbers that are not the ratio of two whole numbers are called *irrational* or 'un-ratio-able', the Latin word *ratio* coming from the Greek for '*logo*' which means ratio. We use the word *logic* to symbolize reason and it is a pejorative statement to describe someone as being irrational.

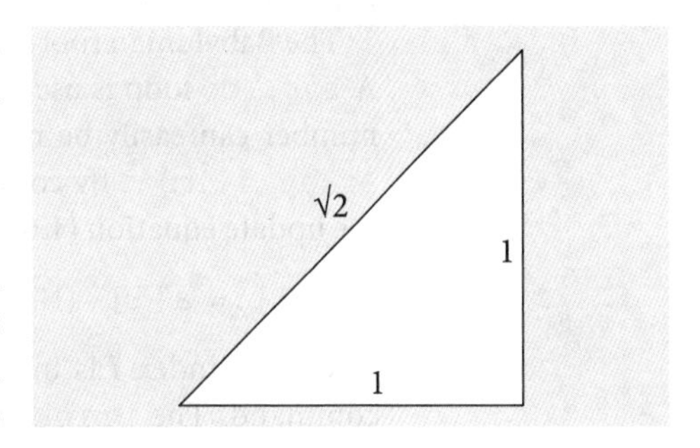

Fig. 1.4 Unit right-angled triangle.

The right-angled triangle does not of itself produce the value of $\sqrt{2}$ very accurately, since the length of the line would have to be measured; it is more of a *geometrical definition*. The first calculation of a square root was made by the Babylonians (modern-day Iraqis), about 4000 years ago, and is inscribed in a clay tablet preserved at Yale University. A geometric method was used that could be converted into an algorithm. The result on the tablet was accurate to five decimal places and the calculation shows the Babylonians knew Pythagoras's theorem about 100 years before Pythagoras discovered it. This result is remarkable, and it is possible that a formula was used similar to what is now called Heron's method after the Alexandrian mathematician and engineer who lived about 2000 years ago; his dates vary from 150 BC to AD 250.

Suppose that N is the number for which the square root is sought, then let a be an estimate of this number and therefore we want an algorithm with which to find a. Suppose that a^2 differs from the true value N by an error ε; hence

$$\varepsilon = N - a^2. \tag{1.4}$$

The next step is to find a better approximation and instead of just randomly guessing a new value for a, a *small* positive or negative number c is added. Now $a + c$ is the new approximation to the square root, therefore

$$\varepsilon_c = N - (a+c)^2. \tag{1.5}$$

Assuming that adding c makes the error ε_c zero, by expanding the bracket, substituting for $N - a^2$ and rearranging produces

$$\varepsilon = 2ac + c^2.$$

However, as c is small, then c^2 is even smaller,[2] and we will suppose we can ignore it and then $c = \varepsilon/2a$. Next we suppose that the approximation to the square root is a_1 where $a_1 = a + c$. Hence

$$a_1 = a + \varepsilon/2a = a + \frac{(N-a^2)}{2a} = \frac{1}{2}\left(a + \frac{N}{a}\right). \qquad (1.6)$$

Repeating the procedure gives $a_2 = (a_1 + N/a_1)/2$ and then $a_3, a_4 \cdots$ and so forth, and a_n will eventually produce $\sqrt{N}$ after n iterations. It is implicitly assumed that this sequence will converge to a real number as the error ε gets smaller.

To calculate $\sqrt{2}$, then $N = 2$, and starting with a poor initial guess of $a = 1/2$, the steps where ε is explicitly calculated are

Step 1: $\varepsilon = N - a^2 = 7/4$, $\quad a_1 = 1/2 + 7/4 = 9/4$
Step 2: $\varepsilon_1 = N - a_1^2 = 2 - 81/16 = -3.0625$, $\quad a_2 = a_1 + \varepsilon_1/2a_1 = 1.569$
Step 3: $\varepsilon_2 = N - a_2^2 = 2 - (1.569)^2 = -0.4618$, $\quad a_3 = a_2 + \varepsilon_2/2a_2 = 1.4218$
Step 4: $\varepsilon_3 = 2 - (1.4218)^2 = -0.0215$, $\quad a_4 = a_3 + \varepsilon_3/2a_3 = 1.4142$

After four steps, a value accurate to three decimal places is produced; the convergence is very rapid because of the a^2 term in the equation.

Heron's formula can also be described as an average, because if a is an estimate of $\sqrt{N}$ then $(a + N/a)/2$ is the average of the number $\sqrt{N}$ plus a small increment a and the value N/a, which is less than $\sqrt{N}$. This result is also produced from Newton's method for approximating the roots of the equation, $N + a^2 = 0$, see Chapter 3.

The Babylonian root estimation process can be calculated as an algorithm using Maple. A `for..do` loop is used to repeat the calculation, in this case seven times, although this number can easily be increased. The estimates are held in the array a, defined as `a:= Array(1..m):`. By combining the equations by putting ε into the new equation for a, the update equation (1.6) for each step is:

```
a[i+1]:= a[i]+(N-a[i]^2)/( 2*a[i] ):
```

where the index i is incremented automatically in the `do i from` to the `end do` command. The `print` statements produce the results and the `evalf(a[i+1],20)` command produces a real number, in this case to 20 decimal places.

Algorithm 1.2 Heron's or the 'Babylonian' algorithm for square roots

```
> restart:
> N:= 2: a0:= 1/2:                    # initial values
  m:= 7:                              # number of repeats
  a:= Array(1..m):                    # define array to hold a's
  print(`square root of `, N, ` initial guess `, a0);
  a[1]:= a0 + (N - a0^2)/(2*a0):
  print( 1, evalf( a[1],20) );        # print 1st value
  for i from 1 to m-1 do              # do iteration, print values
    a[i+1]:= a[i]+( N - a[i]^2 )/( 2*a[i] ):
    print( i+1, evalf( a[i+1],20) );
  end do:
  print( ` using sqrt` , evalf(sqrt(N),20));
```

$$\textit{square root of}, 2, \textit{initial guess}, \frac{1}{2}$$

1, 2.2500000000000000000
2, 1.5694444444444444444
3, 1.4218903638151425762
4, 1.4142342859400733411
5, 1.4142135625249320650
6, 1.4142135623730950488
7, 1.4142135623730950488

using sqrt, 1.4142135623730950488

[2] Making approximations is something that should not be done in the middle of a calculation. In this case its iterative nature effectively corrects for this, i.e. it is justified by leading to a formula that iteratively approximates the correct result.

If you try this algorithm with other values of N, you will soon realize that from a very wide range of initial values, convergence takes only a few iterations; it is far more efficient than the algorithm calculating π. You will need to increase m to get more iterations with larger numbers.

1.2.3 Golden ratio

The golden ratio $1.618033\cdots$ was thought by the ancient Greeks to be the perfect ratio of width to the height of a picture, but to me it appears a little too wide for its height. The golden ratio has been written about extensively and has somewhat of a cult following! In *ca* 300 BC Euclid divided the line below, Fig. 1.5, so that AC is to AB as AB is to BC. This is the ratio of the total length and of the larger section, to the ratio of the larger to small sections. As an equation $\phi=\dfrac{AC}{AB}=\dfrac{AB}{BC}$, then

$$\frac{\phi+1}{\phi}=\phi. \tag{1.7}$$

Rearranging produces $\phi^2-\phi-1=0$, which has the solution $\phi=\dfrac{1\pm\sqrt{5}}{2}=\pm1.6180339\cdots$, and the golden ratio ϕ, is the positive root of the two solutions. Interestingly, $1/\phi = 0.6180339\cdots$. As ϕ is the solution of an equation, it is an algebraic, not transcendental number.

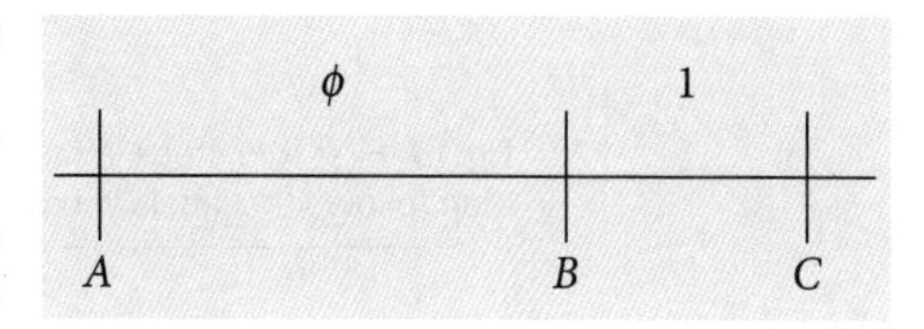

Fig. 1.5 Defining the golden ratio.

The golden ratio appears in unusual places, such as in rectangles and pentagons, as well as the coordinates of the edges of an icosahedron. These can be found from the values $(0, \pm u, \pm v)$, $(\pm u, 0, \pm v)$, $(\pm u, \pm v, 0)$ by taking all possible plus and minus combinations, and where $u^2 = \phi/\sqrt{5}$ and $v^2 = (\phi\sqrt{5})^{-1}$. Boron suboxide, B_6O forms particles of icosahedral symmetry.

Fullerenes, such as C_{60}, and soccer balls, have a truncated icosahedral structure in which the vertices of the icosahedron are cut away to reveal a regular solid with sides of pentagons and hexagons. The truncated icosahedron is one of the Archimedean solids.

A practical use of the golden ratio is that it provides the most efficient way of dividing a curve when trying to find its minimum numerically. This can be used when trying to find the minimum difference between a set of data y_{expt} and its fit to a theoretical expression y_{calc}. The minimum difference thus produces the best fit of the theoretical model to the experimental data. Suppose that the function y_{calc} describes the first-order decay of a chemical species with time t, the theoretical equation is $y_{calc} = e^{-kt}$ where k is the rate constant for the reaction we want to determine by fitting to experimental data.

The parameter whose minimum is sought is the square of the difference between the experimental data points and the theoretical equation calculated at the same t values but with different values of k. This is called the residual and is the sum $R_k = \sum_{i=0}^{n}(y_{expt_i} - y_{calc_i})^2$ where data points are labelled i. The curve whose minimum values is sought is then R vs k

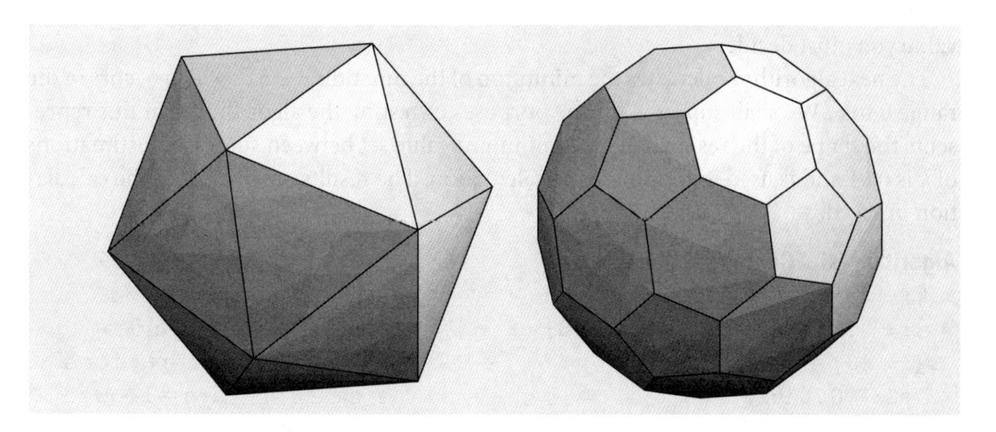

Fig. 1.6 An icosahedron (left) and a truncated icosahedron (right) which is the shape of a football and C_{60} molecules.

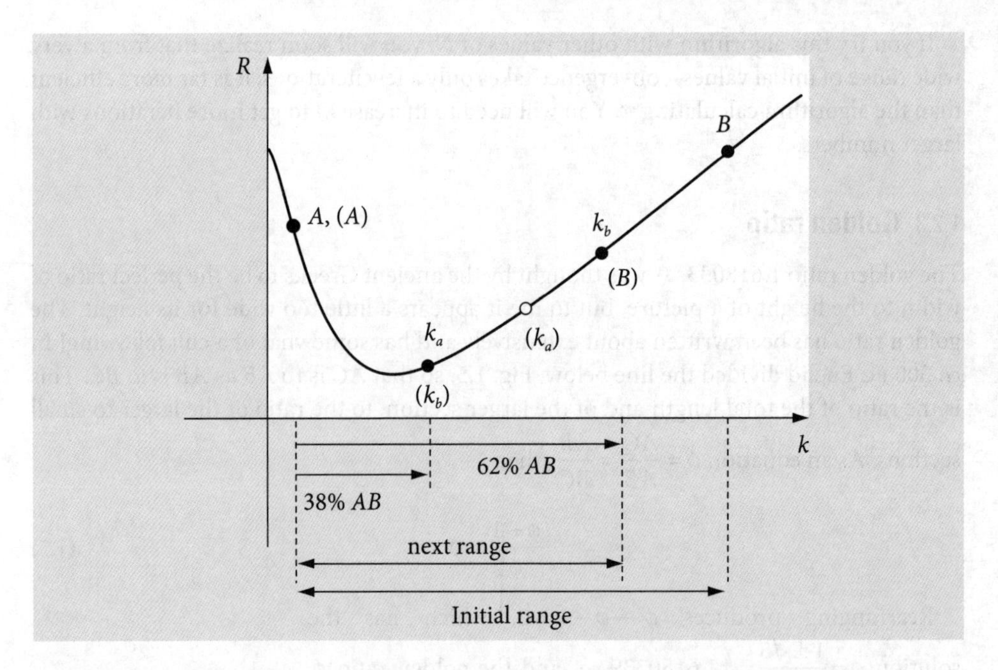

Fig. 1.7 First two steps in finding the minimum of a curve R vs k using the golden ratio. The second step follows the labels in brackets.

for a range of k values; recall that y_{calc} depends on k. One tedious way would be to start with some very large or very small value of k, far away from the true value, and calculate R_k by changing k by small amounts and comparing values of R until a larger value of the residual than its predecessor is found. This method is called an exhaustive search, and this would take a huge number of computations to complete.

A smarter way, Fig. 1.7, chooses two points A and B, known to straddle the true value of k, and places two new points in the interval AB in the ratio $1/\phi$ and $1 - 1/\phi$ as a first step to determining where the minimum might lie. One point k_a is therefore placed at ≈38% along the interval, the point k_b at ≈62%, and the values at k_a and k_b are compared. If, as is the case in Fig. 1.7, the value at k_a is smaller than k_b the minimum is in the region A to k_b, otherwise it is in k_a to B.

The rules of this 'game' define the algorithm and are:

(i) If the value of the function R at $k_a < k_b$, then B is moved to k_b to reduce the interval over which the minimum is to be found; the point k_a becomes k_b and a new k_a is calculated in the original region A to k_b and is placed at $1/\phi$ along this interval labelled in the diagram as (k_a).

(ii) If the values of the curve at $k_1 > k_2$ then labels a and b are swapped around, meaning that A is moved to x_a, to reduce the search interval; k_b becomes k_a and a new k_b is placed at $1 - 1/\phi$ along the original region k_a to B.

These processes are repeated until a fixed number of iterations have occurred, or the minimum difference in two consecutive k values found to within a certain error whose value you must decide.

The next algorithm calculates the minimum of the function $x^3 - 5x^2$, which occurs in the range 0 to 6. We shall suppose, for the purposes of testing the algorithm, that this represents the shape of the residual R. The minimum value set between successive estimations of k is 0.01 and is used to terminate the calculation. The results are printed as the calculation proceeds.

Algorithm 1.3 Golden section search

```
> f:= x-> x^3 - 5*x^2;                          # define a function
> a:= 0.0;           b:= 6.0;                   # set limits a and b
  N:= 20;                                       # number of iterations
  Lmt:= 0.01;                                   # smallest b-a allowed
  g:= (sqrt(5.0)-1)/2.0;                        # golden ratio
  ka:= g*a+(1-g)*b;                             # define start points
  kb:= g*b+(1-g)*a;
  fa:= f(ka);          fb:= f(kb);              # function at start pos'n
```

```
print("i, ka, kb, fa, fb, b-a ");
i:= 0:
while abs(b-a)> Lmt and i < N-1 do
  i:= i+1:
  if fa < fb then                              # rule(i)
    b:= kb:
    kb:= ka:                fb:= fa:
    ka:= g*a+(1-g)*b;       fa:= f(ka)
  else                                         # rule(ii)
    a:= ka;
    ka:= kb:                fa:= fb:
    kb:= g*b+(1-g)*a;       fb:= f(kb)
  end if;
  print(i, ka, kb, fa, fb, b-a);               # print all results
end do:                                        # end of while..do
 i, ka,     kb,       fa,       fb,     b-a
 1 3.7082 4.5836 -17.7632 - 8.7484 3.7082
 2 3.1672 3.7082 -18.3851 -17.7632 2.2918
 3 2.8328 3.1672 -17.3913 -18.3851 1.4164
 4 3.1672 3.3738 -18.3851 -18.5102 0.8754
..
13 3.3295 3.3322 -18.5184 -18.5185 0.0115
14 3.3322 3.3338 -18.5185 -18.5185 0.0071
```

The table of results shows how the search range is reduced as iteration proceeds. The minimum of the function is known by differentiation, which is how we can check that the calculation is correct, and is at $x = 10/3 = 3.33\cdots$ with a value -18.519; effectively the same as found by iteration.

1.3 The exponential *e* and e^x

The final number we consider is the exponential $e = 2.7182818284\cdots$ familiar from Boltzmann's equation, the growth of populations of bacteria, or the decay of a compound by its first-order reaction. It also describes the largest rate of increase of compound interest achievable in your savings account or mortgage.

The number *e* is transcendental, just like π or $\sqrt{2}$, in that it cannot be expressed as the solution to an algebraic equation; it transcends algebra and its value is obtained by expanding an algebraic series or as a limit of the expression;

$$e = \lim_{k\to\infty}(1 + 1/k)^k \qquad \text{or} \qquad e^x = \lim_{k\to\infty}(1 + x/k)^k.$$

The limit is found by trying progressively larger values of *k*; and the first expression tends towards the constant value, *e*.

k	limit
10	2.593742
1000	2.716924
100 000	2.718268
100 000 00	2.712828

In a savings account you receive interest on your capital sum. If this is an amount £*N* and if the interest is at an annual rate of *r*% which is 'compounded' *k* times per year, then at the end of a year the capital has grown to

$$N\left(1 + \frac{0.01r}{k}\right)^k.$$

Starting with £3000, if this is compounded annually at 5%, you should receive $3000\left(1 + \dfrac{0.05}{1}\right)^1 = £3150$ at the end of the first year. However, if the interest is compounded quarterly this rises to $3000\left(1 + \dfrac{0.05}{4}\right)^4 = £3152.84$ also at the end of the first year. If compounded daily ($k = 365$) for a year the interest would only rise to £3153.80 and, eventually, if continuously compounded, k would tend to infinity giving the exponential. The maximum, therefore, that you could ever obtain would be £3000e$^{0.05}$ = £3153.81.

The exponential is *defined* as a series expansion as was first discovered by Euler in 1748. The series converges absolutely for any finite value of x even though it extends to infinity:

$$e^x = 1 + x + \frac{x^2}{2!} + \frac{x^3}{3!} + \frac{x^4}{4!} \cdots = \sum_{n=0}^{\infty} \frac{x^n}{n!}. \tag{1.8}$$

The Σ sign at the end of the series is a shorthand and indicates summation in this case from $n = 0$ to infinity. The factorials 2!, 3!, etc. are products, such that,

$$0! = 1,\ \ 1! = 1,\ \ 2! = 2 \times 1,\ \ 3! = 3 \times 2 \times 1,\ \ n! = n \times (n-1) \times (n-2) \times (n-3) \times \cdots \times 2 \times 1$$

and so on. Factorials are calculated at positive integer values only (see Section 1.7). The value of the exponential series with $x = 1$ is

$$e = 1 + \frac{1}{2!} + \frac{1}{3!} + \frac{1}{4!} \cdots = 1 + 1/2 + 1/6 + 1/24 \cdots = 2.71828 \cdots.$$

Consider now two series,

$$f(x) = 1 + x + \frac{x^2}{2!} + \frac{x^3}{3!} + \frac{x^4}{4!} + \cdots \text{ and } f(y) = 1 + y + \frac{y^2}{2!} + \frac{y^3}{3!} + \frac{y^4}{4!} + \cdots$$

By multiplying term by term and rearranging, it is found that

$$f(x)f(y) = f(x+y),$$

which means that the familiar property that powers of products of numbers add has been confirmed because,

$$e^x e^y = e^{x+y} \quad \text{and} \quad e^a e^b e^c = e^{(a+b+c)}. \tag{1.9}$$

You can try calculating the exponential series by hand and easily see how the accuracy improves term by term. A quick examination of the series shows that you can use a previous value to obtain the next; starting at the second term, the third is calculated by multiplying the second by $x/3$, the fourth by multiplying the third by $x/4$, and so on. This type of process is called *recursion* and the calculation of π and of square roots described in Sections 1.2.1 and 1.2.2 are both recursive. The Maple code to calculate e^x recursively is shown below; you can add more terms in the `for .. do` loop to get a better answer, which is the sum of terms s, and do not forget to give x a value unless you want an algebraic answer.

Algorithm 1.4 Recursive calculation of the exponential function

```
> restart:
  Digits:= 20:            # ensure numerical precision
  p:= 1; s:= 1;
  x:= 1.0:                # omit this line for the algebraic series
  for i from 1 to 30 by 2 do
    p:= p * x/i:
    s:= s + p:
    print(i,s):
  end do:
```

Your hand calculator has to evaluate the exponential series each time a value is required and by calculating more terms in the series a better estimate of the true value is obtained, so a computer working in double precision will produce a value, say, to 15 decimal places, which is more accurate than your calculator. Computer algebra programs, such as Maple, can calculate to almost unlimited decimal places, and obtaining 20 000 decimal places in not a problem. However, I cannot imagine that anyone would ever need to use such a precise number!

There are numerous examples of the exponential function in chemistry and physics. One of the most important is Boltzmann's equation, which relates the population of two energy levels, an upper one n_2 to that of a lower one n_1 and which are separated by energy ΔE at absolute temperature T,

$$\frac{n_2}{n_1} = e^{-\Delta E/k_B T}.$$

Boltzmann's constant is $k_B = 1.38 \times 10^{-23}$ J K^{-1}.

1.4 Logarithms

The 'inverse' of the exponential, is the logarithm, because if $y = e^x$ then a function, the logarithm, can be defined as $\ln(e^x) = x$, and therefore $\ln(y) = x$. More generally put, the logarithm of x is the power by which a base number b must be raised to give x. As an equation this is

$$x = b^{\log_b(x)}.$$

The value of x must be greater than zero, $x > 0$, for the log to be a real number. If $x < 0$, then a complex number is obtained, see Chapter 2.

A series of increasing powers of 2, familiar in computing where 2 kb ≡ 2048 bytes, is shown on the top row of the table; the bottom row is the power with which to raise 2 to obtain the number, e.g. $2^3 = 8$.

n	−2	−1	0	1	2	3	4	5	6	7	8
2^n	1/4	1/2	1	2	4	8	16	32	64	128	256

Calculating 1/4 multiplied by 128 can be done directly of course, but we want instead to avoid multiplication. As 1/4 has a power $n = -2$, and 128 has a power of $n = 7$, adding the powers and looking up the answer under 5, we get 32. This is written as $128/4 = 2^7/2^2 = 2^7 \times 2^{-2} = 2^5 = 32$. The idea behind the logarithm, therefore, is to use addition to solve multiplications and subtraction to perform division, and this is because logs relate to powers of numbers. The calculation 128/4 done with logs is

$$\log_2(1/4) + \log_2(128) = -\log_2(4) + \log_2(128) = 5,$$

and as $5 = \log_2(32)$ the result is 32.

If $a > 0$ and $b > 0$ the first two laws of logs are

$$\log(a) + \log(b) = \log(a \times b) \tag{1.10}$$

$$\log(a) - \log(b) = \log\left(\frac{a}{b}\right) \tag{1.11}$$

and these relationships are true no matter what the base is. Note that this rule does not apply to $\dfrac{\log(a)}{\log(b)}$, which cannot be simplified.

The 'third law' of logs is:

$$\log(x^n) = n\log(x). \tag{1.12}$$

but note that $\log(x)^n$ means that the log of the number is raised to n; for example, $\log(x)^3 = \log(x)\log(x)\log(x)$, whereas $\log(x^3) = 3\log(x)$.

The number $2^8 = 256$ written in logarithmic form is $8 = \log_2(256)$, which means that '8 equals log base 2 of 256' and 'base 2' means that we are raising numbers to powers of 2. The general formulae relating a number N with base b and a power p are:

$$N = b^p \qquad \text{or} \qquad p = \log_b(N) \tag{1.13}$$

Natural (Napierian) logs, usually written as ln, or $\log_e$, use e as the base number with which we raise to a power, and log or $\log_{10}$ use 10 as the base, for example because $10^3 = 1000$ we can write this as $3 = \log_{10}(1000)$. The number $e^3 = 20.086$ can be written as $3 = \ln(20.086)$. The series of powers of 10 and e are shown below:

n	−2	−1	0	1	2	3	4
10^n	0.01	0.1	1	10	$10^2 = 100$	1000	10000
e^n	$e^{-2} = 0.135$	0.368	$e^0 = 1$	$e^1 = 2.718$	$e^2 = 7.389$	$e^3 = 20.086$	54.598

The change in population of a chemical species $c(t)$ decaying by a first-order process follows an exponential law, $c(t) = c_0e^{-kt}$. This is more easily analysed to find the rate constant k by taking logs of both sides to give $\ln(c(t)/c_0) = -kt$ because a plot of the ln vs time is a straight line of slope $-k$. Dividing the concentration $c(t)$ by that initially present c_0 makes the log dimensionless.

When Napier invented logarithms in *c.*1614, he used bones inscribed with numbers. The old-fashioned slide rule uses the principle of addition and subtraction to do multiplication and division with lines marked on rulers that slide past one another. The slide rule is quite easy to use with practice, and not that much slower, but less accurate, than a hand calculator.

1.4.1 Changing the base of logs

By definition

$$x = b^{\log_b(x)}$$

but if the log to a different base a is needed, then taking logs of both sides gives

$$\log_a(x) = \log_a(b)\log_b(x). \tag{1.14}$$

When we want to change the base of logs, such as when plotting $\log_{10}$ graphs but want $\log_e$ answers, which is common in chemical kinetics, the formula to use is

$$\ln(x) = \ln(10)\log(x). \tag{1.15}$$

which is $\ln(x) = 2.30258\log(x)$.

1.4.2 Summary of logs and powers

Definition $x = b^{\log_b(x)}$

$a^0 = 1$	$\log(0) = -\infty$	$\log(1) = 0$	$\log(\infty) = \infty$
$a^{n+m} = a^m a^n$	$\log(a) + \log(b) = \log(a \times b).$		if $a > 0, b > 0$
$a^{n-m} = a^m a^{-n} = a^m/a^n$	$\log(a) - \log(b) = \log\left(\dfrac{a}{b}\right).$		if $a > 0, b > 0$
$(a^m)^n = a^{mn}$	$\log(a^n) = n\log(a)$		if $a > 0, n \neq 0$
	$\log_a(x) = \log_a(b)\log_b(x)$		

1.4.3 Comparison of log and exponential

The graph in Fig. 1.8 shows the e^x, e^{-x}, and $\ln(x)$ functions. Notice how the exponential and log functions are symmetrical about the line $y = x$. The curve that would be symmetrical about e^{-x} is $\ln(|x|)$ where $|x|$ means taking the absolute value, which if x is real means changing $-x$ into x.

The log is normally defined only over the range of positive x values $0 < x < \infty$, $\ln(1) = 0$ and $\ln(0) = -\infty$. The log with negative x values is a complex number; see Chapter 2.8. The exponential is defined over all values of x and has the values $e^0 = 1$, $e^{-\infty} = 0$ and $e^{\infty} = \infty$. It is worth the effort to memorize Fig. 1.8.

1.4.4 Log as a series

Just as with the exponential, the log can be written as a series expansion, but only if $|x| < 1$:

$$\ln(1 + x) = x - \frac{x^2}{2} + \frac{x^3}{3} - \frac{x^4}{4} + \cdots = \sum_{n=1}^{\infty}(-1)^{n+1}\frac{x^n}{n},$$

$$\ln(1 - x) = -x - \frac{x^2}{2} - \frac{x^3}{3} - \frac{x^4}{4} - \cdots = -\sum_{n=1}^{\infty}\frac{x^n}{n},$$

and where the $(-1)^{n+1}$ makes the even-valued terms in n negative. If we want to calculate $\ln(q)$, we can make $q = 1 + x$ and substitute $x = q - 1$. This and other series are described in Chapter 5.

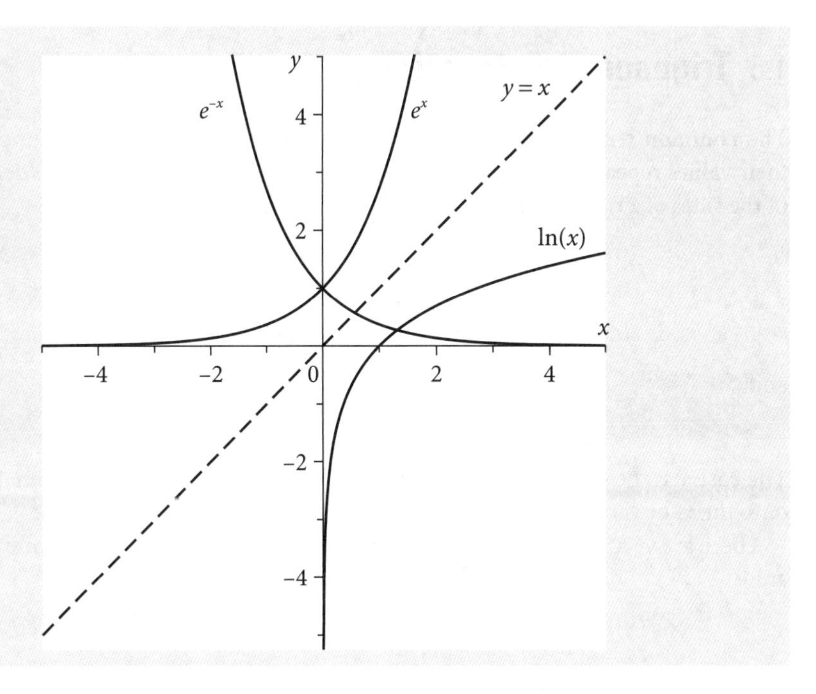

Fig. 1.8 Graph of exponential and log functions with the straight line $y = x$.

1.4.5 Questions

Full solutions are available at **www.oxfordtextbooks.co.uk/orc/beddard**.

Q1.1 Calculate the relative populations of two energies separated by 1 cm^{-1} and then by 1500 cm^{-1}, at both 30 and 300 K. If these energy levels are found in a molecule, what two different types of motion are implicated?

Q1.2 Calculate without using your calculator;

(a) $\log_{10}(100)$, **(b)** $\log_{10}(10)$, **(c)** $\log_e(e)$, **(d)** $\log_b(b)$, **(e)** $\log_b(b^x)$, **(f)** $\log_2(1/8)$, **(g)** Find n in $\log_{10}(n) = -4$, **(h)** Find n in $\log_n(1/16) = -2$, **(i)** Find n in $\log_5(125) = n$.

Strategy: Solve using equation (1.13).

Q1.3 Simplify without using your calculator

(a) $\ln(3^x)$, **(b)** $\ln(a^2/b)$, **(c)** $\ln(a^{1/2}/b^3)$, **(d)** $\ln(3^{1/(x+6)})$, **(e)** $\ln\left(\dfrac{x^{1/2}\sin(x)}{\ln(x^n)}\right)$.

Strategy: Use equations to (1.10) to (1.13).

Q1.4 Express $(1 + s/a)^r$ in exponential form.

Strategy: Take $\log_e$ of the expression.

Q1.5 Evaluate as far as possible without using a calculator:

(a) $\log_{10}(10^6)$, **(b)** $\log_e(e^5)$, **(c)** $\ln(e^2)^3$, **(d)** $\log_3(n^6)$.

Q1.6 A man drinks $1/n^{\text{th}}$ of a pint or beer and replaces this volume with lemonade. Supposing that he could do this an infinite number of times, is any beer left in the glass and if so how much?

Q1.7 The chlorophyll in the leaves on a tree absorbs light but some will still reach the ground, even in a dense forest. The cross section for the absorption by one leaf at a certain wavelength, is $\sigma = \alpha A$ where α is the absorption per unit area and A the area of a leaf. If I_0 is the incident light intensity, and I_n that after n layers of leaves, what fraction of light (I_n/I_0) reaches the ground and how much when $n \to \infty$? Assume that a sufficient number of leaves are present so that no light reaches the ground without passing through a leaf.

Strategy: The light absorbed plus transmitted, equals that incident.

1.5 Trigonometric functions

The common trigonometric functions, sine and cosine, are circular functions because their values repeat as x is increased. Together with the tangent, they are defined as ratios of the sides of a right-angled triangle and are illustrated in Fig. 1.9.

$$\sin(\theta) = opposite/hypotenuse = p/h;$$
$$\cos(\theta) = adjacent/hypotenuse = a/h$$
$$\tan(\theta) = opposite/adjacent = p/a.$$

and

$$\tan(x) = \sin(x)/\cos(x).$$

The cosine is the *projection* of the hypotenuse onto the x-axis, and the sine the projection onto the y- or vertical axis. The tangent is the gradient of the hypotenuse.

There are very many relationships between the trig functions; some of the most useful are

$$\cos^2(x) + \sin^2(x) = 1, \qquad \cos^2(x) - \sin^2(x) = 2\cos^2(x) - 1 = \cos(2x)$$

$$\cos^2(x) = \frac{1}{2}[1 + \cos(2x)], \qquad \sin^2(x) = \frac{1}{2}[1 - \cos(2x)].$$

$$\sin(2x) = 2\sin(x)\cos(x),$$

Many more are known and can easily be produced using Maple; a couple are shown below.

```
> expand( sin(x+y) );
```

$$\sin(x)\cos(y) + \cos(x)\sin(y)$$

```
> expand( cos(x+y) );
```

$$\cos(x)\cos(y) - \sin(x)\sin(y)$$

The sine, cosine, and tangent functions can also be described as series expansions, see Chapter 5, and as complex exponential functions. The equations are given here but described in detail in Chapter 2, which describes complex numbers.

$$\sin(x) = \frac{e^{+ix} - e^{-ix}}{2i}, \qquad \cos(x) = \frac{e^{+x} + e^{-ix}}{2}.$$

The word tangent derives from the Latin *tangere*, which means to touch, and this is also the usage when the tangent is used to define the gradient of a curve at a given point, Fig. 1.10. If the curve is a circle, the tangent is perpendicular to the radius.

Although hand-held calculators usually use degrees to calculate trigonometric functions, this disguises the fact that these functions require the angle to be in radians. A radian is the angle subtended at the centre of a circle by travelling a distance around the circumference that is equal to its radius, thus a complete rotation of $360° \equiv 2\pi$ radians and 1 radian is approximately 57.5°. Radians are not always included as units; the exception is in the study of NMR where the magnetogyric ratio γ, which is the constant that relates the magnetic dipole to the magnetic field, is always quoted in rad T^{-1} s^{-1}, the T here being tesla.

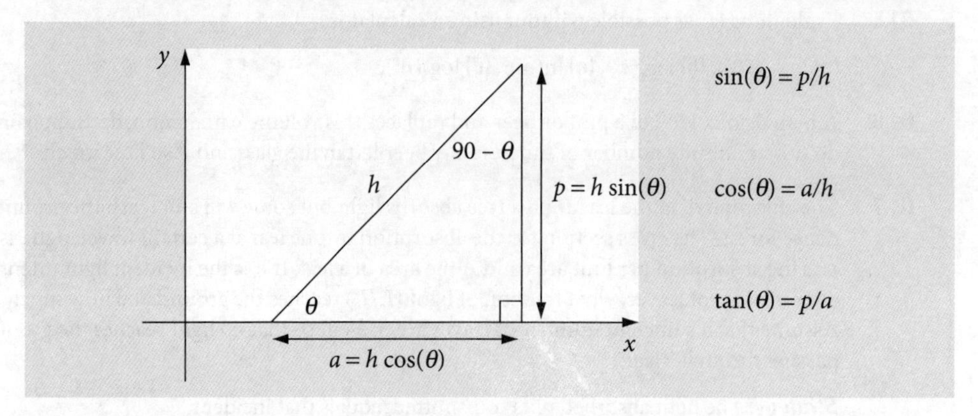

Fig. 1.9 Trigonometric relationships in a right-angled triangle and relationships to the x- and y-axes.

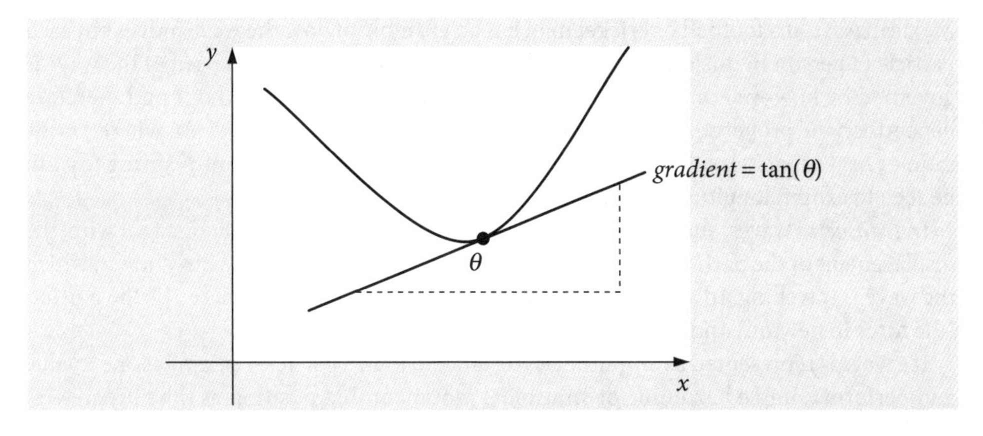

Fig. 1.10 The tangent touches the curve at a point and defines the gradient at that point.

In spectroscopy generally, frequencies are usually quoted either as s^{-1}, with the symbol ν, (nu), or as angular frequencies, ω in rad s^{-1}. These are related as $2\pi\nu = \omega$.

1.5.1 **Trigonometric relationships**

There are many useful identities between sine and cosines. Some of these are given below.

$\sin(-a) = -\sin(a)$ $\qquad$ $\cos(-a) = \cos(a)$

$\cos^2(a) + \sin^2(a) = 1$

$\cos(2a) = \cos^2(a) - \sin^2(a) = 2\cos^2(a) - 1 = 1 + 2\sin^2(a)$

$\sin(2a) = 2\sin(a)\cos(a)$

$\sin(a \pm b) = \sin(a)\cos(b) \pm \cos(a)\sin(b)$ $\qquad$ $\cos(a \pm b) = \cos(a)\cos(b) \mp \sin(a)\sin(b)$

$2\sin(a)\sin(b) = \cos(a - b) - \cos(a + b)$ $\qquad$ $2\cos(a)\cos(b) = \cos(a - b) + \cos(a + b)$

$2\cos(a)\sin(b) = -\sin(a - b) + \sin(a + b)$ $\qquad$ $2\sin(a)\cos(b) = \sin(a - b) + \sin(a + b)$

If you want to generate more relationships use Maple and `expand(..)` or `combine(..)`, for example

```
> expand(tan(2*a)^2);
```

$$\frac{4\tan(a)^2}{(1 - \tan(a)^2)^2}$$

1.5.2 **Waves**

In electromagnetic radiation, such as X-rays or radio waves and, more familiarly, visible light, the periodic displacement of the electric and magnetic fields is described by a sine or cosine function. Similarly, sine and cosine describe the periodic displacement of molecules as a sound wave travels through a gas, liquid, or solid, and describe a molecule's internal vibrational normal modes. Any motion described by a single sine or cosine is called simple harmonic motion (SHM), for example, the motion of a pendulum. A spring is often called a harmonic oscillator and this term is also used to describe molecular vibrations. Mathematically, waves form the basis of the Fourier transforms and Fourier series method described in Chapter 9.

A sine wave is defined by its wavelength, amplitude, and phase. The amplitude is its maximum displacement from zero. The wavelength is one full or two half periods and is the distance from one node to the next but one. A node is any of the points where the wave has zero amplitude.

In a *longitudinal* wave, such as a sound wave, the gas molecules are made to move about their equilibrium position in simple harmonic motion by the disturbance generating the sound. This simple harmonic motion occurs along the direction that the wave travels. The wave moves through the gas because molecules interact with one another causing regions of increased pressure and rarefaction. Even though each molecule only moves about its equilibrium position as it does so, it influences its neighbour and the wave moves. A

Mexican wave at a football or cricket match is an example of how the wave moves but each particle (a person in this case) does not move from equilibrium. The speed of the wave is governed by how well one molecule (person) interacts with its neighbour and therefore depends on the properties of the medium. In a gas, the speed is $v = \sqrt{\gamma RT/M}$ where γ is the ratio of heat capacities at constant pressure and volume. A helical 'slinky' spring toy can be seen to exhibit longitudinal waves.

In *transverse* waves, such as a water wave, electromagnetic radiation, or a taut wire, the displacement of the particles about equilibrium is at right angles to the direction in which the wave is travelling. In a taut wire the wave's velocity is $v = \sqrt{T/\mu}$ where T is the tension (the force in newton) and μ the mass per unit length.

If a wave is represented as amplitude vs distance x, then a particle remains at one x value and performs simple harmonic or sinusoidal motion at that position as time progresses. Its displacement amplitude is

$$y = A\sin(kx + \varphi).$$

The term k is sometimes called the wavevector and is $k = 2\pi/\lambda$ where λ is the wavelength. The phase is φ (radians) which might be zero, is a constant displacement of the wave. At zero x the amplitude is $A\sin(\varphi)$. Sine and cosine always have a fixed phase difference between them of $\pi/2$ or 90°, thus a sine wave is changed into a cosine, and vice versa by a $\pi/2$ phase change.

If the plot is one of amplitude vs time then the wave represents how the particle moves in time at a given position. The wavelength is now the period T of the motion and $T = 1/\nu$ where ν is the frequency in s^{-1}. The *angular* frequency is $\omega = 2\pi\nu$, in radian s^{-1}, hence the following equations are equivalent

$$y = A\sin(\omega t + \varphi) = A\sin(2\pi\nu t + \varphi).$$

The velocity of the wave v is the product of frequency and wavelength; $v = \nu\lambda = \omega/k$.

The general equation for a travelling wave, either transverse or longitudinal, is

$$y = A\sin(\omega t - kx + \varphi). \tag{1.16}$$

In this equation the sign $-kx$ means that the wave is travelling to the right, which is positive x. If the term were written $+kx$, then the wave would be travelling to the left.

Depending upon whether x or t is plotted the similar but different plots in Fig. 1.11 are produced. On the left the displacement x is plotted and therefore as time progresses the particle moves up and down at fixed x. On the right the displacement vs time at a fixed x is plotted.

The phase φ displaces the wave but does not change its frequency or wavelength. The dashed curve, Fig. 1.12 has $\varphi = +2/3$ radians compared the solid curve where the phase is zero. Note that a positive phase displaces the curve to the left compared to zero phase.

When two waves are added they can reinforce or destroy one another depending on their phase. Clearly, when they are in phase, the phase being zero or integer multiples of 2π, the waves add and a double-sized wave is produced. When 180° or integer multiples of π out of phase destructive interference occurs.

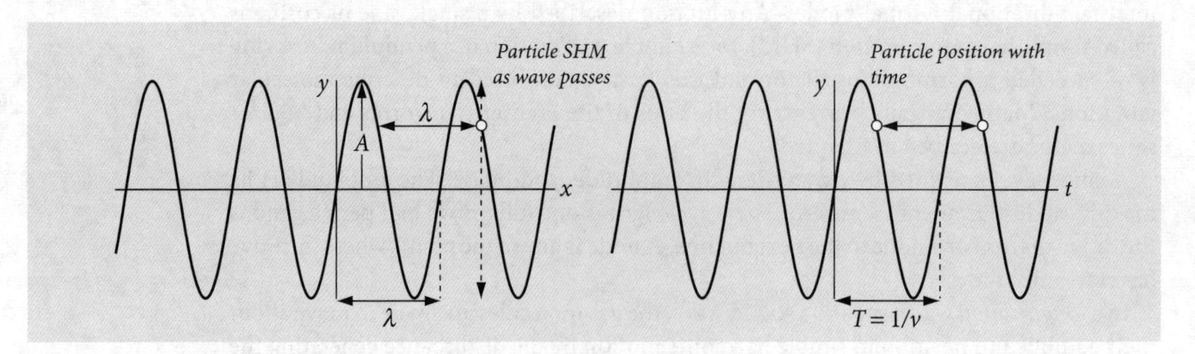

Fig. 1.11 Left: The wave on the left is $y = \sin(2\pi x/\lambda)$. The particle (circle) remains at a fixed x but is displaced and follows SHM motion as the wave passes. After one period, it is at the same displacement. Right: The particle's motion is shown with time; the wave is $y = \sin(2\pi\nu t)$. After one or any whole number of periods the particle (circle) is again at the same displacement y.

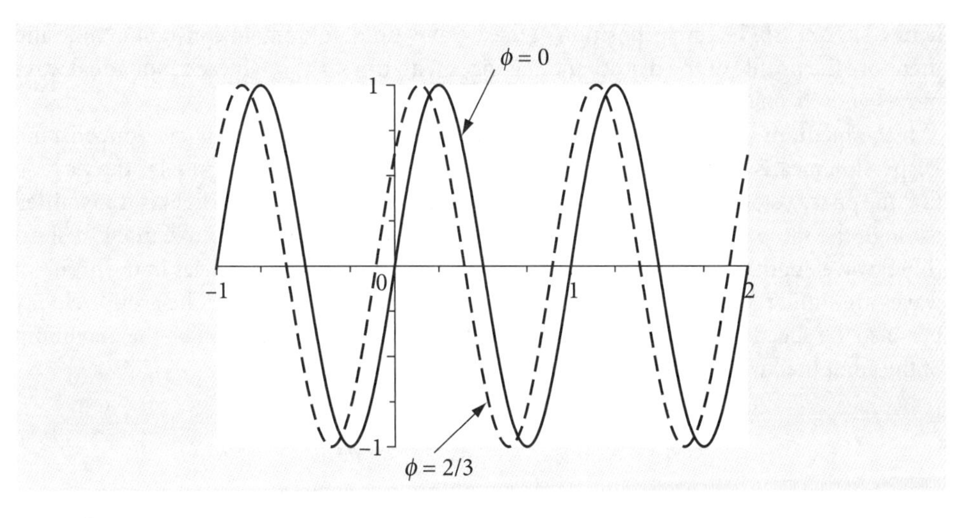

Fig. 1.12 Comparison of $\sin(2\pi\nu t + 2/3)$ with $\nu = 1$ (dashed line) compared to the sine wave with zero phase.

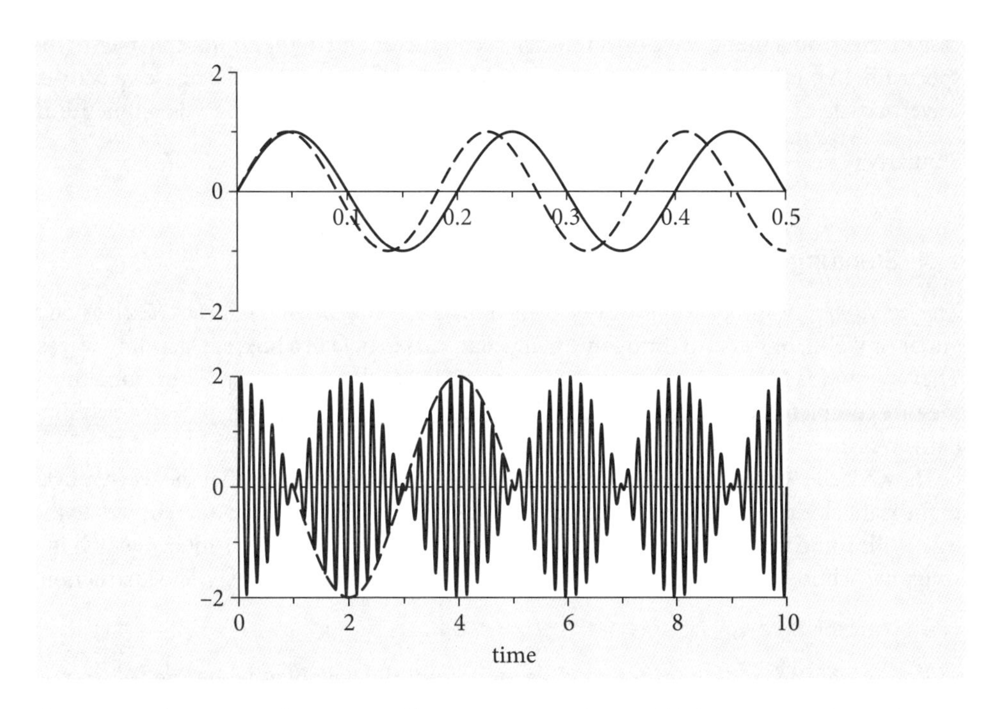

Fig. 1.13 Top: the two travelling waves of frequency $\nu = 5$ and 5.5 quickly become out of step with one another. Bottom: the sum of the wave shows beating with a frequency of 0.5 Hz. (Note that different abscissa are used).

If the waves are of different frequency then they become out step with one another, as shown in Fig. 1.13. However, the two waves will periodically have places where both waves have zero amplitude. The effect is called *beating* and in a sound wave this is heard as a repeating hum, hence 'beats', when the second sound wave is turned on. The beat frequency is the difference in frequency of the two individual waves and is *twice* the lowest frequency of the combined wave. The combined wave has frequencies that are half the sum and half difference of the two fundamental waves. Fig. 1.13 shows this effect on waves with $\nu = 5$ and 5.5 Hz. The beating is clear when the waves are added together.

1.5.3 Phase and group velocity

A wave is defined with an angular frequency $\omega = 2\pi\nu$ and wavevector $k = 2\pi/\lambda$ and has the general form of equation (1.16). The *phase velocity* is what we would colloquially describe as the velocity of a wave and is $v = \omega/k$.

The *group velocity* is also a measure of the speed of a wave-like disturbance. It applies only when a wave is composed of many individual waves of different frequencies and is defined as $\Delta\omega/\Delta k$. The superposition of these waves does not remain constant in time and therefore the profile of the disturbance changes with time, although each individual wave moves with its phase velocity.

If the medium is *dispersive*, as is air or glass to visible light, then ω/k is not constant. Dispersion means that the medium through which the wave is passing affects the wave so that the phase velocity of the wave depends on the frequency; thus, red light is transmitted through the same piece of glass in less time than blue light. The refractive index at short (blue) wavelengths is larger than at longer wavelengths. If the frequencies in the group of waves are similar, then the change $\Delta\omega/\Delta k \to d\omega/dk = v_g$ can be made. The group velocity v_g is also the speed at which energy is transmitted and is the speed at which the maximum of the combined wave moves. As the *phase* velocity is $v = \omega/k$ then

$$v_g = \frac{d\omega}{dk} = v + k\frac{dv}{dk} = v - \lambda\frac{dv}{d\lambda}.$$

If the gradient $\frac{dv}{d\lambda} > 0$ then the medium exhibits what is called *normal dispersion* (e.g., visible light in glass); when $\frac{dv}{d\lambda} < 0$ it is called *anomalous dispersion*. This is normally the case for electromagnetic waves in an electrical conductor and in the far infrared part of the spectrum. In quantum mechanics, it is sometimes useful to calculate the phase velocity of wave/particle as $v = E/p$ because the kinetic energy is $E = p^2/2m$ where p is momentum. Similarly $v_g = \frac{d\omega}{dk} = \frac{dE}{dp}$ is the group velocity.

1.5.4 Standing waves

The *fundamental* and *overtone* waves formed in a taut wire when it is struck, such as on a guitar or violin, or those in the quantum mechanical particle in a box, are standing waves. They rise and fall in time with their nodes remaining at the same place. Thus, sometimes they are completely zero everywhere, and at other times they are maximal. When two similar waves travelling in opposite directions are added, a standing wave is formed. This can be shown if the waves are written as $y_1 = \sin(k(x - vt))$, which means that the wave moves to the right along the x-axis with velocity v, and the wave $y_2 = A\sin(k(x + vt))$ moves to the left. With some manipulation the sum of these waves is $y = 2\cos(kvt)\sin(kx)$, which is a sine wave whose amplitude rises and falls in time with frequency $\omega = kv$. The instruction

```
> animate( plot, [A*sin(2*x), x=-Pi..Pi], A=-1..1 );
```

will show what a standing wave looks like as its amplitude changes with time. When you see the graph, click on it and the animation bar appears at the top of the Maple worksheet.

In practice, if you strike the strings of a violin or piano the motion produced is complex because many fundamental frequencies (called normal modes) are simultaneously excited when the instrument is played. Some calculations that illustrate this are presented in Chapter 7.

1.5.5 Hyperbolic functions

The hyperbolic functions are closely related to the circular functions, sine, cosine, etc. of trigonometry. The equation of a circle is $x^2 + y^2 = 1$ and in parametric form $\cos(\theta) = x$, $\sin(\theta) = y$ and any point has coordinates $\{\cos(\theta), \sin(\theta)\}$. The equation for a hyperbola is $x^2 - y^2 = 1$ or, in parametric form, $\cosh(t) = x$ and $\sinh(t) = y$, but t is not a measure of angle. The two curves produced are shown in Fig. 1.14, and one point plotted on each, at a value of θ or t of $2\pi/3$. The hyperbolic functions are often pronounced, 'shine', 'cosh' and 'than', with emphasis on the 'th'. The shape of a freely hanging cable supported at its ends only, follows the equation $y = \cosh(x)$, and is called a catenary. The arch in St Louis in the United States follows closely that of an inverted cosh.

The equations of the hyperbolic functions are similar to those of normal trig functions, but not identical, for example;

$$\tanh(x) = \frac{\sinh(x)}{\cosh(x)}, \qquad \cosh^2(x) - \sinh^2(x) = 1, \qquad \cosh^2(x) + \sinh^2(x) = \cosh(2x)$$

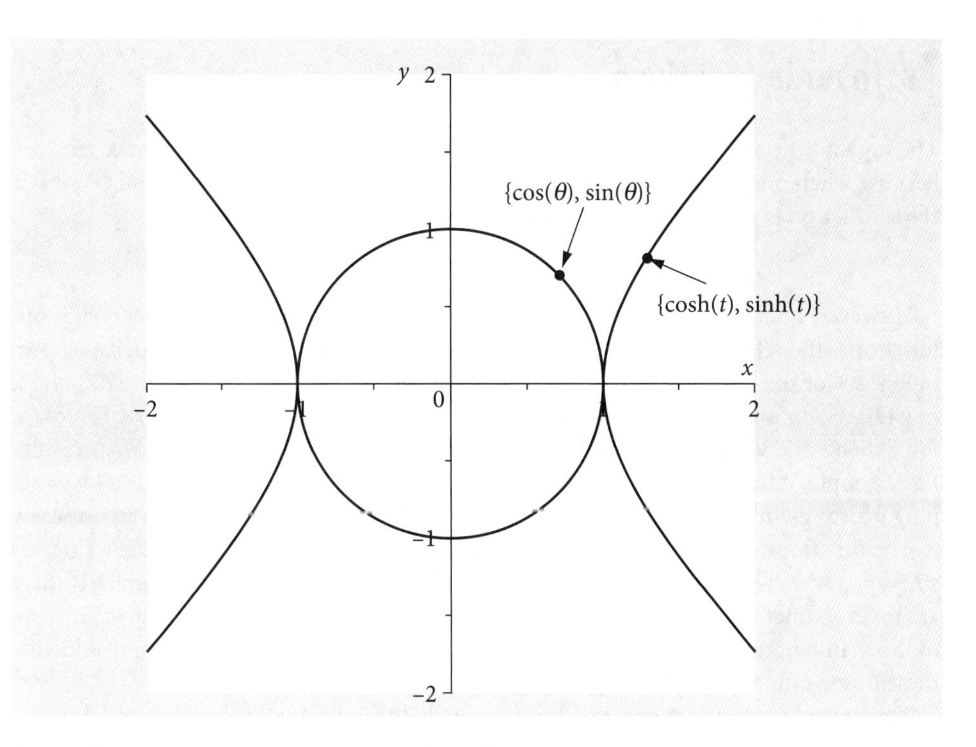

Fig. 1.14 Circle and hyperbola. The point is at θ or t of $\pi/4$.

and the last identity can be checked using Maple

```
> simplify( cosh(x)^2 + sinh(x)^2 - cosh(2*x) );  0
```

The circular functions sine, cosine, and tangent can be represented as real functions involving *complex* exponentials, see Chapter 2. The hyperbolic functions sinh(x), cosh(x), and tanh(x) are also real functions and can also be represented as *real* exponentials. The hyperbolic functions are

$$\sinh(x) = \frac{e^x - e^{-x}}{2} \qquad \cosh(x) = \frac{e^x + e^{-x}}{2} \qquad \tanh(x) = \frac{e^x - e^{-x}}{e^x + e^{-x}}$$

Some values are

$$\begin{aligned} \sinh(0) &= 0, & \sinh(\infty) &= \infty \\ \cosh(0) &= 1, & \cosh(\infty) &= \infty \\ \tanh(0) &= 0, & \tanh(\pm\infty) &= \pm 1 \end{aligned}$$

The functions are plotted in Fig. 1.15 and compared to the normal trig functions.

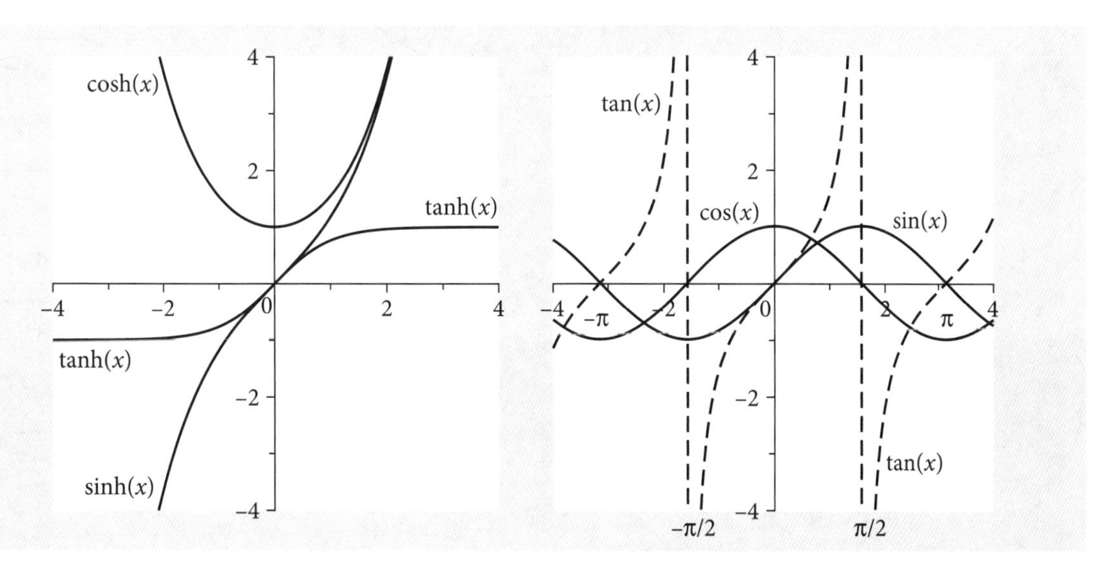

Fig. 1.15 Some of the hyperbolic and normal trigonometric functions. Notice how the tan(x) goes to $\pm\infty$ in multiples of $x = \pm\pi/2$.

1.6 Inverse functions

The log and exponential functions are inverses of one another, and similarly x^n and $x^{1/n}$ because when one is made a function of the other, the result is x; $\ln(e^x) = x$ and $(x^{1/n})^n = x$. If the function is written as $f(x)$ then the inverse is $f^{-1}(x)$, therefore

$$f^{-1}[f(x)] = x.$$

However, not all functions have an inverse. When plotted the observation is that one function is the reflection of the other through a straight line of gradient ±1, such as shown in Fig. 1.8 for the positive exponential and log. More formally, the function and its inverse must have a one-to-one relationship. If an inverse exists when the function $f(x)$ is solved for x, then x and y exchanged and the result substituted into the original equation, x results. For example, if $f(x) \equiv y = (x+2)/3$ then $x = 3y - 2$, the inverse function is found by swapping y for x, giving $f^{-1}(x) = 3x - 2$, and then substituting back $f^{-1}[f(x)] = 3(x+2)/3 - 2 = x$.

The function $y = \sin(x)$ has $y = \sin^{-1}(x)$ as its inverse function, which can be written as $\sin(y) = x$, which is a sine wave moving up the y-axis, as shown in Fig. 1.16. Similarly there are inverse functions $\cos^{-1}(x)$ and $\tan^{-1}(x)$. Notice that the computer notation is different to the mathematical one; for example, $\sin^{-1}(x) \equiv \arcsin(x)$, the prefix 'arc' being added to these inverse functions.

The inverse trig functions are multi-valued; the principal range for $\sin^{-1}(x)$ and $\cos^{-1}(x)$ is $x = \pm 1$ and the function's principal values lie between $y = \pm\pi/2$. The inverse function $\tan^{-1}(x)$ is also multi-valued and ranges from $x = \pm\infty$ and has principal values between $y = \pm\pi/4$.

The equations defining inverse trig and hyperbolic functions are shown below, and were obtained using Maple, where I is used instead of i to represent $\sqrt{-1}$.

$$\arcsin(x) = -\mathrm{I}\ln(\sqrt{1-x^2} + \mathrm{I}\,x)$$

$$\arccos(x) = \frac{1}{2}\pi + \mathrm{I}\ln(\sqrt{1-x^2} + \mathrm{I}\,x)$$

$$\arctan(x) = \frac{1}{2}\mathrm{I}\,(\ln(1 - \mathrm{I}\,x) - \ln(1 + \mathrm{I}\,x))$$

$$\mathrm{arcsinh}(x) = \ln(x + \sqrt{x^2+1})$$

$$\mathrm{arccosh}(x) = \ln(x + \sqrt{x-1}\sqrt{x+1})$$

$$\mathrm{arctanh}(x) = \frac{1}{2}\ln(x+1) - \frac{1}{2}\ln(1-x)$$

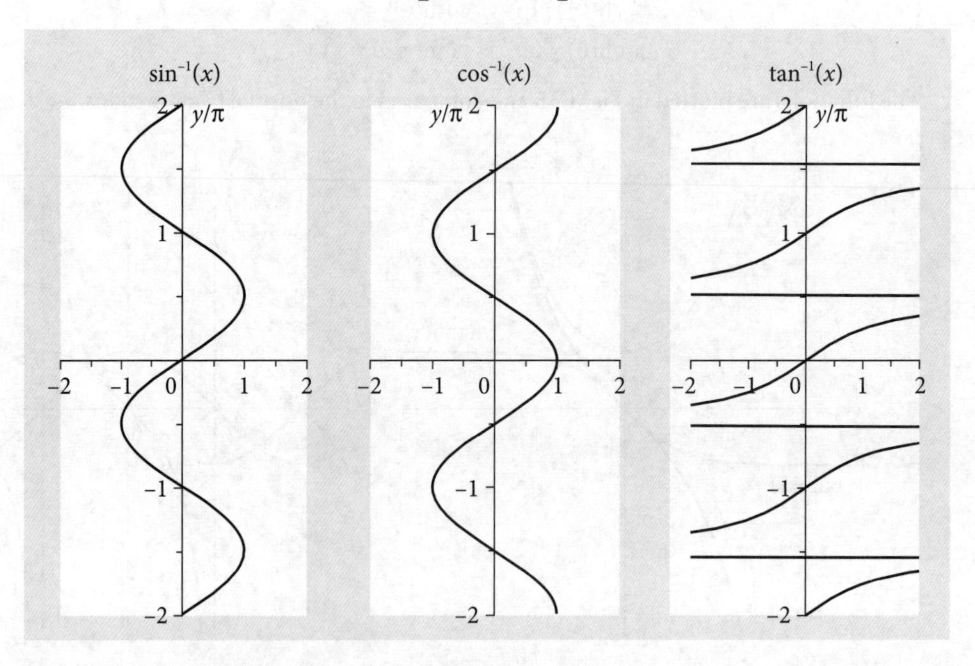

Fig. 1.16 Inverse trig functions with y limited to the range $\pm 2\pi$. The $\tan^{-1}$ functions extend to $x = \pm\infty$ with a limit of integer multiples of $y = \pm\pi/2$.

1.7 Cartesian and polar coordinates

When drawing a graph a right-angled set of axes is normally used. The origin is the point {0, 0} and then any other point is represented by two numbers $\{x, y\}$ in the x–y plane. In three dimensions the coordinates are $\{x, y, z\}$. These coordinates are described as rectilinear when right-angled axes are used and as Cartesian when $\{x, y, z\}$ are used.

There are other ways of defining a point in space and some of them will be used when the situation requires it. In crystallography, for example, it is often sensible to define axes of particular crystal types that are not always at 90° to one another and of unequal length. Monoclinic and trigonal crystal types, among others, have these 'sloping' axes. On the surface of a sphere, such as the earth, there are no straight lines, and latitude and longitude, which are angles, are used to define a location. In other situations, for example defining the three-dimensional shapes of atomic orbitals, *spherical polar* coordinates are used and in these coordinates, two angles and a radius define a point. These coordinates are used because they reflect the underlying symmetry of the problem and equations become easier to solve. In the hydrogen molecule, *prolate spheroidal* coordinates are used. These have two origins, one at each atom, see Fig. 4.30 for a sketch of coordinates. In other situations, *cylindrical* coordinates are used. There are many other coordinate systems, see Arkfen (1970) for a description.

The reason for using different coordinate systems is not to make life difficult, far from it, but is that calculations are often easier done in these coordinates than in Cartesians; the disadvantage of these other coordinate systems is only unfamiliarity.

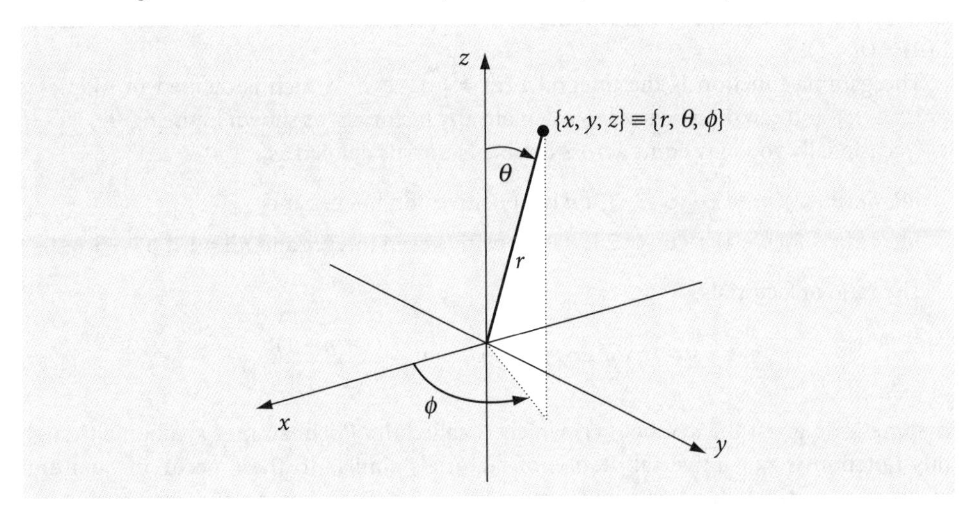

Fig. 1.17 Spherical polar coordinates. The point at $\{x, y, z\}$ in Cartesian coordinates is at $\{r, \theta, \varphi\}$ in spherical polar coordinates. The azimuthal angle φ is in the x–y plane and measures the angle from the x-axis to the line formed by the projection of the point onto this plane, and has a maximum of 360°. θ is the polar angle measured from 0 to 180° from the z-axis.

In each case, it is possible to convert from one coordinate 'system' to another. In spherical polar coordinates the radius r is the distance of a point from the origin, θ is the polar angle measured from the z-axis and φ the azimuthal or equatorial angle measured in the x–y plane from the x- to the y-axis. The conversion between Cartesian and spherical polar coordinates is

$$x = r\sin(\theta)\cos(\varphi), \qquad y = r\sin(\theta)\sin(\varphi),$$
$$z = r\cos(\theta) \qquad \text{and} \qquad r^2 = x^2 + y^2 + z^2.$$

In two dimensions, *plane polar* coordinates are used to define a point in the x–y plane. If z is included then these coordinates become the cylindrical coordinates. The relationship of plane polar to Cartesian coordinates is

$$x = r\cos(\theta), \qquad y = r\sin(\theta), \qquad \text{and} \qquad r^2 = x^2 + y^2.$$

The angle θ is measured from the x- to the y-axis.

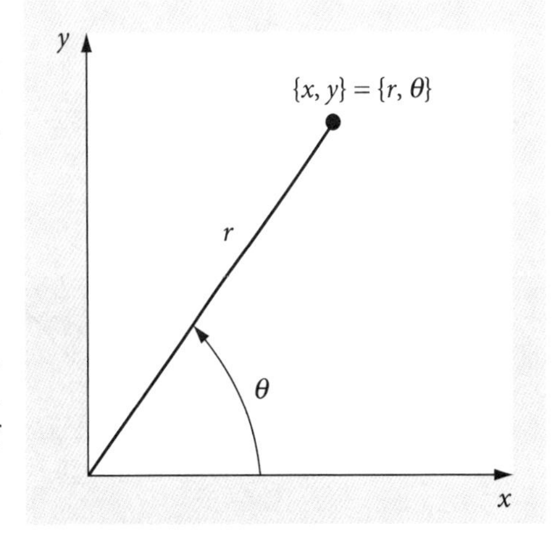

Fig. 1.18 Plane polar coordinates.

A circle of radius α in Cartesian coordinates is the equation $y=\sqrt{\alpha^2-x^2}$; in plane polar coordinates it is $r=\alpha$. To see this use Maple to plot the two curves,

```
plot([ sqrt(4-x^2),-sqrt(4-x^2)], x=-2..2, view=[-2..2,-2..2]);
plot(2,x=0..2*Pi, coords=polar, view=[-2..2,-2..2]);
```

In the first case, both the positive and negative parts of the square root have to be plotted to form a circle. In the second case, the `coords=polar` option is used to ensure that the correct type of coordinates are used.

1.8 Factorials

The factorial function is the product of a set of *positive integers* from 1 up to n;

$$n! = n\times(n-1)\times(n-2)\times\cdots\times 2\times 1$$

with the limits that

$$0! = 1, \qquad (-n)! = \infty.$$

The factorials can also be calculated recursively as $(n+1)! = (n+1)n!$. In Fig. 1.19 notice how rapidly the factorials increase as n increases. The plot is of $\ln(n!)$ vs n, which shows that $n!$ increases more rapidly than does e^x, which would be a line of slope 1 on this graph.

Fractional, negative, and complex factorials are defined by the (complete) gamma function $\Gamma(n)$. Positive integer values of this function produce the factorials, and are related as $\Gamma(n) = (n-1)!$.

The gamma function is the integral $\Gamma(z) = \int_0^\infty t^{z-1}e^{-t}dt$, which is defined in Maple as `GAMMA(x)` and can then be used to calculate any factorials, as shown in the figure.

Occasionally you may come across double factorials defined as

$n!! = n(n-2)(n-4)\cdots 5\cdot 3\cdot 1$ if n is a positive odd integer and

$n!! = n(n-2)(n-4)\cdots 6\cdot 4\cdot 2$ if n is a positive even integer and $n!! = 1$ if n is -1 or 0.

The ratio of factorials

$$x(x+1)(x+2)\cdots(x+n-1) = \frac{(x+n-1)!}{(x-1)!} \tag{1.17}$$

is sometimes given the symbol $(x)_n$ which is called the Pochhammer symbol, although this notation is not universal. Ratios of factorials similar to these occur in quantum mechanics, particularly when angular momentum quantum numbers are involved.

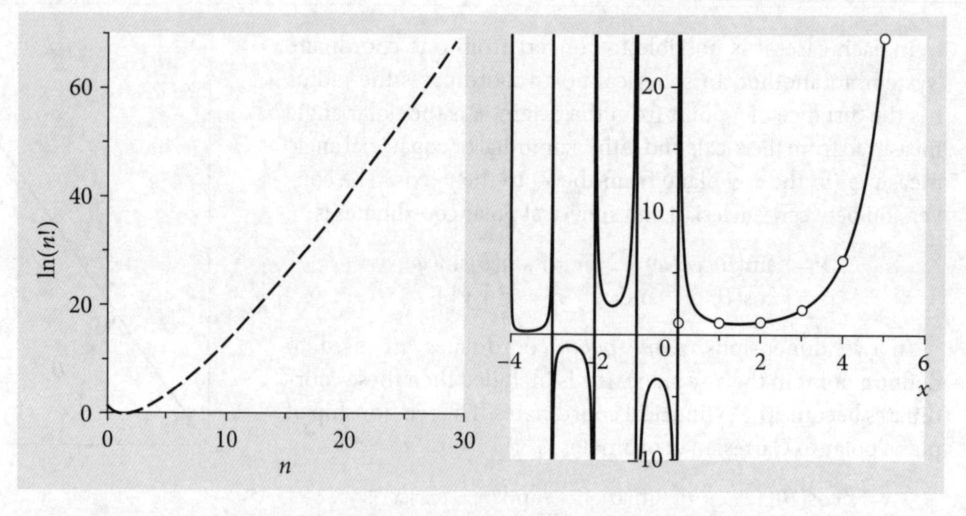

Fig. 1.19 Left: ln(n!) vs n. Right: The gamma function (solid line) and the factorial (circles).

The log of the factorial is often used as a way of calculating factorials and Stirling's formula

$$\ln(n!) = n\ln(n) - n \tag{1.18}$$

is particularly useful for large n.

1.8.1 Recurrence

Recurrence is a general method by which certain functions, such as a factorial or a polynomial, can be calculated by starting with one or two values and using these as a seed to generate further values. The recursion equations are sometimes also called difference equations. The algorithms to calculate the square root, e, and π are each recursive because a previous result is used to calculate the next one. The factorial is the product of numbers separated by unit values, so starting with one value and an increment, all the rest can be calculated up to a given number. The equation is $(n + 1)! = (n + 1)n!$. An algorithm to calculate factorials up to a value of m, is

Algorithm 1.5 Recursive calculation of factorials

```
> m:= 6:
  n:= 1;
  for I from 1 to m-1 do
    n:= n*(i+1);
    print(I, n);
  end do:
```

where n returns the factorial. If you want a 'pretty' output, print using the instruction

```
printf(" %d %g\n", I , n);
```

The algorithm sets an initial value of n and increments this by multiplying by `i+1` each time around the `for .. do` loop. The last calculated value of n is then used in the next calculation. Notice the use of the assignment `n:= n*(n+1);` which means that the new value of `n` is `n*(n+1)`. This is not the way the equation is written, which is $(n + 1)! = (n + 1)n!$, but is this quantity, i.e. `n:=` is $(n + 1)n!$.

The polynomials used to define wavefunctions in quantum mechanics can be calculated by recursion and while there are other ways of calculating them, such as the generating function method, see Chapter 9, recursion is often convenient. The Hermite polynomials $H_n(x)$ describe part of the wavefunctions of the harmonic oscillator; the other parts multiply the Hermite by a normalization constant and by $e^{-x^2/2}$. In the Hermite polynomial, n is an integer $n \ge 0$ representing the quantum number, and x is proportional to the displacement of the oscillator from equilibrium. One recursion formula when $n \ge 1$ is

$$H_{n+1}(x) = 2xH_n(x) - 2nH_{n-1}(x)$$

with starting values $H_0(x) = 1$ and $H_1(x) = 2x$. The next value with $n = 1$, is $H_2(x) = +2xH_1(x) - 2H_0(x) = 4x^2 - 2$.

Writing this in Maple produces

Algorithm 1.6 Recursive calculation of Hermite polynomials

```
> H(0,x):= 1:                                    # initial values
  H(1,x):= 2*x:
  for n from 1 to 6 do                           # calc to H7(x)
    H(n+1,x):= 2*x*H(n,x) - 2*n*H(n-1,x);        # recursion eqn
    print(H_polynom(n+1,x)= simplify(%) );        # % is last value
  end do:
```

$$H\_polynom(2, x) = 4x^2 - 2$$
$$H\_polynom(3, x) = 8x^3 - 12x$$
$$H\_polynom(4, x) = 16x^4 - 48x^2 + 12$$
$$H\_polynom(5, x) = 32x^5 - 160x^3 + 120x$$
$$H\_polynom(6, x) = 64x^6 - 480x^4 + 720x^2 - 120$$
$$H\_polynom(7, x) = 128x^7 - 1344x^5 + 3360x^3 - 1680x$$

The `%` character in the print command means 'use the last calculated value'; `H(n+1,x)` could be used instead.

1.8.2 Questions

Full solutions are available at **www.oxfordtextbooks.co.uk/orc/beddard.**

Q1.8 The particles in a wave have displacement defined as $y = 0.25\sin(125t - 3k + 0.6)$ metres. What is **(a)** the amplitude, **(b)** the frequency in s^{-1}, **(c)** the wavelength in metres, **(d)** the phase, **(e)** the wave velocity? You will need to know how to differentiate to do part **(e)**.

Q1.9 **(a)** If two sound waves of equal amplitude and phase and of frequency ω_1 and ω_2 are travelling in the same direction, show that the frequencies of the resulting waves are $\omega_1 \pm \omega_2$ and hence exhibit beating as shown in Fig. 1.13 when observed at some position x. **(b)** In this figure convince yourself that the beat frequency is 0.5 Hz.

Strategy: Use the appropriate trigonometric identity from Section 1.5. The beat frequency is the difference of the two frequencies but the lower frequency of the summed wave is half the difference in frequency. Because the beating occurs in time, the term $-kx$ and phase φ can be ignored in the general wave equation (1.16).

Q1.10 Calculate $\dfrac{52!}{50!}, \dfrac{10!}{6!4!}, \dfrac{52!}{10!48!}$

Q1.11 For what n does $n!$ first exceed 100, 10^3, 10^6, 10^9, 10^{12}?

Q1.12 Calculate the polynomial functions given by equation (1.17) when $n = 0, 1, 2$ and 3.

Q1.13 The recursion equation for the Legendre polynomials is

$$(n+1)P_{n+1}(x) = (2n+1)xP_n(x) - nP_{n-1}(x) \quad \text{with} \quad P(0,x) = 1, \quad P(1,x) = x$$

and for the Chebychev polynomials,

$$T_n(x) = 2xT_{n-1}(x) - T_{n-2}(x) \quad \text{with} \quad T_0(x) = 1, \quad T_1(x) = x.$$

Calculate the first six polynomials in each case. Use Maple if necessary.

Q1.14 An ancient way to calculate the square root of a number N is to use the recursion formula $r_i = 2kr_{i-1} + (N - k^2)r_{i-2}$ where k is the largest integer such that $k^2 < N$, and then calculate $\dfrac{r_{i+1}}{r_i} - k$ which approximates the square root. Calculate $\sqrt{23}$. The initial two r values can be chosen to be 0 and 1.

Q1.15 **(a)** Use a recursion equation and Maple to calculate the Fibonacci series whose first two values are 1, and all other values are the sum of the previous two. **(b)** Show numerically that the ratio of two adjacent Fibonacci numbers tends to the golden ratio. **(c)** If the recursion expression is $f_n = 2f_{n-1} + f_{n-2}$ show that the ratio of two adjacent numbers tends to $1 + \sqrt{2}$ and if the equation is $f_n = f_{n-1} + 2f_{n-2}$ show that ratio tends to 2. In both these formulae $f_1 = 1$ and $f_2 = 1$.

Strategy: Define a vector to hold the values f_n, f_{n-1}, etc. define the first two values then use a `for..do` loop the increment values.

Q1.16 Pascal's triangle is a mnemonic for binomial coefficients. If a pyramidal triangle is made, the coefficients are placed in rows one above the other and any value is found by adding together the numbers one to the left and one to the right from the row above. If a right-angled triangle is made, then the numbers added are the one above and the one to the left. The binomial coefficients also form the pattern of splitting, in simple AX type, NMR spectra showing the $n : n+1$ rule. For example, a CH_2 next to CH_3 has 4 lines of intensity 1:3:3:1.

(a) Make a Pascal's triangle by adding numbers as described above.

(b) Show that the recurrence formula $\begin{pmatrix} n \\ q+1 \end{pmatrix} = \dfrac{n-q}{q+1}\begin{pmatrix} n \\ q \end{pmatrix}$ where $q = 0, 1, 2, \cdots, n-1$ is true and use this to compute the binomial coefficients for $n = 12$ starting with $\begin{pmatrix} n \\ 0 \end{pmatrix} = 1$. See equation (1.21) for a definition of $\begin{pmatrix} n \\ q \end{pmatrix}$.

1.9 Sophisticated counting: permutations, combinations, and probability

The branch of mathematics dealing with permutations, combinations, and probability is perhaps that most closely related to everyday experience, particularly so if you play cards or do the lottery. There are different ways of counting the number of the arrangements of objects, such as molecules or footballs and these are permutations and combinations.

Permutations count the number of ways of arranging objects so that each permutation is unique. This means that the order is important.

Combinations count the number of ways of selecting objects from a group without considering the order of selecting.

We shall describe these quantities in terms of 'objects' and 'boxes' and let them variously apply to dice, playing cards, electrons, atoms, molecules, and energy levels as the context requires.

In genetics, probabilities are used when calculating the outcome from mixing genes through the generations. In physical science, statistical mechanics uses ideas based on placing particles into energy levels and from their distribution, partition functions can be calculated which in turn lead to thermodynamic qualities.

The probability[3] or *chance* of an event occurring, will be defined as the ratio of the number of successful outcomes to the total number of possible outcomes, and can only have values from 0 to 1. It is implicitly assumed that any one event is just as likely to occur as any other. A particular outcome is not therefore predictable; only that a certain fraction of times the expected result will occur if many trials are carried out. For instance, you would not expect to be able to throw a die so that a 1 is always produced. One might obtain a 1 on the first throw. If a 1 is obtained on the second throw, this is surprising, but if on a third, this suggests, but does not prove, that the die might be biased because we expect a die to produce a 1, or any of its other numbers, *on average* only once in six throws. Probability theory allows the calculation of the exact chance of each possible outcome without having to do the experiment. Because the probability or chance of a successful event p, cannot be greater than 1, the chance of failure is $q = 1 - p$.

1.9.1 Permutations

A permutation is an *ordered* arrangement of objects and if the order is changed then a new permutation is produced. The five letters A to E arranged as $A\ B\ C\ D\ E$ form one permutation; $A\ B\ C\ E\ D$ and $A\ B\ E\ C\ D$ are others. If there are five objects then there are $5! = 5 \times 4 \times 3 \times 2 \times 1 = 120$ permutations. The proof is straightforward: the letter E can be put into five different positions. If D is now moved then it can be placed in four positions, C in three, and so forth; therefore, the number of permutations when all n objects are chosen is $n(n-1)(n-2) \cdots 1$ which is $n!$. This can easily be a huge number, but if there are only three objects then there are only $3! = 6$ ways of doing this which are ABC, ACB, CAB, BAC, BCA, CBA. The permutation of n objects is written as

$$P(n, n) = n!$$

When only some of the objects are chosen, the permutations will be fewer. Suppose that either p objects at a time are chosen out of n, or that p objects are placed into n boxes so that no more than one is in any box, then the number of ways of doing this 'p from n' calculation is

$$P(n, p) = n(n-1)(n-2) \cdots (n-p+1) = \frac{n!}{(n-p)!}. \tag{1.19}$$

Choosing any two letters, $p = 2$, from three, $n = 3$ produces $3!/1! = 6$ permutations. For example, if the letters are ABC, then the six choices are AB, BA, AC, CA, BC, CB. Because each permutation is distinct, if we were to place them into groups or 'boxes' then only one arrangement goes into each box. When $n = p$, because $0! = 1$ this equation equals $P(n, n) = n!$. If each of the permutations is equally likely, then the *probability* of any one

[3] For a discussion of alternative definitions of probability, see Barlow (1989), chapter 7.

occurring is $1/P(n, p)$. Note that the notation $P(n, p)$ is not universal and ${}_nP_p$, ${}^nP_p$ or P_p^n are also commonly used.

1.9.2 Permutation with groups of identical objects

If some of the n objects are identical then clearly the number of choices is going to be reduced. To take an extreme example, if all the objects are identical or indistinguishable from one another, then there is only one way of arranging them and the number of permutations is one. If there are n objects split into two groups and each of v and w are identical objects, the number of permutations is reduced by dividing by the number of ways of separately arranging every identical group, and the result is $n!/(v!w!)$. If the n objects are $A\ A\ B\ B\ B\ E\ C\ D$, then there are $8!/(2!3!) = 8 \times 7 \times 6 \times 5 \times 2$ ways of arranging them or 12 times less than if all the letters were distinct.

The identical grouping permutation can be stated more formally as the number of ways of selecting n objects if these are in r ($\le n$) groups of $m_1, m_2, m_3 \cdots m_r$ objects. The total of all m_r objects must be n. The number of permutations with groups is

$$P_G = \frac{n!}{\prod_{i=1}^{r} m_i!}.$$

the symbol Π indicates the product. This number is also the number of ways of placing n *distinguishable* objects into r distinguishable boxes so that boxes contain $m_1, m_2, m_3 \cdots m_i$ objects each, and each of the objects in any box is alike.

The number of ways of arranging the amino acid residues of even a small protein is astronomically large, but countable. An active protein in a bee's sting is called mellitin. It is a protein with only 26 amino acids and which forms two short α-helical regions, with a bend in between. Two such helices are seen in the crystal structure 2MLT.pdb in the RCSB data base (**www.rcsb.org/pdb/home/home.do**). The sequence of the structure is

```
GLY ILE GLY ALA VAL LEU LYS VAL LEU THR THR GLY LEU
PRO ALA LEU ILE SER TRP ILE LYS ARG LYS ARG GLN GLN
```

Collecting the residues together produces groups of 3 GLY, 3 ILE, 2 ALA, 2 VAL, 4 LEU, 3 LYS, 2 THR, 1 PRO, 1 SER, 1 TRP, 2 ARG, and 2 GLN and this produces

$$\frac{26!}{3!3!2!2!4!3!2!2!2!} \approx 2.4 \times 10^{21}$$

ways of arranging a chain. Nature has had to search in the 'space' of all combinations to find an effective protein (one that causes pain when you are stung) and has had a very long time to do so. However, this number of permutations is still so large that even producing a different sequence at one a minute, supposing that this were possible, would have taken $\approx 5 \times 10^{15}$ years, which is far longer than the age of the earth. This is, of course, a misleading calculation for two reasons at least. One is that it assumes that the protein always had 26 amino acids whereas, in the distant past, it was probably far smaller but nevertheless had enough effect to give the bee's ancient ancestor an evolutionary advantage. Secondly, many permutations of even a few amino acids will not have the stability to form any structure other than a random coil and so could never exist as a functioning protein. Those that do form some stable structure and are effective are then preserved and passed down to the next generation, and by mating and random mutations, improved. Therefore the whole of the possible permutation space is never searched, but the search algorithm that is natural selection very effectively finds a working solution and one that is usually close to the optimum.

1.9.3 Combinations

A combination is really a misnomer, because it is the number of ways of *choosing* p distinguishable objects from a group of n distinguishable objects, and the order of choosing these p objects does not matter. If two letters from ABC are chosen, the number of combinations is three and the choices or combinations are $AB \equiv BA$, $AC \equiv CA$, $BC \equiv CB$

because the order does not matter. If we think of placing objects into boxes, combinations, unlike permutations, allow more than one object to be placed in each box. For example, the letters *ABC* fill three boxes each containing two objects. Removing p of the permutations is equivalent to dividing the objects into two groups, the chosen group of p objects and another group of $n-p$ objects.

In a permutation, there are $n!$ ways of choosing (if all the objects are different) and a combination must be less than this because the ordering of similar objects does not matter, and is less by the factorial of the number chosen, which is $p!$. Four objects *A B C D* produce $4! = 24$ permutations. If any three ($p = 3$) are chosen at a time, the four combinations $C(4, 3)$ are *ABC, ACD, ABD, BCD*. Each of these groups has $p! = 3! = 6$ permutations making $4 \times 6 = 24$ permutations in total. Thus the number of combinations C is

$$C(n, p) = \frac{P(n, p)}{p!} \tag{1.20}$$

Therefore, using (1.19), the number of ways of choosing p objects at a time out of a total of n, is

$$C(n, p) = \frac{n!}{p!(n-p)!} = \binom{n}{p}. \tag{1.21}$$

The second notation, 'n over p', is that used for the coefficients in the binomial expansion. As for permutations, the notations ${}_nC_p$, ${}^nC_p$, C_p^n are also commonly used.

The original context of the word combination is that the number $C(n, p)$ is the number of combinations of n things selected p at a time. Since $\binom{n}{p} = \binom{n}{n-p}$ this is also equal to the combination of n things selected $n-p$ at a time.

The chance of winning a lottery can be found from the number of combinations. For instance, choosing 6 numbers out of 48 produces

$$C(48, 6) = \frac{48!}{6!42!} = \frac{48 \times 47 \times 46 \times 45 \times 44 \times 43}{6 \times 5 \times 4 \times 3 \times 2} = 12\,271\,512$$

possible choices or just under one in 12 million chances of winning, as the chance of any one combination being chosen is just as likely as any other, then its probability is $1/C(n, p)$. If ≈ 12 million people play each week, then on average one might expect one winner each week, assuming that the numbers are equally likely to be chosen by the players.

If you are making choices when two or more conditions apply, then the combinations are multiplied together. For example, suppose that a study is to be conducted in which 25 patients are to be placed into three groups of equal size. The control group must contain 8 persons and therefore so must the experimental groups. These must be selected from $25 - 8 = 17$ and 9 persons each. The number of ways of making this choice is huge even for such small numbers, and is

$$C(25, 8)C(17, 6)C(9, 8) = \frac{25!}{8!(25-8)!}\frac{17!}{6!(17-8)!}\frac{9!}{8!(9-8)!} = 236\,637\,794\,250.$$

1.9.4 Indistinguishable objects

The cards in a pack of playing cards are clearly distinguishable; it would be pointless if they were not. White tennis balls are generally indistinguishable from one another and so would golf balls be if they were not numbered after manufacture to enable players to identify one from another. Atoms or photons are truly indistinguishable; we cannot label them to tell one from another. The number of ways of placing (distributing) n *indistinguishable* objects into p *distinguishable* boxes with $p \ge n$ and with any number of objects being allowed in any one box is,

$$C(p+n-1, n) = \frac{(p+n-1)!}{n!(p-1)!}. \tag{1.22}$$

The proof is involved and given in Margenau & Murphy (1943). We note here only that there are $(n+p-1)!$ permutations when the n objects are placed in $p-1$ boxes. Using only

$p - 1$ boxes is correct because if there is just one box then there are $n!$ permutations if we suppose, for the moment, that the objects are distinguishable. However, if the objects are indistinguishable, the number of permutations has to be divided by $n!$. (There is only one permutation of n indistinguishable objects in one box.) Finally, there are $(p - 1)!$ permutations of the (distinguishable) boxes for a given configuration and therefore equation (1.22) follows. If there are $n = 2$ objects and $p = 3$ boxes in which to place them, then there should be $4!/2!2! = 6$ arrangements. Labelling both objects x these are

$$\begin{bmatrix} xx & - & - \\ - & xx & - \\ - & - & xx \\ x & x & - \\ - & x & x \\ x & - & x \end{bmatrix}$$

Particles with zero or integer spin angular momentum, such as photons and deuterons, obey Bose–Einstein statistics. Any number of them can occupy a quantum state. When dealing with atoms and molecules and distributing particles among their energy levels, the number of boxes p, becomes the *degeneracy* g of any energy level. The degeneracy is the number of states belonging to one energy level. If the angular momentum quantum number is S, then this state is $g = 2S + 1$ degenerate. If there are $i = 1 \cdots N$ energy levels then the total number of ways of distributing particles among the levels is $W = \prod_{i=1}^{N} \dfrac{(n_i + g_i - 1)!}{n_i!(g_i - 1)!}$ where g_i is the degeneracy of level i. (see Margenau & Murphy 1943, Chapter 12). For one level, the number of combinations is $C(n + p - 1, n)$ and the chance of observing any one distinguishable arrangement is considered to be equal in Bose–Einstein statistics and is $1/C(n + p - 1, n)$.

Fermions are half integer spin particles and include electrons, protons, and atoms such as ^{14}N, which are made up of an odd number of fermions. They obey Fermi–Dirac statistics and are restricted so that no more than one of them can be in any quantum state. By the Pauli exclusion principle, a fermion's wavefunction has to be asymmetric to the exchange of coordinates, and each fermion must have a unique set of quantum numbers. In apparent contradiction an orbital can contain zero, one, or two electrons. Two electrons are allowed to be in an orbital if they have different quantum numbers and are therefore different fermions. An electron's spin quantum number is $S = 1/2$ but there is a second quantum number $m_s = \pm 1/2$ value related to the spin's orientation, colloquially spin 'up' or 'down', so that an orbital can have up to two different fermions in it. However, no more than one of them is in any one quantum state. There are only two sets of the electrons' quantum numbers with which to label the electrons (1/2, 1/2) and (1/2, −1/2), and so no more that two electrons can be in any orbital. If an (imaginary) particle had $S = 1$ then m_s would be $0, \pm 1$ and a maximum of three of them could fill any orbital.

With indistinguishable particles there are now only $C(p, n)$ ways of choosing p singly occupied states from the p available, and if these distributions are equally likely then $1/C(p, n)$ is the probability that any one is occupied. Again, if there are $n = 2$ fermions to be placed in $p = 3$ levels then the only possible arrangements are

$$\begin{bmatrix} a & a & - \\ - & a & a \\ a & - & a \end{bmatrix}$$

which is $C(3, 2) = 3!/(2!1!) = 3$.

The number of combinations just described also answers an apparently harder question. If there are n boxes, and $p < n$ indistinguishable objects to be placed in the boxes so that no more than one is in any box, then the number of ways of doing this is $C(n, p)$. The assignment into boxes is the same as selecting p out of n objects. In an atom when calculating the term symbols, the number of microstates in a configuration must be enumerated. If there are two electrons to be placed into the three 2p orbitals, a p^2 configuration, then there are $C(6, 2) = 6!/(2!4!) = 15$ microstates. Why the 6 when there are only three p orbitals? The electrons must each have a unique set of quantum numbers therefore the *spin* states are unique; spin up is different from spin down. See Steinfeld (1981) and also Foote (2005) for a diagrammatic way of calculating term symbols.

1.9.5 Sampling with replacement

If a bag contains n objects and we choose p of them but return each object to the bag before making the next choice, then there are n^p ways of choosing them: there are always n ways of choosing, and this is done p times over. If there are four letters *ABCD*, then choosing three of them produces $4^3 = 64$ possible samples. The number of samples under permutation rules is $4!/1! = 24$ and $4!/(3!1!) = 2$ under combination.

The number of UK car registration plates can be calculated by 'sampling with replacement'. Although the way that number plates are labelled has recently changed, there are still many cars with the form of a letter to identify the year of manufacture, a three digit number and three letters. A plate such as K 446 LPW is typical. In this form, each year there are $3^{10} \times 3^{26} = 150\,094\,635\,296\,999\,121 \approx 10^{17}$ possible registrations; a ridiculously large number even when many are not used for various reasons. Even if only nine numbers and ten letters were used, there would still be more that 10^9 possible registrations.

Braille is a representation of letters and numbers using a pattern that can be felt by the fingertips, and which enables blind people to read. The pattern consists of raised dots and gaps in a rectangular shape whose height is greater than its width. There are six dots and gaps combined making $2^6 = 64$ ways of arranging the patterns and that is enough to encode all the letters, numbers, and punctuation marks commonly used.

The molecular motor ATPase reversibly converts ADP into ATP + phosphate (Pi). The protein crystal structure has been determined to high precision, see the Brookhaven Database (pdb) entry 1E79 and also Gibbons et al. (2000). The protein called F_1 contains the rotor part of the motor, has threefold symmetry, and three sites at which the reaction can occur. The reaction site has four possible states

$$\text{Empty} \leftrightarrow \text{ATP bound} \leftrightarrow (\text{ADP} + \text{Pi}) \text{ bound} \leftrightarrow \text{ADP bound} \leftrightarrow \text{Empty}$$

therefore, there are $4^3 = 64$ possible binding states in the protein at any one time.

1.9.6 Probability

When answering questions about the probability or chance of some event occurring, it is always worth considering whether the question is real. The question 'what is the chance that next Friday is the 13th?' is not a question involving chance, since checking with a calendar will produce the answer. Similarly, asking 'what is the chance that that runner will win this race?', or 'what is the chance that Portsmouth will beat Manchester United?' is not a question that probability theory can answer, since there are factors involved other than pure chance that make the outcome unpredictable. The question 'what is the chance that I will win the lottery?' is a question to which only a probabilistic, not predictable, answer can be given, provided of course that you have bought a ticket! A probabilistic answer is possible because one number is just as likely to be drawn as another is, otherwise the lottery would not be fair. The *just as likely* is important here as it indicates that random chance is involved.

Quite often, some caution is necessary in analysing problems involving chance or random events. For example, if two coins are simultaneously flipped what is the change of observing two heads? The ways the coins can fall is either head *H*, or tail *T*, but to think that there are only three outcomes, *HH*, *HT*, *TT* and the chance 1/3 is wrong. This is because there are *four* outcomes *HH*, *HT*, *TH* and *TT* so the chance of observing *HH* is 1/4. The chance of a head *and* a tail is 1/2.

1.9.7 Calculating probability

The foundations of this subject are based on the ideas of sets and subsets of objects and their properties; for example 'union' and 'intersection', Fig. 1.20. A set is defined as a collection of objects such as the letters of the alphabet. A subset of these could be the vowels, {a, e, i, o, u}. The *sample space* is the total number of arrangements of objects that are possible for any particular problem. Flipping two coins has the sample space of four elements {*HH*, *HT*, *TH*, *TT*}. To determine an *event* or successful outcome is the purpose or object of the calculation, and is a *subset* of this sample space. Suppose that the event we want is that one head is to be produced when two coins are tossed; then this is the subset

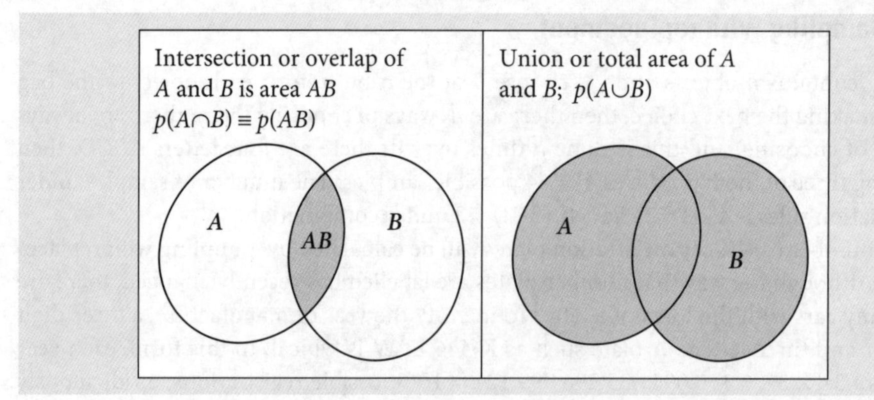

Fig. 1.20 Venn diagrams. Left: *A* and *B* are two overlapping ellipses, *AB* is their overlap and is shaded. The chance of *A* or *B* being observed is the intersection of *A* and *B*, which is $p = p(A) + p(B) - p(A, B)$ and is proportional to the total area within the circles less that area overlapped. If *A* and *B* do not overlap the events are mutually exclusive and $p(AB) = 0$. Right: The union of *A* and *B* is the chance of belonging to at least one of *A* and *B* and is the shaded area.

$A = \{HH, HT, TH\}$. If we want the event, which is two heads, then only one element exists and this is $B = \{HH\}$. An event such as B that contains only one sample point is called a *simple event*. As probability is defined as the ratio of the number of successful outcomes to the total number of possible outcomes, the probability of observing event A, that of one head, is the number of arrangements for this event over the total number in the sample space, making the probability 3/4. Similarly, observing two heads occurs on average 1/4 of the times two coins are thrown.

As an example, consider a die where faces 5 and 6 are black and the other four faces are white. We would like to know the chance p of the top face being white and the chance of it being black. This must be $1 - p$ since there are no other colours. The sample space is $\{1 \cdots 6\}$ and, as usual, it is assumed that each outcome 1 to 6 is equally likely. In the first case, outcome *white* = {1, 2, 3, 4} and occurs with the probability (chance) 4/6; a black face being uppermost occurs with a chance 2/6. See Stewart (1998) and Goldberg (1986) for clear discussions of probability.

1.9.8 Definitions, notation, and some useful formulae

(i) The probability of an event A is $p(A) \ge 0$.

(ii) The certain event S has a probability of 1; $p(S) = 1$.

(iii) The probability of an event $p(A)$ is the sum of simple events in the sample space.

(iv) The word 'or' is used in the inclusive sense. Thus, *A or B* means 'either A or B, or, both'. The notation $p(A + B)$, is the probability that either A or B or both occurs. The notation $p(AB)$ is a joint probability and means that both A and B occur. In set theory, this is the intersection or overlap of A and B, and is usually written as $p(A \cap B)$; Fig. 1.20.

(v) If several *independent* events each of a probabilistic nature occur to produce a successful outcome, then the overall chance of this happening is the product of the probabilities of the individual events. An independent event is one whose outcome does not influence that for any of the others; $p(A\&B) = p(A)p(B)$.

(vi) If two events A and B can occur, their *inclusive probability* $p(A + B)$ means that at least one event occurs, which is to be interpreted as event A or B, or both occur. In Fig. 1.20 (left) the sample spaces are related as

$$n(A) + n(B) = n(A + B) + n(AB).$$

The area $n(A)$ and $n(B)$ is the *whole* of their respective ellipses, $n(A + B)$ is the total area less the overlap $n(AB)$. When divided by the number of arrangements in the sample space these numbers become probabilities. The inclusive probability, either A or B or both, is the probability that A occurs, plus, the chance that B occurs minus the chance that both occur or

$$p(A+B)=p(A)+p(B)-p(AB), \tag{1.23}$$

(vii) A mutually exclusive event is one whose outcome prevents any others occurring. The two events A and B, are *mutually exclusive* if there is no intersection or overlap of A and B, $p(AB) = 0$, therefore the probability of the occurrence of at least one out of two possible events is the sum of the individual probabilities,

$$p(exclusive) = p(A) + p(B).$$

This equation only applies to two events.

(viii) The sample space in tossing a coin is heads and tails, $\{H, T\}$; in tossing a die this is $\{1 \cdots 6\}$, in one set of playing cards $\{1 \cdots 52\}$, and so forth. If three coins are used, the sample space contains 2^3 elements, HHH, HHT, etc. If n_S represents the number of arrangements in the whole sample space, n_A the subset that is the number of ways of arranging events in a successful outcome, A and n_{NA} the subset that is *not* A then, clearly, $n_S = n_A + n_{NA}$. The probability of outcome A is therefore

$$p=\frac{n_A}{n_S}=1-\frac{n_{NA}}{n_S}. \tag{1.24}$$

(ix) If p is the chance that an event occurs, then $1-p$ is the chance that it will not,

$$p(not\ A) = 1 - p(A).$$

This is called the *complement*. On the right of Fig. 1.20, the complement of $p(A \cup B)$ is the area outside the two shaded circles. In the left-hand sketch, the complement of the intersection $p(A \cap B) \equiv p(AB)$, is *all* the area not labelled AB inside the square.

(x) If two objects are placed into two different boxes, hence distinguishable, then the outcomes are $\{AB, -\}$, $\{-, AB\}$, $\{A, B\}$, $\{B, A\}$. As each of these is a simple event, the probability of each occurring is 1/4. If the objects are indistinguishable, then there are three arrangements $\{xx, -\}$, $\{xx, -\}$, $\{x, x\}$, but the last may be considered to be two events and in this case would occur with a probability of 2/4. However, if we take the three outcomes to be equally likely then the probability of the last is 1/3, and this is the case for bosons.

1.9.9 Independent and exclusive events, sample spaces, and conditional probability

(i) If a coin and a die are thrown they are clearly independent events and the chance of observing a head and a 6 is $(1/2) \times (1/6)$. This follows from the fundamental principle of counting; if a job is completed in n ways and another in m ways then both can be completed in $n \times m$ ways. For instance, if there are 6 anions and 8 cations then 48 different salts can be produced. Now, suppose that two dice are thrown and you want to find the chance that the total of their numbers is 10. The two dice are independent, the result of one does not influence the other, and the probabilities therefore multiply. As one die can fall in one of six ways, two can fall in $6 \times 6 = 36$ different ways. The number 10 can be obtained in three different ways, each of which is equally likely to occur: $6 + 4$, $4 + 6$, and $5 + 5$. The probability of observing 10 is therefore 3/36. If a sum of 6 is sought, then this is produced in the combinations $1+5, 5+1, 4+2, 2+4, 3+3$ and so would be expected to be observed 5/36 times the dice are thrown.

(ii) If you want two cards containing the number 7 to be drawn in succession from a pack of 52 playing cards, what is the chance of this happening if the first card chosen is not replaced in the pack? The chance of the first 7 being chosen is 4/52 because there are four 7s in a pack of 52 cards. It is now assumed that a 7 has been removed and therefore the chance that the second card removed is a 7 is 3/51 making the chance $(4/51) \times (3/51) = 1/221$ overall. The second choice is 3/51 because we have only 51 cards left and one 7 is assumed to have been removed in our first try. Had we chosen to find the probability that a 7 *and* a 6 were to be removed in succession, the chance would be $(4/52) \times (4/51)$. If instead we wanted to draw a 7 *or* a 6 from the pack, then the probability would be $4/52 + 4/52$ as these are independent of one another; drawing one card does not depend on the other.

(iii) Independent events can occur in the way molecules react. If a molecule can react to produce two different products A and B with rate constants k_A and k_B respectively, the chance of product A being observed is $p_A = k_A/(k_A + k_B)$ and of B is $1 - p_A$, which is

$p_B = k_B/(k_A + k_B)$. The sum of both events is 1. In chemistry, probability p_A is normally called the yield of A and is often expressed as a percentage. If an excited state of a molecule can fluoresce with rate constant k_f or produce another state such as a triplet by intersystem crossing with rate constant k_i then the fluorescence yield is $k_f/(k_f + k_i)$ and the triplet yield $k_i/(k_f + k_i)$.

(iv) Suppose that two cards are drawn from a pack and the first not replaced, and we want the chance that the second is a 7 of diamonds. This question has two answers. If the first card happens to be the 7, then the chance of the second being this card is obviously zero. If the first card is not the 7 of diamonds, the chance of choosing it a second time is (1/51) making the chance $(1/52) \times (1/51)$ overall.

(v) Suppose a die is thrown n times, what will be the chance that a 3 appears *at least once*? In one throw the 3 appears with a chance 1/6 and there is a 5/6 chance that it does not appear. After n throws, then the chance is $(5/6)^n$ that the 3 does not appear and therefore $1 - (5/6)^n$ that a 3 appears at least once. For two throws, this is 11/36. This calculation can also be described using inclusive probability. Suppose that there are two outcomes A and B, and at least one outcome is required, then the probability is that of event A, plus event B minus that of both A and B or $p = p(A) + p(B) - p(AB)$. The chance that a 3 is thrown, is 1/6 on the first throw (outcome A), and again on the second throw is 1/6, and the chance of both occurring $p(AB) = 1/36$ making $1/6 + 1/6 - 1/36 = 11/36$ overall.

(vi) ***Summary***: If n and m are the numbers 1 to 6 (the sample space) on a die, then the probability of throwing;

(1) any number n is 1/6 and of not throwing any n is $1 - 1/6$.

(2) either n or m is $1/6 + 1/6 = 1/3$.

(3) the same n twice in two throws is 1/36.

(4) n *at least* once in two throws is $1 - (5/6)^2 = 11/36$.

(5) n *exactly* once in two throws is $1/6 + 1/6 \times (1 - 2/6)$.

(vii) To illustrate explicitly the use of a sample space, consider the problem of calculating whether at least two people from a group of 25 have the same birthday (see Goldberg 1986, p. 53). First, to simplify the calculation it is necessary to ignore leap years, and then to assume that there are no twins in the group and that births occur with equal probability throughout the year. None of these may be true in a real sample of people, but we will assume that they are.

The sample space is defined as the total number of arrangements $n_S = n_A + n_{NA}$ split into those in the group we want to determine n_A, and those that we do not, n_{NA}. The sample space is huge, $n_s = 365^{25}$, because this is the number of ways that the birthdays can be arranged. Let n_A be the number of arrangements where at least two people have the same birthday and n_{NA} the number of those with different birthdays, then n_{NA} is the number of ways of selecting 25 different days from 365. The first birthday can be chosen in 365 ways, the second in 365 – 1 and so forth down to 365 – (25 – 1) ways. This makes $n_{NA} = 365 \times 364 \times \cdots \times 342 \times 341$ which is the permutation $P(365, 25) = 365!/(365 - 25)!$.

The number of different ways that 25 people can be selected is therefore $n_A = n_S - n_{NA} = 365^{25} - 365!/(365 - 25)!$. The probability that at least two people have the same birthday is $p = \dfrac{n_S - n_{NA}}{n_S}$, if we assume that each of the 365^{25} outcomes is equally likely. The result is $1 - \dfrac{365 \times 364 \cdots 341}{365^{25}} = 1 - \dfrac{365!}{(365-25)!365^{25}} = 0.568$. It is surprising that in such a small group the chance of two or more people having the same birthday is more likely than not.

(viii) Sometimes *conditional probabilities* are required; for example, tossing three coins and deciding what is the chance that the outcome is *at least* two tails (outcome A) and knowing that the first coin to fall is a head (outcome B). Tossing three coins can only produce the patterns

$$HHH, HHT, HTH, HTT, THH, THT, TTH, TTT$$

and each has a chance of 1/8 of being produced in three throws. The first outcome (A) is the chance of having at least 2 tails and by direct counting this is $p(A) = 4/8$. The chance of having a head as the first coin is $p(B) = 4/8$. The chance that both conditions apply, is calculated $p(A, B) = 1/8$ by inspecting the sequence of coins. The conditional probability $p(A \mid B)$, is the chance that both conditions apply divided by the chance that condition B applies, and this is

$$p(A \mid B) = \frac{p(A, B)}{p(B)} \tag{1.25}$$

which is 1/4. This means that the added knowledge that the first coin to fall must be a head, has reduced the odds of obtaining two tails, which is not unexpected since insisting on a head as the first coin reduces the choices available. The equation is 'symmetrical' and can be rearranged to

$$p(A, B) = p(A)p(B, A) = p(B)p(A, B).$$

1.9.10 The binomial distribution

If n boxes each contain W white balls and B black ones, and if we take one ball from each box, we want to find the chance that *exactly* m of them will be white. The chance of choosing a white ball from a single box is $p = W/(W + B)$ and correspondingly a black one $q = B/(W + B) = 1 - p$. Therefore if $n = 1$, the probabilities are p white balls and q black. If there are two boxes, the probabilities are distributed in the same manner as tossing two coins. The coins land as HH, HT, TH, TT and each has a 50% chance of being H or T, therefore the probabilities are $(1/2)^2$, $2(1/2)(1/2)$, and $(1/2)^2$ or 1/4, 1/2, and 1/4. (HT being the same as TH). Choosing white and black balls from two boxes has the equivalent probability p^2, $2pq$, q^2. Choosing m white balls from n specified boxes, and $n - m$ black ones from the rest, has the probability $q^m p^{n-m}$, but there are a combinatorial number of ways $C(n, m)$ of choosing m boxes from the total of n. Therefore, the total probability of choosing m balls from n boxes, is

$$p(n, m, q) = \frac{n!}{m!(n-m)!} q^m p^{n-m}$$

where $p = 1 - q$, and is called the binomial distribution because the binomial expansion is

$$(p + q)^n = \sum_{m=0}^{n} \frac{n!}{m!(n-m)!} q^m p^{n-m}$$

and the terms in this distribution are just the probability $p(n, m)$. This distribution is normalized $\sum_{m=0}^{n} p(n, m, q) = 1$ because $p + q = 1$ and $(p + q)^n = 1$. The mean of the distribution is nq where it is also a maximum. The binomial distribution, and the Gaussian and Poisson distributions which are derived from it, are examined in more detail in Chapter 13 where descriptive statistics are discussed.

In calculating the probabilities, the values of p or q have to be found from the problem at hand. For instance, to find the chance that no red cards are drawn in a single attempt from each of seven packs of cards, we need to know that half the cards are red. As $p = 1/2 = q$ this chance is

$$p(7, 0, 1/2) = \frac{7!}{0!(7-0)!}\left(\frac{1}{2}\right)^0 \left(1 - \frac{1}{2}\right)^{7-0} = \frac{1}{128}.$$

The chance of picking four red cards is similarly 35/128, and of picking four picture cards (12 in each pack) out of five packs, is $p(5, 4, 12/52) = 4050/371\,298 \approx 0.01$. The chance of obtaining an even number of aces from six packs of cards, is $313\,201/4\,826\,809 \approx 0.065$ and is the sum of choosing 2, 4 and 6 aces, each with a chance $q = 4/52$.

1.9.11 Multinomial distribution

The binomial distribution is concerned with selecting between two classes: success and failure; but the choices need not be restricted to two classes. We have already calculated the chance of observing groups of objects or events, such as the chance of 3 GLY, 3 ILE, etc. in the protein sequence in Section 1.8.2. When there are several choices the multinomial distribution is required and is

$$P_{mG} = \frac{n!}{\prod_{i=1}^{r} m_i!} \prod_{i=1}^{r} p_i^{m_i} \tag{1.26}$$

where p_i is the probability of choosing any object from the n boxes and m_i are the numbers in each of r groups. In the calculation of the Maxwell–Boltzmann distribution, n objects are placed into s identical boxes. The distribution becomes

$$\frac{n!}{\prod_{i=1}^{r} m_i!} \left(\frac{1}{s}\right)^n$$

from which the m_i has to be found and the distribution is maximized subject to the restriction that both the number of particles and the total energy are constant. This is usually done by taking the log of the distribution and, as each m_i is large using Stirling's approximation for $m_i!$. The constraints are dealt with using the method of Lagrange multipliers. The full calculation can be found in textbooks on statistical mechanics.

1.9.12 **Genetics**

Mendel discovered that pure bred but different strains of peas gave rise to offspring that showed different features but did so in definite proportions. His peas had yellow and green seeds, but 75.05% of the offspring in 8023 trials were yellow, and the remainder, green. Others have since found very similar numbers in even larger trials and in other organisms, such as fruit flies or mice (Maynard Smith 1995). The genetic information we now know is encoded in DNA. The regions involved in transmitting genetic information are called genes, and these form part of a chromosome of which several different types are present in an organism. We are familiar with the observation that the probability with which molecules react is exponentially distributed, which is expressed in the Arrhenius rate equation, the rate being probability/time. In genetics, the situation is entirely different. The probability distribution is flat and the chance that genes mix genetic information occurs with the same constant probability as if they were selected out of boxes, each one being equally probable. Although these genes 'mix' during reproduction, they still retain their own identity and are passed down through the generations. An individual needs two genes to express a trait; yellow vs green seeds or stumpy vs normal wings in a fruit fly, and these mix freely from both parents to produce offspring. These traits are usually called *phenotypes*. The traits the offspring show are expected to be those of the parents, but depending on the genetic make up (*genotype*) of the parent, some traits appear less than would be expected from simple numerology. The reason for this is that some genes are recessive to others and therefore a characteristic trait does not manifest itself in the presence of a dominant gene. Only when the genotype is examined can the true make-up of an individual be found.

If the two genes from each parent are described as Y or y and G or g they can be arranged in a diagram. The uppercase letter indicates a dominant gene, the lower case a recessive. Starting with the pure strains YY and yy, which are called *homozygous*, the first generation (F_1) are *heterozygous* Yy. If these are now bred, the homozygous reappear with the heterozygous, Fig. 1.21.

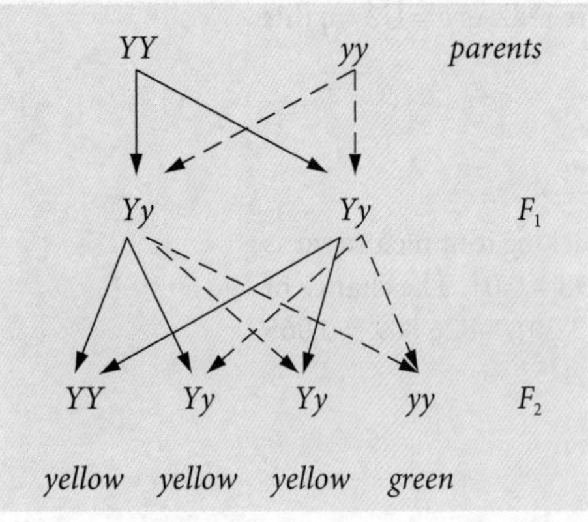

Because the yellow colour in the pea seeds is dominant over the green, although this is not reflected in the genetic make-up (the genotype), the chance of observing yellow peas is therefore 75%, which is very close to that observed experimentally. The chance of observing YY and yy genotypes is 1/4 in each case, and of the hybrid Yy, 1/2, because the genes mix freely but retain their own identity, rather like atoms do in forming a molecule. The yellow seeded plants are present in the ratio 3:1 over the green ones. The phenotype ratio yellow to green is therefore 3:1, or put another way, yellow seeded plants are expected to be observed on average in 3/4 of all experiments.

Fig. 1.21 Genetic make-up after two generations (after Maynard Smith 1995). The yellow colour in pea seeds is dominant over green and occurs in the ratio 3:1, although this is not reflected in the genetic make-up.

If a homozygous plant is crossed with a heterozygous one, the chances that green and yellow seeded plants will be produced are equal.

If, say, 20 experiments are performed it should not be assumed that exactly equal numbers of green and yellow seeded plants will be observed since the number of experiments are small and many factors may influence what the plant looks like. This 'noise' means that accurate and precise ratios are only produced after averaging a large number of experiments. The numbers could range from 0 to 20 but the average should be 10 with the actual probabilities being distributed according to the binomial theorem. The chance (frequency) of observing m green plants is then

$$p(20, m, 1/2) = \frac{20!}{m!(20-m)!}\left(\frac{1}{2}\right)^m\left(\frac{1}{2}\right)^{20-m},$$

which works out to be a very small number, 9.5×10^{-7} of observing 0 or 20 green pea seeds. The chance of there being exactly 10 plants of each type is 46 189/262 144 = 0.1762 or 17.6%. The chance that the actual result differs by ±1 from the average value of 10 is

$$\frac{20!}{10!10!}\left(\frac{1}{2}\right)^{20} + \frac{20!}{9!11!}\left(\frac{1}{2}\right)^{20} + \frac{20!}{11!9!}\left(\frac{1}{2}\right)^{20} = 0.496$$

which is approximately a 50% chance that the numbers of peas produced will be 9, 10, or 11 and a 50% chance that they will be greater or smaller than these numbers. If the number of experiments is increased to 2000, then there is only a 5.3% chance that the average number will not be between 1001 and 999.

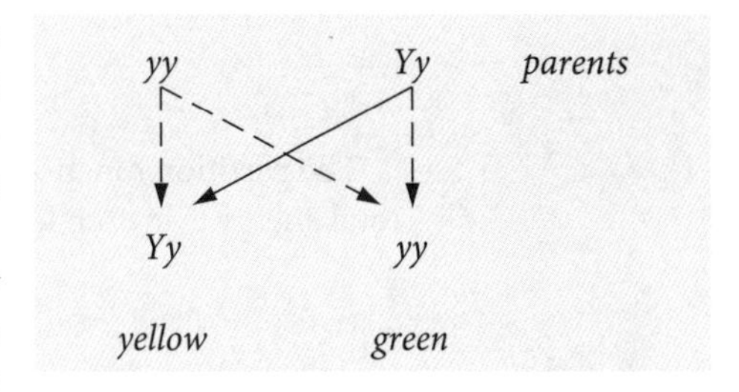

Fig. 1.22 A cross between *yy* and *Yy* plants.

The Punnett square or matrix is a convenient pictorial way of assessing the outcomes of mixing genes. All of the possible genotypes of one parent are placed on the top row, and those of the other parent in the first column and the genotypes of the offspring fill the matrix. The crossing of the F_1 hybrids in Fig. 1.21 is represented as

	Y	y
Y	YY	Yy
y	Yy	yy

which gives the same result as in this figure but in more compact form.

In many cases there are two genes that encode for a phenotype, and we assume that the genes are located on different chromosomes and are therefore independent. We can predict the fraction of the types of offspring of, for example, mice with brown vs white coats, brown being dominant and short vs long tails, short being dominant. The genotypes are labelled as B and b for the colour, and T and t for the tails, so that the genotypes are BT, Bt, bT, bt, the lower case being recessive. If these animals are mated, the offspring genotypes are shown in the table, where each entry is arranged with B/b as the first label and T/t as the second, and the dominant before the recessive.

	BT	Bt	bT	bt
BT	BBTT	BBTt	BbTT	BbTt
Bt	BBTt	BBtt	BbTt	Bbtt
bT	BgTT	BtTt	bbTT	bbTt
bt	BbTt	Bbtt	bbTt	bbtt

An offspring's phenotypes can be characterised using the dash — to mean B/b or T/t as appropriate. The table shows that

(i) 9/16 of genotype B–T– has phenotype brown, long tailed,

(ii) 3/16 as bbT– ≡ white, long tail,

(iii) 3/16 as B–tt ≡ brown, short tail (shown shaded in the square),

(iv) 1/16 as bbtt ≡ white short tail.

The phenotype ratios are 9:3:3:1 and this was observed by Mendel in his studies of peas.

1.9.13 **Partition functions**

The partition function plays a central role in statistical mechanics because all other thermodynamics properties, the average energy or the entropy, for example, can be calculated from it. The partition function is the sum of the statistical weights of the energy levels. The statistical weight of a state i with energy E_i is the Boltzmann factor e^{-E_i/k_BT} multiplied by the degeneracy of that state g_i. The partition function is therefore $Z=\sum_i g_i e^{-E_i/k_BT}$. The constant k_B is Boltzmann's constant, 1.3805×10^{-23} J K^{-1} and T is the temperature in K. A state is degenerate if it has more than one energy level. The vibrational levels in molecules are singly or non-degenerate, the rotational levels have a degeneracy that increases with the quantum number and is $g_J=2J+1$ where J is the quantum number.

As the total probability that a 'particle' has to be in one of possibly many states must be one, $\sum_i p_i=1$ where p_i is the probability of being in state i and is proportional to the statistical weight from that state hence $p_i=\alpha g_i e^{-E_i/k_BT}$ where α is a normalization. The total probability is therefore $\sum_i p_i=\alpha\sum_i g_i e^{-E_i/k_BT}=1$ and $\alpha=1/\sum_i g_i e^{-E_i/k_BT}$. Conventionally we write $\alpha=1/Z$, and so the probability that the particle is in level i is

$$p_i=\frac{g_i e^{-E_i/k_BT}}{Z}.$$

The partition function Z is the normalization term and determines what fraction of the total energy is in each level or how energy is partitioned among the various levels.

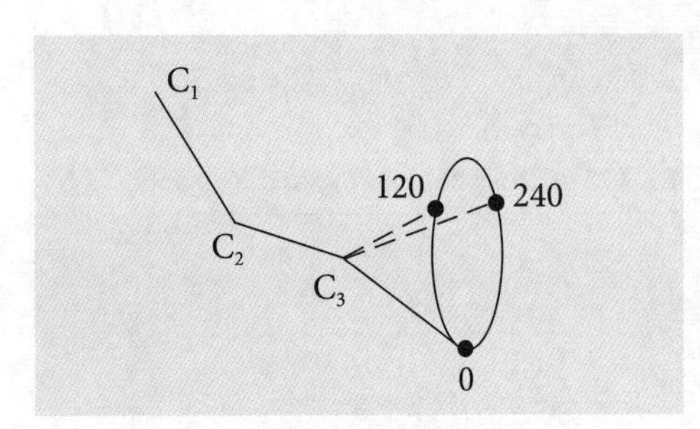

Fig. 1.23 Bond angles in butane

The use of the word 'particle' is rather general, it might represent a molecule in a given vibrational or rotational energy level, a nuclear spin state in an NMR experiment, or the torsional energy of a restricted rotor as in the alkyl chain of hydrocarbons (Jackson 2006). In this last example, consider butane, which has trans and gauche configurations as shown in Fig. 1.23. These have different energies due to interactions between the protons on carbon 1 and 4 as bond C_2 to C_3 rotates.

The energy of the trans state at 0° is taken as zero, $E_0=0$. The energy of the other two conformations, measured relative to this, have minima at 120° and 240° and are E_{120} and E_{240}. The partition function is $Z=1+e^{-E_{120}/k_BT}+e^{-E_{240}/k_BT}$. The probability of being in the trans state is $1/Z$ and of being in the 120 gauche state is $e^{-E_{120}/k_BT}/Z$ which is less but changes with temperature. By symmetry, the energy of the 120 and 240 states are the same and the probability of being in either is $2e^{-E_g/k_BT}/Z$ where E_g is an abbreviation for E_{120} and E_{240}. These two levels are accidentally degenerate. Once the partition function has been found, the average energy follows easily because by definition $\langle E\rangle=\sum_i E_i p_i$ then

$$\langle E\rangle=\frac{2}{Z}E_g e^{-E_g/k_BT}$$

because $E_0=0$. The average length and the average of the square of the length are

$$\langle L\rangle=\sum_i L_i p_i=\frac{1}{Z}(L_0 e^{-E_0/k_BT}+2L_g e^{-E_g/k_BT}),\quad \langle L^2\rangle=\sum_i L_i^2 p_i=\frac{1}{Z}(L_0^2 e^{-E_0/k_BT}+2L_g^2 e^{-E_g/k_BT}).$$

If the alkyl chain is longer then the probability of the final state being a trans or gauche configuration is multiplied by however many repeat units there are, the partition function becoming Z^n for n units.

As an example, if $E_g/k_BT=1$, which would be the case for a small barrier between trans to gauche or at a high temperature, then the partition function $Z=1.736$. If there are 10 repeat units in an alkane then the chance of an all-trans chain is $1/Z^{10}=0.004$ compared to 0.57 for butane. If E_g/k_BT is larger which it will be at a low temperature then the probability of being in the all-trans state is almost 1, even for a long alkyl chain because $Z\to1$ at low a temperature.

1.9.14 Questions

Full solutions are available at **www.oxfordtextbooks.co.uk/orc/beddard.**

Q1.17 The permutation of 1, 2, 3, 4, 5 produces a 5 digit number. If these permutations are listed in order of increasing value, what is the 33rd?

Strategy: Do not enumerate the numbers, but split them into 5 groups of 24, then 4 groups of 3 etc.

Q1.18 Carbon consists of two common isotopes ^{12}C and ^{13}C. ^{12}C is defined to have a relative atomic mass of exactly 12, whereas tables give the value of C as 12.011.

(a) What is the fraction of ^{13}C present in a normal sample of carbon?

(b) What is the chance that two ^{13}C nuclei are present in an ethane molecule, C_2H_6?

(c) An ethane molecule has to have at least one ^{13}C if a ^{13}C NMR spectrum is to be recorded. What is the chance of this occurring? If the magnetogyric ratio of ^{13}C is $\gamma = 6.728 \times 10^7$ and for protons $\gamma = 26.75 \times 10^7$ rad T^{-1} s^{-1} how much smaller is the ^{13}C NMR spectrum than the H spectrum, if the NMR spectrometer's sensitivity is proportional to γ^3?

Q1.19 In the study of genetics, probability can be used to predict the fraction of offspring with various traits. Animals or plants can have dominant A or recessive a genes which can mix to give offspring with genetic make up AA, Aa and aa.

(a) If p is the fraction of a population with gene A and $q = 1 - p$ with a, what is the incidence of individuals with AA, Aa and aa genes?

(b) The trait associated with a recessive gene a only manifests itself in an individual with a genetic makeup aa. What is the chance that an individual *without* this trait has one a gene?

Q1.20 It is possible, but extremely unlikely, that due to random thermal motions all the air molecules in a room will occupy just its top part. Assuming that the molecules do not interact, calculate:

(a) What is the chance a molecule will only occupy V_A if a room is split into two volumes V_A and V_B?

(b) What is the chance that all molecules occupy V_A, if the volumes are equal, and there are 100, and then 1000 gas molecules in the total volume?

(c) If the entropy is defined as $S = k\ln(\Omega)$ where k is Boltzmann's constant and Ω the probability, what is the entropy change upon isothermally expanding N gas molecules from V_A to $V_A + V_B$?

Q1.21 **(a)** In DNA there are four bases labelled G, C, A, T. If three bases encode an amino acid, how many amino acids could there be?

(b) If only 20 amino acids exist, how many 100 amino acid proteins are possible?

Q1.22 List all the distinguishable arrangements of two distinguishable particles in three degenerate energy levels.

Q1.23 A nitrogen atom has the configuration $2s^3 2p^3$ and an Fe atom $2p^6 3d^6$. How many states are expected in each case?

Q1.24 A microstate in an atom comprises each conceivable set of the m_s and m_l quantum numbers for a configuration such as p^2 or d^5 and is therefore the total degeneracy of the configuration. The electron in any of the five degenerate d orbitals has m_l orbital quantum numbers with values −2, −1, 0, 1, 2 but only two m_s values, ±1/2. If there is one electron then there are 10 ways of distributing this in d orbitals producing 10 microstates; colloquially put, the electron can be 'spin up' or 'spin down' in each of five orbitals. The configurations of d orbitals run from d^0 to d^{10}, scandium to copper. How many microstates are their in each configuration?

Q1.25 How many distinguishable states will be produced if five quanta are distributed in a doubly degenerate vibration and how many in a triply degenerate one?

Q1.26 Fig. 1.24 shows a few of the ways in which five identical particles can be distributed among seven energy levels. Work out the ratio of the probabilities with which each of these distinguishable arrangements is produced.

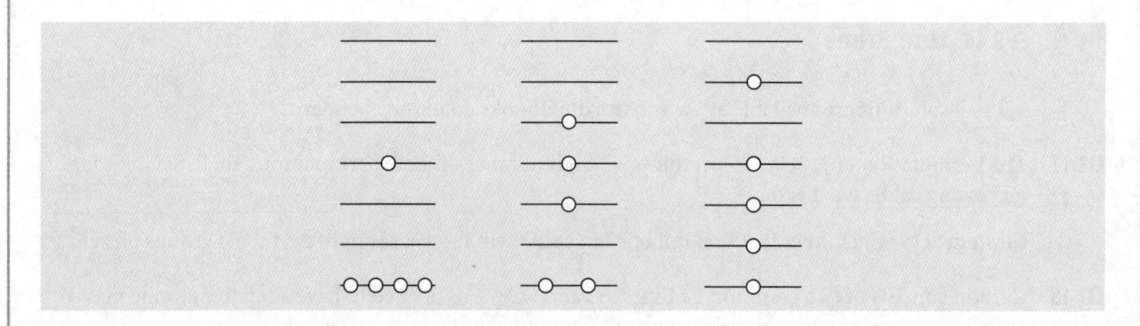

Fig. 1.24 Some of the distributions of five quanta among seven levels.

Q1.27 It is tempting to suppose that the reaction $H_2 + D_2 = 2HD$ has an equilibrium constant $K = 1$, because of the random distribution of H and D isotopes, but this would be wrong, because $K = 4$. If f_H is the fraction of H atoms (protium) and f_D that of deuterium the fraction by random mixing in H_2 is f_H^2, in D_2 is f_D^2 and in HD is $2f_Df_H$. The factor 2 arises because of the two ways of arranging H and D—just as with two coins.

The equilibrium constant is therefore $K = \dfrac{(2f_{HD})^2}{f_H^2 f_D^2} = 4$. As there is no energy change in the reaction, (there is no isotope fractionation), the randomization of the isotopes is the driving force for the reaction, i.e. $\Delta_r H = 0$ and the reaction is caused by an increase in entropy and $\Delta G = -RT\ln(4)$.

Calculate the equilibrium constant K for each reaction, assuming $\Delta_r H = 0$;

(a) HCOOH + DCOOD = HCOOD + DCOOH

(b) $P^{35}Cl_3 + P^{37}Cl_3 = P^{37}Cl_2{}^{35}Cl + P^{35}CL_2{}^{37}Cl$

(c) $C^{35}Cl_4 + C^{37}Cl_4 = 2C^{37}Cl_2{}^{35}Cl_2$

Strategy: Calculate the fraction of each isotopic atom in a molecule, and multiply by the number of ways this can be achieved.

Q1.28 In the eighteenth century the Chevalier de Meré made a fortune by betting on small but favourable odds that at least one six would appear in four throws of a die. He lost his fortune in small favourable odds of throwing at least two sixes in 24 throws. Work out the probabilities in each case and explain what happened.

Q1.29 Two cards are drawn simultaneously from two groups of cards which are {AD, 2H, 1C, 2C, 3C} and {AD, 2H, 3H, 1C} (where AD is the ace of diamonds, 1C the 1 of clubs etc.) **(a)** what is the chance of getting at least one club, **(b)** at least an ace, **(c)** at least one cards less than 4?

Q1.30 The compound phenylthiocarbamide (PTC) has no taste to some individuals but is bitter to others. This trait is controlled by a single gene with the allele for bitter taste being dominant. If the dominant allele is B and the recessive b,

(a) Over a whole population by selecting just those heterozygous tasters who have children, what fraction will be able to taste the PTC?

(b) What is the chance that each of the first four children, and what fraction of all children, will be non-tasters?

(c) Of these children what is the chance that any of their children will be a tasters?

Q1.31 If **(a)** 10, **(b)** 100, **(c)** 1000 random digits, 0–9 are chosen, what is the chance of finding a run of half of the total numbers which are each zero?

Strategy: In **(b)** and **(c)** first calculate the log of the probability by using Stirling's approximation for the factorial in the binomial distribution.

Q1.32 When a whole population is considered, boys and girls are born with equal frequency.

(a) What fraction of families with 6 children is expected to have 4 girls?

(b) How is this fraction changed if the birth ratio is 3:2 in favour of girls?

Q1.33 Prove the following relationships between binomial coefficients

(a) $q\begin{pmatrix} n \\ q \end{pmatrix} = n\begin{pmatrix} n-1 \\ q-1 \end{pmatrix}$, **(b)** $\begin{pmatrix} n \\ q \end{pmatrix} = \begin{pmatrix} n-1 \\ q-1 \end{pmatrix} + \begin{pmatrix} n-1 \\ q \end{pmatrix}$

1.10 **Modulo arithmetic**

Clock time is usually measured to base 60 for minutes, and 12 for hours. For example, if you arrange to meet someone at 10:35 and they are 40 minutes late, then rather than noting the time as 10:75 we say it is 11:15. This is an easy mental calculation expressed more awkwardly and mathematically as $35 + 40 \equiv 15 \bmod 60$. Symbolically, if the difference between two integers a and b is divided by an integer m, or $(a - b)/m$, and if the result is an integer, then a and b are said to be 'congruent modulo m'; hardly an expression that flows naturally off the tongue. As an *identity*, this is written as

$$a \equiv b \bmod m$$

which means that a is *equivalent* to b mod m. More familiarly, it means that integer division a/m produces the *remainder* b; thus $75/60 = 1$ with remainder 15. Integer division measures only how many times one integer goes into another, ignoring any remainder; thus $3/2 = 1$, $23/4 = 5$, and so forth.

However, the usage in most computer languages is different to this and is written as

$$a = b \bmod m,$$

and notice the = sign. The equation means that a is the *remainder* of dividing a positive number b by m a whole number of times. Using Maple produces

```
> 39 mod 10;           9
```

because integer division $39/10 = 3$ with a remainder of 9. When b is a negative integer, the result is the same as calculating $m - (+b \bmod m)$. For example,

```
> -39 mod 10;          1
> 10-(39 mod 10);      1
```

One important example is found in the CAS registry number, which is used to identify uniquely every chemical compound. Its last digit is a check digit, used to confirm the uniqueness of the number. There are at least 35 million compounds known, so uniqueness is important. The check digit is obtained by multiplying each preceding digit by its position in the number; taken in *reverse* order, starting by multiplying the last digit by 1, the second to last by 2, etc. then adding the result and calculating modulo 10 of the number produced. The CAS number for naphthalene is 91 – 20 – 3; the 3 is the check digit. This is obtained with the following sum, $0 \times 1 + 2 \times 2 + 1 \times 3 + 9 \times 4 = 4$ and finally, $43 \bmod 10 = 3$, where the 3 is the remainder of a whole number of divisions of 10 into 43.

Musical scales and clock time use circular arithmetic. The equal temperament scale has 12 frequencies in each octave given by $2^{n/12}$ times the base frequency where $n = 0$ to 12, this being the scale that most people find is most pleasing to their ears. The next octave, above or below, has exactly the same ratio of frequencies and so follows modulo 12 arithmetic.

1.10.1 **Questions**

Full solutions are available at **www.oxfordtextbooks.co.uk/orc/beddard.**

Q1.34 The CAS registry numbers for pheophytin-a (chlorophyll-a with the Mg atom replaced by two H atoms), retinal and trans-stilbene are 100 759 – 86 – x, 116 – 31 – x, and 103 – 30 – x respectively. Work out their check-sum numbers x then look them up to confirm their values.

1.11 **Delta functions**

The Kronecker delta is a function with two indices, and has a value of 1 when these are the same, and 0 when they are different.

$$\delta_{n,m} = 1 \text{ if } n = m, \text{ i.e. } \delta_{n,n} = 1, \text{ otherwise } \delta_{n,m} = 0.$$

This function is met when calculating the orthogonality of many functions with integer arguments, for instance, sin(nx) and cos(mx), see Chapter 9, and is commonly also found in quantum problems where it is often used to pick out one term in a summation;

$$\begin{aligned} s_j &= \sum_i a_i \delta_{ij} \\ &= a_0\delta_{0j} + a_1\delta_{1j} + \cdots a_j\delta_{jj} + \cdots a_n\delta_{nj} \\ &= a_j \end{aligned}$$

The second type of delta function $\delta(x)$ is named after Dirac. This behaves like a normal continuous function but has a positive value only at x and elsewhere is exactly zero. The area under the function, which is its integral, is unity. This means that the function is a spike at position x of infinitesimal width but unit area. Further properties of this function are described in Chapter 9. The Dirac delta can be derived in a number of ways, but the function may be realized by drawing a Gaussian (bell-shaped curve), then making it narrower and narrower until, at the limit, it becomes the δ function.

1.12 Series

A series of numbers is 1, 2, 3, 4 ⋯ or 1, 2, 4, 8, 16 ⋯ . If each term is made from the previous one by adding a constant number, this is an *arithmetical* progression. If, however, each term after the first is multiplied by a constant term, the series is called a *geometrical* progression. Both series may continue to infinity, but the last value may still be finite.

Examples of arithmetic progressions are

$$1+2+3+4+\cdots \qquad \text{or} \qquad 1+3+5+7+9+\cdots$$

or in general

$$a_1 + a_2 + a_3 + a_4 + \cdots = a_1 + (a_1 + d) + (a_1 + 2d) + \cdots + (a_1 + [n-1]d).$$

where the constant additional term is d.

In the geometric progressions below, in G_1, each term after the first is multiplied by 1/2; in G_2 the multiplier is 4 and is x in G_x.

$$G_1 = 1 + 1/2 + 1/4 + 1/8 + \cdots$$

$$G_2 = 1 + 4 + 16 + 64 + \cdots$$

$$G_x = a + ax + ax^2 + ax^3 + \cdots ax^{n-1} + \cdots = a(1 + x + x^2 + x^3 + \cdots x^{n-1} + \cdots)$$

The first of these three series, G_1, converges to a finite value even after an infinite number of terms, and the sum of the terms to infinity is finite.

The spectrum of atomic hydrogen and other atoms with one outer electron show an emission spectrum that converges to a limit. In the ultraviolet part of the spectrum, the lines are called the Lyman series and have the form

$$\tilde{\upsilon} = R\left(1 - \frac{1}{n^2}\right)(\text{cm}^{-1})$$

where n is an integer greater than 1. The positions of the lines are sketched in Fig. 1.25. The symbol $\tilde{\upsilon}$ represents the frequency in wavenumbers or cm^{-1}. These are habitually used by

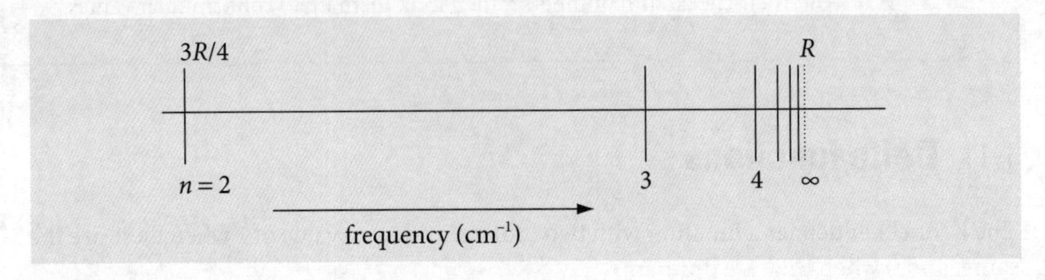

Fig. 1.25 The Lyman series converges to a finite value.

spectroscopists rather than s^{-1} (or Hertz) because the numbers are smaller, a few thousands of wavenumbers correspond to $\sim 10^{13}$–10^{15} s^{-1}; 1 $cm^{-1} \triangleq 3 \times 10^{10}$ s^{-1}.

1.13 Estimation

Estimating quantities is sometimes relatively easy to do and can often be used to help design experiments. An accurate number is not sought; just an order of magnitude estimate is often sufficient. A straightforward example is to estimate the minimum useful concentration of a sample knowing the sensitivity of an instrument. This can be put the other way around. At the maximum allowable sample concentration, we can estimate what the size of the signal will be and whether the method proposed will work or whether another more sensitive instrument should be used. Other forms of estimation can be used to obtain quantities that would otherwise seem impossible, such as calculating the weight of a mountain or the weight of the atmosphere. In these types of problems, we make some simplifying, but not unreasonable assumptions to be able to reach a sensible conclusion; the mountain could reasonably be approximated by a cone or a hemisphere depending on its shape. Several interesting estimation examples are given by Adam (2003).

(i) In an X-ray experiment on crystalline Be, suppose that the detector (e.g. a CCD camera) can respond to single X-ray photons; estimate the least number of electrons that there must be in the crystal's scattering plane for a signal to be detected. What does this imply for the size of the crystal? (Assume that the scattering from n electrons is n times that from a single one and the following data is assumed to be known.)

The intensity, I, of X-ray scattering by a *single electron* is approximately given by

$$I = I_0 \frac{e^4}{(4\pi\varepsilon_0)^2 m^2 c^4} \frac{1}{R^2} \frac{(1+\cos^2(\phi))}{2}$$

where $I_0 = 10^8$ is the initial number of X-ray photons per second, R the distance from the centre of scattering to a detector, and ϕ the angle through which the X-ray is scattered. The detector is 10 cm from the sample and at an angle $\phi = 20°$.

In this example, although calculating I/I_0 is straightforward it is not if you use a calculator because the powers produced are too large. The calculation is done by first adding the powers, then the remaining numbers are easily evaluated.

$$\begin{aligned} \frac{I}{I_0} &= \frac{(1.6 \times 10^{-19})^4}{(4\pi 8.8 \times 10^{-12})^2 (9.1 \times 10^{-31})^2 (3 \times 10^8)^4 100} \left(\frac{1.8}{2}\right) \\ &= \frac{16^4 \times 10^{-4}}{4^2\pi^2 \times 8.8^2 \times 9.1^2 \times 3^4 \times 2} 10^{-76+24+62-32-2} \times 1.8 \\ &\approx 10^{-31} \end{aligned}$$

If the initial X-ray intensity is 10^8 photons/sec then $\approx 10^{-23}$ X-rays are detected from each electron, which means that 10^{23} electrons have to be present to detect one X-ray. As Be contains four electrons, $1/4 \times 10^{23}$ atoms are needed, which is $10^{23}/(4 \times 6 \times 10^{23})$ moles = 0.04 moles. The molar mass is 9 g mol^{-1}. Thus, to detect 1 photon/second, the sample has to weigh at least 0.36 g and as the density of Be is 1.8 g cm^{-3} this is a cube of volume 0.36/1.8 cm^3 which has a side of $\approx$6 mm.

(ii) Sometimes, a particular estimation seems to be very hard but is not. To calculate the mass of the atmosphere would appear to be difficult because of the changing density with increasing altitude. However, if atmospheric pressure is taken at sea level to be 1 atm = 101 325 Pa all that is needed is the surface area of the earth and the fact that a pascal is a N m^{-2} which is force/area. Assuming that the earth is a sphere of radius 6378 km, its surface area is $4\pi \times 6378^2 \times 1000^2 = 5.1 \times 10^{14}$ m^2 and the total force on its surface due to the atmosphere is 101 325(Pa) $\times 5.1 \times 10^{14}$ (m^2) $\approx 5.2 \times 10^{19}$ N (kg m s^{-2}). The atmospheric mass is therefore $\approx 5.18 \times 10^{19}/g = 5.27 \times 10^{18}$ kg where g is the acceleration due to gravity 9.81 m s^{-2}.

(iii) The American Benjamin Franklin is famously remembered for flying a kite with an attached wire into a thunderstorm in an attempt to capture lightning and store it in a

Leyden jar (Bernal 1973). Luckily, he survived. He is less well known for changing the florid opening of the American Constitution to read 'We hold these truths to be self-evident, that all men are created equal'. He travelled to England in 1757 to help with the tax situation of Pennsylvanians, and while sailing noticed that oil on water calmed waves, something that had been noticed since the time of the Greeks. Being interested in surface phenomena, when again in London in 1770, he poured no more than a teaspoon of oil on a pond on Clapham Common and observed that the oil spread of its own volition over the pond, sweeping leaves and other debris out of the way as it did so. He estimated that the oil covered about half an acre of the pond, and he repeated this experiment everywhere he went from then on. He also recorded seeing colours due to thin film interference effects but could not have explained this, nor could he have realized that a monolayer of molecules was formed because the idea of atoms and molecules was not clearly understood at that time. He appears not to have worked out the thickness of the oil layer. To calculate the monolayer thickness, a teaspoon holds approximately 4.5 ml and half an acre is 2023 m^2. From the amount of oil used, the thickness of the layer was 4.5×10^{-6} m^3/2023 m^2 = 2.2 nm. This is approximately the thickness of a close-packed one-molecule thick layer of a long chain fatty acid monolayer, and gives a simple and direct estimate of the length of the oil molecules used. Oleic acid is a major constituent of olive oil and is a by-product of making candles; it has an 18-carbon chain so when fully extended would be about 2.8 nm long, remarkably similar to our estimate.

(iv) If the density of Li is 0.534 g ml^{-1}, what is average size of an atom? A mole contains 6.022×10^{23} molecules, therefore 6.9 g (the molar mass) contains this number of atoms occupying a volume (0.0069 kg mol^{-1})/534 (kg m^{-3}) = 2.17×10^{-29} m^3 or 2.18×10^{-2} nm^3. Assuming spherical atoms, each has a radius 0.17 nm which is not a bad estimation considering the approximate nature of the calculation. The similar calculation for Pb produces 0.19 nm.

1.14 Rounding numbers and units

1.14.1 Significant figures and rounding numbers

A measurement always has two parts: a numerical value and its associated units. It is essential to report numbers to the appropriate number of decimal places. This is done either according to what is possible from the experimental conditions, or from the precision of numbers used in a calculation; and therefore some adjustment, called rounding, of the number is necessary.

There is no single or perfect way of representing a number and so there is room for some personal preference, but scientific notation offers the least ambiguity. To represent a number in this notation, write it with one leading digit followed by the decimal point and then more digits followed by a power of 10; for instance,

$$97\,453.1 \rightarrow 9.74531\times10^4, \qquad 0.0245 \rightarrow 2.45\times10^{-2}$$

and the leading zeros in the second number are seen not to be significant. If a series of numbers are to be compared, then it is best to use the same power of 10 for each; for example, the following three numbers shows how the steady increase in value can easily be recognized. It is even clearer when these numbers are rounded up, say to the nearest 1000, and this gives the values shown on the right.

$$\begin{array}{lllll} 984.7 & \rightarrow & 0.9847\times10^3 & \rightarrow & 1\times10^3 \\ 60\,357 & \rightarrow & 60.357\times10^3 & \rightarrow & 60\times10^3 \\ 124\,560 & \rightarrow & 124.560\times10^3 & \rightarrow & 124\times10^3 \end{array}$$

By the nature of any experiment, a measurement is only known to a certain number of significant digits. These are independent of the size of the number. For instance, 123.4 and 0.00001234, both contain four significant figures and these numbers can be written as 1.234×10^2 and 1.234×10^{-5} respectively. The number 123.40 contains five significant figures, because writing the last zero implies that this digit is known. The limited number of significant figures occurs because of many unknown variations in the way a quantity is measured. Imprecision may occur due to a sloppy experimental technique, but supposing that this is not the case 'noise' may be added to a measurement from any number of sources. By 'noise' is meant the random variability that is seen in any measurement that you would prefer is not there and which often masks the true signal. Perhaps this noise is due to interference from other instruments in the laboratory, from mains voltage fluctuations, or from temperature variations. Noise is present because measurements were at the limit of your instrument's capability. Often data is subsequently analysed to extract the information it contains; the slope of a line, for example, and then the number of significant figures reported as the slope and its uncertainty must be carefully considered. A calculator or computer will happily produce 10 or more digits in an answer; almost invariably far more than is realistic. First, to decide how many significant figures there really are, the original data must be looked at before reaching a conclusion. The result can then be rounded to the correct number of significant figures at the very end of any calculation. Both the result and its associated error will need to be rounded, but this is done with different rules.

In rounding numbers we examine the *last digit to be retained* and

(1) Retain no more digits beyond the first uncertain one.

  (i) Increase this digit by 1 if the residue is greater than 5.

  (ii) Leave this digit unchanged if the residue is less than 5.

  (iii) When the residue is exactly 5, leave this digit unchanged if it is even, or increase it by 1 if it is odd. This is done so that rounding is unbiased.

  Using these rules produces the following values when rounded to 4 significant figures or three decimal places.

  $1.02055 \rightarrow 1.021$; rule (i).

  $1.02345 \rightarrow 1.023$; rule (ii).

  $1.02350 \rightarrow 1.024$; rule (iii) increase by 1, as the last retained digit is 3.

  $1.02450 \rightarrow 1.024$; rule (iii) unchanged as digit 4 is even.

(2) In addition or subtraction, do not retain any more digits in the answer than the number with the smallest number of digits; $21.1 + 2.035 + 6.12 = 29.255 \rightarrow 29.3$.

(3) In multiplication or division, the result should have no more digits than the least precise number, which has the smallest number of significant digits:

  $21.1 \times 0.029 \times 83.2 = 50.91008 \rightarrow 51$, because 0.029 has 2 digits. This result would be better written as 51.0 because 51 implies that the number is an integer which is known exactly.

  $291 \times 272/0.086 = 920\ 372.093 \rightarrow 9.2 \times 10^6$ because 0.086 has 2 digits.

(4) The log of a number should have as many digits to the right of the decimal point (the mantissa) as there are significant digits in the number.

(5) The mean of a number has as many significant figures as the observations upon which it is made; only such a number of significant digits should be retained so that the uncertainty in the mean corresponds approximately to its standard deviation.

Examples of these rules are

$$121.1 + 2.035 + 6.12 = 129.255 \rightarrow 129.3$$

$$291 \times 272/0.086 = 920\ 372.093 \rightarrow 9.2 \times 10^6$$

$$\log(4.000 \times 10^{-5}) = -4.3979$$

$$10^{12.5} = 3.16 \times 10^{12}$$

1.14.2 Experimental results with experimental uncertainties or errors

The error quoted with a measured mean value represents the chance that the data will fall between the error values quoted, usually this is approximately 68%, meaning that by random chance 32% of the times a measurement is made, a value outside the range will be observed. In some cases a 95% limit may be used; this will depend on how the standard deviation or the standard error is defined and this is explained in more detail in Chapter 13. We shall suppose that this has been decided upon, so that just the numbers are examined. To report a result as 7.56 ± 0.03456 kJ/mole would be wrong. It is reasonable to assume that the result is the mean value of several measurements because an error is given. The mean is quoted to only 3 significant figures, 1 part in a 1000, or 2 decimal places, and as the error is usually calculated from the data, it cannot be known to more figures than this. The result should be reported as 7.56 ± 0.03 kJ/mole as the 3 in the quoted error falls in the same decimal place as the 6 in the number. If you want to be cautious, then 7.56 ± 0.04 would be acceptable as would $7.56 \pm 0.03_4$ to indicate the figure that could be rounded.

As a rule of thumb, experimental uncertainties (errors) should be rounded to one significant figure, unless the measurement is *very* precise, then two figures may be used. The error would normally always be rounded up. Once the uncertainty in the measurement is determined, the significant figures in the measured value may need to be revised. This would be the case if the error were determined by some means other than from analysing the data. Reporting $\Delta G = -6051.78 \pm 30$ J/mole is just not reasonable as the error is so large; the result should be $\Delta G = -6050 \pm 30$ J/mole, as an uncertainty of 30 means that the result could be as small as 6020 or as large as 6080, so that the trailing digits, 1, 7, and 8 do not matter.

A second rule of thumb is that the last significant figure in any stated answer should be of the same order of magnitude, i.e. in the same decimal position, as the uncertainty. For example,

$$92.81 \pm 0.3 \text{ should be reported as } 92.8 \pm 0.3;$$

$$92.81 \pm 3 \quad \rightarrow 93 \pm 3.$$

$$92.81 \pm 30 \quad \rightarrow 90 \pm 30.$$

In the last case, the rounded result is a little smaller than the result but the error is so large that this is of no consequence. In other words, the 92.81 is only one of many results that could have been obtained had the experiment been repeated many more times and values from at least 60 to 120 are to be expected. In cases such as 92.5 ± 3.5, you might not want to round up either to 93 or down to 92. In this case the rounded number could be reported as $(0.92_5 \pm 0.04) \times 10^2$.

1.15 SI units and prefixes

The International System of Units (SI) is nowadays used in all textbooks and much of the scientific literature. See Mills et al. (1993) or Cohen et al. (2007) for a full description of all units. The SI system is based on a set of defined units so that a quantity either has one of these units, or is derived from them. There is also a set of named prefixes for numerical values and these are shown below.

1.15.1 Defined units

The defined units are the metre (m), the kilogram (kg), the second (s), the kelvin, the unit of thermodynamics temperature (K), the ampere (A), the mole (mol), and the candela, which is luminous intensity (cd).

1.15.2 Derived units

All other units we use, such as the joule, are derived from these base units. The joule (J) measures energy and $1\text{ J} = 1\text{ kg m}^2\text{ s}^{-2}$. The SI unit for force, the newton (N) is derived from SI base units using the relationship:

$$\text{Force} = \text{mass} \times \text{acceleration} \qquad \text{N} = \text{kg} \times \text{m s}^{-2}$$

One newton is defined as the force required to give a mass of 1 kilogram an acceleration of 1 metre per second per second.

The table lists some derived units and other commonly used quantities and their units

Quantity	Symbol	SI base units
Force (newton)	N	kg m s^{-2}
Energy, (force × distance), work, heat, (joule)	J	kg m^2 s^{-2}
Pressure (force/area) (pascal)	Pa	N m^{-2}
Electrical charge (coulomb)	C	A s
Electrical potential (volt)	V	J C^{-1}
Electrical resistance (ohm)	Ω	V A^{-1}
Electrical capacitance (farad)	F	C V^{-1}
Inductance (henry)	H	V A^{-1} s = m kg s^{-2} A^{-2}
Angular velocity	ω	rad s^{-1} (or s^{-1})
Heat capacity, entropy	C_p, C_V, S	J K^{-1}
Molar heat capacity, molar entropy	C_p, C_V, S	J K^{-1} mol^{-1}
Concentration	[c]	mol m^{-3} (Conventionally mol dm^{-3})
Viscosity*	η	Pa s = kg m^{-1} s^{-1}
Frequency (hertz)	Hz	s^{-1}
Wavenumber	cm^{-1}	m^{-1} but conventionally cm^{-1}

*Viscosity is still commonly quoted in centipoise (cP) where 1 cP = 10^{-3} Pa s and this unit is used because water has a viscosity of ≈1 cP at room temperature, and ethylene glycol about 18 cP.

Powers of 10 can be subsumed into the unit by use of prefixes.

10^{-1}	10^{-2}	10^{-3}	10^{-6}	10^{-9}	10^{-12}	10^{-15}	10^{-18}	10^{-21}	10^{-24}
deci	centi	milli	micro	nano	pico	femto	atto	zepto	yocto
d	c	m	μ	n	p	f	a	z	y
10	10^2	10^3	10^6	10^9	10^{12}	10^{15}	10^{18}	10^{21}	10^{24}
deca	hecto	kilo	mega	giga	tera	peta	exa	zeta	yotta
D	H	K	M	G	T	P	E	Z	Y

The bond length of HCl is 0.00000000012745 m; possible choices for recording this are

$$r = 1.2745 \times 10^{-10}\ \text{m}$$
$$= 0.12745\ \text{nm}$$
$$= 127.45\ \text{pm}$$

of which the final form is becoming common. The angstrom Å is 10^{-10} m and although not an SI unit, is still very frequently used. This bond length is 1.2745 Å.

1.15.3 **Atomic units**

When working at quantum problems, it is sometimes easier to use atomic units. In these units the electron charge e is taken to be 1 unit of charge and its mass m_e also 1 unit of mass. The energies are always electrostatic and hence proportional to $e^2/(4\pi\varepsilon_0)$ where ε_0 is the permittivity of free space and has units F m^{-1} = C^2 J^{-1} m^{-1}. This quantity in SI units is $C^2/(C^2\ J^{-1}\ m^{-1})$ = J m and therefore has dimensions mass × length^3 × time^{-2}. Planck's constant squared has units of $(J\ s)^2$ = mass^2 × length^2 × time^{-1} and using these quantities can now *define* a length as

$$a_0 = \frac{\hbar^2}{m_e(e^2/4\pi\varepsilon_0)} = 0.529177 \times 10^{-10}\ \text{m}$$

This is the Bohr radius and the unit of length in the atomic units. It corresponds to the radius of the 1s orbit of a H atom. The unit of energy is the hartree, which is

$$E_h = \frac{e^2}{4\pi\varepsilon_0}\frac{1}{a_0} = 4.35974 \times 10^{-18}\ \mathrm{J} = 27.2114\ \mathrm{eV}.$$

and is twice the absolute value of the energy of the 1s electron in hydrogen or twice the ionization energy. The unit of time in atomic units is $\hbar/E_h = 2.41888 \times 10^{-17}$ s.

1.15.4 Converting a number to different units

It is often necessary to convert between different sets of units. All non-SI units have a *definition* in terms of the corresponding SI units. An example conversion table for pressure is

$$\begin{aligned}
&1\ \mathrm{Pa} = 1\ \mathrm{N\ m^{-2}} = 1\ \mathrm{kg\ m^{-1}\ s^{-2}}\\
&1\ \mathrm{bar} = 100\ \mathrm{kPa}\\
&1\ \mathrm{atm} = 101\ 325\ \mathrm{Pa}\\
&1\ \mathrm{atm} = 760\ \mathrm{Torr}\\
&1\ \mathrm{psi}^* = 6894.76\ \mathrm{Pa}
\end{aligned}$$

*psi is pounds per square inch.

A good source of conversions is the *CRC Handbook of Chemistry and Physics* (Weast). Maple can also be used to produce a conversion table and the search engine Google allows one to ask questions such as 'calculate the speed of light in furlongs per fortnight'. The answer by the way is 1.8×10^{12} and a furlong is 220 yards. However, although these programs will do this for you it is clearly better to have an idea of what to do yourself.

To convert pressure data from the units of Torr to the SI unit of N m^{-2} (or Pa), one of two methods can be used

Method 1; Direct Substitution

Substitute the value of 1 torr for its equivalent in N m^{-2} as if 'torr' were a variable in the equation:

$$p = 760\ \mathrm{torr} = 760\ (133.322\ \mathrm{N\ m^{-2}}) = 101\ 325\ \mathrm{N\ m^{-2}}$$

Method 2; 'Multiply by 1'

This is the most reliable method to use. The equation is multiplied by 1 by using a unit conversion so that the unit to remove is on the denominator. For example from the definition of a torr,

$$1 = 133.322\ \mathrm{N\ m^{-2}/torr}$$

and substituting makes the equation

$$p = 760\ \mathrm{torr} = 760\ \mathrm{torr}\ (133.322\ \mathrm{N\ m^{-2}})/\mathrm{torr} = 101\ 325\ \mathrm{N\ m^{-2}}$$

and the units cancel giving the result in the new units. To express a pressure of 62 psi in Pascal, using the conversion table above and the 'multiply by 1' method, gives

$$62\ \mathrm{psi} = 62\ \mathrm{psi}\left[\frac{6894.7\ \mathrm{Pa}}{\mathrm{psi}}\right] = 4.27 \times 10^5\ \mathrm{Pa}$$

Planck's constant is $h = 6.626 \times 10^{-34}$ J s, but in some cases it is easier to use this in units of eV ps where 1 ps = 10^{-12} s, and one electron volt, 1 eV = 1.602×10^{-19} J (see table in Section 1.14.5), giving

$$h = 6.626 \times 10^{-34}\ \mathrm{J\ s}\left(\frac{\mathrm{eV}}{1.602 \times 10^{-19}\ \mathrm{J}}\right)\cdot\left(\frac{\mathrm{ps}}{10^{-12}\ \mathrm{s}}\right) = 4.136 \times 10^{-3}\ \mathrm{eV\ ps}.$$

This is even more useful in cm^{-1} ps, which is calculated as

$$h = 6.626 \times 10^{-34}\ \mathrm{J\ s}\left(\frac{\mathrm{cm^{-1}}}{1.986 \times 10^{-23}\ \mathrm{J}}\right)\cdot\left(\frac{\mathrm{ps}}{10^{-12}\ \mathrm{s}}\right) = 33.36\ \mathrm{cm^{-1}\ ps}.$$

1.15.5 Generating conversion tables of scientific units using Maple

Maple can be used to generate conversion tables, lists of scientific constants, and other atomic properties. To obtain a named constant, use

```
> with(ScientificConstants):
> GetConstant(h);
```

$$Planck\_constant,\ symbol = h,\ value = 6.62606876\ 10^{-34},$$
$$uncertainty = 5.2\ 10^{-41},\ units = J\ s$$

```
> hbar:= evalf(Constant(h)/(2*Pi));
```

$$hbar := 1.054571596\ 10^{-34}$$

and note the use of capital letters.

The syntax to generate a table of energy conversions is

```
> with(ScientificConstants):
> alist:= [ J, kJ/mole, eV, cm^(-1) ]:
> convert(alist, conversion_table, energy, output=grid );
```

	To:	J	$\frac{kJ}{mol}$	eV	$\frac{1}{cm}$
Unit Name joules	*Symbol J*	1.	$6.0221367\,10^{20}$	$6.2415064\,10^{18}$	$5.0341125\,10^{22}$
kilogjoules per mole	$\frac{kJ}{mol}$	$1.6605402\,10^{-21}$	1.	0.010364272	83.593461
electronvolts	eV	$1.6021773\,10^{-19}$	96.485309	1.	8065.5409
1 per centimeter	$\frac{1}{cm}$	$1.9864475\,10^{-23}$	0.011962658	0.00012398424	1.

This is read as row to column, for example, 1 eV = 8065 cm^{-1}, a larger list can be made by using the `output=columns` command instead of `output=grid`.

To obtain data on an element, use

```
> alist:= GetElement(Ga):
    for i from 1 to nops([alist]) do print(alist[i]); end do;
```

31
symbol = *Ga*
name = *gallium*
names = { *gallium* }
electronegativity = [*value* = 1.81, *uncertainty* = *undefined*, *units* = 1]
electronaffinity = [*value* = 0.3, *uncertainty* = 0.15, *units* = *eV*]
atomicweight = [*value* = 69.723, *uncertainty* = 0.001, *units* = *u*]
ionizationenergy = [*value* = 5.9993, *uncertainty* = *undefined*, *units* = *eV*]
boilingpoint = [*value* = 2477., *uncertainty* = *undefined*, *units* = *K*]
meltingpoint = [*value* = 302.91, *uncertainty* = *undefined*, *units* = *K*]

$$density = \left[\ value = 5.91,\ uncertainty = undefined,\ units = \frac{g}{cm^3}\right]$$

If the atomic number is added after the name, data for that isotope is produced.

1.15.6 Questions

Full solutions are available at **www.oxfordtextbooks.co.uk/orc/beddard**.

Q1.35 Rewrite and/or correct the following in the most appropriate form.

$5.6789 \times 10^{-7} \pm 4 \times 10^{-9}$ kg	30.01 mm ± 0.001 cm	9.36156 ± 0.312 kJ/mole
$1.2345 \times 10^{4} \pm 3$ m/s	3.45 ± 0.6564 tesla	123.34 ± 20 K
2312.128 ± 0.01 cm^{-1}	19.67346 ± 1.57 kJ/mole	101352 ± 133.322 N m^{-2}

Q1.36 Record the results of the following calculations to an appropriate number of significant figures.

$$30 + 2.167 =$$
$$3.0004/12.2 =$$
$$\log(1.001) =$$

Q1.37 Convert voltage, resistance, and capacitance into base SI units.

Q1.38 **(a)** If a piece of paper is 0.1 mm thick, how many times will it have to be folded to reach the moon? This is hardly realistic, nevertheless, suppose that it were possible.

(b) Graphene is a one carbon thick layer of atoms with the structure of graphite having sp^2 hybridized C-C bonds forming six membered rings. If a sheet of graphene were folded for the same number of times as the paper, how thick would it be?

Q1.39 In natural history programmes, it is common to see pictures of dried up lakes due to intense heat and lack of rainfall.

(a) Calculate how much energy is needed to evaporate the last 1 inch depth of each 1 km^2 of a lake.

(b) How many full car petrol tanks contain this much energy?

(c) If the solar irradiance is 1.3 kw /m^2 how long will it take the sun to evaporate the same volume of water assuming all the incident energy is used to do this?

(d) The water evaporating from the lake will rise, cool and condense into a cloud as it meets colder air. Why do clouds float when they contain so much water?

Complex Numbers

2

2.1 Motivation and concept

Complex numbers arise naturally in mathematics, often when solving quadratic equations such as $x^2 + x + 1 = 0$, which has the solutions $x = \dfrac{1 \pm \sqrt{-4}}{2} = \dfrac{1}{2} \pm \sqrt{-1}$. Because the negative square root cannot be evaluated, as no ordinary number can be negative when squared, a new number conventionally called i (although engineers call this j) was invented with the property $i^2 = -1$ or $i = \sqrt{-1}$. The solution to the equation becomes $x = 1/2 \pm i$. This new number is one of a new class called *complex numbers*. These are not numbers in the elementary sense used in counting or measuring, but constitute new mathematical objects and have an existence of their own. These numbers are called 'complex' only because they contain two parts and can always be written in the form

$$z = a + ib \tag{2.1}$$

where a is called the *real* (*Re*) part and b the *imaginary* (*Im*) part of the number. The complex number $z = i$, if written in the form of equation (2.1), has a real part $a = 0$ and an imaginary part $b = 1$. The latter is rather a misnomer as b is just as 'real' as is a; it is just a number and perhaps, therefore, the best way to view a complex number is to consider it a number in two dimensions with amounts a and b in each of these dimensions. In that case, a complex number can be represented as a point on a graph rather than being a point on a line, as a normal number may be considered to be. The graph is called an Argand diagram, if drawn with the real part a along the conventional x-axis and b along the y; the area defined by a and b is also called the Argand or Gauss plane. The imaginary number i has a real part that is 0 and an imaginary part that is 1, and is represented by the point $\{0, 1\}$ on the y-axis of an Argand diagram.

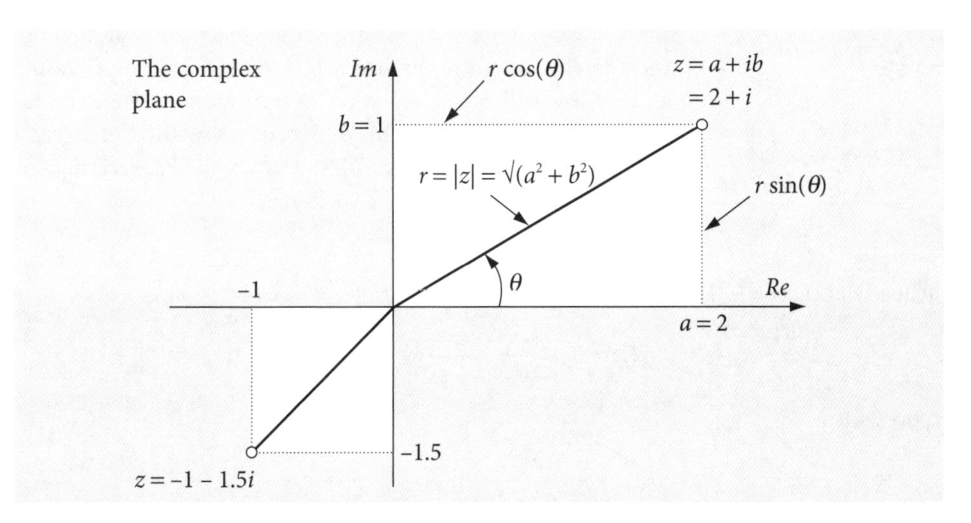

Fig. 2.1 The Argand diagram showing two complex numbers in the form $z = a + ib$, r is the modulus of the complex number z and θ the argument measured anticlockwise from the real axis.

The Argand diagram is not like a normal graph in which a function such as $y = x^3$ is plotted, because the value of y on the graph normally shows how large the function is at a given value of x. The Argand diagram shows one point in the real and imaginary plane for each complex number so is more like a map that locates a place with latitude or longitude.

Performing algebra with complex numbers is no more difficult than with 'normal' numbers, because the prime rule of algebra still applies:

'Whatever I do to one side of an equation I do to the other side.'

The normal rules for addition and multiplication apply but with the additional rule that additions and subtractions are kept separate for the real and imaginary parts, as is done for components of vectors. A complex number can be divided in the usual way by a real number. Dividing by a complex number has the additional step that the top and bottom of the expression are first multiplied by the complex conjugate of the denominator. This is explained below. Although i is a complex number, $i^2 = -1$ and is a real number:

$$i = \sqrt{-1}, \quad -i = -\sqrt{-1}, \quad i^2 = -1, \quad i = \frac{-1}{i}.$$

2.2 Complex conjugate

Complex numbers possess a new property compared to real numbers and this is the complex conjugate. If $z = a + ib$ then the *complex conjugate* is defined as

$$z^* = a - ib, \tag{2.2}$$

where, by convention, an asterisk is added and every i is replaced with $-i$; the result is that z^*z is always a real number;[1]

$$z^*z = (a + ib)^*(a + ib) = (a - ib)(a + ib) = a^2 + b^2. \tag{2.3}$$

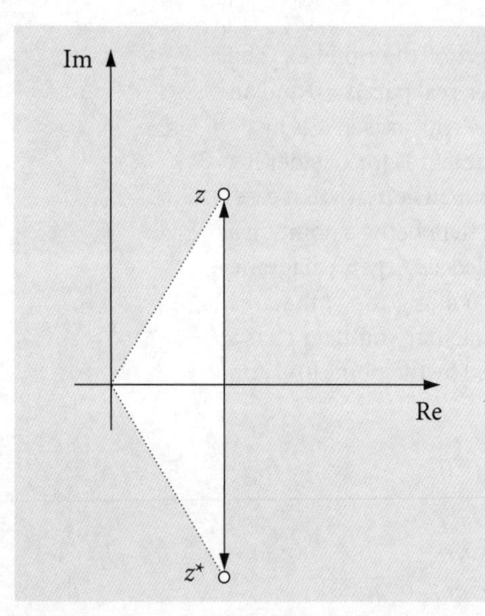

Fig. 2.2 The complex number z and its complex conjugate z^*.

In geometrical terms, forming the complex conjugate is equivalent to a reflection in the real axis because only the imaginary part is inverted.

In quantum mechanics, the wavefunction is often found to be a complex quantity and, therefore, the complex conjugate is always used to calculate expectation values such as $\langle x \rangle = \int \psi^* x \psi dx$ and probabilities $p = \int \psi^* \psi dx$ because only a mathematically real quantity is measured in an experiment, not an imaginary one.

The quantity $z + z^*$ is always a real number equal to $2\text{Re}(z)$ or $2\text{Re}(z^*)$ which is the same. It is worth remembering the rules

$$(z_1 + z_2)^* = z_1^* + z_2^*, \qquad (z_1 z_2)^* = z_1^* z_2^*.$$

In some textbooks and some scientific papers, formulae involving complex numbers are written in a form that does not include the complex conjugate but instead has the notation *+c.c.* at the end of the equation to indicate that the complex conjugate is to be added. This is primarily a method of increasing the readability of formulae. An electric field describing linearly polarized light could be written as

$$E(t, x) = E_0(e^{i(\omega t - kx)} + c.c.)$$

instead of $E(t, x) = E_0(e^{i(\omega t - kx)} + e^{-i(\omega t - kx)})$. Similarly,

$$\chi(t) = E_0\left(\frac{e^{i\omega t}}{\omega_a^2 - \omega^2 + 2i\omega/T} + c.c\right)$$

represents

$$\chi(t) = E_0\left(\frac{e^{i\omega t}}{\omega_a^2 - \omega^2 + 2i\omega/T} + \frac{e^{-i\omega t}}{\omega_a^2 - \omega^2 - 2i\omega/T}\right)$$

[1] Some texts use a bar over the number to represent the complex conjugate although this rare.

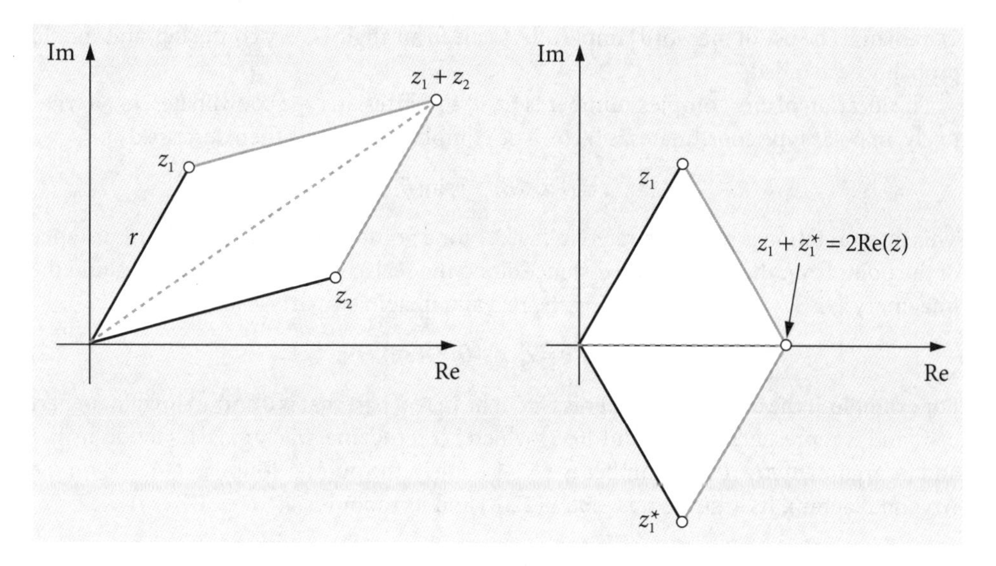

Fig. 2.3 Left: Adding two complex numbers together to form $z_1 + z_2$, dashed line. Right: Adding $z_1 + z_1^*$.

2.2.1 Adding complex numbers

The real and imaginary parts are added separately as shown in Fig. 2.3. This is somewhat like adding two vectors.

2.2.2 Multiplying and dividing complex numbers

Multiplying complex numbers is straightforward using the normal rules of algebra but remembering to use $i^2 = -1$ where necessary.

$$(3 + 5i)(1 - 2i) = 3 - 6i + 5i - 10i^2 = 13 - i.$$

Dividing numbers is a little more difficult. Always multiply top and bottom of the whole expression by the complex conjugate of the denominator, because this makes the denominator a real number, and is equivalent to multiplying by 1. An example makes this clearer.

$$\frac{3+5i}{1-2i} = \left(\frac{3+5i}{1-2i}\right)\frac{(1-2i)^*}{(1-2i)^*} = \left(\frac{3+5i}{1-2i}\right)\left(\frac{1+2i}{1+2i}\right) = \frac{13-i}{5}.$$

2.2.3 Modulus and Argument

The second new property held by complex numbers is variously called the *modulus, magnitude, absolute value*, or *norm* of the complex number. This is calculated in a similar way to that of a vector and is the length of the complex number measured from the origin, Figs 2.1–2.4.

The modulus r of the complex number $z = a + ib$ is

$$r = +\sqrt{a^2 + b^2}. \tag{2.4}$$

It is variously written as

$$r = |z| = |a + ib| = +\sqrt{z^* z} = |z^*|. \tag{2.5}$$

The square of a complex number is the square of the modulus;

$$|a + ib|^2 = (a + ib)^*(a + ib) = (a - ib)(a + ib) = a^2 + b^2 = z^* z = |z|^2$$

and is always a positive number.

In Fig. 2.1 and Fig. 2.4 the line from the origin to the complex number is at an angle θ given by

$$\tan(\theta) = b/a, \qquad \theta = \tan^{-1}(b/a) \tag{2.6}$$

measured anticlockwise from the real axis. This angle θ is called the *argument, amplitude, polar angle*, or *phase* of the complex number and is measured in radians, a full circle being

2π radians. The use of the word 'amplitude' to mean an angle is very confusing, and should probably be avoided.

The location of any complex number is $\{a, b\}$ in Cartesian type coordinates, or alternatively, in polar type coordinates is $\{r, \theta\}$. The complex number is then described as

$$z = r[\cos(\theta) + i\sin(\theta)].$$

This interpretation is also illustrated in Fig. 2.1 for a point $z = a + ib$ where r is the distance of the point from the origin. The distance along the real axis is $a = r\cos(\theta)$ and along the imaginary axis, $b = r\sin(\theta)$. Equating the real and imaginary parts gives

$$z = a + ib = r[\cos(\theta) + i\sin(\theta)]. \tag{2.7}$$

For example, if the complex number is $z = i$, it has a real part that is 0 and an imaginary part of 1, and is represented by a point $\{0, 1\}$ which is on the imaginary axis. Its modulus is 1 and its argument $\pi/2$. If the number is $z = -1 - i$ then the point is found at $\{-1, -1\}$ on the Argand diagram. Its argument is $-5\pi/4$ (225°) and its modulus $\sqrt{(-1-i)(-1+i)} = \sqrt{2}$.

2.3 Summary

If the complex number is $z = a + ib = r[\cos(\theta) + i\sin(\theta)]$ where a and b are real numbers, then

$a = \text{Re}(z)$ is the *real part* of z

$b = \text{Im}(z)$ is the *imaginary part* of z

$r = |z| = \sqrt{z^*z}$ is the *modulus* of z, or *absolute* value, *magnitude* or *norm*.

$\theta = \tan^{-1}(b/a)$ is the *argument* of z, also called the *polar angle* or *phase*.

$z^* = a - ib = r[\cos(\theta) - i\sin(\theta)]$ is the *complex conjugate* of z.

$zz^* = |z|^2 = |z^*|^2$ is the *absolute* value is always a positive real number.

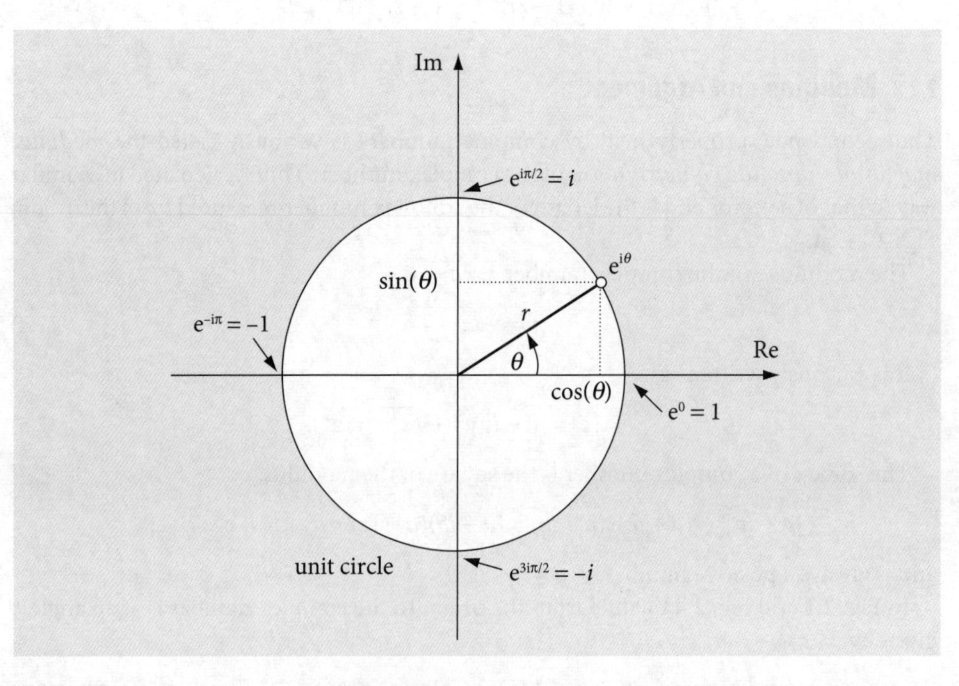

Fig. 2.4 As the angle (argument) θ varies anticlockwise from 0 to 2π, the complex number changes from 1 to i to -1 to $-i$ according to Euler's theorem, equation (2.19). A unit circle has radius of 1.

2.4 Using Maple

When using Maple to perform calculations with complex numbers, a non-conventional notation is used, and `I` (capital i) represents i in mathematical notation. Also, to evaluate expressions with complex numbers, it is necessary to use `evalc(..)` to force a calculation to happen. For example,

```
> evalc( exp(I*Pi));                -1
```

The real `Re(..)` imaginary, `Im(..)` and absolute (modulus) `abs(..)` values are next calculated with a function defined as f. Notice how `evalc(..)` has to be used to force the result.

```
> f:= exp(-I*x/2)*I*Pi/2;
```

$$f := \frac{1}{2}\,\mathrm{I}\,\mathrm{e}^{-\frac{1}{2}\mathrm{I}\,x}\,\pi$$

```
> real_part:= Re(f);                value:= evalc(Re(f));
  Imaginary_part:= Im(f);           value:= evalc(Im(f));
  absolute_value:= abs(f);          value:= evalc(abs(f));
```

$$real\_part := -\frac{1}{2}\,\pi\,\Im\left(\mathrm{e}^{-\frac{1}{2}\mathrm{I}\,x}\right) \qquad value := \frac{1}{2}\,\pi\,\sin\left(\frac{1}{2}\,x\right)$$

$$Imaginary\_part := \frac{1}{2}\,\pi\,\Re\left(\mathrm{e}^{-\frac{1}{2}\mathrm{I}\,x}\right) \qquad value := \frac{1}{2}\,\pi\,\cos\left(\frac{1}{2}\,x\right)$$

$$absolute\_value := \frac{1}{2}\,\mathrm{e}^{\frac{1}{2}\Im(x)}\,\pi \qquad value := \frac{1}{2}\,\pi$$

In Section 2.8, it is shown how easy it is to evaluate these apparently complicated expressions.

2.5 Questions

Full solutions are available at **www.oxfordtextbooks.co.uk/orc/beddard**.

Q2.1 If $z_1 = 2 + i$ and $z_2 = -1 - 3i/2$,

(a) calculate $z_1 + z_2$ and **(b)** $z_1 - z_2$. **(c)** What is $-iz_1z_2$?

Q2.2 **(a)** If $i^2 = -1$, what are i^3, i^4, i^5, and i^6?

(b) What relationship links positive powers of i?

Q2.3 If $z = a - ib$ what is i^3z?

Q2.4 If $z = 3 + 4i$ find z^2 and the modulus and argument of z^2.

Q2.5 Calculate $z = (2 - 5i)(3 + i) + 3i$, and find the modulus and argument of the result.

Q2.6 Express the number $\dfrac{5-i}{2-3i}$ in the form $z = a + ib$ and find its modulus and argument.

Q2.7 Simplify $z = (2 - 5i)(3 + i)/(3 - i)$ and find the modulus and argument of the result.

Q2.8 Find the modulus and argument of

(a) $\cos(\theta) - i\sin(\theta)$, and

(b) $1 - i\tan(\theta)$, where $0 < \theta < \pi/2$ in both cases.

Q2.9 If $w = z^2$ and $z = x + iy$ and $w = u + iv$, find u and v.

2.6 DeMoivre's theorem and powers of complex numbers

A complex number z can be written as

$$z = r[\cos(\theta) + i\sin(\theta)],$$

and if n is any number, what is $z^n = r^n[\cos(\theta) + i\sin(\theta)]^n$? The trigonometric part can be shown to have the simple form,

$$[\cos(\theta) + i\sin(\theta)]^n = \cos(n\theta) + i\sin(n\theta), \tag{2.8}$$

therefore,

$$z^n = r^n[\cos(n\theta) + i\sin(n\theta)] \tag{2.9}$$

which is called DeMoivre's theorem and is essential to calculating powers of complex numbers. One of the unexpected things that can be done is to find the nth root of 1, i, -3 or any other number for that matter.

To demonstrate that DeMoivre's theorem is correct, calculate the product of two complex numbers expressed in angular form, and then let $\theta_1 = \theta_2$. Suppose, for simplicity, that $r_1 = r_2 = 1$, then the product of two numbers is

$$\begin{aligned}&[\cos(\theta_1) + i\sin(\theta_1)][\cos(\theta_2) + i\sin(\theta_2)]\\ &= \cos(\theta_1)\cos(\theta_2) + i\cos(\theta_1)\sin(\theta_2) + i\sin(\theta_1)\cos(\theta_2) - \sin(\theta_1)\sin(\theta_2)\\ &= \cos(\theta_1 + \theta_2) + i\sin(\theta_1 + \theta_2).\end{aligned}$$

The double angle formula (Chapter 1.5.1) was used in the last step, and letting $\theta_1 = \theta_2$ produces

$$[\cos(\theta) + i\sin(\theta)]^2 = \cos(2\theta) + i\sin(2\theta).$$

as predicted by DeMoivre's theorem. This result can be generalized to any power of a real or complex value n.

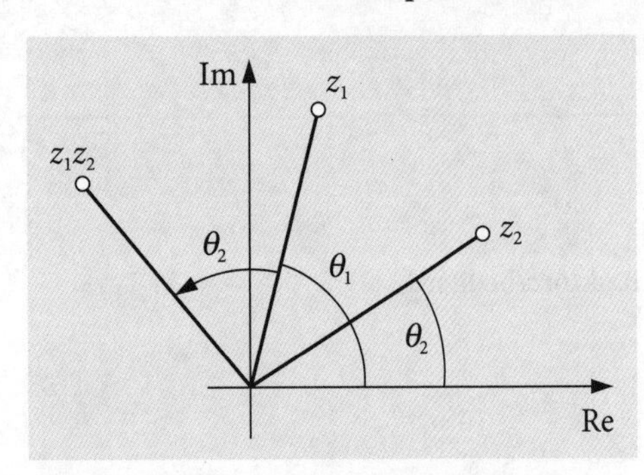

Fig. 2.5 Geometrical interpretation of the multiplication of two complex numbers.

The product z_1z_2 and quotient z_1/z_2 of two complex numbers are written in this form as

$$z_1z_2 = r_1r_2[\cos(\theta_1 + \theta_2) + i\sin(\theta_1 + \theta_2)], \tag{2.10}$$

where the angles add, and provided that $z_2 \ne 0$,

$$\frac{z_1}{z_2} = \frac{r_1}{r_2}[\cos(\theta_1 - \theta_2) + i\sin(\theta_1 - \theta_2)]$$

where the angles subtract. There is a geometrical interpretation to multiplying two complex numbers. If their moduluses are unity, $z_1 = \cos(\theta_1) + i\sin(\theta_1)$ and $z_2 = \cos(\theta_2) + i\sin(\theta_2)$, then multiplication results in rotation about the origin, equation (2.10). Geometrically this is shown in Fig. 2.5.

2.6.1 Roots of a complex number

Suppose that w is a real or complex number whose roots we need to find, then mathematicians have shown that, in general, the answer will be a complex number. If the n roots of a number z are expressed as $w = z^{1/n}$, then the equation to examine is $w^n = z$.

We will let both sides of this equation be different complex numbers. Expressing the left-hand side in angular form using DeMoivre's theorem with a polar angle ϕ gives

$$w^n = R^n[\cos(n\phi) + i\sin(n\phi)]. \tag{2.11}$$

The right-hand side of the equation is

$$z = r[\cos(\theta) + i\sin(\theta)] \tag{2.12}$$

since any complex number can be written in this way. Therefore,

$$R^n = r$$

where both R and r are real numbers. The angles ϕ and θ are related in the most general way as

$$n\phi = \theta + 2\pi k \tag{2.13}$$

where $k = 0, 1, 2, \dots n-1$ because sine and cosine are cyclic functions; $\sin(\theta) = \sin(\theta + 2\pi) = \sin(\theta + 4\pi)$ and so forth, therefore there will be more than one root to the equation. Using $n\phi = \theta$ only allows one root to be found. Using equations (2.11) and (2.13), gives

$$w = R^{1/n}\left[\cos\left(\frac{\theta + 2\pi k}{n}\right) + i\sin\left(\frac{\theta + 2\pi k}{n}\right)\right]. \tag{2.14}$$

In the special case of calculating the nth root of unity, $w^n = 1$ and $z = 1$, then from equation (2.12), $r = 1$, $\theta = 0$ and therefore,

$$w = \left[\cos\left(\frac{2\pi k}{n}\right) + i\sin\left(\frac{2\pi k}{n}\right)\right]. \tag{2.15}$$

There is always one real root and the other roots fall on the vertices of a polygon which is formed inside a circle of unit radius and touches the circle at its vertices.

To illustrate the method, $w^5 = 1$ is solved to find the five fifth roots of unity. The equation to solve is $w^n = z$ with $n = 5$ and $z = 1$. The roots are the solution of equation (2.15) with $n = 5$,

$$z = 1^{1/5} = \cos(2k\pi/5) \pm i\sin(2k\pi/5)$$

where $k = 0, 1, 2, 3, 4$. The *principal value* of the equation is the one solved with $k = 0$. The five roots are then

$$w = 1, \cos(2\pi/5) + i\sin(2\pi/5), \quad \cos(4\pi/5) + i\sin(4\pi/5),$$
$$\cos(6\pi/5) + i\sin(6\pi/5), \quad \cos(8\pi/5) + i\sin(8\pi/5),$$

and as $\sin(2\pi/5) = -\sin(8\pi/5)$ and so forth, only the positive terms need be used. Only one of the roots is not a complex number and as this first root lies on the real axis, the angle to the next root is

$$\theta = \tan^{-1}[\sin(2\pi/5)/\cos(2\pi/5)] \equiv 72°$$

and the other roots are separated from each other by the same angle as expected for a pentagon.

In the Maple calculation, the sequence command `seq` is used to generate pairs of numbers, which are plotted with the `pointplot` procedure. The circle function is used and `display` plots both curves together. To do this, each plot is given a name, `c1` and `p1`. The `constrained` instruction ensures that the x- and y-axes are the same size on the page and so produce a circle. The code here can be used to find and plot any root by defining `n` with `theta:=0`.

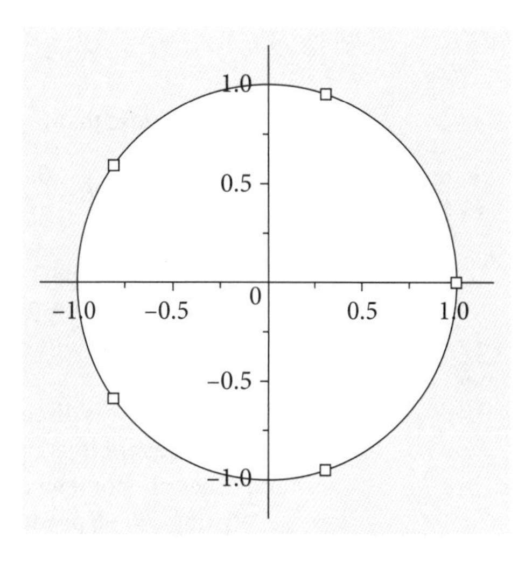

Fig. 2.6 The five roots of the equation $z^5 = 1$. The points form a pentagon.

Algorithm 2.1 *n* roots of unity

```
> with(plottools): with(plots):
  c1:= circle([0,0],1):                # name for plot circle
  n:= 5:   theta:= 0:                  # define values; n>0
  s1:= seq([cos((theta+2*Pi*k)/n ),sin((theta+2*Pi*k)/n )],
               k = 0..n-1):            # pairs of points
  p1:= pointplot([s1], symbol=BOX, scaling = constrained ):
  display([ p1, c1 ]);                 # plot both p1 and c1
```

2.7 Questions

Full solutions are available at **www.oxfordtextbooks.co.uk/orc/beddard.**

Q2.10 Find the four roots of $(-3)^{1/4}$.

Strategy: This problem is the same as solving the equation $w^4 = -3$ and as there are four roots they must form a square on an Argand diagram whose corners lie on a circle of radius $3^{1/4}$. The roots of a negative number are sought so these must all be complex with a zero real part; i.e. with an imaginary part only.

Q2.11 Find the square roots of i, i.e. $w^2 = i$. Find their magnitude and plot them on an Argand diagram.

Q2.12 Solve $w^4 = 16$.

Strategy: Because the equation is fourth order, there are four solutions and not just the two real ones $w = \pm 2$. Use the method of previous questions.

Q2.13 Calculate the modulus and argument of $2 + 3i$ then calculate its square roots. What is the radius of the circle on which the roots lie and at what angles?

Strategy: The complex number $2 + 3i$ is best converted into its trigonometric form to calculate the modulus and argument.

2.8 Euler's theorem

The exponential series is $e^x = 1 + x + \dfrac{x^2}{2!} + \dfrac{x^3}{3!} + \cdots$, and similarly a series can be formed in the complex number w,

$$e^w = 1 + w + \frac{w^2}{2!} + \frac{w^3}{3!} + \cdots.$$

Now suppose that $w = i\theta$, where θ is real, then rearrange into real and imaginary terms;

$$e^{i\theta} = 1 + i\theta - \frac{\theta^2}{2!} - i\frac{\theta^3}{3!} + \frac{\theta^4}{4!} \cdots = \left[1 - \frac{\theta^2}{2!} + \frac{\theta^4}{4!} - \cdots\right] + i\left[\theta - \frac{\theta^3}{3!} + \frac{\theta^5}{5!} - \cdots\right].$$

The real and imaginary parts are expansions of the cosine and sine functions respectively, therefore, if z is a complex number

$$z = e^{i\theta} = \cos(\theta) + i\sin(\theta). \tag{2.16}$$

This equation was discovered in 1748 by the Swiss mathematician Euler, and is extremely important as it crops up everywhere from quantum mechanics to X-ray crystallography and other phenomena connected with waves.

Writing $\theta = -\theta$ produces

$$e^{-i\theta} = \cos(\theta) - i\sin(\theta)$$

and therefore, for a general complex number with (modulus) r as a real number,

$$re^z = re^{i\theta} = r[\cos(\theta) + i\sin(\theta)].$$

DeMoivre's theorem can be derived from these equations: the power of a complex number w is

$$w^n = r^n e^{in\theta} = r^n[\cos(n\theta) + i\sin(n\theta)]. \tag{2.17}$$

Adding and subtracting $e^{i\theta}$ and $e^{-i\theta}$ gives

$$\cos(\theta) = \frac{e^{i\theta} + e^{-i\theta}}{2} \qquad \text{and} \qquad \sin(\theta) = \frac{e^{i\theta} - e^{-i\theta}}{2i} \tag{2.18}$$

which are equations that prove most useful in manipulating trig functions.

Calculating $e^{i\theta}$ with $\theta = \pi$ and $r = 1$ produces

$$e^{i\pi} = -1 \quad \text{or} \quad e^{i\pi} + 1 = 0 \tag{2.19}$$

which some consider the most beautiful equation in mathematics, as it connects the most important numbers of mathematics (0, 1, i, e, and π) and uses the most important operations (multiplication, exponentiation, negation, and addition). Furthermore, an integer is produced by raising an irrational number π times the imaginary unit i to the power of another irrational number, e. It is not at all obvious why this connection exists from an arithmetical standpoint, but from a geometrical one it is clearer. Consider a circle of unit radius on an Argand diagram; as the angle θ increases from 0 to 2π, the modulus (radius) is 1 when $\theta = 0$, and is i when θ is $\pi/2$, and -1 when the angle is π and so on; see Fig. 2.4.

Euler's formula is important in science, because it permits the description of a sinusoidally varying real quantity by means of complex exponentials. This change simplifies equations, because it is far easier to manipulate exponentials than trig functions. For example, the general form of a sinusoidally varying quantity, such as a plane wave, is $f(t) = a_0\cos(\omega t - \theta)$, where a_0 is the amplitude, ω the frequency, and θ the phase. These are all constants, and t is time and is a real variable. The equivalent complex function is

$$g(t) = a_0 e^{i(\theta - \omega t)} = a_0[\cos(\omega t - \theta) - i\sin(\omega t - \theta)]$$

therefore $f(t) = \text{Re}[g(t)]$. Very often in chemistry and physics, the complex form is used without explicitly stating that it is only the real part that represents the waveform. Fig. 2.7 compares various waveforms.

As an example of using Euler's equation, we will evaluate $w = \ln(-1)$ even though it doesn't exist—at least as a pure real number, then $w = \ln(i)$ and $w = \ln(z/3)$ are calculated where z is any complex number. The strategy in problems of this type is to convert the number -1, or i, or whatever it is into an exponential form using Euler's theorem.

(i) In the first example, $w = \ln(-1)$ or $e^w = -1$ and w has to be found to solve this equation. A general complex number can always be written as $z = re^{i\theta}$, therefore to find w, let $w = i\theta$. The absolute value (modulus) r of e^w is $\sqrt{e^{i\theta}e^{-i\theta}} = 1$. Because $e^{i\theta} = \cos(\theta) + i\sin(\theta)$, when $\theta = \pi$, $e^{i\theta} = -1$ making the *principal value* of $\ln(-1) = \ln(1e^{i\pi}) = i\pi$, which is a complex number. Note that there are other values of θ separated by $2k\pi i$, where k is an integer because $e^{i\theta}$ is a cyclic function.

(ii) In this example, $w = \ln(i)$ or $e^w = i$ and let $w = i\theta$. As $e^{i\theta} = \cos(\theta) + i\sin(\theta)$, when $\theta = \pi/2$ this equation produces $e^{i\pi/2} = i$ or $\ln(i) = \dfrac{i\pi}{2}$.

(iii) If $w = \ln(z/3)$, then $3e^w = z$, and if z is any complex number then we look for a value of θ such that $3e^{i\theta} = z$. Generally a complex number is represented by $z = re^{i\theta}$, then in this example $w = \ln(z) = \ln(3e^{i\theta}) = \ln(3) + i(\theta + 2\pi k)$ and $2\pi k$ is added because the function is cyclic and k is any integer; recall that the Euler equation can be put into a cosine and sine form, so it is a repetitive function. The principal value occurs when $k = 0$.

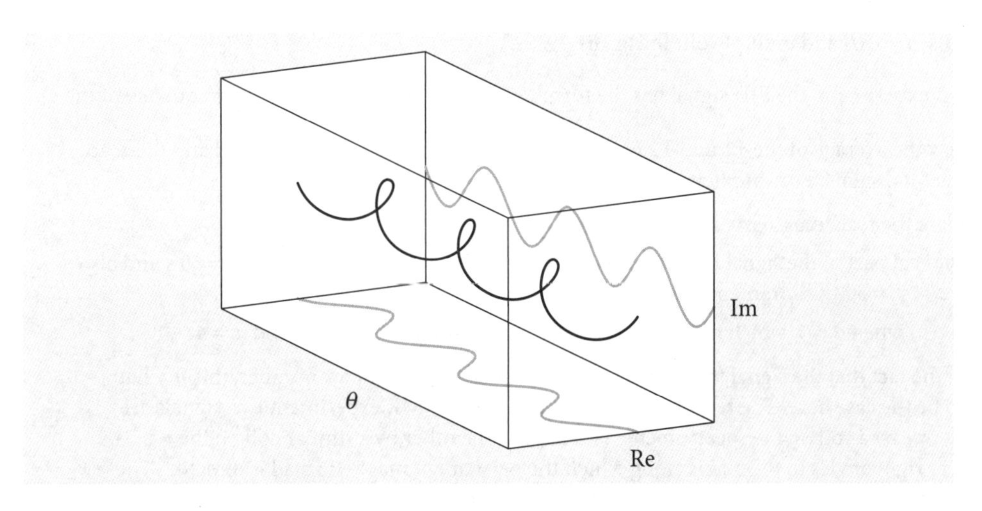

Fig. 2.7 Two visualizations of the complex number $e^{i\theta} = \cos(\theta) + i\sin(\theta)$ illustrate that it has a wavelike form.

Returning to example (i), $w = \ln(-1)$, if the -1 is treated as a complex number with an imaginary part that is zero, then the answer can be written down directly as

$$w = \ln(-1) = \ln(re^{i\theta}) = \ln(1) + i(\pi + 2\pi k)$$

and, since $r = 1$ and $\ln(1) = 0$, this gives the same result as in (i) $\ln(-1) = i\pi$ for the principal value.

2.9 Questions

Full solutions are available at www.oxfordtextbooks.co.uk/orc/beddard.

Q2.14 Calculate e^i.

Q2.15 Find the real and imaginary parts of **(a)** ie^{-ix}, **(b)** $e^{in\pi}$, and **(c)** $e^{in\pi/2}$, where n is an integer.

Q2.16 Calculate **(a)** i^i and **(b)** $i^{1/i}$.

Strategy: Using different bases such as $a^x = e^{x\ln(a)}$ any number can be raised to any power. With complex numbers always try to put the number in terms of Euler's equation.

Q2.17 The cosine function is defined as $\cos(x) = \dfrac{e^{ix} + e^{-ix}}{2}$. What is $\cos^{-1}(x)$?

Strategy: This is a case where x and y are swapped about. If $\cos^{-1}(x)$ then $\cos(y) = x$. It is true also that $\cos(y) = \dfrac{e^{iy} + e^{-iy}}{2}$. Next eliminate the cosine and solve for y and so find $\cos^{-1}(x)$.

Q2.18 Show that the identity $(\cos(x) + \sin(x))^2 = 1 + \sin(2x)$ can be (relatively easily) proved using complex numbers.

Q2.19 Show that $2\sin\left(\dfrac{a+b}{2}\right)\cos\left(\dfrac{a-b}{2}\right) = \sin(a) + \sin(b)$.

Q2.20 Starting with Euler's theorem and letting $\theta = a + b$, calculate $\sin(a + b)$ and $\cos(a + b)$ by equating real and imaginary parts.

Q2.21 Find $\sin(\theta)$ in exponential form then calculate $|\sin(i\theta)|$, and compare it with $|\sin(\theta)|$. Plot values of $|\sin(ix)|$ and $|\sin(x)|$ over the range $x = -4$ to 4.

Q2.22 **(a)** If $z = \cos(x) + i\sin(x)$ show that $\dfrac{dz}{dx} = iz$. **(b)** Integrate this result and prove Euler's theorem.

Q2.23 Calculate the real and imaginary parts of $\dfrac{1}{\sqrt{2\pi}}\left(\dfrac{1 - e^{-i\omega t}}{i\omega}\right)$. This function is the Fourier transform of a square wave of length t; see Chapter 9.5 and 9.6.

Strategy: use $i = -1/i$ and multiply out the terms.

Q2.24 In an NMR experiment, the FID signal has the form $s(t) = \sum_j a_j e^{i\omega_j t - t/\tau_j}$ where ω is the frequency of the transition, τ the average of the T1 and T2 lifetimes, and a the amplitude of each signal and there are j parts to the total signal. For simplicity, assume that τ_j has a constant value τ.

(a) Calculate the real, imaginary, and absolute value of s if $j = 2$.

(b) Plot the real part of the signal if $a_1 = a_2 = 2$ and $2\pi\omega = 1$ Hz and 0.2 Hz and $\tau_1 = \tau_2 = 50$ s and also when $\tau_1 = \tau_2 = 500$ s. Comment on the two results.

(c) Repeat (a) when $a_1 = i$, which means that the initial amplitude is complex, and $a_2 = 1$.

In spite of the fact that the signal from an experiment cannot be a complex number, this is what appears to be the case here. The reason for this is that in a real NMR experiment two signals are measured, one by a coil on the spectrometer's x-axis and the other by a similar coil on the y-axis. These are at right angles to the z-axis along which the permanent magnetic field is directed. These x and y signals are measured in *quadrature*, i.e. 90° out of phase to one another. One signal is taken to be the real component, and one the imaginary. They are then combined to produce $s(t)$ given above.

Strategy: The question asks you to find the components which when combined make $s(t)$. Use the Euler formula to do this and to simplify the complex exponential. As the signal represents the FID from an NMR experiment it oscillating in a sinusoidal way.

Q2.25 Derive the identities **(a)** $4\cos(\theta)\sin^2(\theta) = \cos(\theta) - \cos(3\theta)$ and

(b) $4\sin(\theta)\cos^2(\theta) = \sin(\theta) + \sin(3\theta)$

Strategy: Always use the exponential forms of sine and cosine wherever possible for complicated trig functions. These are

$$\cos(\theta) = \frac{e^{i\theta} + e^{-i\theta}}{2}, \qquad \sin(\theta) = \frac{e^{i\theta} - e^{-i\theta}}{2i}.$$

Q2.26 If C is the series whose n^{th} term is $\cos(nx)/n!$, and S the series $\sin(nx)/n!$, calculate the sum from $n = 1$ to infinity of $C + iS$, and hence find the sum C.

Strategy: Convert to the exponential form using $e^{ix} = \cos(x) + i\sin(x)$, sum the terms then convert back to trig form and separate out the real part of the result.

Q2.27 In the study of the dielectric properties of liquids and in electrochemical techniques that use potentiometry, the response of the solution to different electrical frequencies is studied. The general term for these experiments is impedance spectroscopy. In an experiment where a capacitor C and resistor R are in parallel, the impedance Z, which is a complex quantity, is given by $Z = (R^{-1} + i\omega C)^{-1}$ where ω is the frequency applied to the sample.

(a) Convert Z into the form $Z = Z' - iZ''$.

(b) Plot Z'' as ordinate, and z' as abscissa. Show that the resulting curve is a semicircle. Use $R = 5$ kΩ and C = 1 μF. Decide where high frequency is on the plot. This is not obvious from the graph because ω is not on one of the axes.

Strategy: Multiply top and bottom of the expression by the complex conjugate. Look up the parametric method of plotting graphs in the Maple appendix.

3 Differentiation

The word *calculus* derives from the Latin word for a stone, which, in the form of pebbles, were used in counting boards and the abacus, and led to the Latin verb *calculare* meaning to calculate. In the past, the word calculus was used quite generally to denote any branch of mathematics: arithmetical calculus, algebraic calculus, exponential calculus, and so forth. Nowadays its use is restricted to four areas: integral and differential calculus, which are now called integration and differentiation, differential equations, and the calculus of variations.

Integration, differentiation, and differential equations are used extensivly in science. It is essential for scientists to be familiar with them. With practice, it is a simple thing to differentiate and integrate most of the standard expressions you are likely to meet. Many differential equations can similarly be solved using one of several standard techniques. In addition, computer algebra packages, such as Maple or Mathematica, can help you to check your calculation and to obtain results in the more difficult cases. Although Maple can do almost any differentiation you are likely to meet, there is no substitute for understanding the basics of the calculation. Maple can help to eliminate silly algebraic mistakes, signs missed and so forth, and can not only perform long and tedious calculations as easily as short ones but also do calculations that exceed your current state of mathematical knowledge.

3.1 Concepts

Differentiation allows us to obtain the 'slope', or gradient, of a function, or the rate of change of some quantity with time. Integration allows us to obtain the area under a curve, and this is described fully in the next chapter. Differentiation and integration are shown in pictorial form below for any regular function $f(x)$. By regular is meant a well-behaved function without discontinuities at some value of x.

The differential of the function produces the gradient at the point specified, a in Fig. 3.1, and clearly by moving a along the axis to the right the slope will vary from its negative value, as shown, become zero, when the line is horizontal at the bottom of the curve, and then become positive. The *definite integral* is the area between a and b and clearly grows

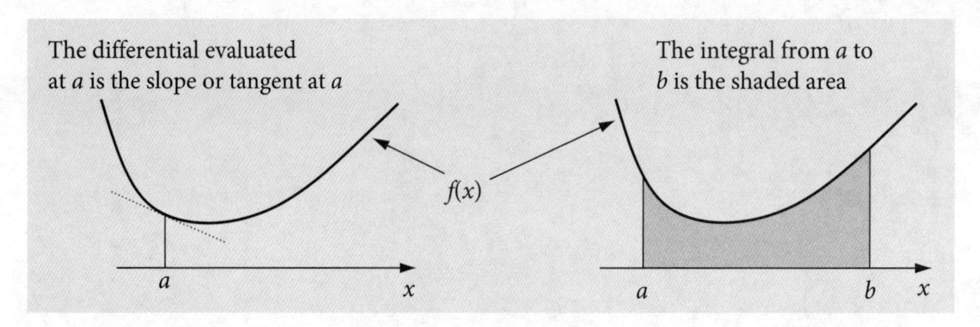

Fig. 3.1 Left: Differentiation determines the slope at point a. Right: Integration from a to b gives the area under the curve.

or shrinks if b is moved along the x-axis with a fixed in place. When evaluated the integral is a number.

The relationship between differentiation, $\dfrac{df}{dx}$, and integration, $\int_a^b f(x)dx$, of an expression $f(x)$ is shown below: moving to the left is differentiation; to the right integration.

$$\frac{df}{dx} \underset{always}{\overset{sometimes}{\rightleftarrows}} f(x) \underset{always}{\overset{sometimes}{\rightleftarrows}} \int_a^b f(x)dx \tag{3.1}$$

Even if we assume that it will always be possible to differentiate our function $f(x)$ it will not always be possible to integrate it algebraically, although this can be done numerically. Integrating df/dx is equivalent to solving a differential equation and these are met primarily in chemical kinetics, quantum mechanics, and dynamics. Before working out the details of differentiation, the effect that a small change in x and y has on an expression is examined, and this will lead naturally into differentiation.

3.2 Differentiation

3.2.1 Gradient calculated from small changes in *x* and *y*

A straight line with equation $y = mx + c$ has a constant slope m, which is the change in y divided by change in x, or $\delta y/\delta x$ for any interval δx. However, most functions are more interesting than the straight line; a parabola for example, $y = mx^2 + c$, has a gradient that changes with x and the value $\delta y/\delta x$ is now clearly going to depend upon how small δx and δy are and where x is on the curve. If δx is made infinitesimally small, the gradient exactly at position x is obtained and this is what is achieved with differentiation.

In Fig. 3.2, the curve is a parabola with the equation $y = x^2 + 1$. The slope in the region P to Q is $\delta y/\delta x$. The point P has coordinates $\{2, 5\}$ and if $\delta x = 0.3$ then the point Q is $\{x, y\} = \{2.3, 2.3^2 + 1\}$, making $\delta y = 1.29$ and $\delta y/\delta x = 4.3$. Similarly, the slope can be calculated starting at any other point on the curve. As δx is small but finite, the gradient is the average over the range δx, and if $f(x)$ is any (normal) function such as $x^2 + 1$, then its slope is given by

$$\frac{\delta y}{\delta x} = \frac{f(x+\delta x) - f(x)}{\delta x}, \tag{3.2}$$

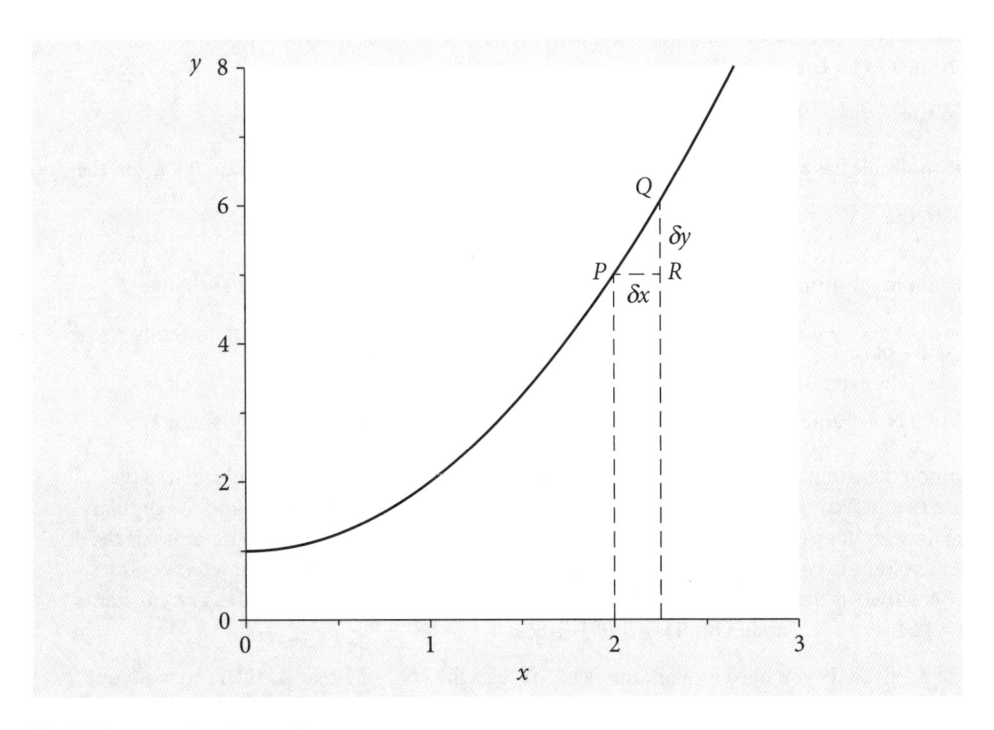

Fig. 3.2 Constructing the gradient on a curve.

but this only approximately measures the true slope, which is obtained when $\delta x \to 0$. The slope can also be interpreted in trigonometric terms because in any right-angled triangle, such as PQR, $\tan(\theta) = \dfrac{QR}{RP}$, where θ is the angle QPR and therefore the gradient is also $\tan(\theta) = \dfrac{f(x+\delta x) - f(x)}{\delta x}$.

As an example, consider calculating the change in volume of a sphere that has a radius r that is increased by a small amount δr. The initial volume is $V = \frac{4}{3}\pi r^3$ and is increased to $V + \delta V = \frac{4}{3}\pi(r+\delta r)^3$. The extra volume is therefore

$$\delta V = \frac{4}{3}\pi(r+\delta r)^3 - \frac{4}{3}\pi r^3 = \frac{4\pi}{3}(\delta r^3 + 3r\delta r^2 + 3r^2\delta r).$$

As δr is, by definition, small compared to r, the δr^2 and δr^3 terms in the bracket can be ignored because they are even smaller than δr. The result for the increase in volume is

$$\delta V = 4\pi r^2 \delta r.$$

Suppose that the initial radius is 10 cm and increases by 1 mm. The change in volume is 40π cm^3. Including all the terms, the volume change is 40.4π, therefore, our approximation is quite good and accurate to approximately 1%.

In the questions below, the strategy is always to take the function given, make a small change, then find the difference with respect to the original value. The subsequent expression may need simplifying and usually powers of small changes can be ignored. For example, δx^2 is far smaller than δx because δx is, by definition, far less than 1.

3.3 Questions

Full solutions are available at **www.oxfordtextbooks.co.uk/orc/beddard.**

Q3.1 A gas occupies a volume V m^3 at pressure p bar and at some temperature $pV = 1000$ joules. Express V in terms of p and δV in terms of p and δp. What is the change in volume when the pressure is increased from 1 to 1.1 bar?

Strategy: Write $V = 1000/p$, increment V and p then calculate $\delta V = (V + \delta V) - V$.

Q3.2 A diatomic molecule has a vibrational frequency of ν s^{-1}, a force constant k Nm^{-1}, mass μ kg, and the vibrational frequency is given by $\nu = \dfrac{1}{2\pi}\sqrt{\dfrac{k}{\mu}}$.

(a) If, by isotopic substitution μ is increased by 1%, show that the relative or fractional change is $\dfrac{\delta\nu}{\nu} = -\sqrt{\dfrac{1}{1+\delta\mu/\mu}} + 1$, which can be approximated as $\dfrac{\delta\nu}{\nu} = -\dfrac{1}{2}\dfrac{\delta\mu}{\mu}$. Justify the approximation you make. The expansion formula needed is $(1+x)^{-1/2} = 1 - x/2 + 3x^2/8 - \cdots$

(b) If $k = 518.0$ N m^{-1} and $\nu = 5889.0$ cm^{-1} what is the absolute change in frequency in s^{-1} or Hz?

Q3.3 In the common thermometer, the thermal expansion of a liquid is measured and calibrated to the temperature rise. Mercury or ethanol is often used. If this is held in a 1 ml reservoir and the capillary of the thermometer has a 0.12 mm diameter, work out the sensitivity of this thermometer if β is the coefficient of volume expansion. The liquid's volume expands as $V = V_0(1 + \beta\delta T)$ for a temperature rise of δT. Sensitivity is the change in length of the liquid for a 1 K rise in temperature. The constants are $\beta(\text{Hg}) = 1.81 \times 10^{-4}$ K^{-1}, and $\beta(\text{EtOH}) = 1.08 \times 10^{-3}$ K^{-1}.

Strategy: The sensitivity you need to work out is $\delta L/\delta T$. Use the volume of the capillary to work out its length.

3.4 The machinery of differentiation

In the previous section, the effect a small change δx on a quantity x was calculated, now δx is made infinitesimally small and the limit found when $\delta x \to 0$. The consequence of making the change $\delta x \to 0$ is profound and leads to differentiation, which is of universal application. If [C] is concentration, differentiation with time gives the rate of a chemical reaction, or if x is position, differentiation gives the speed of the body. A derivative in general describes the rate of change of one quantity with respect to another and 'rate of change' is commonly used not only to mean change with time, its original meaning, but also of any quantity with respect to another; dy/dx for instance.

If a function has a value $y = f(x)$ at x and at a nearby point $x + \delta x$ its value is $y + \delta y = f(x + \delta x)$, then the slope is approximately $\dfrac{\delta y}{\delta x} = \dfrac{f(x+\delta x) - f(x)}{\delta x}$ provided that $\delta x \neq 0$. To find the gradient exactly at the point x, δx must tend to zero; $\delta x \to 0$, which means finding the *limit* of the equation, and at the limit δ is changed to d and the *derivative* of the function is written in mathematical notation as $\dfrac{dy}{dx}$. The derivative can be written in a number of different ways besides dy/dx, although this is the most common. The following forms are equivalent:

$$\begin{aligned}\lim_{\delta x \to 0} \frac{f(x+\delta x) - f(x)}{\delta x} &\equiv \lim_{\delta x \to 0} \frac{\delta y}{\delta x} \\ &\equiv \frac{dy}{dx} = \frac{df}{dx} \equiv \frac{df(x)}{dx} \equiv \frac{d}{dx} f(x) \\ &\equiv y' \equiv f'(x) \equiv f' \equiv \dot{f} \\ &\equiv Dy \equiv Df \equiv Df(x)\end{aligned} \tag{3.3}$$

The derivative df/dx is the *gradient* of the function $f(x)$ at point x. The gradient is also the *tangent* to the curve at the point x; a line at right angles to the tangent is called the *normal* to the curve, Fig. 3.3.

In the next few sections simple functions are differentiated and the methods for dealing with products of these functions and more complex cases are described. Many are illustrated with Maple and while this could be used to perform every calculation, using it as a 'black box' is not advisable; it is important to understand the approach used to reach a result.

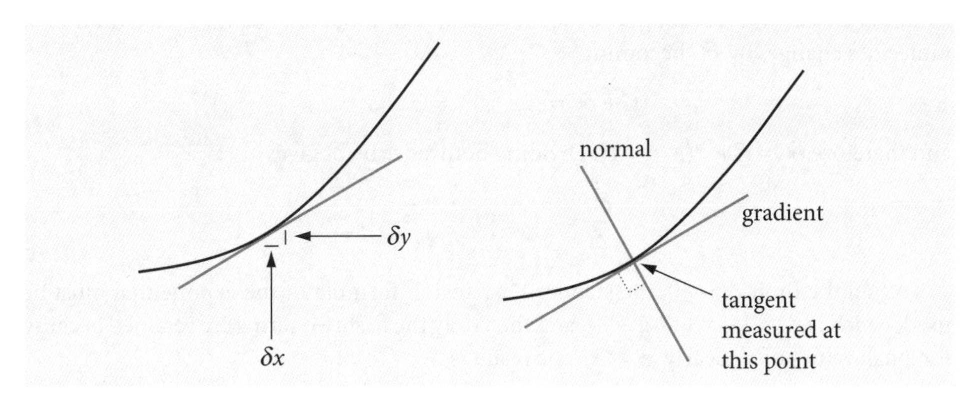

Fig. 3.3 Left: Infinitesimal changes in δx and δy lead to the calculation of the slope at x and to differentiation when $\delta x \to 0$. Right: The gradient and a line normal, i.e. at right angles, to the gradient.

3.4.1 Finding the limit when $\delta x \to 0$

In finding the limit $\delta x \to 0$ although $f(x + \delta x) - f(x)$ and δx tend to zero, the ratio of them does not and is finite. Even though this may not seem to make sense, it is true for ordinary functions, such as $\sin(x)$, x^2, and so forth. For example, choosing two equations $y_1 = 3x$ and $y_2 = 2x$ as $x \to 0$, both y_1 and y_2 tend to zero, but their ratio is $3x/2x = 3/2$ for all values except *exactly* at $x = 0$. The similar effect is seen in a triangle because the ratio of the length of the base to the perpendicular side is constant except at exactly $x = 0$, Fig. 3.4.

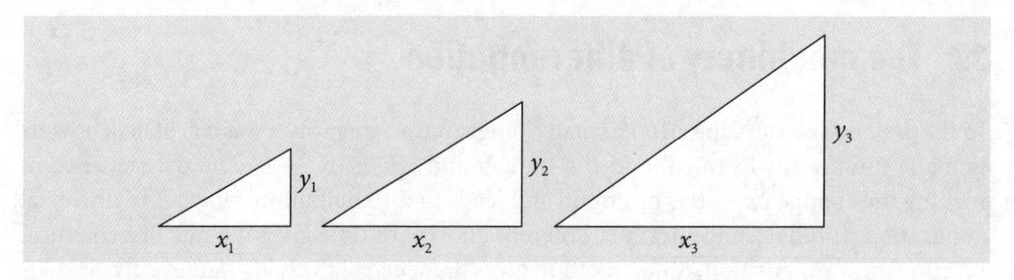

Fig. 3.4 The ratio of y/x is constant for different sized triangles of similar shape.

The gradient of $y = x^2$ will clearly change with the value of x being negative when $x < 0$, zero at $x = 0$, and increasingly positive with increase in x. See Fig. 3.2. Let us consider a point δx larger than x then $y + \delta y = (x + \delta x)^2$ and therefore $\delta y = 2x\delta x + \delta x^2$ so that $\dfrac{\delta y}{\delta x} = 2x + \delta x$, which is the average gradient in the region x to $x + \delta x$. Next, making δx smaller and smaller to evaluate the limit, $\delta x \to 0$ produces the result

$$\left.\frac{\delta y}{\delta x}\right|_{\delta x \to 0} = \frac{dy}{dx} = 2x. \tag{3.4}$$

Sometimes d/dx is called the *differential operator* because it operates on x^2 to form $2x$, which is the gradient at any point x. Now that x^2 has been differentiated, returning to equation (3.2) where an approximate gradient of 4.3 was found for the parabola the true value at $x = 2$ is 4 so our estimate was reasonable but still about 7% out, however, it was more difficult to calculate than the exact value!

3.4.2 **Differentiating exponentials**

The exponential function is very important in science and arises in many situations. For example, in chemical kinetics when a molecule such as $\mathrm{ICN} \xrightarrow{200\,fs} \mathrm{I} + \mathrm{CN}$ dissociates, the rate of decrease of the reactant concentration C is a first-order process, with rate constant $k = 5 \times 10^{12}\ \mathrm{s^{-1}}$ in this case, and

$$C = C_0 e^{-kt}$$

where C_0 is the initial amount present. The differential of the exponential function is very simple and easily derived. Suppose that in a small additional time δt, the concentration of molecules changes by δC becoming

$$C + \delta C = C_0 e^{-k(t+\delta t)} = C_0 e^{-kt} e^{-k\delta t}$$

and therefore $\delta C = C_0 e^{-kt}(e^{-k\delta t} - 1)$. Dividing both sides by δt gives

$$\frac{\delta C}{\delta t} = -C_0 e^{-kt}\left(\frac{e^{-k\delta t} - 1}{\delta t}\right).$$

To find the limit $\delta t \to 0$, the standard expansion formula of the exponential must be used, which is $e^{-x} \approx 1 - x + x^2/2 - \cdots$, and then only the first two terms are retained because for small values of x, clearly $x^2 \ll x$. The result is

$$\left(\frac{e^{-k\delta t} - 1}{\delta t}\right)_{\text{limit}\,\delta t \to 0} = \frac{(1 - k\delta t) - 1}{\delta t} \to -k$$

and

$$\frac{\delta C}{\delta t} \to \frac{dC}{dt} = -kC_0 e^{-kt},$$

producing the differential of the exponential. By substitution of the original equation, this result can be written as

$$\frac{dC}{dt} = -kC,$$

which is the first-order rate equation; the rate of change is proportional to the amount of substance *unreacted at time t*, which is C.

In the general case, if

$$y = e^{-ax+b}$$

then its derivative is the function itself multiplied by the constant multiplier of the x in the argument of the exponential. Far simpler when put as an equation

$$\frac{d}{dx}e^{-ax+b} = e^{b}\frac{d}{dx}e^{-ax} = -ae^{-ax+b}.$$

If $y = e^{x}$, then the value of the gradient is the size of the function itself

$$\frac{d}{dx}e^{x} = e^{x}.$$

3.4.3 Powers of *x*

The similar procedure making $\delta x \to 0$ can be followed for lots of different functions, but need not be carried through because the results and methods are well understood and the results will be quoted. The derivatives of $y = x^{n}$ where n is any positive or negative number, including fractions, have the form

$$\frac{dy}{dx} = nx^{n-1}. \tag{3.5}$$

Some examples are

$$\frac{d}{dx}x^{4} = 4x^{3}; \qquad \frac{d}{dx}x^{-3} = -3x^{-4} = -\frac{3}{x^{4}}$$

$$\frac{d}{dx}(x^{4} - 3x^{2} + 2) = 4x^{3} - 6x \qquad \frac{d}{dx}(x^{4/3} - 3x^{-1/2} + 2) = \frac{4}{3}x^{1/3} + \frac{3}{2}x^{-3/2}$$

3.4.4 Constants

Constants always differentiate to zero because a graph of $y =$ constant has a zero gradient.

3.4.5 *y*

The differential of y is $\frac{dy}{dx}$; that of y^{2} is $2y\frac{dy}{dx}$ where the power of y is differentiated as if it were x then multiplied by the differential of y itself. It is very important to remember to do this. For example, differentiating

$$y^{3} = x^{4} \qquad \text{produces} \qquad 3y^{2}\frac{dy}{dx} = 4x^{3}.$$

The equation could also have been written as,

$$y = x^{4/3} \qquad \text{producing} \qquad \frac{dy}{dx} = \frac{4}{3}x^{1/3},$$

which is the same when worked through. Try it if you are not convinced. If g is a function of y such as y^{3}, the general form for such an equation is

$$g(y) = f(x), \qquad \frac{d}{dy}g(y)\frac{dy}{dx} = \frac{d}{dx}f(x) \tag{3.6}$$

as illustrated in the example but here written formally.

3.4.6 Differentiating with Maple

Differentiating with Maple is very easy. Nevertheless, it is important to be aware of some of the simpler results to get an idea of what to expect to check this against your answers. The syntax is shown in these few examples, those shown in the text follow this form. As with almost all Maple instructions, a pair of brackets is placed immediately after naming the type of calculation you want to do. For example, differentiating is performed by using `diff(..,x);` and a function such as `sin(a*x)` is always placed between the brackets with any additional instructions, viz.:

```
> diff( sin(a*x) , x );
```

which can be put in a more pleasing form using `% = value(%);`

```
> Diff( x^n ,x): % = value(%);
```

$$\frac{\partial}{\partial x}x^n = \frac{x^n n}{x}$$

Sometimes simplifying may be necessary to get a clearer result,

```
> Diff( x^n,x): % = simplify( value(%), symbolic );
```

$$\frac{\partial}{\partial x}x^n = x^{n-1}n$$

The capital `D` in `Diff` prevents evaluation until the `value` instruction is met. This is a common, although not universal feature of Maple, whereby some operations are rendered inert by using a capital first letter in the instruction. This can be useful in checking that the correct equation has been typed. If a lower case first letter is used, the result is calculated immediately.

Note that Maple produces results with the differentials written with a 'curly d', e.g. $\frac{\partial}{\partial x}$, which is the notation for partial differentiation rather than that of the conventional d/dx. This is because other potential variables, such as n or a are present. However, Maple still produces the $\frac{\partial}{\partial x}$ notation even if it is told that n or a is constant.

3.4.7 Repeated differentiation

The result of differentiation is often another function in x, so it is possible to differentiate again. Performing the operation twice is written as $\frac{d^2}{dx^2}f(x)$, and the notation is similar even if you must differentiate three or more times. Sometimes the first derivative is written as f', the second as f'', and so forth. The second derivative means perform the differentiation twice over, the third derivative three times;

$$\frac{d^2}{dx^2}f(x) = \frac{d}{dx}\left(\frac{d}{dx}f(x)\right). \tag{3.7}$$

For example, if $y = ax^2$, differentiating once produces $\frac{dy}{dx} = 2ax$ and differentiating this result gives

$$\frac{d^2y}{dx^2} = \frac{d}{dx}2ax = 2a$$

and again produces

$$\frac{d^3y}{dx^3} = 0$$

because a is a constant and one cannot go any further. Any positive integer power of x repeatedly differentiated will eventually result in zero. Maple performs these differentiations as

```
> Diff(a*x^2,x): % = value(%);
```

$$\frac{\partial}{\partial x}(a\,x^2) = 2\,a\,x$$

```
> # notice x, x to differentiate twice
> Diff(a*x^2,x,x): % = value(%);
```

$$\frac{\partial^2}{\partial x^2}(a\,x^2) = 2\,a$$

3.4.8 Many variables and partial differentiation

Some functions depend upon more than one variable, say x and z, which is written as $y(x, z)$ or $f(x, z)$, it is now possible to differentiate with respect to x then z or vice versa. For example, if

$$y(x, z) = x^3z^4,$$

differentiating by z but keeping x constant produces

$$\frac{dy}{dz} = 4x^3z^3,$$

and similarly,

$$\frac{dy}{dx} = 3x^2z^4$$

if z was constant. These are normally called partial derivatives and the equations would usually be written as $\left(\frac{\partial y}{\partial z}\right)_x = 4x^3z^3$ and $\left(\frac{\partial y}{\partial x}\right)_z = 3x^2z^4$; the curly ∂ is used and the subscript indicates what is held constant. Repeating the differentiation, first in the order x and z then z and x produces the same result,

$$\frac{d}{dz}\left(\frac{dy}{dx}\right) = \frac{d^2y}{dzdx} = 12x^2z^3 \qquad \text{and} \qquad \frac{d}{dx}\left(\frac{dy}{dz}\right) = \frac{d^2y}{dxdz} = 12x^2z^3.$$

In Maple this double differentiation is written as

```
> y:= x^3 * z^4:
> Diff(y, x, z): % = value(%);
```

$$\frac{\partial^2}{\partial z\,\partial x}(x^3\,z^4) = 12\,x^2\,z^3$$

where the `,x ,z` in the `Diff` causes repeated differentiation of y. Partial differentiation is discussed more fully in Section 3.12.

3.4.9 Sine and cosine

Sine and cosine functions are almost mirror images of one another when differentiated,

$$\frac{d}{dx}\sin(x) = \cos(x)$$

$$\frac{d}{dx}\cos(x) = -\sin(x), \tag{3.8}$$

which makes sense if one inspects a graph of these functions, see Fig. 3.5. The *value* of the cosine at any point is the *gradient* of the sine at that point and the gradient of the cosine is −1 times the value of the sine. When the sine has its maximum or minimum, where the gradient is 0, the value of the cosine is also 0. Furthermore, when the gradient of the sine is positive, $x/\pi = 0$ to 1/2 and also 3/2 to 2, the value of the cosine, which is the derivative, is positive. When the gradient of the sine is negative, the value of the cosine is negative. It is therefore not surprising that on differentiating these functions again they turn into one another but with a change of sign depending upon how many times this is done,

$$\frac{d^2}{dx^2}\sin(ax) = \frac{d}{dx}a\cos(ax) = -a^2\sin(ax),$$

$$\frac{d^3}{dx^3}\sin(ax) = -a^3\cos(ax)$$

$$\frac{d^4}{dx^4}\sin(ax) = a^4\sin(ax).$$

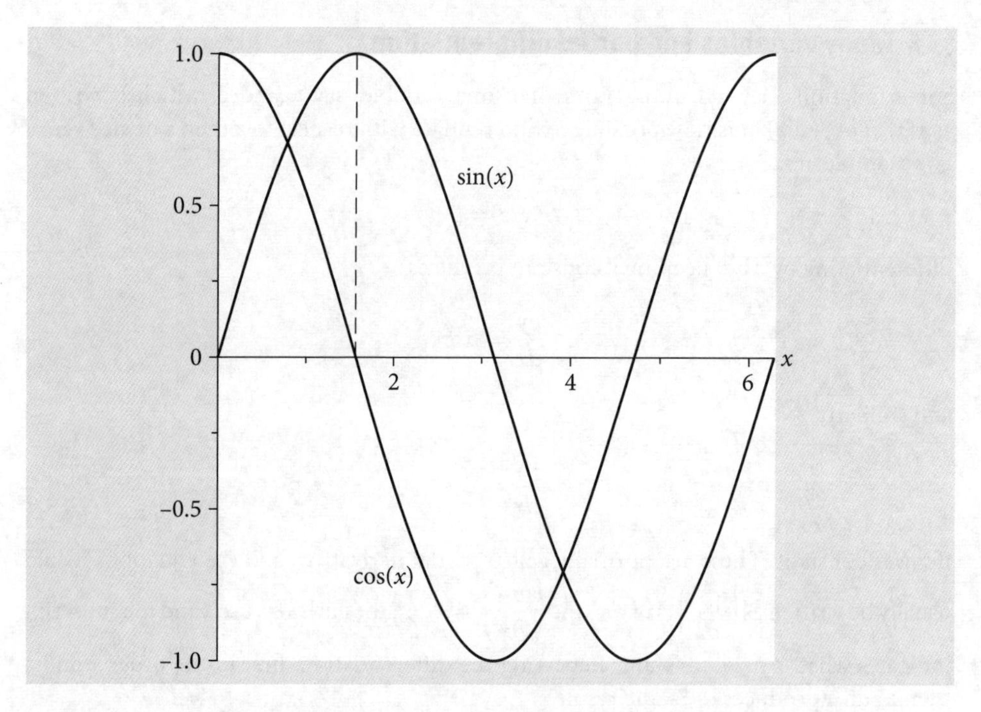

Fig. 3.5 The sine and cosine functions plotted over one period of 0 to 2π.

3.4.10 Other trig functions, tan, sinh, cosh, and their inverse

There are many other trigonometric functions besides sine and cosine, and most can be treated by the differentiation methods described in the next few sections based on basic differentiation. One useful way is to convert to the exponential form first. Sine and cosine can easily be differentiated this way, then converted back to a trig form. Tan, for example, is the ratio of sine/cosine and is treated as a ratio. Others functions such as sec(x) which is 1/cos(x) can be treated as a function of a function, Section 3.5.1, either in the trig or exponential form. The hyperbolic functions cosh, sinh, and tanh also have exponential representations.

The inverse functions $y = \sin^{-1}(x)$, $y = \cosh^{-1}(x)$, and so forth can be treated by rearranging the equation, for instance,

$$y = \sin^{-1}(x^2) \qquad \text{into} \qquad \sin(y) = x^2$$

and differentiating both sides, see equation (3.6). Doing this produces

$$\cos(y)dy/dx = 2x,$$

and y can be substituted into this result using $\cos^2(y) + \sin^2(y) = 1$ giving

$$\cos(y) = \sqrt{1 - \sin^2(y)} = \sqrt{1 - x^4}.$$

The final result is

$$dy/dx = 2x/\sqrt{1 - x^4}.$$

The Pythagoras right-angled triangle can also be used to convert between trig functions. See Chapter 1.5.

3.4.11 Repeated differentiation

The sine, cosine, and exponential functions are capable of endless repeated differentiation, as shown in Section 3.4.9. If $y = e^{-ax}$, then repeated differentiation gives

$$\frac{d^2}{dx^2}\exp(-ax) = -\frac{d}{dx}a\exp(-ax) = a^2\exp(-ax) = a^2y. \tag{3.9}$$

Continuing the differentiation, the powers of a increase and the sign alternates being positive for even powers and negative for odd ones. Therefore the nth derivative is

$$\frac{d^n}{dx^n}\exp(-ax) = (-1)^n a^n \exp(-ax)$$

which, by substitution, can also be put in the form of a differential equation

$$\frac{d^n y}{dx^n} = (-1)^n a^n y.$$

Incidentally, this equation tells us that the solutions of this type of differential equation are exponentials; n can be 1, 2, 3, and so forth.

3.4.12 Logarithms

Differentiating logarithms always has the form: '*derivative divided by the function*'.

$$\frac{d}{dx}\ln[f(x)] = \frac{f'(x)}{f(x)}, \qquad \left(\text{note that } f' = \frac{d}{dx}\right) \qquad (3.10)$$

For example,

$$\frac{d}{dx}\ln(x) = \frac{1}{x}$$

$$\frac{d}{dx}\ln[\sin(x)] = \frac{\cos(x)}{\sin(x)} = \frac{1}{\tan(x)}.$$

Differentiating $\ln(y)$ with respect to x produces a most useful form, which is worth remembering:

$$\frac{d}{dx}\ln(y) = \frac{1}{y}\frac{dy}{dx}. \qquad (3.11)$$

In thermodynamics, the van 't Hoff equation has this last form,

$$\frac{d\ln(K_P)}{dT} = \frac{\Delta_r H^{\ominus}}{RT^2}$$

which describes the change of an equilibrium constant K_P for a reaction carried out at constant pressure with temperature T and is an example of Le Chatelier's principle. $\Delta_r H^{\ominus}$ is the standard molar enthalpy of the reaction.

3.4.13 *x* as a power: $y = a^x$

In cases where there are powers of x it is best to take logs of both sides first. For example, if $y = a^x$, taking logs of both sides gives $\ln(y) = x\ln(a)$ and differentiating produces

$$\frac{1}{y}\frac{dy}{dx} = \ln(a). \qquad (3.12)$$

This can be simplified to

$$\frac{dy}{dx} = \ln(a)y = \ln(a)a^x.$$

In the special case that $a = e$ (e is the exponential constant), then the exponential derivative is retrieved because $\ln(e) = 1$.

3.4.14 Reciprocal derivatives

Occasionally it is necessary, or simpler, to find dx/dy rather than invert the equation to put it in terms of $y = \cdots$ and calculate dy/dx. The derivatives are related as reciprocals;

$$\frac{dy}{dx} = 1/\frac{dx}{dy}. \qquad (3.13)$$

As an example, suppose that $\sin(y^2) = x$, differentiating by y gives the result $dx/dy = 2y\cos(y^2)$. Differentiating by x could mean that a rearrangement must first be done to form $y = \sqrt{\sin^{-1}(x)}$ and then this differentiated, which is quite involved. Instead using equation (3.6), the result is obtained directly $2y\cos(y^2)dy/dx = 1$ and these two results show that equation (3.13) is true.

3.4.15 **Differentiating integrals**

If you are unfamiliar with integration, it will help to know the basic rules; see Chapter 4. It is possible to differentiate integrals; well why not! Integration and differentiation are linked as shown in equation (3.1) and as given by the fundamental theorem of the calculus:

$$\frac{d}{dx}\int_a^x f(u)du = \frac{d}{dx}[F(x) - F(a)] = f(x). \tag{3.14}$$

a being a constant and F is the result of integration. The variable u used in the integration is a dummy variable, any letter could be used instead, but it is better in this instance not to use x as this can be confusing. Differentiating the integral with two limits both of which are constants, i.e. simply numbers a and b, produces a result of 0, because integration with such limits produces a number, e.g. the area under the curve from a to b, see Fig. 3.1, and the differential of a constant is 0,

$$\frac{d}{dx}\int_a^b f(u)du = 0.$$

In the more complex and general cases, where the limits u and v are themselves functions of x, the function of function rule is used (Section 3.5.1),

$$\frac{d}{dx}\int_{v(x)}^{u(x)} f(s)ds = f(u)\frac{du}{dx} - f(v)\frac{dv}{dx}, \tag{3.15}$$

which is also called Leibniz's Rule. An example where x^2 is one limit is

$$\frac{d}{dx}\int_a^{x^2} e^{-au^2}\,du = 2x\,e^{-ax^4}$$

and a contrived example with limits x and x^2 is, using Maple:

```
> Diff(Int(u^3,u = x..x^2),x): % = value(%);
```

$$\frac{\partial}{\partial x}\int_x^{x^2} u^3\,du = 2\,x^7 - x^3$$

and this result can be obtained by hand as $2xx^6 - x^3$ using equation (3.15).

The aim in differentiating integrals is not to work out the integral first, which might not be possible anyway, and then differentiate the result, but to use equation (3.15), which avoids doing this.

A different case, and one to be aware of, involves function in two variables, say y and x. In this example, notice the variable of the integration is y, that of differentiation x, and therefore the differentiation is first performed *inside* the integration.

$$\frac{d}{dx}\int_a^b f(x,y)dy = \int_a^b \frac{\partial f(x,y)}{\partial x}dy.$$

After differentiation with respect to x, x is treated as a constant because integration is in y. For example,

$$\frac{d}{dx}\int_0^1 x^2y\,dy = 2x\int_0^1 y\,dy = xy^2\Big|_0^1 = x$$

where the integral of y is $y^2/2$, see Chapter 4. Another example of this is

$$\frac{d}{da}\int e^{ax}dx = \int \frac{\partial}{\partial a}e^{ax}dx = \int xe^{ax}dx$$

where differentiation is with respect to a and not x. The partial derivative symbol ∂ is used inside the integration sign to specify that only a in this case is to be differentiated.

3.4.16 Fractional derivatives

While it is possible to repeatedly take derivatives of many functions, for instance $\frac{d^3}{dx^3}\sin(x)$, what about the 1/3 or 1/2 or −1 derivative? What would such a thing mean? In the case of 1/2 derivatives we can say that if the function is x^n then the half derivative is such that $\frac{d^{1/2}}{dx^{1/2}}\frac{d^{1/2}}{dx^{1/2}}x^n = nx^{n-1}$. In other words differentiating, or *operating* twice on x^n with $d^{1/2}/dx^{1/2}$, is the same as differentiating once with dy/dx. However, these unusual derivatives need not have more than a curiosity interest for us; they appear in Morse's paper on the anharmonic oscillator (Morse 1929) and hardly anywhere else.

3.4.17 Table of the differentials of simple functions

a, b, and n are treated as constants:

$$\frac{dx^n}{dx} = nx^{n-1}$$

$$\frac{d\sin(a\,x+b)}{dx} = a\cos(a\,x+b)$$

$$\frac{d\cos(a\,x+b)}{dx} = -a\sin(a\,x+b)$$

$$\frac{d\tan(a\,x+b)}{dx} = \frac{2\,a}{\cos(2\,a\,x+2\,b)+1}$$

$$\frac{d\ln(a\,x+b)}{dx} = \frac{a}{a\,x+b}$$

$$\frac{d\,e^{a\,x}}{dx} = a\,e^{a\,x}$$

$$\frac{d\,a^x}{dx} = a^x\ln(a)$$

$$\frac{d\sinh(a\,x+b)}{dx} = a\cosh(a\,x+b)$$

$$\frac{d\cosh(a\,x+b)}{dx} = a\sinh(a\,x+b)$$

$$\frac{d\tanh(a\,x+b)}{dx} = \frac{2\,a}{\cosh(2\,a\,x+2\,b)+1}$$

$$\frac{d\arcsin(a\,x+b)}{dx} = \frac{a}{\sqrt{1-a^2x^2-2\,a\,x\,b-b^2}} \qquad \arcsin(n)\equiv\sin^{-1}(n)$$

$$\frac{d\arccos(a\,x+b)}{dx} = -\frac{a}{\sqrt{1-a^2x^2-2\,a\,x\,b-b^2}} \qquad \arccos(n)\equiv\cos^{-1}(n)$$

Maple can be used to work out other differentials by defining the function such as,

```
> f:= x-> arctan(x):
  Diff(f(x),x): %=value(%);
```

$$\frac{d}{dx}\arctan(x) = \frac{1}{1+x^2}$$

3.4.18 Questions

Full solutions are available at **www.oxfordtextbooks.co.uk/orc/beddard.**

Q3.4 **(a)** Differentiate with respect to x, a being constant.
(i) $3x^4 + \sin(x)$, **(ii)** $10x^{-4} + 5x^2 + a$, **(iii)** $\ln(x) + (ax)^{-1}$, **(iv)** $\exp(ax) + x^2$

(b) Differentiate with respect to a, x being constant.
(i) $3x^4 + \sin(x)$, **(ii)** $e^{ax} + x^2$, **(iii)** $y = ax^2$.

(c) Differentiate $\sin(ax)$ n times with respect to x when n is even and when n is odd. Differentiate up to $n = 5$ before deciding on the pattern of equations, then use the identity $\cos(x) = \sin(x + n\pi/2)$.

Q3.5 Differentiate x^n n times, n being a positive integer, and find $\dfrac{d^n}{dx^n}x^n$.

Strategy: In situations like this, where there are repeated operations and n is undefined, it is best to try to get an answer by induction. Start with $n = 1, 2, \cdots$ and so forth, then build up a pattern. Find the answer for some general or intermediate term, such as the mth, then finally make $m = n$.

Q3.6 Differentiate $y = e^{-ax}$ n times.

Q3.7 A gas of volume V m^3 has a pressure p bar at a constant temperature T K.

(a) Using the ideal gas law, show that $\dfrac{dp}{dV} = cV^{-2}$ where c is a constant.

(b) Find $\dfrac{dV}{dp}$.

(c) What is the relationship between $\dfrac{dV}{dp}$ and $\dfrac{dp}{dV}$?

Q3.8 **(a)** If a ball is thrown with an initial velocity u, the distance it travels in time t is $s = ut + \dfrac{1}{2}at^2$. Find its velocity v at any time t.

(b) What is the meaning of the parameter a?

Q3.9 **(a)** The electric field of a laser or other plane light wave of frequency ω is $E = Ae^{i(\omega t - kx + \varphi)}$, where x is the distance from the source, k the wavevector $(2\pi/\lambda)$, ϕ the phase, and A is a constant and is the amplitude of the wave at $t = 0$ and $x = 0$. Show that the nth derivative with respect to time of E is $\dfrac{d^nE}{dt^n} = (i\omega)^n E$.

(b) Calculate the similar derivative equation for distance x.

Strategy: Start with the first derivatives, look for a pattern and substitute for E into your answer. This equation for the electric field represents a general wave because $e^{i\theta} = \cos(\theta) + i\sin(\theta)$. This equation is described in more detail in Chapter 1.

Q3.10 The rate of diffusion / area / time in a solution of ions is described as

$$D_F\frac{dc}{dx} = cD\frac{d\ln(c\gamma)}{dx},$$

where D_F is the Fick's law diffusion coefficient and D is the (kinematic) diffusion coefficient, c is the concentration of ions in solution, and γ the activity coefficient. This is given by the modified Debye–Hückel expression $\ln(\gamma) = -\dfrac{A\sqrt{c}}{1 + B\sqrt{c}}$ where A and B are constants that depend on the solvent and its temperature and the size of the ions in B.

(a) Show that $D_F = D\left(1 + \dfrac{d\ln(\gamma)}{d\ln(c)}\right)$

(b) Evaluate the derivative.

Strategy: Do not let the unusual form of the differential put you off. Use $d\ln(c) = \dfrac{dc}{c}$.

Q3.11 Evaluate $\dfrac{d}{dx}\displaystyle\int_0^{a/x}\frac{x^2}{e^{-x} - 1}dx$.

Strategy: Use equation (3.15).

3.5 Going beyond simple functions

Many equations have the form where a power of x is multiplied by another, or by an exponential, a sine or cosine and so forth, and you must be able to handle these more complicated equations because they occur so frequently. A couple of simple rules allow us to do this quite easily.

These are the *product rule*, which is used when there is a function that is a product or ratio of other functions e.g. $y = \sin(x)\cos(x)$, and the *function-of-function* rule or *chain rule* used when a function has a structure such that one function is inside another, e.g. $y = \sin(ax^2 + x^{-3})$ or $f(x) = e^{(x^2+\sin(x))}$ or in general $f[g(h)]$ for any variable h.

3.5.1 Function of a function or chain rule

The method when one function is a function of another is to

differentiate the main or 'outside' function,
then multiply this by the differential of the 'inside' function.

This simple but powerful method is most easily explained by examples; (i) to (iv) are typical. The general formula is given in the next section. To differentiate the functions, use the table or various sections above to remind you of the rules. You can of course, always use Maple but then you will not learn how the method works.

$$\text{(i)}\;\; y = \sin(ax^2 + x^{-3}), \qquad \text{(ii)}\;\; y = e^{(ax^2+\sin(x))},$$

$$\text{(iii)}\;\; p = \ln(e^{aq} + q^3) \text{ and find } dp/dq, \qquad \text{(iv)}\;\; y^n = \sin(ax^3).$$

(i) To differentiate $y = \sin(ax^2 + x^{-3})$ treat the sine as the outside function, differentiate it to produce $\cos(\cdots)$, then multiply by the differential of the terms inside the sine bracket

$$y = \sin(\underline{ax^2 + x^{-3}}) \tag{3.16}$$
$$\text{outside}\uparrow \qquad \uparrow\text{inside}$$

$$\frac{dy}{dx} = \cos(ax^2 + x^{-3})\frac{d}{dx}(ax^2 + x^{-3}) = \cos(ax^2 + x^{-3})(2ax - 3x^{-4}).$$

In equation **(ii)**, $y = e^{(ax^2+\sin(x))}$, treat the power of the exponential as the inside function and differentiate the exponential and then multiply by the differential of the inside function to give $\dfrac{dy}{dx} = e^{(ax^2+\sin(x))}(2ax + \cos(x))$.

In **(iii)**, $p = \ln(e^{aq} + q^3)$, the exponential and q^3 are the inside functions then using the function of function rule produces $\dfrac{dp}{dq} = \dfrac{ae^{aq} + 3q^2}{e^{aq} + q^3}$, which can be checked either by using the table of derivatives, or the formula for logs, equation (3.10), or with Maple.

In example **(iv)**, $y^n = \sin(ax^3)$ and the equation can be differentiated either as written, or the nth root taken first. Taking the equation as written, because y^n is a function of x when it is differentiated and using the function of a function method, it produces $\dfrac{dy^n}{dx} = ny^{n-1}\dfrac{dy}{dx}$ because dy/dx is the derivative of y. The result of differentiating both sides is therefore

$$ny^{n-1}\frac{dy}{dx} = 3ax^2\cos(ax^3)$$

and y can be substituted into this result to obtain the equation in x alone. Taking the nth root first gives $y = [\sin(ax^3)]^{1/n}$ making it a double, function-of-function differentiation. The result is

$$\frac{dy}{dx} = \frac{3ax^2}{n}[\sin(ax^3)]^{(1/n)-1}\cos(ax^3),$$

which should be the same result as was obtained by using the first method but some substitution and rearranging is necessary to show this.

Maple produces

```
> Diff( y(x)^n = sin(a*x^3),x): % = simplify(value(%));
```

$$\frac{d}{dx}(y(x)^n = \sin(a\,x^3)) = (y(x)^{(n-1)}n\left(\frac{d}{dx}y(x)\right) = 3\cos(a\,x^3)\,a\,x^2)$$

and y has to be explicitly defined as a function of x for Maple to know what to differentiate with respect to. A Maple method to differentiate functions of y is done elegantly using the `alias` command. See Appendix 1.

3.5.2 Function of a function general formula

The general equation on which to use the function-of-function or chain rule is $f[g(h)]$, which means that f is a function of a second function g, with the generic variable here called h; $\sin(ax^2 + x^{-3})$ has this form with f being $\sin(\cdots)$ and $g(x)$ being $ax^2 + x^{-3}$ and the generic variable h is x in this case.

Differentiation is performed by letting $u = g(h)$ giving

$$\frac{d}{dh}f[g(h)] = \frac{df}{du}\frac{du}{dh}. \tag{3.17}$$

and this formally expressed, is what was done in our previous examples. Notice how du is on the bottom of the derivative on the left, and on the top on the right. To differentiate the equation using the formal equation is harder than following the rule 'differentiate the outside and then the inside', but can be done. Suppose that $f = (a + x^2)^3$ is to be differentiated with respect to x. Letting $x = h$ and $u = 1 + x^2$ gives $du/dx = 2x$ and as $f = u^3$, then $df/du = 3(u)^2$ and the result is $df/du = 6x(1 + x^2)^2$.

In thermodynamics, derivatives often seem to appear out of nowhere. For instance, $\dfrac{dp}{dV} = \dfrac{dp}{dT}\dfrac{dT}{dV}$ and $\dfrac{dS}{dp} = \dfrac{dS}{dT}\dfrac{dT}{dp}$ and these are examples of the chain rule used in a way that you may feel is back to front. The left-hand side is expanded to give the right, rather than starting with an explicit function, of say, S in terms of T and p. The functional form does exist but it is not needed for us to be able to write down the expression; the fact that it exists is enough for further calculations. The reason for using this approach is that the change of entropy S with pressure can be converted into a measurement of entropy with temperature, and a measurement of the change of temperature with pressure. The entropy change with temperature can be accurately measured with an electrochemical cell.

3.5.3 The product rule

When there are products or ratios of functions, differentiation is done in two steps;

First, differentiate one function leaving the other alone,
then add to this the result of differentiating the second leaving the first alone.

An example should make this clear. To differentiate $y = (x^2 + a)(x^3 + x)$ let the first function be $(x^2 + a)$ and differentiate this to $2x$ and multiply by the second function $(x^3 + x)$, then do the reverse and add the two terms together.

$$\frac{dy}{dx} = 2x(x^3 + x) + (x^2 + a)(3x^2 + 1).$$

As a check, expand the brackets to $y = x^5 + (1 + a)x^3 + ax$ and then differentiate giving $\dfrac{dy}{dx} = 5x^4 + 3(1 + a)x^2 + a$, which is the same as the first result. This can be seen by expanding all the brackets in both results.

3.5.4 General formula

The general form of the product rule is rather formidable:

$$\frac{d}{dh}fg = g\left[\frac{d}{dh}f\right] + f\left[\frac{d}{dh}g\right] \equiv gf' + fg' \tag{3.18}$$

where f' is the derivative $\dfrac{df}{dh}$ and the functions f and g are actually functions of variable h and should more properly be written as $f(h)$ and $g(h)$. The formula can be understood by defining $y = fg$, where f and g are functions of x. In full notation, this is written as $y(x) = f(x)g(x)$. First taking logs of both sides

$$\ln(y) = \ln(f) + \ln(g),$$

then differentiating with respect to x,

$$\frac{1}{y}\frac{dy}{dx} = \frac{1}{f}\frac{df}{dx} + \frac{1}{g}\frac{dg}{dx}$$

and multiplying out produces equation (3.18)

$$\frac{dy}{dx} = g\frac{df}{dx} + f\frac{dg}{dx}, \tag{3.19}$$

if the general variable h is x.

3.5.5 Ratios of functions

The product rule and function-of-function rule can be used to calculate quotients. If $y = \dfrac{f(x)}{g(x)}$ then this can be represented as $y = f(x)g(x)^{-1}$ and the derivative is,

$$\frac{d}{dx}f(x)g(x)^{-1} = \frac{1}{g(x)}f'(x) - \frac{f(x)}{g(x)^2}g'(x), \tag{3.20}$$

which, if simplified, can be written as:

$$\frac{d}{dx}fg^{-1} = \frac{gf' - fg'}{g^2} \tag{3.21}$$

where the brackets and x are suppressed for clarity. For example using (3.20)

(a) $y = \dfrac{\sin(x)}{(1+x^2)}$ then $\dfrac{dy}{dx} = \dfrac{\cos(x)}{(1+x^2)} - 2x\dfrac{\sin(x)}{(1+x^2)^2}$.

(b) $y = \tan(x)$, replace the tan as a ratio $y = \dfrac{\sin(x)}{\cos(x)}$ then using (3.21)

$$\frac{dy}{dx} = \frac{\cos^2(x) + \sin^2(x)}{\cos^2(x)} = 1 + \tan^2(x)$$

3.5.6 Differentiation with respect to a function

Sometimes in thermodynamics, but also elsewhere, equations are put in the form $d/d(1/x)$ rather than d/dx. For example, at constant pressure the Gibbs–Helmholtz equation, the derivative of the change of (Gibbs) free energy ΔG with temperature, can be written as $\dfrac{d(\Delta G/T)}{d(1/T)}$ rather than $d\Delta G/dT$. The reason for doing this is largely historical and relates to a time not so long ago when computers were not readily available, and so the equation was simplified making it easier to integrate or plot by hand on graph paper. The Gibbs–Helmholtz equation is now normally written as

$$\left(\frac{\partial \Delta G}{\partial T}\right)_p = \frac{\Delta G - \Delta H}{T} \tag{3.22}$$

where ΔH the enthalpy and T temperature. The curly ∂ indicates a partial derivative, which means only that ΔG, depends on something other than T alone, which here is p, but is otherwise just the same as any other derivative. The p subscript indicates constant pressure, but we will drop this now for clarity. To calculate $d(\Delta G/T)/dT$ the function-of-function method is used

$$\frac{d}{dT}\left(\frac{\Delta G}{T}\right) = \frac{1}{T}\frac{d\Delta G}{dT} - \frac{\Delta G}{T^2}$$

and then substituting for the 'normal' derivative gives

$$\frac{d}{dT}\left(\frac{\Delta G}{T}\right)=\frac{1}{T}\left(\frac{\Delta G-\Delta H}{T}\right)-\frac{\Delta G}{T^2}, \quad \text{and} \quad \frac{d}{dT}\left(\frac{\Delta G}{T}\right)=-\frac{\Delta H}{T^2}.$$

So far, this is what we have done several times in one form or another. The next step is to find the derivative with respect to $1/T$. The best way is to recast the equation with substitution, for example, $u = 1/T$ makes $du = -dT/T^2$. Substituting for dT gives $-\frac{1}{T^2}\frac{d}{du}\left(\frac{\Delta G}{T}\right)=-\frac{\Delta H}{T^2}$ therefore

$$\frac{d(\Delta G/T)}{d(1/T)}=\Delta H.$$

A second commonly met equation is the van 't Hoff isotherm

$$\left[\frac{\partial \ln(K_p)}{\partial T}\right]_p=\frac{\Delta H^\circ}{RT^2}$$

which describes how the equilibrium constant of a reaction at constant pressure varies with the temperature. To put this in the form $\frac{d\ln(K_p)}{d(1/T)}$, the same substitution, $u=1/T$, can be employed, producing

$$\frac{d\ln(K_p)}{d(1/T)}=\frac{-\Delta H^\circ}{R}.$$

Therefore, a plot of the log of the equilibrium constant vs reciprocal temperature is a horizontal line of intercept $-\Delta H^\circ/R$. If the line is not constant, but sloping or varying, then ΔH° depends on temperature.

3.5.7 Implicit differentiation

Quite often, an equation we may want to differentiate contains powers or functions of y as well as x. For example, $\sin(y) + x^2e^x = 1$, and cannot easily be put into the form $y = \cdots$. In this case implicit differentiation is necessary, which is just a grand sounding name for a rather simple procedure and one with which you are already familiar. All that is done is to differentiate any function of y using the function-of-function or product rule or both, as appropriate, and remembering that the differential of y is dy/dx.
Differentiating $\sin(y) + x^2e^x = 1$ with respect to x produces

$$\cos(y)\frac{dy}{dx}+2xe^x+x^2e^x=0.$$

To test how well you understand this, find dx/dy by direct calculation using $\sin(y) + x^2e^x = 1$, and thus show that $\frac{dy}{dx}=\left(\frac{dx}{dy}\right)^{-1}$.

3.5.8 Parametric functions and their differentiation

Sometimes it is easier to describe equations in parametric form as two functions with a common or dummy variable, rather than in the conventional way; there is no fundamental reason for doing so, just convenience. An example is the equation of a circle. If it has a radius a and is centred at the origin,

$$x^2+y^2=a^2.$$

In parametric form, this same equation becomes the *pair* of equations;

$$x=a\cos(t) \quad \text{and} \quad y=a\sin(t),$$

with t as the *parametric variable.* By squaring both x and y and adding, the familiar equation is formed because $\sin^2(t) + \cos^2(t) = 1$. When graphing a parametric equation both y and x are calculated with t varying. The range of t will, naturally, depend on the equations being used, but with a circle need be no more than 0 to 2π.

The parametric equations of an ellipse are $x = a\cos(t)$, $y = b\sin(t)$ where a and b are constants that define the axes of the ellipse. When they are the same, a circle results. The equation of a parabola is $y^2 = 4ax$; in parametric form this can be written as $x = at^2$ and $y = 2at$.

To differentiate the parametric form use the function-of-function approach and write

$$\frac{dy}{dx} = \frac{dy}{dt}\frac{dt}{dx} = \frac{dy}{dt}\bigg/\frac{dx}{dt}, \qquad \text{if} \qquad dx/dt \neq 0. \tag{3.23}$$

Therefore, dy/dx is always the ratio of the two derivatives with respect to t, the parametric variable. The second derivative can be written, using $y' = dy/dt$ and $x' = dx/dt$ for clarity, as;

$$\frac{d^2y}{dx^2} = \frac{d}{dx}\left(\frac{dy}{dx}\right) = \frac{d}{dx}\left(\frac{y'}{x'}\right) = \frac{d}{dt}\left(\frac{y'}{x'}\right)\frac{dt}{dx}.$$

Evaluating the last step further gives,

$$\frac{d}{dt}\left(\frac{y'}{x'}\right)\frac{dt}{dx} = \left(\frac{y''}{x'} - \frac{y'x'}{(x')^2}\right)\frac{1}{x'} = \frac{x'y'' - x''y'}{(x')^3}.$$

The result, in normal notation, is shown below using Maple to confirm this horribly tricky calculation.

```
> Diff(Diff(y(t),t)/Diff(x(t),t),t)/Diff(x(t),t):
  % = simplify( value(%));
```

$$\frac{\frac{d}{dt}\left(\frac{\frac{d}{dt}y(t)}{\frac{d}{dt}x(t)}\right)}{\frac{d}{dt}x(t)} = \frac{\left(\frac{d^2}{dt^2}y(t)\right)\left(\frac{d}{dt}x(t)\right) - \left(\frac{d}{dt}y(t)\right)\left(\frac{d^2}{dt^2}x(t)\right)}{\left(\frac{d}{dt}x(t)\right)^3}$$

Consider finding the gradient of the elegant looking curve shown in Fig. 3.6 at a point such as $x = 2$, and then find the equation of the tangent line at this point and where the tangent is horizontal and where vertical. The curve is described by the parametric equations

$$x = t^3 - t, \qquad y = 2\sin(t^2).$$

The gradient from equation (3.23) is

$$\frac{dy}{dx} = \frac{4t\cos(t^2)}{3t^2 - 1}.$$

To reform this into an equation in y is possible, but to make an equation in x will be messy and there is no reason to do so. At our 'victim' point, $x = 2$, the value of t is the solution of $t^3 - t = 2$ which, using Maple, is

```
> solve(t^3-t-2.0,t);
```

$$1.5214, -0.7607 + 0.8579\,\mathrm{I}, -0.7607 - 0.8579\,\mathrm{I}$$

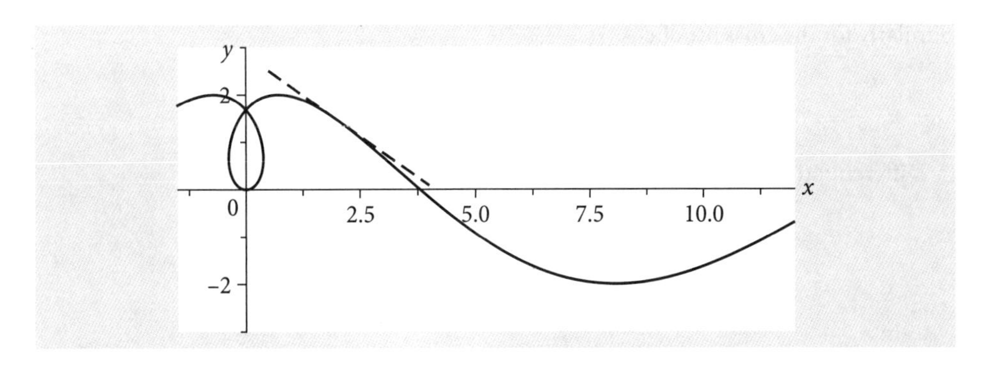

Fig. 3.6 Parametric curve $x = t^3 - t$, $y = 2\sin(t^2)$ and its tangent at $x = 2$.

and only the real value is needed, making the gradient −0.693. The equation of a straight line through points $\{x_0, y_0\}$ is $y - y_0 = m(x - x_0)$, which using the gradient just found and the initial coordinates is $y - 1.47 = -0.693(x - 2)$; this line is also shown on the sketch. The value for y_0 is found from $y = 2\sin(t^2)$.

The horizontal and vertical tangents can be obtained from the individual derivatives when $dy/dt = 0$ and $dx/dt = 0$ respectively; (see also Section 3.9 for maxima and minima). The horizontal tangent is $dy/dt = 4t\cos(t^2) = 0$ where $t = 0$, or $\pm\sqrt{(n\pi/2)}$ where n is a constant with values 1, 3, 5 ⋯. The integers n appear because the function is oscillatory along the x-axis and cosines are zero at multiples of $\pi/2$. The values are

$$t = 0, \qquad \{x, y\} = \{0, 0\}$$

$$t = \sqrt{\pi/2}, \qquad \{x, y\} = \left\{\left(\frac{\pi}{2}\right)^{3/2} \mp \sqrt{\frac{\pi}{2}}, 2\right\},$$

$$t = \sqrt{3\pi/2}, \qquad \{x, y\} = \left\{\frac{3\sqrt{6}}{4}(\pi)^{3/2} \mp \sqrt{\frac{3\pi}{2}}, -2\right\} \quad \text{or} \quad \{\pm 8.06, -2\}$$

and so forth for other tangents. The vertical tangent occurs at $dx/dt = 3t^2 - 1 = 0$ or $t = 1/\sqrt{3}$ with $x = \pm 0.385$ and $y = 0.654$, which are either side of the central loop in the curve.

3.5.9 Differentiation of vectors

To differentiate any vector, each of its elements is treated separately. In the previous section, we saw how to represent a curve in parametric form. The two parametric equations can also be represented in vector form; for example, a parabola is represented by the equation $y^2 = 4ax$, but in parametric form this can be written as $x = at^2$ and $y = 2at$ and in row vector form as $v = [x \quad y] \equiv [at^2 \quad 2at]$. Since all elements are treated separately,

$$\frac{dv}{dt} = \begin{bmatrix} \frac{dx}{dt} & \frac{dy}{dt} \end{bmatrix}.$$

Evaluating each term produces $\dfrac{dv}{dt} = [2at \quad 2a]$. The second derivative is $\dfrac{d^2v}{dt^2} = [2a \quad 0]$. Previously, equation (3.23), it was shown that the gradient was the ratio of the derivative of $y(t)$ over that of $x(t)$. Applying this produces

$$\frac{dy}{dx} = \frac{2a}{2at} = \frac{1}{t}$$

so that the gradient is $\sqrt{a/x}$ which is the same result as starting with $y^2 = 4ax$.

3.5.10 Differentiating dot and cross products

When differentiating dot and cross products, the normal differentiation and the normal vector rules apply. Suppose $\boldsymbol{u}$ and $\boldsymbol{v}$ are vectors in some variable s then, as $\boldsymbol{u}\cdot\boldsymbol{v}$ is the dot product

$$\frac{d}{ds}\boldsymbol{u}\cdot\boldsymbol{v} = \frac{d\boldsymbol{u}}{ds}\cdot\boldsymbol{v} + \boldsymbol{u}\cdot\frac{d\boldsymbol{v}}{ds}.$$

Similarly for the cross product,

$$\frac{d}{ds}\boldsymbol{u}\times\boldsymbol{v} = \frac{d\boldsymbol{u}}{ds}\times\boldsymbol{v} + \boldsymbol{u}\times\frac{d\boldsymbol{v}}{ds}.$$

Differentiating vectors is described in more detail in Section 3.13.

3.6 Summary

Notation: $f' = d/dx$

Exponentials

$$y = e^{f(x)} \quad \rightarrow \quad \frac{dy}{dx} = f'(x)e^{f(x)}$$

$$y = e^{-ax^2} \quad \rightarrow \quad \frac{dy}{dx} = -2axe^{-ax^2}$$

Logarithms

$$\frac{d}{dx}\ln[f(x)] = \frac{f'(x)}{f(x)}$$

Function of a function or chain rule

$$\frac{d}{dh}f[g(h)] = \frac{df}{du}\frac{du}{dh} \qquad \text{where } u = g(h)$$

or

$$\frac{d}{dh}f[g(h)] = f'g'$$

or

$$\frac{dy}{dx} = \frac{dy}{dt}\frac{dt}{dx}$$

e.g.

$$y = \sin(\underline{ax^2 + x^{-3}}) \quad \rightarrow \quad \frac{dy}{dx} = \cos(ax^2 + x^{-3})(2ax - 3x^{-4})$$

f = outside ↑ ↑g = inside

$$y^n = x^2 \quad \rightarrow \quad ny^{n-1}\frac{dy}{dx} = 2x$$

Product rule:

$$\frac{d}{dh}fg = g\left[\frac{d}{dh}f\right] + f\left[\frac{d}{dh}g\right]$$

$$\equiv gf' + fg'$$

or explicitly

$$\frac{d}{dh}f(h)g(h) = f'(h)g(h) + f(h)g'(h);$$

e.g.

$$y = (x^2)e^{-x} \quad \rightarrow \quad \frac{dy}{dx} = 2xe^{-x} - x^2e^{-x}$$

Functions of powers of x

$y = a^x$ then take logs;

$$\ln(y) = x\ln(a) \quad \rightarrow \quad \frac{1}{y}\frac{dy}{dx} = \ln(a)$$

Reciprocal derivatives

$$\frac{dy}{dx} = \left(\frac{dx}{dy}\right)^{-1} \qquad \frac{d^2y}{dx^2} = -\frac{d^2x}{dy^2}\left(\frac{dx}{dy}\right)^{-3}$$

Changing variables

If a function is expressed as $f(x)$ and you want it as $f(t)$ then;

$$\frac{df}{dt} = \frac{df}{dx}\frac{dx}{dt}$$

Parametric equations

$$y = f(t), \qquad x = g(t) \qquad \text{then} \qquad \frac{dy}{dx} = \frac{dy}{dt}\frac{dt}{dx} = \frac{\frac{dy}{dt}}{\frac{dx}{dt}}$$

Integrals

$$\frac{d}{dx}\int f(u)du = f(x) \qquad \frac{d}{dx}\int_a^x f(u)du = f(x)$$

If u and v are functions of x then use Leibniz's rule

$$\frac{d}{dx}\int_{v(x)}^{u(x)} f(s)ds = u'f(u) - v'f(v)$$

If the differentiation variable is not the integration variable

$$\frac{d}{da}\int f(a, s)ds = \int \frac{\partial}{\partial a} f(a, s)ds$$

3.7 Questions

Full solutions are available at **www.oxfordtextbooks.co.uk/orc/beddard**.

Q3.12 Calculate dy/dx if

(a) $y = \sin(ax)e^{-bx}$ **(b)** $y = \tanh(x)e^{-bx}$,
(c) $y = e^{-b\sin(x^2)}$, **(d)** $y = \ln(a + bx)\cos(ax)$,
(e) $y = \ln(a + bx)\tan^{-1}(ax)$, **(f)** $y = \ln(a + b\sin(x))$,
(g) $y = [\cos(e^{x^2} - 1)]^{1/2}$, **(h)** $x = \cosh(y)$
(i) $y = \sin^{-1}(x/a)$, **(j)** $y = \sin(x/a)^{-1}$
(k) If $p = \sin^2(q) + q^2\cos(q)$, find dp/dq.

In each case you could check your answers using Maple.

Q3.13 Differentiate with respect to x without using the computer:

(a) $y = x^y$, **(b)** $x = y^y$, **(c)** $y = x^x$

Strategy: Usually with powers of x and y it is simpler to take logs first then differentiate.

Q3.14 **(a)** Repeatedly differentiate $\int e^{ax}dx$ by a and show that

$$\int x^n e^{ax}dx = \frac{d^n}{da^n}\left(\frac{e^{ax}}{a}\right)$$

(b) Find two similar relationships for $\int \sin(ax)dx$ when n is even and for $\int \cos(ax)dx$ when odd. This is quite a neat way to perform these integrals.

Q3.15 **(a)** Use Maple to plot the parametric function

$$x = \sin(t) - \sin^3(2t), \qquad y = \cos(t) - \cos^3(2t).$$

(b) Calculate dy/dx.

(c) Find the vertical tangents.

This is a rather complex curve that would be hard to define in terms of x and y alone. Try changing the constant 2 to 3 and the powers to see how the curve alters.

Q3.16 **(a)** Calculate the gradient of the cardioid with parametric equations

$$x = 2\cos(t) + \cos(2t) + 1, \qquad y = 2\sin(t) + \sin(2t),$$

at the point t, and where t is not a multiple of π. A plot of a cardioid is shown in the solution to this question.

(b) Calculate where the tangent is zero; only two solutions are sensible, why is this and where is the tangent infinite?

Q3.17 Calculate $\dfrac{d}{dx}\displaystyle\int_{\ln(x)}^{x^2} e^{-s^2}\,ds$ without actually integrating.

Strategy: Use equation (3.15).

Q3.18 If $y = x^2 + 2$, $x = \sqrt{w}$, $w = \cos(u)$, using the chain rule write down dy/du in terms of x, u, and w, then calculate the derivative in terms of u alone. Confirm the result by substituting the equations first.

Q3.19 **(a)** If $y = [\ln(x)]^x$ find dy/dx.

(b) Plot graphs of y and its derivative and explain why the curves have the form they do. What happens when $x < 1$ and why are the curves so similar when x is large?

Q3.20 If $f = ye^{-ax}$ and x and y are both functions of a variable s, what is df/ds if a is a constant?

Strategy: Being a function of s means the functions are $y(s) =$ and $x(s) =$ 'something in s'. In full notation $f(s) = y(s)e^{-ax(s)}$. As both x and y are functions of the independent variable s, then the derivatives of x and y with respect to s must be found. It is here necessary to overcome our prejudice of x as the only differentiation variable. If the equation had different symbols, it might be easier!

Q3.21 If $y = x^n\ln(x)$

(a) Show that $xy' = ny + x^n$.

(b) Calculate y''.

Q3.22 Show that $\dfrac{d}{dx}\left(\dfrac{v^2}{2}\right) = \dfrac{d^2x}{dt^2}$ where x represents the distance travelled by a body in time t when v is its velocity.

Strategy: Notice that the derivatives are in time t and distance x and that they have to be connected; the way to do this is *via* the velocity.

Q3.23 **(a)** Find the gradient dy/dx of the lemniscate of Bernoulli, $(x^2 + y^2)^2 = a^2(y^2 - x^2)$ shown in Fig. 3.7.

(b) Determine where the gradient is zero; it's easy to see on the figure, but show it mathematically.

(c) Repeat the calculation by changing coordinates from Cartesian to Polar (see Chapter 1.7 for the definition of polar coordinates).

Strategy: In part (b), the tangent is horizontal when $dy/dx = 0$, and is infinite when $dx/dy = 0$. You will find $\cos^2(\theta) - \sin^2(\theta) = \cos(2\theta)$ useful.

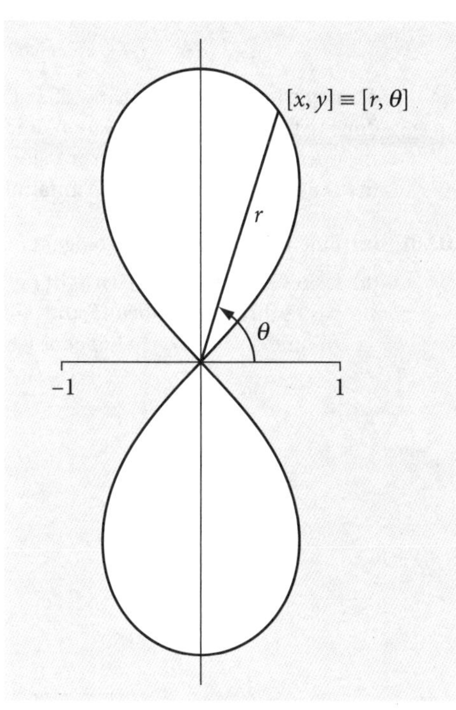

Fig. 3.7 'Lemniscate of Bernoulli' with $a = 2$ and showing the definition of the polar coordinates. The axes' scale is that of the Cartesian coordinates.

Q3.24 Calculate the second and third derivatives of

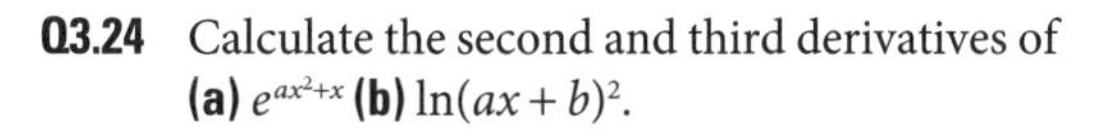

(a) e^{ax^2+x} **(b)** $\ln(ax+b)^2$.

Q3.25 If $y^2 = a^2 - x^2$ prove that $y'' = \dfrac{a^2}{xy^2}y'$ where y'' is the second derivative, y' the first.

Strategy: Start with the equation as written and either differentiate y^2 with respect to x or take the square root first; either way will work its your choice! This equation is that for a circle centred at the origin with radius a.

Q3.26 If D is the differential operator, this acts on a function f to give the same function in return, $Df(x) = f(x)$, and where $f(0) = 1$. Repeatedly differentiate n times and find $f(x)$ by using the Maclaurin series, which is

$$f(x) = f(0) + f'(0)x + \frac{f''(0)}{2!}x^2 + \frac{f'''(0)}{3!}x^3 + \cdots$$

where each derivative is evaluated at $x = 0$.

Strategy: The differential operator is a fancy name for d/dx and therefore the equation is $\dfrac{d}{dx}f(x) = f(x)$.

Q3.27 Show that $udu/dx = du/dt$ if $u = dx/dt$.

Strategy: Perform the differentiations in a rather formal way by using differentiation as an operator such as $\dfrac{d}{dt}(u)$, which is d/dt operating on u. Only at the end of the calculation cancel any terms.

Q3.28 Differentiate the equation $(1 - q^2)f(qN) + qN\ln(q) = 0$

(a) with respect to q and **(b)** to N where $f(\cdots)$ represents any normally differentiable function.

Q3.29 **(a)** if $x^n + y^n = 1$ show that $dy/dx = -(x/y)^{n-1}$

(b) find the first dy/dx, and second derivatives of $\tan(y) = \sinh(x)$.

Q3.30 The Schrödinger equation can be written with certain units so that $\hbar^2/2m = 1$ producing $-D^2\psi + V\psi = E\psi$ where the potential energy V and wavefunction ψ are both functions of x. The energy is E and must always be independent of any coordinate. The operator $D = d/dx$. Verify that

$$\psi D^4\psi = D(\psi^2 DV) + (V - E)^2\psi^2.$$

Strategy: Look at the left-hand side of the equation that is to be verified, and work out what has to be done to the Schrödinger equation. Recall that D means differentiate once, D^2 means twice, and that the operator only works on expressions to its right, and both V and ψ are functions of x. Try to use the D notation, if it proves too unfamiliar revert to using d/dx.

Q3.31 **(a)** Find the equation of the tangent of a parabola with parametric equations $x = at^2$ and $y = 2at$.

(b) Using Fig. 3.8 show that a light ray parallel to the x-axis reflected off a parabolic mirror always passes through the focus F; the point $\{a, 0\}$. Do this by finding the point P and the lengths PF and FT, and consider what happens when point T is varied.

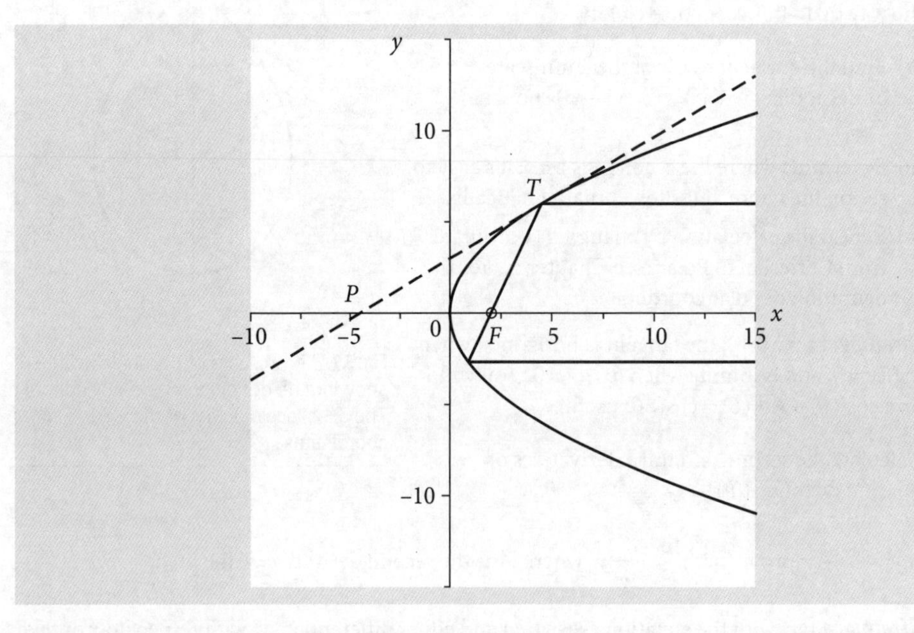

Fig. 3.8 Light reflected off the parabola passes through the focal point $F = \{a, 0\}$. The tangent line at point T is PT.

Strategy: Use the parametric form to calculate the slope at point T in terms of variable t. Calculate the equation of the line PT, then the coordinates of point P and so the distances PF and FT.

Exercise: Show that a spherical surface does not cause all parallel rays entering to be focused to a point. The parametric equations for a circle are $x = \cos(t)$, $y = \sin(t)$.

Q3.32 The internal energy U of a crystal at temperature T is, according to the Debye model,

$$U = \frac{9}{8}Nh\upsilon_D + \frac{9Nh}{\upsilon_D^3}\int_0^{\upsilon_D} \frac{\upsilon^3}{e^{h\upsilon/k_BT} - 1}d\upsilon,$$

where υ is frequency, υ_D is proportional to the highest phonon frequency of the solid, N is Avogadro's number, and k_B Boltzmann's constant.

(a) Confirm the units of the internal energy are correct. What does the first term on the right represent?

(b) Calculate the heat capacity (at constant volume) C_V, which is the rate of change of energy with temperature.

(c) Convert the result to dimensionless quantities using $x = h\upsilon/k_BT$ so that the result is given in terms of temperature with $\theta = h\upsilon_D/k_B$ being the Debye temperature. Why are the units of x and θ dimensionless?

(d) By changing the limits to the integration, calculate C_V as $T \to 0$ and $T \to \infty$ and show that C_V varies as $(T/\theta)^3$ at low temperatures, $C_V \to 0$ at 0K and to $3R$ at very high temperatures in accordance with the empirical law of Dulong and Petit. At high temperatures, expand the terms inside the integral.

Strategy: The heat capacity is the rate of change of energy with temperature and because the temperature is inside the integral, the differentiation can be performed inside the integral, and Leibniz's rule, equation (3.15) does *not* apply.

Q3.33 The buffer capacity β of a weak acid–conjugate base buffer is defined as the number of moles of strong acid needed to change the pH by 1 unit, where

$$\beta = \frac{d[A]}{dpH}$$

and the concentration of acid present is

$$[A] = \frac{K_w}{[H^+]} - [H^+] + \frac{C_BK_a}{[H^+] + K_a}.$$

K_w is the water ionization equilibrium constant $K_w = [H^+][OH^-] = 10^{-14}$, K_a is the acid dissociation constant, and C_B the *total* concentration of buffer.

(a) Calculate β.

(b) Plot a graph of β vs pH for benzoic acid $pK_a = 4.2$ and sodium benzoate each at 0.01 M. $(pK_A = -\log(K_A))$.

(c) Differentiate β. By ignoring the buffer concentration term, find the minimum value of β using $d\beta/dpH = 0$. Next by ignoring terms not involving C_B find the maximum β. Justify these simplifications.

Q3.34 A skydiver jumps from a plane. The distance y that she drops in time t increases as $y = a\ln\left(\frac{(e^{bt} + e^{-bt})}{2}\right)$, which assumes that the air resistance is proportional to the square of her velocity. The constants a and b are $a = \frac{2m}{\rho CA}$ and $b = \sqrt{\frac{g\rho CA}{2m}}$ where g is the acceleration due to gravity, ρ the density of air ~1.29 kg m^{-3}, A the cross-sectional (frontal) area of the falling body of mass m, and C ~ 0.5 is a shape-dependent resistance factor.

(a) Calculate her velocity at time t, and show that this reaches a maximum or terminal velocity of $\sqrt{\frac{2gm}{\rho CA}}$.

(b) What is her terminal velocity if she weighs 60 kg and if $A = 0.7$ m^2? How long does it take to reach 95% of the terminal velocity and how far will she have travelled vertically?

(c) Calculate the terminal velocity for a hailstone weighing approximately 4 g and of 10 mm radius; comment on the relative speeds and energies of the skydiver and hailstone. If there were no air resistance what would happen?

(d) It is an urban myth, apparently, that a cat jumping out of a 10-storey window will survive the fall; you would almost certainly not. The cat reputedly reaches a terminal velocity of only 10 mph, and if it weighs 2 kg what is its surface area? Do you think that the result make sense, and if not, why not?

Q3.35 The Schrödinger equation has the form $H\psi = E\psi$ where H is the Hamiltonian operator describing kinetic and potential energy, E is the energy of the eigenstate and is called the eigenvalue and ψ, the wavefunction, is called the Eigenfunction. $H\psi = E\psi$ is an eigenvalue–eigenvector equation and its characteristic is that operating on ψ produces a constant times ψ. If the operator is $H = \dfrac{d^n}{dx^n}$,

(a) Show that a possible wavefunction is $e^{\alpha x}$ with eigenvalue α^n.

(b) Is $\sin(\alpha x)$ a possible wavefunction?

***Strategy*:** The particular feature of note about eigenvalue–eigenvector equations is that the result of operating on a function is to produce the same function but multiplied by a constant. In the Schrödinger equation, the function is the wavefunction, the constant is the energy, and the operator is the kinetic plus potential energy operator $-\dfrac{\hbar^2}{2m}\dfrac{d^2}{dx^2} + V(x)$ and, for the sake of clarity, the constants in the question are in units such that they $\dfrac{\hbar^2}{2m} = 1$ and also $V(x) = 0$.

Q3.36 The wavefunction for a harmonic oscillator with quantum number $\upsilon = 1$ is

$$\psi = (4\alpha^3/\pi)^{1/4} x e^{-ax^2/2}.$$

The square of the wavefunction describes the chance of finding the nuclei at position x during the vibrational motion of the atoms of a diatomic molecule. The constant α is given by $\alpha = (\mu k/\hbar^2)^{1/2}$, the force constant k, $\hbar$ is Planck's constant divided by 2π, and μ is the reduced mass. Show that ψ is a solution of the Schrödinger equation:

$$-\frac{\hbar^2}{2m}\psi'' + \left(\frac{kx^2}{2}\right)\psi = E\psi,$$

where the energy of the $\upsilon = 1$ level is $E = \dfrac{3}{2}\hbar\left(\dfrac{k}{\mu}\right)^{1/2}$.

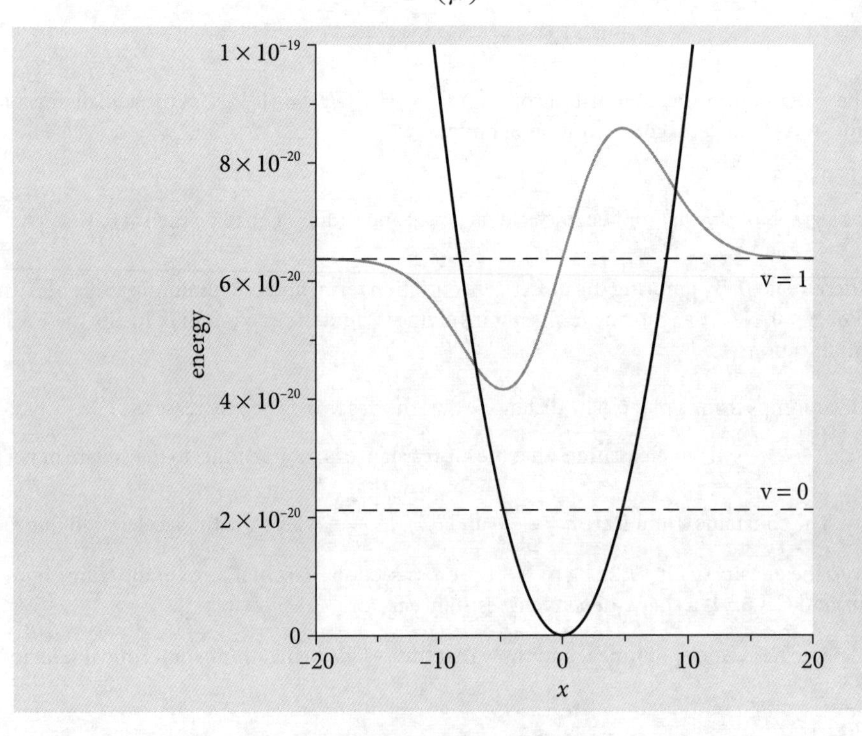

Fig. 3.9 $v = 1$ wavefunction drawn at its energy level in the harmonic potential. The lower line shows the zero-point energy, x is the bond extension. The extension is in picometres, the energy in Joules.

Strategy: To show, rather than prove, that ψ is a solution means demonstrating that the left side of the equation is the same as the right. The equation written more conventionally is, $-\frac{\hbar^2}{2m}\frac{d^2\psi}{dx^2}+\frac{kx^2}{2}\psi=E\psi$ and writing it this way immediately makes the equation look easier. First, substitute for E and rearrange to group terms in ψ then differentiate ψ twice with respect to x and finally do some tidying up. You could also do the calculation by starting by differentiating ψ if you want; the order does not matter.

Q3.37 A wavepacket is the sum of many wavefunctions and because this sum is not an eigenstate, it evolves in time. Generating a wavepacket is easy using femtosecond laser pulses because they have a wide energy spread and many molecular vibrational or rotational energy levels can be (almost) simultaneously excited. The wavepacket has the appearance of a classical particle moving from side to side in the potential, this can be observed using a second femtosecond laser pulse.

The frequency ω of the wavepacket's oscillatory motion can be calculated from $\omega=\frac{1}{\hbar}\frac{\partial E_n}{\partial n}$ where n is the vibrational quantum number and E_n the energy. This formula is obtained from the 'angle-action' or Hamilton–Jacobi formulation of mechanics (see Goldstein 1980, p. 459). This formula enables the frequency of any oscillation to be found even if the equations of motion are not solved. The energy E of a Morse (anharmonic) oscillator of frequency ω_e and anharmonicity x_e is

$$E=\hbar\omega_e(n+1/2)-\hbar x_e\omega_e(n+1/2)^2,$$

(a) Find the wavepacket frequency ω in terms of E instead of n.

(b) Show that this frequency decreases as the energy of an exciting photon increases towards the dissociation energy and is $\omega^2=\omega_e^2-4x_e\omega_e(E_{h\nu}-E_0)$.

Q3.38 A circular cone with a volume $V=\frac{\pi}{3}h^3\tan^2(\alpha)$ has a height h, base radius b, and semi-angle α.

(a) If α is constant what is dV/dx, where x is any value between 0 and h. What does dV/dx represent?

(b) A filter paper is formed into a circular cone and placed in a conical funnel and solvent flows out at a constant rate of 5 $cm^3\ min^{-1}$. Find the rate of decrease of the solvent's level when there is 2 cm of solvent in the funnel if it has a base of 60 mm and depth 90 mm.

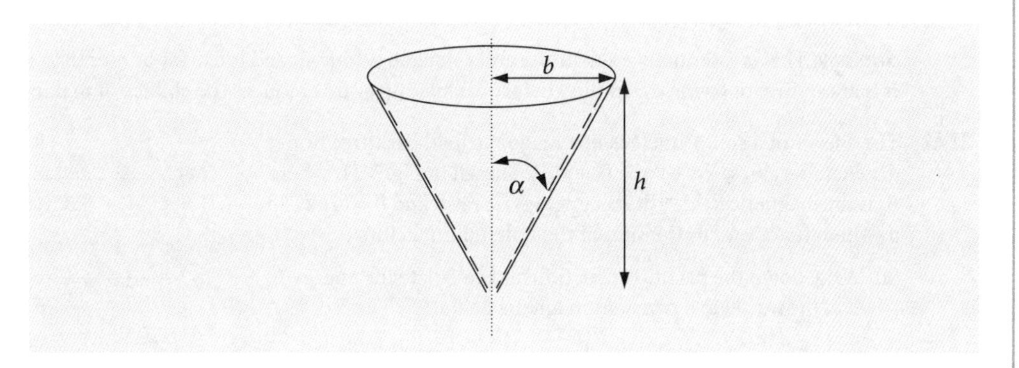

Fig. 3.10 Conical filter paper

Strategy: Part (b) requires the change of volume with time to be calculated, but in (a) the change with x was calculated and these two quantities can be related using the chain rule.

Q3.39 The radius of curvature ρ of a function $f(x)$ is given by

$$\rho=\frac{[1+f'(x)^2]^{3/2}}{f''(x)}. \tag{3.24}$$

where $f'(x)$ and $f''(x)$ are the first and second derivatives respectively. Quantum mechanically the vibrational motion of diatomic molecules can be described by the harmonic oscillator or Morse potentials. Calculate the radius of curvature at the minima of

(a) The harmonic oscillator potential V_{HO} given that $V_{HO}=k(r-r_e)^2$ and

(b) The Morse potential $V_M=D_e(1-e^{-\beta(r-r_e)})^2$.

(c) Assuming that at the minima the anharmonic curve has the same shape as the harmonic, relate k, the force constant, to D_e the dissociation energy and parameter β. The equilibrium separation of the atoms is r_e.

(d) Dimensionally check the results for the radius of curvature and force constant, i.e. check that the units are the same on both sides of your final equation.

Strategy: If you do this calculation by hand, work out the differentiations separately for each potential then put into the curvature equation. The variable to differentiate with respect to is r, which replaces x in the generic formula. The potential energy minimum occurs when $r = r_e$, so substitute this into the result and simplify. However, this is a fiddly rather than hard calculation and it is easier, and mistakes are less likely, if Maple is used, particularly so with the Morse potential.

Q3.40 Force is the derivative of the potential with extension. Hooke's law states that the restoring force is force constant × extension so it follows that the potential energy is proportional to the bond *extension* squared, which is the harmonic potential. Using the Morse potential given in the previous question, show that the force constant k near the bottom of the potential well, where Hooke's law applies, is $k = 2D_e\beta^2$.

Q3.41 The fraction of non-degenerate particles with a given energy level E_v, is given by the Boltzmann distribution $\dfrac{n_v}{N} = \dfrac{e^{E_v/k_BT}}{Z}$, where Z is the partition function, $Z = \sum_{v=0}^{\infty} e^{-E_v/k_BT}$, N the total number of particles, and n_v the number in level v. In the harmonic oscillator model of a diatomic molecule the energy levels are $E_v = \hbar\omega(v + 1/2)$, with quantum number v, and

$$Z = \frac{e^{-\theta/2T}}{1 - e^{-\theta/T}}$$

where $\theta = \hbar\omega/k_B$ is a characteristic temperature for the vibration which is in the range of a few hundred to a few thousand degrees K.

(a) Calculate the internal energy $U = Nk_BT^2\dfrac{d\ln(Z)}{dT}$ and heat capacity $C_V = \dfrac{dU}{dT}$, which is the rate of change of internal energy with temperature.

(b) Determine the low and high temperature limits of U and C_V.

(c) Plot graphs of the functions calculated using dimensionless units; $U/Nk_B\theta$ and C_v/Nk_B vs T/θ. If you want plot with real numbers, $k_B = 0.69$ cm^{-1} K^{-1} and in iodine, $\hbar\omega = 127$ cm^{-1}, and in NO, 1876 cm^{-1}. See McQuarrie & Simon (1997) for other values.

Strategy: This is essentially a substitute and calculate problem, and looks far harder than it really is. It is always worth trying to simplify before calculating as this reduces the chance of making an error.

Q3.42 The energy of a spin 1 nucleus in a magnetic field has three non-degenerate values, $E = -\gamma\hbar B_0, 0, +\gamma\hbar B_0$, shown in Fig. 3.11, where B_0 is the magnetic field, γ the magnetogyric ratio and $\hbar = h/2\pi$. The partition function Z is the sum of the Boltzmann factors e^{-E_i/k_BT}.

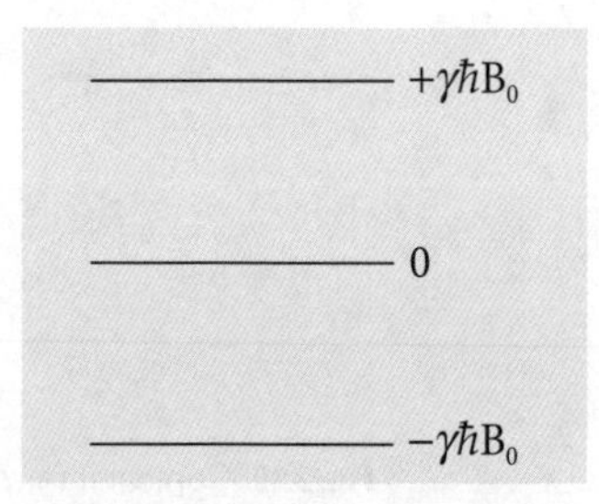

Fig. 3.11 Spin 1 nucleus' energy levels in a magnetic field.

(a) Write down the partition function and calculate the energy U of the nuclei in a constant magnetic field if

$U = -N\dfrac{d\ln(Z)}{d\beta}$ and $\beta = 1/k_BT$.

(b) Calculate the magnetic field contribution to the heat capacity $C_B = \dfrac{dU}{dT}$. (The calculation is easier if the derivative is converted to $\dfrac{dU}{d\beta}$ first.)

(c) Investigate the high and low temperature limits of U and C_v.

(d) Calculate the entropy $S = \dfrac{U}{T} + Nk_B\ln(Z)$ and show that at high temperatures this reaches the limit $S \to k\ln(3^N)$. Comment on the result.

(e) Plot graphs of U and C_B for ^{14}N nuclei if $B_0 = 14$ T and $\gamma = 1.97 \times 10^7$ rad T^{-1}s^{-1}.

Strategy: The partition function summation contains only three terms, which can be easily written down. The rest of the question is a direct calculation but notice that taking logs does not simplify the partition function as only products or quotients can be simplified with logs.

3.8 Limits: l'Hôspital's rule

In many calculations limits are encountered. For example, in the theory of diffraction the function $\dfrac{\sin(x)}{x}$ is met which when $x \to 0$ has the form 0/0 and appears at first sight to be indeterminate. There are other forms similar to this such as ∞/∞, $\infty/0$, 0^∞, 0^0, $\infty - \infty$ and l'Hôspital's rule is a method, sometimes used with a little additional ingenuity, of determining these limits. This topic is discussed here, not only because it requires differentiation, but also because it will be needed in the next section.

The method is

(a) Rearrange the limit required, if necessary, so that it becomes a ratio. This may require some cunning.

(b) Differentiate top and bottom *separately* with respect to the limit variable.

(c) Repeat this until one of the terms becomes constant. At this point, substitute the limiting value of x.

An example should make this clearer; the ratio $\dfrac{\sin(x)}{x}$ appears to be 0/0 as $x \to 0$ but it has a finite value which is determined using l'Hôspital's Rule:

$$\lim_{x\to 0}\frac{\sin(x)}{x} \to \frac{\dfrac{d}{dx}\sin(x)}{\dfrac{d}{dx}x} = \frac{\cos(x)}{1} = 1,$$

and the limiting value of $x = 0$ is only applied in the last step. As a check, one way to determine any limits close to zero is to plot the function. You will see that $\lim_{x\to 0}\dfrac{\sin(x)}{x}$ does indeed have a value of 1 at $x = 0$, see question 3.47 for other examples.

Expressions such as $\dfrac{e^x - 1}{x^2}$ often need to be evaluated when $x \to 0$ or $x \to \infty$. In statistical mechanics, for example, x may be $-E/k_BT$, a ratio of energies where k_B is the Boltzmann constant and T temperature. When $x \to 0$, which corresponds to high temperatures, the function $(e^x - 1)/x^2$ appears to have the indeterminate form 0/0, but the limit by l'Hôspital's is

$$\lim_{x\to 0}\frac{e^x - 1}{x^2} \to \frac{e^x}{2x} \to \frac{e^x}{2} = \frac{1}{2},$$

where differentiation was performed twice over. Notice that the limiting value is only applied when the constant 2 appears on its own. The other limiting case $x \to \infty$ really is infinity:

$$\lim_{x\to \infty}\frac{e^x - 1}{x^2} \to \frac{e^x}{2x} \to \frac{e^x}{5} = \infty$$

and this occurs because e^x increases faster with x than x^2 does. You could plot a graph to see that this is true.

The limit $\lim_{x\to\infty}\dfrac{e^{-x} - 1}{x^2 - 3}$ can perhaps be appreciated by looking at the ratio and noticing that when x is large e^{-x} becomes small but x^2 large so the limit is expected to be zero. Checking this properly gives the same result as intuition:

$$\lim_{x\to \infty}\frac{e^x - 1}{x^2 - 3} \to \frac{-e^{-x}}{2x} \to \frac{e^{-x}}{2} = 0.$$

Limits of products need to be rearranged first, for instance $\lim_{x\to 0} x\ln(x)$ should be rearranged to

$$\lim_{x\to 0}\frac{\ln(x)}{1/x} \to \frac{1}{x}\frac{1}{(-x^{-2})} \equiv \lim_{x\to 0}(-x) = 0.$$

Fractions are treated similarly; the following fraction is nominally undefined as $\infty - \infty$ but is rearranged into a ratio to become undefined as 0/0;

$$\lim_{x\to 0}\frac{1}{x}-\frac{1}{\sin(x)}=\lim_{x\to 0}\frac{\sin(x)-x}{x\sin(x)}$$

$$\to\frac{\cos(x)-1}{x\cos(x)+\sin(x)}\to\frac{-\sin(x)}{-x\sin(x)+2\cos(x)}\to 0$$

In the last step the sine can be divided to give a constant numerator,

$$\frac{-\sin(x)}{-x\sin(x)+2\cos(x)}=\frac{-1}{-x+2\cos(x)/\sin(x)}$$

and when $x \to 0$ the cosine is 1 and the sine zero, making $\cos(x)/\sin(x) \to \infty$ with the result that, because infinity is in the denominator, the limit is zero.

3.9 **Extrema: maxima, minima and inflection points**

One very useful property of derivatives is that they allow us to find the maxima and minima of functions; these are also called stationary points of the function. The extrema might be the maximum or minimum but can also be the limit where the function goes to $\pm\infty$.

The maximum of a curve is a point in whose locality all surrounding points have smaller y values. The minimum is defined similarly but points adjacent to it have larger y values and in both case the gradient is exactly zero. In Fig. 3.12 it is clear that the gradient is zero at the maximum and again at the minimum of the curve. The right-hand graph shows the first f' and second f'' derivatives. An inflexion point can occur when the gradient is zero but y is smaller on one side of the point than it is on the other. An inflexion can also occur when the gradient is not zero in which case the curvature of the line changes from concave to convex or vice versa.

The first derivative is zero at $x = 4/3$ and 3, which are the maximum and minimum respectively. The second derivative, which is the straight line, is negative at the maximum and positive at the minimum, so that the maximum and minimum of a function can be found from knowledge of the first and second derivatives.

The function shown in Fig. 3.12 is the equation $y = 2(x-2)^3 - x^2 + 4$, which has three roots, where $y = 0$ at $x \approx 0.8$, 2, and ≈ 3.7. The first derivative is $y' = 6(x-3)(x-4/3)$, which has roots at $x = 4/3$ and 3, which from the graph (left) are the maximum and minimum. The second derivative $y'' = 12x - 26$ is negative where $x < 13/6$ (≈ 2.2) as can also be seen from the graph.

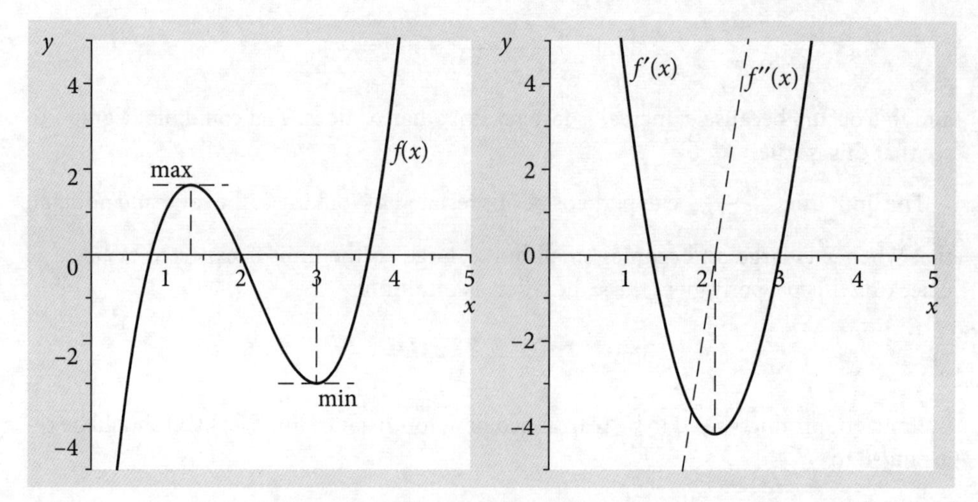

Fig. 3.12 Left: The maximum and minimum of a (cubic) function. Right: Its first and second derivatives.

3.9.1 Summary

A function $f(x)$ has its maximum or minimum when

$$\frac{d}{dx}f(x)=0 \tag{3.25}$$

The maximum occurs when

$$\frac{d^2}{dx^2}f(x)<0 \tag{3.26}$$

and the minimum when

$$\frac{d^2}{dx^2}f(x)>0 \tag{3.27}$$

If both first and second derivatives are 0 at the same value of x, this is a *point of inflexion*

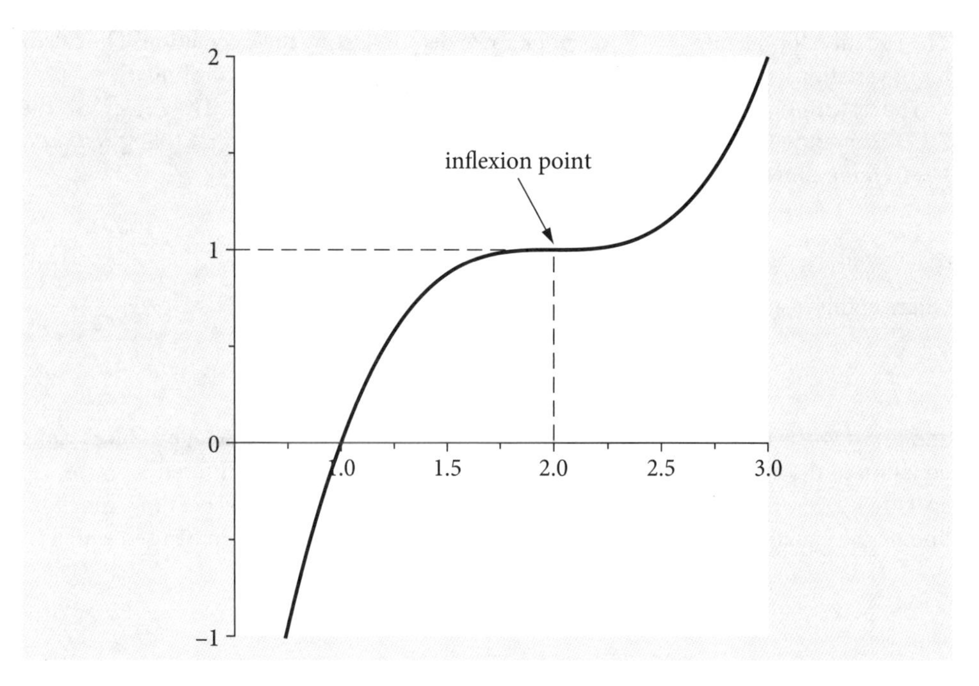

Fig. 3.13 A point of inflexion occurs when both first and second derivatives are 0 at the same point on the curve, in this curve $y=(x-2)^3+1$ this is at the point {2, 1}.

3.9.2 The Calculus of Variations

Instead of finding the maximum or minimum of a curve, consider finding the shortest distance between two points on the earth's surface or on a cone, or finding the equation giving the minimum area of a surface, or the curve of fastest descent between two points. The Calculus of Variations allows us to work out solutions to problems of this type.

You can imagine a graph that shows the many possible different curves that will fit between any two points. The calculus of variations is so named because it varies the path by an amount δy and, as the variation $\delta y \to 0$, the path along y becomes the same as that along $y+\delta y$. Because the whole path is sought, the problem is to find a function f that, provided that it exists, will make an *integral* have a stationary value, also called an *extremal*, which usually means that it has the smallest possible value. This condition is written as

$$I=\int_a^b f(x, y, dy/dx)dx. \tag{3.28}$$

Notice that the function f normally includes terms in the independent variable x, the dependent one y and the derivative dy/dx. Although this equation contains an integral, when solved to find its minimum, differentiation is mainly involved. You may need to consult Chapter 4 on integration to complete the last step in this type of problem.

Consider, for example, finding the equation that describes the minimum value of *all* possible surfaces of revolution. A surface of revolution is the surface obtained by rotating a curve, such as a parabola, about an axis; Fig. 3.8 shows the shape of a parabola and its surface of revolution is shaped somewhat like a bowl. Whatever the equation, $y = \cdots$ is, the surface area is always given by the integral

$$2\pi\int_a^b y ds = 2\pi\int_a^b y\sqrt{1+\left(\frac{dy}{dx}\right)^2}dx. \tag{3.29}$$

and a straightforward integration with the parabola $y = 2\sqrt{ax}$ will produce the parabola's surface area. The term in the square root is the length of a small element of the curve, see Fig. 4.19 (Q4.51) and the integral has the form of equation (3.28). Imagine now a surface film suspended between two similar wire hoops at $x = a$ and b, in practice this could be a soap film, see Fig. 3.27. Now, suppose that the problem is to find that one curve, of all possible curves, that will produce the minimum surface area within the two rings; this minimum area surface is the surface formed by a soap film, and its profile is called the catenary. The equation for the film $y = \cdots$ was not known before starting the calculation. The calculus of variations allows it be found by first finding dy/dx and then integrating it.

The calculus of variations defines a formula, variously called The Euler or the Euler–Lagrange equation, by which it is possible to evaluate the integral (3.28) *so that it has its minimum value*. The Euler equation is

$$\frac{\partial f}{\partial y} - \frac{d}{dx}\frac{\partial f}{\partial y_x} = 0 \tag{3.30}$$

where in this instance the function f is

$$f = y\sqrt{1+\left(\frac{dy}{dx}\right)^2},$$

and $y_x = dy/dx$ is the expression we want to find, which, when integrated, produces y, the equation of the curve required. Notice that the equation tells us to differentiate with respect to dy_x; see Section 3.5.6, which describes differentiation with respect to a function. Should the equation f *not explicitly contain* x then a simpler version can be used which is

$$\frac{\partial f}{\partial x} - \frac{d}{dx}\left(f - y_x\frac{\partial f}{\partial y_x}\right) = 0 \tag{3.31}$$

and the extremal is found from

$$f - y_x\frac{\partial f}{\partial y_x} = const. \tag{3.32}$$

The words 'not explicit' mean that f is a function of y or dy/dx such as $f = y^2 + dy/dx$ which does not explicitly depend on x; the function $f = x^2$ explicitly depends on x.

3.9.3 Minimum surface of revolution

Using the Euler equation is not that difficult. The surface of revolution is given by equation (3.29); if the surface is to be a minimum then (3.30) or (3.32) must apply. The latter is easier to use with $f = y\sqrt{1 + (dy/dx)^2}$ because this does *not explicitly* contain x. By letting $y_n = dy/dn$ and calculating $\partial f/\partial y_n$ equation (3.32) becomes

$$y\left[1+\left(\frac{dy}{dx}\right)^2\right]^{1/2} - y\left(\frac{dy}{dx}\right)^2\left[1+\left(\frac{dy}{dx}\right)^2\right]^{-1/2} = \frac{y}{\left[1+\left(\frac{dy}{dx}\right)^2\right]^{1/2}} = a$$

where a is a constant. The equation to solve is therefore

$$\frac{dy}{dx} = \sqrt{(y/a)^2 - 1},$$

which is a complicated integration giving $y = a\cosh(x/a + b)$, where b is also a constant. This is the catenary or minimum surface of revolution.

3.9.4 The brachistochrone

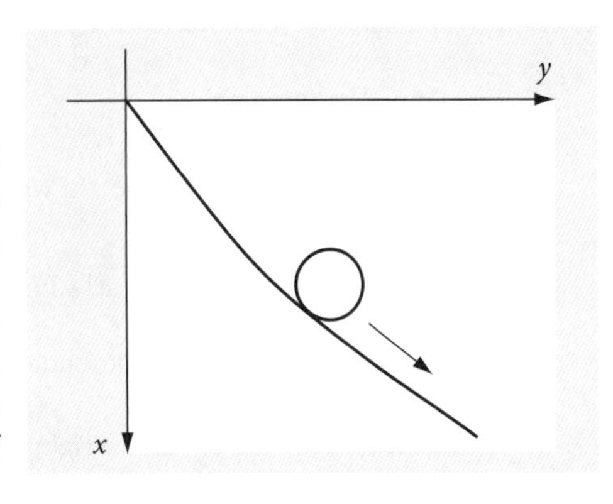

Fig. 3.14 Notice the unconventional axes directions used in the brachistochrone calculation.

The brachistochrone is the name of the curve a frictionless particle will travel along to pass between two points in the shortest time when acted on by a force such as gravity. The time taken is much less than that taken to move down a straight slope or in fact any other slope.

As it will simplify the calculation, we will suppose that a particle of mass m travels downwards in the $+x$ direction and moves to the right as the $+y$ direction after starting at rest at the origin (Margenau & Murphy, 1943). Using conservation of potential and kinetic energy,

$$mgx = mv^2/2$$

where v is the velocity at any point on the path and g the acceleration due to gravity. The velocity is found starting with a small element ds of the path length and using Pythagoras to give

$$v = \frac{ds}{dt} = \frac{\sqrt{dx^2 + dy^2}}{dt}$$

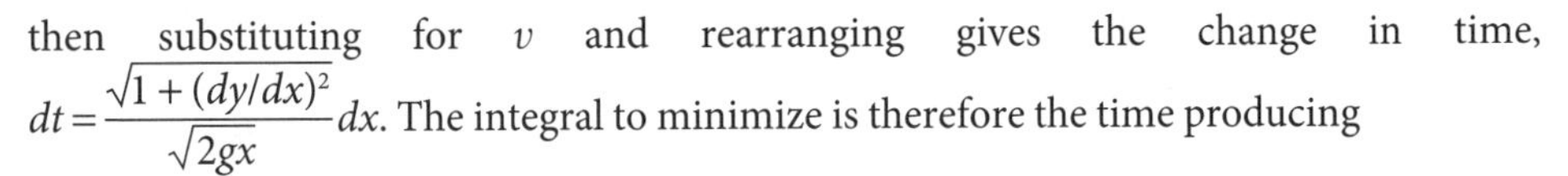

then substituting for v and rearranging gives the change in time, $dt = \dfrac{\sqrt{1 + (dy/dx)^2}}{\sqrt{2gx}} dx$. The integral to minimize is therefore the time producing

$$t = \frac{1}{\sqrt{2g}} \int_{x_0}^{x} \frac{\sqrt{1 + y_x^2}}{\sqrt{x}} dx$$

where $y_x \equiv dy/dx$ and the constant $\sqrt{2g}$ can be ignored for it will not enter into the shape of the curve. The result of the calculation is an equation y vs x that the particle follows. Using the Euler equation (3.30) with

$$f = \frac{\sqrt{1 + y_x^2}}{\sqrt{x}}$$

gives

$$-\frac{d}{dx}\left(\frac{y_x}{\sqrt{x}\sqrt{1 + y_x^2}}\right) = 0$$

because $df/dy = 0$. This is integrated once to produce

$$\frac{y_x}{\sqrt{x}\sqrt{1 + y_x^2}} = \sqrt{c},$$

where c is a constant. (Since c is a constant $\sqrt{c}$ rather than c is chosen to make the following equations simpler.) Squaring both sides, solving for y_x and then multiplying top and bottom inside the square root by $cx - 1$ and rearranging gives

$$\frac{dy}{dx} = \frac{\sqrt{cx(1 - cx)}}{cx - 1}.$$

Using Maple, as this is a complicated integral, produces

```
> Int(sqrt(c*x*(1-c*x))/(c*x-1),x): % = simplify(value(%));
```

$$\int \frac{\sqrt{c\,x(1 - c\,x)}}{-1 + c\,x} dx = -\frac{1}{2}\,\frac{-2\sqrt{c}\,\sqrt{-x(-1 + c\,x)} + \arcsin(-1 + 2\,c\,x)}{c}$$

Simplifying leads to

$$y = \frac{\sin^{-1}(2cx - 1)}{2c} - \sqrt{\frac{x - cx^2}{c}} + c_1$$

where c_1 is also a constant of integration. If the particle starts at the origin then $y = 0$ when $x = 0$ and therefore $c_1 = -\pi/4c$. The solution to this problem was found independently by the Bernoulli brothers, and by Leibniz and Newton towards the end of the seventeenth and beginning of the eighteenth centuries.

3.9.5 Questions

Full solutions are available at www.oxfordtextbooks.co.uk/orc/beddard.

Q3.43 Show that $F_a = \left(\dfrac{\partial f}{\partial y} - \dfrac{d}{dx}\dfrac{\partial f}{\partial y_x}\right) y_x$, where $y_x = dy/dx$ and f is a function of x, y, and y_x, is the same as $F_b = \dfrac{d}{dx}\left(f - y_x \dfrac{\partial f}{\partial y_x}\right)$ if f is independent of x i.e. does not explicitly depend on x.

Strategy: By placing y_x on the right it can be operated on by d/dx. Expand both expressions and only then let f not depend explicitly on x. ($f = y + dy/dx$ does *not explicitly* depend on x, $f = x + y + dy/dx$ does.)

Q3.44 **(a)** Find df/dx if $f = x^2 + y(x)^2 + (dy/dx)^3$ and $y = \exp(x^2)$.

(b) If f is a function of x, of $y(x)$ and $y_x = dy/dx$, written as $f(x, y, y_x)$, calculate df/dx.

(c) Confirm your result of (b) using (a).

Strategy: **(b)** Because y and y_x depend *implicitly* on x, these are not constants. The result will contain partial derivatives because when differentiating with x, y and y_x have to be held constant. Treat y and y_x as you would any other function of x by the function-of-function or chain rule.

Q3.45 Show that the shortest distance between two points on a plane is a straight line. Find the equation if the two points joined are $\{x_0, y_0\}$ and $\{x_1, y_1\}$. A line of shortest distance between two points on a surface is called a *geodesic*.

Strategy: The equation to use is $I = \displaystyle\int \sqrt{1 + \left(\frac{dy}{dx}\right)^2}\, dx$. Complete the calculation then work out the constants at the end using the coordinates given.

Q3.46 Light travels through any medium with a speed given by c/n, where n is the refractive index, and always takes the path of shortest time. Glass can be made with a refractive index that varies with distance. In each case, calculate the path taken by a beam in a two-dimensional medium where,

(a) $n = 1/x$, **(b)** $n = \sqrt{x}$, **(c)** $n = x$.

The integral equation giving the shortest time is $t = c^{-1}\int n(x, y)\sqrt{1 + y_x^2}\, dx$, where c is the speed of light and $y_x \equiv dy/dx$.

Strategy: Using equation (3.30) can be a little tricky. Differentiate with respect to y, which gives 0 in this case, then with respect to y_x, but do not differentiate by x, as this will produce 0 for the whole expression. Instead, integrate the second term with x, but do not forget to add the integration constant. Then solve for y_x and integrate again in x to obtain the final equation.

Q3.47 Find the limits **(a)** $\lim_{x\to 0} \dfrac{\cos(x)}{x}$, **(b)** $\lim_{x\to 0} \dfrac{x}{\sin(x)}$, **(c)** $\lim_{x\to 0} \dfrac{x^2}{\sin(x)}$,

(d) $\lim_{x\to 0} \dfrac{e^{-x} - 1}{x^3}$, **(e)** $\lim_{x\to -\infty} xe^x$, **(f)** $\lim_{x\to\infty} x^{1/x}$, **(g)** $\lim_{\theta\to\pi/2} \sin(\theta)^{\tan(\theta)}$,

(h) $\lim_{x\to 0} \dfrac{e^{\pi x} - e^{-\pi x}}{ix}$, **(i)** $\lim_{\theta\to 0} \dfrac{\sin(n\theta)}{\sin(\theta)}$, **(j)** $\lim_{x\to 0} x^x$, **(k)** $\lim_{n\to\infty} n!^{1/n}$,

(l) $\lim_{n\to\infty} \dfrac{n!^{1/n}}{n}$ (In questions (k) and (l) use Stirling's approximation for large n which is $\ln(n!) = n\ln(n) - n$ for very large n.)

Q3.48 Find **(a)** $\lim_{x\to 0} \dfrac{\sin(x) - x^2}{x}$ **(b)** $\lim_{x\to 0} \dfrac{(\sin(x) + x)^2}{x}$ **(c)** $\lim_{x\to\infty} x - \sqrt{x^2 + x}$ and **(d)** $\lim_{x\to\infty} x + \sqrt{x^2 + x}$.

Q3.49 A base jumper drops from a skyscraper and falls with velocity v, which is limited by the air resistance of the parachute and is proportional to v^2 making his velocity at time t, $v = v_\infty \tanh(gt/v_\infty)$. The corresponding distance dropped is $x = \dfrac{v_\infty^2}{g}\ln[\cosh(gt/v_\infty)]$, where g is the acceleration due to gravity.

(a) By taking the limit $t \to \infty$, what does v_∞ represent?

(b) Find the limits on v and x when $v_\infty \to \infty$, which corresponds to falling in a vacuum.

Strategy: Rewrite the equations in terms of exponentials then take limits. In evaluating the limit $v_\infty \to \infty$ change v_∞ to $u = 1/v_\infty$ to more easily solve this problem.

Q3.50 Show that $\dfrac{x^n}{e^x} \to 0$ for all positive values of n as $x \to \infty$.

Q3.51 The Cauchy function, shown in Fig. 3.15, is $f = e^{-1/x^2}$ and has the unusual property that any derivative d^n/dx^n of this function is zero at $x = 0$. Show that this is true.

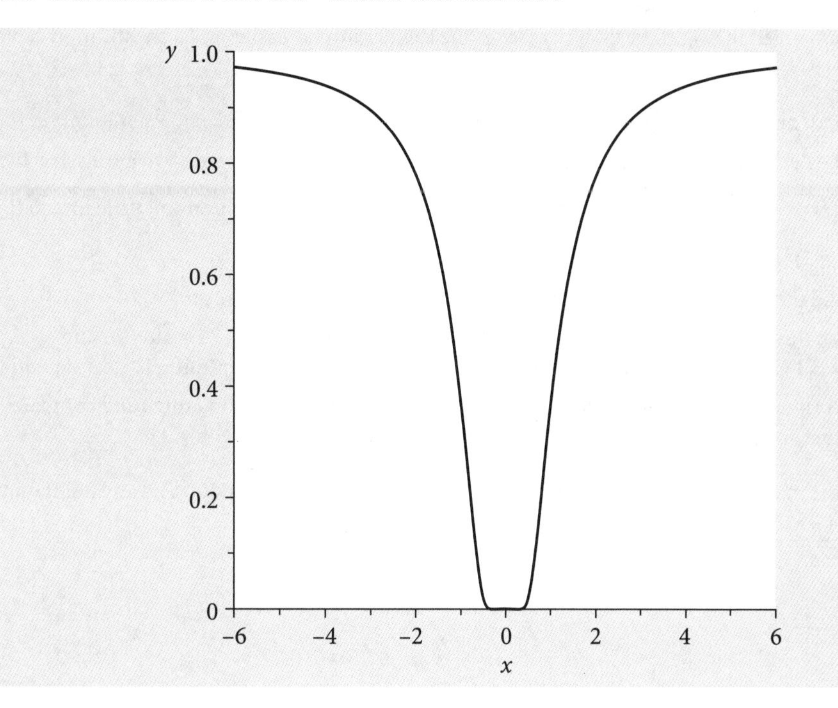

Fig. 3.15 Cauchy function.

Strategy: The reciprocal x^2 makes the calculation difficult when l'Hôspital's rule is used. Instead, substitute $u = 1/x$ and change the limit.

Q3.52 Find the maximum, minimum, and inflexion points of $y = x^2(x-1)^3$. Plot the graph to confirm your calculation:

Q3.53 A cycloid, Fig. 3.16, is the path (locus) drawn out by a point on a tyre as it rolls along a road and can be described with the parametric equations $x = a[t - \sin(t)]$ and $y = a[1 - \cos(t)]$. Find its maximum value. Incidentally, the inverted cycloid is called the brachistochrone, which is the curve along which an object will fall, in the absence of resistance, in the shortest time between two points, see Section 3.9.4.

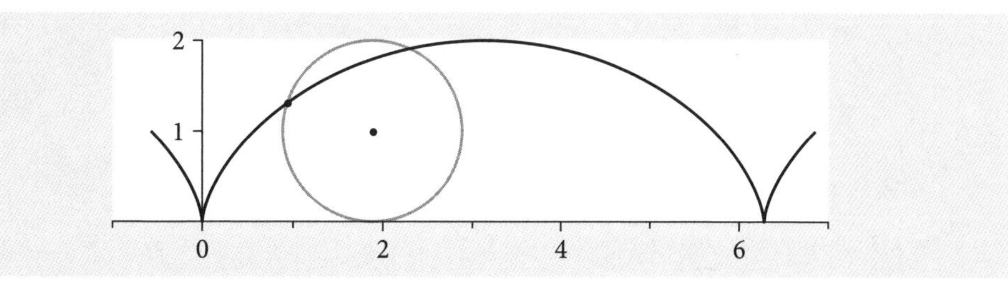

Fig. 3.16 A cycloid with its circle of radius 1. Any fixed point on the circle's circumference draws out a cycloid as the circle rolls along.

Q3.54 **(a)** Find the value of x where $y = x^2e^{-ax^2}$ has its maximum if a is a positive constant.

(b) The Maxwell–Boltzmann distribution $P(u)$ describes the distribution in the speed u of gaseous molecules where

$$P(u) = 4\pi\left(\frac{m}{2\pi k_BT}\right)^{3/2} u^2e^{-mu^2/2k_BT}$$

and m is the mass (in kg) of one molecule, k_B Boltzmann's constant and T temperature (K). Calculate the most probable speed, u_p from the distribution.

(c) Plot a graph of the speed distribution of SO_2 molecules at 100, 300, and 500 K. Indicate u_p for one temperature. In plotting you may find it easier to define P as a function of T and u by `P:=(u,T)->..`*etc.* ($k_B = 1.38 \times 10^{-23}$ JK^{-1}; don't forget to put m in kg.)

(d) Comment on the shape of the distributions.

Strategy: Note the similarity in form between the two equations in (a) and (b). Using the equation of the form given in (a) rather than (b) makes the calculation far easier and a substitution can be made as the last step.

Q3.55 **(a)** Find the minimum value of $y = a/x^2 - b/x^4$.

(b) The Lennard-Jones potential, which describes the attraction between two atoms, has the form $V(r) = -\varepsilon\left[\left(\frac{\sigma}{r}\right)^6 - \left(\frac{\sigma}{r}\right)^{12}\right]$ where σ is approximately the size of the atom, ε the interaction energy, and r the separation of the atoms.

(c) Find the separation of the atoms at which the energy is at a minimum.

(d) Plot the potential using the parameters for neon, which are $\sigma = 2.74 \times 10^{-10}$ m and $\varepsilon = 3.1 \times 10^{-3}$ eV. (Choose the scale carefully in eV so that the minimum is clear on your plot.)

(e) Work out the numerical value of the minimum separation of neon atoms and compare it with the experimental value of 3.13×10^{-10} m. Find the minimum energy.

Q3.56 Find the maximum area of the right-angled triangle OPN when the point P is any point on the curve $y^2 = \frac{(4-x)^3}{x}$.

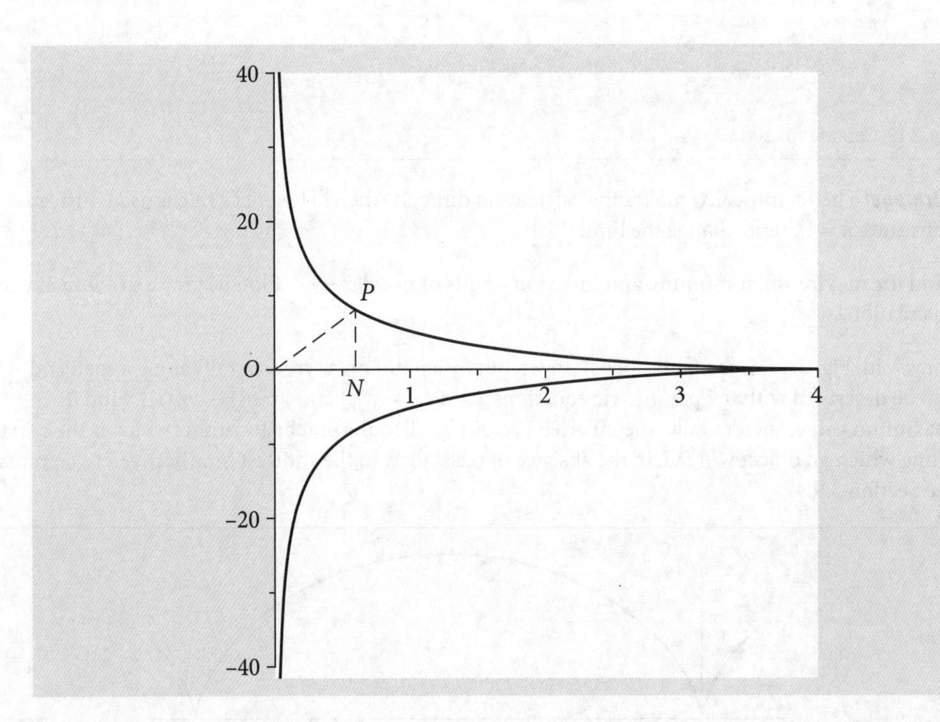

Fig. 3.17 The triangle to maximize is OPN.

Q3.57 Suppose that the power p an aeroplane has to produce to fly at a speed v is given by $p = \frac{aw^2}{v} + bv^3$ where its weight is w and a and b are positive constants depending on the shape and size of the aeroplane. What is the optimum speed when the power is least?

Q3.58 The pear-shaped quartic has the form $(by)^2 = x^3(a - x)$, a and b being constants. Where is the tangent horizontal?

Strategy: A horizontal tangent is the same as zero gradient.

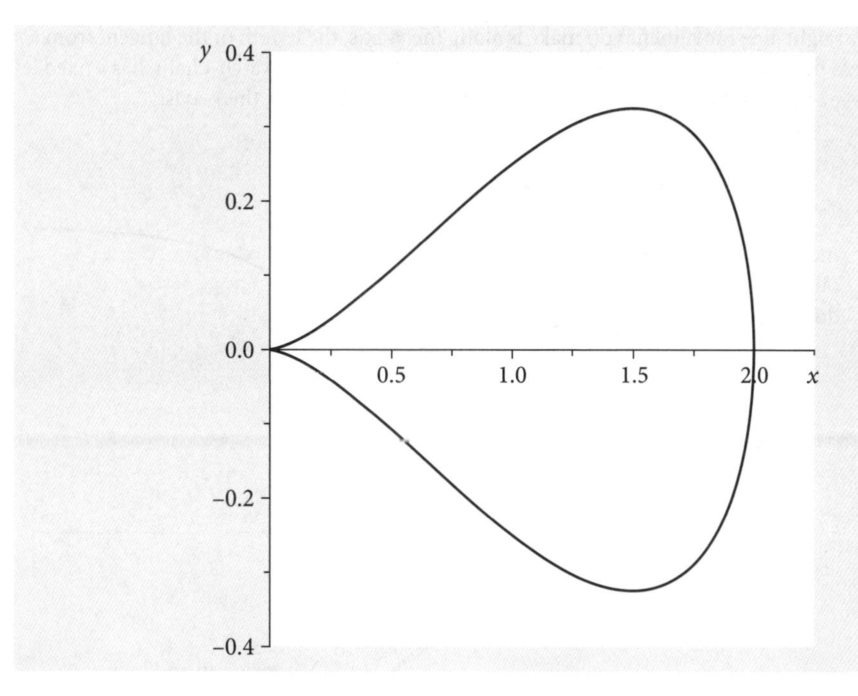

Fig. 3.18 Pear-shaped quartic when $a = 2$ and $b = 4$.

Q3.59 If the inverse sine function is $u = \sin^{-1}(p)$

(a) find $\dfrac{du}{dp}$ and $\dfrac{d^2u}{dp^2}$,

(b) Generalize your result to any inverse function $u = f^{-1}(p)$ and show that

$$\frac{d^2f}{du^2}\left(\frac{du}{dp}\right)^2 + \frac{df}{du}\frac{d^2u}{dp^2} = 0.$$

Strategy: Note that $\sin^{-1}(p)$ is *not* the same as $1/\sin(p)$; use the fact that $\sin(u) = p$ and using a right-angled triangle find $\tan(u)$.

Q3.60 At school you will have learnt that to throw a ball the furthest distance, with a certain amount of effort, it should be launched upwards at 45°. Observing football (soccer) and rugby players throwing balls, they rarely appear to do so at 45° but at a smaller angle. A careful examination of the motion (Linthorne 2006) shows that the equation governing the horizontal distance R travelled is

$$R = \frac{v^2}{2g}\sin(2\theta)\left(1 + \sqrt{1 + \frac{2gh}{[v\sin(\theta)]^2}}\right)$$

where h is the height above ground level that the ball is launched at and with a speed of v m s^{-1}; g is the acceleration due to gravity.

Calculate the maximum distance thrown at a given angle and velocity and calculate how the launching angle varies with initial speed. Plot the *surface* showing the distance R vs angle and initial speed and superimpose the maximum distance vs angle assuming that the ball is launched from head height of about 2 m and with initial speeds of 0 to 15 m s^{-1}.

Strategy: Differentiating this equation with respect to θ gives a very complex equation, which may defeat you. It is easier to differentiate with respect to $\sin(\theta)$ and call this x. The conversion you will also need is $\sin(2\theta) = 2\sin(\theta)\cos(\theta)$ and the cosine is calculated using $\cos^2(\theta) + \sin^2(\theta) = 1$.

Q3.61 The shape of the tractrix or hound-curve is shown in Fig. 3.19. The name arises from the behaviour of a dog that stubbornly resists walking but is nevertheless, pulled along by its lead which has a *fixed* length. Other examples are the sideways path a toy follows when pulled by a child, or a boat pulled from its distant mooring by walking along the side of the quay. Huygens originally studied this problem in 1692, and Leibniz later solved it after a friend described the path taken by a heavy fob watch that was placed on a table and moved by pulling the end of its chain along the table's side.

The straight-line movement you make is along the y-axis, the length of the tangent from a point on the line to the y-axis is always a constant because the dog's lead, or watch-chain, has a fixed length. The curve Fig. 3.19 shows the position of the dog as you move along the y-axis.

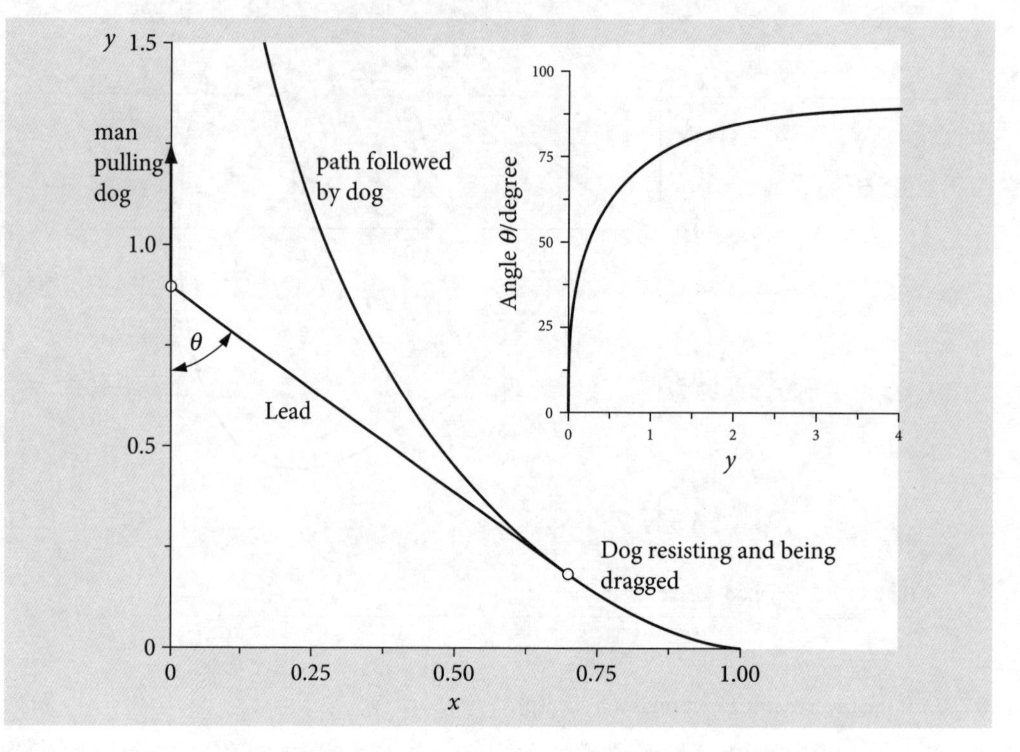

Fig. 3.19 The tractrix. The length of the lead is fixed at $a = 1$ as the dog is pulled. You start at $y = 0$, the dog at {1, 0}. The inset shows the angle θ which you look back at the dog (in degrees) vs the distance walked along the y-axis.

The equation of the curve is $y = a\ln\left(\dfrac{a + \sqrt{a^2 - x^2}}{x}\right) - \sqrt{a^2 - x^2}$, where a is a constant such that $a \le x$ and the man walks up the y-axis.

(a) Find the angle θ the end of the lead makes at any point on the curve.

(b) Find the radius of curvature of the line: equation (3.24) question Q3.39.

(c) Calculate the length of the path the dog takes if the arc length s of a curve between points x_α and x_β on the x-axis is

$$s = \int_{x_\alpha}^{x_\beta} \sqrt{1 + \left(\frac{dy}{dx}\right)^2}\, dx.$$

(d) If the dog has an effective weight of 25 kg (including friction), how much work does the man have to do to move the dog 10 metres assuming that he starts to pull the dog from the x-axis. The dog's lead is 1 metre long.

Q3.62 The bones of animals and birds have to be strong enough to withstand the forces applied to them as the animals run, jump, or fly but not so strong that the extra weight and energy needed to build and maintain them imposes a disproportionate burden; see Alexander (1996) for a full discussion. The strength of a cylinder, which will suffice to describe the shape of a bone, clearly depends upon the material with which it is made as well as its size. A bone has a hard outer region of hydrated calcium phosphate (as calcium apatite), and this is intimately associated with collagen, a protein which gives the bone some elasticity. Bone also has a soft inner core consisting of marrow and this is where new white blood cells are made, Fig. 3.20.

The strength of a material is characterized by an elastic modulus Y that scales how the material responds to an applied force (or strain) as,

$$\text{restoring force} = \text{modulus} \times \text{strain},$$

and by a *moment of the force M*. This moment is the energy or work needed to bend an object; it is the applied force multiplied by the distance from the point of bending to the point of applying that force.

Imagine the effort required trying to bend a length of straight copper pipe by placing in on your knee and pulling on the ends. The further from your knee you can grab hold of the pipe, the easier it is to bend because you can apply more energy or torque. This is called a *moment*, and is defined as

$$\text{moment} = \text{applied force} \times \text{distance from force}.$$

It is intuitive that a long, thin bone will bend and break more easily than a shorter but otherwise similar one.

If a hollow cylinder, see Fig. 3.20, of outer radius r with the central hole having a radius kr is subject to a moment of force, then, by using standard engineering equations, the radius of the bone is related to the strength Y and the moment M as

$r^3 = \dfrac{M}{Y(1-k^4)}$, where $0 < k < 1$ and is a constant depending on the bone.

When k is large ≈ 1, the bone has a thin wall and when $k \approx 0$ the bone is almost solid, therefore it would be interesting to know how to optimize the mass of the bone for a given strength. Bones that are too strong and heavy for any danger that the animal may experience may not offer any evolutionary advantage, but require more effort to move, and more food to grow and maintain. Conversely, weak bones that break under moderate exertion will prove to be deadly, as once crippled an animal is usually easy prey.

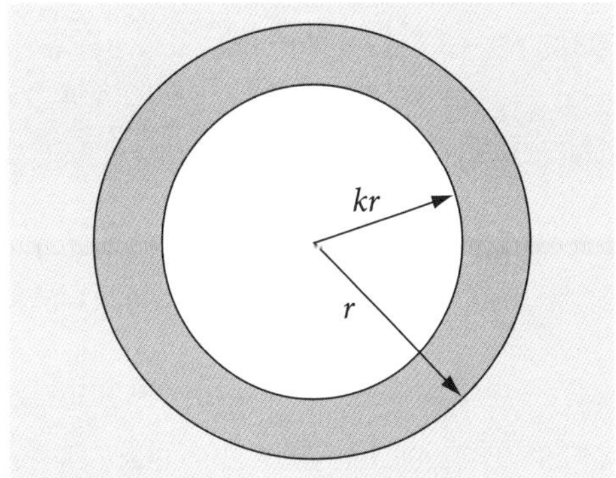

Fig. 3.20 Bone cross section

(a) If the density of the bone is ρ and m the mass per unit length, find the optimum amount of marrow in the bone, i.e. the optimum value of k if the marrow has half the density of bone.

(b) Plot separately the mass of the bone and marrow vs k (assuming M and Y and ρ each equal 1) and also their sum to show that the minimum mass is at about $k = 0.62$ and independent of M and Y. Speculate as to why k appears to be approximately the same in mammal bones. Experimental values for the femur are 0.63 in foxes, 0.56 in lions, and 0.62 camels.

Strategy: Work out the mass m of the bone and minimize this by finding dm/dk. The minimum in the curve of m vs k should be at the optimum value. Use Maple to solve the polynomial equation formed. Inspect the answers to determine what the physically realistic value is.

Q3.63 The intensity of the rotational spectrum of a heteronuclear rigid rotor diatomic molecule, is given approximately by $I = (2J+1)e^{-E_J/k_BT}$, where the rotational energy increases with quantum number J as $E_J = BJ(J+1)$. The rotational constant is B, conventionally this has units of cm^{-1}, and k_B is Boltzmann's constant and T the temperature.

(a) At what quantum number is the maximum line intensity expected?

(b) What is this value for the vibrational ground state of HCl, CO, and iodine at 100, 300, and 500 K if the rotational constants are 10.59, 1.93, and 0.3736 cm^{-1} respectively?

Strategy: Differentiate the equation for the intensity with respect to J, which it is assumed is a continuous variable, not a set of integers, then set the result to zero and solve for J. The maximum will be called J_{max}.

Q3.64 A line in a nuclear magnetic (NMR) experiment on an isolated nucleus, or the spectral lines observed in the absorption or emission of isolated atoms and molecules, are not infinitesimally narrow but have an inherent width, even when Doppler, collisional, and other broadening effects are absent. The width $\Delta\nu$ and lifetime τ of the transition are related by $\Delta\nu\Delta t \approx 1$. Because one or both states involved must have a finite lifetime and decay with an exponential probability distribution, i.e. as $e^{-t/\tau}$, the transition intensity has the Lorentzian line shape

$$g(\omega) = \left(\frac{\tau}{\pi}\right)\frac{1}{1+\tau^2(\omega-\omega_0)^2}$$

where τ is the lifetime of the excited state involved, ω_0 the transition frequency, and ω the radiation's angular frequency; $\omega = 2\pi\nu$ in radians s^{-1}. This line shape is similar to that of the Gaussian or bell shaped curve but is slightly wider in the wings.

(a) Show that the maximum intensity is at the transition frequency, ω_0, and calculate its full width at half-maximum, FWHM.

(b) In EPR spectroscopy the derivative of the line shape is plotted rather than the line shape itself; show that the difference between the maximum and minimum of this derivative lineshape is $2/(\tau\sqrt{3})$.

(c) Plot the lineshape function and its derivative to confirm your calculation. Use values typical for an EPR experiment with a transition at 50 MHz. You should choose a suitable value of τ.

Q3.65 By concluding that radiant energy must be quantized, Planck derived a formula for the radiant energy density of a black body at a wavelength λ and temperature T, which is

$$\rho(\lambda, T) = \frac{8\pi hc}{\lambda^5}\left(\frac{1}{e^{hc/\lambda k_B T} - 1}\right).$$

(a) Show, provided that $hc/\lambda k_B T \gg 1$, that $\lambda_{max}T = \text{constant}$ describes the maximum energy density. Wein discovered empirically that $\lambda_{max}T = \text{constant}$, which is now called the Wein displacement law.

(b) Estimate the constant and limits to λ and T that makes this approximation valid.

Q3.66 The probability of finding the nuclei at a given internuclear separation x in a diatomic molecule is given by $\psi^*(x)\psi(x)$, where $\psi(x)$ is the wavefunction and the * indicates that if ψ has a complex part the complex conjugate should be used. The $v = 1$ vibrational level of a harmonic oscillator has the wavefunction $\psi(x) = \left(\dfrac{\alpha}{4\pi}\right)^{1/4}(2\alpha x^2 - 1)e^{-\alpha x^2/2}$. At what internuclear separation is the maximum and at what the minimum probability?

Q3.67 If a protein or a length of DNA is attached to one end of the cantilever of an atomic force microscope, AFM, the other end being fixed to the microscope's base and then the cantilever retracted, the force produced on the cantilever during extension is described by the *worm-like-chain* (*wlc*) model of extension. The protein acts like a spring but one whose force vs extension varies in a complicated way. At low extension, Hooke's law is obeyed, but at larger extension, this is not the case. Experimentally it is found that, to a good approximation, the force f is related to the extension x as predicted by the *wlc* model,

$$f = \frac{k_B T}{p}\left(\frac{1}{4(1 - x/L_p)^2} - \frac{1}{4} + \frac{x}{L_p}\right),$$

where L_p is the length of the protein (~30 nm) and p is the persistence length (≈0.4 nm), which is a measure of how easily the protein can bend. The measured force is small as would be expected for a single molecule and is measured in piconewton; 1 pN ≡ 10^{-12} N.

The true response of the cantilever also depends upon how easily the cantilever itself bends Fig. 3.21. The compliance h describes both the protein and cantilever behaviour and is $h = \dfrac{1}{k_s} + \dfrac{dx}{df}$, where the cantilever has a force constant k_s and h has units of m/N. The compliance is the reciprocal of the total force constant.

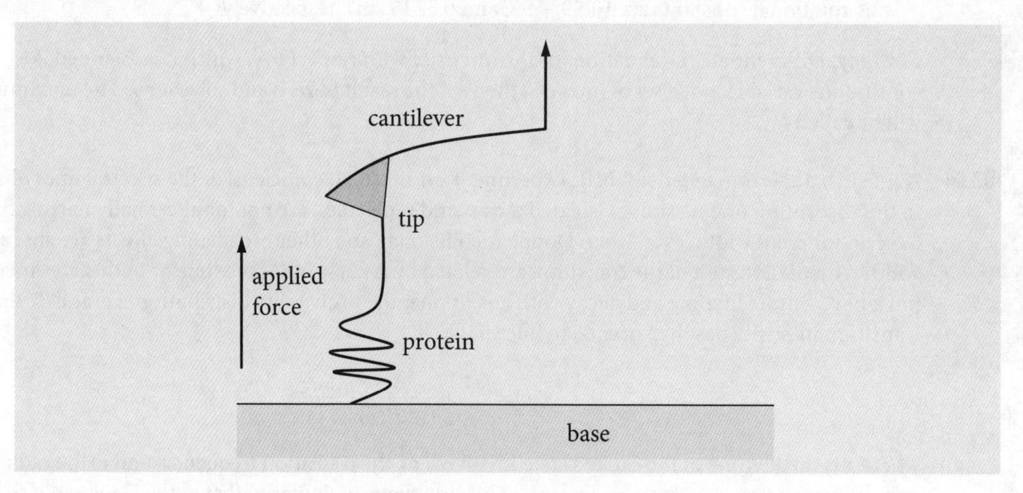

Fig. 3.21 In the AFM, the cantilever is moved upwards relative to the base. The protein partially resists extension, behaving like a stiff spring and bends the cantilever.

(a) Calculate the compliance h in terms of extension x.

(b) Starting with the equation for h, and without substituting for dx/df, show that

$$\frac{dh}{df} = -\frac{\dfrac{d^2f}{dx^2}}{\left(\dfrac{df}{dx}\right)^3}$$

using results in the differentiation summary box (3.6) as necessary.

Strategy: (a) The compliance requires dx/df, but f is a function of x. Therefore the expression to use is that between reciprocal derivatives, $\dfrac{dx}{df} = \left(\dfrac{df}{dx}\right)^{-1}$. (b) The derivative dh/df is sought and instead of substituting for f, because this produces a very complicated expression, simply differentiate $h = \dfrac{1}{k_s} + \dfrac{dx}{df}$ with respect to f with k_s as a constant.

Q3.68 The potential energy V of a diatomic molecule is often more accurately described by a Morse potential than the harmonic oscillator, in which case $V = D_e(1 - e^{-\beta(r-r_e)})^2$, where β is a constant that depends upon the force constant and mass of the molecule and r_e is the equilibrium bond length.

(a) What does D_e represent? What are the units of β?

(b) Evaluate dV/dr and find its minimum using d^2V/dr^2.

(c) Show that the bond extension at the maximum attractive force experienced by the molecule is $r_{\text{max}f} = r_e + \ln(2)/\beta$.

(d) If the H and Cl atoms approach one another with only a small amount of kinetic energy, i.e. essentially zero, what is the closest distance they can approach one another?

(e) Calculate all these quantities for $H^{35}Cl$, using the values $\nu_e = 2989.7\ \text{cm}^{-1}$, $x_e = 0.0174098$ and $r_e = 1.274$ Å and the constants are $\beta = 2\pi c\nu_e\sqrt{\dfrac{\mu}{2D_e}}$ and $D_e = \dfrac{hc\nu_e}{4x_e}$, where μ is the reduced mass and c the speed of light.

Strategy: **(b)** Force is the derivative of potential energy, calculate this and then simplify to find the minimum energy and force. If you are not familiar with the shape of the Morse potential, you can either look up the anharmonic oscillator in a textbook, or start the question by calculating its shape with Maple using the values given in the question.

Q3.69 The energy levels in the Morse potential are given by

$$E_n = \nu_e(n + 1/2) - x_e\nu_e(n + 1/2)^2,$$

where ν_e is the frequency in cm^{-1}, n the vibrational quantum number and x_e the (dimensionless) anharmonicity.

(a) Show that there are a finite number of energy levels in the Morse potential. What is n_{max} for HCl?

(b) Show that the energy with this quantum number is less than D_e.

(c) Calculate the dissociation energy by adding up all the *differences* in energy levels with integer values of n. Use Maple to do this, and then compare the two results you obtain for the maximum quantum number.

(d) Plot a graph of the differences vs quantum number.

(Use data given in the previous question and assume that n is a continuous variable in any differentiations.)

Strategy: By assuming that n is continuous allows us to differentiate the energy, and when the energy difference between levels is zero it is assumed that this occurs just at the dissociation limit. Possibly this will not lead to a very good answer because the dissociation energy must be the sum of all of the energy level differences from $n = 0$ to the dissociation limit.

Q3.70 Gaseous N_2O_4 thermally decomposes into two NO_2 molecules. The reaction is

$$N_2O_{4(g)} \rightleftarrows 2NO_{2(g)}.$$

Starting with 1 mole of N_2O_4 the number of moles reacted is $1 - \xi$ and the number of moles of NO_2 produced will be 2ξ where ξ (pronounced *xi*) is the extent of reaction and has units of moles. The extent of reaction ξ is the number of moles reacted multiplied by the stoicheiometry constant. The total number of moles is $1 - \xi + 2\xi = 1 + \xi$ and the mole fraction of N_2O_4 is $x_{N_2O_4} = \dfrac{1-\xi}{1+\xi}$ and of NO_2 is

$x_{NO_2} = \dfrac{2\xi}{1+\xi}$. If the total pressure is held constant at 1 bar, then $p_{N_2O_4} = x_{N_2O_4}p_0$ and $p_{NO_2} = x_{NO_2}p_0$.

The total Gibbs energy is

$$G(\xi) = (1-\xi)\left[\Delta G^{\ominus}_{N_2O_4} + RT\ln\left(\frac{1-\xi}{1+\xi}\right)\right] + 2\xi\left[\Delta G^{\ominus}_{NO_2} + RT\ln\left(\frac{2\xi}{1+\xi}\right)\right].$$

(a) Calculate the equilibrium extent of reaction ξ_{eq} when $dG/d\xi = 0$.

(b) As $\Delta G^{\ominus}_{N_2O_4} = 97.79$ kJ mole^{-1} and $\Delta G^{\ominus}_{NO_2} = 51.26$ kJ mole^{-1} calculate ξ_{eq} at 320 K.

(c) Confirm your answer by plotting $G(\xi)$ vs ξ at 320 K.

Q3.71 In the protein containing the reaction centre of photosynthetic bacteria, an electron is released from the reaction centre after it receives energy from an antenna bacteriochlorophyll molecule. The path along which the electron moves after leaving the bacteriochlorophyll dimer of the reaction centre is,

$$(\text{BChl})_2 \xrightarrow{2\,ps} \text{BPh} \xrightarrow{200\,ps} \text{Quinone},$$

where $(\text{BChl})_2$ is a dimer of bacteriochlorophyll molecules. BPh is a bacteriopheophytin that lacks the Mg atom at the centre of the chlorophyll and which is replaced by two H atoms. (The actual scheme is slightly more complex than presented here but the essential features are retained.)

In a pump-probe laser experiment using femtosecond-duration light-pulses the BPh has to be detected. If C is the concentration of the $(\text{BChl})_2$, and k_c the decay rate constant then

$$\frac{dC}{dt} = -k_c C \quad \text{and} \quad \frac{dP}{dt} = k_c C + k_P P,$$

where P is the concentration of the pheophytin. At time t, given that the initial amount of $C = C_0$, then $C = C_0 e^{-k_c t}$ and

$$P(t) = C_0 \frac{k_c}{k_c - k_p}\left[e^{-k_p t} - e^{-k_c t}\right].$$

(a) At what time would it be best to observe $P(t)$?

(b) If the spectra of BPh and $(\text{BChl})_2$ were significantly overlapped with one another what would then be the best time to observe the BPh?

Q3.72 Ammonia has a triangular pyramidal structure with the N atom at the apex. One of its vibrational normal modes is inversion of the positions of the atoms to their mirror image. It is a moot point as to whether you consider the N atom or the H atoms as moving during the vibration. However it is viewed, the vibrational potential has two wells separated by a barrier. Experiment shows that the nitrogen atom moves by approximately ±0.38 Å either side of the plane formed by the three H atoms.

(a) If the potential V is given, approximately, by the formula $V = 155(\theta^4 - 7.2\theta^2) + 2010$ cm^{-1} where θ is the bend angle from the planar structure, where are the potential minima and how high is the barrier?

(b) How many inversions/second does the NH_3 molecule undergo in its lowest vibrational level?

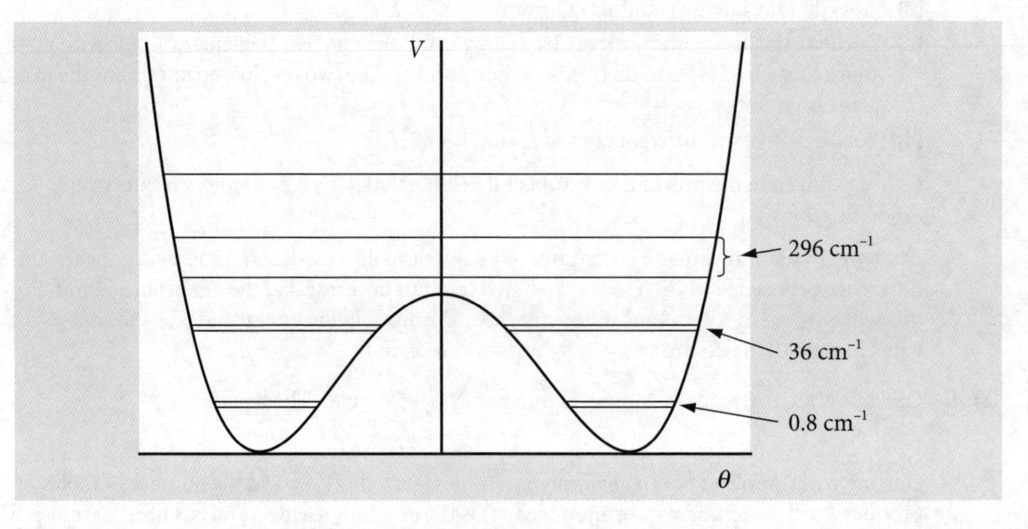

Fig. 3.22 A sketch of the potential energy of the ammonia inversion potential and some vibrational energy levels. Notice that the levels below the barrier are doubled.

The two inversion forms, if considered isolated from one another, have the same energy because they are mirror images of one another. If they are now allowed to interact they will influence one another's energy causing one to increase, and the other to fall, which produces level splitting. The sum of the energy levels remains the same but the population in the lower level will be greater by the Boltzmann distribution causing an overall reduction in energy. The energy levels below the barrier influence one another only slightly, leading to a small splitting of levels. Above the barrier, the interaction is larger, therefore the splitting is larger, and the levels appear as single lines. See Chapter 11.10 for a method with which to calculate the energy levels and wavefunctions.

Q3.73 A ferromagnet is a material that has a spontaneous magnetization in the absence of an external magnetic field and all the magnetic moments are aligned in one direction; e.g. $\cdots\uparrow\uparrow\uparrow\uparrow\cdots$. In an anti-ferromagnet neighbouring magnetic moments are paired as $\cdots\downarrow\uparrow\downarrow\uparrow\cdots$. The compounds of rare earths form crystals containing parallel layers in each of which spins are aligned as in a ferromagnet which can all point in a different direction to that of adjacent layers. However, the spins are coupled between layers with an exchange constant J_1 to neighbouring layers and J_2 to next neighbours. The energy of any plane containing N spins, each having spin quantum number S, can be written as (Blundell 2001)

$$E=-2NS^2(J_1\cos(\theta)+J_2\cos(2\theta)).$$

Find three possible minimum energies if θ is the angle between the overall magnetic moment between planes.

3.10 The Newton–Raphson algorithm: Finding the roots of equations numerically

Two numerical methods are now described for solving equations, which means finding their roots, i.e. the values of x when $y=0$. The Newton–Raphson method uses the derivatives of a function to determine ever-closer approximations to the root of an equation; the related secant and regula-falsi (false-position) methods can be used on real experimental data whose equation is unknown. These methods work by making successive approximations.

The equation used in Q3.62 to work out the optimum strength of bones was given terms of a parameter k; the same equation in x is $3-8x^2+x^4=0$ and this is quite a tough equation to solve because there are four roots which can be found algebraically. Plotting the function $y=3-8x^2+x^4$ shows approximately where the roots are; Fig. 3.23 (left). More difficult equations than this occur in many situations. For instance, a theoretical model might produce a transcendental equation that cannot be solved algebraically when $y=0$; for example $y=-2x+e^{-x^2}-2$ as shown in the right-hand graph of Fig. 3.23 and it is then essential to resort to a numerical method to find the root or roots.

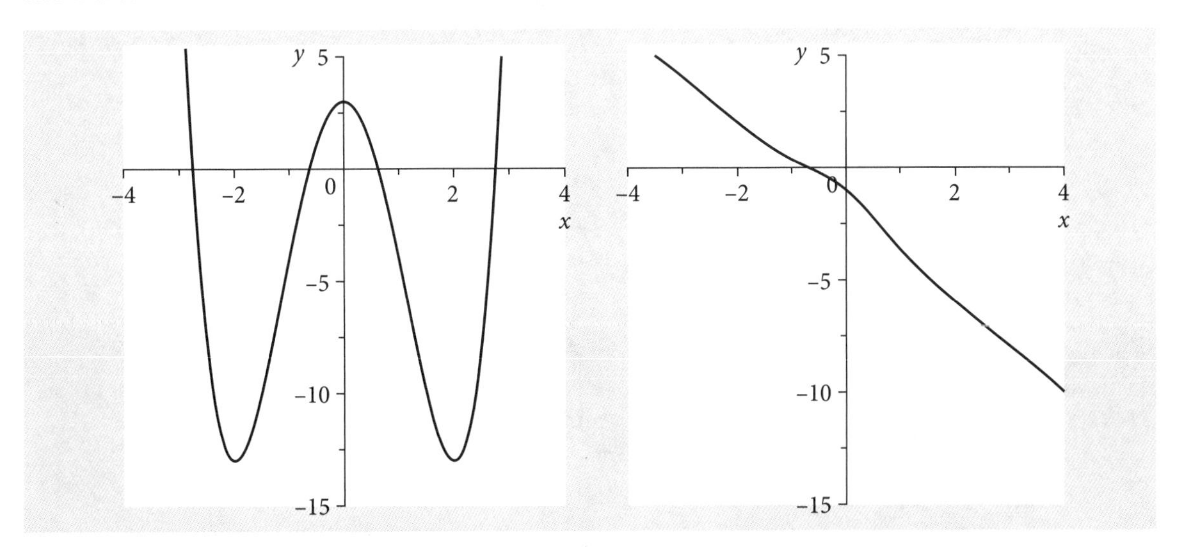

Fig. 3.23 Left: Graph of the polynomial $y=3-8x^2+x^4$ showing roots at approximately ±0.6 and ±2.7. Right: The single root of the transcendental equation $-2x+e^{-x^2}-2=0$ is at about $x=-0.7$.

3.10.1 Derivation of the Newton–Raphson algorithm

Any one of the equations for which a solution is sought will be represented as $f(x)=0$. The equation's y value is made equal to zero because this is where the root will be found. In the derivation of differentiation the equation

$$\Delta y/\Delta x \approx dy/dx$$

occurs and this is now rewritten as

$$\Delta y \approx (dy/dx)\Delta x,$$

where Δy and Δx are small changes in y and x. In terms of a function f at any two points x_i and x_{i+1} this last equation becomes

$$f(x_i) - f(x_{i+1}) + f'(x_i)(x_{i+1} - x_i) = 0, \tag{3.33}$$

because $\Delta y = f(x_i) - f(x_{i+1})$, and $f'(x_i)$ is the derivative dy/dx at point x_i. The subscripts i and $i+1$ are indices indicating successive values of x at which the function is calculated. It can be shown that the root, i.e. solution of the equation, is found when the new value of the function is zero, or $f(x_{i+1}) = 0$. Equation (3.33) can be rearranged to produce the x_{i+1}th value, viz.;

$$x_{i+1} = x_i - \frac{f(x_i)}{f'(x_i)}. \tag{3.34}$$

This is the Newton–Raphson algorithm and only applies where the gradient is not 0. One initial guessed point x_1 is needed to start the calculation and this should be made close to the root if possible; see Fig. 3.24.

A more appealing derivation is to use a Taylor expansion of our function about a point x_0 (see Chapter 5)

$$f(x) = f(x_0) + (x - x_0)f'(x_0) + (x - x_0)^2\frac{f''(x_0)}{2!}.$$

Ignoring the second derivative and making $x_0 \equiv x_i$ to be our initial point, the Newton–Raphson formula is obtained by solving for x where, to be consistent with equation (3.34), $x \equiv x_{i+1}$.

The secant method is effectively a discrete version of the Newton–Raphson method where the gradient is calculated with small but finite differences $f'(x) \to \Delta f(x)/\Delta x = \dfrac{f(x_i) - f(x_{i-1})}{x_i - x_{i-1}}$. The algorithm is

$$x_{i+1} = x_i - f(x_i)\frac{x_i - x_{i-1}}{f(x_i) - f(x_{i-1})} \tag{3.35}$$

This algorithm can be used on real data, i.e. lists of numbers either to find where they accurately cross 0 or by addition or subtraction of a constant to find any other value. The

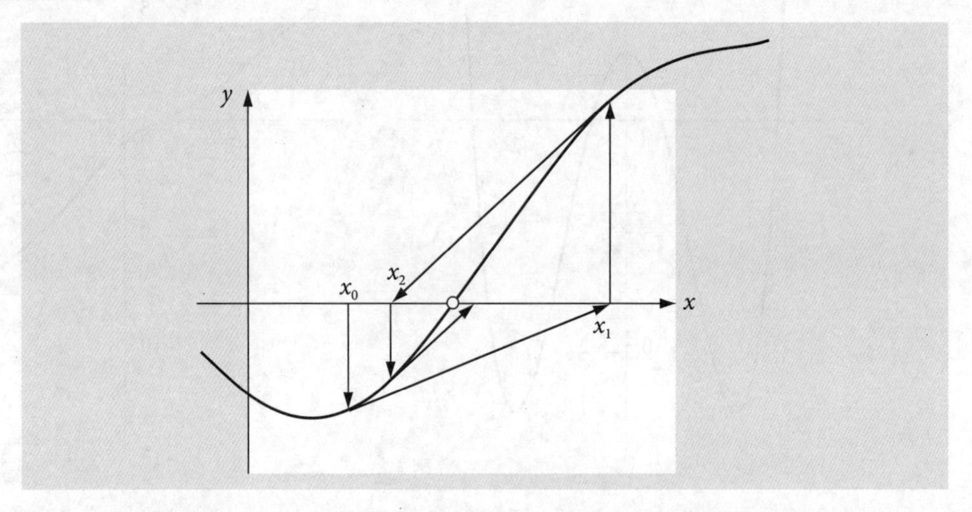

Fig. 3.24 Schematic of the pathways followed by the Newton–Raphson root-finding algorithm, x_0 is the initial value. The tangent at each point is projected along to the x-axis and the next y value and tangent obtained. This process is repeated and should converge on the root.

secant and bisection methods are described further in Chapter 11, which deals with numerical methods.

To use any of these iterative methods, one or more initial guess for x has to be made and then the calculation repeated until successive determinations of the function are sufficiently close to 0. You will have to decide how small a difference in x values is acceptable to stop the calculation and this will vary from case to case. If you want an error of no more than some value ε, then stop the calculation when consecutive results differ by $<\varepsilon/2$.

One Maple way of using the Newton–Raphson method is illustrated using the equation $-2x+e^{-x^2}=2$, which is written as $f(x)=-2x+e^{-x^2}-2$ so that its value is zero when the root is found; see Fig. 3.23. The derivative is $f'(x)=-2-2xe^{-x^2}$.

Algorithm 3.1 Newton–Raphson

```
> f:= x-> -2*x + exp(-x^2) -2:        # function
  df:= x-> -2 -2*x*exp(-x^2):         # first derivative
  x:= 1.0:                            # initial guess
  for i from 1 to 6 do                # recursion eqn
    x:= x - f(x)/df(x);
  end do;
```

$$x := -.3276464466$$
$$x := -.6439952625$$
$$x := -.6887987802$$
$$x := -.6889487343$$
$$x := -.6889487343$$
$$x := -.6889487343$$

The algorithm is contained mainly in the single line of the `for .. do` loop and works by starting with an initial guess, a poor one in this case, and incrementing values using the loop. No more that six iterations is enough to provide convergence to 10 decimal places in this particular example. However, more iterations may be needed to produce an acceptable answer with other equations.

There is no need to store the x_i and x_{i+1} values as the loop automatically does this for us. Remember that `x:= x - f(x)/df(x);` is an *assignment* statement not an equation; it means '*using the current value of x calculate $x-f(x)/f'(x)$ then change x on the left into this value*'. The next time around the loop, this new x value is used in calculating $x-f(x)/f'(x)$, and so on. You should modify the algorithm to include a statement to compare consecutive values if you want the root to be accurate to a fixed number of decimal places rather than performing a fixed number of iterations.

3.10.2 Where the method fails

The Newton–Raphson method will fail when the derivative is 0, but more often it can fail when the next iteration occurs at a point having a gradient that has the opposite sign to that of any previous points. This occurs after a maximum or minimum in the curve is exceeded. Figure 3.25 explains this more simply. The gradient at point x_1, moves the next point crossing the x-axis away from the root sought, which means that it is usually

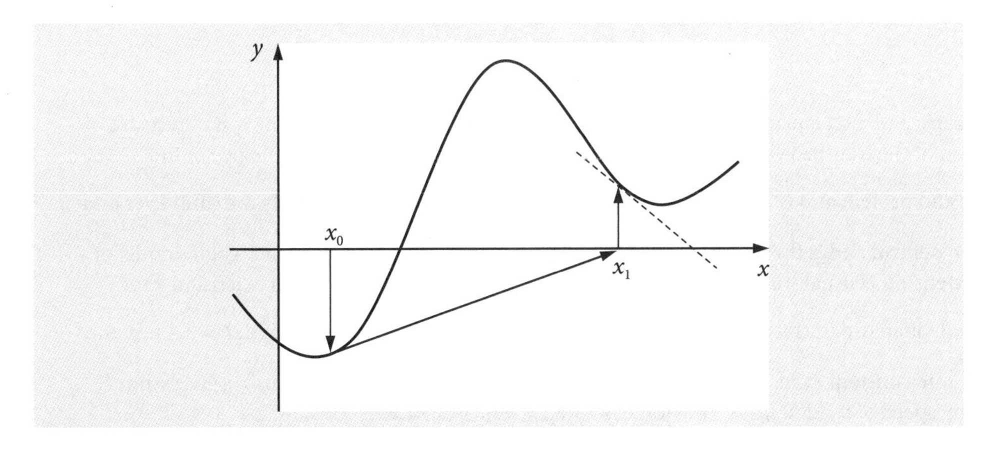

Fig. 3.25 A case where the Newton–Raphson method fails.

sensible to plot the function first, and then decide where to put the initial point. For the function shown in the sketch starting closer to the root will keep the iterations between the maximum and minimum in the curve.

3.10.3 Questions

Full solutions are available at **www.oxfordtextbooks.co.uk/orc/beddard.**

Q3.74 One use of the Newton–Raphson formula is to calculate square, cube, or n^{th} roots of numbers.

(a) Calculate $\sqrt[5]{5}$.

(b) Does this method produce all the roots to this equation?

Strategy: Rewrite the equation as $x^5 = 5$ which is the function $f(x) = x^5 - 5$ and use the Maple method in the script.

Exercise: Write a general Maple procedure to find any real n^{th} real root of a number m, e.g. $\sqrt[n]{m}$.

Q3.75 Look at Fig. 3.23 and find the roots of $3 - 8x^2 + x^4 = 0$. Compare with exact values calculated with Maple.

Strategy: Look at the plot of the function and start close to one of the roots to make sure you get the right one. The function is even and replacing x by $-x$ produces the same value; therefore, only two of the roots need to be calculated because the other two will be -1 times these.

Q3.76 Planck's black-body radiation distribution law is derived as the product of the average energy of the (Bose–Einstein) photons of frequency ν, which is $\langle E\rangle = h\nu[e^{h\nu/k_BT} - 1]^{-1}$, and the density of these states, the number per unit frequency, or $dn/d\nu = 8\pi\nu^2/c^3$. The energy/unit volume/unit frequency is

$$\rho(\nu) = 8\pi h\left(\frac{\nu}{c}\right)^3 [e^{h\nu/k_BT} - 1]^{-1}.$$

This energy density vs frequency can be used to determine the temperature of a hot body such as a lava flow, a furnace, the sun, or another star. The distribution of energy density has a maximum value at a certain frequency at a given temperature.

When $\rho(\nu)$ is at its maximum, the temperature and frequency have a fixed ratio. What is this ratio?

Strategy: Because the distribution $\rho(\nu)$ is never negative, the maximum clearly occurs when $\dfrac{d\rho(\nu)}{d\nu} = 0$ and this equation can be solved using the Newton–Raphson method. If temperature and frequency have a fixed ratio when the maximum condition is satisfied, the frequency and temperature should be present only as their ratio ν/T.

Q3.77 In problem Q3.65 we found that the maximum of the Planck radiation distribution is given by $\lambda_{max}T = \dfrac{hc}{5k_B}\left(\dfrac{1}{1 - e^{-hc/\lambda_{max}k_BT}}\right)$ and made an approximation to find $\lambda_{max}T$. Use the Newton–Raphson method to find an accurate value.

Q3.78 In experiments to measure the unfolding of a protein by pulling on one end with the tip of an AFM, the following equation was used to calculate the force f at a given retraction speed v of the AFM tip,

$$-\frac{x_u}{k_BT} - \frac{d\ln(h)}{df} + \frac{kh}{v}e^{\frac{fx_u}{k_BT}} = 0.$$

This is a transcendental equation and cannot be solved algebraically. The constant x_u is a measure of the position of the transition state on the pathway to unfolding and h is the compliance of the cantilever and protein and is $h = \dfrac{1}{k_s} + \dfrac{1}{df/dx}$. The compliance describes how easily the cantilever and the protein deform and is the inverse of the effective force constant. The worm-like-chain model of force vs extension is usually used to describe the protein's behaviour and at small extension x the force f, is adequately described as $f = \dfrac{3k_BT}{2p}x$. The constants are the persistence length $p = 0.4$ nm, the unfolding rate constant at zero force $k = 0.06\ \text{s}^{-1}$, and the cantilever force constant $k_s = 40\ \text{pN nm}^{-1}$. At a pulling speed of $v = 200\ \text{nm s}^{-1}$ the force is 150 pN. The Boltzmann constant is k_B.

(a) Show that the compliance is a constant independent of force.

(b) Using the values given, calculate x_u. Work in units of pN and nm.

***Strategy*:** Use the Newton–Raphson method to calculate x_u since this is defined in an equation that has no algebraic solution.

Q3.79 **(a)** Show that the concentration of H_3O^+ in a solution of C_A mol dm^{-3} of the acid HA and C_B of the salt NaA, is given by

$$\frac{x}{K_A} = \frac{C_A - x + k_W/x}{C_B + x - k_W/x},$$

where $x = [H_3O^+]$ and the equilibrium constants are $K_A = \dfrac{[H_3O^+][A^-]}{[HA]}$ and $K_w = [H_3O^+][OH^-]$.

(b) Calculate $x = [H_3O^+]$ and hence the pH of the solution if $[HA] = 0.2$ mol dm^{-3}, $[NaA] = 0.01$ mol dm^{-3}, and $K_A = 10^{-3}$.

***Strategy*:** Write down the equations for mass and charge balance, then use the equilibrium constants to eliminate all species other than H_3O^+, because the equation you need only contains H_3O^+. In (b) use the Newton–Raphson method to solve for numerical values. Recall that, by definition, $k_W = 10^{-14}$.

Q3.80 What is the pH of an $c = 0.1$ M solution of ammonia, given that $pK_A(NH_4^+) = 9.25$ and $pK_w = 14$?

***Strategy*:** The species present must be protons, hydroxyl ions, ammonia, and ammonium ions. The equations are determined by mass and charge balance, and the equations defining the equilibrium between protons and OH^- and NH_3/NH_4^+. Recall that if K is an equilibrium constant $pK = -\log_{10}(K)$ and similarly $pH = -\log_{10}([H^+])$.

Q3.81 The energy levels of a particle in a box of *finite* height, or a well of finite depth, cannot be calculated algebraically because the energy is contained within a transcendental equation. The calculation proceeds by specifying three regions, the central one of length L where the potential is $-V_0$, and two outer regions where the potential is *zero*. The Schrödinger equation is solved in each of these three regions and the wavefunction made continuous between them. The equation containing the energy E is $\beta[1 + \tan^2(\beta)] = \alpha$ where $\beta = 2m(E + V_0)L^2/4\hbar^2$ and $\alpha = 2mV_0L^2/4\hbar^2$.

If a quantum well is constructed with a length L of 1 nm, calculate the energy levels, if m is the mass of an electron, and V_0 the potential depth is

(a) -1230 cm^{-1}

(b) 12 times that of (a).

(c) What is the meaning of multiple solutions to the equation?

In each case, comment on any spectra that could be produced by electronic transitions between levels and compare your answer with that of an infinitely deep well.

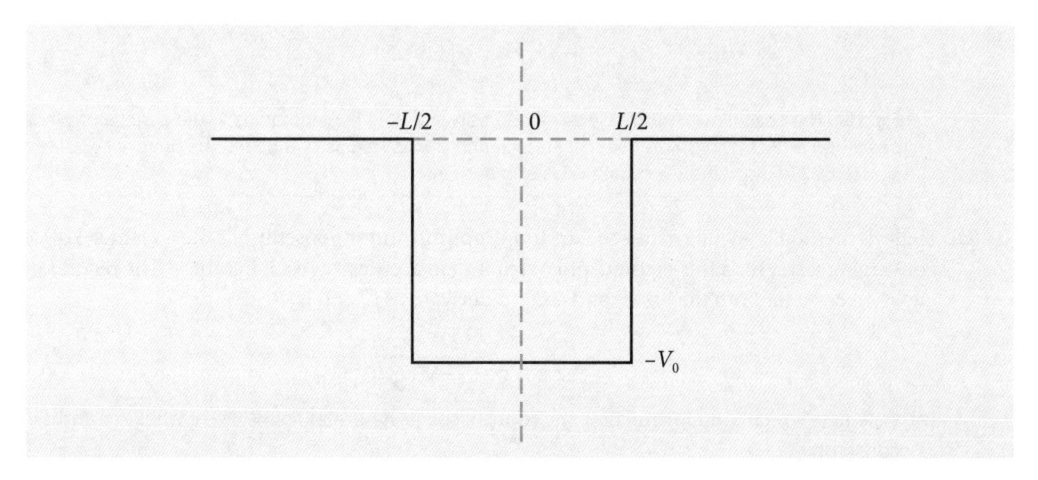

Fig. 3.26 A well of finite depth.

***Strategy*:** Calculate a numerical value for α and plot the function to find where it crosses zero. Alternatively, $[1 + \tan^2(\beta)]$ can be plotted together with α/β each vs β. The solution to the equation occurs when these curves cross.

Q3.82 Adding a surfactant and increasing the viscosity of water by adding one of many compounds such as glycerol lowers its surface tension and makes bubble formation easy. The surface produced has considerable elasticity; the intermolecular forces present must be of longer range than in pure water.

A soap bubble can be formed between two wire hoops of radius 1 positioned at $\pm x$, Fig. 3.27. The equation describing the line the bubble takes between opposite points on the rings placed at $\{-x, 1\}$ and $\{x, 1\}$, is $y = r\cosh(x/r)$ where r is the radius of the bubble's waist where $x = 0$. The surface produced is called a catenoid, and the lowest energy surface is a surface of minimum area. At a given size of the wire rings—in this case their radius is unity—there is a maximum value of $x = x_0$ at which the bubble can exist between the rings. The corresponding waist is r_0.

(a) By rearranging the catenoid equation, plot a graph of x vs r, and then using the Newton–Raphson method show that the maximum x_0, which is just at the point where the film breaks, is 0.6627, and the corresponding value of $r_0 = 0.5524$. Use Maple to plot the graph and to do the differentiation of $\cosh^{-1}$ or look it up in Section 3.4.17.

(b) The surface area of a function y over the range a to b is calculated using the standard formula $S = 4\pi\int_a^b y\sqrt{1+y'^2}\,dx$ (see Section 3.9.2), where y' is the derivative. Work out the surface area for *both* values of r when $x = 0.4$. Decide which surface has a large and which a small waist. Look up the integration or use Maple. Comment on their value relative to 2π. What do you think will happen if the bubble is initially made with the larger volume and smaller waist?

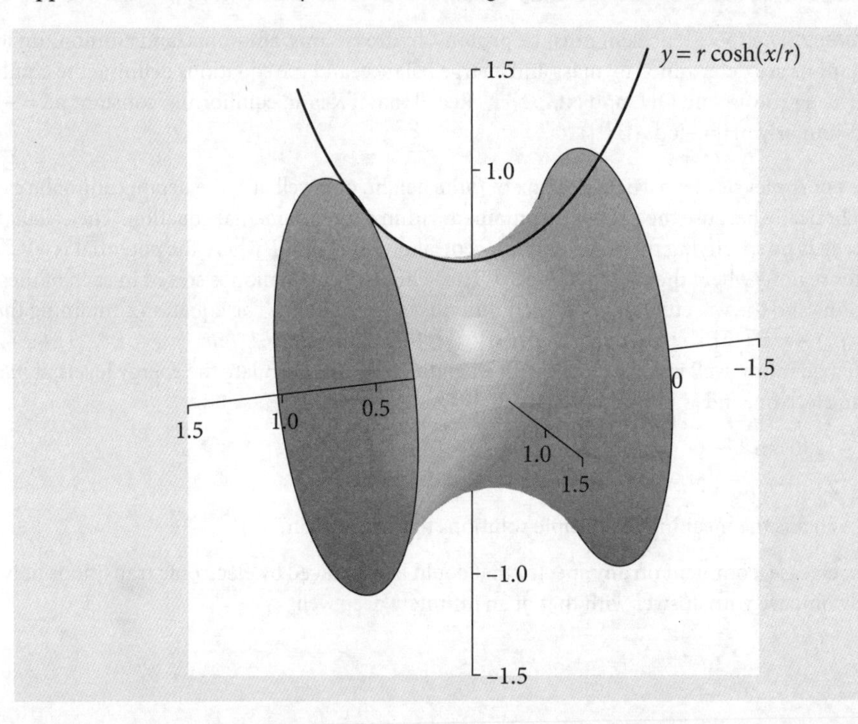

Fig. 3.27 The soap film is maximally stretched between two rings of radius 1, the surface of which is described as a catenoid with minimal surface and a waist radius of 0.5524. The rings are positioned at $x_0 = \pm 0.6627$. The curve $y = r\cosh(x/r)$ is also shown.

Q3.83 Halley's method is an alternative to Newton–Raphson and supposedly has the advantage of converging faster than this method and of tripling in accuracy at each iteration. The recursion equation contains both the first f' and second derivatives f'' and is

$$x_{n+1} = x_n - \frac{2f'(x_n)f(x_n)}{2f'(x_n)^2 - f''(x_n)f(x_n)}$$

(a) Calculate a recursion formula for the roots of the polynomial $f(x) = x^m - c$ where m and c are constants.

(b) Use this to find $79^{1/9}$ and $\left(\dfrac{11}{9971}\right)^{30/91}$ to eight decimal places.

Strategy: To calculate $79^{1/9}$ let $c = 79$ and $m = 9$, and find the root when $x^m - c = 0$.

Q3.84 The planets move in elliptical orbits around the sun. An ellipse has semi-major and semi-minor axes that are the maximum and minimum radii respectively, measured from the centre of the ellipse to its

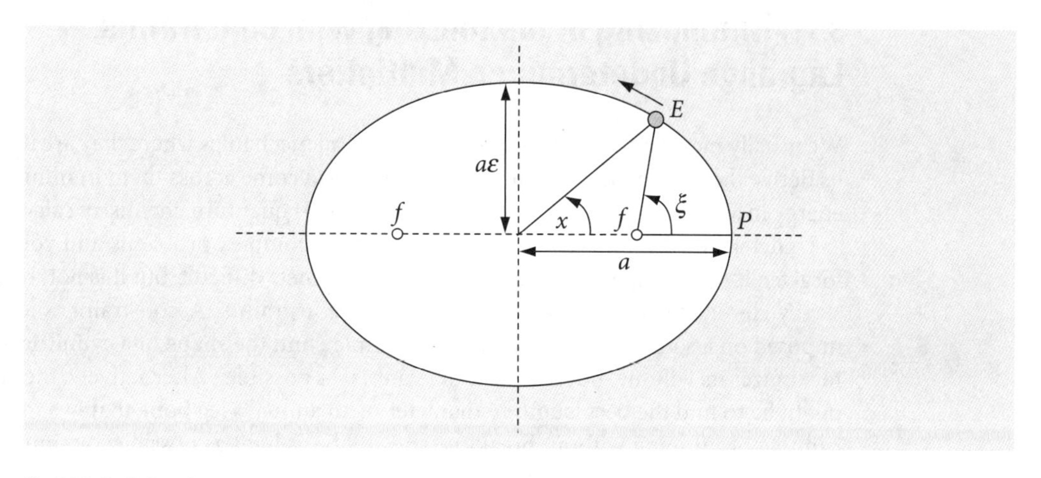

Fig. 3.28 Defining the eccentricity ε and eccentric anomaly x.

circumference. If a is the semi-major axis, then the semi-minor axis is $b = \varepsilon a$ where ε is the eccentricity of the orbit; $0 < \varepsilon \le 1$. For the earth $\varepsilon = 0.01671$, for Mercury 0.2056, and for Pluto 0.2288.

An important problem in celestial mechanics is to know where a planet or its moon was at a given time in the past or will be in the future, when its current position is known. Kepler solved this problem with the equation

$$m = x - \varepsilon \sin(x)$$

where m is called the *mean anomaly*, ε the eccentricity, and x the *eccentric anomaly*. The word *anomaly* as used here is a misnomer and is used nowadays out of historical precedence; anomalies are angles measured in radians, and shown in Fig. 3.28. The mean anomaly is $m = \dfrac{2\pi}{T}(t - t_0)$ radians where T is the period of the orbit, approximately 365 days for the earth, t the present number of the day in the year and t_0 the number of the day at which the body is closest to the centre of its orbit. This is generally called the *periapse* but is called the *perihelion* for the earth–sun system, and occurs on either 4 or 5 January each year. The eccentric anomaly x is the angle from the centre measuring the position of the planet or satellite moon. In polar coordinates the position of the earth is $\{a\cos(x), b\sin(x)\}$ where $a = 149\,600\,000$ km and is the distance of the earth from the sun. The true anomaly is the angle ξ (xi) and is the angle defined by p-f-E, see Fig. 3.28, where the periapse is p, the foci f, and the planet E. The true anomaly ξ and the eccentric anomaly x are related as

$$\cos(\xi) = \frac{\cos(x) - \varepsilon}{1 - \varepsilon\cos(x)}.$$

As Kepler's equation is transcendental, it has either to be solved numerically or approximated algebraically. Halley's method uses the recursion equation

$$x_{n+1} = x_n - \frac{2f'(x_n)f(x_n)}{2f'(x_n)^2 - f''(x_n)f(x_n)}.$$

Use this method to calculate the position of the earth in 2012 at the vernal and autumnal equinoxes, 20 March and 22 September, at the summer and winter solstices, 21 June and 21 December, and at aphelion, 5 July. Perihelion is on 5 January, (see Richards 2002, p. 130 and references therein for further discussion and derivation of Halley's method).

Q3.85 The Fourier transform of a square pulse has the form $g(\omega) = (1 - e^{-i\omega\tau})/i\omega$ where τ is the pulse's duration.

(a) Calculate the half width at half maximum (*hwhm*) of the real part of the transform which is

$$f(\omega) \equiv \mathrm{Re}[g(\omega)] = \frac{\tau}{\sqrt{2\pi}}\frac{\sin(\omega\tau)}{\omega\tau}$$

(b) Evaluate the *hwhm* when $\tau = 3$.

Strategy: (a) Calculate the function at $\omega = 0$ using l'Hôspital's rule, then calculate $f(0) = 1/2$ and obtain an equation for ω. (b) The resulting transcendental equation has to be solved numerically. Plot a graph to convince yourself of the answer.

3.11 Minimizing or maximizing with constraints: Lagrange Undetermined Multipliers

We usually meet Lagrange multipliers in statistical mechanics where they are used to help us derive the Boltzmann distribution; you may also come across them in minimizing the energy in a Hartree–Fock, self-consistent field (SCF) quantum chemistry calculation.

Undetermined multipliers appear in these very complex problems and you would be forgiven if you thought that this method was extremely difficult, but it is not. It is instead, a very clever idea to solve a problem that has constraints. A constraint is a limitation imposed on a solution to a problem for example, 'find the shape of a cylindrical can that maximizes its volume, but uses as little material as possible'. Alternatively, the constraint might be to find the best length to diameter of an animal's leg bone that has to withstand running or jumping without breaking, and still be as light as possible; or simply find the shortest distance from a point to a curved line.

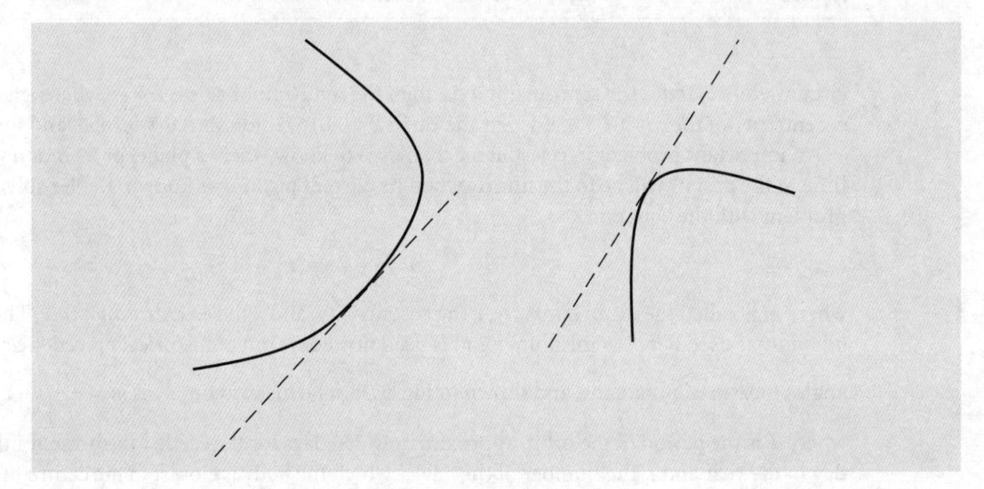

Fig. 3.29 Tangents to two curves, λ scales them to be the same.

Because a maximum or minimum quantity is sought, differentiation is involved. However, in these problems, there are two functions, one is the constraint and the other the function that is to be maximized or minimized. By differentiating both expressions, the function and constraint we find their tangents which will probably not have the same value. The undetermined multiplier, conventionally called λ, scales these tangents to be the same. Therefore,

$$\textit{Differential}\,(\text{function}) + \lambda\, \textit{Differential}\,(\text{constraint}) = 0.$$

In using this method, λ is found first, then the x and y values of the minimum or maximum. Of course, if there is more than one constraint there is an undetermined multiplier for each.

The convention is that $f(x, y)$ represents the function, $g(x, y) = 0$ the constraint, and λ the multiplier. The new function Q we have to solve is

$$Q(x, y, \lambda) = f(x, y) + \lambda g(x, y) \tag{3.36}$$

which is done by taking (partial) derivatives of Q in x, y, and λ and letting $\partial Q/\partial x = \partial Q/\partial y = \partial Q/\partial \lambda = 0$ and solving the simultaneous equations,

$$\frac{df(x,y)}{dx} + \lambda\frac{dg(x,y)}{dx} = 0 \tag{3.37}$$

$$\frac{df(x,y)}{dy} + \lambda\frac{dg(x,y)}{dy} = 0.$$

Notice that differentiation is with respect to x and then with y as variables on both function and constraint. The method or algorithm is

(a) Determine what is the function f and what the constraint g.

(b) Work out the four derivatives shown in (3.37).

(c) Find the multiplier λ by solving the simultaneous equations (3.37).

(d) Solve equations (3.37) for x and y using the constraint $g(x, y) = 0$ if necessary. This is the third derivative $\dfrac{\partial Q}{\partial \lambda} = 0$.

(e) The x and y values calculated are those of the maximum and minimum and are substituted back into the equation describing the *function* to find its value.

An example shows how this algorithm works. Suppose you are set the problem to find the shortest distance from any point $\{x_0, y_0\}$, which could be the origin $\{0, 0\}$, to the Gaussian curve $y = e^{-x^2}$.

(a) The exponential curve defines the constraint, which means that the points we find must lie on this curve, and therefore g is defined as

$$g(x, y) = y - e^{-x^2}$$

The *function* to minimize is the distance from $\{x_0, y_0\}$ to a point $\{x, y\}$ on the curve and this is calculated using Pythagoras' theorem,

$$f(x, y) = \sqrt{(x - x_0)^2 + (y - y_0)^2}. \tag{3.38}$$

The point $\{x_0, y_0\}$ is made the origin $\{0, 0\}$ at the end of the calculation.

(b) Differentiating both f and g with respect to x and to y produces

$$\frac{df(x, y)}{dx} = -\frac{x - x_0}{\sqrt{(x - x_0)^2 + (y - y_0)^2}}, \qquad \frac{dg(x, y)}{dx} = 2xe^{-x^2}$$

$$\frac{df(x, y)}{dy} = -\frac{y - y_0}{\sqrt{(x - x_0)^2 + (y - y_0)^2}}, \qquad \frac{dg(x, y)}{dy} = 1$$

(c) The two simultaneous equations, (3.37) after substituting the square root for $f(x, y)$, equation (3.38), are,

$$-\frac{x - x_0}{f} + \lambda 2xe^{-x^2} = 0, \qquad -\frac{y - y_0}{f} + \lambda = 0$$

and from this last equation, $\lambda = \dfrac{y - y_0}{f}$.

(d) Substituting λ into the first of the two equations in **(c)** gives, after some simplification,

$$-(x - x_0) + (y - y_0)2xe^{-x^2} = 0.$$

This equation defining x, the point at the shortest distance from any point $\{x_0, y_0\}$ to the curve $y = e^{-x^2}$, is after substituting for y,

$$-(x - x_0) + (e^{-x^2} - y_0)2xe^{-x^2} = 0.$$

(e) The quantities x_0 and y_0 are constants in the last equation, which can be solved to find x and then this substituted into our function $y = e^{-x^2}$ to obtain y. The distance from the origin $x_0 = 0$ and $y_0 = 0$ to the curve is found by solving

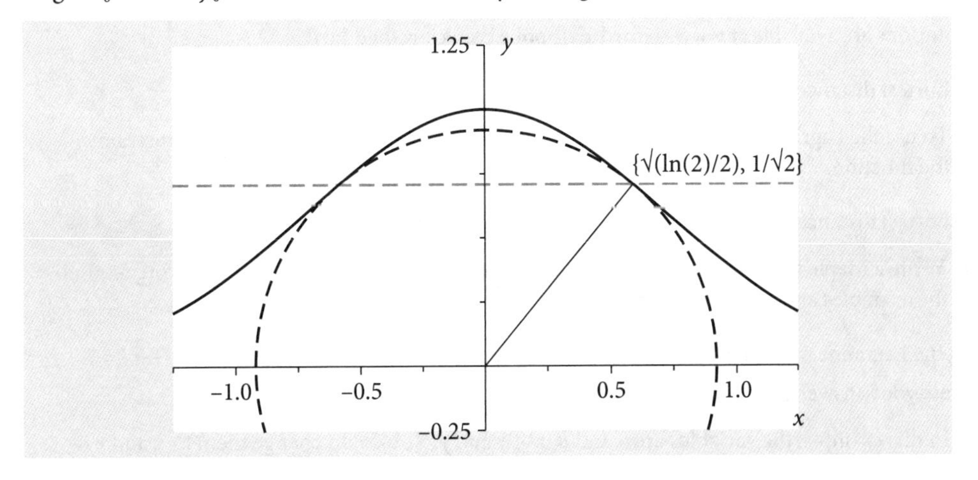

Fig. 3.30 The Gaussian curve e^{-x^2} and the shortest distance from origin (radius) to the curve and the circle with this radius. The shortest distance to the curve is ≈0.92.

$$-1 + 2e^{-2x^2} = 0$$

giving $x = \pm\sqrt{\ln(2)/2} = \pm 0.5887$ and $y = 1/\sqrt{2}$ as the coordinates of the shortest distance, see Fig. 3.30. This distance itself is found by substituting into $f(x, y)$, equation (3.38), and is $\sqrt{(\ln(2) + 1)/2} \approx 0.920$.

In this second example, the closest distance an object can be to its image, when using a thin lens, is found. The lens has a focal length f, the object is at a distance u before, and the image v, after the lens. The lens maker's formula $\dfrac{1}{u} + \dfrac{1}{v} = \dfrac{1}{f}$ relates these distances to the focal length. The total object–image distance is $u + v = c$, where c is a constant, and this distance must be minimized subject to the lens maker's formula. The figure (3.31) shows the object and image distances which are taken to be positive.

Using the Lagrange multiplier method, the function is clearly the lens maker's formula,

$$f(u, v) = 1/u + 1/v - 1/f$$

and the constraint is

$$g(u, v) = u + v - c.$$

Differentiating $f(u, v) + \lambda g(u, v)$ with respect to u and then to v produces the two equations (3.37)

$$-u^{-2} + \lambda = 0, \qquad -v^{-2} + \lambda = 0,$$

from which c has disappeared so that its value was unimportant.
Combining these equations gives $u^{-2} = v^{-2}$ or $u = v$ and consequently from the lens maker's formula $u = v = 2f$ is the closest distance for an image–object relationship to hold as shown in the figure.

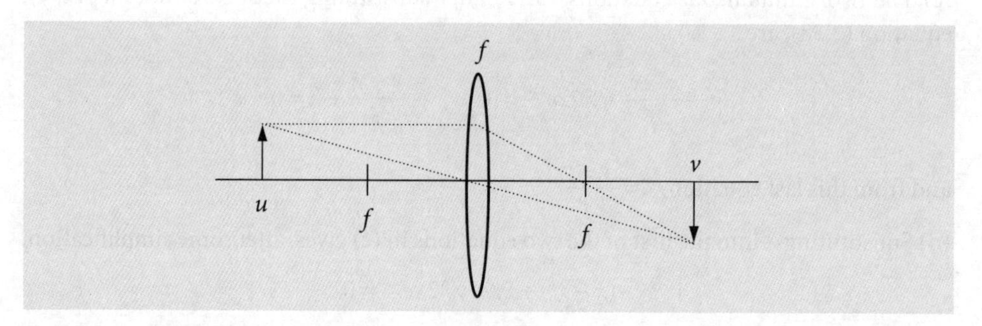

Fig. 3.31 Object and image distances with a thin lens.

3.11.1 Questions

Full solutions are available at **www.oxfordtextbooks.co.uk/orc/beddard.**

Q3.86 Find the shortest distance from the point {2, 2} to the line $y = e^{-x}$.

Strategy: To use the Lagrange undetermined multiplier method, first decide what is the constraint and what the function.

Q3.87 Find the shortest distance from the origin to $y^2 = (4 - x)^3/x$, see Fig. 3.17.

Exercise: Write a Maple worksheet to calculate the distance from the origin to any curve you choose. Test it on the examples and questions above.

Q3.88 **(a)** Using the Lagrange multiplier method, maximize the product ab with the constraint $a^2 + b^2 = 1$.

(b) Illustrate your answer graphically.

Strategy: In this example, the variables are a and b, not x and y. You could change them to x and y if you find this easier.

Q3.89 Photons always take a path through a series of optical elements, of any description whatsoever, that minimizes the total time taken. Prove Snell's law $n_a \sin(\theta_a) = n_b \sin(\theta_b)$, by calculating the minimum time a photon takes to get from A to B when it crosses a boundary from one medium to another with refractive indices n_a and n_b and $n_b < n_a$, as shown in Fig. 3.32. The speed of light in a medium is c/n where n is the medium's refractive index. Use Maple to do the calculation if you wish.

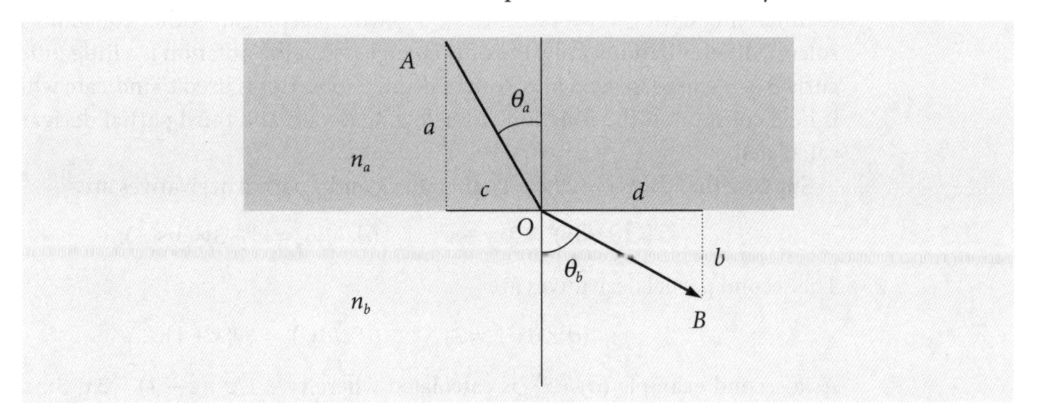

Fig. 3.32 Illustrating Snell's law. The refractive indices $n_a > n_b$.

***Strategy*:** The constraint is not mentioned, but as A and B are fixed, the constraint must be to make the horizontal distance cd constant. It is not necessary to know this constant because when differentiated it becomes 0. If the vertical distance $a + b$ was chosen as the constraint then it is found that λ, the multiplier, disappears from the problem so this cannot be correct.

Q3.90 The entropy of mixing different gases or solutions is $S = -k_B \sum_{i=1}^{n} x_i \ln(x_i)$, where x_i is the mole fraction of component i, k_B is Boltzmann's constant, and n the number of species. Show that the maximum entropy of mixing of n species is $S = k_B \ln(n)$.

***Strategy*:** A thought experiment suggests that the most likely entropy is when each component is present with a mole fraction of $1/n^{\text{th}}$ of the total if there are n species present in the mixture, since this is the most varied way the solution can be divided up into its components.

If there are two components, with mole fraction x_1 and x_2 and therefore $x_2 = 1 - x_1$, the entropy is

$$S = -k_B[x_1 \ln(x_1) + (1 - x_1)\ln(1 - x_1)].$$

The minimum is found by differentiating, which produces

$$\frac{dS}{dx_1} = -k_B[\ln(x_1) - \ln(1 - x_1)] = 0,$$

which has the solution $x_1 = 1/2$ if the sum of the mole fractions add to unity which, by definition, they must. If there were n species this approach would become impossible, or at least very tedious, and therefore the Lagrange method is preferable.

Q3.91 In the gas phase reaction $F + H_2 \rightarrow HF_v + H$, the product HF is produced in a range of vibrational quantum states, only $v = 0, 1, 2, 3$ being energetically possible under the experimental conditions used. The fraction of HF molecules found in vibrational energy level v is x_v but these are subject to a constraint such that $\sum_{v=1}^{n} x_v = 1$ where n is the total number of levels populated, which happens to be four in this experiment but of course could be different with different reactants. The distribution of x_v vs v is expected to maximize the entropy; if this is the case what will the distribution be? This is the same as asking the question 'what then will be the equation for x_i'? The vibrational contribution to the entropy (in units of k_B) is defined as $S = -\sum_{v=0}^{n} x_v \ln(x_v)$.

***Strategy*:** First decide what the function to minimize or maximize is, and because there is a constraint use the Lagrange multiplier method. The Q function, equation (3.36), is $Q = S - \lambda g$ where g is the constraint. The sum of fractional populations, x_v is unity and S is the entropy function.

Q3.92 Find the maximum and minimum of the function $f(x, y, z) = 4x + 3y + 10z$ with the two constraints, $x + y + z = 1$ and $x^2 + z^2 = 1$.

***Strategy*:** As there are two constraints there will be two undetermined multipliers; if these are λ and μ then the Q equation is $Q = f + \lambda g + \mu h$, where the constraints are the functions g and h which are the equations in x, y and z. Differentiation has also to be performed in x, y and z.

3.12 Partial differentiation

3.12.1 Differentiating equations in more than one variable

When an equation depends on more than one variable, x and y perhaps, then it is possible to differentiate with respect to either x or y while keeping the other constant. The normal rules of differentiation apply; the only change is that the notation is a little different with a curly ∂ being used instead of a Roman d and a subscript is used to indicate which variable is held constant. If the function will allow it, second and third partial derivatives can be calculated.

Suppose that $z = x^2y + y^3(x+1)$, then the x and y partial derivatives are

$$(\partial z/\partial x)_y = 2xy + y^3, \qquad (\partial z/\partial y)_x = x^2 + 3y^2(x+1).$$

The second partial derivatives are

$$(\partial^2 z/\partial x^2)_y = 2y, \qquad (\partial^2 z/\partial y^2)_x = 6y(x+1).$$

As a second example $(\partial y/\partial x)_z$ is calculated where $(y + 1/z^2)(z - 1) = 3x$. Since it is indicated by the subscript that z is a constant, differentiating gives $(z-1)\left(\dfrac{\partial y}{\partial x}\right)_z = 3$.

Differentiating the pressure in the ideal gas law $p = nRT/V$ to obtain the rate of change with respect to temperature is expressed as $(\partial p/\partial T)_{n,V}$ if the number of moles, n, and volume are held constant. The result is

$$\left(\frac{\partial p}{\partial T}\right)_{n,V} = \frac{nR}{V}.$$

Similarly, at constant n and T, $\left(\dfrac{\partial p}{\partial V}\right)_{n,T} = -\dfrac{nRT}{V^2}$, and constant V and T gives $\left(\dfrac{\partial p}{\partial n}\right)_{V,T} = \dfrac{RT}{V}$.

Calculating second derivatives has the same notation but, in this example, only $\left(\dfrac{\partial^2 p}{\partial V^2}\right)_{n,T} = 2\dfrac{nRT}{V^3}$ is not 0; the other two derivatives are constants that evaluate to 0:

$$\left(\frac{\partial}{\partial T}\left(\frac{\partial}{\partial T}\frac{nR}{V}\right)\right)_{n,V} = \left(\frac{\partial^2 p}{\partial T^2}\right)_{n,V} = 0.$$

As further examples, consider using the van der Waals equation $(p + a/V^2)(V - b) = RT$ to find $(\partial p/\partial T)_V$. Differentiating with respect to T, and using the usual rules of differentiation with V as a constant gives,

$$\left(\frac{\partial p}{\partial T}\right)_V (V-b) = R.$$

Now find $(\partial p/\partial V)_T$. Differentiating with respect to V at constant T is easier if the equation is expanded first as $p(V-b) = -\dfrac{a}{V} + \dfrac{ab}{V^2} + RT$; then differentiating as a product produces

$$\left(\frac{\partial p}{\partial V}\right)_T (V-b) + p = \frac{a}{V^2} - 2\frac{ab}{V^3}.$$

3.12.2 Geometrical interpretation of partial derivatives

If a function can be differentiated with respect to either of two variables, x and y, then it represents a surface. The derivatives then represent the slope of the surface at any point but in a direction at a fixed x or y depending on which is held constant.

The Sackur–Tetrode equation for the translational entropy S of a perfect gas is

$$S = S_0 + nC_p\ln(T) - nR\ln(p),$$

where C_p is the heat capacity at constant pressure p, S_0 is a constant entropy, n is the number of moles of the gas and R the gas constant. The entropy vs temperature and pressure is shown as the surface in Fig. 3.33. Two partial derivatives are also shown, one at fixed p

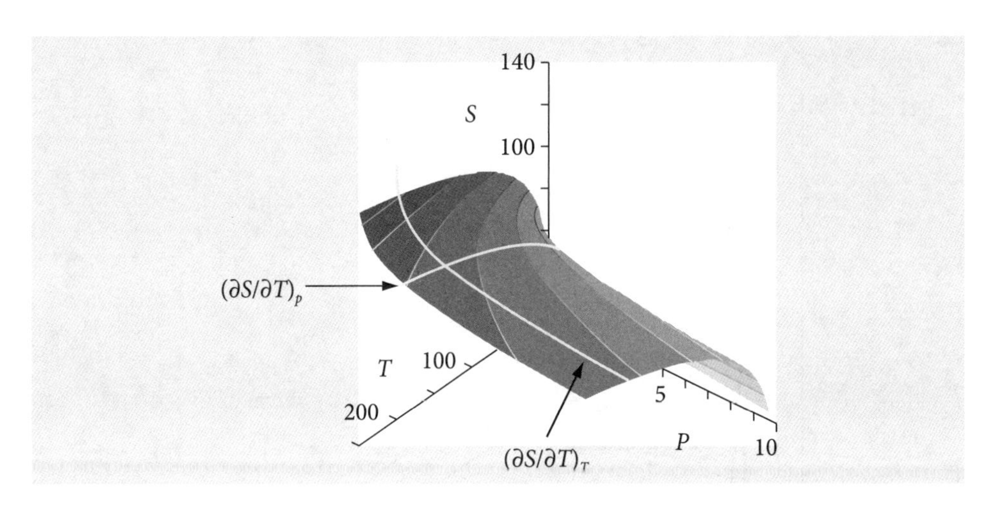

Fig. 3.33 Three-dimensional contour plot of the Sackur–Tetrode equation for 1 mole of an ideal monatomic gas where $C_p = 5R/2$, with pressure in bar and T in kelvin. The contours which show values at constant entropy are at separations of 15 J mol^{-1} K^{-1} starting at 0 and assuming $S_0 = 0$. The thick white lines show how the partial derivatives vary with pressure or temperature when the other variable is held constant.

parallel to the T axis and one at fixed T, parallel to the p axis. The gradients are parallel to the axes because there is no term in both p and T in the partial derivative equations.

The liquefaction of gases was studied by Andrews before 1870 and he discovered the critical point using carbon dioxide. This gas was presumably used as it would be readily available from brewing. The van der Waals equation is nowadays commonly used to study non-ideal, i.e. real, gases. The equation is

$$(p + a/V^2)(V - b) = RT$$

where a is a measure of the attractive forces between molecules and b accounts for their finite size. V is here the molar volume, which is volume divided by number of moles; this is often represented by $\bar{V}$ or V_m. Fig. 3.34 shows the pVT surface and shows gradients at three points with different variables held constant. The partial derivatives are also shown. The contour lines follow constant pressure where $(\partial V/\partial T)_p = constant$. Fig. 3.35, shows the pV or isotherm plot, with the inflexion at the critical temperature of 385 K is where the

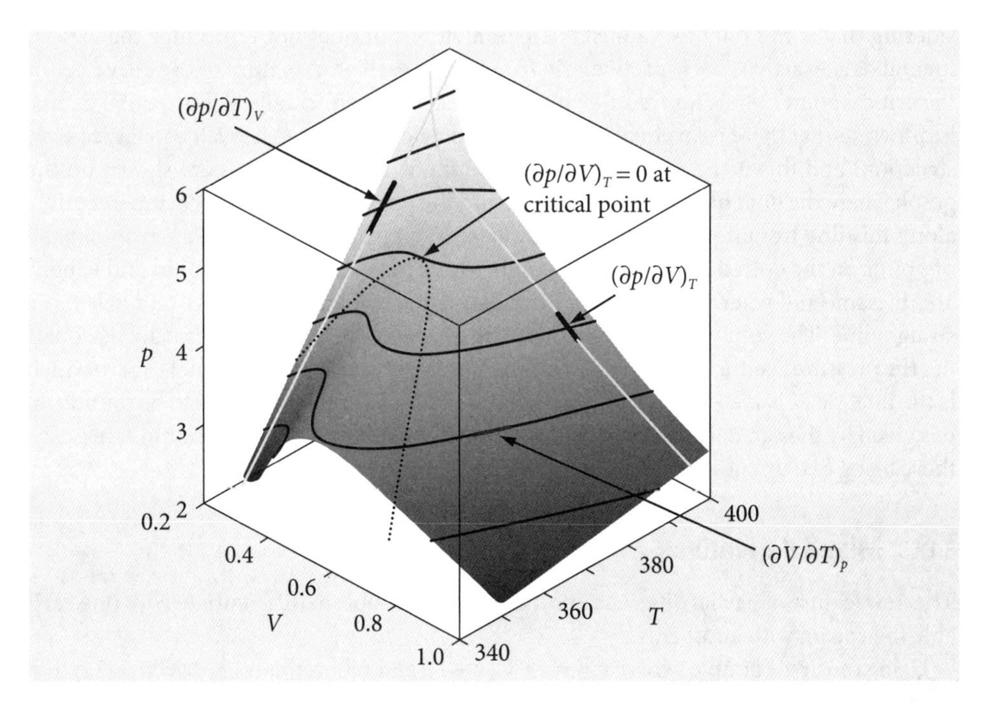

Fig. 3.34 PVT profile for a van der Waals gas (pressure in kbar, molar volume in dm^3). This figure and the next are calculated for freon, CCl_2F_2. The critical temperature is 385 K where there is an inflexion point, where $\partial p/\partial V = \partial^2 p/\partial V^2 = 0$.

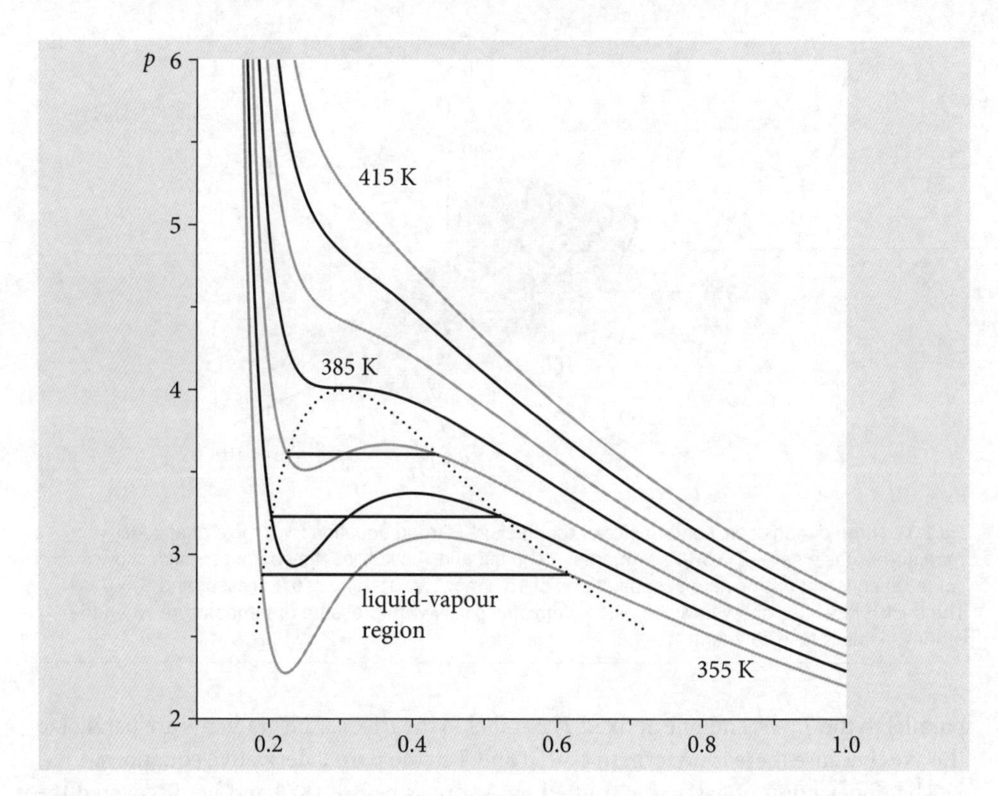

Fig. 3.35 *PV* plot at different temperatures separated by 10K. Pressures are in kbar and volumes in m^3. The parameters for freon, CCl_2F_2 were used. The oscillating lines for pressures at temperatures lower than the critical point and inside the region bounded by the dotted line are not physically realistic. The straight solid lines show the path taken by the fluid in changing from a liquid to vapour. The curves on this graph have a gradient $(\partial p/\partial V)_T$.

gradients, $\partial p/\partial V$ and $\partial^2 p/\partial V^2$ are both zero. The curve $p = a(V-2b)/V^3$ (not shown) is the locus (passes through) of the maximum and minimum of each isotherm and is a maximum at the critical point. The data used to make the plot is that for freon, CCl_2F_2, with $a = 1.078$ m^6 bar mol^{-2} and $b = 9.98 \times 10^{-5}$ m^3 mol^{-1}.

The van der Waals equation provides a surprisingly good description of a real gas, considering that it has only two adjustable parameters, but does not reproduce the experimental data exactly. This is particularly true for the oscillatory nature of the curve below the critical point (385 K for freon), which is not seen experimentally. A horizontal tie line is drawn so that there is an equal area above and below the line (Maxwell equal area construction) and this is the path followed by the fluid. Three such lines are shown on the graph where the fluid has the same pressure and the phase change proceeds imperceptibly along this line from liquid to gas or vice versa. At a given temperature, horizontal pairs of points on the dotted line define the points where the pressure of the liquid and vapour are the same and where the area of the oscillating curve is the same above and below the straight line. The volume at each end of the tie line has to be found numerically by equating the pressure at either end $p_{V_1} = p_{V_2}$ and making the areas mentioned equal to zero which is the integral $\int_{V_1}^{V_2} p_V dV - (V_2 - V_1)p_{V_1} = 0$, where V_1 and V_2 are the volumes to be found and p is given by the van der Waals equation. These two equations are solved simultaneously, there being two unknowns, V_1, V_2 at each temperature.

3.12.3 Mixed derivatives

The new feature of partial differentiation is that it is possible to differentiate with one variable then again with another.

Using the first example, where $z = x^2y + y^3(x+1)$, and differentiating $(\partial z/\partial x)_y = 2xy + y^3$ with y produces

$$\frac{\partial^2 z}{\partial y \partial x} = 2x + 3y^2.$$

Next differentiating $(\partial z/\partial y)_x = x^2 + 3y^2(x+1)$ with x produces $\dfrac{\partial^2 z}{\partial x \partial y} = 2x + 3y^2$ which is the same result. The order of carrying out the differentiation is immaterial, and there is a mathematical theorem that guarantees this. In general for any well behaved function $f(x, y)$,

$$\frac{\partial^2 f}{\partial x \partial y} = \frac{\partial^2 f}{\partial y \partial x}. \tag{3.39}$$

This is also sometimes written in a more formal form. The left side of the equation tells us to differentiate with y keeping x constant, and then differentiate the result with x, keeping y constant,

$$\left[\frac{\partial}{\partial x}\left(\frac{\partial f}{\partial y}\right)_x\right]_y = \left[\frac{\partial}{\partial y}\left(\frac{\partial f}{\partial x}\right)_y\right]_x.$$

This is also frequently written as

$$\frac{\partial}{\partial x}\left(\frac{\partial f}{\partial y}\right) = \frac{\partial}{\partial y}\left(\frac{\partial f}{\partial x}\right)$$

where the subscripts are assumed.

In the ideal gas law, $pV = nRT$, differentiating p with respect to T at constant V and n (number of moles) produces

$$\left(\frac{\partial p}{\partial T}\right)_{n,V} = \frac{nR}{V}.$$

Next, differentiating with respect to V also at constant n produces.

$$\left(\frac{\partial^2 p}{\partial V \partial T}\right)_n = -\frac{nR}{V^2}.$$

Repeating the process but in the opposite order gives

$$\left(\frac{\partial^2 p}{\partial T \partial V}\right)_n = \left(\frac{\partial^2 p}{\partial V \partial T}\right)_n = -\frac{nR}{V^2}.$$

3.12.4 Chain rule

The chain rule can also be used with partial derivatives. If w is some complicated expression in x, y and z the familiar function-of-function rule can be used to calculate for example, $(\partial y/\partial x)_z$ with z held constant. The result is

$$\left(\frac{\partial y}{\partial x}\right)_z = \left(\frac{\partial y}{\partial w}\right)_z \left(\frac{\partial w}{\partial x}\right)_z.$$

Notice the symmetry in the derivatives. A similar equation can be written with x held constant if the differentiation were with respect to z. Using this rule, if $y = \sin(azx^2 + z^2)$ then

$$\left(\frac{\partial y}{\partial x}\right)_z = 2azx\cos(azx^2 + z^2)$$

where $\left(\dfrac{\partial y}{\partial w}\right)_z = \cos(azx^2 + z^2)$. The other derivative is

$$\left(\frac{\partial y}{\partial z}\right)_x = \cos(azx^2 + z^2)(ax^2 + 2z).$$

In thermodynamics, the chain rule is very often used to expand an expression in a new variable w and this proves to be very useful. The reason for doing this is to change an unfamiliar derivative into two expressions each of which is related to something that can be measured.

For example, suppose that the change in enthalpy with temperature at constant pressure is required this is $(\partial H/\partial T)_p$, and is the constant pressure heat capacity C_p, but, for the moment, assume that we do not know this. To find out what this derivative is, expand it in some other thermodynamic variable, entropy for example, then

$$\left(\frac{\partial H}{\partial T}\right)_p = \left(\frac{\partial H}{\partial S}\right)_p \left(\frac{\partial S}{\partial T}\right)_p.$$

Notice again the symmetry of the derivatives when making such an expansion. This procedure does not seem to make much sense until the following derivatives are looked up

$$\left(\frac{\partial H}{\partial S}\right)_p = T, \qquad \left(\frac{\partial S}{\partial T}\right)_p = \frac{C_p}{T}$$

making $\left(\frac{\partial H}{\partial T}\right)_p = C_p$ the heat capacity at constant pressure. Had we chosen some other variable, G or U for example, then although derivatives could be found these may not correspond to anything that could be measured. The second derivative produces the integral used to (experimentally) determine entropy $S = \int \frac{C_p}{T} dT$ at constant pressure, by measuring the heat capacity vs temperature.

3.12.5 Reciprocal derivatives

Reciprocal derivatives follow the same rules as for normal differentiation: take the reciprocal and flip the derivative.

$$\left(\frac{\partial y}{\partial x}\right)_z = 1/\left(\frac{\partial x}{\partial y}\right)_z \tag{3.40}$$

For example if $y = azx^2 + z^2$ then differentiating with respect to x produces $\left(\frac{\partial y}{\partial x}\right)_z = 2azx$ and differentiating with y produces $1 = 2azx\left(\frac{\partial x}{\partial y}\right)_z$ which proves (40).

3.12.6 Total derivatives

Total derivatives are probably used in thermodynamics more than elsewhere. It is common to see expressions such as

$$dp = \left(\frac{\partial p}{\partial T}\right)_V dT + \left(\frac{\partial p}{\partial V}\right)_T dV. \tag{3.41}$$

The notation here is different to that used so far because dp, dT, and dV exist as entities in themselves, rather than as a ratio, such as dp/dV.

In thermodynamics, one needs to know which variable depends upon another because this is rarely stated in the equations. An example is $p \equiv p(T, V)$ where pressure depends on T and V, and therefore temperature depends on p and V and V depends on T and p; $V \equiv V(T, p)$. By definition, the internal energy of an *ideal gas* U, depends only on temperature $U \equiv U(T)$ because there are no interactions between molecules. For example, does the entropy S depend either on p and V, or H and G, or on other quantities? Not apparently an easy question to answer, and in fact any one thermodynamic quantity can depend upon any of the others, but is *defined* when it is a parameter of any other two; which two depends upon what problem you are trying to solve. There is almost too much choice and this can lead to confusion even though there are natural variables for each quantity and these are are generally used.

A total differential can be derived in the same manner as was originally done for a simple derivative, such as dy/dx; we follow here McQuarrie & Simon (1997). Making a small change in pressure Δp caused by a small change in temperature and volume produces

$$\Delta p = p(T + \Delta T, V + \Delta V) - p(T, V),$$

and now subtract a small change in volume at temperature T from the first term of this equation and add the same to the second, so adding zero. The result is

$$\Delta p = p(T + \Delta T, V + \Delta V) - p(T, V + \Delta V) - p(T, V) + p(T, V + \Delta V).$$

Next, multiply both terms by unity, the first pair of terms with $\Delta T/\Delta T$ and the second pair with $\Delta V/\Delta V$ making

$$\Delta p = \left[\frac{p(T+\Delta T, V+\Delta V) - p(T, V+\Delta V)}{\Delta T}\right]\Delta T + \left[\frac{p(T, V+\Delta V) - p(T, V)}{\Delta V}\right]\Delta V.$$

In the first square brackets only T changes; increasing to $T + \Delta T$, the volume is unchanged at $V + \Delta V$ and in the second term, only V changes. Now take the limits $\Delta T \to 0$ and $\Delta V \to 0$ to form the differential

$$dp = \left(\frac{\partial p}{\partial T}\right)_V dT + \left(\frac{\partial p}{\partial V}\right)_T dV \tag{3.42}$$

and again notice the symmetry in the expression. This derivation is quite general: if a function f depends on variables g and h, i.e. $f(g, h)$, then it is *always* possible to write;

$$df = \left(\frac{\partial f}{\partial g}\right)_h dg + \left(\frac{\partial f}{\partial h}\right)_g dh. \tag{3.43}$$

3.12.7 The 'minus 1' rule or Euler's chain rule

Euler's chain rule is a product of three derivatives that is very useful in thermodynamics and elsewhere. The pressure of a gas can be written as a function of temperature and pressure, $p = f(T, V)$. This can be expanded as a total derivative, equation (3.41), and if the pressure is kept constant $dp = 0$ then

$$\left(\frac{\partial p}{\partial T}\right)_V dT + \left(\frac{\partial p}{\partial V}\right)_T dV = 0.$$

The two terms dT and dV can have any value but their ratio dV/dT is fixed when $dp = 0$ and so when dividing by dT the notation should be changed to $(\partial V/\partial T)_p$. Doing this produces

$$\left(\frac{\partial p}{\partial T}\right)_V + \left(\frac{\partial p}{\partial V}\right)_T \left(\frac{\partial V}{\partial T}\right)_p = 0.$$

Because V is constant, by the gas law, p is only a function of temperature. Therefore

$$\left(\frac{\partial p}{\partial T}\right)_V = 1/\left(\frac{\partial T}{\partial p}\right)_V$$

producing

$$\left(\frac{\partial T}{\partial p}\right)_V \left(\frac{\partial p}{\partial V}\right)_T \left(\frac{\partial V}{\partial T}\right)_p = -1 \tag{3.44}$$

Notice the symmetry in the expression; each derivative involves all three parameters in a cyclic fashion. This type of equation is quite general; if a function has the form $f(x, y, z) = 0$ then

$$\left(\frac{\partial x}{\partial y}\right)_z \left(\frac{\partial y}{\partial z}\right)_x \left(\frac{\partial z}{\partial x}\right)_y = -1,$$

and in thermodynamics a function such as $f(p, V, T) = 0$ is usually called an *equation of state*; this means that

$$df = \left(\frac{\partial f}{\partial T}\right)_{V,p} dT + \left(\frac{\partial f}{\partial V}\right)_{p,T} dV + \left(\frac{\partial f}{\partial p}\right)_{V,T} dp = 0.$$

The Euler 'minus 1' equation (3.44) can be obtained in a different way using this last equation. Suppose that there is no change in pressure, which is the case in an isobaric process, then

$$\left(\frac{\partial f}{\partial V}\right)_p dV + \left(\frac{\partial f}{\partial T}\right)_p dT = 0 \quad \text{or} \quad \left(\frac{\partial V}{\partial T}\right)_p = -\left(\frac{\partial f}{\partial T}\right)_p \Big/ \left(\frac{\partial f}{\partial V}\right)_p.$$

This result is not very useful because it still contains the function f. To remove this, the equation can be divided top and bottom by $\partial p/\partial f$ which produces

$$\left(\frac{\partial V}{\partial T}\right)_p = -\frac{\left(\dfrac{\partial f}{\partial T}\right)\left(\dfrac{\partial p}{\partial f}\right)}{\left(\dfrac{\partial f}{\partial V}\right)\left(\dfrac{\partial p}{\partial f}\right)} = -\left(\frac{\partial p}{\partial T}\right)_V \bigg/ \left(\frac{\partial p}{\partial V}\right)_T$$

which is equivalent to equation (3.44) but arrived at in a different way.

3.12.8 Partial Derivatives in Thermodynamics

In the study of thermodynamics, partial, mixed, and total derivatives are commonly used; this adds to the complexity of the subject, particularly if these appear to be 'pulled out of fresh air'. Understanding how these relationships are produced really helps in understanding thermodynamics as it removes the mathematical burden, allowing the subject itself to be better understood. To this end, the parameters used in chemistry are p, V, A, T, H, U, G, S and each one depends upon the others. Besides p, V, T and S, the other parameters used are U the internal energy, H the enthalpy, G the Gibbs free energy and A the Helmholtz free energy.

When one quantity is to be determined, for instance the entropy S, then it is defined provided it can be calculated as a function of any other two parameters. The two parameters chosen are normally determined by what can be measured experimentally; however, it can be shown that there are natural variables for each parameter, for the internal energy U; these are volume and entropy. The defining equations for U, H, G, and A in terms of natural variables are shown in the table. When constructing equations involving thermodynamic derivatives as a rule of thumb, 'x' values are usually T, V, p, and S and 'y' values H, U, G, S, and A.

One example of using Table 3.1 is to calculate the partial derivatives with respect to entropy at constant pressure

$$\left(\frac{dU}{dS}\right)_p = T, \qquad \left(\frac{dH}{dS}\right)_p = T$$

and then derivatives with temperature

$$\left(\frac{dG}{dT}\right)_p = -S, \qquad \left(\frac{dA}{dT}\right)_V = -S.$$

3.12.9 Exact and non-exact differentials and state functions

In an adiabatic change, no heat enters or leaves the thermodynamic system. The first law therefore asserts that

$$dQ = dU + pdV = 0. \tag{3.45}$$

Now, if $U \equiv U(V, T)$ then $dU = \left(\dfrac{\partial U}{\partial V}\right)_T dV + \left(\dfrac{\partial U}{\partial T}\right)_V dT$ and so combining this with the first law produces

$$dQ = \left(\frac{\partial U}{\partial V} + p\right)dV + \frac{\partial U}{\partial T}dT = 0. \tag{3.46}$$

Table 3.1 Thermodynamic relationships and natural variables in a closed system where no matter enters or leaves.

	Equation	Natural variables
Internal energy	$dU = TdS - pdV$	$U(S, V)$
Enthalpy	$dH = TdS + Vdp$	$H(S, p)$
Gibbs energy	$dG = -SdT + Vdp$	$G(T, p)$
Helmholtz energy	$dA = -SdT - pdV$	$A(T, V)$

This differential expression is in two variables and has the same form as the differential the mathematicians call a *Pfaffian*:

$$dQ = A(x, y)dx + B(x, y)dy. \tag{3.47}$$

where $A(x, y)$ and $B(x, y)$ are two functions of x and y. If the derivative of each of these two functions is now taken, but with respect to the other variable, only if $\partial A/\partial y$ and $\partial B/\partial x$ are then found to be equal is the differential said to be *exact* or *perfect*;

$$\frac{\partial A}{\partial y} = \frac{\partial Q}{\partial x \partial y} = \frac{\partial B}{\partial x}. \tag{3.48}$$

The importance of this condition is understood when equation (3.47) is integrated to find Q. If the differential is *exact*, then, provided Q exists,

$$\int_1^2 dQ = Q(2) - Q(1) = \Delta Q$$

and Q is said to be *integrable*. The result of integrating is just the difference in Q at value of 2 compared to that at 1 i.e. *the integral does not depend upon the path* taken from the start to the end but just the values of the function Q at the start and end.

In thermodynamics, Q is called a *state function*: enthalpy, entropy, temperature, and internal energy are examples of state functions. Heat and work, which depend on the path taken from start to end, are not state functions. The internal energy of a molecule consists of energy in rotational and vibrational levels. It does not matter how the energy gets into these levels, i.e. on the path that is taken to reach a certain amount of internal energy. For example, in an experiment, we might use radiation of appropriate frequencies, to excite rotational levels first then vibrational ones or vice versa; the result is that the same amount of internal energy is contained within the molecule. However, the amount of work done to achieve this need not be the same. Another state function is gravitational potential energy, which is the energy gained on climbing a hill. You will have the same potential energy whether you run up, walk up, or parachute there from a plane, but the amount of work done and heat generated to get you there, will be very different in each case.

The equivalence between the derivatives in equation (3.48) may not be true in all cases and then the differential is *not exact*. The integration now depends upon the path taken from 1 to 2 and the integration must be explicitly performed.

Suppose that θ is an expression that depends on x or y, then it can be shown that the differential equation (3.47) can be made exact by dividing the expression by this factor,

$$\frac{dQ}{\theta} = \frac{1}{\theta}[A(x, y)dx + B(x, y)dy]. \tag{3.49}$$

As an example, let us start with equation (3.46) or (3.47) and calculate (3.48) to see if the two derivatives are the same. If, in equation (3.46) we let

$$A = \left(\frac{\partial U}{\partial V} + p\right), \qquad B = \frac{\partial U}{\partial T},$$

differentiating A with respect to T produces

$$\frac{\partial A}{\partial T} = \frac{\partial}{\partial T}\left[\left(\frac{\partial U}{\partial V}\right)_T + p\right] = \frac{\partial^2 U}{\partial V \partial T} + \left(\frac{\partial p}{\partial T}\right)_V$$

and B with V

$$\frac{\partial B}{\partial V} = \frac{\partial^2 U}{\partial V \partial T}.$$

These two derivatives are clearly not the same; for example, in an ideal gas the derivative $\partial p/\partial T \neq 0$ but it should be 0 if equation (3.48) is to be obeyed.

Now do a similar calculation, but instead use equation (3.49), with $\theta = T$ and calculate dQ/T, which is the entropy S. Because T depends upon P and V, by equation (3.49) a perfect differential should result;

$$\frac{dQ}{T} = \frac{1}{T}\left(\frac{\partial U}{\partial V} + p\right)dV + \frac{1}{T}\frac{\partial U}{\partial T}dT.$$

To check this, start with 3.48, calculate the derivatives again using the product rule with A the term preceeding dV and B that preceeding dT in the last equation. The result is

$$\frac{\partial A}{\partial T} = \frac{\partial}{\partial T}\left[\left(\frac{\partial U}{\partial V}\right)_T + p\right]\frac{1}{T}$$

$$= \frac{1}{T}\left[\left(\frac{\partial^2 U}{\partial V \partial T} + \left(\frac{\partial p}{\partial T}\right)_V\right] - \frac{1}{T^2}\left[\left(\frac{\partial U}{\partial V}\right)_T + p\right]$$

and

$$\frac{\partial B}{\partial V} = \frac{1}{T}\frac{\partial^2 U}{\partial T \partial V}$$

If the derivatives of A and B are equal, then with some rearranging the following relationship results

$$\left(\frac{\partial U}{\partial V}\right)_T = T\left(\frac{\partial P}{\partial T}\right)_V - p, \tag{3.50}$$

which must be true if we believe that the original differential is exact. To confirm that this is the case, consider an ideal gas where the internal energy depends only on the temperature, then $dU/dV = 0$, and as $pV = nRT$, then $\left(\frac{\partial p}{\partial T}\right)_V = nR/V$. As a check, substituting into equation (3.50) gives $0 = \frac{TnR}{V} - \frac{nRT}{V}$ which shows that equation (3.50) is correct and so is (3.49).

3.12.10 Questions

Full solutions are available at **www.oxfordtextbooks.co.uk/orc/beddard.**

Q3.93 Find $\partial z/\partial x$, $\partial z/\partial y$ and $\partial^2 z/\partial x \partial y$ if $z(x, y) = (x^2 + y^2)\sin(y/x)$.

Q3.94 **(a)** If $z = \exp(x + cy) + \ln(x - cy)$, c being a constant, show that z is a solution of the wave equation $\frac{\partial^2 z}{\partial y^2} = c^2\frac{\partial^2 z}{\partial x^2}$.

(b) If the function z is a general one made of a linear combination of two arbitrary functions, $z = f(x + cy) + \phi(x - cy)$ where f and ϕ are those two functions. Is the wave equation in (a) also true?

Q3.95 A flux J is defined as the amount of substance passing through an area A of a solution in time t, and is $J = dn/dt$, which means that dn molecules pass through the area in time dt. In a solution of constant temperature, if you do not stir it, this flux is due only to the diffusion of the molecules. If there is no concentration gradient, then clearly the solution has a uniform concentration and the flux from bottom to top through a victim area A is the same as top to bottom, making zero in total. If you make a concentration gradient, for example by heaping two teaspoons of sugar in the bottom of a cup of coffee, then the flux of the sugar one way through our victim area A is greater than the reverse; diffusion acts to equalize the concentration and thus the concentration depends on time t and position in the solution x.

In one dimension, the flux is related to Fick's first law of diffusion as

$$J = -D\frac{\partial c}{\partial x}$$

where D is the diffusion coefficient and c the concentration at position x, and the flux J is therefore now defined in units of mol dm^{-2} s^{-1}. Clearly, the flux is greatest when the concentration gradient is greatest. Partial derivative notation ∂ is used because the concentration depends on position and time.

(a) Diffusion in one dimension may be supposed to be that down a thin, smooth-walled capillary whose length x is far greater than its diameter. In this case the rate of change of concentration is $\dfrac{\partial c}{\partial t} = -\dfrac{\partial J}{\partial x}$. Differentiate J and so find Fick's second law.

(b) Fick's second law can be solved by realizing that the flux J must be zero at the closed ends of our capillary tube; this is the boundary condition, and the resulting concentration is given by $c = \dfrac{c_0}{2\sqrt{\pi D t}}\exp\left(-\dfrac{x^2}{4Dt}\right)$ where c_0 is the total concentration of solution. Show that this equation for the concentration is a solution of Fick's second law.

(c) When does the equation for concentration given above not apply?

(d) Assume that an amount c_0 is placed in a thin disc at $x = 0$ in the centre of an infinitely long capillary; Fig. 3.36. Plot the concentration as a function of distance $-x$ to x and at different times when $D = 2.5 \times 10^{-9}$ m^2 s^{-1}, which is approximately that of water, and assume that $c_0 = 1$ mol dm^{-3}. Choose suitable times, e.g. microseconds. Distances are estimated using the fact that the distance diffused in one dimension in time t is $x = \sqrt{2Dt}$.

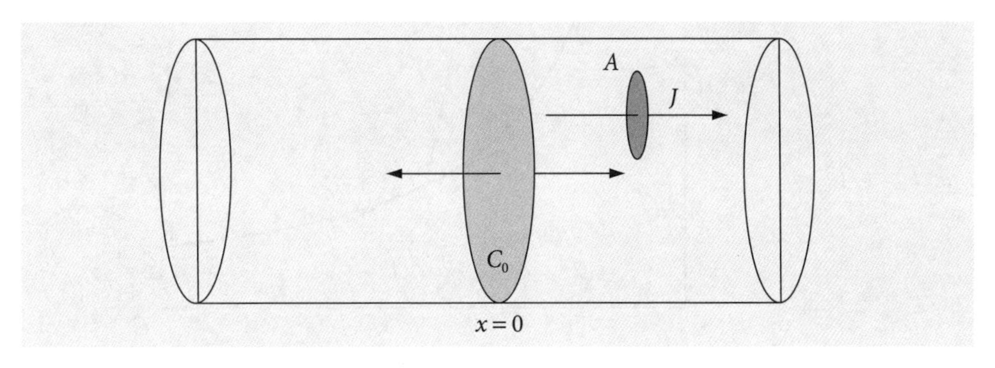

Fig. 3.36 Part of an infinitely long cylinder with initial concentration C_0 at its centre, $x = 0$, and zero elsewhere and shows the flux through a small victim area A.

Q3.96 The van der Waals equation is $\left(p + \dfrac{a}{V^2}\right)(V - b) = RT$, where a and b are constants.

(a) Calculate $\left(\dfrac{\partial V}{\partial T}\right)_p$.

(b) At the critical point, $\left(\dfrac{\partial p}{\partial V}\right)_T = \left(\dfrac{\partial^2 p}{\partial V^2}\right)_T = 0$. Show that the critical parameters are $T_c = \dfrac{8a}{27Rb}$, $V_c = 3b$ and $p_c = \dfrac{a}{27b^2}$.

Q3.97 In the equation $H = U + pV$, H is the enthalpy, U the internal energy, p pressure, and V volume. As each parameter depends upon temperature, i.e. $U \equiv U(p, T)$ and so forth, show that $\left(\dfrac{\partial U}{\partial T}\right)_p = C_p - p\left(\dfrac{\partial V}{\partial T}\right)_p$, where C_p is the heat capacity at constant pressure; $C_p = \left(\dfrac{\partial H}{\partial T}\right)_p$.

Q3.98 The Sackur–Tetrode equation for the translational entropy S of a perfect gas is

$$S = S_0 + nC_p \ln(T) - nR\ln(p),$$

where C_p is the heat capacity at constant pressure p, S_0 is a constant, n is the number of moles of the gas, and R the gas constant.

Calculate the change of entropy with temperature at constant pressure and at constant volume, i.e. $\left(\dfrac{\partial S}{\partial T}\right)_p$ and $\left(\dfrac{\partial S}{\partial T}\right)_V$. (Hint: use the gas law to obtain volume from pressure.)

Q3.99 If U is the internal energy of a gas, use the relationship $\left(\dfrac{\partial U}{\partial V}\right)_T = T\left(\dfrac{\partial p}{\partial T}\right)_V - p$ to evaluate $\left(\dfrac{\partial U}{\partial V}\right)_T$

(a) for an ideal gas,

(b) for a van der Waals gas. Comment on the results obtained.

Q3.100 The compressibility of a substance is defined as $\kappa = -\dfrac{1}{V}\dfrac{\partial V}{\partial p}$ and its inverse is called the bulk modulus. The compressibility is the rate of change of volume with pressure normalized to the volume, and the negative sign is included to make κ positive because the slope of the P–V plot is negative. Calculate the *isothermal* compressibility for

(a) an ideal gas,

(b) a van der Waals gas.

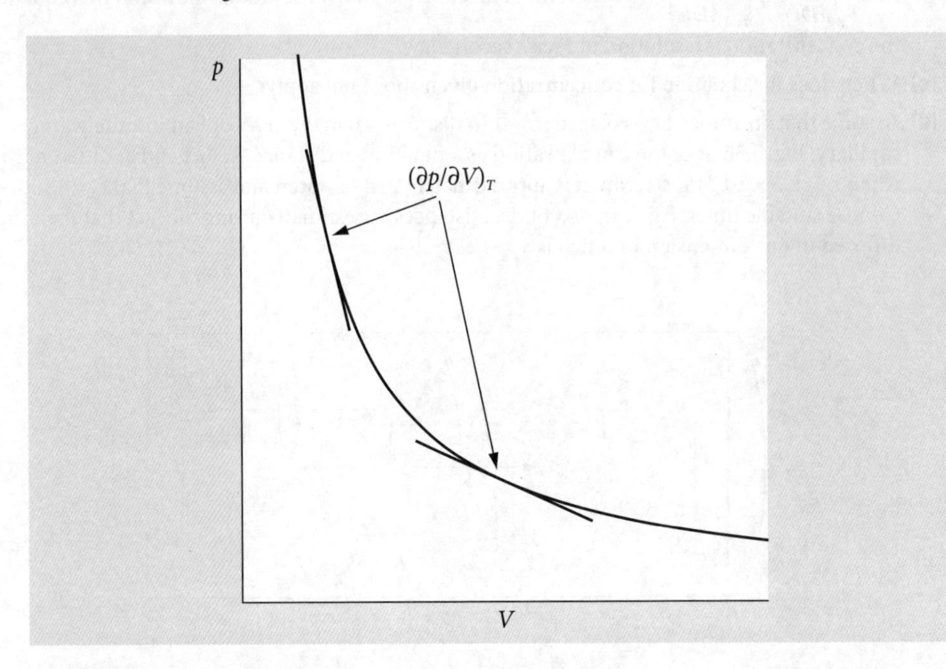

Fig. 3.37 Pressure vs volume at constant temperature for an ideal gas.

Q3.101 The equation $\partial c/\partial t = D\partial^2 c/\partial x^2 - v\partial c/\partial x$ describes the one-dimensional diffusion of molecules or colloids in the presence of a steady force moving the particles at velocity v. D is the diffusion coefficient, c concentration, x position, and t time.

(a) What are the units of the diffusion coefficient D?

(b) Show that $c = c_0 e^{v(x-x_0)/2D} e^{-v^2 t/4D}$ is a solution to this equation if c_0 and x_0 are constants. What do the constants c_0 and x_0 represent?

Q3.102 **(a)** Starting with the first law, show that the heat capacities for 1 mole of an ideal gas at constant pressure and volume are related by $C_P = C_V + R$.

(b) The enthalpy is defined as $H = U + pV$. Show that $C_p = \left(\dfrac{\partial H}{\partial T}\right)_p$.

Notes: Assume that the internal energy U is only a function of the temperature. The heat capacity of a substance is defined as the heat absorbed per unit change in temperature, $C = dQ/dT$.

Strategy: The first law relates heat change dQ to change in internal energy dU and work $-PdV$. Make U a function of temperature; $dU = \dfrac{dU}{dT}dT$ means that the heat capacity has to be found in terms of dU/dT.

Q3.103 Combining the first and second laws produces $dU = TdS - pdV$, where S is the entropy and is a function of T and V, i.e. $S = S(T, V)$.

(a) Calculate $\left(\dfrac{\partial S}{\partial T}\right)_V$ and $\left(\dfrac{\partial S}{\partial V}\right)_T$ by assuming that the internal energy also is a function of T and V, i.e., $U = U(T, V)$.

(b) Next, using $H = U + pV$ state dH as a total derivative and so find $\left(\dfrac{\partial S}{\partial T}\right)_p$, where $H = H(U, p, V)$ and $S = S(p, T)$.

Q3.104 Using equation (3.50) find the expression for the internal energy of a van der Waals gas.

Q3.105 In thermodynamics, equations sometimes seem to appear out of nowhere. This is such a case; find $(\partial V/\partial T)_E$ if $V = f(p, V, E)$ with E constant.

Q3.106 If $p(V - B_T) = RT$ find $(\partial V/\partial p)_T$, $(\partial V/\partial T)_p$, and the mixed derivatives if B_T is a function only of T.

Q3.107 One of the Maxwell relationships, $\left(\dfrac{\partial S}{\partial p}\right)_T = -\left(\dfrac{\partial V}{\partial T}\right)_p$, can be used to calculate an entropy change. The difference $\Delta S_{1\to p}$ between the entropy of a real gas at pressure p and the entropy it would have at 1 bar if it were an ideal gas can be obtained by calculating the entropy change of the real gas from 0 to p and subtracting from this number the entropy change of an ideal gas from 0 to 1 bar. This subtraction has to be done, because the entropy of a gas is infinite at zero pressure. In each case, integrate the Maxwell equation and calculate the entropy change $\Delta S_{1\to p}$ from 1 to p bar if the real gas follows

(a) the van der Waals equation of state $(p - a/V^2)(V - b) = RT$ where a and b are constants depending on the type of gas.

(b) the Berthelot equation of state, $pV = RT + \dfrac{9pRT_c}{128p_c}\left(1 - 6\dfrac{T_c^2}{T^2}\right)$ where T_c and p_c are constants. (This question is based on one in Barrow 1979, Chapter 8.)

Q3.108 Enthalpy is defined as $H = U + pV$. Total differentiation produces $dH = dU + pdV + Vdp$ and since $dU = TdS - pdV$ then $dH = TdS + Vdp$.

(a) Using $\left(\dfrac{\partial S}{\partial p}\right)_T = -\left(\dfrac{\partial V}{\partial T}\right)_p$ and the previous equation for dH, find an equation for $(\partial H/\partial p)_T$ that does not involve the entropy.

(b) Differentiate this equation with respect to T at constant pressure to obtain an equation for $(\partial C_p/\partial p)_T$ where C_p, the heat capacity at constant pressure, is $C_p = dH/dT \equiv (\partial H/\partial T)_p$. Formally integrate the result to obtain an equation showing how C_p *changes* with pressure from 0 to p_1. Calculate the change for a van der Waals and Berthelot gas. (See previous question.)

Q3.109 A gas changes in volume during heating; the heat absorbed q differs from the increase in internal energy U by the amount of work done by the gas, pdV, hence $q = U + pdV$. The heat capacity at constant pressure C_p is the rate of change of enthalpy with temperature and that at constant volume, C_V, is the rate of change of internal energy with temperature. The enthalpy is $H = U + pV$ and the internal energy is a function of p, V, and T, i.e. $U = f(p, V, T)$.

(a) Write down derivative expressions for C_v and C_p and show that

$$C_p = \left(\frac{\partial U}{\partial T}\right)_p + p\left(\frac{\partial V}{\partial T}\right)_p.$$

(b) Starting with the last equation, show that $C_p = C_V + \left(p + \left(\dfrac{\partial U}{\partial V}\right)_T\right)\left(\dfrac{\partial V}{\partial T}\right)_p$.

Q3.110 If entropy is a function of volume and pressure $S = f(T, V)$, write down an equation for dS and then by using one of Maxwell's equations $\left(\dfrac{\partial S}{\partial V}\right)_T = \left(\dfrac{\partial p}{\partial T}\right)_V$ show that $TdS = C_V dT + T\left(\dfrac{\partial p}{\partial T}\right)_V dV$.

Q3.111 **(a)** Calculate the change in internal energy of a van der Waals gas when the volume increases from V_1 to V_2 at a constant temperature. **(b)** Calculate this energy for 1 mole of CO_2 expanding from 10 to 20 dm^3.

Strategy: It is not obvious where to start but the internal energy U is given either by $H = U + pV$ or $A = U - TS$. Differentiation with respect to V at constant T is required, so the second equation describing the Helmholtz free energy A should be tried. Recall that A, U, and S are each functions of V and T. The equation $dA = -SdT - pdV$ is also needed. The van der Waals equation is $(p - n^2a/V^2)(V - nb) = nRT$, where n is the number of moles and a and b are constants, and for the calculation in (b) the value of each constant should be looked up in a textbook such as McQuarrie & Simon (1997).

Q3.112 An alternative starting point to find the internal energy of a non-ideal (monatomic) gas, is to combine the first and second laws as $dU = TdS - pdV$, and then to use $\Delta U = \displaystyle\int\left(\frac{\partial U}{\partial V}\right)_T dV$ to find the internal energy due to expansion of the gas at constant temperature.

(a) Find $(\partial U/\partial V)_T$ as in equation (3.50) and

(b) the *total* energy of the van der Waals gas, which is its internal plus kinetic energy.

(c) One mole of chlorine is compressed to 0.1 dm^3 and has a temperature of 300 K. Assuming it follows van der Waals behaviour, it is freely expanded, so that no work is done and there is no change in heat content until its pressure is 1 bar where it behaves as an ideal gas. Calculate the change in temperature. The van der Waals constant $n^2a = 0.658$ kJ dm^3 mol^{-2}.

Strategy: The internal energy of an ideal gas is zero but it still has kinetic energy. An ideal monatomic gas has energy $3nk_BT/2$, where k_B is Boltzmann's constant. A term $nk_BT/2$ is added for each degree of freedom in the molecule, one for its kinetic energy in three dimensions, or equivalently each squared term in the energy. Therefore a diatomic gas has an extra $nk_BT/2$ added because the bond stretching energy varies in proportion to the square of the extension. In (a) the Maxwell equation $\left(\frac{\partial S}{\partial V}\right)_T = \left(\frac{\partial p}{\partial T}\right)_V$ is required. See Chapter 4 if you are unfamiliar with integration. In (c) calculate the total energy at both pressures and equate them.

Q3.113 Calculate $\left(\frac{\partial k}{\partial T}\right)_p$ if the rate constant $k = Ae^{-\Delta G/(RT)}$ and if the pre-exponential 'constant' A depends on temperature. Use $\left(\frac{\partial H}{\partial T}\right)_p = T\left(\frac{\partial S}{\partial T}\right)_p$ to simplify the result.

Strategy: Rewrite using $\Delta G = \Delta H - T\Delta S$ then take the log of the rate constant and differentiate with T keeping p constant.

Q3.114 In solving the quantum mechanical problem of particle on a ring or the two-dimensional rigid rotor (see Chapter 10.5.9), and in the solution of differential equations such as two-dimensional diffusion or vibrations of a membrane, the conversion of derivatives from Cartesian to plane polar or cylindrical coordinates is necessary. The equations require the derivatives $\partial z/\partial x$ and $\partial^2 z/\partial x^2$ to be found in terms of r and θ using the conversion from rectilinear to cylindrical coordinates; $x = r\cos(\theta)$, $y = r\sin(\theta)$, $z = z$ and $r^2 = x^2 + y^2$.

(a) calculate these derivatives and

(b) show that

$$\frac{\partial^2 z}{\partial x^2} + \frac{\partial^2 z}{\partial y^2} = \frac{\partial^2 z}{\partial r^2} + \frac{1}{r}\frac{\partial z}{\partial r} + \frac{1}{r^2}\frac{\partial^2 z}{\partial \theta^2}.$$

The first derivative calculation is very easy, the second very tough!

Strategy: (a) Use the chain rule to find $\partial z/\partial x$ by expanding this as a total derivative in terms of r and θ. Then differentiate again. To find derivatives such as $\partial r/\partial x$ the equations must be rearranged so that there are none with mixed coordinates on the same side of the equation, thus to find $\partial r/\partial x$ start with $r = \sqrt{x^2 + y^2}$, differentiate and then simplify. In the second derivative calculation, you will need to recall that that r and θ are only functions of x and y.

3.13 Differentiation of vectors

3.13.1 Del, div, grad, Laplacian, and curl

There are four common vector operators involved in differentiation and the *del* operator. They are widely used in describing the physics of electrostatics, magnetism, and flowing liquids, and in the properties of fields in general. Only the briefest outline is given here; see a specialist text such as *Div, Grad, Curl and All That* (Shey 1993) for more details.

3.13.2 Del, ∇

The *vector operator* '*del*' is defined in the (i, j, k) basis as

$$\nabla = \frac{\partial}{\partial x}i + \frac{\partial}{\partial x}j + \frac{\partial}{\partial x}k$$

and the *gradient*, or rate of change of a (scalar) function $f(x, y, z)$ is the vector

$$\nabla f=\frac{\partial f}{\partial x}\boldsymbol{i}+\frac{\partial f}{\partial x}\boldsymbol{j}+\frac{\partial f}{\partial x}\boldsymbol{k}.$$

3.13.3 Grad, ∇f

The *gradient* is a vector that gives the maximum rate of change of a function f and its magnitude in a given direction. For example, if the function is $f=\sin(x)yz^2$ then the rate of change is the vector

$$\nabla f=\frac{\partial f}{\partial x}\boldsymbol{i}+\frac{\partial f}{\partial x}\boldsymbol{j}+\frac{\partial f}{\partial x}\boldsymbol{k}=\cos(x)yz^2\boldsymbol{i}+\sin(x)z^2\boldsymbol{j}+2\sin(x)yz\boldsymbol{k},$$

which can be resolved into components along the base vector's $\boldsymbol{i}, \boldsymbol{j}, \boldsymbol{k}$ directions.

3.13.4 Div, $\nabla\cdot\boldsymbol{v}$

If a vector $\boldsymbol{v}$ is represented in the $(\boldsymbol{i}, \boldsymbol{j}, \boldsymbol{k})$ basis as $\boldsymbol{v}=a\boldsymbol{i}+b\boldsymbol{j}+c\boldsymbol{k}$ then the dot product with the vector ∇ is called the *divergence* and is a scalar defined as

$$\begin{aligned}\nabla\cdot\boldsymbol{v} &= \left(\frac{\partial}{\partial x}\boldsymbol{i}+\frac{\partial}{\partial y}\boldsymbol{j}+\frac{\partial}{\partial z}\boldsymbol{k}\right)\cdot(a\boldsymbol{i}+b\boldsymbol{j}+c\boldsymbol{k})\\ &=\frac{\partial a}{\partial x}+\frac{\partial b}{\partial y}+\frac{\partial c}{\partial z}\end{aligned}$$

and the a, b, and c must depend on x, y, and z otherwise the differential would be zero. Without specifying at what point we want the gradient to be calculated, the calculation cannot be continued further as was the case ∇f calculated above, but suppose the point is $\{\pi/2, 2, 3\}$ then the gradient, or rate of change at this point, is the vector $\nabla f=0\boldsymbol{i}+9\boldsymbol{j}+12\boldsymbol{k}$; its magnitude is the absolute value of the vector which is $\sqrt{81+144}=15$. Furthermore suppose that we want the gradient from our point in the direction towards the origin $\{0, 0, 0\}$. The vector to the origin is $\boldsymbol{v}=-\pi/2\boldsymbol{i}-2\boldsymbol{j}-3z$ which has magnitude $\sqrt{(\pi/2)^2+4+9}$. To make it a unit vector $\boldsymbol{u}$, it is divided by its length

$$u=\frac{1}{\sqrt{\pi^2/4+13}}\left(-\frac{\pi}{2}\boldsymbol{i}-2\boldsymbol{j}-3\boldsymbol{k}\right).$$

The magnitude of the rate of change is therefore

$$\begin{aligned}\nabla f\cdot u &= \frac{1}{\sqrt{\pi^2/4+13}}(0\boldsymbol{i}+9\boldsymbol{j}+12\boldsymbol{k})\cdot\left(-\frac{\pi}{2}\boldsymbol{i}-2\boldsymbol{j}-3\boldsymbol{k}\right)\\ &=\frac{-18-26}{\sqrt{\pi^2/4+13}}=-11.18\end{aligned}$$

and occurs in the direction given by $\nabla f=0\boldsymbol{i}+9\boldsymbol{j}+12\boldsymbol{k}$, and the maximum rate of change is the size of this vector which is $\sqrt{81+144}=15$.

3.13.5 Laplacian, $\nabla\cdot\nabla f$

The dot product of the operator del with itself is a scalar function called the *Laplacian*, which is

$$\nabla^2 f=\nabla\cdot(\nabla f)=\frac{\partial^2 f}{\partial x^2}+\frac{\partial^2 f}{\partial x^2}+\frac{\partial^2 f}{\partial x^2}.$$

3.13.6 Curl, $\nabla\times\boldsymbol{v}$

The cross product of del with a vector $\boldsymbol{v}$ is a vector called the *curl*

$$\nabla\times\boldsymbol{v}=\begin{vmatrix}\boldsymbol{i} & \boldsymbol{j} & \boldsymbol{k}\\ \dfrac{\partial}{\partial x} & \dfrac{\partial}{\partial x} & \dfrac{\partial}{\partial x}\\ a & b & c\end{vmatrix}$$

This vector is best left as a determinant in which form it is easier to remember.

4 Integration

4.1 Basic concepts

Any book on chemical physics, mathematical biology, quantum mechanics, optics, astronomy, classical physics, and material science is peppered with examples of integration and differential equations. It is essential to be familiar with the essence of integration and then to know when a problem is difficult enough to turn to a book or to computer algebra for help. Even when using programs such as Maple or Mathematica, some knowledge of integration is always necessary.

Broadly speaking, integration is used to calculate the amount of some quantity. This might be the total concentration of a chemical product produced up to a certain time, the total amount of work done by a gas when expanding or that done in moving an object against gravity. Consider, for example, calculating the amount of oxygen in a tall column of still air. This is not simply the volume of the column times the gas density because gravity increases the O_2 concentration close to ground level. To calculate the total amount of O_2, the change in density has to be allowed for by integrating from the ground upwards, which means knowing how the density varies with altitude. Bearing in mind that integration will give us the total amount of O_2, the obvious way would be to take the column and divide it into many thin horizontal slabs, work out the amount of O_2 in each, and add them all up. The integral is the result that would be obtained when the slabs are made infinitesimally thin and, as with differentiation, the change is from a finite amount to an infinitesimal one, $\Delta x \to dx$. When this change is made the summation becomes an integral.

Now consider a general curve described by some function $f(x)$, and further suppose that it can be integrated because not all functions can be. Imagine splitting the area under a graph of $f(x)$ into n small rectangles over the range $x = a$ to $x = b$ some of which are shown in the sketch, Fig. 4.1, where there are only four rectangles. The purpose of doing this is to find the total area under the curve as accurately as possible. The total *area* of all the rectangles is

$$\begin{aligned} A_{approx} &= f(s_1)\Delta x_1 + f(s_2)\Delta x_2 + f(s_3)\Delta x_3 + f(s_4)\Delta x_4 \\ &= \sum_{i=1}^{n} f(s_i)\Delta x_i \end{aligned}$$

and the value of the function at position s_1 is $f(s_1)$ and so forth. You can see that the rectangles only very approximately follow the curve since only four are used and a poor approximation to the area under the curve is obtained.

To increase the accuracy of the calculation, it is obvious that the rectangles have to be made narrower and more of them have to be used. When Δx_i is made infinitesimally small, in the limit $\Delta x \to 0$ and $n \to \infty$, the summation A_{approx}, still over the range a to b, becomes the *definite* integral

$$A = \int_a^b f(x)dx$$

which is, without approximation, exactly the *area* under the curve of $f(x)$ from an x value of a to that at b. The integration range a to b is often called the *closed interval* a to b and the definite integral is sometimes called the Riemann integral. The area under the curve $f(x)$

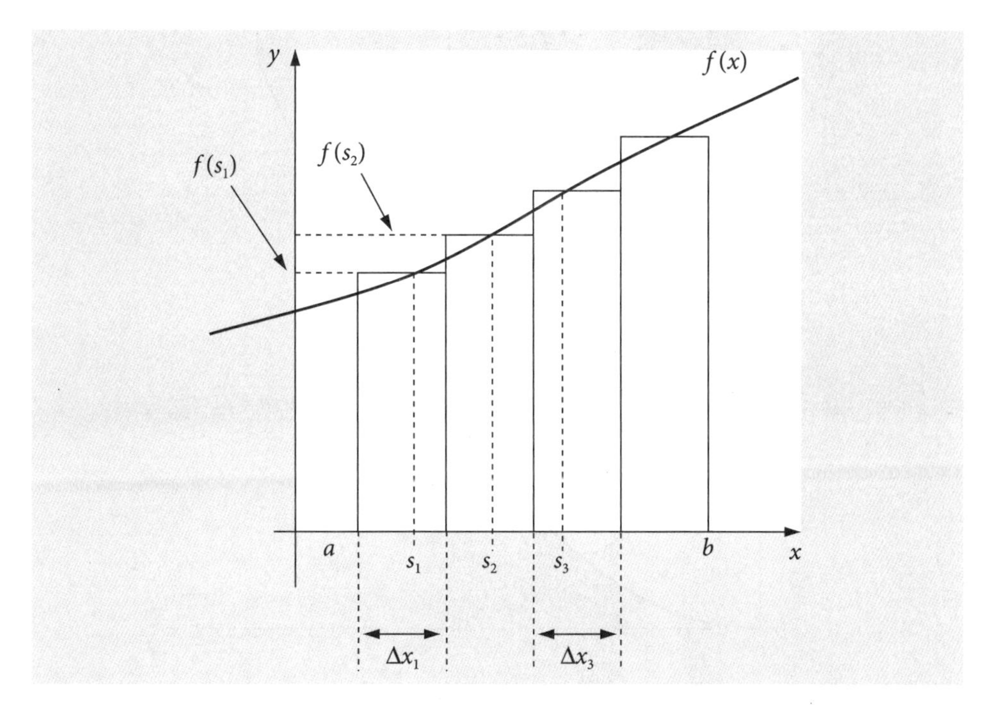

Fig. 4.1 Pictorial representation of the integral as area under a curve shown as a very approximate evaluation of an integral as the sum of a set of narrow rectangles. The width of each rectangle is made to approach zero, and their number increased to infinity, when the integration is performed.

is calculated even though the function depends only on x, and this is because $f(x)dx$ is dimensionally the height $f(x)$ times the width dx. In a general sense, integration increases the dimension of the calculation: a point to a line, a line to an area, an area to volume and so forth.

Often, the curve falls entirely below the x-axis and integration will produce a negative result. Mathematically, there is nothing wrong with a negative area but sometimes this may present a problem in imagining the integration as an area belonging to something physical; however, work and energy can be positive or negative, so the context is all-important in interpreting what the integration means. A result should be checked against intuition, by asking the question: 'am I expecting a positive or a negative result?' You might also want to ask 'is the result going to be large or small?'

4.1.1 A note on notation[1]

The integration operation be written either as $\int f(x)dx$ or $\int dxf(x)$. The latter is often used in physics texts particularly when $f(x)$ is a long and complex expression. In this case the integration is always assumed to extend over the function immediately following dx.

4.1.2 Indefinite and Definite Integrals

Integrating a function, $f(x)$, called the *integrand*, produces a new function $g(x)$, the *integral*, when the *integration operator* $\int \cdots dx$ is used; the notation is

$$\int f(x)dx = g(x) + \text{constant}. \qquad (4.1)$$

This result is an *indefinite integral* and a constant is always added to the answer because the starting and ending values of x were not defined. You can see from Fig. 4.2 that the result of integration is going to be different depending on where the integration starts and ends. The constant effectively allows this to be determined later. Recall that differentiating

[1] The symbol $\int$ was first introduced by Leibniz (1646–1716) is the stylized S from the Latin word *summa*, reminding us that integration and summation are intimately linked. Leibniz named the integral calculus, *calculus summatorius*. An integral is also sometimes called an anti-derivative.

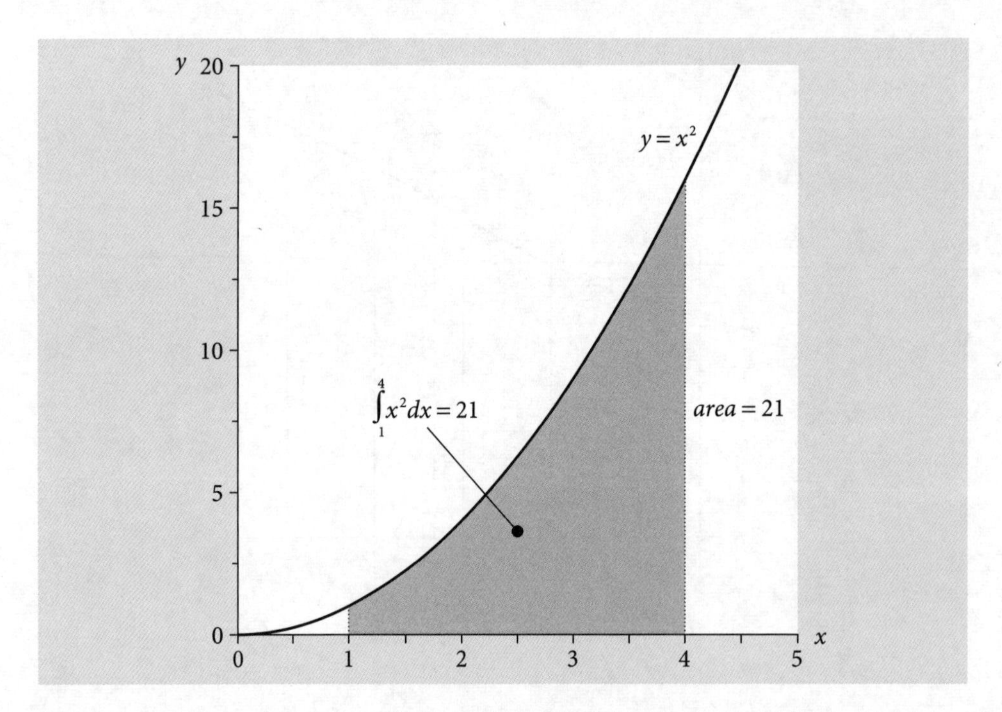

Fig. 4.2 The integral of x^2 from 1 to 4, changing the limits changes the value of the integral.

a constant produces zero, and therefore differentiating ($g(x)$ + constant) reproduces $f(x)$ no matter what value the constant has, i.e. the constant is arbitrary, therefore

$$\frac{d}{dx}\left(\int f(x)dx\right) = \frac{d}{dx}(g(x) + const) = f(x).$$

The integration symbol $\int$ of equation (4.1) has no upper and lower limits on it, and this indicates that it is an indefinite integral. When the limits are added to an integration a definite integral results, and this is a number independent of the variable x, provided that the chosen limits a and b are not functions that themselves contain x;

$$\int_a^b f(x)dx = g(x)\Big|_a^b = g(b) - g(a) \tag{4.2}$$

The symbol $|_a^b$ is the substitution symbol. Because neither $g(b)$ or $g(a)$ are functions of x, $g(b) - g(a)$ is a number and differentiating the result with respect to x would produce zero. The definite integral, equation (4.2), is sometimes called the *fundamental theorem of calculus.*

An example met early on in studying thermodynamics is to calculate the work w done on a gas, as its volume changes. Work is force × distance moved, pressure is force divided by area, so work is also the pressure × change in volume. To account for a series of infinitesimal volume changes, dV, made under reversible conditions, the quantity pdV must be integrated. The integral is

$$w_{rev} = -\int pdV,$$

and, by convention, the negative sign indicates that work is done on the gas. As it stands, this integral cannot yet be solved because we need to know how p and V are related. The next step is to use the ideal gas law, $pV = nRT$, to make an integral in V alone, giving

$$w_{rev} = -nRT\int \frac{1}{V}dV.$$

This equation tells us that the work done on the gas is linearly proportional to the number of moles n present and to the gas constant times the temperature. It also tells us that the work is the area under the curve of $1/V$ as V is changed. Assuming that we already know how to do the integration, which is explained below, the result is

$$w_{rev} = -nRT\ln(V) + const.$$

The constant appears because no upper or lower limit on the volume was defined and so the value of the constant is unknown. Mathematically this equation is fine as it stands, but to add real numbers the log has to be dimensionless, however, volume has dimensions of m^3 therefore V in the equation is made dimensionless by making it volume per m^3. Alternatively, because the constant is arbitrary until some values are added, the equation can be rewritten as $w_{rev} = -nRT\ln(V/V_c)$ where the constant is now changed to V_c and incorporated into the log. V_c is still undefined, but has units of volume, and also reminds us that the log is dimensionless.

4.1.3 Changing limits, odd and even functions

A definite integral is the difference in value of the integral calculated at the two limits, equation (4.2), and changing their the order clearly produces −1 times the result,

$$\int_a^b f(x)dx = -\int_b^a f(x)dx.$$

Furthermore, if there is a point s between limits a and b then the integral can be split into two continuous parts. Using common sense, it is not surprising that the total is the sum of the parts by thinking of the result of integration as an area. Suppose the limits range from a to s and then s to b, then

$$\int_a^b f(x)dx = \int_a^s f(x)dx + \int_s^b f(x)dx \qquad (4.3)$$

which is shown in Fig. 4.3.

When the limits are $\pm a$ and 0 the relationship

$$\int_{-a}^0 f(x)dx = \int_0^a f(-x)dx \qquad (4.4)$$

is sometimes useful. Does this make sense? Try it on equation (4.2) to convince yourself.

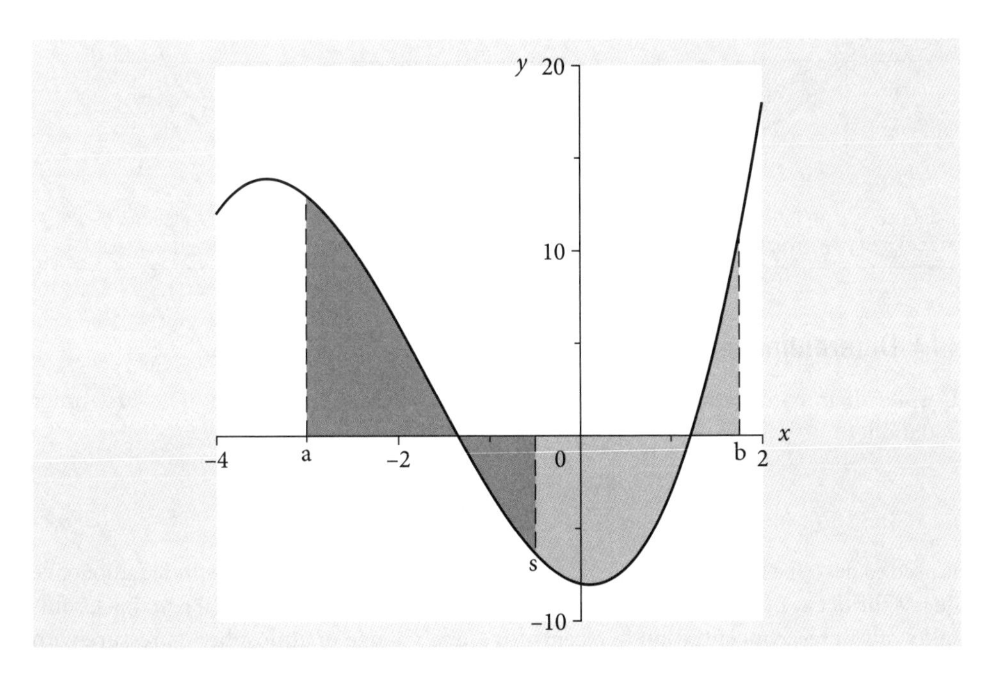

Fig. 4.3 Integration limits a, s and b

An integral with *symmetrical* limits $-a$ to a can be written in different forms

$$I = \int_{-a}^{a} f(x)dx$$

$$= \int_{-a}^{0} f(x)dx + \int_{0}^{a} f(x)dx \tag{4.5}$$

$$= \int_{0}^{a} f(-x)dx + \int_{0}^{a} f(x)dx$$

and this is also very useful in determining whether the integral is zero or not. An *odd* function has the property $f(-x) = -f(x)$, and when integrated about the symmetrical limits $-a$ and $+a$ the integral will be zero. An *even* function has the property $f(-x) = f(x)$; the integral from $-a$ and $+a$ may perhaps be small but is not *exactly* 0. Determining whether integrals are 0 or not is important in the study of quantum mechanics and spectroscopy. However, group theory has to be used to examine more complex functions than is apparent from simple 'odd–even' behaviour.

Summary

If $I = \int_{-a}^{a} f(x)dx$, and $f(x)$ is odd, $f(-x) = -f(x)$, then $I = 0$,

however, if $f(x)$ is even, $f(-x) = f(x)$, then $I \neq 0$.

The left graph Fig. 4.4 shows an odd function, $x^3 - x$. There are many others; for example, $\sin(x)$ over the range $-\pi$ to π, where the area from zero to $-x$ is equal but opposite of that from zero to $+x$ making 0 in total. The integral $\int_{-a}^{a}(x^3 - x)dx = \left.\frac{x^4}{4} - \frac{x^2}{2}\right|_{-a}^{a} = 0$ and this is because $a^4 = (-a)^4$ and $a^2 = (-a)^2$. The right-hand graph shows an even function where the area is not zero but negative.

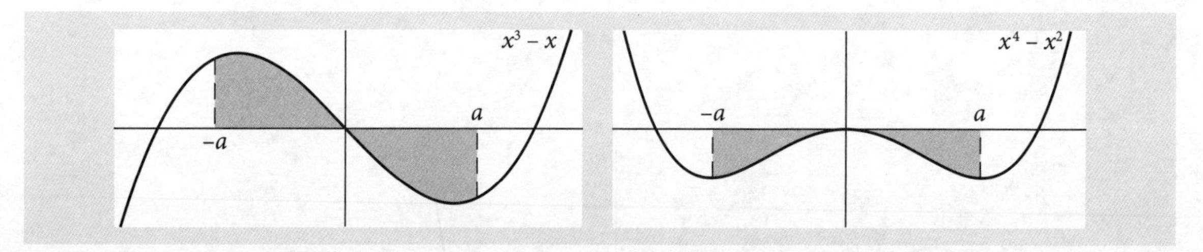

Fig. 4.4 Left: The odd function $x^3 - x$, and the range of the integral (shaded) when $a = 3/4$. Right: The even function $x^4 - x^2$ where the integral is not zero, but negative, over the same range. Both graphs cross or touch the x-axis at −1, 0, and +1.

4.1.4 Differential equations

Despite their name, differential equations are solved by integration. The first-order equation

$$\frac{dc}{dt} = \pm kc$$

is used to describe many phenomena. Depending on the sign, ±, these include radioactive decay, the decay of excited states, growth of bacteria, first-order chemical reactions, solubility of solutes, concentration in a centrifuge, and change of atmospheric pressure with height. If c is a concentration, the equation tells us that the rate of change of the concentration of species c is proportional to the amount of c present at any time, i.e. the amount unreacted, the constant of proportionality being the rate constant k. If the sign is negative, this indicates that c is being lost and is forming another species that is not specified. This differential equation is solved by separating out the two variables c and t and integrating each side separately:

$$\int \frac{1}{c} dc = -k \int dt.$$

This example shows that the integration takes us from describing the rate of a process to the actual quantity; in this case, from the rate of change of c with time, to the amount of c at any time: $\ln(c) = -kt + const$. To make sure that the log is dimensionless, the equation can again be rewritten in an equivalent form as $\ln(c/c_0) = -kt$, where c_0 is a constant equal to the amount of c present at $t = 0$. Solving differential equations is described in detail in Chapter 10.

To show how integration can solve many chemical and physical problems is one aim of this chapter, but first, the mechanics of performing integration must be understood.

4.2 Mechanics of integration

Integration is often a trial and error process, and some experimenting with different options is necessary even when using the computer. Start by simplifying expressions, perhaps by using partial fractions, then look for standard formulas and standard methods such as integration by parts. If none are found suitable, substitutions can be tried, then standard forms and methods looked for again because only with these can an integration be found. If all else fails, numerical methods have to be used; see Chapter 11.

The integration formulae for various functions can be derived in a similar way to that for differentiation by taking small values of dx and dy and then calculating $x + dx$ and $y + dy$. The method is now well established and most results are given below without derivation. Understanding the physical world is hard enough work without having to prove all of mathematics before we can use it!

Several integrations are listed in Section 4.2.13, and Maple is used to do even more complex integrations. Why then, you might ask, should you bother to learn how to do integration? Calculations are done 'by hand' not only because it is often easier and quicker to do so, but also because it encourages a greater understanding of how the result was obtained.

4.2.1 Powers of x, but not x^{-1}

Integrating x^2 produces

$$\int x^2 dx = \frac{x^3}{3} + \text{constant}$$

as can be seen by differentiation of the result. Similarly,

$$\int (x^3 + x) dx = \frac{x^4}{4} + \frac{x^2}{2} + \text{constant}.$$

The general rule for any power of x, other than x^{-1}, should be remembered and is

$$\int x^n dx = \frac{x^{n+1}}{n+1} + \text{constant}. \tag{4.6}$$

Definite integrals have limits that are determined by the problem being solved, for example,

$$\int_1^4 x^2 dx = \left.\frac{x^3}{3}\right|_1^4 = \frac{4^3}{3} - \frac{1^3}{3} = 21,$$

and the definite integral is always evaluated by substituting the two limits into the algebraic result of the integration and therefore there is no constant term, as shown in Fig. 4.2. Notice the vertical line after the calculation and the limits top and bottom to the right of this. The top limit always comes first in the evaluation and the value of the integral with the second limit is always subtracted from the first. In integrations representing physical phenomena, such as the expansion or compression of a gas, the lower limit is always the initial or starting value and the upper limit the final value no matter which is larger.

4.2.2 $\int d\ln(x)$ and similar expressions

In chemistry and physics textbooks it is not uncommon to meet expressions such as $\int dx$, $\int d\ln(x)$, or $\int d\sin(x)$. These cases are the simplest to evaluate because the answer is given in the integral,

$$\int 1dx \equiv \int dx = x + c, \qquad \int 2dx = 2\int dx = 2x + c$$

$$\int d\ln(x) = \ln(x) + c \qquad \int d\sin(x) = \sin(x) + c,$$

and so on; c is an arbitrary constant.

4.2.3 Integrating $1/x$ produces a logarithm

The $x^{n+1}/n+1$ rule is not going to work when $n = -1$ as this will produce 1/0 which is infinity. In the case of reciprocal x the integral is:

$$\int \frac{dx}{x} = \ln(x) + c \qquad \int_a^b \frac{dx}{x} = \ln(b) - \ln(a) = \ln\left(\frac{b}{a}\right) \tag{4.7}$$

where the last result is only possible if a and $b > 0$. If $a = 1$, for example, $\int_1^x \frac{dx}{x} = \ln(x)$ because $\ln(1) = 0$.

Note that to produce a log, the denominator must be linear in x, say $2 + 3x$, but cannot contain anything like x^2 or x^3. The general formula is

$$\int \frac{1}{ax+b}dx = \frac{1}{a}\ln(ax+b). \tag{4.8}$$

4.2.4 Partial fractions

Integrations involving reciprocal powers of x can often be solved by expanding as partial fractions and then integrating each term as a log. The method of obtaining a partial fraction is systematic, as illustrated in the following example. Suppose the integration required is

$$\int \frac{x-3}{x^2-2x-15}dx$$

and this has to be simplified; the hardest part is factoring the denominator. Some experimenting or using `factor(x^2-2*x-15)` with Maple gives

$$\frac{x-3}{x^2-2x-15} = \frac{x-3}{(x+3)\,(x-5)}.$$

The next step is to write this in terms of two fractions with constant A and B that must be determined.

$$\frac{x-3}{(x+3)(x-5)} = \frac{A}{x+3} + \frac{B}{x-5} = \frac{A(x-5)+B(x+3)}{(x+3)(x-5)}.$$

By equating the left- and right-hand expressions,

$$x-3 = A(x-5) + B(x+3),$$

which is true for all values of x. Therefore, if $x = -3$ then $A = 6/8 = 3/4$ and if $x = 5$ then $B = 1/4$. The integration becomes

$$\int \frac{x-3}{x^2-2x-15}dx = \int \frac{3}{4(x+3)} + \frac{1}{4(x-5)}dx = \frac{3}{4}\ln(x+3) + \frac{1}{4}\ln(x-5) + c$$

The same result is produced with Maple

```
> Int((x-3)/(x^2-2*x-15),x): %= value(%);
```

$$\int \frac{x-3}{x^2-2x-15}\mathrm{d}x = \frac{1}{4}\ln(x-5) + \frac{3}{4}\ln(x+3)$$

Notice the Maple syntax. It has the standard form of the name of the operation, `Int` in this case, surrounded by brackets with the function inside the brackets separated by a comma from the integration variable e.g.;

```
> Int( any function in x , x );
```

4.2.5 Exponentials

When differentiating an exponential, the exponential is returned but multiplied by the derivative of the function within the exponential. For example,

$$\frac{d}{dx}e^{-ax+b} = -ae^{-ax+b}.$$

Integrating both sides gives

$$\int \frac{d}{dx}e^{-ax+b} = \int -ae^{-ax+b},$$

therefore the general form, after rearranging and with c as the constant, is

$$\int e^{-ax+b}dx = -\frac{1}{a}e^{-ax+b} + c. \tag{4.9}$$

Note that this formula only applies when the argument to the exponential is to the first power of x and some constants. Integrating e^{-x^2}, for example, cannot be done this way.

A few examples are

$$\int e^{-x}dx = -e^{-x} + c,$$

$$\int e^{3x-6}dx = e^{-6}\int e^{3x}dx = \frac{1}{3}e^{3x-6} + c,$$

$$\int_{-1/3}^{1/3} e^{3x-6}dx = \frac{1}{3}e^{3x-6}\Bigg|_{-1/3}^{1/3} = \frac{1}{3}(e^{-5} - e^{-7}).$$

This last function and its integral from −1/3 to 1/3 is shown in Fig. 4.5.

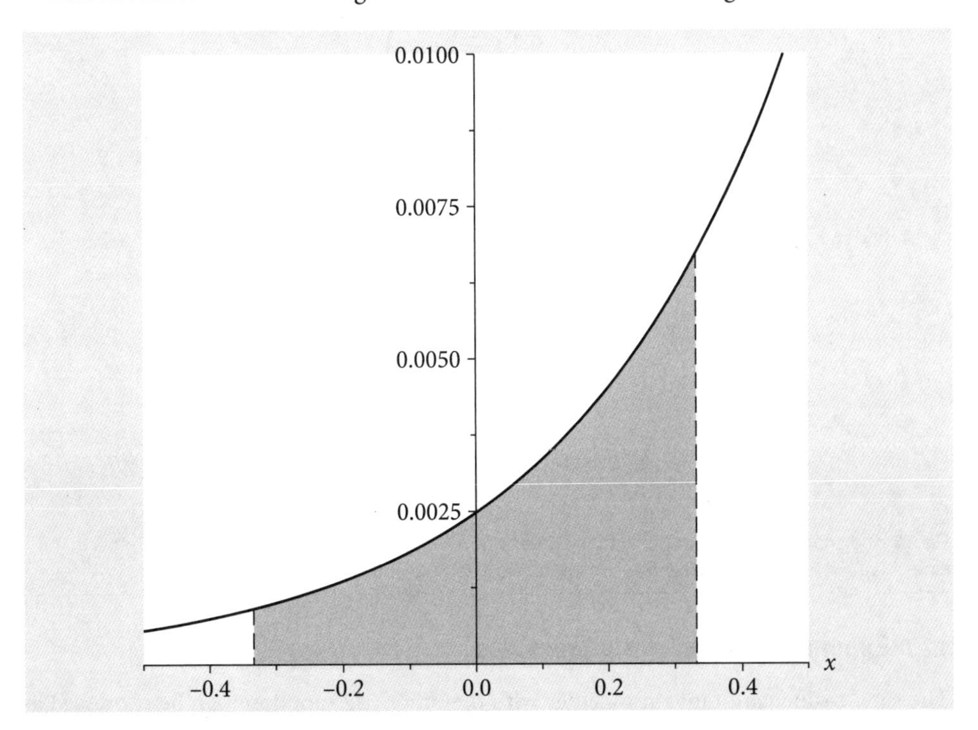

Fig. 4.5 The exponential e^{3x-6} and its integral from −1/3 to 1/3.

4.2.6 Sine, cosine, and tangent

Differentiating the sine or cosine produces one from another and so it should not come as a surprise that the same happens upon integration. For example, if

$$\frac{d}{dx}\cos(3x+2) = -3\sin(3x+2)$$

integrating both sides and rearranging produces,

$$\int \sin(3x+2)dx = -\frac{1}{3}\cos(3x+2) + c.$$

Similarly, the general form is

$$\int \cos(ax+b)dx = \frac{1}{a}\sin(ax+b) + c, \tag{4.10}$$

and

$$\int \sin(ax+b)dx = -\frac{1}{a}\cos(ax+b) + c. \tag{4.11}$$

Note that these integrations are true only for first powers of x in the sine and cosine.

The sine and cosine functions illustrate clearly that integration is the area under the curve. In Fig. 4.6 the cosine and sine curves are plotted. The area under the cosine curve when integrated from 0 to x is the cosine integral, which is the *value* of the sine curve at x, i.e. $\sin(x)$. Putting limits into the integration gives

$$\int_a^x \cos(x)dx = \sin(x)\Big|_a^x = \sin(x) - \sin(a)$$

and when $a = 0$ the integral is $\sin(x)$. When $x = 2\pi$ there is as much of a positive as negative area shaded making the total area zero as is $\sin(2\pi)$. The area under the cosine from 0 to π is also similarly zero and so is $\sin(\pi)$.

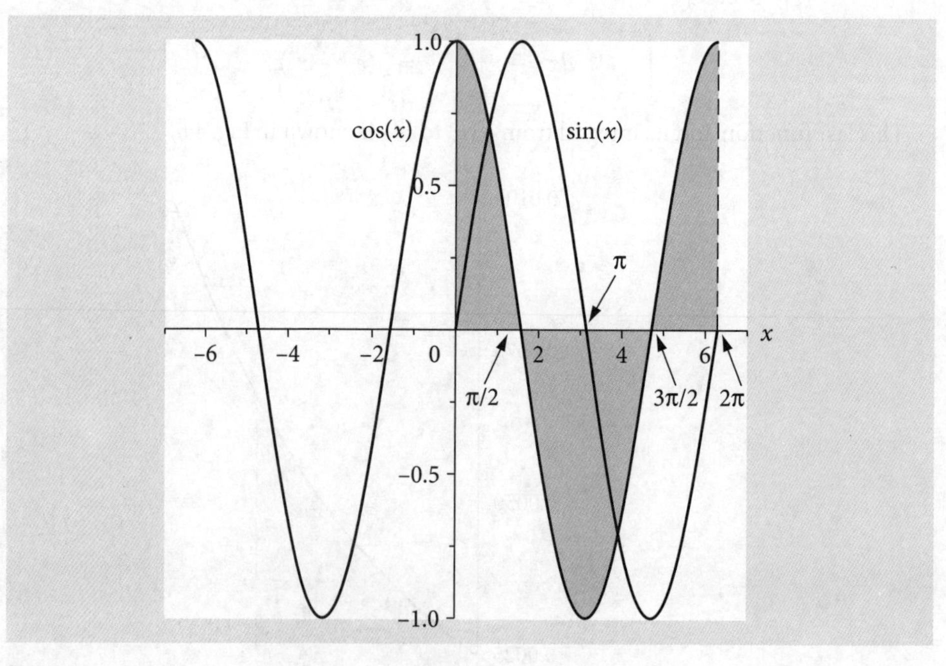

Fig. 4.6 The shaded area under the cosine curve is zero from 0 to 2π; the integral from zero to x is the sine curve, which is also zero at 2π. The integral is also zero at 0 and π.

4.2.7 Exponential form of trig functions

The sine, cosine, and tangent together with the other trig functions can be expressed as complex exponentials and in this form are easy to integrate as are the hyperbolic functions sinh, cosh, and tanh; see Chapter 1.

The square of sines and cosines are also easily integrated using exponentials. For example, $\int \sin^2(x)dx$ is converted into an exponential form using $\sin(x) = \dfrac{e^{ix} - e^{-ix}}{2i}$ where $i = \sqrt{-1}$ or $i^2 = -1$. The integral is

$$\int \sin^2(x)dx = \int \left(\frac{e^{ix} - e^{-ix}}{2i}\right)^2 dx$$

$$= -\frac{1}{4}\int e^{2ix} - 2 + e^{-2ix}dx = -\frac{1}{4}\left(\frac{e^{2ix}}{2i} - 2x - \frac{e^{-2ix}}{2i}\right)$$

$$= \frac{x}{2} - \frac{1}{4}\sin(2x)$$

where the exponentials are converted back to a sine in the last step. Similarly integrations of the form $\int e^x \sin(x)dx$, $\int e^x \cosh(x)d\theta$, and so forth can be evaluated by converting to their exponential form first.

4.2.8 Non-integrability

There are several functions that look quite simple but for which there is no integral. These integrals usually have to be evaluated numerically even though they may be defined with a special function. These integrals include

$$\int \frac{\sin(x)}{x}dx, \qquad \int \frac{\cos(x)}{x}dx, \qquad \int \sin(x^2)dx$$

$$\int \frac{e^x}{x}dx, \qquad \int \frac{1}{\ln(x)}dx, \qquad \int e^{-ax^2}dx.$$

When the limits are 0 and or $\pm\infty$ a result does exist for some of these non-integrable integrals, but it is always a number. The last integral is probably the one you will meet most often; it is the 'bell' shaped curve and produces the error function when integrated,

$$\int e^{-ax^2}dx = \sqrt{\frac{\pi}{4a}}\, erf(x\sqrt{a}).$$

The error function can only be calculated numerically. It is 0 when x is 0 and when x is large it rapidly approaches 1. Therefore,

$$\int_0^\infty e^{-ax^2}dx = \sqrt{\pi/4a}.$$

Other examples are $\displaystyle\int_0^\infty \frac{\sin(x)}{x}dx = \pi/2$ and $\displaystyle\int_0^\infty \sin(x^2)dx = \sqrt{2\pi}/4$, the other integrals listed are either undefined or infinity.

4.2.9 Improper and undefined integrals

A number of functions, when integrated, produce infinity. These are conventionally called *improper* integrals, although there is nothing 'improper' about them, e.g. $\displaystyle\int_0^\infty \frac{e^x}{x}dx = \infty$. The integrand $f(x)$ usually has a point in the range of integration at which it becomes infinite. This can often be spotted, and surprisingly, only sometimes does this produce an infinite result. As an example, in the integral $\displaystyle\int_0^1 \frac{1}{\sqrt{x}}dx$, the reciprocal of the square root becomes infinite at $x = 0$. Ignoring this for the moment, integrating using equation (4.6) then working out the limits produces a finite answer,

$$\int_0^1 \frac{1}{\sqrt{x}}dx = 2\sqrt{x}\,\Big|_0^1 = 2.$$

Conversely the integrals of the form $\int_0^1 \frac{1}{x^2}dx$ or $\int_0^2 \frac{1}{x-1}dx$ and many similar reciprocal functions produce infinity and so are undefined or improper. Not surprisingly, if the range of the integration is changed so that the integrand is not infinite in the range, the integrals behave normally and give a finite result.

4.2.10 Integrals with infinite limits

Even though the limit of an integral may extend to infinity, some integrals are nonetheless finite. One example of this was given in the previous section. Naturally, some may also become infinite: $\int_0^\infty xdx$ is clearly going to be infinite as its result x^2 will become infinite when x is infinity. It is probably true that most polynomial functions in which its largest power is positive will reach $\pm\infty$ with limits 0 to infinity. If a function is odd, such as xe^{-x^2} and the limits extend equally far to the left and right even to infinity, then the integral is always zero, as discussed in Section 4.1.3.

With reciprocal functions, or polynomials with negative powers, as the limit extends to infinity the function becomes smaller and smaller and so may converge to a finite value. For instance, integrating $1/x^2$ from 1 to infinity produces a finite result:

$$\int_1^\infty \frac{1}{x^2}dx = -\frac{1}{x}\Big|_1^\infty = 1$$

Integrating $1/x$ over the same range produces infinity and you can appreciate this if you look at a graph of the log function because this increases for all $x > 1$.

$$\int_1^\infty \frac{1}{x}dx = \ln(x)\Big|_1^\infty = \infty.$$

4.2.11 Discontinuities

Very occasionally a function has a discontinuity such as a vertical step. If this is at point b and the integration is from a to c with b in between, the integral is split into two parts, one part from a to b the other from b to c; see Section 4.1.3.

4.2.12 Mean value theorem for integrals

The *first mean value theorem* states that for a point s in the range of an integration, $a \le s \le b$ then

$$\int_a^b f(x)dx = (b-a)f(s).$$

where $f(s)$ is the average value of f. This is a very convenient way of calculating averages and is described further in Section 4.8. This theorem also proves to be useful when performing numerical integrations with the Monte Carlo method, Chapter 12, where the equation is worked backwards, $f(s)$ is calculated directly to produce the integral.

4.2.13 Common integrals calculated using Maple

Integration constants are not added.

$$\int x^n\, dx = \frac{x^{n+1}}{n+1} \qquad \int e^{ax}\, dx = \frac{e^{ax}}{a}$$

$$\int \frac{1}{ax}dx = \frac{\ln(x)}{a} \qquad \int \frac{1}{ax+b}dx = \frac{\ln(ax+b)}{a}$$

$$\int \sin(ax+b)\, dx = -\frac{\cos(ax+b)}{a} \qquad \int \cos(ax+b)\, dx = \frac{\sin(ax+b)}{a}$$

$$\int \tan(ax)\, dx = -\frac{\ln(\cos(ax))}{a}$$

$$\int \ln(ax)\,\mathrm{d}x = x(\ln(a) + \ln(x) - 1)$$

$$\int \ln(ax+b)\,\mathrm{d}x = \ln(ax+b)\,x - x + \frac{\ln(ax+b)\,b - b}{a}$$

$$\int \frac{1}{(ax+b)^2}\,\mathrm{d}x = -\frac{1}{(ax+b)\,a}$$

$$\int \frac{1}{x^2a^2+b^2}\,\mathrm{d}x = \frac{\arctan\left(\frac{xa}{b}\right)}{ba}$$

$$\int \sqrt{ax+b}\,\mathrm{d}x = \frac{2}{3}\frac{(ax+b)^{3/2}}{a}$$

$$\int \frac{1}{\sqrt{ax+b}}\,\mathrm{d}x = \frac{2\sqrt{ax+b}}{a}$$

$$\int x\,\mathrm{e}^{-ax}\,\mathrm{d}x = \left(-\frac{1}{a^2} - \frac{x}{a}\right)\mathrm{e}^{-ax}$$

$$\int x^2\,\mathrm{e}^{-ax}\,\mathrm{d}x = \left(-\frac{2}{a^3} - \frac{2\,x}{a^2} - \frac{x^2}{a}\right)\mathrm{e}^{-ax}$$

$$\int x^3\,\mathrm{e}^{-ax}\,\mathrm{d}x = \left(-\frac{6}{a^4} - \frac{6\,x}{a^3} - \frac{3\,x^2}{a^2} - \frac{x^3}{a}\right)\mathrm{e}^{-ax}$$

$$\int x^4\,\mathrm{e}^{-ax}\,\mathrm{d}x = \left(-\frac{24}{a^5} - \frac{24\,x}{a^4} - \frac{12\,x^2}{a^3} - \frac{4\,x^3}{a^2} - \frac{x^4}{a}\right)\mathrm{e}^{-ax}$$

$$\int \mathrm{e}^{-ax^2}\,\mathrm{d}x = \frac{1}{2}\frac{\sqrt{\pi}\,\mathrm{erf}(\sqrt{a}\,x)}{\sqrt{a}}$$

$$\int x^2\,\mathrm{e}^{-ax^2}\,\mathrm{d}x = -\frac{1}{2}\frac{x\,\mathrm{e}^{-ax^2}}{a} + \frac{1}{4}\frac{\sqrt{\pi}\,\mathrm{erf}(\sqrt{a}\,x)}{a^{3/2}}$$

$$\int x^4\,\mathrm{e}^{-ax^2}\,\mathrm{d}x = \left(-\frac{1}{2}\frac{x^3}{a} - \frac{3}{4}\frac{x}{a^2}\right)\mathrm{e}^{-a\,x^2} + \frac{3}{8}\frac{\sqrt{\pi}\,\mathrm{erf}(\sqrt{a}\,x)}{a^{5/2}}$$

$$\int x\,\mathrm{e}^{-ax^2}\,\mathrm{d}x = -\frac{1}{2}\frac{\mathrm{e}^{-ax^2}}{a}$$

$$\int x^3\,\mathrm{e}^{-ax^2}\,\mathrm{d}x = \left(-\frac{1}{2\,a^2} - \frac{1}{2}\frac{x^2}{a}\right)\mathrm{e}^{-ax^2}$$

$$\int \sinh(ax+b)\,\mathrm{d}x = \frac{\cosh(ax+b)}{a}$$

$$\int \cosh(ax+b)\,\mathrm{d}x = \frac{\sinh(ax+b)}{a}$$

$$\int \tanh(ax+b)\,\mathrm{d}x = \frac{-\frac{1}{2}\ln(\tanh(ax+b)-1) - \frac{1}{2}\ln(\tanh\,(ax+b)+1)}{a}$$

$$\int \sin(ax+b)^2\,\mathrm{d}x = \frac{1}{2}x + \frac{-\frac{1}{2}\cos(ax+b)\sin(ax+b) + \frac{1}{2}b}{a}$$

$$\int \cos(ax+b)^2\,\mathrm{d}x = \frac{1}{2}x + \frac{\frac{1}{2}\cos(ax+b)\sin(ax+b) + \frac{1}{2}b}{a}$$

$$\int \cos(ax+b)\sin(ax+b)\,\mathrm{d}x = \frac{\frac{1}{2} - \frac{1}{2}\cos(ax+b)^2}{a}$$

To make your own table using Maple, put the function into a list and use a loop to print the results. For example;

```
> ss:=[x^n, exp(a*x), 1/(a*x)]:          # make list
  for i from 1 to 3 do
     Int(ss[i],x):                        # integrate one by one
     print(%= simplify(value(%)));
  end do;
```

4.2.14 Questions

Full solutions are available at **www.oxfordtextbooks.co.uk/orc/beddard.**

Q 4.1 The vibrational motion of HCl is characterized by its force constant, $k = 518$ N m^{-1}, and equilibrium internuclear separation, $r_e = 0.127$ nm.

(a) What is the energy involved in extending the bond to 0.146 nm? This is an extension to $\approx 1.146r_e$ nm and is the maximum (classical) extension of the bond in the $n = 1$ vibrational level from its equilibrium position.

(b) What fraction is this of the dissociation energy, which is 440.2 kJ mol^{-1}? Assume the bond extends as a harmonic oscillator and use Hooke's law; force $\propto$ extension or $f = k(r - r_e)$, where k is the bond force constant. The bond extension is $r - r_e$.

(c) Show that the energy is the same as that given by $E_n = h\nu(n + 1/2)$.

Strategy: Energy is the integral of a force f over distance. At the classical turning point the molecule is instantaneously stationary, therefore the kinetic energy is zero and the total energy E is the same as the potential energy. The vibrational frequency is $\nu = \dfrac{1}{2\pi}\sqrt{\dfrac{k}{\mu}}$ s^{-1} if μ is the reduced mass.

Q4.2 **(a)** Starting with the Arrhenius equation, show that $\dfrac{d\ln(k)}{dT} = \dfrac{E_a}{RT^2}$ and

(b) if k_1 and k_2 are rate coefficients at temperatures T_1 and T_2 respectively show that $\ln\left(\dfrac{k_2}{k_1}\right) = \dfrac{E_a}{R}\left(\dfrac{T_2 - T_1}{T_2T_1}\right)$.

Strategy: You need to know that the Arrhenius equation is usually written as $k = k_0e^{-E_a/RT}$. It is in this form in all the textbooks. Note that k, the rate constant, more properly the rate coefficient, is a function of temperature and can therefore form the derivative dk/dT and that k_0 is a constant and differentiates to zero.

Q4.3 Integrate the first-order rate expression, $\dfrac{dc}{dt} = -kc$ where c is concentration, k the rate constant, and t time. Suppose that at $t = 0$ the concentration is c_0.

Q4.4 The movement of a molecule of mass m in a viscous medium, such as its solvent, is subject to Newton's equation: force equals mass × acceleration, or

$$m\frac{dv}{dt} = F_{frict}$$

where its velocity is v and F_{frict} is the frictional force on the molecule from the solvent and is given by Stokes' law $F_{frict} = -3\pi\delta\eta v$. The friction is negative because it opposes motion. The molecule's diameter is δ, the solvent viscosity η, which has units of kg m^{-1} s^{-1}. More commonly viscosity is measured in centipoise cP, where 1 cP = 10^{-3} kg m^{-1} s^{-1}.

(a) Find the velocity at time t if $v = v_0$ at $t = 0$ and show that the constants are equivalent to a time $\tau = \dfrac{m}{3\pi\delta\eta}$. This time is the time for the molecule to 'lose its memory' of its previous velocity and position and is therefore a measure of how rapidly its kinetic energy is dissipated into the solvent. See Finkelstein & Ptitsyn (2002) page 100 for an interesting discussion relating to proteins.

(b) Estimate this time for a molecule such as benzene and a protein with $\delta = 3$ nm both in water. The density of a typical protein is 1 g cm^{-3}.

Q4.5 Find **(a)** $I=\displaystyle\int\frac{dx}{3x-2}$ **(b)** $I=\displaystyle\int\cosh^2(x)dx$ **(c)** $I=\displaystyle\int\cosh(x)\sinh(x)dx$

(d) $I=\ln\left(\dfrac{2x^2}{1+x}\right)$ **(e)** $I=\displaystyle\int_{-\infty}^{\infty}x^9e^{-ax^2}\,dx$ **(f)** $I=\displaystyle\int_{-1}^{1}\sqrt{x-3}\,dx.$

Q4.6 Show that the integral $\int_0^L\sin^2(xL)dx$ at large L approaches $L/2$.

Strategy: This is a standard integral, and can be looked up in the table or converted into its exponential form and then integrated. The result should be inspected at large L.

Q4.7 The acceleration f of a particle is given by the vector equation,

$$f=2\sin(\omega_0 t)\boldsymbol{i}+\cos(\omega_0 t)\boldsymbol{j}+t\boldsymbol{k}.$$

Calculate its velocity and displacement (position) at time t if the particle is at rest at the origin at time zero. If you are not familiar with the vector notation you may want to look this up in Chapter 6.3.

Strategy: If $\boldsymbol{r}$ is position, velocity is the vector $d\boldsymbol{r}/dt$ and acceleration $f=d^2\boldsymbol{r}/dt^2$. The acceleration has to be integrated once to obtain the velocity and again to find the position vector. The initial conditions are that the velocity and acceleration are both zero at time $t=0$ and the particle is at the origin or $\boldsymbol{r}=0\boldsymbol{i}+0\boldsymbol{j}+0z$. Each integral is really three integrals, one for each of the $\boldsymbol{i}$, $\boldsymbol{j}$ and $\boldsymbol{k}$ base vectors. Treat each integral separately and add the result.

Q4.8 In an ideal gas $pV=nRT$. Show that for an isothermal change

$$\int_{V_1}^{kV_1}pdV=nRT\ln(k).$$

Strategy: As the change is isothermal, T is constant. Substitute for p and integrate.

Q4.9 5.0 m^3 of air at atmospheric pressure (101 325 Pa) is compressed adiabatically in the cylinder of an old steam engine until its volume is 10% of that originally present. In an adiabatic process $pV^\gamma=k$, where γ is the ratio of specific heat capacities C_p to C_V and is measured to be 1.404 for air, and k is a constant. In an adiabatic change no heat enters or leaves during the process, consequently the temperature must change. Also, by the first law, the work done must be equal to the change in internal energy U of the gas.

(a) Calculate the work done on the gas $w=-\int pdv$.

(b) Calculate the instantaneous temperature rise if the engine is initially at 300 K, assuming, somewhat unrealistically, that the cylinder itself does not absorb any heat. Assume also that the air is an ideal diatomic gas with heat capacity per mole $C_V=5R/2$.

You will need to use $dU=C_VdT$ where U is the internal energy and know or look up the first law of thermodynamics.
(Strictly speaking, C_V depends on temperature but you may assume that this change is small and make C_V constant. The increase in C_V of air has been measured to be only ≈ 5 J K^{-1} mole^{-1} on changing the temperature from 400 to 1000 K.)

Strategy: You are given the equation with which to calculate the work done. Note that the engine compresses the gas, so the final volume is less than that of the initial, hence the upper limit of the integration must be smaller than the lower. The lower limit is always taken to be the initial value, V. Because the compression is adiabatic, you must use the equation given, $pV^\gamma=k$, rather than the ideal gas law. It is easier to do the calculation algebraically first and evaluate k afterwards using the values in the question.

In the second part, calculate the work as the increase in internal energy from an initial to final temperature and knowing the work from part (a), find the final temperature.

Q4.10 **(a)** Calculate the work done in increasing the volume of a van der Waals gas where $(p-an^2/V^2)(V_m-nb)=nRT$ in which a and b are constants characteristic of the gas but independent of temperature.

(b) If the gas is chlorine then $a=6.343$ dm^6 bar mol^{-2} and $b=0.05422$ dm^3 mol^{-1} and the initial volume is 20 litres. Calculate the work done in isothermally and reversibly compressing 3 moles to 1 litre.

(c) Look up and calculate the work needed to compress O_2 and H_2 under the same conditions and then comment on the work needed to compress the van der Waals gas compared to an ideal one. The data can be found in Weast (section 6.48, 75th edition).

Q4.11 The heat capacity of α-quartz (SiO_2) is measured to be $C_P = a + bT + c/T^2$ where a, b, and c are experimentally determined constants where $a = 46.0$ J mol^{-1} K^{-1}, $b = 0.00334$ J mol^{-1} K^{-2}, and $c = -8.9 \times 10^{-5}$ J mol^{-1} K.

(a) Find the equations describing the enthalpy H and entropy S changes if

$$H_T^0 - H_{298}^0 = \int_{298}^{T} C_p dT \quad \text{and} \quad S_T^0 - S_{298}^0 = \int_{298}^{T} \frac{C_p}{T} dT.$$

(b) Calculate the change in enthalpy and entropy between 298 and 350 K.

Q4.12 The Clapeyron equation describes the relationship between temperature and pressure of a vapour.

(a) Look up the equation in your textbook and write it in terms of $\Delta_{vap}H$ and ΔV, then integrate the equation from pressure p_1 to p_2 and temperature T_1 to T_2 assuming ΔV is constant.

(b) Show how the Clapeyron equation is modified to become the Clausius–Clapeyron equation.

(c) Integrate the Clausius–Clapeyron equation and show that the ratio of vapour pressures is described by $\ln(p_2/p_1) = \dfrac{-\Delta_{vap}H}{R}\left(\dfrac{1}{T_2} - \dfrac{1}{T_1}\right)$ between temperatures T_1 and T_2. Using this equation, knowing the pressure at one temperature, the pressure can be found at any other. This is illustrated in the next problem.

Q4.13 Given the following data, use Maple to calculate and plot the phase diagram for benzene near the triple point from $T = 250 - 300$ K and $p = 0 - 10\,000$ Pa.
At the triple point $p = 36$ torr and $T = 5.5$°C, $\Delta_{fus}H = +10.6$ kJ mol^{-1}, $\Delta_{vap}H = +30.8$ kJ mol^{-1}, ρ(solid) = 0.891 g cm^{-3}, ρ(liquid) = 0.879 g cm^{-3}.

The coexistence curves are described by the Clapeyron and Clausius–Clapeyron equations calculated in the previous problem. The phase diagram will show regions containing solid, vapour, and liquid separated by the lines you calculate. To obtain a phase diagram similar to that in your textbook, plot three lines on the same graph but starting and ending at different temperatures each one containing the common value that is the triple point.

***Strategy*:** Look up a typical gas–liquid–solid phase diagram in your textbook to see what your calculation should look like. Use the equations derived in the previous question. Note that the enthalpy of sublimation is that due to fusion plus that for vaporization. The common point is the triple point therefore the fusion curve ends at this point and the melting and vaporization curves start there. One atmosphere equals 760 torr which is 101 326 Pa, therefore 1 torr equals 10 1326/760 Pa. You may find it easier to plot the log of the pressure.

Q4.14 A tube that is pinched in the middle and used to accelerate the gases in a steam turbine, jet aircraft, or rocket engine, is called a de Laval or convergent—divergent ('condi') nozzle. In physical science, it is also used as a means of cooling a jet of gas molecules down to a few degrees Kelvin when gas from high pressure is expanded into a vacuum. The profile of a de Laval nozzle is shown in the sketch, Fig. 4.7.

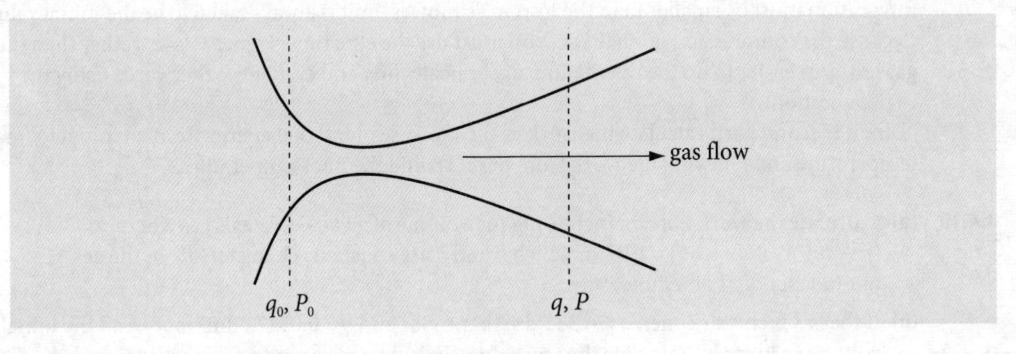

Fig. 4.7 Sketch of the shape of the de Laval or condi nozzle.

The fuel and air heated in a jet engine increases in velocity as it passes through the engine and by reaction on the turbine blades provides the thrust to drive the plane, boat, or train forward. The rocket engine similarly has the shape of a de Laval nozzle whose profile decreases then increases along its length. In any turbine or rocket engine the gas is heated so quickly that this is done adiabatically (and therefore $pV^\gamma = c$, where c is a constant) and also *isentropically* because the gas experiences no change in entropy. Because there is no heat loss, $\delta q = 0$, consequently the entropy change is also 0 because, under reversible conditions, $\delta q_{rev} = TdS$. The enthalpy change (in kJ mole^{-1}) from the front of the engine q_0 to a value q, at some point along the path, is equal to the change in kinetic energy of the gas which is moving at a speed u at that point, therefore for m g mole^{-1} of gas,

$$mu^2/2 = q_0 - q.$$

(a) By integrating the relationship $\left(\dfrac{\partial q}{\partial p}\right)_S = V$ at constant entropy S and between inlet pressure p_0 to pressure p, show that the enthalpy change is

$$q - q_0 = \frac{\gamma}{\gamma - 1} p_0 V_0 \left[1 - \left(\frac{p}{p_0}\right)^{\frac{\gamma-1}{\gamma}}\right]$$

then calculate the gas velocity.

(b) The cross-sectional area of the jet nozzle is σ and therefore mass of gas/second passing through is $\sigma um = V\mu$ where μ is the mass flow rate in kg sec^{-1}. Using the relationship $pV^\gamma = p_0 V_0^\gamma$ rearrange this to an expression for V, then obtain the jet cross section and find its minimum area that is at the throttle point. Plot the nozzle cross section as a function of the ratio of pressure to input pressure, p/p_0, and hence determine the shape of the gas profile in the jet and therefore by inference the internal profile of the engine. This should look something like that in the sketch.

(c) Show that the speed of the gas at the throttle point is that of sound; $u_s = \sqrt{\dfrac{2\gamma}{\gamma+1}\dfrac{RT}{m}}$. Plot the gas velocity relative to u_s assuming constant temperature. Identify the subsonic and supersonic regions. The throttle point occurs at the minimum nozzle cross section.

Typical values of pressures and temperature in a rocket nozzle are given below, but you will only need γ to plot the graphs.

Temperature, 3500 K, input pressure 7 MPa, exhaust pressure 0.1 MPa.

Ratio of specific heats, $\gamma = C_p/C_V = 1.22$, molar mass $m = 22$.

***Strategy*:** The fact that the entropy is constant allows $\left(\dfrac{\partial q}{\partial p}\right)_S = V$ to be integrated but otherwise does not enter into the calculation. The expression for q is first separated by writing $dq = \cdots$ then integrated with limits q, q_0 and p, p_0. The velocity is obtained via the kinetic energy expression and the cross section *via* the mass flow equation.

4.3 Integration by substitution

It may appear at first that if an integral is not in a standard form then it cannot be solved—not so! The purpose of substitution is to convert the integral into a standard and therefore simpler form whose solution is known. Choosing a substitution is rather an art and some trial and error is usually required. However, just as in football, a clever substitution can produce the required result.

There are three parts to the method, which will be illustrated with $\int \sin^2(x)\cos(x)dx$.

(1) Change a complex expression in x into a simpler one in another variable u by a substitution, For example, $\sin^2(x)\cos(x)$ is simplified with $u = \cos(x)$ to $(1 - u^2)u$.

(2) Work out dx as an expression in du. In our example, $du = -\sin(x)dx$.

(3) Change all limits to the new variable.

The usage of du and dx as independent entities in themselves in step (2) appears to be different to that of du/dx, which has previously been used as a ratio. However, it is quite permissible to use du and dx and so forth on their own, provided their ratio could still be made by rearranging the equation.

As a rule of thumb with complicated sine and cosine expressions, try sine and cosine substitutions. If you are lucky, your first guess may be the correct one; if not, try again. Start by guessing that the substitution $u = \cos(x)$ might work, therefore $\dfrac{du}{dx} = -\sin(x)$ or $dx = \dfrac{du}{-\sin(x)}$. Next substituting for $\cos(x)$ and dx gives

$$\int \sin^2(x)\cos(x)dx = -\int \sin(x)udu = -\int \sqrt{(1-u^2)}udu$$

and in the last step $\cos^2(x) + \sin^2(x) = 1$ was used. Notice that no integration has yet been performed, just the substitutions. However, this last integral looks as if it might be possible but it is still a bit complicated, so instead of persevering it is easier to try again and see if a better result is possible. You can always return to this result if necessary. As a second try, substitute $u = \sin(x)$ and then $du = \cos(x)dx$. Therefore,

$$\int \sin^2(x)\cos(x)dx = \int u^2du = \frac{u^3}{3} = \frac{\sin^3(x)}{3}$$

which turns out to be far simpler.

Using the right-angled triangle it is possible to substitute trigonometric functions for algebraic ones and vice versa. The triangles and their corresponding trig relationships are shown in their respective columns in Table 4.1.

Consider the integral $\displaystyle\int \frac{x^2}{\sqrt{4-x^2}}dx$. The substitution $\tan(\theta) = x/\sqrt{a^2-x^2}$ with $a = 2$ is appealing. However, when differentiated to substitute $d\theta$ for dx, it produces a horrible result. Trying $\cos(\theta) = (\sqrt{2^2-x^2})/2$ instead proves to be a good starting place because when differentiated $d\theta$ produces a reciprocal square root that cancels with that in the integral. Differentiating gives $2\sin(\theta)d\theta = x(4-x^2)^{-1/2}dx$ changing the integral into

$$2\int \sin(\theta)xd\theta = 4\int \sin(\theta)\sqrt{1-\cos^2(\theta)}d\theta.$$

$$= 4\int \sin^2(\theta)d\theta$$

This is now a standard form and can be converted to an exponential form (or looked up in Section 4.2.13) and then converted back to x using trig functions.

$$\int \frac{x^2}{\sqrt{4-x^2}}dx = 4\int \sin^2(\theta)d\theta$$

$$= 2\theta - 2\sin(\theta)\cos(\theta) + c$$

$$= 2\sin^{-1}(x/2) - \frac{x}{2}\sqrt{4-x^2} + c$$

Table 4.1 Trig formulas useful for substitutions

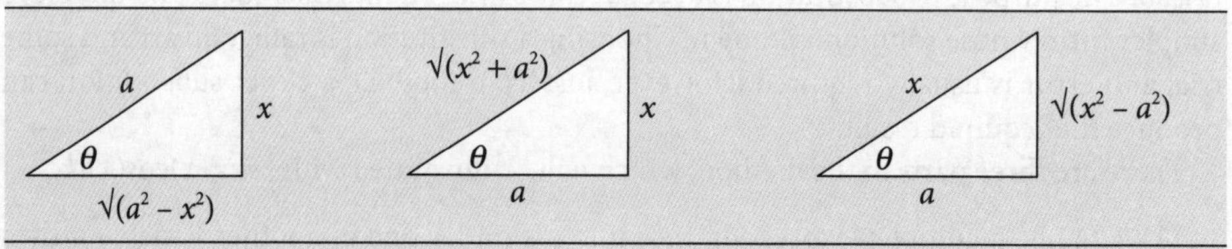

$\sin(\theta) = x/a$	$\sin(\theta) = x/\sqrt{x^2+a^2}$	$\sin(\theta) = \dfrac{\sqrt{x^2-a^2}}{x}$
$\cos(\theta) = \dfrac{\sqrt{a^2-x^2}}{a}$	$\cos(\theta) = a/\sqrt{x^2+a^2}$	$\cos(\theta) = a/x$
$\tan(\theta) = \dfrac{x}{\sqrt{a^2-x^2}}$	$\tan(\theta) = x/a$	$\tan(\theta) = \dfrac{\sqrt{x^2-a^2}}{a}$

This integral was hard to do by hand. Fortunately, Maple gives the same answer;

```
> Int(x^2/sqrt(4-x^2),x): % = value(%);
```

$$\int \frac{x^2}{\sqrt{4-x^2}}\mathrm{d}x = -\frac{1}{2}x\sqrt{4-x^2} + 2\arcsin\left(\frac{1}{2}x\right)$$

4.3.1 Definite integrals with substitution

When calculating a definite integral first evaluate the indefinite integral and then using the limits, work out the final value. If a substitution is made *it is also necessary to change the limits*, thus making it unnecessary to reverse the substitution to obtain the result.

Suppose the integral is $\displaystyle\int_2^8 \frac{3}{(1+2x)^2}dx$, trying the substitution $u = 1 + 2x$ seems an obvious choice and therefore $dx = du/2$. The limits are $x = 2$ and $x = 8$, which become $u = 5$ and $u = 17$ in the new variable, making the integral

$$\int_5^{17} \frac{3}{2u^2}du = -\frac{3}{2u}\bigg|_5^{17} = -\frac{3}{34} + \frac{3}{10} = \frac{18}{85}.$$

Alternatively, calculate the integral leaving out the limits to begin with, then convert back to x and then use the original limits;

$$\int \frac{3}{2u^2}du = -\frac{3}{2u} \equiv -\frac{3}{2(1+2x)}\bigg|_2^8.$$

Maple also does this calculation easily;

```
> Int( 3/(1+2*x)^2, x = 2..8): %= value(%);
```

$$\int_2^8 \frac{3}{(1+2x)^2}\mathrm{d}x = \frac{18}{85}$$

4.4 Three useful results with a function and its derivative

Integrals where the function in the numerator is the *derivative* of the function in the denominator, evaluate to logs. For example,

$$I = \int \frac{f'(x)}{f(x)}dx = \ln(|f(x)|) + c. \tag{4.12}$$

where $f'(x)$ is the derivative of f and c is the constant of integration. The symbols $|\,|$ indicate that the absolute value of the function must be taken because the log of a negative number is not permissible. This result can be demonstrated by substitution of $u = f(x)$ then $du = f'(x)dx$ into equation (4.12) giving

$$I = \int \frac{du}{u} = \ln(u) = \ln(|f(x)|) + c.$$

This result is not surprising when recalling the differential of $\ln(f(x))$. An example is

$$\int \frac{\cos(\theta)}{\sin(\theta)}d\theta = \ln(|\sin(\theta)|) + c.$$

The cosine is the derivative of the sine, and the absolute value is taken because the sine is negative for some values of θ.

Related to equation (4.12) is the integral

$$I = \int \frac{f'(x)}{\sqrt{f(x)}}dx = 2\sqrt{f(x)} + c \tag{4.13}$$

and this can be verified by letting $f(x) = u$ and differentiating $\sqrt{f(x)} + c$. Try this and then integrate both sides of the equation.

The third equation involves a function and its derivative and is the product,

$$I = \int f(x)f'(x)dx = \frac{1}{2}[f(x)]^2 + c. \tag{4.14}$$

Notice that the function is written first, for example,

$$\int \cos(\theta)\sin(\theta)d\theta = -\frac{1}{2}\cos^2(\theta) + c$$

where the function f is the cosine and its derivative f' the sine. Of course, the integral in this particular case could have been written the other way round. The result would then be

$$\int \sin(\theta)\cos(\theta)d\theta = +\frac{1}{2}\sin^2(\theta) + c$$

and because $\cos^2(\theta) + \sin^2(\theta) = 1$ this is the same result if the 1 is now included in the arbitrary constant c.

4.5 Integration by parts

Very often the products of two functions must be integrated and one way of doing this is to use 'integration by parts'. Integration by parts is a major tool to integrate seemingly difficult expressions.

In differentiating products of functions (Chapter 3.5.1), such as u and v where each is itself a function of x, we used the equation,

$$\frac{d}{dx}uv = u\frac{dv}{dx} + v\frac{du}{dx} \tag{4.15}$$

Integrating this equation gives

$$uv = \int u\frac{dv}{dx}dx + \int v\frac{dv}{dx}dx \qquad \text{or} \qquad uv = \int udv + \int vdu$$

and evaluating further gives

$$\int udv = uv - \int vdu \tag{4.16}$$

which is the *integration by parts* formula. This reads:

The integral of udv is u times the integral of dv, which is v, minus the integral of v times the derivative of u.

The trick is to find u that can be differentiated, which is usually easy, and v that can be integrated, which is sometimes much harder. It is worth exchanging u and v if the resulting integration gets more complicated. Some examples should make this important method clearer.

(a) To solve $\int y\cos(y)dy$ using equation (4.16),
(i) Start by trying $u = y$ and $dv = \cos(y)dy$.
(ii) To find uv, integrate $dv = \cos(y)dy$ to $v = \sin(y)$ and then multiply this by $u \equiv y$.
(iii) Form the vdu integral by differentiating u giving $du \equiv dy = 1$ and multiply by $v = \sin(y)$ to give $\int vdu = \int \sin(y)dy$.
(iv) Combining these terms gives

$$\int y\cos(y)dy = y\sin(y) - \int \sin(y)dy = y\sin(y) + \cos(y) + c.$$

Checking with Maple confirms the result;

```
> Int(y*cos(y),y): % = value(%);
```

$$\int y\cos(y)\,dy = \cos(y) + y\sin(y)$$

(b) The integral $\int x\ln(x)dx$ can be evaluated by parts if $u = \ln(x)$ and $dv = xdx$, producing

$$\int x\ln(x)dx = \ln(x)\frac{x^2}{2} - \frac{1}{2}\int \frac{1}{x}x^2dx$$

$$= \ln(x)\frac{x^2}{2} - \frac{x^2}{4} + c$$

with c as the integration constant. Trying this integral the other way round with $u = x$, and $dv = \ln(x)dx$ is more difficult because to obtain v we have to know the integral of $\ln(x)$. This can be looked up and is $x\ln(x) - x$, and has to be integrated again in the next step.

$$\int x\ln(x)dx = x(x\ln(x) - x) - \int (x\ln(x) - x)dx.$$

What has happened here is that the original integral is produced on the right, making the method recursive in this instance. Rearranging gives

$$2\int x\ln(x)dx = x(x\ln(x) - x) + \int xdx$$

$$= x^2\ln(x) - \frac{x^2}{2} + c$$

which is the same result as by the first method when both sides are divided by two.

(c) Sometimes successive integration is necessary, as with $\int e^x\sin(x)dx$. To integrate this function let $u = e^x$ and $dv = \sin(x)dx$. This produces

$$\int e^x\sin(x)dx = -e^x\cos(x) + \int e^x\cos(x)dx$$

It is then necessary to perform the right-hand integration in a similar way, giving

$$\int e^x\cos(x)dx = e^x\sin(x) - \int e^x\sin(x)dx$$

By substituting this into the first result gives

$$\int e^x\sin(x)dx = -e^x\cos(x) + e^x\sin(x) - \int e^x\sin(x)dx$$

which, after rearranging, is

$$\int e^x\sin(x)dx = e^x[\sin(x) - \cos(x)]/2 + c.$$

As a check, differentiate the result:

$$\frac{d}{dx}\frac{e^x}{2}[\sin(x) - \cos(x)] = \frac{e^x}{2}[\cos(x) + \sin(x)] + \frac{e^x}{2}[\sin(x) - \cos(x)]$$

$$= e^x\sin(x).$$

Notice that the integral could more easily be solved by converting the sine to its exponential form. It is often, but not always the case, that more than one method could be used to solve an integral. Which you choose depends on your particular liking for one method over another.

4.5.1 Questions

Full solutions are available at **www.oxfordtextbooks.co.uk/orc/beddard.**

Q4.15 Evaluate **(a)** $I = \int \cos^3(x)dx$, **(b)** $I = \int \dfrac{3x}{(5+3x)^4}dx$, **(c)** $I = \int \dfrac{e^{\sqrt{x}}}{\sqrt{x}}dx$,

(d) $I = \int \cot(ax)dx$, **(e)** $I = \int \dfrac{x^2}{8+x^3}dx$, **(f)** $I = \int \dfrac{e^{ax}}{1-e^{ax}}dx$

Q4.16 Find $I = \int \dfrac{dx}{\sqrt{(a^2-x^2)}}$

***Strategy*:** You might well decide at this point to try using Maple as this integration is looking quite complicated and the substitution to use is not obvious. Evaluation can sometimes be successful by starting with $a^2\cos^2(u) + a^2\sin^2(u) = a^2$ or $a\cos(u) = \sqrt{a^2 - a^2\sin^2(u)}$. Make the substitution $x = a\sin(u)$ and its differential $dx = a\cos(u)du$.

Q4.17 **(a)** Integrate $I = \int xe^{-x^2}dx$ and
(b) Evaluate and plot the function from $x = -3$ to 3 and the result of the integration from 0 to x.

Q4.18 Integrate $I = \int \dfrac{x-2}{x\sqrt{x+3}}dx$ by hand and by Maple,

Q4.19 Evaluate $I = \int_{-1}^{2} \dfrac{x^2}{16+x^6}dx$

Q4.20 Determine by parts $I = \int \sin^{-1}(ax)dx$.

Q4.21 The integral $Ei(x) = -\int_{-x}^{\infty} \dfrac{e^{-t}}{t}dt$ is called the *exponential integral* but cannot be evaluated explicitly but only numerically. It can occur in the calculation of molecular orbitals.

(a) Show that $Ei(-x) = \int_{x}^{\infty} \dfrac{e^{-t}}{t}dt = \int_{-\infty}^{x} \dfrac{e^{t}}{t}dt$

(b) Expand $Ei(x)$ by parts and form a series expansion of the integral.

Q4.22 **(a)** Integrate $I = \int xe^{ax}dx$.
(b) Sometimes integrals can be written in terms of others. Suppose that $I_0 = \int e^{ax}dx$, $I_1 = \int xe^{ax}dx$, and $I_2 = \int x^2e^{ax}dx$. By induction find a formula for $I_n = \int x^n e^{ax}dx$ where n is a positive integer.

***Strategy*:** Calculate I_2 and then I_3, by parts. By 'induction', effectively means look for a pattern in the answers and find the nth term.

Q4.23 **(a)** Integrate $I_1 = \int_0^{\infty} e^{-ax^2}xdx$ by parts by choosing $v = e^{-ax^2}$ and $u = 1$. This is an alternative method to that in Q4.17.
(b) Integrate $I_n = \int_0^{\infty} e^{-ax^2}x^n dx$ by parts to find a recurrence relationship between I_n and I_{n+2}.

***Strategy*:** Look at the general form of the 'by parts' integral, work out what dv is and rewrite the integral.

Q4.24 Integrate $\int \dfrac{\cos(\ln[x])}{x}dx$.

***Strategy*:** Find $d\ln(x)$ and then make a substitution.

Q4.25 Find $\int \sec^2(\sqrt{x})dx$

Q4.26 Integrate by parts **(a)** $I = \int \cosh(x)\sinh(x)dx$, **(b)** $\int x\cos(x)dx$

Q4.27 Integrate $\displaystyle\int \frac{dx}{\sin(ax)}$. Is your answer the same as given by Maple?

Strategy: Convert the sine to its exponential form then try a substitution.

Q4.28 **(a)** Find $\int_{-\pi}^{\pi} dx$ **(b)** show that $\int_{-\pi}^{\pi} e^{-imx}e^{-inx}dx = 2\pi\delta_{nm}$ where m and n are integers. If $m = n$ then $\delta_{nm} = 1$, otherwise it is 0.

Q4.29 Occasionally it is possible to evaluate some definite integrals by engaging in subterfuge, changing the integral and differentiating first. Start with the integral $I = \displaystyle\int_0^{\infty} \frac{\sin(x)}{x}dx$, multiply this by $e^{-\beta x}$ to produce $I_\beta = \displaystyle\int_0^{\infty} \frac{e^{-\beta x}\sin(x)}{x}dx$, and differentiate with respect to β. Next, integrate twice over, first with respect to x and then with respect to β. Use the limits 0 and ∞ for the first integration. Finally calculate what happens when $\beta = 0$, thus making the exponential term unity and so returning the original integral. The other limit is ∞ because this makes $I_\beta = 0$. You will find that the initial differentiation removes the denominator, making the integral simpler. It is an integral met before and one that can be performed by parts; see Section 4.5.

Q4.30 The equation describing the shape of the soap film described in Q3.82 may be derived by considering the surface tension T of the film. Every point on a vertical circle around the film has an equal horizontal surface tension, which is equal to a constant c. Therefore, $2\pi yT\cos(\theta) = c$. By calculating the cosine for small changes of x and y in the limit of dy and dx, find an equation for dy/dx. Separate this into integrals in y and x and integrate both sides to find the equation for y, the film's radius. Finally convert the equation into a hyperbolic trig form.

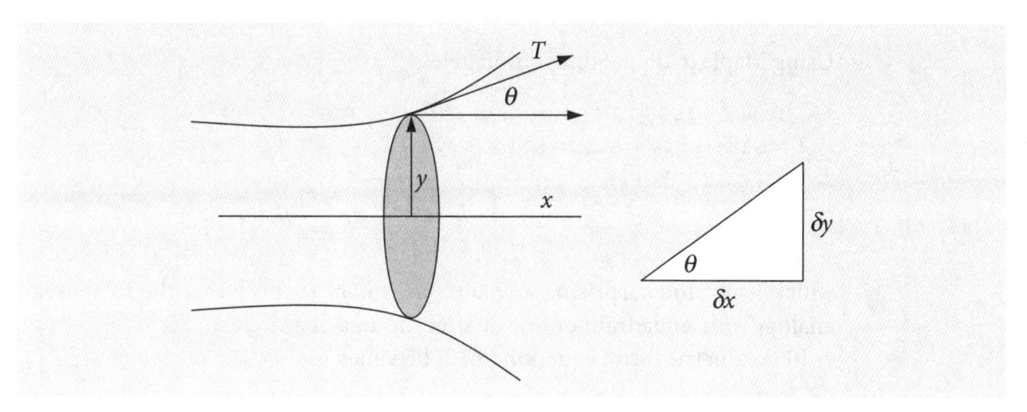

Fig. 4.8 Geometry for Q4.30.

4.6 Integration using parametric equations

Complex curves are often given as a pair of parametric equations. The ellipse $\dfrac{x^2}{a^2} + \dfrac{y^2}{b^2} = 1$, for instance, is represented also as the pair of equations in the *parametric variable t* as

$$x = a\cos(t), \qquad y = b\sin(t).$$

The constants a and b are the semi-major and minor axes of the ellipse. A circle has the similar equation with radius $a = b$. To plot the parametric equations, t is varied and points at $\{x, y\}$ plotted. How large t has to be to complete the curve depends on how t and x and y are related and has to be tested before plotting. Parametric equations are used because many complex equations do not simplify to something that can be readily integrated and it is simply easier to use the parametric form.

In parametric form an integral is always written as

$$A = \int_{\alpha}^{\beta}\left[y(t)\frac{dx(t)}{dt}\right]dt \tag{4.17}$$

where $x(t)$ and $y(t)$ are the parametric equations. Since the variable is t, not only is it necessary to convert dx into dt using $(dx/dt)dt$, but also to convert the integral's limits

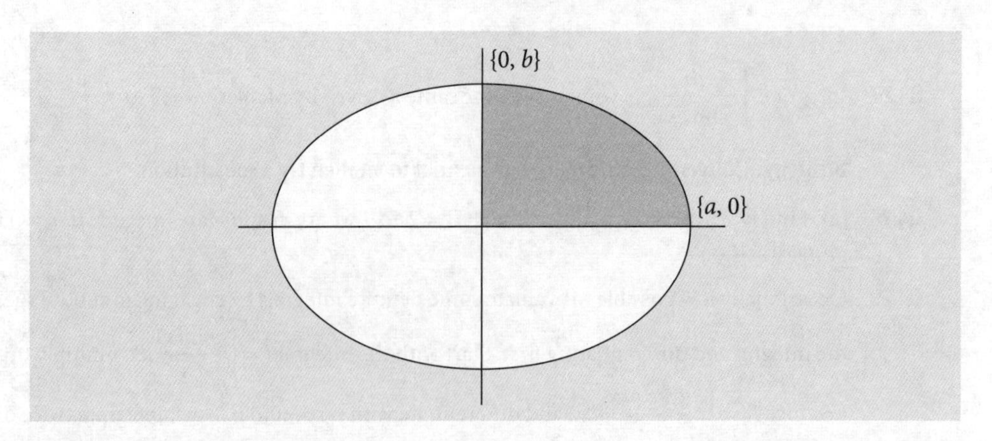

Fig. 4.9 The area of a quadrant is $ab\pi/4$.

α and β. If these are *not* defined in terms of t but as x_α and x_β, it is necessary to define $x_\alpha = f_x(\alpha)$ and $x_\beta = f_x(\beta)$ where f_x is the x parametric function and both these equations will have to be solved to find α and β.

To make this clearer, a comparison is made between calculating the area under the first quadrant of the ellipse from coordinate {0, b} to {a, 0} in the conventional way and in the parametric way. In the conventional way

$$A = \int_0^a y dx = b\int_0^a \sqrt{1 - x^2/a^2} dx.$$

Using Maple to do the integration gives

```
> Int(b*sqrt(1-x^2/a^2),x=0..a):
  % =simplify(value(%),symbolic);
```

$$\int_0^a b\sqrt{1-\frac{x^2}{a^2}}\,dx = \frac{1}{4} b\,\pi\,a$$

which is not too surprising a result. One might surmise that the total area is $\pi ab/4$ by analogy with a quadrant or one quarter the area of a circle.

In parametric form, equation (4.17) becomes

$$A = \int_\alpha^\beta \left[a \sin(t)\frac{d}{dt} b\cos(t)\right] dt = ab\int_\alpha^\beta \sin^2(t)dt.$$

and the new limits α and β must be determined. The integral can be performed by converting the sine to its exponential form first, which makes it much easier, see Section 4.2.7. The new limits α and β are found using $x_1 = 0$ and $x_2 = a$ with $x_1 = a\cos(\alpha)$ and $x_2 = a\cos(\beta)$. If $x_1 = 0$, then $\cos(\beta) = 0$ making the limit $\beta = \pi/2$. When $x_2 = a$, $t = 0$ because $\cos(0) = 1$ making $\alpha = 0$. Using the result of Section 4.2.7 to do the integration gives

$$A = ab\left[\frac{t}{2} - \frac{\sin(2t)}{4}\right]_0^{\pi/2} = \frac{ab\pi}{4}$$

and the order of the limit has been changed here to give a positive area. The calculation in Maple confirms this

```
> Int(a*b*sin(t)^2,t = 0..Pi/2):% = value(%);
```

$$\int_0^{\frac{1}{2}\pi} ab\sin(t)^2\,dt = \frac{1}{4} ab\,\pi$$

4.7 Integration in plane polar coordinates

Sometimes equations are simpler and calculations easier when not done in Cartesian (rectilinear) coordinates but in polar or one of several other coordinate systems. Using Cartesian coordinates to integrate closed curves, such as circles or cardioids and so forth, can also lead to some pitfalls. The equation of a circle of radius 1 and centred at the origin {0, 0} is $x^2 + y^2 = 1$. When integrated, because exactly half the area is above the x-axis, this part should have a positive area and the half below a negative one making the total 0, whereas the area is clearly π. Rearranging to find y and integrating gives $\int_{-1}^{1}\sqrt{1-x^2}dx = \pi/2$ and although the integration was across the diameter $x = -1$ to 1, only half of the area has been evaluated. This is because, by convention, only the positive part of the square root was used, and although it appears that we have integrated over the whole area in fact only half of the area has been covered. If $-\int_{-1}^{1}\sqrt{1-x^2}dx$ is added to account for the other half, the result is zero. The area of the cardioid shown in Fig. 4.11 would similarly be zero because there is as much negative area below the x-axis as there is positive above it. A common way round this problem of false zero areas is to identify a symmetrical curve, calculate part of it and multiply the result according to symmetry, to obtain the whole area.

Two-dimensional curves represented in plane polar coordinates $\{r, \theta\}$ have coordinates that are the radius arm r, the distance from the origin called the *pole*, to a point on the curve at an angle θ, and this angle is conventionally measured anticlockwise upwards from the horizontal. The pole is at $r = 0$ and the horizontal $\theta = 0$. A circle of radius a centred at the origin is $x^2 + y^2 = a^2$ in Cartesians but $r = a$ in plane polar coordinates. If the origin of the coordinates lies on the circumference of a circle and the line $\theta = 0$ passes horizontally through the circle's centre, the equation $r = 2a\cos(\theta)$ describes this circle, see Fig. 4.12.

The cardioid shown in Fig. 4.11 has the form $r = a[1 + \cos(\theta)]$ and shows the definitions of r and θ. The equations to convert to plane polar coordinates from Cartesian ones are described in Chapter 1.6.1 and are

$$r^2 = x^2 + y^2, \qquad x = r\cos(\theta), \qquad y = r\sin(\theta). \tag{4.18}$$

If a general curve described by some function of the angle is $r = f(\theta)$, the area swept out by the radius in moving from the line $\theta = \alpha$ to the line $\theta = \beta$ is the area bounded by the curve and is

$$A = \frac{1}{2}\int_{\alpha}^{\beta} r^2 d\theta \equiv \frac{1}{2}\int_{\alpha}^{\beta} f(\theta)^2 d\theta. \tag{4.19}$$

The integration limits are in the range $0 < \alpha + \beta \le 2\pi$ and therefore cover the whole range of angles at maximum. A derivation of this formula is now given; see Fig. 4.10. At an angle φ up from the horizontal, a sector is defined with a small angle $\delta\theta$. The area of this small segment is $A_{seg} = f(\varphi)\lambda/2$ where λ is the length of the segment at radius $r = f(\varphi)$ on the curve $r = f(\theta)$.

The arc length λ can be calculated using geometry because $\delta\theta$ is so small that the triangle can be considered right-angled, hence $\tan(\delta\theta) = \lambda/f(\varphi)$. As $\delta\theta$ is a small angle, $\tan(\delta\theta) \approx \delta\theta$ making $\lambda \approx f(\varphi)\delta\theta$ and the area of the small segment is $A_{seg} = f(\varphi)^2\delta\theta/2$. If all these areas are added up and the limit $\delta\theta \to 0$ taken, the integral (4.19) is produced.

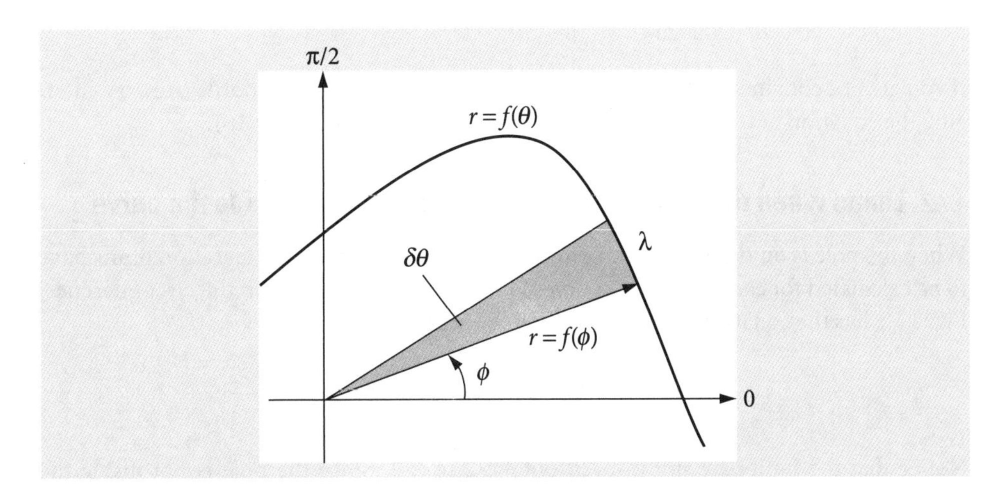

Fig. 4.10 Geometry to define the area swept out by radius arm *r*.

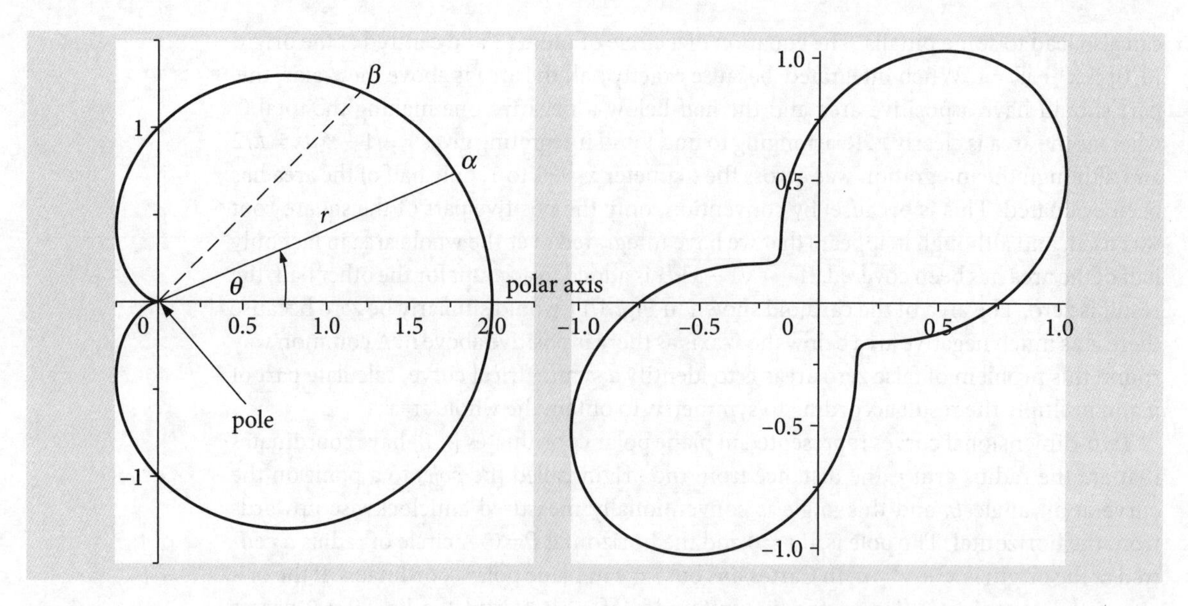

Fig. 4.11 Left: The cardioid $r = a[1 + \cos(\theta)]$ with $a = 1$ drawn in polar coordinates. Right: 'Peanut', $r = 1/4 + \sin^2(\theta + \pi/4)$. The Cartesian axes shown are rather confusing but conventionally drawn, and the axes give the value of r (the radius vector) only when it is either horizontal or vertical and are not the axis on which the plot is drawn. The axes do not show the value of θ.

To test equation (4.19), the area of a circle of radius a, which has the polar equation $r = a$, is

$$A = \frac{a^2}{2}\int_0^{2\pi} d\theta = \pi a^2,$$

which is just as well, otherwise our method would be faulty! Notice that the integration is around the full circle.

Imagine in Fig. 4.11 that the radius arm r extending from 0 to α is moved anticlockwise to point β, a movement that produces the segmental area between the pole and α to β. If β continues all the way round and back again to α then the whole area is calculated. Two cases now have to be distinguished; the first when the pole is within the curve as shown on the right of Fig. 4.11. The second is when the pole is outside the curve or on its circumference and these cases are dealt with next.

4.7.1 Limits when the pole is inside the curve

When the origin of the curve, i.e. the pole, is inside the curve, as it is in the peanut curve, integration is always from 0 to 2π. Integrating around this curve, Fig. 4.11, which has the polar equation $r = 1/4 + \sin^2(\theta + \pi/4)$, is done with limits of 0 to 2π and its area is

$$A = \frac{1}{2}\int_0^{2\pi} [1/4 + \sin^2(\theta + \pi/4)]^2\, d\theta = 3\pi/16.$$

If you are not convinced of the simplicity produced by using polar coordinates, try changing the equation of this curve back to Cartesians using equations (4.18).

4.7.2 Limits when the pole is on the circumference or outside the curve

When the pole is on the curve itself or outside the curve, then the integration limits have to be calculated for each particular curve. The circle with its origin on the circumference, Fig. 4.12, has the equation $r = 2a\cos(\theta)$ and the area is

$$A = \frac{1}{2}\int_{-\pi/2}^{\pi/2} r^2 d\theta = 2a^2\int_{-\pi/2}^{\pi/2} \cos^2(\theta) d\theta = \pi a^2.$$

Notice that the limits are not 0 to 2π but $-\pi/2$ to $\pi/2$. Since the pole is not inside the curve, the angle the tangent makes with the horizontal at the pole, going clockwise and

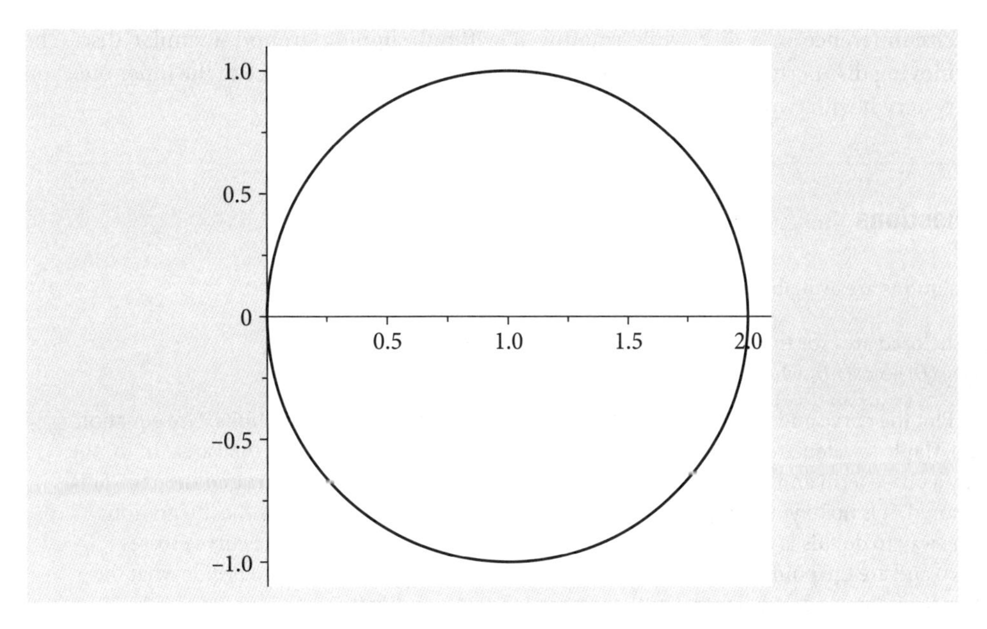

Fig. 4.12 Plot of $r = 2\cos(\theta)$. The angle the curve makes with the horizontal or polar axis, at the pole (origin) is $\pm\pi/2$.

anticlockwise, must be calculated. This is done by making $r = 0$ and solving the equations. For this particular circle $2a\cos(\theta) = 0$, and the angles forming the integration's limits are therefore $\pm\pi/2$ or $\pm k\pi/2$ where k is one of the odd integers $\cdots -3, -1, 1, 3 \cdots$ and so forth. This multiplicity is expected because the cosine function does repeatedly become zero. The limits to the integration are therefore $-\pi/2$ and $\pi/2$ when $k = -1$ and 1, which moves the radius arm only once around the curve during integration. The curve can be plotted using Maple as,

```
> plot(2*cos(x),x= 0..Pi, coords = polar);
```

It is necessary to specify that the coordinates are polar or a normal cosine curve would be produced.

Doing the same calculation for the cardioid, the integration limits are the angles that make $[1 + \cos(\theta)] = 0$, which occur when $\cos(\theta) = -1$ and is satisfied when $\theta = -\pi$ or π and multiples thereof. Therefore, the angle is $\pm k\pi$ and hence the limits are $-\pi$ and π. The total area of the cardioid is

$$A = \frac{1}{2}\int_{-\pi}^{\pi} r^2 d\theta = \frac{a^2}{2}\int_{-\pi}^{\pi} [1 + \cos(\theta)]^2 d\theta$$

$$= \frac{a^2}{2}\int_{-\pi}^{\pi} 1 + 2\cos(\theta) + \cos^2(\theta) d\theta$$

Each of these integrals is standard and can be evaluated by converting to their exponential form, but as a check,

```
> Int( cos(theta)^2, theta ): %= value(%);
```

$$\int \cos(\theta)^2 \, \mathrm{d}\theta = \frac{1}{2}\cos(\theta)\sin(\theta) + \frac{1}{2}\theta$$

The total area is $A = \dfrac{a^2}{2}\left[\theta + 2\sin(\theta) + \dfrac{1}{2}\cos(\theta)\sin(\theta) + \dfrac{\theta}{2}\right]_{-\pi}^{\pi} = 3\pi\dfrac{a^2}{2}$.

The cardioid in Cartesian coordinates is the quartic

$$(x^2 + y^2 - ax)^2 - a^2(x^2 + y^2) = 0,$$

which would have to be solved to obtain $y = \cdots$ before integration. Looking at this equation it is easy to appreciate how much simpler the polar equation is to integrate.

The general form of equation $r = a + b\cos(\theta)$ is called a limaçon. The cardioid is so named after its heart shape and is, incidentally, the curve generated by a point on the circumference of a disc while rotating it without slipping around a similar disc. The moving disc performs two rotations in traversing the circumference of the inner one; you can try it with two similar coins.

4.7.3 Questions

Full solutions are available at **www.oxfordtextbooks.co.uk/orc/beddard.**

Q4.31 Find the enclosed area for the curves **(a)** $r = a\cos(\theta)\sin(\theta)$, **(b)** $r = a\sin(3\theta + \pi/3)$, **(c)** $r = n\cos(\theta) + \cos(n\theta)$ when n is an odd integer.

Strategy: Plot the curve and find the angle when $r = 0$, which will determine the limits. Use equation (4.19) and Maple to calculate the area of one loop, then use symmetry to find the total area. In (b) the $\pi/3$ is only a phase term and does not alter the area—it simply rotates the curve around the axis. This can be ignored, it is up to you, but either way the angle at $r = 0$ has to be found for the integration limits. One way to do this is to plot the curve but not using polar coordinates. The curve crosses zero at $r = 0$ and are the points you are looking for. (c) Try values $n = 1, 3, 5$, etc. to decide what the integration limits are and then generalize the integration and simplify it.

Q4.32 **(a)** Find the area included between the two loops of $r = a\,(2\cos(\theta) + 1)$, **(b)** Show that the area included between the circle of radius a and the curve $r = a\cos(n\theta)$ is three-quarters of the area of the circle where n is an integer but not zero.

Q4.33 Figure 4.13 shows the hyperbolic curve[2] $x^2 - y^2 = 1$. If $OA = \cosh(\theta)$, prove that $AB = \sinh(\theta)$ and show that the area $QAB = \int_0^\theta \sinh^2(\theta)d\theta$. Find the area of the triangle OAB and of OQB.

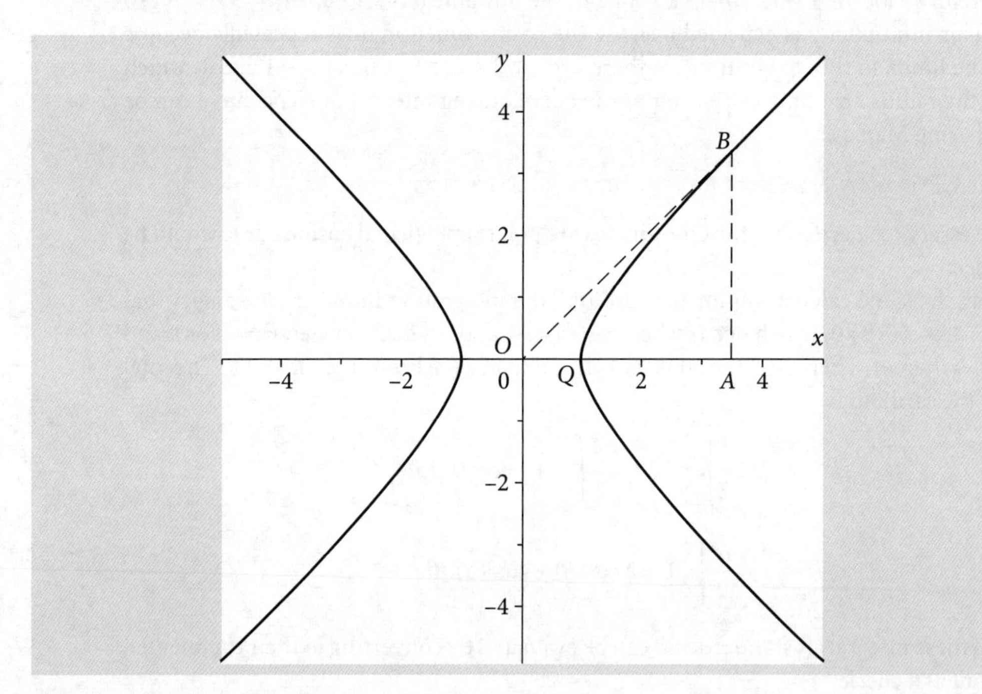

Fig. 4.13 The hyperbolic curve, $x^2 - y^2 = 1$

Strategy: Remind yourself that hyperbolic functions described in Chapter 1 are very similar to, but not the same as, the more familiar sine, cosine, and tangent but are easier to use than these functions because in their exponential form no complex numbers are involved.

Q4.34 Show that the curve $r = 1 + 2\sin(\theta)$ has an inner and outer loop.

(a) Determine what part of the curve, by plotting on Cartesian coordinates, corresponds to the inner loop

[2] The normal euphony is 'cosh theta', and also 'than' for tanh with emphasis on the '*th*', and 'shine' for sinh. An alternative usage is 'sinsh' for sinh and 'tansh' for tanh.

(b) Find the area of the whole curve and

(c) that of the inner loop.

Strategy: Use the area formula equation (4.19). To calculate (*b*) use limits 0 to 2π and (*c*) limits only of the inner loop. The limits are found by solving $1 + 2\sin(\theta) = 0$. There are repeated roots—four in the range 0 to 2π—and this is because the curve is not symmetrical about the *x*-axis. The hardest part of this problem is finding all the roots of the equation $1 + 2\sin(\theta) = 0$.

Q4.35 A radar transmitter has an emission pattern in the shape of the cardioid, $r = a[1 + \cos(\theta)]$. Calculate the ratio of the total transmitted signal to that transmitted backwards which is in the region where $\pi/2 \le \theta \le 3\pi/2$.

Strategy: The ratio is equivalent to the area with an angle $\pi/2 \le \theta \le 3\pi/2$ to that of the complete area. The cardioid is shown in Fig. 4.11 and the area is calculated with equation (4.19).

Q4.36 An *operator* will convert one function into another and this is a very familiar process. For example, the log function converts *x* to ln(*x*) and indeed this is so common that ln() is not thought of as an operator. In quantum mechanics, the Hamiltonian operator is $-\dfrac{\hbar^2}{2m}\dfrac{\partial}{\partial x^2} + V(x)$, which is the differential $-\dfrac{\hbar^2}{2m}\dfrac{\partial}{\partial x^2}$ that describes the kinetic energy, plus $V(x)$ the potential energy. Suppose an operator is instead an integral and has the behaviour

$$L[f(t)] = F(s) = \int_0^\infty f(t)e^{-st}dt,$$

then this is the *Laplace transform operator*. Notice the notation: the function to be operated upon is placed inside the square brackets on the left. The variables *s* and *t* are dummy variables and any other two letters could be used instead. The formula tells us that the operator *L* changes $f(t)$, a function of *t*, into a new function $F(s)$, a function in the variable *s*, and which is the integral.

Follow the 'recipe' given by the formula and assuming that *a* is a constant, show that;

(a) $L[a] = \dfrac{a}{s}$, **(b)** $L[e^{at}] = \dfrac{1}{s-a}$, **(c)** $L[t] = \dfrac{1}{s^2}$,

(d) $L[f(at)] = \dfrac{1}{a}F\left(\dfrac{s}{a}\right)$ **(e)** $L\left[\dfrac{df}{dt}\right] = sF(s) - f(0)$.

(f) Plot some of the functions and their transforms such as **(a)** and **(b)** with $a = 2$.

Q4.37 Find the area under the curve $x^2e^{-\alpha x^2}$ from $x = x_0$ to infinity where $x_0 \ge 0$ and $\alpha > 0$.

Q4.38 Use Maple to investigate integrals of the form $\int x^n e^{-\alpha x^2}dx$.

Q4.39 The first few harmonic oscillator wavefunctions are

$$\psi_0(x) = N_0e^{-\alpha x^2/2}, \qquad \psi_1(x) = \sqrt{2\alpha}N_0xe^{-\alpha x^2/2},$$

$$\psi_2(x) = \frac{N_0}{\sqrt{2}}(2\alpha x^2 - 1)e^{-\alpha x^2/2} \qquad \psi_3(x) = \sqrt{\frac{\alpha}{3}}N_0(2\alpha x^3 - 3x)e^{-ax^2/2},$$

where $N_0 = (\alpha/\pi)^{1/4}$ and $\alpha = \sqrt{k\mu}/\hbar$, *k* being the force constant and μ reduced mass. By inspection, classify the wavefunctions into odd or even functions.

Q4.40 **(a)** Wavefunctions can be described as orthogonal and or normalized. What does this mean? (Look this up in a physical chemistry textbook if you are uncertain.)

Using harmonic oscillator wavefunctions from previous questions show by direct integration that ψ_0 and ψ_1 are **(b)** orthogonal and **(c)** normalized.

Strategy: Two wavefunctions are orthogonal if the product of their integral over all coordinates is zero. If wavefunctions are normalized and orthogonal, they are called *orthonormal*. A normalized wavefunction means that $\int_{-\infty}^{\infty}\psi_1^2dx = 1$, which is interpreted as meaning that the nuclei of the vibrating molecule exists somewhere among all possible values of *x*. This is common sense. Summarizing:

$\int_{-\infty}^{\infty}\psi_i^*\psi_j dx = \delta_{ij}$ where δ_{ij} is the Kronecker delta function defined as $\delta_{ij} = 1$, if $i = j$, and 0 if $i \ne j$.

In general, normalization is interpreted to mean that the probability of a particle being somewhere in all possible space is 1 and the probability of being between *x* and $x + dx$ is ψ_1^2dx.

Q4.41 In an SIS type of disease where an individual is either susceptible S or infected I, then

$$S \xrightarrow{k_1IS} I \xrightarrow{k_2I} S$$

and $N = S + I$ is the total number of individuals (Britton 2003). When there are very few infected individuals, $S \approx N$ and the incidence n of the disease, which is the rate of new infections, is $n = k_1NI$ and can be shown to be the convolution

$$n(\tau) = k_1N\int_0^{\infty} n(\tau - t)I(t)dt.$$

If the number infected is exponentially distributed $I(t) = e^{-k_2t}$ and if $n(t) = n_0e^{rt}$, find r, the initial per capita growth rate of the disease. (Convolution is described in Chapter 9.)

Q4.42 In any molecule the electronic wavefunctions are *orthogonal* to one another. In an electric dipole transition, a photon is absorbed and when this is polarized in the x-direction the probability of absorption is $|\langle\mu_x\rangle|^2$ where

$$\langle\mu_x\rangle = \int_{-\infty}^{\infty} \psi_i^*(x)\mu_x\psi_f(x)dx = \langle i|\mu_x|f\rangle \tag{4.20}$$

and μ_x is the transition dipole moment. This is an 'odd' function of x because it is proportional to the displacement of the charges. The wavefunction has quantum number i in the initial state and f in the final one. The * indicates the complex conjugate.

Show that the symmetry of the initial and final state cannot be the same in an allowed transition.

Q4.43 The transition dipole for a vibrational transition observed in the ground state of a molecule, is $\mu_x = \mu_{0x} + x\left(\dfrac{d\mu_{0x}}{dx}\right)_0$ where x is the bond displacement from the equilibrium bond length, and μ_0 is the permanent molecular dipole that is a constant, but not zero. The term $(d\mu_{0x}/dx)_0$ is the *rate of change* of this dipole with bond displacement x evaluated at the equilibrium position $x = 0$ and is also a constant. If μ_0 is zero as in H_2, O_2, or any other homonuclear molecule, the derivative is zero and no transition occurs. Water has a dipole moment of 1.94 D, ammonia 1.47 D, and HCN 3.00 D and there are similar measured values for other molecules; see the appendix in Gordy et al. (1953). The derivative $(d\mu_{0x}/dx)_0$ has dimensions of electronic charge and typically has values in the range 0.1 to 5×10^{-19} C.

(a) Assuming that $\mu_0 \ne 0$, determine each term in $\langle\mu_x\rangle$, equation (4.20).

(b) Calculate the transition moment for a vibrational transition from level i to f using the wavefunctions in Q4.39. Choose suitable i and f values.

The derivative $(d\mu_{0x}/dx)_0$ means evaluate the gradient of the dipole moment with extension at the equilibrium position $x = 0$, which is the equilibrium bond length at the bottom of the potential well, and assuming that $\mu \sim x$ then $(d\mu_{0x}/dx)_0 = constant$.

Strategy: In any complete set of wavefunctions that describe an atom or molecule, any one wavefunction is always orthogonal to all the others. This means that the product of two different wavefunctions integrated over all the coordinate(s) is always zero. More formally put, this is the 'odd × even' description. If a wavefunction has been multiplied by itself, and then the product is integrated and the result has the value 1, then the wavefunction has been normalized. Always use orthogonality if possible to determine if an integral is zero or not.

Q4.44 If a voltage gradient is imposed on a semi-permeable membrane, such as that in a biological cell, a current made up of ions will flow. The flux J of these ions (mole s^{-1} m^{-2}) through a membrane is given by (Jackson 2006),

$$J = -D\left(\frac{zFc}{RT}\frac{dV}{dx} + \frac{dc}{dx}\right) \tag{4.21}$$

where D is the diffusion coefficient, c the ion concentration, which is a function of position x, z the ion's charge, and F the Faraday constant. The flux equation is the sum of two terms: one is from the flux due to the concentration gradient across the membrane, by Fick's law this is $-Ddc/dx$, and the other term is caused by the applied voltage itself $-(zFc)dV/dx$. Multiplying J by zF converts flux into a current per area (units A m^{-2}),

$$I = -zFD\left(\frac{zFc}{RT}\frac{dV}{dx} + \frac{dc}{dx}\right). \tag{4.22}$$

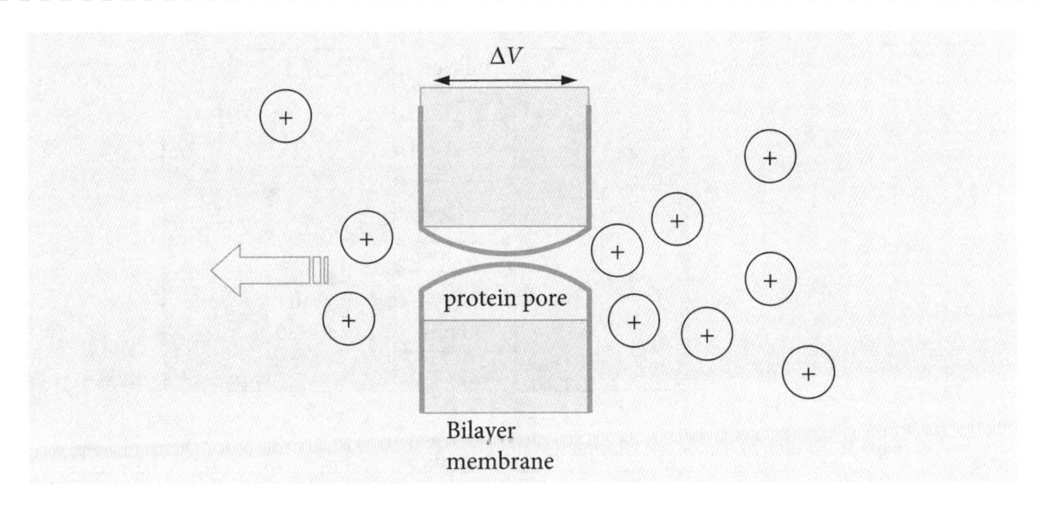

Fig. 4.14 Sketch of a pore in a bilayer membrane, Q4.44.

(a) Show that the units of J are mole s^{-1} m^{-2} and confirm the units of the equation for the current I.

(b) Verify by differentiation that the current I can be written as

$$Ie^{zFV/RT} = -zDF\frac{d}{dx}(ce^{zFV/RT}). \tag{4.23}$$

(c) Integrate both sides of this last equation separately with respect to x from $x = a$ to $x = b$. Do not forget that the concentration depends on x and thus the concentration becomes c_a and c_b. When integrating the left side, use $V = Ex$ where E is the electric field and is a constant.

(d) Simplify the resulting equation to find the current I. The final step produces an expression in E and you need to know that the electric field is proportional to the voltage difference and this can be changed to $E = \alpha\Delta V$ where ΔV is the difference in voltage over the fixed distance a to b.

(e) Calculate the equation giving the limits at large positive and large negative voltages.

(f) Plot a graph of the current vs voltage. Choose $D \approx 10^{-10}$ m^{2}s^{-1}, and voltage difference a to b of a maximum of a few hundred mV and concentrations of mM so that $c_b \approx 10^{-4}$ dm^3 mol^{-1} and $c_b \approx 10c_a$. Choose the positive constant α to suite your plotting scale; typically $\alpha = 10^6$.

***Strategy*:** (a) Look up the SI units of voltage, Faraday constant, and so forth and substitute into the equation. In (b) to 'show' or verify does not mean 'prove' so start from the answer given and work back.

Q4.45 In a medium of relative permittivity or dielectric constant ε, the force between two ions of charges, q_1 and q_2 and separated by R nm is

$$F = -\frac{q_1 q_2}{4\pi\varepsilon_0}\frac{1}{\varepsilon R^2}.$$

(a) Find the work to move a sodium ion and a chloride ion from infinity to within r_0 of one another and evaluate it if their final separation is 1 nm and the medium is water, $\varepsilon = 80$.

(b) What would the energy be if the solvent has a dielectric constant of 2?

(c) Explain in physical terms why there is such a difference in values. What does this imply for dissociation of the salt in a very dilute solution of each solvent?

***Strategy*:** Because dimensionally energy = force × distance, the work done or its equivalent the energy required, is the integral of the force along the path followed by the ions from infinity to some given separation.

Q4.46 A spherical ion of radius r and charge q in a medium of dielectric constant ε can be considered to be a capacitor with capacitance $C = 4\pi\varepsilon_0\varepsilon r$ and its potential is by definition q/C volts. The work required to increase the charge by dq is $dw = \dfrac{q}{C}dq$ and the total work required is the integral when the charge increases from 0 to q.

(a) Calculate the total work to charge the ion.

(b) Suppose that you could remove an ion from the water, $\varepsilon_{H_2O} = 80$, and place it into a lipid bilayer membrane of $\varepsilon_m = 2$. What would be the energy required to do this if the ion is K^+ which has a radius of 2 Å?

(c) Estimate the rate constant for this process if the pre-exponential factor is 10^{12} s^{-1}. Comment on your result and consider whether it is possible to pass an ion directly through a membrane at 300 K.

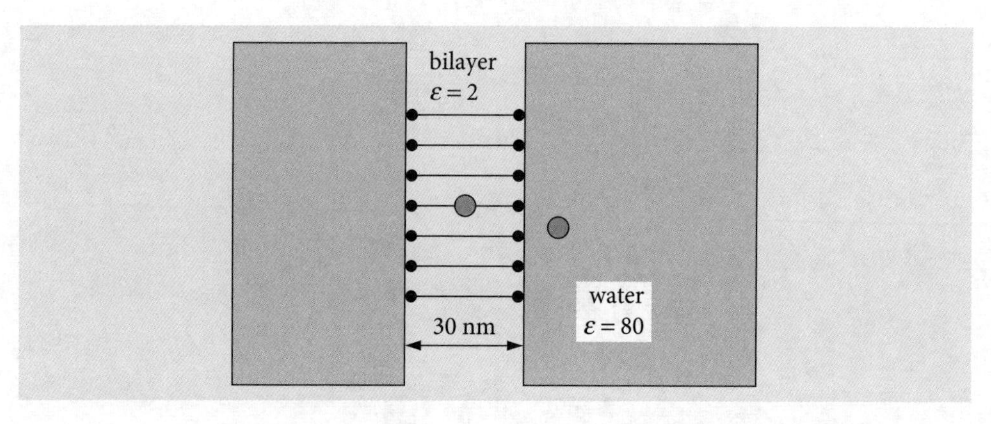

Fig. 4.15 Sketch of bilayer membrane, Q4.46.

Q4.47 The Maxwell–Boltzmann distribution describes the distribution in the *speed*, u, of gaseous molecules in three dimensions as

$$P(u)du = 4\pi\left(\frac{m}{2\pi k_B T}\right)^{3/2} u^2 \exp\left(\frac{-mu^2}{2k_B T}\right)du \tag{4.24}$$

where m is the mass (in kg) of one molecule, k_B Boltzmann's constant, and T temperature (K). Use Maple to plot the equation and to perform integrations.

(a) Show that the speed distribution is normalized.

(b) By integrating the distribution $P(u)$ from u_0 to infinity, find the probability that a molecule has a speed greater than u_0.

(c) Using your result from (b), calculate the fraction of SO_2 molecules which have a speed greater than 350 ms^{-1} at 100, 300 K, and at 500 K. ($k_B = 1.38 \times 10^{-23}$ JK^{-1}. Do not forget to put m in kg.)

(d) Calculate the fraction of molecules with energy $> k_B T$ by converting the equation from velocity to energy and calculate $\int_{k_B T}^{\infty} P(E)dE$ using $E = \frac{1}{2}mu^2$.

Strategy: Normalizing the function means making its total area equal to unity, hence an integration is needed, $\int_0^{\infty} P(u)du = 1$. The integral starts at zero because negative speeds are not allowed. Speed, unlike velocity, is not a vector quantity and is always positive because the direction in which a particle is moving is not important it its calculation. Modify the integration in part (b) to start at speed u_0 instead of zero.

Before doing the integrations, add the following instructions to help Maple do the calculation,

```
> assume(m > 0,kB > 0,T > 0);
```

Q4.48 **(a)** The escape velocity from the surface of the moon is 2380 ms^{-1}. If two cylinders, one containing He and the other N_2 gas, had their gas released on the surface of the moon which is at an average surface temperature of approximately 380 K, what fraction of each gas has enough velocity to escape the effects of gravity?

(b) The escape velocity on earth is 11 200 m s^{-1}. Repeat the calculation in (a) at 1400 K, the approximate temperature of the earth's atmosphere at an altitude of 100 km. (It is arbitrarily assumed that the gas has so few collisions at this altitude than once a molecule has enough speed it will escape gravity.) Is it possible to conclude that this high temperature is the reason that there is so little He in the atmosphere (≈5.2 ppm at sea level) even though He is produced continuously by radioactive decay in rocks?

(c) Calculate the fraction of molecules at an altitude of 100 km where ionization can occur and ions may be constrained by the earth's magnetic fields and re-examine the result of the previous two parts.

(The acceleration due to gravity is $g = GM/R^2$ where G is the gravitational constant, 6.67×10^{-11} N m^2 kg^{-2}, M the mass, and R the radius. You will need to look these up.)

Strategy: (a) and (b) use the formula developed in the previous question for the fraction of molecules having a speed greater than a given value. (c) Use the Boltzmann distribution to calculate the fraction of atoms and molecules at a given height. The potential energy of a mass m is mgh at altitude h, g being the acceleration due to gravity.

4.8 Calculating the average value of an expression

You will be familiar with obtaining the average of a set of numbers obtained in an experiment by adding them all together and dividing by their number. The average value from some theoretical expression, for example, the energy of molecules, may be an important quantity with which to compare with an experimental measurement and therefore a general way of calculating averages is required. To do this, the probability distribution P of the quantity must be known.

4.8.1 The average as a summation

Suppose the numbers q = 10, 13.5, 14, 16.2 have been measured, then their average is clearly $\langle q\rangle = (10 + 13.5 + 14 + 16.2)/4 = 53.7/4$. It has been implicitly assumed that each number is equally likely to have been measured, which is the familiar case because experiments are usually arranged in this way making them easier to do. It is not always true, however, that each value is equally probable, there is often a distribution of values that has to be taken into account, the Boltzmann distribution of energies or the degeneracy of rotational energy levels, for example.

Replacing the numerical example above with a formula makes the calculation appear much more complicated when it is in fact just the same,

$$\langle q\rangle = \frac{\sum_{i=1}^{N} q_i P(q_i)}{\sum_{i=1}^{N} P(q_i)}. \tag{4.25}$$

In this equation the probability of observing the ith observation of q is $P(q_i)$ and if $P(q_i) = 1$ then each measurement has an equal chance of occurring, which is 1, and the probability distribution is uniform or flat. If there are four measurements, then $N = 4$. Figure 4.16 shows two probability distributions. On the left is the uniform distribution, on the right a Gaussian one.

To use equation (4.25), multiply each value q_i by the chance of observing it, which is $P(q_i)$, and then sum all the values together. The denominator tells us to add together all the probabilities. If the distribution is normalized this sum will be 1. Using the values for the numbers and P, then the average, or mean value of q is

$$\langle q\rangle = \frac{\sum_{i=1}^{4} q_i P(q_i)}{\sum_{i=1}^{4} P(q_i)} = \frac{10 + 13.5 + 14 + 16.2}{\sum_{1}^{4} 1} = \frac{53.7}{4}.$$

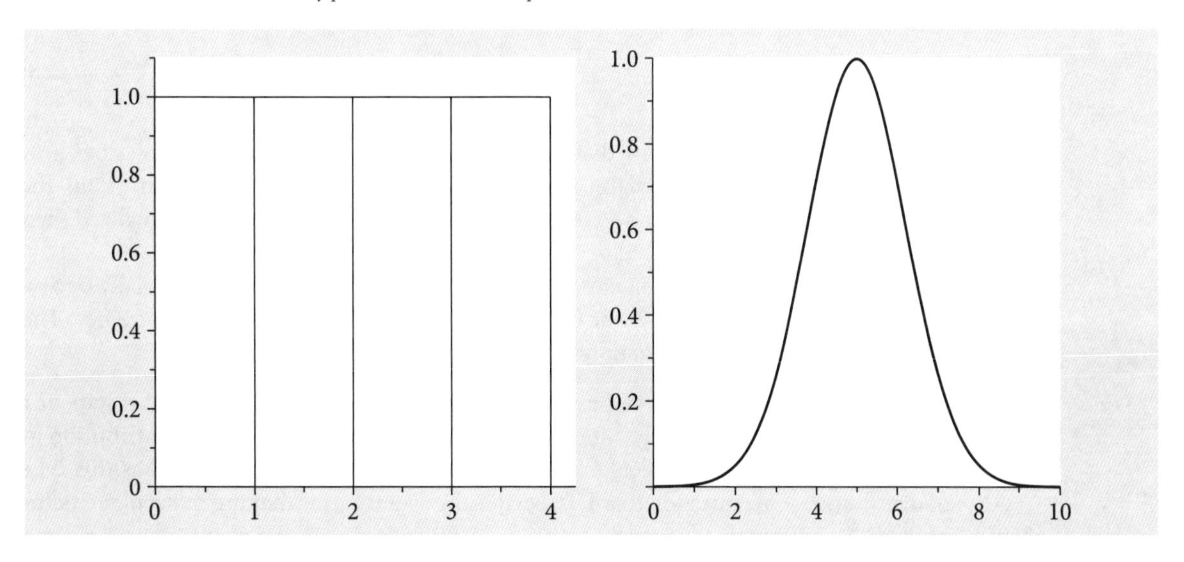

Fig. 4.16 Left: A constant or uniform distribution. Right: A Gaussian distribution. The distributions are not normalized because their area is not unity.

The average of q^2 can also be calculated using the related formula

$$\langle q^2\rangle = \frac{\sum_{i=1}^{4} q_i^2 P(q_i)}{\sum_{i=1}^{4} P(q_i)} = \frac{10^2+13.5^2+14^2+16.2^2}{\sum_1^4 1} = \frac{740.4}{4}.$$

The *variance* σ^2 is a measure of the spread about the mean value. This spread is generally called the *dispersion* and often has a Gaussian or bell-shaped distribution. The variance of a set of numbers is *defined* as

$$\sigma^2 = \langle q^2\rangle - \langle q\rangle^2 = \langle q^2\rangle - \mu^2 \tag{4.26}$$

where μ is the (population) mean, see Chapter 13. The standard deviation, the square root of the variance, is also used as a measure of dispersion and with this particular set of numbers $\sigma = 2.22$. If the numbers q represent energy E then its standard deviation is $\sigma_E = \sqrt{\langle E^2\rangle - \langle E\rangle^2}$ or if position x then $\sigma_x = \sqrt{\langle x^2\rangle - \langle x\rangle^2}$. When analysing sets of data, the average value of a set of repeated measurements is reported with the standard deviation. The result is quoted as a mean value $\mu \pm \sigma$. By blind chance, 68% of observations are to be expected to fall in the range $\mu + \sigma$ to $\mu - \sigma$ and clearly 32% outside this range. Chapter 13 describes this in more detail.

To calculate the average of higher powers of q, the nth, for example, use

$$\langle q^n\rangle = \frac{\sum_{i=1}^{4} q_i^n P(q_i)}{\sum_{i=1}^{4} P(q_i)}. \tag{4.27}$$

In statistical mechanics, the partition function is evaluated as the summation of the terms of an equation. Such sums and others are discussed in Chapter 5.

4.8.2 The average as an integration

The average quantity derived from a theoretical model of a chemical or physical process, and described by an equation, is always used to compare with the experimental value from a set of measurements. Rather than a set of numbers, a calculation similar to that described above is needed to calculate the theoretical average. However, if the function is continuous, then integration rather than summation must be used.

Suppose that x is the quantity whose average is $\langle x\rangle$, then this is calculated as

$$\langle x\rangle = \frac{\int xP(x)dx}{\int P(x)dx} \tag{4.28}$$

where $P(x)dx$ is the probability distribution of x and is the equation that describes how x is distributed and, colloquially speaking, $P(x)$ is its shape. In equation (4.28) the denominator is the normalization term but if the distribution is already normalized then $\int P(x)dx = 1$ and there is no need to calculate this integral.

The limits of the integration cover the whole range of the distribution, usually 0 to ∞, or $-\infty$ to ∞ but may be 0 to 2π or 0 to π if the calculation involves angular values. The integration limits used depend upon the particular problem.

(i) As an example, it is observed experimentally that the electronically excited state of a molecule decays exponentially, which means that there is an exponential distribution of times during which the molecule remains excited. Radioactive decay also has this type of distribution. One question to ask is; what is the average time that the molecule remains excited? To answer this question, equation (4.28) is used with the quantity x = time and the distribution function is

$$P(t)dt = \exp(-t/\tau)dt$$

which is the chance the molecule has of still being excited during a time interval dt from t to $t+dt$, if τ is its lifetime, and it was already excited at time t. The lifetime is defined as the time an ensemble of the excited molecules take to decay to $1/e$, or ~37% or their initial number. The reciprocal of the lifetime is the rate constant. The half-life is the time taken to halve an initial population but half-lives are nowadays rarely used except for radioactive species. Using equation (4.28),

$$\langle t\rangle = \frac{\int tP(t)dt}{\int P(t)dt} = \frac{\int_0^\infty t\exp(-t/\tau)dt}{\int_0^\infty \exp(-t/\tau)dt} \tag{4.29}$$

Note that the integration limits start at $t=0$ and extend to infinity. Both integrals are ones met before, and can be found using Section 4.2.13,

$$\int_0^\infty t\exp(-t/\tau)dt = \tau^2, \qquad \int_0^\infty \exp(-t/\tau)dt = \tau,$$

and their ratio produces $\langle t\rangle = \tau$ so that the average decay time is also the lifetime. Note that in this case the probability distribution is not normalized because the integral $\int P(t)dt = \tau$ and not 1. Because it is often not known beforehand if a distribution is normalized, this is ensured in the calculation by dividing by $\int P(t)dt$ as in equation (4.28).

(ii) Another example is that of calculating the mean time that a person is infected with a disease. Suppose that you are infected at time t, and suppose that the chance of no longer being infected in the next small time interval δt, is $a\delta t$, where a is a constant depending on the type of disease. This is the rate constant with which the disease will die out. If $p(t)$ is the chance that you were infected at time t, during the time $t+\delta t$ this changes to $p(t+\delta t) = p(t)(1-a\delta t) + O(\delta t^2)$, which is the chance of being infected at t multiplied by the chance of remaining infected during the following time δt. The term $O(\delta t^2)$ means that terms in δt^2 and higher are ignored because they are so small (see Chapter 5 for 'big O' notation). Subtracting $p(t)$ from both sides and dividing by δt gives

$$\frac{p(t+\delta t)-p(t)}{\delta t} = -ap(t).$$

Forming the differential with the limit $\delta t \to 0$ gives

$$\frac{dp(t)}{dt} = -ap(t).$$

As only one individual was considered $p(0) = 1$, and when integrated, by separating variables, $p(t) = e^{-at}$. If N individuals were initially infected then $p(t) = Ne^{-at}$. The average or mean time of infected is therefore by (4.29),

$$\langle t\rangle = \frac{\int_0^\infty t\,e^{-at}\,dt}{\int_0^\infty e^{-at}\,dt} = \frac{1}{a}$$

A similar type of argument is followed to find the mean free path of gas molecules or atoms but with the distance between collisions taking the place of time.

(iii) A third example is that of finding the position $\langle s\rangle$ of the centre of gravity or mass of a body that has a weight distribution $w(s)$. The weight distribution describes how the mass resides in the object and therefore describes both its shape and how the mass varies across that shape. In general, there are three values of the centre of mass, one in each of the x-, y-, and z-axes, so it is necessary to specify which one is being considered. The centre of gravity is the average position according to its mass and therefore, for each axis, the appropriate equation to use has the form of equation (4.28),

$$\langle s\rangle = \frac{\int sw(s)ds}{\int w(s)ds} = \frac{M_s}{A} \tag{4.30}$$

and s can represent any one of the x-, y- or z-coordinates but for simplicity, we will consider only two-dimensional shapes and uniform density. If the body has uniform density the coordinates of the centre of mass is often called the *centroid*. The integral $\int sw(s)ds$ is

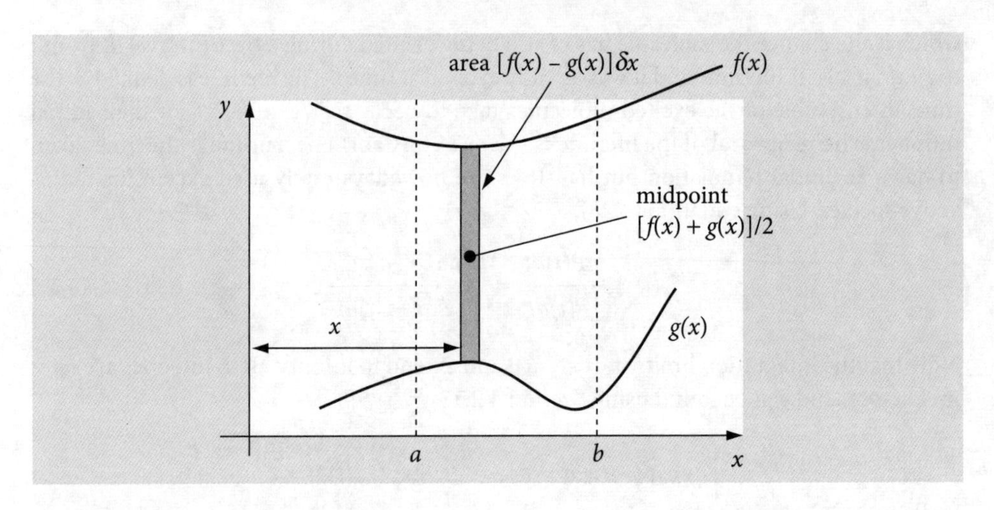

Fig. 4.17 Geometry for calculating centroids.

the *moment* M_s about an axis s and $\int w(s)ds$ the area A. If the density is not uniform, then the function w has to be multiplied by how the density varies.

(iv) Consider calculating the centroids of an area in the x–y plane. The method is to calculate the moment of a strip about each of the x- and y-axes in turn then divide the result by the total area of the object. In this case this is bounded above and below by functions f and g and by a and b along the x-axis. This area is called a *lamina* if it is of uniform density. The area is shown shaded in Fig. 4.17.

The small element between the two curves has an area $[f(x)-g(x)]\delta x$ and when multiplied by the density this would be its mass. The centre of gravity, or mean value of the strip is at y value of $[f(x)+g(x)]/2$ and the moment of the strip about the x-axis is the product of these two terms.

$$\delta M_x = \frac{1}{2}[f(x)-g(x)][f(x)+g(x)]\delta x.$$

Taking the limit to form a true differential, makes the change $\delta x \to 0$ and then dM_x/dx is formed which when integrated to cover the whole area from a to b gives

$$M_x = \frac{1}{2}\int_a^b f(x)^2 - g(x)^2 dx. \tag{4.31}$$

In the special case that the lower curve is the x-axis, then the function g is 0 and

$$M_x = \frac{1}{2}\int_a^b f(x)^2\, dx = \frac{1}{2}\int_a^b y^2 dx. \tag{4.32}$$

The similar calculation about the y-axis produces the moment

$$\delta M_y = x[f(x)-g(x)]\delta x$$

which produces the equation,

$$M_y = \int_a^b xf(x) - xg(x)dx. \tag{4.33}$$

This, when g is the x-axis, reduces to

$$M_y = \int_a^b xf(x)dx = \int_a^b xydx. \tag{4.34}$$

The area is $A = \int_a^b f(x) - g(x)dx$ and the centroids therefore,

$$\langle y\rangle = M_x/A, \qquad \langle x\rangle = M_y/A. \tag{4.35}$$

Notice how the y centroid depends on M_x and vice versa. The centroid of the curve $y = x^{1/3}$ from 0 to 1 is obtained by first finding the area, which is the integral

$$A=\int_0^1 x^{1/3}dx=\frac{3}{4}x^{4/3}\bigg|_0^1=\frac{3}{4}.$$

The moment about the x-axis is $M_x=\frac{1}{2}\int_0^1 x^{2/3}dx=\frac{3}{10}$ and the centroid is at $\langle y\rangle=M_x/A=$ 2/5. The moment about the y-axis is $M_y=\int_0^1 x^{4/3}dx=\frac{3}{7}$ and the centroid is at $\langle x\rangle=M_y/A=$ 4/7.

4.8.3 Average of a function

The average of a function of x, $f(x)$, is evaluated in a similar way as for a variable,

$$\langle f(x)\rangle=\frac{\int f(x)P(x)dx}{\int P(x)dx}. \tag{4.36}$$

If the function is $f(x) = e^{-\alpha x}$ and is distributed according to $P(x) = 1/\cosh(\alpha x)$, its average value is

$$\langle f(x)\rangle=\frac{\int e^{-\alpha x}/\cosh(\alpha x)dx}{\int 1/\cosh(\alpha x)dx}$$

which looks rather complicated. However, cosh, like many trig functions, can be represented in an exponential form as $2\cosh(x) = e^x + e^{-x}$ and this makes this normalization integral simpler;

$$\int_0^\infty 1/\cosh(\alpha x)dx=\int_0^\infty \frac{2}{e^{\alpha x}+e^{-\alpha x}}dx.$$

Trying a substitution $u = e^{\alpha x}$, then $du = \alpha e^{\alpha x}dx = \alpha u dx$ and also changing the limits produces

$$\int_0^\infty \frac{2}{e^{\alpha x}+e^{-\alpha x}}dx=\frac{2}{\alpha}\int_1^\infty \frac{1}{u^2+1}du=\frac{2}{\alpha}\tan^{-1}(u)\bigg|_1^\infty=\frac{\pi}{2\alpha}$$

where the integral in u is a standard one given in Section 4.2.13 or obtained using Maple. The numerator is

$$\int_0^\infty \frac{e^{-\alpha x}}{\cosh(\alpha x)}dx=\int_0^\infty \frac{2}{e^{2\alpha x}+1}dx.$$

Similarly making the substitution $u = e^{2\alpha x}$, using $du = 2\alpha u dx$, changing the limits, and expanding the result by partial fractions gives

$$\int_0^\infty \frac{2}{e^{2\alpha x}+1}dx=\frac{1}{\alpha}\int_1^\infty \frac{1}{u^2+u}du$$

$$=\frac{1}{\alpha}\int_1^\infty \frac{1}{u}-\frac{1}{u+1}du=\frac{1}{\alpha}\ln(u)\bigg|_1^\infty-\frac{1}{\alpha}\ln(u+1)\bigg|_1^\infty=\frac{\ln(2)}{\alpha}$$

making the average $\langle f(x)\rangle=2\ln(2)/\pi$.

4.8.4 Expectation values

In quantum mechanics, if ψ are the normalized wavefunctions and x is the displacement of an atom in a vibrating diatomic molecule from its equilibrium value, then its average position is

$$\langle x_k\rangle=\int_{-\infty}^{\infty}\psi_k^* x\psi_k dx=\langle k|x|k\rangle. \tag{4.37}$$

Expressions of this form are called the *expectation value of the operator*, which is x in this case. The wavefunctions also depend on position x although this is not explicitly shown. The subscript k tells us that we are examining level k; the superscript * shows that this is the complex conjugate of the wavefunction. The $\langle\ \rangle$ around the x indicates an average value. Alternatively this can be written with a bar as $\bar{\mu}_x$. To the right of the equation the compact and much clearer *bra-ket*, or Dirac, notation is used, which uses only the indices of the wavefunction; see Chapter 8. Formally, in equation (4.37), x is the operator and it is always placed between the two wavefunctions. In this particular instance it does not make any difference where x is, but if the operator was d/dx then this has to be followed by ψ and preceded by ψ^*.

Suppose that you are in the teaching lab and are asked to devise an experiment that attempts to measure $\langle x\rangle$ for a harmonic oscillator at several different vibrational energies. Do you rush away and collect lots of equipment or sit down and think about the problem? The former is definitely more exciting—select a laser, several mirrors and lenses, vacuum pump, oscilloscope, detectors, and so forth—but ultimately futile. Thinking about the experiment first is always the better option; it saves you wasting time on pointless experimentation, leaving you time to tackle interesting problems. The resulting average displacement $\langle x\rangle$, as you will have realized for a harmonic oscillator, should be zero.

Why is it concluded that the average displacement is zero just by looking at equation (4.37)? Because the wavefunctions are the same, their product must be an even function; any normal function squared is an even function. Because any displacement such as x, is an odd function: $-x$ is not the same as $+x$, the integral is odd overall and *must* integrate to zero, provided the integration range is symmetrical about zero; Section 4.1.3. This result corresponds to the common sense observation that in the harmonic potential, the bond extension is equal and opposite to the compression, making the average displacement zero for any value of the vibrational quantum number. Figure 4.38 (see online solutions) shows a harmonic potential. It was worth simply thinking about measuring the average displacement first. An experiment will still be needed, although we now know what to expect; it is often a mistake not to do an experiment because you 'know' what the answer is. If the average displacement measured from different vibrational levels were different, what then might you conclude?

The symmetry rules are:

$$(\text{odd}\times\text{odd}) = \text{even} \qquad (\text{even}\times\text{odd}) \equiv (\text{odd}\times\text{even}) = \text{odd}$$

$$(\text{odd}\times\text{odd}\times\text{even}) = \text{even} \qquad (\text{even}\times\text{odd}\times\text{odd}) = \text{even}.$$

If the product of the wavefunctions and operator x is even, there may still be enough symmetry left so that the integral is zero because the positive and negative parts can exactly cancel. With an odd function the integral is always zero. The odd/even classification is very simple; space is split into two parts. If it were split into four parts then, as you can appreciate, things would be more complicated. In general where functions are more complex, as the molecular orbitals or the vibrational normal modes of molecules can be, then group theory (Chapter 7) must be used to decide whether or not the direct product of the irreducible representations are, or are not, non-totally symmetric. The jargon makes this calculation sound difficult but, as with most jargon, it hides simple things.

The general expectation value between levels with different quantum numbers n and m is always calculated as:

$$\langle Q_{nm}\rangle = \frac{\int \psi_n^* Q \psi_m dx}{\int \psi_n^* \psi_m dx} \tag{4.38}$$

where Q is the operator acting on the wavefunction ψ_m for state m. The wavefunction is a function only of x in this example and that is implied by the fact that integration is only with respect to x. You should be aware, however, that it is common to use τ to represent general coordinates which may be x, y, or z or r, θ, and ϕ, and then the integral would be a triple one. Note that if the wavefunctions are normalized then the denominator is unity and

$$\langle Q_{nm}\rangle = \int \psi_n^* Q \psi_m dx. \tag{4.39}$$

Obviously, the operator Q changes depending upon the problem at hand. Often this is the Hamiltonian operator, which represents the total energy, kinetic plus potential, but Q could describe a dipole as in equation (4.37). If H is the Hamiltonian representing energy, the familiar equation

$$E = \int \psi^* H \psi d\tau \tag{4.40}$$

is produced from $H\psi = E\psi$ by multiplying by ψ^* and integrating both sides. The integration differential $d\tau$ is shorthand for whatever the coordinates of the wavefunction are. Note that the coordinates on ψ are assumed, as are the indices for quantum levels. The equation could also be written as $H\psi_n(x) = E_n\psi_n(x)$ for any quantum level n. Also, note that in (4.40) ψ is assumed to be normalized.

It is emphasized again that it is important to note the order of the calculation in equations of the type (4.37) to (4.39). The wavefunction is potentially changed by the operator Q, then it is multiplied by ψ^* and the result integrated. If Q is d/dx the order of calculation really does matter, because $\psi^*\left(\dfrac{d}{dx}\psi\right) \neq \dfrac{d}{dx}(\psi^*\psi)$. For example, if the wavefunction is $\psi(x) = \sin(ax)$, which it might be for a particle in a box, then

$$\sin(ax)\frac{d}{dx}\sin(ax) = a\sin(ax)\cos(ax), \text{ but } \quad \frac{d}{dx}\sin^2(ax) = 2a\sin(ax).$$

In some simple cases when the operator is $Q = x$ or x^2 then the multiplication order does not matter. An important detail is that in quantum mechanics time is a parameter and it is never possible to obtain $\langle t\rangle$ using an equation of the form of (4.38) because the wavefunctions always depend on some coordinate and time is never the integration variable. The same is true of mass.

4.8.5 Higher averages: moments and variance

The average $\langle x\rangle$ is also called the *first moment* of its distribution; $\langle x^2\rangle$ the second moment and so on. The general relationships are, assuming that $P(x)$ is a normalized distribution,

$$\langle m_k\rangle = \int x^k P(x)dx \tag{4.41}$$

$$m_1 \equiv \langle x\rangle = \int xP(x)dx \qquad m_2 \equiv \langle x^2\rangle = \int x^2P(x)dx. \tag{4.42}$$

The variance of a distribution is

$$\sigma_x^2 \equiv \langle x^2\rangle - \langle x\rangle^2 = \int (x - \langle x\rangle)^2 P(x)dx \tag{4.43}$$

4.8.6 Questions

Full solutions are available at **www.oxfordtextbooks.co.uk/orc/beddard.**

Q4.49 **(a)** Find the coordinates of the centroid M_x and M_y and average x and y of the first quadrant of a circle whose equation is $r^2 = x^2 + y^2$, r being the radius.

(b) Find the centroids of the enclosed area between the circle in (a) and $g = x^2$.

Strategy: Use equations (4.31) and (4.33) to calculate the centroids; the averages are given by equation (4.35).

Q4.50 **(a)** Find the mass and centre of gravity measured from the flat side of a quartz 'fisheye' lens of uniform density ρ_0 which has a parabolic profile on one side and is plain on the other. The equation of the parabola is $y = (ax)^{1/2}$ where a is a constant and the lens is h cm thick.

(b) Subsequent experiments show that better results are obtained if the lens is next made from gradient index glass with density changing as $\rho(x) = \rho_0(1 + bx^2)$. What effect does this have on the centre of gravity?

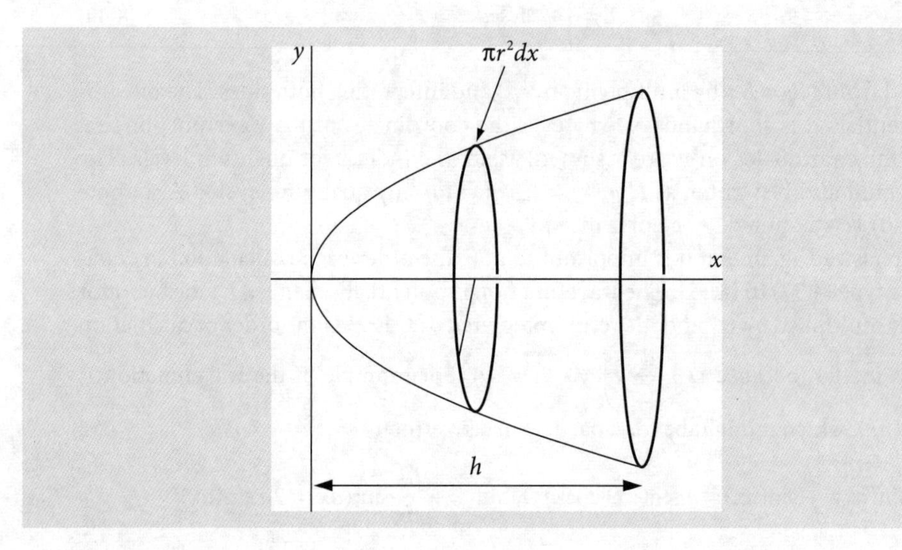

Fig. 4.18 Sketch for Q4.50.

Q4.51 Calculate the volume and surface area formed from the hyperbola $y = 1/x$ from $x = 1$ to ∞. The shape of the volume produced is sometimes called Gabriel's horn. The volume is obtained by calculating the area of a disc and integrating along x. The surface area S is obtained by calculating the circumference of a circle $2\pi y$ at position x then integrating along the whole arc, which has a length element $ds = \sqrt{1 + y'^2}dx$. $S = \int_a^b 2\pi y ds$ and y' is the derivative. Comment on the surface area with respect to the volume. Does this make sense?

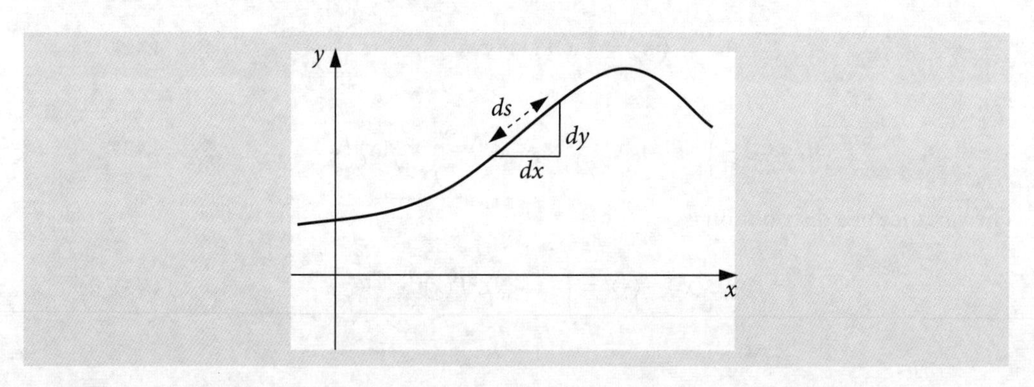

Fig. 4.19 The arc length ds is, by Pythagoras' theorem, $ds^2 = dx^2 + dy^2$, which by dividing by dx^2 can be rearranged into $ds = \sqrt{1 + y'^2}dx$.

Q4.52 In the study of the sequence of base pairs in DNA (Perkus 2001), the occurrence of overlapping sequences of base pairs (clones) can be estimated statistically. The probability of a clone starting at a given site is p and x is its position along the DNA chain taken to be a continuous variable.

(a) Evaluate $F = \dfrac{d}{dp}\displaystyle\int_a^b e^{-px}dx$ by integrating first then differentiating the result, then doing the calculation in reverse order.

(b) Do the same with $F = \dfrac{d}{dp}\ln\left[\displaystyle\int_a^b e^{-px}dx\right]$ and show that this is the average $\langle x\rangle$ of the exponential distribution e^{-px}.

Strategy: Note that, by the way the equations are written, the parameter p is not a variable of the integration and is therefore a constant during integration. In (b) treat the integral as you would any other function.

Q4.53 Evaluate $\dfrac{\partial}{\partial\alpha}\ln\left(\int e^{-E_{\alpha,x}}dx\right)$ where $E_{\alpha,x} = \alpha x^2$.

Strategy: The integral is a function of x, but contains α so differentiate first. This is very similar to part (b) of the previous question. Maple is helpful in obtaining the individual integrals that are also standard.

Q4.54 The probability distribution (or probability density) describing the chance that a molecule has energy in range E to $E + dE$ is given by the Boltzmann equation

$$P(E)dE = \exp\left(-\frac{E}{k_BT}\right)dE$$

where T is the temperature in K, and k_B Boltzmann's constant $= 1.38 \times 10^{-23}$ J K^{-1} and $E \geq 0$.

(a) Calculate the average energy $\langle E\rangle$ and the mean square energy $\langle E^2\rangle$ for an ensemble of N molecules.

(b) Calculate also the dispersion $\sigma_E^2 = \langle E^2\rangle - \langle E\rangle^2$ and comment on the size of the ratio $\sigma_E/\langle E\rangle$.

Q4.55 **(a)** Calculate $\dfrac{d}{dT}\int f(x, T)dx$.

(b) The slope of the log of the rate constant $k(T)$ vs. $1/T$ is used to measure the activation energy E_a of a chemical reaction. The slope is $\dfrac{d\ln(k(T))}{d(1/T)} = -\dfrac{E_a}{k_B}$.

Show that this is $\dfrac{dk(T)}{dT} = \dfrac{E_a k(T)}{k_BT^2}$.

(c) Collision theory defines the rate constant as

$$k(T) = 2\sqrt{\frac{2}{\pi\mu}}\left(\frac{1}{k_BT}\right)^{3/2}\int_0^\infty \sigma(E)Ee^{-E/k_BT}dE$$

where $\sigma(E)$ is the reaction cross section at energy E and E is the relative kinetic energy of the reacting species. Calculate E_a using the result of (b) and explain what each of the terms represents.

(d) Calculate E_a if the cross section varies as predicted by the line of centres geometrical model of collisions where $\sigma(E) = \pi d^2(1 - E_0/E)$ above the threshold energy E_0 and is zero below this. The sum of molecular radii on collision is d. While the integration is straightforward, use Maple to obtain the result and make E_0 the lower range of the integration.

Q4.56 In Q4.47, a plot of the Maxwell–Boltzmann distribution was used to illustrate the probability $P(u)$ of a gaseous molecule having a speed u to $u + du$ where,

$$P(u)du = 4\pi\left(\frac{m}{2\pi k_BT}\right)^{3/2}u^2\exp\left(\frac{-mu^2}{2k_BT}\right)du.$$

(a) Calculate the average speed $\langle u\rangle$ and the root mean square speed (rms) speed $\sqrt{\langle u^2\rangle}$.

(b) Rewrite the probability $P(u)$ in terms of the kinetic energy $E = \dfrac{1}{2}mu^2$ by replacing $P(u)du$ with $P(E)dE$ in your equations, i.e. substitute for u and use $du = dE/(2mE)^{1/2}$.

(c) Calculate $\langle E\rangle$ and $\langle E^2\rangle$ using the fact that the heat capacity $C_V = \dfrac{\partial\langle E\rangle}{\partial T}$ show that $\sigma_E^2 = k_BT^2C_V$ where σ^2 is the variance, equation (4.26).

(d) Using C_V, calculate the ratio $\sigma_E/\langle E\rangle$.

Strategy: This is the type of question that requires direct application of the formulas; use Maple when integration involves square roots and/or terms with e^{-x^2}.

Q4.57 A harmonic oscillator, such as a vibrating spring or a diatomic molecule, has potential energy given by $E_x = 0.5kx^2$ where x is the extension from the equilibrium length at $x = 0$ and therefore x can be both positive and negative; see Fig. 4.38 (see online solutions). The average energy is k_BT.

(a) Calculate the average $\langle x\rangle$ and mean square displacement $\langle x^2\rangle$. Calculate the variance in position. Explain what you understand by $\sqrt{\langle x^2\rangle}$.

(b) The frequency ν of a molecule's vibration (in s^{-1}) is related to the reduced mass μ and the force constant k by $\nu = \frac{1}{2\pi}\sqrt{\frac{k}{\mu}}$. The force constant for I_2 is 173 N m^{-1} and the equilibrium bond length 267 pm. The vibrational frequency is $\approx$213 cm^{-1}. Calculate and comment on the magnitude of the root mean square displacement ($\sqrt{\langle x^2 \rangle}$) *vs.* the equilibrium bond length.

Q4.58 The displacement at time t of a classical harmonic oscillator is $x = x_0 \cos(\omega t)$ where ω is the frequency and x_0 the maximum amplitude.

(a) Find the probability $P(x)dx$ of being at position x to $x + dx$ during the motion, this being proportional to the time taken to move by dx.

(b) Show that your distribution is normalized even though it is infinite at two places.

(c) Calculate $\langle x \rangle$ and $\langle x^2 \rangle$ using this distribution.

Strategy: Assume that the time to move a distance dx is proportional to the time dt taken to do so. Therefore find dx/dt and convert the sine (using Pythagoras) into a function in x. The period of oscillation is $T = 1/\nu = 2\pi/\omega$.

Q4.59 In the previous problem $\langle x \rangle$ and $\langle x^2 \rangle$ were calculated for a classical harmonic oscillator. In this problem, the same calculation is done quantum mechanically. The vibrational wavefunction of a harmonic oscillator for the $n = 0, 1, 2,$ and 3 quantum numbers are given in Q4.39. The probability distribution of the nuclei being at a position between x and $x + dx$ is $\psi^* \psi dx$.

(a) Determine the equations and then calculate $\langle x \rangle$ and $\langle x^2 \rangle$ for each vibrational level.

(b) By induction find a general formula for $\langle x^2 \rangle$ that depends upon n the vibrational quantum number. Find this also in terms of the energy and compare it with the classical value of the previous example.

(c) Using the data below, plot the wavefunctions and the potential energy curve on the same graph, placing the wavefunction at the appropriate energy. To produce a realistic figure, use the data for CO, which is $r_e = 113$ pm and $\omega_e = 2170.2$ cm^{-1}.

(d) Explain the result you obtain for $\langle x \rangle$, which is easy, and that for $\langle x^2 \rangle$, which is harder. You need to consider the shape of the wavefunctions.

(e) Calculate the mean momentum $\langle p \rangle$ and $\langle p^2 \rangle$ and confirm the uncertainty principle as $\sqrt{\langle x^2 \rangle}\sqrt{\langle p^2 \rangle} \ge \frac{\hbar}{2}$.

The energy of a vibration is $E_n = h\nu(n + 1/2)$ where ν is the vibrational frequency in s^{-1} and n the quantum number. The formula $\nu = \frac{1}{2\pi}\sqrt{\frac{k}{\mu}}$ relates the force constant to the vibrational frequency. The quantum operator for momentum is $-i\hbar d/dx$.

(In Maple it will help to use `assume(alpha > 0 );` etc. when evaluating the integrals.)

Strategy: Although this is a long question, it is easily split into manageable parts. The equation to use is (4.28) for part (a) and (4.38) for the momentum in part (e). Equation (4.38) is used because the momentum operator acts on the wavefunction ψ before this is multiplied by ψ^* (the complex conjugate of ψ) and integrated. When the operator is x or x^2 multiplying ψ, ψ^* and x in any order makes no difference to the result and we can use (4.28).

In calculating $\langle x \rangle$ it is sufficient to look up the equations for the wavefunctions and study their odd or even symmetry. Determine which ones are even $\psi(-x) = \psi(x)$ and which odd $\psi(-x) = -\psi(x)$. No integration is needed. The wavefunctions are normalized, therefore the denominator in the equations (4.28) and (4.38) is unity. (You will need to multiply the wavefunction by a very small number to be able to plot on the same scale as the potential.)

Q4.60 The vibrational properties of a molecule can be described by several classical harmonic oscillators. If the zero of energy is placed at the zero-point energy, then the probability distribution of s independent classical harmonic oscillators having total energy E distributed among them is given by

$$P(E)dE = \frac{\beta}{(s-1)!}(\beta E)^{s-1} e^{-\beta E} dE$$

where $\beta = 1/k_B T$ and k_B is the Boltzmann constant and T the temperature. This equation is, of course, not such an accurate model for every molecule but if the vibrational energies are small compared to $k_B T$ then it is a good approximation.

(a) The distribution $P(E)$ has the same form as a well know distribution. What is this called, and under what assumptions is it valid? Show that the distribution is normalized.

(b) Calculate the average energy $\langle E\rangle$ and $\langle E^2\rangle$ and the standard deviation or width of the energy distribution.

(c) Plot the distribution $P(E)$ vs E for 10 and 50 oscillators with $\beta = 1/200\ (\text{cm}^{-1})^{-1}$ and energy up to 20 000 cm^{-1}, and confirm that the average energies are correct. This value of β is approximately that at room temperature.

***Strategy*:** Equation (4.28) gives the formula for an average. Use Maple to perform the calculation or use the table for the general integration result. A gamma function is produced in the result. It is useful to know that if s is a positive integer $\Gamma(s+1) = s!$ See chapter 13 for the distribution.

Q4.61 The particle in a box wavefunction is $\psi_n(x) = N_n \sin(n\pi x/L)$, where N_n is the normalization and L the length of the box and n the quantum number which has values $n = 1, 2, 3, \cdots$. The wavefunction is defined from $0 < x < L$ and is zero outside these limits.

(a) Calculate the normalization constant and

(b) $\langle x\rangle$ and $\langle x^2\rangle$ and σ_x^2 where $\sigma_x^2 = \langle x^2\rangle - \langle x\rangle^2$ for different n.

Q4.62 Calculate the values of the electronic energy levels of a molecule represented as a particle in a box of length L using the wavefunctions from the previous example. The Hamiltonian operator contains only the kinetic energy part as the potential energy is zero. The kinetic energy operator is $-\dfrac{\hbar^2}{2m}\dfrac{d^2}{dx^2}$ where m is the mass of the particle in this case an electron.

***Strategy*:** Use equation (4.38) and place the operator after the first wavefunction. Differentiate twice and then integrate. The wavefunctions are normalized so that the denominator in (4.38) is 1 and can be ignored and becomes (4.39).

Q4.63 Using the normalized particle in a box wavefunctions in the previous problem, calculate the intensity of each of the four electronic absorption transitions in octatetraene originating from the HOMO orbital. Assume that the octatetraene's energy levels can be described by such a simple model and that the transition dipole leading to absorption is $\mu = q(x - L/2)$ where q is the charge on the electron. The absorption probability is $|\langle\mu\rangle|^2$, which is the square of the absolute value of the expectation value μ. The intensity of the transition between the lower level with quantum number n and an upper level with number k is

$$A_{nk} = \beta v_{nk}\, |\langle\mu_{nk}\rangle|^2 \qquad \text{dm}^3\ \text{mol}^{-1}\ \text{s}^{-1}\ \text{cm}^{-1}$$

and where v_{nk} is the transition frequency and $\beta = \dfrac{8\pi^3 N}{3000hc}$. Calculate the absorption spectrum if the molecule has a length 7×1.40 Å.

***Strategy*:** Draw out the energy levels using the energy from the previous question if you want to do this accurately. Octatetraene has eight π electrons and eight orbitals are required. The lowest energy transition must be from $n = 4$ to 5 and the others to levels 6, 7, and 8. The expectation value is equation (4.38) with operator $Q \equiv \mu = q(x - L/2)$. As the wavefunctions are normalized the denominator in this equation is unity and there is no need to calculate it. A particular transition frequency can be calculated from the difference in energy levels.

Q4.64 In electronic spectroscopy an electron is excited from the ground state to one of several excited states but often only to the lowest. These transitions produce the familiar colours we see from the artificial dyes used in our clothes and paints, natural dyes in food and the green colour of leaves (chlorophyll) and varied colours of flowers. Franck–Condon (FC) factors determine the intensity of the vibrational part of these electronic spectra and hence largely their colour, and are the square of the absolute amount by which two vibrational wavefunctions overlap with one another.

The probability of absorbing a photon is proportional to the Franck–Condon integral

$$F_{n,m} = \left|\int_{-\infty}^{\infty} \psi_{an}^{*}(x - x_a)\,\psi_{bm}(x - x_b)\,dx\right|^2 \equiv |\langle n|m\rangle|^2$$

where $\psi_{an}(x - x_a)$ is the vibrational wavefunction of state a, with quantum number n, and $\psi_{bm}(x - x_b)$ that of the other state b with quantum number m. The coordinate of the a and the b wavefunction is different because they are in different electronic states; x_a and x_b are the equilibrium internuclear separation of each of the states. Because they belong to different states, the wavefunctions ψ_a and ψ_b are not orthogonal to one another and the Franck–Condon integral is not identically zero.

The general normalized wavefunctions for the harmonic oscillator are (see Fig. 4.48 of the online solutions for sketch)

$$\psi_n(x-x_0)=\sqrt{\frac{1}{2^n n!}}\left(\frac{\alpha}{\pi}\right)^{1/4}H_n[(x-x_0)\sqrt{\alpha}]e^{-\alpha(x-x_0)^2/2} \qquad (4.44)$$

where $H_n(z)$ is the Hermite polynomial with quantum number n and general coordinate z. The constant $\alpha=\sqrt{mk/\hbar^2}$, where k is the force constant (units N m^{-1}) of the vibration and m the mass.

(a) What does it mean if two wavefunctions are not orthogonal?

(b) Use Maple to calculate the first few Hermite polynomials and check a few wavefunctions with those in Q4.39.

(c) Calculate the Franck–Condon factors originating from the $n=0$ state of a diatomic molecule to the mth vibrational level of one of its excited electronic states. Use Maple to perform the integrations and show by induction, i.e. by trying several values of m and generalizing the result, that

$$F_{0,m}=|\langle 0|m\rangle|^2=\frac{X^m}{2^m m!}e^{-X/2}$$

where $X=\alpha(x_n-x_m)^2$ is the dimensionless displacement between the two electronic states. What are the units of α if X is dimensionless?

(d) Plot the FC factors with quantum numbers from 0 to 10 and with reduced displacement $X=1$ and 5. Comment on the shape of the graph as the displacement increases.

(e) Using values for CO ground state of $x_a=1.1281$ Å, $\upsilon_a=2170.2$ cm^{-1} and the X electronic excited state $x_b=1.235$ Å, $\upsilon_b=1515.6$ cm^{-1} calculate the spectrum from the $n=0$ level of the ground state. The $n=0$ to $m=0$ electronic energy gap is 64746.5 cm^{-1}. You only need to go to $m=10$ in the excited state.

Strategy: (a) Look up how to define Hermite polynomials in Maple; the mathematical notation is different to the one Maple uses. (b) Use the equation given for the wavefunction and list it with $n=0$ and $m=0, 1, 2$, and so forth. Before calculating a FC factor simplify the calculation by removing the constants to outside the integration and substituting for the $n=0$ wavefunction. Calculate the remaining integral and construct the FC factors by multiplying with the constants. (c) Use the numerical values given to calculate actual values for the Franck–Condon factors. Plot a stick graph where the x values are the transition energies and y values are the Franck–Condon factors.

Q4.65 The process of Forster or dipole–dipole energy transfer from an excited donor D^* to an acceptor A is described by the equation $D^*+A\xrightarrow{k_R}D+A^*$. The donor excited state is quenched and its fluorescence is reduced in intensity; conversely the acceptor emission is stimulated. The acceptor emission is usually fluorescence but could be phosphorescence if A is a triplet state of a molecule. The Forster transfer process forms the basis of the FRET techniques now widely used in the biosciences to estimate distances between pairs of chromophores on proteins, DNA or membranes. FRET is a mnemonic for fluorescence, resonance energy transfer. This energy transfer competes with fluorescence; $D^*\xrightarrow[k_f]{h\nu}D$ which has a rate constant k_f.

In solution, the rate constant of energy transfer is not constant but varies with time. This happens because of the R^{-6} dependence of the transfer rate upon separation, R, of the donor to its surrounding and randomly positioned acceptors. Integrating over all these distances, the intensity of the fluorescence emitted by a donor molecule is given by

$$I(t)=I_0\exp\left(-k_f t-\frac{c}{c_0}\sqrt{\pi k_f t}\right)$$

where c the concentration of acceptor molecules $c_0=3\pi N/4R_0^3$ is a constant and R_0 is the characteristic energy transfer distance between donor and acceptor. This is a measure of the overlap of the donor's emission with the acceptor's absorption spectrum. Experimental measurements confirm this theoretical expression.

Using Maple calculate the relative yield, ϕ/ϕ_0, which is the ratio of fluorescence yield at zero quencher ϕ_0 with that at a quencher concentration c. A fluorescence yield is the total amount of fluorescence emitted over all time. Plot the resulting graph after making the substitution $x=\frac{c}{c_0}\frac{\sqrt{\pi}}{2}$ in the result. By definition, the relative fluorescence yield is the ratio of integrals

$$\frac{\phi}{\phi_0}=\int_0^\infty I(t)dt\bigg/\int_0^\infty I_{c=0}(t)dt, \qquad (4.45)$$

where the denominator is the total emission when the quencher concentration is zero.

Q4.66 In Q4.65 the following expression is found and has to be evaluated when x is large.

$$\frac{\phi}{\phi_0} = 1 - x\sqrt{\pi}\,[1 - erf(x)]e^{x^2}$$

Use L'Hôspital's rule to evaluate this when $x \to \infty$ and show that it tends to zero.

Q4.67 Direct spin–spin coupling is observed in solid-state NMR spectroscopy but not in solution. The direct spin–spin coupling between two nuclear spin magnetic moments is dependent on the square of the cosine of the angle between them, the effect being proportional to $\dfrac{3\cos^2(\theta) - 1}{r^3}$ where r is the separation of the spins. Show that the effect of averaging over all angles due to the rotational motion in solution makes the effective coupling zero, i.e. that $\langle \dfrac{3\cos^2(\theta) - 1}{r^3} \rangle = 0$.

Strategy: If the coupling is zero then it must follow that after averaging over both polar and azimuthal angles $\langle \cos^2(\theta) \rangle = 1/3$ because then

$$\langle 3\cos^2(\theta) - 1 \rangle \equiv 3\langle \cos^2(\theta) \rangle - 1 = 0.$$

To average over angles the area element $\sin(\theta)d\theta d\phi$ is used and the average is not only over the polar angle θ but also 'around the equator' with the azimuthal angle ϕ to allow for all possible motions. Again use equation (4.28). It is always necessary in three-dimensional problems such as this, to average over both angles defining a point, i.e. the polar and azimuthal, even though only one is present in the formula. This is a three-dimensional problem because the molecule whose NMR spectrum is measured will undergo rotational diffusion in solution.

Q4.68 The wavefunction of the 1s state of a hydrogenic atom in spherical polar coordinates is

$$\psi(r,\theta,\phi) = \left(\frac{Z}{a_0}\right)^{3/2} \frac{e^{-Zr/a_0}}{\pi^{1/2}},$$

where r is the radial distance from the nucleus, a_0 is the Bohr radius, 52.9 pm, and Z is the atomic number of the hydrogenic atom, 1 for H, 2 for He^+, 3 for Li^{2+}, and so forth.

(a) Show that the *radial distribution* of this 1s wavefunction is normalized. (It is necessary to integrate over all angles as well as r.)

(b) Show that $\langle 1/r \rangle = Z/a_0$ and then calculate $\langle 1/r^2 \rangle$, $\langle 1/r^3 \rangle$, $\langle 1 \rangle$, $\langle r \rangle$, and $\langle r^2 \rangle$. Next use Maple to find $\langle r^n \rangle$ and show that the general formula produces the previous results.

(c) Calculate the standard deviation of the electron probability distribution.

(d) Calculate the probability that that the electron is within a_0 of the nucleus. How does this change as Z increases and why? You may wish to plot the function.

Strategy: The radial parts of the wavefunctions are only normalized when multiplied by r^2; the radial distribution part is therefore $\int_0^\infty \psi^* \psi r^2 dr$. The total wavefunction also has angular parts described by a polar angle θ, which is integrated from 0 to π and an azimuthal or 'round the equator' angle ϕ, which ranges from 0 to 2π. The normalizing equation, as proved in your textbook, is the product of three integrals

$$N = \int_0^{\pi}\int_0^{2\pi}\int_0^{\infty} \psi^* \psi r^2 \sin(\theta) dr d\theta d\phi.$$

The volume element for integrating over all coordinates r, θ, and ϕ is $r^2 \sin(\theta) dr d\theta d\phi$. However, as there are no terms in products $r\theta$, $\theta\phi$, or $r\phi$, such as $\sin(r\theta)$, this allows the integral to be separated into three parts;

$$\int_0^{\pi}\int_0^{2\pi}\int_0^{\infty} \psi^* \psi r^2 \sin(\theta) dr d\theta d\phi = \int_0^{\pi} \sin(\theta) d\theta \times \int_0^{2\pi} d\varphi \times \int_0^{\infty} r^2 \psi(r)^* \psi(r) dr \qquad (4.46)$$

In part (b) of the question the average $\langle 1 \rangle$ is the same as $\langle r^0 \rangle$ and is just the normalization term.

Q4.69 A solution of molecules each of which has a dipole μ is placed in a cuvette across which an electric field of strength ε is applied. Each molecule experiences a torque tending to align it with the direction of the field. However, the alignment is far from complete, because random thermal (Brownian) motion causes the solvent molecules to collide with the dipolar ones and randomizes their orientation.

The interaction energy of a dipole with the electric field is $E_\theta = -\varepsilon\mu\cos(\theta)$ where θ is the angle between the dipole and the field. The Boltzmann distribution of orientational energies is

$$W(E_\theta)dE_\theta = \exp\left(-\frac{E_\theta}{k_BT}\right)dE_\theta.$$

The angular distribution of molecules over the angular range angle θ to $\theta + d\theta$ is $2\pi\sin(\theta)W(E_\theta)d\theta$ on a sphere of unit radius. Note that the azimuthal (round the equator) angle is already included as 2π in the angular distribution. If there are N dipoles, then the fraction that lie in the angular range θ to $\theta + d\theta$ is just $N/4\pi$ times this quantity because the total surface area of the unit sphere is 4π, therefore the fraction at angle θ is $P(\theta)d\theta = N\sin(\theta)W(E_\theta)d\theta/2$. The calculation is shown in the left-hand part of Fig. 4.20.

(a) Calculate the average energy of the dipole $\langle E\rangle$ in the field ε by averaging over all θ angles, 0 to π radians. Simplify the result.

(b) Assume that $\varepsilon\mu/k_BT \ll 1$, which is usually the case, and show that $\langle E\rangle$ is proportional to $(\mu\varepsilon)^2/3k_BT$. Explain whether this result makes physical sense. (Hint: expand the exponentials $e^x \approx 1 + x + x^2/2! + \cdots$ when $x < 1$ or use the Maple `series()` command.) Calculate $\langle E\rangle$ at large field and again comment on the result.

(c) Calculate $\langle E^2\rangle$ and using the fact that the heat capacity $C_V = \dfrac{\partial\langle E\rangle}{\partial T}$, and by assuming that $\varepsilon\mu/k_BT \ll 1$ only after calculating C_v, show that $\sigma_E^2 = k_BT^2C_V$ where $\sigma_E^2 = \langle E^2\rangle - \langle E\rangle^2$.

(d) Calculate $\sigma_E^2 = \langle E^2\rangle - \langle E\rangle^2$ at large field strength or low temperatures when $\varepsilon\mu/k_BT \gg 1$.

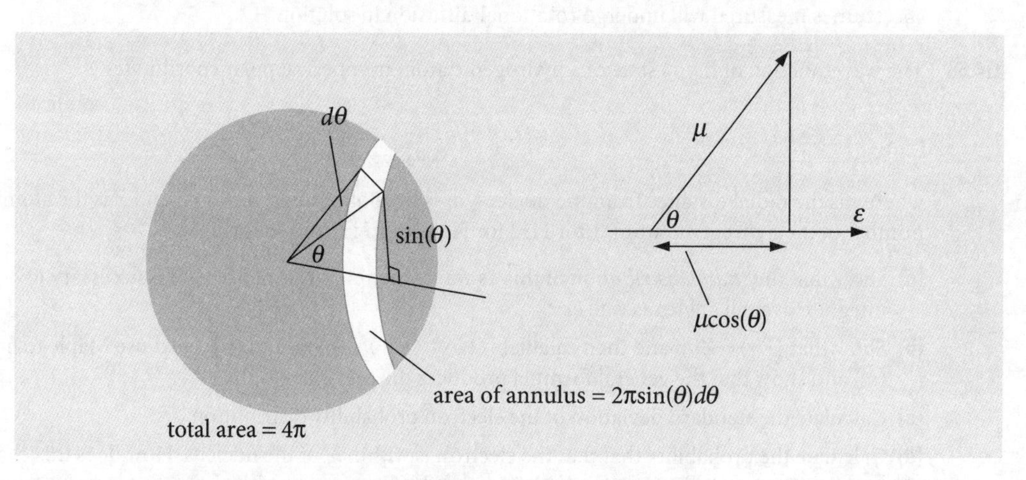

Fig. 4.20 Left: A sphere and the solid angle area element. Right: The projection of the dipole onto the field ε, which has the value $\mu\cos(\theta)$ and therefore the dipole energy is $E_\theta = -\varepsilon\mu\cos(\theta)$.

Strategy: Use the standard equation where the variable for integration is the angle θ. You are given the fraction or probability in the question. The denominator in equation (4.28) is not unity.

Q4.70 The rotational energy E_J of a molecule assumed to behave as a rigid rotor depends upon the quantum number J as $E_J = BJ(J+1)$. B is the rotational constant and J has values $J = 0, 1, 2, 3, \cdots$ The rotational constant depends upon the average bond length r of the molecule and its reduced mass μ as $B = \dfrac{\hbar^2}{2\mu r^2} = \dfrac{\hbar^2}{2I}$ where I is the moment of inertia. The probability distribution describing the chance of a molecule having energy E_J is obtained from the Boltzmann equation $P(E_J)dE_J = g_J\exp\left(-\dfrac{E_J}{k_BT}\right)dE_J$, where T is the temperature (K), k_B Boltzmann's constant (1.38×10^{-23} JK^{-1}), and g_J is the degeneracy of level J; in this case this is $2J + 1$.

(a) Calculate the average rotational quantum number at temperature T and, by assuming that $B < k_BT$, which, except for H_2 and other light molecules is always true, simplify your result. Assume that J; is a continuous variable that can be integrated. Use Maple to approximate the error function produced. Plot the distribution for $^{79}Br_2$ at 30 K, 300 K and 600 K. The equilibrium bond length is 228 pm.

(b) Calculate the average J quantum number and by differentiation of P the quantum number at the maximum of P. Explain your results in physical terms.

Q4.71 The partition function Z of an atomic or molecular ensemble is the sum of the terms in a Boltzmann distribution and is used to calculate the fraction of the total number of particles in any given energy level i, which is n_i/Z. The translational quanta for a molecule are so small that the summation can be replaced by integration. If $g(\varepsilon)$ is the degeneracy of the energy levels at energy ε then

$$Z=\sum_{\varepsilon=0}^{\infty} g(\varepsilon)e^{-\varepsilon/k_BT} \rightarrow \int_0^{\infty} g(\varepsilon)e^{-\varepsilon/k_BT}d\varepsilon,$$

and for translational motion $g(\varepsilon)=\dfrac{4\pi\sqrt{2}}{\hbar^3}Vm^{3/2}\varepsilon^{1/2}$, where m is the particle's mass and V the volume of the container.

Calculate Z using Maple as necessary to perform the integrations.

Strategy: Integrations with exponentials multiplied by fractional powers of x are usually not able to be integrated directly, but produce instead a result that is another function such as the error function. In these cases, use Maple.

Q4.72 A free particle is one that is not influenced by a potential. If it is travelling in the x-direction with momentum p_x it has the wavefunction $\psi=e^{-p_x x/\hbar}$.

(a) Find the expectation $\langle p_x\rangle$ for the operator $p_x=-i\hbar\partial/\partial x$.

(b) Calculate $\Delta p^2=\langle p_x^2\rangle-\langle p_x\rangle^2$.

(c) Is ψ an eigenfunction of both operators p_x and p_x^2?

Strategy: The limits to the integration are not given, so it is unlikely that they will be needed. Eigenfunctions of the same operator commute; see Chapter 7.5.3.

4.9 The Variational Method in Quantum Mechanics

The variational method has been widely used in quantum chemistry to obtain the energy levels of polyatomic molecules and other similarly complicated problems. This method, also called the Rayleigh–Ritz method, is an approximate way of obtaining energies where no exact solution is possible. If an exact quantum calculation has ground state energy E_0^0 and wavefunction φ_0, the Schrödinger equation is $H\varphi_0=E_0^0\varphi_0$. The variational theorem asserts, for any trial wavefunction ψ that is perhaps related to, but is not φ_0, that

$$E_V=\frac{\int\psi^*H\psi d\tau}{\int\psi^*\psi d\tau}\geq E_0^0. \tag{4.47}$$

This means that the energy E_V, calculated with any trial wavefunction ψ, cannot be lower that the ground state energy E_0^0, calculated with the same Hamiltonian H using $H\varphi_0=E_0^0\varphi_0$. The calculation to find E_0^0 may not be possible and this will be unknown, but E_V can still be found as an approximation to E_0^0 even if the integration has to be done numerically. This is important, for in a complicated molecule with many nuclei and even more electrons, solving the Schrödinger equation exactly is not possible in principle even with enormous multiprocessor computers. The variational method, however, allows us to approximate the energy and most importantly, the approximate energy is never lower than the true one; this is what makes this method useful. Initially, the wavefunction ψ is usually a guess based on intuition. Better guesses of ψ will make the left-hand side of equation (4.47) closer to the true value E_0^0, which means in practice that lower and lower values of the energy are calculated until energies asymptotically approaching, but never falling below, the true value are found.

Because the variational principle minimizes the total energy, the sum of the kinetic and potential energies, this is varied to find its minimum and this is done by differentiation. The total energy is reflected in the shape of the wavefunction and so the variational parameter used, a, is a parameter that defines the shape of the wavefunction, for example $\psi=e^{-ar^2}/a^3$ or the amount of one wavefunction vs. another as in a linear combination $\psi=a\varphi_1+\varphi_2$. Although any fantastical function can be chosen as a wavefunction, those leading to a good value of the minimum energy when compared with a known solution are those that represent wavefunctions in some general way. These are single valued, do not have discontinuities and tend to zero at infinity.

The variational calculation is done in five easy steps:

(a) Guess a function to be the wavefunction ψ. This will depend on the problem at hand but starting with a wavefunction from a similar problem is often a good starting point. It must have at least one parameter that is to be varied to find the minimum energy.

(b) Using the total Hamiltonian, $H = -\dfrac{\hbar^2}{2m}\dfrac{\partial^2}{\partial u^2} + V(u)$ where $V(u)$ is the potential energy, which depends upon the problem being studied, calculate $H\psi$. The kinetic energy is always $-\dfrac{\hbar^2}{2m}\dfrac{\partial^2}{\partial u^2}$ where m is the mass and u the appropriate positional coordinate, for example the Cartesians $\{x, y, z\}$ or polar coordinates $\{r, \theta, \phi\}$.

(c) Calculate the integrals in equation (4.47). This is probably the hardest part of the problem.

(d) When the equation for the variational energy E_V has been found in terms of the variational parameter a, differentiate this to find the minimum energy; $dE_V/da = 0$. If there is more than one parameter they must all be minimized.

(e) The value of a found is put back into the equation for E_V to obtain its minimum value. This will be the best approximation to the energy with the type of wavefunction chosen. Better or worse energies can be found with different wavefunctions.

As a one-dimensional example, suppose that an electron is subject to a delta function potential that exists only at the origin (Szabo & Ostlund 1982). This potential is simple to deal with mathematically, but it is somewhat unrealistic: it could perhaps approximate a defect in a nanowire or an oxygen atom in a chain of carbons. The Schrödinger equation in atomic units (see chapters 1.15.3 & 11.10.2) is

$$\left(-\frac{1}{2}\frac{d^2}{dx^2} - \delta(x)\right)\psi = E\psi \tag{4.48}$$

and the delta function has the property that it is unity when x is zero but is zero elsewhere. This simplifies the integration considerably because for any normal function f

$$\int_{-\infty}^{\infty} \delta(x) f(x) dx = f(0),$$

and the delta function can be removed from the integral and replaced by the value of the function f at $x = 0$.

(a) As a trial wavefunction $\psi = e^{-ax^2}$ is tried and the value of a will have to be found that minimizes the energy.

(b) In equation (4.48), the Hamiltonian is the term in brackets and acts on the wavefunction producing

$$\left(-\frac{1}{2}\frac{d^2}{dx^2} - \delta(x)\right)\psi = (a - 2a^2x^2 - \delta(x))e^{-ax^2}$$

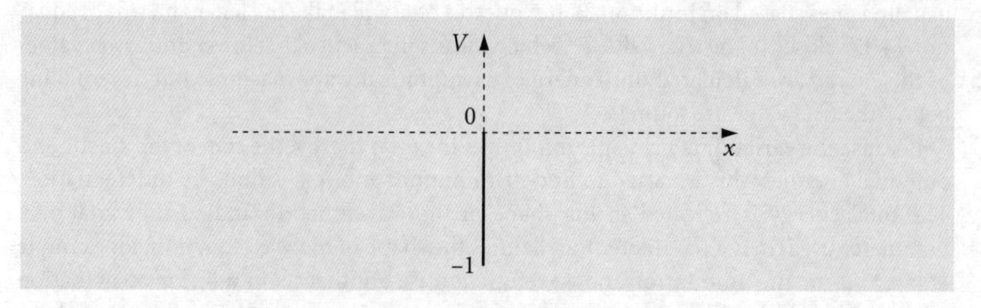

Fig. 4.21 A delta function potential with a value of −1 at the origin.

(c) To use equation (4.47) the integrals have to be evaluated. The denominator, which ensures normalization, is

```
> Int( psi(x)^2,x= -infinity..infinity): %= value(%);
```

$$\int_{-\infty}^{\infty} (\mathrm{e}^{-a\,x^2})^2\,\mathrm{d}x = \frac{1}{2}\frac{\sqrt{2}\,\sqrt{\pi}}{\sqrt{a}}$$

and the expectation value integral $\int \psi H \psi dx$ is

$$\int_{-\infty}^{\infty} (a - 2a^2x^2 - \delta(x))e^{-2ax^2}\,dx = \int_{-\infty}^{\infty} (a - 2a^2x^2)e^{-2ax^2}\,dx - 1.$$

The -1 comes from the integral of the delta function. The remaining integrals are standard ones between the limits of $\pm\infty$. However, Maple can perform the whole integral

```
> Int(psi(x)*(-Diff(psi(x),x,x)/2-psi(x)*Dirac(x)),
        x=-infinity..infinity): %= value(%);
```

$$\int_{-\infty}^{\infty} \mathrm{e}^{-ax^2}\left(-\frac{1}{2}\left(\frac{\partial^2}{\partial x^2}\mathrm{e}^{-ax^2}\right) - \mathrm{e}^{-ax^2}\,\mathrm{Dirac(x)}\right)\mathrm{dx} = -1 + \frac{1}{4}\sqrt{a}\,\sqrt{2}\,\sqrt{\pi}$$

The variational energy is therefore

$$E_v = \frac{\int \psi^* H\psi dx}{\int \psi^* \psi dx} = \frac{\sqrt{2\pi a}/4 - 1}{\sqrt{\pi/2a}} = \frac{a}{2} - \sqrt{\frac{2a}{\pi}}$$

but this gives the energy in terms of a, which is the variational parameter also called the Ritz parameter.

(d, e) To find the minimum energy, E_V has to be minimized. By differentiation with respect to a

$$\frac{d}{da}(a/2 - \sqrt{2a/\pi}) = 1/2 - \sqrt{1/2\pi a} = 0,$$

from which $a = 2/\pi$. Substituting into E_V gives the energy as $-1/\pi$ or -0.318 fairly close to, but greater than, the true value of -0.5.

4.9.1 A semi-proof of the method

A rigorous proof, which would be inappropriate here, rests on the mathematical foundation that an approximate solution to the Schrödinger equation can be written as a linear combination of the wavefunctions of an exact solution. To illustrate the method, the starting point is the exact solution to the Schrödinger equation; $H\varphi_i = E_i^0\varphi_i$ for energy levels $i = 0, 1, 2, 3, \cdots$ The ground state energy is E_0^0. The problem to solve approximately, using the variational method, has the same Hamiltonian and is $H\psi_i = E_i\psi_i$. We start by guessing that the wavefunction $\psi = a\varphi_1 + b\varphi_2$ will be a good solution provided coefficients a and b are optimized. The wavefunction ψ in this form is said to be 'expanded in the basis set' of just two normalized and orthogonal wavefunctions φ.

The variational energy is

$$E_V = \frac{\int (a\varphi_1 + b\varphi_2)^* H\,(a\varphi_1 + b\varphi_2)du}{\int (a\varphi_1 + b\varphi_2)^*\,(a\varphi_1 + b\varphi_2)du}$$

where the * indicates the complex conjugate, which will be ignored from now on by assuming that the wavefunctions are real. The energy E_V has to be shown to be greater than E_0^0. Expanding out the denominator, which is the normalization term, gives

$$\int (a\varphi_1 + b\varphi_2)(a\varphi_1 + b\varphi_2)du = a^2\int \varphi_1^2 du + b^2\int \varphi_2^2 du + 2ab\int \varphi_1\varphi_2 du$$

and similarly for the numerator

$$\int (a\varphi_1 + b\varphi_2)H(a\varphi_1 + b\varphi_2)du$$

$$= a^2\int \varphi_1 H\varphi_1 du + b^2\int \varphi_2 H\varphi_2 du + ab\int \varphi_1 H\varphi_2 du + ab\int \varphi_2 H\varphi_1 du$$

To evaluate these integrals further the wavefunction normalization and orthogonality conditions have to be invoked. As φs are exact solutions of the Schrödinger equation they satisfy the orthonormality condition

$$\int \varphi_1\varphi_2 du = \delta_{12}$$

where δ is the Kronecker delta and has the properties $\delta_{i,j} = 0(i \neq j)$, $\delta_{i,j} = 1(i = j)$. Using these rules and the essential fact that because φ are exact solutions then

$$\int \varphi_i H\varphi_j du = \delta_{ij}E_j^0.$$

After substituting for the integrals the variational energy is

$$E_V = \frac{a^2E_1^0 + b^2E_2^0}{a^2 + b^2}.$$

If E_0^0 is now subtracted from both sides

$$E_V - E_0^0 = \frac{a^2(E_1^0 - E_0^0) + b^2(E_2^0 - E_0^0)}{a^2 + b^2} \geq 0$$

It must be true that $E_V - E_0^0 \geq 0$ because E_0^0 is the ground state energy and E_1^0 and E_2^0 are levels with larger quantum numbers and are therefore higher in energy. Repeating the calculation with wavefunctions φ_0 and φ_1 or with larger linear combinations essentially produces the same result, which is that a linear combination, and by inference any other approximation, is never smaller than the true value for the lowest energy level. Further details on the variational theorem are to be found in most quantum and physical chemistry textbooks such as McQuarrie & Simon (1997), Levine (2001), or Atkins & de Paula (2006).

4.9.2 Questions

Full solutions are available at **www.oxfordtextbooks.co.uk/orc/beddard.**

Q.4.73 For the particle in a one-dimensional box, calculate the variational energy if the trial wavefunction is $\psi = x(L - x)$ where L is the length of the box for which the potential energy $V = 0$. Compare your result with the true energy for the lowest level which is $\dfrac{h^2}{8mL^2}$. The Schrödinger equation is $-\dfrac{\hbar^2}{2m}\dfrac{\partial^2}{\partial x^2}\psi = E\psi$.

Strategy: Calculate the two integrals in equation (4.47) separately; remember that H acts on the wavefunction to its right and the result is then multiplied by ψ before the integration is evaluated.

Q4.74 Suppose that in the quartic oscillator $V(x) = kx^4$, the trial wavefunction is chosen to be $\psi = \dfrac{\alpha^{1/2}}{\pi^{1/4}}e^{-\alpha x^2/2}$ where α is the variable parameter. Calculate the minimum variational energy and compare your result with the exact result $E_0 = 1.060k^{1/3}\left(\dfrac{\hbar^2}{2m}\right)^{2/3}$.

Q4.75 The variational treatment can be used to approximate the ground state energy of the He atom. The energy cannot be calculated exactly because there are three particles, the nucleus and two electrons. Try to find a solution using the normalized wavefunction $\varphi = \frac{1}{\pi}\left(\frac{\zeta}{a_0}\right)^3 e^{-\zeta(r_1-r_2)/a_0}$ with ζ (Greek *zeta*) as a variational parameter to replace the nuclear charge Z; a_0 is the Bohr radius and r_1 and r_2 are the coordinates of the two electrons, which are independent of one another. The parameter ζ should be smaller than Z because one electron is shielded from the nucleus by the other and vice versa.

The energy expectation value is a very complicated integral and evaluates to $E = \int \varphi^* H\varphi du = \frac{e^2}{4\pi\varepsilon_0 a_0}(\zeta^2 - 2Z\zeta + 5\zeta/8)$ where H is the Hamiltonian operator and du represents integration over all coordinates.

(a) Look up the formula for the energy of an H atom. Calculate the (zeroth order) ground state energy of the He atom as twice the energy of an H atom with Z = 2. What interaction terms are missing in this crude model?

(b) Calculate the ionization energy of He^+ and add this to the experimentally measured first ionization energy of 24.5 eV then call this the 'experimental' energy of the He atom. Compare this with the energy from (a).

(c) Find the variational energy of the ground state of the He atom. Compare it with the energy calculated assuming that $\zeta = Z$, and with your 'experimental' value.

Q4.76 Using the trial wavefunction $\psi_x = e^{-ax^2/2}$, calculate the variational energy of the lowest bound state of an electron if the potential has the form of a Gaussian well $V_x = 1 - e^{-\beta x^2}$. Use the Schrödinger equation in atomic units: $\left(-\frac{1}{2}\frac{d^2}{dx^2} + V_x\right)\psi_x = E\psi_x$. Compare the result with the numerical value of 0.5226 when $\beta = 1$ and 0.20473 when $\beta = 10$. (See Chapter 11.10 for details of a numerical method.) Comment on the results obtained.

Strategy: The potential is anharmonic with a value of 1 at large values of $\pm x$ and zero at the origin. The expansion of the potential at small x is

$$1 - e^{-\beta x^2} \approx \beta x^2 - \beta^2 x^4/2 \cdots$$

and therefore the Gaussian wavefunction suggested as a trial wavefunction should be a good approximation as this is the form of the lowest wavefunction for a harmonic potential. To simplify the calculation, use Maple to find the optimum value of α. Do the calculation with $\beta = 1$ then 10, rather than use β in the formula. If you use β rather than a number, solving for a becomes very difficult.

Q4.77 Using atomic units (see Chapter 1.14.3), the Schrödinger equation for the hydrogen atom is

$$-\frac{1}{2}\nabla^2\psi - \frac{\psi}{r} = E\psi$$

where the operator del squared is

$$\nabla^2 = d^2/dx^2 + d^2/dy^2 + d^2/dz^2.$$

Using a trial radial wavefunction $R(r) = e^{-\alpha r^2/2}$ and the variational method, show that the ground state energy is $-4/(3\pi) = -0.42$, which is slightly more than the exact energy which is $-1/2$. You will need to use the relationship $\nabla^2 f(r) = \frac{1}{r}\frac{d^2 rf(r)}{dr^2}$ and the volume element in spherical coordinates is $dxdydz \to r^2 \sin(\theta)drd\theta d\phi$.

Strategy: The Schrödinger equation is given in mixed coordinates Cartesian and radial. This is not unusual because it is assumed that you know how to convert from one to the other; the relationship is given in the question. However, there are three coordinates, not just r to deal with, but the wavefunction is only given in terms of r. The reason for this is that the lowest energy corresponds to an s orbital, which is spherically symmetrical. The integrations have the form

$$\iiint R^*(r)R(r)dxdydz \to \iiint R^*(r)R(r)r^2 \sin(\theta)drd\theta d\phi$$

and because the wavefunction does not depend on θ and ϕ these integrals can be separated out. You must decide what the limits are; the variational parameter is α.

4.10 **Multiple integrals**

Many functions contain two or more variables; for instance, the distance of the electron in a hydrogen atom depends on its x, y, and z-position from the nucleus. Any point on a plane is given by its x- and y-coordinates or on the earth by its latitude and longitude. To evaluate a double or triple integral, each integral is performed with the same rules as a single integral but care must be taken to sort out the integration limits. The use of multiple integrals quite often involves transforming variables, from Cartesian x, y, z to different forms of polar or cylindrical coordinates depending on the problem. The only reason for doing this is to simplify the integration and the only difficulty is unfamiliarity with the new coordinates.

Integrating over a surface such as $f(x, y)$ with the double integral $\iint f(x, y)dydx$ means integrating along the x- and y-axes to obtain the volume between the surface and the x–y plane. The extent of the integration is determined by the limits to the integration in the x–y plane. Figure 4.22 shows the case when the limits are constants and then the integral is written as

$$A = \int_a^b \int_c^d f(x, y)dydx. \tag{4.49}$$

The volume produced is like a rectangular rod with a flat base but a top cut to the shape of the function. The next integral is easily evaluated as two separate ones because the integration limits are constants and the terms can be separated. Note how the integral signs and the variables are written—inside to inside and first to last.

$$\int_0^{\pi}\int_0^{\pi 2} \cos(\theta)\sin(\phi)d\theta d\phi = \int_0^{\pi} \sin(\phi)d\phi \int_0^{\pi/2} \cos(\theta)d\theta$$

$$= -\cos(\phi)\Big|_0^{\pi} \sin(\theta)\Big|_0^{\pi/2} = 2$$

In quantum mechanics, double and triple integrals often have the form where one coordinate's function is unity and integration is made over some range of angles; ϕ ranges from 0 to π in the following integral

$$\int_0^{\pi} d\phi \int_0^{\pi/2} \cos(\theta)\sin(\theta)d\theta = \pi \int_0^{\pi/2} \cos(\theta)\sin(\theta)d\theta = \pi/2.$$

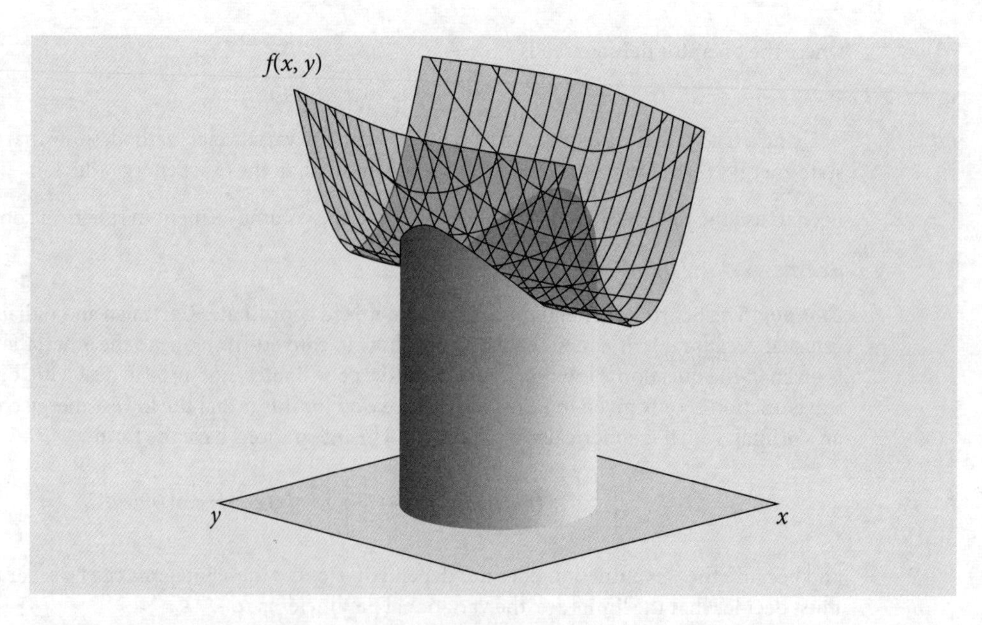

Fig. 4.22 Double integral pictured as the volume between a surface and the x–y plane.

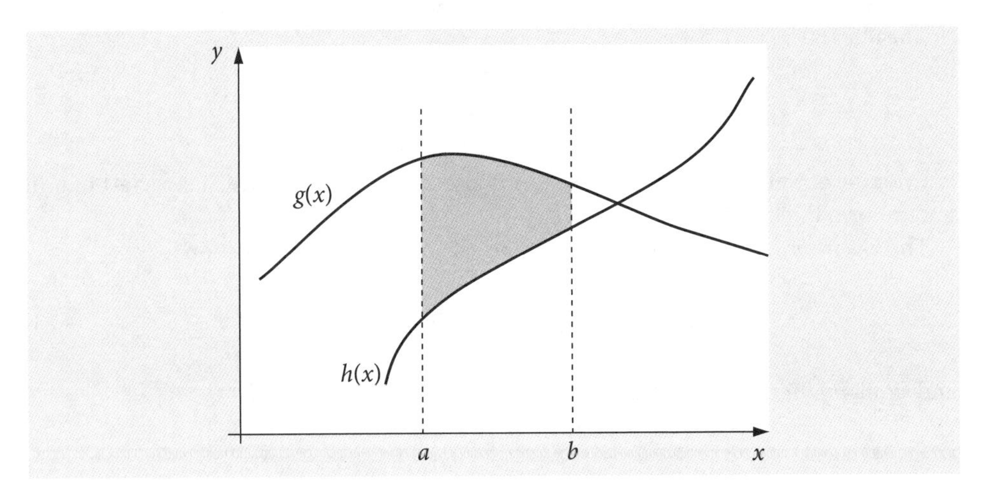

Fig. 4.23 Limits to the double integral in the x–y plane. The integral is the volume above the shaded area and towards the reader and extends to the function $f(x,y)$.

More interesting is the case when the double integral is bounded in the plane by variable limits. As shown in Fig. 4.23, x has limits a and b but y has limits which are the *functions* g and h, which both depend on x. As the y limits are not constant the integral is written as

$$\int_R\int f(x,y)dydx \equiv \int_a^b\left[\int_{g(x)}^{h(x)} f(x,y)dy\right]dx. \tag{4.50}$$

Notice also the order of integration. As written, the integration on y must be performed first and the result is a function of x and is integrated last. The y integration has to have limits depending on x or ones that are constant. It cannot have limits depending on y because the result could not be integrated by x.

Suppose that $f = \ln(x)$ and the two functions g and h are $g = x^2$ and $h = 2 - x^2$ and the x integration range is from -1 to 1, as shown in Fig. 4.24. The double integral is

$$\int_R\int f(x,y)dxdy = \int_{-1}^{1}\int_{x^2}^{2-x^2} \ln(y)dydx \tag{4.51}$$

The inner integral has to be found first which will give a result in x. This is then integrated. The inner integral is a standard one and using Maple to do the calculation and put in limits gives

```
> Int( ln(y), y= x^2..2-x^2 ):%= value(%);
```

$$\int_{x^2}^{2-x^2} \ln(y)\,\mathrm{d}y = -x^2\ln(x^2) + 2x^2 + 2\ln(2-x^2) - \ln(2-x^2)x^2 - 2$$

The next step is to integrate this result from $x = -1$ to 1 and this is clearly more complex. Letting Maple do the whole calculation gives

```
> Int( Int( ln(y),y= x^2..2-x^2 ),x=-1..1 ):%= value(%);
```

$$\int_{-1}^{1}\int_{x^2}^{2-x^2} \ln(y)\,\mathrm{d}y\,\mathrm{d}x = -\frac{64}{9} + \frac{16}{3}\sqrt{2}\,\mathrm{arctanh}\left(\frac{1}{2}\sqrt{2}\right)$$

4.10.1 Mean values, moments of inertia, and centroids

The double integral is useful in obtaining the mean value, centroids and moments of inertia of functions and is an alternative way to that described in Section 4.8.2. In these equations the order of integration is not necessarily implied by the way in which they are written. The integration order is unimportant if the integration limits are constants but is if they depend on x then the y integral must be done first as in equation (4.51).

The integral

$$A = \iint 1dxdy \tag{4.52}$$

is a volume of unit thickness. The result A is also called a lamina, which is a sheet of unit but uniform thickness.

The *x-centroid*, the average position of the x-coordinate of the function is

$$\langle x \rangle = \frac{1}{A}\iint xdxdy \tag{4.53}$$

and similarly for the y-coordinate

$$\langle y \rangle = \frac{1}{A}\iint ydxdy$$

If the density is not constant but is described by a function $f(x, y)$ then the formulae are changed to

$$\langle x \rangle = \frac{1}{A}\iint xf(x, y)dydx, \qquad \langle y \rangle = \frac{1}{A}\iint yf(x, y)dydx, \qquad A = \iint f(x, y)dydx \tag{4.54}$$

which is essentially equation (4.28). The position of the centroids is often called the *centre of gravity* of the object.

By definition, the moment of inertia of an object about an axis is the product of the distance squared from the axis times the mass; $I = mL^2$. If the body is extended, then the integration must be performed over the entire shape. Should the mass is distributed as $f(x, y)$ then the x-direction moment of inertia depends on the distance from the y-axis and is

$$I_x = \iint y^2 f(x, y)dydx. \tag{4.55}$$

The y-direction moment is calculated similarly using x^2,

$$I_y = \iint x^2 f(x, y)dydx$$

If the mass is uniform then the function f is a constant and can be taken outside the integration.

These various calculations are now illustrated. To calculate the lamina's area using equation (4.52) the limits have to be defined. In Fig. 4.24 the closed area is that bounded by $g = x^2$ and $h = 2 - x^2$ from $x = -1$ to 1 and these will determine the integration limits. Notice that the y integral is performed first which produces a result in x that is then integrated. The area is calculated as

$$A = \int_{-1}^{1}\int_{x^2}^{2-x^2} dydx = \int_{-1}^{1} y\Big|_{x^2}^{2-x^2} dx = \int_{-1}^{1} 2 - 2x^2\,dx = \frac{8}{3}$$

This particular result can also be obtained in a simpler way by calculating the integral of both g and h and subtracting them;

$$\int_{-1}^{1}(h - g)dx = 2\int_{-1}^{1}(1 - x^2)dx = 8/3$$

The centroids are

$$\langle x \rangle = \frac{1}{A}\int_{-1}^{1}\int_{x^2}^{2-x^2} xdydx = \int_{-1}^{1} x\left(y\Big|_{x^2}^{2-x^2}\right)dx = \frac{1}{A}\int_{-1}^{1} 2x - 2x^3\,dx = 0$$

$$\langle y \rangle = \frac{1}{A}\int_{-1}^{1}\int_{x^2}^{2-x^2} ydydx = \frac{1}{A}\int_{-1}^{1}\left(\frac{y^2}{2}\Big|_{x^2}^{2-x^2}\right)dx = \frac{1}{A}\int_{-1}^{1} 2 - x^2 - x^4\,dx = 1$$

as would be expected for this symmetrical shape.

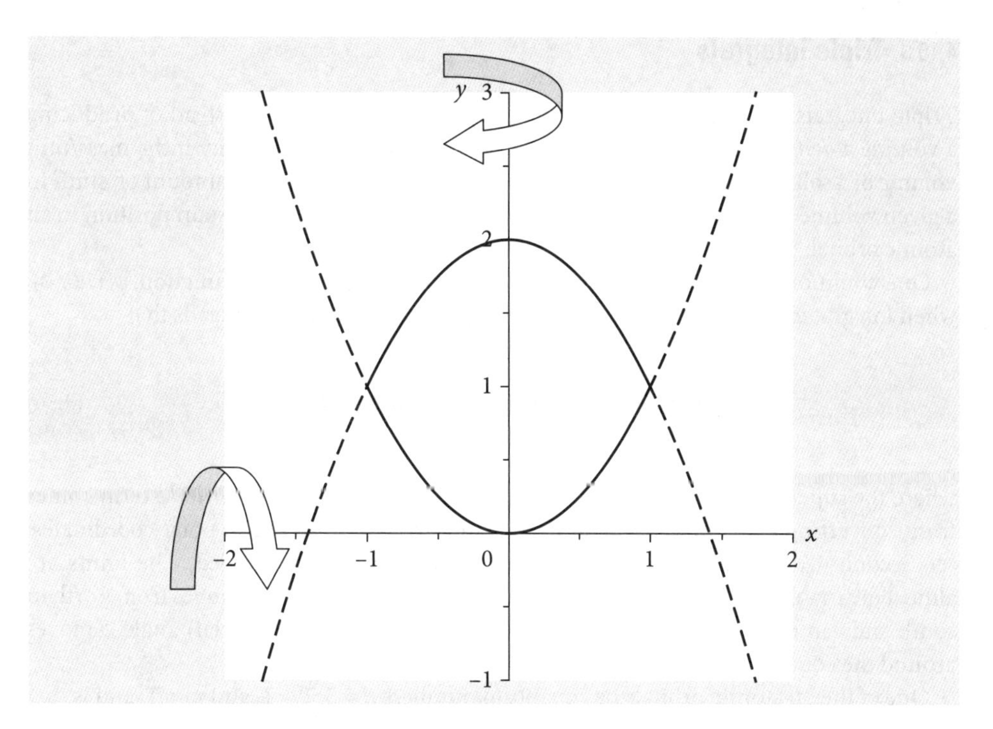

Fig. 4.24 Area bound by x^2 and $2 - x^2$ between −1 and 1 is shown as a solid line. Rotation about x and y is indicated illustrating that the x-axis moment of inertia will be largest.

The moments of inertia of an object, even with some symmetry, are generally quite different along its various axes and this is the case for the x and y moments of inertia of the lamina Fig. 4.24. Using (4.55) with y^2 or x^2 as necessary and assuming uniform density of 1, $f(x, y) = 1$ then the moments of inertia I_y and I_x about the y- and x-axes are respectively,

```
> Int(Int(x^2,y = x^2..2-x^2),x = -1..1): % = value(%);# Iy
```

$$\int_{-1}^{1}\int_{x^2}^{2-x^2} x^2\, dy\, dx = \frac{8}{15}$$

```
> Int(Int(y^2,y = x^2..2-x^2),x = -1..1): % = value(%);# Ix
```

$$\int_{-1}^{1}\int_{x^2}^{2-x^2} y^2\, dy\, dx = \frac{344}{105}$$

The moment of inertia about the x-axis is greater (by approximately 6 times) than that about y because the rotation about x involves the whole body rotating around this axis whereas the body is symmetrically disposed about the y-axis and therefore the moment of inertia is smaller. The moments of inertia of molecules are described in Chapter 7.15.

4.10.2 **Question**

Full solutions are available at **www.oxfordtextbooks.co.uk/orc/beddard**.

Q4.78 The mass of a flat plate varies as $f = \sigma \sin(x + y)$ and the plate is cut by the straight lines $g = x$ and $h = 2x$ and with $x = 0$ to 1, which makes a triangular area in the x–y plane. Calculate the centre of mass and the moments of inertia.

Strategy: Use equations (4.54) to calculate the centre of mass and (4.55) for the moments of inertia.

4.10.3 Triple integrals

Triple integrals are calculated in a similar way to double ones, but instead of producing a volume a *density* function is produced. This might literally be density if the mass/unit volume of a solid is known, but generally, density is taken to mean the amount of 'stuff' in a given volume such as electron density or probability of being at a certain position in an atomic orbital.

One commonly met triple integral is the normalization of a wavefunction $\psi(r, \theta, \phi)$; when integrated over all space the product $\psi^*\psi$ must be unity. This means that

$$\int_0^{2\pi}\int_0^{\pi}\int_0^{\infty} \psi(r,\theta,\phi)^*\psi(r,\theta,\phi)r^2\sin(\theta)drd\theta d\phi = 1 \tag{4.56}$$

where the superscript * indicates a complex conjugate. The $r^2\sin(\theta)drd\theta d\phi$ term comes from converting the volume element $dxdydz$ in Cartesian to spherical polar coordinates; see Section 4.11 where the calculation of this conversion is described. The limits are almost invariably the same for quantum problems. The polar angle θ ranges from north to south and can only have values from 0 to π. The azimuthal (equatorial) angle ϕ moves around the equator so ranges from 0 to 2π.

One of the 3p atomic orbitals has quantum numbers $n = 3$, $\ell = 1$, and $m = 1$, and is

$$\psi_{311} = N\frac{r}{a_0}\left(6 - \frac{r}{a_0}\right)e^{-\frac{r}{3a_0}}\sin(\theta)e^{+i\phi}. \tag{4.57}$$

where N is the normalization constant we want to find. Because r, θ, and ϕ are separate there being no term in the product $\theta\phi$ when calculating equation (4.56), the integrals in r, θ, and ϕ can be treated separately. The integral in r is

$$\frac{N^2}{a_0^2}\int_0^{\infty} r^3\left(6 - \frac{r}{a_0}\right)^2 e^{-\frac{2r}{3a_0}}dr \tag{4.58}$$

which has a standard form, (two terms of the type $x^n e^{-ax}$) which can be integrated by parts; see (4.2.13). Using Maple, because the calculation while straightforward is involved, gives for the radial part equation (4.58),

```
> assume(a0 > 0):
  psir:= r->(r/a0)*(6-r/a0)*exp(-r/(3*a0)); # wavefunction
  Int(r^2*psir(r)^2,r= 0..infinity)*N^2:    # eqn (4.58)
  %= collect( (value(%)),exp );
```

$$N^2\int_0^{\infty}\frac{r^4\left(6-\frac{r}{a_0}\right)^2\left(e^{-\frac{1}{3}\frac{r}{a_0}}\right)^2}{a_0^2}\,dr = \frac{19\,683}{8}N^2 a_0^3$$

The angular parts of the integral $\int_0^{2\pi}\int_0^{\pi}\sin(\theta)e^{i\phi}\sin(\theta)e^{-i\phi}\sin(\theta)d\theta d\phi$ are simplified by separating the θ and ϕ integrals into two then evaluating the complex conjugate first, because $e^{-i\varphi}e^{i\phi} = 1$. The remaining integral is

$$\int_0^{2\pi}d\phi\int_0^{\pi}\sin^3(\theta)d\theta = 2\pi\int_0^{\pi}\sin^3(\theta)d\theta = 8\pi/3$$

and the sine integral was worked out by converting to the exponential form. Multiplying the two results and rearranging gives the normalization as

$$N = \frac{1}{81\sqrt{\pi}}\sqrt{\frac{1}{a_0^3}}.$$

4.10.4 Questions

Full solutions are available at **www.oxfordtextbooks.co.uk/orc/beddard**.

Q4.79 Another 3p orbitals is $\psi_{310} = N\frac{r}{a_0}\left(6 - \frac{r}{a_0}\right)e^{-\frac{r}{3a_0}}\cos(\theta)$, show that this is orthogonal to orbital ψ_{311} (equation (4.57)). The normalization N is the same for both orbitals.

Strategy: If two orbitals y_a, and y_b are orthogonal then $\int_0^{2\pi}\int_0^{\pi}\int_0^{\infty}\psi_a^*\psi_b \sin(\theta)drd\theta d\phi = 0$. Normalization is calculated from equation (4.56).

Q4.80 The angular part of the $d_{x^2-y^2}$ wavefunction is $N\sin^2(\theta)e^{2i\phi}$ and that of the d_{xy} is the complex conjugate of this. Find the normalization N and show that these two orbitals are orthogonal.

4.11 Change of variables in integrals: Jacobians

In Section 4.3 the method of simplifying an integration by a change of variable was described. A commonly used change of variables in multiple integrals is from Cartesian either to *plane polar* or *spherical polar* coordinates. The (plane) polar coordinates are two dimensional and spherical polar are three dimensional; see Chapter 1.6.1. They are used only to simplify a calculation by using those coordinates that reflect the underlying symmetry of the problem being studied, thus the *shapes* of the s, p, d and other atomic wavefunctions (orbitals) are naturally described in terms of three-dimensional spherical polar coordinates with a radius r, a polar θ, and an equatorial (azimuthal) angle φ. However, many two or three or higher dimensional integrations can be simplified by a suitable algebraic substitution, which may also be thought of as a change of coordinates. Fortunately, there is a systematic way of doing this using a determinant of derivatives, called the *Jacobian* and these are described in this section. Determinants are described in Chapter 7.

A one-dimensional example is considered first. An apparently hard integral such as $\int_0^b x\sqrt{a^2 - x^2}\,dx$ can be simplified by substituting $u = a^2 - x^2$, calculating the differential $du = -2xdx$ and changing the limits. The result is

$$-\frac{1}{2}\int_{a^2}^{a^2-b^2}\sqrt{u}\,du = \frac{1}{3}(a^3 - [a^2 - b^2]^{3/2}).$$

Ignoring the limits for clarity, a general integral of a function $f(x)$ and its substitution can be written as

$$\int f(x)dx = \int F(u)\frac{dx}{du}du = \int F(u)J(x, u)du$$

where F is the function f in the new variable u and the new function J contains the terms needed to 'distort' dx into du. This is done by using the differential $dx = J(x, u)du$ where $J(x, u) = \frac{dx}{du} = -\frac{1}{2x} = -\frac{1}{2\sqrt{a^2 - u}}$.

A two-dimensional integral in its general form with a change of coordinates is

$$\iint f(x, y)dxdy = \iint F(u, v)J(x, y, u, v)dudv$$

where f is some normal function of x and y, perhaps $\sqrt{\sin(y)/\sin(x)}$ and u and v are *functions* of x and y. What these are depends on the particular calculation. In three dimensions, the general equation for the transformation is similar but rather formidable,

$$\iiint f(x, y, z)dxdydx = \iiint F(r, \theta, \varphi)J(x, y, z, r, \theta, \varphi)drd\theta d\varphi.$$

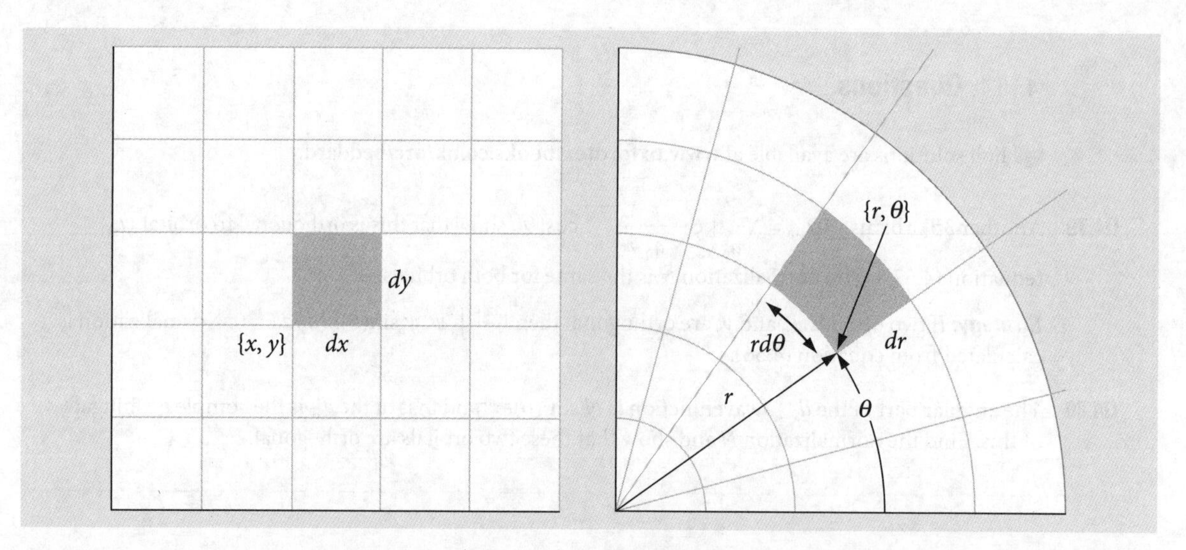

Fig. 4.25 An area in *dxdy* and morphed to an equal value *rdrdθ* in Cartesian and polar coordinates. The figures are not to scale relative to one another, and with infinitesimal lengths *dx* etc. greatly exaggerated relative to the axes.

The change of coordinates means that the volume $dxdydz$ has to be distorted or morphed into an equivalent volume in the new coordinates $drd\theta d\varphi$. Therefore the new function, the Jacobian $J(x, y, z, r, \theta, \varphi)$, has to be found.

Any coordinate change or substitution has three parts

(a) Calculating the Jacobian,
(b) Substituting the new variables into the function f,
(c) Changing any limits on the integration to the new coordinates.

These are best illustrated with examples. A point $\{x, y, z\}$ is equivalently $\{r, \theta, \varphi\}$ in spherical polar coordinates, the connection between the two sets of coordinates is described by geometry and is

$$x = r\sin(\theta)\cos(\varphi), \qquad y = r\sin(\theta)\sin(\varphi), \qquad z = r\cos(\theta).$$

In (plane) polar coordinates the point $\{x, y\}$ is represented as $\{r, \theta\}$ with

$$x = r\cos(\theta), \qquad y = r\sin(\theta).$$

In this case the area element $dxdy$ becomes $rdrd\theta$ and these are shown in Fig. 4.25.

In the case of the polar coordinates the area is relatively easily calculated. The circumference of a circle is $2\pi r$, which is the radius times the angle rotated which is 2π radians. The length of the arc for a small angle is therefore $rd\theta$ for angular change $d\theta$. The radius extends from r to $r + dr$ making the area $rdrd\theta$. For other coordinates, the geometrical calculation is complex and an algebraic method is therefore preferred. This method, presented without proof, is to form the Jacobian, which is the determinant of the derivatives of the equation converting one set coordinates into the other.

Consider now the spherical polar coordinates, the function $J(x, y, z, r, \theta, \varphi)$ is needed and changing to the conventional notation this is the determinant of the partial derivatives of x, y and z with r, θ and φ and is defined as

$$J\left(\frac{x, y, z}{r, \theta, \varphi}\right) \equiv \frac{\partial(x, y, z)}{\partial(r, \theta, \varphi)} = \begin{vmatrix} \dfrac{\partial x}{\partial r} & \dfrac{\partial x}{\partial \theta} & \dfrac{\partial x}{\partial \varphi} \\ \dfrac{\partial y}{\partial r} & \dfrac{\partial y}{\partial \theta} & \dfrac{\partial y}{\partial \varphi} \\ \dfrac{\partial z}{\partial r} & \dfrac{\partial z}{\partial \theta} & \dfrac{\partial z}{\partial \varphi} \end{vmatrix} \tag{4.59}$$

Notice the ordering; the old coordinates x, y, z are on the top of each differentiation. Note also the notation in the brackets with J. Using equation (4.59) the determinant is

$$J\left(\frac{x,y,z}{r,\theta,\varphi}\right)=\begin{vmatrix}\sin(\theta)\cos(\varphi) & r\cos(\theta)\cos(\varphi) & -r\sin(\theta)\sin(\varphi)\\ \sin(\theta)\sin(\varphi) & r\cos(\theta)\sin(\varphi) & r\sin(\theta)\cos(\varphi)\\ \cos(\theta) & -r\sin(\theta) & 0\end{vmatrix}=r^2\sin(\theta)$$

The volume element conversion is then written as

$$dxdydz=r^2\sin(\theta)drd\theta d\varphi. \tag{4.60}$$

In some cases, the determinant may produce a negative answer depending on the order of calculating the derivatives; however, the Jacobian represents an area or volume element so the positive result may legitimately be taken in such cases.

A few examples are now worked through.

(i) The complicated integral $\iint((x-y)^2+2(x+y)+1)^{-1/2}dxdy$ can be simplified markedly with the coordinate change $x=u(1+v)$, $y=v(1+u)$. The general form of the equation is $\iint f(x,y)dxdy=\iint F(u,v)\dfrac{\partial(x,y)}{\partial(u,v)}dudv$. The first two steps in the calculation are necessary because no limits are given. In step (a) the Jacobian is calculated and is

$$\frac{\partial(x,y)}{\partial(u,v)}=\begin{vmatrix}1+v & u\\ v & 1+u\end{vmatrix}=1+u+v$$

Step (b) is substituting into the function to find $F(u,v)$ and this produces

$$\begin{aligned}((x-y)^2+2(x+y)+1)^{-1/2}&=((u-v)^2+2(u+v)+4uv+1)^{-1/2}\\&=(u^2+v^2+2(u+v)+2uv+1)^{-1/2}\\&=\frac{1}{u+v+1}\end{aligned}$$

and multiplying this with the Jacobian makes the integral rather simple:

$$\iint((x-y)^2+2(x+y)+1)^{-1/2}dxdy=\iint 1dudv=uv.$$

(ii) The integral $I=\int_0^1\int_0^{\sqrt{1-x^2}}(x-x^2y^3)dydx$ can be solved by converting to polar coordinates. The limits are converted first. As x extends from 0 to 1 so does r as this is the radius in polar coordinates. As $x=r\cos(\theta)$ and $y=r\sin(\theta)$ then $r^2=x^2+y^2$, which represents a circle. The maximum value r has is 1; therefore $x^2+y^2=1$ and integration is around the first quadrant of a circle of radius 1, in the new coordinates the integration area is a rectangle where r ranges from 0 to 1 and θ from 0 to $\pi/2$.

The angle θ varies from 0 to $\pi/2$ and r from 0 to 1; Fig. 4.26.

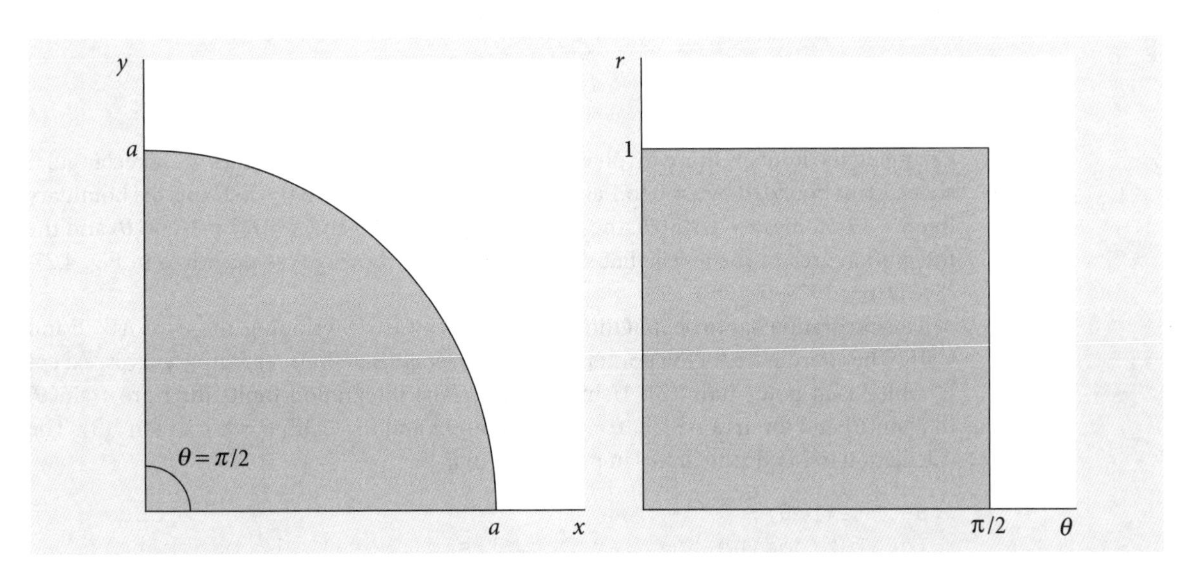

Fig. 4.26 The integration area in the x–y and r–θ planes, r ranges from 0 to 1 and θ from 0 to $\pi/2$.

$$\int_0^1\int_0^{\sqrt{1-x^2}}(x-x^2y^3)dydx=\int_0^1\int_0^{\pi/2}(r\cos(\theta)-r^2\cos^2(\theta)r^3\sin^3(\theta))rd\theta dr$$

Both limits are constant so the order of integration does not matter but the integral in θ is performed first as this is the most difficult.

```
> Int( r^2*cos(theta)-r^6*sin(theta)^3*cos(theta)^2,theta):
  %= value(%);
```

$$\int(r^2\cos(\theta)-r^6\sin(\theta)^3\cos(\theta)^2)\,d\theta=\sin(\theta)r^2-r^6\left(-\frac{1}{5}\sin(\theta)^2\cos(\theta)^3-\frac{2}{15}\cos(\theta)^3\right)$$

The integration in r is easy and the final result is $I=11/35$.

(iii) In this next example the integral $I=\int_0^\infty\int_0^x x^2e^{x^2+y^2}dxdy$ will be solved by transforming to polar coordinates. Using equation (4.59) gives

$$\frac{\partial(x,y)}{\partial(r,\theta)}=\begin{vmatrix}\cos(\theta) & -r\sin(\theta)\\ \sin(\theta) & r\cos(\theta)\end{vmatrix}=r$$

as determined also by the geometrical argument in Fig. 4.25. A limit of x means that this varies as $y=x$ a line with a gradient of one or at 45° to the x-axis which is the same as $\theta=\pi/4$. The integration is therefore in the area from $\theta=0$ to $\pi/4$ and with r extending from 0 to ∞. Substituting into the integral gives $\int_0^\infty\int_0^x x^2e^{-x^2-y^2}dydx=\int_0^\infty\int_0^{\pi/4}r^2\cos^2(\theta)e^{-r^2}rdrd\theta$ and this can be separated into two integrals $\int_0^\infty r^3e^{-r^2}dr\int_0^{\pi/4}\cos^2(\theta)d\theta$ since there is no term in r and θ; and the limits of the integration are constants. The integrals are standard ones, see Section 4.2.13, but can easily be calculated with Maple

```
> Int(r^3*exp(r^2),r): %= value(%);
```

$$\int r^3e^{r^2}\,dr=\frac{1}{2}(-1+r^2)\,e^{r^2}$$

```
> Int(cos(x)^2,x): %= value(%);
```

$$\int\cos(x)^2\,dx=\frac{1}{2}\cos(x)\sin(x)+\frac{1}{2}x$$

Adding the limits produces the answer $I=\left(\frac{\pi}{2}-1\right)\frac{1}{8}$.

(iv) The complicated integral $I=\int_1^3\int_0^1\frac{x^2}{(x^2+y^2)^{5/2}}dxdy$ can be solved by converting to polar coordinates first then changing the integration limits. Converting produces

$$I=\int\int\frac{\cos^2(\theta)}{r^2}drd\theta.$$

Changing the limits is more involved in this example. The initial values are a rectangular shaped area bounded by $x=0$ to 1 and $y=1$ to 3. When x is zero $\theta=\pi/2$, and the boundary line $y=1$ becomes $r=1/\sin(\theta)$, the line $y=3$, $r=3/\sin(\theta)$ and $x=1$, $r=1/\cos(\theta)$ and the integration area in the r–θ is that shape enclosed by these curves as shown in Fig. 4.27. (See Dence 1975, p. 109.)

The integration has to be split into two parts because CD is sloping: the areas ABCE and CDE. The coordinates of the points are the intersections of their respective curves, except E, which is at point $\{\tan^{-1}(3), 1/\sin(\tan^{-1}(3))\}$. The integration limits for r are $1/\sin(\theta)$ to $3/\sin(\theta)$ and for area ABCE, $\theta=\tan^{-1}(3)$ to $\pi/2$ and for CDE, $\theta=\pi/4$ to $\tan^{-1}(3)$. The calculation in Maple integrates in r first then in θ.

```
> assume(r>0);                                    # area CDE
  Int(Int(cos(theta)^2/r^2,r= 1/sin(theta)..1/cos(theta)),
                              theta= Pi/4..arctan(3)):
  % = simplify(value(%));
```

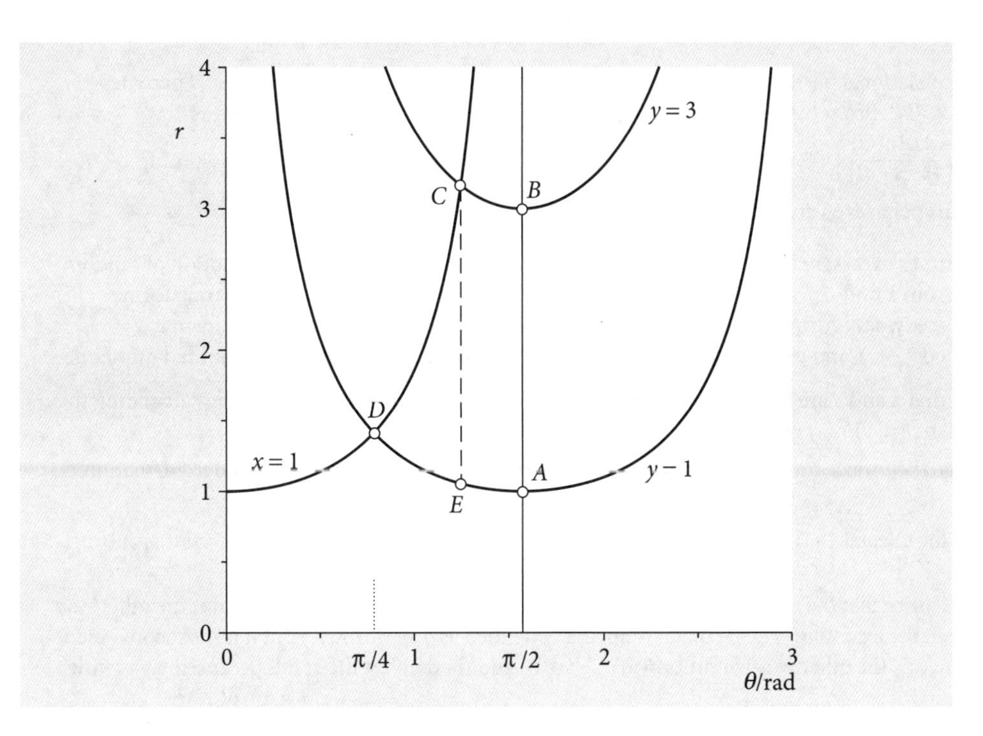

Fig. 4.27 Integration limits in the θ–r plane. (Redrawn based on Fig. 3.4 of Dence 1975.)

$$\int_{\frac{1}{4}\pi}^{\arctan(3)} \int_{\frac{1}{\sin(\theta)}}^{\frac{1}{\cos(\theta)}} \frac{\cos(\theta)^2}{r^2}\, dr\, d\theta = \frac{1}{2}\sqrt{2} - \frac{16}{75}\sqrt{2}\sqrt{5}$$

```
> Int(Int(cos(theta)^2/r^2,r= 1/sin(theta)..3/sin(theta)),
                              theta= arctan(3)..Pi/2):
  % = simplify(value(%));
```

$$\int_{\arctan(3)}^{\frac{1}{2}\pi} \int_{\frac{1}{\sin(\theta)}}^{\frac{3}{\sin(\theta)}} \frac{\cos(\theta)^2}{r^2}\, dr\, d\theta = \frac{1}{450}\sqrt{10}$$

The result, $\dfrac{1}{\sqrt{2}} - \dfrac{19}{90}\sqrt{10} = 0.0395$ is the sum of the two integrals.

4.11.1 **Questions**

Full solutions are available at **www.oxfordtextbooks.co.uk/orc/beddard.**

Q4.81 The integral $I = \int_0^\infty e^{-x^2} dx$ cannot be evaluated in this form. Show that it can be evaluated by forming $I^2 = \int_0^\infty e^{-x^2} dx \int_0^\infty e^{-y^2} dy$ and converting to polar coordinates.

Q4.82 The translational partition function in one dimension is $Z = \sum_{n=1}^{\infty} e^{-E_n/K_B T}$ where $E_n = n^2h^2/(8mL^2)$ in a box of length L. As the energy levels are so closely spaced they may effectively be considered as being continuous and the sum changed into an integral.

(a) Evaluate this integral and using values for benzene vapour at 300 K and 1 bar pressure calculate Z^3—the three-dimensional partition function is the cube of the one-dimensional value—using the mean free path as the size of the 'box'.

(b) Calculate the energy gap between two levels at the average thermal energy $3k_BT/2$. The number you should find is of the order of an NMR transition energy, and much larger than your textbook may indicate because there is no known spectroscopy associated with translational motion the gaps being too small. What is going on here?

The cross-sectional area for benzene is $\sigma = 0.88\ \text{nm}^2$.

(c) The translational partition function when corrected for the indistinguishability of N particles is $Q = Z^N/N!$. From thermodynamics the translational entropy at constant volume is $S = k_B\left(\dfrac{\partial T\ln(Q)}{\partial T}\right)_V$. Estimate S for a mole of mercury vapour at 300 K and 1 bar assuming that this behaves as an ideal gas.

Q4.83 It is sometimes necessary to produce a two-dimensional Gaussian (or normal) distribution of random numbers from a uniform distribution. A commonly used method is the Box–Muller transform. If u and v are two uniformly distributed random numbers between 0 and 1, the transform is $x = \sqrt{-2\ln(u)}\cos(2\pi v)$, and $y = \sqrt{-2\ln(u)}\sin(2\pi v)$ where x and y are two Gaussian distributed numbers.

(a) Prove that x and y are Gaussian distributed with a mean of 0 and a variance of 1 by calculating the Jacobian $\partial(u, v)/\partial(x, y)$. Why is the Jacobian appropriate?

(b) Change the transform equations to produce a distribution with mean μ and standard deviation σ.

Q4.84 Show that the integral $\displaystyle\int_0^{\pi/2}\int_0^{\pi/2}\sqrt{\frac{\sin(\varphi)}{\cos(\theta)}}d\varphi d\theta = \pi$ using the substitution $u = \sin(\varphi)\cos(\theta)$, $v = \sin(\varphi)\sin(\theta)$.

Strategy: Notice that the new coordinates u and v are defined as functions of the old, not the other way around as in the examples. The Jacobian is normally defined as $\partial(old)/\partial(new)$ but it is most convenient to calculate this the other way round as $\partial(u, v)/\partial(\varphi, \theta)$ and then invert the result because just as with normal derivatives, inverting the differentiation order inverts the result or $\dfrac{\partial(u, v)}{\partial(\varphi, \theta)}\dfrac{\partial(\varphi, \theta)}{\partial(u, v)} = 1$.

Q4.85 Using Maple, calculate the Jacobian $\partial(x, y, z)/\partial(u, v, \theta)$ for the confocal elliptical coordinates called 'prolate spheroidal' coordinates. The coordinate transformations are, with α as a constant,

$$x = \alpha\sqrt{(u^2-1)(1-v^2)}\cos(\theta), \qquad y = \alpha\sqrt{(u^2-1)(1-v^2)}\sin(\theta), \qquad z = \alpha uv.$$

The range of values is $1 \le u < \infty$, $-1 \le v \le 1$ and $0 \le \theta \le 2\pi$.
The volume produced by these coordinates is similar to that of a rugby ball, with the z-axis being the long axis. As the coordinates are elliptical there are two origins on the z-axis at $\{0, 0, \pm\alpha\}$. These coordinates are important in quantum calculations of diatomic molecules because each atom is placed at an origin.

Q4.86 When the energies of an atom or molecule are calculated using the Schrödinger equation, the solutions are stationary states and do not change with time; once an atom or molecule is in that state it will remain there unless something happens to perturb it. If an optical transition between two electronic states is to occur, an oscillating electric dipole must exist in the atom or molecule. This will interact with the oscillating electric field of the radiation and absorption (or stimulated emission) can occur if the radiation's energy matches that of the two atomic or molecular orbitals. One unit of angular momentum is also transferred between the radiation and the atom or molecule. The expectation value of a transition with linearly polarized radiation in the z-direction is $\langle z\rangle = \alpha\int\varphi_1^* z\varphi_2 d\tau$ where $\varphi_{1,2}$ are the wavefunctions, α a collection of constants, normalization, etc., and $d\tau$ the area element for the integration. If a transition is to occur then this integral must not be zero and this can be determined by the symmetry properties of the integral.

This picture is, however, a static one; if the orbitals are made time dependent, then the oscillating dipole, if it exists, can be observed by examining the product $\psi_{12}^*\psi_{12}$ where ψ is a *superposition state* $\psi_{1,2} = (a\varphi_1 + b\varphi_2)$. To find the probability of being in a small region z to $z + dz$ at any time t, the integral $P(z, t)dz = \int\psi_1^*(\tau, t)\psi_2(\tau, t)d\tau\, dz$ has to be evaluated. The a and b in the superposition equation are constants that normalize ψ; therefore $a^2 + b^2 = 1$. The integral P can be evaluated using cylindrical coordinates, which then becomes the *double* integral $\int_0^{2\pi}\int_0^{\infty}\psi_1^*(r, \theta, t)\psi_2(r, \theta, t)rdrd\theta$. Cylindrical coordinates are used because the s and p_z orbitals we will use are symmetric about the x- and y-axes. The time dependence of a wavefunction is always given by

$$\varphi_j(t) = \varphi_j e^{-iE_jt/\hbar}$$

where φ_j is the jth wavefunction calculated by solving the normal (time-independent) Schrödinger equation. E_j is the energy. If an atom is considered, the 1s, 2s, and 2p atomic orbitals have wavefunctions

$$\varphi_{1s} = \left(\frac{1}{\pi}\right)^{1/2}\left(\frac{1}{a_0}\right)^{3/2}e^{-r/a_0} \qquad \varphi_{2s} = \frac{1}{8}\left(\frac{2}{\pi}\right)^{1/2}\left(\frac{1}{a}\right)^{3/2}\left(2 - \frac{r}{a_0}\right)e^{-r/(2a_0)}$$

$$\varphi_{2p_z} = \frac{1}{8}\left(\frac{2}{\pi}\right)^{1/2}\left(\frac{1}{a_0}\right)^{3/2}\frac{z}{a_0}e^{-r/(2a_0)},$$

where r is the radial distance from the origin and the substitution $z = r\cos(\theta)$ has been made in φ_{2pz} to convert from spherical polar to cylindrical coordinates, see Fig. 4.28. Note that r and θ in spherical polar coordinates is not the same as r and θ in cylindrical coordinates.

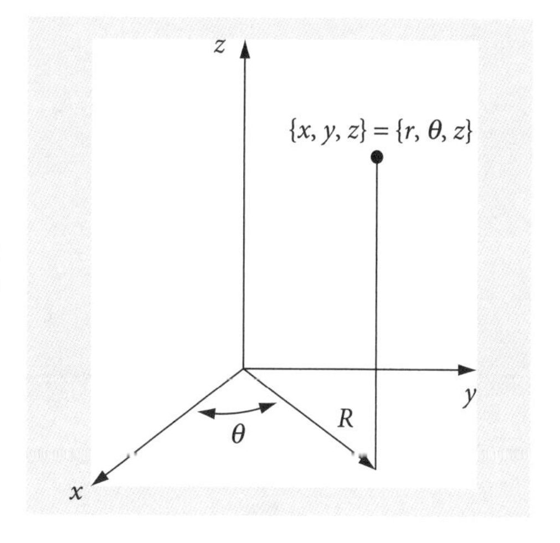

Fig. 4.28 Cylindrical coordinates.

(a) Show that if a wavefunction ψ is a superposition of stationary states that have the same energy, as do each of the 2s and 2p orbitals, then the time dependence cancels out and there can be no fluctuating dipole.

In (b) and (c) use *cylindrical* coordinates to show that

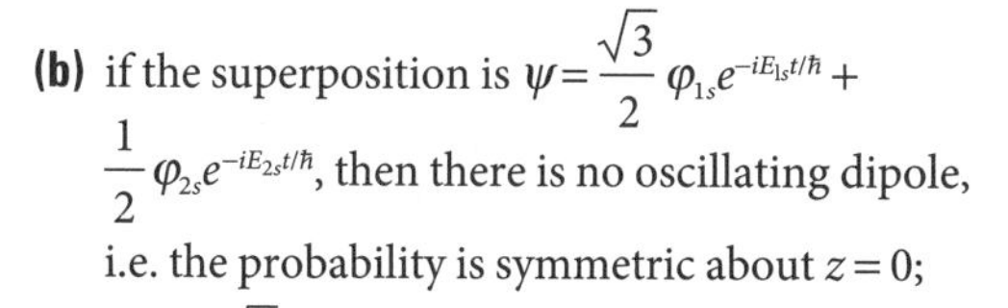

(b) if the superposition is $\psi = \dfrac{\sqrt{3}}{2}\varphi_{1s}e^{-iE_{1s}t/\hbar} + \dfrac{1}{2}\varphi_{2s}e^{-iE_{2s}t/\hbar}$, then there is no oscillating dipole, i.e. the probability is symmetric about $z = 0$;

(c) if $\psi = \dfrac{\sqrt{3}}{2}\varphi_{1s}e^{-iE_{1s}t/\hbar} + \dfrac{1}{2}\varphi_{2p_z}e^{-iE_{2p_0}t/\hbar}$ the atom does have an oscillating dipole.

(d) Plot some example graphs in (b) and (c).

Strategy: (a) No integration is necessary here, the product $\psi^*\psi$ can be shown to have no time dependence. Remember that in the complex conjugate $i \to -i$.

In (b) and (c) expand the terms as in (a) before calculating any integrals. The wavefunctions are written in spherical polar coordinates where the distance to any point $\{x, y, z\}$ from the origin is r where $r = \sqrt{x^2 + y^2 + z^2}$. In cylindrical coordinates the same distance from the origin $r = \sqrt{R^2 + z^2}$ but now R is the *radial distance* in the x–y plane.

4.11.2 Calculating the energy of a chemical bond using molecular orbitals

Molecular orbitals (MO) are orbitals that extend over the whole molecule as opposed to being located on or between particular atoms. In H_2^+, one of the simplest molecules, the MOs are formed by bringing the two lowest energy atomic orbitals (wavefunctions), φ_{sA} and φ_{sB} together. In the situation that the MO is $\varphi_{sA} + \varphi_{sB}$, the atomic orbitals are in phase and a bond is formed; constructive interference has occurred between the two wavefunctions. The other simple combination, $\varphi_{sA} - \varphi_{sB}$, leads to destructive interference and little electron density between the nuclei. Such a combination is known as an anti-bonding MO. The amplitude of the two MOs are shown in Fig. 4.29.

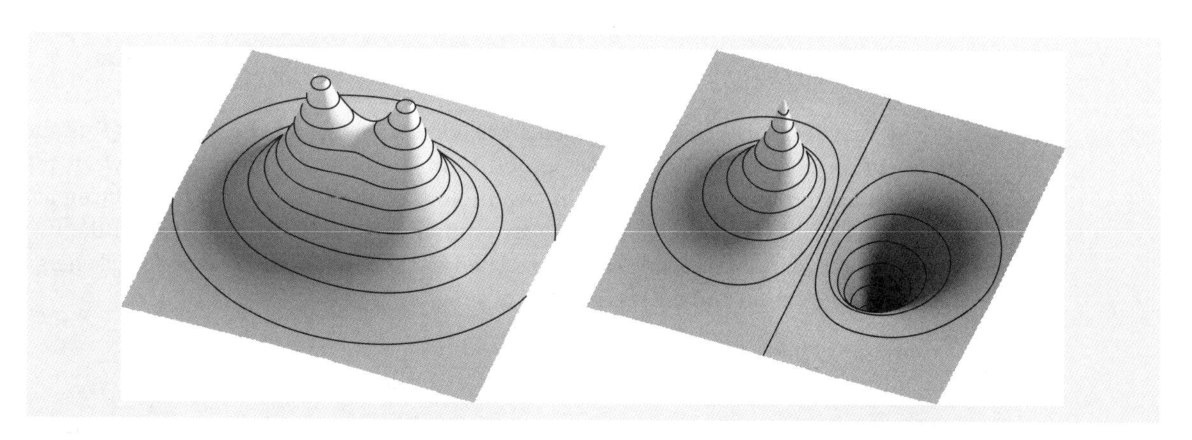

Fig. 4.29 Bonding (left) and anti-bonding combination of orbitals. Note the nodal plane in the anti-bonding orbital where the electron density falls to zero and the also change in sign of the orbital. All of the bonding orbital has the same sign.

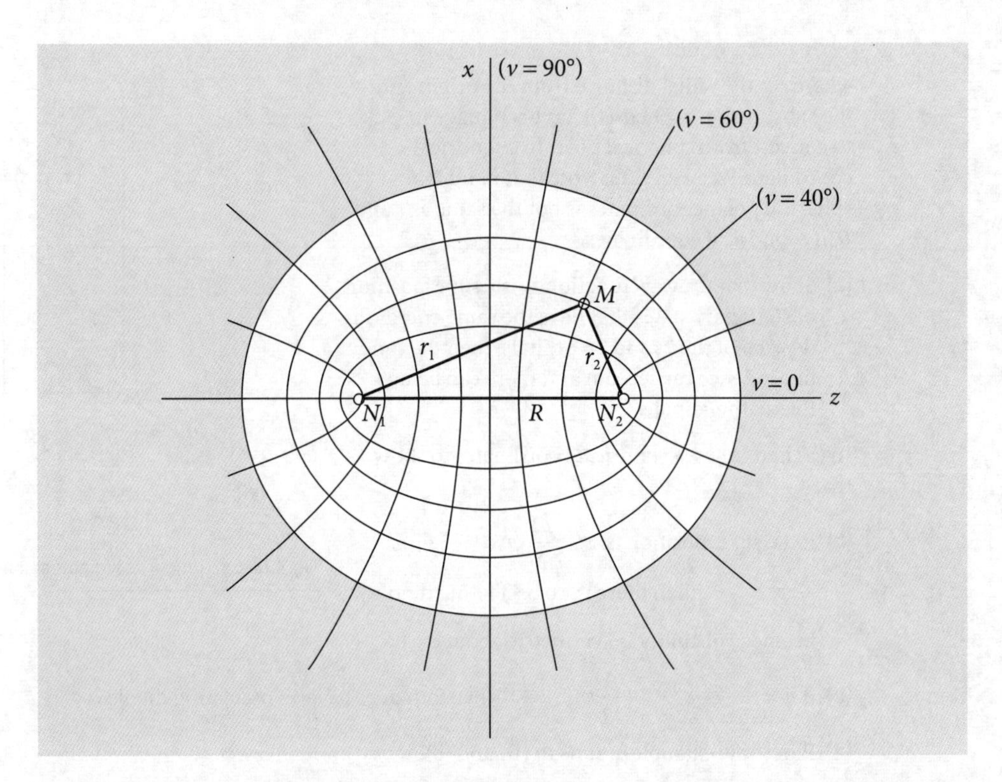

Fig. 4.30 The geometry of the H_2^+ calculation and prolate spherical coordinates. The distance between the two protons (N_1 and N_2) is R, and that between the protons and the electron (M) is r_1 and r_2 respectively. The origin of the coordinates is at $R/2$ so that N_1 is at position $\{x, y, z\} = \{0, 0, -R/2\}$ and N_2 at $\{0, 0, R/2\}$. The electron can take any position in $\{x, y, z\} = \{u, v, \theta\}$ space with a probability determined by the wavefunction. The ellipses and hyperbola show some of the values for the elliptical coordinates used to solve the problem. Specifically these are prolate spheroidal coordinates and are rotationally symmetrical around the z-axis. The parabolas are lines in v the ellipses in u. The equations are $(x/\cosh(u))^2 + (y/\sinh(u))^2 = (R/2)^2$ and $(x/\cos(v))^2 - (y/\sin(v))^2 = (R/2)^2$.

The geometry of the H_2^+ molecular ion is defined as a function of the separation R of the protons, N_1 and N_2, and of the electron M with each proton, Fig. 4.30. One possible separation of the electron from the two protons is shown but the electron can occupy any position in space. The interaction between the electron and the proton is electrostatic and attractive while that between the two protons is repulsive. The electron has kinetic energy as do the nuclei, but to make the calculation easy we assume that the nuclei do not vibrate but have fixed positions on the z-axis and so have no kinetic energy. This approach is called the Born–Oppenheimer approximation and is a good approximation when nuclear motion is far slower than that of electrons. This is due to the difference in their masses; the ratio of mass is at least $m_p/m_e = 1836$.

In the Schrödinger equation, $\boldsymbol{H}\psi = E\psi$, the Hamiltonian operator $\boldsymbol{H}$ represents the kinetic and potential energy of the nuclei and electron (of mass m). The equation is

$$\left(\frac{\boldsymbol{P}^2}{2m} + V(\boldsymbol{r}, R)\right)\psi(\boldsymbol{r}) = E\psi(\boldsymbol{r}).$$

The first term in brackets represents the kinetic energy of the electron ($\boldsymbol{P}$ is the momentum operator $-i\hbar\partial/\partial\tau$), the second term the potential energy V between the electron and protons and between the two protons. The coordinates of the electron are written as $\boldsymbol{r}$ which is a vector defining its position anywhere in space around the two protons. At some juncture, this has to be related to r_1, r_2, and R. The total potential energy between the electron and the two protons is

$$V(\boldsymbol{r}, R) = -\frac{e^2}{r_1} - \frac{e^2}{r_2} + \frac{e^2}{R},$$

and each term in V represents an electrostatic interaction. For clarity, the scaling term $(1/4\pi\varepsilon_0)$, which would multiply V to convert it into SI units, is ignored. The first term in V is the attractive interaction between the electron and proton 1 which are separated by r_1, the second term is the interaction between the electron and proton 2, and the third, the repulsion between the two protons. The full Schrödinger equation is therefore

$$\left[\frac{P^2}{2m} - \frac{e^2}{r_1} - \frac{e^2}{r_2} + \frac{e^2}{R}\right]\psi(r) = E\psi(r).$$

The energy is calculated with the variational principle using a linear combination of atomic orbitals as the basis to form ψ. The orbitals are chosen to be those of the hydrogen atoms based on proton 1 and 2, and are φ_1 and φ_2 respectively and make the linear combination

$$\psi = c_1\varphi_1 + c_2\varphi_2.$$

The atomic wavefunction for a hydrogen atom is

$$\varphi(r) = \frac{1}{\sqrt{\pi a_0^3}} e^{-r/a_0}.$$

and by substituting r_1 and r_2 for r produce two atomic orbitals φ_1 and φ_2.

The energy of the lowest level of the H *atom* can also be calculated from the Schrödinger equation. The equation is $\left[\frac{P^2}{2m} - \frac{e^2}{r}\right]\varphi(r) = E\varphi(r)$ where r is the vectorial (radial) distance of the electron from the proton. The lowest energy is $-e^2/2a_0$ and the ionization energy, being positive, is $E_I = e^2/2a_0$. The constant a_0 is the Bohr radius, for hydrogen $a_0 = \frac{4\pi\varepsilon_0\hbar^2}{me^2} = 52.92$ pm. In SI units, $R_H = \frac{1}{hc}\left(\frac{1}{4\pi\varepsilon_0}\right)\frac{e^2}{2a_0} = E_I$, where R_H is the Rydberg constant in cm^{-1}.

We shall use equations (4.61) to find the energies of the H_2^+ ion. From the variational calculation (see Section 4.9) the bonding E_+ and anti-bonding energies E_- are

$$E_+ = \frac{H_{11} + H_{12}}{1 + S} \qquad E_- = \frac{H_{11} - H_{12}}{1 - S} \tag{4.61}$$

where S is the overlap integral, and H_{11} and H_{12} are the energy expectation value integrals. These are calculated next. These energies are functions of the internuclear separation. This may seem contradictory, as it was stated that R was constant; however, this is only for the purpose of calculating the energy in the Born–Oppenheimer approximation. In this calculation, the electrons are allowed to move to find their minimum energy at any internuclear separation R. This is then changed and the energy recalculated. This makes the energy a function of nuclear position R, but not the electron's position r, and thereby the potential energy curve is obtained. If the motion of the electrons and nuclei could not be separated then it would not be possible to draw a potential energy curve because the energy would depend on both r and R.

In calculating the energy, equation (4.61), three integrals have to be evaluated. The overlap integral S is easy to understand, its name describes what it is, which is the extent to which the two atomic orbitals occupy the same region of space: In symbolic form this is

$$S = \int \varphi_1^* \varphi_2 d\tau = \int \varphi_2^* \varphi_1 d\tau. \tag{4.62}$$

and from now on, the complex conjugate will be ignored because the wavefunctions are real. Thinking of the integral in simple terms as an area, the 'area' is that of the product of the two wavefunctions and is obviously large only where they overlap. Integration is written with in symbolic coordinates $d\tau$ that represents a volume element of three-dimensional space. When the coordinates are included properly, S becomes a triple integral.

The H_{11}, H_{22}, and H_{12} integrals are different as they contain the energy (Hamiltonian) operator $\boldsymbol{H}$. In symbolic form these integrals are

$$H_{11} = \int \phi_1 \boldsymbol{H} \phi_1 d\tau \tag{4.63}$$

$$\boldsymbol{H} = \left[\frac{\boldsymbol{P}^2}{2m} + V_{en} + V_{nm}\right] = \left[\frac{\boldsymbol{P}^2}{2m} - \frac{q^2}{r_1} - \frac{q^2}{r_2} + \frac{q^2}{R}\right]$$

which contains both kinetic ($P^2/2m$) and potential energy terms which are the electron–proton interaction, $V_{en} = -\dfrac{q^2}{r_1} - \dfrac{q^2}{r_2}$ and proton–proton interaction $V_{nn} = +\dfrac{q^2}{R_1}$. The electronic charge is q and is used instead of the conventional e because of possible confusion when exponentials are used later on. The other integrals, H_{22} and H_{12}, are formed in a similar way.

The calculation now involves evaluating these integrals and to calculate the energies E_+ and E_- and plot them vs internuclear separation R. To make the equations simpler, two things are done. First, the coordinates are changed to prolate spheroidal ones, second, reduced distances are used, and these are defined in terms of a_0 the Bohr radius:

$$\rho_1 = r_1/a_0, \qquad \rho_2 = r_2/a_0, \qquad \rho = R/a_0. \tag{4.64}$$

In reduced distance the wavefunctions are

$$\varphi_1 = \frac{1}{\sqrt{\pi a_0^3}} e^{-\rho_1} \qquad \varphi_2 = \frac{1}{\sqrt{\pi a_0^3}} e^{-\rho_2}. \tag{4.65}$$

4.11.3 Calculation of the overlap integral *S*

Substituting for the atomic wavefunctions and simplifying gives

$$S = \int \frac{1}{\sqrt{\pi a_0^3}} e^{-r_1/a_0} \frac{1}{\sqrt{\pi a_0^3}} e^{-r_2/a_0}\, d\tau$$
$$= \frac{1}{\pi a_0^3} \int e^{-\rho_1-\rho_2}\, d\tau \tag{4.66}$$

In this equation, r_1 and r_2 have been converted into reduced distances ρ_1 and ρ_2. To integrate over $d\tau$ we must know how $d\tau$ and ρ_1 and ρ_2 are related. The protons and electron are placed in real space so that $d\tau$ is a volume element,

$$d\tau = dxdydx,$$

and therefore the integral is three dimensional; a triple integral. The integral has coordinates in distances r_1 and r_2 and R which are in Cartesian (x, y, z) coordinates which makes the calculation very difficult. Fortunately, in this problem the maths works out far easier if elliptic coordinates are used, which means transforming $\{x, y, z\} \to \{\mu, v, \theta\}$. These new coordinates are used because they have an origin on each atom, see Fig. 4.30. In this new coordinate system the volume element, or the Jacobian is,

$$d\tau = dxdydz = \frac{\rho^3 a_0^3}{8}(u^2 - v^2)dudvd\theta \tag{4.67}$$

and the distances ρ_1 and ρ_2 are defined in terms of u and v as,

$$\rho_1 = \frac{u+v}{2}\rho \qquad \rho_2 = \frac{u-v}{2}\rho \tag{4.68}$$

and θ is the angle around the z-axis. So now, all the elements with which to calculate the solution are assembled.

To calculate the integral, first substitute for $d\tau$ into the S integral equation (4.66) and then substitute for ρ_1 and ρ_2 to give

$$S = \frac{1}{\pi a_0^3} \iiint \exp\left(-\frac{u+v}{2}\rho\right) \exp\left(-\frac{u-v}{2}\rho\right) \frac{\rho^3 a_0^3}{8}(u^2 - v^2)dudvd\theta$$

The next step is to put in the limits to the integration. These come from the new coordinate system and are:

$$u \to 1 \text{ to } \infty; \qquad v \to -1 \text{ to } +1; \qquad \theta \to 0 \text{ to } 2\pi.$$

Therefore, adding limits and with some simplification to the exponential and constant terms,

$$S = \frac{\rho^3}{8\pi} \int_0^{2\pi} \int_{-1}^{1} \int_1^{\infty} e^{-u\rho}(u^2 - v^2)dudvd\theta.$$

Note the order of the limits with respect to the order of integration; this is u, v, then θ. The integral in θ can be separated out, because this does not appear in the integral, which becomes

$$S=\frac{\rho^3}{4}\int_{-1}^{1}\int_{-1}^{\infty}e^{-u\rho}(u^2-v^2)dudv.$$

This is a standard integral and using Maple produces

```
> rho^3/(4)*
  Int(Int(exp(-u*rho)*(u^2-v^2),u=1..infinity),v=-1..1):
  %= simplify(value(%)); S:= rhs(%);
```

$$\frac{1}{4}\rho^3\left(\int_{-1}^{1}\int_{1}^{\infty}e^{-u\rho}(u^2-v^2)\,du\,dv\right)=\frac{1}{3}e^{-\rho}(\rho^2+3+3\rho)$$

$$S:=\frac{1}{3}e^{-\rho}(\rho^2+3+3\rho)$$

where $\rho=R/a_0$. The result shows that the overlap decreases with separation of the nuclei because the exponential decreases more rapidly than ρ^2 increases.

4.11.4 Calculation of the self-energy integrals H_{11} and H_{22}

The self-energy H_{11} or H_{22} is the energy the electron and proton 1 will have as if they formed an atom whose energy is influenced by a nearby positive charge, which is that of the other proton. Most of the calculation of this energy is done before actually calculating an integral but is complicated by the fact that the momentum $\boldsymbol{P}$ is included in the Hamiltonian and this has to be dealt with. The equation is

$$\begin{aligned}H_{11}&=\int\varphi_1\boldsymbol{H}\varphi_1d\tau\\&=\int\varphi_1\left(\frac{\boldsymbol{P}^2}{2m}-\frac{q^2}{r_1}-\frac{q^2}{r_2}+\frac{q^2}{R}\right)\varphi_1d\tau\\&=\int\varphi_1\frac{\boldsymbol{P}^2}{2m}\varphi_1d\tau-\int\varphi_1\frac{q^2}{r_1}\varphi_1d\tau-\int\varphi_1\frac{q^2}{r_2}\varphi_1d\tau+\int\varphi_1\frac{q^2}{R}\varphi_1d\tau\end{aligned}\tag{4.69}$$

Understanding what each integral represents helps in their solution. Remember that as H_{11} is calculated only the electron and proton 1 are involved. The $\boldsymbol{P}^2/2m$ operator in the first integral represents the electron's kinetic energy and the second integral its potential energy with respect to proton 1. The first two integrals must therefore represent the lowest energy of a hydrogen atom, or minus one times the ionization energy, $-E_I$, where $E_I=q^2/2a_0$.

The fourth integral is also easy to solve; the electron's atomic wavefunction φ (equation (4.65)) does not depend upon the separation R of the nuclei, but only on the electron's position $\boldsymbol{r}$. This is because we have defined the nuclei to be fixed in space and so R can be taken out of the integral. Therefore the integral is

$$\int\varphi_1\frac{q^2}{R}\varphi_1d\tau=\frac{q^2}{R}\int\varphi_1\varphi_1d\tau=\frac{q^2}{R}=\frac{q^2}{\rho a_0}.$$

So far the integral is

$$H_{11}=-E_I+e^2/R-C\tag{4.70}$$

where C is the third integral,

$$C=\int\varphi_1\frac{q^2}{r_2}\varphi_1d\tau\tag{4.71}$$

and is called the *Coulomb Integral.* This is calculated in a similar manner to the overlap integral. The Coulomb integral describes the electrostatic potential energy between proton 2 and the charge distribution of the electron when it is associated with the 1s orbital around proton 1. Substituting for the wavefunctions gives

$$C=\int \frac{1}{\sqrt{\pi a_0^3}}e^{-\rho_1}\frac{q^2}{a_0\rho_2}\frac{1}{\sqrt{\pi a_0^3}}e^{-\rho_1}d\tau=\frac{q^2}{\pi a_0^4}\int\frac{e^{-2\rho_1}}{\rho_2}d\tau$$

Converting now to prolate spherical coordinates produces

$$\begin{aligned}C&=\frac{q^2\rho^3}{8\pi a_0}\int_0^{2\pi}\int_{-1}^{1}\int_1^{\infty}e^{-(u+v)\rho}\frac{(u^2-v^2)}{(u-v)\rho/2}du\,dv\,d\theta\\&=\frac{q^2\rho^2}{2a_0}\int_{-1}^{1}\int_1^{\infty}e^{-(u+v)\rho}(u+v)du\,dv\end{aligned}$$

Using Maple produces

```
> q^2*rho^2/(2*a0)*
  Int(Int(exp(-(u+v)*rho)*(u+v),u= 1..infinity),v= -1..1):
  %= simplify(value(%)); C:=rhs(%);
```

$$\frac{1}{2}\frac{q^2\rho^2\left(\int_{-1}^{1}\int_1^{\infty}e^{-(u+v)\rho}(u+v)du\,dv\right)}{a_0}=-\frac{q^2(-1+e^{-2\rho}+e^{-2\rho}\rho)}{\rho\,a_0}$$

$$C:=-\frac{q^2(-1+e^{-2\rho}+e^{-2\rho}\rho)}{\rho\,a_0}$$

The coulomb C integral is large at small internuclear separation, where $1/\rho$ is large, and decreases to zero at large ρ, when $1/\rho\to 0$ and the exponential term is small. Plotting the energy H_{11} as a function of ρ shows that H_{11} is large at small separation because C is large, but constantly decreases and then reaches a constant value of $-E_I$ at large proton separation.

4.11.5 Calculation of interaction energy integral H_{12}

Integrals such as H_{12} are said to 'cause' the interaction between the electron and each nucleus and lead to the formation of the bond. This is because this integral measures the interaction of the electron on one atom with the other atom.

Following the procedure for H_{11} the integral to be evaluated is

$$\begin{aligned}H_{12}&=\int\varphi_1 H\varphi_2 d\tau=\int\varphi_1\left(\frac{P^2}{2m}-\frac{q^2}{r_2}-\frac{q^2}{r_1}+\frac{q^2}{R}\right)\varphi_2 d\tau\\&=\int\varphi_1\left(\frac{P^2}{2m}-\frac{q^2}{r_2}\right)\varphi_2 d\tau-\int\varphi_1\left(\frac{q^2}{r_1}\right)\varphi_2 d\tau+\int\varphi_1\left(\frac{q^2}{R}\right)\varphi_2 d\tau\end{aligned}\tag{4.72}$$

Because the atomic orbitals φ_1 and φ_2 form the basis set for our calculation the first of the three separate integrals is related to the ionization energy of the atom just as in the integral H_{11}. However, the calculation is not the same because the second orbital is φ_2. To evaluate the integral, it is multiplied by 1 in the form of the wavefunction normalization, $\int\varphi_1\varphi_1 d\tau=1$. This cunning 'trick' is valid! The integral is rearranged as follows,

$$\begin{aligned}I_1&=\int\varphi_1(P^2/2m-q^2/r_2)\varphi_2 d\tau\int\varphi_1\varphi_1 d\tau\\&=\int\varphi_1(P^2/2m-q^2/r_2)\varphi_2\varphi_1\varphi_1 d\tau\\&=\int\varphi_1(P^2/2m-q^2/r_2)\varphi_1 d\tau\int\varphi_1\varphi_2 d\tau\\&=-E_I S\end{aligned}$$

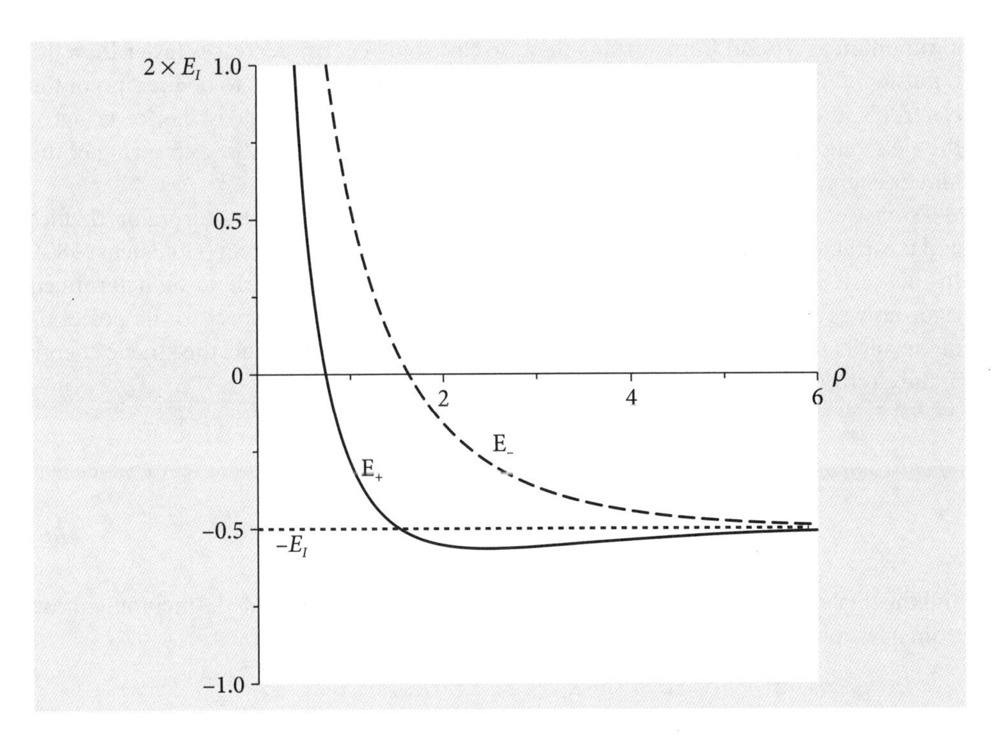

Fig. 4.31 Energies E_+ and E_- (dashed line) vs reduced internuclear separation $\rho = a_0 R$ and energy in units of ionization energy E_1.

In the second step, the integrals are merged as the integration is over the same coordinates, represented as $d\tau$. In the third step, the integrals are separated after rearranging terms and in the last step, they are identified as integrals already calculated. The first integral is the same as in (4.69) and the second is the overlap S.

The second integral of H_{12} has the form

$$Z = \int \varphi_1 \frac{q^2}{r_1} \varphi_2 d\tau, \tag{4.73}$$

and is called the *resonance* or *exchange integral* and is not zero. It expresses the possibility of the electron moving from the neighbourhood of one proton to that of the other and is a purely quantum effect: it has no classical counterpart. If at some time we could start the electron in orbital φ_1 and compel its wavefunction at later times to have the form $c_1\varphi_1 + c_2\varphi_2$, so it is 'shared' between two orbitals, then the values of c_1 and c_2 would oscillate with time. This means that after a certain time interval the wavefunction would be φ_2 and shortly after this, it would return to φ_1 and so on. This quantum mechanical oscillation is, by analogy with the mechanical oscillation, called resonance. The total integral is therefore

$$H_{12} = -E_I S + q^2 S/R - Z. \tag{4.74}$$

where R is the internuclear separation. If the energy H_{12} is plotted it has a minimum energy at a distance somewhat shorter than that measured for the chemical bond, and is zero at large internuclear separation. It is zero because the overlap integral is zero at large ρ as is the resonance integral, see Q4.87 for further calculations.

Plotting the graphs of the energies E_+ and E_- produces Fig. 4.31. It is clear from the shape of the curves that a minimum in the energy exists for the E_+ and not for the energy E_-. The minimum energy calculated and hence bond length occurs at $\approx 2.5a_0$ or 130 pm. The experimental value is 106 pm (Herzberg 1950), so this simple model produces a reasonably good result.

4.11.6 The origin of the chemical bond and the virial theorem

To determine the origin of the stability of the chemical bond, the virial theorem can be used. When the separation between the protons is reduced, the energy of their mutual (electrostatic) repulsion increases. The fact that the total energy of H_2^+ passes through a minimum as a bond forms means that the electronic energy decreases faster than the repulsion q^2/R increases. At any bond length, the total energy E has to be the sum of the electron's kinetic and potential energies, hence the question is: does the lowering of the electronic energy arise from a reduction in the potential energy or a lowering of the kinetic energy or both?

To answer this, the virial theorem can be used. This is a general result from mechanics and is extremely powerful as it can be applied to any mechanical system (Goldstein 1980). In this case, it allows the kinetic and potential energy of a molecule to be determined rigorously. To do this the variation of the total energy E, with respect to the positions of the nuclei must be known. In the special case of a diatomic molecule the kinetic energy of the electron $\langle T_e \rangle$ and its potential energy $\langle V_e \rangle$ are given by

$$\langle T_e \rangle = -E - R\frac{dE}{dR}, \tag{4.75}$$

$$\langle V_e \rangle = 2E + R\frac{dE}{dR}, \tag{4.76}$$

and the total energy $E = \langle V_e \rangle + \langle T_e \rangle$. Using the results already calculated, (including those from question Q4.87) the potential energy is calculated using

```
> Te:= -Eplus - rho*diff( Eplus, rho):
  Ve:= 2*Eplus + rho*diff( Eplus, rho):
  plot([Ve,Te],rho= 0..10);
```

The plot shows that both the kinetic and potential energy have minima in the region of the bond but that the decrease in kinetic energy is very small compared to that in the potential energy. The minimum in total energy when the bond is formed is at $\rho \approx 2.5$, see dotted line on Fig. 4.32, and is E_+ as in Fig. 4.31. The decrease in potential energy when the two nuclei come together is therefore what dominates the stabilization of the chemical bond because the kinetic energy is rising slightly at the bond minima but the potential energy is falling rapidly. If the kinetic and potential contributions to the energy of the antibonding orbital E_- are calculated, the potential energy still shows a small minimum but this is dominated by a large positive kinetic energy contribution, leading to no overall minimum in the energy.

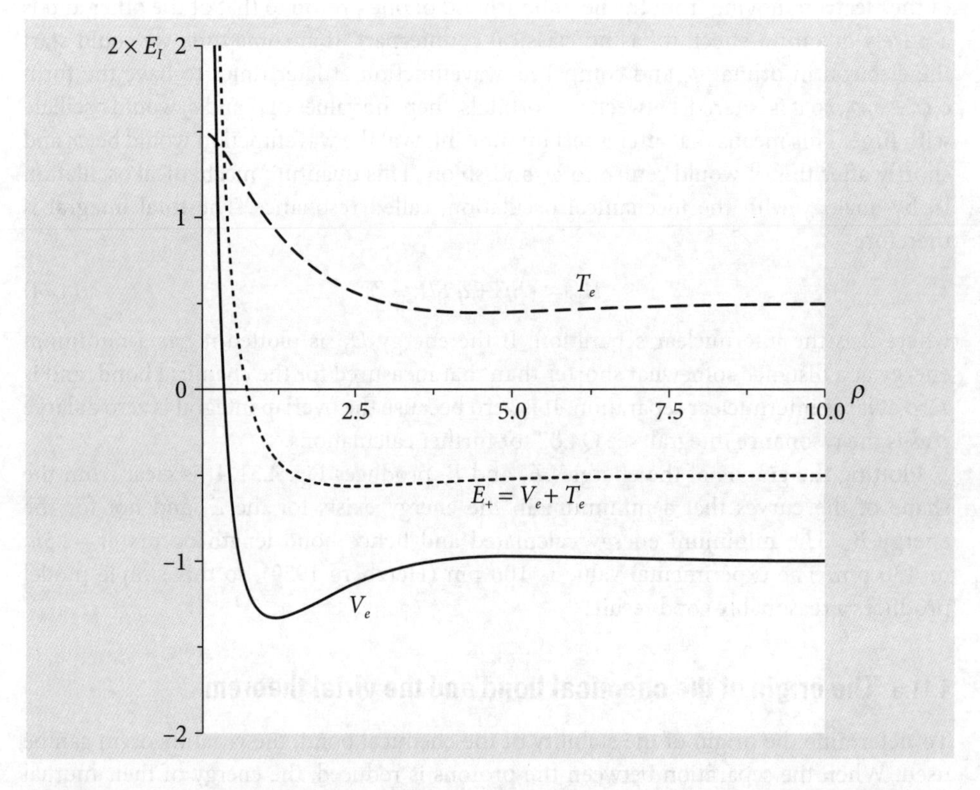

Fig. 4.32 The kinetic energy T_e and potential energy V_e of the electron in H_2^+ vs internuclear separation ρ. The total energy is the dotted curve.

4.11.7 **Question**

Full solutions are available at **www.oxfordtextbooks.co.uk/orc/beddard.**

Q4.87 This question is about the energy of H_2^+ molecular orbitals. The definitions in 4.11.2 are used. If φ_1 and φ_2 are H atom 1S orbitals show that

(a) $\int \varphi_1 \frac{q^2}{R}\varphi_2 d\tau = \frac{q^2 S}{R}$ where R is the internuclear separation.

(b) Calculate the resonance integral $Z = \int \varphi_1 \frac{q^2}{r_2}\varphi_2 d\tau$ belonging to H_{12} and thereby calculate the energy H_{12}.

(c) Use the results from the text and those just calculated to plot the two energies $E_+ = \frac{H_{11}+H_{12}}{1+S}$ and $E_- = \frac{H_{11}-H_{12}}{1-S}$ as a function of reduced internuclear separation and reproduce Fig. 4.31. Use $\rho = a_0 R$, $a_0 = 1$ and electronic charge $q = 1$.

4.12 Line integrals

The integrals so far studied have the form $\int f(x)dx$, which means that integration proceeds uniformly along the x-axis. Now suppose instead that we wish to integrate along some other direction, such as any curved line that is in the x–y plane. This is not as obscure as it may sound. When you ride a bike, the force needed to keep you moving will obviously vary depending on which way or how fast you want to go. Similarly, the amount of energy you consume getting from one place to another depends on the path taken such as walking around a hill instead of over the top. In thermodynamics, *line* or *path* integrals are very important and explain why some integrals are represented as the difference between starting and ending values while other integrals have to be evaluated explicitly.

Three things are needed to calculate a line integral:

(a) The path to be followed, C.

(b) Its starting and ending points, A and B.

(c) The 'line' or surface function itself, $P(x, y)$.

The line is often a two-dimensional function of x and y. A typical example would be to calculate the line integral of $P(x, y) = x + y$ along a path or curve given by $y = x^2/2$ from $x = 0$ to $x = 2$. This is written in the form

$$\int_C (x+y)dx \text{ where the path } C \text{ is } y = x^2/2 \text{ with limits } x = 0 \text{ to } x = 2.$$

The C subscript on the integral is the conventional notation to indicate a line integral, although this is not usually used in thermodynamics. The integral $\int (x + y)dx$ cannot be evaluated as it stands, because we do not yet know how y is related to x and integration is with respect to x. The path taken determines the relationship between y and x and mathematically, this can take almost any form we choose. In practice, of course, the specific problem being studied will determine this relationship. Therefore integration of $\int (x+y)dx$ could just as easily be over the path $y = e^x$ from $x = -2$ to $x = 2$ or $y = \sin(3x)$ from 0 to $\pi/2$ and so forth and consequently there is no limit to the number of examples that can be devised even with a single surface. Figure 4.33 shows the path of a circle of radius 3 projected onto the surface $P(x, y) = x^3y + xy^3 + 50$. The line integral is the curtain-like area between the lower circle C in the x–y plane and its projection hugging the surface. If this seems rather esoteric, consider instead a prosaic example of finding the weight of a piece of string. This is path C (Fig. 4.33). Now suppose that the string has a weight that varies with its length and the weight changes as described by $P(x, y)$. The total weight must be the integral along the length of the string each little element of which is accounted for in proportion to P by integrating. The weight at each point is the height of the line draped on P above the x–y plane as shown in Fig. 4.33 and the whole weight is the area between curve C in the x–y plane and its projection on P.

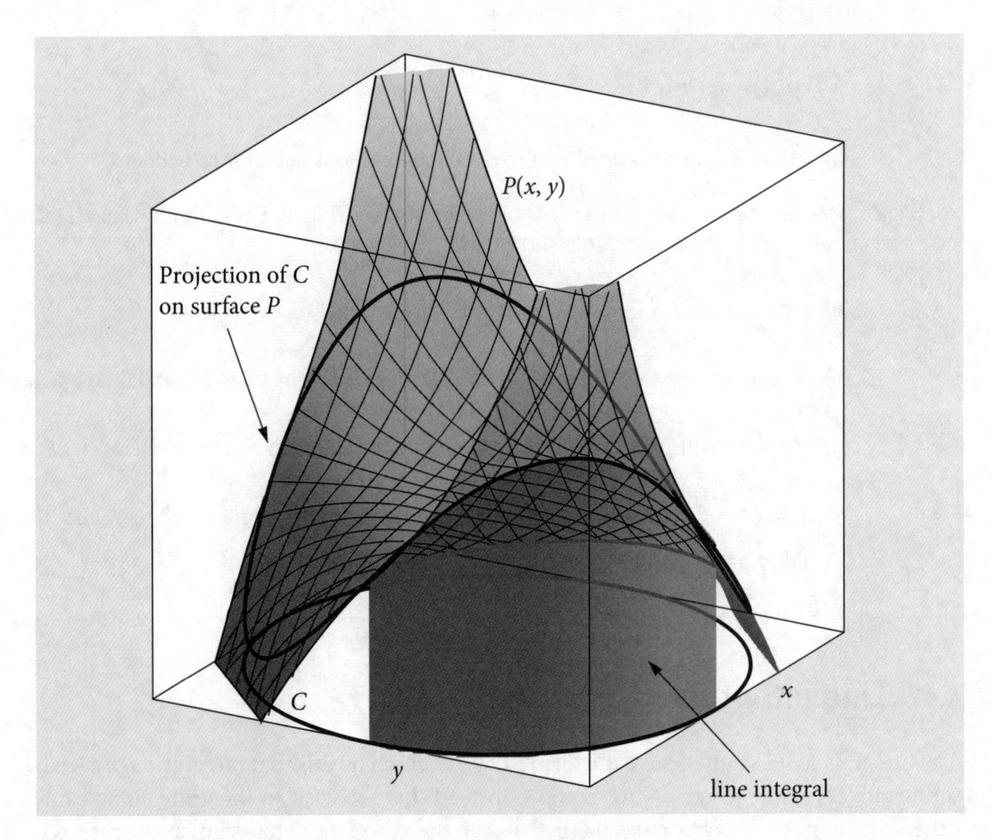

Fig. 4.33 The surface $P(x, y) = x^3y + xy^3 + 50$ with the curve (circle) C, $x^2 + y^2 = 9$. The dark line is the path of the circle projected onto the surface; the circle is shown below. The line integral is the curtain-like area shown between the circle and its projection onto the surface. In this figure this ranges from 0 to $\pi/2$ around the circle.

Returning to the original example, to calculate the integral is easier than it may seem. All that is necessary is to substitute y into the function $P = x + y$ and integrate as normal. If the path C is $y = x^2/2$ then

$$\int_C (x+y)dx = \int_0^2 (x + x^2/2)dx = [x^2/2 + x^3/6]_0^2 = 10/3$$

and if path C is $y = e^x$ from $x = -2$ to $x = 2$ then

$$\int_C (x+y)dx = \int_{-2}^2 (x + e^x)dx = \left[\frac{x^2}{2} + e^x\right]_{-2}^2 = e^2 - e^{-2}.$$

More formally, the line $P(x, y)$ is defined to be a surface or function of x and y, and assumed to be single valued. C is the curve in the plane defined as $y = f(x)$ going from points $A = \{x_1, y_1\}$ to $B = \{x_2, y_2\}$ as sketched in Fig. 4.34. The integral could therefore be written as

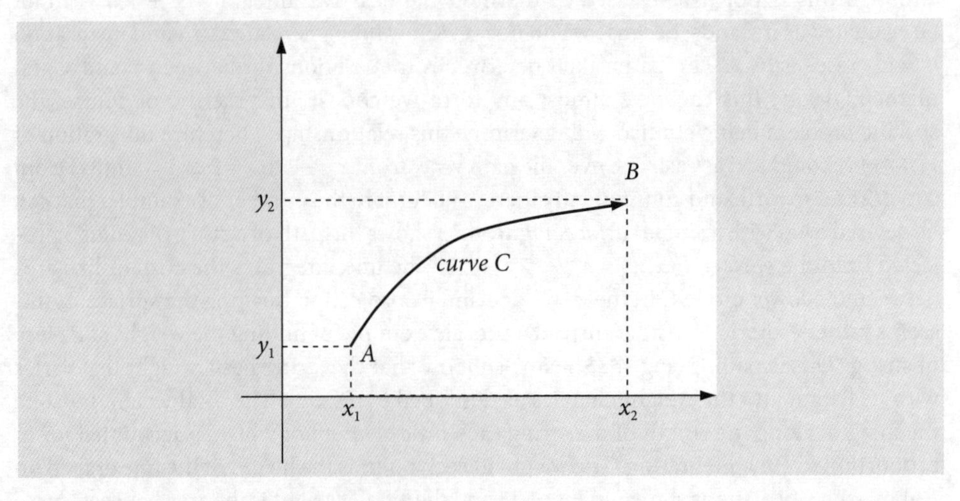

Fig. 4.34 A line integral is made along curve C.

$$\int_C P(x, y)dx = \int_{x_1}^{x_2} P(x, f(x))dx \tag{4.77}$$

which does make the integral look rather ordinary.

4.13 Definitions of some different forms of line integrals

4.13.1 Normal, i.e. continuous and single value functions

The line integral of a surface $P(x, y)$ with curve $C \equiv f(x)$ and limits x_1 and x_2 is

$$\int_C P(x, y)dx = \int_{x_1}^{x_2} P(x, f(x))dx. \tag{4.78}$$

A similar integral could be defined in terms of an integral in y; for example, if $Q(x, y)$ is the line, then

$$\int_C Q(x, y)dy = \int_{x_1}^{x_2} Q(x, f(x))\frac{dy}{dx}dx \tag{4.79}$$

but note that it is necessary to multiply by dy/dx and integrate in dx.

4.13.2 Parametric forms

Parametric equations are very useful in defining complex curves; a circle of radius 2 about the origin has the form $x^2 + y^2 = 4$ in Cartesian coordinates, and in parametric form, $x = 2\cos(t)$, $y = 2\sin(t)$; see Section 4.6. A curve defined in parametric form is written as $[x(t), y(t)]$. Integration is obtained by substituting equation (4.77) for x and y in terms of t and integrating in t,

$$\int_C P(x, y)dx = \int_{x_1}^{x_2} P(x(t), y(t))\frac{dx}{dt}dt \tag{4.80}$$

and differentiation is needed here to change dx into dt. If the line is again $x + y$ and the integration is around the circle $x = 2\cos(t)$, $y = 2\sin(t)$, choosing limits from 0 to π gives

$$\int_C (x + y)dx = -4\int_0^{\pi} [\cos(t) + \sin(t)]\sin(t)dt$$

$$= -4\int_0^{\pi} [\cos(t)\sin(t)dt - 4\int_0^{\pi} \sin^2(t)dt$$

Using Maple to do the integrations produces

```
> Int(sin(t)*cos(t),t):%= value(%);
```

$$\int \sin(t)\cos(t)\,dt = \frac{1}{2}\sin(t)^2$$

```
> Int(sin(t)^2,t):%= value(%);
```

$$\int \sin(t)^2\,dt = -\frac{1}{2}\sin(t)\cos(t) + \frac{1}{2}t$$

and adding the limits produces 0 for the first integral and π/2 for the second. The result is $\int_C (x + y)dx = -2\pi$.

4.13.3 Arc length on a surface *P*(*x*, *y*)

Suppose that the length of a small section of the curve C is ds then a line integral can be defined in terms of this arc length. See the sketch, Fig. 4.35,

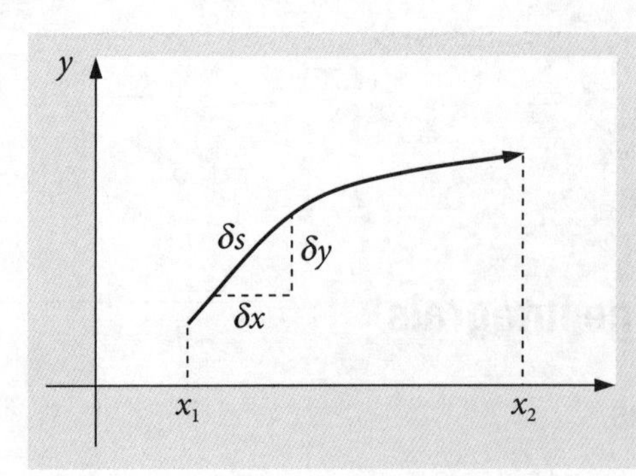

Fig. 4.35 Calculating a small length of a curve δs.

$$S=\int_C P(x,y)ds=\int_{x_1}^{x_2} P(x,y)\frac{ds}{dx}dx. \tag{4.81}$$

By Pythagoras's theorem, $\delta s^2 = \delta x^2 + \delta y^2$, and rearranging and taking the limit to form the differential $\dfrac{ds}{dx}=\sqrt{1+\left(\dfrac{dy}{dx}\right)^2}$ leads to the result:

$$S=\int_{x_1}^{x_2} P(x,y)\sqrt{1+\left(\frac{dy}{dx}\right)^2}\,dx, \tag{4.82}$$

where $y=f(x)$. Suppose that the surface is $P(x,y)=x+y$ then the arc length from $x=0$ to 4 along a curve C that we choose to be the parabola $y=1+x^2$ is

$$\int_C (x+y)ds=\int_{x_1}^{x_2}(x+y)\frac{ds}{dx}dx=\int_0^4 (x+1+x^2)\sqrt{1+4x^2}\,dx=190.36.$$

If the equation is in parametric form then ds/dt is used instead of ds/dx in equation (4.82), where,

$$\frac{ds}{dt}=\sqrt{\left(\frac{dx}{dt}\right)^2+\left(\frac{dy}{dt}\right)^2}.$$

4.13.4 Length of a curve

When only the length of a curve C is required, all that is necessary is to make $P(x,y)=1$ but still use the equation of the curve to define dy/dx;

$$S=\int_C ds=\int_{x_1}^{x_2}\frac{ds}{dx}dx=\int_{x_1}^{x_2}\sqrt{1+\left(\frac{dy}{dx}\right)^2}\,dx \tag{4.83}$$

As an example, consider finding the length of the curve of the same parabola and limits as in Section 4.13.3. The integral is

$$S=\int_{x_1}^{x_2}\frac{ds}{dx}dx=\int_0^4\sqrt{1+4x^2}\,dx$$

Using Maple to do the calculation gives

```
> Int( sqrt(1+ (Diff(1+x^2,x)^2)),x): % = value(%);
```

$$\int\sqrt{1+\left(\frac{d}{dx}(1+x^2)\right)^2}\,dx=\frac{1}{2}x\sqrt{1+4x^2}+\frac{1}{4}\operatorname{arcsinh}(2x)$$

and adding limits gives a value of 16.8, approximately.

If a curve is defined in plane polar coordinates, the arc length is

$$S=\int_{x_1}^{x_2}\sqrt{\left(\frac{dx}{d\theta}\right)^2+\left(\frac{dy}{d\theta}\right)^2}\,d\theta$$

which after substituting for x and y becomes

$$S=\int_{x_1}^{x_2}\sqrt{r^2+\left(\frac{dr}{d\theta}\right)^2}\,d\theta \tag{4.84}$$

The Archimedean spiral has the form $r=a\theta$ where a is a constant. The length of this curve is $a\int_0^{5\pi}\sqrt{1+\theta^2}d\theta$ if the limits are 0 to 5π. Using Maple to perform the integral produces

```
> Int(sqrt(1+theta^2),theta): %= value(%);
```

$$\int\sqrt{1+\theta^2}\,d\theta=\frac{1}{2}\theta\sqrt{1+\theta^2}+\frac{1}{2}\operatorname{arcsinh}(\theta)$$

Putting in the limits produces $S=123.34a$

4.13.5 Questions

Full solutions are available at **www.oxfordtextbooks.co.uk/orc/beddard**.

Q4.88 Calculate the arc length for **(a)** a circle of radius R, **(b)** the logarithmic or equiangular spiral $r = e^{-\theta/a}$ from 0 to 2π, **(c)** the catenary $y = \cosh(x)$ from $x = 0$ to x_0, and **(d)** the Archimedean spiral $r = a\theta$ from 0 to 2π

Strategy: Use equation (83) or (84).

4.13.6 A surface defined by two functions

Suppose that the surface is represented by two functions M and N, which is often the case in thermodynamics, then the line integral can be written as two integrals in dx and dy,

$$\begin{aligned} f &= \int_C M(x,y)dx + N(x,y)dy \\ &= \int_C \left[M(x,y) + N(x,y)\frac{dy}{dx} \right] dx \end{aligned} \tag{4.85}$$

As an example, the integral

$$\int_C y^2dx + 2x^2ydy \tag{4.86}$$

will be calculated over the upper half of the semicircle of a unit disc. The curve C is the equation of a unit circle, $x^2 + y^2 = 1$, and the integration proceeds from $\{1, 0\}$ to $\{-1, 0\}$, which is from the positive x-axis anticlockwise to the negative x-axis. The integral is, with the derivative in brackets,

$$\begin{aligned} \int_C y^2dx + 2x^2ydy &= \int_C 1 - x^2 + 2x^2\sqrt{1-x^2}\left(\frac{-2x}{2\sqrt{1-x^2}} \right) dx \\ &= \int_1^{-1} 1 - x^2 - 2x^3dx = x - \frac{x^3}{3} - \frac{x^4}{2}\bigg|_1^{-1} = -\frac{4}{3} \end{aligned}$$

4.13.7 Path independent integrals

As an example of an integral that is independent of the path, consider calculating the integral $\int_C xdx + ydy$ first along the sinusoidal path $y = \sin(x)$ and then along the straight path $y = 2x/\pi$ both from $\{0, 0\}$ to $\{0, \pi/2\}$. Using (4.85) the first curve produces

```
> Int( x + sin(x)*cos(x),x ): %= value(%);
```

$$\int (x + \sin(x)\cos(x))\, dx = \frac{1}{2}x^2 + \frac{1}{2}\sin(x)^2$$

which has the value $\pi^2/8 + 1/2$ with limits 0 and $\pi/2$. The second, with the curve $y = 2x/\pi$ produces

```
> Int(x + x*(2/Pi)^2, x=0..Pi/2):%= value(%);
```

$$\int_0^{\frac{1}{2}\pi} \left(x + \frac{4x}{\pi^2} \right) dx = \frac{1}{8}\left(1 + \frac{4}{\pi^2} \right)\pi^2$$

which simplifies to the same result. Therefore, this integral is independent of the path *provided both paths pass through the same end points.* There are clearly many other curves that pass through the points $\{0, 0\}$ to $\{0, \pi/2\}$ and all of these will return the same value of the integral.

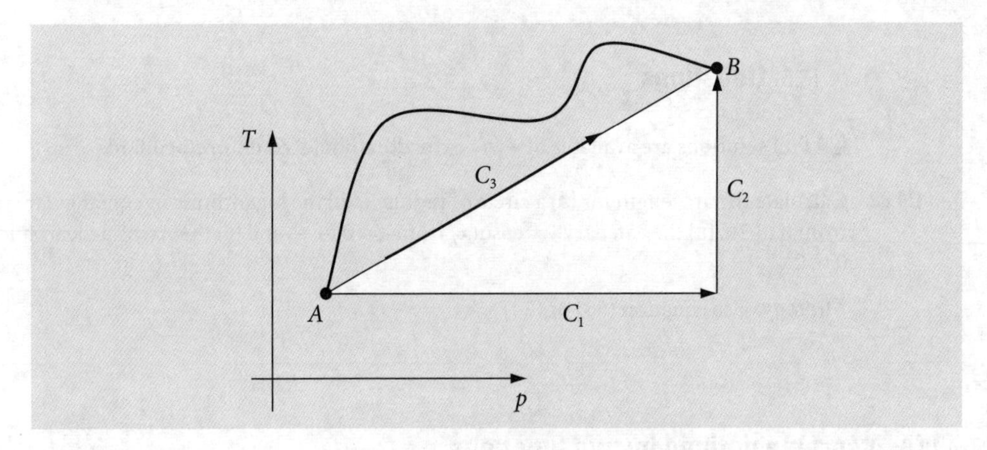

Fig. 4.36 The integral of a state function is independent of path and depends only on starting and ending positions *A* and *B*.

The surface $x + y$ would be called a *state function* if x and y had a meaning in thermodynamics. The value of a state function only depends on the starting and ending conditions and not on the means of getting from one to the other. The condition for the integral $f = \int M(x, y)dx + N(x, y)dy$ to depend only on its end-points is

$$\frac{\partial N}{\partial x} = \frac{\partial M}{\partial y}, \tag{4.87}$$

a relationship which has been met in the discussion on partial derivatives; see Chapter 3.12.

4.14 Path integrals in thermodynamics

In thermodynamics, all the integrals you are likely to meet are line functions in 8-dimensional space. Thermodynamics can consequently appear to be so very complicated because the choice of variables is so large. Normally, 'natural' variables are chosen depending upon which can easily be measured, usually these are p, V and T. However, any two independent variables can be chosen from the state functions p, V, A, T, H, U, G, S.

The first law states that the heat gained by a system q and the work done on the system w, define the change in internal energy ΔU as

$$\Delta U = q + w$$

or in differential form

$$dU = \delta q + \delta w.$$

The δ notation reminds us that q and w are not state functions but depend on the path. The integral therefore has to be performed as per normal and not obtained from the difference in starting and ending values, as is the case for a state function such as U. Some authors write $dU = đq + đw$ instead. The δ or $đ$ notation also means that q and w are not perfect differentials and are calculated by path integrals. Other authors just use $dU = dq + dw$ and leave it for you to decide which the perfect differentials are.

Suppose the heat absorbed on expansion of a gas is taken to be a function of internal energy U that changes from U_0 to U_1 and volume V that changes from V_0 to V_1. Since U is a state function and work is not, integrating q as $q = \int dq = \int dU - \int dw$ and replacing work with $dw = -pdV$ gives

$$q = U_1 - U_0 + \int_{V_0}^{V_1} pdV.$$

It is implied by choosing U and V that p is a function of both, i.e. $p(U, V)$, but it is never written like that. In a general thermodynamic system, it is unknown how the pressure depends on both the volume and internal energy, and how they vary along the path chosen. Therefore, the integral cannot be evaluated unless these relationships can be found. If the ideal gas law or van der Waals equation applies, then the relationship between p and V is known and the work integral can be calculated.

Suppose now that T and V are chosen as independent variables. By definition, $C_V = (\partial U/\partial T)_V$, then $\int dU = \int C_V dT$ and the heat change $q = \int dq = \int dU - \int dw$ becomes

$$q = \int C_V dT + T(\partial p/\partial T)_V dV.$$

This can be evaluated because the heat capacity and change of pressure with temperature and volume can be measured. Similarly, if p and V are the independent variables, then

$$q = \int C_V(\partial T/\partial p)_V dp + C_p(\partial T/\partial V)_p dV.$$

If gas expansion is carried out at constant pressure, then the first term is zero; and the second term is zero if carried out at constant volume; if neither is the case then the path equation must be known and dV/dp calculated to make the integral one in p or V alone.

As an example of a thermodynamic calculation, consider calculating the entropy, which is a state function and can be defined as

$$dS = \frac{C_P}{T}dT - \left(\frac{\partial V}{\partial T}\right)_p dp \tag{4.88}$$

and in an integrated form as

$$S = \int \frac{C_P}{T}dT - \left(\frac{\partial V}{\partial T}\right)_p dp. \tag{4.89}$$

if variables T and p are chosen. It has been stated that S is a state function: to prove that this is true, equation (4.87) must hold true for the derivatives in equation (4.88). To illustrate this, it will be necessary to relate T, V, and p and for which the ideal gas equation will be used. The first step is to form equation (4.87), which is done by looking at the 'symmetry' of the first of equations (4.85), $Mdx + Ndy$ and comparing it with (4.87), $\partial N/\partial x = \partial M/\partial y$. In this equation M is a derivative with respect to y, and N to x. Therefore, choosing $M = C_p/T$ and $N = (\partial V/\partial T)_P$ these must be functions of the other variable used in (4.88) or (4.89). This means that equation (4.87) becomes

$$\frac{\partial}{\partial p}\left(\frac{C_P}{T}\right) = \frac{\partial}{\partial T}\left(\frac{\partial V}{\partial T}\right)_p \tag{4.90}$$

Next, it is shown that this relationship is true. As the ideal gas law is $pV = RT$ then $(\partial V/\partial T)_p = R/p$ and so

$$\frac{1}{T}\frac{\partial C_P}{\partial p} = \frac{\partial}{\partial T}\left(\frac{R}{p}\right) \tag{4.91}$$

The heat capacity is by definition, $(dU/dT)_p$ making the left-hand side of equation (4.91)

$$\frac{1}{T}\frac{\partial C_P}{\partial p} = \frac{1}{T}\frac{\partial^2 U}{\partial p \partial T} = 0$$

because the internal energy U of an ideal gas depends only on the temperature. The right-hand side of (4.91) is also 0 as $\dfrac{\partial}{\partial T}\left(\dfrac{R}{p}\right) = 0$, verifying that dS is a state function.

4.14.1 State functions do not depend on the path

To show that S does not depend on the path, the entropy S of 1 mole of an ideal gas, expanded reversibly, will be calculated using equation (4.89). The starting pressure and temperature are $\{p_0, T_0\}$ (point A, Fig. 4.36) and ending $\{p_1, T_1\}$ (point B) and the calculation follows each of the two paths shown in (a) lines $C_1 + C_2$ and (b) line C_3.

Path (a) consists of two straight lines C_1 and C_2 at right angles to one another. C_1 is the path P_0 to P_1 at T_0, then C_2 from T_0 to T_1 at P_1 and path (b) is along a straight line C_3 from start to finish.

The method to follow is to:

(1) Substitute for any partial derivatives using (in this example) the ideal gas law to make an equation in p or T as necessary.
(2) Use equation (4.85) to make equations in dp and dT.
(3) Work out the remaining derivative dp/dT or dT/dp depending on the path taken.
(4) Integrate the resulting equation.

In step (1), starting with equation (4.89), substituting the partial derivative using the ideal gas law produces

$$S=\int \frac{C_P}{T}dT-\frac{R}{p}dp. \tag{4.92}$$

Step (2): making the integral a function of p and of T alone gives

$$S_T=\int_{p_0}^{p_1}\left(\frac{C_P}{T}\frac{dT}{dp}-\frac{R}{p}\right)dp, \qquad S_p=\int_{T_0}^{T_1}\left(\frac{C_P}{T}-\frac{R}{p}\frac{dp}{dT}\right)dT,$$

as illustrated in (4.85).

Steps (3) and (4): The first part of the path (a) is at constant $T = T_0$ and from p_0 to p_1; therefore S_T, is used and as $dT/dp = 0$

$$S_{T_0}=\int_{p_0}^{p_1}\frac{R}{p}dp=-R\ln(p_1/p_0).$$

This is also obvious because T is constant along this part of path (a) and equation (4.92) could have been used directly. In the second step at constant P_1 with the temperature changing from T_0 to T_1,

$$S_{p_1}=\int_{T_0}^{T_1}\frac{C_P}{T}dT=C_p\ln(T_1/T_0)$$

and the total change is the sum of these last two integrals.

The same method is used for path (b). As this path goes directly from start to finish it is the equation of a straight line;

$$T=\frac{T_1-T_0}{p_1-p_0}(p-p_0)+T_0.$$

The gradient $\dfrac{dT}{dp}=\dfrac{T_1-T_0}{p_1-p_0}=m$ and substituting this and T into the equation for S_T and integrating produces

$$\begin{aligned}S&=\int_{p_0}^{p_1}\left(\frac{C_P m}{m(p-p_0)+T_0}-\frac{R}{p}\right)dp\\&=C_p\ln\left(\frac{T_1-T_0}{p_1-p_0}(p-p_0)+T_0\right)\Bigg|_{P_0}^{P_1}-R\ln(p_1/p_0),\\&=C_p\ln(T_1/T_0)-R\ln(p_1/p_0)\end{aligned}$$

which is the same as along the path (a); therefore entropy is a state function. (The first part of the last integration is a standard one: $\int dx/(ax+b)=\ln(ax+b)/a$.)

4.14.2 Green's function

A line integral of two functions is by Green's theorem equal to the surface integral around a *closed area*, or

$$\int_C Mdx+Ndy=\iint\left(\frac{dN}{dx}-\frac{dM}{dy}\right)dxdy \tag{4.93}$$

where integration is taken in an anticlockwise direction. It can be evaluated either as two or more line integrals, one for each part of the loop, or from the double integral depending which is easier, but is always around a closed area. The notation for this is often given as $\int_C Mdx + Ndy$. If the curve is given parametrically, the integral is

$$A = \frac{1}{2}\int\left(x\frac{dN}{dt} - y\frac{dM}{dt}\right)dt.$$

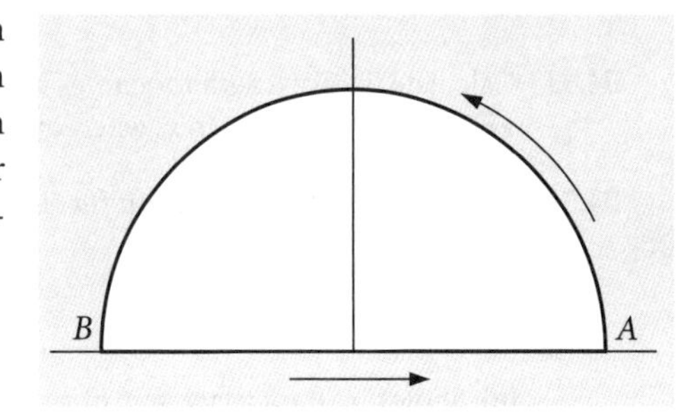

Fig. 4.37 Integrating around the curve A to B to A.

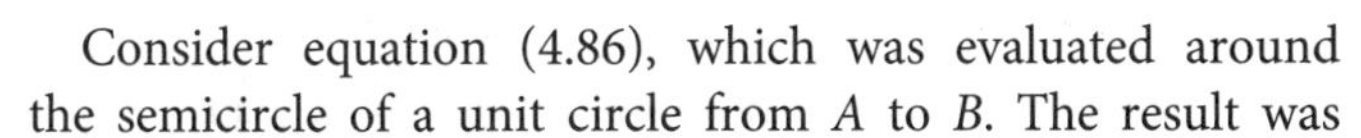

Consider equation (4.86), which was evaluated around the semicircle of a unit circle from A to B. The result was $\int_C y^2dx + 2x^2ydy = -\frac{4}{3}$.

The second part of the calculation of the closed loop must also contain the line integral from B to A, the curve for this is $y = 0$ and therefore the integral B to A is zero because y is the same, zero, in both limits of the integral. The total line integral from A to B then straight to A again is therefore also $-4/3$.

Now consider the Green's function integral, equation (4.93) with $M \equiv y^2$ and $N \equiv 2x^2y$ then

$$\begin{aligned}\int_C y^2dx + 2x^2ydy &= \iint\left(\frac{dN}{dx} - \frac{dM}{dy}\right)dxdy \\ &= \int_{-1}^{1}\int_{0}^{\sqrt{1-x^2}}(4xy - 2y)dydx \\ &= \int_{-1}^{1}(2xy^2 - y^2)\Big|_0^{\sqrt{1-x^2}}dx = \int_{-1}^{1}(2x-1)(1-x^2)dx \\ &= \left(x^2 - \frac{x^4}{2} - x + \frac{x^3}{3}\right)\Bigg|_{-1}^{1} = -\frac{4}{3}\end{aligned}$$

and note the change of order of integration in the second line which ensures that an integral in x with constants as limits is the final integration. The result is the same that obtained before.

4.14.3 **Questions**

Full solutions are available at **www.oxfordtextbooks.co.uk/orc/beddard**.

Q4.89 Find the area, the x and y centroids and moments of inertia I_x, I_y, and I_z of the ellipse $\frac{x^2}{a^2} + \frac{y^2}{b^2} = 1$.

Q4.90 Calculate the mean value of $r^2 = x^2 + y^2$ over the ellipse defined in the previous question.

Q4.91 If C is a line joining $\{0, 0\}$ to $\{a, b\}$ calculate

$$\int_C e^x\sin(y)dx + e^x\cos(y)dy$$

Strategy: Use the two function formula and convert dy into dy/dx where y is determined by the limits on the line, in this case a straight line from the origin to $\{a, b\}$.

Q4.92 **(a)** Find the area under one arch of the cycloid that is described by the parametric equations $x = a[t - \sin(t)]$ and $y = a[1 - \cos(t)]$. A description and sketch of the cycloid is given in Fig. 3.16.

(b) Find the length of the arch.

Q4.93 Calculate the arc length for curves **(a)** $r = 1$ and **(b)** $r = e^{-\theta}$ from 0 to 2π, and **(c)** the catenary $y = \cosh(x)$ from $x = 0$ to x_0 where $x_0 > 0$.

Q4.94 The surface area of a function $f(x)$ is given by

$$A = 2\pi \int_a^b f(x)\sqrt{1 + [f'(x)]^2}\, dx$$

(a) Show that the surface area of a sphere is $4\pi r^2$ starting with a circle of radius r, in which case $f(x) = \sqrt{r^2 - x^2}$, and effectively rotating this to form the surface. The integration limits are $\pm r$.

(b) Work out what fraction of the earth's surface is north of the seaside town of Dunbar, Scotland that is situated at exactly lat 56°.00 N.

Note: $f'(x)$ is the first derivative. Latitude is the angle from the equator to the pole.

Strategy: In (a) substitute, simplify, and find a very simple integral. In (b) take the south to north axis of the earth to be the x-axis and work out the x integration limits.

Q4.95 **(a)** In thermodynamics, what is a state variable?

(b) The work required to expand a gas is the line integral $w = -\int p dV$. If T and p are the variables to be used, this equation can be written as $w = \int p\left(\dfrac{\partial V}{\partial T}\right)_p dT + p\left(\dfrac{\partial V}{\partial p}\right)_T dp$. For 1 mole of an ideal gas calculate w along each of the two paths used in the example in Section 4.13.9 and Fig. 4.36 and hence show that w is not a state function.

Strategy: Follow the example and make the integral into one in dp and then dT alone. Substitute for the partial derivatives and use the gas law to substitute variables to make an equation in p or T as necessary. Only then, work out the remaining derivative, dp/dT or dT/dp depending on the path taken.

Summations, Series, and Expansion of Functions

5

5.1 Motivation

When calculating a partition function, working out the dissociation energy of a molecule by counting energy levels, or calculating a Madelung constant, a series of terms must be summed. Conversely, a theoretical expression may perhaps describe some molecular behaviour at a particular concentration or temperature and we shall need to discover what happens at other concentrations and temperatures. The assumption is made in all cases that it is possible to expand an expression as a Taylor or Maclaurin series and thereby learn something about regions in which we have no information. This assumption is based on the understanding that any function used varies in a slow and predictable way and that the extrapolation is not taken too far. In this chapter, both approaches are described.

5.2 Power series

The series expansions of exponential, sine, and cosine functions have already been met and there is nothing unusual or special about these. Any regular function can be expanded in a similar manner in powers of x, to produce its unique power series, which is in general

$$f = a_0 + a_1x + a_2x^2 + a_3x^3 + \cdots$$

where the as are positive or negative constants. In the reverse process, some series can be summed to form a function. The power series

$$1 + x + x^2 + x^3 + x^4 + \cdots + x^n +$$

can be summed to the algebraic expression,

$$\frac{1}{1-x} = 1 + x + x^2 + x^3 + x^4 + \cdots + x^n + \cdots \tag{5.1}$$

provided that $|x| < 1$ where $|x|$ is the absolute values of x. If this condition is not true then the expansion tends to infinity and gives the wrong answer. The summation notation for this series is

$$\frac{1}{1-x} = \sum_{n=0}^{\infty} x^n \tag{5.2}$$

where the Σ (sigma) is the summation symbol with limits $n = 0$ to $n = \infty$. The function $f(x) = \dfrac{1}{1-x}$ can then be represented by the polynomial series equation (5.1), but this is only true if $|x| < 1$. The series is calculated as

$$f(0) = 1, \qquad f(1/2) = 1 + 1/2 + 1/4 + 1/8 + \cdots$$

and so forth, which looks as though it will converge as terms are becoming increasingly smaller as the power of x increases. This always needs to be checked. If x is greater than 1 then the series obviously increases to infinity, which is clear to see from

$$f(2) = 1 + 2 + 4 + 8 + \cdots$$

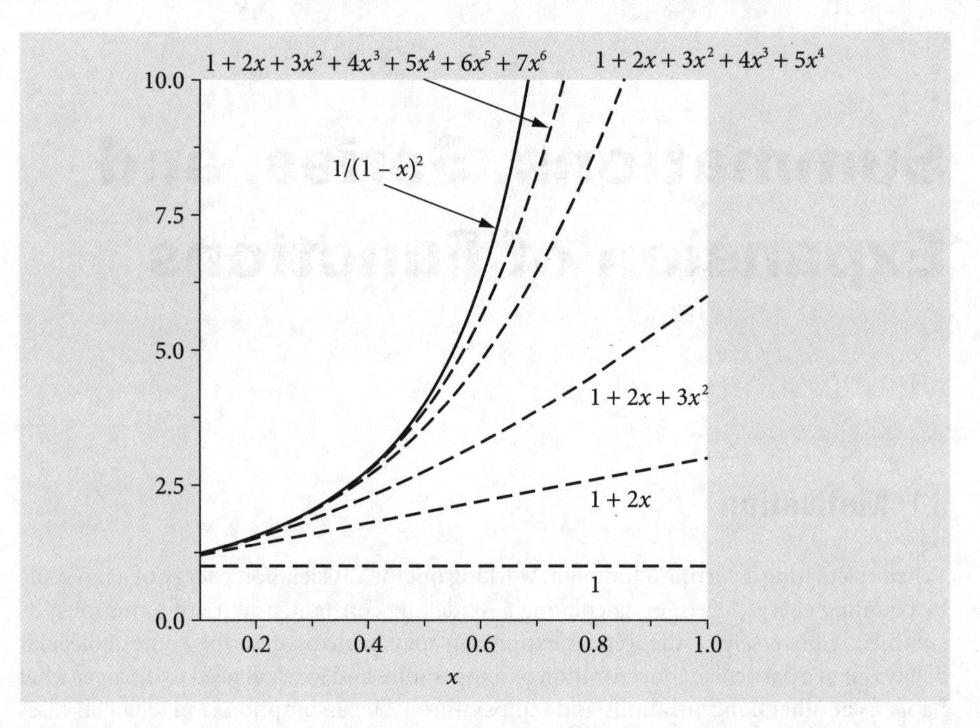

Fig. 5.1 The function $f'(x) = (1 - x)^{-2}$ and its approximation as series of increasing length over the range $x = 0$ to 1.

The expansion of $1/(1 + x)$ and of $1/(1 + ax)$, for example, follows immediately by substitution of $x \to -x$ and $x \to ax$ into the series of equation (5.1). In the latter case for the series to be valid $|ax| < 1$, where a is a constant.

Suppose each term in the series of $f(x)$ is differentiated, then

$$f'(x) = \frac{1}{(1-x)^2} = 1 + 2x + 3x^2 + 4x^3 + \cdots + nx^{n-1} + \cdots \tag{5.3}$$

and if $|x| < 1$ this is the expansion of $1/(1 - x)^2$, which can be written as

$$\sum_{n=0}^{\infty} nx^{n-1} = \frac{1}{(1-x)^2}. \tag{5.4}$$

Substituting $x \to -x$ gives the series

$$\frac{1}{(1+x)^2} = 1 - 2x + 3x^2 - 4x^3 + \cdots. \tag{5.5}$$

The summation expression is $\sum_{n=0}^{\infty}(-1)^n nx^{n-1} = \frac{1}{(1+x)^2}$ and the $(-1)^n$ ensures that alternate terms are positive and negative.

Fig. 5.1 shows the function of equation (5.3) and several approximations which are ever longer series with increasing powers of x. Expanding the series as far as x^6 only matches the $f'(x)$ well up to about $x = 0.5$ indicating that many terms may be needed to accurately reproduce this and similar functions.

5.2.1 Convergence

In summing a series, it is important to ensure that it converges to some sensible expression such as $1/(1 - x)$, and is not going to be infinite or undefined. With many series, the summation is infinite and cannot therefore be expressed in a simple form. There are a number of convergence tests to ensure that a series has a finite result; many of these are complex, but a ratio test is a good way of determining if the series will be finite.

In the ratio test, the ratio of any term w in the series to its preceding term is calculated

$$r = \lim_{n\to\infty}\left|\frac{w_{n+1}}{w_n}\right|,$$

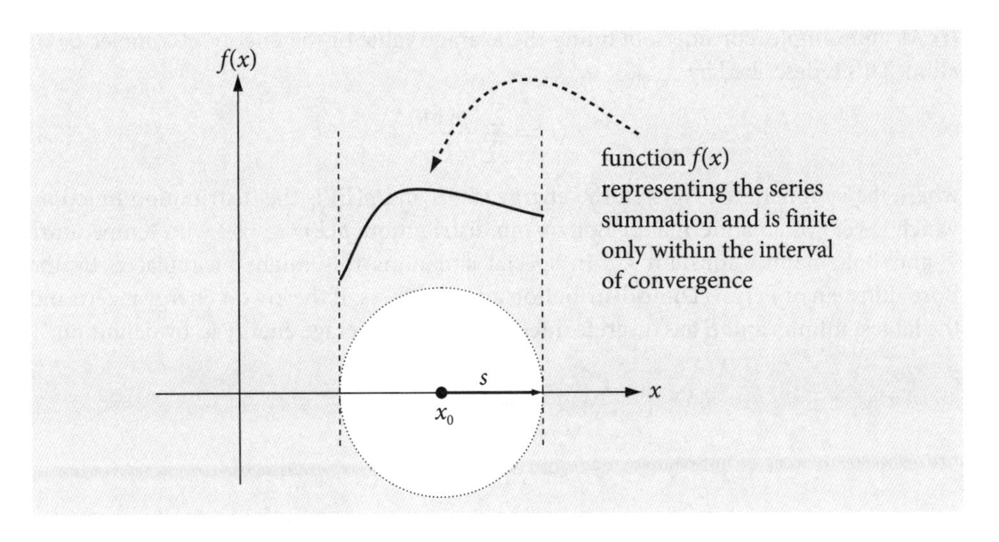

Fig. 5.2 The series converges only for values $x = x_0 \pm s$, elsewhere it is infinite.

and then the limit is taken as n tends to infinity. If the $(n+1)^{\text{th}}$ term is smaller than the nth, then the series is converging. The second step is to check whether $r < 1$, as n goes to infinity.

In the series expansion of $(1-x)^{-1} = 1 + x + x^2 \cdots$ any individual term in the series is x^n so the ratio $r = \lim_{n\to\infty}\left|\dfrac{x^{n+1}}{x^n}\right| = |x|$ is less than 1 only if $|x| < 1$; therefore when this is true the series converges. The interval of convergence for this sum is $0 \le x < 1$, and the radius of convergence s is 1. For any other values this series diverges. Should the radius s be infinity the series is convergent for all values of x; for instance, the exponential series is convergent for all x. Testing for convergence properly is far more complicated than this example leads us to believe and a maths textbook should be consulted to understand how to do this rigorously.

In general, a series converges if x is smaller than some number s, so that $|x_0 - x| < s$ where x_0 is a displacement about which the series converges; see Section 5.6 describing Taylor series for examples of this. The number s can be determined from the convergence test, Fig. 5.2. A useful property is that if a power series converges to some number and if each term of it is differentiated, the resulting series can be shown also to converge.

5.3 **Average quantities**

Any quantity x, such as a set of measurements of some phenomenon or of the discrete energy levels of an atom or molecule has an average value

$$\langle x \rangle = \frac{\sum xp(x)}{\sum p(x)} \tag{5.6}$$

which is also called the *first moment* of the distribution $p(x)$; the second moment is $\langle x^2 \rangle$, the fourth $\langle x^4 \rangle$, etc. The average of x^2 is

$$\langle x^2 \rangle = \frac{\sum x^2 p(x)}{\sum p(x)} \tag{5.7}$$

The variance or standard deviation squared of x is

$$\sigma^2 = \langle x^2 \rangle - \langle x \rangle^2$$

A more meaningful description is that σ is the *dispersion* in the value of x, meaning it is the spread in its value; experimental values are often quoted as $\langle x \rangle \pm \sigma$. The chapter on integration has examples of averaging using integrals rather than summations, and variance is described in Chapter 13.

(i) As an example, consider obtaining the average value of the energy of a molecule or atom. This is described by

$$\langle E\rangle = \frac{\sum Ep(E)}{\sum p(E)} \tag{5.8}$$

where the summation covers all the energy states and $p(E)$ is the distribution function, which is very often a thermal or Boltzmann distribution, $p(E) = e^{-E/k_BT}$ with temperature T and Boltzmann's constant k_B. In special situations this might be replaced by the Bose–Einstein or Fermi–Dirac distribution among others. If the lowest energy is zero and the largest infinity and E has discrete integer values the average energy is, by definition,

$$\langle E\rangle = \frac{\sum_{E=0}^{\infty} Ee^{-E/k_BT}}{\sum_{E=0}^{\infty} e^{-E/k_BT}}.$$

The denominator is called the partition function (see Section 5.4) and using the property of exponentials $e^{nx} = (e^x)^n$ is written as

$$Z = \sum_{E=0}^{\infty} (e^{-1/k_BT})^E = \frac{1}{1-e^{-1/k_BT}}$$

based on the summation $\sum_{n=0}^{\infty} x^n = \dfrac{1}{1-x}$ and by substituting $x = e^{-1/k_BT}$. Some guile is now needed in writing the numerator, which becomes $e^{-1/k_BT}\sum_{E=0}^{\infty} E(e^{-1/k_BT})^{E-1}$. Notice that e^{-1/k_BT} is put outside the summation to put the summation into the mathematical form $\sum_{E=0}^{\infty} Ex^{E-1} = \dfrac{1}{(1-x)^2}$, see equation (5.4), which when x is substituted back gives

$$\sum_{E=0}^{\infty} Ee^{-E/k_BT} = \frac{e^{-1/k_BT}}{(1-e^{-1/k_BT})^2}.$$

Dividing by the partition function Z produces the average energy as

$$\langle E\rangle = \frac{e^{-1/k_BT}}{(1-e^{-1/k_BT})}.$$

(ii) In synthesizing polymers, the random nature of the chemistry that adds monomers to an already growing chain dictates that a range of polymer lengths is usually produced; the polymer is poly-disperse. Two different averages are frequently taken to characterize the polymer; one is the number average mass, the other the weight average mass. The number average mass is

$$\langle m_n\rangle = \frac{\sum_k m_k p(n_k)}{\sum_k p(n_k)} = \frac{n_1m_1 + n_2m_2 + n_3m_3 + \cdots}{n_1 + n_2 + n_3 + \cdots} \tag{5.9}$$

where p is the distribution of the number of polymers of length k and m_k is the mass of each polymer k. The sum $\sum_k p(n_k) = N$ the total number of molecules. The weight (mass) average mass is

$$\langle m_w\rangle = \frac{\sum_k m_k \times m_k p(n_k)}{\sum_k m_k p(n_k)} = \frac{n_1m_1^2 + n_2m_2^2 + n_3m_3^2 + \cdots}{n_1m_1 + n_2m_2 + n_3m_3 + \cdots} \tag{5.10}$$

If the number average and mass (weight) average is the same, the polymer is mono-disperse. However, this is not usually the case. To illustrate that these two averages are different, suppose the number distribution, which may not necessarily relate to a particular polymer but has approximately the right shape, is $p(n_k) = n_k e^{-an_k}$ where a is a positive constant roughly describing the width of the distribution. If $n_k = k$ then, by definition, the number of moles of polymers with length 1 is 1 and the number of length 2 is 2 and so forth. The mass of the polymer with k segments is $m_k = km_0$, if m_0 is the mass of a mole of monomer units, then the number average, if k is taken to infinity, is

$$\langle m_n\rangle = m_0\frac{\sum_{k=1} n_k^2 e^{-an_k}}{\sum_{k=1} n_k e^{-an_k}} = m_0\frac{e^a+1}{e^a-1}.$$

The weight average is

$$\langle m_w \rangle = m_0 \frac{\sum_{k=1} n_k^3 e^{-an_k}}{\sum_{k=1} n_k^2 e^{-an_k}} = m_0 \frac{4e^a + e^{2a} + 1}{e^{2a} - 1}$$

which is clearly not the same as $\langle m_n \rangle$ unless $a \to \infty$ which is not the case.

5.4 Partition functions

One of the commonest uses of the summation of a series of terms is in the calculation of the partition function. These are met formally in statistical mechanics but are also used in kinetic theory and in spectroscopy.

The partition function is the sum over states of the distribution of energy levels usually assumed to follow Boltzmann's exponential law (or distribution), which describes the probability $pdE = ge^{-E/k_BT}dE$ that an atom or molecule has energy in the range E to $E + dE$ and where g is the degeneracy of the energy level, k_B is the Boltzmann constant and T the temperature. This distribution is obtained by considering the most probable way of arranging a fixed number of particles among all the available energy levels in a constant volume and with a constant total energy. Related distributions, but applicable only in certain circumstances, are the Fermi–Dirac and Bose–Einstein.

The German word for the partition function, *Zustandssumme*, translates as 'sum over states'. In English, it is called the partition function because it describes how much energy is in one level compared to the total, i.e. how energy is partitioned when the system, an ensemble of molecules for example, is in contact with a heat bath. A distribution based on being in thermal equilibrium with a heat bath is also called a canonical distribution. Once the partition function is calculated all the various thermodynamic properties, such as heat capacity, entropy, and free energy, can be calculated. The partition function is the summation

$$Z = \sum_{i=0}^{m} g_i e^{-E_i/k_BT}. \tag{5.11}$$

which extends over all energy levels from $i = 0$ to m. The energy of the ith level is E_i, its degeneracy g_i. In some situations, such as the harmonic oscillator, there is an infinite number of levels to sum in the partition function and $m = \infty$; in others, such as when an atom is in a magnetic field, there is a finite number.

The fractional population in any one level is the ratio

$$f_i \equiv \frac{n_i}{N} = \frac{g_i e^{-E_i/k_BT}}{\sum_{i=0}^{m} g_i e^{-E_i/k_BT}} \tag{5.12}$$

and the partition function normalizes the expression. N is the total number of particles. The partition function assumes that the energy is based on a thermal scale starting at zero, if this is not the case plausible answers can be obtained, which on scrutiny may prove to be wrong. The way to avoid this is always to make the energy scale start at zero; Q5.10 illustrates this effect.

5.5 Questions

Full solutions are available at **www.oxfordtextbooks.co.uk/orc/beddard**.

Q5.1 The exponential series is $e^x = 1 + x + \frac{x^2}{2!} + \frac{x^3}{3!} + \frac{x^4}{4!} + \cdots$, find values of x for which this series is convergent.

Q5.2 What is the expansion of $1/(1 + ax)^3$?

Strategy: Seek the answer by starting with an expression for which a series is known and repeatedly differentiate this expression if necessary to make the function in the question.

Q5.3 Calculate the atomic weight of lead and its standard deviation using the data below, which is taken from the NIST data handbook (www.physics.nist.gov/PhysRefData).

Isotope	Mass (amu)	Abundance
^{204}Pb	203.973020	1.40%
^{206}Pb	205.974440	24.1%
^{207}Pb	206.975872	22.1%
^{208}Pb	207.976627	52.4%

Q5.4 If, in one experiment, a die is repeatedly thrown and then in another experiment two are repeatedly thrown, calculate the expected (average) value of the number on the die face $\langle x\rangle$, and also $\langle x^2\rangle$ and σ in both experiments. Compare the averages and standard deviations in both cases. In the second case, write down in mathematical notation a summation formula for $\langle x\rangle$ and calculate its value. Using Maple makes this much easier.

Q5.5 Repeat the previous calculation for 10-sided dice with numbers 0 to 9. This is effectively estimating the average value of a random digit.

Q5.6 Assuming that Boltzmann's distribution applies, calculate the partition function Z,

(a) for a beam of n photons of frequency v s^{-1} and energy $E_n = nhv$;

(b) for a harmonic oscillator where $E_n = hv(n + 1/2)$. The vibrational quantum number is n, the vibrational frequency v s^{-1}.

(c) Calculate the relative population of the first five levels for N_2 and I_2 given that the vibrational temperatures $\theta = hv/k_B$ are 3394.6 and 308.7 K respectively.

Q5.7 Using the data in the previous question, work out at a given temperature

(a) The average *number* of photons $\langle n\rangle$.

(b) The average *number* of quanta $\langle n\rangle$ in a harmonic oscillator.

(c) Use data for N_2 and I_2 from the previous question to calculate the average number of vibrational levels populated at room temperature. Give a reasoned argument to explain what you think will be the effect on $\langle n\rangle$ of making the potentials for these molecules anharmonic, such as Morse potentials.

Strategy: Starting with equation (5.6), work out what x is and then the probability distribution $p(x)$. Write down the partition function and then $xp(x)$ and simplify or rearrange before trying to sum.

Q5.8 In this question, an unknown vibrational frequency v_4 will be found by combining spectroscopic and thermodynamic data. The classical energy of a molecule comprises at term $k_BT/2$ for each 'squared' or quadratic term in the calculation of the energy—this is known as the *equipartition theorem.* Therefore, for translation in each of the x-, y- or z-directions and each of the rotations about these axes and amount $k_BT/2$ is added to the energy. (Note that a linear molecule only has moments of inertia about 2 axes.) This limit is true provided the temperature $T \ge hv/k_B$ where v is the quantum and k_B is Boltzmann's constant, and applies to translations at all temperatures and rotations for most molecules above $\approx$10 K. Classically, each vibration also contributes k_BT, which is a term $k_BT/2$ for the vibrational potential energy, as this varies as bond extension squared. However, at room temperature, the classical limit is not reached for vibrations and instead the total quantized vibrational energy for all n normal modes contributes $k_B\sum_{i=1}^{n}\left[\dfrac{\theta_i/2T}{\sinh(\theta_i/2T)}\right]^2$ to the heat capacity. The 'vibrational temperature' for mode i is $\theta_i = hv_i/k_B$ and v_i the vibrational frequency.

Ethylene $CH_2{=}CH_2$ has one twisting vibrational normal mode v_4 of A_u symmetry species that is neither infrared nor Raman active, and is therefore invisible to spectroscopic methods. This occurs because the molecule has a centre of symmetry and hence no symmetric vibration can be IR active (no changing dipole) and no anti-symmetric vibration Raman active (no change in polarizability). This is true for any molecule in a point group containing a centre of inversion.

The heat capacity at *constant pressure* has been measured calorimetrically to be $4.37k_BT$ at 211 K.

(a) Explain why there are 12 normal mode vibrations.

(b) Show that ethylene belongs to the D_{2h} point group and that only normal modes belonging to the A_u symmetry species are both IR and Raman inactive.

(c) Calculate the vibrational contribution to the heat capacity from the partition function, and show that it has the form given above for each mode. The vibrational partition function Z for a harmonic oscillator is calculated in Q5.6 and $E_{vib} = k_B T^2 \left(\dfrac{\partial \ln(Z)}{\partial T}\right)_V$ and $C_V = \dfrac{dE_{vib}}{dT}$.

(d) How are C_V and C_p related?

(e) Calculate the frequency ν_4 using the following data.

Mode frequencies (cm^{-1}) for ethylene C_2H_4.

A_g	A_u	B_{1g}	B_{1u}	B_{2g}	B_{2u}	B_{3u}
3019	ν_4	3075	949	943	3105	2989
1623						
1342		1236			810	1443

Strategy: Find C_V for the vibrations and add terms for the rotations and translations at their classical limit. Look up how C_V and C_p differ. Compare the calculated C_p with the experimental value. Work out the summation using Maple, then isolate the term due to the unknown vibrational frequency and solve this equation for the A_u frequency.

Q5.9 The average energy of an ensemble (i.e. group) of molecules is given by $\langle E\rangle$. Each molecule is in an energy level with value E and these levels are discrete because the molecules obey the laws of quantum mechanics. The energies are $E = 0, 1, 2, \cdots, \infty$.

(a) Calculate $\langle E\rangle$ and show that at low temperatures it has a value of zero but at high temperatures $\langle E\rangle = k_B T$. What is a high temperature in this context?

(b) Calculate $\langle E^2\rangle$ and find its value at high and low temperatures.

(c) Obtain a formula for the ratio $\sigma_E/\langle E\rangle$ where σ_E is the standard deviation of the energy, its square is the variance defined as $\sigma_E^2 = \langle E^2\rangle - \langle E\rangle^2$.

(d) Comment on the formula obtained as the temperature is changed.

(e) Using the fact that the heat capacity $C_V = \dfrac{\partial\langle E\rangle}{\partial T}$, show that $\sigma_E^2 = k_B T^2 C_V$.

Strategy: This is a long question, so split it into smaller parts. Start by writing down the formulae needed and look up the summations in the table, or use Maple to do them for you. The answer to part of the first question is given in the text.

Q5.10 A sample of $CHCl_3$ is placed in an NMR spectrometer with a magnetic field of intensity B tesla. The energy (in joules) of the ^1H nuclei is given by $E_m = -\gamma m\hbar B$ where γ is the magnetogyric ratio, which is positive and describes how strongly the nuclei interact with the field, and m is the spin quantum number along the direction of the magnetic field which for a spin $1/2$ nucleus has only the values $m = -1/2$ and $+1/2$. The lowest energy state is that which aligns itself with the magnetic field and has $m = 1/2$ and is the state in which the molecule is more likely to be found.

(a) What is the energy gap ΔE responsible for the NMR transition and how does it compare to thermal energy at room temperature?

(b) Using Boltzmann's distribution, derive equations for $\langle E\rangle$ and $\langle E^2\rangle$. Derive the low and high temperature limits. Do they make sense? Move the zero of energy to the lowest level (see Fig. 5.20 of the online solutions) and repeat the calculation.

(c) Calculate also exact values at 10 K and 300 K using the values $\gamma = 26.7 \times 10^7$ rad tesla^{-1} s^{-1} (for the ^1H nuclei) and $B = 10$ tesla and room temperature.

This problem illustrates a subtle and possibly misleading effect when summing quantum numbers and energies not originating at zero.

Q5.11 In a homonuclear diatomic molecule, nuclear spin plays an important part in determining which molecular rotational energy levels are present because the nuclei are indistinguishable. Because of this indistinguishability, the nuclear wavefunctions must consist of linear combinations of individual wavefunctions that are either symmetric or anti-symmetric to exchange of coordinates. If the nuclear spin is an integer or zero, the nuclear wavefunction has symmetrical nuclear states and the molecule is therefore a boson. However, the total wavefunction must be asymmetrical overall (Pauli Principle) since it is the product of nuclear spin and rotational parts. If the nuclear spin has a half-integer value the molecule is a fermion and the total wavefunction must be *asymmetrical* to exchange.

Molecular hydrogen H_2 has two protons each with nuclear spin quantum number of 1/2 and exists as two forms; ortho- and para-hydrogen. In ortho-hydrogen molecules, the nuclear spins are parallel, giving total spin quantum number of $S=1$ and magnetic spin numbers of 1, 0, and −1. Because the total quantum number is one, there are $2S+1=3$ *symmetric* nuclear spin combinations and the molecule therefore only exists in *anti-symmetric* rotational states, such as those with odd J quantum numbers. The three nuclear levels are degenerate so the molecule exists in three ground states.

Para-hydrogen has opposed nuclear spins, with total spin quantum number $S=0$, and magnetic spin quantum number of zero, and an asymmetrical nuclear spin wavefunction with symmetric rotational spin wavefunctions, $J=0, 2, 4, \cdots$ and so forth.

If S is the nuclear spin quantum number, $g=2S+1$ is the nuclear spin degeneracy. The total partition function for half-integer spin states is $Z=Z_{para}+Z_{ortho}$ where

$$Z_{para}=\frac{g(g-1)}{2}\sum_{J=0,2,4..}(2J+1)e^{-BJ(J+1)/k_BT}$$

and

$$Z_{ortho}=\frac{g(g+1)}{2}\sum_{J=1,3,5..}(2J+1)e^{-BJ(J+1)/k_BT}.$$

The fraction $g(g-1)/2$ represents the number of asymmetrical nuclear spins states and $g(g+1)/2$ those of the symmetrical spin states. The total number of spin states is g^2.

(a) Calculate the total degeneracy of the Jth rotational level with ortho- and with para- nuclei. Use Maple to calculate and plot the equilibrium fraction of ortho molecules present in H_2 with temperature. The rotational constant is 60.8 cm^{-1}. Calculate the composition at room temperature, say 300 K, and at the normal boiling temperature of 20.4 K.

(b) Calculate how the rotational contribution to the total heat capacity varies with temperature. Explain the shape of the heat capacity curve.

(c) The heat of vaporization of liquid H_2 is 0.904 kJ mole^{-1}. What would be the consequence of spontaneous conversion of a mole of liquid H_2 at its room temperature composition of ortho- to para- molecules but at its boiling point? Consequently, explain why it is necessary to catalyse the conversion from ortho- to para- before it is cooled and put into cryostats as liquid H_2.

Strategy: Calculate the ortho- and para- series separately but with only alternate odd or even terms. At equilibrium the fraction of ortho-hydrogen present is $Z_{ortho}/(Z_{ortho}+Z_{para})$. The heat capacity is the rate of change of internal energy with temperature, so it is necessary to find the relationship between the partition function and internal energy. Look this up in your textbook.

Q5.12 Type II restriction enzymes are used by bacteria to inactivate viral DNA by identifying then cutting out specific sequences usually $n=4$, 6, or 8 base pairs long; these are called 'words'. The bacteria's own DNA is protected from this enzyme by methylation. The resulting distribution of shorter segments of DNA can be used as a DNA fingerprint if the fragments are cloned so that there is enough material to be characterized, for example, by gel electrophoresis.

Assuming that the four base pairs occur with equal probability, the fraction $1/4^n$ is the probability of reading the correct n letter word that leads to DNA scission. The enzyme then moves along the DNA to read the next word.

(a) Convince yourself that the probability of an enzyme *not* starting a new word after $L-1$ moves along the DNA but starting one at move L is the (geometric) distribution

$$p[L]=\frac{(1-1/4^n)^{L-1}}{4^n}$$

(b) Is this distribution normalized?

(c) Show that the *mean* distance to the next word is approximately 4^n base pairs.

Strategy: Think about the chance that one word is read and then not read and recall that probabilities multiply.

Q5.13 A ball that is initially stationary is dropped from a height h. Its position y at time t is $y = -gt^2/2 + h$ if its motion is not impeded by friction due to the viscosity of the air. The mass of the ball is m, g the acceleration due to gravity, and h the initial height. The equation describing the balance of forces is $md^2y/dt^2 + mg = 0$. The similar equation including friction from the air, which is proportional to velocity, is $md^2y/dt^2 + mg + mcdy/dt = 0$, where c is a constant. The solution to this equation is $y = \dfrac{hc^2 + g(1 - e^{-ct})}{c^2} - \dfrac{gt}{c}$. Show that this solution reduces to the first when $c = 0$.

5.6 Maclaurin and Taylor series expansions

Many normal functions can be expanded into a series by the method devised by Maclaurin and by Taylor. Series are often used to simplify a function at some value of its argument, for example at small x, $\sin(x) \approx x$, and this may enable an equation to be solved, as in the case of a pendulum's motion. The equation of motion is greatly simplified by assuming that the pendulum is restricted to move only to small angles from the vertical. The other use of a series expansion is to extrapolate from some simple law where dependence may be linear in x, into more realistic conditions by adding terms in x^2 and x^3 and so on. This approach extends our knowledge into unknown regions by using what is presently known about the behaviour of the equations, and assumes that they change only in a predictable and gradual way. This is the approach made when the virial coefficients of a real gas are calculated. The equation has the form $pV/(nRT) = c_0 + c_1(n/V) + c_2(n/V)^2 + \cdots$ with virial coefficients c_0, c_1, etc. and is clearly seen to be based on the ideal gas law.

5.6.1 Maclaurin series

A power series has the general form

$$f(x) = a_0 + a_1x + a_2x^2 + a_3x^3 + \cdots + a_nx^n + \cdots \tag{5.13}$$

where the coefficients a_0, a_1, etc. are independent of the parameter x. We shall clearly want to be able to find the coefficients a_0, a_1, and so forth, and to do this shall rely on the fact that by differentiating a *converging* power series the resulting series can be shown also to converge.

The coefficients are calculated by repeatedly differentiating $f(x)$ and then substituting $x = 0$. This will lead to the *Maclaurin* series. (It is implicitly assumed that the function can be differentiated, functions such as $|x|$ generally have no series as the differential is undetermined at $x = 0$.) If the notation $f'(x)$ represents the first derivative, $f''(x)$ the second, etc., the derivatives of $f(x)$ are

$$f'(x) = a_1 + 2a_2x + 3a_3x^2 + 4a_4x^3 + \cdots + na_nx^{n-1} + \cdots$$

$$f''(x) = 2a_2 + 6a_3x + 12a_4x^2 + \cdots + n(n-1)a_nx^{n-2} + \cdots$$

$$f'''(x) = 2 + 6a_3 + 24a_4x + \cdots + n(n-1)(n-2)a_nx^{n-3} + \cdots$$

The coefficients are found by substituting $x = 0$ into equation (5.13) and into the derivatives. The first four are

$$a_0 = f(0), \qquad a_1 = f'(0), \qquad a_2 = f''(0)/2! \qquad a_3 = f'''(0)/3!,$$

and the nth one is $a_n = f^n(0)/n!$.
By reassembling the parts, the Maclaurin series for $f(x)$ is produced:

$$f(x) = f(0) + xf'(0) + \frac{x^2}{2!}f''(0) + \frac{x^3}{3!}f'''(0) + \cdots + \frac{x^n}{n!}f^n(0) + \cdots \tag{5.14}$$

Sometimes the equation is written in the equivalent form

$$f(x) = f(0) + x\left(\frac{df}{dx}\right)_0 + \frac{x^2}{2!}\left(\frac{d^2f}{dx^2}\right)_0 + \frac{x^3}{3!}\left(\frac{d^3f}{dx^3}\right)_0 + \cdots + \frac{x^n}{n!}\left(\frac{d^nf}{dx^n}\right)_0 + \cdots \tag{5.15}$$

where the subscript 0 means evaluate the derivative at $x = 0$.

5.6.2 Taylor series

In the Maclaurin series, the derivative is calculated at $x = 0$ and the function is described as having been 'expanded about zero'. However, a function can be expanded about any other value of x, say x_0; this will generate the *Taylor* series which is the 'expansion about x_0'. This series is

$$f(x) = f(x_0) + (x - x_0)f'(x_0) + \frac{(x - x_0)^2}{2!}f''(x_0) + \cdots + \frac{(x - x_0)^n}{n!}f^n(x_0) + \cdots \tag{5.16}$$

or equivalently,

$$f(x) = f(x_0) + (x - x_0)\left(\frac{df}{dx}\right)_{x_0} + \frac{(x - x_0)^2}{2!}\left(\frac{d^2f}{dx^2}\right)_{x_0} + \cdots + \frac{(x - x_0)^n}{n!}\left(\frac{d^nf}{dx^n}\right)_{x_0} + \cdots \tag{5.17}$$

where the subscript x_0 means evaluate the derivative at $x = x_0$. The Taylor series evaluated at zero is the Maclaurin series.

These series are extremely useful, for example, to calculate $\sin(1.3)$, $\cos(\pi/4)$, $\ln(1 + x)$, and so forth the function can be expanded as a Taylor series and terms added until the required numerical accuracy is needed. This is tedious to do by hand rather than being difficult, but is ideally suited for computer calculation where repetitive calculations are easily performed.

5.6.3 O(*x*) and ~ notations

In a series expansion of a function $f(x)$ it is often possible to state something general perhaps that that $f(x) \to \infty$ as $x \to \infty$. This is, however, often too general and if a series is to be used only to a finite number of terms we would like to know how quickly the term after the last one used is varying as x changes; does it change as x^2 or e^x for example? The 'big-O' notation, for example, $O(x^2)$, $O(e^x)$, and so forth, is used for this purpose and means that the function whose series we are examining grows at a rate x^2 or e^x at the point that the function is being evaluated. An example is the series $x^{-2} - x^{-4} + O(x^{-6})$ when $x > 1$ which means that the next term in the series is vastly smaller than the x^{-4} term and can be safely ignored.

The symbol ~ is generally interpreted as meaning 'is proportional to' and as used in $\sin(x) \sim x$ when $x \to 0$ means that the function grows at the same rate as x.

5.6.4 Useful series expansion formulae

Several expansions calculated using Maple are listed below. If you want to calculate, for example $1/(1 - ax)$, replace each x in the first series below with $-ax$ and evaluate the result; $(1 - ax)^{-1} = 1 + ax + (ax)^2 + (ax)^3 + \cdots$. This substitution can be performed with any series and means that fewer need to be remembered. The $O(x^6)$ means that the next term in the series has power of order of x^6. You must check on the value of x used determine if this is an acceptable approximation.

$$\frac{1}{1+x} = 1 - x + x^2 - x^3 + x^4 - x^5 + O(x^6)$$

$$\frac{1}{(1+x)^2} = 1 - 2x + 3x^2 - 4x^3 + 5x^4 - 6x^5 + O(x^6)$$

$$\sqrt{1+x} = 1 + \frac{1}{2}x - \frac{1}{8}x^2 + \frac{1}{16}x^3 - \frac{5}{128}x^4 + \frac{7}{256}x^5 + O(x^6)$$

$$\frac{1}{\sqrt{1+x}} = 1 - \frac{1}{2}x + \frac{3}{8}x^2 - \frac{5}{16}x^3 + \frac{35}{128}x^4 - \frac{63}{256}x^5 + O(x^6)$$

$$\sin(x) = x - \frac{1}{6}x^3 + \frac{1}{120}x^5 + O(x^6)$$

$$\cos(x) = 1 - \frac{1}{2}x^2 + \frac{1}{24}x^4 + O(x^6)$$

$$\tan(x) = x + \frac{1}{3}x^3 + \frac{2}{15}x^5 + O(x^6)$$

$$\ln(1+x) = x - \frac{1}{2}x^2 + \frac{1}{3}x^3 - \frac{1}{4}x^4 + \frac{1}{5}x^5 + O(x^6)$$

$$e^x = 1 + x + \frac{1}{2}x^2 + \frac{1}{6}x^3 + \frac{1}{24}x^4 + \frac{1}{120}x^5 + O(x^6)$$

$$(a+x)^n = a^n + \frac{a^n n}{a}x + \frac{1}{2}\frac{a^n n(n-1)}{a^2}x^2 + \frac{1}{6}\frac{a^n n(n-1)(n-2)}{a^3}x^3 + O(x^4)$$

5.6.5 Trig functions

The Maclaurin series of $f(x) = \sin(x)$ is equivalent to a Taylor series expanded about zero. By definition the expansion is

$$\sin(x) = \sin(0) + x\left(\frac{d\sin(x)}{dx}\right)_0 + \frac{x^2}{2!}\left(\frac{d^2\sin(x)}{dx^2}\right)_0 + \frac{x^3}{3!}\left(\frac{d^3\sin(x)}{dx^3}\right)_0 + \cdots.$$

The derivatives of the function are

$$f'(x) = \cos(x) \qquad f''(x) = -\sin(x) \qquad f'''(x) = -\cos(x)$$

$$f^4(x) = \sin(x) \qquad f^5(x) = \cos(x), \cdots,$$

and when $x = 0$ they follow a repeating pattern with a period of four terms,

$$f'(x) = 1 \qquad f''(x) = 0 \qquad f'''(x) = -1 \qquad f^4(x) = 0 \qquad f^5(x) = 1$$

where each even power derivative is zero and the odd power derivatives alternate between 1 and −1. With this in mind, the expansion for $\sin(x)$, which must contain only odd powers of x, is

$$\sin(x) = x - \frac{x^3}{3!} + \frac{x^5}{5!} - \frac{x^7}{7!} + \cdots = \sum_{n=0}^{\infty}(-1)^n \frac{x^{2n+1}}{(2n+1)!}. \tag{5.18}$$

The summation formula on the right was obtained 'by inspection', which in practice can involve quite a bit of 'trial and error' to get correct. For example the $(-1)^n$ ensures that terms are negative when n is odd, n being the *position* of a term in the series. The first, x, is at position 0.

The series for $\cos(x)$ can be found by differentiating the series for $\sin(x)$ because $f'(\sin(x)) = \cos(x)$, therefore the series only contains even powers of x and is,

$$\cos(x) = 1 - \frac{x^2}{2!} + \frac{x^4}{4!} - \frac{x^6}{6!} + \cdots = \sum_{n=0}^{\infty}(-1)^n \frac{x^{2n}}{(2n)!} \tag{5.19}$$

The series expansion of $\sin(a+x)$ contains terms in all powers of x,

$$\sin(a+x) = \sin(a) + \cos(a)x - \sin(a)\frac{x^2}{2!} - \cos(a)\frac{x^3}{3!} + \sin(a)\frac{x^4}{4!} + \cdots$$

as does the cosine series

$$\cos(a+x) = \cos(a) - \sin(a)x - \cos(a)\frac{x^2}{2!} + \sin(a)\frac{x^3}{3!} + \cos(a)\frac{x^4}{4!} - \cdots.$$

and when $a = 0 \pm k\pi$, where k is an integer, this reverts to equation (5.19) and the sine expansion to (5.18). The graph in Fig. 5.3 shows the Taylor series approximation to $\sin(x)$ taken up to x^{15}. The dashed curve is the approximation which is almost exact up to about $x = \pm 6$. The Maple code is shown below.

This code is not the most sophisticated way of doing this calculation but by being straight forward should be clear. The `for deg... do` loop generates a series of calculations with different number of powers in the expansion. The series expansion is `series(fun(x), x, deg)` where `fun(x)` is the function to be expanded and `deg` the power to end the expansion. `convert(..,polynom)` makes a polynomial out of the result, this removes the order $(O)^n$ at the end of the output and `unapply(.. x)` makes

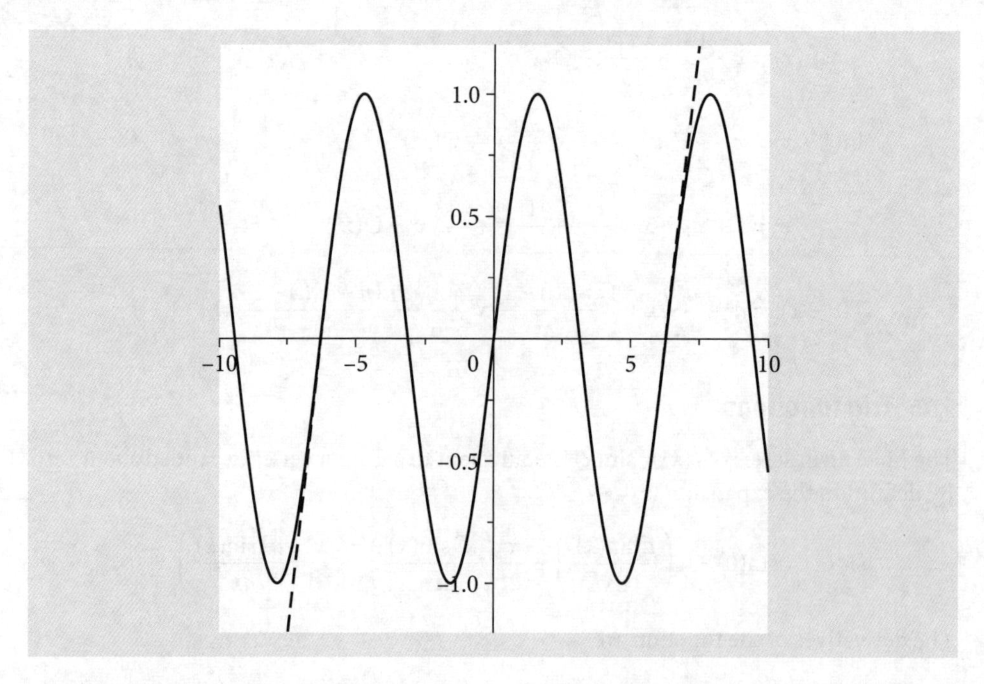

Fig. 5.3 Taylor expansion of sin(x) up to power x^{15} (dashed line) together with the function itself. The expansion is good to about ±3.

the series a function of x. The `display` instruction is used to plot the initial curve and the series expansion curves on the same axes and `seq(..)` makes a list of plots so you can animate them.

Algorithm 5.1 Making a Taylor series

```
> with(plots):
  fun:= x-> sin(x):                    # define function
  n:= 20:                              # maximum degree of polynom
  plt:= array(1..n):                   # make array to hold plots
  i:= 0:
  for deg from 1 by 2 to n do:
     s:= series(fun(x),x, deg);              # make series
     P:= convert(s, polynom):                # make into polynomial
     f:= unapply(P,x);                       # make function in x
     i:= i + 1:
     plt[i]:= plot( [fun(x),f(x)],x = -10..10):# plot array
  end do:
> display([seq( plt[deg],deg= 1..n/2)],insequence =true);
```

When the graph is plotted, click on it and the animation icons appear on the top of the Maple worksheet. The animation shows the effect of adding terms into the expansion, but you will probably need to slow it down to see this clearly.

This routine can easily be customized for other functions by replacing the sin(x) with something else, for example exp($-x$). The Taylor series are not necessarily very good at approximating all functions, and very large powers are often needed. Mathematicians have developed other power expansions, such as the Fourier series and Lagrange, Hermite and Chebychev polynomials, which are more efficient in many cases. Fourier series are described in Chapter 9.

5.6.6 Exponential expansion

The expansion of the exponential is easy to achieve with $f(x) = e^{ax}$ and a as a constant. The derivatives are

$$f'(x) = ae^{ax}, \qquad f''(x) = a^2e^{ax}, \qquad f'''(x) = a^3e^{ax}, \qquad f^4(x) = a^4e^{ax},$$

then, when $x = 0$, the derivatives can be written as $f^n(0) = a^n$ and therefore the coefficients in the series are $\dfrac{a^n}{n!}$. Substituting using equation (5.16) produces

$$e^{ax} = 1 + ax + a^2\frac{x^2}{2!} + a^3\frac{x^2}{3!} + \cdots + a^n\frac{x^n}{n!} + \cdots = \sum_{n=0}^{\infty}\frac{(ax)^n}{n!}. \tag{5.20}$$

This expansion can be shown to be true for any value of x. When the constant $a = 1$ the summation is $e^x = \sum_{n=0}^{\infty}\dfrac{x^n}{n!}$, which is worth remembering as is the expansion for *small ax* which is

$$e^{ax} \approx 1 + ax, \qquad \text{and} \qquad e^{-ax} \approx 1 - ax.$$

These approximations are perfectly acceptable provided that $ax < 1$, and therefore $(ax)^2 \lll ax$ and so forth. The indication of 'less than' means ten times less rather than just slightly less for this approximation to be a good one. You could investigate this for yourself.

5.6.7 Log expansion

Finding the expansion of log functions is also easy, but in this case it is restricted to the limits $-1 < x \le 1$ when $f(x) = \ln(1 + x)$. Differentiating produces

$$f'(x) = \frac{1}{1+x}, \qquad f''(x) = \frac{-1}{(1+x)^2}, \qquad f'''(x) = \frac{2}{(1+x)^3}, \qquad f^4(x) = -\frac{6}{(1+x)^4}$$

and so forth and the series when $x = 0$, has coefficients

$$f(0) = 0, \qquad f'(0) = 1, \qquad f''(0) = -1, \qquad f'''(0) = 2, \qquad f^4(0) = -6, \cdots$$

therefore

$$\ln(1 + x) = x - \frac{x^2}{2} + \frac{x^3}{3} - \frac{x^4}{4} + \cdots \tag{5.21}$$

This series is restricted to the limits $-1 < x \le 1$. To understand why this is, suppose that $x = -1$, then this would mean calculating $\ln(0)$, which is $-\infty$. The series would be $-1 - \dfrac{1^2}{2} - \dfrac{1^3}{3} - \dfrac{1^4}{4} + \cdots$ or $-\left(1 + \dfrac{1}{2} + \dfrac{1}{3} + \dfrac{1}{4} + \cdots\right)$. The series within the brackets is called the harmonic series which diverges, i.e. continuously increases, so the log series expansion is not valid when $x = -1$. To check that the harmonic series grows as terms are added, try the calculation using Maple. Superficially, it appears that the harmonic series will converge to a finite limit, but the reciprocal $1/n$ does not decrease fast enough as n increases to ensure this. When $x = 1$, however, the series is $1 - \dfrac{1}{2} + \dfrac{1}{3} - \dfrac{1}{4} + \cdots$, which is known to *converge*; therefore $\ln(2) = 1 - \dfrac{1}{2} + \dfrac{1}{3} - \dfrac{1}{4} + \cdots$. This has very slow convergence and many thousands of terms are needed to obtain an accurate result; consequently this is not a very practical way of calculating $\ln(2)$.

When the value of x is small and $x^2 < x$ and higher powers of x are also less than x then expansion can be limited to just one term: $\ln(1 + x) \sim x$.

Interestingly, if the standard integral $\displaystyle\int_0^x \frac{dx}{1+x} = \ln(1 + x)$ is expanded using the series formula for $1/(1 + x)$ this produces

$$\int_0^x \frac{dx}{1+x} = \int_0^x [1 - x + x^2 - x^3 + x^4 \cdots]dx.$$

Integrating term by term gives

$$\int_0^x \frac{dx}{1+x} = x - \frac{x^2}{2} + \frac{x^3}{3} - \cdots$$

which is just the expansion for $\ln(1 + x)$ as it should be.

This series expansion method does give us a way of evaluating integrals numerically. The method is to keep on adding terms to the series until the answer does not vary to within a sufficiently large number of decimal places or the answer is considered accurate enough. An example is given below. There are several well known methods for numerical integration such as the Euler–Maclaurin method or Simpson's rule and generally, these should be used. Numerical integration is described in Chapter 10 and the Euler-Maclaurin in Section 5.7.

5.6.8 Binomial expansion $(1+x)^m$

One of the most general formulas is obtained by expanding $f(x)=(1+x)^m$ where m can be positive, negative, an integer, or a fraction. Starting with the expansion formula

$$f(x)=f(0)+f'(0)x+\frac{x^2}{2!}f''(0)+\cdots$$

the first term $f(0)=1$, and the function has derivative

$$\begin{aligned}
f'(x)&=m(1+x)^{m-1} && \rightarrow f'(0)=m\\
f''(x)&=m(m-1)(1+x)^{m-2} && \rightarrow f''(0)=m(m-1)\\
f'''(x)&=m(m-1)(m-2)(1+x)^{m-3} && \rightarrow f'''(0)=m(m-1)(m-2).
\end{aligned}$$

The general series is

$$(1+x)^m=1+mx+\frac{m(m-1)}{2!}x^2+\frac{m(m-1)(m-2)}{3!}x^3\cdots. \tag{5.22}$$

After some trial and error this can also be written as,

$$(1+x)^m=\sum_{n=0}^{\infty}\frac{m!}{n!(m-n)!}x^n \tag{5.23}$$

and is valid only when $|x|<1$. The *binomial* coefficients are often written as

$$\frac{m!}{n!(m-n)!}=\binom{m}{n}. \tag{5.24}$$

This ratio is usually used to describe the number of ways of placing n objects into one of two bins if there are a total number of m objects.

Provided that $|x|<1$ the expansion of $(1+x)^{1/2}$ is

$$(1+x)^{1/2}=1+\frac{x}{2}-\frac{1}{8}x^2+\frac{1}{16}x^3\cdots \tag{5.25}$$

and similarly

$$(1+x)^{-1/2}=1-\frac{x}{2}+\frac{3}{8}x^2-\frac{5}{16}x^3\cdots. \tag{5.26}$$

As an example, consider calculating $\sqrt{3/2}$, using equation (5.25). Letting $x=1/2$ the series is $(1+x)^{1/2}=1+\frac{1}{4}-\frac{1}{32}+\frac{1}{128}\cdots$ which evaluates to 1.2265. The answer to four decimal places is 1.2247 and the series has to be extended to nine terms, the highest power being x^8 to reach this level of accuracy. Trying to calculate $\sqrt{3}$ using this series gives a divergent result; why is this? Try it for yourself. The ancient Algorithm 1.2 could, however, be used.

In common with other expansions when $x \ll 1$, higher powers of x are even smaller and the series can be truncated at the second term.

5.6.9 Derivatives of series

Sometimes it is convenient to obtain a series by taking the derivative of each term in a known series rather than work out the series from scratch; for example, by differentiating term by term

$$\frac{d}{dx}(1+x)^{1/2} = \frac{1}{2\sqrt{1+x}} = \frac{1}{2} - \frac{1}{4}x + \frac{3}{16}x^2 - \frac{5}{32}x^3 + \cdots$$

a reciprocal square root series is found. Differentiating the log series gives the result.

$$\frac{d}{dx}\ln(1-x) = \frac{-1}{1-x} = 1 + x + x^2 + x^3 + \cdots$$

Reversing the equation, the summation is $\sum_n x^n = x + x^2 + x^3 + x^4 + \cdots = \dfrac{1}{x-1}$ and if differentiated is

$$\sum_n nx^{n-1} = 1 + 2x + 3x^2 + 4x^3 + \cdots = \frac{1}{(1-x)^2}.$$

5.7 **Euler–Maclaurin formula**

Sometimes a series cannot be expanded to simple formulae but its integral can be evaluated. If the terms in a series can each be differentiated, the Euler–Maclaurin formula, first developed in 1732–3, connects the summation of a series with its integral and provides a series of correction terms to make one the same as the other.

A common form of the Euler–Maclaurin equation puts the summation on the left-hand side starting at $k = 1$ and going to $n - 1$, which is an integer. On the right-hand side, the integral starts at zero and runs to n and the other terms are each evaluated at zero and n,

$$\sum_{k=1}^{n-1} f_k = \int_0^n f(x)dx + \frac{1}{2}[f(n) + f(0)] + \frac{1}{12}[f'(n) - f'(0)] \\ - \frac{1}{720}[f^3(n) - f^3(0)] + \frac{1}{30\,240}[f^5(n) - f^5(0)] - \cdots. \tag{5.27}$$

The superscripts on f indicate derivatives with respect to x. The lower limit on the integral can be 1 rather than 0, and then the function and its derivatives are evaluated at 1. The lower limit on the summation may be zero. The odd numbered derivatives of the function are indicated by the superscript dash or numbers. The numerical coefficients and derivatives are given more generally by $\sum_{k=1}^{\infty} \dfrac{B_{2k}}{(2k)!}[f^{2k-1}(n) - f^{2k-1}(0)]$ where B_{2k} are the Bernoulli numbers (see Abramowicz & Stegun 1965), some of which are $B_0 = 1$, $B_1 = -1/2$, $B_2 = 1/6$, $B_4 = 1/42$, $B_8 = -1/30$. The missing odd indexed numbered values are each zero. These numbers can be found using the `bernoulli(i)` function in Maple for index i. The general form of the Euler–Maclaurin equation, where the function is evaluated from $x = a$ to b, is

$$\sum_{k=0}^{m} f(a+kh) = \frac{1}{h}\int_a^b f(x)dx + \frac{1}{2}[f(b) + f(a)] + \sum_{k=1}^{\infty} \frac{h^{2k-1}B_{2k}}{(2k)!}(f^{2k-1}(b) - f^{2k-1}(a))$$

where m is an integer and h the element $h = (b - a)/m$ and the summation goes from $f(a)$ to $f(b)$ in m steps. The summation to infinity has only as many terms as necessary to obtain a convergence to the required number of decimal places. There is an error term associated with the summation in this form or that given in equation (5.27), but this is complicated to evaluate and is not given. It is not necessary to calculate it, however, because using the computer the results can be listed until sufficient precision is reached. To obtain equation (5.27) from the general equation the substitutions $n = m + 1$, $b = n$ and $a = h = 1$ are used. The summation containing the Bernoulli numbers is usually limited to just a few terms because this summation eventually diverges becoming very large at large k, but converges for smaller k. It seems rather strange that it can be used when it has this property but it does give accurate results when just a few terms are used.

(i) As an example of using the Euler–Maclaurin equation, some series will be calculated, starting with that for k^2. As only the first derivative is not zero the series is short,

$$\sum_{k=1}^{n} k^2 = \int_0^n x^2 dx + \frac{n^2}{2} + \frac{n}{6}$$
$$= \frac{n^3}{3} + \frac{n^2}{2} + \frac{n}{6}$$

and this result is the same as the series summation for x^2 as may be confirmed using Maple, `sum( x^2, x = 1..n );`.

(ii) The series for $\ln(k)$ is

$$\begin{aligned}\sum_{k=1}^{n} \ln(k) &= \int_1^n \ln(x)dx + \frac{[\ln(n)+\ln(1)]}{2} + \frac{[1/n-1]}{12} - \frac{2/n^3-2}{720} \\ &= n\ln(n) - n + 1 + \frac{\ln(n)}{2} + \frac{[1/n-1]}{12} - \frac{2/n^3-2}{720}\end{aligned} \tag{5.28}$$

where the integral is $\int \ln(x)dx = x\ln(x) - x$. Evaluating the series directly (with Maple `add()`) and with the Euler–Maclaurin formula produces similar results; the summation has a value of 148.4778 if $n = 50$, and 148.4781 if the Euler–Maclaurin formula is used, which is a very close match. You have little hope of evaluating 50 terms in the series expansion with your calculator but could manage the Euler–Maclaurin calculation.

(iii) A Maple algorithm is now described because the Euler–Maclaurin formula is often a very awkward one to evaluate because the third, fifth, and perhaps higher odd powers of derivatives are needed. A procedure is made and the function f is passed into it with limits m and n. The integral is evaluated and saved as `s1` and the derivative parts saved as `s3`. This is then made into a function s4 using `unapply` to evaluate the derivatives.

Algorithm 5.2 Euler–Maclaurin formula, equation (5.27)

```
> Euler_Maclaurin:= proc(f,m,n)           # eqn (5.27)
    local ff,s1,s2,s3,func_s3;
    ff:= unapply(f,x);                    # make function ff
    s1:= int(f, x = m..n);                # int term
    s2:= 1/2*( ff(n) + ff(m));            # 1st sum
    s3:= +1/12*diff(f,x)          -1/720*diff(f,x$3)
       +1/30240*diff(f,x$5) -1/1209600*diff(f,x$7):
    func_s3:= unapply(s3, x);      #use for f'(n) & f'(m)
    evalf( s1 + s2 + func_s3(n) - func_s3(m) );
  end proc:
> m:= 1:       n:= 50:             # terms in sum
  f:= ln(x):                       # expression to sum: ln(x)
  Euler_Maclaurin(f,m,n);          # calculate result
```

148.4781

5.7.1 Questions

Full solutions are available at www.oxfordtextbooks.co.uk/orc/beddard.

Q5.14 Expand $e^x \tan(x)$ using Maclaurin's theorem to powers up to x^4.

Q5.15 Show that $x\dfrac{(1+x)}{(x-1)^2} = x + 3x^2 + 5x^3 + 7x^4 + \cdots$.

Strategy**:** The expansion formulae all have the form $(1 \pm x)$ not $(x-1)$ so the original equation must be put into the correct form. This can be done by multiplying both the top and bottom by -1.

Q5.16 The relativistic energy of a particle with mass m travelling at speed v is $E = mc^2\left(1 - \dfrac{v^2}{c^2}\right)^{-1/2}$ where c is the speed of light. Compare this energy with the classical energy of the particle.

Strategy: Expand the square root using the binomial expansion and in terms of $v/c = x$.

Q5.17 Solve $ax^2 + bx + c = 0$ and show that when $b^2 \gg ac$, one root of the equation is small and tending to zero, the other to b/a.

Q5.18 Use a series expansion to find the limit of $\dfrac{x}{e^x - 1}$ as $x \to 0$ and show that this gives the same answer as l'Hôspital's rule.

Q5.19 **(a)** Expand the summation $\sum_{k=1}^{\infty} e^{ikx}$ where $i = \sqrt{-1}$. Rearrange the terms and re-sum to find a neat expression.

(b) Find an expression for $\sum_{k=1}^{\infty} \dfrac{e^{ikx}}{k}$.

Q5.20 In condensation polymerization end-group analysis, which assumes that the rate of reaction is independent of the polymer size, show that if k monomers are joined together, this has a probability of $p(k) = (1-p)p^{k-1}$ where p is the fraction of groups that have reacted. To form an '*n-mer*' it is necessary for $k - 1$ links to have been made with probability p^{k-1}. What remains is $1 - p$, the fraction that has not reacted.

(a) Show that $\sum_{k=1}^{\infty} p(k) = 1$.

(b) Calculate $\langle k \rangle$, the number average, and $\langle k^2 \rangle$ and σ. To calculate $\langle k^2 \rangle$ start by taking the derivative of $\langle k \rangle$.

Q5.21 The *sinc*^{2} function $\sin^2(x)/x^2$, crops up frequently with Fourier transforms, diffraction, absorption of radiation, and the theory of short laser pulses.

(a) Expand the function as a series and determine the limit when $x \to 0$.

(b) Plot the function from -4π to 4π to see if your answer is correct.

Strategy: Expand the sine and take the limit when x is small.

Q5.22 The error function $erf(x) = \dfrac{2}{\sqrt{\pi}}\displaystyle\int_0^x e^{-s^2}\,ds$ is the integral of the 'bell-shaped' or Gaussian probability distribution and is used to obtain the probability of an event with value x.

(a) The integral cannot be evaluated algebraically. Therefore, derive an expansion formula for this integral up to powers of x^9.

(b) Compare your answer with an accurate value computed by Maple, for $x = 0.1$, 0.5, and 1.0.

Strategy: Expand e^{-s^2} as a Taylor series and integrate each term.

Q5.23 If $\tanh^{-1}(x) = \displaystyle\int_0^x \frac{dt}{1-t^2}$ find $\tanh^{-1}(x)$ by expanding term by term and integrating. Compare your answer at $x = 0.5$ using `arctanh(0.5)` with Maple.

Q5.24 In numerically integrating the one-dimensional Schrödinger equation $-D^2\psi + V\psi = E\psi$, where both the potential V and the wavefunction ψ are functions of x, and the operator is $D = d/dx$, the finite difference equation

$$h^2D^2\psi \approx \psi(x+h) + \psi(x-h) - 2\psi(x)$$

could be used if h is a small increment in x. Show that the Taylor expansion of

$$\psi(x+h) + \psi(x-h) - 2\psi(x)$$

to lowest order produces $h^2D^2\psi$.

Strategy: We are not told what point to expand the function about so try a general point x_0 and see what happens. In the Taylor expansion equation (5.17), substitute $x \to x \pm h$ as appropriate.

Exercise: Show that the next term in the series is $\dfrac{h^4}{12}D^4\psi(x)$

Q5.25 Expand the integral $\tan^{-1}(x) = \displaystyle\int_0^x \frac{1}{1+u^2}\,du$ as a series and find a formula for π.

Strategy: Expand the argument of the integral and integrate term by term. You may need to try some values of $\tan^{-1}(x)$ or plot a graph to see what value of x is likely to produce terms in π.

Q5.26 Evaluate $\displaystyle\int_0^2 \cos(x^3)dx$ using a series expansion.

Strategy: Evaluating this integral with Maple, a Fresnel integral is produced which has itself to be evaluated numerically. Instead, estimate the integral with a Taylor series expanded inside the integral with limits 0 and x and then replace x with 2.

Q5.27 Evaluate $\displaystyle\int_0^{20} e^{-x}(3x^2 - x^3)dx$ with Maple by using a series expansion.

Strategy: The numerical limits to the integral mean that the result is a number not a formula. Expanding the function produces terms that do not obviously lend themselves to being coerced into a neat summation expression, so calculate 'by gosh and by golly' i.e. bash your way through to an answer!

Q5.28 Find the Maclaurin expansion of $(a + x)^{1/2}$ and use this to find the square root of 5 to three reliable decimal places.

Q5.29 The factorial of n is written $n! = n(n-1)(n-2)\cdots 1$. The Stirling approximation is often used for calculating factorials of large integer numbers and is frequently used in statistical mechanics. It has the form $\ln(n!) = n\ln(n) - n$. To appreciate the simplification this formula produces, imagine calculating directly $10^{23}!$ or even $10^6!$ compared to using Stirling's formula; it would be impossible to multiply together a million terms. Using the Euler–Maclaurin formula, derive the Stirling formula and plot the approximation vs the true value on a log–log scale.

Strategy: Use the fact that the log of a product of numbers is the same as the sum of their logs, for example $\ln(3!) = \ln(1) + \ln(2) + \ln(3)$.

Q5.30 The partition function for the rigid rotor model of molecular rotation is often evaluated as an integral rather than a summation because the summation has no closed form. This approximation leads to a *systematic* error in its value and this is justified by saying that the energy gaps are small and therefore the error in this approximation is small.

(a) Why is this approximation better for heavy molecules?

(b) Use the Euler–Maclaurin formula to calculate the partition function for a molecule with rotational constant B at different temperatures.

(c) Compare your results with the integral and summation for HD and I_2 molecules at 30 and 300 K. The bond length for HD is 0.7413 Å and for I_2, 2.67 Å. Boltzmann's constant $k_B = 0.695\ \text{cm}^{-1}\ \text{K}^{-1}$.

Strategy: Look up the formula for the rigid rotor energy levels in your textbook. The rotational quantum number is J, which starts at zero when the molecule has no rotational energy and hence is not rotating. Notice that the partition function also starts with quantum number $J = 0$ but the Euler–Maclaurin formula starts the summation at 1 so there is an extra term with $J = 0$ to add to the result. Alternatively, the formula can be modified to start at zero and many textbooks quote this.

Q5.31 In an NMR experiment, the fractional population between two nuclear spin states α and β determines the size of the measured signal. Calculate the fractional population $f = \dfrac{n_\beta - n_\alpha}{N}$ when each level is populated according to a Boltzmann distribution with energy $E_{mz} = -\gamma\hbar B_z m_z$. Spin state β has an azimuthal (or magnetic or projection) quantum number along the z-axis of $m_z = 1/2$ and state α has $m_z = -1/2$. N is the total population, γ the positive magnetogyric ratio, B_z the magnetic field in the direction of the z-axis. The upper level is labelled α.

(a) Why does the upper level have $m_z = -1/2$?

(b) Show that because the energy gap between levels is very small relative to thermal energy k_BT that $f = \dfrac{\Delta E}{2k_BT}$ and $n_\beta - n_\alpha = \dfrac{N\gamma\hbar}{2}\dfrac{B_z}{k_BT}$.

Strategy: The populations are always given by the Boltzmann distribution as the molecules are in contact with a 'heat bath', which is primarily the solvent as this is in vast excess. The total population is $N = n_\alpha + n_\beta$.

Q5.32 A protein or length of DNA can be extended and straightened either by pulling with an atomic force microscope, AFM or by inserting it into a flowing viscous liquid. The change in internal energy U is given by the thermodynamic relationship $dU = TdS - PdV + Fdx$ where F is the applied force and x the extension. The entropy of conformation of the protein, considering it to be a freely jointed chain that contains a total of N segments each of length L is

$$S = -\frac{1}{2}k_B N[(1+\alpha)\ln(1+\alpha) + (1-\alpha)\ln(1-\alpha)],$$

where $\alpha = x/NL$, x is any extension of the chain such that $0 < x \le NL$ and NL is its full length. A conformation is any one of the possible arrangements of the atoms of a molecule that can be reached without breaking bonds or moving atoms through one another.

(a) State Hooke's law. How is (potential) energy related to force?

(b) Show that at small extension Hooke's law is obeyed with the force given by $F = \frac{x}{NL^2}k_B T$. This means that the protein is perfectly elastic and that the internal energy is independent of extension, provided this occurs at a constant temperature and volume. (See Atkins 2001 for a fuller discussion.)

Strategy: To obtain the force it is necessary to differentiate the equation for dU with x, because energy equals force × distance and force is the derivative of potential energy with distance; $F = \frac{dU}{dx}$. Note that the entropy is a function of extension and T and V are constant. The requirement to calculate at small extension indicates that a series expansion will be required.

Q5.33 The worm-like chain model of extension is normally used to describe the forced extension of a protein or DNA with an AFM or optical tweezers. The force is given by $F = \frac{k_B T}{p}\left(\frac{1}{4(1-\alpha)^2} - \frac{1}{4} + \alpha\right)$, where $\alpha = x/L$ is the fractional extension, x is the end-to-end distance and L the contour (total) length. The constant p is the persistence length and represents the smallest rigid length the polymer exhibits. The persistence length is a measure of bending resistance or *flexural rigidity*, and this is typically ≈0.4 nm for a protein and ≈40 nm for micrometer long polymers such as DNA. The force resisting extension exists because of the elastic properties of the polymer and this is contributed to by entropy, and by intermolecular forces and hydrogen bonds which resist extension. The graph of force vs extension is similar in general shape to that in Fig. 5.29 (see online solutions).

(a) Show that at small extension Hooke's law is obeyed.

(b) At 300 K calculate the force constant for a small protein with ≈80 residues and total length $L = 28$ nm and $p = 0.4$ nm and also calculate that for a long piece of DNA with $p = 40$ nm and $L = 16$ μm. Comment on their relative magnitudes and compare these with the energy to break a typical chemical bond. Calculate the energy needed to unfold, assuming Hooke's law applies at large extension, and compare with that of a typical chemical bond.

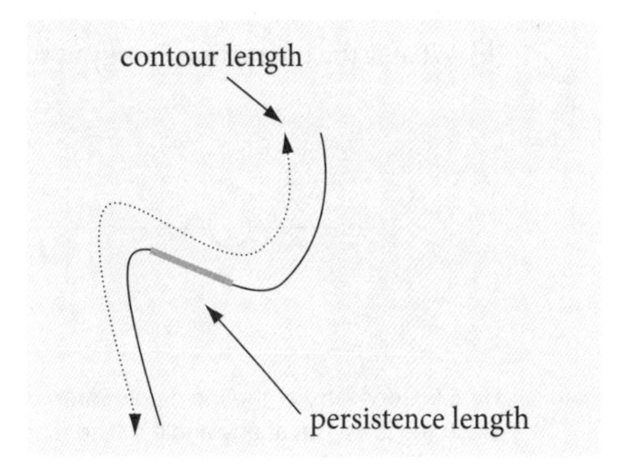

Fig. 5.4 Contour length.

Strategy: To calculate the force for small extensions in α, it is necessary to expand $(1-\alpha)^{-2}$ using the binomial expansion. A small extension implies $\alpha < 1$.

Q5.34 The mean square end-to-end separation of a DNA or other polymer chain is given by $\langle r^2 \rangle = 2pL - 2p^2[1 - e^{-L/p}]$, where L is the contour length and p the persistence length. Calculate $\langle r^2 \rangle$ **(a)** if the polymer is long compared to the persistence length $L \gg p$ and **(b)** If the polymer is short $L \ll p$.

Q5.35 The Lennard-Jones 6–12 potential, which describes the intermolecular energy between a pair of molecules, has a minimum at $r_e = 2^{1/6}\sigma$. The potential is

$$U(r) = -4\varepsilon\left[\left(\frac{\sigma}{r}\right)^6 - \left(\frac{\sigma}{r}\right)^{12}\right]$$

where ε is the depth of the energy well and σ the diameter of a molecule.

(a) Show that the minimum energy is $-\varepsilon$.

(b) Expand the potential about the minimum energy at intermolecular separation r_e using a Taylor series.

(c) Calculate the approximate Hooke's law force constant, k, around the minimum energy. This is the slope of the derivative of the potential with extension x, i.e. $dU/dx = -kx$. Calculate the approximate classical vibrational frequency in the bottom of the potential using parameters for Xe; $\varepsilon = 20.0$ meV and $\sigma = 398$ pm.

(d) To check your calculation, plot the equation and the expansion to the second power of $(r - r_e)$.

Strategy: (b) Use a Taylor expansion about the minimum separation r_e, and then ignore terms in higher powers of $(r - r_e)$ as the change from the equilibrium position is small. Because you have to take the derivative of the potential to find the force constant, expand the potential at least to quadratic terms.

Q5.36 A dipole $q^+ - q^-$ will interact with an ion in solution because the electric field of the ion will extend through the solution and so cause a force to exist between them. The electric field strength E around a charge is the force / unit charge, or $E = \dfrac{f}{q}$. Because there is a force between the charges, energy is needed to place the dipole and ion at any given separation. Your textbook will state that this force varies as the inverse *cube* of the separation from the dipole when the separation is larger than the size of the dipole itself. However, two isolated point charges, q_1 and q_2, will interact with a *force* given by the inverse *square* of their separation,

$$f = \frac{q_1 q_2}{(4\pi\varepsilon_0)} \frac{1}{\varepsilon r^2}, \tag{5.29}$$

where ε_0 is the permittivity of free space and ε is the relative permittivity (dielectric constant) of the intervening medium, such as the solvent. Force written in this way has SI units of J m^{-1}. The interaction energy in joules between two *point charges* q_1 and q_2 at separation r is

$$U = \frac{q_1 q_2}{(4\pi\varepsilon_0)} \frac{1}{\varepsilon r}$$

and is obtained by integrating the force over the distance from infinity to r.

(a) By calculating the electric field at the ion situated along the x-axis with charge $+z$, show that the ion-dipole interaction varies with separation as $1/x^3$; Fig. 5.5 illustrates the geometry.

(b) What is the interaction energy at separation x? Determine that it has the correct units.

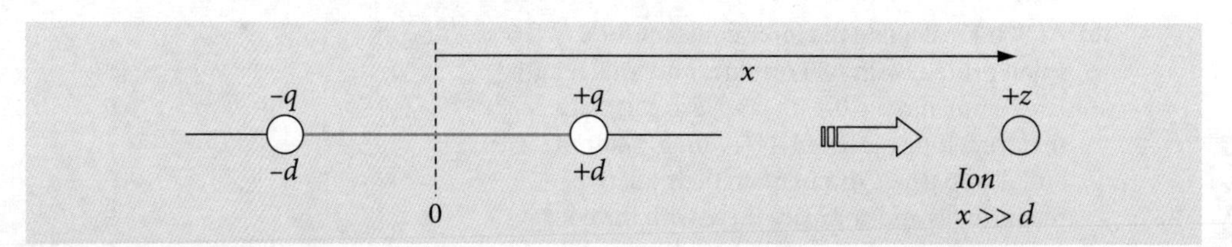

Fig. 5.5 Geometry of the ion-dipole interaction. A more complete, and more complicated calculation, would allow the ion to be at any angle to the dipole and the results averaged, but the result is qualitatively the same. The dipole length is 2*d*.

Strategy: Calculate the electric field using charges $+q$ and $-q$ then calculate the energy. In electrostatic calculations the field and energy is always calculated as the sum of the individual contributions between each pair of charges. As the separation is large compared to the size of the ion or dipole expand the field in terms of the fractional separation. Using the diagram, the dipole has charges $+q$ and $-q$ and the ion $+z$. Note that E is used to represent the *electric field*.

Q5.37 The energy of two interacting dipoles with the geometry shown in Fig. 5.6 is

$$U = \frac{q^2}{4\pi\varepsilon_0}\left[\frac{1}{x} + \frac{1}{(x + d_2 - d_1)} - \frac{1}{(x - d_1)} - \frac{1}{(x + d_2)}\right].$$

(a) Explain how this equation is obtained.

(b) Show that if $x \gg d_1$ and d_2, the energy varies as $\mu_1\mu_2/x^3$ if μ_1 and μ_2 are the dipole moments equal to qd_1 and qd_2 respectively.

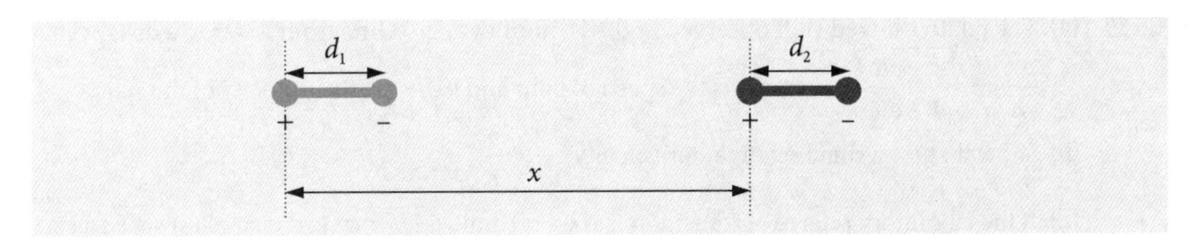

Figure 5.6 Definition of dipole's geometry.

(c) Calculate the interaction energy if two dipoles each of 5 D are separated by 2 nm, as shown in Fig. 5.6. Compare this to thermal energy at room temperature.

***Strategy*:** Because the interaction is electrostatic (or Coulomb) in nature, the energy is always calculated by adding together the interaction between pairs of charges; one charge each end of the dipole on one molecule with each of the charges on the other. The energy is inversely proportional to the separation of each pair of charges so there are four terms to consider.

Q5.38 The pitch of an ambulance's siren sounds higher as it speeds towards you and lower as it recedes. This is caused by the Doppler effect. Because the source is moving, the separation between the sound waves becomes smaller as the source approaches and longer as it recedes.

In approaching you, the sound frequency appears to be up-shifted from f_0 to $f = f_0\left(\dfrac{s+v_0}{s-v}\right)$ where s is the speed of sound in air, approximately 331 ms^{-1} or 740 m.p.h., f_0 the true frequency of the siren (440 Hz), v (60 m.p.h.) the velocity of the ambulance and v_0 the speed of you the observer. When the vehicle moves away from you the perceived frequency is lower as now $f = f_0\left(\dfrac{s+v_0}{s+v}\right)$; notice the sign change.

(a) Sketch how the sound frequency perceived by a stationary observer positioned, as shown in Fig. 5.7, would change as the vehicle passes.

(b) Show that the perceived *frequency shift* $\dfrac{f-f_0}{f}$ is proportional to v, the speed of the ambulance.

Assume that your speed v_0 is small compared to the speed of sound.

***Strategy*:** (a) The frequency heard is higher than normal when the ambulance is approaching and coming directly towards us, but is at exactly frequency f_0 when it is right in front of us, and falls as it departs. (b) If we were to assume that both v_0 and v are small compared to s, the speed of sound in air, and simply ignore them, then $f = f_0$ and the frequency would not change. Experience tells us that the perceived frequency does change, so this assumption cannot be correct because it is too crude. Instead, rearrange the frequency equation into two parts, and ratio the speeds to produce terms such as $(1 - v/s)^{-1}$ and then expand into a series.

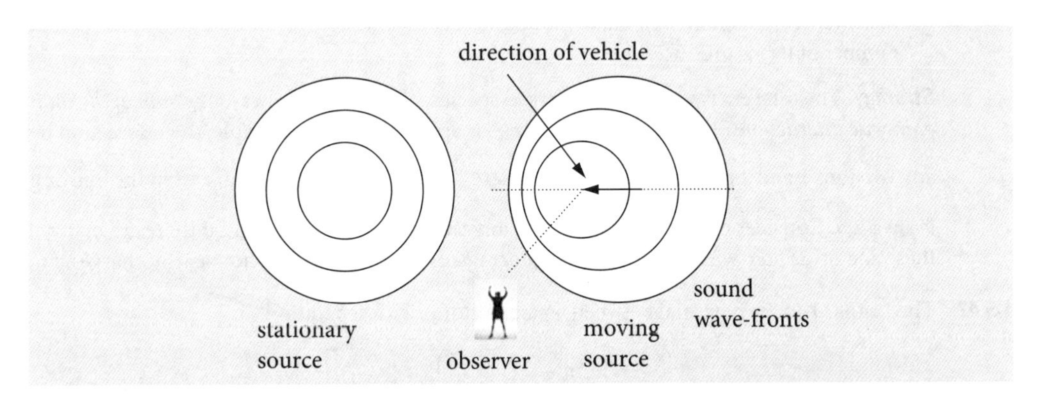

Fig. 5.7 The stationary vehicle produces sound wave-fronts that are equally spaced from one another in all directions if measured at equal time intervals. The moving vehicle causes the sound waves to appear to close up in the direction of travel, and to move apart in the opposite direction. If you are at the side of the road, the sound is that component of the forward motion in your direction. If you are in the vehicle, the pitch of the sound appears to be the same whether you are moving or stationary because the sound waves are always generated at the same frequency and because they are moving much faster than the vehicle.

Q5.39 **(a)** The relativistic red shift observed in the H atom Lyman-α line from a star in a distant galaxy is $\dfrac{\Delta\lambda}{\lambda}=\sqrt{\dfrac{1+v/c}{1-v/c}}-1$ where c is the speed of light and v the relative velocity of the star.

(b) Show that for a small relative star velocity $\dfrac{\Delta\lambda}{\lambda}=\dfrac{v}{c}$.

(c) If the laboratory reference transition is $\Delta\lambda = 0.1$ nm wide, what is the smallest speed a star must be receding by to separate it from the reference line, assuming that a separation of $2\Delta\lambda$ is needed?

Q5.40 Two molecular energy levels of energy E_1 and E_2 and separation ΔE interact with a 'coupling energy' V. Perturbation theory applied to quantum mechanics allows us to calculate how these levels are shifted in energy as a result of this interaction. One level rises, the other falls and their new energies are,

$$E_{\pm}=\frac{E_1}{2}+\frac{E_2}{2}\pm\frac{1}{2}\sqrt{\Delta E^2+4V^2} \tag{5.30}$$

although the total energy remains the same, see Fig. 5.8.

(a) Calculate the total energy before and after the interaction and show that they are the same.

(b) Calculate the two energies when $V \ll \Delta E$, both being positive, and when $V \gg \Delta E$.

(c) Plot the correct energies if $E_1 = 2$, $E_2 = 3$ and V varies from 0 to 1, and compare them with the approximations from (b).

Strategy: In (b) when $V \ll \Delta E$ expand the square root in $E_{\pm}$. Do this by rearranging to get a term in $\sqrt{1+\dfrac{4V^2}{\Delta E^2}}$.

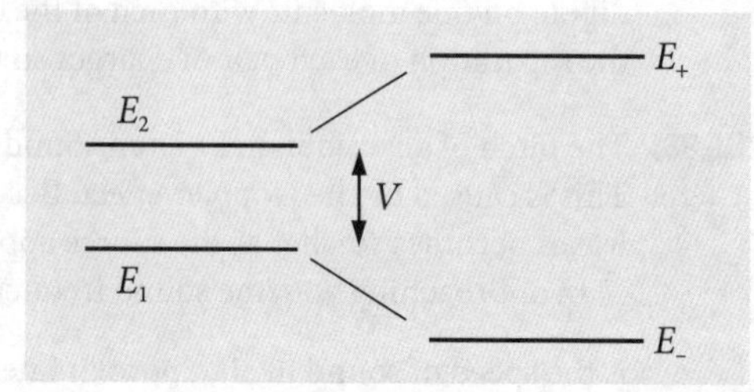

Fig. 5.8 Initial levels (left) interact with coupling V to produce two new levels (right).

Q5.41 Crystals of simple salts consist of ordered lattices of anions and cations where the interactions are predominantly due to the Coulomb electrostatic force. As there are many ions, the total interaction acting upon any one of them is due to the effect of all the others. The energy between any two ions 1 and 2 separated by a distance d is $U_{12}=\dfrac{q_1q_2}{4\pi\varepsilon_0}\dfrac{1}{d}$ where the charge on an ion is $q = eZ$, and e is the electronic charge 1.6022×10^{-19} C. The charge number Z can be positive or negative.

(a) Find the total energy of a positively charged ion in a linear chain of alternating positive and negatively charged ions with charges Z and $-Z$. Find the Madelung constant M, which is the numerical factor that contributes to the energy and is due solely to the positions and charges of the ions. The total energy is $U_M=\dfrac{q_1q_2}{4\pi\varepsilon_0}\dfrac{M}{d}$.

(b) Repeat the calculation on a square grid of alternating charges. Now the summation has to be evaluated numerically. Use Maple's `add` instruction to do this, but take care; you will need very many terms (thousands) to make the addition converge. The result is -1.612 but a reasonable number of terms produce -1.6.

Strategy: The total energy of several charged species, of any sort, is always the *sum of the individual pair-wise interactions*, + to +, – to – and + to – as appropriate. For example, the interaction between any two ions 1 and 2 is $U_{12}=\dfrac{q_1q_2}{4\pi\varepsilon_0}\dfrac{1}{d_1-d_2}$ where $d_1 - d_2$ is their separation, and in the line or grid, this is always d. Consider only the interaction of any two species at a time, and if there are many charges these add up as pair-wise contributions ignoring any intervening or other nearby charges, Fig. 5.9.

Q5.42 The Lennard-Jones potential between a pair of atoms with separation r is

$$U=-4\varepsilon\left[\left(\frac{\sigma}{r}\right)^6-\left(\frac{\sigma}{r}\right)^{12}\right].$$

The potential acts mainly at short range and ε is the strength of the intermolecular interaction and σ scales the interaction and is approximately 0.3 nm for solids of the noble gases. The interaction energy ε is 0.0031 eV for Ne and 0.020 eV for Xe. When there are many atoms in a solid the *cohesive energy* U_c is calculated as the sum of the *pair-wise* interactions between atoms i and j

$$U_c=-4\varepsilon\sum_{j\neq i}\left[\left(\frac{\sigma}{r_{ij}}\right)^6-\left(\frac{\sigma}{r_{ij}}\right)^{12}\right].$$

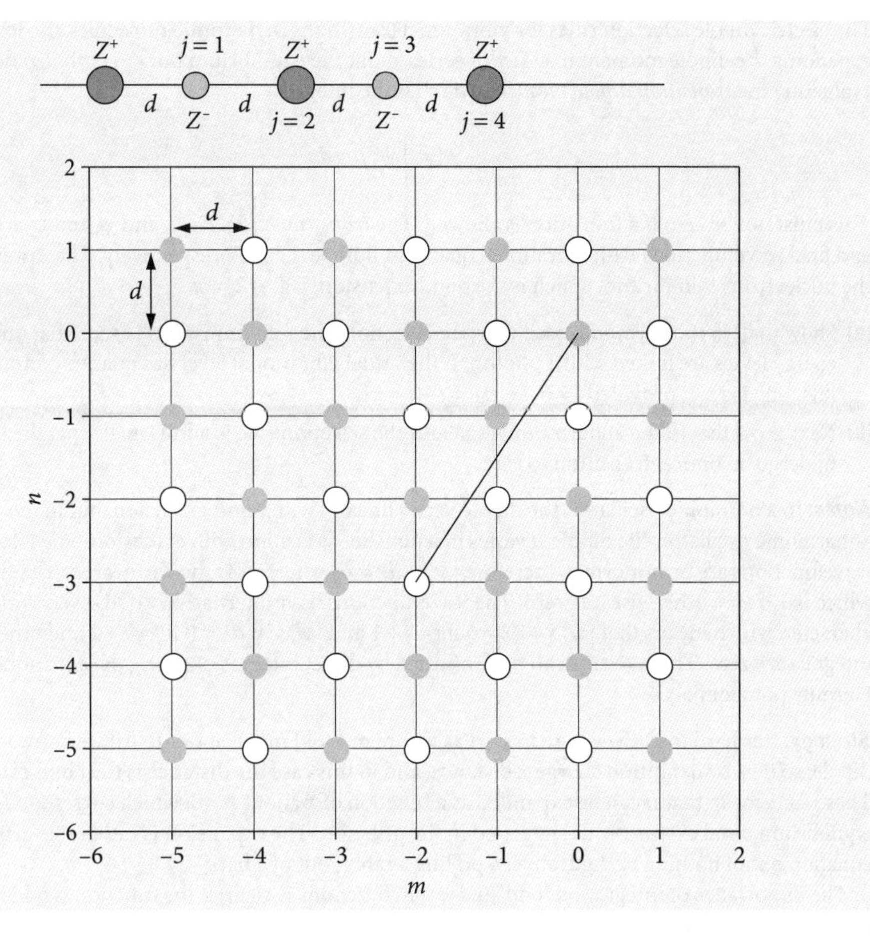

Fig. 5.9 Top: Ions placed in a line. Bottom: Some of the ions whose charges alternate on a square lattice of atoms with grid spacing d. The diagonal shown has length $d\sqrt{13}$.

In a cubic crystal the separation of any pair of atoms is represented in terms of multiples of the near-neighbour separation, R, where $r_{ij} = \alpha_{ij}R$. The number α, which need not be an integer, clearly depends on the crystal geometry. The summation becomes

$$U_c = -4\varepsilon \sum_{j\neq i}\left[\left(\frac{\sigma}{\alpha_{ij}R}\right)^6 - \left(\frac{\sigma}{\alpha_{ij}R}\right)^{12}\right].$$

Calculate the lattice sums $A_6 = \sum_{j\neq i}\frac{1}{\alpha_{ij}^6}$ and $A_{12} = \sum_{j\neq i}\frac{1}{\alpha_{ij}^{12}}$ for a simple cubic crystal lattice using the diagram shown in Fig. 5.10. Calculate the value for a unit cell then use Maple to calculate the sum over as many cells as necessary to achieve two decimal places of accuracy.

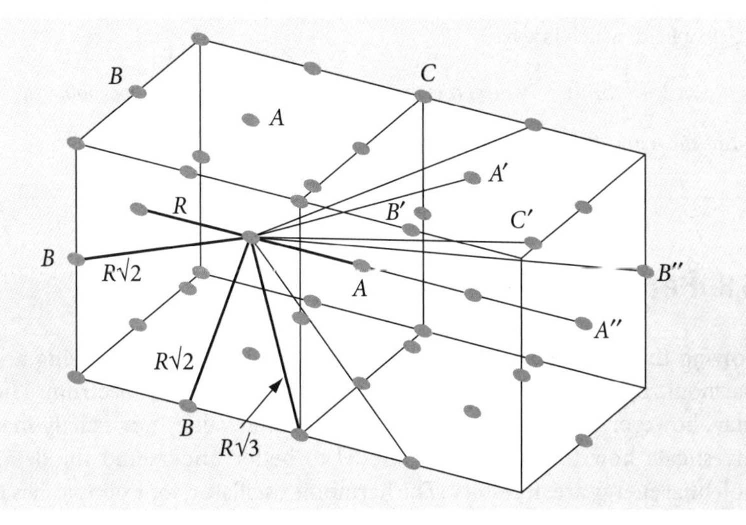

Fig. 5.10 A simple cubic structure with near neighbours (A), and some of the next near neighbours (B) and (C). Atoms in the other adjacent unit cells, which are not shown, will also contribute to the summation.

Q5.43 The electric dipole selection rules for vibrational transitions in diatomic molecules are described by expanding the dipole moment in a Taylor series about the equilibrium bond length R_e, and then evaluating the *transition dipole moment*, which is the integral

$$M = \int \psi_f^* \mu \psi_i dx.$$

This must not be zero if a transition is allowed. The *transition dipole* is μ, and ψ_i and ψ_f are the initial and final wavefunctions with vibrational quantum numbers f and i respectively. The displacement of the nuclei from equilibrium, which is the bond extension, is $R - R_e = x$.

(a) Show that, in the harmonic oscillator, the selection rule for a transition is such that only adjacent energy levels are linked with a photon. If the initial vibrational level has quantum number i the final one is $i \pm 1$, provided $i \neq 0$, i.e. $\Delta v = f - i = \pm 1$.

(b) Next show that in the anharmonic oscillator the selection rule is additionally that the level i can undergo an optical transition to $i \pm 2$.

Notes: In a harmonic oscillator, the dipole varies linearly with bond extension, but in the anharmonic oscillator, the dipole μ varies in a non-linear fashion with extension. The vibrational wavefunctions are orthonormal, therefore $\int \psi_f^* \psi_i dx = \delta_{if}$ where δ_{if} is the Kronecker delta function which is 1 if $i = f$, otherwise it is zero. The wavefunctions have alternatively odd–even symmetry character, which means that $\int \psi_f^* x \psi_i dx \neq 0$ if $f = i + 1$ and $\int \psi_f^* x^2 \psi_i dx \neq 0$ if $f = i + 2$ and otherwise the integrals are zero. These results can be confirmed by direct integration using the equations for the Hermite polynomials.

Strategy: It is hard to know where to start as we are not told much about μ. All we know is that it is a dipole, so it is, by definition charge × distance, and in this case the distance is the bond extension x. These facts mean that μ can be expanded as a function of x about $x = 0$, which corresponds to the equilibrium bond extension, as suggested in the question. The expansion is rather like generating an equation out of nothing or, figuratively, pulling a rabbit out of a hat!

The importance of integration 'odd' and 'even' functions is clear; if the function is odd the integral over all space is *always* zero, if even the integral generally is not zero. In the more general sense, group theory should be used to determine if the integral belongs to the *totally symmetric representation* of the point group of the molecule, which, if it does, the integral is finite. See Chapter 7 for a fuller discussion.

Q5.44 The Hellmann–Feynman theorem states that for a property q the energy of a molecule U and its Hamiltonian H are related as $\dfrac{dU}{dq} = \langle \dfrac{dH}{dq} \rangle$. The angle brackets indicate an average value is measured. Suppose the property q is an external electric field E then $q \equiv E$, and in the presence of this field, the Hamiltonian is $H = -\boldsymbol{\mu} \cdot \boldsymbol{E}$ where $\boldsymbol{\mu}$ and $\boldsymbol{E}$ are vector quantities. To simplify matters suppose that the field only exists along the z-axis then $H = -\mu_z E$.

(a) Calculate $\dfrac{dU}{dE}$.

(b) Use a Taylor series to expand the molecular energy U in terms of the electric field E about the energy U_0 in a field, which is zero.

(c) If $\langle \mu_z \rangle = \mu_{z0} + \alpha E + \dfrac{1}{2}\beta E^2 + \cdots$ where α is the *polarizability* and β the *hyperpolarizability*, find expressions for α and β as derivatives of the energy with field strength.

5.8 Perturbation theory

Solving the Schrödinger for the energy levels of a molecule, using a model such as a harmonic oscillator or rigid rotor, allows us to predict the spectrum. These spectral lines may, however, only be close to the experimental values not exactly matching them. To investigate how to improve our model to better understand the data, changes to the potential energy are necessary. The harmonic oscillator, for example, has the Hamiltonian,

$$H^0 = -\frac{\hbar^2}{2m}\frac{d^2}{dx^2} + \frac{k}{2}x^2.$$

We might try to improve on this, perhaps by adding cubic or quartic terms, such as $ax^3 + bx^4$, to describe the effect of the bond stretching. Similarly, we might consider adding terms in J^4 to the rigid rotor model to allow for centrifugal distortion during rotation. Alternatively, the harmonic potential model may be satisfactory, but the experiments report on the effects of small changes to the spectrum caused by external electric or magnetic fields. In this case, if the electric field ε is along the x-axis a term εex could be added to the Hamiltonian to allow for this. The use of perturbation theory allows for the incorporation of these and many other *small* potential energy terms to extend our knowledge a little into unfamiliar territory by using what is already known as a starting point.

5.8.1 **Formal derivation**

First, the formal derivation is presented in most of the important details and the rather simple results, equations (5.41) and (5.43), that describe the changes in energy and wavefunctions are produced. These equations are then applied to some examples. If you want to try the problems and are not interested in the derivation details, then these equations are the results you will need.

The Schrödinger equation whose solutions are known, is defined as,

$$H^0\psi_n^0 = E_n^0\psi_n^0. \tag{5.31}$$

The superscripts 0 indicate that the wavefunctions and energies are known for every quantum number $n = 0, 1, 2, \cdots$ The wavefunctions form an orthonormal set so that

$$\int \psi_n^{0*}\psi_m^0 d\tau = \delta_{nm} \tag{5.32}$$

and formally the equation is solved by left multiplying equation (5.31) by a wavefunction and integrating. The result is

$$\int \psi_n^{0*}H^0\psi_m^0 d\tau = E_n^0\int \psi_n^{0*}\psi_m^0 d\tau = \delta_{nm}E_n^0, \tag{5.33}$$

where δ_{nm} is the Kronecker delta function which is zero unless $n = m$ when it is unity. (Note that n and m are both dummy indices labelling energy levels, it does not matter if they are exchanged; equation (5.31) could be labelled with ms rather than ns.)

An extra potential term is now added to the Hamiltonian operator, changing it from $H^0 = -\dfrac{\hbar^2}{2m}\dfrac{d^2}{dx^2} + V^0$, whose solution is known to

$$H = H^0 + \lambda V, \tag{5.34}$$

whose solution is sought. It is assumed that the potential term $V \ll V^0$ produces only a small perturbation to the system. The parameter λ is a dimensionless coefficient that is allowed to vary from 0, which is no perturbation, to 1, the full perturbation. It is not necessary to know what λ is; it is only as a way of reaching an answer. To make the calculation simpler, the energy levels E_n^0 are always assumed to be non-degenerate. Degenerate levels can be dealt with in a similar manner; see Atkins & Friedmann (1997),

The perturbed Schrödinger equation is

$$H\varphi_n = E_n\varphi_n \qquad \text{or} \qquad (H^0 + \lambda V)\varphi_n = E_n\varphi_n \tag{5.35}$$

where all the new wavefunctions φ_n and energies E_n are unknown so far. Both the wavefunction φ and energy are functions of the parameter λ. Our aim is to be able to express the perturbed energy levels E_n in terms of the unperturbed energies E_n^0 and their wavefunctions ψ_n^0 both of which are already known.

The assumption is now made that perturbed energies and wavefunctions can be expanded as Taylor series in λ; see equation (5.15). This assumption is justified by the validity of the final result. Expanding E and φ gives

$$E_n = E_n|_{\lambda=0} + \lambda\left.\frac{dE_n}{d\lambda}\right|_{\lambda=0} + \left.\frac{\lambda^2}{2!}\frac{d^2E_n}{d\lambda^2}\right|_{\lambda=0} + \cdots$$

$$\varphi_n = \varphi_n|_{\lambda=0} + \lambda\left.\frac{d\varphi_n}{d\lambda}\right|_{\lambda=0} + \left.\frac{\lambda^2}{2!}\frac{d^2\varphi_n}{d\lambda^2}\right|_{\lambda=0} + \cdots.$$

A new notation is now introduced to make the equations simpler to read,

$$E_n^{(j)} \equiv \frac{1}{j!}\frac{d^jE_n}{d\lambda^j}\bigg|_{\lambda=0} \qquad \varphi_n^{(j)} \equiv \frac{1}{j!}\frac{d^j\varphi_n}{d\lambda^j}\bigg|_{\lambda=0}$$

therefore

$$E_n = E_n^0 + \lambda E_n^{(1)} + \lambda^2 E_n^{(2)} + \cdots \tag{5.36}$$

$$\varphi_n = \psi_n^0 + \lambda\varphi_n^{(1)} + \lambda^2\varphi_n^{(2)} + \cdots \Lambda \tag{5.37}$$

Notice that $E^{(0)} \equiv E^0$ is the unperturbed energy and $\varphi^{(0)} \equiv \psi^0$ the unperturbed wavefunction. Next, E and φ are substituted into the perturbed Schrödinger equation (5.35) and an apparently very complicated equation results;

$$\begin{aligned}&(H^0 + \lambda V)(\psi_n^0 + \lambda\varphi_n^{(1)} + \lambda^2\varphi_n^{(2)} + \cdots)\\ &= (E_n^0 + \lambda E_n^{(1)} + \lambda^2 E_n^{(2)} + \cdots)(\psi_n^0 + \lambda\varphi_n^{(1)} + \lambda^2\varphi_n^{(2)} + \cdots)\end{aligned} \tag{5.38}$$

However, grouping terms with similar powers of λ allows equation (5.38) to be rearranged in a far clearer way;

$$\begin{aligned}&(H^0 - E_n^0)\psi_n^0 + && \leftarrow \text{zeroth order}\\ &\lambda[(V - E_n^{(1)})\psi_n^0 + (H^0 - E_n^{(0)})\varphi_n^{(1)}] + && \leftarrow \text{first order.}\\ &\lambda^2[(V - E_n^{(1)})\varphi_n^{(1)} + (H^0 - E_n^{(0)})\varphi_n^{(2)}] + \lambda^3[\cdots] + \cdots = 0\end{aligned}$$

The zeroth-order term is the initial unperturbed equation. It should contain λ^0 but as this is 1, it is not written explicitly. To obtain the *first-order* correction it is assumed that λ^2 is smaller than λ and similarly for higher powers and therefore these can be ignored. The first-order energy correction is obtained by extracting just the terms in λ and is

$$(V - E_n^{(1)})\psi_n^0 + (H^0 - E_n^{(0)})\varphi_n^{(1)} = 0.$$

The perturbed energies $E_n^{(1)}$ are calculated from this last equation by left multiplying it by the complex conjugate of ψ_n^0 and integrating. This is what is always done to form the expectation value to calculate the energy, see equation (5.33) and recall that H^0 is an operator and must always act on terms to its right. It cannot be removed from the integration as can the energy, which is a constant. Integrating as indicated produces the result

$$\int \psi_n^{0*}(V - E_n^{(1)})\psi_n^0 d\tau + \int \psi_n^{0*}(H^0 - E_n^{(0)})\varphi_n^{(1)} d\tau = 0.$$

Expanding the brackets gives

$$\int \psi_n^{0*} V\psi_n^0 d\tau - E_n^{(1)}\int \psi_n^{0*}\psi_n^0 d\tau + \int \psi_n^{0*} H^0\varphi_n^{(1)} d\tau - E_n^{(0)}\int \psi_n^{0*}\varphi_n^{(1)} d\tau = 0.$$

Using the fact that the wavefunctions are orthogonal and normalized allows a little simplification to the second term giving

$$\int \psi_n^{0*} V\psi_n^0 d\tau - E_n^{(1)} + \int \psi_n^{0*} H^0\varphi_n^{(1)} d\tau - E_n^{(0)}\int \psi_n^{0*}\varphi_n^{(1)} d\tau = 0. \tag{5.39}$$

At this point we can go no further because we do not know how to deal with the $\varphi_n^{(1)}$ wavefunctions; they are unknown. An important step is now taken and this is to expand the unknown wavefunctions φ in the *basis set of our original wavefunctions* ψ^0 because these are a complete orthonormal set; this is a standard mathematical procedure taken from a study of linear algebra and often used in quantum mechanics (see chapters 6 and 8). Using the basis set ψ, the wavefunctions $\varphi_n^{(1)}$ are defined by, and expanded in an infinite series as

$$\varphi_n^{(1)} = \sum_k a_k\psi_k^0, \tag{5.40}$$

where a_k are the expansion coefficients. These have to be calculated if the new wavefunction is required but do not enter into the final equation for the perturbed energy. By substitution for φ the integrals in equation (5.39) become expressible in known wavefunctions and coefficients a. The result is clearly going to be complicated but will simplify greatly if each term is examined separately. The complete equation is

$$\int \psi_n^{0*} V \psi_n^0 d\tau - E_n^{(1)} + \int \psi_n^{0*} H^0 \sum_k a_k \psi_k^0 d\tau - E_n^{(0)} \int \psi_n^{0*} \sum_k a_k \psi_k^0 d\tau = 0.$$

Expanding out the third term produces

$$\int \psi_n^{0*} H^0 \sum_k a_k \psi_k^0 d\tau = \int \psi_n^{0*} H^0 (a_1 \psi_1^0 + a_2 \psi_2^0 + a_3 \psi_3^0 + \cdots a_n \psi_n^0 + \cdots) d\tau,$$

However, because the ψ's are orthonormal (equation (5.32)), only the nth term remains after integration and all the others are zero. An individual n to k integral is

$$\int \psi_n^{0*} H^0 \psi_k^0 d\tau = E_n^0 \int \psi_n^{0*} \psi_k^0 d\tau = \delta_{nk} E_n^0$$

(see equation (5.33)); therefore only the term with index n remains producing $E_n^0 \int \psi_n^{0*} \psi_n^0 d\tau = E_n^0$.

Using the same procedure and arguments the second of the two summations evaluates to $-E_n^0$ and therefore the sum of these two integrals is zero. The outcome is a simple and elegant equation for the first-order energy correction

$$E_n^{(1)} = \int \psi_n^{0*} V \psi_n^0 d\tau. \tag{5.41}$$

The total energy is

$$\begin{aligned} E_n &\approx E_n^0 + E_n^{(1)} \\ &\approx E_n^0 + \int \psi_n^{0*} V \psi_n^0 d\tau \end{aligned} \tag{5.42}$$

This equation tells us that to obtain the energy change to level n caused by the small perturbation V, only the wavefunctions belonging to quantum level n of the initial *unperturbed* Hamiltonian H^0, equation (5.31), need to be known. The energy change itself is the average of the perturbation potential V with the unperturbed wavefunctions.

The correction to wavefunctions for each level are $\varphi_n^{(1)}$, and are the summation of the basis set wavefunctions ψ^0 weighted by the ratio of the perturbed energy to energy gap; the result for the nth level is,

$$\varphi_n^{(1)} = \sum_{n \neq k} a_k \psi_k^0 = \sum_{n \neq k} \frac{\int \psi_k^{0*} V \psi_n^0 d\tau}{E_n^0 - E_k^0} \psi_k^0 \tag{5.43}$$

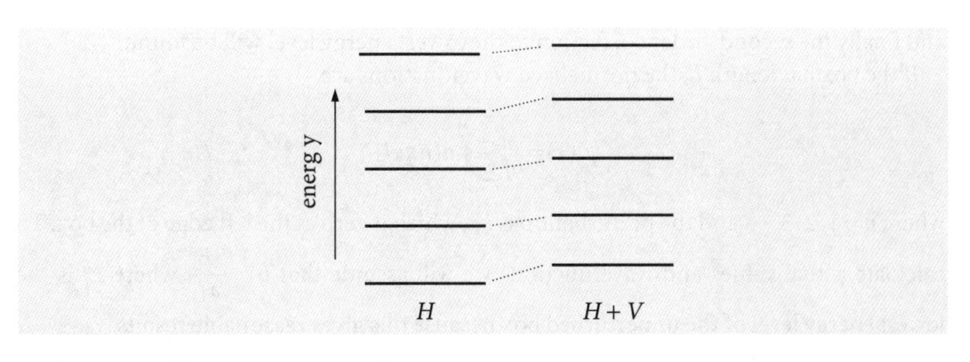

Fig. 5.11 The perturbation V alters the energy levels. The exact amount depends on how V changes.

where the $n=k$ term is removed; otherwise this would lead to division by zero. This result represents the correction to the wavefunction as a linear combination of the original ones with a_k coefficients, which are numbers, given by

$$a_k = \frac{\int \psi_k^{0*} V \psi_n^0 d\tau}{E_n^0 - E_k^0} \equiv \frac{\langle \psi_k^0 | V | \psi_n^0 \rangle}{E_n^0 - E_k^0}.$$

where the right-hand expression is in bra-ket notation. Notice that the conjugate is not indicated, it is assumed by virtue of being on the left of the expression. The integration limits are not included either because they depend on the problem but for a harmonic oscillator they would be ± infinity. The size of the coefficient belonging to a level n is inversely proportional to how far away in energy every other level (index k) is. Clearly the closer in energy level k is to level n the bigger the effect it has primarily due to the reciprocal energy difference; if this is small the coefficient will be large and vice versa. Equation (5.43) shows how the new potential effectively mixes the original wavefunctions ψ to produce the new ones. Notice that this equation produces the *correction* to the wavefunction and ψ_n^0 has to be added to obtain the perturbed wavefunction.

The calculation can be repeated for the second-order perturbation and so on; the general observation is that if the wavefunction to order m is known then the energy to order $2m+1$ can be calculated. The zeroth-order states ψ^0 therefore allow us to find the first order energy correction and so forth.
Extending the calculation to second order produces the energy correction, (see Atkins & Friedmann 1997 for details),

$$E_n^{(2)} = \sum_{n \neq k} \frac{|\langle \psi_k^0 | V | \psi_n^0 \rangle|^2}{E_n^0 - E_k^0} \tag{5.44}$$

This term is usually negative for the lowest energy as the energy for quantum numbers larger than n is usually greater than that for n. The total energy to second order is therefore

$$E_n = E_n^0 + E_n^{(1)} + E_n^{(2)} \tag{5.45}$$

When calculating the perturbed energy how do we know that our calculation is valid? The potential added V might be too big for this method to work. Fortunately, this has also been worked out; the result is that the expectation values divided by the energy gap must be much less that 1. This means checking terms with indices $k=n\pm 1$ because the denominator is usually smallest in this case. The limiting condition for our victim state n is thus

$$\left| \frac{\langle \psi_k^0 | V | \psi_n^0 \rangle}{E_n^0 - E_k^0} \right| \ll 1, \tag{5.46}$$

for each of the nearby states k where $k \neq n$.

5.8.2 Calculation of the energy of a perturbed particle in a box

Consider now a particle in a one-dimensional box that is subjected to a small linear potential ramp; the potential is bx where b is a constant. Such a model might represent an electron in a deep quantum well experiencing an applied electric field. Using the perturbation method, the first-order correction to the energies and wavefunctions will be calculated. The validity of the result will also be checked according to equation (5.46), and finally the second-order correction to the lowest energy level will be found.

If the box has length L, the normalized wavefunctions are

$$\psi_n(x) = \sqrt{\frac{2}{L}} \sin(n\pi x/L)$$

where $n = 1, 2, 3, \cdots$ and the perturbation is bx, which is zero at the left edge of the box. To calculate actual values and wavefunctions we will assume that $b = \dfrac{3E_1^0}{4L}$ where E_1^0 is the lowest energy level of the unperturbed box because this gives reasonable results.

To start the calculation, the unperturbed energy levels of the particle in a box are found by using the Schrödinger equation with a zero potential $V^0 = 0$,

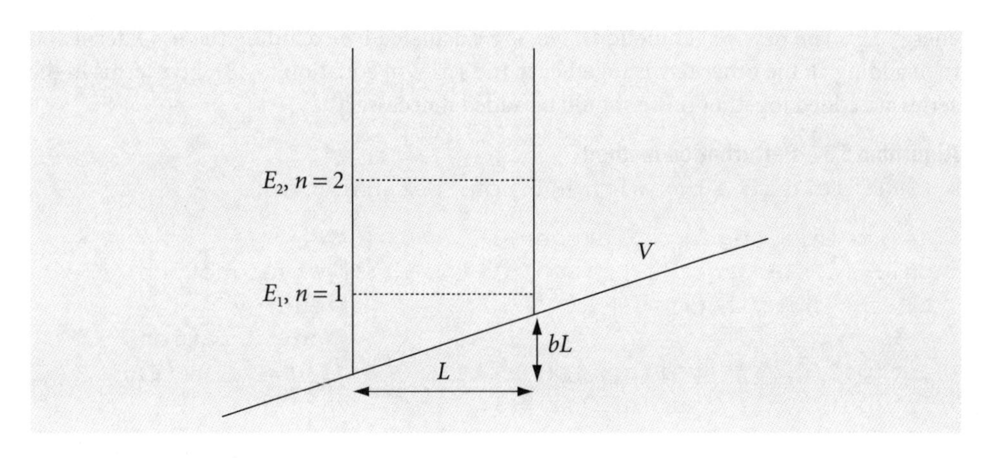

Fig. 5.12 Square well with a linear applied potential *V*.

$$-\frac{\hbar^2}{2m}\frac{d^2\psi_n(x)}{dx^2} = E_n\psi_n(x).$$

Multiplying both sides by the wavefunction, and integrating, produces

$$-\int_0^L \psi_n(x)\frac{\hbar^2}{2m}\frac{d^2\psi_n(x)}{dx^2}dx = E_n\int_0^L \psi_n(x)\psi_n(x)dx.$$

The complex conjugate is not indicated because the wavefunctions are real. The wavefunctions are normalized so that $\int_0^L\psi_n(x)\psi_n(x)dx = 1$ and the unperturbed energies after substituting for the wavefunctions and performing the differentiation and integration are

$$E_n = \frac{2}{L}\frac{\hbar^2}{2m}\left(\frac{n\pi}{L}\right)^2\int_0^L \sin(n\pi x/L)^2 dx = \frac{\hbar^2}{2m}\left(\frac{n\pi}{L}\right)^2. \tag{5.47}$$

If the mass is that of the electron and the box has a length of 1 nm, the energies of the levels with quantum number n are $6.023 \times 10^{-20} \times n^2$ J or $3032 \times n^2$ cm^{-1}.

The sloping potential increases linearly across the box and has the value $V = bx$. Using equation (5.41) the energy change to first order of the nth level is

$$E_n^{(1)} = \frac{2}{L}\int_0^L \sin(n\pi x/L)V\sin(n\pi x/L)dx = \frac{bL}{2} \tag{5.48}$$

Using Maple to check this integral gives

```
> assume(n,integer, k,integer):
  psi:=(n,x)-> sqrt(2/L)*sin(n*Pi*x/L):
  b*Int(psi(n,x)*x*psi(n,x),x=0..L): % = value(%);
```

$$b\int_0^L \frac{2\sin\left(\frac{n\pi x}{L}\right)^2 x}{L}\,dx = \frac{1}{2}bL$$

The result is that the energy shift to first order $E_1^{(1)} = bL/2$, is half the final value of the applied potential, which makes sense. Furthermore, it is independent of the quantum number n, which is not always to be expected; if the potential was bx^2 the first-order shift would depend on the quantum number. Using the values for the constant the energy shift is 2.258×10^{-20} J or 1137 cm^{-1}, the first, unperturbed, level is at an energy of 3032 cm^{-1}, so the shift is substantial. The validity of the calculation is checked below.

The wavefunctions are calculated from the sum over all adjacent levels, see equation (5.43), and the calculation continued until subsequent terms are small. This calculation is the *correction* to the wavefunction and this is then added to ψ_n^0 for each level as shown in equation (5.37). This calculation is clearly going to be simpler using a computer and if the expectation values $\langle\psi_k^0|V|\psi_n^0\rangle$ are known before hand, the summation will be easier to calculate. The integrals in the algorithm are $\langle\psi_k^0|V|\psi_n^0\rangle$, the first has $n = k$ and produces energy $E_n^{(1)}$. The new wavefunctions, *phi*, are calculated by excluding the $n = k$ term and then adding all the other terms together in the series of equation (5.43). Five terms in the series are added together more should be added if necessary.

Algorithm 5.3 Perturbation method

```
> # Particle in a box with a linear perturbation

> assume(n,integer, k,integer):
  psi:=(n,x)-> sqrt(2/L)*sin(n*Pi*x/L): # wavefunction
  E:= n-> hbar^2/(2*mu)*(n*Pi/L)^2:      # energy
  V:= b*x:                               # perturbation
  Int(psi(n,x)*V*psi(n,x),x=0..L):   'En(1)'= value(%);
  Int(psi(k,x)*V*psi(n,x),x = 0..L): % = value(%);

  ak:= unapply( rhs(%),n, k, b ):        # ak equation (5.43)
```

$$\int_0^L \frac{2\sin\left(\frac{k\pi x}{L}\right) bx \sin\left(\frac{n\pi x}{L}\right)}{L}\,dx = -\frac{4Lbkn(1+(-1)^{1+k+n})}{\pi^2(k^4 - 2k^2n^2 + n^4)}$$

```
  num:= 20:          # large to ensure convergence of series
  s:=( k, n, x)->
     if n <> k then

       ak( n, k, b)* psi(k, x)/( E(n) - E(k) ) else 0

     end if;
  phi:=(n, x)-> add( s(k, n, x), k = 1..num ); # eqn (5.43)
```

To complete the calculation as shown in Fig. 5.13 the energies, have to be added and the constants evaluated.

The correction to the basic (unperturbed) wavefunction has the opposite parity to the unperturbed wavefunction. The correction to $n = 1$ is mainly the $n = 2$ wavefunction scaled by the constant a_k, and this is not so surprising, the adjacent wavefunctions have the biggest correcting effect due to the inverse dependence of the energy gap; equation (5.43).

To check on the validity of the calculation use equation (5.46), and ensure that the ratio is less than unity. The values $L = 1$ nm and $b = \frac{3E_1}{4L}$ are used and the corrections for the $n \pm 1$ levels are calculated with,

```
> evalf( abs( ak(n,n+1,b)/(E(n)-E(n+1)) ) );

> evalf( abs( ak(n,n-1,b)/(E(n)-E(n-1)) ) );
```

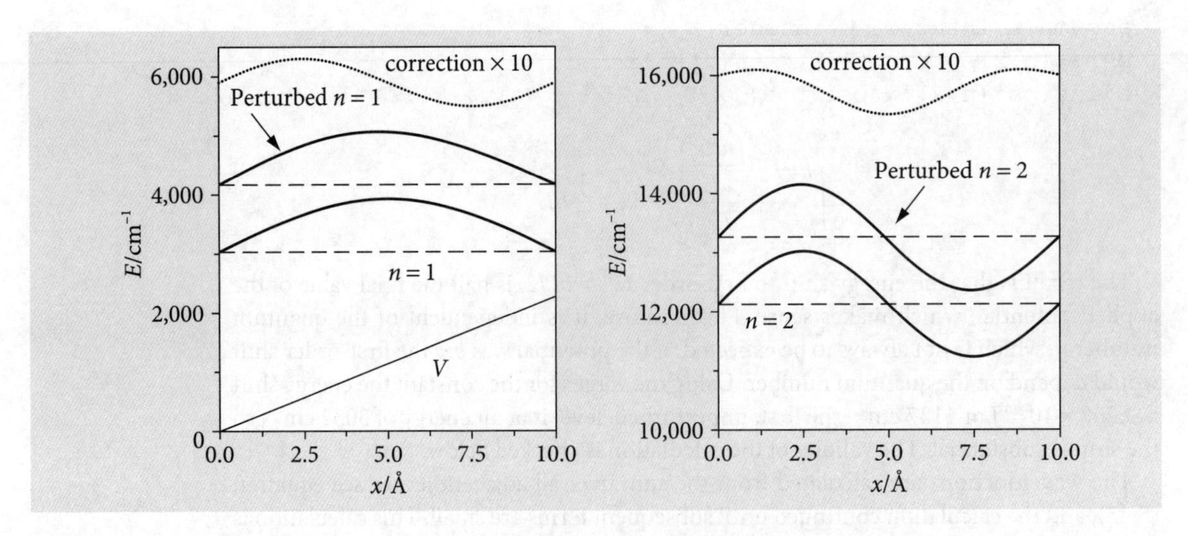

Fig. 5.13 Perturbed and non-perturbed energy levels for the particle in a box. The left-hand figure shows the case when $n = 1$ and on the right-hand side when $n = 2$. (Note the change in energy). The perturbation correction to the wavefunction is shown at the top of the figure (displaced and x10 for clarity) and the perturbed wavefunction and energy level is in both cases at greater energy.

For the first level this check produces a value 0.045, which is clearly less than 1 and also for the second level, 0.029 and 0.045 and the perturbation is small enough to be in the region where it is valid even though the shift in energy levels is large.

The second-order energy correction to the lowest level is shown in equation (5.44), which, using the previous result for the expectation value `ak` given above, is for $k = 2, 3, \cdots$ the term

```
> En2:=(n, k, b)-> ak(n,k,b)^2 / ( E(n)-E(k) );
```

The first value with $n = 1$, and $k = 2$ is

$$E_1^{(2)} = -\frac{512}{243}\frac{L^4 b^2 \mu}{\pi^6 \hbar^2}$$

and has the relatively small value of -3.664×10^{-22} J or -18.44 cm^{-1}. The next and each term where k is odd is zero. The term with $k = 4$ is also negative and far smaller (-0.02 cm^{-1}) than that with $k = 2$. The zero valued terms when k is an odd number can be seen from the integral producing a_k. The numerator is zero when $(-1)^{k+1+n}$ is odd.

5.8.3 Questions

Full solutions are available at **www.oxfordtextbooks.co.uk/orc/beddard**.

Q5.45 Repeat the particle in box example calculation, but with the box extending from $-L/2$ to $L/2$, and with a perturbing potential of the form $V = bx^2$ making it zero in the centre of the box. The constant is $b = 1/4$. Use Maple, based on the code in the example or otherwise, to calculate some of the energy corrections $E_1^{(1)}$ and $E_1^{(2)}$ and so forth.

Q5.46 A heteronuclear diatomic molecule, which can be adequately described as a harmonic oscillator, is placed in an electric field aligned with the molecule's long axis and so experiences an additional and linear potential of magnitude ax. Calculate the change in the energy levels and the resulting spectrum. The harmonic oscillator has vibrational frequency ω and reduced mass μ and orthonormal wavefunctions,

$$\psi(n, x) = \frac{1}{\sqrt{2^n n!}}\left(\frac{\alpha}{\pi}\right)^{1/4} H(n, x\sqrt{\alpha})e^{-\alpha x^2/2}$$

where $\alpha = \sqrt{k\mu}/\hbar$.

***Strategy*:** Use the perturbation method to calculate the change in energy. In each case use the harmonic oscillator wavefunctions. The Hamiltonian is $H = H^0 + ax$ where H^0 solves the normal harmonic oscillator with energy $E_n = \hbar\omega(n + 1/2)$.

Q5.47 Suppose that a harmonic potential is modified by a perturbing cubic term of magnitude bx^3, the oscillator now becomes anharmonic. Calculate the energy levels and spectrum.

Q5.48 The particle on a ring can approximate the energy levels of a cyclic polyene. The potential energy is zero and the Schrödinger equation $-\dfrac{\hbar^2}{2\mu}\dfrac{d^2\psi}{d\phi^2} = E\psi$ where the angle ϕ has values from $-\pi$ to π radians. The wavefunction is $\psi_n = \dfrac{1}{\sqrt{2\pi}}e^{in\phi}$ and the quantum numbers have values $n = 0, \pm 1, \pm 2, \cdots$.

(a) Calculate the unperturbed energies E_n.

(b) Calculate the perturbed energy of the lowest level ($n = 0$) to second order, when the potential has the value V from $-a\pi$ to $a\pi$ where a is a fraction < 1. If we were to suppose that our ring was pyridine then the nitrogen would have a different potential to that of the carbons. Call this value V, and then a could be 1/6. Find the energy if $V = 0.1E_1$.

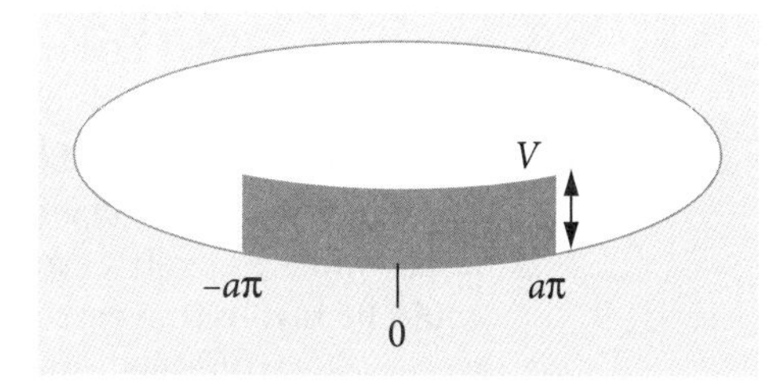

Fig. 5.14 Particle on a ring with a small region of perturbation.

5.9 Quantum superposition and wavepackets

In calculating the energy and wavefunctions of the harmonic oscillator, hydrogen atoms, the rigid rotor, and many similar problems, the *time-independent* Schrödinger equation, $H\psi = E\psi$, is used. However, many processes evolve in time and then a time-dependent form of the equation must be used and this is

$$H\psi = i\hbar\frac{d\psi}{dt} \tag{5.49}$$

where H is still the Hamiltonian operator. The wavefunction must now depend upon time or its derivative would be zero. In some experiments a superposition of different wavefunctions is unavoidable and a *non-stationary* state is produced which must evolve with time if the energy levels differ. An example is an electronically excited state that is produced by absorbing photons from a femtosecond-duration laser pulse. Many vibrational levels are simultaneously excited, because a narrow laser pulse must have a wide energy spread, and the wavepacket produced evolves with time. This can be observed by pump-probe spectroscopy (Zewail 1994). The amount of each level populated depends on the intensity of the light at each absorption frequency, the Franck–Condon factors involved in the absorption and the shape of the laser pulse.

Suppose that $\psi_1^0(x)$ is the time-independent wavefunction which is the eigenfunction of $H\psi_1 = E_1\psi_1$. This wavefunction could be that of the harmonic oscillator, of s or p-orbitals, or any other quantum mechanical model of an atom or molecule, with energy eigenvalue E_1. The time-dependent version of this wavefunction is

$$\psi_1(x,t) = \psi_1^0(x)e^{-iE_1t/\hbar} \tag{5.50}$$

where t is time and x is a spatial coordinate, which may be the bond extension of a vibrating molecule or the position of an electron in an atom. The probability of observing this single wavefunction in a region x to $x + dx$ is $P(x, t)dx = \psi(x)^*\psi(x)dx$, which is constant in time because the complex conjugate removes any time dependence;

$$\psi_1^0(x)e^{+iE_1t/\hbar}\psi_1^0(x)e^{-iE_1t/\hbar} = \psi_1^0(x)\psi_1^0(x)$$

A second wavefunction describing energy E_2 is $\psi_2(x, t) = \psi_2^0(x)e^{-iE_2t/\hbar}$; then a *normalized superposition* of these two wavefunctions could be

$$\Psi = \frac{\psi_1 + \psi_2}{\sqrt{2}}$$

if equal contributions are made from each wavefunction. The x and t labels are suppressed for clarity. In general, with two wavefunctions $\Psi = \dfrac{\alpha\psi_1 + \beta\psi_2}{\sqrt{\alpha^2+\beta^2}}$, where α and β are the amounts of ψ_1 and ψ_2. These are limited by the normalization condition $|\alpha|^2 + |\beta|^2 = 1$. The modulus ($|\alpha| = \sqrt{\alpha^*\alpha}$) is used here as α may contain a complex quantity. If many wavefunctions are in the superposition or *wavepacket*, then

$$\Psi = \sum_{n=1}^{N}\alpha_n\psi_n$$

and if normalized the condition is $\sum_{n=1}^{N}|\alpha|_n^2 = 1$. In each of these cases, Ψ evolves in time, as does the probability $P(x, t)$, and this is explained next.

The stationary-state wavefunction is $\psi^0(x)$ and depends only on position. A superposition or wavepacket might be formed by absorbing photons with an energy spread covering only two different energy levels E_1 and E_2. However, a wavepacket containing only two levels, while formally correct, is hardly a wavepacket at all and in practice many levels could be involved. Nevertheless, two will be used here to illustrate what happens. The wavepacket with equal amounts of the two wavefunctions and added in phase is

$$\Psi(x,t) = \psi_1^0(x)e^{-iE_1t/\hbar} + \psi_2^0(x)e^{-iE_2t/\hbar},$$

and the amount of each wavefunction varies as time proceeds because E_1 and E_2 are different. The exponential terms can be considered as phase factors and, since they oscillate in time and because E_1 and E_2 are different, they cause the wavefunctions to be added together in different proportions. Recall the identity $e^{-ix} = \cos(x) - i\sin(x)$; this can be used to illustrate how the wavefunctions oscillate in time. The two-wavefunction wavepacket could be written as,

$$\Psi(x,t) = \psi_1^0(x)\cos(E_1t/\hbar) + \psi_2^0(x)\cos(E_2t/\hbar)$$
$$-i[\psi_1^0(x)\sin(E_1t/\hbar) + \psi_2^0(x)\sin(E_2t/\hbar)],$$

but this is rather long and the exponential form is preferred. Any measurement of the wavepacket is always proportional to $\Psi^* \times \Psi$ so the imaginary part is lost in the product via the complex conjugate leaving only real terms. This can be seen by calculating the probability of the molecule being at a position x at a time t which is

$$P(x,t) = \Psi^*\Psi = (\psi_1 + \psi_2)^*(\psi_1 + \psi_2),$$

where the * indicated the complex conjugate; the labels x and t on the wavefunction are removed for clarity. In any classical calculation, terms are squared then added; in quantum mechanics always do the opposite, therefore,

always add the wavefunctions then square,

because this leads to the interference or cross terms such as $\psi_1\psi_2$ that are absent from any classical calculation. Hence,

$$(\psi_1 + \psi_2)^*(\psi_1 + \psi_2) = (\psi_1^* + \psi_2^*)(\psi_1 + \psi_2)$$
$$= \psi_1^*\psi_1 + \psi_2^*\psi_2 + \psi_1^*\psi_2 + \psi_2^*\psi_1.$$

Substituting for the wavefunctions and taking complex conjugates gives

$$P(x,t) = [\psi_1^0(x)]^2 + [\psi_2^0(x)]^2 + \psi_1^0(x)\psi_2^0(x)[e^{i(E_1-E_2)t/\hbar} + e^{-i(E_1-E_2)t/\hbar}], \quad (5.51)$$

of which the last term is the most interesting as it represents the interference of the two wavefunctions with one another and is a purely quantum effect. The first two constant terms are those produced by a classical calculation. Defining a frequency as $\omega = \dfrac{E_1 - E_2}{\hbar}$ and using the definition of cosine equation (2–18) gives,

$$P(x,t) = [\psi_1^0(x)]^2 + [\psi_2^0(x)]^2 + 2\psi_1^0(x)\psi_2^0(x)\cos(\omega t) \quad (5.52)$$

which shows that the probability oscillates with frequency ω in addition to a constant term. This means that the amount of each wavefunction in the superposition changes in time if the energy difference $E_1 - E_2$ is not zero. Note that the probability oscillates only because there is an energy difference between two levels.

If the wavefunctions are added in phase, this means that there are terms where time origin is the same in each, for example $e^{-iE_1t/\hbar}$ and $e^{-iE_2t/\hbar}$. However, if we delay the generation of one of the wavefunctions, as is possible when using finite-duration laser pulses to form wavepackets, then we would produce terms such as $e^{-iE_1t/\hbar}$ and $e^{-iE_2(t-t_0)/\hbar}$ where t_0 is a constant time difference between exciting each energy level and hence forming the wavepacket. This constant time delay is equivalent to a phase delay ϕ because in a sine or cosine a constant term is equivalent to a phase delay,

$$\cos(E_1t/\hbar + E_1t_0/\hbar) \equiv \cos(E_1t/\hbar + \phi).$$

Any laser pulse exciting molecules has a finite duration, therefore there is a range of phases produced during excitation because different parts of the wavepacket start at different times. The effect is that of convoluting in time the laser pulse with $P(x, t)$. Convolution is described in Chapter 9.7.

The effect of adding together some harmonic oscillator wavefunctions is shown in Fig. 5.15. The wavefunctions are added in equal amounts with vibrational quantum numbers $n = 3$ to 8. Below this are shown the wavepackets formed at $t = 0$, 1/3, and 1/2 period. At $t = 0$ it is easy to see that the addition will produce amplitude at the right of the potential. At 1/2 a period the wavepacket is the mirror image of that at $t = 0$ as the phase of the wavefunctions, the $e^{i\omega t/\hbar}$ part, has reversed and therefore reinforcement occurs on the left part of the potential. At intermediate times, the wavepacket is some arbitrary shape depending on how the exponential terms make the wavefunctions add together. The wavepacket is placed at its average energy on the potential.

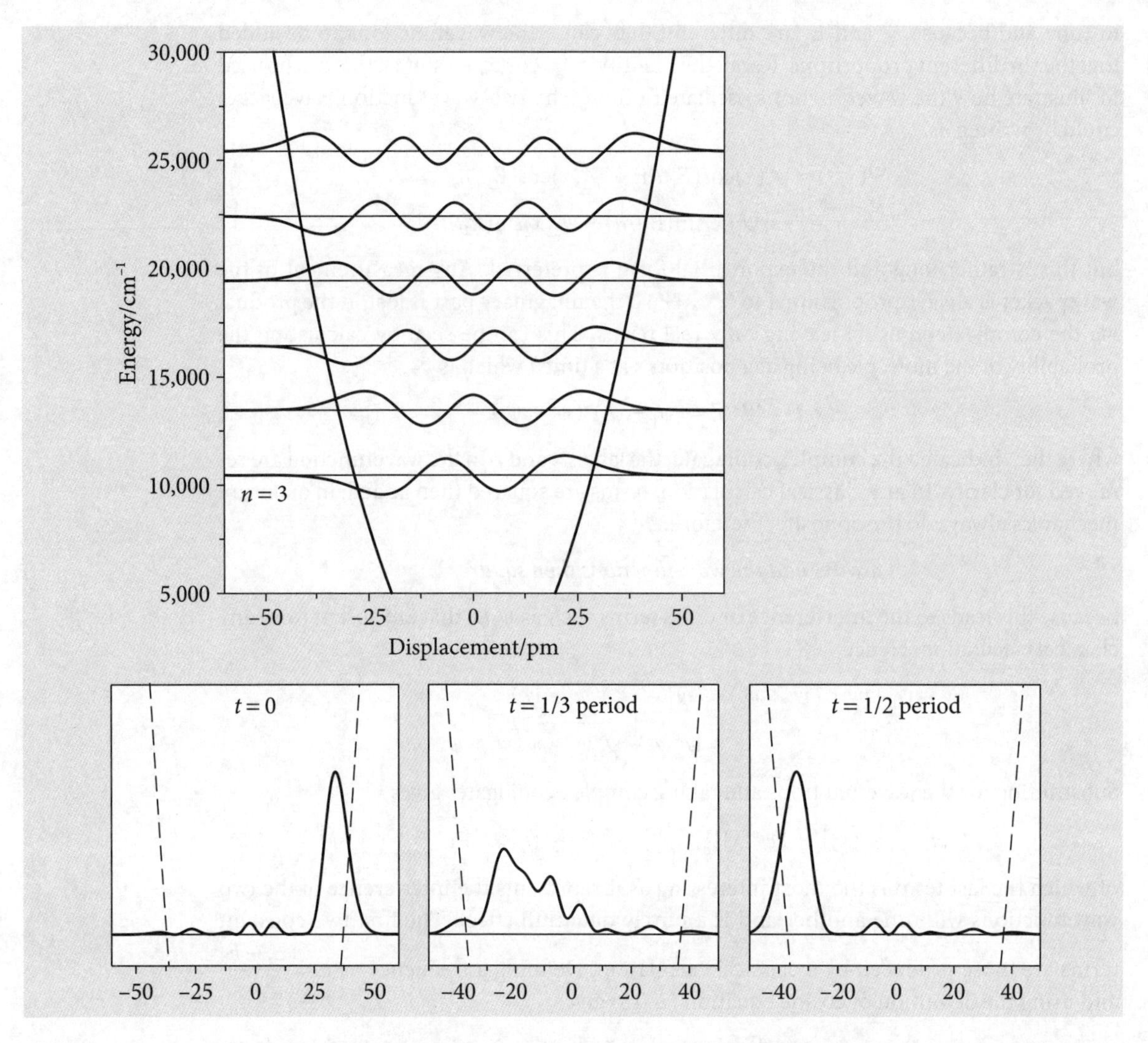

Fig. 5.15 Harmonic oscillator wavefunctions for the $n = 3$ to 8 levels of the ground state of HCl drawn with the potential energy and placed at the energy of each level. On the bottom is shown the probability density $\Psi^*\Psi$ for three wavepackets; at $t = 0$, at 1/3, and at 1/2 a period. The wavefunctions are added equally to make the wavepacket.

The initial wavepacket is formed on the right of the potential where the wavefunctions are all positive, and although not exactly in phase they are sufficiently so that at this bond displacement, the resultant wavepacket is positive. On the left of the potential, the wavefunctions are alternately positive and negative so add approximately to zero. As time progresses the phases associated with the wavefunctions change; imagine this as the sine part becoming cosine and vice versa and this leads to the superposition being large on the left side of the potential when a time of half a period is reached. As more time elapses, the wavepacket continues to change and will reform after one period and then repeats the cycle ad infinitum.

The harmonic oscillator is unusual in that all the energy levels are separated by an equal amount; this means that all the components of the wavepacket are changing at the same single frequency. In an anharmonic oscillator, rigid rotor, or H atoms, for example, the energy levels are not equally spaced; so many frequencies must be present in the wavepacket. Initially suppose that the wavepacket has an approximately Gaussian or bell shape in its spatial coordinate. The different frequencies present mean that as time progresses the wavepacket spreads out and appears to become random, but it will nevertheless reform itself after a certain period, governed by integer multiples of the lowest frequency present. This period is often a large number of roundtrips across the potential, unlike the harmonic potential where the wavepacket reforms after each round trip. The process of many oscillators reforming the original wavepacket is called *recurrence*. After recurring, the wavepacket appears to become random again, although it is not, and then recurs and repeats the sequence unless something happens to stop this. Usually collisions with other molecules or energy flowing into vibrational modes not excited by the laser will

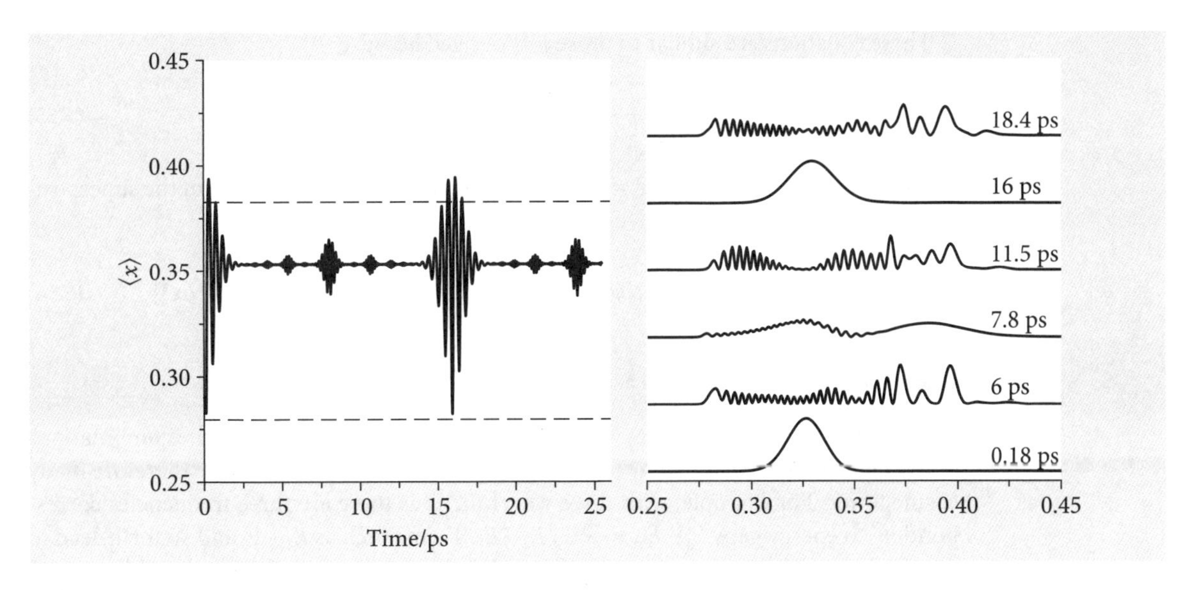

Fig. 5.16 Left: The average position of I_2 wavepackets $\langle x\rangle$ vs time. Right: The wavepackets at various times showing recurrence. (The size of the wavepackets is scaled to fit the figure so that the area under each is not the same.)

destroy the wavepacket. Children playing on swings can be seen to 'recur' when at some time after starting, they all swing in phase for an instance, then become out of phase. In fact any system consisting of several oscillators with different frequencies will show recurrences governed by integer multiples of the lowest frequency. It is even possible also to observe fractional recurrences (Vrakking et al. 1996).

One way of detecting recurrences is to calculate the mean position $\langle x\rangle$ of the wavepacket. The wavepacket is initially formed at one side of the potential (by absorbing a photon from the ground state) and then propagates back and forth across the potential, as shown in Fig. 5.15. If the wavepacket is spread out its mean value is not necessarily that of the equilibrium internuclear separation because any real molecule has an anharmonic potential, such as the Morse potential and the mean value is at longer internuclear separation than $r_e = 0.308$ nm for this I_2 potential. The mean wavepacket position is shown in Fig. 5.16 for the B excited state of iodine excited by a 40 fs Gaussian shaped pulse centred at 550 nm. The recurrences at about 15.7 ps and partial recurrences at other times are clear. The mean value is $\langle x\rangle_t = \int xP(x,t)dx$ which is usually replaced by a summation in practice, because discrete values of position are normally calculated. The figure also shows how the wavepacket behaves like a particle oscillating in the potential as the bond stretches and compresses during the first 2 ps at $\approx$16 ps and less so at $\approx$8 ps. But at other times there is no apparent motion because the *average* bond extension is constant in time. What happened here is that the wavepacket has split into parts that counter propagate and/or interfere with one another in such a way that for a while the average is constant. The figure on the right shows a few wavepackets measured at approximately 0.325 nm, which is about halfway across the potential. The recurrence at about 16 ps is clear and partial recurrence at $\approx$ 8 ps is also seen, with apparently random motion at other times.

When a pump-probe experiment is performed, it is possible to observe wavepacket motion, but not the wavepacket itself because there is no known microscope with which this could be done. What this method does is to excite the molecule and then probe the excited state later and a signal is produced as the wavepacket moves through the region covered by the probe (Atkins & de Paula 2006, pp. 892–3 and Zewail 1994). The experimental observation is a signal oscillating in time.

5.9.1 Superposition of many wavefunctions

The general superposition where each stationary-state wavefunction is weighted with its contribution α or β is

$$\Psi = \frac{\alpha\psi_1 + \beta\psi_2}{\sqrt{\alpha^2+\beta^2}}.$$

These equations are similar to those just derived because

$$\Psi(x,t)=\frac{1}{\sqrt{\alpha^2+\beta^2}}[\alpha\psi_1^0(x)e^{iE_1t/\hbar}+\beta\psi_2^0(x)e^{iE_2t/\hbar}]. \tag{5.53}$$

Calculating the probability as $P=\Psi^*\Psi$ by taking the complex conjugate of the superposition wavefunction and multiplying gives

$$P(x,t)=\frac{1}{\alpha^2+\beta^2}[[\alpha\psi_1^0(x)]^2+[\beta\psi_2^0(x)]^2+2\alpha\beta\psi_1^0(x)\psi_2^0(x)\cos(\omega t)] \tag{5.54}$$

where the frequency is defined as $\omega=\dfrac{E_1-E_2}{\hbar}$.

When there are several, say S, wavefunctions in the superposition, all the combinations of frequencies as determined by the energy gaps are present and there are therefore many cosine terms. For example, with three wavefunctions there are three frequencies corresponding to energy gaps E_3-E_2, E_3-E_1, E_2-E_1. Terms such as E_3-E_3 and so forth, lead to a constant term because cos(0) = 1. The probability in the general case where there are S levels in total is this monstrous looking equation:

$$P(x,t)=N^2\sum_{n=1}^{S}\sum_{m=1}^{S}c_nc_m\psi_n^0\psi_m^0\cos\left(\frac{E_n-E_m}{\hbar}t\right)$$

where and c_n and c_m are the coefficients which, above, were called α and β when there were just two terms; N is the normalization constant.

In its most general form, where the coefficients may also be complex numbers, as may the wavefunctions, the wavepacket has the form

$$\left[\sum_{i=1}^{n}c_i^*(\psi_i^0(x))^*e^{-iE_it/\hbar}\right]\left[\sum_{i=1}^{n}c_i\psi_i^0(x)e^{iE_it/\hbar}\right],$$

and when calculating with Maple it is often convenient to make the sum first then take its complex conjugate and multiply them together. This is the form of equation used to produce Fig. 5.16, when convoluted with the laser pulse shape. The wavefunctions for the Morse potential were used (Morse 1929), but in situations where the wavefunctions are not known because the Schrödinger equation cannot be solved algebraically, the numerical method known as the split-time operator method can be used (see Kosloff 1988).

To illustrate how to use the cosine equation $P(x, t)$, let $S=2$, then taking the terms in the summation in order $(n, m)=(1, 1), (1, 2), (2, 1), (2, 2)$ gives

$$\begin{aligned}P(x,t)&=N^2\sum_{n=1}^{2}\sum_{m=1}^{2}c_nc_m\psi_n^0\psi_m^0\cos\left(\frac{E_n-E_m}{\hbar}t\right)\\&=\frac{1}{c_1^2+c_2^2}\left[\begin{matrix}c_1c_1\psi_1^0\psi_1^0\cos\left(\dfrac{E_1-E_1}{\hbar}t\right)+c_1c_2\psi_1^0\psi_2^0\cos\left(\dfrac{E_1-E_2}{\hbar}t\right)+\\c_2c_1\psi_2^0\psi_1^0\cos\left(\dfrac{E_2-E_1}{\hbar}t\right)+[c_2\psi_2^0]^2\end{matrix}\right]\end{aligned}$$

which is the same result worked out in detail in equation (5.52) after performing some more simplifications and $\alpha=c_1$, $\beta=c_2$ and substituting $\omega=(E_1-E_2)/\hbar$, the energy gap always being taken as positive. The probability of observing the wavepacket is

$$P(x,t)=\frac{1}{c_1^2+c_2^2}[[c_1\psi_1^0]^2+[c_2\psi_2^0]^2+2c_1c_2\psi_1^0\psi_2^0\cos(\omega t)]$$

Finally, note that sometimes the equations are written as

$$\Psi(x,t)=\frac{1}{\sqrt{|\alpha|^2+|\beta|^2}}[\alpha(t)\psi_1^0(x)+\beta(t)\psi_2^0(x)]$$

where the time dependence is put into each constant as $\alpha e^{iE_1t/\hbar}\equiv\alpha(t)$ and the normalization involves calculating the modulus or absolute value of α and β not just their square. The final result is the same; it is only a matter of choice how the equation is written.

5.9.2 Questions

Full solutions are available at **www.oxfordtextbooks.co.uk/orc/beddard**.

Q5.49 Make a superposition of the third and fourth wavefunctions of a particle in a one-dimensional box where the amount of ψ_3 is 1/2 and ψ_4 is 2/5. Use the wavefunctions $\psi_n(x) = \sqrt{\frac{2}{L}}\sin\left(\frac{n\pi x}{L}\right)$ where $n = 1, 2, 3, 4, \cdots$ and L is the length of the box. The energy of the nth level is $E_n = \frac{1}{8m}\left(\frac{hn}{L}\right)^2$.

(a) Calculate the period of the oscillation.

(b) The change in probability $P(x, t)$ over a period of one cycle.

(c) Check your results with Maple and plot the probability at several times over a period if the box is 1 nm in length and m is the mass of an electron.

Q5.50 A vibrational wavepacket is the coherent sum of many wavefunctions, and can be made by exciting a molecule into an excited state from the ground state with a laser pulse of femtosecond duration. If the pulse is short ~ 10 fs, it excites many of the vibrational energy levels at effectively the same time. The amount excited depends on the central wavelength, the wavelength spread of the laser pulse, and the Franck–Condon factors. As an electronic transition occurs in a time far shorter than a vibrational period, the molecules must be excited at a similar internuclear extension as determined by the uncertainty limit. The wavepacket is created coherently in two senses; all the molecules are excited at approximately the same time, limited by laser pulse duration, and at the same position on the potential.

The wavepacket Ψ depends on x the bond extension and time t. It depends on extension because the wavefunctions ψ depend on x and on time because a non-stationary state is produced, which is not an eigenstate and this must evolve in time. The wavepacket is the summation of the amplitudes of each wavefunction multiplied by the time or phase factor, which is the exponential term, and is

$$\Psi(x, t) = \sum_n a_n \psi(x) e^{-iE_n t/\hbar} \tag{5.55}$$

where E_n is the energy of the nth vibrational level and a_n is the amount of each wavefunction present in the superposition and is the product of the Franck–Condon factor for excitation from the ground state and the amplitude of the electric field of the laser pulse at a given energy.

(a) If the molecule is a harmonic oscillator show that the wavepacket recurs, i.e. returns periodically to the same place in the potential, at times $t = mT$ where $m = 0, 1, 2, \cdots$ and $T = 1/\nu = 2\pi/\omega$ is the period of the vibration with frequency ν sec^{-1}. (The energy levels can be written in different, but equivalent ways, as $E_n = \hbar\omega(n + 1/2) = h\nu(n + 1/2) = hc\bar{\nu}(n + 1/2)$ with $\omega = 2\pi\nu$ where ν has units of s^{-1}, ω of radians s^{-1} and $\bar{\nu}$ wavenumbers.)

(b) Using the definitions of the harmonic oscillator wavefunctions and using vibrational levels $n = 0, 1, 2, 3, 4$ with $a_n = [0.1, 0.2, 0.25, 0.2, 0.1]$ use Maple to illustrate the wavepacket motion and confirm the recurrences.

Use numerical values for HI, which has vibrational frequency $\bar{\nu} = 2309.5$ cm^{-1}. The harmonic oscillator wavefunction ψ at extension x and quantum number n are $\psi_n(x) = N_n H_n(x\sqrt{\alpha})e^{-\alpha x^2/2}$, where the normalization is $N_n = \left(\frac{\alpha}{\pi}\right)^{1/4}\left(\frac{1}{2^n n!}\right)^{1/2}$, $\alpha = \frac{\sqrt{k\mu}}{\hbar}$ and H is the Hermite polynomial, k the force constant and μ the reduced mass.

Strategy: In (a), do not try to solve the wavepacket equation but show that it is the same at times mT as it was at time 0. A good starting point would therefore seem to be to substitute for the energy and then time into equation (5.55) and rearrange the resulting equation.

Q5.51 As the energy of an H atom increases, the energy levels become closer together because $E_n = R/n^2$ where R is the Rydberg constant and n the principal quantum number. Exciting an H atom from the ground state into a level close to the dissociation limit with a picosecond laser excites many levels virtually simultaneously compared to the classical orbital period of these levels, and a wavepacket is produced. If the atom is optically excited from the ground state s orbital with $\ell = 0$, the wavepacket is a sum of the radial distributions of the p or $\ell = 1$ electronic wavefunctions with different principal quantum numbers. The electron is promoted to a higher energy orbital, which has a large orbital radius; the many similar wavefunctions add up to generate a wavepacket whose amplitude increases and decreases, and gives the appearance that the electron orbits the nucleus.

(a) Calculate and plot the radial distribution of the wavepacket produced as time vs position.

(b) Comment on the size of the wavepacket and its period of the motion compared to that with $n = 1$. The oscillation frequency is given by $\omega = \dfrac{1}{\hbar}\left|\dfrac{dE_n}{dn}\right|$, and the absolute value is taken only to make the frequency positive.

Assume that the laser's energy is such that the central wavelength excites the levels with $n = 40$, and five levels are excited either side of this, and the laser excites these levels with amounts 0.1, 0.2, 0.3, 0.5, 0.8, 1, 0.8, 0.5, 0.3, 0.2, 0.1. (Different amounts would be used depending on an actual laser's pulse duration and shape but this list corresponds very approximately to a Gaussian shaped pulse of $^1/_2$ ps duration.)

Strategy: Look up the equations for the radial distribution functions of the H atom. The associated Laguerre polynomial is involved which can be calculated explicitly using Maple, but there is a built-in function `LaguerreL`, which provides a rapid and accurate way of calculating this function.

Q5.52 When two or more atomic or molecular eigenstates are coherently excited into an electronically excited state, for example by a laser pulse, and their total emission to a *common* ground state measured, *quantum beats* may be observed as oscillations in the total fluorescence. In the experiment three states are involved, see Fig. 5.17. The beat frequencies directly measure the energy splitting ΔE in pairs of excited levels. This experiment is a very good way of measuring small energy gaps between levels, because small gaps produce long beat periods that are easily measured in a time-resolved experiment and this, therefore, can be a good way of doing high-resolution molecular spectroscopy.

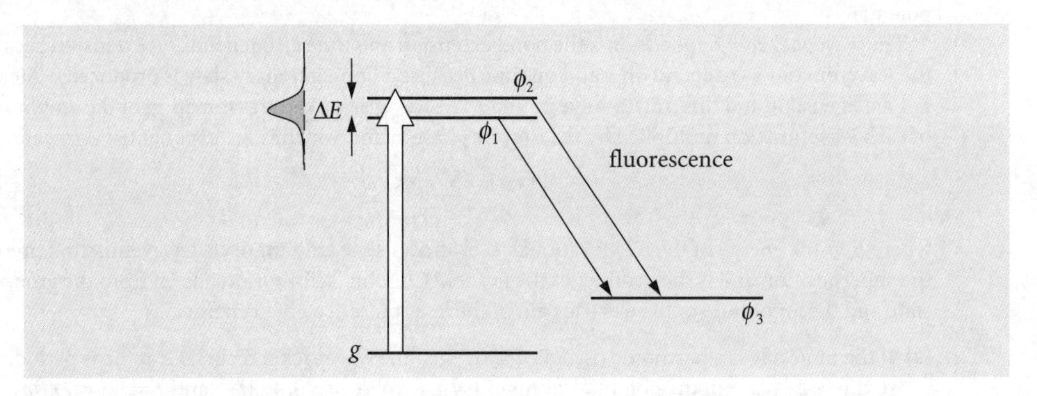

Fig. 5.17 Quantum beat spectroscopy. Two levels are excited from a vibrational level g in the ground state to the first excited electronic state and their *total* fluorescence decay from excited levels 1–3 and 2–3 to a common state is measured.

The fluorescence signal is given by

$$F(t) = \beta\left|\int \varphi_3 \mu \psi(t) dq\right|^2 \tag{5.56}$$

where φ_3 is the wavefunction of the final state and $\psi(t)$ the superposition state (wavepacket) excited by the laser. This is made up from the two states φ_1 and φ_2, and μ is the dipole moment operator coupling excited state levels 1 and 2 to final state ground state level 3. The integral is over a spatial dimension q of the wavefunction, such as a bond extension. The constant β consists of instrumental parameters. The laser pulse exciting the states is long compared to the vibrational period, so that position is averaged out in these experiments; this is why there is the integration over bond extension q. The interference leading to beats arises from the fact that two pathways, 1–3 and 2–3, see Fig. 5.17, are simultaneously measured and in quantum processes one path cannot be distinguished from the other when both fluorescence wavelengths are measured at the same time. The signal is always the square of the sum of the amplitudes of each contribution. The superposition of the two states populated with the laser is the wavepacket

$$\Psi(t) = a_1\varphi_1^0 e^{-iE_1t/\hbar} + a_2\varphi_2^0 e^{-iE_2t/\hbar}$$

and as both levels fluoresce with the same rate constant k, then this equation can be changed to

$$\Psi(t) = a_1\varphi_1^0 e^{-iE_1t/\hbar - kt/2} + a_2\varphi_2^0 e^{-iE_2t/\hbar - kt/2}.$$

(a) Calculate the decaying fluorescence signal if the decay rate constant has the same value k from each level excited. Assume that the integrals are $\int \varphi_3 \mu \varphi_1^0 dq = B_{31}$ and $\int \varphi_3 \mu \varphi_2^0 dq = B_{32}$.

(b) Normalize the signal at $t = 0$.

(c) Plot a graph of the normalized signal if $k = 1\ \text{ns}^{-1}$ and $\Delta E = 1\ \text{cm}^{-1}$ and the laser produces levels with amounts $a_1 = 0.3$ and $a_2 = 0.7$ and $B_{32} = B_{31} = 2$.

6 Vectors

6.1 Motivation and concept

Mass, temperature, energy, entropy, pressure, volume, and speed among many other quantities can be represented by a scalar, which is a simple number expressing the size or amount of something. Velocity, force, electric and magnetic fields, and acceleration each have magnitude and direction and therefore require two quantities to describe them, and they are called vectors.[1] At school, you will probably have met a vector as an arrow drawn on a graph and which, therefore, has magnitude and direction. In biology, the idea is a related one, a vector being 'something that carries' and an infection is described as having a vector that causes its transmission from one creature to another.

Although vectors, such as force or velocity, have two 'dimensions'—magnitude and direction—mathematically, they can have any number of dimensions. A vector's components need not be objects in any physical space; this is not as bizarre as it sounds, for example, your car can be described by its make, colour, manufacture date, engine-capacity, and model. These values can form a vector such as [*Ford, yellow, 2005, 1900cc, estate*]. However, a vector is more than just a list; it is constructed from a *basis set*. The basis set contains the primitive components and the vector is constructed as a linear combination of these elements; more of this later. The basis set is not always apparent, for example, when drawing arrows on a graph, but in all cases, including topics as diverse as quantum mechanics and geometry, the basis set must be explicitly defined before a calculation is started. However, in apparent contradiction to this, many vector properties and formulae can be obtained by symbolically, which means algebraically, $\boldsymbol{A} + \boldsymbol{B}$, etc., without a thought about the basis set, and some of these calculations are described in Section 6.2.

6.1.1 Notation

It is important to distinguish vectors from scalars. Scalars are printed in normal typeface, 1, 2, 3, etc. Symbols printed in bold, italic typeface $\boldsymbol{A}$, $\boldsymbol{B}$, and so forth will represent vectors. Square brackets $[1 \quad 3 \quad -4]$ are used to represent a vector's components 1, 3, and -4, so a vector is written as $\boldsymbol{A} = [1 \quad 3 \quad -4]$. If $\boldsymbol{A}$ ends at a point on a graph, it is defined to have a basis set (x, y, z) which, when it is necessary to be explicitly stated, will have round brackets. The coordinates of a point in space are surrounded by curly brackets $\{4, 0, 2\}$. Some authors put a caret above a letter, e.g. $\hat{B}$, to indicate a vector, others place an arrow over the letter to describe its start and end; for example, vector A to B is written as $\overrightarrow{AB}$, but we will use $\boldsymbol{AB}$ to indicate the same vector. The vector's magnitude or length is enclosed in a pair of vertical lines as $|\boldsymbol{A}|$; however, vertical lines are also used to indicate the absolute value of a quantity and represent a determinant, but the context should make the meaning clear.

[1] W. Hamilton, of Hamiltonian fame, first seems to have used the word 'vector'. J. Willard Gibbs who developed much of thermodynamics also largely developed vector analysis.

6.1.2 **Basis sets**

In describing many phenomena, such as the static magnetic field in an NMR machine, right-angled (rectilinear) or Cartesian axes are used. A vector can be represented in this space as a linear combination or multiples of three vectors of *unit length* each one pointing from the origin along each of the x-, y-, or z-axes. These three *unit vectors* form the *basis set*, and in three dimensions are normally labelled $\boldsymbol{i}, \boldsymbol{j}$, and $\boldsymbol{k}$. In quantum mechanics, for instance, the basis set needed is not always that of spatial coordinates, and often comprises the set of quantum numbers a given atom or molecule possesses. The basis set for a car with the vector [*Ford yellow 2005 1900cc estate*] would be (*make, colour, manufacture date, engine size, model*).

Before basis sets are considered, some general vector properties are defined, which still apply whether we specifically use them in the form of the components of a basis set or not.

6.1.3 **Position vectors**

Conventionally, positional vectors are drawn as arrows to indicate which way they point, and a set of axes is not usually shown. Some of the infinite number that could be drawn is shown in Fig. 6.1. Notice that in the top left of the figure, vectors $\boldsymbol{R}$ and $\boldsymbol{S}$ are the same even though they are displaced from one another because they have the same length and direction. Consequently, on the top right, $\boldsymbol{R} \neq \boldsymbol{S}$ because their directions differ even though their lengths are identical. The coordinate origin of each vector is always at its base, so $\boldsymbol{R}$ and $\boldsymbol{S}$ have their own local coordinate origin whether or not they are equal to one another.

To add two vectors, form a triangle; to add three, form a rectangle; and form a polygon if there are more, so that the sum of all the vectors is zero. In Fig. 6.1 (iii), the vectors $\boldsymbol{A}$, $\boldsymbol{B}$, and C are added by placing the head (or tip) of one to the tail of the other and completing the triangle, the equation is therefore

$$\boldsymbol{A}+\boldsymbol{B}+\boldsymbol{C}=\boldsymbol{0}.$$

where $\boldsymbol{0}$ is the null or zero vector. It follows that $\boldsymbol{A}+\boldsymbol{B}=-\boldsymbol{C}$ with $\boldsymbol{C}$ as drawn in the figure with its head at the foot of $\boldsymbol{A}$. The vector $-\boldsymbol{C}$ has the same direction as $\boldsymbol{C}$ but points in the opposite direction, i.e. its arrow-head would be at the head of $\boldsymbol{B}$. Alternatively, the equation could be *defined* as $\boldsymbol{A}+\boldsymbol{B}=\boldsymbol{C}$ where $\boldsymbol{C}$ is the resultant of $\boldsymbol{A}+\boldsymbol{B}$. In this case, $\boldsymbol{C}$ would have been drawn so that it starts at the foot of $\boldsymbol{A}$ and points to the head of $\boldsymbol{B}$. It is easy to get muddled with vector directions and the simplest thing is to make $\boldsymbol{A}+\boldsymbol{B}+\boldsymbol{C}=\boldsymbol{0}$, where the arrows are always head to tail, but note that in this case the resultant of adding $\boldsymbol{A}$ and $\boldsymbol{B}$ is $-\boldsymbol{C}$.

To subtract the two vectors shown in (iii), to obtain $\boldsymbol{A}-\boldsymbol{B}$, reverse $\boldsymbol{B}$, which means placing the arrow tip on the other end of the line, see Fig. 6.1 (iv). Then place the tip of $\boldsymbol{A}$ to the base of the reversed $\boldsymbol{B}$; the result we shall call $\boldsymbol{D}$. The subtraction $\boldsymbol{B}-\boldsymbol{A}$ produces the vector $-\boldsymbol{D}$.

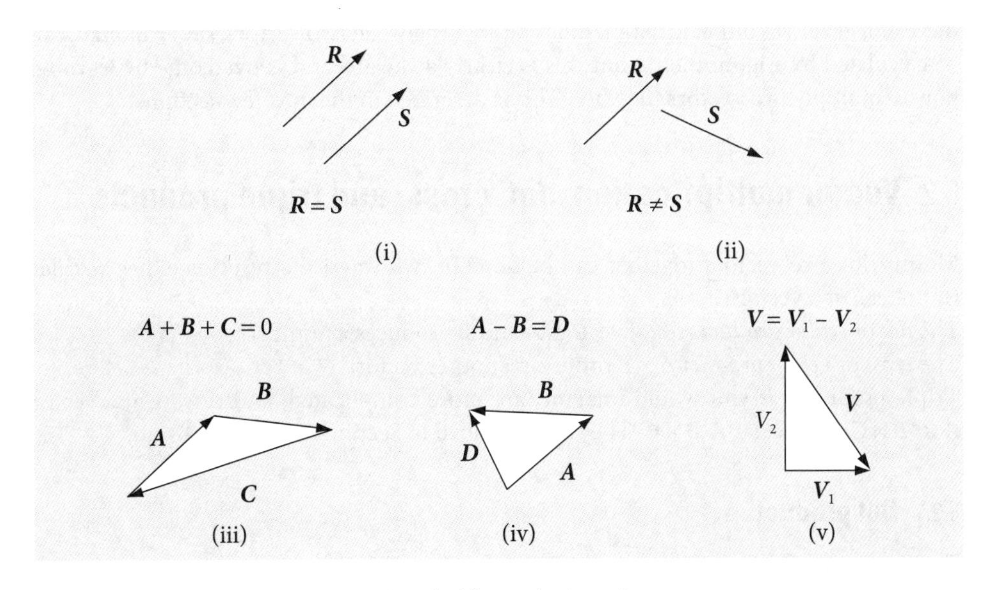

Fig. 6.1 Diagram showing some examples of adding and subtracting vectors.

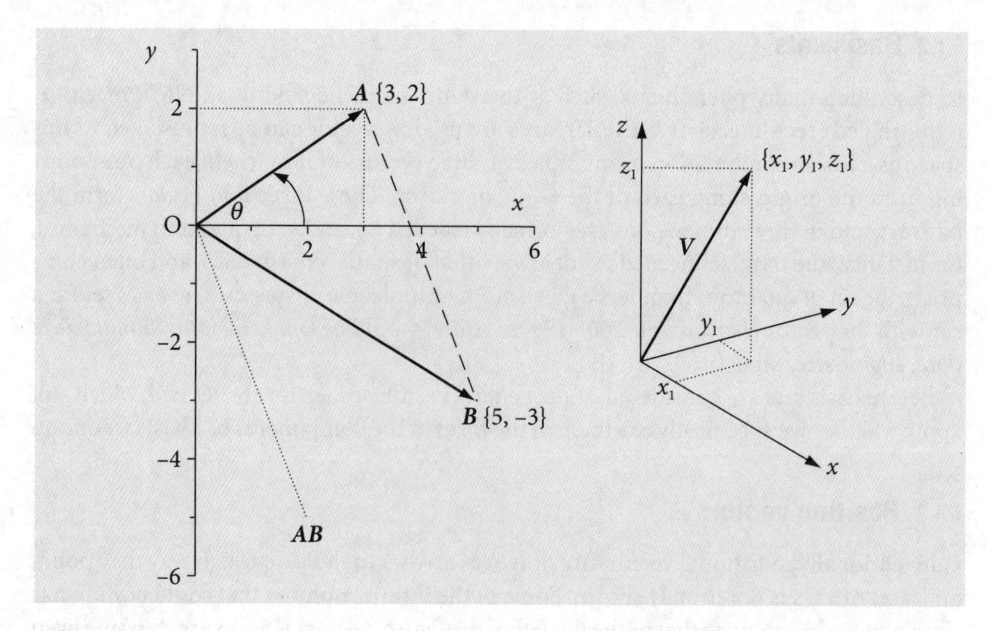

Fig. 6.2 Two- and three-dimensional Cartesian coordinates. Right: A vector ***V*** is shown pointing from the origin to point $\{x_1, y_1, z_1\}$.

In Fig. 6.2, ***OA*** and ***OB*** are vectors and because both start at the origin, they are called *position vectors*. The basis set for ***OA*** and ***OB*** are unit vectors along the *x*- and *y*-axes only, because they are two dimensional. Vector ***OA*** comprises 3 units of the *x*-axis base vector and 2 units of the *y*-axis base vector. Later on, we will find that to solve some problems this basis set will have to be written explicitly as $(\boldsymbol{i}, \boldsymbol{j})$ where $\boldsymbol{i}$ and $\boldsymbol{j}$ are the unit vectors along *x* and *y*, but before doing this the properties of positional vectors are described and some problems solved.

The length of a position vector is, by Pythagoras' theorem,

$$|A| = \sqrt{a_1^2 + a_2^2 + a_3^2}$$

where a_1, a_2, and a_3 are the components; if V (Fig. 6.2, right) ends at $\{x_1, y_1, z_1\} = \{3, 4, 5\}$ then $|V| = \sqrt{9+16+25} = 5\sqrt{2}$. In vector A, $a_1 = 3$ and $a_2 = 2$ or $A = [32]$ and there is no a_3 because A is two dimensional; its length is $\sqrt{13}$. The length or projection of A along the *x*-axis is $A_x = |A|\cos(\theta) = 3$ and along the *y*-axis $A_y = |A|\sin(\theta) = 2$ and the angle $\theta = \tan^{-1}(2/3)$.

The vector AB is defined as $\mathrm{AB} = B - A$ (see Fig. 6.2) and the individual coordinates are subtracted, therefore $AB = [2, -5]$ and has length $|AB| = \sqrt{29} = 5.385$. This vector is not the same as the dashed line in Fig. 6.2 between A and B, but a parallel one starting at the origin and ending at $\{2, -5\}$. This illustrates that each vector has its own set of axes from which it springs; to make AB start at the end of A the origin of its axes have to be placed there, not at zero as our calculation did. The angle between two vectors, say A and AB, can be calculated by trigonometry, but this is clearly awkward and is overcome by learning how to multiply two vectors together. This is described in the next few sections.

6.2 Vector multiplication: dot, cross, and triple products

Multiplying two vectors together can be done in two ways; the result is either a scalar (number), or a vector.

The *dot* or *scalar product* $A{\cdot}B$ of vectors A and B produces a number.

The *cross* or *vector product* $A \times B$ produces another vector.

Triple products, as you would imagine, are more complicated, and can form a vector $A \times (B \times C)$ or a scalar $A{\cdot}B \times C$. They are described in Sections 6.17 and 6.18.

6.2.1 Dot product

The dot product of two vectors A and B is defined as

$$A{\cdot}B = |A||B|\cos(\theta) \tag{6.1}$$

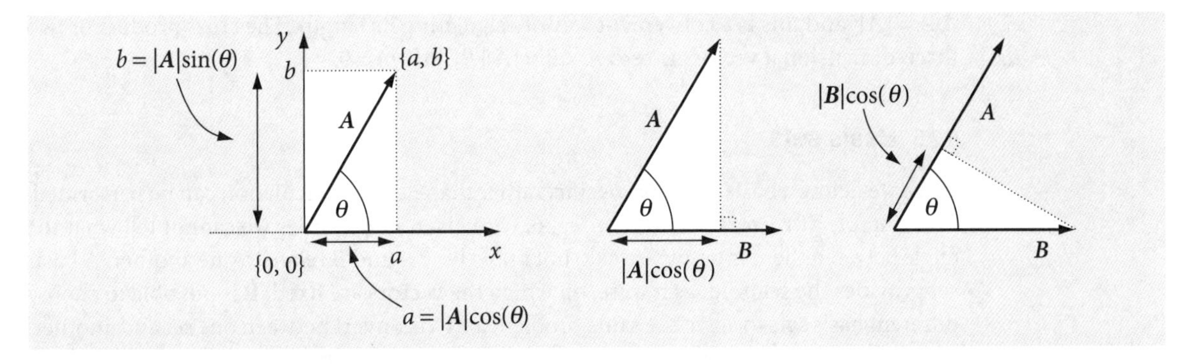

Fig. 6.3 Left: Vector $\boldsymbol{A}$ with components a and b on the x and y axes; $a = |\boldsymbol{A}|\cos(\theta)$, $b = |\boldsymbol{A}|\sin(\theta)$. Right and bottom: Projection of vector $\boldsymbol{A}$ on $\boldsymbol{B}$ which is $|\boldsymbol{A}|\cos(\theta)$ and of $\boldsymbol{B}$ on $\boldsymbol{A}$ which is $|\boldsymbol{B}|\cos(\theta)$.

where $|A|$ is the absolute value of the vector A, which is its length, and θ is the angle between the vectors. Fig. 6.3 shows a two-dimensional vector. The dot product is a number not a vector because the absolute values $|A||B|$ are numbers and as shown in Fig. 6.3, it is the length of A times $|B|\cos(\theta)$, which is the same as the length of B times $|A|\cos(\theta)$.

We can find the angle between any two vectors by rearranging equation (6.1);

$$\cos(\theta) = \frac{A\cdot B}{|A||B|}. \tag{6.2}$$

Before this equation can be evaluated, the vectors A and B have to be represented in a *basis set*, that is, as components of basis vectors. Usually, and for simplicity, we choose a basis set that runs along the x-, y-, and z-axis.

6.2.2 Cross product

The dot product is not the only way we can multiply two vectors, this can be done so that a vector rather than the scalar is produced. The cross product is

$$A \times B = |A||B|\sin(\theta)\vec{n}$$

where $\vec{n}$ is a unit vector perpendicular to A and B. More details are given in Section 6.16. This product vector is at right angles to the other two, consequently the cross product is *unique* to vectors in a three-dimensional space. The cross product, also called vector product, is used to simplify geometric calculations, such as calculating torsion angles between bonds in proteins or other molecules, or the distance of an atom from the plane of a molecule; for instance, an oxygen molecule above the plane of the haem in haemoglobin, see Q6.52. Cross products have other geometric uses, such as calculating the torque when tightening a screw thread, or the angular momentum and velocity of rotating bodies. Cross products also enable us to calculate the area of a triangle or parallelogram.

6.2.3 Perpendicular vectors

If the angle θ between two vectors is 90°, then the dot product is zero

$$A\cdot B = |A||B|\cos(90°) = 0.$$

Because the two vectors are at right angles to one another, the *projection* of A onto B and vice versa is zero, but this does not mean that either $A = 0$ or $B = 0$. If two vectors A and B satisfy the condition, $A\cdot B = 0$ they are said to be *orthogonal*. If the vectors both have a length of unity $|A| = |B| = 1$, they are normalized, and if $A\cdot B = 0$ then they are also orthogonal and together these two conditions make the vectors *orthonormal*.

The cross product of two perpendicular, three-dimensional vectors is not zero, but $A \times B = |A||B|\vec{n}$ and the unit vector $\vec{n}$ is perpendicular to both A and B.

6.2.4 Parallel vectors

If two vectors are parallel, $A\cdot B = |A||B|\cos(0) = |A||B|$, because the angle between them is zero. Therefore, the dot product of a vector with itself is the square of its magnitude,

$A \cdot A = |A|^2$ and this is a convenient way of calculating its length. The cross product of two three-dimensional vectors is zero $A \times B = |A||B|\sin(0)\vec{n} = 0$.

6.2.5 Basis sets

Once we define a basis set then specific, rather than general calculation can be performed. For example, if vectors A and B of Fig. 6.3 are given magnitudes we cannot tell what the angle between them is because we do not know how A and B relate to one another. A basis set provides the scaffold, as it were, on which the vectors are fixed. It is possible to choose different basis sets to solve the same problem and to convert between one set and another. The following three properties of vectors and basis sets are important:

(i) Every vector is always made from a *linear combination of the components* of its *basis set.*

(ii) Because a basis set is the minimal possible set, it cannot be decomposed any further or made any simpler.

(iii) Any vector is unchanged by changing its basis set and therefore its coefficients must change between bases.

For example, the vector A in Fig. 6.3 has components of a units on the x-axis and b on the y and is therefore described $A = ai + bj$. In Cartesian coordinates, the basis set most often used is made of *orthogonal, unit vectors* along each of the x-, y- and z-axes; conventionally we represent this as the basis set (i, j, k) where i, j and k are *orthogonal unit vectors.* Fig. 6.4 shows the basis set vectors and a three-dimensional vector separated into it basis set components. A unit vector has length of 1, so a vector can be made from multiples of this. The (i, j, k) basis set is described more detail in Section 6.3.

Any vector can always be represented in several different but equivalent ways depending on which is easiest. We can choose to manipulate a three-dimensional vector in multiples of unit basis vectors i, j, and k or transform it into column or row matrices. The three-dimensional i, j, and k basis set written as three single column matrices and in standard form is

$$(i, j, k) \equiv \left(\begin{bmatrix} 1 \\ 0 \\ 0 \end{bmatrix}, \begin{bmatrix} 0 \\ 1 \\ 0 \end{bmatrix}, \begin{bmatrix} 0 \\ 0 \\ 1 \end{bmatrix} \right),$$

where the i, j, and k unit vectors are $i = \begin{bmatrix} 1 \\ 0 \\ 0 \end{bmatrix}, j = \begin{bmatrix} 0 \\ 1 \\ 0 \end{bmatrix}, k = \begin{bmatrix} 0 \\ 0 \\ 1 \end{bmatrix}$.

A vector $2i + 3j - k$ put into the column matrix form of the basis set is represented as $\left(\begin{bmatrix} 2 \\ 0 \\ 0 \end{bmatrix}, \begin{bmatrix} 0 \\ 3 \\ 0 \end{bmatrix}, \begin{bmatrix} 0 \\ 0 \\ -1 \end{bmatrix} \right)$ or more commonly and more concisely $\begin{bmatrix} 2 \\ 3 \\ -1 \end{bmatrix}$. If we choose the basis set to be row matrices then the same vector is $[2 \quad 3 \quad -1]$.

The matrix formulation of vectors is very adaptable and not restricted to three dimensions; in fact, any number of dimensions can be used. However, the (i, j, k) basis set can only be used up to three dimensions.

The elements of basis set can be numbers, or they could be a vector, a matrix, or a function or a set of functions. Some examples are considered more fully after some basic properties have been described. Using functions as a basis set is described in Chapter 9. For the present, we shall assume that a suitable basis set always exists and perform some calculations.

6.2.6 Origin

If it is not otherwise specified, it is assumed that the origin of a vector is at the centre of the coordinates, i.e. at {0, 0, 0}. If this is not the case and the origin is at some other place, $\{a, b, c\}$ then we must subtract a, b, and c from each of the three components and use $[x - a \quad y - b \quad z - c]$ as our vector. This has the effect of moving the origin to the base of our vector. In addition, when subtracting two sets of coordinates to make a vector, its base is now at the origin of the new set of coordinates, i.e. the origin has again been shifted. This is important; a vector always starts at the origin of its own local coordinates, which may not be the same as initially drawn on a diagram, so some care is needed; vectors are only a direction and length in space; they carry around their own coordinate system or basis set with them but this is hardly ever drawn.

6.2.7 Magnitude

The magnitude of a vector A, which may be imagined as its length, is used as $|A|$. If $A = [a \quad b \quad c]$ where A has components a, b, and c in the 3-dimensional basis set (x, y, z), then

$$|A| = \sqrt{a^2 + b^2 + c^2}. \tag{6.3}$$

We can also calculate the magnitude as,

$$|A| = \sqrt{A \cdot A}, \tag{6.4}$$

because any vector is parallel to itself and therefore $\theta = 0$ in the dot product. This latter equation is very useful indeed when using the computer to do calculations, because in Maple and other programs the dot product is an in-built function.

6.2.8 Vector addition

Addition follows the familiar rules, *except* that each element in the vector adds separately; for example, the three-dimensional vectors represented by the two sets of coordinates is

$$\{1, \quad -3, \quad 4\} + \{-2, \quad 4, \quad -5\} = \{-1, \quad 1, \quad -1\}$$

and subtracting these two vectors produces

$$\{1, \quad -3, \quad 4\} - \{-2, \quad 4, \quad -5\} = \{3, \quad -7, \quad 9\},$$

which, in effect, is what is done when drawing out a vector triangle. The same rule applies when adding matrices. In row and column matrix-vector notation, the same vector addition is written as

$$[1 \quad -3 \quad 4] + [-2 \quad 4 \quad -5] = [-1 \quad 1 \quad -1]$$

or

$$\begin{bmatrix} 1 \\ -3 \\ 4 \end{bmatrix} + \begin{bmatrix} -2 \\ 4 \\ -5 \end{bmatrix} = \begin{bmatrix} -1 \\ 1 \\ -1 \end{bmatrix};$$

Addition is commutative,

$$A + B = B + A,$$

and associative,

$$A + (B + C) = (A + B) + C.$$

This means that vectors can be added or subtracted in any order.

6.2.9 Dot products or inner products in matrix-vector form

The dot product using vectors in a three-dimensional basis set is calculated by defining the vectors as one-dimensional matrices, multiplying element by element and summing the result,[2]

$$A \cdot B = [a \quad b \quad c] \begin{bmatrix} d \\ e \\ f \end{bmatrix} = ad + be + cf. \tag{6.5}$$

[2] Chapter 7 on matrices explains this and other matrix multiplications.

In a dot product, the left-hand of the two vectors is always written as a row and the right always a column. The vector dot product is also called the vector *inner product*.

Changing a row into a column vector is called *transposing*, and this operation is indicated by a superscript T. In quantum mechanics, the complex conjugate of each element in the row is always made when the dot product is formed from column vectors. The transpose is

$$[a \quad b \quad c]^T = \begin{bmatrix} a \\ b \\ c \end{bmatrix}$$

The angle between two vectors is $\cos(\theta) = (\boldsymbol{A}\cdot\boldsymbol{B})/|\boldsymbol{A}||\boldsymbol{B}|$ equation (6.2), which can be *expanded in the basis set* giving

$$\cos(\theta) = \frac{(ad + be + cf)}{\sqrt{a^2 + b^2 + c^2}\sqrt{d^2 + e^2 + f^2}}. \tag{6.6}$$

The normalization terms on the bottom of this equation are just the length of each vector as calculated by Pythagoras' theorem. The angle between two vectors $\boldsymbol{A} = [3 \quad 2 \quad 1]$ and $\boldsymbol{B} = [-1 \quad 0 \quad 1]$ is calculated as follows: the length of each vector is $|\boldsymbol{A}| = \sqrt{9+4+1} = \sqrt{14}$ and $|\boldsymbol{B}| = \sqrt{1+0+1} = \sqrt{2}$, and the dot product is $[3 \quad 2 \quad 1]\begin{bmatrix} -1 \\ 0 \\ 1 \end{bmatrix} = -2$ and the angle $\cos(\theta) = \dfrac{-2}{\sqrt{14}\sqrt{2}}$, which is 1.958 radians or 112.2°.

6.2.10 Unit vectors and normalizing vectors

Unit vectors, as their name suggests, have a length of 1. A vector can be made into a unit vector by dividing by its length. If a vector is $\boldsymbol{v} = [a \quad b \quad c \quad d]$ the normalizing condition is

$$N^2(a^2 + b^2 + c^2 + d^2) = 1, \tag{6.7}$$

where N is the normalization constant. If, for example, $a = 2$, $b = -4$, $c = 5$, and $d = 1$, then $N^2(4 + 16 + 25 + 1) = 1$ or $N = 1/\sqrt{46}$ and the normalized or unit vector is $[0.295 \quad 0.589 \quad 0.737 \quad 0.147]$.

A second method is to use the dot product formula on the same vector, $\boldsymbol{v}\cdot\boldsymbol{v}$, and because these must be parallel, $\cos(\theta) = 1$, then,

$$\boldsymbol{v}_n = \frac{\boldsymbol{v}}{|\boldsymbol{v}|} = \frac{\boldsymbol{v}}{\sqrt{\boldsymbol{v}\cdot\boldsymbol{v}}}, \tag{6.8}$$

where $|\boldsymbol{v}|$ is the absolute value or length of the vector,

$$|\boldsymbol{v}| = \sqrt{\boldsymbol{v}\cdot\boldsymbol{v}} = \sqrt{a^2 + b^2 + c^2 + d^2}. \tag{6.9}$$

The angle θ between any two *normalized* vectors $\boldsymbol{v}_{n1}$ and $\boldsymbol{v}_{n2}$ is

$$\cos(\theta) = \boldsymbol{v}_{n1}\cdot\boldsymbol{v}_{n2}. \tag{6.10}$$

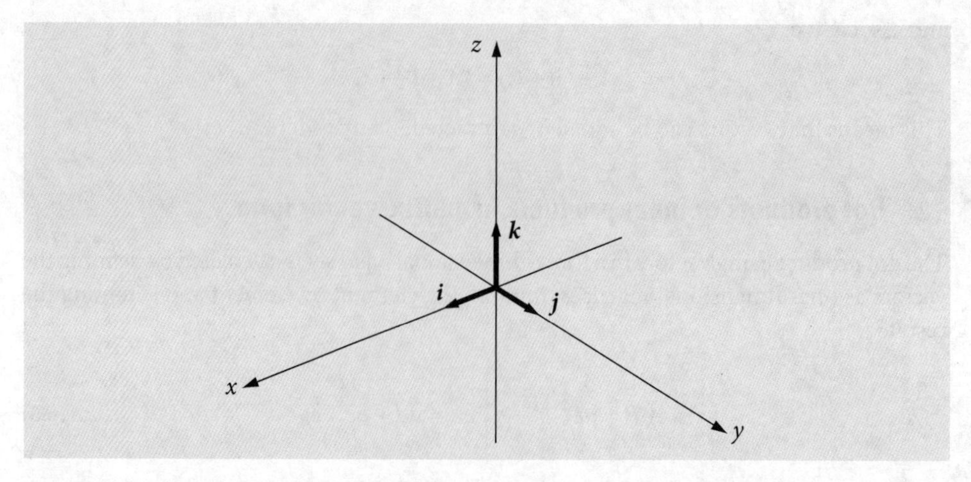

Fig. 6.4 The orthogonal and normalized base vectors (***i***, ***j***, ***k***).

6.3 The orthonormal *i, j, k* base vectors

A general method for describing a three-vector is to express it in terms of vectors along a set of axes. If the mutually perpendicular set of x-, y-, and z-axes is chosen, the basis set is orthogonal set. If the base vectors are also normalized to 1, each base vector is a unit vector and the set is orthonormal. Conventionally, orthonormal base vectors are labelled $\boldsymbol{i}, \boldsymbol{j}$, and $\boldsymbol{k}$ and any vector in this basis is written as,

$$\boldsymbol{v} = a\boldsymbol{i} + b\boldsymbol{j} + c\boldsymbol{k}$$

magnitude ↙ ↘ direction

which is interpreted as having a length (magnitude) of a along the x-axis and of b and c along the y- and z-axes respectively, as shown in Fig. 6.4 and Fig. 6.5 the magnitude, or length, of the vector is $\sqrt{a^2 + b^2 + c^2}$.

To be clear: any vector such as $\boldsymbol{v}$ is the *linear combination of base vectors* $\boldsymbol{i}, \boldsymbol{j}$, and $\boldsymbol{k}$ taken in proportions a, b, and c. This is why the vectors $\boldsymbol{i}, \boldsymbol{j}$, and $\boldsymbol{k}$ are called the *base* vectors or the *basis set* for describing other vectors.

If two vectors are

$$\boldsymbol{v} = 3\boldsymbol{i} - 4\boldsymbol{j} + \boldsymbol{k} \qquad \text{and} \qquad \boldsymbol{w} = 5\boldsymbol{i} + 8\boldsymbol{j} - 10\boldsymbol{k},$$

their dot product is $\boldsymbol{v}\cdot\boldsymbol{w} = -27$. By multiplication, the following sum results:

$$\boldsymbol{v}\cdot\boldsymbol{w} = 15\boldsymbol{i}\cdot\boldsymbol{i} - 32\boldsymbol{j}\cdot\boldsymbol{j} - 10\boldsymbol{k}\cdot\boldsymbol{k} + 24\boldsymbol{i}\cdot\boldsymbol{j} - 30\boldsymbol{k}\cdot\boldsymbol{j} - 20\boldsymbol{j}\cdot\boldsymbol{i} + 40\boldsymbol{j}\cdot\boldsymbol{k} + 5\boldsymbol{k}\cdot\boldsymbol{i} + 8\boldsymbol{k}\cdot\boldsymbol{j}$$

At first sight, this seems a very complicated way of finding the dot product. However, the full multiplication is not necessary for two reasons; first, because $\boldsymbol{i}, \boldsymbol{j}$, and $\boldsymbol{k}$ are themselves orthogonal vectors of length 1, i.e. normalized so the dot product is 1,

$$\boldsymbol{i}\cdot\boldsymbol{i} = \boldsymbol{j}\cdot\boldsymbol{j} = \boldsymbol{k}\cdot\boldsymbol{k} = 1$$

because any vector is clearly parallel to itself and $\cos(0) = 1$. Secondly, the base vectors $\boldsymbol{i}, \boldsymbol{j}$, and $\boldsymbol{k}$ are orthogonal, because they lie along axes mutually set at 90°; therefore $\boldsymbol{i}\cdot\boldsymbol{j} = \boldsymbol{j}\cdot\boldsymbol{k} = \boldsymbol{i}\cdot\boldsymbol{k} = 0$. We can calculate a dot product fairly simply with the following rules:

$$\boldsymbol{i}\cdot\boldsymbol{i} = \boldsymbol{j}\cdot\boldsymbol{j} = \boldsymbol{k}\cdot\boldsymbol{k} = 1, \tag{6.11}$$

$$\boldsymbol{i}\cdot\boldsymbol{j} = \boldsymbol{j}\cdot\boldsymbol{k} = \boldsymbol{i}\cdot\boldsymbol{k} = 0. \tag{6.12}$$

Stated in words:

'Calculate the dot product by multiplying both of the $\boldsymbol{i}$ vector terms together, making $\boldsymbol{i}\cdot\boldsymbol{i} = 1$, do the same for the $\boldsymbol{j}$ and $\boldsymbol{k}$ terms. Ignore, or make zero, all other cross terms, such as, $\boldsymbol{i}\cdot\boldsymbol{j}$, $\boldsymbol{i}\cdot\boldsymbol{k}$ and $\boldsymbol{j}\cdot\boldsymbol{k}$.'

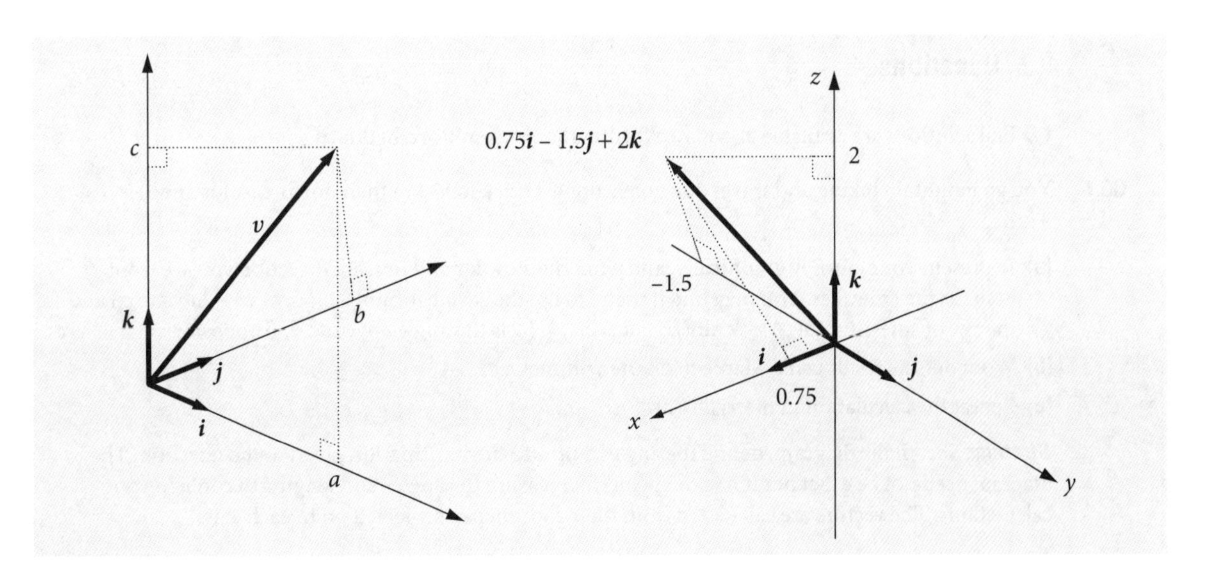

Fig. 6.5 Unit vectors $\boldsymbol{i}, \boldsymbol{j}, \boldsymbol{k}$, and the vector, $\boldsymbol{v} = a\boldsymbol{i} + b\boldsymbol{j} + c\boldsymbol{k}$, which has length a, b and c along each of the axes shown and which are the projections of $\boldsymbol{v}$ on the axes. Right: the vector $0.75\boldsymbol{i} - 1.5\boldsymbol{j} + 2\boldsymbol{k}$.

Therefore, for our example

$$v \cdot w = 15 - 32 - 10 = -27,$$

and this is the same result as would be obtained from the vector dot product equation (6.5)

$$[3 \quad -4 \quad 1]\begin{bmatrix} 5 \\ 8 \\ -10 \end{bmatrix} = -27.$$

This is an important point as it means we have two ways of calculating vectors. When vectors have more than three dimensions, the matrix methods become far simpler, particularly when using the computer to do calculations.

6.4 Summary

Dot product $A \cdot B = |A||B|\cos(\theta)$. Magnitude $|A| = \sqrt{A \cdot A}$

Vector defined as $A = [a \quad b \quad c \quad d]$ with base vectors $\left(\begin{pmatrix} 1 \\ 0 \\ 0 \\ 0 \end{pmatrix}, \begin{pmatrix} 0 \\ 1 \\ 0 \\ 0 \end{pmatrix}, \begin{pmatrix} 0 \\ 0 \\ 1 \\ 0 \end{pmatrix}, \begin{pmatrix} 0 \\ 0 \\ 0 \\ 1 \end{pmatrix}\right)$

has dot product $A \cdot B = [a \quad b \quad c \quad d]\begin{bmatrix} e \\ f \\ g \\ h \end{bmatrix} = ae + df + cg + dh.$

In three dimensions only, in addition to the matrix method, orthonormal base vectors $\boldsymbol{i}, \boldsymbol{j}, \boldsymbol{k}$. can be used, e.g. $A = a\boldsymbol{i} + b\boldsymbol{j} + c\boldsymbol{k}$.

$$A \cdot B = (a\boldsymbol{i} + b\boldsymbol{j} + c\boldsymbol{k}) \cdot (c\boldsymbol{i} + d\boldsymbol{j} + e\boldsymbol{k}) = ac + bd + ce$$

where $\boldsymbol{i} \cdot \boldsymbol{i} = \boldsymbol{j} \cdot \boldsymbol{j} = \boldsymbol{k} \cdot \boldsymbol{k} = 1 \qquad \boldsymbol{i} \cdot \boldsymbol{j} = \boldsymbol{j} \cdot \boldsymbol{k} = \boldsymbol{i} \cdot \boldsymbol{k} = 0.$

Cross product exists only in three dimensions

$A \times B = |A||B|\sin(\theta)\tilde{n}$, where $\tilde{n}$ is a unit vector perpendicular to A and B.

6.5 Questions

Full solutions are available at **www.oxfordtextbooks.co.uk/orc/beddard.**

Q6.1 You go mountain biking and travel due north along a track for 2 km then north-east for another 3 km.

(a) Represent your journey graphically, and work out how far and in what direction you are from home as the crow flies. You may find it useful to use the cosine formula, $c^2 = a^2 + b^2 - 2ab\cos(\alpha)$ and the law of sines, $ab\sin(\alpha) = bc\sin(\beta) = ac\sin(\chi)$. ($\alpha$ is the angle opposite a, β opposite b).

(b) Work out the result using Maple to do the arithmetic.

(c) Repeat the calculation in matrix form.

Strategy: Sketch the diagram, define the angles, and add the information given in the question. The diagram need not be exact but if it is then you can measure the angles and lengths to confirm your calculations. The vectors are calculated from their coordinates projected on to each axis.

Q6.2 Six forces $f_1 \cdots f_6$, act simultaneously on a body. What force must be applied to prevent the body from moving?

Strategy: If a body does not move, then the resultant of all forces acting on it must be zero. The resultant force must be the equal and opposite of the six applied forces.

Q6.3 On a windless day, an aeroplane travels at 200 mph to its destination R in a north-westerly direction. On another day, the wind is blowing from the west at 50 mph. In what direction and how fast should the plane now fly to reach its destination?

Strategy: Draw out the vector diagram and sum the vectors. Let O be the origin. The plane travels along the line $\boldsymbol{OR}$ over the ground with its nose pointing somewhat into the wind. To compensate for the wind, the plane thus points vectorially along the course $\boldsymbol{OP}$ which has a larger angle from the north and so is to the west of $\boldsymbol{OR}$.

Q6.4 Using the triangle in Fig. 6.6, prove the cosine formula $c^2 = a^2 + b^2 - 2ab\cos(\chi)$, where a, b, and c are the lengths of the vectors $\boldsymbol{A}$, $\boldsymbol{B}$, and $\boldsymbol{C}$ respectively.

Strategy: The simplest way of finding a^2, b^2, and c^2 is with a dot product.

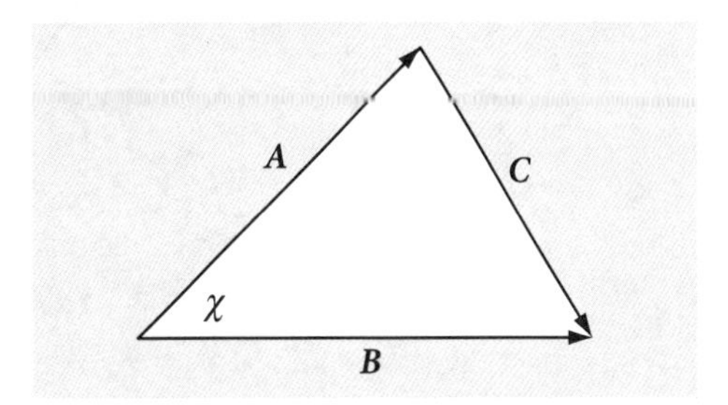

Fig. 6.6 Construction for cosine formula.

Q6.5 Use vectors to show that the internal angle of a triangle ($\angle ACB$), Fig. 6.7, whose hypotenuse is the diameter of a semicircle, is 90°. The equation of a circle centred at the origin is $x^2 + y^2 = r^2$.

Strategy: Let the triangle be ABC. If the angle ACB is a right angle, then the dot product of vector $\boldsymbol{AC}$ and $\boldsymbol{CB}$ is zero; it is necessary, therefore, to find these vectors with the restriction that C always lies on the circumference of the circle. Define the origin of the coordinates as the centre of the circle; O then OC, OA, OB are radii of length r. Use two-dimensional vectors in the basis set $\boldsymbol{i}$ and $\boldsymbol{j}$ or in matrix-vector form. Find $\boldsymbol{CA}\cdot\boldsymbol{CB}$, it is zero if they are at 90°.

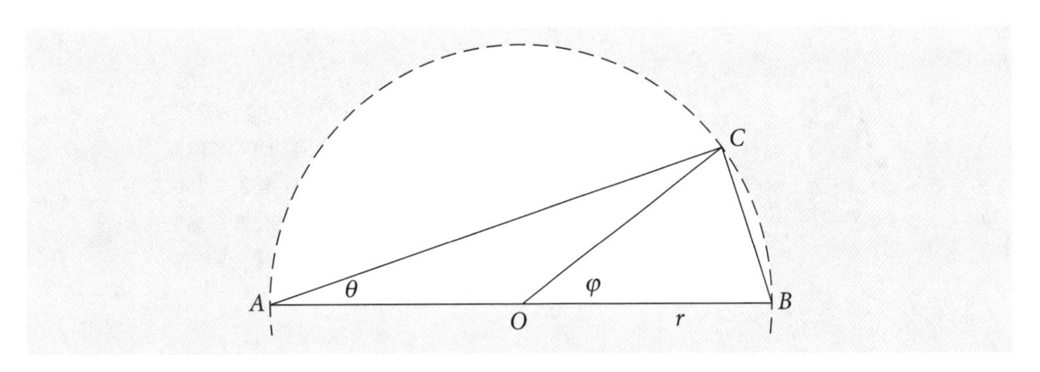

Fig. 6.7 Construction to show that an internal angle is twice that at circumference.

Q6.6 Use vectors to prove that that the angle at the centre of the circle φ, (see Q6.5) is twice that at the circumference θ.

Strategy: Calculate both φ and θ using dot products, then use the identity $\cos(2\alpha) = 2\cos^2(\alpha) - 1$, that is true for any angle α, to prove that $\varphi = 2\theta$.

Q6.7 **(a)** For what value of a is $\boldsymbol{A} = a\boldsymbol{i} + 2\boldsymbol{j} - \boldsymbol{k}$ perpendicular to $\boldsymbol{B} = 3a\boldsymbol{i} + a\boldsymbol{j} + 11\boldsymbol{k}$?

(b) Can these vectors be made parallel?

Q6.8 The sides of a triangle, Fig. 6.8 are given by $A = [3 \quad 2 \quad -1]$ and $B = [3 \quad -5 \quad 6]$. Determine its angles. This shows how easy a calculation is when basis sets are used.

Strategy: Find the vector of the third side and calculate two dot products; the third angle can be determined because the sum of the internal angles is 180°.

Q6.9 Find the angle between vectors $A = [2 \quad 3 \quad 4]$ and $B = [4 \quad 3 \quad 2]$.

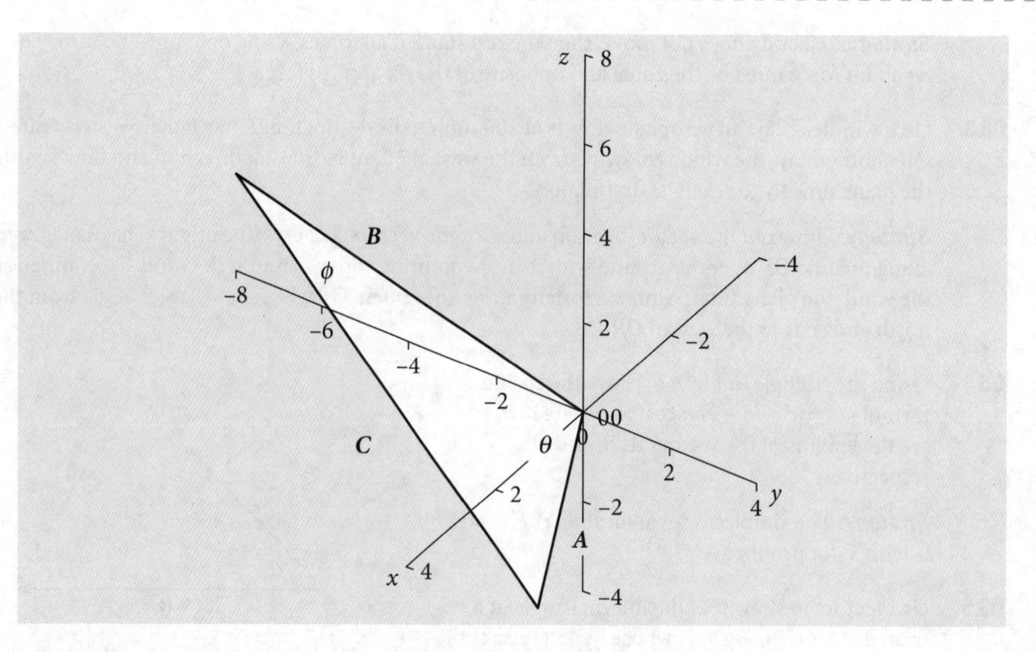

Fig. 6.8 Triangle with sides ABC.

Q6.10 The atomic coordinates in Fig. 6.9 form part of a protein backbone. Calculate the angles O-C-C_A, C-C_A-N and bond lengths CO and C_AN.

Strategy: For each angle, convert the three sets of coordinates into two vectors with the start of each vector at the central atom, then calculate the dot product.

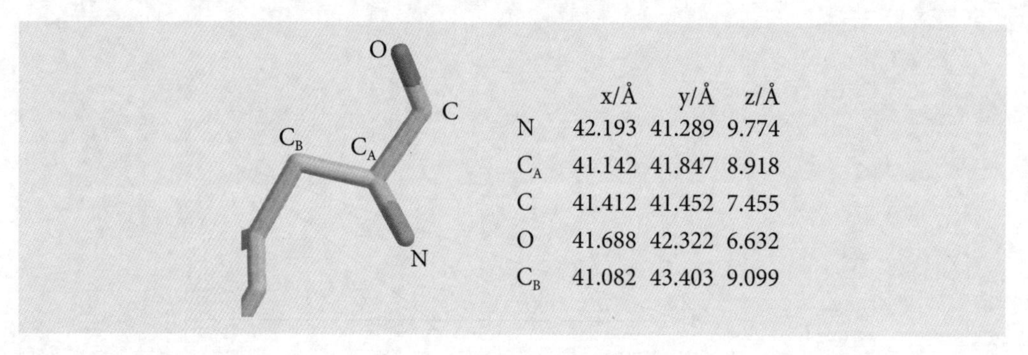

	x/Å	y/Å	z/Å
N	42.193	41.289	9.774
C_A	41.142	41.847	8.918
C	41.412	41.452	7.455
O	41.688	42.322	6.632
C_B	41.082	43.403	9.099

Fig. 6.9 Part of protein backbone.

Q6.11 A straight line joins point {2, 3, −6} with {3, −2, 4}; find the acute angles this line makes with the axes. The cosines of these angles are called *direction cosines*, and they are the dot product of the vector with each axis considered to be a vector.

Strategy: The points are made into a vector defined as the difference of one from another. Let the x-, y-, and z-axes have unit vectors pointing along them, for example $x = [1 \quad 0 \quad 0]$.

Q6.12 Find the acute angle θ between the two diagonals of a cube. Show that the obtuse angle is 109.47°, which is that for tetrahedral sp^3 hybrid orbitals and found in saturated hydrocarbons.

Strategy: Draw a diagram and add coordinates to the cube. Use the dot product between two vectors to calculate the angle.

Q6.13 Fig. 6.10 shows part of the backbone of a carbon polymer chain drawn within two cubes, and with each bond of length d. Calculate the angles θ, ϕ, and χ the angle to the y-axis, and the bond angle α.

Strategy: Define the sides and diagonals as three-dimensional vectors. Choose O as the origin to begin with. The basis set is that of three orthogonal vectors along the x-, y-, and z-axes. To calculate α move this carbon atom to the origin and recalculate vectors ***go*** and ***gf***.

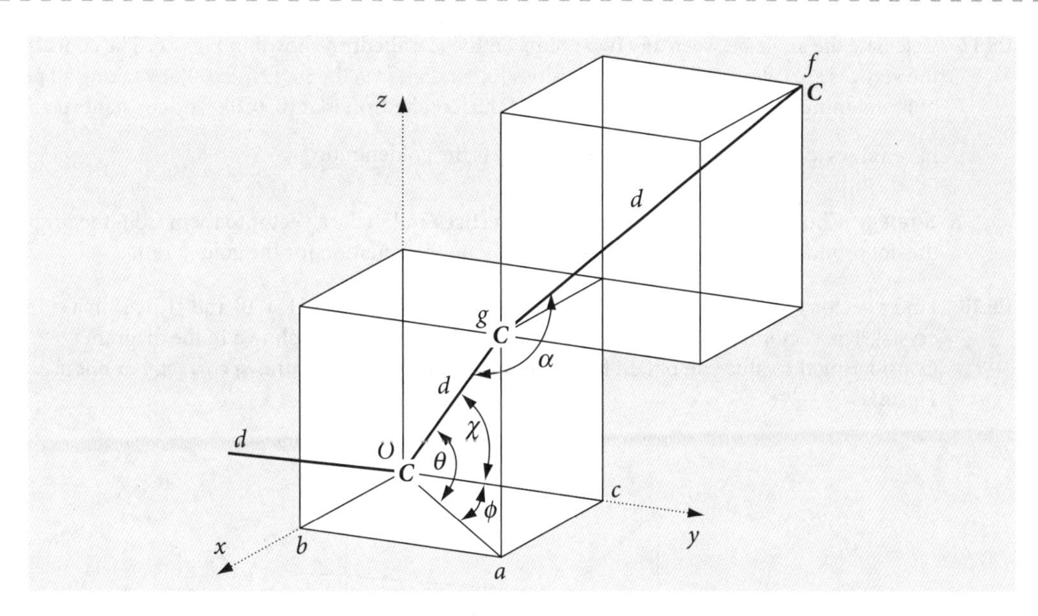

Fig. 6.10 Polymer backbone of tetrahedral carbons atoms and superimposed cubes.

Q6.14 All methane molecules have tetrahedral symmetry. The angle between each bond is ≈ 109°. A tetrahedron is drawn by placing the carbon atom at the centre of a cube, and the H atoms at opposite corners. The side of the cube is chosen to have length $2a$ and the C atom is chosen to be the origin.

(a) If the CH bond length is 1.57Å, what is the distance between any two H atoms?

(b) Calculate the *exact* bond angles.

Strategy: The C atom should be placed at the origin {0, 0, 0}, then the H atoms can be placed at the corners of a cube of side $2a$. (Choosing the length of a side to be $2a$ makes the arithmetic easier.) Using the information given, work out the other coordinates of the H atoms and use Pythagoras' theorem to calculate distances. Use dot products to calculate angles.

Q6.15 Using the data below, calculate the bond lengths and angles in the water molecule. Round the answers to the appropriate number of significant figures. The coordinates in Å are

	x	y	z
O	2.0317	−1.1893	−1.0464
H_1	1.1374	−1.0385	−0.7880
H_2	2.5201	−0.4437	−0.7383

Q6.16 The atomic coordinates of ammonia are listed below.

(a) Calculate the bond lengths and the HNH angles.

(b) Calculate how far out of the plane of the H atoms the N atom is in ammonia, and hence the size of this atom's motion in its vibrational inversion or umbrella mode. Assume that the molecule has threefold symmetry. The coordinates are given below, but are only accurate to three significant figures.

	N	H_1	H_2	H_3
x	−0.0200	0.5390	0.1972	−0.9788
y	−0.9149	−0.0887	−1.5122	−0.6416
z	0.1781	0.0957	−0.6032	0.0954

Strategy: (a) The bond lengths and angles are calculated as in the previous problem. Because the molecule has threefold symmetry, the height that the N atom is above the plane of the H atoms, is calculated by finding the mid-point of the plane, making a vector to this point from the N atom, and calculating the distance.

Q6.17 Calculate the angle between any two points on the icosahedron shown in Fig. 1.6. The coordinates of the vertices are calculated to have the values $\{0, \pm u, \pm v\}$, $\{\pm u, 0, \pm v\}$, $\{\pm u, \pm v, 0\}$, by taking all possible plus and minus combinations. The centre of the icosahedron is at $\{0, 0, 0\}$. The constants u and v have values $u^2 = \gamma/\sqrt{5}$ and $v^2 = \dfrac{1}{\gamma\sqrt{5}}$, where γ is the golden ratio.

Strategy: Choose two points from the set of vertices and make a vector to them from the origin. Use the dot product formula to find the angle. Look up the equation for the golden ratio.

Q6.18 Find a vector $\boldsymbol{p}$, perpendicular to the vector $\boldsymbol{v}$, connecting points $\{1, 1, 0\}$ and $\{1, 1, 1\}$ in a cubic crystal. The vector $\boldsymbol{v}$ and *one* of many perpendicular vectors $\boldsymbol{p}$ are shown in the diagram. Considering the value you obtain for $\boldsymbol{p}$, comment on whether $\boldsymbol{p}$ is drawn correctly or not in Fig. 6.11.

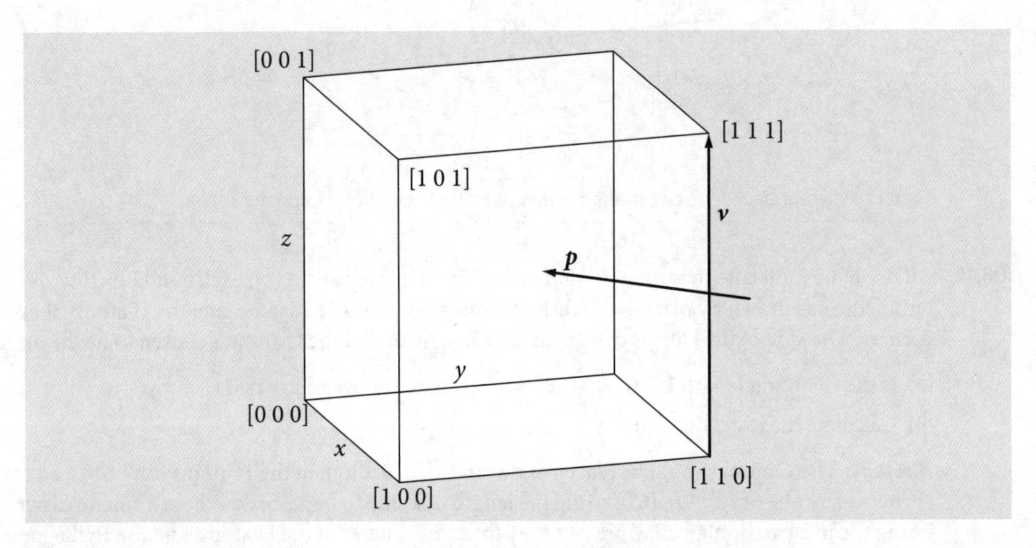

Fig. 6.11 Perpendicular vectors.

Strategy: If one vector is perpendicular to another, then the angle between them is 90° and their dot product is zero. Choose the vector $\boldsymbol{p}$ to have any general component values $[p_1\, p_2\, p_3]$, then work out what these are as far as is possible.

Q6.19 NaCl normally grows as a cubic crystal but crystals of the shape of a truncated cube, Fig. 6.12, are formed as the 'habit' if NaCl crystals are grown in the presence of urea. Triangular pyramidal shapes are missing from the cube; the remaining shapes and similar ones when corners or edges are missing, are called coigns, and can sometimes be seen on the sides of old houses. Vertex A is at a distance 1 from the centre, so that a sphere of unit radius touches each point equivalent to A.

(a) If the three vertices A, B, C of the triangle in the centre of the figure are $\{u, v, w\}$, $\{u, w, v\}$, and $\{w, v, u\}$, respectively, where $u = v = 0.6786$ and $w = 0.2811$, show that the angle subtended from the centre at $\{0, 0, 0\}$ is 32.6° between each vertex A, B, C.

(b) Calculate the coordinates of the point p at the centre of the triangle ABC, and show that the plane ABC is perpendicular to the line from the centre to p.

Strategy: (a) Draw out and label a diagram of triangle OAB to help you visualize the problem, and then use the dot product formula to find the angles.

Q6.20 Suppose that X-rays of wavelength λ are diffracted off a line of atoms equally spaced by d Å. Diffraction occurs when the paths taken by two rays scattered off the atom's electrons, are an integer multiple of the wavelength. i.e., $n\lambda$ where $n = 0, \pm 1, \pm 2 \cdots$ and so forth. One possible geometry is shown in Fig. 6.13; the lines representing the diffracted X-rays are the limits of a cone of diffracted rays whose axis is the line of atoms, as shown on the right.

(a) Calculate the Laue equation $d[\cos(\theta) - \cos(\phi)] = n\lambda$, by considering the path difference between AB and CD, and therefore the angle θ diffracted from the line of atoms for the zero and ± 1 and ± 2 orders.

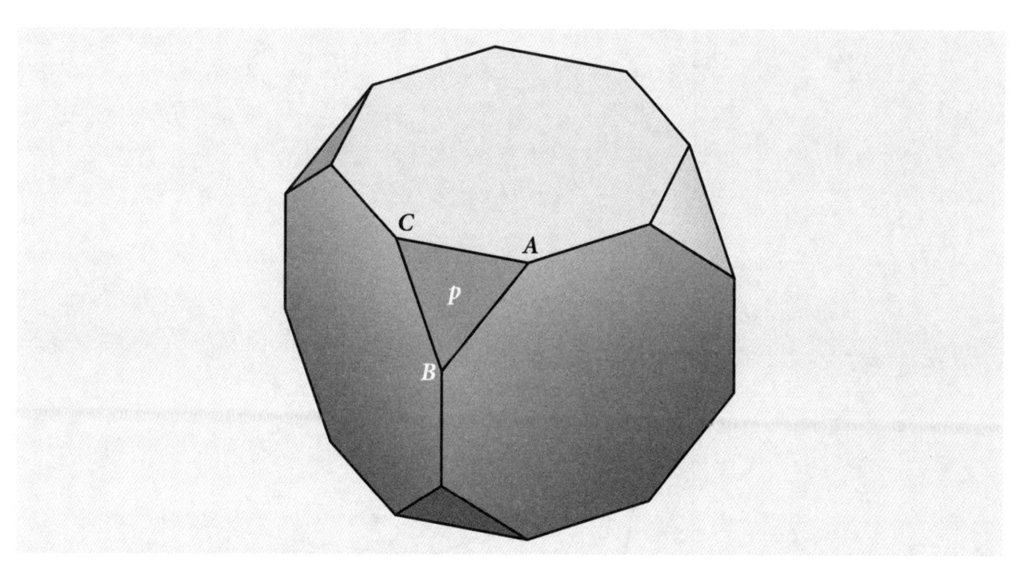

Fig. 6.12 Truncated cube or hexahedron.

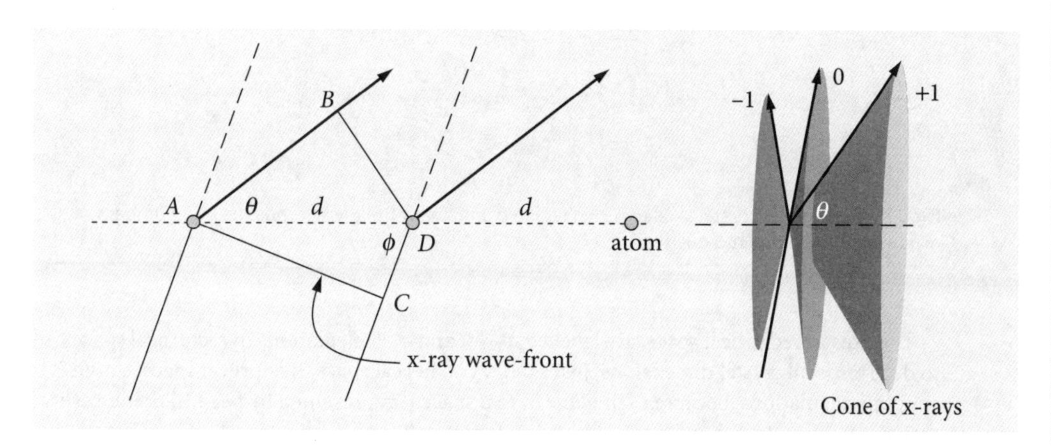

Fig. 6.13 Geometrical construction showing X-rays diffracted off a row of atoms, and right the cones of diffracted X-rays generated for the −1, 0, and +1 diffraction orders.

(b) Calculate all possible input angles ϕ, and diffraction angles θ, by plotting a graph. Assume that $d = 4$ Å, and that Cu Kα1 radiation is used, which has a wavelength $\lambda = 1.54060$ Å.

***Strategy*:** (a) Use the definition of the cosine to calculate the distances AB and CD; their difference must be $n\lambda$. (b) Not all angles ϕ and θ will satisfy the Laue equation, therefore plot a graph to find out what these are.

Q6.21 The diagram Fig. 6.14 shows the lattice planes of a orthorhombic crystal and their Miller indices $(h\,k\,l)$. The crystal's *unit cell* axes are $\boldsymbol{a}$, $\boldsymbol{b}$, and $\boldsymbol{c}$. The Miller index of the $\boldsymbol{a}$ axis is

$$h = a/a'$$

where a' is the intercept in units of $\boldsymbol{a}$ (i.e. fraction of $\boldsymbol{a}$) where the plane intersects the axis; similar equations apply for the k and l axes. The lower plane has indices

$$h = a/(a/2) = 2, \qquad k = b/(b/2) = 2, \qquad \text{and} \qquad l = c/(c/4) = 4,$$

or $(h\,k\,l) = (2\,2\,4)$ which is a plane parallel to the plane (1 1 2) and (4 4 8).

If the intercept is infinity, the plane is parallel to one or more of the axes and the index is zero. By convention, negative indices are represented with a bar over the number in the Miller index; e.g. $(\bar{2}\ \ \bar{1}\ \ 0)$ has fractional intercepts of −1/2, −1, and infinity. If any fractions remain after the calculation, they are cleared to generate Miller indices with the lowest whole numbers.

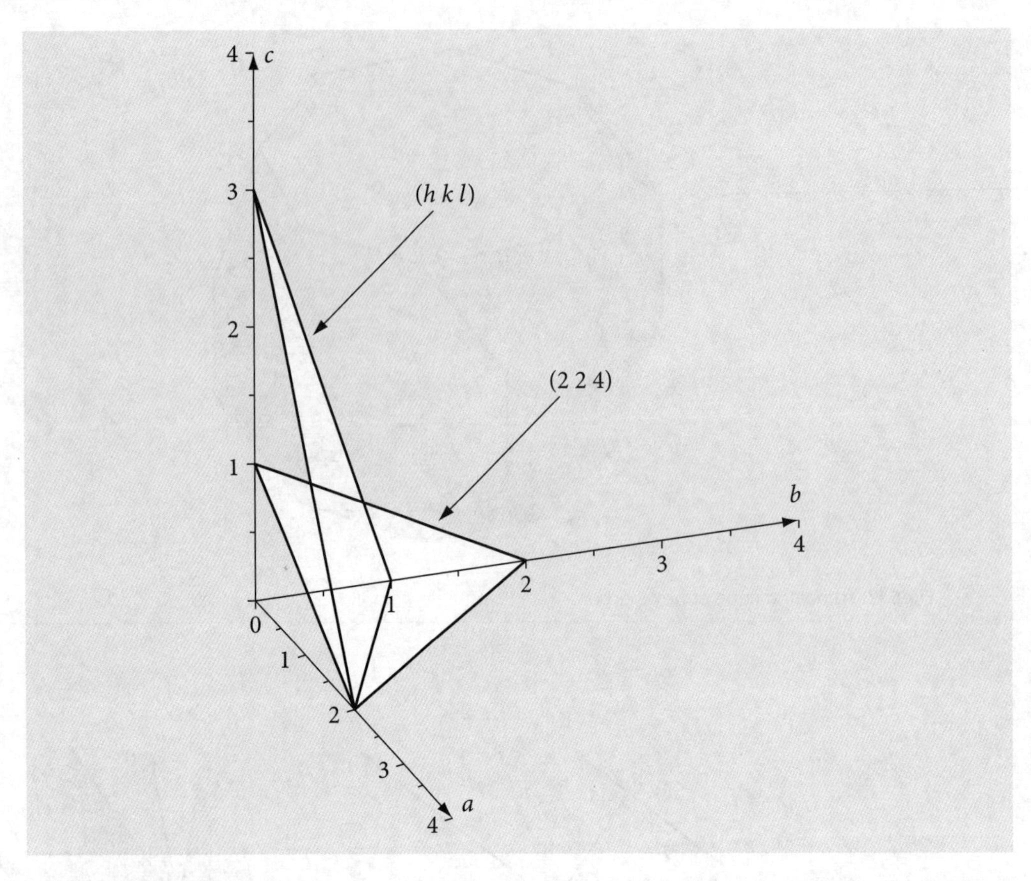

Fig. 6.14 Orthogonal crystal axes and two lattice planes with their Miller indices. The numbers on the axes divide each total axis into quarters.

The axes in reciprocal space, are labelled a^*, b^*, and c^*. Calculations in reciprocal space can be used to determine all crystal dimensions, just as they can in real space. Reciprocal space is used is to simplify calculations, because each *plane* in real space is represented by *one* Miller index $(h\,k\,l)$ in reciprocal space. In cubic crystals, reciprocal lattices offer no particular advantage but, of course, not all crystals are cubic and so reciprocal space comes into its own.

(a) Draw the planes with Miller indices (1 1 2), (3 3 6), and (4 4 8) and show that they are parallel to (2 2 4).

(b) Derive the general formula using indices $(h\,k\,l)$ and $(h'\,k'\,l')$ to describe the two planes.

(c) Calculate the Miller indices for the plane labelled $(h\,k\,l)$ in the figure and show that it is parallel to (3 6 2).

(d) Is it possible to find the equation for the plane perpendicular to $(h\,k\,l)$?

Strategy: (*a*) When the planes are found in terms of Miller indices, which is in reciprocal space, make vectors along the ***a***, ***b***, and ***c*** axis as necessary, and form the cross product which is zero when the vectors chosen are parallel. (*b*) instead of using numbers, make vectors based on *h*, *k*, *l* for the general lattice.

Q6.22 **(a)** Calculate the distance from the origin to planes with Miller indices, $(h\,k\,l)$ in an orthorhombic crystal.

(b) Calculate the distance *d* between consecutive parallel planes; for example, with Miller indices (1 1 0) and (2 2 0).

(c) Find the formula for the distance between any two adjacent parallel planes.

You may wish to look at Section 6.16.4 before attempting this question.

Strategy: An orthorhombic crystal has a unit cell with orthogonal axes of different length, ***a***, ***b***, and ***c***. The distance from the origin to a plane is given by equation (6.47), but in the case that the intercepts are known, the plane then passes through points $\{p, 0, 0\}$, $\{0, q, 0\}$ and $\{0, 0, r\}$. The equation is $\dfrac{x}{p}+\dfrac{y}{q}+\dfrac{z}{r}=1$ where the axes are $x \equiv a$, $y \equiv b$ and $z \equiv c$.

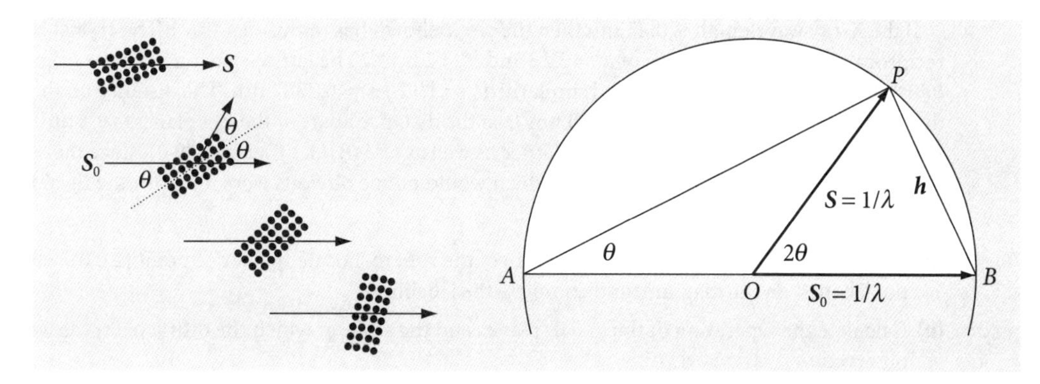

Fig. 6.15 Left: Diffraction off planes of atoms. Right: Ewald Sphere construction

Q6.23 Figure 6.15 (left) shows the geometry for diffraction off a lattice of atoms. When the Bragg condition is obeyed, $\lambda = 2d\sin(\theta)$, where the X-ray wavelength is λ, d the spacing between layers of atoms, and θ the diffraction angle. The Ewald construction used in X-ray crystallography is shown on the right and crystallographers have used this diagram to identify whether or not a lattice point P will diffract the X-rays to an angle θ.

The initial X-ray travels along vector S_0, and the diffracted X-rays along vector S. In constructing the Ewald sphere, the two vectors S_0 and S are made $1/\lambda$ long, centred at the origin and a sphere of radius $1/\lambda$ drawn around them. The diagram is therefore one of reciprocal space, in which each *set of planes* in the crystal is represented by a *single point*. Suppose lattice points have a separation d, and it is found that the point P touches the sphere when the diffracted X-ray is at an angle θ, then the vector $S - S_0$, labelled $\boldsymbol{h}$ on the diagram, has a length $1/d$. If $|\boldsymbol{h}| = 1/d$ the Bragg condition has been satisfied.

Show that $\boldsymbol{h} = S - S_0$ and that $\boldsymbol{h}\cdot\boldsymbol{h} = |S - S_0|^2$ and hence find the Bragg law $\lambda = 2d\sin(\theta)$. (Use the identity $2\sin^2(\theta) = 1 - \cos(2\theta)$.)

Strategy: Using the diagram, the vector $\boldsymbol{h}$ is the difference between S and S_0 therefore $\boldsymbol{h} = S - S_0$ and since the angle between a vector and itself is zero then $\boldsymbol{h}\cdot\boldsymbol{h} = |S - S_0|^2$. You must show this by calculating actual values.

Q6.24 Miller indices define points in reciprocal space which represent planes in real space and are labelled $(h\,k\,l)$. Figure 6.16 shows how a cut through the Ewald sphere at $l = 0$, locates reflection planes. The origin of the reciprocal lattice points, given in units of the Miller indices $(h\,k\,l) = (0\,0\,0)$, is placed where the incident X-ray *exits* the sphere. The planes generating a diffraction spot are those that intersect with the sphere. Any lattice point with a negative index is shown with a line above the number. The reciprocal of the separation of any point from the origin gives the separation of that lattice plane. Its angle from the centre of the sphere is twice the diffracted angle.

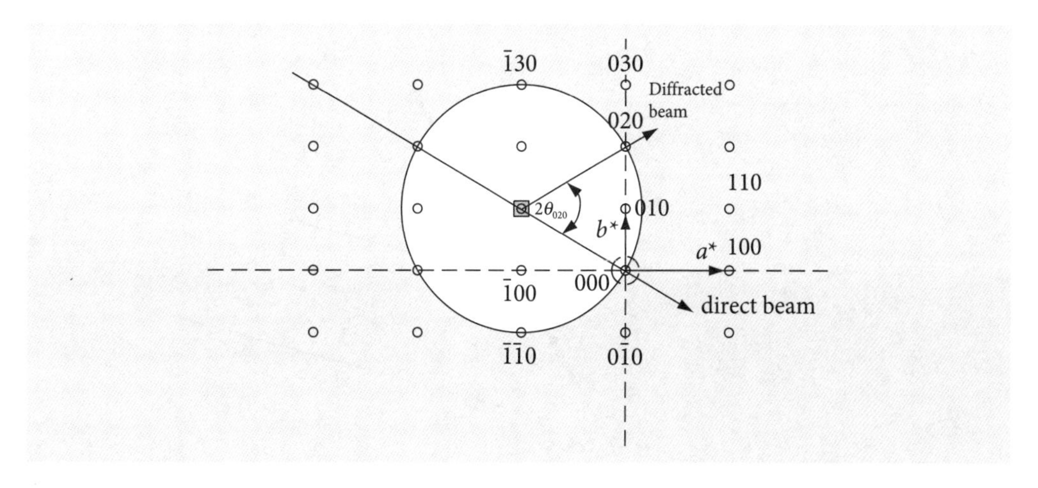

Fig. 6.16 Ewald sphere and lattice points.

If the X-ray wavelength is 0.25 nm then the circle shown has radius of 4 nm^{-1}. The crystal has reciprocal unit cell dimensions of $a^* = 2\sqrt{3}$ and $b^* = 2$ nm^{-1}. The lattice spacing from the origin (0 0 0) to (0 1 0) is $d_{010} = 0.5$ nm, and similarly $d_{100} = 1/(2\sqrt{3}) = 0.2887$ nm. The distance on the Ewald diagram from the origin to (0 1 0) is 10 nm^{-1}, so the distance apart of the d_{010} planes is 0.1 nm. Diffraction is observed from the (0 2 0) lattice as well as ($\bar{1}$ 3 0), ($\bar{1}$ $\bar{1}$ 0), and ($\bar{2}$ 0 0), since these reciprocal lattice points lie on the circle, which would not be obvious were it not for the Ewald sphere construction.

(a) Show that the point ($\bar{1}$ 1 0) is at the centre of the sphere. Locate the lattice point ($\bar{2}$ 0 0), which is not labelled on the diagram, and complete the labelling.

(b) Calculate the separation of the ($\bar{1}$ 3 0) planes and the angle at which the diffracted spots would be observed.

(c) What X-ray wavelength would be needed to observe diffraction from the (1 2 0) and (0 1 0) planes?

(d) It is far easier to rotate the crystal than change the X-ray wavelength. What angle must the crystal be rotated by to observe the (1 2 0) and (0 1 0) planes?

6.6 Projections and components

To find the *projection* or *component* of a vector A that lies along an axis, the dot product with one of the vector's *orthonormal* base vectors is calculated. For example, to calculate the component along the x-axis the base vector $\boldsymbol{i}$ is used;

$$i{\cdot}A = i{\cdot}(ai + bj + ck) = ai{\cdot}i = a. \qquad (6.13)$$

This shows that only the x component is extracted and you can see how to obtain the b and c components using $\boldsymbol{j}$ and $\boldsymbol{k}$ instead of $\boldsymbol{i}$. More usefully, since the dot product can be written as

$$i{\cdot}A = |A|\cos(\theta),$$

then

$$a = |A|\cos(\theta)$$

is the projection on component of vector A along the x-axis, which is the same result as obtained by trigonometry. Recall that $|A|$ is the magnitude of A.

The *projection* of a vector $\boldsymbol{w}$ onto another $\boldsymbol{v}$, produces a new vector $\boldsymbol{P}$, which is calculated in a similar way to the projection onto a base vector. The diagram Fig. 6.17 shows two vectors and their projection, $\boldsymbol{P}$.

The projection of $\boldsymbol{w}$ onto $\boldsymbol{v}$ is made in three steps: the length of the projection is the size of the vector $\boldsymbol{w}$ multiplied by the cosine of the angle between them, as in Fig. 6.3.

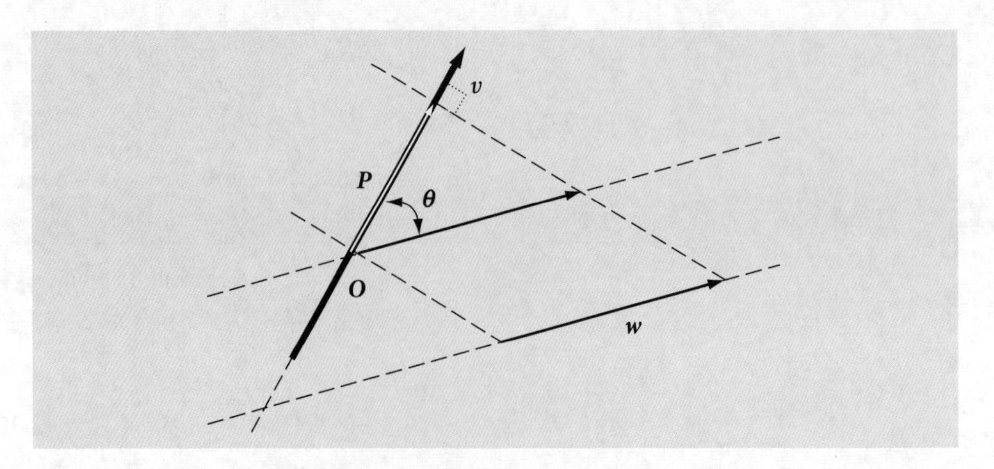

Fig. 6.17 The projection of vector ***w*** onto ***v***. Even if ***w*** and ***v*** are separated, they can be translated to meet at one point O and the projection is unchanged. The projection is vector ***P***.

(a) Convert $\boldsymbol{v}$ and $\boldsymbol{w}$ into unit vectors (i.e. normalize them) and calculate their dot product, which is $\cos(\theta)$.

(b) Multiply by the *magnitude* of $\boldsymbol{w}$ to get the actual projection. This produces a scalar.

(c) Multiply by the unit vector of $\boldsymbol{v}$. All this does is to make the projection vector lie in the direction of $\boldsymbol{v}$.

Steps (a) and (b): the normalized dot product of the two vectors is $\dfrac{\boldsymbol{w}\cdot\boldsymbol{v}}{|\boldsymbol{w}||\boldsymbol{v}|}$, where $|\boldsymbol{w}|$ and $|\boldsymbol{v}|$ are the lengths of the vectors. This dot product is also the cosine of the angle between them.

Step (c): Multiplying by the length (magnitude) of $\boldsymbol{w}$ and by the unit vector $\boldsymbol{v}/|\boldsymbol{v}|$ produces the projection of $\boldsymbol{w}$ onto $\boldsymbol{v}$

$$\begin{aligned} \boldsymbol{P}_{w\to v} &= \left[|\boldsymbol{w}|\frac{\boldsymbol{w}\cdot\boldsymbol{v}}{|\boldsymbol{w}||\boldsymbol{v}|}\right]\frac{\boldsymbol{v}}{|\boldsymbol{v}|} \\ &= \left[\frac{\boldsymbol{w}\cdot\boldsymbol{v}}{|\boldsymbol{v}|^2}\right]\boldsymbol{v} \end{aligned} \tag{6.14}$$

The term in square brackets is a scalar number. The length of the projection and of vector $\boldsymbol{P}$ is a scalar and is

$$L = \frac{|\boldsymbol{w}\cdot\boldsymbol{v}|}{|\boldsymbol{v}|} \tag{6.15}$$

and the absolute value of the dot product makes the length positive. You can see, perhaps, that the projection of $\boldsymbol{v}$ onto $\boldsymbol{w}$ is not the same as $\boldsymbol{w}$ onto $\boldsymbol{v}$. If the matrix-vector notation is used, then the projection of n-dimensional vectors onto one another can be worked out and this type of calculation is common in quantum mechanics.

As an example, consider the work done W, on moving an object. This is force × distance, but if the force is not applied in the direction that the object is moved, but at an angle θ, then only a component of the force is effective. The total work done is $W = |\boldsymbol{F}||\boldsymbol{d}|\cos(\theta) = \boldsymbol{F}\cdot\boldsymbol{d}$, which is a scalar number. If force is given by the vector $\boldsymbol{F} = [5 \quad -3 \quad 1]$ and the direction of motion by $\boldsymbol{d} = [2 \quad 3 \quad 4]$, then the work done is,

$$W = \boldsymbol{F}\cdot\boldsymbol{d} = (5\boldsymbol{i} - 3\boldsymbol{j} + \boldsymbol{k})\cdot(2\boldsymbol{i} + 3\boldsymbol{j} + 4\boldsymbol{k}) = 10 - 9 + 4 = 5.$$

In vector-matrix form this is $W = [5 \quad -3 \quad 1]\begin{bmatrix}2\\3\\4\end{bmatrix} = 5.$

The same calculation, as a *projection* of the force onto the direction of motion, is the vector $\boldsymbol{P}$, equation (6.14),

$$\left[(5\boldsymbol{i} - 3\boldsymbol{j} + \boldsymbol{k})\cdot\frac{(2\boldsymbol{i} + 3\boldsymbol{j} + 4\boldsymbol{k})}{\sqrt{4+9+16}}\right]\frac{(2\boldsymbol{i} + 3\boldsymbol{j} + 4\boldsymbol{k})}{\sqrt{4+9+16}} = \frac{5}{29}(2\boldsymbol{i} + 3\boldsymbol{j} + 4\boldsymbol{k}),$$

where the term in square brackets is a scalar. The magnitude of the projection is L, equation (6.15),

$$L = (5\boldsymbol{i} - 3\boldsymbol{j} + \boldsymbol{k})\cdot\frac{(2\boldsymbol{i} + 3\boldsymbol{j} + 4\boldsymbol{k})}{\sqrt{29}} = \frac{5}{\sqrt{29}}.$$

To calculate the total work, this projected length has to be multiplied by the distance moved which is the *length* of vector $\boldsymbol{d} = (2\boldsymbol{i} + 3\boldsymbol{j} + 4\boldsymbol{k})$ which is $\sqrt{29}$, therefore the work is 5 as calculated above.

If the vectors are greater than three dimensional then the $\boldsymbol{i}, \boldsymbol{j}, \boldsymbol{k}$ basis cannot be used and matrices are used instead. For example if $\boldsymbol{w} = [1 \quad 3 \quad 5 \quad 7]$ and $\boldsymbol{v} = [2 \quad 4 \quad 6 \quad 8]$ the projection of $\boldsymbol{w}$ onto $\boldsymbol{v}$ is $\boldsymbol{P}_{w\to v} = \left[\dfrac{\boldsymbol{w}\cdot\boldsymbol{v}}{|\boldsymbol{v}|^2}\right]\boldsymbol{v}$. The square of the magnitude $\boldsymbol{v}$ is $\boldsymbol{v}\cdot\boldsymbol{v} = [2 \quad 4 \quad 6 \quad 8]\begin{bmatrix}2\\4\\6\\8\end{bmatrix} = 4 + 16 + 36 + 64 = 120$ and therefore $|\boldsymbol{v}|^2 = 120$. The dot

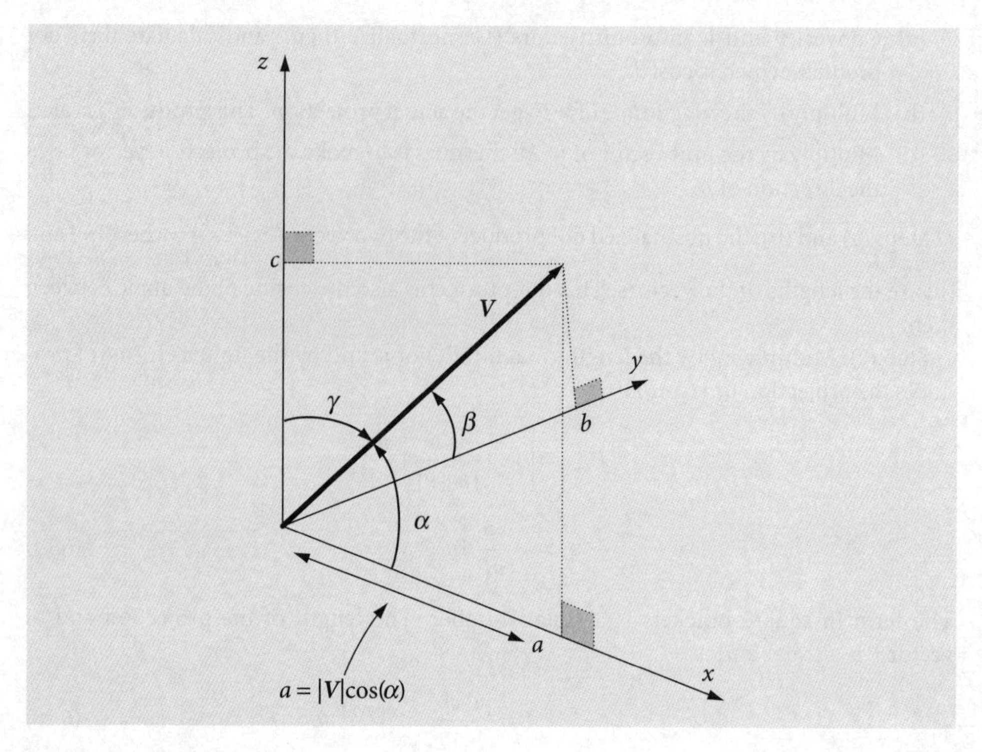

Fig. 6.18 Vector $\boldsymbol{v}$ defined by angles, α, β, and γ.

product is $\boldsymbol{w}{\cdot}\boldsymbol{v} = 100$. Therefore the projection is $\boldsymbol{P}_{w\rightarrow v} = [2 \quad 4 \quad 6 \quad 8] \times 5/6$ and the length of the projection $50/\sqrt{30}$. The projection of $\boldsymbol{v}$ onto $\boldsymbol{w}$ is $\boldsymbol{P}_{v\rightarrow w} = [1 \quad 3 \quad 5 \quad 7] \times 25/21$ of length $50/\sqrt{21}$.

6.6.1 Direction cosines

A three-dimensional vector is often defined as $V = a\boldsymbol{i} + b\boldsymbol{j} + c\boldsymbol{k}$. However, the vector can equally well be described by the angles it makes to the x, y and z-axes, when these angles are α, β, and γ, as shown in Fig. 6.18, where the angles are measured away from the axes to the vector.

The vector $\boldsymbol{v}$ is now described by

$$\boldsymbol{v} = |\boldsymbol{v}|[\cos(\alpha)\boldsymbol{i} + \cos(\beta)\boldsymbol{j} + \cos(\gamma)\boldsymbol{k}] \tag{6.16}$$

and the *components* $\cos(\alpha)$, $\cos(\beta)$, $\cos(\gamma)$ are each *direction cosines*. The direction cosines are, therefore, the projections the normalized vector $\boldsymbol{v}$ makes onto each of the x-, y- and z-axes. They are also the dot product of vector $\boldsymbol{v}$ with each base vector $\boldsymbol{i}, \boldsymbol{j}, \boldsymbol{k}$ in turn. Taking the dot product with base vector $\boldsymbol{i}$ gives

$$\boldsymbol{v}{\cdot}\boldsymbol{i} = |\boldsymbol{v}|\cos(\alpha) \qquad \text{or} \qquad \cos(\alpha) = \frac{\boldsymbol{v}{\cdot}\boldsymbol{i}}{|\boldsymbol{v}|}.$$

Because the vector can also be written as $\boldsymbol{v} = a\boldsymbol{i} + b\boldsymbol{j} + c\boldsymbol{k}$, the direction cosine can also be written as

$$\cos(\alpha) = \frac{a}{|\boldsymbol{v}|} = \frac{a}{\sqrt{a^2 + b^2 + c^2}}.$$

In matrix-vector notation, this calculation is

$$\cos(\alpha) = \frac{1}{|\boldsymbol{v}|}[a \quad b \quad c]\begin{bmatrix} 1 \\ 0 \\ 0 \end{bmatrix} = \frac{a}{|\boldsymbol{v}|}$$

and $|\boldsymbol{v}| = \sqrt{\boldsymbol{v}{\cdot}\boldsymbol{v}}$ where $\boldsymbol{v}{\cdot}\boldsymbol{v} = [a \quad b \quad c]\begin{bmatrix} a \\ b \\ c \end{bmatrix} = a^2 + b^2 + c^2$.

If we take the dot product of $\boldsymbol{v}$ with itself, then the sum rule for direction cosines is found,

$$\cos(\alpha)^2 + \cos(\beta)^2 + \cos(\gamma)^2 = 1. \qquad (6.17)$$

This equation shows that the angles are not independent because knowing two of them determines the third. Note that direction cosines describe the vector by angles rather than components of $\{x, y, z\}$ and hence can be used to find these angles; alternatively, if the angles are given, they can be used to find the x, y, z components.

As an example, let us find the angles that the vector $\boldsymbol{R} = 3\boldsymbol{i} + 6\boldsymbol{j} - 2\boldsymbol{k}$ makes with the x-axis and calculate the direction cosines for this vector. First, assume that the angles are α, β, and γ with respect to the positive x-, y-, and z-axes, as in Fig. 6.18, as this is the convention. The dot product with unit vector $\boldsymbol{i}$ produces

$$\boldsymbol{i}\cdot\boldsymbol{R} = |\boldsymbol{i}||\boldsymbol{R}|\cos(\alpha) = 7\cos(\alpha),$$

because $|\boldsymbol{i}| = 1$ and the length of the vector is $\sqrt{3^2 + 6^2 + 2^2} = 7$. Next, using the vector components

$$\boldsymbol{i}\cdot\boldsymbol{R} = \boldsymbol{i}\cdot(3\boldsymbol{i} + 6\boldsymbol{j} - 2\boldsymbol{k}) = 3,$$

and combining these two equations gives the direction cosine as $\cos(\alpha) = 3/7$, and the angle $\alpha = \cos^{-1}(3/7)$ or $\alpha = 64.62°$. A similar calculation gives the other direction cosines as

$$\cos(\beta) = 6/7, \qquad \cos(\gamma) = -2/7,$$

and angles $\beta = \cos^{-1}(6/7)$ and $\gamma = \cos^{-1}(-2/7)$. As a check, the rule for direction cosines is $\cos(\alpha)^2 + \cos(\beta)^2 + \cos(\gamma)^2 = 1$. In this example this is shown to be correct because $(9 + 36 + 4)/49 = 1$.

6.7 Questions

Full solutions are available at **www.oxfordtextbooks.co.uk/orc/beddard**.

Q6.25 Two unit vectors, $\boldsymbol{a}$ and $\boldsymbol{b}$ in the x–y plane, make angles α and β with the x-axis respectively.

(a) Show that $\boldsymbol{a} = \cos(\alpha)\boldsymbol{i} + \sin(\alpha)\boldsymbol{j}$ and $\boldsymbol{b} = \cos(\beta)\boldsymbol{i} + \sin(\beta)\boldsymbol{j}$ and then by calculating $\boldsymbol{a}\cdot\boldsymbol{b}$ find the trigonometric formulae for $\cos(\alpha - \beta)$.

Start with angles α and β in the first quadrant 0 to $\pi/2$, and then make β negative in the range 0 to $-\pi/2$.

(b) Find $\cos(\alpha + \beta)$.

(c) Repeat the dot product calculation in matrix form.

Strategy: Draw out the vector diagram using only the first quadrant, as this makes the calculation easier. Label the diagram and add the projections on to the axes. Calculate the dot product in *two* ways, one in terms of vector components and the other from the angle between the vectors.

Q6.26 The 2s and three 2p orbitals in a one electron atom, such as hydrogen, are degenerate in energy if all angular momentum is ignored. The basis set orbitals for the angular component of the 2p wavefunctions are, $\psi_{2p_{-1}}, \psi_{2p_0}, \psi_{2p_{+1}}$ where the 0 and ±1 represent the m quantum numbers. These orbitals are orthogonal because they are the eigenvectors of the Schrödinger equation and we assume that they are normalized. Mathematically, these orbitals are spherical harmonic polynomials. In the $\psi_{2p_{-1}}, \psi_{2p_0}, \psi_{2p_{+1}}$ form, the ±1 orbitals cannot be plotted as they contain complex numbers and instead, linear combinations are made which are called ψ_{2p_x} and ψ_{2p_y} and which point in the x and y directions respectively. The $2p_0$ orbital is real and a linear combination is not necessary and points in the z direction $\psi_{2p_z} = \psi_{2p_0}$. The x, y, and z orbitals have the familiar double lobed p orbital shape.

(a) Show that the orbitals

$$\psi_{2p_x} = \frac{1}{\sqrt{2}}(\psi_{2p_{-1}} - \psi_{2p_{+1}}) \text{ and } \psi_{2p_y} = \frac{i}{\sqrt{2}}(\psi_{2p_{-1}} + \psi_{2p_{+1}})$$

are orthogonal and normalized.

(b) Are the hybrid orbitals

$$\psi_1 = \sqrt{\frac{1}{3}}\psi_{2s} - \sqrt{\frac{1}{2}}\psi_{2p_{+1}} - \sqrt{\frac{1}{6}}\psi_{2p_{-1}},$$

$$\psi_2 = \sqrt{\frac{1}{3}}\psi_{2s} + \sqrt{\frac{1}{2}}\psi_{2p_{+1}} - \sqrt{\frac{1}{6}}\psi_{2p_{-1}}$$

orthogonal with each other or with the $2p_x$ and $2p_y$ orbitals? (ψ_s is the 2s orbital which is normalized and orthogonal to each 2p orbital).

(c) Are the hybrid orbitals

$$\psi_{sp^2(+xy)} = \sqrt{\frac{1}{3}}\psi_{2s} + \sqrt{\frac{1}{2}}\psi_{2p_x} + \sqrt{\frac{1}{6}}\psi_{2p_y},$$

$$\psi_{sp^2(-xy)} = \sqrt{\frac{1}{3}}\psi_{2s} + \sqrt{\frac{1}{2}}\psi_{2p_x} - \sqrt{\frac{1}{6}}\psi_{2p_y},$$

$$\psi_{sp^2(x)} = \sqrt{\frac{1}{3}}\psi_{2s} - \sqrt{\frac{2}{3}}\psi_{2p_x}$$

orthogonal to one another and with the $2p_x$, $2p_x$, and $2p_y$ orbitals?

(d) Plot the angular parts of the p_x, p_y and p_z and the sp^2 hybrid orbitals.

The basis set wavefunctions angular functions are $\psi_{2p_0} = \left(\dfrac{3}{4\pi}\right)^{1/2}\cos(\theta)$ and $\psi_{2p_{\pm 1}} = \mp\left(\dfrac{3}{8\pi}\right)^{1/2}\sin(\theta)e^{\pm i\varphi}$. The angular part of the s orbital is $\psi_s = 1/\sqrt{4\pi}$. You should plot the absolute value of the function, because some wavefunctions can contain complex numbers. Use Maple's `plot3d` function with `cords = spherical` to convert between angular and x, y, z coordinates automatically. The Maple code to plot an orbital ψ is

```
plot3d(abs(psi(theta, phi)), phi= 0..2*Pi, theta=0..Pi,
                              coords=spherical, axes=boxed);
```

Notice that the angle phi (ϕ) comes before theta (θ) which is the mathematicians' notation and is the opposite way round to that used in science.

(e) Show, by direct integration, that the $\psi_{2p_{0,\pm 1}}$ wavefunctions are orthogonal and normalized.

Strategy: (a–c) Treat the orbitals as vectors in the 0, ±1 orbital basis set. If these are orthonormal, the dot product of any base vector with itself will be unity and with any other vector, it will be zero. In matrix vector form, the base vectors can be written as

$$\psi_{2p_0} = \begin{bmatrix}1\\0\\0\end{bmatrix} \; \psi_{2p_{+1}} = \begin{bmatrix}0\\1\\0\end{bmatrix} \text{ and } \psi_{2p_{-1}} = \begin{bmatrix}0\\0\\1\end{bmatrix}.$$

(e) In the integration, check for normalization and orthogonality. In symbolic form, $\int \psi_m^* \psi_n d\tau = \delta_{n,m}$. Recall that in polar coordinates a volume element is necessary when integrating, see Chapter 4.7 and 4.11. The polar angle varies from 0 to π and the azimuthal from 0 to 2π.

Q6.27 Show that **(a)** the orbitals represented by $\psi_1 = \frac{1}{2}(\psi_s + \psi_{px} + \psi_{py} + \psi_{pz})$ and $\psi_2 = \frac{1}{2}(\psi_s + \psi_{px} - \psi_{py} - \psi_{pz})$ are orthogonal and **(b)** the angle between the orbitals is ≈109°. Assume that the s, p_x, p_y, and p_z orbitals are orthogonal.

Strategy: If the vectors are three dimensional, the cosine of the angle found from the dot product has some meaning. In four or higher dimensional vectors, the dot product still produces the cosine of an angle although this angle does not have any geometric meaning; instead, if the cosine is 0 the vectors are orthogonal, if 1, parallel. Because the s orbital is spherically symmetrical, it cannot contribute to the orbital's direction and can be ignored. Form a basis set out of the four s and p orbitals, as they are orthogonal.

Q6.28 An sp^n hybrid orbital may be written as a linear combination of atomic orbitals (LCAO) $\psi = a\psi_s + b\psi_{px} + c\psi_{py} + d\psi_{pz}$ where, a, b, c, and d are the amounts of the respective s and p atomic orbitals ψ_s and ψ_p, and n can be 1, 2, or 3 depending upon the type of hybrid.

(a) Find the angles between normalized orbitals represented by

$$\psi_1 = \frac{1}{\sqrt{6}}(\sqrt{2}\psi_s - \psi_{px} + \sqrt{3}\psi_{py}) \text{ and } \psi_2 = \frac{1}{\sqrt{6}}(\sqrt{2}\psi_s - \psi_{px} - \sqrt{3}\psi_{py}).$$

Assume that the s, p_x, and p_y orbitals are orthogonal.

(b) What hybridization does your answer imply?

(c) In (a) we assumed that the orbitals were orthogonal, prove this repeating (a) using the definitions of ψ_{2p_x} and ψ_{2p_y} given in Q6.26 using a basis set of s, p_0, p_{+1}, and p_{-1} orbitals.

(d) The basis set orbitals s, p_0, p_{+1}, and p_{-1} each have a set of well-defined quantum numbers. The principal quantum number is 1, 2, 3 for s orbitals, 2, 3, 4, $\cdots$ for p orbitals, 3, 4, 5 for d orbitals, and so forth. The orbital angular momentum is $l = 0$ for s orbitals, $l = 1$ for p orbitals, and 2 for d orbitals. The projection or magnetic quantum numbers m are limited to $\pm l$ and integer values in between, and are 0 and ± 1 for p orbitals. The electron has a quantum number of $S = 1/2$ and projection quantum numbers of $m_s = \pm 1/2$. When hybrid orbitals such as p_x, and p_y are formed, some quantum numbers may not be defined. Make a table of the quantum numbers of the basis set, and p_x, p_y, p_z, and ψ_1 and ψ_2 orbitals.

Strategy*:** Think about the geometry of the *s* and *p* orbitals and more particularly the directions in which they can point. Relate ψ to ψ_1 and ψ_2 using the ***i, ***j***, ***k*** basis set vectors.

Q6.29 Some atoms are shown in Fig. 6.19 positioned in the arrangement of the body-centred and face-centred unit cells. The length of one side of the unit cell is q.

(a) Using the vectors shown, calculate the distances from atom a to b, c, d, and e and the angles (i) bac, (ii) bad, and (iii) bae. The atom labelled a in the body-centred cell (left) is described by the vector $\frac{q}{2}(i + j + k)$, and the atom a in the rear of the face-centred cube at $\frac{q}{2}(i + 2j + k)$, where $\boldsymbol{i}, \boldsymbol{j}, \boldsymbol{k}$ form an orthonormal set of vectors.

(b) Use Maple to do a similar calculation on the face-centred cell. In this calculation, it will be easier to convert from the ***i***, ***j***, ***k*** to matrix vector method.

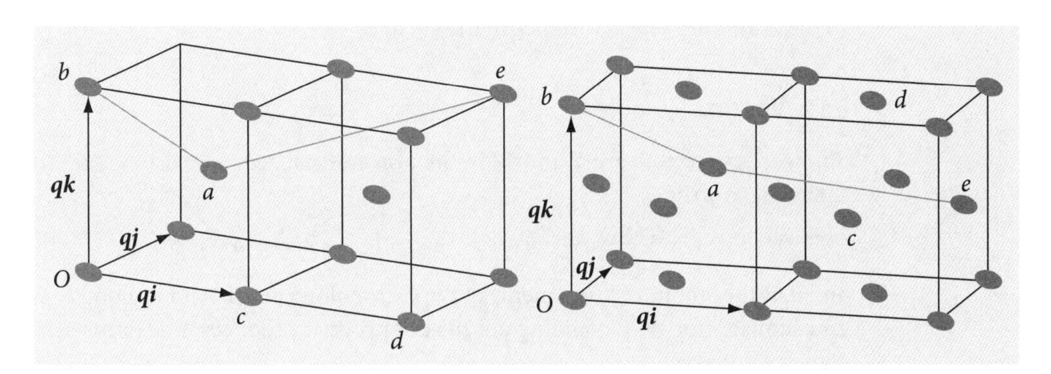

Fig. 6.19 Body-centred (left) and face-centred cubes. The angle *bae* is shown with dotted lines.

***Strategy*:** First, work out the coordinates of the points and then the vectors from the origin O. As q is a common factor to all the coordinates, it is easier to work by defining its numerical value to be 1 and finally multiply distances by q.

Q6.30 Derive equation (6.17).

Q6.31 Find $A + B$, $|A|$, $|B|$, $A \cdot B$ and $\cos(\theta)$ for the following vectors:

(a) $A = i - j + k$, $\quad B = i + 3j + 5k$.

(b) $A = 3i + j + 4k$, $\quad B = 2i - 5j + 3k$.

(c) $A = i + j$, $\quad B = i + k$.

Q6.32 The dipole moments of a molecule may be constructed, at least approximately, from the vectorial sum of individual bond dipole moments which are tabulated. Using the values in the table, calculate the dipole moments for **(a)** meta-dichloro-benzene, **(b)** methylene chloride, and **(c)** chloroform.

Bond dipole moments are given in units of debye, see Weast for values for other bonds. A positive value means the dipole direction is towards its positive end and to the right-hand atom of the pair. (The bond angle in tetrahedral molecules is $\cos^{-1}(-1/3) = 109.47°$.)

bond	μ/D	bond	μ/D
C–H	−0.4	C–Cl	−1.56
C–I	−1.29	C=O	−2.40
N≡C	3.6	N–H	1.31

6.8 Not all axes are right-angled or of equal length

6.8.1 Basis set

In crystallography the coordinates of the unit cell are often reported in terms of the crystal type, monoclinic, triclinic, and so forth, whose axes are not mutually at right angles or always of the same length. The base vectors are no longer the orthonormal ***i***, ***j***, ***k*** set we have been using and are now labelled as ***a***, ***b***, ***c*** to indicate that they are non-orthogonal. By convention, the angle α lies between ***b*** and ***c***, β between ***a*** and ***c***, and γ between ***a*** and ***b***. This is shown in Fig. 6.20.

As the base vectors are not at 90° to one another, when a dot product is calculated, the terms $\boldsymbol{a}\cdot\boldsymbol{b}$, $\boldsymbol{a}\cdot\boldsymbol{c}$, and $\boldsymbol{c}\cdot\boldsymbol{b}$ are not zero as are the corresponding cross terms $\boldsymbol{i}\cdot\boldsymbol{k}$, $\boldsymbol{i}\cdot\boldsymbol{j}$, and $\boldsymbol{j}\cdot\boldsymbol{k}$ in the right-angled basis set. Additionally, the base vectors are often not of unit length, but have the length, and relative angles, of the unit cell, which means that they are not normalized and not orthogonal—just about as bad as it can get! However, the method to calculate a dot product is, in principle, no different to that already described. Although ***a***, ***b***, and ***c*** represent the length of the unit cell, these are not unit vectors; the 'unit' here refers to the minimum repeat distance in the crystal.

6.8.2 Vectors

Suppose two vectors are V and W in the non-normalized, non-orthogonal (***a***, ***b***, ***c***) basis then we can write

$$V = v_1\boldsymbol{a} + v_2\boldsymbol{b} + v_3\boldsymbol{c} \qquad \text{and} \qquad W = w_1\boldsymbol{a} + w_2\boldsymbol{b} + w_3\boldsymbol{c}$$

and v_1 is the amount (component) of the vector along axes ***a***, and v_2 along ***b***, and so forth, and similarly for W. Expanding out the dot product produces nine terms, which do have some symmetry;

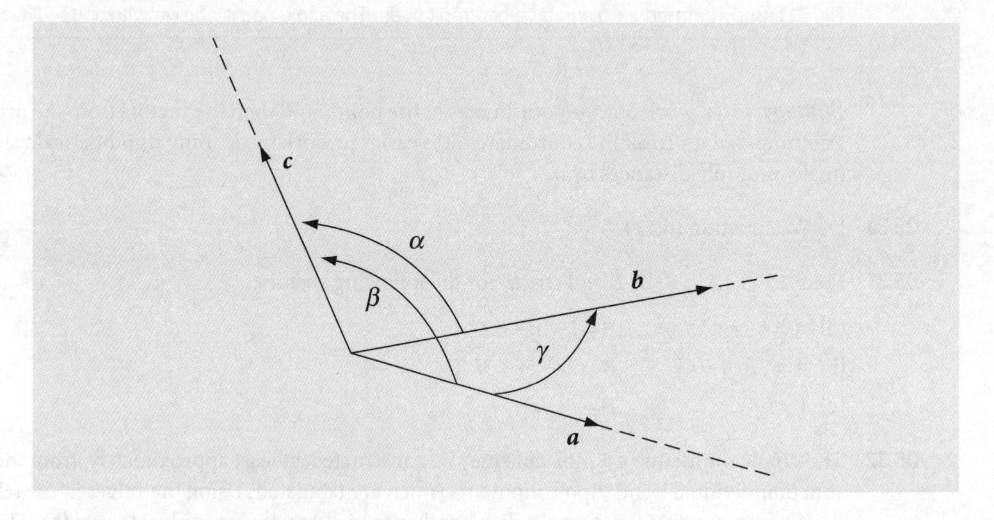

Fig. 6.20 Non-orthogonal axes and non-normalized base vectors ***a***, ***b*** and ***c***.

$$\begin{aligned} \boldsymbol{V}\cdot\boldsymbol{W} &= (v_1\boldsymbol{a}+v_2\boldsymbol{b}+v_3\boldsymbol{c})\cdot(w_1\boldsymbol{a}+w_2\boldsymbol{b}+w_3\boldsymbol{c}) \\ &= v_1w_1\boldsymbol{a}\cdot\boldsymbol{a}+v_2w_2\boldsymbol{b}\cdot\boldsymbol{b}+v_3w_3\boldsymbol{c}\cdot\boldsymbol{c} \\ &\quad +(v_1w_2+v_2w_1)\boldsymbol{a}\cdot\boldsymbol{b} \\ &\quad +(v_1w_3+v_3w_1)\boldsymbol{a}\cdot\boldsymbol{c} \\ &\quad +(v_2w_3+v_3w_2)\boldsymbol{b}\cdot\boldsymbol{c} \end{aligned} \tag{6.18}$$

Keeping our nerve, we can simplify, since $\boldsymbol{a}\cdot\boldsymbol{a}=a^2$, if a is the length of vector $\boldsymbol{a}$ and similarly for $\boldsymbol{b}$ and $\boldsymbol{c}$. The dot product gives the cosine of the angle between any two vectors, just as it does for orthogonal vectors, equation (6.2). From Fig. 6.20 it can be seen that

$$\begin{aligned} \boldsymbol{a}\cdot\boldsymbol{b} &= ab\cos(\gamma), \\ \boldsymbol{a}\cdot\boldsymbol{c} &= ac\cos(\beta), \\ \boldsymbol{b}\cdot\boldsymbol{c} &= bc\cos(\alpha). \end{aligned} \tag{6.19}$$

The cosines remain because the axes are not at 90° so that the contribution a component along each axis makes to the others is not zero. Substituting into equation (6.18) gives,

$$\begin{aligned} \boldsymbol{V}\cdot\boldsymbol{W} = v_1w_1a^2+v_2w_2b^2+v_3w_3c^2 \\ +(v_1w_2+v_2w_1)ab\cos(\gamma) \\ +(v_1w_3+v_3w_1)ac\cos(\beta) \\ +(v_2w_3+v_3w_2)bc\cos(\alpha). \end{aligned} \tag{6.20}$$

The magnitude of V is calculated in a similar way

$$\begin{aligned} |\boldsymbol{V}|^2=\boldsymbol{V}\cdot\boldsymbol{V}=v_1^2a^2+v_2^2b^2+v_3^2c^2 \\ +2v_1v_2ab\cos(\gamma) \\ +2v_1v_3ac\cos(\beta) \\ +2v_2v_3bc\cos(\alpha). \end{aligned} \tag{6.21}$$

as is $|\boldsymbol{W}|$ by substituting w's for v's.

6.8.3 Calculating bond lengths

In X-ray crystallography, an atom's coordinates are often given in terms of the unit cell dimensions and these are not always at right angles to one another. Suppose that the vector of atom C_A's coordinates, given in unit cell dimensions, is $\boldsymbol{A}$ and that for atom C_B is $\boldsymbol{B}$. The bond length is calculated from the difference of two vectors $\boldsymbol{V}=\boldsymbol{A}-\boldsymbol{B}$ and then by calculating the dot product of this vector with itself, equation (6.21)

$$\boldsymbol{V}\cdot\boldsymbol{V}=|\boldsymbol{V}|^2. \tag{6.22}$$

To illustrate the method described in Section 6.8.2, (i) the C–N_1 bond length and (ii) the angle between the N_1CN_2 atoms in s-tetrazine, $C_2N_4H_2$, will be calculated; its geometry is shown in Fig. 6.22. The crystal structure was determined using Cu radiation with a wavelength of 1.542 Å. The unit cell has dimensions of $a=0.523$ nm, $b=0.579$ nm, and $c=0.663$ nm (Bertinotti & Giacomello 1956). The angle between the long and either short axis is 90°, and the angle between the two short axes is 115°30′. The crystal is therefore monoclinic and this crystal type is characterized by one twofold symmetry axis and one mirror plane. The *fractional* coordinates of some of the atoms in the crystal are shown in the table. These are in units of the length of each base vector or unit cell dimension.

	v_1/a	v_2/b	v_3/c
C	0.2546	0.0153	0.1380
N_1	0.1834	−0.1669	−0.0015
N_2	0.0849	0.1795	0.1486
N_3	−0.0849	−0.1795	−0.1486

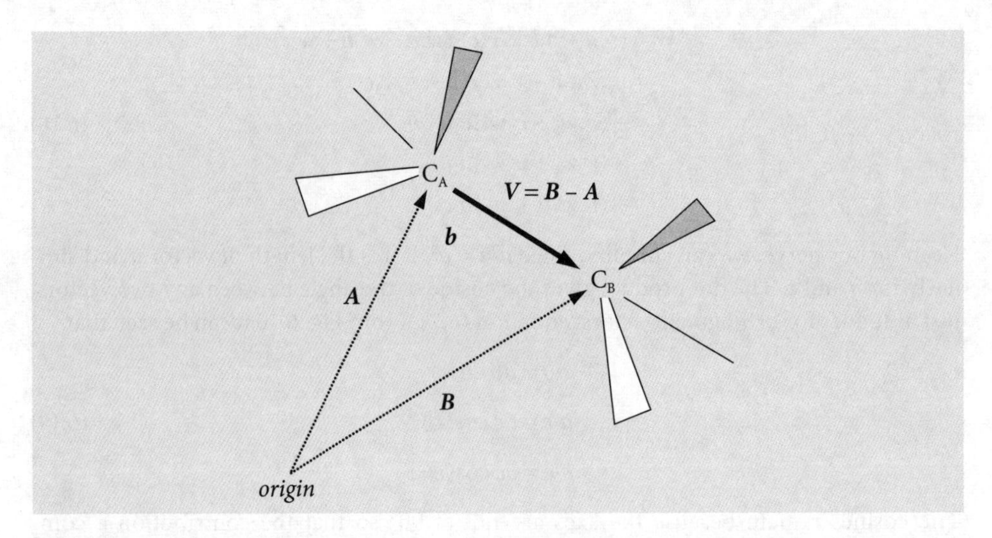

Fig. 6.21 Figure to show how to calculate a bond length given vectors ***A*** and ***B*** of the two atom's coordinates.

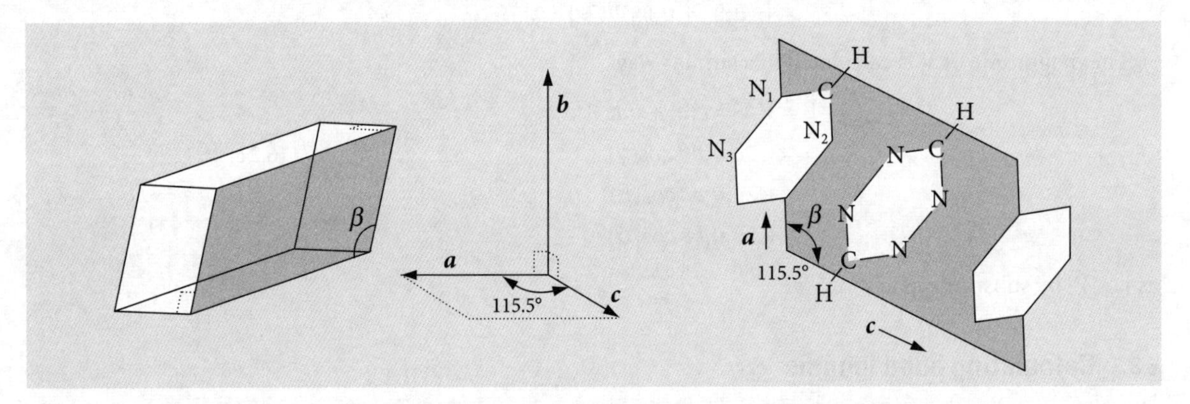

Fig. 6.22 A monoclinic crystal, the axes, and a sketch (right) of the projection of the tetrazine structure looking down the ***b*** axis. The plane of the *s*-tetrazine molecule lies in the ***a*** – ***c*** plane. (Not all atoms are explicitly shown.)

The vectors $\boldsymbol{a}$, $\boldsymbol{b}$, and $\boldsymbol{c}$ form the basis-set axes with lengths a, b, and c, respectively, which are the same as the unit cell. The origin of coordinates is assumed to be $\boldsymbol{O} = [0 \quad 0 \quad 0]$ and no H atoms are resolved.
Using the data, the magnitude of the base vectors are $|\boldsymbol{a}| = a = 0.523$ nm, $|\boldsymbol{b}| = b = 0.579$ nm, and $|\boldsymbol{c}| = c = 0.663$ nm. The angle β between $\boldsymbol{a}$ and $c = 115°30' \equiv 115.5°$ and the other angles are each 90°.

(i) To find the C–N_1 bond length, let the vector for the *position* of the N_1 atom be $\boldsymbol{W}_{N1} = w_1\boldsymbol{a} + w_2\boldsymbol{b} + w_3\boldsymbol{c}$. Using data in the table, this vector is $\boldsymbol{W}_{N1} = 0.1834a\boldsymbol{a} - 0.1669b\boldsymbol{b} - 0.0015c\boldsymbol{c}$ where $w_1 = 0.1834a$, $w_2 = -0.1669b$, $w_3 = -0.0015c$ and the constants a, b, c are present because the lengths are given as fractions of the unit cell length. These could be multiplied out at the outset to give the vector

$$\boldsymbol{W}_{N1} = 0.095918\boldsymbol{a} - 0.096635\boldsymbol{b} - 0.0009945\boldsymbol{c}$$

but can be left and done at the end of the calculation. The vector for the C atom is

$$\boldsymbol{V}_C = 0.2456a\boldsymbol{a} + 0.0153b\boldsymbol{b} + 0.1380c\boldsymbol{c}$$

Define a new vector Δ to be the bond vector $\boldsymbol{V}_C - \boldsymbol{W}_{N1}$, then

$$\Delta = \Delta_1\boldsymbol{a} + \Delta_2\boldsymbol{b} + \Delta_3\boldsymbol{c}$$

where $\Delta_1 = (v_1 - w_1)$, $\Delta_2 = (v_2 - w_3)$, $\Delta_3 = (v_3 - w_3)$ and as $|\Delta| = d$ is the bond length, using the definition of the dot product $\boldsymbol{\Delta}\cdot\boldsymbol{\Delta}$, given in equation (6.21), and substituting for d will give

$$d^2 = \Delta_1^2 a^2 + \Delta_2^2 b^2 + \Delta_3^2 c^2 + 2\Delta_1\Delta_2 ab\cos(\gamma) + 2\Delta_1\Delta_3 ac\cos(\beta) + 2\Delta_2\Delta_3 bc\cos(\alpha) \tag{6.23}$$

where a, b, and c are the magnitudes of their corresponding vectors $\boldsymbol{a}$, $\boldsymbol{b}$, and $\boldsymbol{c}$. In this particular example, $\beta = 115.5°$ and the angles γ and α are both 90° and their cosines zero; therefore,

$$d^2 = \Delta_1^2 a^2 + \Delta_2^2 b^2 + \Delta_3^2 c^2 + 2\Delta_1\Delta_3 ac\cos(\beta). \tag{6.24}$$

Substituting for numerical values gives $\Delta_1 = 0.1834 - 0.2546 = -0.07128$, $\Delta_2 = -0.1822$, etc. Therefore

$$d^2 = (-0.07128a)^2 + (-0.1822b)^2 + (-0.1395c)^2 + 2(-0.07128a)(-0.1395b)\cos(115.5\pi/180)$$

and using the values for a, b, and c produces a C–N_1 bond length of $d = 0.1346$ nm.

(ii) The angle N_1–C–N_2 (or any other) is found by defining difference vectors Δ_{CN2} and Δ_{CN1} for the two bonds as just described, then calculating their dot product

$$\cos(\theta) = \frac{\Delta_{CN_2}\cdot\Delta_{CN_1}}{d_{CN_1}d_{CN_2}}, \tag{6.25}$$

The values d_{CN1} and d_{CN2} are the bond lengths and using the calculated bond lengths the N_1–C–N_2 angle is 114.5°. To make sure you understand the calculation, calculate the C–N_2 and N_1–N_2 bond lengths. You should find them to be 0.1323 and 0.1321 nm respectively.

6.9 Conversion from one basis set to another

It is sometimes essential to be able to convert from one basis set to another, and if you were involved in computer graphics this would be something that you would have to do all the time as the viewing point and perspective of the scene changes. In this case, the basis set would probably be called a reference frame. In geography or astronomy, converting from earth-based coordinates to celestial ones is commonplace, in chemistry and physics converting rectangular to polar coordinates is often necessary when dealing with atomic or molecular orbitals. An X-ray crystallographic analysis of a molecule usually results in coordinates that are quoted in terms of the unit cell parameters the axes of which may be neither at right angles to one another nor of equal length. Although a molecule's bond lengths and angles can be calculated in these non-orthogonal coordinates, as illustrated in Section 6.8.3, it is often practical to convert to normal right-angled coordinates, particularly for drawing structures.

The method of converting from triclinic to orthonormal coordinates is now described and is used to illustrate the general method of converting one basis set into another. We use triclinic coordinate axes because these axes are set at different angles to one another and are each of different length. This conversion will lead to a related but more useful method of calculating geometries from X-ray data than that of Section 6.8. The two sets of axes shown in Fig. 6.23 are the general triclinic a, b, c and the orthogonal or conventional Cartesian x, y, and z.

Any vector is unchanged by changing its basis set and therefore its coefficients must change between bases. If V is any vector and $g_{x,y,z}$ is the set of its coefficients in an $\boldsymbol{xyz}$ basis and $g_{A,B,C}$ in another basis $\boldsymbol{ABC}$, this equality is written as

$$V = g_x\boldsymbol{x} + g_y\boldsymbol{y} + g_z\boldsymbol{z} = g_A\boldsymbol{A} + g_B\boldsymbol{B} + g_C\boldsymbol{C}. \tag{6.26}$$

To transform from the $\boldsymbol{ABC}$ basis with known coefficients g_A, g_C, g_C to another $\boldsymbol{xyz}$ basis a new set of coefficients g_x, g_y, g_Z need to be calculated.

To effect the conversion of a vector, the axes must be converted first, and then the vector's new coefficients calculated. These coefficients will be calculated by expanding each of the $\boldsymbol{A}$, $\boldsymbol{B}$, and $\boldsymbol{C}$ vectors, equations (6.26), in turn with equations like those of (6.27).

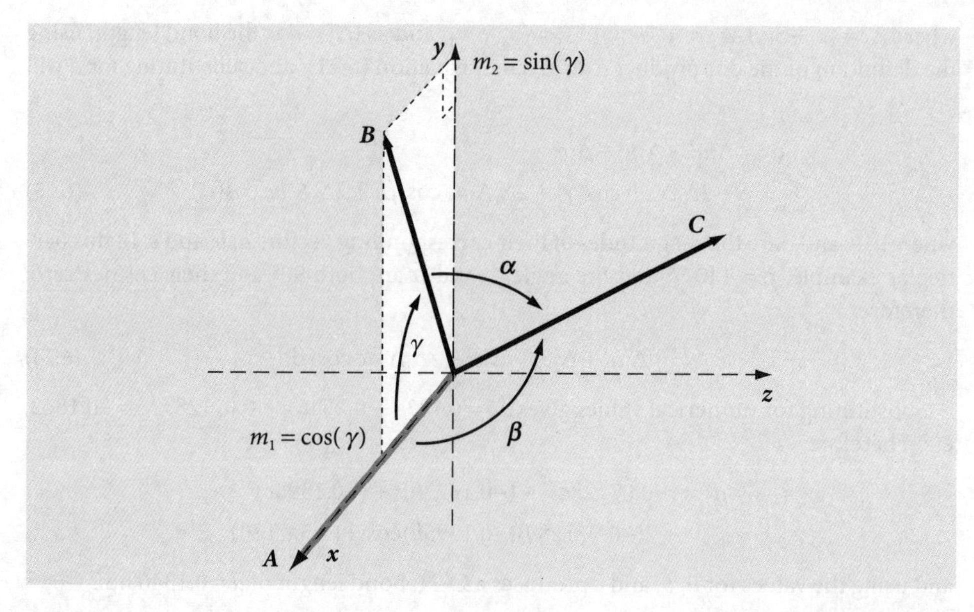

Fig. 6.23 Two sets of unit vectors; ***xyz*** are orthogonal, ***ABC*** are not. Vector ***A*** lies along ***x*** and ***B*** is in the ***xy*** plane.

6.9.1 Converting axes

We will assume that the ***xyz*** basis set[3] is orthogonal, as is the convention, therefore the basis vectors ***x***, ***y***, and ***z*** are orthogonal to one another, for instance, $\boldsymbol{x}\cdot\boldsymbol{y} = \boldsymbol{x}\cdot\boldsymbol{z} = \boldsymbol{z}\cdot\boldsymbol{y} = 0$ and as the vectors are normalized $\boldsymbol{x}\cdot\boldsymbol{x} = \boldsymbol{y}\cdot\boldsymbol{y} = \boldsymbol{z}\cdot\boldsymbol{z} = 1$. The non-orthogonal axes have unit vectors ***A***, ***B***, ***C*** lying along them and which are projected in turn onto the right-angled ***xyz*** unit vectors, in which case they are not a basis set, and must therefore be represented as linear combinations of ***x***, ***y***, and ***z*** unit vectors

$$\begin{aligned} \boldsymbol{A} &= k_1\boldsymbol{x} + k_2\boldsymbol{y} + k_3\boldsymbol{z} \\ \boldsymbol{B} &= m_1\boldsymbol{x} + m_2\boldsymbol{y} + m_3\boldsymbol{z} \\ \boldsymbol{C} &= n_1\boldsymbol{x} + n_2\boldsymbol{y} + n_3\boldsymbol{z} \end{aligned} \tag{6.27}$$

The base vectors ***A***, ***B***, and ***C*** have unit lengths, therefore, $k_1^2 + k_2^2 + k_3^2 = 1$ and similarly for the m and n components, $m_1^2 + m_2^2 + m_3^2 = 1$ and $n_1^2 + n_2^2 + n_3^2 = 1$. Changing a basis does not change the vector at all, it merely moves the components around so that they have different values in a different basis set. The equations (6.27) represent the axes conversion from the ***ABC*** basis set to the ***xyz***. These three equations can be put into a matrix form that shows the pattern of indices nicely. You should consult Chapter 7 if you are not familiar with matrices and their multiplication.

$$\begin{bmatrix} \boldsymbol{A} \\ \boldsymbol{B} \\ \boldsymbol{C} \end{bmatrix} = \begin{bmatrix} k_1 & k_2 & k_3 \\ m_1 & m_2 & m_3 \\ n_1 & n_2 & n_3 \end{bmatrix} \begin{bmatrix} \boldsymbol{x} \\ \boldsymbol{y} \\ \boldsymbol{z} \end{bmatrix} \tag{6.28}$$

The two sets of axes could be placed at any orientation with respect to one another but to make things easier, let vectors ***A*** and ***x*** coincide and let ***B*** be in the x–y plane, see Fig. 6.23. If this orientation is calculated in some other way then the basis set (***x y z***) may need to be reordered to get everything correct. The way to check that everything is all right is to make all the angles 90° then a diagonal matrix should result.

6.9.2 Calculating the new coefficients

Vector ***A*** lies along the x-axis, therefore its y and z components k_2 and k_3 are zero. The remaining component k_1 has a length 1, which is the length of vector ***A***, therefore the first of equations (6.27) representing the vector ***A*** in the ***xyz*** basis set is simply;

[3] The ***x***, ***y***, ***z*** unit vectors are the same as the ***i***, ***j***, ***k*** unit vectors used before, but here this notation would be confusing.

$$A = \boldsymbol{x}.$$

Similarly, vector $\boldsymbol{B}$ which is in the x–y plane, has a component on the x-axis of $m_1 = \cos(\gamma)$, its y component is $m_2 = \cos(\pi/2 - \gamma) = \sin(\gamma)$. The component on the z-axis is zero, $m_3 = 0$, because the vector $\boldsymbol{B}$ is in the x–y plane. The second vector is therefore

$$\boldsymbol{B} = \cos(\gamma)\boldsymbol{x} + \sin(\gamma)\boldsymbol{y}.$$

Calculating $\boldsymbol{C}$ is a little more complex. The component (projection) onto the x-axis is $n_1 = \cos(\beta)$. Calculating the other components is harder because the angle α only connects vector $\boldsymbol{C}$ to $\boldsymbol{B}$ and not to a y- or x-axis. However, with a little cunning, starting with the dot product the coefficients can be found. The dot product is

$$\boldsymbol{B}\cdot\boldsymbol{C} = \cos(\alpha),$$

because $\boldsymbol{B}$ and $\boldsymbol{C}$ are unit vectors. Expanding this equation in the new basis set given by equation (6.27), and substituting for values already determined, gives

$$\cos(\alpha) = \boldsymbol{B}\cdot\boldsymbol{C} = (\cos(\gamma)\boldsymbol{x} + \sin(\gamma)\boldsymbol{y})(\cos(\beta)\boldsymbol{x} + n_2\boldsymbol{y} + n_3\boldsymbol{z}).$$

Using the orthogonality rules $\boldsymbol{x}\cdot\boldsymbol{x} = \boldsymbol{y}\cdot\boldsymbol{y} = 1$ together with the rule that all 'cross terms' such as $\boldsymbol{x}\cdot\boldsymbol{y} = 0$ produces

$$\cos(\alpha) = \boldsymbol{B}\cdot\boldsymbol{C} = \cos(\gamma)\cos(\beta) + n_2\sin(\gamma)$$

therefore

$$n_2 = \frac{\cos(\alpha) - \cos(\gamma)\cos(\beta)}{\sin(\gamma)}. \tag{6.29}$$

The coefficient n_3 remains to be determined. This can be found from the normalization condition $n_1^2 + n_2^2 + n_3^2 = 1$ and, again, using values already determined, is

$$n_3^2 = 1 - \cos^2(\beta) - \frac{[\cos(\alpha) - \cos(\gamma)\cos(\beta)]^2}{\sin^2(\gamma)},$$

which is not worth simplifying. The final equation for vector $\boldsymbol{C}$ is

$$\boldsymbol{C} = \cos(\beta)\boldsymbol{x} + n_2\boldsymbol{y} + \sqrt{\sin^2(\beta) - n_2^2}\,\boldsymbol{z}.$$

The derivation so far has involved *unit* vectors, but a real crystal has vectors along the crystallographic axes that are not necessarily all of the same length or even of unit length. The notation usually used is to define the axes as $\boldsymbol{a}$, $\boldsymbol{b}$, and $\boldsymbol{c}$ of length a, b, and c respectively and the unit vectors then can be written as $\boldsymbol{A} = \boldsymbol{a}/a$, and so forth. The bold lower case letter represents the actual crystal vector $\boldsymbol{a}$. Making this change means that each of the vector equations can be rewritten by multiplying by the size of each vector, for instance as,

$$\boldsymbol{b} = b(\cos(\gamma)\boldsymbol{x} + \sin(\gamma)\boldsymbol{y})$$

The three equations written in matrix form are clearer,

$$\begin{bmatrix} \boldsymbol{a} \\ \boldsymbol{b} \\ \boldsymbol{c} \end{bmatrix} = \begin{bmatrix} a & 0 & 0 \\ b\cos(\gamma) & b\sin(\gamma) & 0 \\ c\cos(\beta) & cn_2 & c\sqrt{\sin^2(\beta) - n_2^2} \end{bmatrix} \begin{bmatrix} \boldsymbol{x} \\ \boldsymbol{y} \\ \boldsymbol{z} \end{bmatrix}. \tag{6.30}$$

This equation can be written in shorthand as

$$\boldsymbol{D} = \boldsymbol{M}\boldsymbol{X}$$

where $\boldsymbol{D}$ is the vector $\boldsymbol{abc}$ and $\boldsymbol{X}$ that of unit vectors $\boldsymbol{xyz}$ and the matrix is $\boldsymbol{M}$. This matrix represents the conversion from one set of axes into another; as written converting from the $\boldsymbol{xyz}$ axes to the $\boldsymbol{abc}$. To effect the opposite conversion, the equation to use is

$$\boldsymbol{X} = \boldsymbol{M}^{-1}\boldsymbol{D}$$

where $\boldsymbol{M}^{-1}$ is the inverse of $\boldsymbol{M}$, see Chapter 7. Finally, and as a check when all the angles are 90°, matrix $\boldsymbol{M}$ reduces to a diagonal matrix and this tells us that ordering the axes as $\boldsymbol{ABC}$ gives the resulting order $\boldsymbol{xyz}$.

6.10 Transformation of basis vectors

6.10.1 Calculating bond lengths and angles in non right-angled crystals

Normally, we are not so interested in converting axes but wish to find an atom's coordinates in Cartesian (***xyz***) axes, but which are given in crystallographic tables in terms of the unit cell ***abc*** axes. The convention in crystallographic tables is to quote the coordinates as fractions of the unit cell lengths in columns labelled as x/a, y/b, and z/c, or even as x, y, z. This is initially confusing because xyz axes are not involved.

The method now described is not limited to the specific example considered, but will apply to any two basis sets. The coordinates of an atom describe a vector from the origin and equation (6.26) is used to describe the vector. The same vector has coefficients $g_{x,y,z}$ in the xyz basis set (axes) and $g_{A,B,C}$ in the abc basis set. The right-hand side of equation (6.26) can be expanded by replacing the base vectors $\boldsymbol{A}$, $\boldsymbol{B}$, and $\boldsymbol{C}$ by equation (6.27), which gives

$$\begin{aligned} g_x\boldsymbol{x}+g_y\boldsymbol{y}+g_z\boldsymbol{z} &= g_A(k_1\boldsymbol{x}+k_2\boldsymbol{y}+k_3\boldsymbol{z}) \\ &\quad + g_B(m_1\boldsymbol{x}+m_2\boldsymbol{y}+m_3\boldsymbol{z}) \\ &\quad + g_C(n_1\boldsymbol{x}+n_2\boldsymbol{y}+n_3\boldsymbol{z}) \end{aligned}$$

This is really three equations and, although difficult to understand in this form, becomes clearer when written as matrices:

$$[g_x \quad g_y \quad g_z]\begin{bmatrix} \boldsymbol{x} \\ \boldsymbol{y} \\ \boldsymbol{z} \end{bmatrix} = [g_A \quad g_B \quad g_C]\begin{bmatrix} k_1 & k_2 & k_3 \\ m_1 & m_2 & m_3 \\ n_1 & n_2 & n_3 \end{bmatrix}\begin{bmatrix} \boldsymbol{x} \\ \boldsymbol{y} \\ \boldsymbol{z} \end{bmatrix}$$

which is simplified to

$$[g_x \quad g_y \quad g_x] = [g_A \quad g_B \quad g_C]\begin{bmatrix} k_1 & k_2 & k_3 \\ m_1 & m_2 & m_3 \\ n_1 & n_2 & n_3 \end{bmatrix}$$

and the square matrix is the matrix, $\boldsymbol{M}$ of equation (6.30). This last equation can be summarized as the matrix equation

$$g_{xyz} = g_{ABC}\boldsymbol{M} \tag{6.31}$$

where the **g**s are vectors of the *coefficients.* With X-ray data, the g_{ABC} are the x/a, y/b, z/c data given in crystallographic tables. This equation shows us how to calculate the coordinates in ***xyz*** axes given those in ***abc***. To calculate ***abc*** coordinates given ***xyz***, the equation to use is

$$g_{ABC} = g_{xyz}\boldsymbol{M}^{-1}$$

which is obtained by right multiplying both sides of (6.31) by the inverse of $\boldsymbol{M}$.

If a vector $\boldsymbol{W}$ between two atoms represents the bond vector, then its coordinates are conventionally given in terms of ***abc*** and its length can be calculated from equation (6.21). More conveniently, the dot product $d^2 = |\boldsymbol{W}\cdot\boldsymbol{W}| = |g_{xyz}\cdot g_{xyz}|$ can be used, and with equation (6.31) is

$$d^2 = (g_{ABC}\boldsymbol{M})\cdot(g^T_{ABC}\boldsymbol{M}^T), \tag{6.32}$$

which is the same as equation (6.21) when it is expanded out. The superscript T means that the transpose is taken, see Chapter 7. This equation looks a little different from those dot products calculated previously because the transpose is in a different place. This is because g is a row vector and the left vector has to be a row, the right-hand side as a column, hence the transpose.

The bond angle can be calculated using the dot product of two bond vectors after converting into ***xyz*** coordinates. The angle is therefore

$$\theta_{12} = \cos^{-1}\left(\frac{g_{1xyz}\cdot g_{2xyz}}{|g_{1xyz}||g_{2xyz}|}\right) = \cos^{-1}\left(\frac{g_{1xyz}\cdot g_{2xyz}}{d_1 d_2}\right). \tag{6.33}$$

In a monoclinic crystal, which is perhaps the most common crystal type, the equations are considerably simplified because $\alpha = \gamma = 90^\circ$ then the matrix M (equation (6.30)) becomes

$$M = \begin{bmatrix} a & 0 & 0 \\ 0 & b & 0 \\ c\cos(\beta) & 0 & c\sin(\beta) \end{bmatrix}.$$

and the new coordinates are

$$[g_x \quad g_y \quad g_z] = [ag_A + c\cos(\beta)g_C \quad bg_B \quad c\sin(\beta)g_C]$$

In Section 6.8.3 some data from s-tetrazine was used to calculate bond lengths and angles. The unit cell parameters are $a = 0.523$ nm, $b = 0.579$ nm, and $c = 0.663$ nm and angle $\beta = 115.5^\circ$ making $\cos(\beta) = -0.4305$, $\sin(\beta) = 0.9026$. Using equation (6.31) the $g_{x,y,z}$ coefficients for the carbon atom (in nm) are

$$[g_x \quad g_y \quad g_z] = [0.2546 \quad 0.0153 \quad 0.1380]\begin{bmatrix} a & 0 & 0 \\ 0 & b & 0 \\ -0.4305c & 0 & 0.9026c \end{bmatrix}$$

$$= [0.09606 \quad 0.008859 \quad 0.08258].$$

The C–N_1 bond length can be calculated either by converting the C or N_1 atom's coordinates separately, or by forming the g_{ABC} values of a bond vector as the difference in C and N_1 values, and then transforming with the M matrix. The result is 0.1346 nm, which is the same found in Section 6.8.3 but arrived at far more simply.

A more complex example of using these formulae is to calculate the volume of the unit cell, (see Section 6.10.2) bond distances C_1-C_2 and C_2-C_3 and the C_1-C_2 to C_2-C_3 bond angle in the benzene ring of o-nitrobenzoic acid. Maple will be used to do the arithmetic. The crystal structure is triclinic with the following cell parameters (Wyckoff 1969, p. 158),

$$a = 7.55\text{ Å},\ b = 4.99\text{ Å},\ c = 12.50\text{ Å},\ \alpha = 122^\circ 30',\ \beta = 95^\circ 18',\ \gamma = 118^\circ 54'.$$

	x/a	y/b	z/c
C_1	−0.2182	−0.0628	0.1928
C_2	−0.2308	−0.09721	0.2931
C_3	−0.3639	−0.1913	0.3521

3 2 1 COOH NO_2

The first step is to define the matrix M, and then equation (6.31) where the row vector is the vector of bond coordinates. When M was defined, it was done so in terms of fractional coordinates, as shown in the table, so that these can be used directly in the calculation. The angles are quoted in degrees and minutes and must be converted into fractions of degrees and then radians.

Algorithm 6.1

```
> with(LinearAlgebra):

> a:= 7.55; b:= 4.99; c:= 12.50;
  alpha:= 122.5*Pi/180;          # angle in radians
  beta:= (95+18/60.)*Pi/180;     # convert minutes to decimal
  gama:= (108+54/60.)*Pi/180; # use gama as gamma is reserved
  C1:= < -0.2812 | -0.0628 | 0.1928 >: # x/a, y/a, z/a
  C2:= < -0.2308 | -0.0972 | 0.2931 >:
  C3:= < -0.3639 | -0.1913 | 0.3521 >:
  n2:=(cos(alpha)-cos(gama)*cos(beta))/sin(gama):
  M:= < < a | 0 | 0 >,
        < b*cos(gama)| b*sin(gama)| 0 >,
        < c*cos(beta)| c*n2| c*sqrt(sin(beta)^2-n2^2)> >:
evalf(M);
```

$$\begin{bmatrix} 7.5500000000 & 0.0000000000 & 0.0000000000 \\ -1.6163479190 & 4.7209659390 & 0.0000000000 \\ -1.1546323410 & -7.4943033100 & 9.9374162660 \end{bmatrix}$$

```
volume:= evalf(Determinant(M));                    # in Å³
```

$$volume := 354.2022$$

```
# bond coordinates are vector C2-C1, C2-C3.
dcm1:= evalf((C2-C1).M):
bondC2C1:= sqrt(dcm1 . dcm1);                      # in nm
```

$$bondC2C1 := 1.3898$$

```
dcm3:= evalf((C2-C3).M):
bondC2C3:= sqrt(dcm3 . dcm3);
```

$$bondC2C3 := 1.4063$$

```
angle12_23:= evalf(
        arccos((dcm1.dcm3)/(bondC2C1*bondC2C3))*180/Pi);
```

$$angle12\_23 := 124.2383$$

The volume is in Å^3, the bond lengths in Å, and the angle in degrees. In calculating the angles, it is important to make the central atom either the first or the second atom in both vectors; if it is not then 180° minus the bond angle will be calculated.

6.10.2 Unit cell volume

The absolute value of the determinant of M is numerically equal to the volume of the unit cell, (Giacovazzo et al. 1992, p. 68). When all the angles are 90° then this matrix reduces to a diagonal form, and the determinant is abc, which is equal to the volume. Consider a monoclinic crystal, such as benzoic acid, where only one crystal angle is not 90°, and where $a = 0.552$ nm, $b = 0.514$ nm, and $c = 2.190$ nm and the angle $\beta = 97°$. The volume of the crystal will only be a little smaller than that of a solid rectangle of sides $abc = 0.621$ nm^3 because the angle β is close to 90°. The M matrix for a monoclinic crystal where $\alpha = 90°$ and $\gamma = 90°$ is

$$\begin{bmatrix} a \\ b \\ c \end{bmatrix} = \begin{bmatrix} a & 0 & 0 \\ 0 & b & 0 \\ c\cos(\beta) & 0 & c\sin(\beta) \end{bmatrix} \begin{bmatrix} x \\ y \\ z \end{bmatrix} \tag{6.34}$$

Using Maple to calculate the determinant is easy, although it is very simple to do by hand in this particular example because the terms that are zero mean that the determinant is the product of the diagonal terms.

```
> restart: with(LinearAlgebra): assume(beta > 0);
> M:=< <a|0|0>,<0|b|0>,< c*cos(beta)|0| c*sin(beta) > >;
```

$$M := \begin{bmatrix} a & 0 & 0 \\ 0 & b & 0 \\ c\cos(\beta) & 0 & c\sin(\beta) \end{bmatrix}$$

```
> V:= Determinant(M);
```

$$V := a\,b\,c\sin(\beta)$$

This result makes sense viewing the geometry of a monoclinic crystal, Fig. 6.22, which is that of a rectangular box pushed along one side. The term $\sin(\beta)$ is the amount the height is reduced compared to a rectangular box. Using the cell parameters, the volume of the benzoic acid unit cell is 0.617 nm^3, which is slightly smaller than the rectangular 0.621 nm^3.

6.11 Questions

Full solutions are available at **www.oxfordtextbooks.co.uk/orc/beddard.**

Q6.33 Find the general formula for the volume of a unit cell using equation (6.30). Simplify the answer as far as possible.

Q6.34 Determine what Bravais lattices make $n_2 = \dfrac{\cos(\alpha) - \cos(\gamma)\cos(\beta)}{\sin(\gamma)} = 0$; see equation (6.29).

Q6.35 If a space group is monoclinic then $\alpha = \gamma = 90°$ and $\beta \neq 90°$ and the unit cell dimensions are a, b, and c. By convention, β is the angle between sides a and c, see Fig. 6.20. Show that the bond distance between two atoms is the same using equation (6.21) as equation (6.34).

Q6.36 **(a)** Write down a basis set to define the position of atoms in a cubic crystal with sides a, b, and c, and calculate the angle θ if point 4 is $a/3$ along the side.

(b) For a two-dimensional hexagonal structure such as graphite, as shown in the figure, the unit cell axis are at 60°. The axes can be defined with unit vectors $\boldsymbol{u}$ and $\boldsymbol{v}$. Calculate the lengths 1-2, 1-3, 1-4, 1-5, and angles 2-1-3, 2-1-4, and 2-1-5.

***Strategy*: (b)** The natural basis set should lie along the sides of the hexagonal unit cell and then this is labelled with vectors $\boldsymbol{u}$ and $\boldsymbol{v}$. If the basis set is written as $[u, 0]$, $[0, v]$ then this would be an orthogonal set, but the angle between the vectors is 60° not 90° so this is cannot be right. It is better to transform the vectors into an orthogonal x-y set using the transformation matrix described in the text; equation (6.31). As the structure is two dimensional, then the axis c is zero and the matrix becomes two dimensional. Taking point 1 to be the origin, point 2 is at {3a, 3a}, 3 at $\{2a, 4a\}$, 4 at $\{1a, 4a\}$, and 5 at $\{4a, 1a\}$ in $\boldsymbol{u}$ and $\boldsymbol{v}$ unit vectors. This can be seen by counting the number of diamond shapes defined by the $\boldsymbol{u}$-$\boldsymbol{v}$ basis set needed to cross the hexagons to a given point.

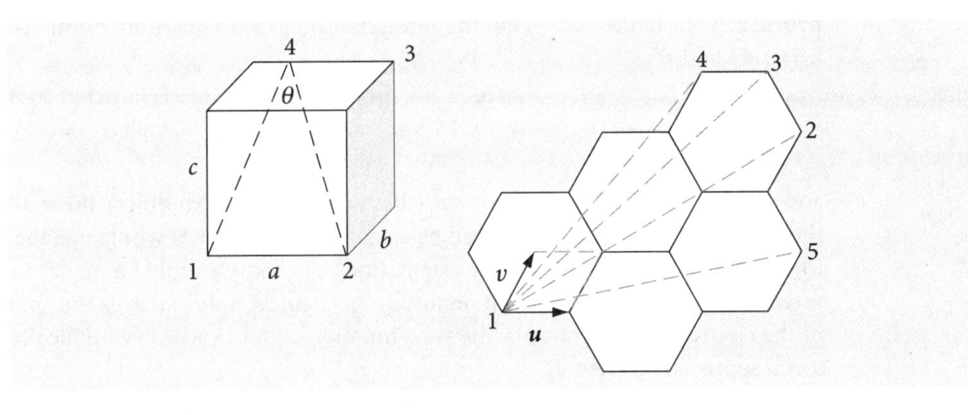

Fig. 6.24 Geometry for a cube and hexagonal structure such as graphite.

Q6.37 Using Maple, repeat the tetrazine example in the text then calculate the C–N_2, N_1–N_3 bond lengths and CN_1N_3 bond angle.

Q6.38 Recalculate the previous question using matrices and Maple.

6.12 Basis sets with more than three dimensions

The x, y, and z-axes at 90° to one another are very familiar and the basis set of unit vectors $\boldsymbol{i}$, $\boldsymbol{j}$, and $\boldsymbol{k}$ can be used to describe any vector in these axes. The basis vectors $\{\boldsymbol{i}, \boldsymbol{j}, \boldsymbol{k}\}$ are orthogonal; geometrically they are each at 90° to one another and their dot product is zero. We have seen that $\boldsymbol{i}\cdot\boldsymbol{j} = \boldsymbol{j}\cdot\boldsymbol{k} = \boldsymbol{i}\cdot\boldsymbol{k} = 0$ and that the vectors are normalized, $\boldsymbol{i}\cdot\boldsymbol{i} = \boldsymbol{j}\cdot\boldsymbol{j} = \boldsymbol{k}\cdot\boldsymbol{k} = 1$; however, not all of the properties of molecules are represented in three dimensions. For instance, molecular orbitals need large basis sets. Benzene for instance, would need at least a six element basis set just to describe the π electrons alone. At the other extreme, we could have an infinite basis set, which is also quite common in quantum mechanical problems and would be needed to handle the infinite number of vibrational levels in a harmonic oscillator. Fortunately, a large basis set is handled in the same way as a small one.

Suppose a four-dimensional set of unit vectors is made each one lying along one of four orthogonal axes. We cannot represent this geometrically in any easy way, but we can do so algebraically. The simplest way is to make a vector, row or column, containing the four elements and then to normalize it if necessary. In question Q6.28 we used a wavefunction of the form $\psi = a\psi_s + b\psi_{px} + c\psi_{py} + d\psi_{pz}$. This wavefunction is expressed in the basis set of one 2s and three 2p atomic orbitals ψ_s, ψ_{px}, ψ_{py}, ψ_{pz}. We assume that each of these orbitals are already orthogonal and normalized; i.e. that they form an *orthonormal basis set*. If they are not orthonormal then we have to perform a Gram–Schmidt orthogonalization procedure, which we need not delve into, but is described in more advanced texts such as Arkfen (1970).

The four-dimensional basis set can be written either as a column or as a row vector

$$[\psi_s \quad \psi_{px} \quad \psi_{py} \quad \psi_{pz}].$$

It does not matter which, but the rules for normalization and orthogonality are here assumed to be understood and are not stated explicitly. In standard column vector notation the basis set is

$$(\psi_s, \psi_{px}, \psi_{py}, \psi_{pz}) = \left(\begin{bmatrix} 1 \\ 0 \\ 0 \\ 0 \end{bmatrix}, \begin{bmatrix} 0 \\ 1 \\ 0 \\ 0 \end{bmatrix}, \begin{bmatrix} 0 \\ 0 \\ 1 \\ 0 \end{bmatrix}, \begin{bmatrix} 0 \\ 0 \\ 0 \\ 1 \end{bmatrix} \right)$$

and the normalization and orthogonality is clear. Wavefunctions describing the eigenstates of a molecule or atom are orthogonal, but not necessarily normalized, and are used to calculate the expectation values of operators. The orbitals ψ_{2px}, ψ_{2py}, and ψ_{2pz} are hybrid orbitals in the basis set of the three 2p orbitals with quantum numbers $n = 2$, $l = 1$, $m = 0, \pm 1$ and are

$$\psi_{2p_x} = \frac{1}{\sqrt{2}}(\psi_{2p_{-1}} + \psi_{2p_{+1}}), \qquad \psi_{2p_y} = \frac{i}{\sqrt{2}}(\psi_{2p_{-1}} - \psi_{2p_{+1}}), \qquad \psi_{2p_z} = \psi_{2p_0},$$

and as the $\psi_{2p0,\pm 1}$ functions are known, (they are spherical harmonics), the spatial extent of the wavefunctions, for instance, can be calculated. However, if we change the calculation to one using vectors, many other calculations can more simply be made. In fact, if the wavefunctions were to represent quantum mechanical 'spin', such as the spin properties of the electron or of a nucleus, the wavefunction is only known symbolically and a vectorial approach is essential.

Returning to the three 2p orbitals these can equivalently be written as

$$\boldsymbol{v}_{2p_x} = \frac{1}{\sqrt{2}}(\boldsymbol{i} + \boldsymbol{j}), \qquad \boldsymbol{v}_{2p_y} = \frac{1}{\sqrt{2}}(\boldsymbol{i} - \boldsymbol{j}), \qquad \boldsymbol{v}_{2p_z} = \boldsymbol{k}$$

in the three-dimensional $\boldsymbol{i}, \boldsymbol{j}, \boldsymbol{k}$ basis set. In vector form the equations are

$$\boldsymbol{v}_{2p_x} = \frac{1}{\sqrt{2}}[1 \quad 1 \quad 0], \qquad \boldsymbol{v}_{2p_y} = \frac{1}{\sqrt{2}}[1 \quad -1 \quad 0], \qquad \boldsymbol{v}_{2p_z} = \frac{1}{\sqrt{2}}[0 \quad 0 \quad \sqrt{2}],$$

and note that column vectors would be equally acceptable. A basis set for the *s* and three 2p orbitals cannot be written using the $\boldsymbol{i}, \boldsymbol{j}, \boldsymbol{k}$ basis set because there are four elements. A vector must be used and in row-vector form a hybrid orbital of s and a mixture of $2p_x$, $2p_y$, and $2p_z$ orbitals making $2sp^2$ hybrid orbitals have combinations

$$\boldsymbol{v}_1 = \frac{1}{\sqrt{6}}[\sqrt{2} \quad -1 \quad \sqrt{3} \quad 0] \quad \text{or} \quad \boldsymbol{v}_2 = \frac{1}{\sqrt{6}}[\sqrt{2} \quad -1 \quad -\sqrt{3} \quad 0]$$

where the *basis set ordering* is s, p_x, p_y, p_z. If we are dealing with a conjugated molecule, conventionally the *z* direction or $2p_z$ orbital is used to form any π bonds, the s and other two 2p orbitals form the sp^2 hybrid, the last entry of the $\boldsymbol{v}_1$ and $\boldsymbol{v}_2$ vectors is thus zero. To test if the orbitals are orthogonal, calculate their dot product. (The superscript *T* means make the transpose that converts rows into columns or vice versa.)

$$\boldsymbol{v}_1 \cdot \boldsymbol{v}_2 = [\boldsymbol{v}_1][\boldsymbol{v}_2]^T = \frac{1}{6}[\sqrt{2} \quad -1 \quad -\sqrt{3} \quad 0]\begin{bmatrix} \sqrt{2} \\ -1 \\ \sqrt{3} \\ 0 \end{bmatrix} = (2+1-3+0)/6 = 0$$

which shows that the two sp^2 orbitals are orthogonal in the basis of the *s* and the three 2p-atomic orbitals. Each orbital is normalized as $\boldsymbol{v}_1 \cdot \boldsymbol{v}_1 = 1$ and similarly for $\boldsymbol{v}_2$. (See Chapter 7 for matrix row-column multiplication rules)

6.13 Large and infinite basis sets

Large basis sets are often, but not always, needed when solving quantum mechanical problems. The basis set will contain *n* vectors starting with $\boldsymbol{x}_1$ and extending up to $\boldsymbol{x}_n$ instead of just three such as $\boldsymbol{i}, \boldsymbol{j}, \boldsymbol{k}$. Any vector is written in the usual way as a linear combination of the $\boldsymbol{x}$ base vectors multiplied with constants or coefficients $\boldsymbol{v}_1$ to $\boldsymbol{v}_n$, and we shall assume, for simplicity, that these are scalars, i.e. simple numbers and not complex numbers as they can sometimes be in quantum problems. The basis set vectors are, by definition, orthonormal.

A vector V is always written as a linear combination of base vectors $\boldsymbol{x}$

$$V = v_1\boldsymbol{x}_1 + v_2\boldsymbol{x}_2 + \cdots + v_n\boldsymbol{x}_n = \sum_{q=1}^{n} v_q\boldsymbol{x}_q$$

where the symbol Σ is the shorthand for summation, and the dummy variable q takes the values from 1 to n. Suppose that there is another vector W then in the same fashion,

$$W = \sum_{p=1}^{n} w_p\boldsymbol{x}_p$$

and the dot product in summation and matrix form is

$$W \cdot V = \left(\sum_{p=1}^{n} w_p\boldsymbol{x}_p\right)\left(\sum_{q=1}^{n} v_q\boldsymbol{x}_q\right) \equiv [w_1 \quad w_2 \quad \cdots]\begin{bmatrix} v_1 \\ v_2 \\ \vdots \end{bmatrix}. \tag{6.35}$$

in this case, also using dummy indices p and q. This equation is interpreted to mean that we first expand out both the summations then multiply term by term but keeping the dot operator $\cdot$ between the left and right hand of each pair of values. The subscripts p or q are only an index to each term; they have no physical meaning and so are dummy subscripts. Let us do the summation with three terms to show how the double summation and dot product works. Notice that we have to use subscripts for the coefficients w_1 etc. if we are going to use this method. Let

$$V = v_1\boldsymbol{x}_1 + v_2\boldsymbol{x}_2 + v_3\boldsymbol{x}_3, \qquad W = w_1\boldsymbol{x}_1 + w_2\boldsymbol{x}_2 + w_3\boldsymbol{x}_3$$

and calculate the product directly by expanding the terms but keeping the base vectors in order,

$$\begin{aligned} W \cdot V &= (w_1\boldsymbol{x}_1 + w_2\boldsymbol{x}_2 + w_3\boldsymbol{x}_3)\cdot(v_1\boldsymbol{x}_1 + v_2\boldsymbol{x}_2 + v_3\boldsymbol{x}_3) \\ &= w_1v_1\boldsymbol{x}_1\cdot\boldsymbol{x}_1 + w_1v_2\boldsymbol{x}_1\cdot\boldsymbol{x}_2 + w_1v_3\boldsymbol{x}_1\cdot\boldsymbol{x}_3 + \cdots. \end{aligned}$$

The basis set vectors are orthonormal, therefore, for instance, $\boldsymbol{x}_1 \cdot \boldsymbol{x}_1 = 1$ and $\boldsymbol{x}_1 \cdot \boldsymbol{x}_2 = 0$. The Kronecker delta $\delta_{p,q}$ is conveniently used to retain terms where $p = q$, when $\delta_{p,p} = 1$, and removes 'cross terms' where $p \ne q$ and $\delta_{p,q} = 0$, making, for example, terms such as $x_2 \cdot x_3 = 0$. The Kronecker delta $\delta_{p,q}$ greatly simplifies the calculation because it is always zero, unless $p = q$, then it is unity. It does exactly what was done earlier with equations of the form $\boldsymbol{i}\cdot\boldsymbol{i} = 1$ and $\boldsymbol{i}\cdot\boldsymbol{k} = 0$ in the $\boldsymbol{i}, \boldsymbol{j}, \boldsymbol{k}$ basis but now for any two terms of our basis set vector x. Using the δ function gives

$$\begin{aligned} W \cdot V = &\, w_1v_1\delta_{11} + w_1v_2\delta_{12} + w_1v_3\delta_{13} \\ &+ w_2v_1\delta_{21} + w_2v_2\delta_{12} + w_2v_3\delta_{23} \\ &+ w_3v_1\delta_{31} + w_3v_2\delta_{32} + w_3v_3\delta_{33} \end{aligned}$$

which in fact hardly need be written down because it becomes

$$W\cdot V = w_1v_1 + w_2v_2 + w_3v_3$$

by evaluating the δ function. This is the familiar dot product equation and in matrix form is $W\cdot V = [w_1 \quad w_2 \quad w_3]\begin{bmatrix} v_1 \\ v_2 \\ v_3 \end{bmatrix} = w_1v_1 + w_2v_2 + w_3v_3$.

Starting with the general formula for n terms, the expansion can be simplified by moving the summation to the front of the calculation, which gives after some lengthy but not difficult algebra,

$$W\cdot V = \sum_{p=1}^{n}\sum_{q=1}^{n} w_pv_q x_p\cdot x_q = \sum_{p=1}^{n}\sum_{q=1}^{n} w_pv_q\delta_{p,q}. \tag{6.36}$$

If $n = 3$, the expansion is

$$\begin{aligned} W\cdot V &= \sum_{p=1}^{3}\sum_{q=1}^{3} w_pv_q\delta_{p,q} \\ &= \underbrace{w_1(v_1\delta_{11} + v_2\delta_{12} + v_3\delta_{13})}_{p=1} + \underbrace{w_2(v_1\delta_{21} + v_2\delta_{22} + v_3\delta_{23})}_{p=2} + \underbrace{w_3(v_1\delta_{31} + v_2\delta_{32} + v_3\delta_{33})}_{p=3} \end{aligned}$$

and when the δ's are evaluated again produces $W\cdot V = w_1v_1 + w_2v_2 + w_3v_3$.

The projection of one vector onto another is calculated with equation (6.14) and this can be expressed also as the summation formula,

$$P_{w\to v} = \left[\frac{W\cdot V}{|V|^2}\right]V = \left[\frac{\sum_{p=1}^{n}\sum_{q=1}^{n} w_pv_q\delta_{p,q}}{\sum_{q}^{n} v_p^2}\right]\sum_{q}^{n} v_q x_q$$

where x are the base vectors. The term in the square brackets is a (scalar) number and is zero except when $p = q$.

Although the summation has been described in detail, the simplest way to evaluate large basis sets is to use row-column matrix multiplication; see equation (6.35), using a computer because all languages now have built in and optimized routines to do this. In practice, when calculating energy levels numerically, as big a basis set as possible is used to ensure an accurate calculation.

Calculating the energy levels of a harmonic oscillator can be done this way, and as the basis sets get bigger, the more accurate the energies become. In the particular case of the harmonic oscillator, we know what the exact energies are, but this is not always so. If two potentials are coupled, or one perturbed by an electric or magnetic field, or in numerous other situations in which the Schrödinger equation cannot be solved algebraically but only numerically, then a large basis set is needed. Some examples of how to solve the Schrödinger equation numerically are illustrated in Chapters 8 and 10.

6.14 Basis sets in molecules

In a molecule with π electrons, such as butadiene, we naturally want to form molecular orbitals in the basis of the carbon atom's 2p π orbitals. However, it is important to realize that as the π orbitals exist on different atoms this has implications for our basis set.

Using chemical intuition, we assume that all the orbitals are pointing in the same direction as shown in Fig. 6.25. As there are 4 $p\pi$ orbitals in butadiene, we must form four MOs, and as there are also 4π electrons, two orbitals will be bonding and two, anti-bonding, Fig. 6.25, but only the lower two orbitals are filled. The π orbitals point out of the plane formed by the atoms in the chemical structure, and out of the page as drawn, but they are turned through a right angle and viewed side-on in the lower part of the figure.

If we try to represent the molecular orbitals with a basis set $(\pi_1, \pi_2, \pi_3, \pi_4)$, where the π's represent *vectors* on the π orbitals on each atom, then this would simply be wrong. This

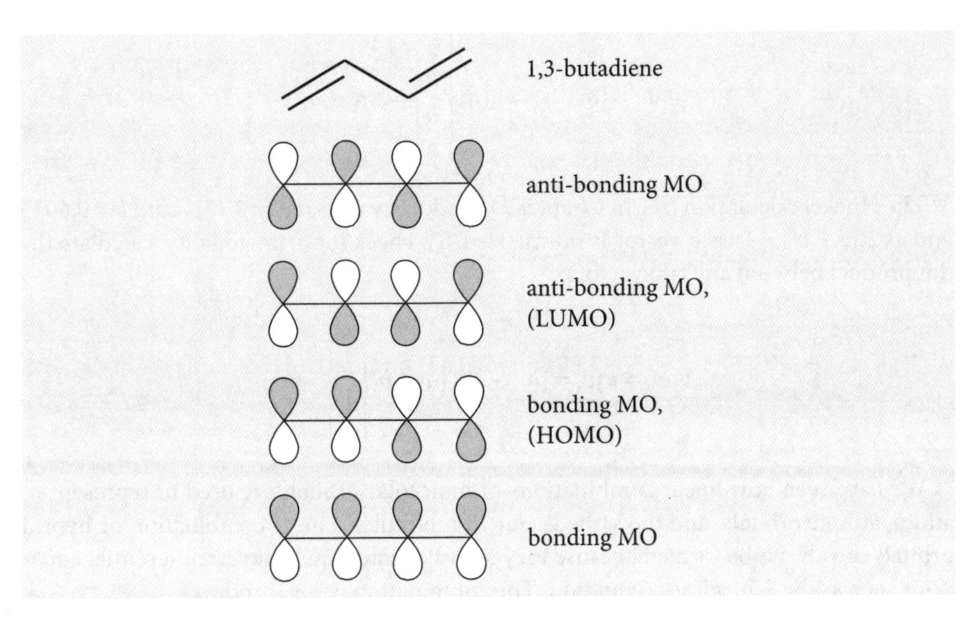

Fig. 6.25 The bonding and anti-bonding molecular pπ orbitals in butadiene are shown in order of increasing energy. The shading represents the relative phase component of the orbital often represented by + and –.

is because the orbitals are on different atoms and so not based on the same axes but axes displaced from one another by the length and angle of the bonds. We would somehow need to relate π_2 to π_1 by allowing for each displacement and angle, and this would be very difficult and ultimately futile. We could still decide that the π orbitals are normalized by insisting that $\pi_m \cdot \pi_m = 1$, where m and n are index numbers 1 to 4, but we cannot expect that $\pi_m \cdot \pi_n = 0$ because of the different coordinate origin of each orbital. It is difficult to know what $\pi_n \cdot \pi_m$ will be, so we cannot very easily do the calculation in this basis set.

An orthonormal basis set can, however, be found based on the π atomic orbital set above but now as *linear combinations* of these π orbitals. Suppose we make a set of four new vectors each of which is a linear combination of the π orbital vectors and where the first vector v_1 is $v_1 = a\pi_1 + b\pi_2 + c\pi_3 + d\pi_4$. We will need to calculate the components a, b, c, d, but as each π vector is now based on its own coordinate on each atom $\pi_m \cdot \pi_n = 0$ and $\pi_m \cdot \pi_m = 1$, which confirms what is already known because atomic orbitals are orthonormal. The general and systematic way to find the coefficients is to use molecular group theory, but for small molecules we can often guess the linear combinations.

In butadiene there are only two classes of atoms, those at the end and those in the middle; therefore, by symmetry, only two constants a and b are needed, the vectors must have the form

$$v = a\pi_1 \pm b\pi_2 \pm b\pi_3 \pm a\pi_4,$$

and the ± signs ensure orthogonality. The *set of the base vectors* is

$$(v_1 \quad v_2 \quad v_3 \quad v_4) = \left(\begin{bmatrix} a \\ b \\ b \\ a \end{bmatrix}, \begin{bmatrix} b \\ a \\ -a \\ -b \end{bmatrix}, \begin{bmatrix} b \\ -a \\ -a \\ b \end{bmatrix}, \begin{bmatrix} a \\ -b \\ b \\ a \end{bmatrix} \right)$$

where each member of the basis set is itself a (column) vector and is an eigenvector from a Hückel calculation. The plus and minus pattern of the vector components is the same as the phase of the π-orbitals. In our previous calculations $\boldsymbol{i}$, $\boldsymbol{j}$, and $\boldsymbol{k}$ base vectors were used, but when there are four or more base vectors, each with four or more terms, a matrix representation has to be used. We could, if so inclined, define a set of four orthogonal unit vectors, say, $\boldsymbol{i}$, $\boldsymbol{j}$, $\boldsymbol{k}$, $\boldsymbol{m}$ and define their properties as for the three-dimensional vectors $\boldsymbol{i}$, $\boldsymbol{j}$, $\boldsymbol{k}$, and use these instead of the matrices but this would be very cumbersome.

As a check that the elements of our butadiene basis set are normalized, the dot product $v_1 \cdot v_1$ is

$$v_1 \cdot v_1 = [a \quad b \quad b \quad a]\begin{bmatrix} a \\ b \\ b \\ a \end{bmatrix} = 2(a^2 + b^2)$$

The Hückel calculation (see in Chapter 7) produces values of $a = 0.3717$ and $b = 0.6015$ and as $2(a^2 + b^2) = 1$ each vector is normalized. To check for orthogonality, calculate the dot product between any two vectors;

$$v_3 \cdot v_1 = v_3^T v_1 = [b \quad -a \quad -a \quad b]\begin{bmatrix} a \\ b \\ b \\ a \end{bmatrix} = 0.$$

We have seen how linear combinations of molecular orbitals are used to represent p_x and p_y atomic orbitals, and the same is true in molecules. Linear combination or hybrid orbitals can always be formed because they are valid solutions of linear differential equations such as the Schrödinger equation. The combination $v_1 + v_3$ produces

$$v_{1+3} = \begin{bmatrix} a+b \\ b-a \\ b-a \\ a+b \end{bmatrix}$$

with a dot product with itself of $4(a^2 + b^2) = 2$, so this vector is normalized by dividing with $1/\sqrt{2}$. The same result can be found by calculating with the base vectors $(v_1 + v_3)\cdot(v_1 + v_3) = 2$. The dot product with v_2 is

$$v_2 \cdot v_{1+3} = [v_2]^T[v_{1+3}] = [b \quad a \quad -a \quad -b]\begin{bmatrix} a+b \\ b-a \\ b-a \\ a+b \end{bmatrix} = 0,$$

showing that this orbital is also orthogonal to the others that are not part of the linear combination.

When orbitals are degenerate, forming linear combinations can be a more convenient way of viewing the orbitals. This is true in benzene where some of the familiar shapes of the Hückel MOs are linear combinations; see Q7.7 and Q7.50. To make a more accurate calculation of other orbitals besides the pπ, perhaps higher s or d orbitals must be added and then the basis set will have to expand. In molecular orbital calculations, a choice of predetermined basis sets of different complexity such as 631G can be used. Very often these basis sets consist of Gaussian functions (ae^{-bx^2}) parameterized with constants a and b to fit to Slater type atomic orbitals. Gaussian functions are used as they can be integrated easily. Huge basis sets containing millions of terms are necessary to represent molecular orbitals accurately and computer calculations of this type are very time consuming and among the most challenging of all calculations requiring super-computers.

? 6.15 Questions

 Full solutions are available at www.oxfordtextbooks.co.uk/orc/beddard.

Q6.39 The wavefunction in a 2-dimensional orthonormal vector basis set $\{\varphi_1, \varphi_2\}$ for a particular molecular orbital (eigenstate), is $\psi_1 = a_1\varphi_1 + a_2\varphi_2$, and for another is $\psi_2 = b_1\varphi_1 - b_2\varphi_2$. The amount of each basis set in each vector is a_1, a_2 and b_1, b_2.

(a) Calculate the dot product of ψ_1 and ψ_2 directly, using the summation formula and then using matrices.

(b) Normalize each wavefunction.

Q6.40 Repeat the last example with $a_1 = 2$, $a_2 = 2$, $b_1 = 2a_1$, $b_2 = -2a_2$. Normalize the wavefunctions and calculate their dot product. What can you conclude about the wavefunctions?

Q6.41 In the s and p-orbital basis set are the vectors $v_1 = [2 \quad 2 \quad 2 \quad 2]$ and $v_2 = \frac{1}{2}[1 \quad 1 \quad -1 \quad -1]$ orthogonal? Check if they are normalized.

Q6.42 Solving the Hückel MOs for benzene produces the following un-normalized vectors in a 2p π orbital basis:

$$v_1 = [1 \quad 1 \quad 1 \quad 1 \quad 1 \quad 1], \qquad v_4 = [-1 \quad 1 \quad -1 \quad 1 \quad -1 \quad 1],$$

$$v_2 = [-1 \quad 1 \quad 0 \quad -1 \quad 1 \quad 0], \qquad v_5 = [-1 \quad -1 \quad 0 \quad 1 \quad 1 \quad 0],$$

$$v_3 = [-1 \quad 0 \quad 1 \quad -1 \quad 0 \quad 1], \qquad v_6 = [1 \quad 0 \quad -1 \quad -1 \quad 0 \quad 1].$$

(a) Show that v_2 and v_3 are orthogonal to all others as are v_5 and v_6 but not to one another.

(b) Make a linear combination $v_2 \pm v_3$ and $v_5 \pm v_6$ and show that they are orthogonal to other vectors. Normalize the new vectors.

(c) If $\pm$ signs represent orbitals phases, plot these new orbitals on a figure of benzene's structure.

Q6.43 Molecules without a centre of inversion possess a dipole. The dipole μ is the difference in positive and negative charges Q multiplied by their separation a. Therefore, the dipole is $\mu = Qa$ which has units of coulomb metres. The coulomb is the SI unit of charge. (In Fig. 6.26 Q = 2δ.)

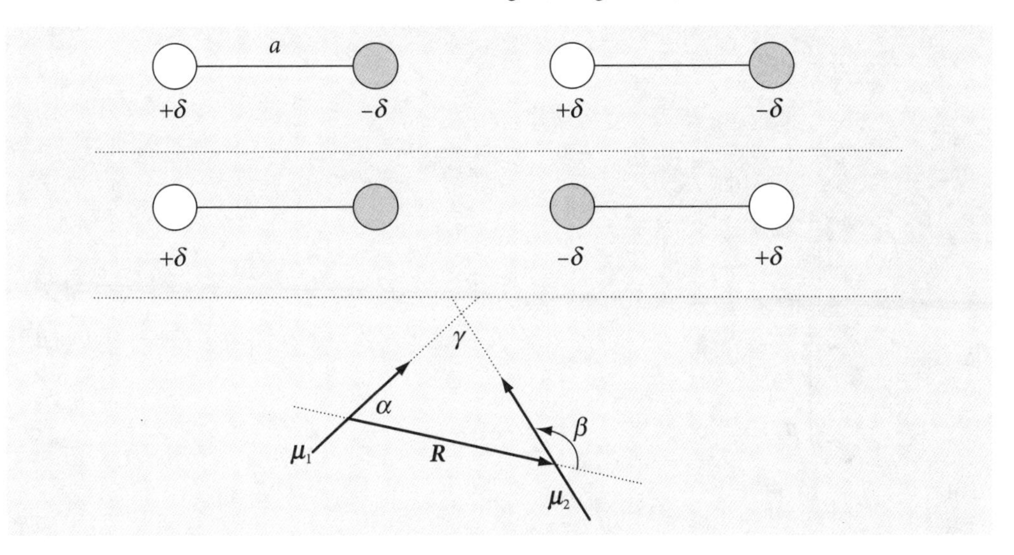

Fig. 6.26 Any two dipoles interact. In the top figure, the interaction is attractive with negative interaction energy, but this is repulsive in the geometry of the middle figure. The lower figure shows definitions of the relative angles of the three vectors μ_1 and μ_2 and $\boldsymbol{R}$ all in the plane of the figure.

Any two dipoles μ_1 and μ_2 will interact with one another, and this depends upon their relative orientation and separation R. If we perform a calculation of the interaction energy from first principles and assume that the dipoles are very small compared with their separation, then this energy in joules is

$$E = \frac{1}{4\pi\varepsilon_0}\left\{\frac{\mu_1 \cdot \mu_2}{R^3} - 3\frac{(\mu_1 \cdot R)(\mu_2 \cdot R)}{R^5}\right\} \tag{6.37}$$

where $\boldsymbol{R}$ is the vector along the line joining the centres of the two dipoles. In deriving this formula, the assumption has been made that the dipoles are small relative to their separation $a \ll R$, and not as shown in the figure.

Assuming for simplicity, that the dipoles are arranged in the plane of the diagram;

(a) What is the largest (attractive and repulsive) and

(b) the smallest interaction energy between any two dipoles? Look for two different solutions here.

(c) In Fig. 6.27 where the dipoles are parallel, show that

$$E = \frac{1}{4\pi\varepsilon_0}\frac{|\mu_1||\mu_2|}{R^3}(1 - 3\cos^2(\theta)).$$

(d) If the pairs of dipoles (i), (ii), and (iii) lie in the plane of Fig. 6.28, what is their interaction energy if $|\mu_1| = \mu_1 = 5$ D and $\mu_2 = 3$ D and $R = 4$ nm? Calculate your answers in $D^2\ nm^{-3}$ and in joules. Compare this energy to thermal energy at room temperature k_BT, and comment on the result. (1 debye, D = 3.335610^{-30} coulomb metres (C m), ε_o, the permittivity of free space = $8.85418\ 10^{-12}\ C^2\ J^{-1}\ m^{-1}$.)

Strategy: Intuitively we can see that the maximum attractive force must occur when the positive and negative ends of the dipole are adjacent to one another and the dipoles are in line, as in the top of Fig. 6.26. Notice how the angles are defined and particularly angle β. The maximum interaction must be equal and opposite to that of the minimum.

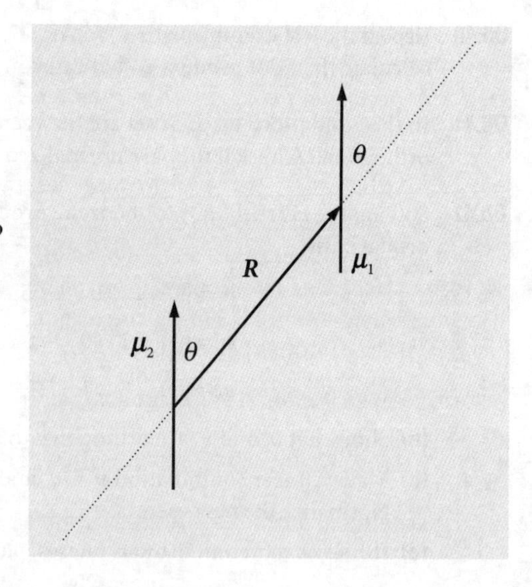

Fig. 6.27 Parallel dipoles.

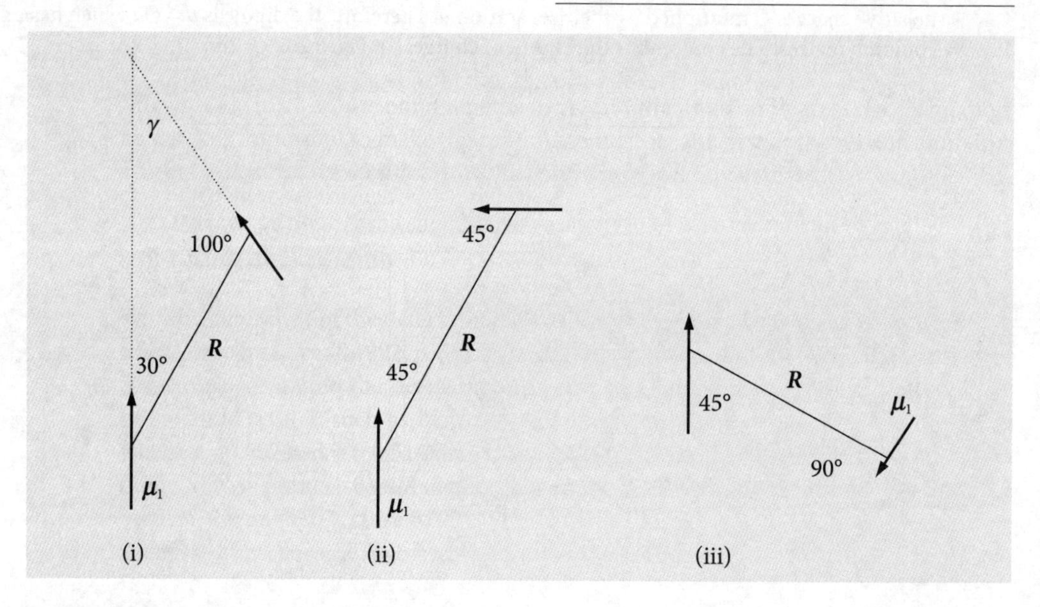

Fig. 6.28 Dipole pairs.

Q6.44 In the previous question the two dipoles were constrained to lie in a plane, the dipoles would have to lie flat on a surface to do this. Now suppose that two dipoles have arbitrary angles, as is normally the case in solution, and are separated by a distance R. The dipolar energy can be written as

$$E = \frac{\boldsymbol{\mu}_1 \cdot \boldsymbol{\mu}_2}{R^3} - 3\frac{(\boldsymbol{\mu}_1 \cdot \boldsymbol{R})(\boldsymbol{\mu}_2 \cdot \boldsymbol{R})}{R^5}$$

$$= -\frac{\mu_1 \cdot \mu_2}{R^3}[2\cos(\theta_1)\cos(\theta_2) - \cos(\psi)\sin(\theta_1)\sin(\theta_2)].$$

To put the energy into SI, units it must be divided by $4\pi\varepsilon_0$. The angles are shown in the diagram with angle $\psi = \varphi_2 - \varphi_1$ between the planes containing $\boldsymbol{\mu}_1$-$\boldsymbol{R}$ and $\boldsymbol{\mu}_2$-$\boldsymbol{R}$. Derive the trig formula from the vector one and so show that the two forms of the equation are the same. The identity

$$\cos(\varphi_2 - \varphi_1) = \cos(\varphi_1)\cos(\varphi_2) + \sin(\varphi_1)\sin(\varphi_2)$$

is needed.

Strategy: This problem becomes easier to understand by redrawing the diagram. By rotating by approximately 90°, it becomes more understandable when the axis is labelled as z. The θ are then identified as polar angles, and φ as the azimuthal angles in spherical polar coordinates, i.e. the question is defined in spherical polar coordinates and, although this is not stated, it is implied by the use of angles. To solve the problem, the x-, y-, and z-values of the tip of the μ_1 and μ_2 vectors have to be found.

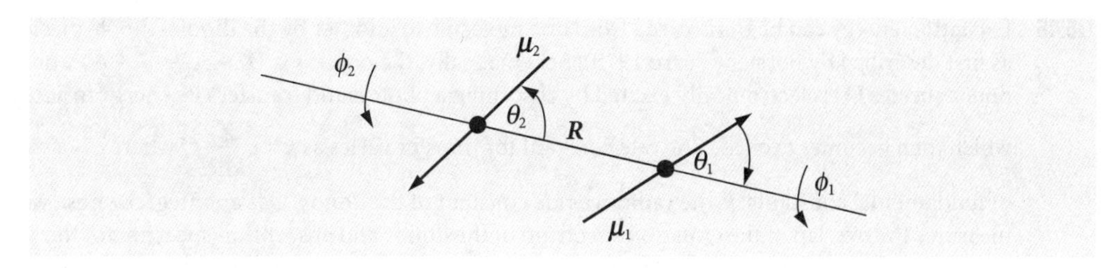

Fig. 6.29 Two dipoles with arbitrary angles.

The dot products with $\boldsymbol{R}$ are easily calculated in terms of the angles. The $\mu_1 \cdot \mu_2$ dot product is more involved. Define a set of x-, y-, and z-axes and work out the components of the μ_1 and μ_2 vectors on these axes then use the $\boldsymbol{i}, \boldsymbol{j}, \boldsymbol{k}$ or vector basis set to calculate the dot product.

Q6.45 In substituted cyclohexanes among other molecules, equatorial substitution is preferred because steric repulsion is minimized. In saturated ring molecules containing an oxygen atom, e.g. substituted tetrahydropyrans, the axial position is favoured in what is called the *anomeric* effect, which is an example of a stereoelectronic effect. The figure shows the most stable (preferred) geometry for each type of molecule.

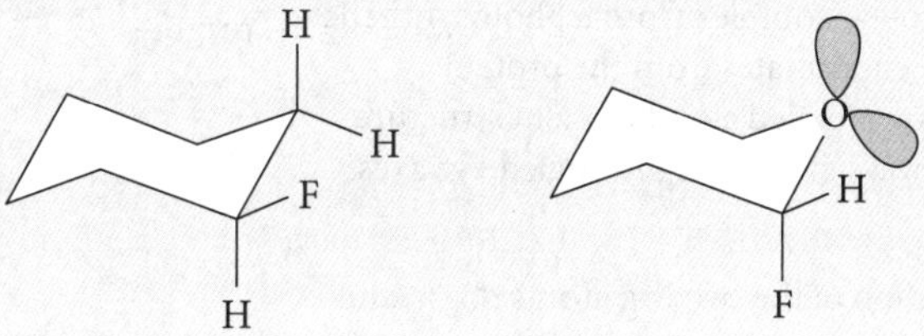

Fig. 6.30 Left: Equatorially substituted fluorocyclohexane. Right: An axial substituted fluorotetrahydropyran, which is the most stable. This is rationalized as due to the anomeric effect.

The anomeric effect is thought to be caused by the O atom's non-bonding electrons partially transferring their charge into an anti-bonding (π^*) orbital on the carbon atom, which is bonded to an electrophyllic substituent such as a halogen, X; this is sometimes called a hyper-conjugation resonance effect. The effect this has is to shorten the ring CO bond about 0.1 Å and lengthen the substituent CX bond by a similar amount. However, solvent polarity also has an effect on the ratio of equatorial to axial substituents, which suggests that the dipole–dipole (or possibly induced dipole–dipole) interaction may contribute. For example, in the tetrahydropyran shown above when X = OMe the percentage of axial substituted molecules changes from 83 in CCl_4 with a dielectric constant of 2.2, to 52 in water, with a dielectric constant of 78.5 (Pialy & Lemieux 1987).

Supposing that the lone pair electrons have an effective dipole μ_O pointing along the putative sp^3 (tetrahedral) bond direction and that the CX bond has a dipole μ_x you will need to calculate and compare the total dipole–dipole interaction when the CX substituent is in the axial and then equatorial position.

(a) Perform a thought experiment to determine whether axial or equatorial dipole-dipole interaction is favoured. The interaction energy calculated according to equation (6.37) must be multiplied by $1/\varepsilon$ where ε is the dielectric constant. Use this to explain the solvent effect.

(b) Calculate energies assuming that the CO bond length is d, and the dipoles are $\mu_x = \beta\mu_{nx}$ and $\mu_O = \alpha\mu_{nO}$ for the substituent and on the oxygen atom respectively. The unit dipoles are μ_{nx} and μ_{nO} and α and β are their respective sizes. Let the vector $\boldsymbol{R}$ in the dipole equation (6.37) be between the C and O atoms rather than the centre of the dipoles, because this simplifies the calculation; this should not alter the nature of the result.

Strategy: In (a) it serves to sketch the dipole directions. It is assumed that the non-bonding orbitals have dipoles along the normal bond directions; they could be combined vectorially into one dipole but this offers little advantage to us. In (b) the geometry of the molecule has to be determined, the bonds are tetrahedral, which means that if a carbon atom is at the centre of a cube its bonds end at $\{1, 1, 1\}, \{1, -1, -1\}, \{-1, 1, -1\}, \{-1, -1, 1\}$. Assume that the O atom is sp^3 hybridized.

Q6.46 Excitation energy can be transferred from one molecule to another by the dipole–dipole mechanism as first described by Förster (Turro 1978). Schematically, the process is $D^* + A \rightarrow D + A^*$ where donor species D is electronically excited by absorbing a photon and transfers its energy to acceptor A which then becomes excited. The rate constant for transfer varies as $k = c\dfrac{\chi^2}{k_f R^6}Q$ where c is a collection of fundamental constants, k_f the radiative rate constant of the donor, Q is an integral whose value measures the overlap of the emission spectrum of the donor and absorption spectrum of the acceptor and ensures energy conservation. R is the separation of donor and acceptor and χ^2 measures the orientation of the two transition dipoles involved and

$$\chi = \vec{\mu}_1 \cdot \vec{\mu}_2 - 3(\vec{\mu}_1 \cdot \vec{r})(\vec{\mu}_2 \cdot \vec{r}),$$

where $\vec{r}$ is the *unit* vector along the line separating the dipoles and $\vec{\mu}_1$ and $\vec{\mu}_2$ are *unit* vectors along the dipole directions. Unit vectors must be used since χ^2 is only a measure of relative orientation.

The X-ray coordinates for two of chlorophyll molecules are listed below in Å and are from the RCSB Protein Data Bank (www.rcsb.org/pdb/home.do 1VCR). The chlorophylls are part of the structure of a chlorophyll protein complex from a photosynthetic organism. The coordinates from the protein database are downloaded as .pdb or .ent structure files and are always given in right-angled *xyz* axes.

Calculate:

(a) The separation of the two Mg atoms $d_{\text{Mg-Mg}}$ and normalize the Mg to Mg atom vector.

(b) The orientation parameter χ^2 for energy transfer between chlorophyll molecules assuming the dipoles lie on the line from the Mg to nitrogen atom ND on each molecule. For the purpose of this calculation, you may assume that the unit vector $\vec{r}$ lies along the line of the Mg–Mg atoms, rather than from the mid-points of the dipole vectors. Take Mg 498 as belonging to the donor molecule.

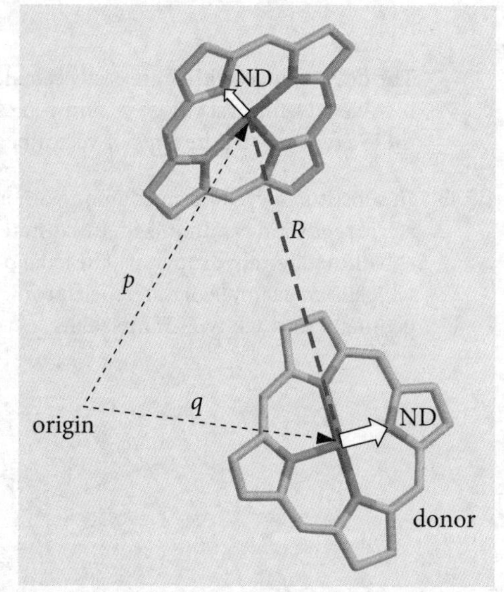

Fig. 6.31 A partial structure of the chlorophyll molecules in a photosynthetic pigment protein complex. A Mg atom is situated in the centre of each molecule. Chlorophyll belongs to a class of molecules called chlorins, but the structure shown here is of a porphyrin as ring 5 was omitted from the X-ray structure. The surrounding protein is present in the X-ray data but is not shown for clarity. The origin of coordinates is only shown diagrammatically and does not represent the true origin.

atom					x	y	z
HETATM	498	MG	CLA	251	114.289	22.722	26.723
HETATM	503	N A	CLA	251	112.811	24.082	27.586
HETATM	508	N B	CLA	251	112.900	21.809	25.675
HETATM	513	N C	CLA	251	115.826	21.842	25.694
HETATM	518	N D	CLA	251	115.789	24.007	27.556
HETATM	523	MG	CLA	252	106.145	30.029	31.708
HETATM	528	N A	CLA	252	105.539	28.845	29.965
HETATM	533	N B	CLA	252	107.913	30.363	30.917
HETATM	538	N C	CLA	252	106.695	30.719	33.562
HETATM	543	N D	CLA	252	104.380	29.292	32.675

Strategy: Locate the coordinates in the table for each atom. Calculate the distance vector $R = p - q$ using the coordinates and its magnitude $|R|$ that is the separation of the Mg atoms. Assume that the origin of the coordinates is at $\{0, 0, 0\}$.

6.16 Cross product or vector product

The cross product of two vectors produces a vector rather than a scalar. This vector is at right angles to the other two; consequently, the cross product is *unique* to vectors in three-dimensional space. The symbol × is used to indicate a cross product; some authors use ∧ although this less is common nowadays. In Maple the cross product command is `&x`.

If $\boldsymbol{u}$ and $\boldsymbol{v}$ are two vectors at an angle θ to one another, the cross product is

$$\boldsymbol{u} \times \boldsymbol{v} = |\boldsymbol{u}||\boldsymbol{v}|\sin(\theta)\vec{\boldsymbol{n}}, \tag{6.38}$$

where $\vec{\boldsymbol{n}}$ is a unit vector at right angles to $\boldsymbol{u}$ and $\boldsymbol{v}$, and shows that the result is a vector. It is not at all obvious why the cross product produces a vector; the product $|\boldsymbol{u}||\boldsymbol{v}|\sin(\theta)$ is a scalar quantity, i.e. a simple number possibly with units, but mathematically $\boldsymbol{u} \times \boldsymbol{v}$ behaves like a vector perpendicular to the plane containing $\boldsymbol{u}$ and $\boldsymbol{v}$, and hence we multiply by $\vec{\boldsymbol{n}}$. It is easy to forget to do this. Because the cross product of a vector is perpendicular to either vector, the cross product of any vector with itself is always zero $\boldsymbol{u} \times \boldsymbol{u} = 0$.

The *magnitude* or absolute value of the resultant vector is a scalar;

$$|\boldsymbol{u} \times \boldsymbol{v}| = |\boldsymbol{u}||\boldsymbol{v}|\sin(\theta). \tag{6.39}$$

The cross product is *anti-commutative*, which means that

$$\boldsymbol{u} \times \boldsymbol{v} = -\boldsymbol{v} \times \boldsymbol{u} \tag{6.40}$$

and this is shown in Fig. 6.32 where the two resultant vectors point in opposite directions. Calculating the numerical value of a cross product can only be done by expanding the vectors in the basis set as components of unit vectors $\boldsymbol{i}, \boldsymbol{j}$, and $\boldsymbol{k}$, or a column or row vectors of the components.

Relationships between combinations of dot and cross products can be calculated symbolically; for example, the expression $|A \times B|^2 + |A{\cdot}B|^2$ can be expanded using the definition of the cross and dot products.

$$|A \times B|^2 + |A{\cdot}B|^2 = |A|^2|B|^2\sin^2(\theta) + |A|^2|B|^2\cos^2(\theta) = |A|^2|B|^2$$

and in the last step $\sin^2(\theta) + \cos^2(\theta) = 1$ is used. It is not surprising that this result is a number because the dot product is always a number and so is the absolute value of the cross product.

As a second example let us calculate $(A + B) \times (A - B)$ if A and B are vectors. Expanding the multiplication but keeping the order of the terms the same gives

$$(A + B) \times (A - B) = A \times A - A \times B + B \times A - B \times B,$$

and using the definition of cross products, $A \times A = B \times B = 0$, gives

$$(A + B) \times (A - B) = -A \times B + B \times A = 2(B \times A) = -2(A \times B)$$

because $-A \times B = B \times A$.

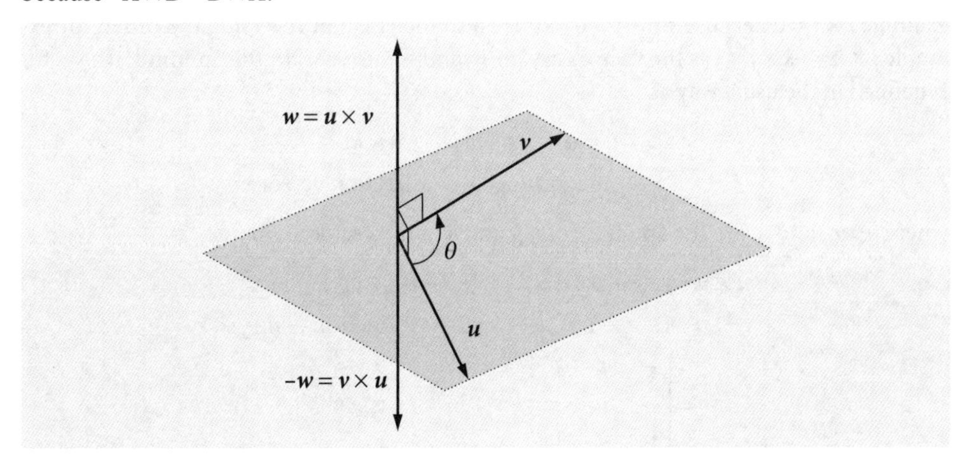

Fig. 6.32 Definition of cross products $\boldsymbol{u} \times \boldsymbol{v}$ and $\boldsymbol{v} \times \boldsymbol{u}$. Notice that $\boldsymbol{u}$ and $\boldsymbol{v}$ are in the same plane and that the two resultant vectors point in opposite directions. Notice also the relative orientation of the vectors.

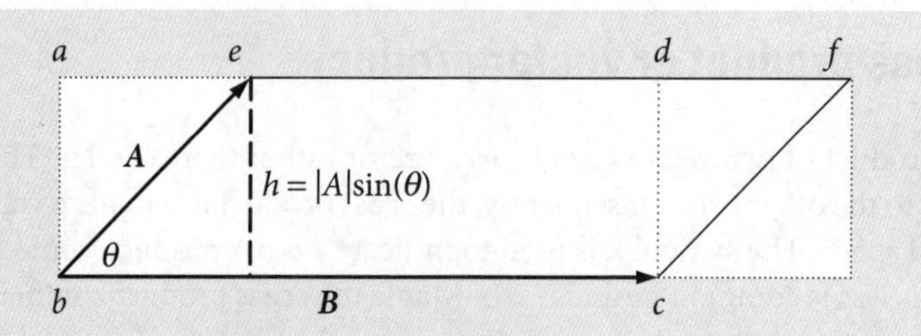

Fig. 6.33 Parallelogram

The area of the parallelogram with sides of length A and B is $AB\sin(\theta)$, where θ is the enclosed angle. A parallelogram is shown in Fig. 6.33. The strategy used to prove this result is to calculate the area of the rectangle *abcd* and to convince yourself that this is the same as the area of the parallelogram by moving triangle *cdf* onto *abe*.

By definition, $\sin(\theta)$ = *opposite/hypotenuse*, therefore the height of the rectangle $h = |A|\sin(\theta)$ and, as the area of the rectangle is $h|B|$, it follows that the area of the parallelogram is $|A||B|\sin(\theta) \equiv |A\times B|$ and the last identity follows by definition. The area of any triangle with sides A and B and enclosed angle θ is therefore

$$area = \frac{1}{2}|A||B|\sin(\theta) = \frac{1}{2}|A\times B|$$

This relationship can be understood by realizing that the area of triangle *bce* is half that of the rectangle. To calculate the area of a triangle we need to use a basis set and this is done next.

6.16.1 Cross products using the *i, j, k* basis set

Returning to equation (6.38) we can see that the cross product of two vectors can be zero; $A\times B = 0$ even when neither A nor B are zero because the angle between them is zero. It is clear that, if A and B are parallel to one another the angle between them being zero, $\sin(0) = 0$; the same is true if the vectors are anti-parallel because $\sin(180°) = 0$, see Fig. 6.32.

In the description of dot products we used the right-angled unit vector (orthonormal) basis set (i, j, k) to describe each vector, and found that in some circumstances calculations were more easily performed in this way. Naturally, we can do the same for the vectors here but we have to learn the rules for calculating cross products of the unit vectors.

The rules are easy to remember as the *indices rotate* about the equations always being in the order, $i \to j \to k$;

$$i\times j = k, \qquad k\times i = j, \qquad j\times k = i.$$

Additionally, we use the rule that the cross product of *any* vector with itself is zero, for example $i\times i = 0$ because $\sin(0) = 0$. Notice also the effect of reversing the order, for example $j\times k = -k\times j = i$ as the vectors are *anti-commutative*. With this in mind, if a vector is defined in the usual way as

$$u = u_x i \qquad +u_y j \qquad +u_z k$$

magnitude ↙ ↘ direction vector

where u_x, u_y, and u_z are the amounts of i, j, and k in the vector, then

$$\begin{aligned} u\times v &= (u_x i + u_y j + u_z k)\times(v_x i + v_y j + v_z k) \\ &= (u_y v_z - u_z v_y)i + (u_z v_x - u_x v_z)j + (u_x v_y - u_y v_x)k \\ &= \begin{vmatrix} i & j & k \\ u_x & u_y & u_z \\ v_x & v_y & v_z \end{vmatrix} \end{aligned} \tag{6.41}$$

where the last equality is a *determinant* and is by far the simplest way of remembering the cross product. Determinants are described in more detail in Chapter 7 and multiplication illustrated in Chapter 7.2.2. In the multiplication, notice that the second (middle) term is pre-multiplied by −1 and the first term starts at the top left of the four terms to be multiplied.

$$\begin{vmatrix} i & j & k \\ u_x & u_y & u_z \\ v_x & v_y & v_z \end{vmatrix} - \begin{vmatrix} i & j & k \\ u_x & u_y & u_z \\ v_x & v_y & v_z \end{vmatrix} + \begin{vmatrix} i & j & k \\ u_x & u_y & u_z \\ v_x & v_y & v_z \end{vmatrix}$$

$$= (u_y v_z - u_z v_y)i \quad - (u_x v_z - u_z v_x)j \quad + (u_x v_y - u_y v_x)k.$$

The area of any triangle, such as in Fig. 6.33, can now be found. Suppose a triangle is enclosed by three vertices {2, −1, 6}, {8, 3, 10}, and {10, −2, 16}. To calculate the area, make the triangle's sides into vectors and calculate half the absolute value of the cross product. The ***i***, ***j***, ***k*** basis set should be used. Three vectors ***a***, ***b***, and ***c*** will form a triangle if $a + b + c = 0$. Using the coordinates, let ***a*** be the difference between the first and second $a = (8-2)i + (3+1)j + (10-6)k$, ***b*** the difference between the first and third $b - 8i - j + 10k$, and then $c = -14i - 7j - 14k$ although it is not needed. The vectors must form a triangle unless they lie on the same straight line, in which case the area would be zero. The cross product is

$$a \times b = \begin{vmatrix} i & j & k \\ 6 & 4 & 4 \\ 8 & 3 & 10 \end{vmatrix} = 28i - 28j - 14k.$$

The area is therefore

$$\frac{1}{2}|a \times b| = \frac{1}{2}|28i - 28j - 14k| = \frac{1}{2}\sqrt{28^2 + 28^2 + 14^2} = 21.$$

6.16.2 Cross product with a vector basis set

If a basis set of vectors such as (1, 0, 0), (0, 1, 0), and (0, 0, 1) is used instead of ***i***, ***j***, ***k*** then the cross product is written slightly differently. If the vectors are $v = [3 \quad 2 \quad 5]$ and $u = [4 \quad 3 \quad 6]$ the cross product determinant gives the vector

$$u \times v \equiv \begin{vmatrix} 1 & 1 & 1 \\ 4 & 3 & 6 \\ 3 & 2 & 5 \end{vmatrix} = [3 \quad -2 \quad -1]$$

where the determinant multiplication is performed in the normal way. The length of the vector is its absolute value, or $\sqrt{14}$.

6.16.3 Distance from a point to a line and between two skew lines

Cross products are useful in calculating the distance between a point and a line or plane and between two skew lines; a calculation that is very hard to do with coordinate geometry. If p is our victim point, Fig. 6.34, and a line goes from point A to B, then the perpendicular (shortest) distance is d. The cross product of vector ***a*** with ***b*** is $a \times b = |a||b|\sin(\theta)\bar{n}$, where

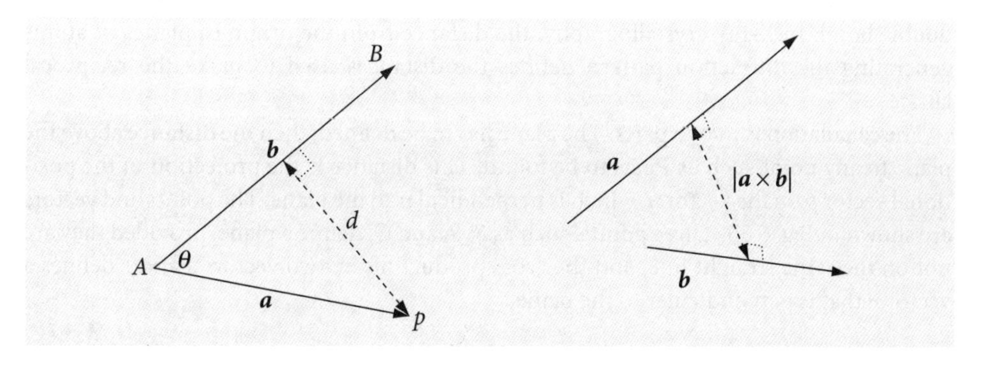

Fig. 6.34 Left: Distance d of point p from line A–B. Right: Distance between two skew lines represented in three dimensions as vectors ***a*** and ***b***.

$\bar{n}$ is a unit vector. The magnitude of this cross product is $|a \times b| = |a||b| \sin(\theta)$ but by trigonometry, $\sin(\theta) = d/a$ where a is the length of vector $\boldsymbol{a}$ or $a = |\boldsymbol{a}|$, then

$$d = |d| = \frac{|a \times b|}{|b|}. \tag{6.42}$$

Notice that the length of the line AB, which is b, goes into the denominator.

Suppose that two points on the same line have coordinates $A = \{1, -2, 3\}$, $B = \{4, 6, 0\}$ and another point p, which is not on the line, has coordinates $p = \{1, 2, 3\}$. The length b is $b = |\boldsymbol{b}| = \sqrt{9+64+9} = \sqrt{82}$ and is that of vector A to B making $\boldsymbol{b} = [3 \;\; 8 \;\; -3]$. Alternatively, $\boldsymbol{b} = 3\boldsymbol{i} + 8\boldsymbol{j} - 3\boldsymbol{k}$. The cross product of the vectors $\boldsymbol{b}$ and $\boldsymbol{a}$ is

$$\boldsymbol{a} \times \boldsymbol{b} = \begin{vmatrix} \boldsymbol{i} & \boldsymbol{j} & \boldsymbol{k} \\ 0 & 4 & 0 \\ 3 & 8 & -3 \end{vmatrix} = -12\boldsymbol{i} - 0\boldsymbol{j} - 12\boldsymbol{k}$$

and the magnitude of this vector is $12\sqrt{2}$. The distance of p from the line AB is therefore $\dfrac{12\sqrt{2}}{\sqrt{82}} = \dfrac{12}{\sqrt{41}}$. Using Maple the calculation, while specific to points A, B, and p, is easily made general and will produce results for any line and a point by changing the coordinate values.

Algorithm 6.2 Perpendicular distance from point to line

```
> with(LinearAlgebra):
> A:= < 1, -2, 3 >: B:= < 4, 6, 0 >: p:= < 1, 2, 3 >:
  b:= B - A;
  a:= p - A;
  ab:= a &x b;                                     # cross product
```

$$ab := \begin{bmatrix} -12 \\ 0 \\ -12 \end{bmatrix}$$

```
> d:= sqrt(ab.ab)/sqrt(b.b);               # perpendicular dist
```

$$d := \frac{6}{41}\sqrt{2}\sqrt{82}$$

This result, with some simplification, is the same as calculated by hand.

Skew lines are straight lines in three dimensions that do not cross because they are displaced from one another. Aircraft trajectories generally follow skew lines, the closest distance of approach permitted is approximately 3 miles. This distance is the absolute value of the cross product of the two vectors defining the trajectories because the shortest approach vector is at right angles to both trajectories, as shown in Fig. 6.34.

6.16.4 Equation of a plane and distance from a point to a plane

Often when studying molecules, the distance of an atom to the bond formed by two other atoms, or to the plane formed by several others, is an important quantity; for example, to calculate the $\pi\pi$ interaction between an atom and an aromatic ring or double bond. In X-ray crystallography, the distance from the origin to planes of atoms generating the diffraction pattern defines the distances used to make the reciprocal lattice.

The calculation is in two parts. The plane has to be defined, then the distance above the plane to any point such as P has to be found. This distance is the projection of the positional vector $\boldsymbol{p}$ on the vector $\boldsymbol{n}$, which is perpendicular to the plane. The points and vectors are shown in Fig. 6.35. Three points, such as S, R, and Q, define a plane, provided they are not on the same straight line, and the cross product of the two vectors $\boldsymbol{s}$ and $\boldsymbol{r}$ defines a vector $\boldsymbol{n}$ that is perpendicular to the plane,

$$\begin{aligned} n &= s \times r \\ &= ai + bj + ck \end{aligned} \tag{6.43}$$

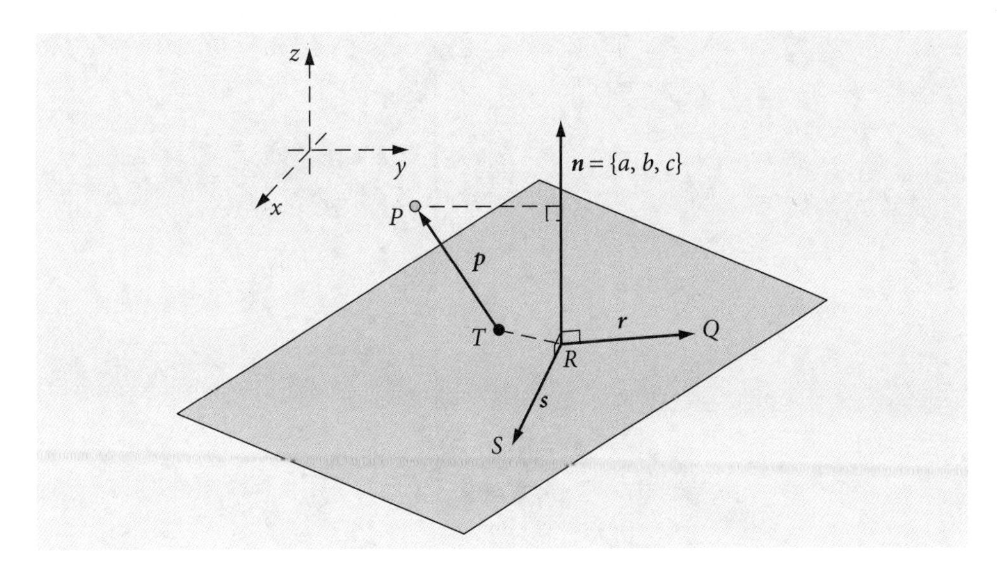

Fig. 6.35 Calculating the distance from a point p to a plane.

where a, b, and c are the coefficients of $\boldsymbol{n}$. These are found by evaluating the cross product as a determinant. Point T is any point in the plane with coordinates $T = \{x_0, y_0, z_0\}$ and the equation of the plane is, by definition,

$$a(x - x_0) + b(y - y_0) + c(z - z_0) = 0$$

with a, b, and c being the coefficients of $\boldsymbol{n}$. Recasting this equation in vector form it becomes rather neat and is

$$\boldsymbol{n}\cdot(\boldsymbol{X} - \boldsymbol{T}) = 0 \tag{6.44}$$

where $\boldsymbol{X}$ is the vector to any point $\{x, y, z\}$ and $\boldsymbol{T}$ the vector to point T which must be in the plane. Because T is not unique, it can be any other point such as one of the points, S, R, or Q that define the plane, and is therefore known. The length of the projection d of the vector $\boldsymbol{p}$ on to $\boldsymbol{n}$ is from equation (6.15)

$$d = \frac{\boldsymbol{n}\cdot(\boldsymbol{p} - \boldsymbol{T})}{\sqrt{\boldsymbol{n}\cdot\boldsymbol{n}}} \tag{6.45}$$

because the vector $\boldsymbol{p}$ joins point T to P. This equation can be solved because the coefficients of $\boldsymbol{n}$ are known *via* the cross product equation (6.43) and because points defining vectors $\boldsymbol{s}$ and $\boldsymbol{r}$ are known. Vector $\boldsymbol{p}$ is known (see Fig. 6.35) because T can be any point such as S, R, or Q and P is known because this is the point whose distance above the plane is sought. If point P has coordinates $\{p_x, p_y, p_z\}$ and T has $\{t_x, t_y, t_z\}$ substituting the values into equation (6.45) gives the formula for the distance of p from the plane.

$$d = \frac{a(p_x - t_x) + b(p_y - t_y) + c(p_z - t_z)}{\sqrt{a^2 + b^2 + c^2}} \tag{6.46}$$

If the point T is made to be at the origin, then the plane passes through the origin and the perpendicular distance to p is therefore

$$d_0 = \frac{ap_x + bp_y + cp_z}{\sqrt{a^2 + b^2 + c^2}} \tag{6.47}$$

Conversely, point P could be placed at the origin and p_x etc. made zero. In either case, the resulting distance could be negative or positive; if you are not interested in whether the point is above or below the plane then the absolute value of the distance is what you will want. If you want to determine which side of a molecule another atom is, then the sign of the distance may help you determine this, but in this case there is no up or down so the sign on the distance really means the 'same side' or 'opposite side'.

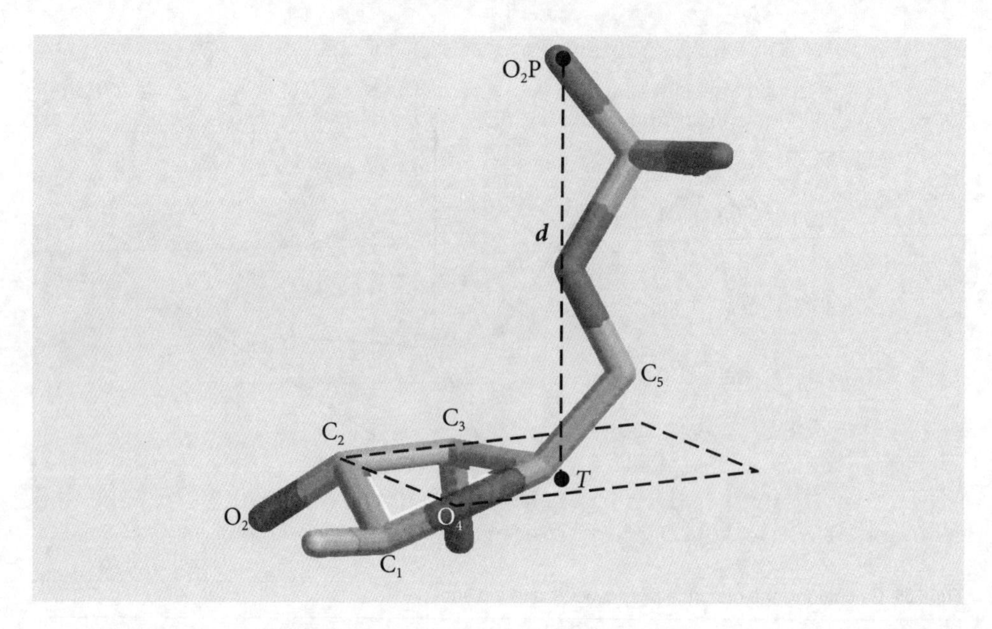

Fig. 6.36 Calculating the perpendicular distance from an atom to the plane.

6.16.5 Plane defined by its intercepts

The distance from the origin to a plane is given by equation (6.47) but in the special case that the intercepts a, b, c are known, and the plane goes through the points $\{a, 0\ 0\}$, $\{0, b, 0\}$, and $\{0, 0, c\}$, the equation is

$$\frac{x}{a}+\frac{y}{b}+\frac{z}{c}=1.$$

The equation for the perpendicular (shortest) distance from the origin to the plane is

$$d_0=\frac{abc}{\sqrt{(ab)^2+(bc)^2+(ac)^2}}=\frac{1}{\sqrt{(1/a)^2+(1/b)^2+(1/c)^2}} \tag{6.48}$$

and *only* applies when the plane is defined by its intercepts; notice that the intercept values only are used. This equation is particularly useful in crystallography to calculate the reciprocal lattice distance from the origin, and hence the inter-lattice spacing of crystals with orthorhombic (orthogonal) axes.

As an example, suppose that you need to know the distance from the oxygen on the phosphate to the plane of the ring defined as atoms C_2, O_4, and C_3 in the ribose phosphate shown in Fig. 6.36. The coordinates are known from the crystal structure and are

Atom	x	y	z
O_2P	115.394	41.169	129.137
O_4	120.546	41.818	127.822
C_3	119.237	43.428	126.672
C_2	119.664	42.262	125.771

To use the method described above, let points S, R, and Q represent atoms O_4, C_3, and C_2 respectively, which will define the plane, see Fig. 6.36. Let P be atom O_2P which is not in the plane and the vector $\boldsymbol{n}$ is then $\boldsymbol{n} = (S - R) \times (Q - R)$. Any point in the plane can be chosen to be T, see Figs 6.35 and 6.36, so it might as well be C_3 or point R. The equation of the plane is then $\boldsymbol{n}\cdot(X - T)$ where X is any point $\{x, y, z\}$ and the result is an equation in x, y and z. Next, choose any point in the plane; T can be used again making length of the normal from the plane to P,

$$d=\left|\frac{\boldsymbol{n}\cdot(P-T)}{\sqrt{\boldsymbol{n}\cdot\boldsymbol{n}}}\right|.$$

The calculation using the coordinates from the structure is as follows.

Algorithm 6.3 Distance from point to a plane

```
> with(LinearAlgebra):
> P:= < 115.394, 41.169, 129.137 >: # O2P in Å
  S:= < 120.546, 41.818, 127.822 >: # O4
  R:= < 119.237, 43.428, 126.672 >: # C3
  Q:= < 119.664, 42.262, 125.771 >: # C2
   n:=( S - R ) &x ( Q - R );         # n is normal vector
```

$$n := \begin{bmatrix} 2.7915 \\ 1.6705 \\ -.8388 \end{bmatrix}$$

```
# choose any point in the plane and call it T
  T:= < R[1], R[2], R[3] >:
  X:= < x, y, z >:                    # any point x y z.
> plane:= n . ( X - T );              # eqn of plane
```

$$plane := 2.7915x - 299.1405 + 1.6705\,y - 0.8388\,z$$

```
> d:= abs(n.(P-T)/sqrt(n.n));         # distance in Å
```

$$d := 4.9319$$

The perpendicular distance of this plane to any other point is calculated with different values of P; for example, the perpendicular distance from the origin is calculated with point $T = \{0, 0, 0\}$ and is

```
> P:= < 0, 0, 0 >;
  d:= abs(n.(P-T)/sqrt(n.n));         # distance in Å
```

$$d := 84.1099$$

6.16.6 Best plane through a set of points

In haemoglobin, the porphyrin has four N atoms roughly, but not exactly, in a plane surrounding the Fe, and similarly in chlorophyll-containing proteins found in photosynthetic organisms, the four N atoms surround the Mg. We often need to know how far an atom is from the plane of another molecule. A plane is determined by choosing three atoms not four. However, four atoms defining the plane would be more useful and, in this case, a best-fit plane to these four atoms is needed. This plane can be calculated using a matrix method similar to that used to determine moments of inertia, and this is described in Chapter 7.13. When this plane is found, the equations developed in Sections 6.16.4 and 6.16.5 can be used to find distances.

6.17 Scalar triple products are numbers

If $\boldsymbol{u}$, $\boldsymbol{v}$, and $\boldsymbol{w}$ are vectors, the expression $\boldsymbol{w}\cdot\boldsymbol{u}\times\boldsymbol{v}$ is a triple product. Using the results from Section 6.16.1, this is equal to

$$\begin{aligned}\boldsymbol{w}\cdot\boldsymbol{u}\times\boldsymbol{v} &= (w_x\boldsymbol{i} + w_y\boldsymbol{j} + w_z\boldsymbol{k})\cdot\begin{vmatrix} \boldsymbol{i} & \boldsymbol{j} & \boldsymbol{k} \\ u_x & u_y & u_z \\ v_x & v_y & v_z \end{vmatrix} \\ &= (w_x\boldsymbol{i} + w_y\boldsymbol{j} + w_z\boldsymbol{k})\cdot[(u_yv_z - u_zv_y)\boldsymbol{i} + (u_zv_x - u_xv_z)\boldsymbol{j} + (u_xv_y - u_yv_x)\boldsymbol{k}] \\ &= \begin{vmatrix} w_x & w_y & w_z \\ u_x & u_y & u_z \\ v_x & v_y & v_z \end{vmatrix}\end{aligned} \tag{6.49}$$

This result shows that the triple product is a number, hence the prefix 'scalar', because it is the determinant of the coefficients of the vectors and a determinant is equivalent to a number not a vector. It must follow that it does not matter in what order vectors occur in this triple product because the result is a number; $\boldsymbol{w}\cdot\boldsymbol{u}\times\boldsymbol{v}$ is the same as $\boldsymbol{u}\cdot\boldsymbol{v}\times\boldsymbol{w}$ etc.

One important application of this product is to calculate the volume of a solid body that has the shape of a parallelepiped; this is a prism whose faces are all parallelograms. If the body is right angled and with vectors $\boldsymbol{a}$, $\boldsymbol{b}$, and $\boldsymbol{c}$ along its edges, then the determinant is diagonal and its value is abc, which is equal to the volume. If the axes are not orthogonal, such as the $\boldsymbol{a}, \boldsymbol{b}, \boldsymbol{c}$ unit cell vectors of crystals, then the dot products are neither 1 nor 0 but have to be calculated as described in equation (6.34) for a monoclinic crystal. In general, for a parallelepiped the volume $= |\boldsymbol{w}\cdot\boldsymbol{u}\times\boldsymbol{v}|$ and the absolute value is used to ensure that this is positive.

The triple scalar product is used in the formation of reciprocal lattices used in crystallography. In a crystal, the axes are represented as vectors labelled $\boldsymbol{a}$, $\boldsymbol{b}$, and $\boldsymbol{c}$, which need not necessarily be orthogonal; the reciprocal lattices are $\boldsymbol{a}^*$, $\boldsymbol{b}^*$, and $\boldsymbol{c}^*$ and are defined so that

$$\boldsymbol{a}\cdot\boldsymbol{a}^* = \boldsymbol{b}\cdot\boldsymbol{b}^* = \boldsymbol{c}\cdot\boldsymbol{c}^* = 1$$

whereas all of the 'cross' terms are 0, i.e. $\boldsymbol{a}\cdot\boldsymbol{b}^* = 0$ and so forth. The new reciprocal vectors are

$$\boldsymbol{a}^* = \frac{\boldsymbol{b}\times\boldsymbol{c}}{\boldsymbol{a}\cdot\boldsymbol{b}\times\boldsymbol{c}}, \qquad \boldsymbol{b}^* = \frac{\boldsymbol{c}\times\boldsymbol{a}}{\boldsymbol{a}\cdot\boldsymbol{b}\times\boldsymbol{c}}, \qquad \boldsymbol{c}^* = \frac{\boldsymbol{a}\times\boldsymbol{b}}{\boldsymbol{a}\cdot\boldsymbol{b}\times\boldsymbol{c}},$$

and the triple product denominator is the volume of the unit cell. From these relationships it is easy to see that $\boldsymbol{a}\cdot\boldsymbol{a}^* = \boldsymbol{b}\cdot\boldsymbol{b}^* = \boldsymbol{c}\cdot\boldsymbol{c}^* = 1$. The vector $\boldsymbol{a}^*$ is at right angles to the plane of $\boldsymbol{b}$ and $\boldsymbol{c}$ and similarly for the other reciprocal basis vectors.

The meaning of a reciprocal lattice can be seen if a*, b^*, and c^* are calculated for an orthogonal unit cell. The simplest way is to define a basis set in three dimensions and if the unit cell is $a = 5/2$, $b = 3/2$ and $c = 2$ then the basis vectors are

$$\boldsymbol{a} = \frac{5}{2}[1 \quad 0 \quad 0], \qquad \boldsymbol{b} = \frac{3}{2}[0 \quad 1 \quad 0], \qquad \boldsymbol{c} = 2[0 \quad 0 \quad 1].$$

The triple product is the determinant

$$\boldsymbol{a}\cdot(\boldsymbol{b}\times\boldsymbol{c}) = \begin{vmatrix} 5/2 & 0 & 0 \\ 0 & 3/2 & 0 \\ 0 & 0 & 2 \end{vmatrix} = \frac{5}{2}\frac{3}{2}2 = \frac{15}{2}.$$

To calculate b^* the cross product needed is $\boldsymbol{c}\times\boldsymbol{a} = \begin{vmatrix} 1 & 1 & 1 \\ 0 & 0 & 2 \\ 5/2 & 0 & 0 \end{vmatrix} = 5$ making $b^* = \dfrac{5\times 2}{15} = \dfrac{2}{3}$ or the reciprocal of b. A similar result is found for a^* and c^*. The calculation follows in the same way if the $\boldsymbol{i}, \boldsymbol{j}, \boldsymbol{k}$ basis set is used.

6.18 Vector triple product

The vector triple product is the identity

$$\boldsymbol{A}\times(\boldsymbol{B}\times\boldsymbol{C}) = (\boldsymbol{A}\cdot\boldsymbol{C})\boldsymbol{B} - (\boldsymbol{A}\cdot\boldsymbol{B})\boldsymbol{C} \tag{6.50}$$

which is the difference between two vectors $\boldsymbol{B}$ and $\boldsymbol{C}$ scaled with a dot product, which is a number. The triple product is therefore a vector, as its name implies. This vector is perpendicular to $\boldsymbol{A}$ and to $\boldsymbol{B}\times\boldsymbol{C}$, which means that it lies in the plane of $\boldsymbol{B}$ and $\boldsymbol{C}$ and is a linear combination of these vectors.

If the brackets are placed differently then a different vector is obtained

$$(\boldsymbol{A}\times\boldsymbol{B})\times\boldsymbol{C} = (\boldsymbol{A}\cdot\boldsymbol{C})\boldsymbol{B} - (\boldsymbol{B}\cdot\boldsymbol{C})\boldsymbol{A}$$

There are other higher vector products, but you will only infrequently meet them.

6.19 **Questions**

Full solutions are available at **www.oxfordtextbooks.co.uk/orc/beddard**.

Q6.47 Show that $\sin(\alpha-\beta)=\sin(\alpha)\cos(\beta)-\cos(\alpha)\sin(\beta)$. See problem Q6.25.

Strategy: As the problem involves cosines we could try to solve it by calculating the cross product of $\boldsymbol{a}$ and $\boldsymbol{b}$ as two *unit* vectors which have an angle $\alpha-\beta$ between them. The cross product is $\boldsymbol{a}\times\boldsymbol{b}=|\boldsymbol{a}||\boldsymbol{b}|\sin(\alpha-\beta)\boldsymbol{k}$ and we calculate this in two ways and compare the results.

Q6.48 If the sides of a triangle have lengths a, b, and c prove the 'law of sines' for plane triangles, $\dfrac{\sin(\alpha)}{a}=\dfrac{\sin(\beta)}{b}=\dfrac{\sin(\chi)}{c}$.

Strategy: Half of the cross product of any two sides of a triangle, taken as vectors, is equal to half its area. Let the sides be described by vectors $\boldsymbol{A}$, $\boldsymbol{B}$ and $\boldsymbol{C}$ and calculate the cross products using different pairs of sides.

Q6.49 Show that $\boldsymbol{A}\times(\boldsymbol{B}\times\boldsymbol{C})=\boldsymbol{B}(\boldsymbol{A}\cdot\boldsymbol{C})-\boldsymbol{C}(\boldsymbol{A}\cdot\boldsymbol{B})$ by defining vectors in a three-dimensional basis set. Use Maple to show that one side of the equation is the same as the other as this is simpler than doing it by hand.

Q6.50 Using figure Fig. 6.34 **(a)** calculate the distance of point B from the line Ap, and **(b)** the distance of A from the line pB.

Q6.51 Show that **(a)** the length of any perpendicular inside an equilateral triangle, of side r, to any vertex is $\dfrac{r\sqrt{3}}{2}$ and

(b) the lengths of the perpendiculars from any point p inside the triangle to the sides add to give $\dfrac{r\sqrt{3}}{2}$.

Strategy: All sides have the same length because the triangle is equilateral and the internal angles are 60°. **(a)** Determine the coordinates of the vertices then calculate one perpendicular distance. **(b)** choose a point p inside the triangle with coordinates $\{x, y, 0\}$ and calculate the distance from p to each side.

Q6.52 The protein haemoglobin binds molecular oxygen to the Fe atom of the porphyrin chromophore of which there are four in the protein. This chromophore gives blood its characteristic purple venous and red arterial colours in its deoxygenated and oxygenated forms respectively. The binding of O_2 causes the iron atom's d-orbital electronic energy levels to shift, causing a change in the haemoglobin's visible absorption spectrum. These changes are sufficiently quantitative that they are used as a non-invasive, optical monitor of the extent of blood oxygenation of patients in hospital.

In the haem protein, the Fe is sixfold coordinated, and the sixth position is taken by a nearby histidine residue whose bonding to the Fe pulls it out of the plane of the porphyrin. When the Fe atom releases the oxygen molecule, the porphyrin changes shape which results in the Fe atom moving more into the plane of the ring. The resulting force on the histidine is sufficient to trigger a shape change in the protein allowing the O_2 molecule to escape.

The following data is taken from the Brookhaven protein data bank (pdb) entry 1THB recorded at 1.5 Å resolution (Waller & Liddington 1990). The data below contains only some of the coordinates, in Å, of the histidine and Fe porphyrin.

(a) Calculate the O-O, Fe-O, and Fe-His nitrogen bond lengths. Look up the Fe-O and Fe-N bond lengths from other compounds and comment on the bond lengths in this protein.

(b) Compared to the plane set by the NA-NB-ND atoms, calculate how much out of the plane, histidine Nitrogen, Fe, and O atoms are.

```
ATOM     648 CD2 HIS A 87   16.821 13.125 16.996 1.00 18.39     1THB 802
ATOM     649 CE1 HIS A 87   18.447 14.161 15.857 1.00 19.70     1THB 803
ATOM     650 NE2 HIS A 87   17.137 14.178 16.171 1.00 21.42     1THB 804
HETATM 1071 FE   HEM A  1   15.828 15.692 15.076 1.00 22.68     1THB1225
HETATM 1076 N A HEM A  1   17.420 16.837 14.530 1.00 19.46     1THB1230
HETATM 1087 N B HEM A  1   15.899 14.779 13.232 1.00 17.39     1THB1241
HETATM 1095 N C HEM A  1   13.910 14.869 15.365 1.00 19.78     1THB1249
HETATM 1103 N D HEM A  1   15.422 17.006 16.640 1.00 22.50     1THB1257
HETATM 1114 O1   HEM A  1   14.753 17.065 14.136 0.40 25.20     1THB1268
HETATM 1115 O2   HEM A  1   13.534 17.296 13.735 1.00 36.05     1THB1269
```

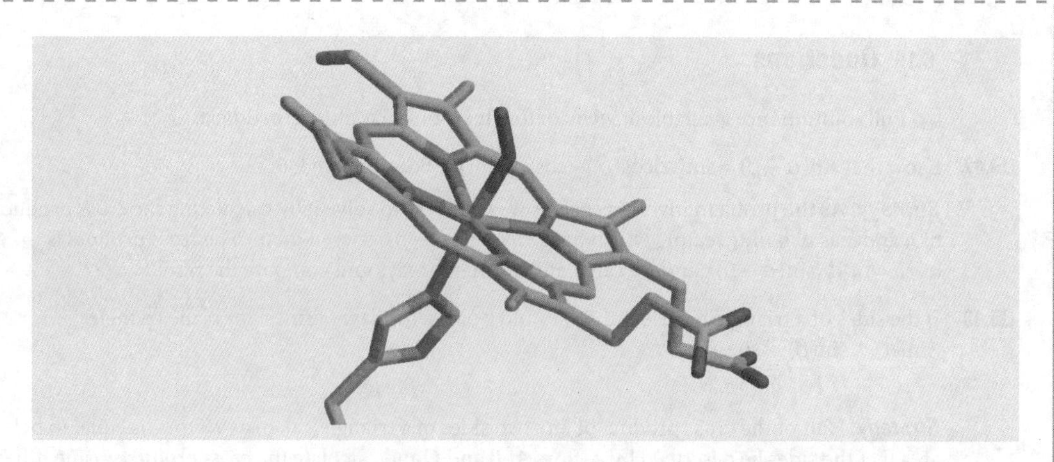

Fig. 6.37 Part of .pdb 1THB showing the porphyrin, O_2, and histidine molecules attached to the Fe at the centre of the porphyrin's ring. The distortion of the porphyrin is clear, showing that the Fe is not in the plane of the four N atoms.

Strategy: The atom positions should be made into vectors. The equation of the plane is given by equation (6.44) and the perpendicular distance to it by equation (6.45).

6.20 Torsion or dihedral angles

The torsion or dihedral angle ψ between two bonds is something we might want to calculate from a chemical structure. A dihedral angle is the angle between two planes, see Fig. 6.41. Two such angles, ψ and ϕ in proteins are used to generate Ramachandran plots as shown in Fig. 6.38. A regular or ideal protein helix would have angles $\phi = -60°$ and $\psi = -50°$ and real proteins with a large amount of α-helical character, such as bacteriorhodopsin, show clusters of points in this area. The other main feature of proteins is the β-sheet and this is a more loosely defined structure and exhibits angles in the top left corner bounded by approximately $\phi = -60°$ and $\psi = +40°$. The exact areas describing the structure are limited by steric repulsion between the atoms as bonds rotate, and are described in most biochemistry, although few chemistry, textbooks. Ramachandran plots

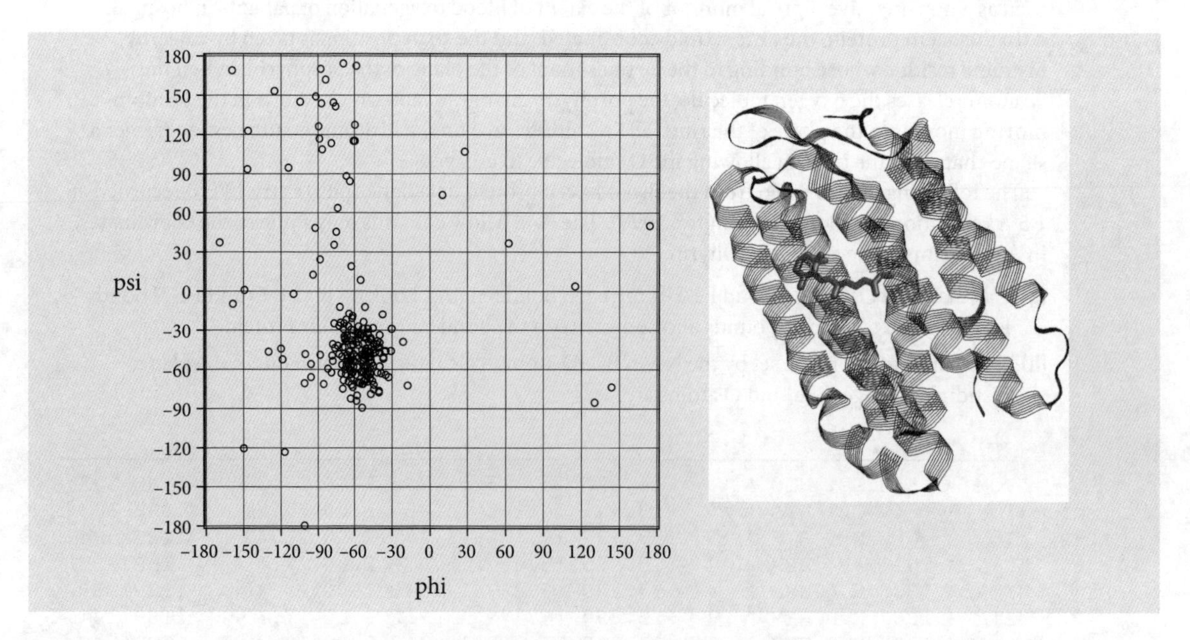

Fig. 6.38 Left: Ramachandran plot of the ϕ and ψ torsion angles in the protein bacteriorhodopsin (pdb 1FBB), which contains extensive α-helices. Most angles cluster around the values $\phi = -56°$ and $\psi = -53°$ typical of an α-helix. A regular or ideal helix would have angles −60° and −50°. The area of β-sheet structure, of which there is very little in this protein, is in the top left corner bounded by approximately −55 and +60 degrees. Right: The structure of the protein shows extensive helical structure. The retinal chromophore, which is positioned in the centre of the column of helices, is shown also.

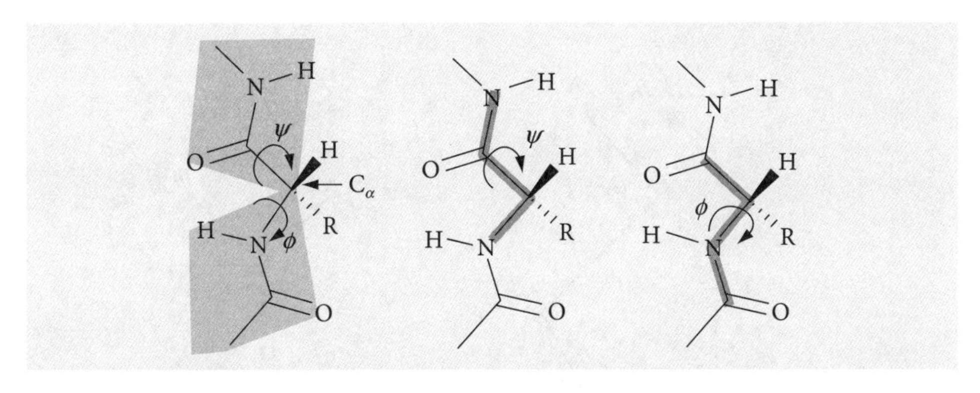

Fig. 6.39 Part of a peptide chain. The four atoms defining the ψ (N to N) and ϕ (CO to CO) torsion angles are shown on the right. As these chains are drawn as if completely extended, the angles are 180°. When the main chain atoms are eclipsed, the angles are zero.

can be produced as part of the web interface of sites such as the RCSB Protein Data Bank (www.rcsb.org/pdb/) as well as in textbooks. You could also write your own program that can read '.pdb' or '.ent' files and produce a plot.

Part of a peptide chain is shown in Fig. 6.39. The backbone forms a repeating sequence $-C_\alpha$-C-N-C_α-C-N-. The peptide unit is the CONH part of the structure, linked together by C_α atoms containing the amino acid side-chain *R*. The C_α-C-O-(N-H) atoms lie in a plane, shown shaded in the top of the figure, as do the atoms O-C-(N-H)-C_α shown shaded in the lower part. There are two main torsion angles, Fig. 6.39, which are ϕ, the torsion angle of the planes containing the CN–C_αC atoms which runs from the C=O to C=O groups, and ψ, the angle between planes containing the NC_α–CN atoms which starts and ends on the nitrogen atoms. A third angle ω is also defined as C_αC–NC_α but in a regular protein, this is very close to 180°. The peptide chain runs from the N terminal to the C. The N terminal of the protein occurs towards the end of the chain with the lowest indexed α-carbon, therefore the angle ϕ is towards the N terminal, the ψ towards the C. Torsion angles for side-chains can also be defined; these are given the symbol χ_1, χ_2 etc. depending on the type of amino acid.

The picture Fig. 6.40 shows that the N to N torsion angles ψ between planes containing N1 and N2, and N2 and N3, are both very small, $\psi_1 \approx -6$ and $\psi_2 \approx -14°$, which is the angle of the plane containing the bond N2CA2 to that containing the bond C2N3. In this notation, which is similar to that in the .pdb data, CA is the alpha carbon. The ϕ_1 angle C1N2-CA2C2 is approximately −90°; the ϕ_2 angle bond, C2N3-CA3C3, is −140° approximately. The angles defined for proteins must range from −180 to +180° and this means

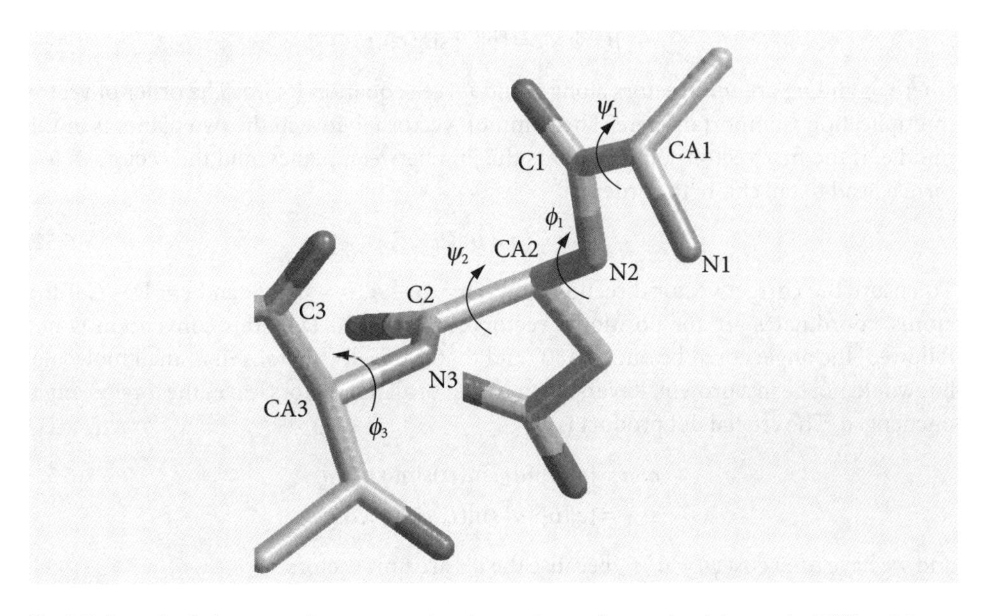

Fig. 6.40 Part of a β-sheet protein structure showing angles ψ of approximately zero (< ±10°) and ϕ angles of approx −100°.

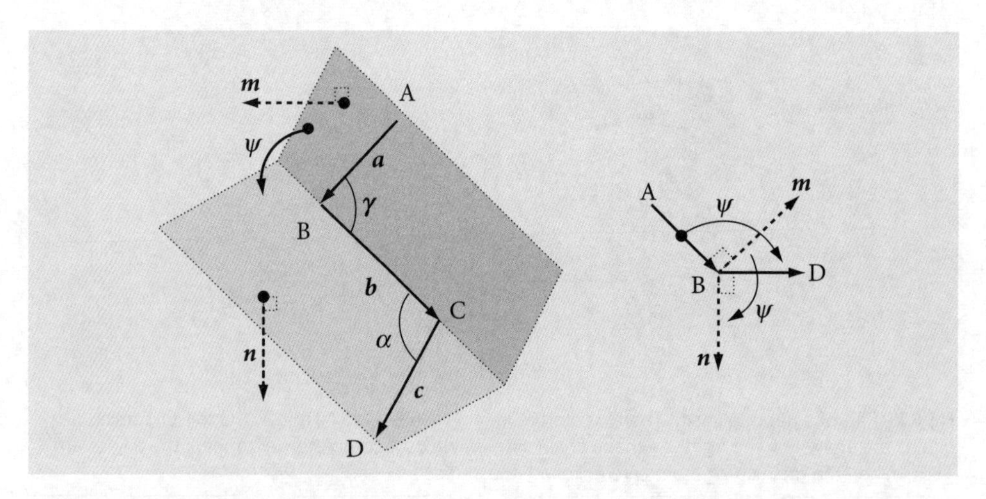

Fig. 6.41 Left: Geometry showing the torsion or dihedral angle ψ. Right: End-view looking down *B* towards *C* and showing the positive dihedral angles ψ between the cross products *n* and *m* in the planes ABC and BCD.

that the formula developed below has to have a condition applied to it and this is described at the end of this section. The calculation of the cosine of the dihedral (torsion) angle is developed next.

The geometry of the dihedral or torsion angle ψ is shown in Fig. 6.41, where A, B, C, and D are atoms. The angle is taken as the rotation of the AB bond (vector $\boldsymbol{a}$) needed to eclipse the CD bond (vector $\boldsymbol{c}$) when viewed down BC, vector $\boldsymbol{b}$. The angles α and γ are in the planes, BCD and ABC respectively, and represent the bond angles, typically 100 to 130° in molecules depending upon bonding type, sp^2 or sp^3. However, these angles are unimportant because the torsion angle is between the *planes* that the atoms lie in.

The torsion or dihedral angle ψ is defined as the angle between the plane ABC and the plane BCD. ABC has a vector $\boldsymbol{m}$ normal (perpendicular) to $\boldsymbol{a} \times \boldsymbol{b}$ and BCD has $\boldsymbol{n}$ normal to $\boldsymbol{b} \times \boldsymbol{c}$; Fig. 6.41. The torsion angle must be independent of the size of the vectors, which means that in the final formula normalized vectors will be used. The angle between vectors $\boldsymbol{m}$ and $\boldsymbol{n}$, the dihedral angle ψ, is found by first calculating their dot product.

$$\boldsymbol{m}\cdot\boldsymbol{n} = |\boldsymbol{m}||\boldsymbol{n}|\cos(\psi)$$

Because the vectors $\boldsymbol{m}$ and $\boldsymbol{n}$ are perpendicular to $\boldsymbol{a}$ and $\boldsymbol{b}$ and $\boldsymbol{d}$ and $\boldsymbol{c}$ respectively then,

$$\begin{aligned} \boldsymbol{m} &= \boldsymbol{a} \times \boldsymbol{b} = |\boldsymbol{a}||\boldsymbol{b}|\sin(\gamma)\boldsymbol{u}_m \\ \boldsymbol{n} &= \boldsymbol{b} \times \boldsymbol{c} = |\boldsymbol{b}||\boldsymbol{c}|\sin(\alpha)\boldsymbol{u}_n \end{aligned} \tag{6.51}$$

where $\boldsymbol{u}_n$ and $\boldsymbol{u}_m$ are *unit* vectors along $\boldsymbol{n}$ and $\boldsymbol{m}$; see equation (6.38). The order of vector multiplication is important here. The common vector $\boldsymbol{b}$ between the two planes is in the middle. If the first vector points towards the join between planes, and the vectors $\boldsymbol{a}$, $\boldsymbol{b}$, $\boldsymbol{c}$ form a head to tail chain, the order is

$$\boldsymbol{m}\cdot\boldsymbol{n} = (\boldsymbol{a} \times \boldsymbol{b})\cdot(\boldsymbol{b} \times \boldsymbol{c}). \tag{6.52}$$

To ensure the correct vector directions, then $\boldsymbol{a} = \boldsymbol{B} - \boldsymbol{A}$, $\boldsymbol{b} = \boldsymbol{C} - \boldsymbol{B}$, and $\boldsymbol{c} = \boldsymbol{D} - \boldsymbol{C}$ if the atoms' coordinates are the positional vectors $\boldsymbol{A}$, $\boldsymbol{B}$, $\boldsymbol{C}$, and $\boldsymbol{D}$. If this convention is not followed, the angle could be out by 180° and that is usually obvious in a small molecule, but would not be in a protein. Reversing the order of all the vectors leaves the torsion angle unchanged. The crucial dot product is

$$\begin{aligned} \boldsymbol{m}\cdot\boldsymbol{n} &= |\boldsymbol{c}||\boldsymbol{b}|^2|\boldsymbol{a}|\sin(\alpha)\sin(\gamma)\boldsymbol{u}_m\cdot\boldsymbol{u}_n \\ &= |\boldsymbol{c}||\boldsymbol{b}|^2|\boldsymbol{a}|\sin(\alpha)\sin(\gamma)\cos(\psi) \end{aligned} \tag{6.53}$$

and we have used $\cos(\psi) = \boldsymbol{u}_m\cdot\boldsymbol{u}_n$ because the $\boldsymbol{u}$'s are unit vectors.

Using equations (6.51) the magnitude (absolute value) of vector $\boldsymbol{n}$ is $|\boldsymbol{n}| = |\boldsymbol{c}||\boldsymbol{b}|\sin(\alpha)$, and $|\boldsymbol{m}| = |\boldsymbol{b}||\boldsymbol{a}|\sin(\gamma)$. This means that the dot product equation (6.53) can be rearranged to give the torsion angle as

$$\cos(\psi) = \frac{\boldsymbol{m}\cdot\boldsymbol{n}}{|\boldsymbol{m}||\boldsymbol{n}|} = \frac{\boldsymbol{m}\cdot\boldsymbol{n}}{\sqrt{\boldsymbol{m}\cdot\boldsymbol{m}}\sqrt{\boldsymbol{n}\cdot\boldsymbol{n}}}. \tag{6.54}$$

To evaluate these equations each vector has to be defined in a basis set. The protein data bank crystallographic and NMR data use a basis set of (orthogonal) Cartesian $\{x, y, z\}$ coordinates, as will those from a molecular dynamics simulation. The $\boldsymbol{n}$ and $\boldsymbol{m}$ vectors are defined as differences in these values, for example, in the $\{\boldsymbol{i}, \boldsymbol{j}, \boldsymbol{k}\}$ basis set $\boldsymbol{n} = p_1\boldsymbol{i} + p_2\boldsymbol{j} + p_3\boldsymbol{k}$ and $\boldsymbol{m} = q_1\boldsymbol{i} + q_2\boldsymbol{j} + q_3\boldsymbol{k}$ where the ps and qs are coefficients or coordinates along each axis. In this case, the angle is calculated using

$$\cos(\psi) = \frac{(p_1q_1 + p_2q_2 + p_3q_3)}{\sqrt{p_1^2 + p_2^2 + p_3^2}\sqrt{q_1^2 + q_2^2 + q_3^2}}.$$

In vector-matrix form where $\boldsymbol{n} = [p_1 \quad p_2 \quad p_3]$ and $\boldsymbol{m} = [q_1 \quad q_2 \quad q_3]$, the equation becomes

$$\cos(\psi) = \frac{\boldsymbol{m}\boldsymbol{n}^T}{\sqrt{(\boldsymbol{m}\boldsymbol{m}^T)}\sqrt{(\boldsymbol{n}\boldsymbol{n}^T)}}$$

where the transpose (superscript T) changes the row vector into a column.

In the special case of proteins, the angles are defined to be between −180° and +180° and a check has to be made because the cosine can be positive between −90° to 90° and negative between −180° to 90° and again from 90° to 180° and one cannot tell from the cosine alone what the angle should be. For example, a cosine of −0.77 could be 140° or −140°. By calculating the *sign* of the dot product $\boldsymbol{a}\cdot\boldsymbol{n}$, the quadrant in which the cosine lies can be determined by multiplying the angle by this sign which is −1 or +1. In Maple there is a function to do determine the sign, which is `sign(a.n)`.

Figure 6.42 shows two residues of strand B of the structure of insulin (www.rcsb.org/pdf/home.od 2INS.pdb). The torsion angle between the aromatic planes of the phenylalanine and tyrosine residues can be calculated since the coordinates of the atoms are known. It is necessary to choose which atoms are to be involved and to order them to form head to tail vectors. These are shown in the figure.

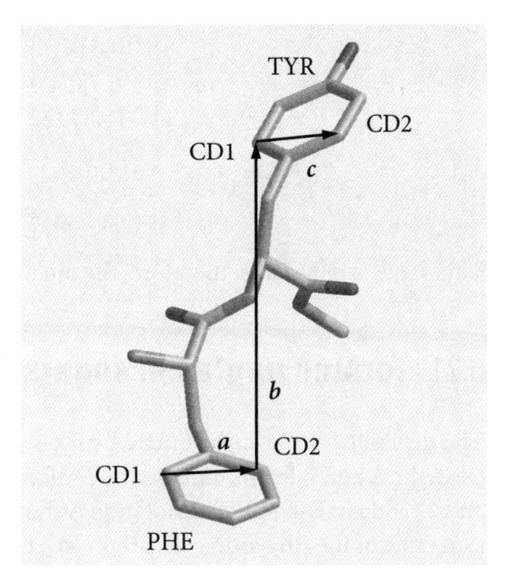

Fig. 6.42 Two residues of strand B of insulin.

The data taken from the .pdb is

						x	y	z			
..											
ATOM	349	CD1	PHE	B	25	-7.497	20.877	-0.865	1.00	45.00	C
ATOM	350	CD2	PHE	B	25	-5.644	21.663	0.493	1.00	23.22	C
..											
ATOM	360	CD1	TYR	B	26	-4.106	13.747	5.282	1.00	16.37	C
ATOM	361	CD2	TYR	B	26	-1.869	14.466	4.723	1.00	12.17	C
..											

Algorithm 6.4 Dihedral angles

```
> with(LinearAlgebra):
> phecd1:= < -7.497, 20.877, -0.865 >:         # from pdb 2INS
  phecd2:= < -5.644, 21.663,  0.493 >:
  tyrcd1:= < -4.106, 13.747,  5.282 >:
  tyrcd2:= < -1.869, 14.466,  4.723 >:
  vec_a:= phecd2-phecd1:                       # define vector a
  vec_b:= tyrcd1-phecd2:
  vec_c:= tyrcd2-tyrcd1:
```

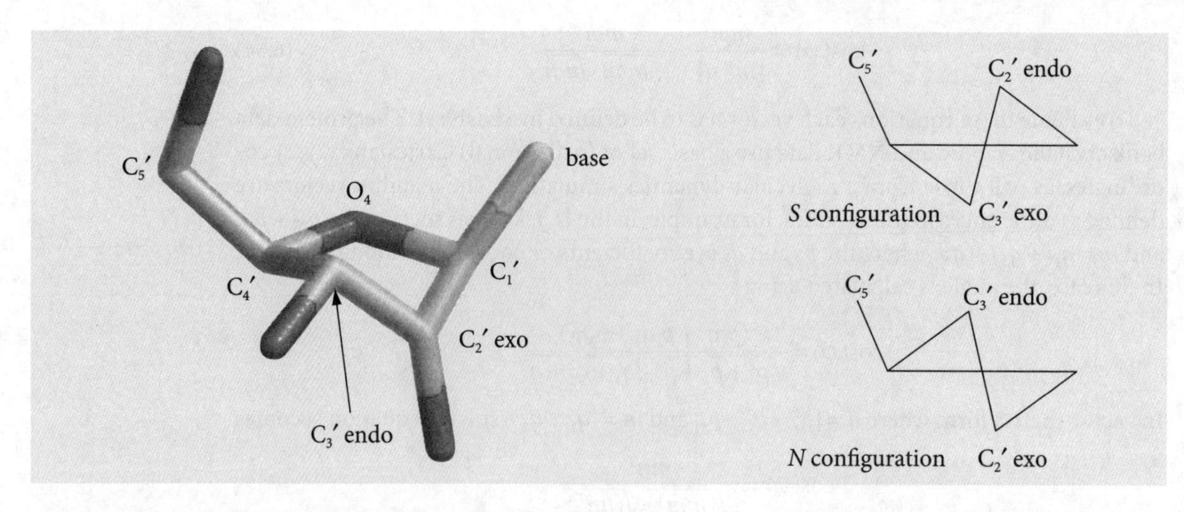

Fig. 6.43 A C_3' endo sugar. The C_4'-O_4'-C_1' are in one plane, the C_3' above the plane on the side of C_5' and C_2' below the plane. The cartoon structures on the right show this also.

```
m:= vec_a &x vec_b;                          # m, eqn. (6.51)
n:= vec_b &x vec_c;
```

$$m := \begin{bmatrix} 14.5141 \\ -6.7854 \\ -15.8772 \end{bmatrix} \quad n := \begin{bmatrix} 0.9818 \\ 11.5727 \\ 18.8139 \end{bmatrix}$$

```
> psi:= sign(vec_a.n)*arccos(n.m/(sqrt(n.n)*sqrt(m.m)));
```

$$\psi := 2.3860$$

This value is in radians; therefore, the dihedral angle ψ is 136.7°.

6.21 Torsion angles in sugars and DNA

The flexibility of the sugar in DNA makes it possible for more than one type of DNA to exist; the A and B forms differ in the conformation of the sugars being C_3' endo in A-DNA and C_2' endo in B-DNA. The notation is shown in Fig. 6.43; *endo* is defined as being on the same side of the ring as the C_5' carbon, *exo* on the opposite side.

A sugar (ribose, for example) is considered to have a twisted (*T*) conformation if three of its five ring atoms lie in one plane; these are the C_1', O_4', and C_4' atoms, as shown in Fig. 6.43, whereas the sugar has the envelope (*E*) form if four of its atoms are in a plane; the C_2' or C_3' carbon atom is then out of the plane.

The conformation of the sugar is often represented in terms of its pseudo-rotation angle *P*, which is the torsion (dihedral) angle $\upsilon_2 = C_1'C_2' - C_3'C_4'$. The configurations are shown in the next diagram, showing very approximate angles; the configuration with $0° < P < 90°$ are C_3' endo and $-90° < P < 0°$ are C_2' endo. Configurations on the right half of the diagram, C_3' endo and C_2' exo, are called north (N), the others are south (S).

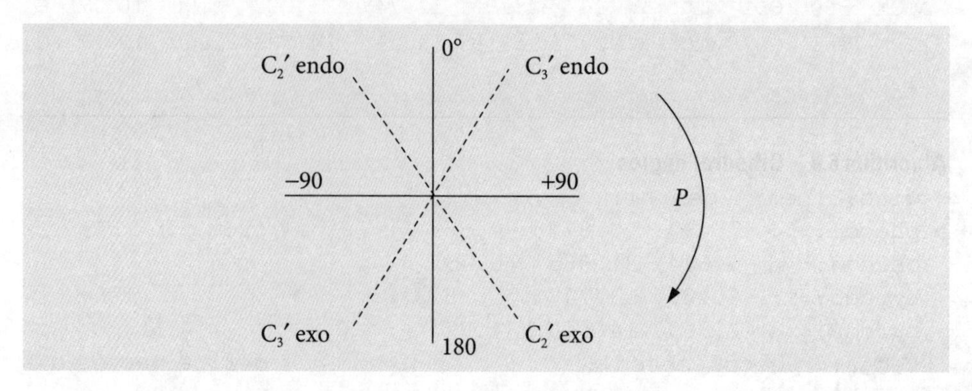

Fig. 6.44 Pseudo-rotation angles.

6.21.1 Pseudo-rotation angles

The torsion angles for the phosphate and sugar in the DNA are shown below, Fig. 6.45; the torsion angle γ is calculated with atoms $O'_5C'_5$-$C'_4C'_3$, the pucker or pseudo-rotation angle $\upsilon_2 = C'_1C'_2 - C'_3C'_4$. The angle χ is calculated with atoms $O'_4C'_1$–N_9C_4 for purines, and $O'_4C'_1$–N_1C_2 for pyrimidines. When $\chi = 0°$, the O'_4–C'_1 bond is eclipsed with the N_9–C_4 bond in purines and with the N_1–C_2 bond in pyrimidines. The C_4 atom in purines is the carbon next to the nitrogen and which joins both rings, and in pyrimidines, it is also next to the nitrogen but is attached to an oxygen atom. A χ torsion angle of about 60° means that in the syn configuration the base lies symmetrically over the sugar and when $\chi = -120°$ the base points symmetrically away from the sugar in the anti configuration. The syn-anti configuration is defined as,

$$\text{syn: } \chi = < |90°| \qquad \text{anti: } \chi = > |90°| \text{ and } < |180°|.$$

A and B type DNA have the torsion angle γ in the range 30 to 90°, and angle $\chi \equiv$ anti, which means that the base does not lie over the sugar. A-DNA has ring pucker with C'_3 endo, and B-DNA, C'_2 endo.

	γ	υ_2	χ
A-DNA	30 to 90°	0 to +90° C'_3 endo, typically 40° 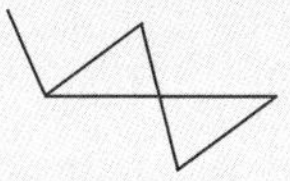	anti 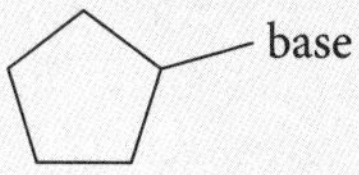
B-DNA	30 to 90°	−90 to 0° C'_2 endo, typically −38 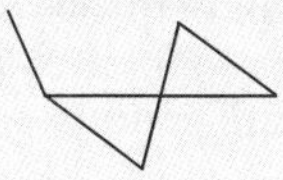	anti 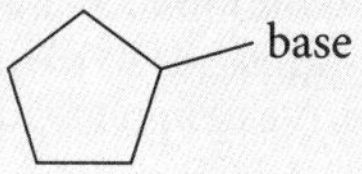
Z-DNA	150 to 210°	0 to +40° C'_3 endo, typically 25° 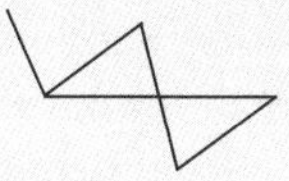	syn 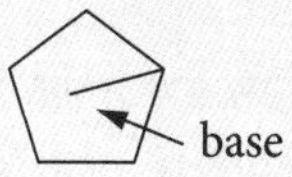

6.21.2 Reading protein data bank data

A protein data bank listing contains many lines of introductory information, with prefixes such as TITLE, REMARK, or SEQRES; most of this self-explanatory. The data on the crystal's space group is also to be found here. After these lines, the atom coordinates start with the word ATOM or HETATM if the atom does not belong to an amino acid, for example a porphyrin. The HETATM lines always follow the ATOM lines. At the end of the list, a TER termination statement occurs and either CONECT index numbers, to help drawing the structure or another ATOM line for the next amino acid chain. Details about the data format can be found on the protein data bank website, www.rcsb.org/pdb/home.do. A summary is given below once the coordinate data starts.

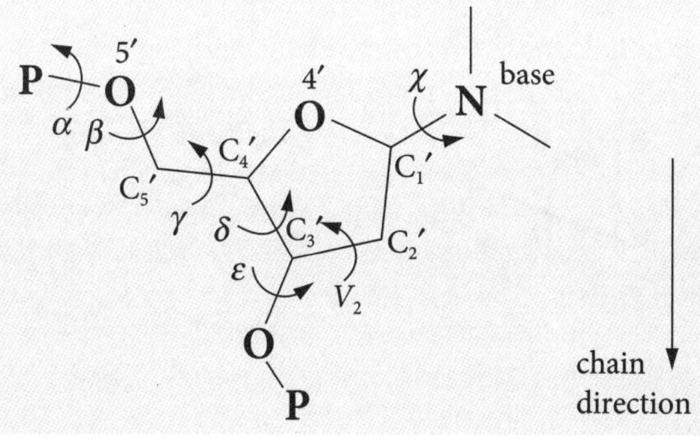

Fig. 6.45 Definition of some torsion angles in DNA. The N in the base is N_9 (purines) or N_1.

```
12345678901234567890123456789012345678901234567890123456789012345678901234567890
   1-10   |  10-20   |  20-30   |  30-40   |  40-50   |  50-60   |  60-70   |  70-80
ATOM    1221  CD1 LEU A 162      35.215  21.713  -1.469  1.00  0.00          1TIMC 65
```

The format of all the lines of ATOM and HETATM information is the same.

Columns 1–6 identifies type as ATOM or HETATOM.

Columns 8–11 contains the index number of the atom in the protein.

Columns 14–16 is the atom type.

CA is carbon alpha; C the backbone carbon attached to the oxygen labelled O, and N is the backbone nitrogen. Therefore, only atoms with these labels are important in calculating the ϕ and ψ torsion angles. The amino acid side chain atoms are labelled CB, CD, CZ, NH1, O1, N1 etc. depending upon the amino acid.

In DNA, the sugar atoms are identified by a star in the order O5*, C5*, C4*, O4*, C3*, O3*, C2*, C1*. The bases are listed next and are identified by unstarred atoms. Phosphorus P is next, followed by its two oxygen atoms O1P and O2P.

Columns 18–20 is a three-letter code identifying the amino acid or HOH for water. In DNA just one letter identifies the base C, G, A or T.

Column 22 gives the letter indicating the amino acid chain A, B, C etc. In DNA, one chain is listed first then the other.

Columns 24–26 gives the number of the amino acid or nucleic acid base in the list.

Columns 31–28, 39–46, and 47–54 are the x-, y-, and z-coordinates in Å. The x data starts on Column 31 and can be up to 8 characters including any minus sign and the decimal point. These values are usually less than ±40 Å because of the relatively small size of proteins.

Columns 55 to 70 can be ignored.

Columns 70–80. This varies but can be protein type or identification and/or data row number.

6.22 Questions

Full solutions are available at **www.oxfordtextbooks.co.uk/orc/beddard**.

Q6.53 Show that in normalizing equation (6.51) the unit vectors disappear.

Q6.54 In the chair and boat form of cyclohexane C_6H_{12} calculate;

(a) the coordinates of each of the atoms,

(b) the distance between the C_1 and C_4 atoms in both chair and boat forms.

Calculate the dihedral angle ψ between the following planes of cyclohexane.

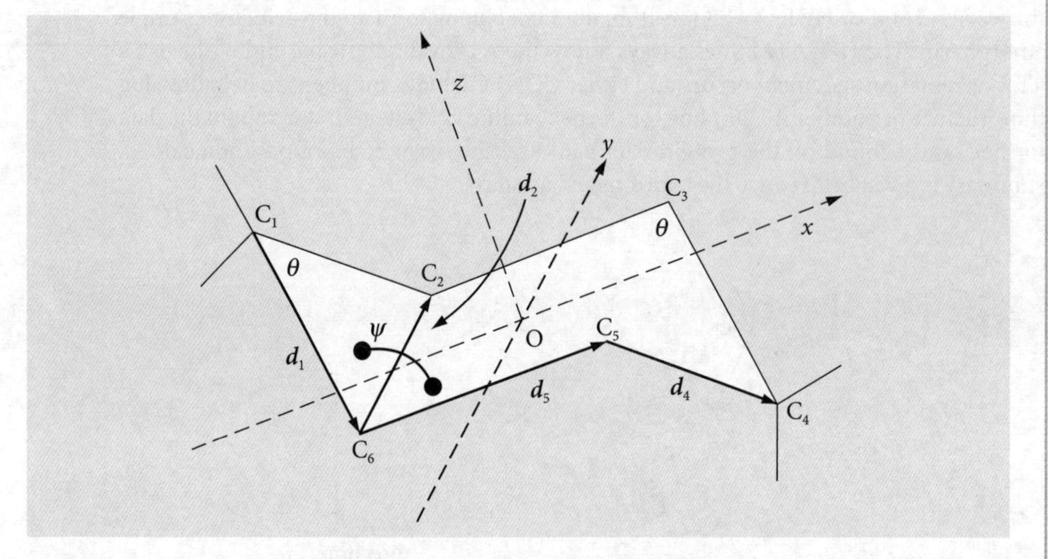

Fig. 6.46 Cyclohexane geometry, labels, and vectors. The dihedral angle ψ for part (a) is shown.

(c) C_1–C_6–C_2 and C_2–C_6–C_5, which is the same in the chair or boat form.

(d) C_1–C_6–C_2 and C_1–C_6–C_5.

(e) C_1–C_6–C_2 and C_5–C_4–C_3 in the boat form.

Assume standard bond lengths d and sp^3 tetrahedral bond angles θ. The bond vectors and axes are labelled in the diagram.

Strategy: (a–b) The C-C bonds are all the same length and the angles are also the same. The chair shaped molecule has a centre of symmetry and the boat has a mirror plane. Use vector notation in the x, y, z frame so that any vector representing a bond, is $a\boldsymbol{i} + b\boldsymbol{j} + c\boldsymbol{k}$ with coefficients a, b, and c which must be determined for each vector. Choose a right-angled axes set with the origin at the centre of inversion. The most difficult part of this problem is determining the coordinates of the atoms. The tetrahedral bond angles are given by $\cos(\theta) = -1/3$ with θ in radians, equivalent to 109.47°. All the bond lengths are d. (c–e) Dihedral or torsion angles are the angles between planes. The dihedral angle is given by equation (6.54). If the vectors $\boldsymbol{n}$ and $\boldsymbol{m}$ are normal to the two planes, the cross and dot product should have the order $\boldsymbol{m}\cdot\boldsymbol{n} = (\boldsymbol{d}_1 \times \boldsymbol{d}_2)\cdot(\boldsymbol{d}_2 \times \boldsymbol{d}_5)$. The vectors are all defined to be in the same direction as shown on the figure.

Q6.55 Figure 6.47 shows just the C, N, and O atoms of part of a peptide backbone from pdb entry 1VCR and some of the coordinates are listed below. Calculate the torsion angles ψ and ϕ using Maple to do so. Write a general procedure that accepts the four atoms coordinates as input. Does this part of the backbone forms an α-helix or β-sheet?

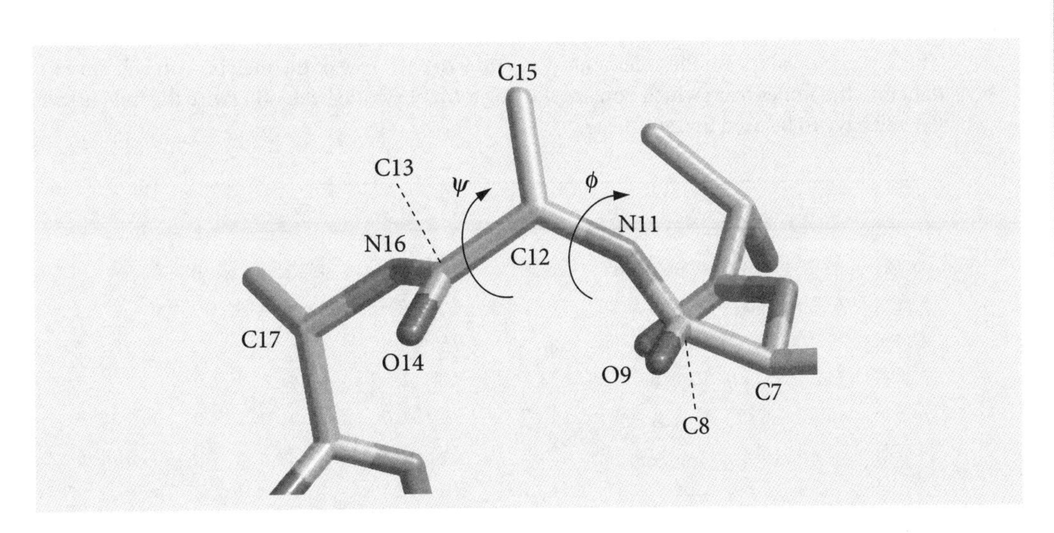

Fig. 6.47 Part of the X-ray structure of a peptide chain excluding H atoms (from pdb 1VCR).

						x	y	z		
ATOM	8	C	GLU	A	56	120.627	4.607	38.990	1.00296.52	C
ATOM	9	O	GLU	A	56	119.838	4.104	38.191	1.00296.52	O
ATOM	10	CB	GLU	A	56	122.694	3.628	38.000	1.00296.52	C
ATOM	11	N	THR	A	57	120.292	4.951	40.228	1.00296.52	N
ATOM	12	CA	THR	A	57	118.955	4.753	40.750	1.00296.52	C
ATOM	13	C	THR	A	57	117.996	5.868	40.333	1.00296.52	C
ATOM	14	O	THR	A	57	117.192	5.699	39.415	1.00296.52	O
ATOM	15	CB	THR	A	57	119.013	4.640	42.275	1.00296.52	C
ATOM	16	N	PHE	A	58	118.093	7.008	41.010	1.00296.52	N
ATOM	17	CA	PHE	A	58	117.229	8.153	40.736	1.00296.52	C
ATOM	18	C	PHE	A	58	117.620	8.930	39.485	1.00296.52	C
ATOM	19	O	PHE	A	58	116.760	9.339	38.702	1.00296.52	O

Strategy: The first step is to determine which atoms are involved and this means consulting Fig. 6.39. The torsion ψ belongs to the atoms including and between the N atoms, and is group N11–C12–C13–N16. The angle ϕ starts and ends at the C=O groups and is grouping C8–N11–C12–C13.

Q6.56 Triose phosphate (pdb 1TIM) has been studied by X-ray crystallography and has a β-barrel structure.

(a) Find the first ψ torsion angle of the part of the β-sheet part of the structure shown, starting from N36. Do this semi-automatically by searching along the data until you find the atoms you want. Look at Fig. 6.39 to decide which type of atoms to use for each angle.

(b) Calculate the remainder of the ψ angles, then modify the Maple code to calculate the ϕ angles.

You may want to read the data into Maple after extracting this part of the data from the 1TIM.pdb file. The Maple instruction to read a block of lines of data, that all have the same format from this .pdb file and put it into a matrix `MAT` is

```
> dataname:='part of 1TIM.pdb':

> MAT:= readdata(dataname,

[string,integer,string,string,string,integer,float,float,float,
float,float,string]):
```

The x, y, and z values are then elements 7, 8, and 9 of each row of the matrix. You will have to put the data into the folder from which you are starting a Maple worksheet, otherwise the full path starting at C:\ will have to be used in `dataname`.

```
                                  x        y        z
ATOM  2128  CB  SER B  34   50.330   75.964  -10.921 1.00  0.00   1TIM2273
ATOM  2129  OG  SER B  34   49.244   75.039  -10.804 1.00  0.00   1TIM2274
ATOM  2130  N   ALA B  35   52.355   75.699   -8.836 1.00  0.00   1TIM2275
ATOM  2131  CA  ALA B  35   52.796   75.402   -7.466 1.00  0.00   1TIM2276
ATOM  2132  C   ALA B  35   51.779   75.906   -6.429 1.00  0.00   1TIM2277
ATOM  2133  O   ALA B  35   51.910   75.645   -5.232 1.00  0.00   1TIM2278
ATOM  2134  CB  ALA B  35   54.179   76.024   -7.200 1.00  0.00   1TIM2279
ATOM  2135  N   ASP B  36   50.780   76.625   -6.939 1.00  0.00   1TIM2280
ATOM  2136  CA  ASP B  36   49.728   77.116   -6.027 1.00  0.00   1TIM2281
ATOM  2137  C   ASP B  36   48.681   75.986   -5.894 1.00  0.00   1TIM2282
ATOM  2138  O   ASP B  36   47.846   76.003   -4.993 1.00  0.00   1TIM2283
ATOM  2139  CB  ASP B  36   49.144   78.437   -6.534 1.00  0.00   1TIM2284
ATOM  2140  CG  ASP B  36   47.765   78.743   -5.929 1.00  0.00   1TIM2285
ATOM  2141  OD1 ASP B  36   47.140   79.720   -6.427 1.00  0.00   1TIM2286
ATOM  2142  OD2 ASP B  36   47.390   77.990   -4.991 1.00  0.00   1TIM2287
ATOM  2143  N   THR B  37   48.800   75.039   -6.817 1.00  0.00   1TIM2288
ATOM  2144  CA  THR B  37   47.914   73.870   -6.899 1.00  0.00   1TIM2289
ATOM  2145  C   THR B  37   48.627   72.644   -6.271 1.00  0.00   1TIM2290
ATOM  2146  O   THR B  37   49.830   72.468   -6.467 1.00  0.00   1TIM2291
ATOM  2147  CB  THR B  37   47.634   73.522   -8.376 1.00  0.00   1TIM2292
```

Strategy: Use the procedure developed in the previous problem, or your own version of it. Use the values in `MAT` to obtain the coordinates of each atom.

Q6.57 The following data is that for ideal A type DNA, should such a molecule exist (data from website of L. Williams. http://web.chemistry.gatech.edu/~williams/).

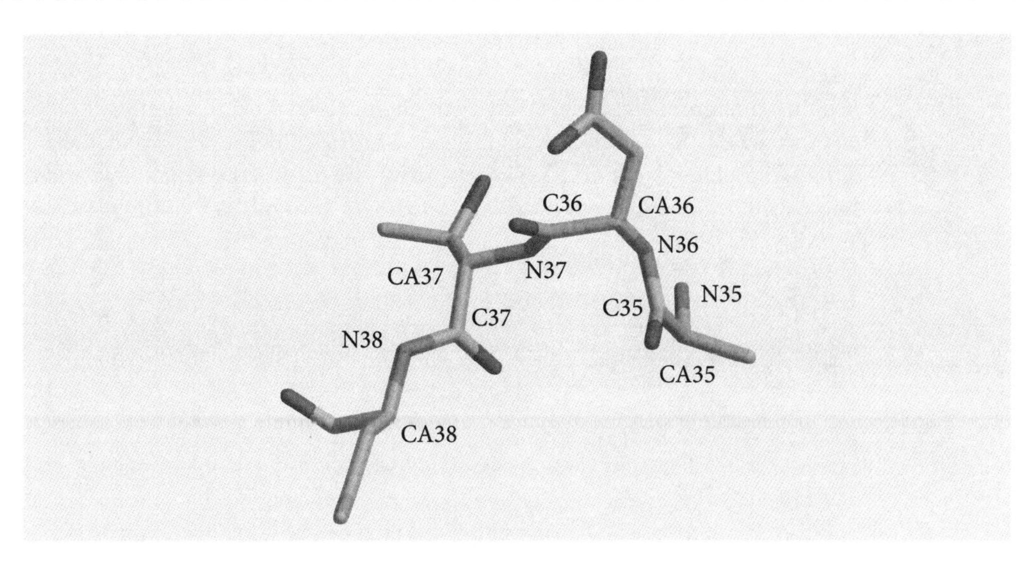

Fig. 6.48 Part of triose phosphate .pdb

						x	y	z		
ATOM	20	P	G	A	2	6.900	5.093	-6.694	1.00	25.00
ATOM	21	O1P	G	A	2	5.428	4.956	-6.641	1.00	25.00
ATOM	22	O2P	G	A	2	7.451	5.729	-7.911	1.00	25.00
ATOM	23	O5*	G	A	2	7.581	3.662	-6.483	1.00	25.00
ATOM	24	C5*	G	A	2	8.989	3.594	-6.187	1.00	25.00
ATOM	25	C4*	G	A	2	9.378	2.172	-5.835	1.00	25.00
ATOM	26	O3*	G	A	2	9.415	0.945	-7.948	1.00	25.00
ATOM	27	C3*	G	A	2	8.751	1.078	-6.697	1.00	25.00
ATOM	28	C2*	G	A	2	8.902	-0.134	-5.779	1.00	25.00
ATOM	29	C1*	G	A	2	8.500	0.483	-4.443	1.00	25.00
ATOM	30	O4*	G	A	2	8.917	1.839	-4.493	1.00	25.00
ATOM	31	N9	G	A	2	7.034	0.441	-4.182	1.00	25.00

(a) Confirm that the data is consistent with A type DNA.

(b) Check the conformation by calculating the distance of the C_3' and C_2' atoms from the plane of C_4'-O_4'-C_1' atoms.

Q6.58 Figure 6.49 shows part of the DNA found in the zinc finger protein (PDB entry 1AAY).

(a) What base is shown in Fig. 6.49 and is this a purine or pyrimidine base?

(b) Calculate the ring pucker υ_2, and γ and χ torsion angles.

(c) What type of DNA is this?

The .pdb coordinates for the top sugar and base in the picture are

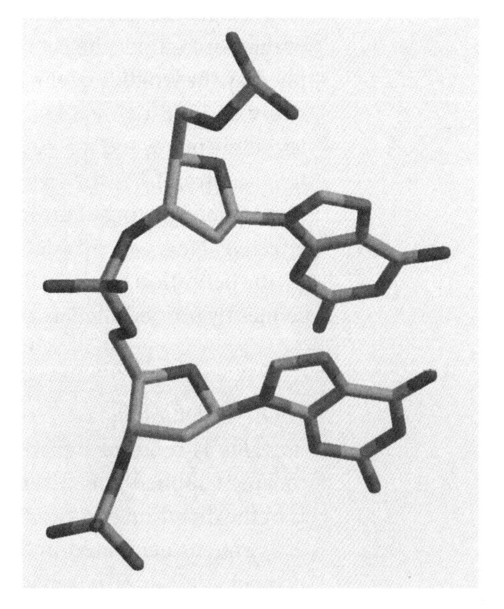

Fig. 6.49 Part of the zinc finger protein pdb 1AAY.

ATOM	837	P	G	B	6	5.282	8.838	56.055	1.00	36.20	DNA1	P
ATOM	838	O1P	G	B	6	6.249	8.553	57.139	1.00	35.84	DNA1	O
ATOM	839	O2P	G	B	6	4.927	7.788	55.059	1.00	35.58	DNA1	O
ATOM	840	O5*	G	B	6	3.903	9.424	56.660	1.00	32.80	DNA1	O
ATOM	841	C5*	G	B	6	3.829	10.592	57.472	1.00	28.09	DNA1	C
ATOM	842	C4*	G	B	6	2.610	11.430	57.099	1.00	27.15	DNA1	C
ATOM	843	O4*	G	B	6	2.772	12.046	55.805	1.00	25.05	DNA1	O
ATOM	844	C3*	G	B	6	1.382	10.524	57.025	1.00	28.06	DNA1	C
ATOM	845	O3*	G	B	6	0.315	11.094	57.776	1.00	30.82	DNA1	O
ATOM	846	C2*	G	B	6	1.034	10.517	55.563	1.00	26.27	DNA1	C
ATOM	847	C1*	G	B	6	1.598	11.813	55.022	1.00	25.13	DNA1	C
ATOM	848	N9	G	B	6	1.981	11.653	53.605	1.00	23.54	DNA1	N
ATOM	849	C8	G	B	6	2.908	10.792	53.071	1.00	23.68	DNA1	C
ATOM	850	N7	G	B	6	3.051	10.912	51.784	1.00	23.73	DNA1	N
ATOM	851	C5	G	B	6	2.155	11.922	51.440	1.00	22.27	DNA1	C
ATOM	852	C6	G	B	6	1.874	12.481	50.170	1.00	20.87	DNA1	C
ATOM	853	O6	G	B	6	2.415	12.213	49.105	1.00	19.80	DNA1	O
ATOM	854	N1	G	B	6	0.895	13.472	50.251	1.00	21.65	DNA1	N
ATOM	855	C2	G	B	6	0.275	13.875	51.422	1.00	21.79	DNA1	C
ATOM	856	N2	G	B	6	-0.637	14.838	51.323	1.00	22.30	DNA1	N
ATOM	857	N3	G	B	6	0.551	13.346	52.614	1.00	20.28	DNA1	N
ATOM	858	C4	G	B	6	1.493	12.380	52.547	1.00	21.68	DNA1	C
ATOM	859	P	G	B	7	-1.054	10.303	58.105	1.00	31.62	DNA1	P

***Strategy*:** First, identify the atoms about which the angles are to be calculated; the sugar atoms are starred so this is easy! By definition, (see Section 6.21), the χ torsion angle is between N_9 and C_4 in this type of base.

6.23 Torque and angular momentum

If two objects are joined with a bolt and this is tightened with a spanner, the clockwise motion of tightening causes a force to be applied to the bolt that generates a *torque* or *moment* of that force about the rotation axis. The torque has a magnitude that is the force F multiplied by the length r of the lever. It is well known that a longer spanner is needed to more easily undo or tighten a stiff bolt. Similarly, if you push on a seesaw, torque is generated about the pivot, and it is easier to move the seesaw if it is pushed at its end rather than close to its centre. If F is the size of the force that is applied at an angle θ at a distance r from a pivot, then the torque has magnitude $Fr\sin(\theta)$. Thus, only a component of the force is effective. Next we consider the force and distance to be vectors $\boldsymbol{F}$ and $\boldsymbol{r}$ respectively, then by the definition of a cross product $|\boldsymbol{F}\times\boldsymbol{r}| = |\boldsymbol{F}||\boldsymbol{r}|\sin(\theta)\vec{n} = Fr\sin(\theta)\vec{n}$. Torque can consequently be identified as a vector in the direction $\vec{n}$ and defined as the cross product,

$$\boldsymbol{T} = \boldsymbol{r}\times\boldsymbol{F}.$$

On the left of Fig. 6.51, the force is applied at right angles, on the right at an angle θ, so its value is reduced by $\sin(\theta)$ in accordance with the cross product. The direction the torque is applied in is determined by the forces and the cross product rule, Fig. 6.50, and is in the direction of a right-handed screw.

Torque is generated by molecular motors and has been measured in the rotor of the protein ATPase (Cherepanov & Junge 2001; Pänke et al. 2001) as well as in everyday objects such as electric motors and petrol engines. A torque wrench is a common tool for a mechanic to use to tighten engine and wheel bolts, and manufacturers often specify the torque to be applied to particular bolts. Power tools used to tighten wheel bolts often have

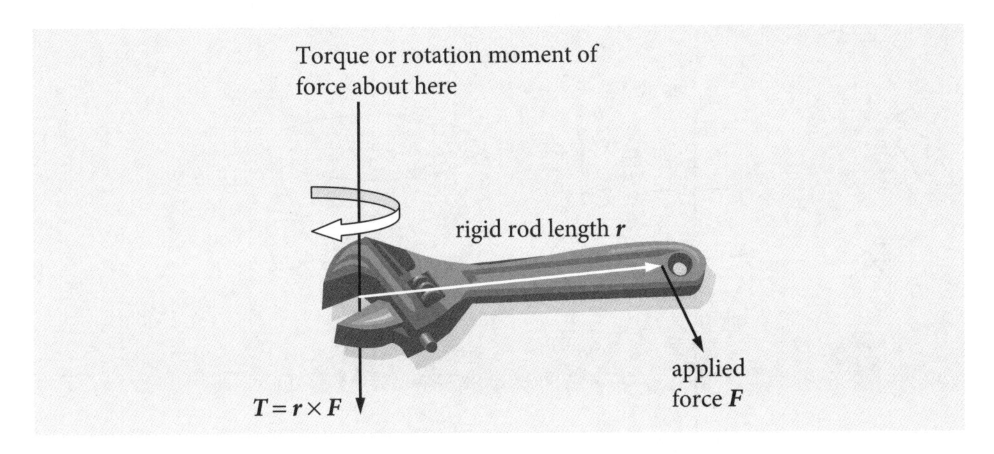

Fig. 6.50 Torque is the cross product, applied force × length of rod.

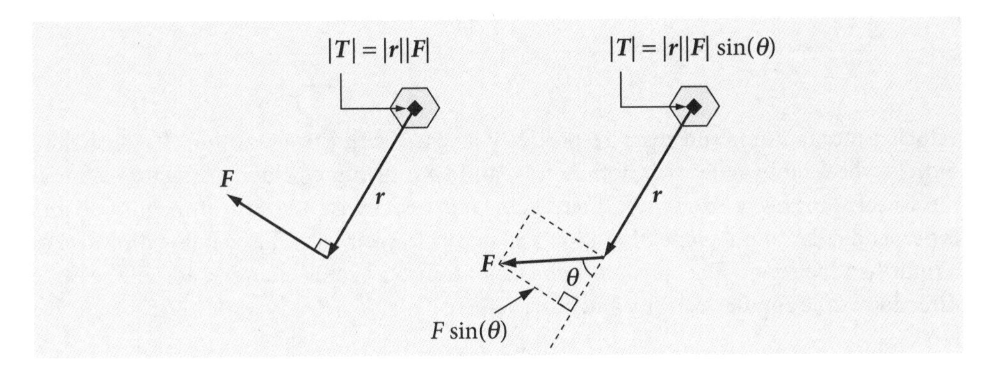

Fig. 6.51 Generating torque by applying a force to cause a rotation. Torque acts into the figure.

a clutch that slips at a certain torque to prevent over-tightening. The units of torque are those of energy, but it is a vector and not a scalar, so also has direction.

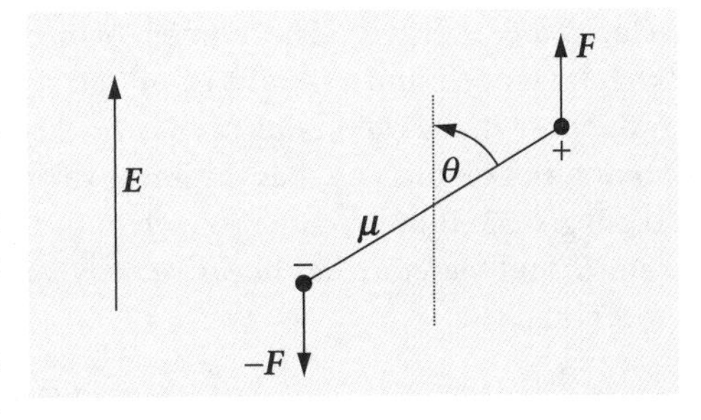

Fig. 6.52 Dipole in a field.

A polar molecule in a non-polar solvent that is placed in a homogeneous electric field of strength E will produce a torque. The molecular dipole produces the torque and the molecule will try to turn to minimize its energy by aligning itself with the field. However, this motion will be resisted by the surrounding solvent molecules and randomized by thermal motion. Suppose that the molecule's dipole moment is $\boldsymbol{\mu} = q\boldsymbol{r}$, where $\boldsymbol{r}$ is a vector of length r and q charge, then the torque is $\boldsymbol{\mu} \times \boldsymbol{E}$ and this is in the direction that is perpendicular to $\boldsymbol{\mu}$ and $\boldsymbol{E}$. The torque operates to minimize a dipole's energy in the field. At some angle θ to the field the dipole has the energy $\mu E \sin(\theta)$ and by convention $\theta = 90°$ is chosen to be zero so that the energy (or work done) to reach angle θ is the integral

$$W = \mu E \int_{\pi/2}^{\theta} \sin(\theta) d\theta = -\mu E \cos(\theta).$$

If the dipole and field are made into vectors the energy is $W = -\boldsymbol{\mu}\cdot\boldsymbol{E}$. A similar calculation would apply if a magnetic dipole were placed in a magnetic field.

6.23.1 Angular momentum

When studying the motion of rotating bodies, nuclei, molecules, or footballs we must define angular velocity $\boldsymbol{\omega}$, angular momentum $\boldsymbol{L}$, and moment of inertia I, which replace

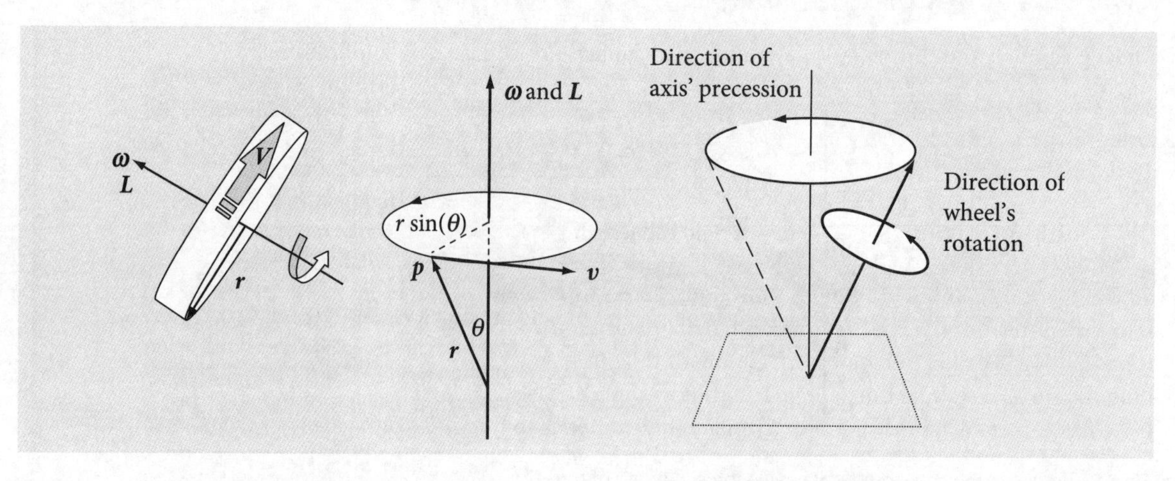

Fig. 6.53 A wheel of radius ***r*** set spinning with a linear velocity ***v*** produces angular velocity ω and has angular momentum $\boldsymbol{L} = I\boldsymbol{\omega}$ in the direction shown. Middle: a rigid body rotating about its axis. Right: Precession of the axis of a wheel in a force such as gravity. Precession causes the axis to rotate about the applied field at the same time as the wheel is rotating about its own axis.

velocity, momentum, and mass respectively as used with linear motion. If you hold a bicycle wheel by its axle with both hands, and spin it in a clockwise direction, so that the wheel spins away from you, then its *angular velocity* $\boldsymbol{\omega}$, a vector of magnitude $|\boldsymbol{\omega}|$, is perpendicular to the plane of motion and points to your left. This is in the direction of a right-handed screw. The *angular momentum* $\boldsymbol{L}$, also a vector quantity, acts in the same direction as the angular velocity and has a value

$$\boldsymbol{L} = \boldsymbol{I\omega}$$

where $\boldsymbol{I}$ is the moment of inertia of the wheel. Linear momentum is the quantity $\boldsymbol{p} = m\boldsymbol{v}$ and therefore, in angular motion, the moment of inertia takes the place of mass and angular velocity that of velocity used in linear motion. Velocity has units of m s^{-1} and angular velocity ω of radian s^{-1}. The moment of inertia has units of mass × distance^2 usually kg m^2. Angular momentum has units kg m^2 rad s^{-1} or rad J s.

Suppose that a rigid body such as a wheel, rotates about a fixed (say z) axis, then any point p, not on this axis, has the linear velocity $\boldsymbol{v}$ in the x–y plane and so is perpendicular to z Fig. 6.53. If the vector $\boldsymbol{r}$ joins p to the axis, then p moves through a circle of radius $r\sin(\theta)$ and therefore has linear velocity $v = \omega r\sin(\theta)$. This is equal to the magnitude of $|\boldsymbol{\omega}\times\boldsymbol{r}|$ and so

$$\boldsymbol{v} = \boldsymbol{\omega}\times\boldsymbol{r}.$$

Angular momentum is by definition

$$\boldsymbol{L} = \boldsymbol{r}\times m\boldsymbol{v} = m\boldsymbol{r}\times(\boldsymbol{\omega}\times\boldsymbol{r})$$

and in the second step the definition of linear velocity is used. An elementary approach to angular momentum does not treat velocity as a vector, but gives the equation $L = mvr$. The limited applicability of this equation can be discovered if the triple product is expanded; see Section 6.18. The result is

$$\begin{aligned}\boldsymbol{L} &= m\boldsymbol{r}\times(\boldsymbol{\omega}\times\boldsymbol{r})\\ &= m(\boldsymbol{r}\cdot\boldsymbol{r})\boldsymbol{\omega} - m(\boldsymbol{r}\cdot\boldsymbol{\omega})\boldsymbol{r}\\ &= m|\boldsymbol{r}|^2\,\boldsymbol{\omega} - m|\boldsymbol{r}||\boldsymbol{\omega}|\cos(\theta)\boldsymbol{r}\end{aligned}$$

In the special case that $\boldsymbol{r}$ and $\boldsymbol{v}$ are in the same plane, then $\boldsymbol{r}$ is perpendicular to the angular velocity making $\theta = 90°$ and the cosine zero. The magnitude of the angular momentum is therefore $|\boldsymbol{L}| = L = mr^2|\boldsymbol{\omega}| = mvr$ where r and v are the magnitudes of their respective vectors.

To illustrate the application of cross products, a basic NMR experiment is briefly described. You will recall that in these experiments the sample is placed in a large static and homogeneous magnetic field $\boldsymbol{B}$, and then irradiated with radio-frequency (RF) radiation at right angles to $\boldsymbol{B}$. The energy of a nucleus with non-zero nuclear spin in a static magnetic field is $E=-\boldsymbol{\mu}\cdot\boldsymbol{B}$, where $\boldsymbol{\mu}$ is the magnetic dipole moment vector and $\boldsymbol{B}$ the static magnetic field vector. Because the nuclear spin is subject to quantum mechanics, the magnetic dipole moment is written as $\boldsymbol{\mu}=\gamma\boldsymbol{I}$ where $\boldsymbol{I}$ is the quantum spin angular momentum and γ the magnetogyric ratio. This is a constant, but different for each type of nucleus, for protons $\gamma=26.7\times10^{7}$ rad $\mathrm{T^{-1}\,s^{-1}}$. The proton has two spin states defined with quantum numbers $\{I, m_z\}=\{1/2, \pm1/2\}$ where I is the spin quantum number and m_z the z component also called the magnetic, azimuthal or projection quantum number. The nuclear spin energies are therefore $E=-\gamma IB=\pm\dfrac{1}{2}\hbar\gamma B_0$ and the dot product was removed because the applied magnetic field B_0 is only along the z-axis, $B_0=|\boldsymbol{B}|$. When the nuclear spin angular momentum is in the same direction as the magnetic field the lower energy of the two energy levels results and has $m_z=1/2$. An NMR transition occurs when one unit of angular momentum is absorbed from, or emitted to, the RF field. The transition energy is always the difference in two energy levels, provided they have a difference of one unit of angular momentum and this energy is always $\Delta E=\hbar\gamma B$. The corresponding transition frequency is $\omega=\gamma B_0$ rad $\mathrm{s^{-1}}$ and is called the Larmor frequency.

It is also necessary to consider the ensemble of spin states called the magnetization $\boldsymbol{M}$. This is the vectorial sum of all the individual spin magnetic dipoles, from states pointing with and against the magnetic field $\boldsymbol{B}$. Consequently, $\boldsymbol{M}$ will lie exactly parallel to $\boldsymbol{B}$. As the magnetization is a macroscopic quantity, it can be treated classically in most instances. This means that it is normally subject to the rules of classical mechanics. When a circularly polarized radio frequency field is applied for a short while, and at 90° to the static magnetic field, it induces nuclear spins to flip from one state to another. The consequence of this is that magnetization experiences a torque and is rotated away from the direction of field $\boldsymbol{B}$ and starts to precess, rather as a spinning top does. Precession is the *rotation of the axis* of a spinning body about the axis of an applied force. The precessing magnetization produces a rotating magnetic field that induces a current in a coil of wire, which, when Fourier transformed, becomes the NMR signal. The precession is caused because a torque is applied to the nuclear spins, and hence magnetization, by the static magnetic field, and is analogous to that observed by a spinning top (gyroscope) in a gravitational field.

The magnetization's angular frequency has the same value ω as that of the NMR transition, but its angular velocity points in the opposite direction to the field if γ is positive, i.e. $\boldsymbol{\omega}=-\gamma\boldsymbol{B}$. To show this, the direction of the torque has to be calculated. The torque $\boldsymbol{T}$ on the magnetization is $\boldsymbol{T}=\boldsymbol{M}\times\boldsymbol{B}$ and is therefore at 90° to $\boldsymbol{M}$ and $\boldsymbol{B}$ as shown in Fig. 6.54 and which corresponds to the definition given in Section 6.16. The linear velocity $\boldsymbol{v}$ must be parallel to the torque and is defined as $\boldsymbol{v}=\boldsymbol{\omega}\times\boldsymbol{M}$. Consequently, when the magnetogyric ratio is positive the angular velocity $\boldsymbol{\omega}$ points downwards and the spin precession is anticlockwise when viewed from below, see Fig. 6.54.

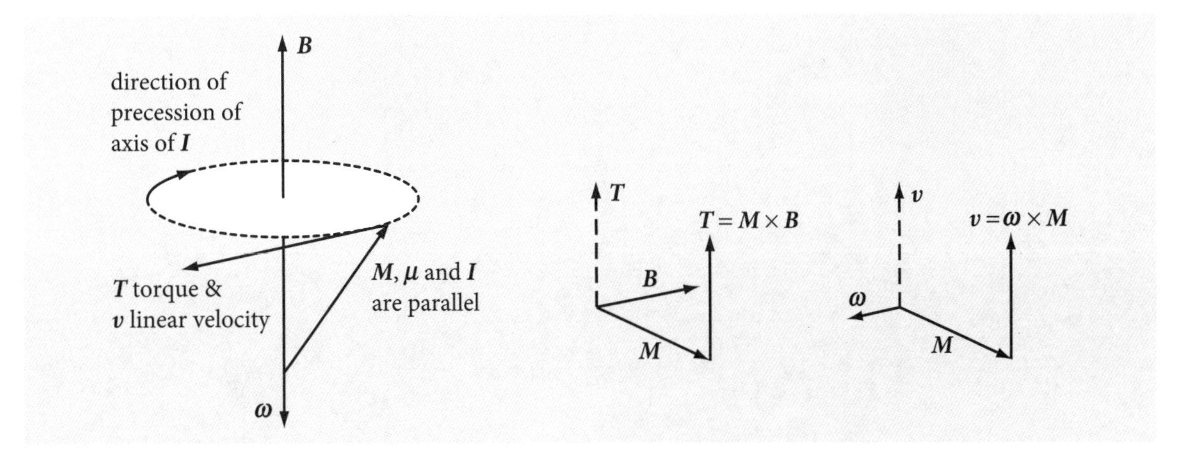

Fig. 6.54 Left: precession of magnetisation of nuclei with positive magnetogyric ratio in a field $\boldsymbol{B}$. Right: sketches of the relative orientation of vectors. These two diagrams need to be rotated to fit that on the left.

6.23.2 Questions

Full solutions are available at **www.oxfordtextbooks.co.uk/orc/beddard.**

Q6.59 The differential of a cross product is $\dfrac{d}{dt}(A\times B) = A\times\dfrac{dB}{dt} + \dfrac{dA}{dt}\times B$, calculate, dL/dt if $L = mv\times\dfrac{dr}{dt}$.

Q6.60 A particle with orbital motion has a centripetal acceleration of $\boldsymbol{a} = \boldsymbol{\omega}\times(\boldsymbol{\omega}\times\boldsymbol{r})$, where ω is angular velocity and $\boldsymbol{r}$ a positional vector. Find $|\boldsymbol{a}|$ if the motion is in a circle where $\boldsymbol{r}$ and ω are perpendicular.

Q6.61 Derive the equation for the linear velocity of a disc, and by inference any rigid body, using the $(\boldsymbol{i}, \boldsymbol{j}, \boldsymbol{k})$ basis set of vectors. Although these are fixed in space they may be used by allowing the amplitude of the vector's components to change with rate α which is also the magnitude of the angular velocity ω,

$$\boldsymbol{r}(t) = r\cos(\alpha t)\boldsymbol{i} + r\sin(\alpha t)\boldsymbol{j}.$$

Calculate the cross product $\boldsymbol{\omega}\times\boldsymbol{r}$ and the position vector $\boldsymbol{r}$. Next, differentiate $\boldsymbol{r}$ to obtain the linear velocity $\boldsymbol{v}$ and equate the two equations. Figure 6.55 shows the definition of vectors. Let r be the magnitude of vector $\boldsymbol{r}$.

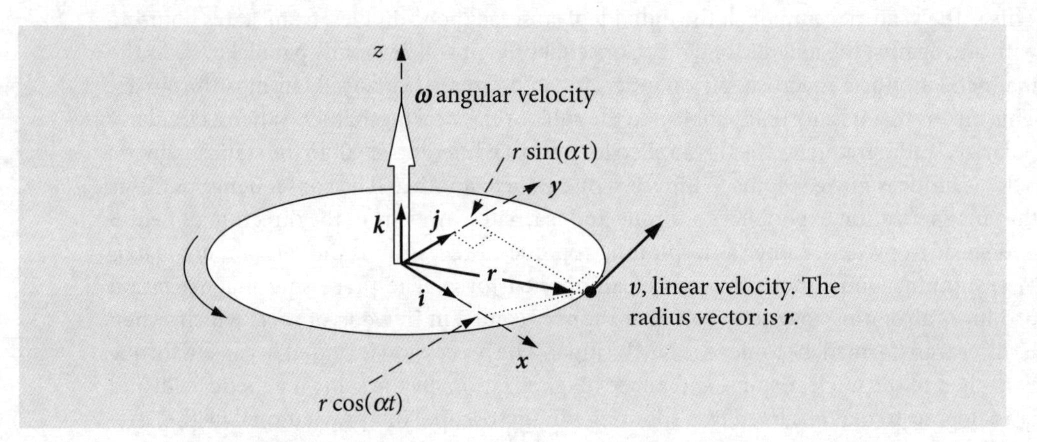

Fig. 6.55 Angular velocity ω, linear velocity $\boldsymbol{v}$ of a disc of radius r with radius vector $\boldsymbol{r}$.

Matrices

7

7.1 Motivation and concept

A *matrix* is a two-dimensional, rectangular or square array of numbers or functions. A *determinant* is always a square block of numbers or functions and is used in solving systems of simultaneous equations and is really an alternative way of representing a number (a scalar) or a polynomial and is evaluated following a single rule.

A matrix, which may not necessarily be square, is a way of representing and handling *linear transformations*, which means, for example, if we want to rotate a picture of a protein or other molecule, a rotation matrix is used to take us from one orientation to another. The transformation is linear as each atom is moved (we could also say mapped) to a new position, but the molecular structure remains the same because the distance and angle between any two atoms is unchanged.

Matrices and determinants are used when calculating electronic energies, such as by the Hückel method, and in quantum mechanics generally. Expectation values, e.g. average energies, positions, momenta, etc., which are the quantities measured in an experiment, are, in quantum mechanics, also called matrix elements. When group theory is used to calculate the normal vibrational modes of a molecule or even the symmetry of a molecular orbital, matrices are involved.

Matrices are represented by square [ ] or round ( ) brackets surrounding the block of numbers; determinants always by straight lines | | and sometimes double straight lines || ||. The following matrix has $m = 4$ rows and $n = 3$ columns. The elements of any matrix or determinant are referenced with two subscripts, so, if the matrix A has m rows and n columns, i.e. an $m \times n$ matrix, the mn^{th} element is written as A_{mn} and the indexing is always row by column. The diagonal is only present in a square matrix or determinant and is the series of terms from top left to bottom right that have the same indices, e.g., a_{22}. The anti-diagonal passes from top right to bottom left.

$$\overset{n=3 \text{ columns}}{m = 4 \text{ rows} \begin{bmatrix} a_{11} & a_{12} & a_{13} \\ a_{21} & a_{22} & a_{23} \\ a_{31} & a_{32} & a_{33} \\ a_{41} & a_{42} & a_{43} \end{bmatrix}} \qquad \overset{n=2}{m = 2 \begin{bmatrix} b_{11} & b_{12} \\ b_{21} & b_{22} \end{bmatrix}}$$

Fig. 7.1 4×3 matrix and, right, a 2×2 square matrix. Notice the subscript numbering scheme; the first index is the row the second the column. The diagonal elements always have the same two indices.

The square matrix has the same number of rows as columns, therefore $m = n$ otherwise the matrix is rectangular. A one-dimensional matrix, Fig. 7.2 is also equivalent to either a row or a column vector; see Chapter 6.

The properties of determinants and some examples of their use are described first. Next, the various properties of matrices are described, followed by sections illustrating how matrices are used to solve a wide range of problems ranging from designing laser cavities, to calculating the symmetry of molecules to the selective breeding of plants.

$$[a_1 \quad a_2 \quad \ldots] \qquad \begin{bmatrix} b_1 \\ b_2 \\ \cdot \\ \cdot \\ \cdot \end{bmatrix}$$

Fig. 7.2 A row and column matrix or vectors.

7.2 Determinants

The value of a determinant is always a number or a polynomial equation. Suppose a determinant consisting of the squares of the numbers one to four is

$$\begin{vmatrix} 1 & 4 \\ 9 & 16 \end{vmatrix}$$

then its value is $1 \times 16 - 4 \times 9 = -20$, and in symbols the calculation is the identity

$$\begin{vmatrix} a & b \\ c & d \end{vmatrix} \equiv ad - cb. \tag{7.1}$$

7.2.1 The characteristic equation

In many situations, the determinant is made equal to zero; for example, suppose that the determinant is

$$\begin{vmatrix} x & 1 \\ 1 & x \end{vmatrix} = 0$$

then this is an alternative way of writing the equation $x^2 - 1 = 0$, which is called the *characteristic* equation or characteristic polynomial.

7.2.2 Evaluating 3 × 3 determinants

One way of writing a 3 × 3 determinant of the squares of integers is

$$\begin{vmatrix} 1 & 4 & 9 \\ 16 & 25 & 36 \\ 49 & 64 & 81 \end{vmatrix}$$

Its value can be determined in a similar way to the 2 × 2 determinant. A pictorial method is shown below and is the same as used to calculate cross products, Chapter 6. The rules are

(a) Each number in the top row is multiplied by the 2 × 2 matrix formed excluding the values in the column underneath that number.

(b) the terms are summed but with *alternate* values being multiplied by −1.

The first term is diagrammatically represented as shown below,

$$\begin{vmatrix} 1 & 4 & 9 \\ 16 & 25 & 36 \\ 49 & 64 & 81 \end{vmatrix} \tag{7.2}$$

which evaluates to $1 \times (25 \times 81 - 64 \times 36)$. The *cofactor* of the top left (1, 1) element of the determinant is $(25 \times 81 - 64 \times 36)$. The second term is $-4 \times (16 \times 81 - 49 \times 36)$, with cofactor $-(16 \times 81 - 49 \times 36)$. Notice the extra minus sign with this term. Graphically this term is

$$\begin{vmatrix} 1 & 4 & 9 \\ 16 & 25 & 36 \\ 49 & 64 & 81 \end{vmatrix}$$

and the third term is $9 \times (16 \times 64 - 49 \times 25)$. It is worth remembering the order of doing the calculation for small determinants even though most calculations will be done using a computer. In expanded form the calculation above is,

$$\begin{vmatrix} 1 & 4 & 9 \\ 16 & 25 & 36 \\ 49 & 64 & 81 \end{vmatrix} = 1\begin{vmatrix} 25 & 36 \\ 64 & 81 \end{vmatrix} - 4\begin{vmatrix} 16 & 36 \\ 49 & 81 \end{vmatrix} + 9\begin{vmatrix} 16 & 25 \\ 49 & 64 \end{vmatrix}$$

Although the top row was used to lead us in factoring the determinant, we could just as easily use any one of the columns or rows with the appropriate cofactors.

The theory of determinants allows us, somewhat unexpectedly, to calculate a determinant from the number of ways of arranging the *symbols* of the determinant's indices; the top left element is labelled with indices 11, the next in the same row 12, and so on. The next row starts with indices 21, 22, and so forth as in Fig. 7.1. In a 3×3 determinant the first index of each number of the triple number is *always* in the order 1, 2, 3; the second indices, which are also 1, 2 or 3, are used in all possible arrangements viz.:

$$+123, \quad +231, \quad +312, \quad -321, \quad -132, \quad -213.$$

All terms have *first* indices whose numbers are always in the order 123 and second indices in this order also, if rotated around one another. In the negative terms, the order of each of the second indices of a positive term is reversed. The determinant is then

$$\begin{vmatrix} a_{11} & a_{12} & a_{13} \\ a_{21} & a_{22} & a_{23} \\ a_{31} & a_{32} & a_{33} \end{vmatrix} = a_{11}a_{22}a_{33} + a_{12}a_{23}a_{31} + a_{13}a_{21}a_{32} - a_{13}a_{22}a_{31} - a_{11}a_{23}a_{32} - a_{12}a_{21}a_{33}$$

7.2.3 Evaluating determinants using Maple

To calculate a determinant the `LinearAlgebra` package must be included to tell Maple that matrices are being used; notice here the odd use of capitals; this package has most if not all its instructions starting with a capital letter.

```
> restart: with(LinearAlgebra):
> M:= < < 1|4|9 > , < 16|25|36 > , < 49|64|81 > > ;
```

$$M := \begin{bmatrix} 1 & 4 & 9 \\ 16 & 25 & 36 \\ 49 & 64 & 81 \end{bmatrix}$$

```
> Determinant(M);
```

$$-216$$

Notice how the matrix is constructed, the vertical bars | separate elements in columns, the commas separate rows. Alternatively, the matrix could be defined as

```
> Matrix(3,3,[ [ 1,4,9] , [16,25,36] , [49,64,81] ] );
```

and which is used is a matter of preference. Notice that Maple does not distinguish in its formatting, a matrix and a determinant; the word `Determinant` is used to obtain a value not to define a determinant.

Using Maple, the calculation of a 3×3 determinant is not a lot shorter than that by hand as we have to set up the calculation. A far larger calculation, however, takes no more effort but would be very tedious if not practically impossible by hand without making an error. You will see this if you try working out the values of a 5×5 determinant by hand.

7.2.4 Determinants have the following properties

(i) A common multiplier can be factored out of a determinant.

(ii) If any two rows or two columns are equal, or if a row or column is zero, the determinant is zero.

(iii) If any two row or columns are interchanged, then the value of the determinant changes sign.

(iv) If any row or column is multiplied by a constant then the whole determinant is multiplied by that factor.

(v) The determinant's value is unaltered if a row or column is multiplied by a constant and then added to another row or column.

7.2.5 Examples

(i) To work out the determinant of the matrix

$$\begin{bmatrix} \cos(x) & \sin(x) \\ -\sin(x) & \cos(x) \end{bmatrix}$$

use the rule in equation (7.1) which produces

$$\begin{vmatrix} \cos(x) & \sin(x) \\ -\sin(x) & \cos(x) \end{vmatrix} = \cos^2(x) + \sin^2(x),$$

and by the well-known trigonometric identity, $\cos^2(x) + \sin^2(x) = 1$. The determinant is therefore unity. This matrix is a rotation matrix and its use is described in Section 7.7. Because the determinant is 1 the matrix does not distort the object rotated.

(ii) The determinant

$$\begin{vmatrix} x & 1 & 1 \\ 1 & x & 1 \\ 1 & 1 & x \end{vmatrix} = 0$$

is another way of writing a polynomial. Since there are three occurrences of x the determinant should produce a cubic equation. Expanding by using cofactors, as in equation (7.2), produces

$$\begin{vmatrix} x & 1 & 1 \\ 1 & x & 1 \\ 1 & 1 & x \end{vmatrix} = x(x^2 - 1) - (x - 1) + 1 - x = x^3 - 3x + 2 = 0$$

and simplifying gives $x(x^2 - 3) + 2 = 0$. The solution of this *characteristic polynomial* may be seen by inspection, which often means by trial and error; there are two roots with $x = 1$ and one with $x = -2$. If you are not sure, substitute these answers into the equation, or use Maple:

```
> solve( x^3 - 3*x + 2, x);       # solve equation for x
```

(iii) Hückel MO method. In the Hückel molecular orbital approximation π bonding energies of molecules, ions, and radicals can be calculated. This method reflects the topology of the structure because only adjacent atoms interact with one another, and the change in energy with the length or angle of bonds is not included in the calculation. In this model of π bonding, each electron in a π orbital interacts only with π electrons on the nearest atoms. This interaction energy is calculated from the resonance (exchange) Coulomb energy integral, is conventionally given the symbol β, and is negative since interaction is attractive. It has a value ≈ -300 kJ/mole, although estimates vary widely. The π electrons on any atom also have their own energy, which is the Coulomb self-energy integral, and this is labelled α. The interaction energies can be made into a determinant, which is called the secular determinant[1] and which is then solved to find the energy. This determinant is formed by a set of simple rules; why it works is explained later on in Section 7.12.3 that describes eigenvalue–eigenvector equations.

The determinant is formed from the rules:

(i) Label the atoms in 1 to n, and $n = 4$ in the case of 1,3-butadiene. This produces an $n \times n$ determinant. Labelling the atoms in this way effectively forms a basis set for the problem, see Section 7.12.8.

(ii) Each diagonal element has the energy $\alpha - E$.

(iii) Each π electron when connected to a nearest neighbour has interaction energy β. The basis set labelling determines where these go in the determinant. Using the labels starting as indicated in Fig. 7.3, atoms 1-2, 2-3, 3-4 only have interaction energy β, which means that β is entered at these positions in the determinant and

[1] The word 'secular' usually means 'not religious', but here it means 'long-term' in the sense of being time independent or unchanging as used in mechanics.

the corresponding positions to make the determinant symmetrical. The reason for this is that interaction 1-2 is the same as 2-1 and this makes the determinant Hermitian; see Section 7.4.9.

(iv) Any remaining entries in the determinant are zero and the determinant is made equal to zero.

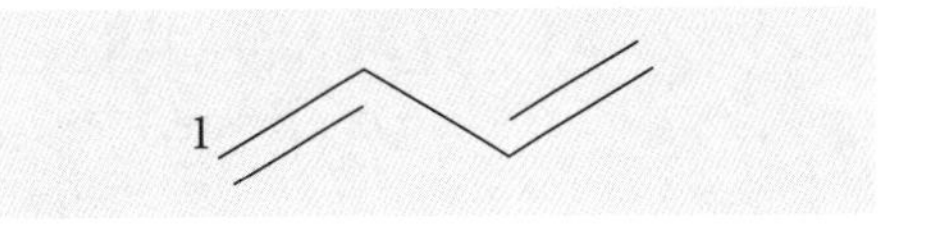

Fig. 7.3 1,3-butadiene.

The determinant describing the interactions of the four π electrons of 1,3-butadiene has a tri-diagonal form. The numbers above the determinant refer to the position in the determinant and to the atom positions in Fig. 7.3.

$$\begin{array}{c|cccc|}
 & 1 & 2 & 3 & 4 \\
1 & \alpha - E & \beta & 0 & 0 \\
2 & \beta & \alpha - E & \beta & 0 \\
3 & 0 & \beta & \alpha - E & \beta \\
4 & 0 & 0 & \beta & \alpha - E
\end{array} = 0$$

Note that the conventional π bonds are ignored; all that matters is where the π electrons are situated and because there are four electrons, there will be four energies and four molecular orbitals.

To solve the determinant each term is divided by β, and letting $x = \dfrac{\alpha - E}{\beta}$, the determinant can be written in a clearer form as

$$\begin{vmatrix} x & 1 & 0 & 0 \\ 1 & x & 1 & 0 \\ 0 & 1 & x & 1 \\ 0 & 0 & 1 & x \end{vmatrix} = 0.$$

We could solve the determinant by hand and then attempt to solve the quartic equation also by hand by factoring out terms or perhaps by plotting the function and looking for its roots where $x = 0$. The Newton–Raphson method could also be used to find roots numerically and until recently, this would have been done. Instead, with Maple the construction and solution of the Hückel determinant is straightforward. The matrix can be constructed manually or the scheme below can be used which will make the Hückel determinant for any linear polyene. The `if .. then .. end if` statement selects off-diagonal terms and puts values into the +1 and −1 diagonals.

Algorithm 7.1 Hückel determinant

```
> restart: with(LinearAlgebra):
  n:= 4:
  M:= Matrix( n, n ):                  # create a 4 x 4 matrix
  for i from 1 to n do                 # fill matrix
    M[i,i]:= x;                        # diag indices are i,i
    if (i > 0 ) and (i < n) then
      M[i,i+1]:= 1:
      M[i+1,i]:= 1:
    end if                             # add +1, -1 diags
  end do:                              # end of matrix filling
  M;                                   # show matrix to check
```

$$\begin{bmatrix} x & 1 & 0 & 0 \\ 1 & x & 1 & 0 \\ 0 & 1 & x & 1 \\ 0 & 0 & 1 & x \end{bmatrix}$$

```
> char_eqn:= Determinant(M);           # calculate and solve
```

$$char\_eqn := x^4 - 3x^2 + 1$$

```
> x:= solve( char_eqn, x );
```

$$x := -\frac{1}{2} + \frac{1}{2}\sqrt{5},\ -\frac{1}{2} - \frac{1}{2}\sqrt{5},\ \frac{1}{2} + \frac{1}{2}\sqrt{5},\ \frac{1}{2} - \frac{1}{2}\sqrt{5}$$

The solutions give each of the values of x and as $x = \dfrac{\alpha - E}{\beta}$ the four energies given by $E = \alpha - \beta x$ are;

$$E = \alpha - \frac{\beta}{2}(1 + \sqrt{5}) = \alpha - 1.618\beta, \qquad E = \alpha - \frac{\beta}{2}(1 - \sqrt{5}) = \alpha + 0.618\beta,$$

$$E = \alpha + \frac{\beta}{2}(1 - \sqrt{5}) = \alpha - 0.618\beta, \qquad E = \alpha + \frac{\beta}{2}(1 + \sqrt{5}) = \alpha + 1.618\beta.$$

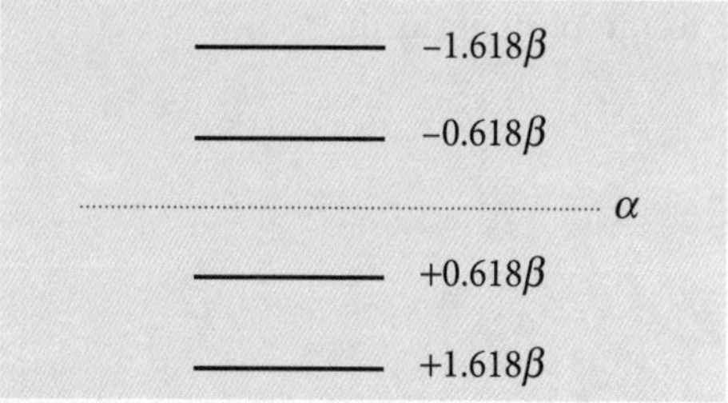

-1.618β

-0.618β

α

$+0.618\beta$

$+1.618\beta$

Fig. 7.4 Butadiene Hückel energy levels measured as changes away from α. (The integral β is negative.)

Notice that when the energies are plotted out, Fig. 7.4, they symmetrically span the energy α, which is the self-energy of the π electrons. If the interaction was zero because $\beta = 0$, then the determinant would be diagonal and all the electrons would be degenerate with energy $E = \alpha$. The initial energy, or rather the energy of 4π electrons is 4α; after the molecular orbital has formed the energy is $2(\alpha + 1.618\beta) + 2(\alpha + 0.618\beta)$, a difference of $2(1.618 + 0.618)\beta$ or 4.472β and this is the energy gained or *stabilization* energy, β being negative. This can be seen in the figure also. However, this is not really what is of interest because if the MO covers all the atoms, as it must, then the *delocalization* energy is of interest as it is the energy saved compared to two individual double bonds; these are equivalent to the π energy of two ethylene molecules which is $4(\alpha + \beta)$. The delocalization energy saving is then

$$4(\alpha + \beta) - 2(\alpha + 1.618\beta) - 2(\alpha + 0.618\beta) = 4\beta - 2(1.618 + 0.618)\beta$$

or 0.472β.

The characteristic polynomial can be plotted, as shown in Fig. 7.5, to convince you that the solutions are correct. The roots of the equation ($y = 0$) can be seen to be at just greater than ± 0.6 and ± 1.6. A more detailed plot would give more accurate answers but not as exactly as the algebraic solution.

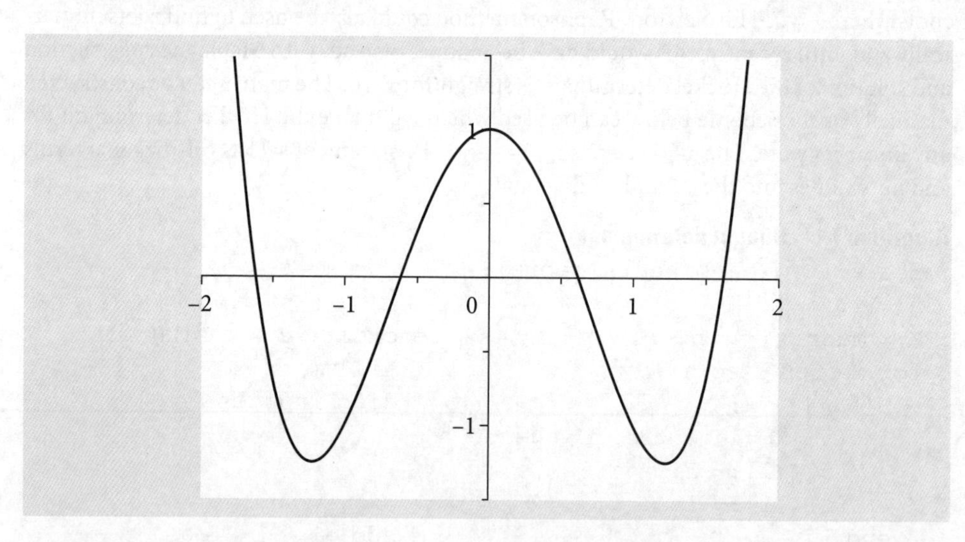

Fig. 7.5 Graphical solution of the characteristic polynomial for butadiene.

7.3 Questions

Full solutions are available at **www.oxfordtextbooks.co.uk/orc/beddard.**

Q7.1 Evaluate the Lorentz determinant

$$d\begin{vmatrix} 1 & -v \\ -\dfrac{v}{c^2} & 1 \end{vmatrix},$$

where c is the speed of light and v the velocity of a body and $d = c/\sqrt{c^2 - v^2}$.

Q7.2 Show that $\begin{vmatrix} 0 & y & z \\ x & 0 & -z \\ x & y & 0 \end{vmatrix} = 0.$

Q7.3 Determine the characteristic polynomial of $\begin{vmatrix} x & a & b \\ a & x & b \\ a & b & x \end{vmatrix}$ and solve it for x.

Q7.4 Calculate the cofactor of the (2, 3) element of $\begin{vmatrix} a & b & x \\ c & d & 1 \\ e & f & y \end{vmatrix}$.

Q7.5 **(a)** Use Maple to calculate the determinant of the first 16 numbers starting at 1. Explain why the determinant is zero. Construct the matrix explicitly then repeat the calculation using a Maple procedure containing two `for..do` loops.

(b) Change the matrix, using your procedure, to calculate the determinant of the cube of each of the first 49 numbers starting at one.

Strategy: Use one of the Maple methods in the text to define a matrix then use Maple 'help' or the Maple Crib appendix to find the syntax for the `for..do` loop. In part (a) the matrix is small enough to be written by hand. In (b) we could also do this by hand but it is easier to use two loops, one for rows the other for columns which enables us to make a matrix of any size.

Q7.6 Use Maple to calculate the determinant of the first 16 prime numbers.

Strategy: First, find out how to obtain prime numbers by using the Help in Maple. The Maple `seq` instruction makes a 'list' and is described in the Appendix.

Q7.7 In a linear polyene, the only interactions lie on the diagonal and 1 element to the right or left of this, a tri-diagonal form, but in the cyclic polyenes the first atom has to interact with the last in the ring, atom 1 with atom 6 in benzene for example.

Modify Algorithm 7.1 and construct the Hückel matrix for any cyclic polyene.

(a) Calculate the MO energies for benzene and cyclo-octatetraene assuming that $x = \dfrac{\alpha - E}{\beta}$.

(b) Sketch and label the energy levels.

Strategy: Work out what it means to form a ring by adding extra matrix elements to the Hückel MO matrix.

Q7.8 Fulvalene has the structure

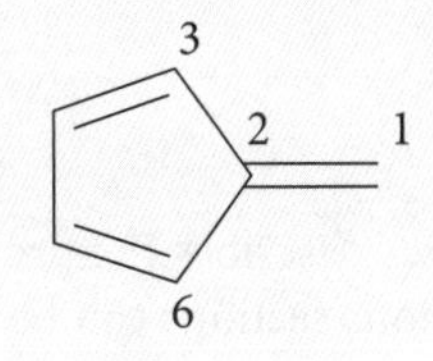

Write down the Hückel matrix in terms of x and calculate the energies. Recall that only adjacent atoms are connected in the matrix. What is the stabilization energy on forming π bonds and is this molecule more or less stable than benzene?

Strategy: The only difference here compared to previous calculations is that the ring joins atoms 6 and 2, not 6 and 1 as in benzene. The matrix elements corresponding to these atoms must each contain 1.

Q7.9 The molecule Na_3 exists as a covalently bound molecule in the vapour phase and its spectroscopy has been studied.

(a) Using Hückel theory, work out if the linear or triangular arrangement of atoms is the most stable.

(b) which arrangement of atoms is most stable in Na_3^+?

(c) Calculate the energy, as the bond is opened from a triangular to straight form, assuming a very crude model in which the overlap on the outside two atoms varies as $\beta e^{-r/d}$ where r is their separation assuming that the bond length is d. The overlap β accounts for the interaction at the normal bond length, and the exponential describes just how the overlap decreases as the atoms separate.

Strategy: The molecule has three s electrons that can become involved in bonding. The reference energy, without forming an MO, is the self-energy α of the three electrons. In (c) the β is changed due to the separation of two of the atoms as they move round in a circle from the bent to straight form. Use the cosine rule to work out how the distance r varies with bond angle θ. It is not obvious which energy levels are lowest, so plot all three to determine this and then calculate the total energy. Use values $\alpha = 0$ and $\beta = -1$ to observe the trend with bond angle. The value of α does not matter, because every energy contains a term $+\alpha$. The value of β should be negative; the exact value does not matter as only trends are sought in this crude model.

Q7.10 **(a)** Show that the characteristic equation for the Hückel energy of the nth linear polyene's molecular orbitals can be written as the recursion formula $f_n(x) = xf_{n-1}(x) - f_{n-2}(x)$ where f_{n-1} is the equation for the $n - 1^{\text{th}}$ polyene. The nth polyene has a characteristic equation found evaluating the $n \times n$ determinant

$$f_n(x) = \begin{vmatrix} x & 1 & 0 & \\ 1 & x & 1 & \\ 0 & 1 & \ddots & \cdots \\ & & \vdots & \ddots \end{vmatrix}$$

and when $n = 1$, a single carbon atom, let $f_1(x) = x$.

(b) Write a recursion equation in Maple to calculate the characteristic equation of the tenth linear polyene.

Strategy: A recursive function is one that depends upon previous functions of the same type, and by knowing only a few initial values, perhaps only one or two, all the others can be determined. Many types of polynomials, such as Hermites, can be defined recursively; see Chapter 1. It is necessary, therefore, to work out a few examples, in this case two initial values, and then by induction find a formula relating f_n to f_{n-1} and f_{n-2} from which all the remaining formulae can be obtained. The subscript n indicates the index of the function.

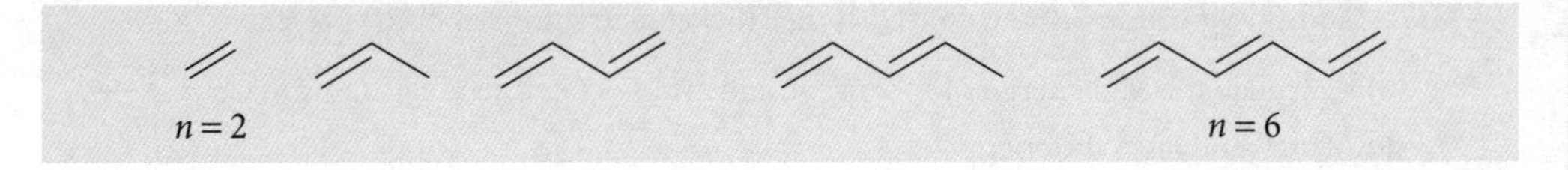

Fig. 7.6 Linear polyenes.

7.4 Matrices

The rest of this chapter describes the properties and uses of matrices. The determinant described in the previous sections can be considered as one of the many properties of a matrix. In several places, reference is made to eigenvalues and eigenvectors before they have been fully explained, which is done in Section 7.12.3. The meaning of these words is outlined below but how they are obtained is left unanswered for the moment.

The *eigenvalue-eigenvector* equation has the form

$$\textit{operator} \times \textit{function} = \textit{constant} \times \textit{same function},$$

provided the function is not zero. In matrix form, the eigenvector–eigenvalue equation is

$$Ax = \lambda x$$

where A is an $n \times n$ square matrix, x is one of n, one-dimensional column matrices (column vectors) and each one is called an eigenvector, λ represents one of n numbers, and each λ is called an eigenvalue.

7.4.1 Basic properties of matrices

To add or subtract two matrices they must have the same shape. To add or subtract a constant or another matrix their individual elements are added or subtracted. Similarly, to multiply or divide by a constant number (a scalar), each element is also multiplied or divided by this value. Multiplying two matrices together is more complicated and depends on the *order* with which this is done and the *shapes* of the two matrices; sometimes it is just not possible.

The notation used, is that bold italic letters indicate a matrix and the numbers ***0*** and ***1*** for the null and unit matrix respectively.

Suppose two matrices are $\boldsymbol{A} = \begin{vmatrix} 2 & 1 & 5 \\ -5 & 2 & 6 \end{vmatrix}$ and $\boldsymbol{B} = \begin{vmatrix} 1 & 10 & 8 \\ -3 & 3 & 9 \end{vmatrix}$ calculating

$\boldsymbol{A} + \boldsymbol{B}$, $\boldsymbol{A} - \boldsymbol{B}$, $3\boldsymbol{A} + \boldsymbol{B}/2$ is easy. Taking elements individually, the answers are

$$\boldsymbol{A} + \boldsymbol{B} = \begin{vmatrix} 3 & 11 & 13 \\ -8 & 5 & 15 \end{vmatrix}, \qquad \boldsymbol{A} - \boldsymbol{B} = \begin{vmatrix} 1 & -9 & -3 \\ -8 & -1 & -3 \end{vmatrix},$$

$$3\boldsymbol{A} + \boldsymbol{B}/2 = \begin{vmatrix} 9.5 & 8.5 & 19 \\ -16.5 & 7.5 & 22.5 \end{vmatrix}.$$

One matrix cannot be divided by another; the *inverse* matrix of the divisor is formed instead and then these matrices are multiplied together. The inverse of a matrix $\boldsymbol{M}$ is always written as $\boldsymbol{M}^{-1}$. Generating the inverse of a matrix is difficult unless the matrix is small, and one would normally use Maple to do this.

Matrices can be multiplied together. The multiplication order is always important; ***AB*** is not necessarily the same as ***BA***. With three or more matrices ***ABC***, the multiplication sequence does not matter as long as the *ordering* is the same. This is the *associative* property

$$\boldsymbol{ABC} \equiv \boldsymbol{A}(\boldsymbol{BC}) \equiv (\boldsymbol{AB})\boldsymbol{C}.$$

If ***BC*** are multiplied first, then the result left multiplied by ***A***, this is the same as multiplying ***AB*** to form a new matrix, ***D*** for example, then performing the multiplication ***DC***. Habitually, we start at the right and multiply the ***BC*** pair first then left multiply by ***A***. The details are described in Section 7.5.

There are a few special matrices and several types operations that can be performed on matrices and these are describe next.

7.4.2 The zero or null matrix

As you would expect it is possible to have a matrix of zeros; adding this to another matrix changes nothing, but multiplying by it annihilates the other matrix leaving only the zero matrix as you would expect.

$$\boldsymbol{0} = \begin{bmatrix} 0 & 0 & \cdots \\ 0 & 0 & \cdots \\ \vdots & \vdots & \ddots \end{bmatrix} \qquad \begin{matrix} \boldsymbol{A} + \boldsymbol{0} = \boldsymbol{A} \\ \boldsymbol{A0} = \boldsymbol{0} \\ \boldsymbol{0A} = \boldsymbol{0} \end{matrix} .$$

7.4.3 The unit or identity matrix *1*

This is a matrix of unit diagonals with all other elements being zero. The unit matrix is also called the *multiplicative identity*.

$$\boldsymbol{1} = \begin{bmatrix} 1 & 0 & 0 & \cdots \\ 0 & 1 & 0 & \cdots \\ 0 & 0 & 1 & \cdots \\ \vdots & \vdots & & \ddots \end{bmatrix} \qquad \boldsymbol{A} + \boldsymbol{1} = \begin{bmatrix} a_{11}+1 & a_{12} & \cdots \\ a_{21} & a_{22}+1 & \cdots \\ \vdots & \ddots & \ddots \end{bmatrix}.$$

$$\boldsymbol{A1} = \boldsymbol{A}$$
$$\boldsymbol{1A} = \boldsymbol{A}$$

7.4.4 Diagonal matrix

This is a matrix whose diagonals can take any value, but all other elements are zero. The unit matrix is a special case of this matrix. A 3×3 diagonal matrix is

$$\begin{bmatrix} a & 0 & 0 \\ 0 & b & 0 \\ 0 & 0 & c \end{bmatrix}.$$

7.4.5 Trace or character of a matrix

The trace, spur, or character of any matrix M, is the sum of its diagonal elements only, $Tr(M) = \sum_{i=1} a_{ii}$. If the matrix is $\begin{bmatrix} 1 & 0 & 2 \\ 3 & -1 & 0 \\ 0 & 4 & -1 \end{bmatrix}$ its character is -1. It is rather tedious to do, but by multiplying out the product of matrices, the trace of the product is found to be unchanged if they are permuted cyclically:

$$Tr(ABC) = Tr(CAB) = Tr(BCA).$$

This fact is very useful in group theory because the trace of a matrix forms a representation of a molecule's symmetry. Because of this rule, the trace, and hence representation, is independent of the basis set used to describe the symmetry.

7.4.6 Matrix transpose, M^T

The transpose operation replaces rows and columns with one another. For example;

$$\begin{bmatrix} a & b & c \\ d & e & f \end{bmatrix}^T = \begin{bmatrix} a & d \\ b & e \\ c & f \end{bmatrix}$$

and for a square matrix this operation leaves the diagonal unchanged

$$\begin{bmatrix} a & b & c \\ d & e & f \\ g & h & i \end{bmatrix}^T = \begin{bmatrix} a & d & g \\ b & e & h \\ c & f & i \end{bmatrix}.$$

The transpose operation exchanges element (1, 2) with (2, 1), and (1, 3) with (3, 1) and so on. You can appreciate that two transposes reproduce the initial matrix $(M^T)^T = M$. A *symmetric* matrix is equal to its transpose and therefore must also be square and have off-diagonal elements $a_{ij} = a_{ji}$. An *antisymmetric* matrix satisfies the identity $A = -A^T$ and must therefore have zeros on its diagonal and have components $a_{ij} = -a_{ji}$; for example, such a matrix and its transpose is

$$\begin{bmatrix} 0 & 2 & 3 \\ -2 & 0 & 4 \\ -3 & -4 & 0 \end{bmatrix}^T = \begin{bmatrix} 0 & -2 & -3 \\ 2 & 0 & -4 \\ 3 & 4 & 0 \end{bmatrix} = -\begin{bmatrix} 0 & 2 & 3 \\ -2 & 0 & 4 \\ -3 & -4 & 0 \end{bmatrix}.$$

When two matrices are multiplied, then the transpose reorders the matrix multiplication because rows are converted into columns and vice versa. This means that

$$(AB)^T = B^T A^T.$$

and an example is given when matrix multiplication has been described.

7.4.7 Complex conjugate M^*

This changes each matrix element with its complex conjugate, assuming there are complex numbers in the matrix; if not, it has no effect. The conjugate of a matrix M is labelled M^*. In making a complex conjugate each i is replaced by $-i$ where $i = \sqrt{-1}$, for example $[3 \quad i \quad 4]^* = [3 \quad -i \quad 4]$.
Some properties are

$$(M^*)^* = M,$$
$$(AB)^* = A^* B^*$$
$$|M^*| = |M|^*,$$

where the last property describes the effect on the determinant of the matrix.

7.4.8 Adjoint, $M^\dagger$

This grand sounding name means

'Form the complex conjugate, then transpose the matrix or vice versa.'

The special symbol † is conventionally used as a superscript. The effect of the adjoint operation is

$$M^\dagger = (M^*)^T = (M^T)^*.$$

For example $\begin{bmatrix} i \\ -3i \\ 4 \end{bmatrix}^\dagger = [-i \quad +3i \quad 4].$

In quantum mechanics, the adjoint is always used to convert a ket into a bra and vice versa, Chapter 8.

7.4.9 Hermitian or self-adjoint matrix $M^\dagger = M$

In these matrices the diagonals are real and the off-diagonals $a_{ij} = a_{ji}^*$. The Pauli spin matrices in quantum mechanics are Hermitian; for example

$$\sigma_2 = \begin{vmatrix} 0 & i \\ -i & 0 \end{vmatrix}.$$

Taking the transpose then the complex conjugate produces a matrix that is clearly the same as σ_2. If the Hermitian matrix is real, which means that it contains only real numbers, then it must be symmetric. In quantum mechanics, a matrix containing the integrals that evaluate to expectation values is a real Hermitian matrix whose eigenvalues are real and eigenvectors orthogonal, but not necessarily normalized.

An *anti-Hermitian* matrix is defined as $M^\dagger = -M$.

7.4.10 Inverse of a square matrix, M^{-1}

One matrix cannot be divided into another or into a constant so the operation $\dfrac{1}{M}$ is not allowed; instead the matrix inverse must be formed. Furthermore the inverse is defined only for a square matrix and then only if its determinant is not zero, $|M| \ne 0$. The inverse is another matrix labelled M^{-1}, such that

$$MM^{-1} = \boldsymbol{1} = M^{-1}M$$

where $\boldsymbol{1}$ is the unit matrix. Formally, we can find the inverse matrix element by element with the ij^{th} element given by

$$(M^{-1})_{ij} = \frac{(-1)^{i+j}}{|M|}|C_{ji}|$$

where $|M| \ne 0$ is the determinant and $|C_{ji}|$ is the cofactor matrix of row j and column i. Note the change in ordering of the indices. Cofactors were described in Section 7.2.2. Generally, it is best to use Maple to calculate the inverse, because, besides being tedious, the chance of making an error is very high. The matrix inverse is met again when solving equations.

If the matrix is diagonal then its inverse comprises the reciprocal of each term.

$$\begin{bmatrix} a & 0 & 0 \\ 0 & b & 0 \\ 0 & 0 & c \end{bmatrix}^{-1} = \begin{bmatrix} 1/a & 0 & 0 \\ 0 & 1/b & 0 \\ 0 & 0 & 1/c \end{bmatrix}$$

and this is used in the calculation of vibrational normal modes.

7.4.11 Singular matrix, determinant $|M| = 0$

The determinant is zero and the matrix has no inverse. This is not the same as the null matrix, $\boldsymbol{0}$. Odd $n \times n$ sized anti-symmetric matrices are singular.

7.4.12 Unitary: $M^\dagger = M^{-1}$ or $M^\dagger M = 1$ and determinant $|M| = 1$

Two examples of unitary matrices are $\begin{bmatrix} 0 & 1 & 0 \\ 0 & 0 & 1 \\ 1 & 0 & 0 \end{bmatrix}$, which has determinant expansion of $0(0-1) - 1(0-1) + 0(0-0) = 1$, and the rotation matrix

$$\begin{bmatrix} \cos(\theta) & \sin(\theta) \\ -\sin(\theta) & \cos(\theta) \end{bmatrix} \tag{7.3}$$

whose determinant is also 1 since $\cos^2(\theta) + \sin^2(\theta) = 1$.

7.4.13 Orthogonal matrices

Orthogonal matrices are always square and have the properties that if M is a matrix then

$$M^T = M^{-1} \qquad \text{or equivalently} \qquad M^T M = 1$$

which means that the matrix transpose is its inverse and when the matrix and its transpose are multiplied together a unit diagonal matrix results. The orthogonal matrix has a determinant that is ± 1; the eigenvalues are all $+1$. However, a determinant of ± 1 does not mean that a matrix is orthogonal. The product of two orthogonal matrices is also an orthogonal matrix.

The rotation matrix (7.3) is orthogonal, as is the matrix $\frac{1}{\sqrt{2}}\begin{bmatrix} 1 & 1 \\ 1 & -1 \end{bmatrix}$, which shows that a real orthogonal matrix is the same as a unitary matrix.

In an orthogonal matrix, rows and columns form an orthonormal basis and each row has a length of 1. When an orthogonal matrix operates on (linearly transforms) a vector or a matrix representing an object, such as a molecule, the molecule is unchanged, meaning that relative bond angles and lengths remain the same but the coordinate axes moved so the object appears to have been rotated on the computer screen. Orthogonal matrices are used to represent reflections, rotations, and inversions in molecular group theory. Matrix decomposition methods make use of orthogonal matrices. One matrix method called singular value decomposition is used in data analysis to separate out overlapping spectra into their constituent parts.

7.4.14 Determinants

Evaluating determinants was described in Section 7.2. The determinants of different types of square matrices have the following properties that do not involve evaluating the determinant.

$$\begin{gathered} |ABC| = |A||B||C| \\ |A^T| = |A| \\ |A^{-1}| = \frac{1}{|A|} \\ |-A| = (-1)^n |A| \end{gathered} \tag{7.4}$$

where n is the size of the matrix and in general $|aA| = a^n|A|$.

Consider calculating the determinant of the inverse of the matrix $A = \begin{bmatrix} 1 & 2 \\ 3 & 4 \end{bmatrix}$ as in equation (7.4). A has a determinant that is -2. The equation for its inverse is given in Section 7.5.7 and is

$$A^{-1} = \begin{bmatrix} 1 & 2 \\ 3 & 4 \end{bmatrix}^{-1} = \frac{1}{|A|}\begin{bmatrix} 4 & -2 \\ -3 & 1 \end{bmatrix} = -\frac{1}{2}\begin{bmatrix} 4 & -2 \\ -3 & 1 \end{bmatrix}$$

producing a determinant $|A^{-1}|$ of $\left| -\frac{1}{2}\begin{bmatrix} 4 & -2 \\ -3 & 1 \end{bmatrix} \right| = \frac{1}{4}\begin{vmatrix} 4 & -2 \\ -3 & 1 \end{vmatrix} = -\frac{1}{2}$. The 1/4 arises using $|aA| = a^n|A|$ where a is $-1/2$ and the matrix size is 2.

7.5 Matrix multiplication

The *shape* of the matrices and their multiplication *order* must both be respected. If multiplication is possible then, in general, multiplication $\boldsymbol{AB}$ does not give the same result as $\boldsymbol{BA}$. The matrices

$$A = \begin{vmatrix} 2 & 1 & 5 \\ -5 & 2 & 6 \end{vmatrix} \quad \text{and} \quad B = \begin{vmatrix} 1 & 10 & 8 \\ -3 & 3 & 9 \end{vmatrix}$$

cannot be multiplied together, because multiplication is only possible if the number of columns of the left-hand matrix is equal to the number of rows of the right-hand one. The matrices are then *commensurate* or *conformable*. The result is a matrix whose size is determined by the number of rows of the left hand matrix and the number of columns of the right hand matrix, e.g.

$$\begin{array}{ccc} \boldsymbol{A} & \boldsymbol{B} & \boldsymbol{C} \\ (n \times m) & (m \times r) \rightarrow & (n \times r) \end{array}$$

Same number of columns in matrix $\boldsymbol{A}$ as rows in $\boldsymbol{B}$

The calculation below shows the multiplication $\boldsymbol{C} = \boldsymbol{AB}$; the arrows show how the top element in $\boldsymbol{C}$ is calculated as the product of element 1 of row 1 with element 1 of column 1. To this value is added the product of element 2 of row 1 and element 2 of column 1 and so on for all the elements in a row. This is why the number of rows and columns must be the same.

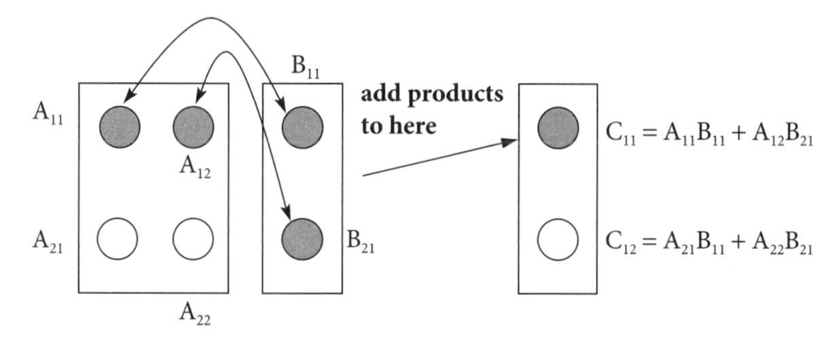

Multiplying two 2×2 matrices produces

$$\begin{bmatrix} a_{11} & a_{12} \\ a_{21} & a_{22} \end{bmatrix} \times \begin{bmatrix} b_{11} & b_{12} \\ b_{21} & b_{22} \end{bmatrix} = \begin{bmatrix} a_{11}b_{11} + a_{12}b_{21} & a_{11}b_{12} + a_{12}b_{22} \\ a_{21}b_{11} + a_{22}b_{21} & a_{21}b_{12} + a_{22}b_{22} \end{bmatrix}$$

and it is useful to look at the pattern of elements where you can see the $\boldsymbol{A}$ matrix in the pattern of a's and columns in $\boldsymbol{B}$ appear as rows in the sums. If the matrices are not square, multiplying a $n \times m$ matrix by a $m \times r$ one results in a $n \times r$ matrix as shown above. The equation with which to calculate the (row–column) ij^{th} element of any matrix multiplication is

$$C_{ij} = \sum_{k=1}^{m} a_{ik}b_{kj} \tag{7.5}$$

where the sum index k runs from 1 to m, the number of columns in matrix $\boldsymbol{A}$ or rows in $\boldsymbol{B}$ and i takes values 1 to n and j from 1 to r. Multiplying a 2 row $\times$ 3 column matrix with a 3×2, the number of columns in the left matrix is the same as the number of rows in the right-hand one and therefore the result is a 2×2 square matrix,

$$\begin{array}{ccccc} \begin{bmatrix} a_{11} & a_{12} & a_{13} \\ a_{21} & a_{22} & a_{23} \end{bmatrix} & \times & \begin{bmatrix} b_{11} & b_{12} \\ b_{21} & b_{22} \\ b_{31} & b_{32} \end{bmatrix} & = & \begin{bmatrix} P & Q \\ R & S \end{bmatrix} \\ \uparrow & & \uparrow & & \uparrow \\ \text{3 columns} & & \text{3 rows} & & 2 \times 2 \text{ matrix} \\ \text{2 rows} & & \text{2 columns} & & \end{array} \tag{7.6}$$

where element R is $a_{21}b_{11} + a_{22}b_{21} + a_{23}b_{31}$ and element S is $a_{21}b_{12} + a_{22}b_{22} + a_{23}b_{32}$. A 2×3 and a 3×1 matrix produce a 2×1 column matrix

$$\begin{bmatrix} a_{11} & a_{12} & a_{13} \\ a_{21} & a_{22} & a_{23} \end{bmatrix} \times \begin{bmatrix} b_1 \\ b_2 \\ b_3 \end{bmatrix} = \begin{bmatrix} P \\ R \end{bmatrix} \tag{7.7}$$

where $P = a_{11}b_1 + a_{12}b_2 + a_{13}b_3$ and $R = a_{21}b_1 + a_{22}b_2 + a_{23}b_3$.

Suppose there are three matrices $\boldsymbol{ABC}$, then the safest rule to follow is to left-multiply $\boldsymbol{C}$ by $\boldsymbol{B}$ first, then to left-multiply the result by $\boldsymbol{A}$. The same rule is applied to several matrices; *start at the right and work to the left*. However, by the associative rule, Section 7.4.1, as long as the order $\boldsymbol{ABCD}$ is maintained, this product can be multiplied in any order.

The diagrams in Fig. 7.7 show, diagrammatically, the result of multiplying differently shaped matrices. The bra-ket notation is shown also. As examples, the following matrices are multiplied together as pairs in every way possible.

$$A = \begin{bmatrix} a \\ b \\ c \end{bmatrix} \qquad B = [p \quad q \quad r] \qquad C = \begin{bmatrix} d & e & f \\ g & h & i \\ j & k & l \end{bmatrix}$$

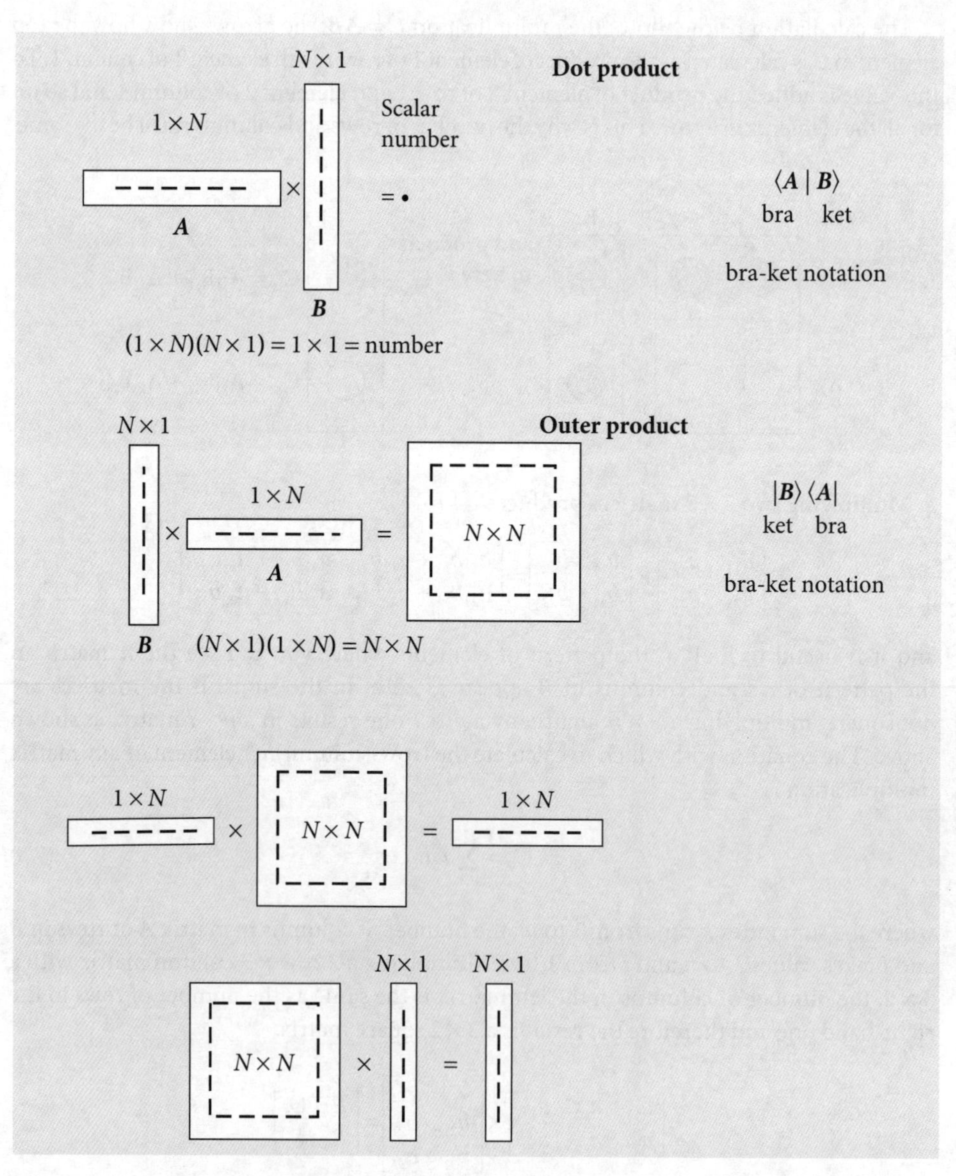

Fig. 7.7 Pictorial representation of matrix multiplication. The bra-ket notation used in quantum mechanics is shown on the right.

giving

$BA = ap + bq + cr$, which is the *dot product* or *inner product* of $B{\cdot}A$.

$$AB = \begin{bmatrix} ap & aq & ar \\ bp & bq & br \\ cp & cq & cr \end{bmatrix}.$$ This is also called the *outer product*.

$$BC = [pd + qg + rj \quad pe + qh + rk \quad pf + qi + rl]$$

$$CA = \begin{bmatrix} ad + be + cf \\ ag + bh + ci \\ aj + bk + cl \end{bmatrix}$$

$$CC = \begin{bmatrix} d^2 + eg + fj & de + eh + fk & df + ei + fl \\ gd + hg + ij & ge + h^2 + ik & gf + hi + il \\ jd + kg + lj & je + kh + lk & jf + ki + l^2 \end{bmatrix}$$

and AC and CB products do not exist and neither does AA and BB.

As a numerical example, if $A = \begin{bmatrix} 1 & 2 \\ 3 & 4 \end{bmatrix}$ and $B = \begin{bmatrix} 5 & 6 \\ 7 & 8 \end{bmatrix}$ then the product AB is

$$\begin{bmatrix} 1 & 2 \\ 3 & 4 \end{bmatrix} \times \begin{bmatrix} 5 & 6 \\ 7 & 8 \end{bmatrix} = \begin{bmatrix} 1\times 5 + 2\times 7 & 6 + 2\times 8 \\ 3\times 5 + 4\times 7 & 3\times 6 + 4\times 8 \end{bmatrix} = \begin{bmatrix} 19 & 22 \\ 43 & 50 \end{bmatrix}$$

and BA is

$$\begin{bmatrix} 5 & 6 \\ 7 & 8 \end{bmatrix} \times \begin{bmatrix} 1 & 2 \\ 3 & 4 \end{bmatrix} = \begin{bmatrix} 5 + 18 & 10 + 24 \\ 7 + 24 & 14 + 32 \end{bmatrix} = \begin{bmatrix} 23 & 34 \\ 31 & 46 \end{bmatrix}.$$

In Section 7.4.6 it was stated that, when a pair of matrices are multiplied then transposed, this is the same as transposing both matrices and multiplying in *reverse* order or $(AB)^T = B^TA^T$. As an example, if $M = \begin{bmatrix} a & b \\ c & d \end{bmatrix}$, $N = \begin{bmatrix} 1 & 2 \\ 3 & 4 \end{bmatrix}$ then

$$(MN)^T = \begin{bmatrix} 1a + 3b & 2a + 4b \\ 1c + 3d & 2c + 4d \end{bmatrix}^T = \begin{bmatrix} 1a + 3b & 1c + 3d \\ 2a + 4b & 2c + 4d \end{bmatrix}$$

Performing the transpose then multiplying in reverse order gives the same result.

$$N^TM^T = \begin{bmatrix} 1 & 3 \\ 2 & 4 \end{bmatrix}\begin{bmatrix} a & c \\ b & d \end{bmatrix} = \begin{bmatrix} 1a + 3b & 1c + 3d \\ 2a + 4b & 2c + 4d \end{bmatrix}$$

7.5.1 Matrix sum

To sum the terms in a square matrix left-multiply by a unit row vector and right-multiply by a unit column vector. For example

$$[1 \;\; 1]\begin{bmatrix} a & b \\ c & d \end{bmatrix}\begin{bmatrix} 1 \\ 1 \end{bmatrix} = a + b + c + d$$

7.5.2 bra-ket notation

Often the *bra-ket* notation is used in quantum mechanics. The *ket* $|k\rangle$ is a single column matrix (a column vector) $|k\rangle = \begin{bmatrix} a \\ b \\ c \end{bmatrix}$, the *bra* is the single row matrix or a row vector, $\langle k| \equiv [a^* \quad b^* \quad c^*]$ and is always the complex transpose of the ket, i.e. the Hermitian transpose; these vectors may be of infinite length. These objects are discussed in more detail in chapters 6 and 8 but are mentioned as they provide a shorthand way of visualizing matrix multiplication; see Fig. 7.7. There is no specific symbol for a scalar or a square matrix; they have to be represented by the bra-ket pair.

7.5.3 Commutation

If matrices are not square then there is only one way to multiply them: the left-hand matrix's number of columns must equal the right-hand matrix's number of rows. When both matrices are square, to be multiplied they must be of the same size and we can then choose which is to be the left-hand and which the right-hand side of the multiplication, and now the order of multiplication does matter as illustrated on the previous page. Sometimes the result is the same, i.e. $\boldsymbol{AB} = \boldsymbol{BA}$ then the matrices are said to *commute*, but generally they do not and, therefore, $\boldsymbol{AB} \neq \boldsymbol{BA}$.

The *commutator* of two square matrices $\boldsymbol{A}$ and $\boldsymbol{B}$ is also a matrix and is defined as

$$[A,B] = AB - BA. \tag{7.8}$$

If the result is the *null* matrix $\boldsymbol{0}$, which is full of zeros, $\boldsymbol{A}$ and $\boldsymbol{B}$ are said to *commute*: $[A,B] = AB - BA = \boldsymbol{0}$. Sometimes expressions such as $[C,[A,B]]$ may be met. This means: work out the inside commutation first, then the commutation with each term that this produces: $[C,[A,B]] = [C, AB - BA] = [C, AB] - [C, -BA]$.

Commutation is very important in quantum mechanics; only observables that commute can be observed simultaneously. Position and momentum, which do not commute, or those components of angular momentum, which also do not commute, have values that are restricted by an amount determined by the Heisenberg uncertainty principle when being observed simultaneously.

Commutation is common of operators in general. The commutator relationship applies to any two operators $\boldsymbol{P}$ and $\boldsymbol{Q}$, which need not be matrices, but could be the differential operator such as d/dx or x itself. This parallel means that we can consider matrices as operators. The commutator will in general operate on a function, for example,

$$[P, Q]f = PQf - QPf$$

and f can be any normal function, $\ln(x)$, $\sin(x)$, and so on. In molecular group theory, the five types of operators that are the identity, rotation, reflection, inversion, and improper rotation (combined rotation and reflection) sometimes commute with one another although this depends on the point group. Section 7.6 describes these operations.

7.5.4 Integral powers of square matrices

Only integral powers of matrices are defined and are calculated by repeated multiplication, for example

$$M^0 = 1, \qquad M^1 = M, \qquad M^2 = MM, \qquad M^3 = MMM,$$

and so on. For example,

$$\begin{bmatrix} 1 & 2 \\ 3 & 4 \end{bmatrix}^2 = \begin{bmatrix} 1 & 2 \\ 3 & 4 \end{bmatrix}\begin{bmatrix} 1 & 2 \\ 3 & 4 \end{bmatrix} = \begin{bmatrix} 7 & 10 \\ 15 & 22 \end{bmatrix}.$$

(Repeated operation to obtain a power is familiar from differentiation as for example $\dfrac{d^2y}{dx^2} = \dfrac{d}{dx}\dfrac{dy}{dx}$.) The *similarity transform*, see Section 7.13.4, can be used to obtain high integer powers of matrices, however, if the matrix is diagonal then taking the power is easy because the diagonal elements are raised to the power. The power need not be positive; therefore, the inverse of a diagonal matrix is easily obtained.

$$\begin{bmatrix} a & 0 & 0 \\ 0 & b & 0 \\ 0 & 0 & c \end{bmatrix}^n = \begin{bmatrix} a^n & 0 & 0 \\ 0 & b^n & 0 \\ 0 & 0 & c^n \end{bmatrix}.$$

7.5.5 Functions of matrices

In the study of the theory of nuclear magnetic resonance, NMR, a quantum mechanical description of nuclear spin is essential. To explore the rotation of the magnetization as used in inversion recovery, spin-echo, or a complicated two-dimensional experiment, such as COSY, requires the exponentiation of matrices (Levitt 2001). Exponentiation can only be performed by expanding the exponential and evaluating the terms in the series one by one as

$$e^M = 1 + M + \frac{M^2}{2!} + \cdots. \qquad \text{and} \qquad e^{-M} = 1 - M + \frac{M^2}{2!} - \cdots.$$

and if x is a variable $e^{-xM} = 1 - xM + x^2\dfrac{M^2}{2!} - \cdots$.

The matrix 1 is a unit diagonal matrix. Similar expressions are formed with trig and log functions according to their expansion formula and with Taylor or Maclaurin series for other functions. The familiar relationship $e^Ae^B = e^{A+B}$ is only true if the matrices A and B commute. In Section 7.13.2 and 13.3 a transformation is describe which enables the exponential of a matrix to be found.

7.5.6 Block diagonal matrices

In many instances, a matrix can be blocked into smaller ones symmetrically disposed along the diagonal. The result of this is that the problem reduces to the smaller one of solving several matrices where each is much smaller than the whole and is therefore more easily solved. Why should we bother with this if the computer can diagonalize any matrix we give it to do? The reason is that eigenvalues can more easily be identified within the basis set by doing the calculation this way. Recall that the basis set you choose to use, for example in a quantum problem, determines the ordering of elements in a matrix. Why does this matter? It matters because when the spectrum from a molecule is observed, which measures only the difference in energy levels, we would like to know what quantum numbers give rise to what spectral lines. If the matrix is a block diagonal one, then this is made somewhat easier because we know what parts of the basis set elements are involved because each block when diagonalized contains only that part of the basis set that was in it in the first place. If the whole matrix is diagonalized blind, as it were, and without thinking about the problem beforehand, this information can be lost because all the elements and hence eigenvalues can be mixed up. The elements in the basis set can be ordered in any way you want, and different basis sets can be chosen for the same problem. By trying different ordering, it is sometimes possible to discover a block diagonal form for a matrix and so aid its solution. In the study of group theory, blocking matrices proves to be a powerful way of determining the irreducible representation; see Section 7.6. The following matrix has a 2×2, a 3×3, and a 1×1 block.

$$\begin{bmatrix} 1 & 2 & 0 & 0 & 0 & 0 \\ 2 & 1 & 0 & 0 & 0 & 0 \\ 0 & 0 & 2 & 3 & 1 & 0 \\ 0 & 0 & 3 & 2 & 1 & 0 \\ 0 & 0 & 1 & 1 & 1 & 0 \\ 0 & 0 & 0 & 0 & 0 & 1 \end{bmatrix}$$

7.5.7 The special case of the 2 × 2 matrix

If the matrix is $M = \begin{bmatrix} A & B \\ C & D \end{bmatrix}$, then the determinant is $|M| = AD - BC$ and the trace $Tr(M) = A + D$.

The inverse is $M^{-1} = \dfrac{1}{|M|}\begin{bmatrix} D & -B \\ -C & A \end{bmatrix}$.

Eigenvalues λ_1 and λ_2 (see Section 7.12.3) are solutions of $\lambda^2 - \lambda(A + D) + (AD - BC) = 0$ or equivalently $\lambda^2 + \lambda Tr(M) + |M| = 0$, which are

$$\lambda_{1,2} = \frac{A + D \pm \sqrt{(A + D)^2 - 4(AD - BC)}}{2}.$$

As a check $\lambda_1 + \lambda_2 = Tr(M)$ and $\lambda_1\lambda_2 = |M|$.

The eigenvectors are $v_1 = k\begin{bmatrix} B \\ \lambda_1 - A \end{bmatrix}$ and $v_2 = k\begin{bmatrix} B \\ \lambda_2 - A \end{bmatrix}$ where k is an arbitrary constant.

7.5.8 Using Maple

Evaluating matrix products and calculating inverse matrices is simple with Maple. Some syntax is shown in Section 7.2.3 and below, and in Appendix 1. There are different forms that can be used, but the linear algebra package must be loaded first.

Notice that we can use a dot or `Multiply` to effect matrix multiplication. When using the bracket notation, elements in a row are separated by the vertical line | and columns by a comma.

Algorithm 7.2 Using Maple to manipulate matrices

```
> with(LinearAlgebra):                # load package
> A:= < <1|1>,< 0|1> >;               # define matrices
> B:= < <1|0>,< 1|1> >;
> Multiply(A,B);                      # matrix multiplication
```

$$\begin{bmatrix} 2 & 1 \\ 1 & 1 \end{bmatrix}$$

```
> A.B;                                # also multiplication
```

$$\begin{bmatrix} 2 & 1 \\ 1 & 1 \end{bmatrix}$$

```
> B.A;                                # A and B do not commute
```

$$\begin{bmatrix} 1 & 1 \\ 1 & 2 \end{bmatrix}$$

```
> # The commutation [ A²B, BA²] = A²B³A - B²A³B
> A.A.B.B.B.A - B.B.A.A.A.B;
```

$$\begin{bmatrix} 3 & 6 \\ -6 & -3 \end{bmatrix}$$

```
> # therefore A²B and B²A do not commute.
> A^(2).B^(3).A - B^(2).A^(3).B; # alternative, note brackets
> B^(-1);                             # inverse of B
```

$$\begin{bmatrix} 1 & 0 \\ -1 & 1 \end{bmatrix}$$

```
> C:= < < 1|2|3> , <4|5|6> , <7|8|9> >;          # 3x3 matrix
```

$$C := \begin{bmatrix} 1 & 2 & 3 \\ 4 & 5 & 6 \\ 7 & 8 & 9 \end{bmatrix}$$

```
> E:= < 10,11,12 >;                   # column matrix
```

$$E := \begin{bmatrix} 10 \\ 11 \\ 12 \end{bmatrix}$$

```
> C.E;                                # only allowed product
```

$$\begin{bmatrix} 68 \\ 167 \\ 266 \end{bmatrix}$$

```
> F:= < 13|14|15 >;                   # row matrix
```

$$F := [\,13 \quad 14 \quad 15\,]$$

```
> F.E;                                # dot product is a scalar
```

$$464$$

```
> E.F;                                #outer product is a matrix
```

$$\begin{bmatrix} 130 & 140 & 150 \\ 143 & 154 & 165 \\ 156 & 168 & 180 \end{bmatrix}$$

7.5.9 Questions

Full solutions are available at **www.oxfordtextbooks.co.uk/orc/beddard**.

Q7.11 Find AB and BA if $A = \begin{bmatrix} 1 & 2 \\ 3 & 4 \end{bmatrix}$ and $B = \begin{bmatrix} 5 & 6 \\ 7 & 8 \end{bmatrix}$. Do these matrices commute and if not what is $[A, B]$?

Q7.12 **(a)** Find A^2, B^2, AB, BA, A^2B, and B^2A if $A = \begin{bmatrix} 1 & 1 \\ 0 & 1 \end{bmatrix}$ and $B = \begin{bmatrix} 1 & 0 \\ 1 & 1 \end{bmatrix}$.

(b) Do A^2B and B^2A commute?

Q7.13 If $\boldsymbol{x} = [x_1 \quad x_2]$, $\boldsymbol{y} = \begin{bmatrix} y_1 \\ y_2 \end{bmatrix}$, and $\boldsymbol{Q} = \begin{bmatrix} 1 & -4 \\ -9 & 16 \end{bmatrix}$, find $\boldsymbol{xQy}$ and $\boldsymbol{yQx}$.

Q7.14 **(a)** Explain why if j is a constant factor that $|jM| = j^n|M|$ for an $n \times n$ matrix.

(b) Confirm equation (7.4) if $A = \begin{bmatrix} a & b \\ c & d \end{bmatrix}$ by calculating A^{-1} and $|A^{-1}|$

(c) Using Maple (or by hand) find A^{-1} and $|A^{-1}|$ for the 3×3 matrix $A = \begin{bmatrix} 0 & b & c \\ b & 0 & d \\ c & d & 0 \end{bmatrix}$.

***Strategy*:** See Section 7.4.14, but now the problem is algebraic not numerical.

Q7.15 **(a)** Show that if $\boldsymbol{P}$ and $\boldsymbol{Q}$ are linear operators, not necessarily matrices, $[\boldsymbol{P}, \boldsymbol{Q}] = -[\boldsymbol{Q}, \boldsymbol{P}]$

(b) (i) find $[\frac{d}{dx}, x]\sin(x)$, (ii) $[\frac{d}{dx}, x]f(x)$, and (iii) $[\frac{d}{dx}, x]$ and

(c) Show that operators d/dx and x do not commute.

(d) Write a Maple commutator function and test your results.

(e) Do $\dfrac{df(x)}{dx}$ and $\displaystyle\int_0^a f(x)dx$ commute?

(f) Does the operator df/dx and a displacement operator $\Delta f = f(x + c)$ commute?

(g) Does the operator xx which means multiply twice by x, commute with the inversion operator, $\text{Inv}(f(x)) = f(-x)$.

Q7.16 $\boldsymbol{B}$ and $\boldsymbol{C}$ are commuting square matrices and the matrix A is defined as $A \equiv e^B = \boldsymbol{1} + \boldsymbol{B} + \dfrac{\boldsymbol{B}^2}{2!} + \cdots$ show that:

(a) $e^B e^C = e^{B+C}$,

(b) $A^{-1} = e^{-B}$,

(c) $e^{CBC^{-1}} = CAC^{-1}$.

***Strategy*:** (a) Expand out the exponentials as shown in Section 7.5.5 and collect terms and try to reform an exponential series in $\boldsymbol{B} + \boldsymbol{C}$. (b) Use the fact that for an inverse matrix $AA^{-1} = \boldsymbol{1}$. (c) The expression CAC^{-1} is a similarity transform and is itself a square matrix. To prove this result, expand the exponential on both sides of the equation.

7.6 Molecular group theory

7.6.1 Motivation and Concepts

When we learn to draw molecules and molecular orbitals their inherent symmetry becomes clear; benzene or perhaps tetrahedral methane spring to mind and possibly the geometry of sp^2 and sp^3 hybridizations. To characterize exactly what such symmetry means is the role of molecular group theory. This can also be used to determine the selection rules of spectroscopic transitions, to characterize the 'shapes' of normal mode vibrations and simplify molecular orbital calculations. In chemistry, the word symmetry, while retaining its colloquial meaning also has a technical meaning and this might appear to be rather abstract and divorced from other topics such as quantum mechanics and spectroscopy. Group theory's jargon does not help in learning the subject mainly because it appears to be so abstract. In fact, it is quite the opposite: it is intensely practical, and expresses a complicated set of rules and ideas in a few symbols; D_{6h}, for example, encapsulates all the many symmetry properties of benzene. The jargon we shall have to understand will lead us to be able to distinguish between symmetry elements, symmetry operations, irreducible and reducible representations, characters, classes, basis sets, similarity transforms, and Mulliken labels.

This section can only give a brief introduction to the subject to act as a basis for further study. There are many books on this topic, but Vincent (2001) follows a tutorial approach; Molloy (2004) has many molecular examples; Atkins & Friedman (1997) has a chapter giving a thorough mathematical approach; and Cotton (1990) and Bishop (1993) discuss the subject fully. The organization of this section is as follows: first, the geometrical properties of symmetry elements and operators are introduced. These are then used to identify a point group. It is then shown how two combined operations lead to the formation of an operator multiplication table, and how a symmetry operation can be represented by a matrix. Next, the pertinent properties of a mathematical group are described, and it is demonstrated how symmetry operations can form such a group. The character table is described next, and it is shown how the symmetry operations can be represented by a row of numbers rather than as a matrix or a symmetry label. Understanding how to use the information presented in the point group to characterize molecular vibrations and orbitals is our ending point.

7.6.2 **Essential jargon**

Symmetry is defined as the relationship between parts of an object or between groups of objects in space. To identify the *symmetry elements* inherent in a molecule, we look at those geometrical operations that can be performed that will make the molecule *indistinguishable* from its initial state. Rotations and reflections are two of five such operations and are described further in Section 7.6.3. Group theory shows how symmetry properties can be represented by a set of numbers, called characters when collected into a table called a *point group* or *character* table, see Fig. 7.15 for an example. It is quite remarkable that in the context of group theory, a symmetry operation such as rotating a molecule can be *represented* by a number, usually 0 or ± 1 although other values are possible depending on the point group. These numbers are the *characters* of the point group and a row of characters is called an *irreducible representation*. The character table uniquely defines the symmetry of a molecule and properties, such as how different molecular orbitals behave and whether visible, infrared, and Raman transitions can occur, and, if they do, the orientation of the transition dipoles in the molecule. All the *symmetry operations* belonging to each point group are listed along the top row of a point group table; see Fig. 7.15. The characters are in the body of the table. In the next paragraphs, these concepts are expanded upon.

Most molecules contain very little symmetry, cholesterol or $SOCl_2$ are examples, but water, ammonia, chlorobenzene, ferrocene, and particularly benzene each have many symmetry elements. The more 'regular' the molecular structure is, the larger the number of *symmetry elements* it contains, which increases the number of *symmetry operations* that can be performed on these elements. To work out what point group a molecule belongs to and to determine its properties, the victim molecule is subjected to a given set of symmetry operations about each of the symmetry elements that may be present. You look at the molecule and then decide, by intuition, experience, or trial and error, which symmetry elements are present. The point group is then identified by the set of operations that leaves the molecule indistinguishable from its starting condition. The starting point is thus to know what the symmetry elements and operations are, and then to learn how to determine which ones are present in a molecule. Usually a simple three-dimensional model will help when doing this; it is sometimes difficult to 'see' the symmetries present from a sketch even if it is in perspective. It is a skill that improves with practice, and, as with riding a bicycle, if it is not done for a while, you can be a bit 'wobbly' to begin with.

7.6.3 Symmetry operations and symmetry elements

Symmetry operations act via those symmetry elements that the molecule contains, which may be an axis, mirror plane or centre of inversion. The operation moves the molecule in space to a new, perhaps indistinguishable position, but the symmetry elements remain fixed. The words 'operation' and 'element' are often interchangeable, but technically the operation can only occur about a symmetry element. For example, with a water molecule which has the same symmetry as ClO_2, Fig. 7.11, you will usually see a mirror plane (the element) and the reflection (the operation) at the same time. The operations and elements are linked because certain operations can only act on certain elements. If an operation does not leave the molecule indistinguishable, then it is not present.

The effect of any valid symmetry operation is always to leave the molecule in an indistinguishable state,[2] not necessarily an identical state. To understand this important distinction further, consider the square shown in Fig. 7.8, where the label is used only to identify one corner. Rotating the square by 90° (in 'math speak' operating with a $+\pi/2$ rotation operator) makes the right-hand square indistinguishable from the left-hand one. Only after four similar rotation operations are the two squares *identical.*

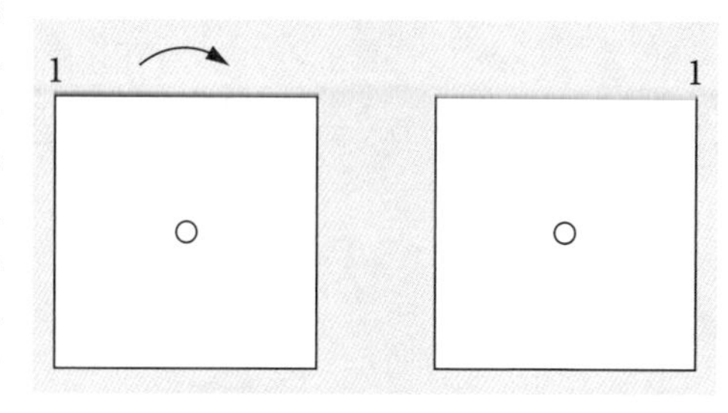

Fig. 7.8 Indistinguishable squares.

A study of group theory shows that there are only five types of operators that could leave a molecule in an indistinguishable state. However, before operating on a molecule, a principal axis must first be chosen. All the symmetry operations are referenced with respect to the principal axis, which is the axis of highest rotational symmetry in the molecule. If two or more axes are the same, then one victim must be chosen. Fig. 7.9 shows some examples. It is normal also to choose the principal axis as the z-axis and then to define x- and y-axes at right angles in the usual way. If the molecule is planar, as is naphthalene, the z-axis is usually chosen to project out of the plane.

The operators are:

(i) The Identity, labelled E. No atoms change position with this operation and all molecules possess the Identity.

(ii) Rotation about an axis, labelled C_n. This symmetry element is an axis that often coincides with an x-, y-, or z-axis, but might be in any other direction depending upon the molecule. Rotation by 180° is labelled C_2, by 120° C_3, etc. The subscript defines how many times the operation that makes the molecule identical has to occur. A molecule may have rotation about more than one axis. The label C is a shorthand for cyclic.

(iii) Reflection in a mirror plane σ.[3] The symmetry element is a plane. There are three types of mirror planes: vertical if the mirror edge runs along the principal axis, horizontal if this axis is at 90° to the mirror, and dihedral if the mirror divides two axes; see Fig. 7.10. There may be more than one of each type of mirror plane in a molecule. For example, water or ClO_2 has two vertical planes labelled σ and σ' or σ_V and σ'_V, see Fig. 7.11. A horizontal mirror plane is labelled σ_h, a dihedral plane σ_d. One or more superscript dashes ′ are added if more than one of a type of mirror plane is present. In cases where a mirror plane falls on two axes this may alternatively be labelled $\sigma(x, z)$ etc.

(iv) Inversion through a centre i. The element is the centre of inversion the operation always changes coordinates from $\{x, y, z\}$ to $\{-x, -y, -z\}$. See Fig. 7.12.

(v) A combined rotation–reflection operation, S_n, also called an improper rotation. There may be more than one of these. The axis subscript n is defined as in (ii). The operation is rotation followed by reflection in a plane perpendicular to the rotation, $\sigma_n C_n$. Fig. 7.13 shows an example of an S_4 operation (S is from the word *Sphenoidisch*).

[2] An operation that 'takes it into itself' is used in some texts to mean indistinguishable.

[3] The Greek letter sigma, σ, is used to represent the initial letter of the word *Spiegelung* or reflection.

Three examples of rotation operations are shown in Fig. 7.9. The ammonia molecule when rotated about the principal axis by 120° becomes indistinguishable. It is also indistinguishable if rotated by twice this amount, but if rotated three times it is more than indistinguishable; it is *identical* to the starting state of the molecule. The molecule is also indistinguishable if rotated by −120° or −240°. The label C_3 represents one rotation operation as the molecule is moved by 120° = 360°/3 of a turn; two rotations are labelled C_3^2 and three $C_3^3 \equiv E$. If you are unclear about this, label the H atoms 1, 2, 3 and draw out the pictures or make a model; it really does help to do this yourself. The water molecule has only to be rotated by ±180° to become indistinguishable, which is a C_2 operation. The fluorinated acetylene can have any angle of rotation to become indistinguishable and this is labelled C_∞.

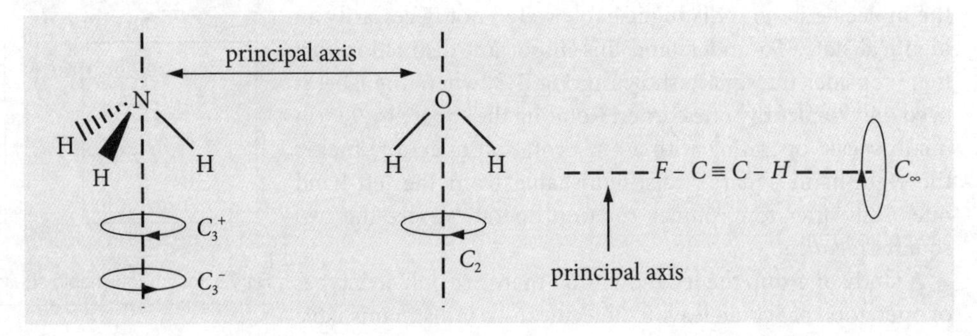

Fig. 7.9 Principal axes (dotted) and rotation operators about this symmetry element that make the molecules indistinguishable.

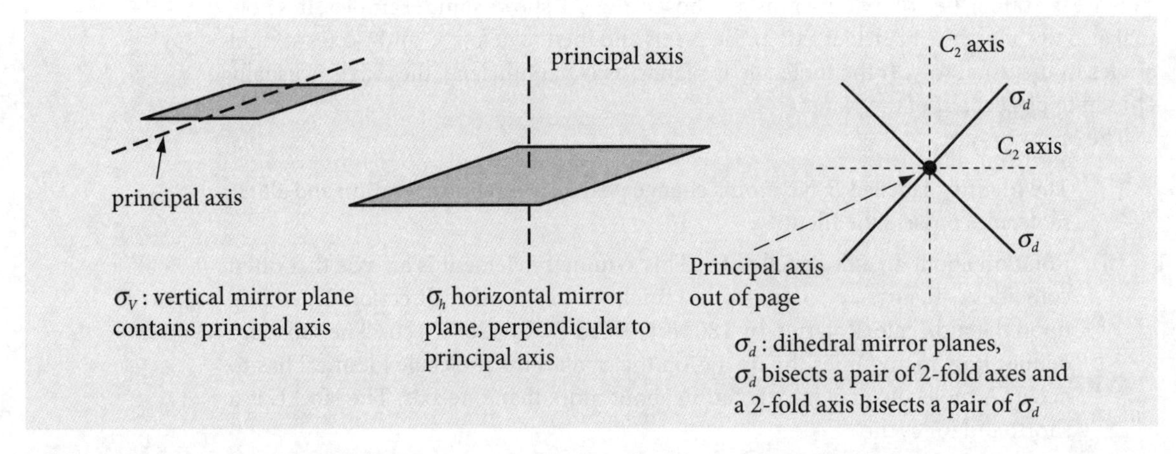

Fig. 7.10 Different types of mirror planes.

Now consider symmetry or mirror planes. There are three types, as shown in Fig. 7.10, defined relative to the principal and other axes. The molecule $SOCl_2$, Fig. 7.11, has one chlorine atom that is in front of the plane and one that is behind it and the SO atoms are in the plane. This molecule has only one vertical mirror plane and no other symmetry operations are valid, other than the identity. Only the mirror plane makes the molecule indistinguishable. The symmetry operations for chloride dioxide, which has a C_{2V} point group label, as do water, SO_2, and pyridine among many others, are also shown in Fig. 7.11, except the identity, which is always present but changes nothing. The first step is to define the principal axis and as ClO_2 is bent into a V shape, the axis has the direction shown in the figure. (The molecule could be also drawn the other way up.) There are only four types of operations in the C_{2V} point group, which are (i) the identity E, (all molecules have this); (ii) rotation by 180° around the principal axis labelled C_2; and (iii) there are two vertical mirror planes σ_V and σ'_V. The superscript dash is only used only to distinguish one axis from the other. If a set of x-, y-, z-axes are drawn on the molecule with z as the principal axis, then the mirror planes could alternatively be labelled $\sigma(xz)$ and $\sigma(yz)$. There is a possible ambiguity because some authors place the x-axis in the plane of the molecule and some place the y-axis here. You need to check this when looking at different point group tables. The mirror planes are vertical because their edge runs along the principal

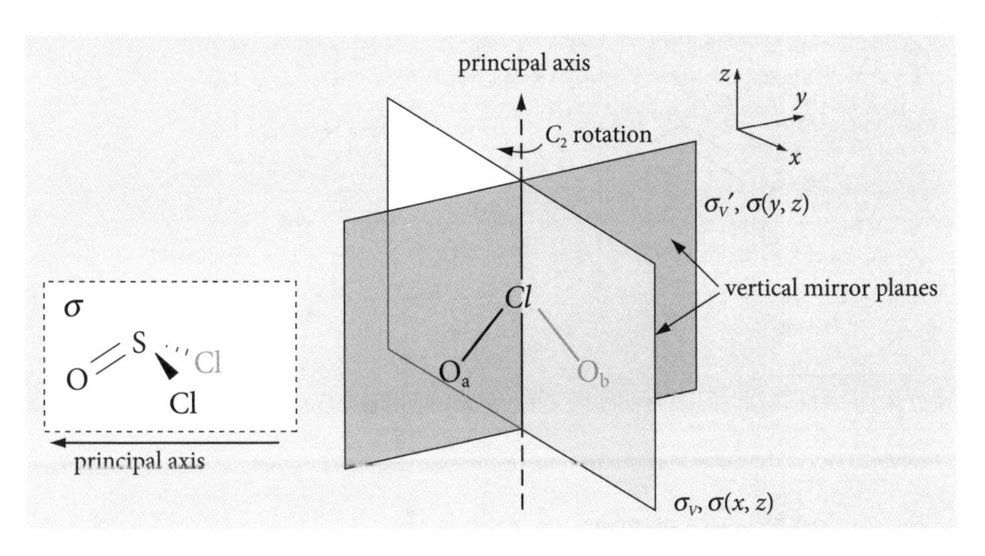

Fig. 7.11 Left: The one mirror plane in a C_s point group. Right: Symmetry operations in a C_{2V} molecule. The C_2 operation is rotation by 180°, σ and σ' operations are reflections in mirror planes as shown. Oxygen atom labels *a* and *b* help when performing operations; the atoms are identical.

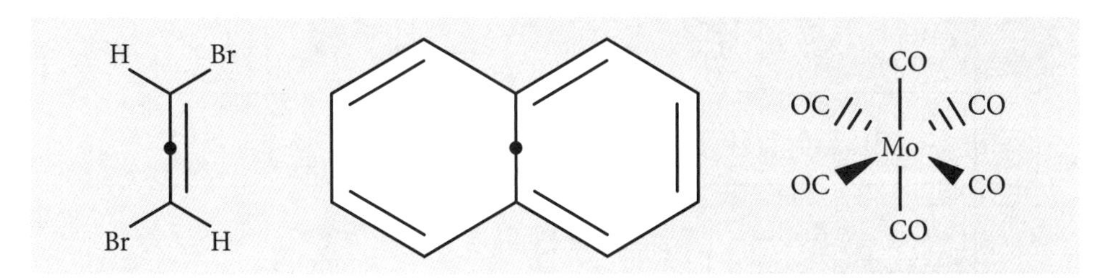

Fig. 7.12 Each of these molecules has an inversion centre. This is shown with a dot but is at the Mo atom in $Mo(CO)_6$. Every atom can be moved through the inversion centre to an equivalent point on the opposite side of the molecule leaving it indistinguishable.

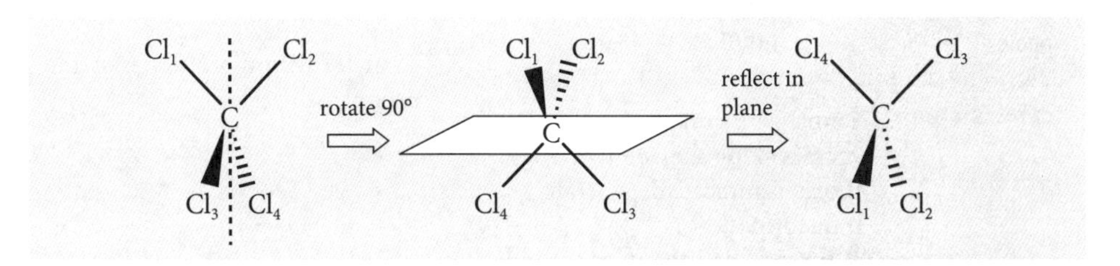

Fig. 7.13 The S_4 rotation–reflection operation applied to the tetrahedral molecule CCl_4. The atoms are labelled only to allow the operations to be followed they are otherwise identical.

axis. If the axis passed perpendicularly through the middle of a mirror plane, this would be a horizontal mirror plane, Fig. 7.10.

To summarize: in the C_{2V} point group, the only symmetry operations present are the identity, whose symbol is E, and is present whatever the point group, one rotation C_2, and two mirror planes, σ_V and $\sigma_{V'}'$. A C_3 or C_4 rotation, which would be rotation by 120° or 90° respectively, cannot make the molecule indistinguishable from its starting position and neither can an inversion or any type of improper rotation operation S, and thus they are not present.

Consider next the naphthalene molecule in Fig. 7.12, which has many more symmetry elements than the centre of inversion. These are indicated in Fig. 7.14. The principal axis could be any of the three C_2 axes because they are all equal; the coordinates drawn show that the out of plane direction is chosen to be z and so this will be chosen as the principal axis. You could choose another orientation of axes if you wanted to, but however the axes are chosen the molecule still has three C_2 axes, three mirror planes running along these axes, a centre of inversion and a mirror plane in the plane of the molecule, labelled σ_h because it is perpendicular to the principal axis. The other element is the identity E. With this information the point group can be identified. The next section shows how this may be done.

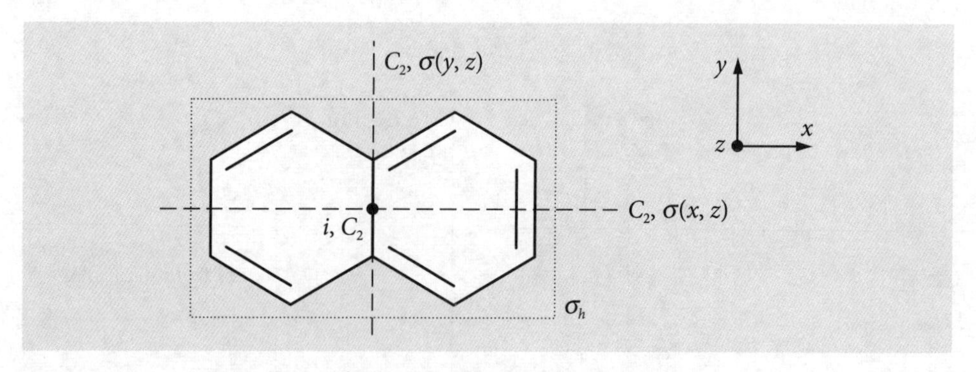

Fig. 7.14 Symmetry elements in naphthalene. The principal axis is out of plane along z.

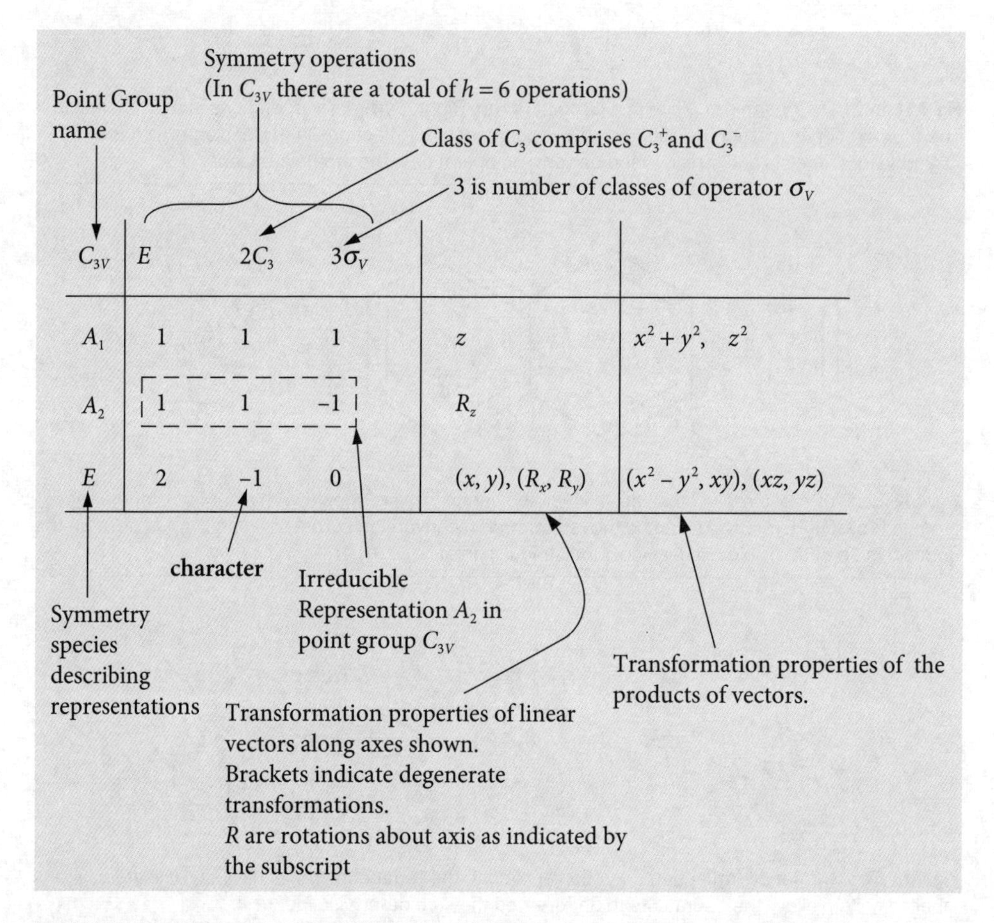

C_{3V}	E	$2C_3$	$3\sigma_V$		
A_1	1	1	1	z	$x^2 + y^2$, z^2
A_2	1	1	−1	R_z	
E	2	−1	0	(x, y), (R_x, R_y)	$(x^2 - y^2, xy)$, (xz, yz)

Fig. 7.15 Navigating the point group character table.

7.6.4 A strategy to identify molecular point groups

When trying to assign a point group, first see if the molecule is a 'special case', that is tetrahedral, octahedral, icosahedral (football shaped), or is 'cylindrical' such as O_2, HCl or FCCF, FCCH, etc. and identify it on this basis alone; see Section 7.6.5 for examples. You can always check later to see if you have guessed correctly by comparing with the point group (character) table. Next, look for any obvious overall rotational symmetry; for example, benzene clearly has sixfold symmetry and pyridine twofold, and this often indicates the principal axis direction and the symmetry label for the highest rotation operator. If a centre of inversion is present, this severely limits the choices of point groups. Particular axes or mirror planes can now be hunted down. Usually these will be enough to restrict your search to one or two point groups. At this point, you will have a list of some rotations and mirror planes and perhaps an inversion. The next step could be to look at tables of point groups and to see how best to match them with your findings so far. The table you choose may suggest the presence of some feature that you have missed.

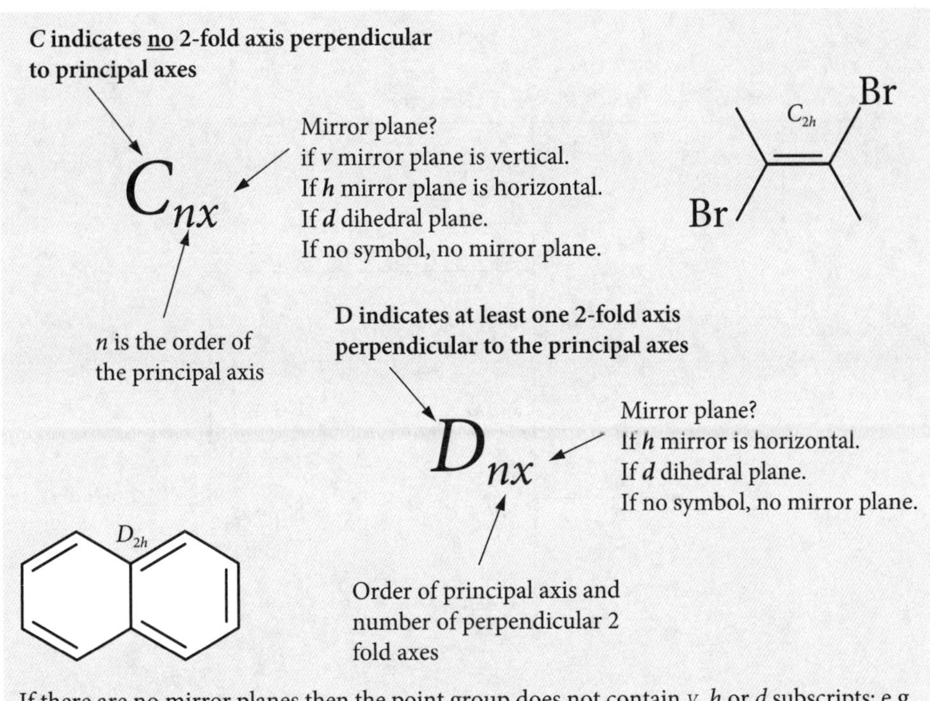

If there are no mirror planes then the point group does not contain *v*, *h* or *d* subscripts; e.g. D_3. which has a propeller shape.
If at least one horizontal mirror plane *h* is present as well as *d* or *v* mirror planes, or both, then *h* takes precedence; *e.g.*, D_{2h}, D_{4h}
If only one type mirror plane is present then this defines the subscript *e.g.* C_{2V}.

Fig. 7.16 Notation for *C* and *D* point groups.

If the highest (principal) axis is twofold symmetric, the point group will be restricted to those with a 2 in their subscript, C_{2V}, D_{2h}, etc. The groups with C_2, C_3, etc. labels are single axis groups, meaning that only one rotation axis is present. Molecules with more than one rotation axis are labelled *D*. Three groups have low symmetry and are C_1, asymmetric, C_s, e.g. $SOCl_2$, with only one mirror plane, and C_i, which has only a centre of inversion. At the other end of the scale, the cubic groups, tetrahedral, octahedral, and icosahedral molecules, have very many symmetry elements and are easily identified.

Better than guessing is to use a systematic way of finding the point group and a 'route map' algorithm is shown in Fig. 7.17. The route map follows roughly the same method as just outlined. Sometimes a shortcut can be made by using the point group labels because they contain a shorthand version of the symmetry operations. The *C* and *D* groups are very common and the meaning of the labels is shown in Fig. 7.16. Assigning point groups is a strange skill; you can become very proficient quite quickly, but lose this skill equally quickly if it is not practised. However, with a little revision, this soon returns.

Summary

(i) Check for special cases; diatomic, octahedral and tetrahedral molecules.
(ii) Look for rotation axes; highest order axis is the principal axis, this gives the first subscript, *n*.
(iii) Determine orientation of any C_2 *axes* perpendicular to principal axis. If none is perpendicular then letter is *C* (or *S*) else *D*.
(iv) Determine the orientation of any mirror planes relative to principal axis; this gives subscripts *v*, *h*, *d*.
(v) Identify all remaining symmetry elements/operation and check with point group tables.

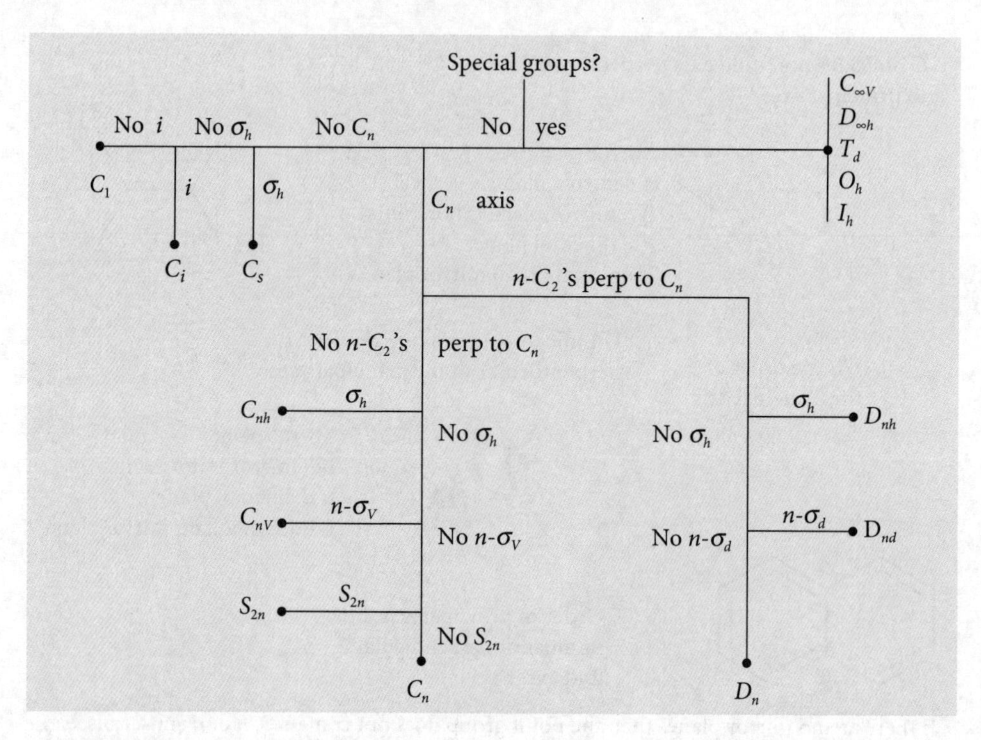

Fig. 7.17 'Road map' to help assign point groups.

7.6.5 Examples of point groups

Most of the H atoms are not included on the structures.

C_1 CHClFBr, cholesterol	C_i staggered 1,2-bibromo-1,2-dichloroethane	C_s meta-chlorophenol, HN_3
C_2 H_2O_2	C_3	S_4 all trans tetrafluorocyclobutane,
C_{2h} trans ClHC=CHCl	C_{3h} 1,3,5-trivinylbenzene (if planar)	C_{4h} tetra fluoroZn-porphyrin
C_{6h} hexa-vinylbenzene (if planar)	C_{2V} H_2O, $COCl_2$, CH_2F_2, $CF_2C{=}CH_2$, pyridine (C_5H_5N)	C_{3V} NH_3, $CHCl_3$, OPF_3

C_{4V} B_5H_9, SF_5Cl 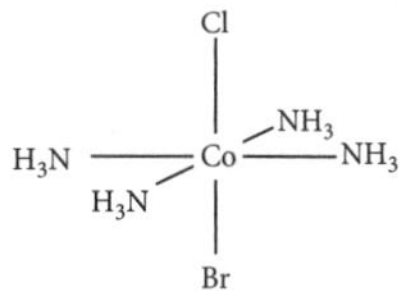	$C_{\infty v}$ HF and all other heteronuclear diatomics, HCN, F–C≡C–H	D_2 twisted biphenyl,  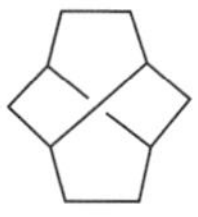 twistane.
D_3 triphenylmethane$^+$ ion (propeller shaped) 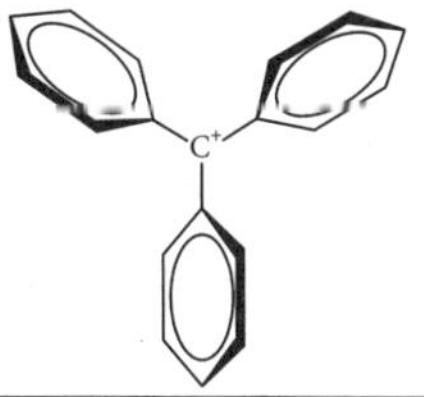	D_{2h} naphthalene, CH_2=CH_2. 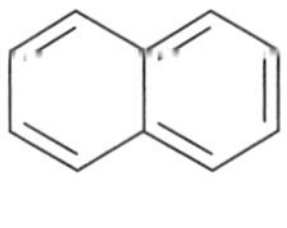	D_{3h} BF_3, C_3H_3, 1,3, 5-trifluorobenzene 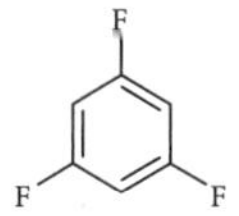
D_{4h} $PtCl_4^{2-}$ 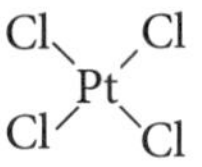 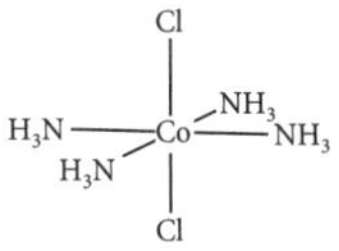	D_{5h} $C_5H_5^-$, IF_7, $Ru(C_5H_5)_2$ 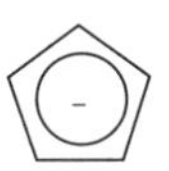	D_{6h} benzene 
D_{7h} tropylium ion $C_7H_7^-$. 	D_{2d} CH_2=C=CCH_2, (allene) 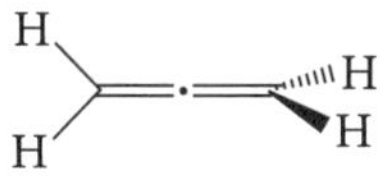	D_{3d} staggered ethane, chair cyclohexane 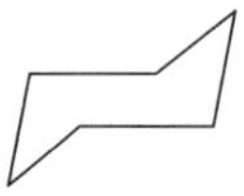
D_{4d} crown S_8 $B_{10}H_{10}^{2-}$, (no H are shown, B atom at each apex) 	D_{5d} ferrocene 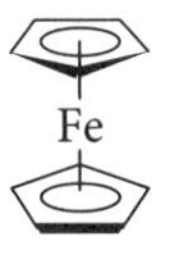	$D_{\infty h}$ H_2 and all other homonuclear diatomics, F–C≡C–F, C_2N_2, CO_2
T_d (tetrahedral) CH_4, CF_4 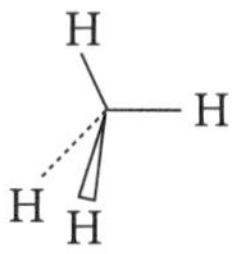	O_h (octahedral) SF_6, cubane, $Fe(CN)_6^{4-}$ 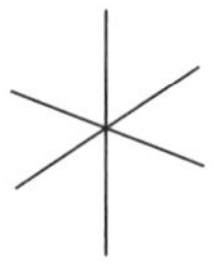	I_h (icosahedral) $B_{12}H_{12}^{2-}$, C_{60}, football, 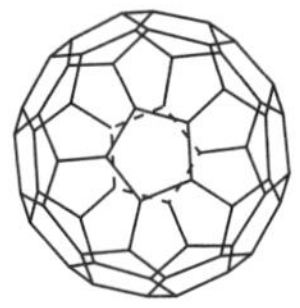

7.6.6 **Products of operators**

To determine what a group multiplication table is, it is necessary to examine the products of two or more symmetry operations. These are then compared with the properties of a mathematical group, and this set of operations may then be associated with a particular group. How this is done is explained in the next few sections.

Using the symmetry operations shown in Figs 7.11–16 and Fig. 7.19, or the matrix representation of the next section, the group multiplication table will be constructed for the C_{2V} point group. The operations are E, C_2, σ_V, σ'_V and a molecule of this point group is shown in Fig. 7.11. The product table is made by multiplying every operation by every other one, both ways round—for example, $\sigma_V C_2$ and $C_2\sigma_V$—and then determining if the product is also one of the operations, which it must be if σ_V and C_2 both belong to the group. The rules of this 'game' are given in Section 7.6.7. Fig. 7.19 shows one C_2 rotation;

C_{2V}	E	C_2	σ_V	σ_V'
E	E	C_2	σ_V	σ_V'
C_2	C_2	E	σ_V'	σ_V
σ_V	σ_V	σ_V'	E	C_2
σ_V'	σ_V'	σ_V	C_2	E

Fig. 7.18 Multiplication table for the C_{2V} point group.

two of them will make the molecule *identical* or $C_2C_2 = C_2^2 = E$ and similarly using Fig. 7.11 shows that reflecting in either of the mirror planes twice each in succession, also produces an identical molecule; therefore, $\sigma_V^2 = E$ and $\sigma_V'^2 = E$. Multiplying an operator by itself produces the diagonal terms in Fig. 7.18. The identity multiplied by itself is still the identity; similarly, the identity multiplied by any other operator leaves the operator unchanged so this produces the left-hand column and top row. What remains are the other off-diagonal terms, such as σ_VC_2 and $C_2\sigma_V$ and these are left for you to confirm. The result for this point group is a symmetrical product table meaning, that in the C_{2V} point group all the operators commute with one another, which is called an Abelian group. This is not always true; for example, see the C_{3V} table produced in Q7.17.

The C_{2V} table shows that the operations form a group, because they conform to the rules of a group as described in Section 7.6.7 and each row in the body of the table is therefore a *representation* of the group. It is not a very convenient representation however, because these can hardly be distinguished from the operators. Another representation can be imagined where all the entries in the table would be 1. This would follow the rules for forming a group but would be useless, as one operation could not be distinguished from another. The clever part was the development of a representation of each point group, such as C_{2V} or D_{2h}, in a meaningful and practically useful way, and to this end, matrices can be used.

7.6.7 Pertinent properties of a mathematical group

The word 'group' in the context of molecular point groups has a precise mathematical definition. The group consists of a set of members that are the symmetry operations and follow four rules.

(i) There must be an identity operator that commutes with all others in the group and leaves them unchanged. The identity is always labelled E.

(ii) The product of two operators A and B is also an operator and member of the group, i.e. AB belongs to the group as does AA and BB.

(iii) The operators follow the associative product rule $(AB)C = A(BC)$.

(iv) Every operator A has an inverse A^{-1} that is also an operator and member of the group. Any operator A that operates on its inverse produces the identity $AA^{-1} = E$. Therefore, by this rule, the operator A^{-1} must be a member of the group.

The members of the C_{2v} group are E, C_2, σ_V, σ_V'. The multiplication table, Fig. 7.18, shows that rule (ii) is followed, because each of the entries in the table is a member of the group. In symmetry operations on molecules, it is common for the inverse and the operator to be identical, i.e. an operator can be its own inverse; for example, $C_2C_2 = C_2^2 = E$ meaning that $C_2^{-1} = C_2$. This shows that rule (iv) is followed.

7.6.8 Symmetry operations as matrices

Although we can perform symmetry operations in a geometrical sense, as was done to produce the C_{2V} product table, these can be rather awkward to use. It turns out that a symmetry operation can be performed as a matrix multiplication using as a basis either a molecule's atoms, or its orbitals or bonds. The trace of the matrix, the sum of its diagonal terms, will form a representation of each operation and so form a representation of

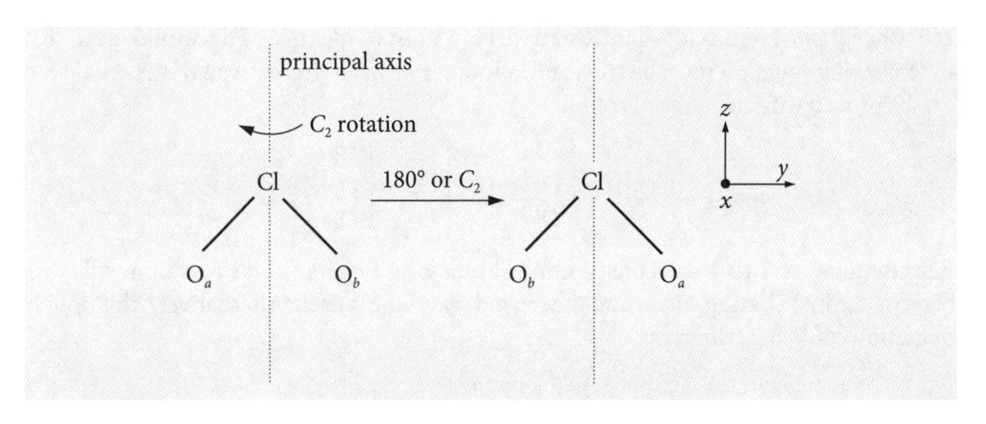

Fig. 7.19 C_2 rotation. The positive *x*-axis projects out of the page.

the group. The matrix used for each symmetry operation used must be a unitary matrix, $|M| = 1$, because bond angles and lengths must be unchanged to maintain a molecule's symmetry. As an example, the symmetry properties of ClO_2 will be examined, using as a basis set the three atoms and with the oxygen atoms labelled *a* and *b* for convenience, this basis is {Cl, O_a, O_b}, Fig. 7.19. The oxygen atoms are labelled only to keep track of them; they are otherwise identical. The identity operation is represented by the matrix equation where the vector containing the atoms does not change,

$$[E]\begin{bmatrix} Cl \\ O_a \\ O_b \end{bmatrix} = \begin{bmatrix} Cl \\ O_a \\ O_b \end{bmatrix}$$

and, by the rules of matrix multiplication, the matrix $[E]$ must be a 3×3 matrix that keeps the left- and right-hand column matrices equal. Note that the two column matrices are identical; *E* is, after all, the identity! Because the identity matrix leaves the column vector unchanged, it must be the unit matrix ***1***,

$$\begin{bmatrix} 1 & 0 & 0 \\ 0 & 1 & 0 \\ 0 & 0 & 1 \end{bmatrix}\begin{bmatrix} Cl \\ O_a \\ O_b \end{bmatrix} = \begin{bmatrix} Cl \\ O_a \\ O_b \end{bmatrix}$$

A C_2 operation or 180° rotation exchanges the oxygen atom positions *a* and *b* but leaves the Cl atom unchanged;

$$\mathrm{Cl} \rightarrow \mathrm{Cl}, \qquad \mathrm{O_a} \rightarrow \mathrm{O_b} \qquad \mathrm{O_b} \rightarrow \mathrm{O_a} \qquad C_2$$

The matrix equation must therefore have the form

$$[C_2]\begin{bmatrix} Cl \\ O_a \\ O_b \end{bmatrix} = \begin{bmatrix} Cl \\ O_b \\ O_a \end{bmatrix}$$

To find out what matrix is needed, a little trial and error is required. In the matrix equation, $\begin{bmatrix} 0 & 1 \\ 1 & 0 \end{bmatrix}\begin{bmatrix} A \\ B \end{bmatrix} = \begin{bmatrix} B \\ A \end{bmatrix}$ the positions of *A* and *B* are swapped by the off-diagonal terms. Using this idea the C_2 rotation matrix is easily guessed as

$$\begin{bmatrix} 1 & 0 & 0 \\ 0 & 0 & 1 \\ 0 & 1 & 0 \end{bmatrix}\begin{bmatrix} Cl \\ O_a \\ O_b \end{bmatrix} = \begin{bmatrix} Cl \\ O_b \\ O_a \end{bmatrix}.$$

Checking that this is correct, the multiplication produces

$$\begin{bmatrix} 1 & 0 & 0 \\ 0 & 0 & 1 \\ 0 & 1 & 0 \end{bmatrix}\begin{bmatrix} Cl \\ O_a \\ O_b \end{bmatrix} = \begin{bmatrix} Cl + 0 \times O_a + 0 \times O_b \\ 0 \times Cl + 0 \times O_a + O_b \\ 0 \times Cl + O_a + 0 \times O_b \end{bmatrix} = \begin{bmatrix} Cl \\ O_b \\ O_a \end{bmatrix}.$$

Now clearly this procedure is a complex business in a molecule even with only three atoms, but, in fact, you effectively do this operation in your head when you look at a picture of the molecule and reflect or rotate it. Suppose that the molecule is again rotated by

180°, then it must return to where it started, i.e. it must be identical. This would mean that the following equation must be true, and, although we have already seen this is the case, it can be proved with the multiplication,

$$[C_2][C_2]\begin{bmatrix} Cl \\ O_a \\ O_b \end{bmatrix} = [C_2]\begin{bmatrix} Cl \\ O_b \\ O_a \end{bmatrix} = \begin{bmatrix} Cl \\ O_a \\ O_b \end{bmatrix}.$$

Because the C_2 matrix swaps the positions of the a and b atoms, this equation will work because the first C_2 swaps them and the second swaps them back. Alternatively, this double operation could be written as

$$[C_2]^2\begin{bmatrix} Cl \\ O_a \\ O_b \end{bmatrix} = \begin{bmatrix} Cl \\ O_a \\ O_b \end{bmatrix} \equiv [E]\begin{bmatrix} Cl \\ O_a \\ O_b \end{bmatrix},$$

where $[C_2]^2 \equiv [E]$. Notice that in this last calculation, the associative product rule of matrices, and of a group (rule (iii)), was used because C_2C_2 was worked out first. To show that this last result is true, work out the direct multiplication,

$$\begin{bmatrix} 1 & 0 & 0 \\ 0 & 0 & 1 \\ 0 & 1 & 0 \end{bmatrix}\begin{bmatrix} 1 & 0 & 0 \\ 0 & 0 & 1 \\ 0 & 1 & 0 \end{bmatrix} = \begin{bmatrix} 1\times1+0+0 & 0 & 0 \\ 0\times1+0+0 & 0+0+1\times1 & 0 \\ 0\times1+1\times0+0 & 0 & 1 \end{bmatrix} = \begin{bmatrix} 1 & 0 & 0 \\ 0 & 1 & 0 \\ 0 & 0 & 1 \end{bmatrix}.$$

Repeating the calculation for the reflections produces the matrices

$$\sigma(x,z) = \begin{bmatrix} 1 & 0 & 0 \\ 0 & 0 & 1 \\ 0 & 1 & 0 \end{bmatrix} \qquad \sigma(y,z) = \begin{bmatrix} 1 & 0 & 0 \\ 0 & 1 & 0 \\ 0 & 0 & 1 \end{bmatrix}.$$

If we perform operations on any pair of matrices, say rotation and reflection, a group multiplication table can be built. With this, and by applying methods from group theory, a character table that describes all the symmetry properties of a molecule can be produced. Instead of using the atoms as a basis, the same matrices would be produced by three unit length vectors, one along each bond and one along the principal axis, and each originating at the Cl atom.

The matrices just calculated represent the operations in the C_{2V} point group with three atoms or unit vectors along the principal axis and bonds and can be collected together as

$$\begin{array}{c|cccc} C_{2V} & E & C_2 & \sigma(x,z) & \sigma'(y,z) \\ \hline \Gamma_R & \begin{bmatrix} 1 & 0 & 0 \\ 0 & 1 & 0 \\ 0 & 0 & 1 \end{bmatrix} & \begin{bmatrix} 1 & 0 & 0 \\ 0 & 0 & 1 \\ 0 & 1 & 0 \end{bmatrix} & \begin{bmatrix} 1 & 0 & 0 \\ 0 & 0 & 1 \\ 0 & 1 & 0 \end{bmatrix} & \begin{bmatrix} 1 & 0 & 0 \\ 0 & 1 & 0 \\ 0 & 0 & 1 \end{bmatrix} \end{array}, \qquad (7.9)$$

which is one form of a *reducible representation* Γ_R, and is reducible because it is not in its simplest form since some matrices have off-diagonal terms, therefore the characters in the point group cannot be determined directly. The trace of each matrix produces the table;

C_{2V} atom basis set	E	C_2	$\sigma(x,z)$	$\sigma'(y,z)$
Γ_R	3	1	1	3

If any other basis set covering the same 'space' were used then the trace of the matrices giving rise to the reducible representation would be the same. If a basis in a different 'space' were used, such as x, y, z unit vectors, then a different reducible representation would be produced, as described in the next section. Two similar spaces could be an atoms' p orbitals in the form p_x, p_y, and p_z or as p_0, p_{-1}, and p_{+1} where the numbers represent the m quantum numbers. These two forms of orbitals can be transformed into one another; $p_z = p_0$, $p_x = (p_1 + p_{-1})/\sqrt{2}$, and $p_y = -i(p_1 - p_{-1})/2$ hence their 'space' is the same.

The process of working out the effect of each operator is not complicated, but can prove tedious; however, it is important because if all the C_{2V} operations can be identified with just one molecule of this point group the same rules must apply to *all molecules of the same point group* no matter how many atoms it has. As these have all been worked out; all we usually need to do is to identify the point group. Fig. 7.20 shows a few molecules belonging to the C_{2V} point group.

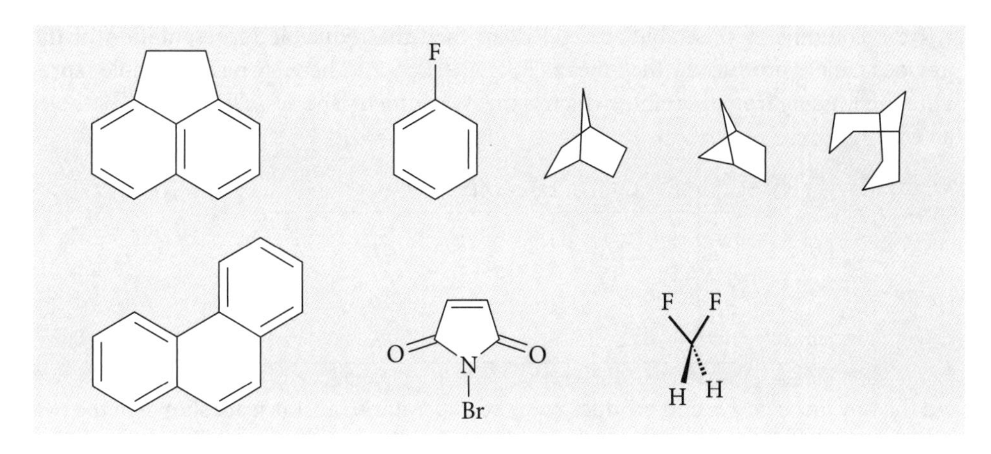

Fig. 7.20 Some molecules belonging to the C_{2v} point group. Most of the H atoms are ignored The structures are only approximately to scale relative to one another.

7.6.9 Representations based on matrices

Choosing a basis set of unit vectors along the x-, y-, and z-axes, and following the method used above, will produce three diagonal matrices for the C_{2V} point group. The vectors can be imagined along the axes shown in Fig. 7.19. A reflection in the z–y plane will leave the z and y vectors unchanged, but invert the x, therefore the matrix equation is

$$\sigma'(y, z): \qquad \begin{bmatrix} -1 & 0 & 0 \\ 0 & 1 & 0 \\ 0 & 0 & 1 \end{bmatrix} \begin{bmatrix} x \\ y \\ z \end{bmatrix} = \begin{bmatrix} -x \\ y \\ z \end{bmatrix}.$$

Fig. 7.21 Reflection of the x unit vector in the y–z plane swaps x coordinates only.

A similar calculation for the x–z plane changes only the y coordinate giving

$$\sigma(x, z): \qquad \begin{bmatrix} 1 & 0 & 0 \\ 0 & -1 & 0 \\ 0 & 0 & 1 \end{bmatrix} \begin{bmatrix} x \\ y \\ z \end{bmatrix} = \begin{bmatrix} x \\ -y \\ z \end{bmatrix}.$$

and the C_2 rotation matrix swaps x with $-x$ and y with $-y$.

The four matrices form a reducible representation, which is now diagonal

C_{2V}	E	C_2	$\sigma(x, z)$	$\sigma'(y, z)$
Γ_R	$\begin{bmatrix} 1 & 0 & 0 \\ 0 & 1 & 0 \\ 0 & 0 & 1 \end{bmatrix}$	$\begin{bmatrix} -1 & 0 & 0 \\ 0 & -1 & 0 \\ 0 & 0 & 1 \end{bmatrix}$	$\begin{bmatrix} 1 & 0 & 0 \\ 0 & -1 & 0 \\ 0 & 0 & 1 \end{bmatrix}$	$\begin{bmatrix} -1 & 0 & 0 \\ 0 & 1 & 0 \\ 0 & 0 & 1 \end{bmatrix}$

The trace of each matrix is the same as for the other basis set, and for any other basis set, which means that the *trace* of the matrix in any basis set (the trace is the sum of the diagonal elements) can form the reducible representation and, clearly, this is far more practical than using the matrix itself.

C_{2V} x, y, z basis set	E	C_2	$\sigma(xz)$	$\sigma'(yz)$
Γ_R	3	−1	1	1

In this particular case, each matrix is diagonal so that the irreducible representation (*irreps*) can be extracted directly, but the matrices are only 3×3 and not 4×4, which they would need to be to produce all four irreps; obviously one will be missing. If there are n, basis set elements then n irreps will be produced. The irreps are:

C_{2V}	E	C_2	$\sigma(xz)$	$\sigma'(yz)$
A_1	1	1	1	1
B_1	1	−1	1	−1
B_2	1	−1	−1	1

If the columns in this table are added up then the reducible representation of the previous table is produced which means $\Gamma_R = A_1 + B_1 + B_2$. There is one irreducible representation missing from this table, which is the A_2 symmetry species. The full C_{2V} character table is

C_{2V}	E	C_2	$\sigma(xz)$	$\sigma'(yz)$		
A_1	1	1	1	1	z	x^2, y^2, z^2
A_2	1	1	−1	−1	R_z	xy
B_1	1	−1	1	−1	x, R_y	xz
B_2	1	−1	−1	1	y, R_x	yz

and the way linear, x, y, z and product xz, z^2, etc. operators transform are shown in the two right-hand most columns. The entries in the body of the table are called the characters, and hence the name *character* table. The symbols (Mulliken labels) in the left-hand column are the labels of the irreducible representations but usually these are called the *symmetry species*. The top row has the symbol A_1, A_g (or Σ_g for $C_{\infty V}$ and Σ_g^+ for $D_{\infty h}$) and is always the totally symmetric representation; the lowest row is the 'least symmetric'. The symmetry species B_2 in the C_{2V} point group has the properties that it is unchanged by the identity E; is changed by 180° rotation about the C_2 axis and by reflection in the mirror plane σ, but unchanged by reflection in mirror plane σ'.

The diagram Fig. 7.15 shows the various properties contained within the point group table, in this case for C_{3V}. The Mulliken label 'E' in the bottom left-hand column in the table means that this irreducible representation is doubly degenerate. This should not be confused with E the identity operation.

7.6.10 Similarity and classes

If symmetry elements in a group C_3^+, C_3^-, σ, etc. are equivalent they satisfy the *similarity* transformation. For example, if $\boldsymbol{A}$, $\boldsymbol{B}$, and $\boldsymbol{C}$ are elements of a group then $\boldsymbol{A}$ and $\boldsymbol{B}$ are equivalent, and are said to be *conjugate*, only if they satisfy the similarity transform

$$A = C^{-1}BC.$$

Equivalent members of a group form a class, a class being a column of characters in the point group table. The number before the symmetry operation is the number of operations in the class, see Fig. 7.15. In C_{2V} there is only one member of each class, in C_{3V} there are two classes of C_3 operations and three of σ_V.

Looking at the multiplication table for C_{2V}, Fig. 7.18, the product $C_2\sigma_V C_2 = C_2\sigma'_V = \sigma_V$, and as C_2 is its own inverse or $C_2 = C_2^{-1}$, this equation has the form of a similarity transform, $\sigma_V = C_2^{-1}\sigma_V C_2$. As each class is one dimensional in this case, the result of the transformation of σ_V has to be σ_V. In C_{3V} we have not yet worked out the direct product table and matrices, this is done in Q7.17–19, but suppose that we want to see if the rotations C_3^+ and C_3^- (Fig. 7.9) are related by a similarity transform involving a mirror plane and if they are whether they belong to the same class. The matrices are

$$C_3^+ \equiv \begin{bmatrix} 0 & 0 & 1 \\ 1 & 0 & 0 \\ 0 & 1 & 0 \end{bmatrix}, \qquad \sigma_V \equiv \begin{bmatrix} 0 & 1 & 0 \\ 1 & 0 & 0 \\ 0 & 0 & 1 \end{bmatrix}.$$

and the transform is $A = \sigma_V^{-1}C_3\sigma_V$. By direct calculation, we find that $\sigma_V^{-1} = \sigma_V$, i.e. σ_V is its own inverse, making the matrix product,

$$A = \begin{bmatrix} 0 & 1 & 0 \\ 1 & 0 & 0 \\ 0 & 0 & 1 \end{bmatrix}\begin{bmatrix} 0 & 1 & 0 \\ 1 & 0 & 0 \\ 0 & 0 & 1 \end{bmatrix}\begin{bmatrix} 0 & 0 & 1 \\ 1 & 0 & 0 \\ 0 & 1 & 0 \end{bmatrix} = \begin{bmatrix} 0 & 1 & 0 \\ 0 & 0 & 1 \\ 0 & 1 & 0 \end{bmatrix} = C_3^-,$$

which indicates that C_3^+ and C_3^- belong to the same class for all point groups that contain a σ_V mirror plane. In C_{3V} there is one column for C_3 operations and has two members in its class, C_3^+ and C_3^-. These are not usually expressed individually in the point group but instead $2C_3$ is used a column heading because the characters are the same for both operators.

7.6.11 **Direct products**

Using characters, it is simple to form a direct product of two or more symmetry species. If operators are A and B, the direct product is written as $A \otimes B$. The symbol $\otimes$ means calculate the direct product by multiplying pairs of characters together column-wise. One of the other symmetry species of the point group must be produced. If either A or B is the totally symmetric representation, the top line in the table, the result is always the other symmetry species B or A respectively.

In the C_{2V} table (see Section 7.6.9), the product $B_1 \otimes B_2 = A_2$ as may be seen by multiplying the two elements of symmetry species B_1 and B_2 column by column and identifying the pattern of characters produced. If two species that are doubly or triply degenerate (E or T Mulliken labels) form a direct product, this has to be reduced in the normal way to a sum of irreducible representations. For example, in C_{3V}, Fig. 7.15, the direct product $E \otimes E$ produces the result $E \otimes E = 4E \oplus C_3 \oplus 0\sigma_V$. The symbol $\oplus$ means the symmetry species are added or, more properly, that $4E$, C_3 but no σ_V are included in $E \otimes E$. It is common in many texts just to use + instead of $\oplus$ and + is used from now on. Reducing direct products is explained in Section 7.6.13, but first one important use of them is illustrated.

7.6.12 **Selection rules**

In the spectroscopy of molecules, symmetry and point group table can be used to determine whether an electronic, vibrational, or rotational transition is going to appear in the spectrum. The probability of a transition is proportional to the expectation value of the operator for that type of transition. This can be written for two wavefunctions for states a and b and an operator μ as $\int \psi_a \mu \psi_b d\tau$ where τ covers all the coordinates the wavefunctions may have. Symmetry can be used to determine if this integral is finite or not, i.e. whether the integral vanishes or not. Only a knowledge of the 'shape' of the wavefunctions and operator are required and no integration is involved and is a sophisticated extension of the odd–even rules to determine if an integral is exactly zero or not.

The absorption or emission of radiation involves an electric-dipole operator. This will transform as a linear vector in the x-, y-, or z-direction because it depends linearly on the change in charge distribution that occurs with the transition. The dipole moment is a vector $\mu = q{\cdot}r$ where q is the charge distribution and r the displacement vector during a transition. Raman scattering depends on having a change in the *polarizability* of the molecule. This is a measure of how easily the electron 'cloud' forming the molecular orbitals changes shape in the presence of the electric field of the radiation. This change is proportional to operators in two dimensions xy, $x^2–y^2$, etc. and these are shown in the last column of the point group table. If a transition is allowed between two states S_1, S_2 with symmetry species ΓS_1 and ΓS_2, then the direct product $\Gamma S_1 \otimes \Gamma\mu \otimes \Gamma S_2$ has to include the totally symmetric representation of the molecule's point group, which is always the top row of the character table. If the transition is of the electric dipole type then the operator's symmetry species, $(\Gamma\mu)$ must be that of x, y or z as shown in the third major column of the point group table. For Raman transitions, the symmetry species corresponding to products xy, yz, etc. are used. For example, in a molecule with C_{2V} symmetry species such as SO_2, a Raman transition from a state with symmetry species A_2 to that with species A_1 will be allowed because an operator $\Gamma\mu$ transforming as xy belongs to the A_2 symmetry species, making

$$\Gamma S_1 \otimes \Gamma\mu \otimes \Gamma S_2 = A_1 \otimes A_2 \otimes A_2 = A_1,$$

the totally symmetric species. An electric dipole transition would not be allowed because no x, y or z operator belongs to the A_2 symmetry species necessary to make the product A_1. A transition between the states with B_2 and B_1 symmetry is not allowed with an electric dipole operator even though both x and y operators have these symmetries. The reason is that the direct product is not totally symmetric. For the y-direction transition, $\Gamma S_1 \otimes \Gamma\mu \otimes \Gamma S_2 = B_2 \otimes B_2 \otimes B_1 = B_1$, similarly the x-direction operator produces the direct product $\Gamma S_1 \otimes \Gamma\mu \otimes \Gamma S_2 = B_2 \otimes B_1 \otimes B_1 = B_2$. The z-direction operator is also no good, producing an A_2 direct product.

This product of symmetry species approach can be extended to cover spin–orbit coupling which allows singlet to triplet transitions between the excited states of molecules and gives rise to the 'El-Sayed' rules, (Turro 1978). The spin–orbit operator transforms as $R_{x,y,z}$

and appears in column 3 in the point group table. Similarly, molecular vibrations can distort the molecule making forbidden transitions formally allowed and intensity for the transition will be borrowed from a nearly allowed transition if such exists. This 'intensity stealing' or Herzberg–Teller coupling gives rise to the vibrational features in the absorption spectrum of benzene among many other molecules (Steinfeld 1981; Atkins & Friedmann 1997). In benzene the ground state to first excited singlet state is formally symmetry forbidden but this transition becomes allowed because a vibration of e_g symmetry species can change the ground or excited state geometry. In this case, the symmetry species of the ground (or excited) state is multiplied by the symmetry of the vibration. This then forms the new symmetry of the ground state and the method outlined above is used to determine if the transition is allowed. The intensity of the transition is borrowed, or 'stolen' from an allowed transition nearby in energy and with the same symmetry as the vibrationally modified state.

7.6.13 Reducible representations

When, for instance, the orbitals in a molecule are operated on with rotations or reflections, a reducible representation Γ_R is produced. This can be decomposed into some irreducible representations, which are the rows that appear in the character table. The effect that symmetry operations have, were calculated in Section 7.6.8 by setting up a matrix for each atom/orbital and working out the effect of each operation. A matrix has the following general form with unit Cartesian vectors, x_1, y_1, z_1 and so forth on each atom.

$$C_2\begin{bmatrix} x_1 \\ y_1 \\ z_1 \\ x_2 \\ y_2 \\ \vdots \end{bmatrix} = \begin{bmatrix} 0 & 0 & -1 & \cdots & \cdots \\ 0 & 1 & & & \\ 0 & & \ddots & & \\ -1 & & & & \\ \vdots & & & & \\ \vdots & & & & \end{bmatrix}\begin{bmatrix} x_1 \\ y_1 \\ z_1 \\ x_2 \\ y_2 \\ \vdots \end{bmatrix}$$

The dimensions of the matrix depend on the number of basis vectors used to describe how the orbitals or vibrations change. The trace of each matrix is then calculated for each class of symmetry operation and then collected together to form the reducible representation. However, such an elaborate approach is not necessary, and the trace can be found more easily by following a small set of rules and so obtaining the reduced representation directly.

First, the atom positions or orbitals are drawn and labelled, then the molecule is subject to each of the symmetry operations of the point group in turn. A table is started and the first row is the list of symmetry operations. The second row is filled in according to four rules for each symmetry operation operating on each base vector.

(i) If unchanged, a value of 1 is entered in the table,
(ii) If changed in sign, −1 is entered.
(iii) If moved in position, 0 is entered.
(iv) Add up all the numbers and enter the result under the symmetry operations column.

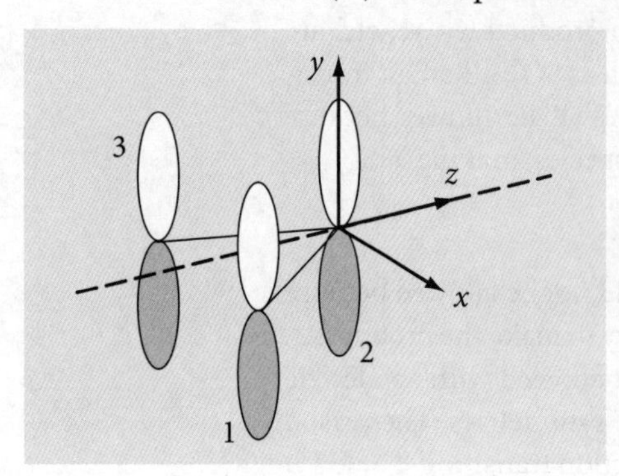

Fig. 7.22 p orbitals in C_{2V} symmetry. The shading indicated +/− phases and are used to identify changes under the symmetry operations.

Consider now the p orbitals on SO_2, as shown in Fig. 7.22, we will use these as a basis for the calculation. They could also be envisaged as unit vectors pointing up from the atoms. The figure tries to show a perspective view; the principal axis, z, is equally placed between atoms 1 and 3 and through atom 2. The molecule has C_{2V} symmetry and the point group, which is shown above, has four operations E, C_2 and two reflections $\sigma(xz)$ and $\sigma(yz)$.

Operating on the molecule according to the rules (i) to (iii) produces the reduced representation Γ_R.

	E	C_2	$\sigma_V(xz)$	$\sigma_V(yz)$
Γ_R	3	−1	−3	1

The 3 in the identity operator column is produced because each orbital is unchanged with this operator. Rotation about the C_2 or z-axis moves vectors on orbitals 1 and 3, so they count zero, and inverts orbital 2; therefore −1 results. The reflections are calculated similarly.

To reduce the representation, a tabular method due to Carter (1997) is very convenient although a formula can be used. This method starts with the table just produced, and multiplies each term it contains by the corresponding element in the character table for each of the symmetry species, and then by the number of operations in each class. The number of operations in a class is shown in the heading of each column in the point group; see Fig. 7.15. In the C_{3V} point group this is two for the C_3 operation. The sum of each row is made and divided by the total number of symmetry operations h.

The C_{2V} point group only has one operation in each class, and the number of symmetry operations $h=4$. The table produced for the reduced representation of the C_{2V} molecule is,

C_{2V}	E	C_2	$\sigma(xz)$	$\sigma'(yz)$	sum/h
Γ_R	3	−1	−3	1	
A_1	3	−1	−3	1	0/4 − 0
A_2	3	−1	$(-1)\times(-3)\times 1=3$	−1	4/4 = 1
B_1	3	1	−3	−1	0/4 = 0
B_2	3	1	3	1	8/4 = 2

The ratio of the sum and h must always be an integer. If not, a mistake has been made. This reduced representation of the p orbitals in SO_2 is therefore composed of an A_1 and two B_2 irreducible representations; $\Gamma_R \equiv A_1 + 2B_2$. We therefore expect molecular orbitals based on p orbitals to have these symmetries.

In the C_{3V} character table, Fig. 7.15, the electric dipole transition from A_1 to E states is potentially allowed under x or y polarized transitions, because the x and y operators belong to the E symmetry species. However the product $A_1 \otimes E \otimes E = E \otimes \mathrm{E}$ is not an irreducible representation because the direct product is $E=4$, $C_3=1$, $\sigma_V=0$. Reducing the $E \otimes E$ direct product is done as follows using the tabular method. Each entry in the table is the number from the reduced representation multiplied by the character and then multiplied by the number in each class. The total number of classes $h=6$ and the $E \otimes E$ table is

	E	$2C_2$	$3\sigma_V$	sum/h
Γ_R	4	1	0	
A_1	4	2	0	6/6 = 1
A_2	4	2	0	6/6 = 1
E	8	−2	0	6/6 = 1

Therefore the product $A_1 \otimes E \otimes E \equiv A_1 \otimes (A_1 + A_2 + E) = A_1 + A_2 + E$. The product $E \otimes E$ therefore contains one of each of the other species in the point group so that the transition would be allowed because A_1 is present.

7.6.14 **Normal mode vibrations**

A normal mode is one of a set of the elementary vibrations of any vibrating object and is fundamental to understanding vibrations of any kind, whether in the engine or suspension of a car, in a washing machine, in a guitar string or in a molecule. If you were able to see a molecule, other than a diatomic, and watch the motion of the atoms, they would appear to be moving rather chaotically. However, to a good approximation it is always possible to assume that the complicated vibrations of any body can be broken down into a set of normal modes and then each of the individual stretching and bending normal motions of a molecule would be apparent. Rather than concentrating on the motion of one atom at a time, the motion of several together is considered and if we get this right, the motion will be that of one of the normal modes. This would, in effect, be a transformation from viewing the molecule in lab xyz coordinates, to viewing internal coordinates based on the motions of these several atoms and there is, fortunately, a systematic way of finding them. In any molecule, all the atoms are coupled through their bonds and in a normal mode, all the atoms move with the same frequency, have fixed amplitude ratios and fixed phase relationships between them. Unless the normal mode is, by symmetry, degenerate with another, each normal mode has a unique frequency.

The symmetry of a normal mode is described by one of the irreducible representations in the molecule's point group. The fact that a symmetry label can be attached to a normal mode encapsulates the idea of collective synchronized motion; some examples are given below for a planar molecule such as HBF_2. Notice that it is possible to have different normal mode displacements, and consequently vibrational frequencies, but with the same overall symmetry.

The first normal mode shown in Fig. 7.23 belongs to symmetry species A_1 and each atom is moved in such a way as to stretch each bond in phase with the other. If the molecule were HBF_2, the centre of mass is below the atom labelled x on a line from z to x. In the bottom right of the figure, the B_1 mode has atoms moving together, but the central atom moves into the plane of the figure while the other three atoms move out, and vice versa.

No external force can act on a molecule due solely to its vibrational motion; therefore, no extra displacement or rotation of the molecule can occur during normal mode vibrations and the centre of mass must remain in the same place. If there are N masses (atoms) connected by forces (chemical bonds), then there are $3N$ modes in a three-dimensional object. However, because we want vibrational normal modes, we have to remove three modes due to translation and three for rotation (two if the molecule is linear) as we are not interested in the whole body's rotation or translation. This leaves $3N - 6$ vibrational normal modes for polyatomic molecules and $3N - 5$ for linear ones.

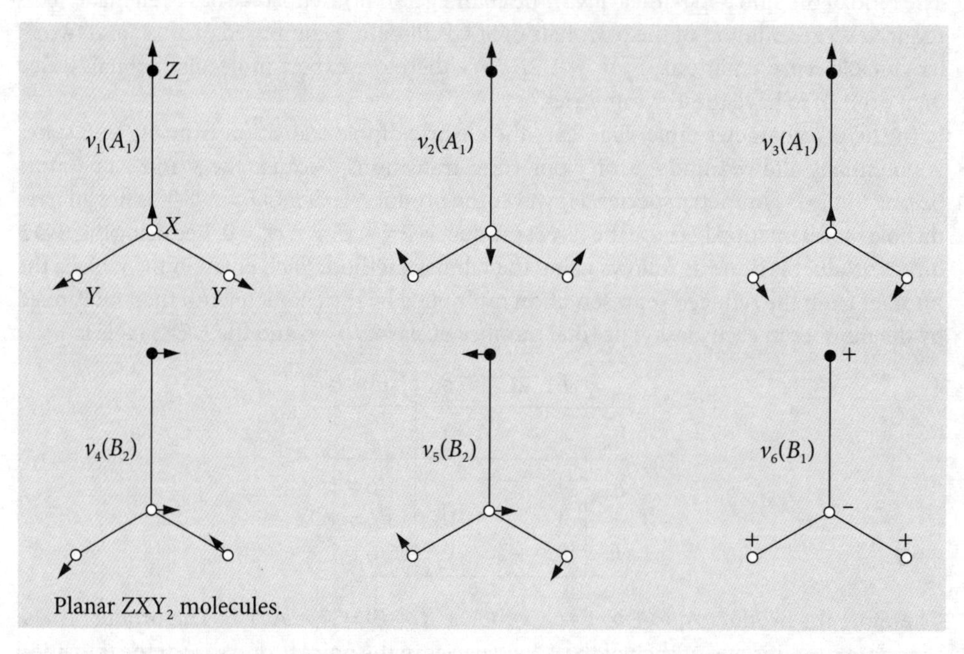

Fig. 7.23 Normal mode displacements (reproduced from Carter 1997). The arrows indicate the direction the atoms move in, but not relatively how far they move; + and – show motion out of the plane of the paper. Typical bond displacements at room temperature are 1% of the bond length. The symmetry labels are also shown and each mode has its own unique frequency ν_1, ν_2 etc.

7.6.15 **Applications to molecular vibrations**

Group theory plays an important part in working out what spectroscopic transitions can occur in a molecule, and this was outlined in Section 7.6.12. In addition, it can be used to work out what the normal mode symmetry species are going to be. Recall that a normal mode is the collective motion of the atoms in a molecule such that a constant phase relationship is maintained between them; see Figs 7.27, 7.78, and 7.79. Often these are characterized as symmetric stretch, asymmetric stretch, and bending vibrations, which is fine in a small molecule, such as SO_2 or water. However, in larger molecule the vibrational motion becomes more complex and cannot be so simply described, but is instead characterized as a symmetry species of the molecule. The normal mode vibrations are given by the Mulliken labels listed in the left-hand column of the point group but usually with a lower case letter, e.g. b_{2g} rather than B_{2g}.

The method outlined for working out orbital symmetry species can be adapted for use here. Again, a table is made up with a top row as the symmetry operations and entries filled in according to rules very similar for those for orbitals.

(a) Choose, either a set of three orthogonal vectors on each atom, or a single vector along each bond to be the basis set for the calculation. Which basis you choose will depend on the problem. The components of these vectors are added up to form the characters needed. This is be done with a set of rules.

(b) Move the molecule according to the symmetry operations in the point group.

(c) For a mirror plane and inversion, add 1 to the character under the symmetry operation for each vector unchanged and −1 for each vector inverted and zero for each atom moved. Do the same for rotations if by 90° or 180°, i.e. for C_2 and C_4 axes.

(d) For other rotations C_3 for example, (or rotation-inversion) a rotation matrix has to be used to calculate the components of the character because the operation may mix the x, y and z components. The trace of the (rotation) matrix gives the character needed.

(e) Reduce the representation produced. If the three orthogonal vectors on each atom were used as the basis set, remove the symmetry species due to three translations and three rotations from the final list as these are implied by the orthogonal vectors.

The case of a C_{2V} molecule such as water is easy to follow; the operators are E, C_2 and $\sigma_V(x,z)$ and $\sigma_V(y,z)$. There are nine vectors, so the identity has a character $E = 9$. The effect of the 180° rotation is shown in Fig. 7.24. The vectors add up to 0 for the H atoms as they move, +1 for the z_1 vector, and −1 for both the x_1 and y_1 producing −1 overall, which is the character for the reducible representation for the C_2 operator. The two reflections are worked out similarly to rotation. Reflection in the x–z plane leaves x and z unchanged making 6 in total but each y is inverted therefore the character for the $\sigma(x,z)$ is 3. The character for $\sigma(y,z)$ is 1 because the H atoms move under this operation so count zero and only x_1 becomes −1 and z_1 and y_1 are unchanged.

The reducible representation is therefore

C_{2V}	E	C_2	$\sigma(xz)$	$\sigma(yz)$
Γ_R	9	−1	3	1

The table to reduce this is

C_{2V}	E	C_2	$\sigma(xz)$	$\sigma'(yz)$	sum/h
Γ_R	9	−1	3	1	
A_1	9	−1	3	1	12/4 = 3
A_2	9	−1	−3	−1	4/4 = 1
B_1	9	1	3	−1	12/4 = 3
B_2	9	1	−3	1	8/4 = 2

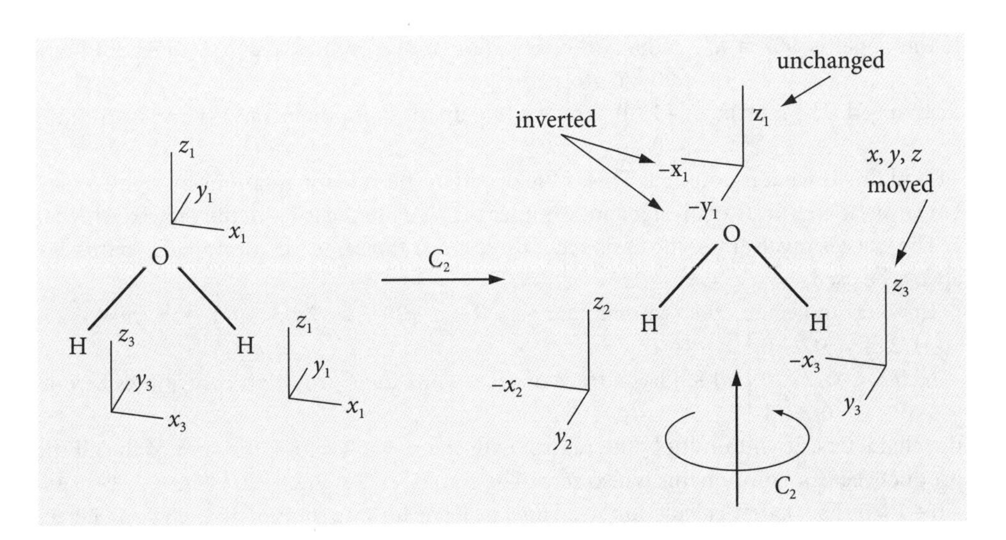

Fig. 7.24 Unit vectors with which to work out vibrational symmetry species.

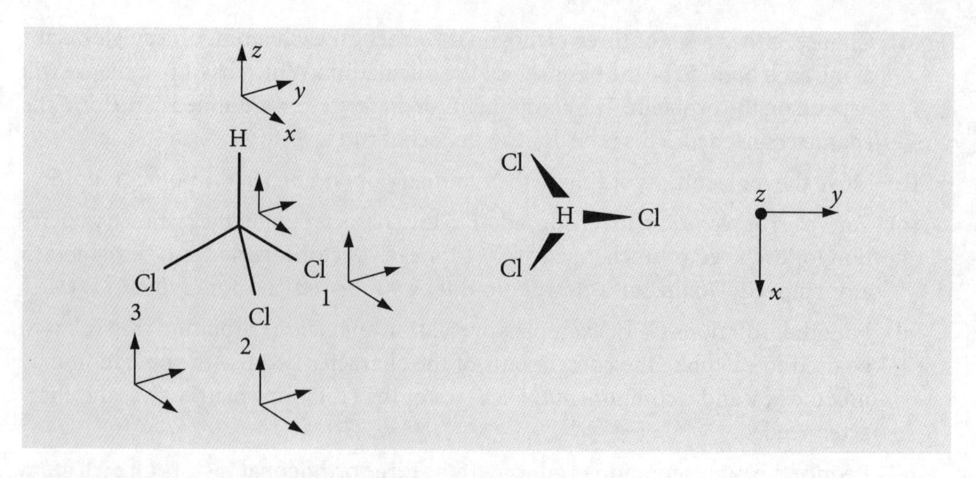

Fig. 7.25 $CHCl_3$ with orthogonal Cartesian sets of unit vectors.

The result is $\Gamma_R = 3A_1 + A_2 + 3B_1 + 2B_2$. There are $3N = 9$ modes in total, $3N - 6$, three of which are vibrations, the others displacements and rotations. These transform as x, y and x and $R_{x,y,z}$ and amount to $A_1 + A_2 + 2B_1 + 2B_2$ leaving $2A_1 + B_1$ as the vibrations. These species correspond to one totally symmetrical stretch and bend, and an asymmetrical stretch, see Fig. 7.26.

In $CHCl_3$ if we place vectors on each atom this makes a 15-dimensional basis set and the C_3 axis is then going to mix the x and y components if z is along the principal axis, Fig. 7.25. Chlorine atom 1 is on the y-axis and the carbon is at the origin. The symmetry operations in C_{3V} are E, $2C_3$, and $3\sigma_V$. The reducible representation is again found by applying the operators in turn to each vector. The identity E scores 15, because each vector is unchanged. The mirror planes run along the principal axis and between each pair of Cl atoms. The mirror plane between Cl atoms 2 and 3 also runs along the H-C-Cl bonds containing Cl atom 1. The Cl-2 and Cl-3 atoms are moved by reflection so count zero for all vectors. The y and z coordinates on the C, Cl(1) and C atoms are unchanged and count 1 each, making 6 in total. The x-direction vectors on these atoms are inverted and count −1 each. Do this for one mirror plane only the effect of the number of classes is accounted for when the representation is reduced. The reflections thus count $6 - 3 = 3$.

The rotations are a little more complicated, but it is clear that the Cl atoms are each shifted by rotation and their vectors count zero. The H and C atoms each have their x- and y-directions mixed. The matrices to rotate a molecule are explained in Section 7.7, and we use this result. The new positions are x_θ etc. after rotation by an angle θ.

$$\begin{bmatrix} x_\theta \\ y_\theta \\ z_\theta \end{bmatrix} = \begin{bmatrix} \cos(\theta) & \sin(\theta) & 0 \\ -\sin(\theta) & \cos(\theta) & 0 \\ 0 & 0 & 1 \end{bmatrix} \begin{bmatrix} x \\ y \\ z \end{bmatrix}$$

If the angle is 90° ≡ π/2 radians then rotation moves y into x and x to $-y$ and z is unchanged. The matrix is $\begin{bmatrix} 0 & 1 & 0 \\ -1 & 0 & 0 \\ 0 & 0 & 1 \end{bmatrix}$. Adding up the components in x, y and z produces +1 and this is what is counted. This is the origin of the rule for rotations by right angles. However, it is entirely equivalent and simpler to calculate the trace of the matrix, which is 1. The trace is invariant of the basis set, if these cover the same function space, so this can always be used.

In our molecule the rotation angle θ is 120° ≡ 2π/3 and the matrix is $\begin{bmatrix} -1/2 & \sqrt{3}/2 & 0 \\ -\sqrt{3}/2 & -1/2 & 0 \\ 0 & 0 & 1 \end{bmatrix}$ which has a trace of zero. Thus the C_3 rotation contributes zero to the characters. If you multiply the matrix with the x, y, z vectors, the sum of the all the product vectors' components is also zero. The sum is $-1/2 + \sqrt{3}/2 + 0 - \sqrt{3}/2 - 1/2 + 0 + 0 + 0 + 1 = 0$. The matrix calculation need only be done for one rotation in each class, i.e. we do not need to do both C_3^+ and C_3^- as this is taken into account in reducing the representation. The result of the calculation is

C_{3V}	E	$2C_3$	$3\sigma_V$
Γ_R	15	0	3

This is reduced using the group table Fig. 7.15 to give

C_{3V}	E	$2C_3$	$3\sigma_V$	*sum*	*sum*/6
Γ_R	15	0	3		
A_1	15	0	9	24	4
A_2	15	0	−9	6	1
E	30	0	0	30	5

The reduced representation of $CHCl_3$ vibrations in the Cartesian basis set is therefore composed of $\Gamma_R = 4A_1 + A_2 + 5E$ making 15 species in total. There are six degrees of freedom, three for translations and three for rotations of the molecule to be subtracted. These add to $A_1 + A_2 + 4E$ (see x, y, z and R_x, R_y, R_z in the point group table), which leaves $3A_1 + 3E$ as the vibrational normal mode symmetry species. The number of normal modes remaining is $3N - 6 = 9$, which is $3A_1$ singly and 3 doubly degenerate E vibrations. The shapes of the normal modes for CH_3Cl are illustrated in Herzberg (1964, vol II p. 314).

Rotation axes C_n are numbered as fractions of 360° or $2\pi/n$ making angle in the rotation matrix $\theta = 2\pi/n$ and the trace of the rotation matrix for rotations with different n is $T_{C_n} = 2\cos(2\pi/n) + 1$. The trace for the first few C_n is

C_2	C_3	C_4	C_5	C_6
−1	0	1	$\frac{1+\sqrt{5}}{2}$	2

The similar calculation for rotation-reflection or S_n axes produces $T_{S_n} = -1 + 2\cos(2\pi/n)$, see Q7.26.

7.6.16 **Determining the shapes of normal mode vibrations and molecular orbitals using projection operators**

Having obtained the symmetry species of vibrations or molecular orbitals it is natural to want to see what these look like and to do this *projection operators* are used. What these operators do is to extract the symmetry adapted functions from the basis functions used to form them and so produce a linear combination L of the basis functions. Many basis functions can be chosen, but the easiest to use are usually vectors pointing along bonds or those representing a pπ orbital. To form a linear combination, a 'victim' vector is chosen and operated on with each symmetry operation in the point group, C_2, $\sigma(xz)$, etc. to find what vector this turns into. This vector's *name* (the basis function) is then multiplied with the character from the point group corresponding to that symmetry operation and symmetry species. Finally, the names are added together by moving from one symmetry operation to the next along the top row of the point group. We expect the result to be the sum or difference of the displacement vectors for a molecular vibration, such as $2v_1 + v_2 - v_3 \cdots$ and so forth. The same formula or algorithm is used if the combinations of atomic orbitals that make up molecular orbitals are sought.

The equation to make a linear combination is

$$L_M = \frac{d}{h} \sum_{j=1..h} c_j S_j(v),$$

where M is the symmetry species label, A_g, B_{3g}, and so forth, h is the order of the group, d the dimension of the irreducible representation, and the sum is over all the classes. Because the resulting vector will be normalized d/h can be ignored. One further point is important. In the C_{3V} point group, for example, there is a heading $2C_3$ as there are two members in this class, and this must be split in the summation into C_3^+ and C_3^-. This is because moving a vector 120° to the right, say, will turn it into a different vector than turning 120° to the left, similarly, any mirror planes must be separated out. The character for

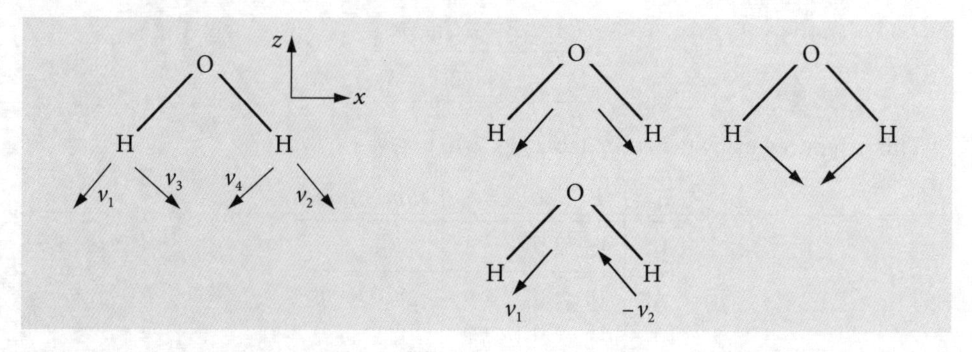

Fig. 7.26 Left: Vectors which are the basis functions used to determine normal modes. Right: The two symmetric A_1 (top) and one asymmetric normal mode B_1 as $v_1 - v_2$.

symmetry species M is c_j and $S_j(v)$ is the effect that operator S in column j of the point group has on our victim vector v. The result of this operation is to produce a vector whose *name* is recorded. For example, if S is the identity then the result is $E_1(v) = v$, other operations may leave v unchanged or change it into *another vector*. Finally, note that if the reduced representation of the vibrations or molecular orbitals contains two or more symmetry species of the same type, e.g. $2A_1$, then two or more different victim vectors will have to be chosen to obtain all the linear combinations.

In H_2O, SO_2, and other triatomics with C_{2V} symmetry, it is clear that two of the vibrational modes stretch the bonds and one changes the bond angle. To work out the normal mode vectors, some representative vectors are placed along the bonds as in Fig. 7.26. You can choose where to put these but physically realistic choices will usually make the calculation simpler.

In C_{2V}, the normal mode vibrations have been found to comprise the symmetry species $2A_1$ and B_1. The L_M formula is shown as a table with v_1 as the victim vector. The table also shows the characters for each symmetry species.

C_{2V}	E	C_2	$\sigma(x,z)$	$\sigma'(y,z)$	L_M	$(d=4, h=4)$
A_1	1	1	1	1		
L_{A1}	v_1	v_2	v_1	v_2	$2(v_1+v_2)$	
A_2	1	1	−1	−1		
L_{A2}	v_1	v_2	$-v_1$	$-v_2$	0	
B_1	1	−1	1	−1		
L_{B1}	v_1	$-v_2$	v_1	$-v_2$	$2(v_1-v_2)$	
B_2	1	−1	−1	1		
L_{B2}	v_1	$-v_2$	$-v_1$	v_2	0	

Notice that what the vector changes into is (obviously) the same for each symmetry species as this is determined by the symmetry operations. Only the value of the character changes in front of each term The final L_M is not multiplied by d/h because the resulting vectors are instead normalized giving, $L_{A_1} = (v_1 + v_2)/\sqrt{2}$ and $L_{B_1} = (v_1 - v_2)/\sqrt{2}$. Note also that v_3 and v_4 do not enter into this table, they are at 90° to v_1 and v_2 and no operation in this point group can interconvert them.

There are two A_1 species produced from the reducible representation but only one has been found so far. This is because the vectors v_1 and v_2 cannot produce a bend, which is the other normal mode. Using vector v_3 as the victim this species appears. As this mode has A_1 symmetry all the characters c_j in the L_M formula are 1, making the calculation easy and the combination is shown below.

C_{2V}	E	C_2	$\sigma(xz)$	$\sigma'(yz)$	L_M
L_{A1}	v_3	v_4	v_3	v_4	$2(v_3+v_4)$

which describes the bending mode shown in Fig. 7.26 in terms of its unit vectors. When normalized this mode is $(v_3 + v_4)/\sqrt{2}$.

As the centre of mass cannot move during a vibration, there being no force to make it do so, therefore the O atom has to move a small distance away from the H atoms. Vectors could be added onto the O atom to show this and L_M terms calculated. The atoms' displacements can also be found when the equations of motion are solved as shown in Section 7.14.

In the case of degenerate symmetry species as occurs in the C_{3V} and most other point groups the results of the summation L_M is to produce two or more vector sums that are not orthogonal. Either these have to be made orthogonal, by choosing another victim for the degenerate term and then using the Gram–Schmidt method to make the two vectors orthogonal, or using the method outlined by Carter (1997) and by Vincent (2001), which involves producing a second vector at 90° to the victim one and repeating the calculation. (The Gram–Schmidt method can be found in Maple and is described by Arkfen 1970.)

The calculation to produce molecular orbitals is essentially the same as for vibrations if the basis vectors are made 'parallel' to the atomic orbitals. For example, for aromatic or other conjugated molecules, a vector pointing in the direction of each π atomic orbital is used as a basis.

Once the symmetry of the molecular orbitals have been found, a secular equation can be set up to calculate their energies. The matrix elements (or expectation value) will have the form $\langle \psi_1 | H | \psi_2 \rangle$ between two molecular orbitals labelled 1 and 2. If more than one orbital is present with a particular symmetry, the orbitals will mix and a secular equation is set up to find their energies, provided that the orbitals are made orthogonal first. This can be done with a Gram–Schmidt method and the results normalized. The molecular orbitals each have the form $\psi = a_1\varphi_1 + a_2\varphi_2 + b_3\varphi_3 + \cdots + b_n\varphi_n$ where φ are the atomic orbitals and a are the coefficients found using the projection operator method just described. We can replace the orbital by the base vectors v and still work out the coefficients. If for example the Hückel Hamiltonian is used, this operates only on adjacent orbitals and is not zero only for terms with $k = j$ and $j \pm 1$ when it produces a constant value. However, whatever the Hamiltonian, an expectation value is calculated with the rather complicated looking formula

$$\langle \psi_1 | H | \psi_2 \rangle = \sum_{k=1}^{n} \sum_{j=1}^{n} a_{1k} a_{2j} \langle v_k | H | v_j \rangle,$$

but many of the terms may be zero, either because of the type of Hamiltonian, e.g. Hückel, or because one of the coefficients a_{1k} or a_{2j} is zero. The expectation values are placed in a secular determinant,

$$\begin{bmatrix} \langle 1|H|1\rangle & \langle 1|H|2\rangle & \cdots & \\ \langle 2|H|1\rangle & \langle 2|H|2\rangle & & \\ \vdots & & \ddots & \\ & & & \langle n|H|n\rangle \end{bmatrix}$$

which is solved to find the n eigenvalues, which are the energies, and the eigenvectors, see 7.2.5.

7.6.17 **Questions**

Full solutions are available at **www.oxfordtextbooks.co.uk/orc/beddard**.

Q7.17 Ammonia belongs to the C_{3V} point group and has the following symmetry elements: the identity, a C_3 axes, and three vertical mirror planes. The rotation about the principal axes is 120° rotation clockwise C_3^+, and either a rotation by −120° (anticlockwise) labelled C_3^- or a rotation by +240°. The three mirror planes are at 120° to one another and are labelled σ_V, σ_V' and σ_V''. Work out the 6×6 group multiplication table.

Strategy: The number of calculations can be reduced from 36 to 19 because the product of a symmetry operation with the identity E leaves the operation unchanged. However, multiplication of an operation with itself does not necessarily produces the identity, which means that only the first row and column can be written down directly. Completing the table is rather like doing a Sudoku problem; each product must be a member of the group and each row and column contains all the elements in the group. This means that as the table approaches completion the missing terms can be inferred from those already in a row or column.

Q7.18 **(a)** Repeat the same calculation as Q7.17 but using the NH *bonds* as a basis set to form 3×3 matrices. Multiply these as appropriate to produce the group multiplication table.

(b) Write a Maple algorithm to do this.

Q7.19 Find the group multiplication table using the symmetry operations from the group E, B, C where E is the identity but B and C are arbitrary.

***Strategy*:** Use the rule for group multiplication that a product is also a member of the group. Also each row and column must contain each member of the group.

Q7.20 Using the similarity transforms, $A_1 = \sigma_V^{-1} C_3^- \sigma_V$ and $A_2 = \sigma_V^{-1} C_3^+ \sigma_V$ show that the σ_V operations belong to the same class in C_{3V} as C_3^+ and C_3^-. Use the matrices from the pervious question and Maple to perform the matrix multiplication if necessary.

Q7.21 The molecule XeF_4 is square planar and because of its high symmetry, it has many elements. Assign its point group and also sketch and list all the symmetry elements. To confirm positively the point group you will need a set of tables. These can be found in most physical chemistry textbooks and in any book on molecular group theory.

***Strategy*:** Not all the symmetry elements need be to found to assign a point group. Sketching the molecule immediately suggests fourfold symmetry. Do not forget to look for S symmetry elements and a centre of inversion. Use the point group 'road map' to identify the point group if necessary; see Fig. 7.17.

Q7.22 Work out the 3×3 matrix for the reflections σ and σ' in C_{2V} as defined in Fig. 6.11 with the basis set $\{Cl, O_a, O_b\}$.

***Strategy*:** Define a symbolic matrix equation to describe what a reflection does then find a matrix that performs the same operation.

Q7.23 **(a)** Work out the effect of a rotation followed by a reflection σ' in the C_{2V} point group. Then reverse the order so that a reflection is followed by a rotation. Use the matrices calculated in the previous question.

(b) Show that operators C_2 and σ' commute.

Q7.24 Using the molecule $H_2C{=}CF_2$, which has C_{2V} symmetry, generate matrices for each operation, C_2, σ and σ'. Use the basis set $\{H_a, H_b, C, C, F_a, F_b\}$. Swap the order of atoms in the C_2 basis set in any way you choose and show that although the matrix formed is different, the matrix multiplication has the same result on the molecule. The fluorine and hydrogen atoms are in the σ plane. (Use Maple to perform the matrix multiplications.)

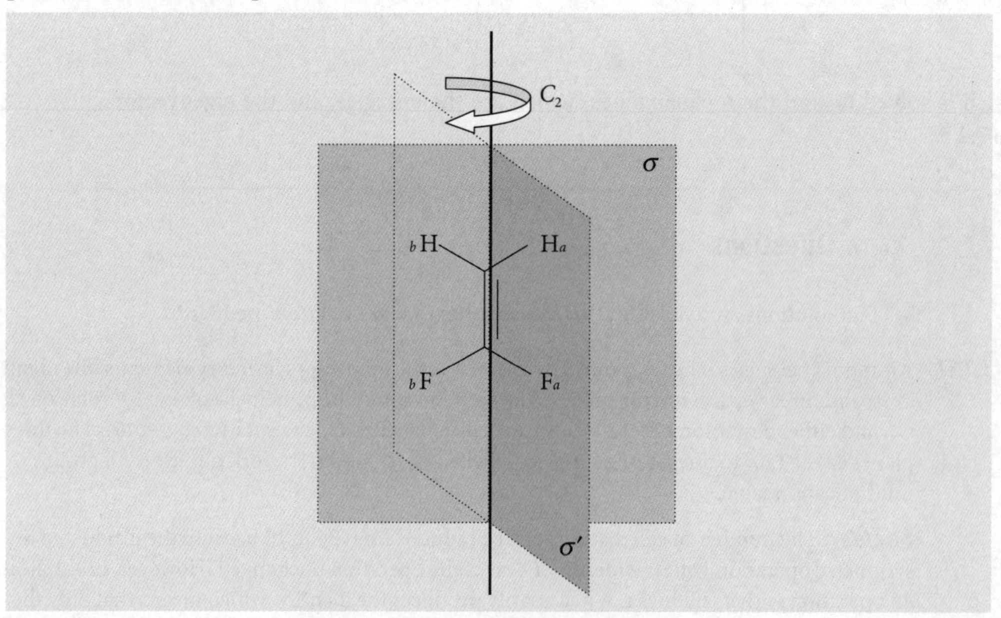

Fig. 7.27 F_2CCH_2 which belongs to the C_{2V} point group as does H_2O, SO_2, NO_2, $COCl_2$ and many other molecules.

Q7.25 The pyramidal molecule NH_3 belongs to the C_{3V} point group, which has operations E, $2C_3$, and $3\sigma_V$. Using an atom basis set of $\{N, H_1, H_2, H_3\}$,

(a) Calculate the effect of a C_3 symmetry operation. Identify the rotations and vertical mirror planes. (The C_3 operation is rotation by $\pm 360/3 = \pm 120°$).

(b) Work out the matrix for reflection.

Strategy: Start by drawing a diagram of the molecule and then identify the rotation and reflection axes. Next, rotate the atoms about the principal axis. This is the axis of highest symmetry and is out of the plane of the page if you draw the molecule 'flat' on the page. The basis set $\{N, H_1, H_2, H_3\}$ means form a vector and matrix in only these atoms as has been done in previous questions.

Q7.26 **(a)** Calculate the matrix for a rotation C_n about the principal axis then reflection in a perpendicular plane and so produce the matrix for the S_n operation. **(b)** Calculate the trace of the matrix.

Q7.27 The molecular ion $C_3H_3^+$ has an equilateral triangular shape.

(a) Sketch the symmetry elements in the point group which is D_{3h}.

(b) The point group can be reduced to C_3 when only p orbitals are considered because they remove the horizontal plane of symmetry. Transform the p orbitals under operations from C_3 and from the reduced representation produced find the irreps and therefore the symmetry species of the π molecular orbitals.

(c) The molecular orbitals (wavefunctions) will contain complex numbers but making a linear combination will make them real. Show that this is true.

The C_3 point group character table is

C_3	E	C_3	C_3^2
A	1	1	1
E	1	ε	ε^*
	1	ε^*	ε

where the labels are $\varepsilon = e^{2\pi i/3}$ and ε^* is the complex conjugate. The E species is doubly degenerate and two rows are listed. A fuller account of calculating the MO's in this molecule can be found in Ratner & Schatz (2001, §13).

Q7.28 Naphthalene has 10 pπ orbitals. **(a)** What is the point group if the phase of the orbitals is taken into account in determining the symmetry? Take the short axis of the molecule to be x the long axis y.

(b) Find the reduced and then irreducible representations (irreps) of these orbitals.

Q7.29 Naphthalene really belongs to the D_{2h} point group. Work out the shapes (symmetry species) of its π molecular orbitals. The point group is

D_{2h}	E	$C_2(z)$	$C_2(y)$	$C_2(x)$	i	$\sigma(xy)$	$\sigma(xz)$	$\sigma(yz)$
A_g	1	1	1	1	1	1	1	1
B_{1g}	1	1	−1	−1	1	1	−1	−1
B_{2g}	1	−1	1	−1	1	−1	1	−1
B_{3g}	1	−1	−1	1	1	−1	−1	1
A_u	1	1	1	1	−1	−1	−1	−1
B_{1u}	1	1	−1	−1	−1	−1	1	1
B_{2u}	1	−1	1	−1	−1	1	−1	1
B_{3u}	1	−1	−1	1	−1	1	1	−1

The atom numbering is that conventionally taken and starts by labeling around the outside of this planar molecule. The axes should be imagined as being centred on the molecule.

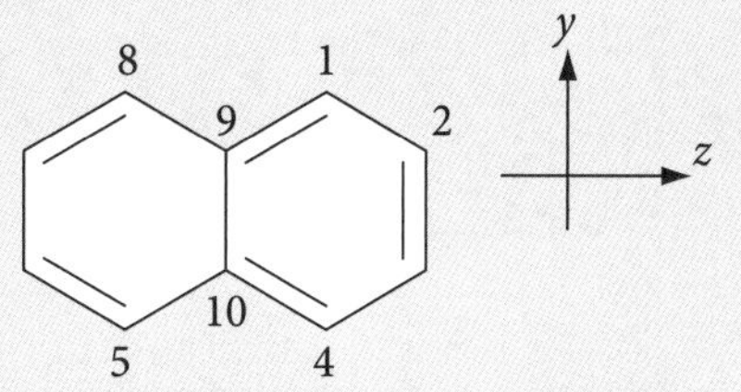

Fig. 7.28 Naphthalene

***Strategy*:** Find the reduced representation first, then place vectors to represent the p orbitals rising out of the plane of the rings. Choosing a victim orbital, work out how this is transformed under each of the symmetry operations in turn, and build up the orbitals comprising a molecular orbital. You will have to choose 3 different orbitals as victims, because more than one of each type of some symmetry species is present. For example, orbitals on atom 1, 2, and 9 do not interconvert so can be chosen as victims. The method is explained in Sections 7.6.13 and 7.6.16.

Q7.30 The molecule BF_3 belongs to the point group D_{3h}. Work out the symmetry of the displacements describing its in plane stretching vibrational normal modes. The point group is

D_{3h}	E	$2C_3$	$3C_2$	σ_h	$2S_3$	$3\sigma_v$
A_1'	1	1	1	1	1	1
A_2'	1	1	−1	1	1	−1
E'	2	−1	0	2	−1	0
A_1''	1	1	1	−1	−1	−1
A_2''	1	1	−1	−1	−1	1
E''	2	−1	0	−2	1	0

(One normal mode is doubly degenerate so use Maple's GramSchmidt function to make the vectors orthogonal.)

***Strategy*:** Place vectors along the bonds to find the geometry of the in-plane stretching modes. First, find the *irreps* that these vectors produce then use the projection operator method to find the modes. There are $3N - 6 = 6$ vibrational normal modes.

7.7 Rotation matrices: moving molecules

Rotating the image of a molecular structure viewed on the computer screen can be easily achieved with a matrix. In two dimensions, the rotation can be defined as moving the co-ordinates of the molecule about an axis perpendicular to the plane of the page (x–y plane) by θ radians, relative to the original axes as shown in Fig. 7.29. Alternatively, the axes themselves can be rotated.

The atom a, on the benzene molecule is shown at $\{x_1, y_1\}$ relative to the initial axes labelled x and y, but in the new axes (labelled x_R and y_R) it is moved to coordinate $\{x_{R1}, y_{R1}\}$ by rotating the axes anticlockwise by θ. From the viewpoint of the new axes, seen by rotating them anticlockwise so that axis x_{R1} appears horizontal, the molecule appears to have rotated clockwise by θ so that atoms a and b now appear to be horizontal. Alternatively, the axes can be fixed and the molecule rotated clockwise about the origin of the axes, as in the right-hand figure, which seems more natural.

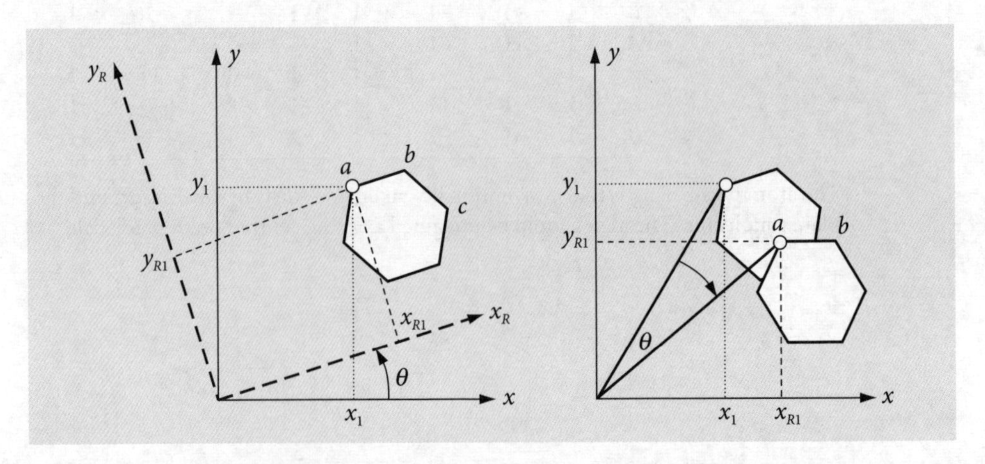

Fig. 7.29 Two equivalent views of rotating a molecule. Left: The axes rotate θ anticlockwise. The point 'a' has coordinates $[x_1, y_1] \equiv [x_{R1} = y_{R2}]$ in the two sets of axes. Right: The molecule is rotated θ clockwise.

If the atom a, has coordinates represented as a vector $\begin{bmatrix} x_1 \\ y_1 \end{bmatrix}$ in the two-dimensional basis set of $\{x, y\}$ then the new coordinates when rotating the molecule clockwise are

$$x_{R1} = x_1 \cos(\theta) + y_1 \sin(\theta) \qquad \text{and} \qquad y_{R1} = -x_1 \sin(\theta) + y_1 \cos(\theta)$$

which is, in matrix form,

$$\begin{bmatrix} x_{R1} \\ y_{R1} \end{bmatrix} = R_\theta \begin{bmatrix} x_1 \\ y_1 \end{bmatrix} \tag{7.10}$$

where

$$R_\theta = \begin{bmatrix} \cos(\theta) & \sin(\theta) \\ -\sin(\theta) & \cos(\theta) \end{bmatrix} \tag{7.11}$$

is the two-dimensional rotation matrix, which is a unitary and orthogonal matrix. Does this rotation matrix make sense? Let us try some with numbers. If $\theta = 0$ then nothing should happen and x_1 should be the same as x_{R1}, and similarly for y values, because $\sin(0) = 0$ and $\cos(0) = 1$. If $\theta = \pi$ radians (180°) then $x_{R1} = -x_1$ and $y_{R1} = -y_1$ so the molecule is inverted by rotating by 180°.

If rotation is by 90° then $x \rightarrow y$ and $y \rightarrow -x$ because $\cos(\pi/2) = 0$ and $\sin(\pi/2) = 1$. The rotation matrices for 180° or π radian, 90° ($\pi/2$ radian), and 45° rotation are

$$R_{180^\circ} = \begin{bmatrix} -1 & 0 \\ 0 & -1 \end{bmatrix} \qquad R_{90^\circ} = \begin{bmatrix} 0 & 1 \\ -1 & 0 \end{bmatrix} \qquad R_{45^\circ} = \frac{1}{\sqrt{2}} \begin{bmatrix} 1 & 1 \\ -1 & 1 \end{bmatrix} \tag{7.12}$$

The effect we observe on rotation is that the molecule will appear to move in an arc around $\{0, 0\}$ or the centre of the axes; the arc atom 'a' follows has a radius $\sqrt{x_1^2 + x_2^2} \equiv \sqrt{x_{R1}^2 + x_{R2}^2}$. If we wanted the molecule to rotate about its own centre then we would have to place the centre of the axes at the centre of the molecule; this can be done by subtracting the coordinates of the centre from each atom, rotating the molecule then replacing these coordinates onto every atom.

Using Maple we define a line (or vector), then draw this as an arrow using the `arrow` instruction `a1:= arrow( base, tip..):` and define a plot object to be drawn later with `display`. When rotating the arrow both base and tip coordinates must be rotated, if you only change one of the pair, the line is only rotated about the other point, and not the origin, and this may not have been your intention.

Algorithm 7.3 Rotating a vector

```
> with(plots):
  Rot:= < < cos(theta)| sin(theta) > ,
          < - sin(theta)|cos(theta) > > ; # rotation matrix
```

$$Rot := \begin{bmatrix} \cos(\theta) & \sin(\theta) \\ -\sin(\theta) & \cos(\theta) \end{bmatrix}$$

```
  tip0:=  < 3.0, 3.0 >:          # define tip of line
  base0:= < 1.0, 1.0 >:          # define base of line
  a1:=arrow( base0,tip0, scaling=constrained, color=red):
  theta:= Pi/3:                  # rotation angle, 60°
  Rot;                           # look at matrix
  tip1: = Rot . tip0;            # rotate tip or arrow
  base1:= Rot . base0;           # rotate base
  a2:=arrow( base1,tip1,scaling=constrained, color=blue):
  display( [a1,a2], view = [0..6,-2..4] ); # plot data
```

$$\begin{bmatrix} \frac{1}{2} & \frac{1}{2}\sqrt{3} \\ -\frac{1}{2}\sqrt{3} & \frac{1}{2} \end{bmatrix}$$

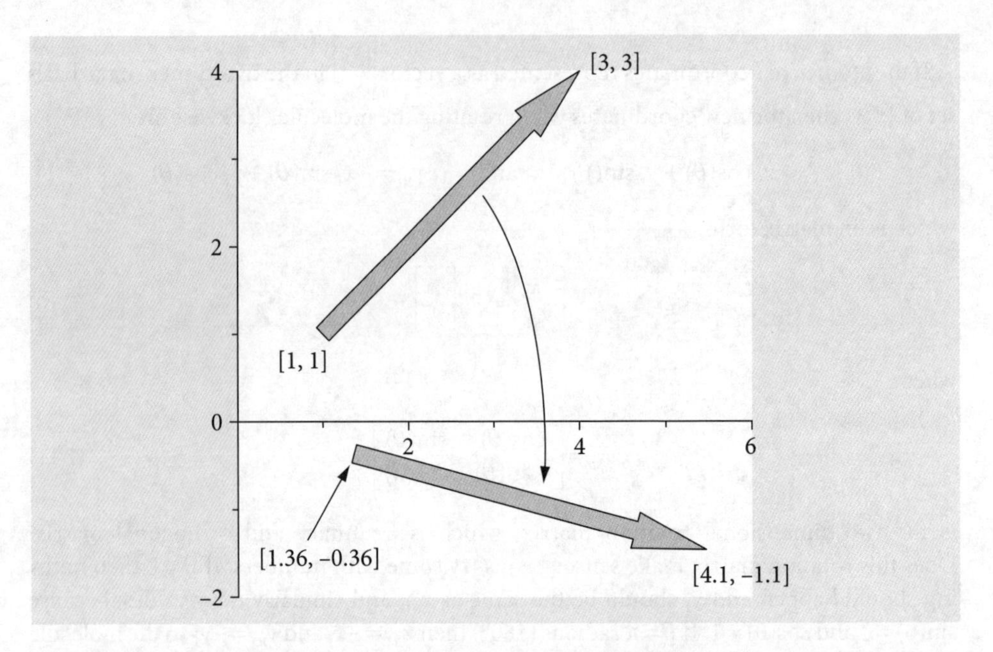

Fig. 7.30 Rotating a line or vector (shown as an arrow) by 60° clockwise about the origin.

To rotate in three directions, three 3 × 3 matrices are needed because rotation about each axis can be performed separately. By multiplication we can combine these into a single matrix, but the elements are complicated and it is easier to use them separately as each has the simple sine/cosine pattern similar to equation (7.11). Next, the axes and rotation angles have to be defined; however, different authors use different labelling systems therefore when doing three-dimensional rotations it is important to notice how the axes are defined. It does not matter which labelling you use as long as you are consistent and know what they are. We use the Euler angles defined in the same manner as Goldstein (1980), the first rotation is anticlockwise looking down the z-axis, the next two rotations are anticlockwise about the *new* axes B then C that are formed as shown in Fig. 7.31. The original axes are indicated in the figure.

The first rotation is around the axis A though angle ϕ. The matrix is

$$A(\phi) = \begin{bmatrix} \cos(\phi) & \sin(\phi) & 0 \\ -\sin(\phi) & \cos(\phi) & 0 \\ 0 & 0 & 1 \end{bmatrix} \tag{7.13}$$

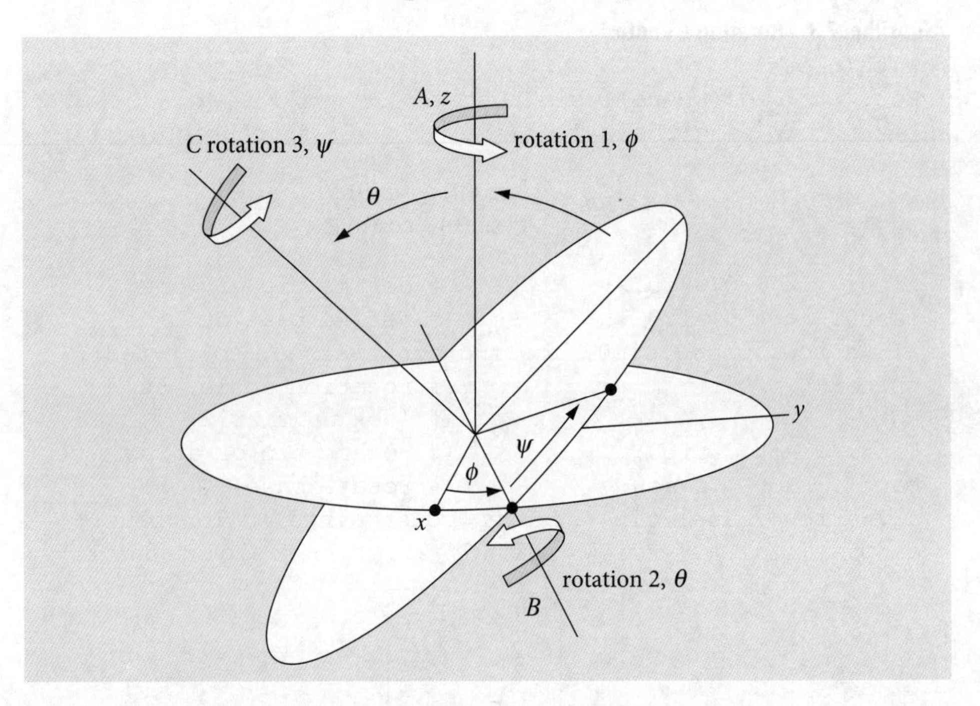

Fig. 7.31 Ordering of rotations. (Fig. based on Goldstein 1980, chapter 4.)

The second is round the axis labelled *B* and is by θ also as shown in Fig. 7.31. The matrix is

$$B(\theta) = \begin{bmatrix} 1 & 0 & 0 \\ 0 & \cos(\theta) & \sin(\theta) \\ 0 & -\sin(\theta) & \cos(\theta) \end{bmatrix} \qquad (7.14)$$

The final rotation by angle ψ is about the axis labelled *C*, giving

$$C(\psi) = \begin{bmatrix} \cos(\psi) & \sin(\psi) & 0 \\ -\sin(\psi) & \cos(\psi) & 0 \\ 0 & 0 & 1 \end{bmatrix}. \qquad (7.15)$$

and the full rotation matrix is the triple multiplication $R = ABC$.

A point $\{x, y, z\}$ whose coordinates are defined by a column matrix after anticlockwise rotation by ψ, then θ, and then ϕ radians is now at

$$\begin{bmatrix} x_R \\ y_R \\ z_R \end{bmatrix} = A(\psi)B(\theta)C(\phi) \begin{bmatrix} x_1 \\ y_1 \\ z_1 \end{bmatrix} \qquad (7.16)$$

and this is just about all you need to know about rotating molecules!

An algorithm in Maple using rotation matrices to rotate a molecules image starts by defining the coordinates, in this example of ethanol. These coordinates are

```
C1:= [ -0.968, -0.008, -0.167]  : #   carbon of CH2OH
Ox:= [ -0.953, 1.395, -0.142 ]  : #   O ATOM
H1:= [ 0.094, -0.344, -0.200 ]  : #   H on C1
C2:= [ -1.683, -0.523, 1.084 ]  : #   Carbon OF CH3,
H2:= [ -1.490, -0.319, -1.102 ] : #   H on C1
H3:= [ -1.842, 1.688, -0.250 ]  : #   H on O atom
H4:= [ -1.698, -1.638, 1.101 ]  : #   H on C2
H5:= [ -1.171, -0.174, 2.011 ]  : #   H on C2
H6:= [ -2.738, -0.167, 1.117 ]  : #   H on C2
```

The input to the procedure must have the three rotation angles and the coordinates of the molecule. A graph of the molecule will be plotted in the usual chemical way after rotation. The labelled structure is shown in Fig. 7.32.

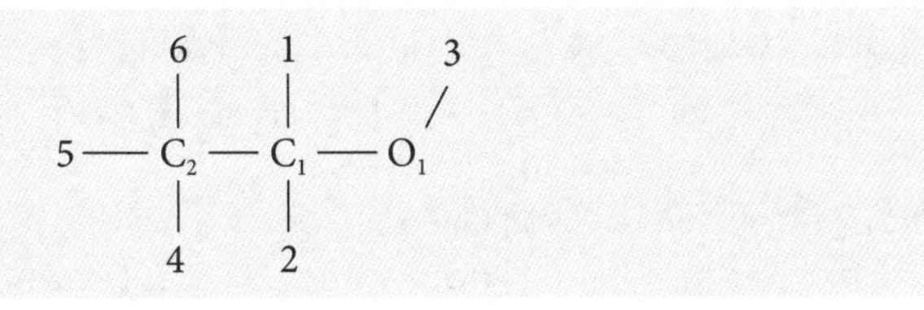

Fig. 7.32 Ethanol, with H atoms replaced by their labels.

If a molecule is plotted only as points showing where the atoms are, it will usually be impossible to decide what its structure is; therefore bonds must be drawn, and this means joining up the atoms in a sensible way with lines. A 'connectivity' list showing which atoms are joined to which has to be made, and this might be done automatically by examining the distances between atoms and deciding which atoms are connected to which depending upon their separation. Ethanol is sufficiently small that the list of atoms and lines between them can be made directly so that the 'pen' is not lifted from the 'paper' when drawing the structure. Clearly, this means going over the same bonds twice and back and forth between some atoms many times. There are several ways of doing this depending on where you start, but the list of atoms below will draw the molecule. Notice how some atoms have to be included more than once.

```
> ethanol:=[H3,Ox,C1,H2,C1,H1,C1,C2,H4,C2,H5,C2,H6];
```

To check, sketch this on paper using this order and keep the pen down.

A Maple procedure is a set of instructions contained within the

```
proc(..) ... end proc;
```

pair of commands and is made to work by calling it by name; for example `rotate_molecule`, which is listed below. (See Appendix 1.)

The parameters used are defined to be valid only within the procedure; this is what the `local` instruction does. The three values of the angles, *a*, *b*, and *c*, in degrees, are passed into the procedure and the name of the list of atoms coordinates (`ethanol`). Next, the angles are converted to radians and called ψ, θ and ϕ. The rotation matrices are calculated and multiplied together to make the matrix `rot`. Next each atom in turn is taken from the list and with a `for..do` loop each set of *x*, *y* and *z* coordinates rotated and put into `new_xyz`. Finally, the result is plotted with `pointplot3d(..)`. To see what has happened the axes are plotted on the graph. If you drag the mouse over the graph with the button down the image will rotate including axes, so these are required to illustrate that the molecule has been rotated. If you want the new coordinates, you can print them from inside the procedure; this is disabled in the listing below.

The 'dot' method of multiplying matrices is used; brackets are added to indicate to us the order of multiplication although they are not necessary, e.g.,

```
rot:= rot_psi . ( rot_theta . rot_phi) ;
```

Algorithm 7.4 Rotating a molecule in three dimensions

```
> restart: with(LinearAlgebra):with(plots):
> # put the list of coordinates given above here.
> # ethanol is the name of atom list to draw molecule
> ethanol:=[H3,Ox,C1,H2,C1,H1,C1,C2,H4,C2,H5,C2,H6];
> pointplot3d( ethanol, symbol = CIRCLE, symbolsize = 20,
  connect = true, color = blue, scaling = constrained,
  thickness = 6, orientation =[65,-100],axes=boxed
  view =[-2..0, -2..2, -2..2], labels =["x", "y", "z"] ):
> ## procedure follows ##
> # input order is psi, theta, phi, name of molecule
> rotate_molecule:= proc(a, b, c, molecule)
    local psi, theta, phi, i, new_xyz, rot, col_vect,
          rot_theta, rot_psi, rot_phi, n;
    n:= nops(molecule):                  # n = number of atoms
    psi:= a*Pi/180.0; theta:= b*Pi/180.0;
    phi:= c*Pi/180.0;                    # make into radians
    # eqns (13) to (15), follow
    rot_psi:= < < cos(psi)|sin(psi)|0 >,
                < sin(psi)|cos(psi)|0 >,
                < 0|0|1>    >;
    rot_theta:= < < 1|0|0>,
                  < 0|cos(theta)|sin(theta) >,
                  < 0|sin(theta)|cos(theta) > >;
    rot_phi:=   < < cos(phi)|sin(phi)|0 >,
                  < -sin(phi)|cos(phi)|0 >,
                  < 0|0|1 >        >;
    rot:= rot_psi . ( rot_theta . rot_phi ) ;
    new_xyz:=[ seq([0,0,0], i = 1..n) ]; # make new coords
    for i from 1 to n do                 # change coords
      col_vect:= rot . convert(molecule[i],Vector) ;
      new_xyz[i]:= Transpose(col_vect); # calc values
      # print( i, new_xyz[i] );          # print coords
    end do;
    pointplot3d( new_xyz, symbol = CIRCLE,
    symbolsize = 20, connect = true, color = blue,
    scaling = constrained, thickness = 6,
    orientation = [65,-100], axes= boxed
    view = [-2..0, -2..2, -2..2],labels=["x", "y", "z"]):
  end proc:                              # end of procedure
> psi:= 0 : theta := 0: Phi : = -40:     # set new angles
> rotate_molecule( psi, theta, phi ,ethanol);# calc & plot
```

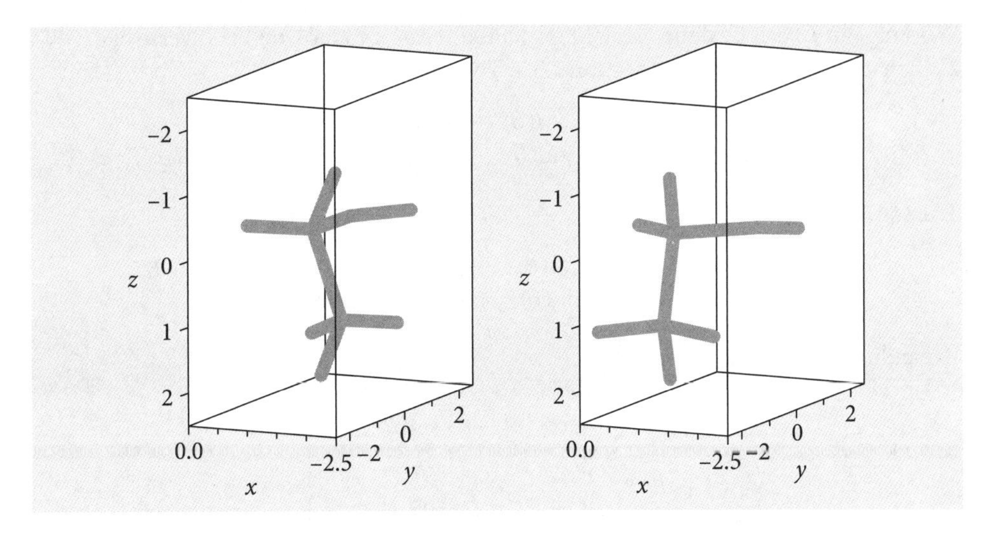

Fig. 7.33 The left structure shows the original orientation {0, 0, 0}, and right after rotation by $\psi = 0$, $\theta = 0$ and $\varphi = -40°$ using the procedure in Algorithm 7.4.

7.8 **Using Jacobians to calculate derivatives in polar coordinates**

Another form of matrix 'rotation' is a coordinate transformation, and to do this the Jacobian, which was introduced in Chapter 4, can be used. In that chapter, it was used to simplify integrations by changing variables and a similar process can be followed when solving differential equations: it is often convenient to change from Cartesian coordinates to polar ones. For example, in two dimensions the Schrödinger equation for a particle on a ring, which is $-\dfrac{\hbar^2}{2m}\left(\dfrac{\partial^2\psi}{\partial x^2}+\dfrac{\partial^2\psi}{\partial y^2}\right)=E\psi$, is difficult to solve using Cartesian coordinates but easier in plane polar coordinates r, θ, see Chapter 10.5.9. Similarly the three-dimensional equation, such as is required to calculate the energy levels and wavefunctions for the H atom, is easier in spherical polar coordinates than in x, y, and z.

In plane polar coordinates the conversion from d^2/dx^2 and d^2/dy^2 to d^2/dr^2 and $d^2/d\theta^2$ can be worked out starting with the conversions $x = r\cos(\theta)$, $y = r\sin(\theta)$ and $r^2 = x^2 + y^2$. However, the calculation of the second derivative is hard; see question Q3.114. Using matrices, the calculation of $\partial/\partial x$ and $\partial/\partial y$ in terms of derivatives of r and θ uses a Jacobian matrix. This first derivative is then used to obtain the second. The method is:

(i) Form the matrix equation

Polar derivatives in (r, θ) = Jacobian × Cartesian derivatives in (x, y).

which can be represented as $\boldsymbol{P} = \boldsymbol{JC}$, where $\boldsymbol{J}$ is the Jacobian matrix, $\boldsymbol{P}$ a column vector of derivatives in polar coordinates and $\boldsymbol{C}$ a column vectors of Cartesian derivatives.

(ii) Invert the Jacobian matrix and left multiply both sides of the equation by $\boldsymbol{J}^{-1}$ to obtain the derivatives in (x, y). This step is

$$\boldsymbol{J}^{-1}\boldsymbol{P} = \boldsymbol{J}^{-1}\boldsymbol{JC} = \boldsymbol{C} \qquad \text{or} \qquad \boldsymbol{C} = \boldsymbol{J}^{-1}\boldsymbol{P},$$

In plane polar coordinates there are only two derivatives, therefore the matrix equation is $\boldsymbol{P} = \boldsymbol{JC}$ and if this acts on a function f then

$$\begin{bmatrix}\partial f/\partial r\\ \partial f/\partial\theta\end{bmatrix}=\begin{bmatrix}\partial x/\partial r & \partial y/\partial r\\ \partial x/\partial\theta & \partial y/\partial\theta\end{bmatrix}\begin{bmatrix}\partial f/\partial x\\ \partial f/\partial y\end{bmatrix}$$

and the Jacobian $\boldsymbol{J}$ is the square matrix. Calculating this gives

$$J=\begin{bmatrix}\cos(\theta) & \sin(\theta)\\ -r\sin(\theta) & r\cos(\theta)\end{bmatrix}.$$

Working out J^{-1} can be done using Maple, but it is easier to look up the inverse of a 2×2 matrix which defined in Section 7.5.7. This gives

$$J^{-1}=\begin{bmatrix}\cos(\theta) & -\sin(\theta)/r\\ \sin(\theta) & \cos(\theta)/r\end{bmatrix},$$

The equation $\boldsymbol{C}=J^{-1}\boldsymbol{P}$ is

$$\begin{bmatrix}\partial f/\partial x\\ \partial f/\partial y\end{bmatrix}=\begin{bmatrix}\cos(\theta) & -\sin(\theta)/r\\ \sin(\theta) & \cos(\theta)/r\end{bmatrix}\begin{bmatrix}\partial f/\partial r\\ \partial f/\partial \theta\end{bmatrix}.$$

Expanding produces

$$\begin{bmatrix}\partial f/\partial x\\ \partial f/\partial y\end{bmatrix}=\begin{bmatrix}\cos(\theta)\dfrac{\partial}{\partial r}f-\sin(\theta)\dfrac{1}{r}\dfrac{\partial}{\partial\theta}f\\ \sin(\theta)\dfrac{\partial}{\partial r}f+\cos(\theta)\dfrac{1}{r}\dfrac{\partial}{\partial\theta}f\end{bmatrix}\tag{7.17}$$

which gives the individual first derivatives as

$$\frac{\partial}{\partial x}f=\cos(\theta)\frac{\partial}{\partial r}f-\frac{\sin(\theta)}{r}\frac{\partial}{\partial\theta}f;\qquad \frac{\partial}{\partial y}f=\sin(\theta)\frac{\partial}{\partial r}f+\frac{\cos(\theta)}{r}\frac{\partial}{\partial\theta}f.$$

If the function f is defined, the gradients of a curve in r and θ can be found in x and y. Suppose, therefore, that the curve is $r=\sin(\theta)$, which is a circle centred at $x=0, y=1$, then the function is $f=r/\sin(\theta)-1$ and the derivatives are found by putting f into equation (7.17) and differentiating. The result, after some simplification, is

$$\frac{\partial}{\partial x}f=\frac{2\cos(\theta)}{\sin(\theta)}\text{ and }\frac{\partial}{\partial y}f=\frac{1-2\cos^2(\theta)}{\sin^2(\theta)}.$$

Returning to the calculation of the second derivatives $\dfrac{\partial^2 f}{\partial x^2}+\dfrac{\partial^2 f}{\partial y^2}$, these are found by taking each derivative to be a function f and substituting into equation (7.17). Thus

$$f_x=\cos(\theta)\frac{\partial}{\partial r}f-\sin(\theta)\frac{1}{r}\frac{\partial}{\partial\theta}f$$

and

$$f_y=\sin(\theta)\frac{\partial}{\partial r}f+\cos(\theta)\frac{1}{r}\frac{\partial}{\partial\theta}f.$$

and substituting each of these into (7.17); $f_x\to f$. For example, the first term of the top row is

$$\cos(\theta)\frac{\partial}{\partial r}\left[\cos(\theta)\frac{\partial}{\partial r}f-\sin(\theta)\frac{1}{r}\frac{\partial}{\partial\theta}f\right]=\cos^2(\theta)\frac{\partial^2}{\partial r^2}f+\frac{\cos(\theta)\sin(\theta)}{r^2}\frac{\partial^2}{\partial r\partial\theta}f.$$

Completing the derivatives, simplifying, and finally summing the result gives

$$\frac{\partial^2}{\partial x^2}+\frac{\partial^2}{\partial y^2}=\frac{\partial^2}{\partial r^2}+\frac{1}{r}\frac{\partial}{\partial r}+\frac{1}{r^2}\frac{\partial^2}{\partial\theta^2},$$

which completes the conversion from Cartesian to plane polar coordinates. The calculation in Maple can be done as follows.

Algorithm 7.5 Derivative calculation using a Jacobian

```
> restart: with(LinearAlgebra):
> fx:=r*cos(theta):
  fy:=r*sin(theta):
   J:= < < diff(fx,r) | diff(fy,r) >,
       < diff(fx,theta)| diff(fy,theta) > >;
```

$$J:=\begin{bmatrix}\cos(\theta) & \sin(\theta)\\ -r\sin(\theta) & r\cos(\theta)\end{bmatrix}$$

```
P:= < diff(f(r,theta),r), diff(f(r,theta),theta) >:
C:=J^(-1).P:
expand( simplify( C[1] ,symbolic));
expand( simplify( C[2] ,symbolic));
```

$$\cos(\theta)\left(\frac{\partial}{\partial r}f(r,\theta)\right)-\frac{\sin(\theta)\left(\frac{\partial}{\partial\theta}f(r,\theta)\right)}{r}$$

$$\sin(\theta)\left(\frac{\partial}{\partial r}f(r,\theta)\right)+\frac{\cos(\theta)\left(\frac{\partial}{\partial\theta}f(r,\theta)\right)}{r}$$

```
fxx:= simplify(
  diff(C[1],r)*cos(theta)-
diff(C[1],theta)/r*sin(theta)):
fyy:= simplify(
diff(C[2],r)*sin(theta)+diff(C[2],theta)/r*cos(theta)):
expand( fxx + fyy );
```

$$\frac{\frac{\partial}{\partial r}f(r,\theta)}{r}+\frac{\frac{\partial^2}{\partial\theta^2}f(r,\theta)}{r^2}+\frac{\partial^2}{\partial r^2}f(r,\theta)$$

7.9 Questions

Full solutions are available at **www.oxfordtextbooks.co.uk/orc/beddard**.

Q7.31 Derive the equations (7.10) and matrix (7.11) for rotating the tip of a vector clockwise using the definitions in Fig. 7.34. The two vectors defined in terms of their polar coordinates, are $\{r, \cos(90-\alpha)\}$ and $\{r, \cos(90-\alpha+\theta)\}$ and their Cartesian equivalents, V_1 and V_2.

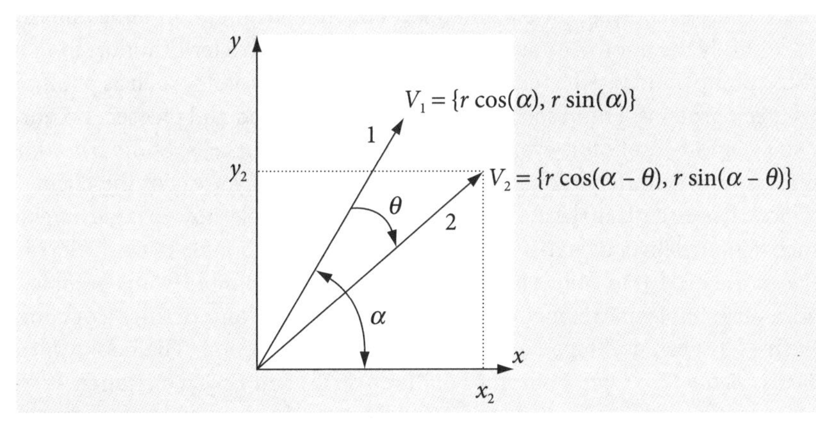

Fig. 7.34 Rotating vectors.

Strategy: Define the x and y coordinates of both vectors in terms of sines and cosines. Expand the double-angle sine and cosine produced using the trig formulas.

Q7.32 Show that by rotating a molecule by θ using the rotation matrix $R(\theta)$, equation (7.11), and then again by θ, the result is the same as the operation $R(\theta)^2$.

Strategy: The total angle moved is 2θ, so we need to show that the rotation matrix R with angle 2θ is the same as R^2 with angle θ so we are likely to need the double angle trig formulas which are $\sin(2\theta) = 2\sin(\theta)\cos(\theta)$ and $\cos(2\theta) = \cos^2(\theta) - \sin^2(\theta)$.

Q7.33 Draw and rotate part of a protein structure you choose; this could be based on algorithm 7.4. Use part of the peptide chain from a .pdb file taken from the Brookhaven database. The size is not so important and will be limited by how easily you can make a connectivity list. Alternatively, you could try to generate this list by measuring the distance between atoms using known bond lengths. These can also be obtained from the coordinates in the .pdb file.

Q7.34 The molecule CCl_4 has tetrahedral symmetry where the bond angles are each 109.7°. The atoms labelled 1 and 2 are in the z–y plane and the other two in the z–x plane.

(a) Using a matrix representation for each symmetry operation, show that a rotation by 90° about the z-axis, then inversion followed by an S_4 rotation–reflection operation, leaves the molecule in an indistinguishable state.

(b) Does operating in the reverse order produce the same result?

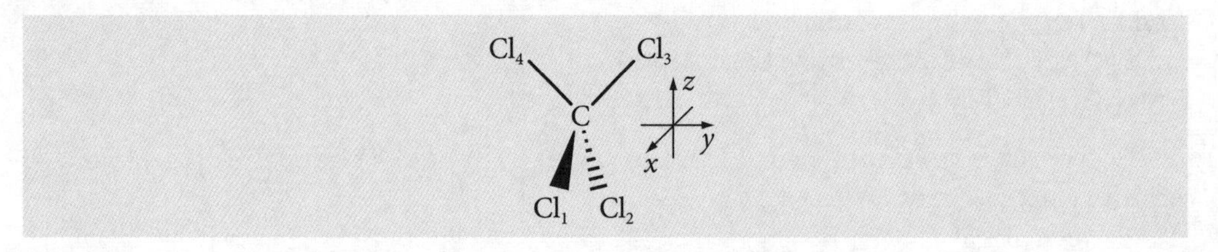

Fig. 7.35 CCl_4.

Strategy: If the angle rotated about the z-axis is θ the rotation matrix when $\theta = 90°$ contains elements of 0 and ±1 only if the other rotation angles are zero. Inversion inverts all the atoms coordinates.

7.10 Matrices in optics and lasers

Chemists are among the largest group of laser users and it is not often appreciated that, nowadays, not only is nearly all spectroscopy performed with lasers, but also that most chemical kinetics measurements use lasers to initiate and follow reactions. Lasers are also widely used to produce fluorescence from dye molecules bound to DNA or proteins and to generate intrinsic fluorescence or Raman scattering from tryptophan and tyrosine residues in proteins. Ultra-short femtosecond duration lasers are used to probe the excited state reactions of the chromophores in photosynthetic antennas and reaction centres, to study the photo-isomerization in vision and in bacteriorhodopsin (a light-driven proton pump), and in other chromophore containing proteins such as photoactive yellow and green proteins. Fluorescence excited by femtosecond and picosecond duration lasers in single and two photon events is used to image living cells. Some basic knowledge of optics is now definitely required to do research in many areas of the chemical and biosciences. It is essential to be able to understand for example, how to build a stable laser, how much light to deliver as watts/cm^2 into a sample, and to be able to work out what lenses or mirrors are needed to collect as much fluorescence as possible from a sample. And to this end, a very simple matrix method of analysing the behaviour of simple or complex optical systems has been developed and is described in this section. The books, *Introduction to Matrix Methods in Optics* (Gerrard & Burch 1975) and *Lasers* (Siegman 1986) provide a detailed description of these methods and the latter book considerably more than this.

The matrix method is based on following a two-component vector through an optical system. The first element of this vector represents the height of a light ray above the (optic) axis and the other element is the angle the ray makes to this axis. The optic axis is the direction the ray takes through the centre of all the optical elements. The angle the ray makes must be small enough, < 6°, to allow the approximations $\tan(\theta) \approx \theta$ and $\sin(\theta) \approx \theta$ to be true, and this is good enough for most applications. This is known as the *paraxial* approximation. The basis set for the vector is $\{y, \theta\}$. The calculation is always contained between pairs of reference planes and you have to decide where to put these based on the problem at hand.

The input ray enters at a height y_1 above the optic axis at reference plane 1 and moves at angle θ_1 to the optic axis as shown in Fig. 7.36. Using the paraxial approximation, the

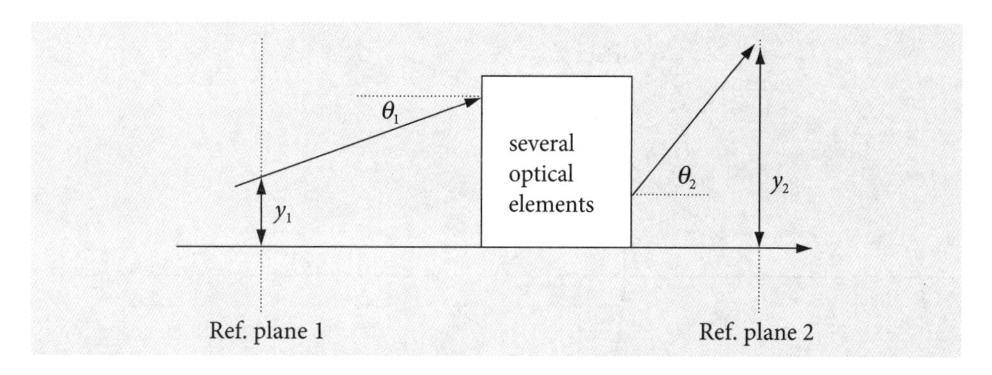

Fig. 7.36 Definitions of input and output rays and reference planes.

output ray at reference plane 2 having passed through one or more optical elements is related by a linear transformation, hence a matrix, to the input ray at the input reference plane as

$$\begin{bmatrix} y_2 \\ \theta_2 \end{bmatrix} = \begin{bmatrix} A & B \\ C & D \end{bmatrix} \begin{bmatrix} y_1 \\ \theta_1 \end{bmatrix} \tag{7.18}$$

where the ***ABCD*** matrix is defined always to have a unit determinant

$$\begin{vmatrix} A & B \\ C & D \end{vmatrix} = AD - BC = 1. \tag{7.19}$$

The values of the elements A, B, C, and D determine the optical properties of a single element, such as a thin lens, an air gap, or a curved surface. Although two reference planes determine where to start and end the calculation, when several optical elements are stacked together, additional reference planes can be added to determine where one ***ABCD*** matrix starts and the other finishes. When many elements are present, the ***ABCD*** matrices are multiplied together to produce a final 2×2 matrix whose *ABCD* values are those of the whole optical assembly. The form of the individual ***ABCD*** matrix is tabulated for various individual optical elements; see Gerrard & Burch (1975) or Siegman (1986). The *ABCD* matrices for three optical elements are:

$$\underset{\text{thin lens}}{\begin{bmatrix} 1 & 0 \\ -1/f & 1 \end{bmatrix}} \qquad \underset{\text{gap } d}{\begin{bmatrix} 1 & d/n \\ 0 & 1 \end{bmatrix}} \qquad \underset{+ve\ mirror}{\begin{bmatrix} 1 & 0 \\ -2n/r & 1 \end{bmatrix}} \tag{7.20}$$

The gap is just that: an air gap or a piece of glass of length d with refractive index n. The refractive index of air is practically 1, and can be taken to be 1 for the purpose of almost all calculations. The refractive index of different glasses depends on their composition but ≈ 1.52 is a typical value for crown glass. Quartz has a smaller value of ≈ 1.48 and glasses containing rare earth elements, values of ≈ 1.8. Water has a refractive index of 1.33 at room temperature, and liquids with more polarizable atoms such as CCl_4 have higher values ≈ 1.6. The reflection matrix shown is for a positive, concave, or focusing mirror with radius of curvature $-r$. A positive radius of curvature $+r$, by convention, applies to a negative mirror, which is defined when the ray approaches from the left and meets a convex mirror surface. This is the convention shown in Fig. 7.37, where the surface is convex to the incoming ray starting at point Q. The radius $+r$ is shown as the line SR, see also Fig. 7.43. The focal length f of a mirror is half the radius of curvature and it is often easier to use this rather than r.

The derivation of each ***ABCD*** matrix is not difficult, but it is unnecessary to know how this is done for every optical element. Therefore the principle is illustrated by calculating the matrix for a single lens surface, as shown in Fig. 7.37. For each optical element, such as a surface or air gap, one reference plane at some position on the x-axis is chosen at which the input beam must be defined, and a second reference plane positioned where the properties of the output beam are required. In the example below, these two planes, rp_1, rp_2 are coincident at the lens surface.

We aim to find equations describing y_2 and angle θ_2 in terms of y_1 and θ_1 and do so with a reference plane where the beam intersects the lens and follow the derivation in Gerrard

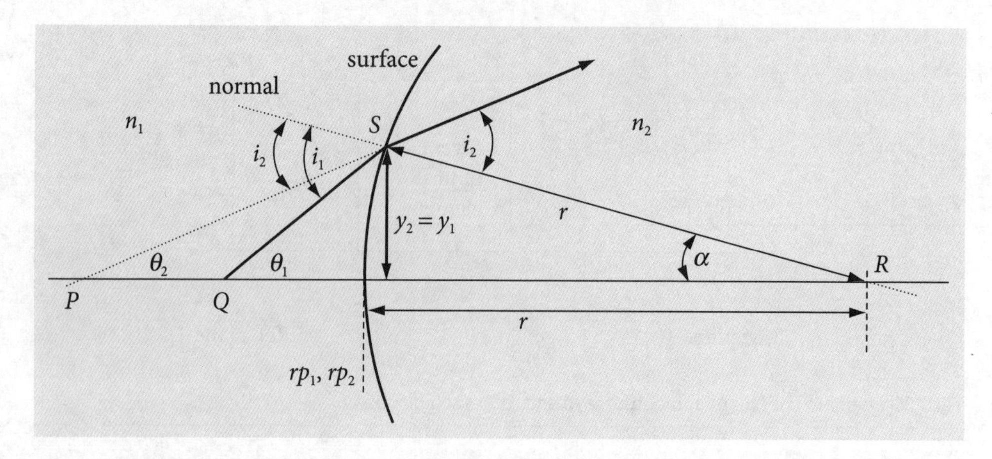

Fig. 7.37 Ray diagram showing refraction at a lens surface with a radius r made of glass with refractive index n_2; the medium outside the lens, say air, has a refractive index n_1. The angles and surface curvature are greatly exaggerated in the diagram. (Based on a figure in Gerrard & Burch 1975.)

& Burch (1975). By Snell's law of refraction $n_1\sin(i_1) = n_2\sin(i_2)$, where i_1 and i_2 are, respectively, the angles of incidence and refraction measured from the normal to the surface. At small angles, $n_1 i_1 = n_2 i_2$. The refractive indices are n_1 outside the lens, and n_2 for the lens' glass. The angles of a triangle add up to 180°; therefore, by the external angle theorem on triangle QRS,

$$i_1 = \theta_1 + \alpha = \theta_1 + y_1/r$$

and on triangle PRS,

$$i_2 = \theta_2 + \alpha = \theta_2 + y_1/r.$$

Combining equations gives the equation for y_1,

$$n_1\theta_1 + n_1 y_1/r = n_2\theta_2 + n_2 y_1/r \qquad \text{or} \qquad y_1\frac{(n_1 - n_2)}{r} = n_2\theta_2 - n_1\theta_1$$

and using the fact that at the surface the equation for y_2 is $y_2 = y_1$ the matrix equation for the surface is

$$\begin{bmatrix} y_2 \\ n_2\theta_2 \end{bmatrix} = \begin{bmatrix} 1 & 0 \\ -\left(\dfrac{n_2 - n_1}{r}\right) & 1 \end{bmatrix} \begin{bmatrix} y_1 \\ n_1\theta_1 \end{bmatrix} \tag{7.21}$$

Multiplying the angle by the refractive index $n\theta$ in the vector makes all matrices unitary; often the notation $V = n\theta$ is used and this is done because, by Snell's law, V remains unchanged as the ray crosses a boundary from one medium to another. The ***ABCD*** matrix for a surface of radius r is

$$\begin{bmatrix} 1 & 0 \\ -\left(\dfrac{n_2 - n_1}{r}\right) & 1 \end{bmatrix}$$

with refractive index n_1 on the left of the surface and n_2 on the right.

If there are several optical elements, the effect of a ray passing through several of them is calculated by multiplying their ***ABCD*** matrices in order, as shown in Fig. 7.38. The reference planes are placed around each optical element but the first and last, which are often gaps, are determined by where the object (ray input) distance and image (ray output) distances are required to be. If a thick lens is present, three thin lens matrices would have to be used, which are those of two surfaces separated by a gap containing the glass. The multiplication order is always that in which the light passes through the elements; for example if M_1 is the matrix next to the input reference plane (plane 1), it is the right-hand matrix in the multiplication; similarly, the output matrix is always placed on the left-hand side. In Fig. 7.38 the right-hand matrix is an air gap and the properties of the beam are those that exist at the reference plane, in this case some distance from the last lens. This might be the focal point of the lens arrangement if the light entering the first lens is parallel.

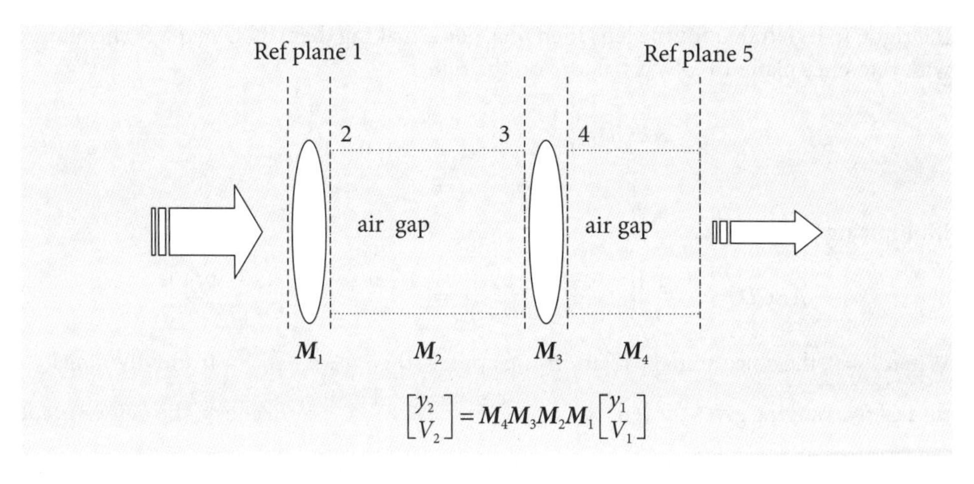

Fig. 7.38 Matrices for lens—gap—lens—gap arrangement.

The different properties of the *ABCD* elements, determine what optical properties we observe. This is explained in the next section.

7.10.1 Understanding the ABCD matrix

The elements of the *ABCD* matrix have the following meaning when each in turn is made equal to zero. The other elements are unchanged.

(i) Making $D = 0$ means that the first (left-most) reference plane must be at the focal point so that rays leave the optical elements parallel to one another.

(ii) Making $A = 0$ puts the final reference plane at the focal point so the rays enter the optical arrangement parallel to one another.

(iii) If $B = 0$, this gives the *object–image* relationship, Fig. 7.39. The object is magnified by amount A and brought to a focus at the final reference plane. The number $1/D$ is also the magnification.

(iv) Making $C = 0$ ensures that both input and output rays are parallel but with angular magnification D. This is the telescopic relationship and the system is called *afocal*.

As an example of using the matrix methods, the lens maker's formula $\dfrac{1}{v} + \dfrac{1}{u} = \dfrac{1}{f}$ will be derived, where u is the object distance from a lens, v the image distance, and f the focal length. The ***ABCD*** matrix for a thin lens of focal length f is $\begin{bmatrix} 1 & 0 \\ -1/f & 1 \end{bmatrix}$ and for an air gap of length d is $\begin{bmatrix} 1 & d \\ 0 & 1 \end{bmatrix}$. The total ***ABCD*** matrix is calculated and by making element $B = 0$ produces the thin lens (or lens maker's) formula. A diagram of the arrangement is shown below.

The matrices are written down in the order of rays passing through the elements and are in the reverse order to that drawn on the diagram; place the matrix of the first gap on

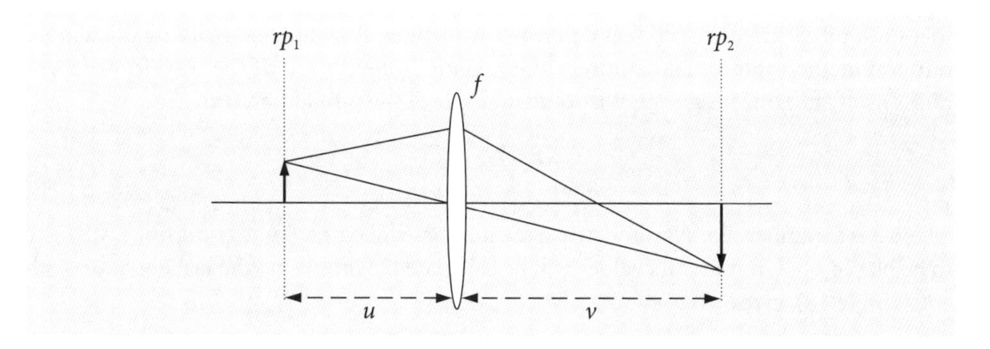

Fig. 7.39 Arrangement for calculating object–image relationship for a thin lens.

the right, to its left that for the lens then final gap v making the ABCD matrix. The matrix with reference plane 1 is always placed on the right.

$$ABCD = \underset{\text{gap after lens}}{\begin{bmatrix} 1 & v \\ 0 & 1 \end{bmatrix}} \underset{\text{lens}}{\begin{bmatrix} 1 & 0 \\ -1/f & 1 \end{bmatrix}} \underset{\text{gap before lens}}{\begin{bmatrix} 1 & u \\ 0 & 1 \end{bmatrix}}.$$

Multiplying by hand gives

$$ABCD = \begin{bmatrix} 1 & v \\ 0 & 1 \end{bmatrix} \begin{bmatrix} 1 & u \\ -1/f & -u/f+1 \end{bmatrix} = \begin{bmatrix} 1-v/f & u+v(1-u/f) \\ -1/f & 1-u/f \end{bmatrix}.$$

When $B = 0$, the object–image relationship applies, then $u + v - uv/f = 0$ and dividing by uv and rearranging gives the lens maker's formula $\dfrac{1}{v} + \dfrac{1}{u} = \dfrac{1}{f}$. As a check, the determinant should be 1. The calculation is $(1 - u/f)(1 - v/f) + uf + vf(1 - uf) = 1$.

7.10.2 **The laser cavity**

A laser cavity consists of two or more mirrors with a gain medium sandwiched between them, Fig. 7.40. Usually one of the end mirrors is 100% reflecting and the other mirror, the output coupler, is partially transmitting. This is typically in the range of 1 to 10% transmitting depending on the type of laser. The gain medium can be a gas, a liquid containing a highly fluorescent dye, such as Rhodamine-6, or a solid, usually a glass doped with a small percentage of luminescent ions, such as Nd^{3+} or Ti^{+}. The gain medium is excited either electrically, forming a gas discharge, or from another laser or flash lamp. Most of the fluorescence is lost because it is emitted over all angles, but the laser's cavity mirrors capture some of it, which is then fed back into the gain medium where amplification caused by stimulated emission occurs. For the laser to work, the photons have to pass back and forth in the cavity indefinitely and can only do so if they follow the same path. This is equivalent to saying that the wave front reproduces itself at the mirrors after each round trip. The wavefront is the shape of the wave inside the cavity and it has the same curvature as that of the mirror as it reaches it because light must be reflected normally off any mirror if it is to return along the same path. The curvature changes inside the cavity to satisfy this normality condition at the end mirrors. As the beam waist is approached, the curved wavefront becomes a plane wave and then changes to the opposite curvature as it moves towards the other mirror. When the laser beam leaves the cavity, its longitudinal profile is that determined by the cavity and it continuously diverges. The transverse profile of the laser intensity is ideally Gaussian, although lasers with other profiles can be made.

If the cavity is represented by an ***ABCD*** matrix ***M***, and we will define exactly what this will be shortly, then, as the input and output rays must reproduce one another, we can suppose that the input and output can only be related by some constant factor. Furthermore, this value must be the same for one round trip as for N round trips of the cavity. For one round trip, from equation (7.18), $\begin{bmatrix} y_2 \\ \theta_2 \end{bmatrix} = M \begin{bmatrix} y_1 \\ \theta_1 \end{bmatrix}$. It is also true that for N round trips

$$\begin{bmatrix} y_N \\ \theta_N \end{bmatrix} = M^N \begin{bmatrix} y_1 \\ \theta_1 \end{bmatrix}, \tag{7.22}$$

therefore some way of finding the Nth power of the matrix has to be used. The method is to use a similarity matrix, which is described in Section 7.13.4 but before doing this a more physical method rather than a purely mathematical one is used. If the wavefront reproduces itself after one pass, then it is some multiple of the input, therefore,

$$\begin{bmatrix} y_2 \\ \theta_2 \end{bmatrix} = \lambda \begin{bmatrix} y_1 \\ \theta_1 \end{bmatrix}$$

where λ is a number. For N round-trips, the number will be λ^N. Some conditions will have to be placed on λ, but first its values will be calculated. Combining the last equation with equation (6.18), gives

$$M \begin{bmatrix} y_1 \\ \theta_1 \end{bmatrix} = \lambda \begin{bmatrix} y_1 \\ \theta_1 \end{bmatrix}$$

and this has the form of an eigenvalue–eigenvector equation see 7.12.13. Rearranging gives

$$(M-\lambda I)\begin{bmatrix} y_1 \\ \theta_1 \end{bmatrix}=0$$

where I is the 2×2 unit or identity matrix. To find λ the determinant has to be solved;

$$|M-\lambda I|=0 \qquad \text{or} \qquad \begin{vmatrix} A-\lambda & B \\ C & D-\lambda \end{vmatrix}=0$$

producing the characteristic equation $(A - \lambda)(D - \lambda) - BC = 0$ or $\lambda^2 - (A + D)\lambda + 1 = 0$ because

$$AD-BC=1$$

giving

$$\lambda=\frac{A+D}{2}\pm\frac{1}{2}\sqrt{(A+D)^2-4}. \tag{7.23}$$

Because the wavefronts must replicate themselves, λ cannot be a real number unless it is 1, which is a trivial result; therefore it follows that λ is purely imaginary. In this case $(A + D)^2 < 4$, which means that

$$\left|\frac{A+D}{2}\right|<1. \tag{7.24}$$

Now this is surprising; we do not need to know the values of λ because only equation (7.24) needs to be obeyed. If it is obeyed, the wavefronts in the cavity replicate themselves and the cavity is stable; if not it is unstable. To design a laser cavity, the first step is to calculate the values of A and D that make a cavity stable with the mirror radius of curvature that have been chosen and to plot a graph of the stable region. The next step is therefore to work out an ***ABCD*** matrix for a cavity.

The simplest cavity has two mirrors, as shown in Figs 7.40 and 7.41. The cavity is split into several regions by defining a reference plane at each mirror, at the edges of each gap

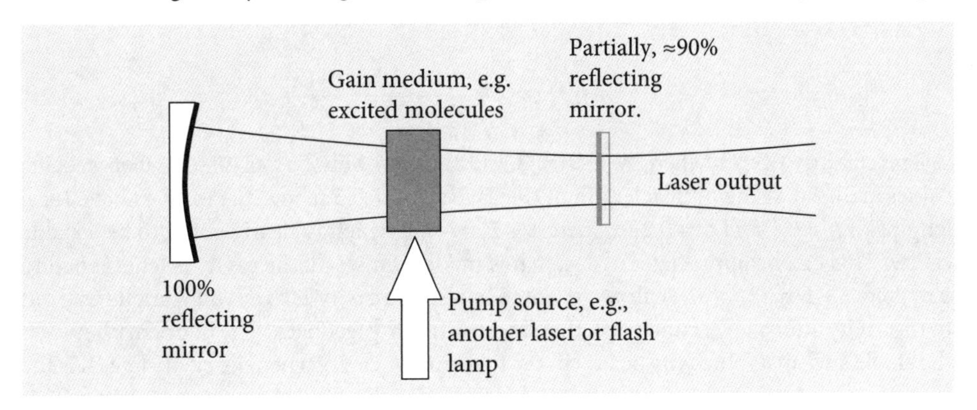

Fig. 7.40 Basic laser cavity.

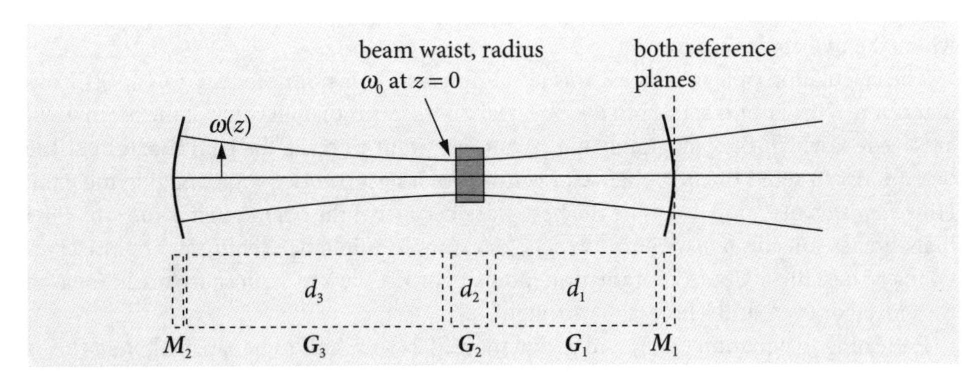

Fig. 7.41 ***ABCD*** matrices for a two-mirror cavity.

and in the gain medium, which is assumed to have flat surfaces, and can be treated as a gap with a different refractive index. The matrices must allow the photons to travel from the edge of the output coupler to the end mirror and back again, but note that the output mirror is only counted once. The scheme is shown in Fig. 7.41.

The *ABCD* matrix *M* is made by following a ray from the output mirror and back again; the sequence of matrices is

$$M = G_1G_2G_3M_2G_3G_2G_1M_1.$$

Notice that mirror 1 is on the right and that each mirror is included only once. The gaps are indicated by the *G* matrices. G_2 is the gap produced by the gain medium. To make the calculation simpler, the three gaps can be combined into one. If their optical lengths are d_1, d_2/n_2, and d_3, the total optical length, assuming the refractive index of air is 1, is $L = d_1 + d_2/n_2 + d_3$ and the combined gap matrix $\begin{bmatrix} 1 & L \\ 0 & 1 \end{bmatrix}$. This is the same as multiplying together the three gap matrices. The matrices for a gap also commute so that $G_1G_2G_3 = G_3G_2G_1$. The overall matrix for the cavity is, therefore,

$$M = \begin{bmatrix} 1 & L \\ 0 & 1 \end{bmatrix}\begin{bmatrix} 1 & 0 \\ -1/f_2 & 1 \end{bmatrix}\begin{bmatrix} 1 & L \\ 0 & 1 \end{bmatrix}\begin{bmatrix} 1 & 0 \\ -1/f_1 & 1 \end{bmatrix}$$

where the focal length of the output coupler mirror is f_1 and that of the high reflector f_2. This can easily be multiplied out by hand, but we will use Maple.

```
> with(LinearAlgebra):
> G:=  < <1|L>,<0|1> >;            # gap
  M1:= < <1|0>,<-1/f1|1> >;        # output coupler
  M2:= < <1|0>,<-1/f2|1> >;        # high reflector
  M:= simplify( G.M2.G.M1 );
```

$$M := \begin{bmatrix} \dfrac{f2\,f1 - L\,f1 - 2\,L\,f2 + L^2}{f2\,f1} & -\dfrac{L(-2\,f2 + L)}{f2} \\ \dfrac{-f1 - f2 + L}{f2\,f1} & -\dfrac{-f2 + L}{f2} \end{bmatrix}$$

```
> 'A+D':= expand( M[1,1] + M[2,2] );
```

$$A + D := 2 - \frac{2\,L}{f2} - \frac{2\,L}{f1} + \frac{L^2}{f2\,f1} \tag{7.25}$$

The stability occurs when $|A + D| < 2$ and this can only be calculated when specific values are used. If the optical length of the cavity is $L = 0.5$ m and the mirrors have focal lengths of $f_1 = 0.1$ and $f_2 = 0.2$ m, then $A + D = -0.5$, which is clearly going to be a stable cavity. This can be appreciated if a sketch is made. If instead, the cavity is lengthened to 1 m then $A + D = 22$, which will prove not to be a stable cavity as a ray will work its way out of the cavity after a few round trips, thereby limiting any feedback. The boundary between the stable and unstable cavity occurs in the two-mirror cavity when the cavity length is the sum of the mirrors' radii of curvature, which is twice the sum of their focal lengths. This can be seen from equation (7.25). A quick way to see if a cavity is stable and to work out where the beam waist is going to be is to draw two circles each with each mirror's radius of curvature and with the centres of the circles separated by $L - r_1 - r_2$. The beam waist is where the two circles cross.

The calculation of a value for λ was left unfinished; it is not necessary to know this to determine if the laser is stable but it is necessary if the position and value of the beam waist are needed, which they invariably are. We might want to place the gain medium at the beam waist to stand the best chance of getting the laser to work by maximizing the gain. However, this may not always be the best place because if the beam waist is too small the high intensity inside gain medium, which, for example, might be a titanium sapphire crystal may cause this to 'burn'. A radiation intensity of $>10^{10}$ w cm^{-2} can cause breakdown in the crystal or even drill a hole right through it.

Returning to equation (7.23) and as equation (7.24) also has to be satisfied, then this is possible if the two eigenvalues λ are

$$\lambda_{1,2} = e^{\pm i\theta} \tag{7.26}$$

where θ is an angle such that $\cos(\theta) = (A + D)/2$ will satisfy the eigenvalues λ. From a study of the Gaussian beam properties of laser cavities (Gerrard & Burch 1975), the following properties are obtained:

Radius of curvature of laser beam at reference plane	$R = \dfrac{2B}{D-A}$
Beam radius at reference plane, (λ is wavelength)	$\omega = \sqrt{\pm\lambda B/\pi\sin(\theta)}$
Location of neck to left of reference plane	$z = (A - D)/2C$
Beam waist at neck	$\omega_0 = \sqrt{\pm\lambda\sin(\theta)/\pi C}$
Confocal beam length	$z_0 = \pi\omega_0^2/\lambda$

The last parameter, the confocal beam length, is the length of the region over which the laser is focused in the sense that it is the length either side of the beam waist where the beam increases by $\sqrt{2}$. The beam waist at z, a position either side of the focus, is given by

$$\omega_z^2 = \omega_0^2\left(1 + \left[\frac{\lambda z}{\pi\omega_0^2}\right]^2\right).$$

In calculating the beam radii ω_z and ω_0 the $\pm$ sign is chosen to ensure that the number is real. When cavities that are more complicated are used, it is necessary to use the complex beam parameter to determine the properties of the cavity. This is explained in specialized texts (Siegman (1986); Gerrard & Burch (1975); Svelto (1982); Yariv (1975)).

7.10.3 Questions

Full solutions are available at **www.oxfordtextbooks.co.uk/orc/beddard**.

Q7.35 Two thin lenses, $f = 10$ cm and $f = -20$ cm are positioned in air as shown in the diagram. If an object is placed at 0.25 m from the first (positive) lens and the lenses are separated by 0.05 m, where is the image to be found, and what is its magnification?

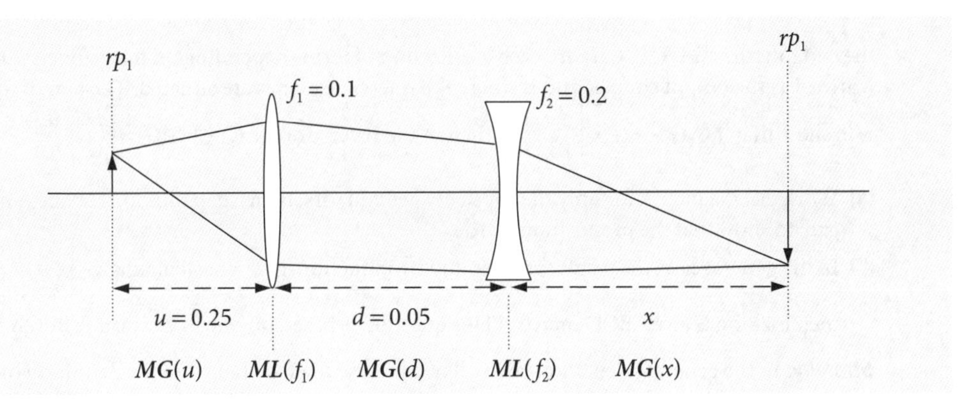

Fig. 7.42 Two thin lenses producing an object–image relationship.

Strategy: Label the known distances as u and d, and focal lengths as f_1 and f_2, and substitute the numbers into the matrices at the end of the calculation. To find the image distance x use the object–image relationship of the $ABCD$ matrix which is $B = 0$. Take care to write matrices down in the correct order, the reverse of that on the diagram. Use Maple to avoid matrix multiplication errors.

Q7.36 A thick lens is to be used to focus a laser beam. For the best focusing with minimal distortion the most curved surface is presented to the laser. The lens thickness is $d_L = 6$ mm and it is made of quartz with a refractive index of 1.48 at the laser wavelength and has surfaces with radii of curvature of 100 mm and 300 mm. Where is the focal point of the lens if the input beam is effectively parallel?

The lens is shown below with surfaces exaggerated very greatly. Use the matrix in equation (7.21) for the lens surfaces between the reference planes shown. Use Maple to perform the matrix multiplications.

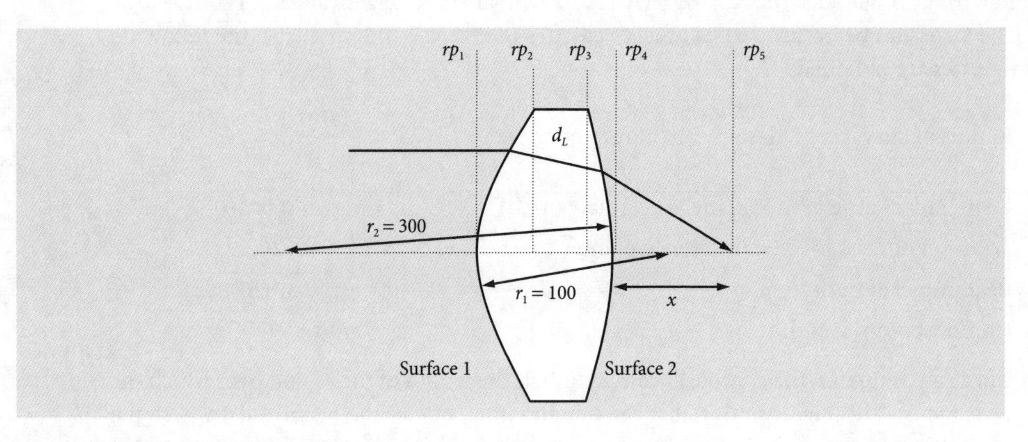

Fig. 7.43 Geometry for a thick lens.

Strategy: The matrices are set up as before between the reference planes. The radius of the second surface r_2 is made negative because it is concave towards the incoming photons. As the input beam is parallel to the optical axis, make the element $A = 0$ in the ***ABCD*** matrix to calculate the focal point. The first reference plane is placed at the lens surface, because the input is parallel, i.e. the object distance is at infinity and it would not make sense to have an infinite term in a matrix. Recall that the gap matrices have to have their distances divided by the refractive index. The gap from the lens (rp_4) to the focal point at reference plane 5, (rp_5) is x.

Q7.37 A laser operates at 600 nm. Using the values $L = 0.5$ m, $f_1 = 0.1$, and $f_2 = 0.2$ m, calculate the position of the beam waist its value and that of the confocal parameter. Plot a graph of the beam profile between the mirrors making the beam waist at $z = 0$. The beam waist will need to be multiplied by 100 to fit on the same scale as the mirrors.

Q7.38 A laser cavity can contain several mirrors; a common design for a continuously pumped dye laser has three mirrors, a cavity dumped dye laser four; see Fig. 7.44. In such cavities, the sequence of matrices and gaps can become very long. The general equation is

$$M = G_1G_2G_3 \cdots G_nM_2G_n \cdots G_3G_2G_1M_1.$$

The whole sequence can also be written as $M = \begin{bmatrix} A & B \\ C & D \end{bmatrix} M_2 \begin{bmatrix} D & B \\ C & A \end{bmatrix} M_1$ where $M_{1,2}$ are the matrices for the end mirrors and A, B, C, D are each a collection of terms depending on the geometry of the cavity. Mirror 1 is the output coupler and is where the reference planes are situated. The aim of this question is to show that if $G_1G_2 \cdots G_n = \begin{bmatrix} A & B \\ C & D \end{bmatrix}$, then in the reverse order $G_n \cdots G_2G_1 = \begin{bmatrix} D & B \\ C & A \end{bmatrix}$.

(a) Work out the matrix for only the gaps between mirrors, then a gap plus mirror and a mirror plus gap, to show that the proposition is true.

(b) In the general case, the result can be found by induction, by first calculating $G_n \begin{bmatrix} -1 & 0 \\ 1 & 0 \end{bmatrix} G_n$ and then making G any ABCD matrix. (This question is based on one in Gerrard & Burch 1975.)

Strategy: In the general calculation you will need to use the fact that from the definition of the matrices for any *individual* gap or mirror matrix $A = D = 1$, and also for any individual or product of matrices $AD - BC = 1$.

Q7.39 Show that the equation developed in the previous question applies to the cavity shown in the sketch; Fig. 7.44. The jet of dye in ethylene glycol, which is the gain medium, could be at the focus between mirrors 1 and 2 and the cavity dumper, a crystal with an attached piezoelectric element, at the focus of mirrors 3 and 4. The cavity dumper crystal has an acoustic field produced in it via an RF source applied to the piezo element and this diffracts, or 'dumps', the light out of the cavity.

Mirrors M_1 and M_2 have focal lengths 60 and 100 cm and 3 and 4 lengths of 30 and 100. Gap 1 is 50, gap 2 is 200 and gap 3 is 35 cm. Is this cavity stable? Use Maple to do the calculation.

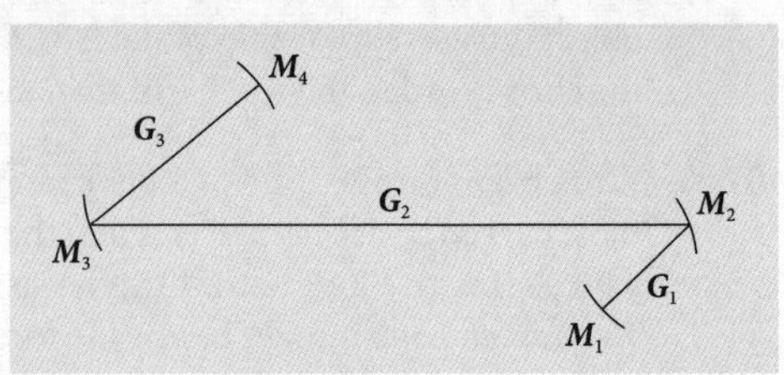

Fig. 7.44 Four mirror cavity.

7.11 Polarizing optics

Using the polarization properties of light can lead to a very sensitive method of detecting spectroscopic features. If light is linearly polarized, its electric field oscillates in one plane only, which is perpendicular to the direction the light travels in. This is shown in Fig. 7.45, where the polarization is in the x–z plane, and this plane is defined as 'vertical'. Horizontally polarized light would have its electric field in the y–z plane. Light can be linearly polarized at any other angle, perhaps 45°, in which case it would have equal components of the horizontal and vertical. Any amplitude of vertical and horizontal linear polarizations can be added vectorially to make linear polarization at any other angle. In this sense, vertical polarization is the addition of vertical with a zero horizontal component and vice versa.

In the most general case, components in the y- and x-directions have a phase difference between them. When the phase lag is zero, linear polarization results; when ±90°, right and left circular polarization results and the electric field rotates about the direction of travel, see Fig. 2.7. Linear polarized light is a special case of elliptical polarization. The description of polarisation states is quite technical and elliptical polarized light somewhat difficult to visualize properly; however, by accepting that such polarization exists, the matrix methods to calculate the complicated changes caused by wave-plates and polarizers are surprisingly simple.

In a typical experiment, a sample is placed between two crossed polarizers the second of which is adjusted usually to 90° compared to the first to extinguish any light passing through the first polarizer and the sample. If a small perturbation is now made to the sample, such as exciting it with a second laser or applying an electric or magnetic field, then its optical properties can be changed. Some light now leaks through the second polarizer and is detected; the amount of light is related to the perturbation and hence can convey some information about the molecules in the sample. The scheme is shown in Fig. 7.46. This is a zero-background type of experiment, fluorescence is another, and

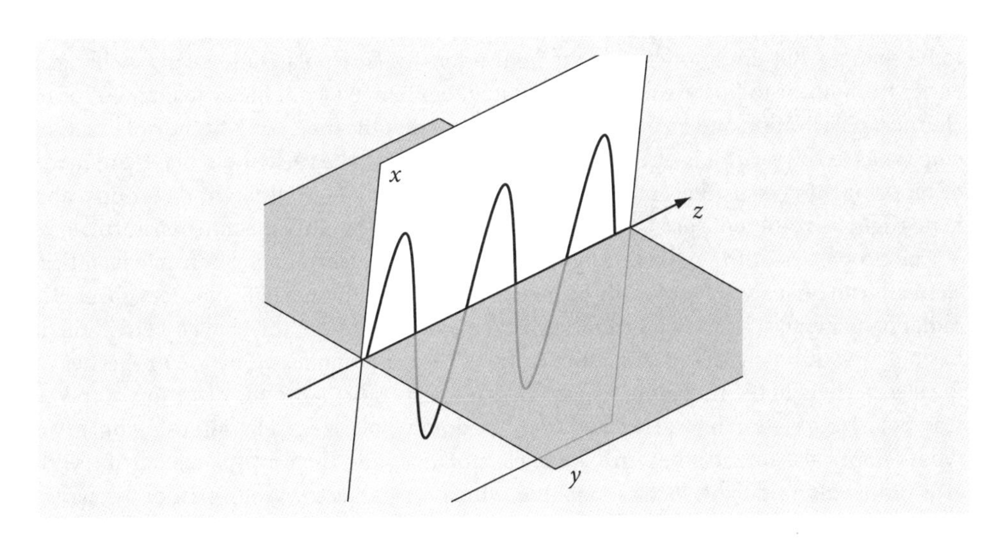

Fig. 7.45 Linear, vertically polarized light. Vertical polarized light has its electric field oscillating in the x-direction.

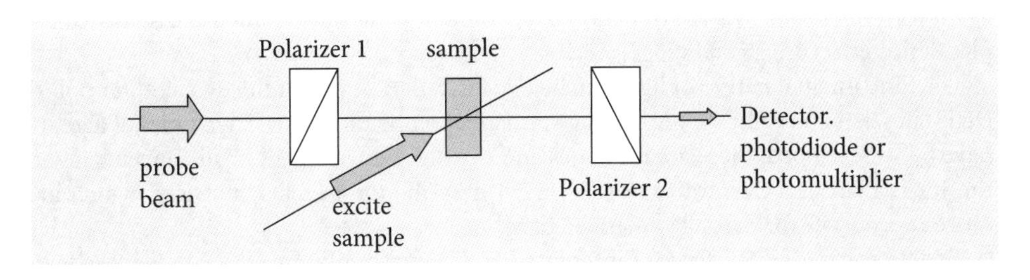

Fig. 7.46 Polarization experiment; polarizer 1 and 2 are crossed and ideally, in the absence of any perturbation, do not transmit any light.

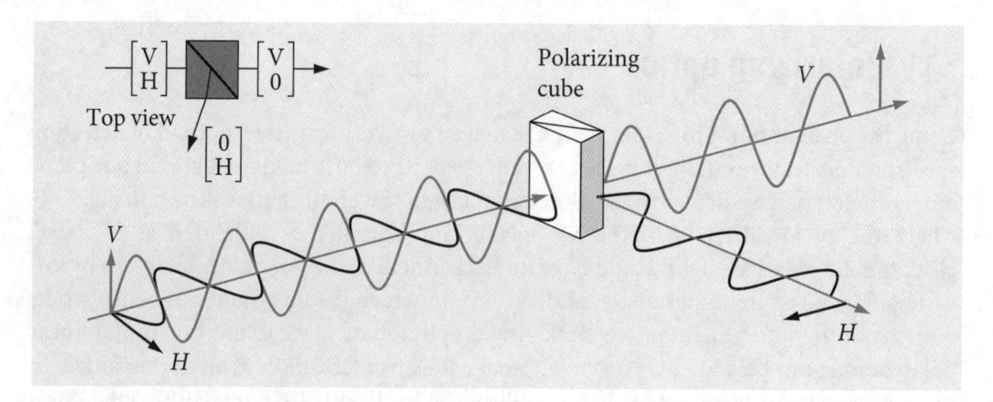

Fig. 7.47 Light composed of equal amounts of horizontal and vertically linearly polarized light is incident on a polarizer. The polarizer is set to pass the vertical component only; the horizontal component is reflected out of the beam. If a film of Polaroid aligned in the same manner was used, the horizontal component would be absorbed by the polarizer.

because no signal is seen before the perturbation occurs, all zero-background types of experiments are very sensitive.

The polarizers could be made from sheets of Polaroid, such as in sunglasses. However, far better ones are made from the transparent and birefringent mineral calcite ($CaCO_3$). The crystal is cut at a certain angle and then made into two similarly shaped prisms or wedges, which, when placed at right angles to one another, form a cube. Variations of these polarizers are called Glan-Thomson or Glan-laser and when crossed only allow 1 part in $\approx 10^6$ of the initial radiation to be passed. A Glan-Thomson polarizer set to pass vertical polarized light rejects horizontally polarized light by total internal reflection at the internal surface of the wedge, and vice versa. The angle to some fixed axis, say the vertical, is sometimes called the 'pass-plane' or 'vibration plane' of the polarized radiation. The effect is illustrated in Fig. 7.47 where the matrices describing the polarization are also shown as an inset.

A second type of polarizing element is a *wave-plate*, which causes the polarization state to be changed but does not change the light intensity. The polarization can be changed either from linear to linear with its axis rotated, or changed from linear to circular, or to elliptical polarization and vice versa. The direction that the major axis of the polarization ellipse has to the vertical is often called the *fast axis*; this is because wave-plates are made of birefringent crystals that have different refractive indices in different directions, and hence light has different speeds. Although one axis is called 'fast', the other is 'normal'.

The properties of the polarized light and polarizing, dichroic, or birefringent optical elements can be analysed very easily by matrix methods. Although light or electromagnetic radiation generally can have a vast range of frequencies, and can travel in any direction, it has only two fundamental polarization states. These are manifested by the properties of the electric field of the radiation, which is always perpendicular to the direction of travel. Any polarizing state can be generated from randomly polarized radiation using only two types of optical elements; these are the linear polarizer and the wave-plate. Starting with arbitrarily polarized light, such as sunshine, which has its electric vector randomly changing and so takes up any angle, passing it through a polarizer produces linear polarized light. If this light is then passed through a wave-plate, depending on the type of wave-plate and the direction of its polarization axis either the linear polarization is rotated, or it becomes circularly polarized. In the most general case, elliptically polarized light results, again with the major axis is a direction determined by the wave-plate used. Some examples of this effect are given in Fig. 7.48.

The fundamental nature of light can be considered to be one of the two circular polarized states, left and right. To allow analysis of any polarization states, two types of algebra have been developed; these are the Jones and Mueller matrices. We shall concentrate on the Jones matrices. Gerrard & Burch (1975) describe the Mueller matrices, which are more appropriate for partially polarized light.

The state of the electric field is represented as a Maxwell column; equation (7.27) shows equivalent forms,

$$\begin{bmatrix} E_V \\ E_H \end{bmatrix} = \begin{bmatrix} E_V^0 \cos(\psi) e^{i\delta V} \\ E_H^0 \sin(\psi) e^{i\delta H} \end{bmatrix} \equiv \begin{bmatrix} E_V^0 \cos(\psi) \\ E_H^0 \sin(\psi) e^{i\Delta} \end{bmatrix} \equiv \begin{bmatrix} A_V \\ A_H e^{i\Delta} \end{bmatrix}, \tag{7.27}$$

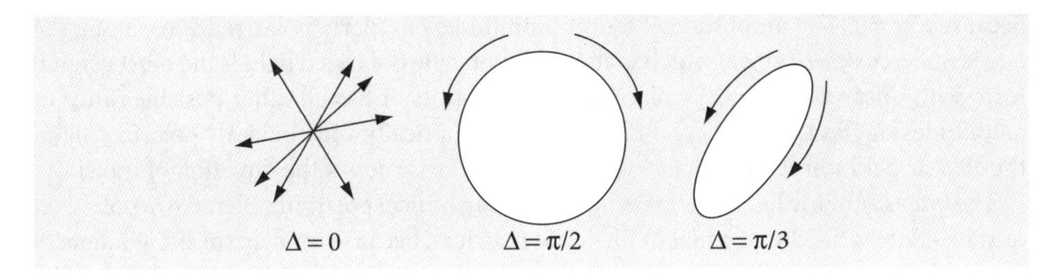

Fig. 7.48 Left: Linearly polarized light, the phase between the *x* and *y* components is zero. This polarization can be at any angle. Middle: Left and right circular polarized light has equal components of *x* and *y* polarizations but 90° out of phase. Right: Elliptically polarized light can be left or right polarized but with different amounts of *x* and *y* components and some arbitrary phase between them.

where Δ is the phase difference between vertical and horizontal components, and ψ is the angle of the electric vector of the light from the laboratory y-axis. A_V is the *amplitude* of the vertical or y, component of the electric field and A_H that of the horizontal component; note that *intensity* (J/cm^2) is amplitude squared and this is what is measured on a photodiode, a CCD detector, or your eye. This is most important; working with the amplitude allows us to calculate correctly the effect of the phase. If we were to work only with the intensity then effects based on phase differences would not be correct.

Linear polarized light is a linear combination of two circular polarization states but with zero phase difference between them ($\Delta = 0$), consequently the orientation of its electric vector lies in one plane only and this may take any angle compared to the lab y-axis. The vectors representing linear polarization with *unity amplitude* in the vertical and horizontal directions respectively, are $V = \begin{bmatrix} 1 \\ 0 \end{bmatrix}$ and $H = \begin{bmatrix} 0 \\ 1 \end{bmatrix}$ because $\psi = 0$ and $\Delta = 0$.

Linearly polarized light can be formed at $\psi = 45°$ to the vertical and is then described by the normalized vector $L_{45} = \dfrac{1}{\sqrt{2}}\begin{bmatrix} 1 \\ 1 \end{bmatrix}$ and has equal components in the vertical and horizontal directions. Linearly polarized light at any angle ψ has vertical and horizontal components of $\sin(\psi)$ and $\cos(\psi)$. The angle ψ is sometimes called the pass-plane or vibration plane of the polarized radiation. Linearly polarized light at an angle ψ to the vertical is shown below left.

Circularly polarized light always has equal contributions of the two base states but they have a phase difference of exactly $\Delta = +90°$ or $-90°$ so that the electric field draws out a left-handed or right-handed helical pattern about the direction of travel. If the angle $\psi = 45°$ the normalized vectors, using equation (7.27), are

$$C_{left} = \frac{1}{\sqrt{2}}\begin{bmatrix} 1 \\ -i \end{bmatrix} \quad \text{and} \quad C_{right} = \frac{1}{\sqrt{2}}\begin{bmatrix} 1 \\ i \end{bmatrix}$$

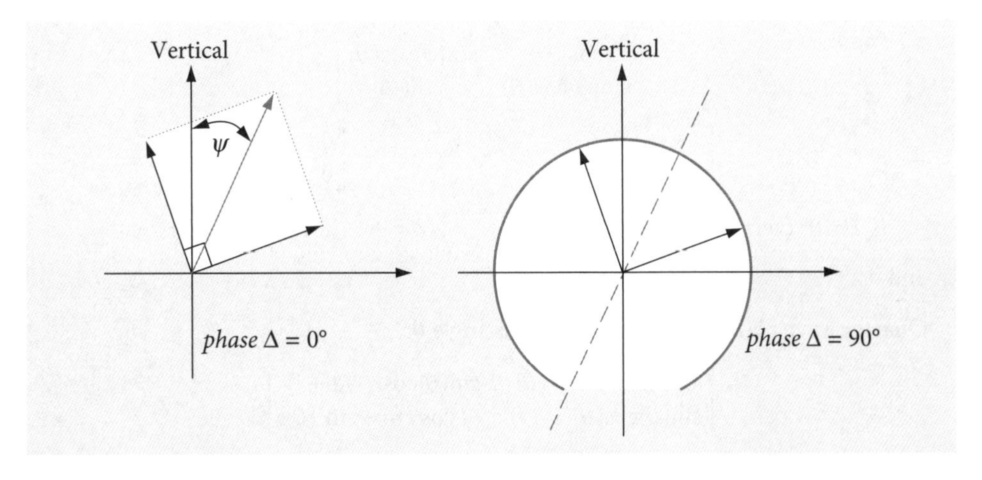

Fig. 7.49 Left: Linearly polarized light, the linear components are at 90° to one another but rotated to give a linearly polarized beam at ψ to the vertical. Right: circular polarized light with equal components but 90° out of phase with one another, the axis of the circle is shown dotted and is at ψ to the vertical as in the left figure.

because $e^{i\pi/2} = i$. The amplitude of a unit-amplitude, circularly polarized wave along the z-axis is $A = \cos(z) + i\sin(z)$ which is also e^{iz}. Elliptically polarized light is the most general form with linear and circular polarization as its limits; it has all other possible ratios of amplitudes, angles ψ, and phase differences Δ. In elliptically and circularly polarized light, the electric field spirals either clockwise or anticlockwise about the direction of travel.

The *intensity* of the light after passing through an optical polarizing element, a polarizer, or wave-plate is the dot product of the Jones matrices, because crudely speaking intensity is amplitude squared. The *Hermitian* transpose must be used to form the row from the column, which means taking the complex conjugate as well as transposing. Thus, for unit amplitude linear V-polarized light the intensity is

$$V^T \cdot V = [1 \quad 0]\begin{bmatrix}1\\0\end{bmatrix} = 1$$

and for left circular polarized light

$$C_L^T \cdot C_L = \frac{1}{2}[1 \quad i]\begin{bmatrix}1\\-i\end{bmatrix} = 1.$$

Calculating the intensities is the last step in the process; the first step is to pass the light trough a number of optical elements, and as the response is linear the output Maxwell column matrix is related to the input by a 2×2 matrix. Each optic has the effect

$$\begin{bmatrix}out_x\\out_y\end{bmatrix} = \begin{bmatrix}a & b\\c & d\end{bmatrix}\begin{bmatrix}in_x\\in_y\end{bmatrix}$$

and the matrices for different elements are tabulated below. When several optical elements are used, the square matrices are multiplied together. Note that vertical polarization is in the x-direction in the way we have defined the matrices. The input light column matrix is always on the right. The other matrices to the left of this are placed in the order of the experimental arrangement; just as with the ***ABCD*** matrices for geometric optical calculations, see Section 7.10. Thus the matrix arrangement for an experimental arrangement such as;

$$\textit{input beam} \rightarrow \textit{polarizer}_1 \rightarrow \textit{wave-plate} \rightarrow \textit{polarizer}_2 \rightarrow \textit{output}$$

is to order the matrices with the input on the right,

$$\begin{bmatrix}\textit{output}\\\textit{column}\end{bmatrix} = [\textit{polarizer}_2][\textit{waveplate}][\textit{polarizer}_1]\begin{bmatrix}\textit{input}\\\textit{column}\end{bmatrix}.$$

The angle θ rotated is calculated from the column output vector as $\tan(\theta) = \dfrac{H_{output}}{V_{output}}$. Now we are in business to do almost any real calculation when we know the form of the 2×2 matrices! These have been worked out and are:

Linear polarizer: polarizer at any angle θ, where θ is rotation from vertical;

$$\begin{bmatrix}\cos^2(\theta) & \sin(\theta)\cos(\theta)\\\sin(\theta)\cos(\theta) & \sin^2(\theta)\end{bmatrix}$$

$$\begin{bmatrix}1 & 0\\0 & 0\end{bmatrix} \qquad \begin{bmatrix}0 & 0\\0 & 1\end{bmatrix}$$

$\theta = 0°$ (vertical polarizer) $\qquad$ $\theta = 90°$ (horizontal polarizer)

and

Quarter wave plate $\Delta = \pi/2$, with fast axis angle θ,

$$\begin{bmatrix}\cos^2(\theta) - i\sin^2(\theta) & \sin(\theta)\cos(\theta)(1+i)\\\sin(\theta)\cos(\theta)(1+i) & -i\cos^2(\theta) + \sin^2(\theta)\end{bmatrix}$$

$$\begin{bmatrix}1 & 0\\0 & -i\end{bmatrix} \qquad \begin{bmatrix}-i & 0\\0 & 1\end{bmatrix}$$

fast axis $\theta = 0°$, vertical. $\qquad$ fast axis $\theta = 90°$, horizontal.

Half wave plate $\Delta = \pi$ with fast axis at θ,

$$\begin{bmatrix} 2\cos^2(\theta)-1 & 2\cos(\theta)\sin(\theta) \\ 2\cos(\theta)\sin(\theta) & 1-2\cos^2(\theta) \end{bmatrix}$$

$$\begin{bmatrix} 1 & 0 \\ 0 & -1 \end{bmatrix} \qquad \begin{bmatrix} -1 & 0 \\ 0 & 1 \end{bmatrix}$$

fast axis $\theta = 0°$ vertical. fast axis $\theta = 90°$, horizontal.

General Linear Retarder or wave-plate:

Rotation angle of fast axis θ from vertical, and phase retardation Δ.

$$\begin{bmatrix} \cos^2(\theta)+\sin^2(\theta)e^{-i\Delta} & \sin(\theta)\cos(\theta)(1-e^{-i\Delta}) \\ \sin(\theta)\cos(\theta)(1-e^{-i\Delta}) & \cos^2(\theta)e^{-i\Delta}+\sin^2(\theta) \end{bmatrix}.$$

The linear retarder or wave-plate can cause the vertical and horizontal components to be delayed from one another by an amount, Δ but can also rotate the axes so that if, say, elliptical light is being formed, its major and so also its minor axes, can be rotated. The amount of retardation is set by the optical properties of the wave-plate's material. Crystalline quartz and calcite are both birefringent materials and therefore each has two refractive indices, an 'ordinary' and an 'extraordinary' one. The crystal has to be orientated and then cut so that the two refractive indices are directed at right angles to one another in the face of the wave-plate. Depending upon the rotation angle that the crystal presents to the incoming light, different amounts of these two refractive indices are used to change the incoming light. One polarization is always delayed with respect to the other in a wave-plate, because the speed of light at a given wavelength is different for the two polarizations because their refractive indices are different. The total phase delay caused by this time delay in passing through the crystal, depends upon cutting the crystal wave-plate to a certain thickness; for example, 100 μm.

As an example, the effect of a wave-plate is examined. Vertically polarized light of amplitude V is passed through a *half-wave* plate with its fast axis at $\theta = 0°$, and in a separate experiment at $\theta = 45°$ and the nature of the resultant beam examined. The arrangement is sketched in Fig. 7.51. Because the wave-plate only rotates the polarization and does not

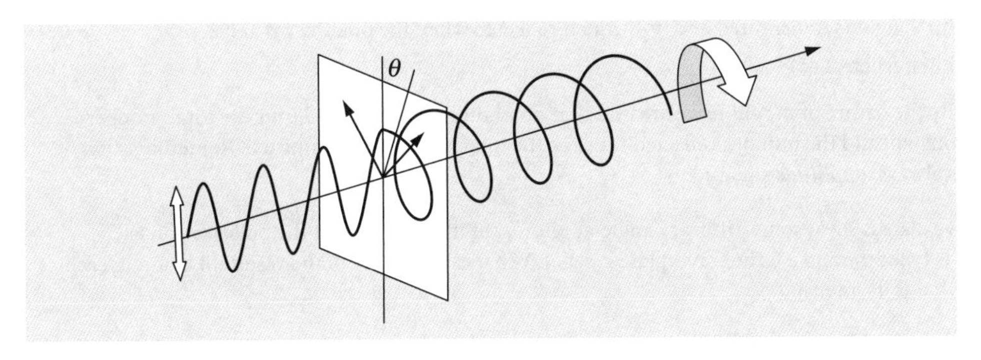

Fig. 7.50 In the general case, a wave-plate converts linearly polarized light into elliptically polarized light with axis depending on the angle θ, the wave-plate and principal axis make to the polarization, and on the crystal thickness, which determines the phase delay Δ.

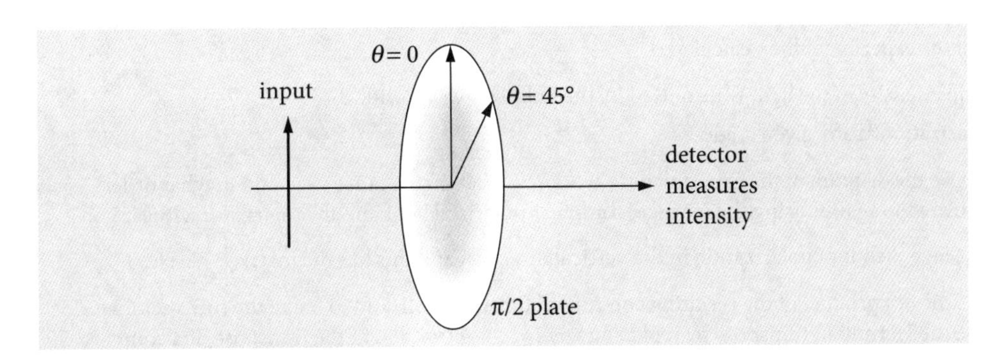

Fig. 7.51 Polarized light passing through a half-wave-plate.

absorb or reflect, the beam's intensity should be unchanged. This should be checked to see that it is true.

Using the matrices in the table, the matrix equation for the wave-plate with its fast axis at 0° is

$$\begin{bmatrix} 1 & 0 \\ 0 & -1 \end{bmatrix}\begin{bmatrix} V \\ 0 \end{bmatrix} = V\begin{bmatrix} 1 \\ 0 \end{bmatrix},$$

which has no effect on the beam. Therefore its intensity is unchanged at V^2 because the dot product of the column vector with itself is unity; $V^2\,[1 \quad 0]\begin{bmatrix} 1 \\ 0 \end{bmatrix} = V^2$

When the wave-plate is rotated to 45° or $\pi/4$ radians, the matrix is

$$\begin{bmatrix} 2\cos^2(\pi/4) - 1 & 2\cos(\pi/4)\sin(\pi/4) \\ 2\cos(\pi/4)\sin(\pi/4) & 1 - 2\cos^2(\pi/4) \end{bmatrix} = \begin{bmatrix} 0 & 1 \\ 1 & 0 \end{bmatrix}$$

and the matrix equation $\begin{bmatrix} 0 & 1 \\ 1 & 0 \end{bmatrix}\begin{bmatrix} V \\ 0 \end{bmatrix} = V\begin{bmatrix} 0 \\ 1 \end{bmatrix}$

This shows that the half wave plate at 45° rotates the vertical polarization to horizontal. This is extremely useful in practice to manipulate laser beams to the correct polarization for lots of different types of experiments. The rotation angle is always twice the wave-plate angle. The output intensity is unchanged at V^2.

7.11.1 Questions

Full solutions are available at www.oxfordtextbooks.co.uk/orc/beddard.

Q7.40 A beam of linearly polarized light with amplitude components V and H is passed through a polarizing cube set to pass vertical polarized light, $\theta = 0$. The light is either reflected or transmitted by the cube.

(a) How much light is transmitted and how much reflected when the polarizer is at 0°?

(b) How much when at 45°?

Strategy: By the nature of the cube, we are told that no photons are absorbed thus the total number remains constant and the number reflected must be the total less those transmitted. Remember that *intensity* is always *amplitude* squared.

Q7.41 A half-wave plate is set at some arbitrary angle φ. Show that the angle ψ that the polarization is rotated to is twice the angle φ the wave-plate is set at. What is the nature of the resultant beam in each case and what is its intensity?

Q7.42 Vertically polarized light of amplitude V is passed through a *quarter-wave* plate **(a)** with its fast axis at $\theta = 0$; **(b)** at $\theta = 45°$, and **(c)** at some arbitrary angle θ. What is the nature of the resultant beam in each case?

Q7.43 A beam of vertical, linearly polarized laser light of unit amplitude is passed into a linear polarizer whose angle θ is rotated from 0 to 180° and the transmitted light measured on a photodiode. Draw a diagram of the experiment then calculate:

(a) how the intensity of the light transmitted by the polarizer varies with θ

(b) its polarization at any given angle.

Strategy: Use the diagram of the experiment to work out what matrices are used and in what order. Next, calculate the answer using the matrices for linear polarized light and the matrix for a linear polarizer. Make the input polarization in the vertical direction and this has the matrix $\begin{bmatrix} 1 \\ 0 \end{bmatrix}$. The intensity is the dot product of the resulting column vector with itself, and to make the row vector we must make the Hermitian transpose by replacing any $i = \sqrt{(-1)}$ with $-i$ in the transpose, if a complex quantity is present.

Q7.44 A half-wave plate is placed between a pair of crossed linear polarizers. A beam of light of unit amplitude is transmitted by the first polarizer set at 0° and is rotated by a half wave plate set at θ.

(a) What is the intensity transmitted by the *second* polarizer for any angle θ of the half wave plate?

(b) Show that the maximum transmitted intensity occurs when the half-wave plate is set at ±45°.

(c) What is the intensity variation if the wave-plate is restricted to small angles from the vertical?

(d) Generalize part (a) to any linear polarizer angles α and β, and wave-plate θ angles and arbitrary input intensities V and H. Calculate the result of part (a) again.

Use Maple to do the matrix multiplications if you wish.

Strategy: Draw out the arrangement first to know the order of the optical elements and the polarization directions. As the beam is passed through the linear polarizer set at 0°, the output from this must be vertically polarized light with matrix $\begin{bmatrix} 1 \\ 0 \end{bmatrix}$ so we can use this to start the calculation. As the polarizers are crossed, the second is at 90° to the first.

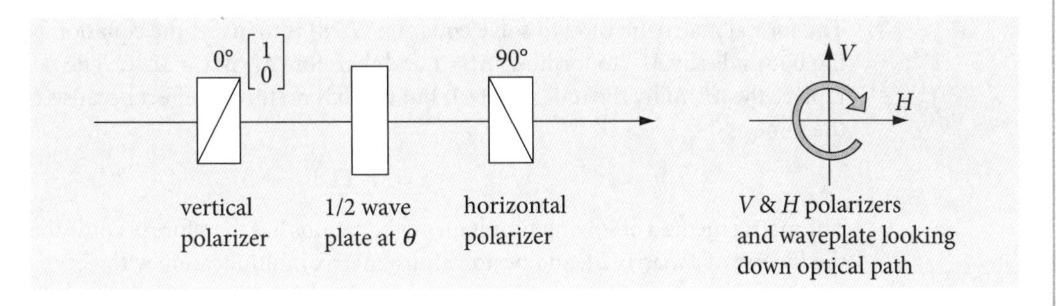

Fig. 7.52 Scheme for Q7.44.

7.12 Solving equations using matrices

7.12.1 Simultaneous equations

A matrix can be of great use in solving simultaneous equations provided that its inverse exists. We consider two types of equations. First is the general case

$$\boldsymbol{M}\boldsymbol{x} = \boldsymbol{C}, \tag{7.28}$$

where $\boldsymbol{M}$ is a matrix and $\boldsymbol{x}$ and $\boldsymbol{C}$ are column vectors (one-dimensional matrices). Later on, we consider eigenvalue–eigenvector equations, which are more important and useful to us.

A set of simultaneous equations can conventionally be written as shown below, but the matrix equation is equivalent and clearer.

$$\begin{aligned} 3x + 4y + z &= 6 \\ x - 3y + 6z &= -3 \\ 2x - y + 4z &= 0 \end{aligned} \quad \text{or} \quad \underset{\boldsymbol{M}}{\begin{bmatrix} 3 & 4 & 1 \\ 1 & -3 & 6 \\ 2 & -1 & 4 \end{bmatrix}} \underset{\boldsymbol{x}}{\begin{bmatrix} x \\ y \\ z \end{bmatrix}} \underset{=}{=} \underset{\boldsymbol{C}}{\begin{bmatrix} 6 \\ -3 \\ 0 \end{bmatrix}} \tag{7.29}$$

To solve equation (7.29), multiply the first equation by 2, the second by 6, and the third by 3, then subtract to eliminate the x and so on, until y and z are found. Clearly, this is a somewhat complicated business especially if there are, for example, ten equations and ten unknowns to solve for. At school, you may have used this method or Cramer's Rule, which is really a matrix method. This method has the following algorithm:

(a) Calculate the determinant of the matrix.

(b) Replace one column of the matrix by the column vector $\boldsymbol{C}$ to form a new matrix.

(c) Divide the determinant of this by the determinant of the original matrix.

(d) Repeat (b, c) choosing different columns.

It is easier to see as an equation, and the result for component i of the vector $\boldsymbol{x}$ is

$$x_i = \frac{|M_{C\to i}|}{|M|}.$$

The value $|M_{C\to i}|$ is the determinant formed with column C replacing column i. The first component $i = 1$ gives the unknown x, the second y the third z. The value of x is

$$\frac{\begin{vmatrix} 6 & 4 & 1 \\ -3 & -3 & 6 \\ 0 & -1 & 4 \end{vmatrix}}{\begin{vmatrix} 3 & 4 & 1 \\ 1 & -3 & 6 \\ 2 & -1 & 4 \end{vmatrix}} = \frac{6(-12+6) - 4(-12) + 1(3)}{3(-12+6) - 4(4-12) + 1(-1+6)} = \frac{15}{19}.$$

7.12.2 A matrix method for simultaneous equations

The formal matrix method to solve equation (7.28) is to invert the equation by multiplying both sides by M^{-1} to form $M^{-1}M = 1$ and therefore $M^{-1}Mx = M^{-1}C$. The next step is to replace the $M^{-1}M$ by the unit matrix 1, but this has no further effect because $1x = M^{-1}C$ is the same as

$$x = M^{-1}C. \tag{7.30}$$

The matrix method of solving simultaneous equations has therefore become that of finding the inverse of a matrix M, and performing a matrix multiplication with C.

In practice, you will generally want to use the computer to solve simultaneous equations and this is illustrated in two ways below using Maple. In the matrix method, C is a column matrix and M has rows of the equation coefficients.

```
> with(LinearAlgebra):
> M:= < < 3 | 4 | 1 > , < 1 | -3 | 6 > , < 2 | -1 | 4 >;
  C:= < 6,-3, 0 >;
  answers:= M^(-1) . C;                    # equation (7.30)
```

$$M := \begin{bmatrix} 3 & 4 & 1 \\ 1 & -3 & 6 \\ 2 & -1 & 4 \end{bmatrix} \quad C := \begin{bmatrix} 6 \\ -3 \\ 0 \end{bmatrix} \quad answers := \begin{bmatrix} \frac{15}{19} \\ \frac{18}{19} \\ -\frac{3}{19} \end{bmatrix}$$

Notice that the order of the results is not the same as that of used to form the matrix M defining x, y and z, and therefore we cannot tell which answer is x, y, or z. A second method, uses the `solve` routine, which does tells us which is x, y, and z. Notice the use of the equal sign in the equations.

```
> eq1:= 3*x + 4*y + z = 6;
  eq2:=   x - 3*y + 6*z = -3;
  eq3:= 2*x - y + 4*x = 0;
  ans:= solve( { eq1, eq2, eq3 },{ x, y, z } ):
```

$$z = -\frac{3}{19} \quad x = \frac{15}{19} \quad y = \frac{18}{19}$$

Notice how in `solve` the equations are normally placed between *curly* brackets `{ }` and similarly the parameters we want to solve for.

A second example concerns rotational spectroscopy. Raman spectroscopy has been used to measure the rotational energy levels of benzene vapour, which are then used to obtain accurate CH bond lengths. The rotational constants obtained from C_6H_6 and C_6D_6, are $B_H = 0.18960 \pm 0.00005$ and $B_D = 0.15681 \pm 0.00008$ cm^{-1}, which describe the motion *perpendicular* to benzene's sixfold axis. The bond lengths will be calculated assuming that the molecule has D_{6h} symmetry, i.e. is a regular hexagon and behaves as a rigid rotor.

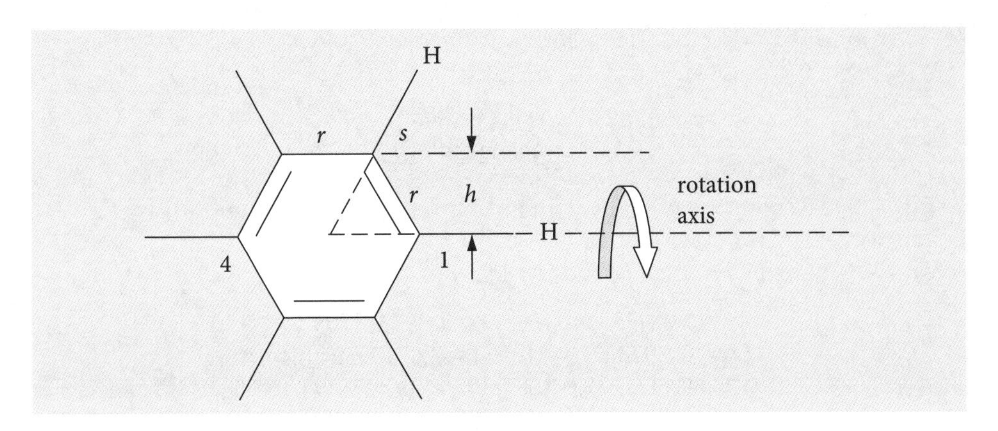

Fig. 7.53 The Raman spectrum has lines measuring the rotational motion in benzene about the twofold (C_2) axis shown.

This calculation shows how isotopic substitution makes it possible to calculate very accurate bond lengths even in large molecules; in fact, far more accurately than by X-ray crystallography.

There does not initially appear to be enough information to do the calculation. However, the rotational constants are related to the moment of inertia of the molecule by $B = \dfrac{\hbar}{4\pi I}\,\mathrm{s}^{-1}$ where I is the moment of inertia (kg m^2). The moment of inertia is related to bond lengths as $I = \sum_i m_i r_i^2$ where the sum is taken over all the distances r_i the i atoms have from the rotation axis. The mass of atom i in kg is m_i. The rotational motion measured is perpendicular to the sixfold axis, which means that the rotation axis must pass through opposite carbon atoms, for instance, 1 and 4, Fig. 7.53. It is necessary to form two equations for the moments of inertia of the two molecules, and then to solve them simultaneously. Notice that the B values are in wavenumbers, so it is necessary also to change B to s^{-1} by multiplying by the speed of light in cm s^{-1}. Moments of inertia calculations are described more fully in Section 7.15.6.

To calculate the moments of inertia, the distances of the atoms from the rotation axis are needed and can be calculated using Fig. 7.53. The CC bond length is r the CH is s. The distance of each C atom from the axis, is either zero, which will not change the moment of inertia, or $h = r\sin(60\pi/180)$. Similarly, each H atom contributes either zero or $(r+s)\sin(60\pi/180)$. The moment of inertia is therefore due to four C and four H atoms,

$$\begin{aligned} I &= 4m_C r^2\sin^2(60\pi/180) + 4m_H(r+s)^2\sin^2(60\pi/180) \\ &= 3[m_C r^2 + m_H(r+s)^2] \end{aligned}$$

Because two rotational constants are measured and these give moments of inertia $I = \dfrac{\hbar}{4\pi B}$, two simultaneous equations are produced for the H and D derivatives. These are

$$(m_C + m_H)r^2 + m_H s^2 + 2m_H rs - \frac{\hbar}{12\pi B_H} = 0$$

$$(m_C + m_D)r^2 + m_D s^2 + 2m_D rs - \frac{\hbar}{12\pi B_D} = 0$$

Using numerical values of the rotational constants and Maple to obtain accurate molecular masses with the `ScientificConstants` package, solving the two equations for r and s gives,

```
> restart: with(ScientificConstants):
> eqH:= (mC + mH)*r^2 + mH*s^2 + 2*mH*r*s-hbar/(12*Pi*BH);
  eqD:= (mC + mD)*r^2 + mD*s^2 + 2*mD*r*s-hbar/(12*Pi*BD);
  mC:= GetValue(Element( C[12], atomicmass ));     # in kg
  mH:= GetValue(Element(  H[1], atomicmass ));
  mD:= GetValue(Element(  H[2], atomicmass ));
```

$$mC := 1.992648240\ 10^{-26}\quad mH := 1.673533980\ 10^{-27}\quad mD := 3.344496969\ 10^{-27}$$

```
hbar:= evalf(Constant(Planck_constant_over_2pi) );
c:= evalf(Constant(speed_of_light_in_vacuum) )*100;
```

$$hbar := 1.054571596\ 10^{-34}$$
$$c := 2.997924580\ 10^{10}$$

```
BH:= 0.18960*c:  BD:= 0.15681*c: # rotation constants
ans:= solve( {eqH, eqD},{r, s}):
ans[1]; ans[2]; ans[3]; ans[4];
```

$$\{s = 1.084338541\ 10^{-10}, r = 1.397327704\ 10^{-10}\}$$
$$\{r = -1.397327704\ 10^{-10}, s = 3.878993949\ 10^{-10}\}$$
$$\{r = -1.397327704\ 10^{-10}, s = -1.084338541\ 10^{-10}\}$$
$$\{r = 1.397327704\ 10^{-10}, s = -3.878993949\ 10^{-10}\}$$

The physically meaningful roots of these equations are r = 1.3973 Å for the CC bond length and s = 1.0843 Å for the CH bond length. The results are quoted to 5 figures, as this is the precision of the initial data.

7.12.3 Eigenvalue–eigenvector equations

A most important and interesting type of equation is the *eigenvalue–eigenvector* equation which has the form

$$operator \times function = constant \times same\ function.$$

provided the function is not zero.[4] The Schrödinger equation $H\psi = E\psi$ is of this type, and is used to calculate vibrational and rotational spectra and virtually all other aspects of molecular quantum mechanics. In this equation, H is the Hamiltonian operator; ψ is the wavefunction and E the energy. In this section, we will deal with operators that are matrices and functions that are vectors.

In matrix form, the eigenvector–eigenvalue equation is

$$Ax = \lambda x, \tag{7.31}$$

where the operator A, which is a square matrix, alters column vector, x to produce itself multiplied by a constant λ. The constant λ is called either an *eigenvalue*, a proper value, or a characteristic value. The *eigenvector* is x. The action of the operator is to magnify or reduce the eigenvector x by an amount λ. In contrast, operators, such as rotation matrices, or those used in group theory, would leave the vector x unchanged in size but pointing in another direction.

Equation (7.31) can be written as

$$Ax_i = \lambda_i x_i,$$

which is preferable, because it indicates that there is one column vector x_i for each eigenvalue λ_i. If the matrix is $n \times n$ then there are n eigenvalues and n eigenvectors. The column eigenvectors x_i are often stacked into a square matrix (modal matrix) and the eigenvalues are usually formed into the single column vector λ. When solving an eigenvalue–eigenvector equation, the values of all the eigenvectors and eigenvalues will be determined; for example in solving the Schrödinger equation, the Hamiltonian, the sum of kinetic and potential energy is defined by the problem at hand. Solving the equations produces not only the energy levels but also the wavefunctions.

7.12.4 Kinetics, dynamics, and quantum mechanics

Before embarking on the details of solving eigenvalue equations, it is worth considering the difference and similarities in the types of problems that can be studied. We shall meet problems in quantum mechanics, chemical kinetics, and dynamics and although there are great differences, the language to describe various problems—that is the maths—follows the same path. In kinetics, eigenvalue equations are used when several *linear* reactions

[4] The equation $d^2e^{-ax}/dx^2 = a^2e^{-ax}$, where a is a constant, is also called an eigenvalue–eigenvector equation, even though no vectors or matrices are involved; e^{-ax} is called the eigenvector and a^2 the eigenvalue.

have to be modelled. If there are five reactions, then a 5×5 matrix of rate constants is constructed to describe the inter-conversion of species. In quantum mechanics, the Schrödinger equation is solved by expanding the new, and so far unknown, wavefunctions in a basis set of other wavefunctions. Interactions are added between various states and these are usually called couplings; mathematically they take the place of the rate constants used in kinetics. The *order* of the *basis set* determines where the couplings between states occur in the matrix. Reordering the basis set moves these positions about in the matrix, but does not affect the outcome. In kinetics, basis sets are usually not talked about; the order of the species occurring in a reaction scheme is, in effect, the basis set. Reorder the scheme, and the rate constants appear in different places in the matrix, but the result is unchanged.

7.12.5 Interpreting eigenvalues and eigenvectors

The eigenvalue and its associated eigenvector provide us with different kinds of information. Eigenvalues are numbers that produce energies, frequencies, and timescales; eigenvectors on the other hand give geometrical type properties, such as normal mode displacements or the composition of a wavefunction leading to the shape of a molecular orbital. Each eigenvalue is always associated with its eigenvector, which as a vector is a series of numbers.

7.12.6 The secular equation and determinant. The evaluation of eigenvalues and eigenvectors

In an eigenvalue–eigenvector equation, the operator A is always an $n \times n$ square matrix and x is an, as yet, unknown eigenvector column matrix, and similarly for the eigenvalues λ which also form an unknown column matrix. Rearranging equation (7.31) produces what is called the *secular equation*,

$$(A - \lambda \mathbf{1})x = \mathbf{0}, \tag{7.32}$$

where $\mathbf{1}$ is the unit matrix and $\mathbf{0}$ is the null matrix. This equation is, in fact, just a set of simultaneous equations in n unknowns, because there are n rows and columns. The eigenvalues are calculated from the *secular determinant* by setting it to zero

$$|A - \lambda \mathbf{1}| = \mathbf{0}$$

and solving the resulting *characteristic* equation, which is a polynomial in λ. The λ's only appear on the diagonal in the matrix because the matrix $\mathbf{1}$ is a unit diagonal matrix with all non-diagonal elements equal to zero. Writing the secular determinant in full gives

$$|A - \lambda \mathbf{1}| = \begin{bmatrix} a_{11} - \lambda & a_{12} & a_{13} & \cdots & a_{1n} \\ a_{21} & a_{22} - \lambda & a_{23} & \cdots & a_{2n} \\ a_{31} & a_{32} & a_{33} - \lambda & \cdots & a_{3n} \\ \vdots & \vdots & \vdots & \ddots & \vdots \\ a_{n1} & a_{n2} & a_{n3} & \cdots & a_{nn} - \lambda \end{bmatrix} = 0. \tag{7.33}$$

The n solutions of the polynomial are the eigenvalues λ_1 to λ_n normally formed into the vector $\boldsymbol{\lambda}$. There are n eigenvectors, x_1 to x_n, each of length n, and to find each of these eigenvectors, one belonging to each eigenvalue, λ, each eigenvalue in turn is substituted back into the secular equation and this is then solved for each x. The jth solution, for example, is obtained using $(A - \lambda_j \mathbf{1})x_j = \mathbf{0}$, and this is the part that takes the computer (or you) most time to evaluate. This is done by computing the ith component of the jth eigenvector, x_j, which is a multiple of

$$x_{ij} = (-1)^{i+k}|(A - \lambda_j \mathbf{1})_{ki}| \tag{7.34}$$

where $(-1)^{i+k}|(A - \lambda_j \mathbf{1})_{ki}|$ is the *cofactor* of the matrix of element ki; note that j is the index of the eigenvalue and k the index of a column. Collecting all the eigenvectors together, they can be placed into a matrix, each column of which is an eigenvector, the first column belonging to the first eigenvalue and so forth; this is why equation (7.34) has two subscripts; the second j identifies the eigenvalue. The matrix of eigenvectors a sometimes called the *modal* matrix.

We shall use Maple to do most of the matrix diagonalization. When the matrix is formed, this is easy, and both eigenvectors and eigenvalues can be produced together. The position of any eigenvalue–eigenvector pair that the computer produces is arbitrary, and this order does not, unless accidentally, correspond to the ordering of the basis set, but the eigenvalue and its corresponding eigenvector are always produced in the same relative positions. The syntax is;

```
> with(LinearAlgebra):
> (eigval, eigvec):= Eigenvectors(M):
```

To illustrate the method, the eigenvalues λ_1 and λ_2 of $A = \begin{bmatrix} 2 & 4 \\ 3 & 1 \end{bmatrix}$ and the corresponding eigenvectors will be found. First, the calculation will be done by hand, then using Maple. Our strategy will be to convert the matrix to the form of equation (7.33), expand the secular determinant, and solve for λ. Equation (7.34) will be used to calculate the eigenvectors. As the calculation is relatively simple for this small matrix, it can be done by hand. The equation to solve is $\begin{bmatrix} 2-\lambda & 4 \\ 3 & 1-\lambda \end{bmatrix} = 0$ or $(2-\lambda)(1-\lambda) - 12 = 0$, which is the characteristic equation $\lambda^2 - 3\lambda - 10 = 0$. Solving the quadratic, produces eigenvalues $\lambda_1 = -2$ and $\lambda_2 = 5$.

The eigenvectors are evaluated following equation (7.34). The cofactor of the top left matrix element is $1 - \lambda$; the cofactor of 4 is 3 and so on. We have to calculate x_{11}, x_{12}, x_{21}, and x_{22} and these terms form a matrix with the eigenvectors as columns. Three subscripts are needed to use equation (7.34); i, j, and k. The eigenvalue index is j and k is any column index, which is 1 or 2 in this example. Suppose we start with $k = 1$ and $i = 1$ then

$$x_{1j} = (-1)^{1+1}|(A - \lambda_j \boldsymbol{1})_{11}| = 1 - \lambda_j.$$

As the cofactor of element (1, 1) is $1 - \lambda$, using $j = 1$ then 2, for λ_1 then λ_2, we obtain $x_{11} = 3$ and $x_{12} = -4$ respectively, because $\lambda_1 = -2$ and $\lambda_2 = 5$.

Next, choose $i = 2$ to obtain the second row of the same eigenvector and again chose $k = 1$ then

$$x_{2j} = (-1)^{2+k}|(A - \lambda_j \boldsymbol{1})_{k2}| = -3$$

therefore, $x_{21} = -3$ and $x_{22} = -3$. The first eigenvector is

$$\boldsymbol{x}_1 = \begin{bmatrix} x_{11} \\ x_{21} \end{bmatrix} = \begin{bmatrix} 3 \\ -3 \end{bmatrix}$$

and following the same procedure with $k = 2$, the second eigenvector is

$$\boldsymbol{x}_2 = \begin{bmatrix} -4 \\ -3 \end{bmatrix}.$$

The eigenvectors can be normalized to give

$$\boldsymbol{x}_1 = \frac{1}{\sqrt{3^2+3^2}}\begin{bmatrix} 3 \\ -3 \end{bmatrix} = \frac{1}{\sqrt{2}}\begin{bmatrix} 1 \\ -1 \end{bmatrix} \quad \text{and} \quad \boldsymbol{x}_2 = \frac{1}{\sqrt{4^2+3^2}}\begin{bmatrix} 4 \\ 3 \end{bmatrix} = \frac{1}{5}\begin{bmatrix} 4 \\ 3 \end{bmatrix}.$$

The negative sign on $\boldsymbol{x}_2$ was ignored because both terms are negative and it does not make any difference to the result if both are negative or both positive. As a check, put these eigenvector values back into the original equation

$$A\boldsymbol{x}_1 \equiv \frac{1}{\sqrt{2}}\begin{bmatrix} 2 & 4 \\ 3 & 1 \end{bmatrix}\begin{bmatrix} 1 \\ -1 \end{bmatrix} = \frac{1}{\sqrt{2}}\begin{bmatrix} -2 \\ 2 \end{bmatrix} = -2\boldsymbol{x}_1, \text{ which is } \lambda_1\boldsymbol{x}_1$$

$$A\boldsymbol{x}_2 \equiv \frac{1}{5}\begin{bmatrix} 2 & 4 \\ 3 & 1 \end{bmatrix}\begin{bmatrix} 4 \\ 3 \end{bmatrix} = \frac{1}{5}\begin{bmatrix} 20 \\ 15 \end{bmatrix} = 5\boldsymbol{x}_2, \text{ which is } \lambda_2\boldsymbol{x}_2.$$

The same calculation in Maple is rather easier and you will undoubtedly always want to do the calculation this way. The eigenvectors are normalized by dividing by the square root of the dot product. The `LinearAlgebra` package has to be loaded first. The eigenvalues and eigenvectors are called `eigval` and `eigvec`. The `Column` instruction extracts the column vector from the `eigvec` matrix.

```
> with(LinearAlgebra):
> M:= < < 2 | 4 >,< 3 | 1 > >;                # define matrix
  (eigval, eigvec):= Eigenvectors(M):
  `eigval`:= eigval;
  `eigvec`:= eigvec;
```

$$eigval := \begin{bmatrix} 5 \\ -2 \end{bmatrix} \qquad eigvec := \begin{bmatrix} \frac{4}{3} & -1 \\ 1 & 1 \end{bmatrix}$$

```
> x:= Column(eigvec,1): x1:= x/sqrt(x.x); # normalize
```

$$x1 := \begin{bmatrix} \frac{4}{5} \\ \frac{3}{5} \end{bmatrix}$$

```
> x:= Column(eigvec,2): x2:= x/sqrt(x.x); # normalize
```

$$x2 := \begin{bmatrix} -\frac{1}{2}\sqrt{2} \\ \frac{1}{2}\sqrt{2} \end{bmatrix}$$

The eigenvector–eigenvalue pair is the same as in our calculation, but the eigenvalues are given in a different order. If you repeat this calculation, the computer may reorder the eigenvalues, but they are always paired with the eigenvectors properly. When using Maple if the vector has any imaginary parts, then use the `DotProduct(V,V)` instruction which automatically takes the complex conjugate.

7.12.7 Properties of eigenvalues and eigenvectors

Some of the more important properties of eigenvalues and eigenvectors are listed here.

(a) Any scalar multiple of an eigenvector is also an eigenvector, because the operator A is linear; e.g. if an eigenvector is $\boldsymbol{x}$ then $3\boldsymbol{x}$ is also an eigenvector. If the operator A is raised to an integer power, $p = 0, \pm 1, \pm 2 \cdots$ then the eigenvalues are λ^p. The eigenvectors are the same as those of A.

(b) Eigenvectors associated with different eigenvalues are linearly independent. If more than one linearly independent eigenvector belongs to the same eigenvalue, then the eigenvalue is degenerate.

(c) When the operator matrix A is Hermitian or symmetric, as in quantum mechanical problems, then the eigenvectors corresponding to different eigenvalues are more than linearly independent; they are mutually orthogonal to one another. For example, if $\boldsymbol{x}_a$ and $\boldsymbol{x}_b$ are two eigenvectors corresponding to eigenvalues a and b, they are orthogonal if $\boldsymbol{x}_a \cdot \boldsymbol{x}_b = 0$. As the eigenvectors are orthogonal, the matrix A will be diagonalizable.

(d) In cases where eigenvectors are not orthogonal, there are strategies that can be used to achieve this if necessary; for example, Gram-Schmidt orthogonalization (Arkfen 1970).

(e) To normalize an eigenvector $\boldsymbol{x}_a$, as with any other vector, divide it by $\sqrt{\boldsymbol{x}_a \cdot \boldsymbol{x}_a}$.

(f) Given eigenvectors $\boldsymbol{x}_a$ and $\boldsymbol{x}_b$ where $A\boldsymbol{x}_a = \lambda_a \boldsymbol{x}_a$ and $A\boldsymbol{x}_b = \lambda_b \boldsymbol{x}_b$, if α and β are numbers (constants), then $A(\alpha \boldsymbol{x}_a + \beta \boldsymbol{x}_b) = \lambda(\alpha \boldsymbol{x}_a + \beta \boldsymbol{x}_b)$.

(g) A graph of the size of $\boldsymbol{\lambda}$ vs their index number, 1 to n for n equations, is sometimes called the *eigenvalue spectrum*.

In case (c), to determine whether the eigenvectors are orthogonal in a symmetrical matrix A, compare the two equations $A\boldsymbol{v} = \lambda_v \boldsymbol{v}$ and $A^T\boldsymbol{u} = \lambda_u \boldsymbol{u}$. The eigenvalues λ_u and λ_v must be different, but because the matrix is symmetrical $A^T = A$. Next, there is a lemma that proposes that the dot product $\boldsymbol{v} \cdot (A^T\boldsymbol{u})$ and $\boldsymbol{u} \cdot (A\boldsymbol{v})$ are equal. Substituting for $A\boldsymbol{v}$ and $A^T\boldsymbol{u}$ gives the equations

$$v\cdot(A^T u) = \lambda_u v\cdot u \qquad u\cdot(Av) = \lambda_v u\cdot v$$

and if they are equal then it must follow that $\lambda_u v\cdot u = \lambda_v u\cdot v$. The only way that this can be true, since $\lambda_u \neq \lambda_v$, is if $u\cdot v = 0$, hence the eigenvectors must be orthogonal.

7.12.8 Basis sets: a reminder

To be able to solve an equation such as (7.31), it is essential to choose a basis set for the vectors. Normally, one would like to choose an orthonormal set, such as the standard basis, and in three dimensions we could use the set $\{x, y, z\}$ with the basis vectors $\{\boldsymbol{i}, \boldsymbol{j}, \boldsymbol{k}\}$ (see Chapter 6) or equivalently in column matrix form $\left\{\begin{bmatrix}1\\0\\0\end{bmatrix}, \begin{bmatrix}0\\1\\0\end{bmatrix}, \begin{bmatrix}0\\0\\1\end{bmatrix}\right\}$ both forms of which are normalized and orthonormal. Any vector can be expanded in components of its basis set; if a vector is V, and if b are the expansion coefficients, then $V = \sum b_j x_j$.

For example, in the three-dimensional basis $\{\boldsymbol{i}, \boldsymbol{j}, \boldsymbol{k}\}$ this vector could be $2\boldsymbol{i} + 3\boldsymbol{j} - 4\boldsymbol{k}$ where the b_js are 2, 3, and −4. The vector could also be represented as the dot product of the basis vector and individual vectors $\begin{bmatrix}2\\0\\0\end{bmatrix}, \begin{bmatrix}0\\3\\0\end{bmatrix}, \begin{bmatrix}0\\0\\-4\end{bmatrix}$ which gives $[1\ 0\ 0]\begin{bmatrix}2\\0\\0\end{bmatrix} = 2$ for the first term of V. All three terms produce the vector $V = \begin{bmatrix}2\\3\\-4\end{bmatrix}$. This calculation has used the fact that the summation of the product of pairs of terms with the same index, is the same as the dot product of two vectors;

$$\sum_{j=1}^{n} b_j x_j \equiv [b_1 \quad \cdots \quad \cdots \quad b_n]\begin{bmatrix}x_1\\ \vdots\\ \vdots\\ x_n\end{bmatrix}.$$

7.12.9 Interpreting the secular determinant

The nature of the secular determinant, and hence the matrix from which it is generated, needs explaining particularly in quantum mechanical problems. In these problems, the secular determinant is always symmetrical and contains only real numbers; it is Hermitian. The order of the element in both rows and columns is the same as the ordering of the basis set. If the secular determinant is diagonal, that is contains only the diagonal elements, and all the rest are zero, then the eigenvalues are the diagonal terms and are the solutions to the Schrödinger equation. If there are non-zero, off-diagonal terms, then these mix the basis set wavefunctions and energies together when the matrix is diagonalized. The off-diagonal terms are often referred to as *coupling* terms, because they 'cause interaction' between the energy levels in the diagonal of that row and column in which that matrix element is present. The type of interaction, Coulombic or dipole–dipole for instance, varies depending on the problem at hand.

In the matrix below (left) the energies E_1^0, E_2^0, and so forth are the solutions to the Schrödinger equation $H^0\psi^0 = E^0\psi^0$ because all the off-diagonal terms are zero. Suppose now that another potential energy term V is added to represent a new interaction, an electric field is applied for instance, this changes the Schrödinger equation to a new one $(H^0 + V)\psi = E\psi$ with different energies (eigenvalues) and wavefunctions (eigenvectors), which will be *linear combinations* of the basis set wavefunctions ψ^0.

$$\begin{bmatrix} E_1^0 & 0 & 0 & 0\\ 0 & E_2^0 & 0 & 0\\ 0 & 0 & E_3^0 & 0\\ 0 & 0 & 0 & E_4^0 \end{bmatrix} \qquad \begin{bmatrix} E_1^0 & 0 & L & 0\\ 0 & E_2^0 & K & 0\\ L & K & E_3^0 & 0\\ 0 & 0 & 0 & E_4^0 \end{bmatrix}$$

$$H^0\psi^0 = E^0\psi^0 \qquad\qquad (H^0 + V)\psi = E\psi$$

The off-diagonal matrix element L indicates that level 1 interacts with level 3, and of course, 3 with 1, and K shows that levels 2 and 3 also interact. The effect of these interactions

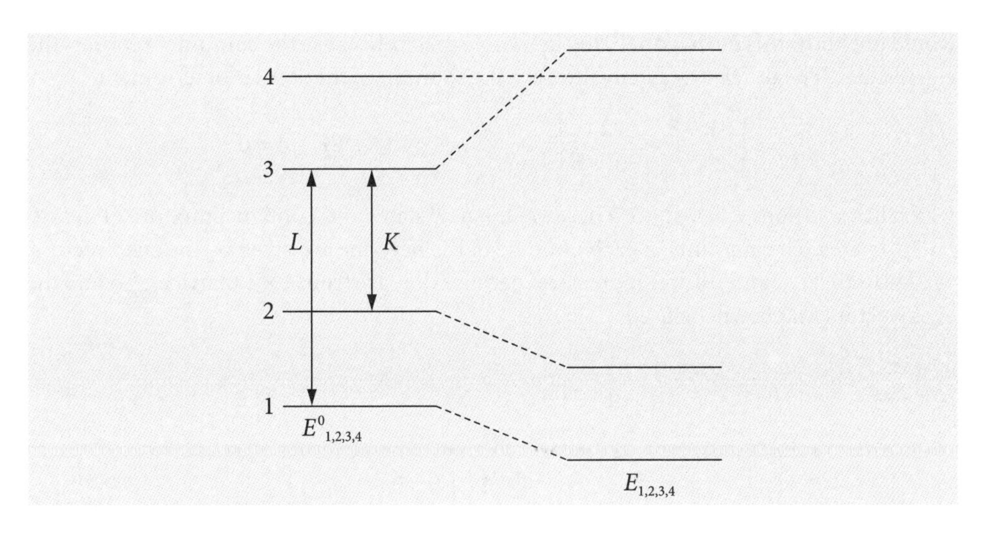

Fig. 7.54 The diagonal matrix produces energy levels (left). The interactions 1–3 and 2–3 move the energy levels about as sketched (right).

is to form new energies E_1, E_2 and E_3, which are functions of the base state energies E_1^0, E_2^0, and E_3^0 and the coupling energies L and K. E_4^0 is unchanged because it is not coupled to any other level. The total energy with these new interactions remains the same which means that some energies must rise and others fall to keep the total energy constant at $E_1^0+E_2^0+E_3^0+E_4^0$.

As an example of using the secular determinant and block diagonal matrices, the coupling between two spins will be calculated. The magnetic moments of electrons and nuclei are coupled via the so-called contact interaction, introduced by Fermi to account for the hyperfine interactions in atomic spectra. The nucleus has spin angular momentum I and the electron S. The interaction represents the energy of the nuclear magnetic moment in the magnetic field at the nucleus, produced by the 'spinning' electrons. The coupling matrix with nuclear spin $S_n = 1/2$ and electron spin $S_e = 1/2$ produces a 4×4 matrix because terms arise from spin combinations $\alpha\alpha$, $\alpha\beta$, $\beta\alpha$, $\beta\beta$. The first letter describes the electron spin, the second nuclear spin. The label α represents either the electron or nuclear spin wavefunction with spin magnetic quantum number $m_s = {}^1/_2$ and β that with $m_s = -1/2$ (m_s is also called the projection or azimuthal quantum number). With the basis set $\{\alpha\alpha, \alpha\beta, \beta\alpha, \beta\beta\}$ determining the order of terms in the rows and columns of the interaction matrix is

$$I{\cdot}S = \frac{a}{4}\begin{bmatrix} 1 & 0 & 0 & 0 \\ 0 & -1 & 2 & 0 \\ 0 & 2 & -1 & 0 \\ 0 & 0 & 0 & 1 \end{bmatrix}$$

where a is a constant. This matrix is essentially the same as that derived in equation (8.25), which describes an NMR spectrum but here with the magnetic field B set to zero. The interaction of state $\beta\alpha$ with $\alpha\beta$, is matrix element (3, 2) and is the same as the interaction of element (2, 3) of state $\alpha\beta$ with $\beta\alpha$. The matrix is Hermitian real and symmetrical, meaning that the eigenvalues will be real and eigenvectors orthogonal to one another. As the matrix is block diagonal containing two 1×1 and one 2×2 matrix, the eigenvalues can easily be found by hand. The same solution is produced by Maple, which will also be used to find the eigenvectors. The wavefunctions for states where there is interaction between the basis states, is in general,

$$\psi = c_1|\alpha\alpha\rangle + c_2|\alpha\beta\rangle + c_3|\beta\alpha\rangle + c_4|\beta\beta\rangle$$

where c_1 to c_4 are the eigenvector columns corresponding to each state. The notation means $|\alpha\alpha\rangle \equiv \psi_{\alpha\alpha}$.

The two 1×1 blocks, which correspond to states $\alpha\alpha$ and $\beta\beta$, each have an eigenvalue of 1 in units of $a/4$. Notice that in this instance we know which states are $\alpha\alpha$ and $\beta\beta$. We would not know this by diagonalizing the whole matrix because the computer reorders the eigenvalues. The other two eigenvalues are the solution of the secular determinant

$$\begin{vmatrix} -1-E & 2 \\ 2 & -1-E \end{vmatrix} = 0 \qquad \text{or} \qquad (1+E)^2 - 4 = 0$$

which has solutions $E = 1$ or -3. These are the new states and contain a mixture of the $\alpha\alpha$ and $\beta\beta$ states. The amounts of each old state in the new one are given by the eigenvectors. We know that $\alpha\alpha$ and $\beta\beta$ are single states because they are from 1×1 matrices. To find the eigenvectors Maple will be used.

```
> with(LinearAlgebra):
> M:=< <1|0|0|0> , <0|-1|2|0> , <0|2|-1|0> , <0|0|0|1> >;
  (eigval, eigvec ):= Eigenvectors(M)> ;
```

$$eigval, eigvec := \begin{bmatrix} 1 \\ 1 \\ 1 \\ -3 \end{bmatrix}, \begin{bmatrix} 0 & 0 & 1 & 0 \\ 0 & 1 & 0 & -1 \\ 0 & 1 & 0 & 1 \\ 1 & 0 & 0 & 0 \end{bmatrix}$$

The eigenvalues confirm the previous calculation. Eigenvector 1 belongs to state $\alpha\alpha$, and 3 to state $\beta\beta$ or vice versa; it makes no difference. The wavefunctions are $\psi = |\alpha\alpha\rangle$ and $\psi = |\beta\beta\rangle$, and both have energy $E = 1$ but are accidentally degenerate. The other two wavefunctions are constructed as $\alpha\beta + \beta\alpha$ and $\alpha\beta - \beta\alpha$ because their eigenvectors are $[0 \pm 1\ 1\ 0]$, which is in the same order as the basis set. The normalization constant for this pair of eigenvectors is $1/\sqrt{2}$, making the normalized wavefunctions for the mixed states $\psi = \dfrac{|\alpha\beta\rangle \pm |\beta\alpha\rangle}{\sqrt{2}}$ with energies $E = 1$ and -3.

7.12.10 Questions

Full solutions are available at **www.oxfordtextbooks.co.uk/orc/beddard.**

Q7.45 Write down the characteristic equations and find the eigenvectors and *normalized* eigenvalues of the following two matrices. Check whether the eigenvectors are orthogonal.

(a) $\begin{bmatrix} 4 & -i \\ i & 2 \end{bmatrix}$, **(b)** $\begin{bmatrix} 1 & -i \\ -i & 1 \end{bmatrix}$.

Strategy: Expand the determinants of the matrices to find the characteristic equations, then use Maple, as shown in the text, or, if you wish, do the calculation by hand. The normalization term is $\sqrt{v^* \cdot v}$ and the Maple instruction to form a dot product is `DotProduct(..,..);` because the vector is complex.

Q7.46 Find the eigenvalues of the matrix by hand

$$\begin{bmatrix} 1 & 0 & 0 & 0 & 0 & 0 \\ 0 & 2 & 4 & 0 & 0 & 0 \\ 0 & 5 & 3 & 0 & 0 & 0 \\ 0 & 0 & 0 & 2 & 3 & 0 \\ 0 & 0 & 0 & 3 & 4 & 0 \\ 0 & 0 & 0 & 0 & 0 & 2 \end{bmatrix}$$

and confirm your result with Maple.

Strategy: This is a block diagonal matrix so use this property to make smaller matrices.

Q7.47 Find the characteristic equation, eigenvalues, and eigenvectors of $\begin{bmatrix} 0 & -u & v \\ u & 0 & 0 \\ -v & 0 & 0 \end{bmatrix}$. Simplify your answers using Maple, and $z^2 = u^2 + v^2$.

Strategy: Use Maple to solve the characteristic equation. The pattern of cofactors are shown in eqn (7.2).

Q7.48 In Section 7.2.5(iii) the MO energies of butadiene were calculated by the Hückel method, using as a basis set the atomic wavefunctions $\{\psi_1, \psi_2, \psi_3, \psi_4\}$, where the subscript labels the $n = 4$ atoms. The Hückel matrix is

$$\begin{vmatrix} x & 1 & 0 & 0 \\ 1 & x & 1 & 0 \\ 0 & 1 & x & 1 \\ 0 & 0 & 1 & x \end{vmatrix} = 0 \qquad \text{where} \qquad x = \frac{\alpha - E}{\beta}.$$

(a) Using the eigenvalue–eigenvector method, calculate not only the energies, but also the orbital coefficients, which are the eigenvectors.

(b) Calculate the delocalization energy, which is the Hückel energy less $n(\alpha + \beta)$ where α is the Coulomb self-energy of a π electron and β the overlap energy.

(c) Calculate the bond order, charge density, and dipole moment. The bond order is $\rho_{ab} = \sum_i^N m_i c_{ai} c_{bi}$ where c_{ai} is the coefficient on carbon atom a and of orbital i and m_i is 0, 1, or 2 and is the number of electrons in orbital i. The total bond order is larger by 1 when the σ bond order is added. The charge density is $q_a = \sum_i m_i |c_{ia}|^2$ and dipole moment $d_\pi = \sum_a (1 - q_a) r_a$ where r_a is the coordinate of atom a. The dipole of a CH bond is 0.3 D.

Strategy: As this is a large matrix and not block diagonal, use Maple to perform the calculation. Each MO with index i is the wavefunction $\Psi_i = c_{1i}\psi_1 + c_{2i}\psi_2 + c_{3i}\psi_1 + c_{4i}\psi_1$ where the cs are the elements of the ith eigenvector; the second (column) index identifies the eigenvalue i, the first the atom.

The full spatial dependence of the orbitals would involve calculating the ψ in three dimensions; instead, and just as effectively, the coefficients c are used to represent the π electron density on each atom which allows us to find the MO's pattern and hence the number of nodes. The node pattern can be used to determine the energy ordering; as a rule of thumb, the larger the number of nodes the higher the energy.

Q7.49 Repeat the calculation of the previous question, Q7.48, for fulvalene; see Q7.8 for the numbering of the atoms. Confirm that the dipole is $-0.711eL$ or 0.48 debye where e is the charge on the electron (in coulombs) and L is the bond length, which is ≈140 pm. The experimentally measured dipole is 0.4 D. Confirm also that the π bond order between atoms is as shown below:

Bond number	Bond order
1 − 2	0.499
2 − 3	0.788
3 − 4	0.520

The remainder of the bond orders follow by symmetry.

Q7.50 Repeat the Hückel MO calculation for benzene; the matrix is worked out in Q6.7. Calculate the eigenvectors and plot out the MO coefficients.

(a) Is the pattern what you expect? You should find that some MOs are simply rotations of others. What distinguishes these?

(b) Make linear combinations of these MOs to form new MOs with the usual orbital shapes. Draw out the results you obtain.

Strategy: for the first part, use Maple to obtain the eigenvectors.

7.13 Rate equations and chemical kinetics

7.13.1 Linear kinetic schemes

The equations of chemical kinetics can quite easily become complicated even with simple schemes. The sequential scheme $A \xrightarrow{k_1} B \xrightarrow{k_2} C$ is quite difficult to solve by direct integration of the equations involved, and if a reversible step is introduced, such as $A \underset{k_{-1}}{\overset{k_1}{\rightleftarrows}} B \xrightarrow{k_2} C$, then solving the rate equations becomes a difficult task. When performing transient kinetics experiments using, for instance, the stopped-flow, flash-photolysis, or femtosecond pump-probe methods, the time profiles of the species present have to be calculated, so that they can be fitted to data to obtain rate coefficients. The transient species present, such as B above, are identified both by their time profile and by spectra.

A kinetic scheme is always written in terms of the rate of change of the species present; for example, the first-order decay of species A is

$$\frac{dA}{dt} = -k_1 A$$

with rate constant k_1, which is in units of s^{-1}. k_1 is more properly called the *rate coefficient*. Conventionally we write the concentration of species A as [A] but for clarity we use A. Note that A is a function of time, this is implied in the equations but almost never explicitly written down, which means that we rarely need to write $A(t)$.

The first-order equation indicates that the rate of change of A is negative; therefore, A is decaying away and being changed into something else and it is doing so at a *rate* of $-k_1 A$ mol dm^{-3} s^{-1}. The rate at any time is thus proportional to how much of A remains unreacted.

In the scheme,

$$A \xrightarrow{k_1} B \xrightarrow{k_2} C,$$

as A decays, B is formed at rate $k_1 A$ and as this decays with rate $k_2 B$, C is formed. The whole scheme is

$$\begin{aligned} \frac{dA}{dt} &= -k_1 A \\ \frac{dB}{dt} &= k_1 B - k_2 B \\ \frac{dC}{dt} &= k_2 B. \end{aligned} \tag{7.35}$$

In solving these differential equations the initial conditions must always be defined. This means defining the concentration of all possible species present in the reaction scheme at time zero. Often it is assumed that only some amount of A present is present, say A_0, and that $B_0 = C_0 = 0$. You will notice that, since we cannot create or destroy atoms in a chemical process, the total amount of all species is constant at all time, so that $A_0 + B_0 + C_0 = A + B + C$, which means that C can always be calculated as the difference $C = A_0 + B_0 + C_0 - A - B$ so there are really only two differential equations. However, if C decomposes with time to another species, we cannot use this last relationship.

We know from direct integration that $A = A_0 e^{-k_1 t}$, and from the theory of coupled linear differential equations that the solution in general for all species are *sums of exponential* terms, see Chapter 10.7. If the initial concentration of A is A_0 and B and C are initially zero, then the concentration of B is

$$B(t) = \frac{k_1 A_0}{k_1 - k_2}(e^{-k_2 t} - e^{-k_1 t}), \tag{7.36}$$

which is of the form expected, with an exponential decay preceded by a grow-in, see Fig. 7.55. Notice that k_1 cannot have the same value as k_2 as then the equation for the population would be 0/0, which is undefined; but in fact there is no reason why these rate constants should not be equal, that is to say that B is formed with the same rate constant as it decays. This special case is dealt with in Chapter 10.7(ii).

7.13.2 Matrix solutions

Using matrix methods changes the way the problem is solved into that of finding eigenvalues and eigenvectors, thereby avoiding the difficulty of integration, but it is only applicable to first order or linear equations; i.e. product terms such as $k_2 AB$ are not allowed. For very complex schemes, consisting of hundreds of equations, the master equation approach is used and is usually solved numerically. A master equation is defined as a phenomenological, first-order differential equation, describing the time evolution of the probability of a 'system' to occupy any one of a discrete set of states.

The matrix method is described first and justified in the next section. Returning to the sequential scheme (7.35), the three equations can be reproduced in matrix form as

$$dM/dt = kM,$$

the solution of which is formally

$$M = M_0 e^{kt}. \tag{7.37}$$

where M is a column matrix of concentrations at time t, M_0 the matrix of their initial values, and k is a square matrix of rate constants. The rate equations (7.35) are rewritten in matrix form as

$$\begin{bmatrix} \dfrac{dA}{dt} \\ \dfrac{dB}{dt} \\ \dfrac{dC}{dt} \end{bmatrix} = \begin{bmatrix} -k_1 & 0 & 0 \\ k_1 & -k_2 & 0 \\ 0 & k_2 & 0 \end{bmatrix} \begin{bmatrix} A \\ B \\ C \end{bmatrix}. \tag{7.38}$$

Notice how the decay rate constant of each species is on the diagonal, and the grow-in or decay of species C from B and B from A, on the off-diagonal. Notice also that the matrix is not Hermitian, i.e. is not symmetrical, although each term is real. This means that when the equation is solved the eigenvectors x are not orthogonal.

Solving the matrix equation (7.37) is done in two steps. First the eigenvalues λ are obtained from the secular determinant of the rate constants, then equation (7.39) is used to obtain the populations with time. The justification for this is given in the next section, 7.13.3; we use it first. The secular determinant of matrix k is

$$\begin{vmatrix} -k_1 - \lambda & 0 & 0 \\ k_1 & -k_2 - \lambda & 0 \\ 0 & k_2 & 0 - \lambda \end{vmatrix} = 0,$$

whose characteristic equation is $(k_1 + \lambda)(k_2 + \lambda)\lambda = 0$ and from which, by inspection, $\lambda_1 = -k_1$, $\lambda_2 = -k_2$, and $\lambda_3 = 0$.

To calculate the populations, or concentrations, the matrix equation

$$M(t) = x[e^{\lambda t}]x^{-1}M_0 \tag{7.39}$$

is used. This produces a column vector $M(t)$ of the populations of each species at time t, M_0 being a column vector of the initial populations. The eigenvectors of matrix k are formed into a (modal) matrix x and x^{-1} is its inverse. The exponential matrix is the diagonal matrix of the exponential of eigenvalues multiplied by time:

$$[e^{\lambda t}] = \begin{bmatrix} e^{\lambda_1 t} & 0 & 0 & \cdots \\ 0 & e^{\lambda_2 t} & 0 & \cdots \\ 0 & 0 & e^{\lambda_3 t} & \cdots \\ \vdots & \vdots & \vdots & \ddots \end{bmatrix}, \tag{7.40}$$

which in this particular example is $[e^{\lambda t}] = \begin{bmatrix} e^{-k_1 t} & 0 & 0 \\ 0 & e^{-k_2 t} & 0 \\ 0 & 0 & 1 \end{bmatrix}$.

The whole equation is, in diagrammatic form,

$$\begin{bmatrix} M_1(t) \\ M_2(t) \\ \vdots \\ \vdots \end{bmatrix} = \begin{bmatrix} x_{11} & x_{12} & \vdots & x_{1n} \\ x_{21} & \vdots & \vdots & \vdots \\ \vdots & \vdots & \vdots & \vdots \\ \vdots & x_{n2} & \vdots & \vdots \end{bmatrix} \begin{bmatrix} e^{\lambda_1 t} & 0 & 0 & \cdots \\ 0 & e^{\lambda_2 t} & 0 & \cdots \\ 0 & 0 & e^{\lambda_3 t} & 0\cdots \\ \vdots & \vdots & 0 & \ddots \end{bmatrix} \begin{bmatrix} x_{11} & x_{12} & \vdots & x_{1n} \\ x_{21} & \vdots & \vdots & \vdots \\ \vdots & \vdots & \vdots & \vdots \\ \vdots & x_{n2} & \vdots & \vdots \end{bmatrix}^{-1} \begin{bmatrix} M_{0,1} \\ M_{0,2} \\ \vdots \end{bmatrix} \tag{7.41}$$

where each eigenvector is a column in the x matrix. The populations of each species are the rows of the $M(t)$ column vector.

The calculation, using Maple, is shown in Algorithm 7.6. It is assumed that the rate constants are k_1 and k_2, and that, at time zero, the amount of A present $A_0 = 1$, and that $B_0 = C_0 = 0$; `k_matrix` is the matrix of rate constants. This code will calculate any $A \rightleftharpoons B \rightleftharpoons C$

scheme by suitable addition of rate constants to the `k_matrix`. Similarly, any initial amounts of A, B, or C can be made by altering vector `M0`.

Algorithm 7.6 Scheme $A \xrightarrow{k_1} B \xrightarrow{k_2} C$

```
> with(LinearAlgebra):
> k_matrix:= < < -k1|0|0 > , < k1|-k2|0 > , < 0|k2|0 > >;
```

$$k\_matrix := \begin{bmatrix} -k1 & 0 & 0 \\ k1 & -k2 & 0 \\ 0 & k2 & 0 \end{bmatrix}$$

```
(eigval, eigvec):= Eigenvectors( k_matrix );
```

$$eigval, eigvec := \begin{bmatrix} -k2 \\ -k1 \\ 0 \end{bmatrix}, \begin{bmatrix} 0 & \dfrac{-k2+k1}{k2} & 0 \\ -1 & -\dfrac{k1}{k2} & 0 \\ 1 & 1 & 1 \end{bmatrix}$$

```
lambda1:= eigval[1]:
lambda2:= eigval[2]:
lambda3:= eigval[3]:                    # eigenvalues
exp_matrix:= < < exp(lambda1*t) | 0 | 0 >,
               < 0 | exp(lambda2*t)| 0 >,
               < 0 | 0 | exp(lambda3*t) > >;
```

$$exp\_matrix := \begin{bmatrix} e^{-k2\,t} & 0 & 0 \\ 0 & e^{-k1\,t} & 0 \\ 0 & 0 & 1 \end{bmatrix}$$

```
> M0:= < A0, 0, 0 >;                      # initial popl'ns
  Pop:= eigvec. exp_matrix . eigvec^(-1). M0; # eqn (7.39)
```

$$Pop := \begin{bmatrix} e^{-k1\,t}\, A0 \\ \left(\dfrac{e^{-k2\,t}\, k1}{-k2+k1} - \dfrac{k1\, e^{-k1\,t}}{-k2+k1}\right) A0 \\ \left(-\dfrac{e^{-k2\,t}\, k1}{-k2+k1} + \dfrac{e^{-k1\,t}\, k2}{-k2+k1} + 1\right) A0 \end{bmatrix}$$

```
> # make functions to plot A, B and C populations
  A:= unapply(Pop[1],t,k1,k2):
  B:= unapply(Pop[2],t,k1,k2):
  C:= unapply(Pop[3],t,k1,k2):
  k1:= 1: k2:= 1.5: A0:=1;      # choose values but not k1=k2.
  plot([A(t,k1,k2),B(t,k1,k2),C(t,k1,k2)],t = 0..5,0..1);
```

The populations of the three species are shown in Fig. 7.55. Intuitively, they have the correct form: A decays to zero, B rises and falls also to zero, and C continuously rises to a constant value, which is A_0, as all molecules must eventually be converted into C. Notice, that even though B decays with a rate constant greater than it forms, it is still possible to see the population of B. You might like to calculate this figure and compare it with that when k_1 and k_2 are swapped in value.

7.13.3 Justification for the matrix method to solve coupled linear differential equations

In using equation (7.38) to solve the scheme, the assumption is made that if M is a vector and k a matrix, we can write $dM/dt = kM$ and that its solution is $M = M_0e^{kt}$. To show that this is valid, we work backwards and start with M, which is differentiated term by term, after first expanding the exponential as a series. In Section 7.5.5 it was shown that the exponential of a matrix can only be manipulated by first expanding.

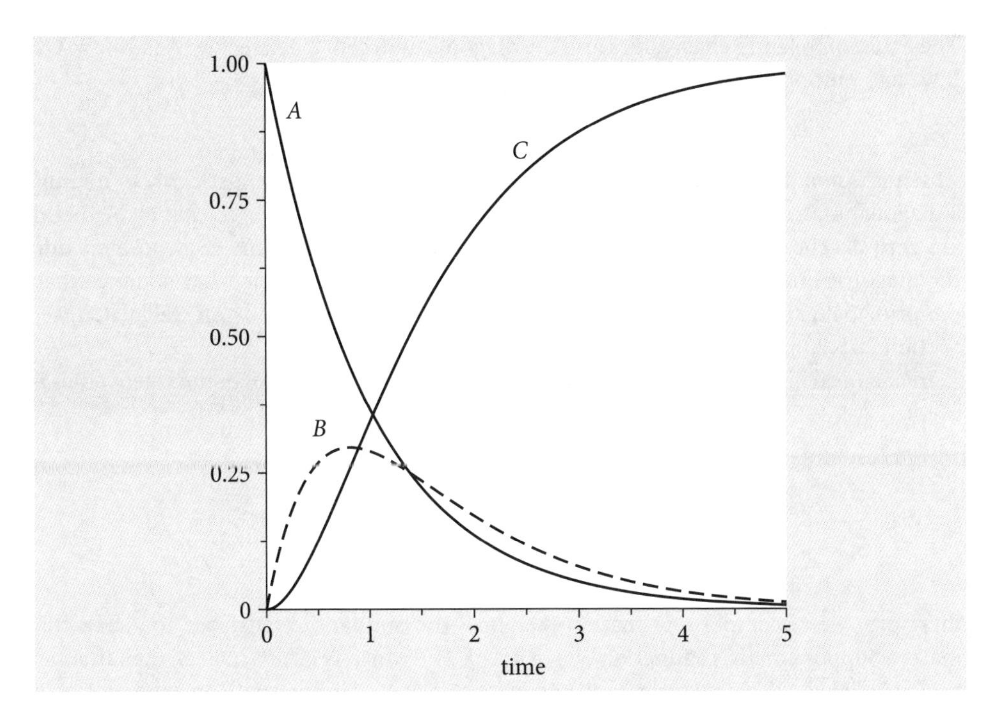

Fig. 7.55 Populations of species A, B, and C with time when $k_1 = 1$ and $k_2 = 1.5$ with initial concentration of $A_0 = 1$ and $B_0 = C_0 = 0$. The scheme is $A \xrightarrow{k_1} B \xrightarrow{k_2} C$.

$$\frac{dM}{dt} = M_0\frac{d}{dt}e^{kt} = M_0\frac{d}{dt}\left(1 + kt + \frac{k^2}{2!}t^2 + \frac{k^3}{3!}t^3 + \cdots\right) = M_0\left(k + k^2t + \frac{k^3}{2!}t^2 \cdots\right)$$

$$= kM_0\left(1 + kt + \frac{k^2}{2!}t^2 + \cdots\right) = kM_0e^{-kt} = kM.$$

In the last two steps, k has been separated out, and the remaining series identified as the first few terms of the exponential expansion.

The second equation, (7.39), used in the calculation, is $M(t) = X[e^{\lambda t}]X^{-1}M_0$ and the form of this needs explaining. First, each value in the column vector $M(t)$ represents the concentration of each species at time t, the column vector M_0 holds the initial concentrations of the same species. The equation we start with is $M(t) = M_0e^{-kt}$. To solve this it has to be converted into $M(t) = X[e^{\lambda t}]X^{-1}M_0$, and to do this a *similarity transform* is needed.

7.13.4 Similarity transforms

In our study of molecular group theory, *similar* matrices were introduced. However, similar matrices are used more generally. A square matrix W is described as being *similar* or *conjugate* to another square matrix N, if there exists a third, non-singular, square matrix X such that

$$W = X^{-1}NX, \qquad (7.42)$$

then $X^{-1}NX$ is a *similarity transform*; XNX^{-1} is also a similarity transform. A similarity transform is equivalent to changing the basis set of one matrix into another basis; this might also be considered a coordinate transformation. If matrix W operates on an object, this is the equivalent of transforming the object (this could be a rotation), then operating on it by N, then undoing the initial transformation with X^{-1}. Thus, W performs the same operation as N but uses a different set of axes.

The similarity transform is most useful in diagonalizing matrices, because, if a matrix W can be diagonalized and its matrix of eigenvectors X calculated, then W is *similar* to the *diagonal matrix of its eigenvalues* Λ, where the eigenvectors X act as the similarity matrices. The diagonal matrix is

$$\Lambda = \begin{bmatrix} \lambda_1 & 0 & 0 & \cdots \\ 0 & \lambda_2 & 0 & \cdots \\ 0 & 0 & \lambda_3 & \cdots \\ \vdots & \vdots & \vdots & \ddots \end{bmatrix}$$

The equation relating eigenvalues and the diagonal matrix of eigenvectors Λ is $WX = X\Lambda$. If we left multiply by the inverse eigenvector matrix X^{-1} then

$$X^{-1}WX = \Lambda \tag{7.43}$$

which is a *similarity transform*. Of course, the inverse matrix X^{-1} must exist. In making a diagonal matrix, the original matrix has been 'rotated', as it were, so that all elements are zero except on the diagonal, the off-diagonal elements become incorporated into the diagonal ones. This 'rotation' or coordinate change analogy, is what is done when the principal axis are found when the moments of inertia of a body are calculated, see Section 7.15.2.

In chemical kinetics, the diagonal matrix of the exponential values of the eigenvalues is used:

$$[e^{\lambda t}] = \begin{bmatrix} e^{\lambda_1 t} & 0 & 0 & \cdots \\ 0 & e^{\lambda_2 t} & 0 & \cdots \\ 0 & 0 & e^{\lambda_3 t} & \cdots \\ \vdots & \vdots & \vdots & \ddots \end{bmatrix},$$

therefore, we need to show that we can use the similarity transform to make this matrix. Suppose there is a function of a matrix $f(W)$, which is itself a matrix, then this can be expanded as a (Taylor) series with coefficients a_i, just as is done with an exponential or other normal function. A general power series is written as

$$f(W) = a_0\mathbf{1} + a_1W + a_2W^2 + \cdots$$

Because $f(W)$ must commute with W, then both matrices can be diagonalized simultaneously with a similarity transform. Using the eigenvector matrix X of matrix W, then the similarity transform is

$$X^{-1}f(W)X = f(\Lambda) \tag{7.44}$$

and $f(\Lambda)$ is a diagonal matrix of function f. For instance, if $f(W) = e^W$, then

$$x^{-1}e^W x = e^\Lambda$$

which is used to solve equations in chemical kinetics.

Suppose that a matrix M is to be raised to its nth power, the function $f(W)$ is then M^n. The similarity transform becomes $X^{-1}M^nX = \Lambda^n$, which can be rearranged to $M^n = X\Lambda^nX^{-1}$ and so the equation can be used both ways round. This last expression is a very convenient way of raising a matrix to a large power, if M can be diagonalized and its eigenvalues X determined. The alternative method, is the repeated multiplication $MMMM\cdots$, which is impracticable should n be large, 100 for example.

To show that a similarity transform on M produces a diagonal matrix of eigenvalues, consider the matrix

$$M = \begin{bmatrix} -2 & 10 \\ 2 & -3 \end{bmatrix}.$$

The eigenvalues are obtained using the secular determinant

$$\begin{vmatrix} -2-\lambda & 10 \\ 2 & -3-\lambda \end{vmatrix} = 0$$

and solving the characteristic equation, which is $(\lambda + 2)(\lambda + 3) - 20 = 0$. This produces eigenvalues of $\lambda_1 = 2$ and $\lambda_2 = -7$. The eigenvectors are

$$x_1 = \begin{bmatrix} 5/2 \\ 1 \end{bmatrix} \text{ and } x_2 = \begin{bmatrix} -2 \\ 1 \end{bmatrix},$$

as shown below using Maple. The dot product of the eigenvectors is -4 so they are not orthogonal, which is expected as the matrix is not Hermitian. The (modal) matrix of eigenvector column vectors is

$$x = \begin{bmatrix} 5/2 & -2 \\ 1 & 1 \end{bmatrix}$$

which has an inverse,

$$x^{-1} = \frac{1}{9}\begin{bmatrix} 2 & 4 \\ -2 & 5 \end{bmatrix}.$$

(The eigenvalues, eigenvectors, and inverse of any 2×2 matrix are given in Section 7.5.6.) The similarity transform is therefore

$$\Lambda = x^{-1}Mx = \frac{1}{9}\begin{bmatrix} 2 & 4 \\ -2 & 5 \end{bmatrix}\begin{bmatrix} -2 & 10 \\ 2 & -3 \end{bmatrix}\begin{bmatrix} 5/2 & -2 \\ 1 & 1 \end{bmatrix} = \frac{1}{9}\begin{bmatrix} 2 & 4 \\ -2 & 5 \end{bmatrix}\begin{bmatrix} 5 & 14 \\ 2 & -7 \end{bmatrix} = \begin{bmatrix} 2 & 0 \\ 0 & -7 \end{bmatrix},$$

and which proves equation (7.43) because a diagonal matrix of eigenvalues is obtained. Using Maple,

```
> with(LinearAlgebra):
> M:= < < -2 | 10 > , < 2 | -3 > >;
  (eigval,eigvec):= Eigenvectors(M);
  Lambda:= eigvec^(-1). M . eigvec; # similarity transform
```

$$eigval,\ eigvec := \begin{bmatrix} 2 \\ -7 \end{bmatrix}, \begin{bmatrix} \frac{5}{2} & -2 \\ 1 & 1 \end{bmatrix}$$

$$\Lambda := \begin{bmatrix} 2 & 0 \\ 0 & -7 \end{bmatrix}$$

To raise M to the 5th power the calculation is $M^5 = X\Lambda^5X^{-1}$ or

$$\begin{aligned} M^5 &= \begin{bmatrix} 5/2 & -2 \\ 1 & 1 \end{bmatrix}\begin{bmatrix} 2 & 0 \\ 0 & -7 \end{bmatrix}^5\begin{bmatrix} 2 & 4 \\ -2 & 5 \end{bmatrix} \\ &= \begin{bmatrix} 5/2 & -2 \\ 1 & 1 \end{bmatrix}\begin{bmatrix} 2^5 & 0 \\ 0 & -7^5 \end{bmatrix}\begin{bmatrix} 2 & 4 \\ -2 & 5 \end{bmatrix} \\ &= \begin{bmatrix} -7452 & 18710 \\ 3472 & -9323 \end{bmatrix} \end{aligned}$$

Using Maple again, this last calculation is written as

```
> eigvec. Lambda^(5).eigvec^(-1);
```

7.13.5 A similarity transform used to evaluate a partition function. The helix-coil transition

A polypeptide or a protein can change from a helix to a random coil structure or vice versa, over a small range of temperatures. This sudden change is suggestive of a cooperative process and Zimm & Bragg (1959) used a statistical model and calculated the partition function for a peptide chain consisting of two types of residues, either those that form a helix or those that do not. A section of protein will be described as having an unstructured coil if it has c type amino acids and h type if they are helix forming such as alanine. A portion of the chain could be $\cdots cchcchhhcchh\cdots$. The statistical weight of a $\cdots ch$ boundary and of continuing to grow a helix $\cdots chhh$ once formed is calculated and this statistical weight is the Boltzmann contribution to the partition function or $e^{-\Delta G/k_BT}$, where ΔG is the change in free energy when the next residue is encountered along the chain. The model is successful despite the assumption of nearest neighbour interactions whereas a helix strictly involves interactions with residues that are not just nearest neighbours.

The model has the following rules.

(i) The energy of the randomly coiled chain is taken as the baseline energy and it has a value of zero and a given a statistical weight of 1.

(ii) A section of helix is energetically favourable as hydrogen bonds are formed. The statistical weight given to a helix forming residue is s, if the helix is already formed, i.e. this is the statistical weight to add an h if the preceding residue is also h. This is sometimes called a helix continuation parameter.

(iii) The first helix forming residue must be at a $\cdots ch$ boundary. There is an entropy term σ due to restricting the rotational motion of the residue on forming a helix from a coil, making the statistical weight σs. The value of σ is always less than 1.

The results of the calculation show that there is cooperativity in the helix-coil transition. When $\sigma \ll 1$ it is hard to form a helix, but once formed it is energetically favourable to continue to add only h residues and the helix-coil 'phase transition' is sharp. Conversely, when $\sigma \sim 1$ there is no cooperativity in the transition and it occurs over a relatively wide range of temperature, which effectively means a large range of s because s is the Boltzmann contribution $e^{-\Delta G/k_BT}$ to the partition function.

Zimm & Bragg (1959) presented a matrix method to calculate the partition function sum, and they showed that the equation $A_{n+1} = MA_n$ can be used where M is a transition matrix and A_n is a column vector containing two terms, one for adding a c residue and one for adding an h residue. The transition matrix represents the statistical weight of adding the next residue but also takes into account the weights of all the preceding ones. This matrix equation is

$$\begin{bmatrix} A^c_{n+1} \\ A^h_{n+1} \end{bmatrix} = \begin{bmatrix} 1 & 1 \\ \sigma s & s \end{bmatrix} \begin{bmatrix} A^c_n \\ A^h_n \end{bmatrix}$$

which can be split into two equations $A^c_{n+1} = A^h_n + A^c_n$ and $A^h_n = \sigma s A^c_n + s A^h_n$ and, written this way, the weights of the preceding residues is made clear. If N residues are considered, the transition equation becomes

$$\begin{bmatrix} A^c_n \\ A^h_n \end{bmatrix} = \begin{bmatrix} 1 & 1 \\ \sigma s & s \end{bmatrix}^N \begin{bmatrix} A^c_0 \\ A^h_0 \end{bmatrix}$$

where A^h_0 and A^c_0 is the initial state of the protein. If this is a single c residue then the equation is

$$\begin{bmatrix} A^h_n \\ A^c_n \end{bmatrix} = \begin{bmatrix} 1 & 1 \\ \sigma s & s \end{bmatrix}^N \begin{bmatrix} 1 \\ 0 \end{bmatrix} = M^N \begin{bmatrix} 1 \\ 0 \end{bmatrix}$$

The partition function is the sum of the two parts A_h and A_c and can be formed by left-multiplying by a unit row vector,

$$Z = (1 \quad 1) M^N \begin{pmatrix} 0 \\ 1 \end{pmatrix}.$$

To complete the calculation the statistical weight matrix has to be solved using a similarity transform. The eigenvalues $\lambda_{1,2}$ are found by expanding the determinant of M in the usual way, and are the solution to the characteristic equation $(1 - \lambda)(s - \lambda) - \sigma s = 0$ which are

$$\lambda_{1,2} = \frac{s - 1 \pm \sqrt{(1 - s)^2 + 4\sigma s}}{2}.$$

The eigenvector modal matrix is

$$X = \begin{bmatrix} \dfrac{1}{\lambda_1 - 1} & \dfrac{1}{\lambda_2 - 1} \\ 1 & 1 \end{bmatrix}$$

and its inverse

$$X^{-1} = \frac{1}{\lambda_1 - \lambda_2} \begin{bmatrix} -(\lambda_1 - 1)(\lambda_2 - 1) & \lambda_1 - 1 \\ (\lambda_1 - 1)(\lambda_2 - 1) & -\lambda_2 + 1 \end{bmatrix}.$$

The matrix is raised to the power N with

$$M^N = X\Lambda X^{-1} = \frac{1}{\lambda_1 - \lambda_2} \begin{bmatrix} \dfrac{1}{\lambda_1 - 1} & \dfrac{1}{\lambda_2 - 1} \\ 1 & 1 \end{bmatrix} \begin{bmatrix} \lambda_1^N & 0 \\ 0 & \lambda_2^N \end{bmatrix} \begin{bmatrix} -(\lambda_1 - 1)(\lambda_2 - 1) & \lambda_1 - 1 \\ (\lambda_1 - 1)(\lambda_2 - 1) & -\lambda_2 + 1 \end{bmatrix}.$$

The expression $Z = [1 \quad 1] M^N \begin{bmatrix} 1 \\ 0 \end{bmatrix}$ completes the calculation and produces the partition function. This calculation is most easily achieved using Maple.

```
> with(LinearAlgebra):
> M:= < < 1 | 1 > , <sigma*s | s > >;
  (eigval,eigvec):= Eigenvectors(M):
  eigval;                                   # lambda1 and lambda2
```

$$\begin{bmatrix} \frac{1}{2}+\frac{1}{2}s+\frac{1}{2}\sqrt{1-2s+s^2+4\sigma s} \\ \frac{1}{2}+\frac{1}{2}s-\frac{1}{2}\sqrt{1-2s+s^2+4\sigma s} \end{bmatrix}$$

```
  eigvec;
```

$$\begin{bmatrix} \dfrac{1}{-\frac{1}{2}+\frac{1}{2}s+\frac{1}{2}\sqrt{1-2s+s^2+4\sigma s}} & \dfrac{1}{-\frac{1}{2}+\frac{1}{2}s-\frac{1}{2}\sqrt{1-2s+s^2+4\sigma s}} \\ 1 & 1 \end{bmatrix}$$

```
  X:= < < 1/(lambda1-1) | 1/(lambda2-1)>,
        < 1             | 1             > >;#eigenvector matrix
  Lambda:= < < lambda1^N | 0         >,
             < 0         | lambda2^N > >; # diagonal matrix
  Z:= simplify( <1|1> . X . Lambda . X^(-1) . <1|0> );
```

$$Z := \frac{-\lambda 1^{1+N}\lambda 2+\lambda 1^{1+N}+\lambda 2^{1+N}\lambda 1-\lambda 2^{1+N}}{-\lambda 2+\lambda 1}$$

The average number of helical residues can be calculated with $\langle h\rangle = \frac{s}{Z}\frac{dZ}{ds}$ assuming constant σ and N, and the number of helical stretches with $\langle\sigma\rangle = \frac{\sigma}{Z}\frac{dZ}{d\sigma}$ at constant s and N. Typical values of s are 0 to 3 and of $\sigma = 10^{-3}$ to 1/2. N can range from 10 to 1000. For further details, see Daune (1999) or Jackson (2006).

7.13.6 **Questions**

Full solutions are available at **www.oxfordtextbooks.co.uk/orc/beddard**.

Q7.51 Calculate the time profile of species A, B and C for the scheme

$$A \underset{k_{21}}{\overset{k_{12}}{\rightleftarrows}} B \xrightarrow{k_2} C \xrightarrow{k_3},$$

where rate constants are k_{12}, k_{21}, k_2, and k_3. Use Maple to investigate the population of each species, as outlined after equation (7.41), with, for example, $A_0 = 1$ and B_0 and $C_0 = 0$ and rate constants $k_{12} = 5$, $k_{21} = 6$, $k_2 = 1$, $k_3 = 0.2$. Try other values of rate constants and initial populations for yourself. (You will find that the algebraic equations produced are very large, so use numerical values from the start.) Convince yourself that the populations you graph make sense.

Strategy: Write down the rate equations and then translate them into the matrix form. Use Maple to do the integration by the matrix method outlined in the text, Algorithm 7.6.

Q7.52 Overwhelmingly, life on earth depends upon photosynthesis. In this process, photons are captured in *antennae* that always consist of many pigments, which are usually, but not exclusively, chlorophyll held in a protein. The energy from the lowest excited state of these pigments is passed to a special pair in the *reaction centre* where electron transfer occurs. The electrons are eventually used to reduce carbon dioxide to carbohydrates and eventually to biomass. The antenna, pigment-protein complex of *Prosthecochloris aestuarii* has seven bacteriochlorophyll (BChl) molecules between which Förster energy transfer occurs. The structure of this membrane bound protein has been determined by X-ray crystallography and is shown in Fig. 7.57. The coordinates are in the Brookhaven data bank .pdb entry 3BCHL.

Energy transfer occurs from the excited state of one BChl molecule to another caused by the dipole–dipole interaction between the two molecules. However, energy can leave one molecule and be transferred to a third or back to the second or on to a fourth and so forth. In this way, the excitation energy diffuses around the antenna and the population of each pigment changes until (dynamic) equilibrium is established between the molecular excited states. Energy transfer to the special pair and fluorescence both compete with this process. The reaction centres, which quench the excited states, are found in another nearby protein and with this quenching and by fluorescence, the total excited state population is eventually lost but not until long after this equilibrium between excited states is established.

The rate constant of energy transfer can be calculated between any pair of molecules and is $k = \chi^2 k_f \left(\dfrac{R_0}{R}\right)^6$ where R is the separation of the centres of the chlorophyll molecules at the Mg atoms. R_0 is a characteristic distance which depends upon the overlap of the donor emission spectrum with the absorption spectrum of the donor molecule and which has a constant value for any donor–acceptor pair. The parameter χ is an orientational factor that we shall ignore for simplicity. The donor and acceptor can be the same type of molecule because the absorption and emission spectra of chlorophyll overlap with one another. This type of energy transfer was first predicted by Förster and normally carries his name (Förster 1959; Bennett & Kellogg 1967). It is a very successful way of describing energy transfer in photosynthetic organisms and is the basis of the technique called FRET, or fluorescence, resonance, energy transfer.

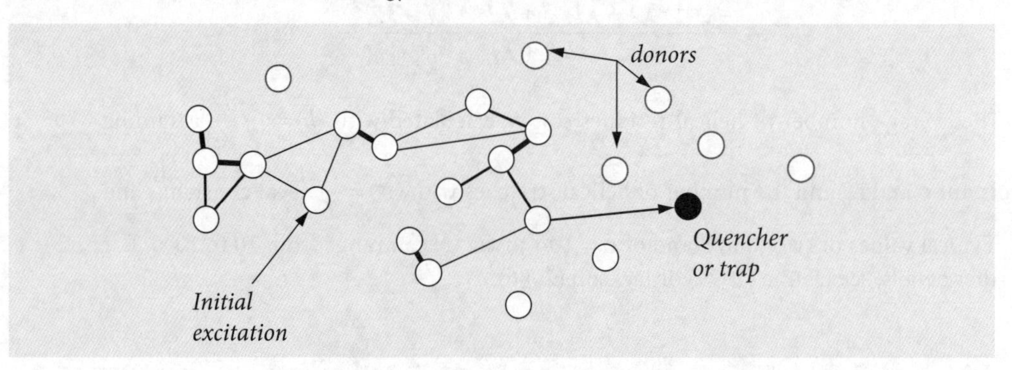

Fig. 7.56 Schematic of energy diffusion (repeated energy transfers) between molecules by the Förster dipole–dipole mechanism. The circles represent the molecules whose positions are fixed. Some possible transfer steps are shown as lines; the thin lines represent just a single transfer, and the thicker lines, repeated transfer back and forth between a pair of molecules. The R^{-6} dependence of the rate of separation ensures that the energy makes many rapid transfers between close pairs of molecules, which tend to slow energy diffusion, but occasional large steps counteract this effect. The diffusion ends when either the molecule fluoresces, or undergoes intersystem crossing (to form a triplet state), or the energy reaches the trap.

If in the antenna protein, molecule 1 is excited, Fig. 7.57, use Maple to calculate the populations on each of the other molecules as a function of time when only fluorescence occurs to deplete the excited state. Assume that any one molecule can transfer to any other, with a rate constant depending on its distance away, and any molecule can therefore receive energy from every other molecule; very democratic!

(a) Why do the calculated rate constants have such a wide range?

(b) Plot the population of molecule 1 and 7 vs time, and any other you want to observe, and estimate how long it takes the energy to equilibrate around the antenna.

(c) Comment on the shape of the calculated populations. Why do they all reach a finite value rather than zero?

(d) Add a quenching term, a rate constant with value 10^{12} s^{-1}, so that energy is lost irreversibly but only from molecule 1 to a nearby quencher, such as a reaction centre.

Strategy: Use the data from the .pdb given below to calculate the separation between each pair of Mg atoms. Then write down the rate constant matrix for the transfer from any one molecule to each of the others. This is a 7×7 matrix and every element will be filled. Because all the molecules are the same, the transfer rate constant from molecule 1 to 2 is the same as the reverse 2 to 1; there is no Boltzmann factor to take into account and therefore the matrix is symmetrical. Assume that the BChl molecules decay with a lifetime of 5 ns so that the fluorescence rate constant is $k_f = 1/5 \times 10^9\ \text{s}^{-1}$ and $R_0 = 70$ Å.

Data extracted from .pdb 3BCHL (Tronrud et al. 1986, b. 443).

	x	y	z Å		
HETATM 2509 MG BCL 1	53.113	58.877	20.553	1.00	16.86
HETATM 2575 MG BCL 2	56.307	55.372	32.462	1.00	14.36
HETATM 2641 MG BCL 3	49.690	44.549	44.813	1.00	10.22
HETATM 2707 MG BCL 4	39.128	41.873	42.544	1.00	10.61
HETATM 2773 MG BCL 5	34.082	47.474	30.966	1.00	11.28
HETATM 2839 MG BCL 6	41.714	47.644	22.127	1.00	12.74
HETATM 2905 MG BCL 7	47.867	43.555	32.934	1.00	10.85

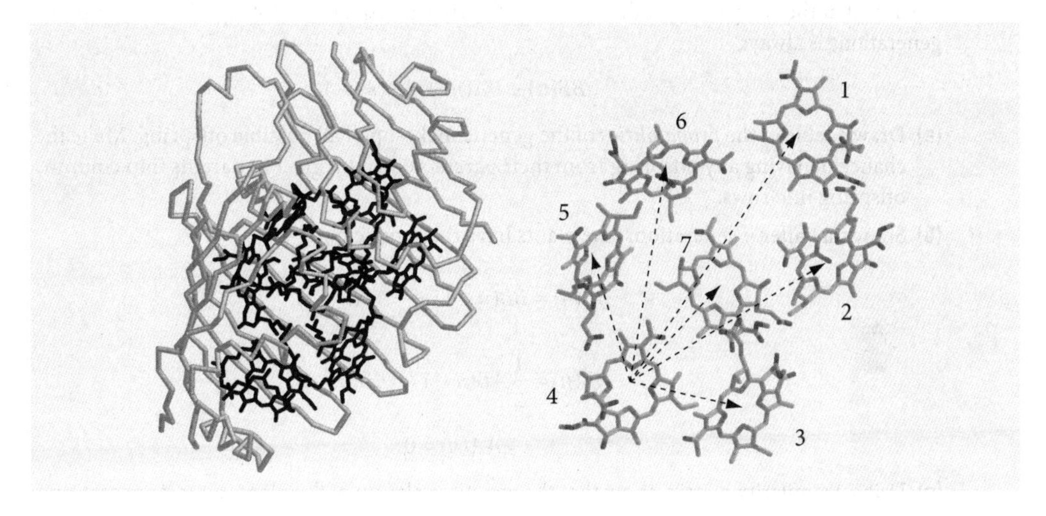

Fig. 7.57 FMO protein with peptide surrounding the BChl molecule, and right, slightly magnified and at a different orientation, to show the bacteriochlorophyll molecules but with their phytyl tails removed for clarity of display. The arrows show transfer from molecule 4 to each of the others. (Structures drawn with Rasmol from .pdb 3BCHL)

Strategy: Although this calculation looks as though it might be very difficult, it is just a bigger version of what was already done in the previous question. The hard part is defining the matrix, so remember that each molecule can transfer to any other. This means that each diagonal term has a rate constant for transfer to *every* other molecule as well as the rate of fluorescence decay. Each molecule also receives energy from every other one, which means that each off-diagonal has one rate constant from each molecule. First write down the rate equation for one molecule, and then use this as a template to work out the rates for the others. It will be easiest to label the rate constants k_{12}, k_{34} and so on where the subscripts label the molecules, k_{12} corresponding to transfer from 1 to 2. Unless you want dozens of pages of algebra, put numerical values into the matrix before calculating the eigenvalues and eigenvectors.

Q7.53 You are presented with a problem in which the matrix $M = \begin{bmatrix} -2a & a \\ a & -2a \end{bmatrix}$ is to be raised to the 50th power and then its determinant evaluated. Direct calculation will inevitably lead to numerical errors. Algebraically, it would be beyond any direct calculation. Therefore, how could this calculation be achieved?

Strategy: A similarity transform is the method to use as this enables us to make a diagonal matrix Λ, *via* $X^{-1}MX = \Lambda$. This can then be manipulated to make M^n with $M^{50} = X\Lambda^{50}X^{-1}$.

Q7.54 In autosomal inheritance, two genes are exchanged during reproduction, but these do not determine in the sex of the organism. Inheriting these genes could mean, in humans for example, producing brown eyes vs blue eyes or, in a flower, red petals vs blue. If an organism is *autosomal dominant*, a gene on one of the non-sex chromosomes is always expressed, even if only one copy is present. If the genes are *AB* with *A* being the dominant mutant, when offspring are produced with a *BB* partner, the result is as shown below in the left most column, Fig. 7.58. The makeup of offspring, by crossing two plants with *AB* or *AA* genes, is also shown in Fig. 7.58, except for *AA* with *AA* and *BB* with *BB* that are unchanged.

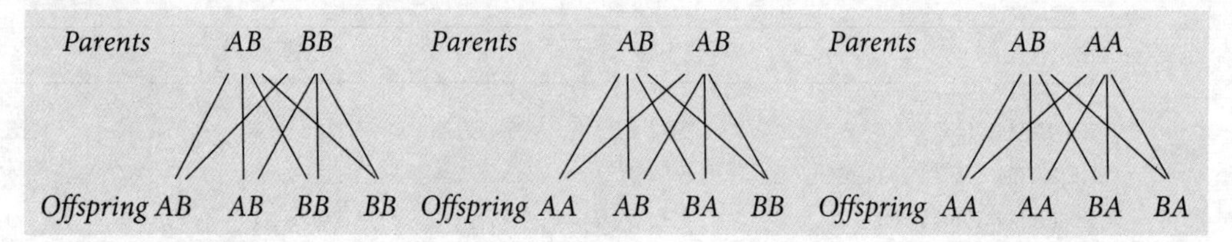

Fig. 7.58 Mixing genes.

A horticulturalist is fertilizing plants *only* with a strain known to be *BB*, which has traits she wants to encourage. When the plants, which are an equal mixture of all possible combinations of *A* and *B* genes, are fertilized they are then replaced by their offspring, and the process repeated *n* times. Assume that the chances of producing each offspring are equal. The total probability after *n* generations is always

$$BB(n) + BA(n) + AA(n) = 1. \tag{7.46}$$

(a) Draw a table of the *probabilities* of the genetic make-up of all possible offspring. Make the total chance of having any offspring from their parent's equal to one. Put parents into columns and offspring into rows.

(b) Show that after *n* generations, the plants have the genetic make-up

$$BB(n) = BB(n-1) + \frac{1}{2}AB(n-1),$$

$$AB(n) = \frac{1}{2}AB(n-1) + BB(n-1),$$

$$AA(n) = 0.$$

(c) Using a similarity matrix, show that the genetic make-up of the plants after *n* generations is predominantly *BB*. How many generations does it take to produce at least 90% of plants of this type?

***Strategy*:** Use the diagrams above to work out what the effect of fertilization will be on individuals with genetic make up AA, BA ≡ AB, and BB. In biologists' jargon, three *genotypes* can be produced, which are *BB*, *AB*, and *AA*.

7.14 Molecular vibrations and pendulums

Matrix eigenvalue–eigenvector problems can also be used to solve a number of related problems such as molecular vibrations, coupled pendulums, and damped oscillators. When the displacements from equilibrium are small the equations produced are linear, and this is always assumed in solving these equations with matrix methods. The numerical methods of Chapter 11 allow the fully non-linear equations to be solved. In this section, the concept of normal modes is described first, then the motion of linked pendulums, and finally the vibrational motion of molecules is described and some examples worked through.

7.14.1 Vibrational normal modes

If the initial displacements made to the atoms of a molecule correspond to that expected of any one of the normal modes, Fig. 7.23, then the oscillations of all atoms continue only in that normal mode. The first reason is because each atom is connected to all others via chemical bonds, and second because each normal mode is orthogonal to every other one. This means that, ideally, no energy flows from one normal mode to any other, and although this happens in practice, for example due to an harmonicity, it need not concern us here. If an arbitrary initial displacement is made that does not correspond to that expected of a normal mode, a complicated motion results. However, this can always be decomposed to a linear combination or sum of the normal modes, taken in various proportions as necessary to describe the motion. The overall motion, when many modes are simultaneously excited, causes beating; this is particularly clear in a coupled pendulum as the two pendulums come into and go out of phase with one another. This is seen when one of the pendulum bobs becomes stationary for an instance while its partner swings maximally; this then loses energy and becomes instantaneously stationary while the other has gained maximum displacement and which continues by swapping energy ad infinitum.

7.14.2 **Calculation of normal modes**

In a mechanical or molecular vibration, we usually try to find the equation relating forces. In simple harmonic motion, such as stretching a chemical bond, the potential energy is by Hooke's law, $V = ks^2/2$, where k is the force constant and s the amount the bond extends away from its equilibrium length. The force f to stretch the bond, is obtained by differentiation $f = -\dfrac{dV}{ds} = -ks$ and is also given by Newton's second law; force = mass × acceleration, $f = m\dfrac{d^2s}{dt^2}$, which provides the connection to time. Equating the two terms produces the equation of motion for small displacements from equilibrium, viz.

$$\frac{d^2s}{dt^2} + \frac{k}{m}s = 0. \tag{7.46}$$

The next step is to find out what k/m represents. By knowing that the motion of the pendulum is periodic, we may assume that the displacement at any time is described by $s = u\cos(\omega t + \phi)$, where ω is an *angular* frequency, ϕ is the phase, t is time, and u is the amplitude of the motion. The phase is arbitrary, and can be set to zero. Differentiating s twice by time and multiplying by m to obtain the force gives

$$m\frac{d^2s}{dt^2} = -m\omega^2 u\cos(\omega t + \phi) = -m\omega^2 s \qquad \text{or} \qquad \frac{d^2s}{dt^2} + \omega^2 s = 0$$

By comparison with equation (7.46), the (angular) oscillation frequency is $\omega = \sqrt{k/m}$, and is in radians s^{-1}. A similar equation describes the oscillation of a pendulum, but now the angular frequency is $\omega = \sqrt{g/L}$ if L is the length of the pendulum and g is the acceleration due to gravity.

When there is more than one oscillator—for instance, two coupled pendulums or the many vibrations of a polyatomic molecule—there is one force equation for each mass in the molecule. The potential V for atom 1 now has terms not only describing its own position, but also of all others connected to it, and could contain terms with each of $s_1, s_2, \cdots$ etc., where s is the displacement of any atom. The equations are therefore coupled and when placed into a matrix form there are off-diagonal force-terms that are not zero;

$$\begin{bmatrix} m_1\dfrac{d^2s_1}{dt^2} \\ m_2\dfrac{d^2s_2}{dt^2} \\ \vdots \end{bmatrix} = \begin{bmatrix} k_{11} & k_{12} & \cdots \\ k_{21} & k_{22} & \\ \vdots & & \ddots \end{bmatrix}\begin{bmatrix} s_1 \\ s_2 \\ \vdots \end{bmatrix} \tag{7.47}$$

The ks in the square matrix are the forces or negative derivatives of the potential energy with respect to each displacement $s_1, s_2, \cdots$. For example, in the coupled pendulum example that will shortly be worked through, we find that $-\dfrac{\partial V}{\partial s_1} = -\dfrac{mgs_1}{L} + k(s_2 - s_1)$ then $k_{11} = -\dfrac{mg}{L} - k$ and $k_{12} = k_{21} = k$.

The next step in the calculation is to write down the equation of motion of the normal modes. Each normal mode has, by definition, its own frequency $\omega_1, \omega_2, \cdots$ and so forth. Occasionally modes are degenerate, but nevertheless, by analogy with a single oscillator, each normal mode has an equation of the form $\dfrac{d^2s_n}{dt^2} + \omega^2 s_n = 0$, where s_n is the coordinate describing the displacement of the normal mode and is not the same as s_1 or s_2 etc. used above. If the equations for the several normal modes are put into matrix form, this is diagonal, because each normal mode is unique and does not depend on any other;

$$\begin{bmatrix} \dfrac{d^2s_{n_1}}{dt^2} \\ \dfrac{d^2s_{n_2}}{dt^2} \\ \vdots \end{bmatrix} = \begin{bmatrix} \omega_1^2 & 0 & 0 \cdots \\ 0 & \omega_2^2 & 0 \cdots \\ 0 & 0 & \ddots \\ \vdots & \vdots & \end{bmatrix} \begin{bmatrix} s_{n_1} \\ s_{n_2} \\ \vdots \end{bmatrix} \tag{7.48}$$

The equations (7.47) and (7.48) describe the same motion but are on different sets of axes. The former, is in terms of displacements of atoms, and the latter, in displacements of groups of atoms in a normal mode. If equation (7.47) is diagonalized, this has the effect of 'rotating' or more generally transforming the atoms' coordinates by mixing them up in such a way that they become the normal mode coordinates. The eigenvalues obtained can then be identified with the squared frequencies. Taking equation (7.47) and making an eigenvalue eigenvector equation gives

$$\begin{bmatrix} k_{11} & k_{12} & \cdots \\ k_{21} & k_{22} & \\ \vdots & & \ddots \end{bmatrix} \begin{bmatrix} s_1 \\ s_2 \\ \vdots \end{bmatrix} = \lambda \begin{bmatrix} s_1 \\ s_2 \\ \vdots \end{bmatrix}$$

which can be solved, based on equation (7.31), by writing it as $\boldsymbol{ks} = \lambda \boldsymbol{s}$, where $\boldsymbol{k}$ is the matrix of force constants, and s the eigenvalues. Once the eigenvalues are obtained, they form a diagonal matrix and, term by term, are equal to the diagonal matrix of squared frequencies. If the eigenvectors are also obtained, their *symmetry* gives the pattern of displacements, by which is meant that, if all the values in an eigenvector are the same, the motion is all in one direction and, if different, then some masses are moving in one direction and other masses in other directions as determined by the sign.

Summary: The matrix of forces is constructed and made into an eigenvalue–eigenvector equation and this is diagonalized. The eigenvalues λ are then equated to the squared frequencies of the normal modes. The eigenvectors indicate the pattern of normal mode displacements.

Using the matrix method to solve problems of this type is best described with an example. The most difficult part is always obtaining the equation for the energy and therefore forces. In this example, the motion of two identical pendulums will be calculated. They have length L and mass m at the end of rigid, mass-less rods, and the masses are coupled together with a spring with force constant k, Fig. 7.59. The pendulums are initially at equilibrium separated by the un-extended length of the spring, and then energy is added by moving one or both of the individual pendulums by a very small amount so that Hooke's law for the spring is obeyed, and the change in potential, due to height, is directly proportional to the change in the vertical angle. There are two normal modes, see Fig. 7.60. Either the pendulums swing together in phase both moving in the same direction at constant separation, or they move together and apart stretching and compressing the spring. These are clearly the only possible motions of the two pendulums for small angular motion and we shall want to find out their frequency (eigenvalues) and the normal mode displacements using the eigenvectors.

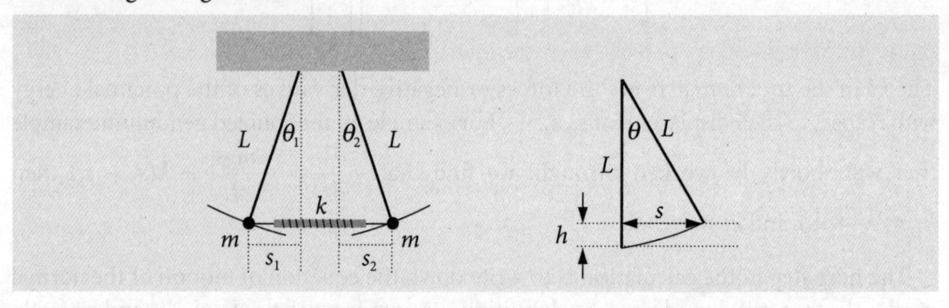

Fig. 7.59 Left: Two coupled pendulums of length L and mass m connected by a spring with force constant k. Right: Calculation of height h.

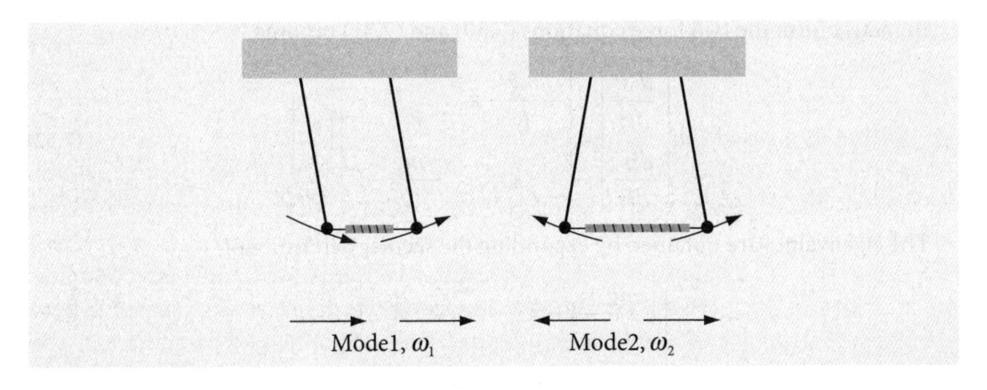

Fig. 7.60 Normal modes of a coupled pendulum.

The first step is to work out the potential energy V, which is the sum of the increase in height as each bob swings and the energy stored due to the change in length of the spring. The potential due to a pendulum's height change is mgh, where h is the vertical displacement when the pendulum swings, defined in the right-hand pane of Fig. 7.59 and g is the acceleration due to gravity and m is mass. Next, the energy due to compressing or extending the spring is found then, using the total potential energy, the forces are calculated. The potential energy of extending a spring of force constant k by distance s, is ks. As the restoring force is always equal to the negative derivative of the potential and is also mass times acceleration, then,

$$force = -\frac{d}{ds}V = m\frac{d^2s}{dt^2}. \tag{7.49}$$

It remains, therefore, to calculate the potential V in terms of the displacement, and to solve the resulting eigenvalue–eigenvector equations.

Using the right-hand part of Fig. 7.59, by definition, $\sin(\theta) = s/L$ and $\cos(\theta) = (L-h)/L$. The differential equations have to be linear to use the eigenvalue–eigenvector method. Thus the displacements and hence angle θ is small. Rearranging to find h, and assuming small angles means that the cosine can be expanding as $\cos(\theta) = 1 - \theta^2/2 + \cdots$ giving $h = L[1-\cos(\theta)] \approx \frac{L}{2}\theta^2$. The potential energy of each of the two bobs due to their increase in height, is $mgh = mgL\theta^2/2$, but the energy is needed in terms of the displacement s not the angle θ and using trigonometry again, $\sin(\theta) = s/L \approx \theta$, gives $\frac{mgs^2}{2L}$ for the potential energy.

The potential energy due to the spring's extension or compression is, by Hooke's law, $\frac{k}{2}\Delta s^2 = \frac{k}{2}(s_2 - s_1)^2$ and the total potential energy is

$$V = \frac{mg}{2L}s_1^2 + \frac{mg}{2L}s_2^2 + \frac{k}{2}(s_2 - s_1)^2$$

where the displacement of the pendulums is s_1 and s_2. This, in effect, means that we have chosen the basis set for the calculation to be the two horizontal displacements. The vector of displacements, which we shall need later, is $\mathbf{s} = \begin{bmatrix} s_1 \\ s_2 \end{bmatrix}$.

The force on pendulum 1 is, by definition, mass × acceleration,

$$m\frac{d^2s_1}{dt^2} = -\frac{\partial V}{\partial s_1} = -\frac{mgs_1}{L} + k(s_2 - s_1) \tag{7.50}$$

and on pendulum 2

$$m\frac{d^2s_2}{dt^2} = -\frac{\partial V}{\partial s_2} = -\frac{mgs_2}{L} - k(s_2 - s_1). \tag{7.51}$$

As a check, the total force on the pendulums should be due to gravity alone, therefore it should not depend on the force constant k, viz.

$$m\frac{d^2s_1}{dt_2} + m\frac{d^2s_2}{dt_2} = -\frac{mgs_1}{L} + k(s_2 - s_1) - \frac{mgs_2}{L} - k(s_2 - s_1) = -\frac{mg(s_1 + s_2)}{L}.$$

In matrix form the two force equations (7.50) and (7.51) become

$$m\begin{bmatrix}\dfrac{d^2s_1}{dt^2}\\[2mm] \dfrac{d^2s_2}{dt^2}\end{bmatrix}=\begin{bmatrix}-\dfrac{mg}{L}-k & k\\[2mm] k & -\dfrac{mg}{L}-k\end{bmatrix}\begin{bmatrix}s_1\\ s_2\end{bmatrix} \tag{7.52}$$

The eigenvalues are obtained by expanding the secular determinant

$$\begin{vmatrix}-\dfrac{mg}{L}-k-\lambda & k\\[2mm] k & -\dfrac{mg}{L}-k-\lambda\end{vmatrix}=0$$

and solving the characteristic equation $\left(-\dfrac{mg}{L}-k-\lambda\right)^2-k^2=0$ for λ. The eigenvalues λ and eigenvectors $\boldsymbol{x}$ are

$$\lambda_1=-\frac{mg}{L}\quad \text{with}\quad x_1=\begin{bmatrix}1\\1\end{bmatrix}$$

and

$$\lambda_2=-\frac{mg}{L}-2k\quad \text{with}\quad x_2=\begin{bmatrix}-1\\1\end{bmatrix},$$

and the eigenvector (modal) matrix is $\begin{bmatrix}1 & -1\\ 1 & 1\end{bmatrix}$. Maple was used to confirm the eigenvalues and eigenvectors. The first eigenvalue corresponds to the first eigenvector, and must be the motion where both pendulums move in the same direction since both values in x_1 have the same sign. The second eigenvector shows that the spring is stretched because the displacements are in opposite directions. We may surmise this also, because only the second eigenvalue involves k, the spring's force constant.

The matrix equation of forces for the normal modes is

$$m\begin{bmatrix}\dfrac{d^2s_{n_1}}{dt^2}\\[2mm] \dfrac{d^2s_{n_2}}{dt^2}\end{bmatrix}=\begin{bmatrix}-m\omega_1^2 & 0\\ 0 & -m\omega_2^2\end{bmatrix}\begin{bmatrix}s_{n_1}\\ s_{n_2}\end{bmatrix} \tag{7.53}$$

and as the matrix is diagonal it can be equated to the (diagonalized) eigenvalue matrix. Therefore, using λ_1 calculated above gives $-\lambda_1=m\omega_1^2=\dfrac{mg}{L}$ and similarly for λ_2, gives the normal mode frequencies

$$\omega_1^2=\frac{g}{L}\quad \text{and}\quad \omega_2^2=\frac{g}{L}+\frac{2k}{m}$$

which can also be written as $\omega_2^2=\omega_1^2+\dfrac{2k}{m}$. These results tell us that one motion of the pendulums has a frequency $\sqrt{g/L}$, which is that of single free pendulum of the same length and mass, and this motion must be that of both pendulums moving in phase with one another in the same direction and without changing the length of the spring because k is not in the formula. The other, out-of-phase frequency is higher, $\sqrt{g/L+2k/m}$, since k is positive, and is the motion when the two pendulums are opposed to one another, and therefore, the spring is continually compressed and extended.

It is interesting to consider what happens when the force constant is very small. In this case, both pendulums tend towards the same frequency because they are hardly coupled; $\sqrt{g/L+2k/m}\approx\sqrt{g/L}\left(1+\dfrac{Lk}{mg}\right)$. You may know that pendulum clocks fastened to the same wall will become synchronized to one another. This is because there is a coupling between them through the wall, and although very weak, it acts like the spring in this problem, and forces the pendulums to come into phase in a normal mode. In the limit $k=0$, then the two pendulums are independent of one another, because their coupling is zero and although both oscillation frequencies are the same they can never become synchronized unless they are started off that way.

When the spring constant is very large, or rather the ratio k/m is large, gravity becomes unimportant and the out-of-phase vibrational normal mode frequency becomes much larger than the in-phase one and approaches $\omega_2 \to \sqrt{2k/m}$. If gravity were 'turned off', $g = 0$, the pendulums would only move if the relative separation of the bobs was changed from the equilibrium separation by compressing or extending the spring. Only one mode is now possible which is the out-of-phase mode 2 with frequency $\omega_2 = \sqrt{2k/m}$.

The two modes with frequencies ω_1 and ω_2 are the *normal modes* of the vibrations; we know this by noting that the eigenvectors are orthogonal as the matrix is Hermitian, but as a check the dot product is $[-1 \quad 1]\begin{bmatrix} 1 \\ 1 \end{bmatrix} = 0$.

Suppose that the normal modes displacements that form the column matrix $\mathbf{Q}$ are Q_1 and Q_2. Their value in terms of the pendulum's displacements are calculated via $\mathbf{Q} = \boldsymbol{x}^{\mathrm{T}}\boldsymbol{s}$ (see next section) where $\boldsymbol{x}^{\mathrm{T}}$ is the transpose of the eigenvector matrix and $\boldsymbol{s}$ the vector of displacements, consequently

$$\mathbf{Q} = \begin{bmatrix} Q_1 \\ Q_2 \end{bmatrix} = \begin{bmatrix} 1 & 1 \\ -1 & 1 \end{bmatrix}\begin{bmatrix} s_1 \\ s_2 \end{bmatrix} = \begin{bmatrix} s_1 + s_2 \\ -s_1 + s_2 \end{bmatrix}$$

or

$$Q_1 = s_1 + s_2 \qquad Q_2 = -s_1 + s_2.$$

To isolate one mode, set $Q_1 = 0$, then the displacements of the bobs are equal and opposite $s_1 = -s_2$. This corresponds to moving one bob of the pendulum to the left, and the other to the right by the same amount; therefore, the motion is at the same frequency and in phase. This is the out-of-phase normal mode with frequency ω_2. The displacements of the bobs follow $s_1 = u\cos(\omega t + \phi)$ and $s_2 = -u\cos(\omega t + \phi)$. The total displacement is zero at all times. Next, to isolate the other mode, set $Q_2 = 0$ then $s_2 = s_1$ and the displacements are always equal to one another during the oscillation, which corresponds to both bobs moving to the right by the same amount, and therefore, they move at the same frequency and with constant phase; this is the in-phase normal mode with frequency ω_1. The motion is $s_1 = s_2 = u\cos(\omega t + \phi)$.

7.14.3 How to calculate vibrational normal modes in a molecule using the *GFG* matrix method

The basic premises are the same as for the pendulum just calculated.

(i) The vibrations undergo simple harmonic motion. In a molecule, this means that each bond is a harmonic oscillator.

(ii) Assume that the energy of each potential energy term, which in this case is the 'spring' connecting the atoms adds separately to the total.

If a molecule has potential energy V, then this energy can be expressed as a Maclaurin expansion about the minimum, or equilibrium, configuration, using displacements of each of the atoms in each of x-, y-, and z-directions. Clearly, this is going to be a complicated function. If there are n atoms with $3n$ coordinates, the displacement from equilibrium is s_i for atom coordinate i. The potential energy is

$$V = V_0 + \sum_{i=1}^{3n}\left(\frac{\partial V}{\partial s_i}\right)_0 s_i + \frac{1}{2}\sum_{i=1}^{3n}\sum_{j=1}^{3n}\left(\frac{\partial^2 V}{\partial s_i \partial s_j}\right)_0 s_i s_j + \cdots. \tag{7.54}$$

The zero subscripts mean that the gradient is evaluated about the equilibrium position at each coordinate, where the extension $s_i = 0$. The summation index covers all coordinates 1 to $3n$. The indexing is up to us to choose, but could be 1 for atom 1's x-coordinate, 2 for y, 3 for z, then 4 for atom 2's x-coordinate, 5 for y and so on. The second derivative has two summations and contains $(3n)^2$ terms. The reason for all these terms is easiest to see with just two dimensions. Imagine the potential is shaped like a bowl; the height is V and x and y are the two displacements, s. The first derivative dV/dx gives the slope (force) in the x-direction at constant y, and similarly dV/dy the force in the y-direction at constant x. When $i = j$ the second derivative d^2V/dx^2 describes how the force varies along the x-axis at constant y. The 'cross terms', when $i \neq j$, measures these values between the axes.

The origin of the potential has a value V_0, which is arbitrary, and because potential energy is a relative measure, V_0 can be set to zero. The first derivative of the potential is the force, and as this is expanding about the equilibrium position, it is zero. The potential is therefore,

$$V \approx \frac{1}{2}\sum_{i=1}^{3n}\sum_{j=1}^{3n}\left(\frac{\partial^2 V}{\partial s_i \partial s_j}\right)_0 s_i s_j \tag{7.55}$$

which is analogous to the quadratic potential based on Hooke's law.

The kinetic energy is by definition $mv^2/2$ for each mass or

$$T = \frac{1}{2}\sum_i^{3n} m_i\left(\frac{ds_i}{dt}\right)^2 \tag{7.56}$$

for all masses. We now have to assume a great deal and jump almost to the answer by using Lagrange's formulation of Newton's laws of motion. Furthermore, we are going straight to the matrix form by defining

$$F \equiv \frac{\partial^2 V}{\partial s_i \partial s_j} = k_{ij}$$

which is a matrix of force constants k. Next, a matrix M is defined whose inverse is the matrix of second derivatives of the kinetic energy with respect to velocities $\dot{s}$, in which terms are of the form $\dfrac{\partial^2 T}{\partial \dot{s}_i \partial \dot{s}_j}$, and this turns out to be a diagonal matrix of the masses,

$$M^{-1} = diag(m_1, m_2 \cdots m_n) \tag{7.57}$$

and therefore

$$M = diag(1/m_1, 1/m_2 \cdots 1/m_n) \tag{7.58}$$

is the diagonal matrix of *reciprocal* masses; see Section 7.4.10.

Writing out the Lagrange equation gives

$$M^{-1}\frac{d^2 s}{dt^2} + Fs = 0 \qquad \text{or} \qquad \frac{d^2 s}{dt^2} + MFs = 0 \tag{7.59}$$

The second of these equations looks rather like that for a single oscillator, equation (7.47), if we identify MF with k/m, which, dimensionally, is correct.

To solve this last equation, it is necessary to show that it can be turned into an eigenvalue–eigenvector equation. Supposing that the motion is harmonic with frequency ω, a solution for the ith atom is $s_i = a_i \cos(\omega t + \phi)$ and by differentiating this displacement twice with time, it is seen that this is a solution of equation (7.59) because the result is

$$\frac{d^2 s}{dt^2} + \omega^2 s = 0. \tag{7.60}$$

Combining the equations (7.59) and (7.60) produces the eigenvalue–eigenvector equation

$$MFs = \omega^2 s \qquad \text{or} \qquad [MF - \omega^2 \boldsymbol{1}]s = \boldsymbol{0} \tag{7.61}$$

with eigenvalues ω^2 as roots of the secular determinant $|MF - \omega^2\boldsymbol{1}| = 0$. The eigenvectors will be labelled $\boldsymbol{x}$. There is one difficulty with this which is that M and F do not always commute and, although it is not obvious, this is easily overcome by using the square root of matrix M instead of M itself. In the form shown below, we *define* a matrix G as

$$G = diag(1/\sqrt{m_1}, 1/\sqrt{m_2}, 1/\sqrt{m_n}) \text{ and } G^{-1} = diag(\sqrt{m_1}, \sqrt{m_2}, \sqrt{m_n}) \tag{7.62}$$

and define the displacement in terms of *reduced* mass

$$q = G^{-1}s, \tag{7.63}$$

therefore, each q matrix element takes the form $q_1 = s_1\sqrt{m_1}$ and so forth.[5] With these changes, equation (7.61) becomes

$$GFGq = \omega^2 q \tag{7.64}$$

[5] Note, however, that some authors prefer to define our G^{-1} as $G^{-1/2}$.

and the secular determinant

$$|GFG - \omega^2 1| = 0. \tag{7.65}$$

The GFG matrix product has the form

$$\begin{bmatrix} 1/\sqrt{m_1} & 0 & \cdots & 0 \\ 0 & 1/\sqrt{m_2} & 0\cdots & 0 \\ \vdots & \vdots & \ddots & \vdots \\ 0 & 0 & \vdots & 1/\sqrt{m_n} \end{bmatrix} \begin{bmatrix} k_{11} & k_{12} & \cdots & k_{in} \\ k_{21} & k_{22} & \cdots & \vdots \\ \vdots & \vdots & \ddots & \vdots \\ k_{n1} & \vdots & \vdots & k_{nn} \end{bmatrix} \begin{bmatrix} 1/\sqrt{m_1} & 0 & \cdots & 0 \\ 0 & 1/\sqrt{m_2} & 0\cdots & 0 \\ \vdots & \vdots & \ddots & \vdots \\ 0 & 0 & \vdots & 1/\sqrt{m_n} \end{bmatrix}$$

$$= \begin{bmatrix} \dfrac{k_{11}}{m_1} & \dfrac{k_{12}}{\sqrt{m_1 m_2}} & \dfrac{k_{13}}{\sqrt{m_1 m_3}} & \cdots \\ \dfrac{k_{21}}{\sqrt{m_1 m_2}} & \dfrac{k_{22}}{m_2} & \cdots & \cdots \\ \vdots & \vdots & \ddots & \end{bmatrix} \tag{7.66}$$

so that each diagonal term is divided by the mass appropriate to that position in the matrix, and the off-diagonal terms at positions i, j are divided by the square root of the product of the corresponding masses in the basis set; i.e. by $\sqrt{m_i m_j}$. The k_{ij} are the force constants. This matrix of reduced mass is the GFG matrix.

A fuller analysis of the problem leads to the conclusion that the eigenvectors x can be used to produce the normal mode displacements Q via

$$Q = x^T q. \tag{7.67}$$

and the coordinate displacements s are given by

$$s = GxQ. \tag{7.68}$$

Summary

G is the diagonal matrix of $m^{-1/2}$ and F is the matrix of force constants. The eigenvalues are the squared frequencies of the molecular normal modes Q, and are calculated from the secular determinant $|GFG - \omega^2 1| = 0$. The normal mode and coordinate displacements are calculated using equations (7.67) and (7.68).

This method is rather abstract but can be made less so by working through an example. We will calculate some of the normal modes of a linear triatomic molecule, such as CO_2, by considering just the motion along the line of atoms, as shown in Fig. 7.61. A linear molecule is chosen to make the calculation simpler; a bent molecule such as H_2O or SO_2 can be calculated in a similar manner, but the algebra in calculating the potential is more complex. Herzberg gives several examples of this including different possible models of the potentials that can be used; Herzberg (1964, vol II, Chapter 2).

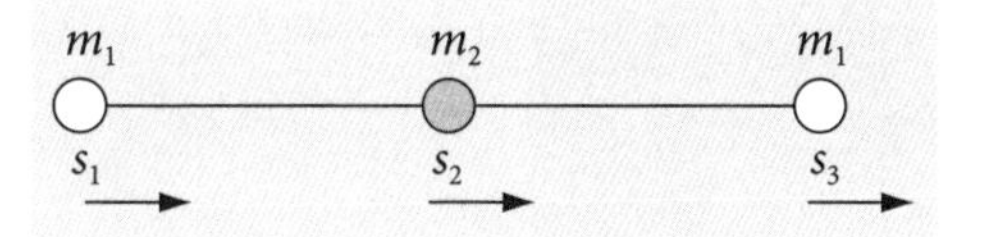

Fig. 7.61 A linear triatomic molecule. The displacements s are measured from the equilibrium positions of the atoms. They define the basis set for the calculation and matrices.

It is normally assumed that the motion of the atoms is harmonic and we additionally assume that the only force constants are those between adjacent atoms; this is equivalent to assuming that the valence bond model describes the bonding. A molecular orbital model would consider forces constants between one atom and every other atom. The calculation will be done in two ways. First directly, as in Section 7.14.2 but using mass weighting from the start; then secondly, using Maple with the GFG matrix method which introduces mass weighting in the GFG product.

The first calculation uses mass-weighted coordinates and force constants. This means that the displacements s of atom i are modified to become $q_i = s_i\sqrt{m_i}$, and force constants between atoms i and j become $K_{ij} = \dfrac{k_{ij}}{\sqrt{m_i m_j}}$ as in equation (7.66).

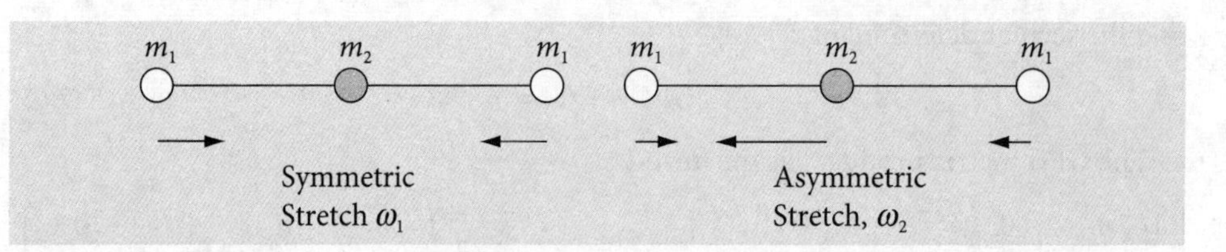

Fig. 7.62 Some normal mode vectors in the linear, triatomic molecule CO_2. The sizes of the displacements are arbitrary but relatively correct in each mode. The displacements are the same in the symmetric stretch but in the asymmetric stretch the central atom, the carbon moves $2m_o/m_c$ as far as the oxygen atoms.

The starting point is to calculate the potential energy and, from this, the forces on the atoms. As the molecule is symmetrical, there is only one force constant, but the mass-weighted force constants will be different. The normal modes for a simple molecule can easily be sketched, and two of them are shown in Fig. 7.62. The molecule is linear giving $3N-5$ or 4 modes in total, the other two are the degenerate bending motion in the plane of the page. One where the carbon atom moves down and the two oxygen atoms move up and then vice versa. The other bending normal mode is the similar motion but perpendicular to the plane of the page.

The potential energy is the sum of terms for the stretching of each bond; therefore,

$$V(s_1, s_2, s_3) = \frac{k}{2}(s_1 - s_2)^2 + \frac{k}{2}(s_2 - s_3)^2.$$

The forces are the negative derivatives with respect to the displacements

$$-\frac{dV}{ds_1} = -k(s_1 - s_2); \quad -\frac{dV}{ds_2} = k(s_1 - s_2) - k(s_2 - s_3); \quad -\frac{dV}{ds_3} = k(s_2 - s_3).$$

With f as the force, placing these equations into matrix form gives

$$\begin{bmatrix} f_1 \\ f_2 \\ f_3 \end{bmatrix} = \begin{bmatrix} -k & k & 0 \\ k & -2k & k \\ 0 & k & -k \end{bmatrix} \begin{bmatrix} s_1 \\ s_2 \\ s_3 \end{bmatrix}$$

but, because the masses are different, we must change to mass weighted force constants, using the formula $K_{ij} = k_{ij}/\sqrt{m_i m_j}$, where i and j are the atom indices; for example $K_{12} = k/\sqrt{m_o m_c}$ is the mass-weighted force constant between atoms 1 and 2, if m_o is the oxygen mass and m_c that of the carbon. The matrix of force constants becomes

$$K = \begin{bmatrix} \dfrac{-k}{m_o} & \dfrac{k}{\sqrt{m_o m_c}} & 0 \\ \dfrac{k}{\sqrt{m_o m_c}} & \dfrac{-2k}{m_c} & \dfrac{k}{\sqrt{m_o m_c}} \\ 0 & \dfrac{k}{\sqrt{m_o m_c}} & \dfrac{-k}{m_o} \end{bmatrix}$$

which is solved as a secular determinant with eigenvalues,

$$\lambda_1 = -k/m_o, \quad \lambda_2 = -k\frac{(2m_o + m_c)}{m_o m_c}, \quad \text{and } \lambda_3 = 0.$$

The frequency of each vibration is $\omega^2 = -\lambda$ so the square of the normal mode frequencies are

$$\omega_1^2 = k/m_o, \quad \omega_2^2 = k\frac{(2m_o + m_c)}{m_o m_c} \quad \text{and } \omega_3 = 0.$$

The normalized[6] eigenvector modal matrix is, after some rearranging and with $M = 2m_o + m_c$ as the total mass,

[6] If a vector is $v = [a \quad b \quad c]$ the normalization is $1/\sqrt{(a^2 + b^2 + c^2)}$ or $\sqrt{v \cdot v}$.

$$x = \frac{1}{\sqrt{2M}}\begin{bmatrix} -\sqrt{M} & \sqrt{m_c} & \sqrt{2m_o} \\ 0 & -2\sqrt{m_o} & \sqrt{2m_c} \\ \sqrt{M} & \sqrt{m_c} & \sqrt{2m_o} \end{bmatrix}.$$

Notice that the eigenvectors do not depend on the force constants; this matrix is used to produce the geometry that is related to the symmetry of the vibrations, and cannot depend on the value of the force constants. The normal modes depend only on the geometry because the 'springs' connecting the atoms can only vibrate in certain patterns governed by the geometry or symmetry of the molecule; the *frequency* and *size* of extension depend on the force constants.

The normal mode coordinates are calculated using $\boldsymbol{Q} = \boldsymbol{x}^T\boldsymbol{q}$ (equation 7.67) where $\boldsymbol{x}^T$ is the transpose of the eigenvector matrix and $\boldsymbol{q}$ the vector of mass weighted coordinates with $q - s\sqrt{m}$.

$$\begin{bmatrix} Q_1 \\ Q_2 \\ Q_3 \end{bmatrix} = \frac{1}{\sqrt{2M}}\begin{bmatrix} -\sqrt{M} & 0 & \sqrt{M} \\ \sqrt{m_c} & -2\sqrt{m_o} & \sqrt{m_c} \\ \sqrt{2m_o} & \sqrt{2m_c} & \sqrt{2m_o} \end{bmatrix}\begin{bmatrix} s_1\sqrt{m_o} \\ s_2\sqrt{m_c} \\ s_3\sqrt{m_o} \end{bmatrix}$$

and the individual modes are

$$Q_1 = \sqrt{m_o/2}(-s_1 + s_3),$$

$$Q_2 = \sqrt{\frac{m_o m_c}{2M}}(s_1 - 2s_2 + s_3),$$

$$Q_3 = (m_o s_1 + m_c s_2 + m_o s_3)/\sqrt{M}.$$

To calculate the individual atom displacements s_1 to s_3 directly, these three equations have to be solved to get s in terms of the Qs. The coordinate displacements are calculated with the matrix equation $\boldsymbol{s} = \boldsymbol{GxQ}$ (equation 7.68), where $\boldsymbol{x}$ is the eigenvector matrix, $\boldsymbol{Q}$ is the column vector $Q = \begin{bmatrix} Q_1 \\ Q_2 \\ Q_3 \end{bmatrix}$ and $\boldsymbol{G}$ is the diagonal matrix of $m^{-1/2}$ equation (7.62).

The result is

$$s_1 = -\frac{1}{\sqrt{2m_o}}Q_1 + \sqrt{\frac{m_c}{2m_o}}\frac{1}{\sqrt{M}}Q_2 + \frac{1}{\sqrt{M}}Q_3$$

$$s_2 = -\sqrt{\frac{2m_o}{m_c}}\frac{1}{\sqrt{M}}Q_2 + \frac{1}{\sqrt{M}}Q_3$$

$$s_3 = \frac{1}{\sqrt{2m_o}}Q_1 + \sqrt{\frac{m_c}{2m_o}}\frac{1}{\sqrt{M}}Q_2 + \frac{1}{\sqrt{M}}Q_3.$$

Looking at these equations, we can choose to make any of the Qs zero to isolate normal mode motion. The three cases are now considered.

(*i*) Suppose we choose $Q_1 = Q_2 = 0$, then all the atom displacements are the same, $s_1 = s_2 = s_3$ and $Q_3/\sqrt{M}$, and each atom is moving in the same direction. This cannot be a vibrational normal mode as the bonds are neither stretched nor compressed, but represents a translation, and clearly, this corresponds to the zero frequency eigenvalue $\omega_3 = 0$. Note that if $Q_3 = 0$ this must mean that the centre of mass does not change, no translation occurs, and so we expect to produce normal modes with this condition.

(*ii*) If $Q_1 = Q_3 = 0$ then $s_1 = s_3 = \sqrt{\dfrac{m_c}{2m_o M}}Q_2$ and $s_2 = -\sqrt{\dfrac{2m_o}{m_c M}}Q_2$. In this case this is the asymmetric stretch with the oxygen atoms moving in the same direction and opposed to that of the carbon, and this is ω_2. The relative motion is $s_1 = s_3 = 0.115Q_2$ to $s_2 = -0.308Q_2$ so the carbon atom moves further than the oxygen atoms do, which is not surprising, because the centre of mass has to be held constant and the C atom has to compensate for the motion of two O atoms. The ratio of the displacements is $2m_o/m_c$. The total centre of gravity does not move because the total moment, found by summing the product of displacement times mass, is zero, viz.

$$2m_o\sqrt{\frac{m_c}{2m_oM}}Q_2 - m_c\sqrt{\frac{2m_o}{m_cM}}Q_2 = \sqrt{2m_cm_o} - \sqrt{2m_om_c} = 0$$

(*iii*) Finally, if $Q_2 = Q_3 = 0$ then the equations are solved when $s_2 = 0$ and $s_1 = -s_3$, which is the symmetric stretch ω_1 with displacements $\pm Q_1/\sqrt{2m_o}$ or $0.176Q_1$.

The full calculation using the *G* and *F* matrix method is given next. You can modify this method to solve many similar normal mode problems with different dimensions and geometries provided the potential energy equation can be worked out and the force constant matrix obtained. If you repeat this calculation the order of the eigenvalues may be different to those shown here, which means that the labels $Q_{1,2,3}$ may be different to those shown here. The following labels are exchanged with respect to those on p 409, $Q_2 \rightarrow Q_1$, $Q_1 \rightarrow Q$.

Algorithm 7.7 Normal mode calculation: linear CO_2

```
> with(LinearAlgebra):
> assume( mo > 0, mc > 0);                                # mass >0
  G:= < <1/sqrt(mo)| 0 | 0          >,
        < 0       | 1/sqrt(mc) | 0    >,
        < 0       | 0    | 1/sqrt(mo) >  >;           #diag sqrt mass
  F:=   < <-k|      k| 0  >,
          < k| -2*k   | k  >,
          < 0|    k   | -k > >:              # force constant matrix
> K:= G.F.G;                          # mass weighted force constant
```

$$K := \begin{bmatrix} -\frac{k}{mo} & \frac{k}{\sqrt{mo}\sqrt{mc}} & 0 \\ \frac{k}{\sqrt{mo}\sqrt{mc}} & -\frac{2k}{mc} & \frac{k}{\sqrt{mo}\sqrt{mc}} \\ 0 & \frac{k}{\sqrt{mo}\sqrt{mc}} & -\frac{k}{mo} \end{bmatrix}$$

```
> (eigval, eigvec):= Eigenvectors( K );
```

$$eigval, eigvec := \begin{bmatrix} -\frac{k(2\,mo + mc)}{mo\,mc} \\ -\frac{k}{mo} \\ 0 \end{bmatrix}, \begin{bmatrix} 1 & -1 & 1 \\ -\frac{2\sqrt{mo}}{\sqrt{mc}} & 0 & \frac{\sqrt{mc}}{\sqrt{mo}} \\ 1 & 1 & 1 \end{bmatrix}$$

```
# normalise eigenvects, first extract column into w
  w:= eigvec[1..3,1]: v1:= w/sqrt(w.w):
  w:= eigvec[1..3,2]: v2:= w/sqrt(w.w):
  w:= eigvec[1..3,3]: v3:= w/sqrt(w.w):
  < v1| v2 | v3 >: x:= simplify( % );                 # x=eigenvectors
```

$$x := \begin{bmatrix} \frac{1}{2}\frac{\sqrt{2}\sqrt{mc}}{\sqrt{2\,mo + mc}} & -\frac{1}{2}\sqrt{2} & \frac{\sqrt{mo}}{\sqrt{2\,mo + mc}} \\ -\frac{\sqrt{2}\sqrt{mo}}{\sqrt{2\,mo + mc}} & 0 & \frac{\sqrt{mc}}{\sqrt{2\,mo + mc}} \\ \frac{1}{2}\frac{\sqrt{2}\sqrt{mc}}{\sqrt{2\,mo + mc}} & \frac{1}{2}\sqrt{2} & \frac{\sqrt{mo}}{\sqrt{2\,mo + mc}} \end{bmatrix}$$

```
# define mass weighted coords equation (7.64)
> s:= < s1, s2, s3 >: q:=G^(-1).s;
```

$$q := \begin{bmatrix} \sqrt{mo}\,s1 \\ \sqrt{mc}\,s2 \\ \sqrt{mo}\,s3 \end{bmatrix}.$$

```
> Q:= Transpose(x).q;                    # Q are normal modes in s
```

$$Q := \begin{bmatrix} \frac{1}{2}\frac{\sqrt{2}\sqrt{mc}\sqrt{mo}\,s1}{\sqrt{2\,mo+mc}} - \frac{\sqrt{2}\sqrt{mo}\sqrt{mc}\,s2}{\sqrt{2\,mo+mc}} + \frac{1}{2}\frac{\sqrt{2}\sqrt{mc}\sqrt{mo}\,s3}{\sqrt{2\,mo+mc}} \\ -\frac{1}{2}\sqrt{2}\sqrt{mo}\,s1 + \frac{1}{2}\sqrt{2}\sqrt{mo}\,s3 \\ \frac{mo\,s1}{\sqrt{2\,mo+mc}} + \frac{mo\,s2}{\sqrt{2\,mo+mc}} + \frac{mo\,s3}{\sqrt{2\,mo+mc}} \end{bmatrix}$$

```
# these modes can be identified with ref to fig. 7.62
# calculate extension s in normal coords Q1, Q2, Q3
> QQ:= < Q1, Q2, Q3 >:
  extns:= G.x.QQ ;                         # s values
```

$$extns := \begin{bmatrix} \frac{1}{2}\frac{\sqrt{2}\sqrt{mc}\,Q1}{\sqrt{mo}\sqrt{2\,mo+mc}} - \frac{1}{2}\frac{\sqrt{2}Q2}{\sqrt{mo}} + \frac{Q3}{\sqrt{2\,mo+mc}} \\ -\frac{\sqrt{2}\sqrt{mo}\,Q1}{\sqrt{mc}\sqrt{2\,mo+mc}} + \frac{Q3}{\sqrt{2\,mo+mc}} \\ \frac{1}{2}\frac{\sqrt{2}\sqrt{mc}\,Q1}{\sqrt{mo}\sqrt{2\,mo+mc}} + \frac{1}{2}\frac{\sqrt{2}Q2}{\sqrt{mo}} + \frac{Q3}{\sqrt{2\,mo+mc}} \end{bmatrix}$$

7.14.4 **Questions**

Full solutions are available at **www.oxfordtextbooks.co.uk/orc/beddard**.

Q7.55 **(a)** Assuming that all angles are small during the motion, find the secular determinant for two rigid linked pendulums as shown in the diagram. The pendulums have lengths L_1 and L_2 and masses m_1 and m_2 at their ends. The rods are stiff but weightless and the bearing connecting the upper rod to its mount and that between the rods is frictionless. Because the equations produced are complex, in the last step of the calculation, assume that the pendulums have the same length L.

(b) Use Maple to solve the determinant and show that the normal mode frequencies are given by $\omega^2 = (1 \pm \sqrt{M})g/L$ where $M = m_2/(m_1 + m_2)$.

(c) Suppose the linkage is a spring with force constant k whose potential energy varies as $k\theta_2^2/2$; recalculate the oscillation frequencies.

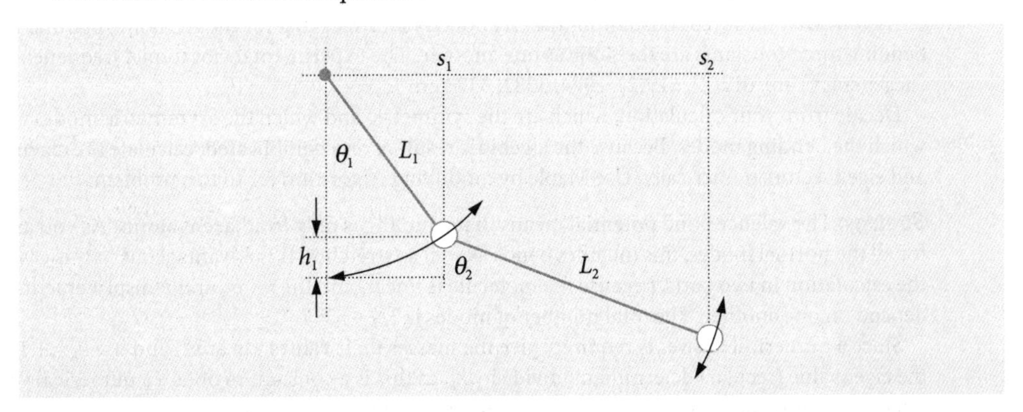

Fig. 7.63 Double linked pendulum. The angles are greatly exaggerated, for in this calculation, only small angle motion is accurately described.

Strategy: Work out the potential energy and differentiate this to find the force. The potential energy against gravity is that due to the vertical height raised from the stationary pendulums. Take them in turn and make s_1 and s_2 the horizontal displacements. The mass needed in working out the potential energy of the upper pendulum is the total mass $m_1 + m_2$. To see this, assume that the lower pendulum hangs vertically, then, clearly, lifting the upper pendulum involves lifting both masses.

Q7.56 Two similar masses are joined together by three springs fixed at either end to rigid walls, as in the figure, and are free to slide on a frictionless surface. Calculate the normal mode frequencies and sketch their geometry if the force constant of the middle spring is n times that of the outer two. Assume Hooke's law applies to the springs.

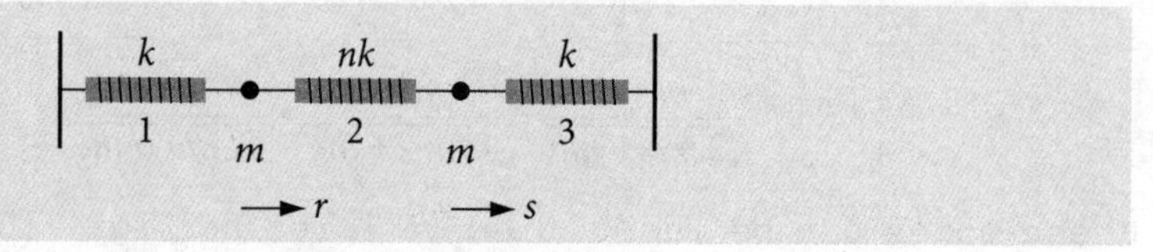

Fig. 7.64 Two masses fixed between three springs. The masses are displaced by r and s.

Strategy: Calculate the potential energy by defining displacements of each mass, then work out the force equation as in equation (7.53).

Q7.57 Calculate the bending normal mode for a linear molecule, in a related manner to that used in the example of the stretching modes. Use the displacement vectors shown in the figure, which are in the plane of the figure. Assume that the bending motion obeys Hooke's law. The force constant is that to bend the molecule rather than to extend its bonds.

(a) Sketch the bending normal modes.

(b) Calculate the normal mode frequencies of the molecule, such as CO_2, by considering the displacement of each atom.

(c) Calculate the correct relative displacements of the atoms.

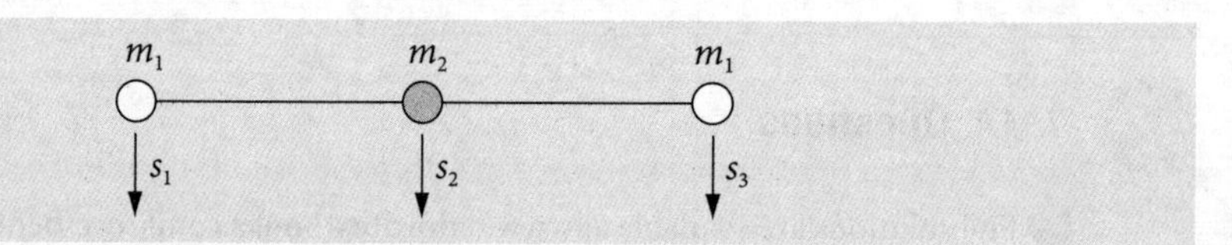

Fig. 7.65 Basis set vectors for a bending mode.

Strategy: To determine the potential energy, use only force constants for bending between atoms 1 and 2, and between 2 and 3. Once the energy is calculated, the force matrix can be set up and solved. There will be two degenerate bending modes in the molecule, giving $3N - 5 = 4$ in total. Two have been identified in the example in the text, which leaves two remaining; one of them is bending in the plane of the figure and the other one is perpendicular to this.

Q7.58 Calculate the frequencies and normal mode vibrations of the linear molecule HCCH, assuming a valence bond potential.

Assume that the force constant for the CH bond is 1.1 times that for the CC bond, but that the bending force constants are the same as one another. The experimental vibrational frequencies (degeneracy) are, 612(2), 619(2), 1974, 3282, 3373 cm^{-1}.

Decide from your calculation, which are the symmetric and which the asymmetric modes and which the bending modes. Because the algebraic result is very complicated, calculate the eigenvalues and eigenvectors numerically. Use Maple by modifying Algorithm 7.7 to this problem.

Strategy: The valence bond potential means that coupling is only to adjacent atoms. As you are asked for all the normal modes, this includes bends as well as stretches. It is advantageous in this case to do the calculation in two parts, because the molecule is linear, and the x-, y-, and z-displacements do not depend on one another. The total number of modes is $3N - 5 = 7$.

Since a numerical answer is required, give the masses their values, (in amu) and use $k_h = 1.1k_c$. In the eigenvalue (secular) determinant, divide by k_h, as this is a constant to obtain a numerical value, but do not forget then to multiply the resulting eigenvalues by k_h. The eigenvectors are independent of k_h. Use the pattern of these to determine the nature of the normal modes.

7.15 Moments of inertia

Angular momentum plays a central role in chemistry, but this is usually in the context of nuclear and electron spin; for example, nuclear spin gives rise to NMR and electron spin (EPR) spectroscopies, as well as being essential for the formation of the chemical bond. The rotation of whole molecules is a natural consequence of thermal motion and is observed with microwave, infrared, Terahertz, or Raman spectroscopy. These techniques are used to measure the spacing between a molecule's rotational energy levels. This is described next, followed by a brief description of angular momentum and moments of inertia. Finally, the calculation of bond lengths is described.

7.15.1 The rotational motion of molecules

The collision between molecules in the gas phase and at the prevailing temperature equilibrates the energy among translation, vibration, and rotational motions. A molecule can only have discrete values of vibrational or rotational energy. The lowest rotational energy is zero, and the energy levels become more widely spaced as the quantum number J is increased. The equation $E_J = \dfrac{\hbar^2}{2I}J(J+1)$ gives the rotational energy, in joules, of a (rigid rotor) diatomic or linear molecule, with quantum number $J = 0, 1, 2, \cdots$. The degeneracy of a rotational level with quantum number J, is $g_J = 2J + 1$. The constant I is the moment of inertia, and is typically $\approx 10^{-45}$ kg m^2 for a molecule. It is this small, because a molecule's mass is small, and its bond lengths short. An apple has a value $I \approx 10^{-4}$ kg m^2 and a lorry's wheel ≈ 10 kg m^2. As we shall see the moment of inertia is related to a bond length.

A rotational constant is defined as $B = \dfrac{\hbar^2}{2I}$ making the energy $E_J = BJ(J+1)$. Remembering that a joule is a unit of energy with base units of mass $\times$ velocity squared, the constant B has units

$$B = \frac{\hbar^2}{2I} \equiv \frac{J^2s^2}{kgm^2} = \frac{kg^2m^4s^{-4}s^2}{kgm^2} = kgm^2s^{-2} = J.$$

Usually, however, B is expressed in units of wavenumbers (cm^{-1}) produced by dividing B by $100hc$; the factor $100c$ is used because c has units of m s^{-1} and we want units in cm^{-1}; $B = \dfrac{1}{100hc}\dfrac{\hbar^2}{2I}$ in cm^{-1}.

The rotational constant B has a value that is typically less than a wavenumber. Its largest value, ≈ 64 cm^{-1}, is for H_2, but clearly this is not typical, as this is the lightest molecule. Once the rotational spectrum is measured, it is easy to measure B, because the lines are spaced by $2B$. In real molecules, things are more complex, because of centrifugal distortion in the rotating molecule, but this effect is small for low J quantum numbers and does not fundamentally change our analysis. The next step is to calculate the bond length from I. At one time, this was the main purpose of microwave spectroscopy, but nowadays, bond lengths of very many small molecules have been accurately measured. The technique is now more often used analytically, for example, to identify species in interstellar dust clouds, or to monitor the ethene produced by ripening fruit.

7.15.2 Angular momentum and moment of inertia

The angular momentum of a solid body is the momentum caused by virtue of its rotational motion and is a vector quantity. In an isolated molecule, this motion is about the centre of mass, also called the centre of gravity. The angular momentum produced has a fixed direction in space. In general, it is just as possible to cause the rotational motion of an object to be about its end as it is about its middle. A rod can be spun about its centre, along its axis, or about its end, and each angular momentum will be different. If the body rotates with angular frequency $\boldsymbol{\omega}$, the angular momentum $\mathbf{L}$ is given by $\mathbf{L} = \mathbf{I}\boldsymbol{\omega}$, where $\mathbf{I}$ is a matrix of moment of inertia values and $\mathbf{L}$ and $\boldsymbol{\omega}$ are one-dimensional matrices or vectors. The reason for the vector quantities is that a body can move in three dimensions and the motion can be split into components along three axes. The moment of inertia of (rigidly connected) masses, where each has mass m_i, is $I=\sum_i m_i r_i^2$ where r_i is the *perpendicular* distance of mass i from an axis. In fact, we can choose the axes to be anywhere we want so the moment of inertia depends on where these axes are placed; it is therefore not a fundamental property of an object. This presents a problem if we are to try to calculate bond lengths, because different values would be obtained depending on where the axes are placed. The obvious place is to locate the axes at the centre of mass since a freely rotating object will rotate about this point. However, in what directions the axes point relative to the molecule has still to be chosen. Very often, we may choose to align the axes with some symmetry axis of the molecule. The molecule will always rotate about its own inertial axes and if these two sets of axes do not coincide, and they usually do not, the moments of inertia will have the form of a matrix with terms that depend on where an atom is with respect to any two axes, I_{xy}, I_{yz} etc. All is not lost, however, because there is a simple way to rotate our chosen axes onto the molecule's inertial axes, making the moment of inertia matrix, diagonal. This will be described in Section 7.15.8.

The total kinetic energy of a rigid body is that due to its linear plus rotational energy. The linear kinetic energy is $E_{lin}=\frac{1}{2}mv^2$ and the rotational kinetic energy $E_{rot}=\frac{1}{2}I\omega^2$; thus, the moment of inertia I, takes the place of the mass used to describe linear motion and angular velocity $\boldsymbol{\omega}$ (radians s^{-1}), replaces linear velocity $\boldsymbol{v}$ in m s^{-1}. Similarly, linear momentum $\boldsymbol{p}=m\boldsymbol{v}$, is replaced by angular momentum $\boldsymbol{L}=I\boldsymbol{\omega}$, which is also vector quantity. The angular momentum relative to the centre of mass about an axis at the centre of a rigid disc of mass m, is shown in Fig. 6.53 and is $\boldsymbol{L}=m\boldsymbol{r}\times\boldsymbol{v}$ where $\boldsymbol{v}$ is the velocity of the edge of the disc and $\boldsymbol{r}$ its radius vector. The angular momentum points in the direction away from you, if the disc rotates clockwise when you look at its underside.

Angular velocity is, by definition, the rate of change of the angle θ the rigid body moves through, and has units of radians / second. It can be shown that the velocity vector $\boldsymbol{v}$ in the centre-of-mass coordinate system is related to the angular velocity $\boldsymbol{\omega}$ as

$$\boldsymbol{v}=\boldsymbol{\omega}\times\boldsymbol{r}.$$

The angular momentum becomes $\boldsymbol{L}=m\boldsymbol{r}\times\boldsymbol{\omega}\times\boldsymbol{r}$, and evaluating the cross product the angular momentum is

$$\boldsymbol{L}=\boldsymbol{I}\boldsymbol{\omega}.$$

where $\boldsymbol{I}$ is a 3×3 matrix of the moment of inertia components. This can be written as

$$\begin{bmatrix} L_x \\ L_y \\ L_z \end{bmatrix}=\begin{bmatrix} I_{xx} & I_{xy} & I_{xz} \\ I_{xy} & I_{yy} & I_{yz} \\ I_{xz} & I_{yz} & I_{zz} \end{bmatrix}\begin{bmatrix} \omega_x \\ \omega_y \\ \omega_z \end{bmatrix}$$

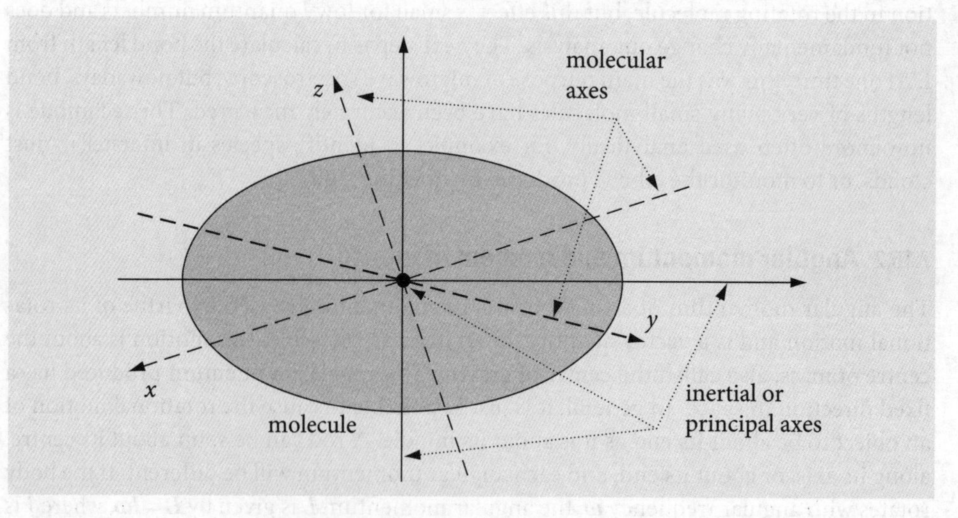

Fig. 7.66 The dashed lines represent the molecular axes; the solid lines the principal axes obtained after diagonalizing the moment of inertia matrix, the third axis of which is perpendicular to the page. The ellipse represents a molecule.

Calculating each of these terms is described in Section 7.13.4. The angular momentum is always proportional to the angular velocity; the proportionality is the moment of inertia, and hence the emphasis falls on calculating this. Sometimes, the angular momentum matrix is called the angular momentum *tensor*.

To summarize: A molecule or other solid object will rotate about its own inertial (principal) axes that are fixed in space. The geometrical axes we chose to place on the molecule need not be coincident with the inertial axes, in fact, they are usually not, and therefore the inertial matrix has off-diagonal elements. Should the inertial and geometrical axes coincide, only the diagonal values exist. By diagonalizing the inertial matrix, the two sets of axes are made coincident, and the diagonal moments of inertia of the principal or inertial axes can be calculated.

7.15.3 Solid bodies

If the rotating body is not made of discrete parts such as a molecule, then integration over the mass and distances must be done instead of summation. The summation is replaced by an integral; some examples are given in Chapter 4.8.2 and 4.10.1. The moments of inertia have been worked out for very many geometrical objects, cones, cylinders and so forth and lists can usually be found in engineering textbooks.

7.15.4 Discrete bodies: molecules

The moment of inertia is not an intrinsic property of a body but depends upon the axis about which the moment is taken. In chemical physics, it is usual to make this axis pass through the centre of mass. In engineering, the moment of inertia about some remote axis may be needed instead. By the *principle of parallel axes*, if the moment of inertia with reference to an axis through its centre of mass $\boldsymbol{I}_{cm}$ is known, it can easily be changed to a value about a *parallel axis* $\boldsymbol{I}_p$ if the parallel axis is a distance d away. The result is

$$I_p = I_{cm} + Md^2 \tag{7.69}$$

and M is the total mass.

The moment of inertia $\boldsymbol{I}$ of a collection of atoms is defined as the summation of the product of the mass and the distance squared from an axis α through the centre of mass,

$$I_\alpha = \sum_i m_i r_i^2 \tag{7.70}$$

where r_i is the distance of mass i from the axis α passing through the centre of mass. As the mass (atom) has coordinates $\{x, y, z\}$, and r is the distance given by Pythagoras, for example, from the x-axis, $r = \sqrt{(z - z_{cm})^2 + (y - y_{cm})^2}$ where x_{cm} and y_{cm} are the coordinate of the centre of mass, and similarly, for the y- and z-axes. The *centre of mass* of i masses is defined as

$$q_{cm} = \frac{\sum_i m_i q_i}{\sum_i m_i} = \frac{\sum_i m_i q_i}{M} \tag{7.71}$$

where q can be x, y, or z and M is the total mass. The moment of inertia about the x-, y-, or z-axes whose origin is at the centre of mass, is

$$I_x = \sum_i m_i(y_i^2 + z_i^2), \qquad I_y = \sum_i m_i(x_i^2 + z_i^2), \qquad I_z = \sum_i m_i(x_i^2 + y_i^2).$$

The total moment of inertia is

$$I_x + I_y + I_z = 2\sum_i m(x_i^2 + y_i^2 + z_i^2) = 2\sum_i m_i r_i^2,$$

where r_i is the distance of mass m_i from the centre of mass.

A planar object, such as a sheet or loop of wire in the plane of x and y, and z is perpendicular to these, has moments of $I_x = \sum_i m_i y_i^2$, $I_y = \sum_i m_i x_i^2$ and $I_z = \sum_i m_i(x_i^2 + y_i^2)$ and therefore

$$I_z = I_x + I_y. \tag{7.72}$$

This is true only for planar bodies or uniform composition (lamina) and is called the *perpendicular axis* theorem.

The moment of inertia can also be defined as

$$I_\alpha = Mk^2 \tag{7.73}$$

where $k = \sqrt{I_\alpha / M}$ is called the *radius of gyration* of the body about axis α; and k is the root mean square radius.

7.15.5 Questions

 Full solutions are available at **www.oxfordtextbooks.co.uk/orc/beddard.**

Q7.59 Calculate the moments of inertia of the planar molecule XeF_4 **(a)** about the C_4 symmetry axis, **(b)** about a diagonal, and **(c)** about an edge. Assume each bond length is 2Å.

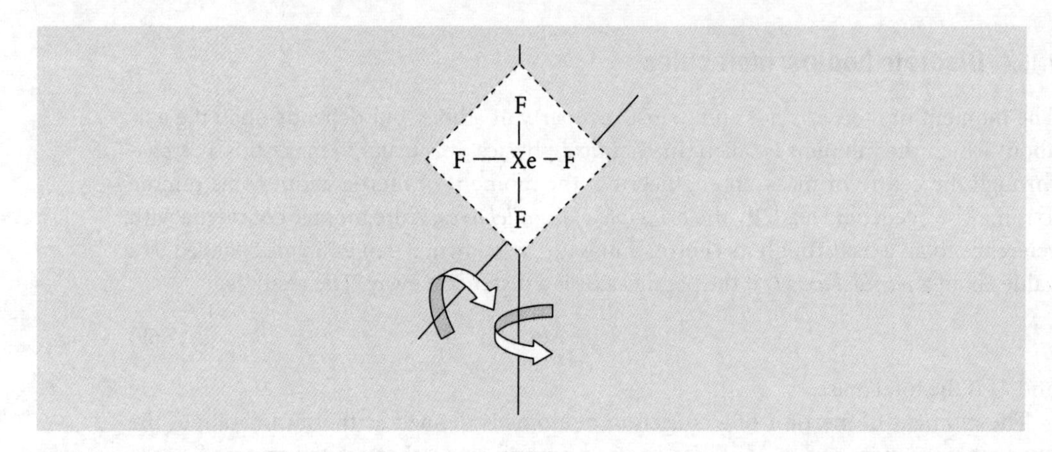

Fig. 7.67 XeF_4 and two rotation axes.

Q7.60 A circular loop of wire is formed by n rigidly connected masses with radius R of total mass M. Explain the different values obtained.

(a) About an axis z, through the centre of the loop,

(b) About a diameter.

(c) About an edge and parallel to z.

The difference in moments of inertia, is illustrated in Fig. 7.68, and in the right-hand diagram where the loop rotates about an edge.

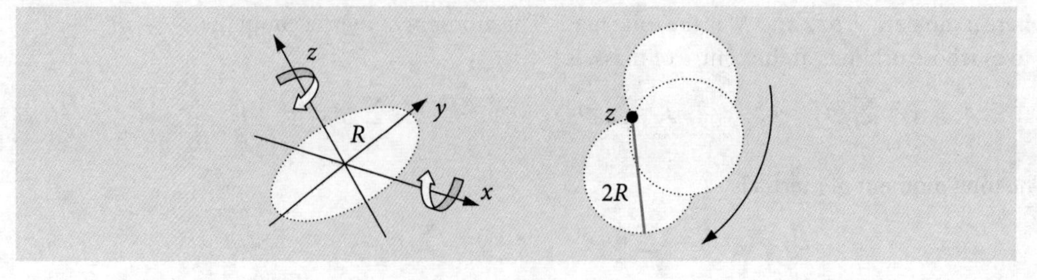

Fig. 7.68 Rotation of a loop about different axes.

Strategy: For a planar body to simplify the calculation, use the perpendicular axes theorem $I_z = I_x + I_y$ and by symmetry $I_x = I_y$.

7.15.6 Calculating bond lengths using moments of inertia

In calculating a molecule's moment or inertia, or other properties, first decide where the coordinate origin is going to be, and in what direction the axes are going to point. Data from an X-ray structure already has coordinates defined for us and usually these would be used. With an arbitrary molecule, for instance CO_2, we have to decide which atom is going to be at zero coordinate or perhaps we want to define zero between atoms; it really does not matter as long as each of the atom's coordinates are relative to zero. Calculating the moment of inertia of a molecule is equivalent to finding its angular momentum, and this could point in any direction but it always passes through the centre of mass. This might therefore be used as the coordinate zero but it is often not convenient to do this unless the molecule is highly symmetrical.

The angular momentum vector $L = I\omega$ of a typical diatomic molecule is shown in Fig. 7.69, where mass B is heavier than A, as the centre of mass indicates, and r is the bond length and is equal to $r = r_A + r_B$. As we have defined the distances r_A and r_B from the centre of mass, we have implicitly assumed that this is at zero. The angular momentum passes through the centre of mass, and points in the direction shown if the molecule is rotating perpendicularly to the page, with atom B going into and A coming out of the page. The molecule, if we could view it, would appear to be oscillating as it rotates around an invisible point in space, not centred exactly in the middle of the atoms. The same effect would be seen if you threw a heavy club hammer by its handle to make it spin; the handle rotates around the massive head that hardly appears to move. Any object, with sufficient effort, can be made to spin in any arbitrary direction, but we can always reduce this motion to a combination of contributions on three orthogonal directions we call the *principal axes*. A molecule or atom, expresses the laws of quantum mechanics far more than macroscopic bodies do and so have only limited values and directions in which the angular momentum can exist. The angular momentum properties are dealt with in detail in most textbooks on quantum mechanics. We will assume that our molecules behave classically.

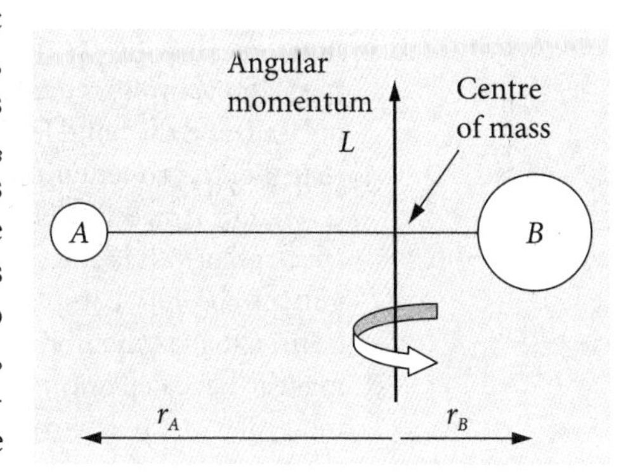

Fig. 7.69 A sketch showing the direction of one component of the angular momentum of a diatomic molecule. The angular momentum on the internuclear axis is zero.

The angular momentum along the line of the atoms, Fig. 7.69, is essentially zero, because the molecule is linear and the atoms have infinitesimal, effectively zero, dimensions. The molecule also has clockwise or anticlockwise rotation in the plane of the figure with angular momentum pointing directly into, or directly out of the page, respectively. By symmetry, this motion has the same angular momentum as shown in the figure.

To calculate bond length r starting with the moment of inertia, we use the centre of mass formula, equation (7.71). Put the bond along the x-axis, and the centre of mass at $x = 0$, then

$$q_x = \frac{-r_A m_A + m_B r_B}{m_A + m_B} = 0,$$

which produces $m_A r_A = m_B r_B$ and relates masses and distances. Equivalently, the turning moments about the centre of mass must be equal, also giving $m_A r_A = m_B r_B$. The moment of inertia is by definition,

$$I = m_A r_A^2 + m_B r_B^2$$

and from these equations, r_A and r_B must be removed and replaced with the bond length $r = r_A + r_B$. Substituting for r_A into $r = r_A + r_B$ gives $r = (m_B/m_A + 1)r_B$, and for r_B gives $r = (m_A/m_B + 1)r_A$.

Next replace r_A and r_B and calculate the moment of inertia

$$I = \frac{m_A r^2}{(m_A/m_B + 1)^2} + \frac{m_B r^2}{(m_B/m_A + 1)^2}.$$

Rearranging and simplifying produces

$$I = \left[\frac{m_B m_B m_A}{(m_A + m_B)^2} + \frac{m_A m_A m_B}{(m_B + m_A)^2} \right] r^2 = \frac{m_B m_A}{m_A + m_B} r^2 = \mu r^2$$

where $\mu = \dfrac{m_A m_B}{m_A + m_B}$ is the *reduced mass*.

The masses of the atoms are known, hence the bond lengths can be calculated using a measured spectroscopic rotational constant B, and the formula $B = \dfrac{1}{100hc} \dfrac{\hbar^2}{2I}$ cm^{-1}. In small molecules, by using laser or microwave spectroscopy, bond lengths can be obtained with extraordinary precision, typically to 0.01 Å. In CO_2 the bond length for the lowest vibrational level is 1.162 Å, in HCN the CH bond is 1.066 and the CN bond 1.153 Å long. Note, however, that the experimental result is more complex, but therefore also more interesting. For example, centrifugal effects cause the bond to stretch as rotational energy is increased, and the bond length also increases with vibrational energy. More subtle is the fact that the experimental measurement is B and what this actually measures is the average $\langle 1/r^2 \rangle$, not r directly. These effects and their resolution are discussed in many books on spectroscopy. It is clear also that for larger molecules, the moment of inertia calculation is going to be very complicated with lots of simultaneous equations, and for these calculations, a matrix method is required. This method, while a little complicated to start with, makes the calculation of the moments of inertia hardly any more difficult for any molecule, irrespective of size, than that just worked through for the diatomic.

7.15.7 Questions

Full solutions are available at **www.oxfordtextbooks.co.uk/orc/beddard.**

Q7.61 **(a)** Calculate the moment of inertia of the linear molecule HCN, using masses m_N, m_H and m_C for the most common isotopes of the atoms and bond lengths r_{HC} and r_{CN} given in the text. Explain why isotopic substitution is necessary to obtain the bond lengths. The parameters needed are defined in Fig. 7.70.

(b) Calculate the numerical value of the com displacement δ, Fig. 7.70, then the moment of inertia I using the values of the bond length given in the text for the isotope containing ^{1}H, ^{14}N and ^{12}C atoms.

(c) Assuming rigid-rotor behaviour, calculate by how much a rotational transition will shift if ^{14}N and then ^{13}C isotopes are used.

Strategy: On the diagram of the molecule choose a point to represent the centre of mass (com) and find its x coordinate. Write down the equation for the moment of inertia based on this point, and the equation for the centre of mass and assume that this is at $x = 0$.

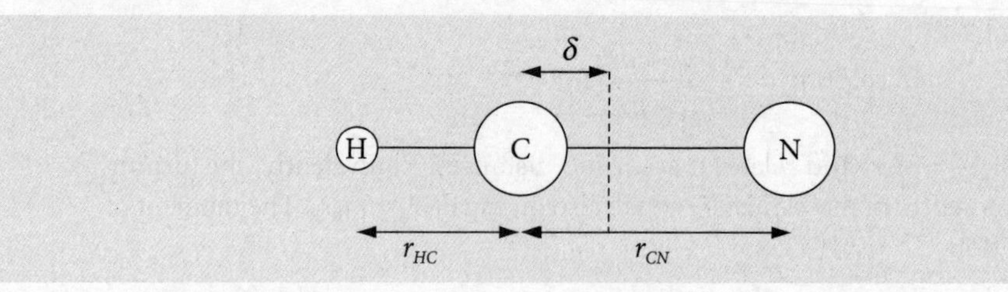

Fig. 7.70 A linear triatomic molecule.

7.15.8 Principal axes

We can cause any mass, such as a ball or spanner, to spin in any direction whatsoever with respect to itself; spin up, spin down, left, right, or any combination. If we calculate the moment of inertia about each of the axes the mass is spinning, we find that it is always possible to represent the motion and moments as a linear combination of three principal axes, which are intrinsic to the body and are defined by the shape and mass of the body itself. It is not so surprising that we can find a unique set of axes because we know that we can decompose any vector into its basis vectors, for example along the x-, y-, or z-axes for the unit vectors $\boldsymbol{i}, \boldsymbol{j}, \boldsymbol{k}$. When decomposing the motion or moments of inertia, if each does not contain components from any of the others, then the axes must be orthogonal and they are the principal axes.

The principal axes about which the body will rotate, are shown in Fig. 7.71; the moment of inertia about axis A will be relatively small, as the girder is long and thin. The moments of inertia about B and C, will be larger than about axis A, but approximately equal to one another because of symmetry. The same arguments apply to a molecule.

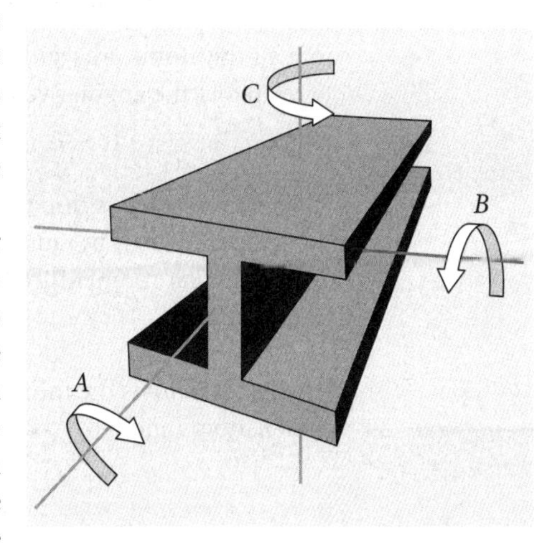

Fig. 7.71 Principal rotation axes about which moments of inertia are calculated.

It is clear from Fig. 7.71 how to choose the principal axes for the girder, but for an oddly shaped body, which means in practice most molecules this is hard to decide per se. If we were to choose another set of axes at some angle to those shown on the girder, then the moments of inertia would contain terms with contributions from the principal axes to a lesser or greater extent, depending on exactly how these other axes are placed with respect to the body. This would be rather awkward because everyone would have calculated different values for the same body, depending on exactly where the axes are chosen to be. By calculating the principal axes, which are unique to the body considered because angular momentum points in a fixed direction for a given rotational motion, then this ambiguity is removed.

Molecules have discrete masses and Fig. 7.72 shows the approximate location of the centre of mass of the chloro-mesiylene and propynal molecules together with two of the *principal rotation axes*; the third is 90° to these two and out of the page. In the vapour phase, the molecule rotates about its centre of mass (or gravity) and this may not be situated on an atom. Most of our calculations are aimed at finding these principal axes because molecular properties can be referenced to these. Principal axes are used in different ways by different authors, but only in the sense that x, y, and z labels become interchanged, therefore, a convention is adopted to label axes with the largest moment of inertia as the axis C and the smallest as A.

In Fig. 7.72 the moment about the axis through the chlorine atom will be small as this heavy atom is on the axis and for this atom, and the other two carbon atoms on the axis the product mr^2 is zero. Rotation about the other two axes will be larger but different to one another.

In calculating the moment of inertia, we are not interested in the exact formula for a particular molecule, which is often very complicated and of no intrinsic interest. We are interested, however, in the numerical values of the moment of inertia and their directions with respect to the molecule, because we ultimately want to calculate bond lengths. A very elegant eigenvalue–eigenvector matrix method can automatically find the principal axes

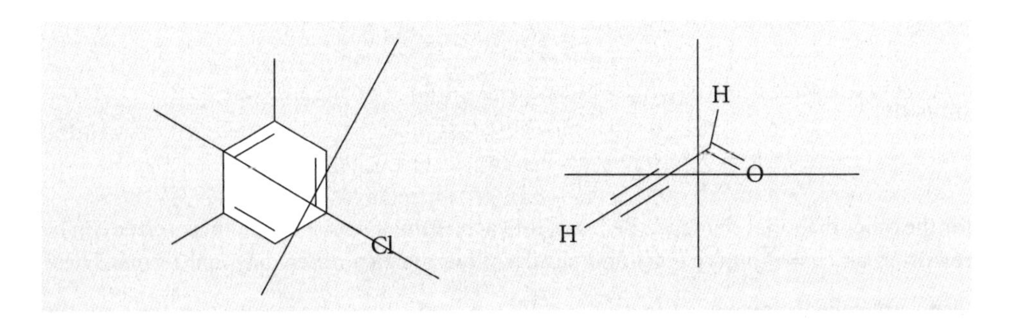

Fig. 7.72 Approximate location of the centre of mass together with two of the principal rotation axes. The third, describing rotational motion in the plane of the figure, is perpendicular to the other axes and points out of the plane of the figure. The molecule rotates about the centre of mass or gravity. This need not be situated on an atom.

and the moments of inertia of very complex molecules. The algorithm with which to do this is shown below. The eigenvalues produced are the moments of inertia; the eigenvectors are used to produce the principal or inertial axes; eigenvectors always produce geometry! The method is described next; equation (7.75) is the one we will use.

7.15.9 Formal description of the method

If all the atoms are rigidly connected together, the kth atom and its velocity vector $\boldsymbol{v}_k$, are related to the angular velocity of the molecule $\boldsymbol{\omega}$ about the centre of mass as

$$\boldsymbol{v}_k = \boldsymbol{\omega} \times \boldsymbol{r}_k$$

where $\boldsymbol{r}_k$ is the position vector from the centre of mass and $\boldsymbol{\omega}$ is a vector but does not carry an index. This is because in a rigid body, all atoms move with the same angular velocity. The angular momentum for the kth atom is defined as the vector cross product

$$J_k = \boldsymbol{r}_k \times \boldsymbol{p}_k$$

where $\boldsymbol{p} = m\boldsymbol{v}$ is the momentum vector and the total *angular* momentum is the sum over all n atoms and is

$$J = \sum_{k=1}^{n} m_k(\boldsymbol{r}_k \times \boldsymbol{v}_k) = \sum_{k=1}^{n} m_k(\boldsymbol{r}_k \times (\boldsymbol{\omega} \times \boldsymbol{v}_k))$$

where we have substituted for $\boldsymbol{p}$ and then $\boldsymbol{v}$. The cross product of a cross product is called a triple product (see Chapter 6.18) and is a vector;

$$\boldsymbol{a} \times (\boldsymbol{b} \times \boldsymbol{c}) = (\boldsymbol{a}\cdot\boldsymbol{c})\boldsymbol{b} - (\boldsymbol{a}\cdot\boldsymbol{b})\boldsymbol{c}.$$

The two dot products each produce a number, and these multiply the vectors $\boldsymbol{b}$ and $\boldsymbol{c}$. We can now write, remembering that $\boldsymbol{r}_k$ is a vector,

$$J = \sum_{k=1}^{n} m_k[(\boldsymbol{r}_k\cdot\boldsymbol{r}_k)\boldsymbol{\omega} - (\boldsymbol{r}_k\cdot\boldsymbol{\omega})\boldsymbol{r}_k] = \sum_{k=1}^{n} m_k(r_k^2\boldsymbol{\omega} - (\boldsymbol{r}_k\cdot\boldsymbol{\omega})\boldsymbol{r}_k).$$

The vector J has components x, y, and z so it represents three equations. This can be written as a matrix equation but note that r_k^2 is a number; it is the perpendicular distance of atom k from an axis, x, y or z, but $\boldsymbol{r}_k$ is the vector $\boldsymbol{r}_k = [x_k \;\; y_k \;\; z_k]$ describing the position of atom k.

$$J_{(x,y,z),k} = m_k r_k^2 \begin{bmatrix} \omega_x \\ \omega_y \\ \omega_z \end{bmatrix} - m_k \left([x_k \;\; y_k \;\; z_k] \cdot \begin{bmatrix} \omega_x \\ \omega_y \\ \omega_z \end{bmatrix} \right) \begin{bmatrix} x_k \\ y_k \\ z_k \end{bmatrix}$$

The x component for the kth atom is found by expanding the dot product as $x_k\omega_x + y_k\omega_y + z_k\omega_z$ and then multiplying by $x_k m_k$ and rearranging a little

$$J_{x,k} = m_k(r_k^2 - x_k^2)\omega_x - m_k x_k y_k \omega_y - m_k x_k z_k \omega_z. \tag{7.74}$$

There are similar equations for the y and z direction components.

By comparing coefficients of ω_x, ω_y, and ω_z equations (7.74) and (7.75), the diagonal terms in this matrix are

$$I_{xx} = \sum_k m_k(r_k^2 - x_k^2),$$

similarly,

$$I_{yy} = \sum_k m_k(r_k^2 - y_k^2) \qquad \text{and} \qquad I_{zz} = \sum_k m_k(r_k^2 - z_k^2)$$

for the other diagonal elements. Because, for each atom, $r^2 = x^2 + y^2 + z^2$, these terms can be rewritten as $I_{xx} = \sum_k m_k(y_k^2 + z_k^2)$ and similarly for the two other diagonal terms. These terms are called *moments of inertia* coefficients and cannot be negative as they are the sum of squared terms. The cross terms I_{xy}, for example, are called *products of inertia* coefficients, and are

$$I_{xy} = -\sum_k m_k x_k y_k, \qquad I_{xz} = -\sum_k m_k x_k z_k, \qquad I_{yz} = -\sum_k m_k y_k z_k.$$

Equation (7.74) can be rewritten for each atom k using the inertial coefficients

$$J_x = I_{xx}\omega_x + I_{xy}\omega_y + I_{xz}\omega_z$$
$$J_y = I_{xy}\omega_x + I_{yy}\omega_y + I_{yz}\omega_z$$
$$J_z = I_{xz}\omega_x + I_{yz}\omega_y + I_{zz}\omega_z$$

and, of the nine coefficients, only six are different because of symmetry; $I_{xy} = I_{yx}$ and so forth. In matrix form these equations are

$$J = I\omega \equiv \begin{bmatrix} I_{xx} & I_{xy} & I_{xz} \\ I_{xy} & I_{yy} & I_{yz} \\ I_{xz} & I_{yz} & I_{zz} \end{bmatrix} \begin{bmatrix} \omega_x \\ \omega_y \\ \omega_z \end{bmatrix}. \tag{7.75}$$

The matrix I, is also sometimes either called the *moment of inertia dyadic* or the *inertia tensor*, but, more importantly, it is symmetrical and Hermitian so has real eigenvalues and orthogonal eigenvectors.

The next step in the calculation is to perform a *principal axis transform*, which we can view as a rotation of the inertia matrix to remove all the off-diagonal terms that become zero on forming a diagonal matrix. The methods of matrix algebra enable us to find for any molecule, or any body in general, the set of Cartesian axes for which the inertia I matrix will be diagonal. The result of this transformation is to produce moments of inertia about the principal axes.

The eigenvalues λ, are found for each atom k, using the secular determinant

$$\begin{vmatrix} I_{xx}-\lambda & I_{xy} & I_{xz} \\ I_{xy} & I_{yy}-\lambda & I_{yz} \\ I_{xz} & I_{yz} & I_{zz}-\lambda \end{vmatrix} = 0.$$

The expressions for the diagonal and off diagonal terms are given above. Because the moments of inertia coefficients contain squared terms we can pictorially view then as an ellipse. The rotation to principal axes is then akin to rotating the ellipse, as shown in Fig. 7.66.

Finally, the kinetic energy relative to the centre of mass is also calculated in a straightforward way in matrix form and is

$$T = \frac{1}{2}\omega\cdot I\cdot\omega = \frac{1}{2}[\omega_x \;\; \omega_y \;\; \omega_z] \begin{bmatrix} I_{xx} & I_{xy} & I_{xz} \\ I_{xy} & I_{yy} & I_{yz} \\ I_{xz} & I_{yz} & I_{zz} \end{bmatrix} \begin{bmatrix} \omega_x \\ \omega_y \\ \omega_z \end{bmatrix}.$$

An example is easier to understand than this complex theory; the moments of inertia of ethanol will now be calculated. The X-ray coordinates give the atoms' coordinates and Maple is used to do the algebra. The rotational constants are then easily calculated and compared with experimental values, which are $A = 1.18$, $B = 0.318$ and $C = 0.277\ \text{cm}^{-1}$ (Senent et al. 2000).

This problem is transferable to any molecule, although inputting data will be tedious for large molecules; only the coordinates C1, Ox, etc. will need to be changed and the order of array `molec` and the `mass`. The numerical diagonalization will normally produce complex numbers as the eigenvalues and eigenvectors. As the determinant is Hermitian, the eigenvalues must be real, and any complex part should be small because it is caused by the method used to numerically solve of the equations, and should be made zero. To force Maple to use real values, use `Re(eigenval[1])` etc. Note that distances are in angstroms, and the masses of the common isotopes, ^{16}O and ^{12}C are in atomic mass units, therefore, we take the mass to be 16 and 12 respectively. The units of the moment of inertia can be changed to kg m^2 units at the end of the calculation. Note that the coordinates are each adjusted to the centre of mass before the calculation of the moments of inertia begins. The centre of mass is labelled `com` and the new coordinates called `Cmolec`.

Algorithm 7.8 Calculating the moments of inertia of a molecule

```
> with(LinearAlgebra):
> # ethanol coordinates in order x, y, z and in angstrom.
> C1:= [  -0.968, -0.008, -0.167 ]:
  Ox:= [  -0.953, 1.395, -0.142 ]:
  H1:= [  0.094, -0.344, -0.200 ]:
```

```
  C2:= [ -1.683,-0.523, 1.084 ]:
  H2:= [  -1.490, -0.319, -1.102 ]:
  H3:= [  -1.842, 1.688, -0.250 ]:
  H4:= [  -1.698, -1.638, 1.101 ]:
  H5:= [  -1.171, -0.174, 2.011 ]:
  H6:= [  -2.738, -0.167, 1.117 ]:
> # molec is list to hold coordinates,
> # the order does not matter provided mass is in same order
  molec:= [ H1, H2, H3, H4, H5, H6, C1, C2, Ox ]:
  mass:=  [ 1, 1, 1, 1, 1, 1, 12, 12, 16 ]: #in amu
  n:= nops(molec);          # get nr of atoms in list molec
  Cmolec:= molec:  Ri:= molec:  # make lists to hold data
  com:= < 0,0,0 >:              # Com is centre of mass
  M:= add( mass[i],i = 1..n);              # find total mass
  # next calculate centre of mass, com
> com:= add( molec[i] * mass[i],i = 1..n ) / M;
```

$$com := [-1.21533, 0.32596, 0.24802]$$

```
  # adjust coordinates to center of mass
  for i from 1 to n do
       Cmolec[i]:= molec[i] - com
  end do:
  # calculate distance from origin of each atom
  for i from 1 to n do
    Ri[i]:=sqrt(Cmolec[i,1]^2+Cmolec[i,2]^2+Cmolec[i,3]^2)
  end do:
  # next define 6 moment of inertia parameters. see p 420
  Ixx:= add( mass[i]*(Ri[i]^2-Cmolec[i,1]^2),i = 1..n) ;
  Iyy:= add( mass[i]*(Ri[i]^2-Cmolec[i,2]^2),i = 1..n) ;
  Izz:= add( mass[i]*(Ri[i]^2-Cmolec[i,3]^2),i = 1..n) ;
  Ixy:= add(-mass[i]*(Cmolec[i,1]*Cmolec[i,2]),i =1..n);
  Ixz:= add(-mass[i]*(Cmolec[i,1]*Cmolec[i,3]),i =1..n) ;
  Iyz:= add(-mass[i]*(Cmolec[i,2]*Cmolec[i,3]),i =1..n) ;
  # put into inertia matrix M and evaluate
  M:= << Ixx|Ixy|Ixz >,< Ixy|Iyy|Iyz >,< Ixz|Iyz|Izz > >;
```

$$M := \begin{bmatrix} 55.09197 & -8.38308 & 9.12077 \\ -8.38308 & 28.94488 & 16.01698 \\ 9.12077 & 16.01698 & 44.53846 \end{bmatrix}$$

```
> (eigval, eigvec):= Eigenvectors( M ):       # diagonalise
  # label Ic largest, Ia smallest, do this by sorting.
  Isorted:=
    sort([ Re(eigval[1]), Re(eigval[2]), Re(eigval[3]) ]);
  Ia:= Isorted[1];
  Ib:= Isorted[2];
  Ic:= Isorted[3];
```

$$Ia := 15.31633 \quad Ib := 52.87277 \quad Ic := 60.38671$$

```
> # make each value of eigenvector real, print 5 places.
  evecs:= map( x-> Re( evalf(x, 5) ), eigvec );
```

$$evecs := \begin{bmatrix} 0.28935 & 0.84488 & .44996 \\ 0.79846 & 0.04624 & 0.60027 \\ -.52796 & 0.53296 & 0.66122 \end{bmatrix}$$

```
> # Rotational constants B1, B2, B3 in wavenumbers.
  hbar:= 1.054e-34: c:= 2.9979e10: amu:= 1.667e-27:
  Angst:= 1.0e-10:  pi:= evalf(Pi):
  A:= hbar/(4*pi*Ia*amu*Anst^2*c);
  B:= hbar/(4*pi*Ib*amu*Anst^2*c);
  C:= hbar/(4*pi*Ic*amu*Anst^2*c);
```

$$A := 1.09578 \quad B := 0.31743 \quad C := 0.27793$$

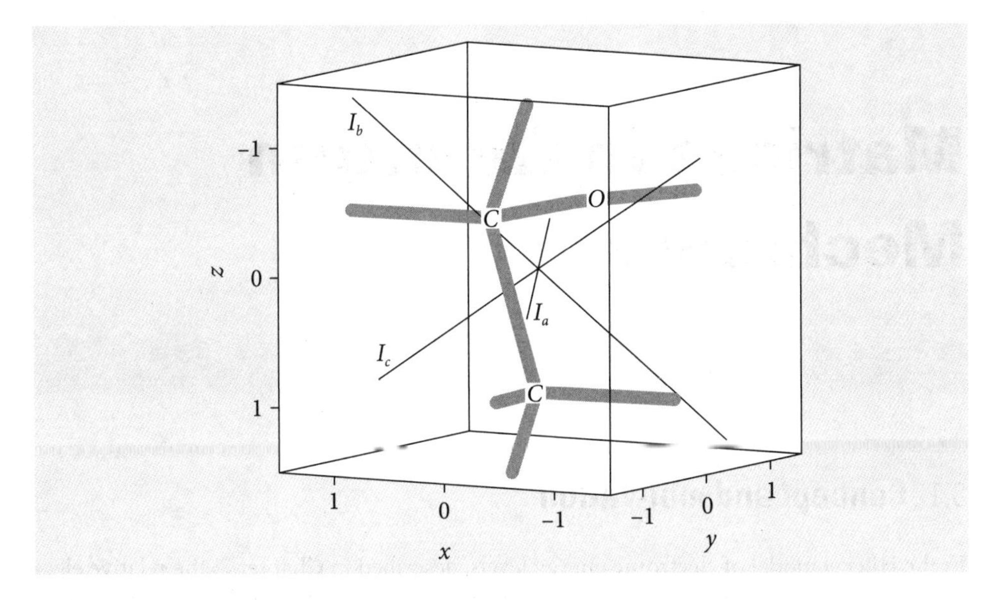

Fig. 7.73 X-ray structure of ethanol and its new inertial axes drawn to scale with respect to one another using the eigenvalues. The lengths are in units of amu A^2 or 1.667×10^{-47} kgm^2, $I_a = 15.2$, $I_b = 52.9$ and $I_c = 60.4$. The H atoms are not labelled.

Note that these results should be rounded to four figures, as this is the precision of the data. The rotational constants compare well with experimentally measured values which are, $A = 1.18$, $B = 0.318$, and $C = 0.277$ cm^{-1}. The inertial axes can now be drawn on top of the molecular structure; see Algorithm 7.4 (p 360) for an example of drawing a molecule. By convention, I_c labels the largest moment of inertia and I_a the smallest. Note that the units we use are in atomic mass units × Å^2, which are equivalent to 1.667×10^{-47} kgm^2.

As might have been anticipated, the centre of mass is between the heavier atoms and is almost in the plane of these atoms. The smallest moment of inertia is about an axis in the plane of the OCC atoms and it points in the OC direction as shown approximately along the 'line' of the CCO atoms. The two largest moments of inertia, which are similar in value, describe motion perpendicular to this axis and are larger because the atoms are further from the axes. The new inertial axes, which are parallel to the moments of inertia, are drawn in proportion to the size of each eigenvector component from the initial molecular x, y, and z-axes; the I_a axis has x, y, z components of approximately 0.29, 0.8, −0.53 as shown in the modal matrix evecs.

7.15.10 **Question**

Full solutions are available at **www.oxfordtextbooks.co.uk/orc/beddard**.

Q7.62 Use an X-ray database to find coordinates for CH_3F, CH_2Cl_2, chlororethylene (CH_2CHCl), and any other molecules you want. Calculate the moments of inertia and plot the inertia axes on top of the molecular structure. The values you obtain should be approximately as given in the table, which are in units of 10^{-47} kg m^2:

	I_a	I_b	I_c
CH_3F	5.3	I_c	33
CH_2Cl_2	26	255	276
CH2CHCl	15	139	154

8 Matrices in Quantum Mechanics

8.1 Concept and motivation

In the Hückel model of electronic energy levels, described in Chapter 7, the relative electronic energies of π orbitals are calculated on the assumption that each π orbital interacts only with it nearest neighbours. A matrix of interactions is constructed and the eigenvalues and eigenvectors found. In this chapter, problems that are more general are explored.

The Schrödinger equation (8.1) can be solved, in principle, by integration in the same manner as other similar differential equations and the energies E and wavefunctions φ determined.

$$\left[-\frac{\hbar^2}{2m}\frac{d^2}{dx^2}+V(x)\right]\varphi=E\varphi. \tag{8.1}$$

However, an algebraic solution turns out to be possible only for a few potentials V. Among these are the harmonic and Morse oscillator, the particle in a box or on a ring, the rigid rotor and the hydrogenic type atoms with a single electron. Flugge (1999) gives several more one dimensional examples. For other problems, for example, to predict an NMR spectrum or that of a non-rigid rotor the equation is solved by recasting it in a basis set and using the properties of eigenvalue–eigenvector matrices. This is in essence an algebraic method although numerical values can be calculated. The Schrödinger equation can also be solved numerically for an arbitrary potential, by using one of several integration methods; see Chapter 11. However, numerical integration does not easily lend itself to predicting trends and identifying important parameters and limits, which an algebraic solution does.

This chapter has two parts, the first assumes some knowledge of basis sets and several problems are solved using them, and, in the second part, basis sets and bra-ket algebra are described, and how these objects are manipulated is illustrated. If you are unfamiliar with these topics, it may be worth looking at the second part first.

8.1.1 Notation

Writing the Schrödinger equation as $H^0\varphi=E\varphi$ is the conventional shorthand that represents the following set of simultaneous equations

$$H^0\varphi_0=E_0\varphi_0,\qquad H^0\varphi_1=E_1\varphi_1,\qquad\cdots\qquad H^0\varphi_i=E_i\varphi_i, \tag{8.2}$$

one for each of the possibly infinite number of energy levels. The superscript zero, e.g. H^0 indicates that this operator exactly solves the Schrödinger equation with wavefunctions φ_i. These wavefunctions φ_i are orthogonal to one another because each describes an eigenstate of the particular potential used and are assumed to have been normalized. The orthonormality condition is $\int\varphi_n^*\varphi_m dx=\delta_{n,m}$ and in Dirac 'bra-ket' notation the orthogonality and normalization of the wavefunctions is written as $\langle\varphi_n|\varphi_m\rangle=\delta_{nm}$, where δ is the Kronecker delta such that $\delta_{nn}=\delta_{mm}=1$ and if $n\neq m$, then δ is zero. The subscripts $i=0,1,2,\cdots$ are used to label the whole set of energy levels but the subscripts n and m are reserved to represent individual levels should that be necessary and are also the quantum numbers. If the harmonic oscillator were being described, i would extend to infinity. In contrast, spin only has a finite set of quantum numbers; two for the spin state of the electron $i=-1/2, 1/2$ and three for the nuclear spin state of ^{14}N or ^{2}H, $0, \pm1$.

8.2 Expectation Values

The *expectation* or average value[1] for the lowest level 0 is the energy E_0. It is calculated by left multiplying equation (8.2) by φ_0^* and integrating over all space x,

$$E_0 = \frac{\int_{-\infty}^{\infty} \varphi_0^* H^0 \varphi_0 dx}{\int_{-\infty}^{\infty} \varphi_0^* \varphi_0 dx} = \int_{-\infty}^{\infty} \varphi_0^* H^0 \varphi_0 dx, \tag{8.3}$$

and because the wavefunction is normalized, $\int_{-\infty}^{\infty} \varphi_0^* \varphi_0 dx = 1$. The same procedure can be followed for every other level to calculate E_n. Any energy can be written in equivalent ways as

$$E_n = \int_{-\infty}^{\infty} \varphi_n^* H^0 \varphi_n dx \equiv \langle \varphi_n^* | H^0 | \varphi_n \rangle = \langle n | H^0 | n \rangle. \tag{8.4}$$

the latter two expressions being in Dirac's bra-ket notation, explained in Section 8.2 together with basis sets. Note that the bra $\langle \varphi_n^* |$ is always contains the complex conjugate of the wavefunction φ, but it is not common to label it as such, therefore the bra can be written as $\langle \varphi_n |$. The ket is $| \varphi_n \rangle$. In general, if S is an operator, its expectation is

$$S_{n,n} = \int_{-\infty}^{\infty} \varphi_n^* S \varphi_n dx \equiv \langle \varphi_n^* | S | \varphi_n \rangle = \langle n | S | n \rangle. \tag{8.5}$$

If the operator S is position x, the expectation value is the average position, if momentum, whose operator is $-i\hbar d/dx$, then S is the average momentum and so forth. When two different states n and m interact under the action of the operator S, the notation used is similar

$$S_{n,m} = \int_{-\infty}^{\infty} \varphi_n^* S \varphi_m dx \equiv \langle \varphi_n^* | S | \varphi_m \rangle = \langle n | S | m \rangle, \tag{8.6}$$

Interactions between different levels n and m lead to interesting effects, and this chapter describes how these may be calculated. No interaction occurs if $S = H^0$ because H^0 exactly solves the Schrödinger equation with wavefunctions φ_i, which are orthogonal,

$$H_{nm} = \int \varphi_n^* H^0 \varphi_m dx = E_0 \int \varphi_n^* \varphi_m dx = 0. \qquad n \neq m \tag{8.7}$$

Expectation values are also called *matrix elements* and these are explained next. Suppose that the (quantum) harmonic oscillator is to be solved. The wavefunction is a function of position, the bond extension, and the operator is

$$H^0 \equiv \left[-\frac{\hbar^2}{2m} \frac{d^2}{dx^2} + V(x) \right]$$

and, for the harmonic oscillator, $V(x) = kx^2/2$. To obtain the energy, H^0 operates on $\varphi(x)$, the result is then multiplied by $\varphi^*(x)$ and this result integrated. The matrix element describing the interaction of one state n to another m based on equation (8.6), has the form $H_{nm} = \int \varphi_n^* H^0 \varphi_m dx$ where H_{nm} is an energy. Because the quantum numbers are discrete, a secular equation can be formed and eigenvalue–eigenvector methods used to solve for the energies. See Chapter 7.12 for the method. The general equation is

$$\boldsymbol{A v} = \lambda \boldsymbol{v}$$

where $\boldsymbol{A}$ is a matrix, $\boldsymbol{v}$ the eigenvectors and λ the eigenvalues. In this context, the matrix $\boldsymbol{A}$ is the matrix of expectation values H_{nm}, the eigenvector's elements are the wavefunctions φ_n and eigenvalues λ_n are the energies E_n. Rewriting in this way produces

[1] The average can also be written as <E> or $\bar{E}$.

$$\begin{bmatrix} H_{00} & H_{01} & \cdots & \cdots \\ H_{10} & H_{11} & H_{12} & \cdots \\ H_{20} & H_{21} & \ddots & \cdots \\ \vdots & \vdots & \vdots & H_{nn} \end{bmatrix}\begin{bmatrix} \varphi_0 \\ \varphi_1 \\ \vdots \\ \varphi_n \end{bmatrix} = \begin{bmatrix} E_0 \\ E_1 \\ \vdots \\ E_n \end{bmatrix}\begin{bmatrix} \varphi_0 \\ \varphi_1 \\ \vdots \\ \varphi_n \end{bmatrix} \tag{8.8}$$

but because each φ is an exact solution to the Schrödinger equation the φ are orthogonal to one another and all values that are non-diagonal with $n \neq m$ equation (8.7) are zero, giving

$$\begin{bmatrix} H_{00} & 0 & \cdots & \cdots \\ 0 & H_{11} & 0 & \cdots \\ 0 & 0 & H_{22} & \cdots \\ \vdots & \vdots & \vdots & \ddots \end{bmatrix}\begin{bmatrix} \varphi_0 \\ \varphi_1 \\ \varphi_2 \\ \vdots \end{bmatrix} = \begin{bmatrix} E_0 \\ E_1 \\ E_2 \\ \vdots \end{bmatrix}\begin{bmatrix} \varphi_0 \\ \varphi_1 \\ \varphi_2 \\ \vdots \end{bmatrix} \tag{8.9}$$

and therefore a set of equations, such as (8.4) is obtained, one for each quantum number n. The matrix (8.9) is already diagonalized, therefore $E_0 = H_{00}$, $E_1 = H_{11}$, and so forth and these are the eigenvalues or energies.

The effect of using a different potential is calculated next. Hamiltonian H has a new term in the potential energy, and this manifests itself as off-diagonal terms in the H_{nm} matrix, equation (8.8). The Hamiltonian can be written as $H = -\dfrac{\hbar^2}{2m}\dfrac{d^2}{dx^2} + V^1(x)$ where V^1 is the new potential. Each integral $H_{nm} = \int \varphi_n^* H \varphi_m d\tau$ has to be solved first then the matrix diagonalized. In some problems, such as the NMR nuclear spin example given below, the wavefunctions are never known, only their integrals with the angular momentum operator, which means that H_{nm} is known directly. In problems that have a large number of states, such as an anharmonic oscillator, a huge matrix will be needed to calculate accurate values of the energies because there are many non-zero, off-diagonal terms. In this case, the larger the matrix is, the more accurate the answers become. In examples involving electron or nuclear spin, the matrix has a finite size that completely determines the problem, and accurate results are obtained from small matrices.

8.2.1 The effect of new potential energy

The energies and wavefunctions of the harmonic oscillator are well known, the Hamiltonian is $H^0 = -\dfrac{\hbar^2}{2m}\dfrac{d^2}{dx^2} + \dfrac{1}{2}kx^2 \equiv H^k + V$. Suppose that to make the potential more like that of a real molecule, a term V^1 replaces the potential energy making it quartic, i.e. $V^1 = k'x^4/2$, but any similar potential, such as the Morse potential, could be used. The harmonic oscillator wavefunctions φ would no longer be 'diagonal in the eigenstates' meaning that the wavefunctions φ that are solutions with H^0 are not solutions of the Schrödinger equation with the new potential. With the new potential, some, if not all of the off-diagonal expectation values (matrix elements) H_{23}, H_{53}, etc. will no longer be zero. The new energies are no longer given by the diagonal elements of the matrix but by an equation of the form of (8.8), which has to be diagonalized.

The wavefunctions of the quartic oscillator will be labelled ψ, however, the Schrödinger equation cannot yet be solved with this potential because the equation $H_{nm} = \int \psi_n H \psi_m dx$ cannot be worked out because the ψ are unknown. One solution to this problem is to solve the Schrödinger equation with the quartic potential, V^1, using the harmonic wavefunctions φ as a basis, and so obtain the quartic eigenvalues (energies) and eigenvectors. The eigenvectors v are used to *expand* the unknown wavefunctions ψ in terms of the known ones φ. A basis set to describe the new wavefunctions has to be constructed. This basis set has to be orthogonal and although it does not need be normalized, it is usually easier to use if it is. A basis set is used because *any* vector can be described as a linear combination of basis vectors (Chapter 6) and it is proposed that any wavefunction can be described as a weighted sum of basis wavefunctions. The basis set used is the set of harmonic oscillator wavefunctions (eigenvectors) φ which solve the equation $H^0\varphi = E\varphi$. The new wavefunctions then have the form

$$\psi = a_1\varphi_1 + a_2\varphi_2 + a_3\varphi_3 + \cdots \tag{8.10}$$

where a_1, a_2 and so forth, are the amounts of each φ needed to make the ψ that solves—which means, will diagonalize the Schrödinger equation $(H^k + V^1)\psi = E\psi$. As there are many levels 0 to n, each ψ has its own expansion and there are, therefore, n^2 coefficients in total. Each wavefunction has its own set of coefficients and it is better to give them two subscripts and label the coefficients $v_{1k}, v_{2k}, \cdots$ as they are elements of the kth (column) eigenvector of the matrix of v's. The kth of a total of n wavefunctions is

$$\psi_k = v_{1k}\varphi_1 + v_{2k}\varphi_2 + v_{3k}\varphi_3 + \cdots + v_{nk}\varphi_n \tag{8.11}$$

The columns of eigenvectors are calculated as the eigenvalue-eigenvector equation (8.8) is solved.

Summary of the method

(i) The wavefunctions ψ are the solutions to Schrödinger's equation with Hamiltonian $H + V^1$, and φ is the *exact* solution to a simpler problem with Hamiltonian H^0, such as the harmonic oscillator or particle in a box, from whose wavefunctions a linear combination is made to find ψ.

(ii) The φ are orthogonal to one another; to form the complete basis set they must be and it is simpler to start with each φ normalized.

(iii) If the number of levels is infinite then clearly the calculation cannot be completed exactly. In this case, the ψ and the corresponding energies are approximate but improve as more energy levels are added. This improvement occurs because the energy levels are coupled one to another and including more of them allow more couplings to be added.

(iv) If the number of levels is finite as in spin problems such as NMR, then the result is exact.

(v) A basis set of any orthogonal set of functions could be chosen for the functions φ as shown in the generalized Fourier method, Chapter 9.4. However, it is natural to try to choose a set of basis functions that have a similar shape to those the new wavefunctions are expected to look like. A good guess will make the off-diagonal terms relatively small, and when diagonalized, which has the effect of mixing all the matrix elements together, the resulting diagonal terms will only be slightly different from the un-diagonalized ones.

(vi) This is not a perturbation method; the more terms that are added, the more accurate the result becomes.

Returning to the quartic oscillator, using the basis set wavefunctions φ the matrix of the eigenvalue equation now has the form

$$\begin{array}{c} \text{ordering of basis set} \rightarrow \\ \begin{array}{c|cccc} & b_1 & b_2 & b_3 & b_4 \cdots \\ b_1 & H_{11} & H_{12} & \cdots & \cdots \\ b_2 & H_{21} & H_{22} & H_{23} & \cdots \\ b_3 & H_{31} & H_{32} & H_{33} & \cdots \\ \vdots & \vdots & \vdots & \vdots & \ddots \end{array} \end{array} \tag{8.12}$$

where b_1, b_2 are only the *indices* of the basis set; for the vibrational oscillator they are labelled 1, 2, 3, 4, $\cdots$ The matrix elements, the integrals H_{11}, H_{12}, $\cdots$ are in the form of equations (8.5) and (8.6). The basis set indices can be put in any order but this must be the same in each row and column. To simplify the calculation, the Hamiltonian operator is split into H and a potential term H^1, as

$$H = H + H^1. \tag{8.13}$$

and $H^1 = V^1 = k'x^4/2$ if the new potential were quartic. The matrix elements formed in the new basis set have the form

$$\begin{aligned} H_{nm} &= \int \varphi_n^* H \varphi_m dx \\ &= \int \varphi_n^* H \varphi_m dx + \int \varphi_n^* H^1 \varphi_m dx = \int \varphi_n^* H^1 \varphi_m dx \end{aligned}$$

Which is not necessarily zero because the operator has been changed to H^1. Notice that the wavefunctions are φ_n and φ_m and are those of the chosen basis set and not the wavefunctions ψ that will solve the new Schrödinger equation; these are not known yet.

Diagonalizing the matrix produces the eigenvalues that are the energies. The eigen vectors v are also obtained and from these the wavefunctions that approximate to the true ones can be calculated using equation (8.11). Eigenvectors are usually formed into a matrix; each column is the eigenvector v for a wavefunction and the order of the columns is the same as the order of the eigenvalues. This means that if v_1, v_2, v_3 are the first three entries for the *column* eigenvector of eigenvalue 1, then the wavefunction is

$$\psi_1 \approx v_{11}\varphi_1 + v_{21}\varphi_2 + v_{31}\varphi_3 + \cdots + v_{n1}\varphi_n$$

and $\approx$ is used as the basis set is finite so this is an approximation, albeit a very good one, if an accurate calculation has been performed with a sufficiently large basis set. The second wavefunction might be $\psi_2 \approx v_{12}\varphi_1 + v_{22}\varphi_2 + v_{32}\varphi_3 + \cdots + v_{n2}\varphi_n$ where the vs are the elements of the second column of eigenvectors.

Finally, note that this matrix/basis set method is an entirely different approach to solving the Schrödinger equation compared to other numerical methods. In those methods (see chapter 10) the equation is solved on a large grid of points using recursive equations that approximate the integral. In the matrix method, the amounts of each of the basis wavefunctions only are calculated. A grid of points is needed only to plot the wavefunction not to do numerical calculation.

Although the example described is a harmonic oscillator becoming a quartic oscillator, it is simpler mathematically as an illustration of this method, to calculate the vibrational energy levels and wavefunctions of a harmonic oscillator, using the values for the HCl molecule and the basis set of the 'particle in a box' wavefunctions. The length of the box will be L, where (arbitrarily) $L/2 = 0.128$ nm, the HCl bond length. The reduced mass $\mu = 35/36 \times 1.6710^{-27}$ kg, and force constant $k = 516$ N m^{-1}. The Schrödinger equation is $H\psi = E\psi$ where $H = -\dfrac{\hbar^2}{2m}\dfrac{d^2}{dx^2} + kx^2/2$ and the basis set φ are the solutions of $H^0\varphi = E\varphi$ where $H^0 = -\dfrac{\hbar^2}{2m}\dfrac{d^2}{dx^2}$ because the potential for a particle in a box is zero, and $H^1 = kx^2/2$.

The particle in a box wavefunctions are sine functions and are similar in shape to the wavefunctions of the harmonic oscillator, which oscillate in a sort of damped sinusoidal manner. Gaussian functions $e^{-a(x-x_i)^2}$ might also do quite well, if one were displaced from another by an amount x_i, and positioned along the bond extension of the molecule. Choosing $\exp(-ax)$ as the basis functions with different α would be a poor choice since adding exponential functions to make an oscillating wavefunction is always going to be difficult. Note that, although we use single functions as the basis set, each element could be several functions added together; you can decide what you want to do, but the basis set functions must be made orthogonal. The Gram–Schmidt method (Arkfen 1970) can be used to do this. Maple also has a function for this operation.

The chosen basis set is the set of 'particle in a box' wavefunctions

$$\varphi_n \equiv \sqrt{\frac{2}{L}}\sin(n\pi x/L), \tag{8.14}$$

with $n = 1, 2, 3, \cdots$ and are defined over the range $0, \cdots, L$, and are zero elsewhere. These wavefunctions are normalized and orthogonal, the first few members of the basis set are $[\sqrt{\frac{2}{L}}\sin(\pi x/L),\ \sqrt{\frac{2}{L}}\sin(2\pi x/L),\ \sqrt{\frac{2}{L}}\sin(3\pi x/L),\ \cdots]$ and x has to range from 0 to L and L has to be slightly bigger than the full bond extension in HCl at the maximum energy to be calculated. The equilibrium extension is at $L/2$. The potential energy on this scale is $V^1(x) = k(x - L/2)^2/2$ where k is the bond force constant and x the extension about the equilibrium position, $L/2$.

Summary of calculation

(i) Use the particle in a box wavefunctions to approximate the harmonic oscillator. The more of these wavefunctions that are added together to make ψ, the better is the approximation. The eigenvectors of the Hamiltonian matrix determine the amount of each 'particle in a box' wavefunction to add together; these are v's of equation (8.11).

(ii) The expectation values (matrix elements) are formed into a matrix that will be diagonalized.

(iii) To form the matrix the order of the basis set is the same order as the elements in the matrix.

(iv) The wavefunctions φ, have indices 1, 2, 3, and so forth, which form the order of the rows and columns. Element $H_{34} = \langle\varphi_3|H|\varphi_4\rangle$ for example, appears at row 3 column 4 in the matrix. The matrix could be ordered with basis set elements as 3, 1, 2, 5, 4, 10, ⋯ etc.; it would not make any difference as long as rows and columns are ordered in the same way and the matrix element entered at the correct position according to the indices. The ordering of the basis set and hence the matrix and corresponding quantum numbers is direct and is

Basis set index 1 2 3 4 5 ⋯ row/column order in matrix

Quantum number 1 2 3 4 5 ⋯ (for particle in a box)

The basis set index is chosen to start at 1 because Maple will be used to diagonalize the matrix and its indices start at 1. The particle in a box quantum numbers start at 1.

8.2.2 Matrix elements

The individual matrix elements are $\langle n|H|m\rangle = \int \varphi_n^* H \varphi_m dx$ where

$$\begin{aligned}\langle n|H|m\rangle &= \langle n|H^0|m\rangle + \langle n|H^1|m\rangle \\ &= \int_0^L \varphi_n^*\left(-\frac{\hbar^2}{2\mu}\frac{d^2}{dx^2}\varphi_m + \frac{k}{2}\left(x-\frac{L}{2}\right)^2\varphi_m\right)d_x \qquad (8.15)\\ &= -\frac{\hbar^2}{2\mu}\int_0^L \varphi_n^*\frac{d^2}{dx^2}\varphi_m dx + \frac{k}{2}\int_0^L \varphi_n^*\left(x-\frac{L}{2}\right)^2\varphi_m dx\end{aligned}$$

where μ is the reduced mass and n and m the quantum numbers. This equation can be simplified because the basis set functions φ are the known solutions to the particle in a box problem. In a particle in a box, the potential is everywhere zero except at 0 and L, where it is infinite. In the first of the two integrals in equation (8.15), the diagonal terms, when $n = m$, have the values

$$\langle n|H^0|n\rangle = -\frac{\hbar^2}{2\mu}\int_0^L \varphi_n^*\left(\frac{d^2}{dx^2}\varphi_n\right)dx = \frac{\hbar^2}{2\mu}\left(\frac{\pi n}{L}\right)^2.$$

With the same reasoning (and shown by direct calculation), when $n \neq m$ these integrals are all zero, they correspond to 'off-diagonal' terms in the matrix. The second integral is

$$\langle n|H^1|n\rangle = \frac{k}{L}\int_0^L \left(x-\frac{L}{2}\right)^2 \sin^2(n\pi \times lL)dx = (n^2\pi^2 - 6)\frac{kL^2}{24n^2\pi^2}.$$

Therefore the diagonal terms in the matrix are

$$\langle n|H|n\rangle = \frac{\hbar^2}{2\mu}\left(\frac{\pi n}{L}\right)^2 + (n^2\pi^2 - 6)\frac{kL^2}{24n^2\pi^2}$$

and the off-diagonal terms

$$\langle n|H|m\rangle = \frac{k}{2}\int_0^L \varphi_n^*\left(x-\frac{L}{2}\right)^2\varphi_m dx.$$

These results can be confirmed using Maple (a concise result is produced by letting Maple know that n and m are integers). The first integral is

Algorithm 8.1 Matrix elements $<\varphi_n|H|\varphi_m>$

```
> assume(n, integer, m, integer);
  phi:= (x,n)-> sqrt(2/L)*sin(n*Pi*x/L);          # φn
# diagonal terms with H⁰,                          <n|H⁰|n>
> -hbar^2/(2*mu))*Int(phi(x,n)*Diff(phi(x,n),x,x ),x=0..L):
  %= value(%);
```

$$-\frac{1}{2}\frac{\sqrt{2}\sqrt{\frac{1}{L}}hbar^2\int_0^L \sin\left(\frac{n\pi x}{L}\right)\left(\frac{\partial^2}{\partial x^2}\left(\sqrt{2}\sqrt{\frac{1}{L}}\sin\left(\frac{n\pi x}{L}\right)\right)\right)dx}{\mu}$$

$$=\frac{1}{2}\frac{hbar^2n^2\pi^2}{L^2\mu}$$

```
# off-diagonal n, m terms with H⁰ are all zero; <n|H⁰|m> = 0
> (-hbar^2/(2*mu)*Int(phi(x,n)* Diff(phi(x,m),x,x),x=0..L):
  %= value(%);
```

$$\int_0^L\left(-\frac{1}{2}\frac{\sqrt{2}\sqrt{\frac{1}{L}}\sin\left(\frac{n\pi x}{L}\right)hbar^2\left(\frac{\partial^2}{\partial x^2}\left(\sqrt{2}\sqrt{\frac{1}{L}}\sin\left(\frac{m\pi x}{L}\right)\right)\right)}{\mu}\right)dx=0$$

```
# Hamiltonian H¹. The second integral equation (8.15)
# The diagonal terms;                              <n|H¹|n>
> Int(phi(x,n)^2*(k/2*(x-L/2)^2), x=0..L): %= value(%);
```

$$\int_0^L\frac{\sin\left(\frac{n\pi x}{L}\right)^2k\left(x-\frac{1}{2}L\right)^2}{L}dx=\frac{1}{24}\frac{L^2k(-6+n^2\pi^2)}{n^2\pi^2}$$

```
# and the off-diagonal terms when n ≠ m,           <n|H¹|m>
> Int(phi(x,n)*phi(x,m)*(k/2*(x-L/2)^2), x=0..L):
  %= value(%);
```

$$\int_0^L\frac{\sin\left(\frac{n\pi x}{L}\right)\sin\left(\frac{m\pi x}{L}\right)k\left(x-\frac{1}{2}L\right)^2}{L}dx=-\frac{2L^2\,knm\,(-1+(-1)^{1+m+n})}{\pi^2(n^4-2n^2m^2+m^4)}$$

Constructing the matrix, equation (8.12), using results from these integrals, but adding just a few of the terms produces

$$\begin{bmatrix}
\frac{\hbar^2}{2\mu}\left(\frac{\pi}{L}\right)^2+k\left(\frac{L}{\pi}\right)^2\frac{(6-\pi^2)}{24} & 0 & \frac{3}{16}k\left(\frac{L}{\pi}\right)^2 & 0\\
0 & \frac{\hbar^2}{2\mu}\left(\frac{2\pi}{L}\right)^2+k\left(\frac{L}{2\pi}\right)^2\frac{(6-(2\pi)^2)}{24} & 0 & \frac{2}{9}k\left(\frac{L}{\pi}\right)^2\\
\frac{3}{16}k\left(\frac{L}{\pi}\right)^2 & 0 & \frac{\hbar^2}{2\mu}\left(\frac{3\pi}{L}\right)^2+k\left(\frac{L}{3\pi}\right)^2\frac{(6-(3\pi)^2)}{24} & 0\\
0 & \frac{2}{9}k\left(\frac{L}{\pi}\right)^2 & 0 & \frac{\hbar^2}{2\mu}\left(\frac{4\pi}{L}\right)^2+k\left(\frac{L}{4\pi}\right)^2\frac{(6-(4\pi)^2)}{24}
\end{bmatrix}$$

Notice that the matrix is Hermitian, real, and symmetrical about the diagonal. The matrix looks rather complicated, but it can be split into two parts just as was done with H, equation (8.13). The first part contains the diagonal terms, the second the couplings between states introduced by the new potential energy terms,

$$\frac{\hbar^2}{2\mu}\begin{bmatrix} \left(\frac{\pi}{L}\right)^2 & 0 & 0 & 0 \\ 0 & \left(\frac{2\pi}{L}\right)^2 & 0 & 0 \\ 0 & 0 & \left(\frac{3\pi}{L}\right)^2 & 0 \\ 0 & 0 & 0 & \left(\frac{4\pi}{L}\right)^2 \end{bmatrix} + k\left(\frac{L}{\pi}\right)^2 \begin{bmatrix} \frac{6-\pi^2}{24} & 0 & \frac{3}{16} & 0 \\ 0 & \frac{6-(2\pi)^2}{24\times 2^2} & 0 & \frac{2}{9} \\ \frac{3}{16} & 0 & \frac{6-(3\pi)^2}{24\times 3^2} & 0 \\ 0 & \frac{2}{9} & 0 & \frac{6-(4\pi)^2}{24\times 4^2} \end{bmatrix}$$

The whole matrix must be diagonalized to obtain an approximate algebraic result, and a numerical solution is required so that the energies can be compared with experiment. One way of using Maple for this calculation is shown below. A procedure is used to evaluate the matrix elements, using the integrals already calculated, then the matrix is filled, eigenvalues and eigenvectors calculated and sorted in pairs in order of increasing energy. Finally, the calculated wavefunctions are plotted. Several lines of the calculation are concerned with sorting the eigenvalues and eigenvectors into energy order. To make the calculation faster the results of the integration already calculated are used again. A matrix of 45×45 elements is necessary to obtain an accurate result for the first ten or so energy levels. A 65×65 matrix will allow the first 25 levels to be accurately found. Some complicated Maple is needed using `map(x->...` to make the eigenvalues real and to sort the eigenvalues and eigenvectors, see Appendix 1. The wavefunctions are calculated according to equations similar to (8.11).

Algorithm 8.2 Calculating energies (eigenvalues) and wavefunctions

```
> with(LinearAlgebra):
> Integs:= proc(n, m)      # procedure evaluate integrals
    if n = m then                          # diagonal terms
        1/2*hbar^2*n^2*Pi^2/( mu*L^2 )+
        1/24*L^2*k*( -6 + n^2*Pi^2 )/( n^2*Pi^2 )
    else                                   # off-diagonals
  -2*L^2*k*n*m*((-1)^(1+m+n)-1)/(Pi^2*(n^4-2*n^2*m^2+m^4));
    end if;
  end proc: # end of calculating matrix elements integrals
# parameters k, mu, L, are for HCl.
  hbar:= 1.054*10^(-34):                   # constants etc
  L:= 2*128.0*10^(-12):
  mu:= 35/36*1.67*10^(-27):
  k:= 516.0;
# define matrix Hnm and fill with matrix elements
  num:= 45:                                # size of matrix
  Hnm:= Matrix(num, num):                  # define Hnm
  for n from 1 to num do                   # fill matrix
    for m from 1 to num do
        Hnm[n,m]:= evalf( Integs(n, m) ):
    end do;
  end do;
# calc eigenvectors & eigenvalues. Make into real numbers
  ( eigval, eigvec ):= Eigenvectors( Hnm ):
  evec:= map( x-> Re(x),eigvec ):            # make real
  evals:= map( x-> Re(x),eigval ):
# sort evec's and evals: keep as pairs by using index
  indx := [ seq( i, i= 1..num ) ]:           # make index
  indx:= sort(indx,(i,j)-> evalb( evals[i] < evals[j] ));
  evals := Array([seq(evals[i],i=indx)]): # sort eigenvals
  Sevec:= Matrix(1..num,1..num):
  for i from 1 to num do      # sort eigenvectors using index
    for j from 1 to num do
        Sevec[i,j]:= evec[i,indx[j]];
```

```
      end do
    end do;
  # end of sort eigenvectors on same index as eigenvalues.
  # note that sorted eigenvectors are now called Sevec
  # print energies
    for i from 1 to num do
      print(i-1, evals[i], hbar*sqrt(k/mu)*(i-1+1/2))
    end do;
  # construct new wavefunctions using particle in
  # box wavefunctions φ with equation (8.11).
  # define procedure psi
    psi:= proc(qn,x)         # make new wavefunctions ψ = Σvφ
      local s, i, n;
      n:= qn + 1:            # qn starts at 0, basis set n at 1
      s:= 0:
      for i from 1 to num do          # i is row index, n column
        s:= s + Sevec[i,n]*sqrt(2/L)*sin(i*Pi*x/L);
      end do;
    end proc:
  # call procedure, quantum number eg 10 and x values in pm.
  > plot(psi(10,x*1e-12)/1e5,x=0..L*1e12); # plot wavefunction
```

The `print` line lists the energies calculated by this method and compares it with the exact values $\hbar\sqrt{k/\mu}(n-1+1/2)$. This is not the usual formula for the harmonic oscillator as $n-1$ is used instead of n. This is because $n=1$ is the lowest index in the basis set because the particle in box has a lowest quantum number of 1, not $n=0$. With a small 4×4 matrix the calculated and exact energy eigenvalues are in poor agreement;

Q*n*.	Matrix method.	Exact values.
0	1.87900e-19	2.97094e-20
1	5.14374e-19	8.91283e-20
2	1.68354e-18	1.48547e-19
3	2.04630e-18	2.07966e-19

With a 45×45 matrix, the agreement is far better to at least four decimal places for the first nine levels.

Qn,	Matrix method.	Exact values.
0	2.97094e-20	2.97094e-20
1	8.91283e-20	8.91283e-20
2	1.48547e-19	1.48547e-19
3	2.07966e-19	2.07966e-19
4	2.67385e-19	2.67385e-19
5	3.26804e-19	3.26804e-19
6	3.86223e-19	3.86223e-19
7	4.45642e-19	4.45642e-19
8	5.05061e-19	5.05060e-19

Figure 8.1 shows the approximate wavefunctions built out of a particle in a box basis set using the procedure `psi(n,x)`. The wavefunction may be negative; this will depend on the sign of eigenvectors that are only defined to within ±1. Negative values are just as correct as positive ones. The wavefunction ψ is constructed according to equation (8.11); the eigenvector values are the constants v. The wavefunctions are very close to the true ones as may be seen by looking at Fig. 4.48 of the online solutions. The reason that the match between the basis set expansion and the true values is better with a bigger basis set is because there are more sine functions and more variables v to match to the harmonic oscillator wavefunctions. The same is true for a series expansion; the more terms the better the expansion will match the function.

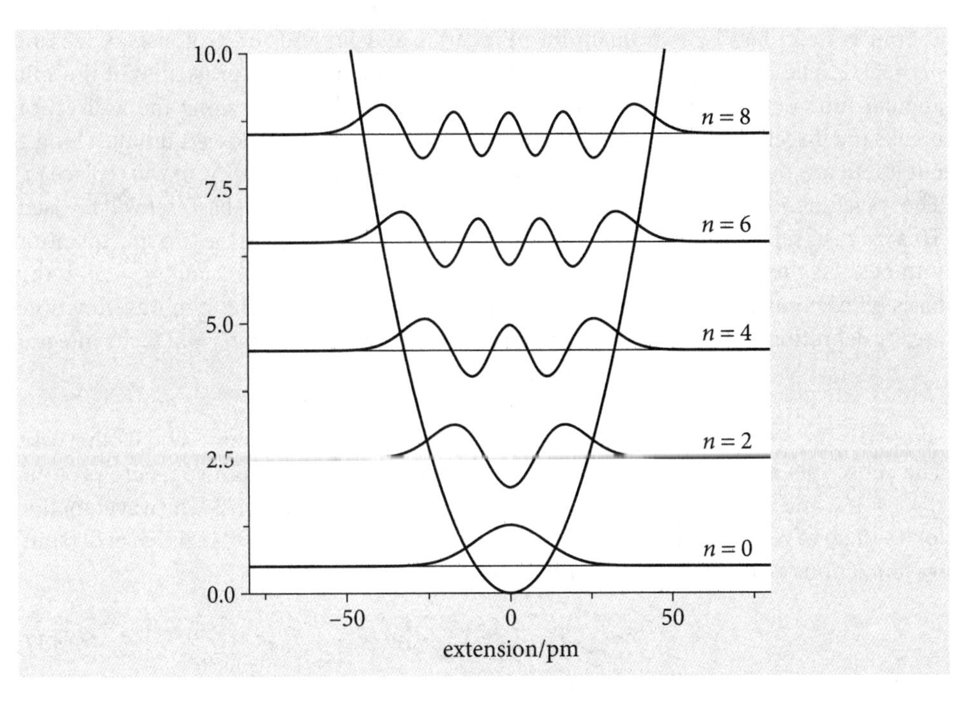

Fig. 8.1 Approximate wavefunctions for $n = 0$, 2, 6 and $n = 8$ levels of HCl as a harmonic oscillator using the particle in a box basis set. The potential is also shown, and the energy is in units of $h\nu$. The wavefunctions are scaled to a convenient amplitude for display and are essentially identical to those calculated using the equations derived directly from Schrödinger's equation, Fig. 4.48 of the online solutions.

You may wish to try adding a quartic term to the quadratic potential, still using the particle in a box wavefunctions. The integrations will have to be recalculated using Maple, the 'force constant' k_Q will have to be about 10^{24} in $k_Q(x - L/2)^4$, and at least 60 terms will be needed in the matrix. The potential becomes steeper and the energy levels farther apart.

8.3 NMR spectrum with two spins

In this second example, the pure nuclear Zeeman effect, which gives rise to the NMR spectrum, will be calculated. In this calculation, and every other involving spin, the matrix element integrals H_{nm} are evaluated directly from the properties of the angular momentum operators, and the spin wavefunctions can never be known.[2] NMR is extremely widely used in chemistry and biochemistry because the spectrum is a measure of the structure of the (diamagnetic) molecule. The many lines in the typical NMR spectrum have two main causes. One is caused by the local magnetic field of the molecule's electrons shielding the nuclei from the external field and is called the chemical shift. The other, J coupling, is due to the local field caused by indirect spin–spin or indirect dipole–dipole interaction mediated by the electrons in the chemical bond. Because both effects are local and this enables the structure to be determined. Enough detailed information is generally present to allow this, but not so much that the spectrum is too complicated to be interpreted.

The Hamiltonian S describing the interactions between i nuclei in the strong external magnetic field $\boldsymbol{B}$ and with the local induced fields, is,

$$S = -\frac{1}{\hbar}\sum_i \gamma_i(1-\sigma_i)\boldsymbol{I}_i \cdot \boldsymbol{B} + \frac{1}{\hbar^2}\sum_{i>j}\sum_j J_{ij}\boldsymbol{I}_i \cdot \boldsymbol{I}_j. \tag{8.16}$$

The applied magnetic field is always in the z-direction with a magnitude B_0 tesla, and is zero in other directions. $\boldsymbol{I}$ is the nuclear spin angular momentum vector with components in the x-, y-, and z-directions of I_x, I_y, and I_z respectively. Angular momentum has the same units as $\hbar$; J s rad^{-1}. The constant γ_i is the magnetogyric ratio (rad T^{-1} s^{-1}) of each atom i, σ_i the shielding constant (dimensionless) leading to the chemical shift and J_{ij} the spin–spin coupling constant (rad s^{-1}) between any two atoms i and j.

[2] Spin eigenfunctions unlike angular momentum do not depend on spatial coordinates; as a result, they do not have to satisfy a periodicity condition and can have half integer quantum numbers. Angular momentum, such as possessed by a rigid rotor molecule has integer quantum numbers.

Spin is described by two quantum numbers s and m_z and for protons $s = 1/2$ and $m_z = \pm 1/2$. The latter is the quantum number associated with the projection of the spin angular momentum on the unique z-axis. The magnetic field $\boldsymbol{B}$ along the z-direction breaks the threefold degeneracy of the angular momentum by making it unique along z, but it remains degenerate in x and y. From now on, and only for clarity, m will replace m_z. The wavefunction is written symbolically as $\psi_{sm} \equiv |sm\rangle$ and the s-m basis set will be used. This is a basis set made up of the spin wavefunctions but labelled with the two spin quantum numbers. As there are two spins, each with quantum numbers $s = 1/2$ and $m_z = \pm 1/2$, this basis set has only four elements. It is described in detail shortly. The spin wavefunctions are, by definition, orthonormal, which means that in general for spin j and k, the integral

$$\langle s_j m_j | s_k m_k \rangle = \delta_{s_j s_k} \delta_{m_j m_k}$$

where δ is the Kronecker delta, which is 1 if the subscripts are the same, and 0 otherwise. The subscripts are the quantum numbers of the two spins. As both spins are protons, $s_j = s_k = 1/2$, and each spin has an m quantum number that can be $\pm 1/2$. The wavefunction for a system of two spins labelled a and b, is made up in the usual way as a basis set of (spin) wavefunctions ψ, multiplied by coefficients c;

$$\psi = \sum_{m_1} \sum_{m_2} c_{m_1,m_2} \psi_a(s_1 m_1) \psi_b(s_2 m_2). \tag{8.17}$$

Because there are two spins, the expectation value integrals are the product of the spin wavefunctions and the operator. For example,

$$H_{1,2} = \langle \psi_a(s_1 m_1) \psi_b(s_1 m_1) | S | \psi_a(s_2 m_2) \psi_b(s_2 m_2) \rangle$$

is the first row and second column expectation value of the Hamiltonian matrix equation (8.12). In such equations, S is the (spin) operator, ψ_a and ψ_b are the spin wavefunctions on atom a and b respectively, and s and m are the spin quantum numbers. The subscript numbers refer to the *indices* of the matrix and are also the indices of the basis set. The ordering used here is

$$(m_a, m_b) = (1/2, 1/2) \quad (1/2, -1/2) \quad (-1/2, 1/2) \quad (-1/2, -1/2)$$

and is shown in equation (8.24). The numbers $\pm 1/2$ are only *symbols*, α and β or any other symbols could be used instead.

A clear notation is obviously of great help in this type of problem. Each nucleus is identified with subscript a or b and the position in the matrix with a number 1 to 4. The matrix element H_{12} can be written with its indices alone; there is no need to write ψ,

$$H_{1,2} = \langle s_{a1} m_{a1} s_{b1} m_{b1} | S | s_{a2} m_{a2} s_{b2} m_{b2} \rangle,$$

and furthermore, as the nuclear spin quantum number is the same on each nucleus, the subscripts on quantum number s can be dropped in the calculation. This produces

$$H_{1,2} = \langle s m_{a1} s m_{b1} | S | s m_{a2}, s m_{b2} \rangle$$

and is used from now on. The spin quantum number s should not be removed from the notation, as it is needed when the matrix elements are worked out. The matrix of expectation values has next to be constructed and diagonalized. To work out the expectation values, the interactions between two spins and between a spin and the external magnetic field have to be examined, equation (8.16). These calculations are done in the following steps.

(i) The spin angular momentum $\boldsymbol{I}$ can be treated as if it were a three-dimensional vector with components in the x-, y-, and z-directions. In the Hamiltonian $\boldsymbol{S}$, the terms $\boldsymbol{I}\cdot\boldsymbol{B}$ and $I_a \cdot I_b = I_{ax} I_{bx} + I_{ay} I_{by} + I_{az} I_{bz}$ for spins a and b have to be evaluated. The magnetic field $\boldsymbol{B}$ only has a z component by definition, and in practice in the NMR machine, so that $\boldsymbol{I}\cdot\boldsymbol{B} = I_z B_0$

(ii) Decide on the s-m basis set order and then fill out the Hamiltonian matrix; see equation (8.12).

(iii) Evaluate each of the elements of the 4×4 matrix using the operator equations given below, and calculate the energy levels algebraically by diagonalising. Maple will be used to do this.

(iv) The energy levels will be found in the limit that the spin-spin coupling J is small compared to the chemical shift; this is called the AX limit.

The integrals between states with angular momentum operators are well known, and can be rigorously derived using the properties of the spherical harmonic functions; see Atkins & Friedmann (1997) or Flygare (1978) for a derivation. The angular momentum operator is conventionally labelled $\boldsymbol{L}$, and the operators are $\boldsymbol{L}^2$, $\boldsymbol{L}_x$, $\boldsymbol{L}_y$ and $\boldsymbol{L}_z$. The shorthand notation of labelling the states with their quantum numbers rather than with a symbol for the wavefunction and its associated quantum numbers is the convention. The operators have the following properties:

$$\langle sm|L^2|sm\rangle = \hbar^2 s(s+1) \tag{8.18}$$

$$\begin{aligned}\langle sm+1|L_x|sm\rangle &= \frac{\hbar}{2}\sqrt{s(s+1)-m(m+1)}\\ \langle sm-1|L_x|sm\rangle &= \frac{\hbar}{2}\sqrt{s(s+1)-m(m-1)}\end{aligned} \tag{8.19}$$

$$\begin{aligned}\langle sm+1|L_y|sm\rangle &= \frac{\hbar}{2i}\sqrt{s(s+1)-m(m+1)}\\ \langle sm-1|L_y|sm\rangle &= -\frac{\hbar}{2i}\sqrt{s(s+1)-m(m-1)}\end{aligned} \tag{8.20}$$

$$\langle sm|L_z|sm\rangle = \hbar m \tag{8.21}$$

and all other combinations of s and m have zero valued integrals. Note that only states in which the m quantum number changes by zero or ±1 are not zero. It is assumed also that sm refers to the *same spin a or b on both sides* of the bra-ket, if not the result is zero. The integral of the product of the wavefunctions is

$$\langle sm_a|sm_a\rangle = \langle sm_b|sm_b\rangle = 1, \qquad \langle sm_a|sm_b\rangle = \langle sm_b|sm_a\rangle = 0$$

which means that the wavefunctions are normalized and orthogonal. (The L_x, and L_y are derived form the raising and lowering operators, see Q8.11.)

In the particular situation of nuclear spin, the angular momentum is labelled $\boldsymbol{I}$ and this will be used from now on. It is vital to realize that the spin operator $\boldsymbol{I}_b$ only works on nucleus b, and $\boldsymbol{I}_a$ only on nucleus a. The operator $\boldsymbol{I}_b$ will treat all the wavefunctions sm_a as constants and *vice versa*. The calculation proceeds as follows:

(i) The summation in (8.16) is the double summation $\sum_{i=1}\sum_{j=1} J_{ij}\boldsymbol{I}_i\cdot\boldsymbol{I}_j$, with the restriction that $i > j$ for i and $j = a$ and b. There are four terms without this restriction because both i and j have two values; $i = a,b$ and $j = a,b$.

$$\sum_{i=a}\sum_{j=b} J_{ij}\boldsymbol{I}_i\cdot\boldsymbol{I}_j = J_{aa}\boldsymbol{I}_a\cdot\boldsymbol{I}_a + J_{ab}\boldsymbol{I}_a\cdot\boldsymbol{I}_b + J_{ba}\boldsymbol{I}_b\cdot\boldsymbol{I}_a + J_{bb}\boldsymbol{I}_b\cdot\boldsymbol{I}_b,$$

but only $J_{ba}\boldsymbol{I}_b\cdot\boldsymbol{I}_a$ remains when the restriction $i > j$ if $b > a$ is applied. The dot product is clearer when written in matrix form and is

$$\boldsymbol{I}_a\cdot\boldsymbol{I}_b = [I_{ax} \quad I_{ay} \quad I_{az}]\begin{bmatrix} I_{bx}\\ I_{by}\\ I_{bz}\end{bmatrix} = I_{ax}I_{bx} + I_{ay}I_{by} + I_{az}I_{bz},$$

and this is the same result as $\boldsymbol{I}_b\cdot\boldsymbol{I}_a$ because operators with an a subscript only act on spin a, and I_b only on spin b, therefore, operating on wavefunctions in the order $I_{ax}I_{bx}$ produces the same result as operating $I_{bx}I_{ax}$.

(ii) Using this last result, equation (8.16) simplifies to

$$S = -\frac{q_a I_{az}}{\hbar} - \frac{q_b I_{bz}}{\hbar} + \frac{J}{\hbar^2}(I_{ax}I_{bx} + I_{ay}I_{by} + I_{az}I_{bz}) \tag{8.22}$$

with the abbreviation $q = \gamma(1-\sigma)B_0$ for each nuclei a and b. The first two terms contain only the z component of the angular momentum, because the magnetic field of the NMR spectrometer points only in the z direction. Planck's constant is present in the equation to give S units of radian s^{-1}, matrix elements are in radian s^{-1} also. The magnetogyric ratio γ has units of radian tesla^{-1} s^{-1} and B_0 has units of tesla, I has the same units as $\hbar \equiv$ joule second radian^{-1}, and σ is dimensionless. Overall the units of S are rad s^{-1} or angular frequency. The constant J also has units of rad s^{-1}.

(iii) A basis set of four vectors can be used, however, the numbers are to be treated as *symbols* to represent the spin state; no arithmetic is done with them.

$$\left\{\begin{bmatrix} s_a \\ m_a \\ s_b \\ m_b \end{bmatrix}\right\} \equiv \left\{\begin{bmatrix} 1/2 \\ 1/2 \\ 1/2 \\ 1/2 \end{bmatrix}, \begin{bmatrix} 1/2 \\ -1/2 \\ 1/2 \\ 1/2 \end{bmatrix}, \begin{bmatrix} 1/2 \\ 1/2 \\ 1/2 \\ -1/2 \end{bmatrix}, \begin{bmatrix} 1/2 \\ -1/2 \\ 1/2 \\ -1/2 \end{bmatrix}\right\} \tag{8.23}$$

Symbols could have been chosen instead of numbers. For example, if $\alpha = 1/2$, $\beta = -1/2$ then the basis set would be

$$\left\{\begin{bmatrix} s_a \\ m_a \\ s_b \\ m_b \end{bmatrix}\right\} \equiv \left\{\begin{bmatrix} \alpha \\ \alpha \\ \alpha \\ \alpha \end{bmatrix}, \begin{bmatrix} \alpha \\ \beta \\ \alpha \\ \alpha \end{bmatrix}, \begin{bmatrix} \alpha \\ \alpha \\ \alpha \\ \beta \end{bmatrix}, \begin{bmatrix} \alpha \\ \beta \\ \alpha \\ \beta \end{bmatrix}\right\}$$

which may be preferable. The matrix of the expectation values is a 4×4 matrix with rows and columns determined only by the s and m combinations, the two m values are listed along the top and down the side. The s_a and s_b quantum numbers in the basis set are not listed because they are always 1/2. The top row and left-hand column show the matrix indices and the indices for the basis set. The Hamiltonian S is

$$\begin{array}{ccc|cccc} & & & 1 & 2 & 3 & 4 \\ & & m_a & 1/2 & 1/2 & -1/2 & -1/2 \\ & m_a & m_b & 1/2 & -1/2 & 1/2 & -1/2 \\ \hline 1 & 1/2 & 1/2 & H_{11} & H_{12} & H_{13} & \vdots \\ 2 & 1/2 & -1/2 & H_{12} & H_{22} & & \vdots \\ 3 & -1/2 & 1/2 & \vdots & & \ddots & \\ 4 & -1/2 & -1/2 & \vdots & & & \ddots \end{array} \tag{8.24}$$

The matrix element $H_{i,j}$ integral written in terms of the quantum numbers is

$$H_{i,j} = \langle sm_{ai}sm_{bi}|S|sm_{aj}sm_{bj}\rangle,$$

where the subscript i refers to rows and j to columns. Looking at the matrix, m_{a1} is 1/2, m_{b2} is −1/2, and S is the operator equation (8.22). Recall that operating on nucleus a with an 'a' nuclear operator does not affect nucleus b and vice versa, and this is why it is necessary to label m with nucleus a or b. (One could remember that the second entry is always nucleus b, which makes the equations less messy, but the risk of making an error is far greater.)

The two nuclei form a basis set of four elements so that there are 16 integrals to evaluate, but looking at the integrals, equations (8.18) to (8.21), only those with the same s and m or those with one m changing by +1, the other by −1, are not zero. This means that most of the matrix contains zero. The matrix, when all terms are evaluated, is

$$\begin{array}{cc|cccc} & m_a & 1/2 & 1/2 & -1/2 & -1/2 \\ m_a & m_b & 1/2 & -1/2 & 1/2 & -1/2 \\ \hline 1/2 & 1/2 & -q_a/2 - q_b/2 + J/4 & 0 & 0 & 0 \\ 1/2 & -1/2 & 0 & -q_a/2 + q_b/2 - J/4 & J/2 & 0 \\ -1/2 & 1/2 & 0 & J/2 & +q_a/2 - q_b/2 - J/4 & 0 \\ -1/2 & -1/2 & 0 & 0 & 0 & +q_a/2 + q_b/2 + J/4 \end{array} \tag{8.25}$$

This matrix shows how the states $|1/2, -1/2\rangle$ and $|-1/2, 1/2\rangle$ are mixed by the J spin–spin interaction because these are the only states with off-diagonal terms. The matrix can be diagonalized by blocking it into two 1×1 matrices and a 2×2 matrix; therefore, two of the energies can be read directly from the matrix.

The terms are calculated as follows: the first integral is

$$H_{11} = \langle sm_{a1}, sm_{b1}| -\frac{q_a I_{az}}{\hbar} - \frac{q_b I_{bz}}{\hbar} + \frac{J}{\hbar^2}(I_{ax}I_{bx} + I_{ay}I_{by} + I_{az}I_{bz})|sm_{a1}, sm_{b1}\rangle \tag{8.26}$$

which has five terms each with quantum numbers $s_{a1} = ½$, $s_{b1} = ½$, $m_{a1} = ½$, $m_{b1} = ½$. The s and m quantum numbers only operate on spins on the same nuclei, i.e. a with a and b

with b. The first term in equation (8.26) with operator I_{az} can be factored into two terms because I_{az} does not operate on nucleus b.

$$-\frac{q_a}{\hbar}\langle sm_{a1}, sm_{b1}|I_{az}|sm_{a1}, sm_{b1}\rangle = -\frac{q_a}{\hbar}\langle sm_{a1}|I_{az}|sm_{a1}\rangle\langle sm_{b1}|sm_{b1}\rangle$$

$$= -\frac{q_a}{\hbar}\langle sm_{a1}|I_{az}|sm_{a1}\rangle$$

The second integral is $\langle sm_{b1}|sm_{b1}\rangle = 1$ because the wavefunctions are normalized. The angular momentum operator I_{az} acts on nucleus a only, the integral has the form $\langle sm|I_z|sm\rangle = \hbar m$, equation (8.21), and as $m_a = 1/2$ therefore,

$$\langle sm_{a1}|I_{az}|sm_{a1}\rangle = \frac{\hbar}{2}.$$

and the second term in equation (8.26) follows in a similar way with operator I_{bz} to produce $-q_b/2$.

The third to fifth terms involve two spin terms, such as I_xI_x, which are evaluated as the product of two terms, one for spin a the other for spin b, giving for the z terms by equation (8.21)

$$\frac{J}{\hbar^2}\langle sm_{a1}, sm_{b1}|I_{az}I_{bz}|sm_{a1}, sm_{b1}\rangle =$$

$$\frac{J}{\hbar^2}\langle sm_{a1}|I_{az}|sm_{a1}\rangle\langle sm_{b1}|I_{bz}|sm_{b1}\rangle = \frac{J}{4}$$

The x and y components I_x and I_y are zero because $m_a = m_b = 1/2$, as required by equations (8.19) and (8.20); all diagonal terms in the matrix are zero for I_x and I_y operators for the same reason. The first diagonal term is therefore $H_{11} = -q_a/2 - q_b/2 + J/4$.

The other diagonal terms in the matrix equation (8.25) are evaluated in a similar manner, the sign changes being due to the different signs that m has.

The two non-zero off-diagonal terms are $H_{3,2}$ and $H_{2,3}$, (column 2, row 3), has m_a decreasing by 1 and m_b increasing by 1. On the off-diagonal, m changes therefore each of the I_z terms in the integral are zero, see equation (8.21). The remaining two integrals with I_x and I_y are factored into two pairs in x and y as follows

$$\frac{J}{\hbar^2}\langle sm_{a2}, sm_{b2}|I_{ax}I_{bx} + I_{ay}I_{by}|sm_{a3}, sm_{b3}\rangle$$

$$= \frac{J}{\hbar^2}\langle sm_{a2}|I_{ax}|sm_{a3}\rangle\langle sm_{b2}|I_{bx}|sm_{b3}\rangle$$

$$+ \frac{J}{\hbar^2}\langle sm_{a2}|I_{ay}|sm_{a3}\rangle\langle sm_{b2}|I_{by}|sm_{b3}\rangle$$

Each of these four integrals is tackled separately. The first is $\frac{J}{\hbar^2}\langle sm_{a2}|I_{ax}|sm_{a3}\rangle$ and equation (8.19) is used to evaluate it. Looking at the matrix, (8.24), the m_a value in row 2 is $m_{a2} = m_{a3} + 1$, with m_{a3} in column 3, therefore

$$\langle sm_{a2}|I_{ax}|sm_{a3}\rangle = \langle sm_{a3} + 1|I_{ax}|sm_{a3}\rangle$$

$$= \frac{\hbar}{2}\sqrt{\frac{1}{2}\frac{3}{2} + \frac{1}{2}\frac{1}{2}} = \frac{\hbar}{2}$$

The I_{bx} integral evaluates similarly so that

$$\frac{J}{\hbar^2}\langle sm_{a2}|I_{ax}|sm_{a3}\rangle\langle sm_{b2}|I_{bx}|sm_{b3}\rangle = \frac{J}{4}.$$

Each y component integral is complex, equation (8.20), but their product is real because $i^2 = -1$;

$$\frac{J}{\hbar^2}\langle sm_{a2}|I_{ay}|sm_{a3}\rangle\langle sm_{b2}|I_{by}|sm_{b3}\rangle$$

$$= -i^2\frac{J}{\hbar^2}\frac{\hbar^2}{4}\sqrt{\left[\frac{1}{2}\frac{3}{2} - \frac{-1}{2}\left(\frac{1}{2}\right)\right]\left[\frac{1}{2}\frac{3}{2} - \frac{1}{2}\left(\frac{-1}{2}\right)\right]} = \frac{J}{4}$$

In total the $H_{2,3}$ term has a matrix entry of $J/2$. Since the matrix is Hermitian, as are all quantum matrices, it is real and symmetrical and therefore, $H_{2,3} = H_{3,2}$.

You can see how tricky it is to evaluate these integrals, but only because the notation is very complicated. Maple can help with the notation and do the calculation. First, define the integrals, which are called `Ix`, `Iy`, and `Iz`, and are equations (8.19) to (8.21), then make them functions of the four s and m quantum numbers, and use `if .. then.. elif .. end if` to check for careless mistakes. The `sa, ma, sb,` and `mb` are defined only as parameters to the functions and they do not exist outside the definitions. Other names could be used.

Algorithm 8.3 Matrix elements equations (8.19) to (8.21)

```
> Iz:= (sa,ma,sb,mb)->
        if (ma = mb) then hbar*ma else 0 end if: # <sm|Iz|sm>
  Iy:= (sa,ma,sb,mb)->
        if (mb = ma + 1) then
           I*hbar/2*sqrt(sa*(sa + 1) - ma*(ma + 1))
        elif (mb = ma - 1) then
           -I*hbar/2*sqrt(sa*(sa + 1) - ma*(ma - 1)) else 0
        end if:                                # -i<sm|Iy|sm+-1>
  Ix:= (sa,ma,sb,mb)->
        if (mb = ma + 1) then
           hbar/2*sqrt(sa*(sa + 1)- ma*(ma + 1))
        elif (mb = ma - 1) then
           hbar/2*sqrt(sa*(sa + 1)- ma*(ma - 1)) else 0
        end if:                                # <sm|Ix|sm+-1>
  delta:= (p, q)->
        if (p = q) then 1 else 0 end if: # Kroneker delta
```

In the next algorithm, the matrix H is calculated. Lists are defined to hold the m quantum numbers, and because the s quantum numbers are always ½ they are ignored. The `m1`, `m11`, `m2`, and `m22` parameters are the quantum number pairs needed to calculate the integrals, and are selected out of the basis set. The delta function is needed to calculate integrals of the form $\langle sm|1|s'm'\rangle \equiv \langle sm|s'm'\rangle = \delta(s,s')\delta(m,m')$ where the delta function is only one if $m = m'$, or $s = s'$, otherwise it is zero. These integrals evaluate this way because the isolated spin wavefunctions are orthogonal and normalized.

Algorithm 8.4 Hamiltonian matrix: two spin ½ system

```
> with(LinearAlgebra):
  n:= 4:                              # size of basis
  H:= Matrix(1..n,1..n):              # define matrix
  ma:= [1/2,1/2,-1/2,-1/2]:           # basis set ma; see (8.25)
  mb:= [1/2,-1/2,1/2,-1/2]:           # basis set mb
  for i from 1 to n do                # i is row, k column
    for k from 1 to n do
    H[i,k]:=0:
    m1:= ma[i]: m11:= ma[k]:          # get quantum number
    m2:= mb[i]: m22:= mb[k]:
    Iaz:= -qa/hbar*Iz(1/2,m1,1/2,m11)*delta(m2,m22);
    Ibz:= -qb/hbar*Iz(1/2,m2,1/2,m22)*delta(m1,m11);
    Ixyz:= J/hbar^2*
      (   Ix(1/2,m1,1/2,m11) * Ix(1/2,m2,1/2,m22)
       +  Iy(1/2,m1,1/2,m11) * Iy(1/2,m2,1/2,m22)
       +  Iz(1/2,m1,1/2,m11) * Iz(1/2,m2,1/2,m22)    );
    H[i,k]:= Iaz  + Ibz + Ixyz;
  end do
end do:
> H;                                  # list matrix
```

This matrix is the same as in equation (8.25). The matrix is real and symmetrical (Hermitian) so the eigenvalues are real and the eigenvectors are orthogonal. The eigenvalues and vector are calculated using

```
> (eigvals, eigvecs):= Eigenvectors(H):
  eigvals;
  eigvecs;
```

The eigenvalues produce four energy levels, which are

$$E_1 = -\frac{J}{4} + \frac{\sqrt{J^2 + (q_a - q_b)^2 B_0^2}}{2}$$

$$E_2 = -\frac{J}{4} - \frac{\sqrt{J^2 + (q_a - q_b)^2 B_0^2}}{2}$$

$$E_3 = \frac{J}{4} - \frac{q_a + q_b}{2} B_0$$

$$E_4 = \frac{J}{4} + \frac{q_a + q_b}{2} B_0$$

These results show how the energy levels (in units of Hz) depend on the applied magnetic field B_0 in tesla. The corresponding eigenvectors are

$$\begin{bmatrix} 0 \\ c_1 \\ 1 \\ 0 \end{bmatrix}, \begin{bmatrix} 0 \\ c_2 \\ 1 \\ 0 \end{bmatrix}, \begin{bmatrix} 1 \\ 0 \\ 0 \\ 0 \end{bmatrix}, \begin{bmatrix} 0 \\ 0 \\ 0 \\ 1 \end{bmatrix}$$

where c_1 and c_2 are complicated terms in q_a, q_b, and J. The eigenvectors are orthogonal because, when evaluated it is found that $c_1c_2 + 1 = 0$. Examining the basis set and the eigenvectors, the third eigenvector with 1 as its first element and energy E_3 corresponds to state $\{1/2 \quad 1/2 \quad 1/2 \quad 1/2\}$ with $m_a = m_b = 1/2$, the fourth to $m_a = m_b = -1/2$. The other two eigenvectors correspond to states with $m_a = ½$ or $m_b = -½$ and vice versa. The NMR transition is allowed only if $\Delta m = \pm 1$; therefore, the transition can occur only between levels 4–2, 4–1 and 2–3, 1–3. If you repeat the calculation, your computer may produce a different eigenvalue and eigenvector pair ordering, but the eigenvalues and vectors are always paired up in the correct way.

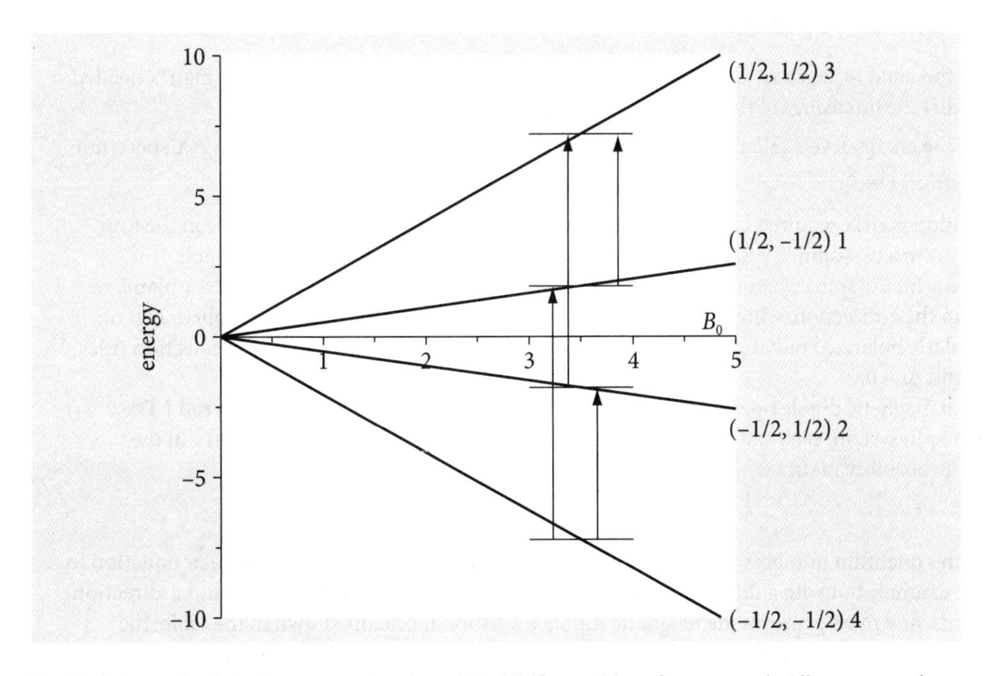

Fig. 8.2 Schematic of the four energy levels and the NMR transitions for a two spin 1/2 system at the same fixed value of the magnetic field and in the AX limit where J is small compared to the chemical shift. The number in brackets after the spin quantum numbers is the eigenvalue index.

8.3.1 AX spectra

If the coupling J is small compared to the difference in chemical shift, the AX limit is produced. The condition is expressed as $J/2 \ll \gamma_a(1-\sigma_a)-\gamma_b(1-\sigma_b)$, and some simplification of the equations is possible:

$$E_2 = -\frac{J}{4} - \frac{\sqrt{J^2+(\gamma_a(1-\sigma_a)-\gamma_b(1-\sigma_b))^2B_0^2}}{2} \rightarrow -\frac{J}{4} - \frac{\gamma_a(1-\sigma_a)-\gamma_b(1-\sigma_b)}{2}B_0$$

and

$$E_1 = -\frac{J}{4} + \frac{\gamma_a(1-\sigma_a)-\gamma_b(1-\sigma_b)}{2}B_0$$

and the other levels are unchanged. This arrangement of levels produces the familiar four level structure and two double peaked NMR spectra, the smaller spacing being the spin–spin coupling J.

Although the equation for the exact energy has been simplified, by looking at the matrix equation (8.25), the same result is obtained if the off-diagonal terms are ignored.

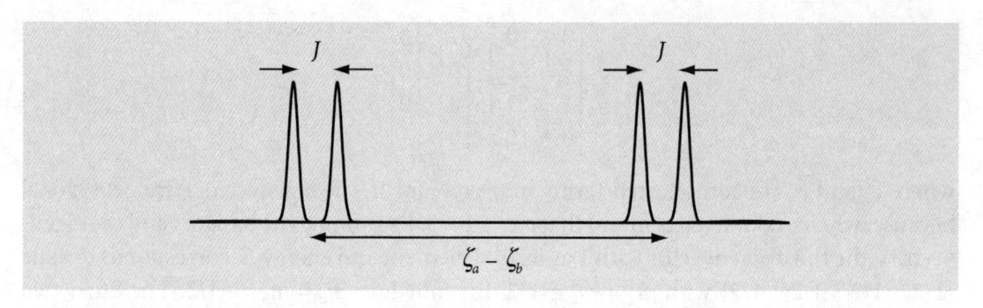

Fig. 8.3 AX type spectrum. The spin–spin coupling is smaller than the difference in chemical shift, $\zeta_a = \gamma_a(1-\sigma_a)/2$ and similarly for ζ_2.

8.3.2 Questions

Full solutions are available at **www.oxfordtextbooks.co.uk/orc/beddard.**

Q8.1 **(a)** Using the s and m_z basis set of the NMR example, work out the transition moment matrix needed to predict the intensities of the lines in the two-spin NMR spectrum.

(b) Using the energy levels calculated in the example show that the transitions form an AX spectrum as shown in Fig. 8.3.

The transition matrix required is the 4×4 matrix of all possible NMR transitions between the four energy levels in a two-spin ½ system. The transitions occur because of the magnetic-dipole interaction with the spins when using right or left circularly polarized radiation in the x–y plane, but none in the z-direction, which is the direction in which the magnetic field B_0 is applied. Left or right circularly polarized radiation can only couple transitions between m states, with selection rules $\Delta m = \pm 1$ and $\Delta s = 0$.

The total magnetic dipole operator for two spins is $\mu = \gamma_a I_a + \gamma_b I_b$, which has units of rad J T^{-1}. In the two-spin system, each state is a product of two wavefunctions and so the i-j^{th} entry in the transition probability matrix is

$$T_{ij} = \langle sm_{ai}, sm_{bi}|\mu|sm_{aj}, sm_{bj}\rangle$$

and as both s quantum numbers are ½, the subscripts have been dropped. This is a similar equation to that in the example but with a different operator. The angular momentum I has x-, y- and z-direction components, and these generate the magnetic dipole transition moments shown in the table; the dipole $\mu_\pm = \mu_x \pm \mu_y$ describes left and right circularly polarized light, and $\mu_z = \mu m$, linearly polarized light. All other combinations of quantum number changes, other than those shown in the table, have zero transition moments. It is only necessary to work with the $\mu_\pm$ operators and the transition moments they produce. There is one $\mu_{\pm}$ term for each spin a and b so $\mu = \gamma_a\mu_{\pm a} + \gamma_b\mu_{\pm b}$ and each $\mu_{\pm}$ is two terms as shown in the table.

Magnetic dipole μ	Transition moment non-zero only when	Transition moment
Linearly polarized light along z	μ_z, $\Delta s = 0$, $\Delta m = 0$	μm
Right and left circularly polarized light in x–y plane	$\mu_{\pm}$, $\Delta s = 0$, $\Delta m = \pm 1$ $\mu_{\pm} = \mu_x \pm \mu_y$	$\mu_{\pm} = \mu\sqrt{s(s+1) - m(m \pm 1)}$

The basis set can be the same as in the example, equation (8.24), which is

$$\begin{array}{cc|cccc} & m_a & 1/2 & 1/2 & 1/2 & 1/2 \\ m_a & m_b & 1/2 & -1/2 & 1/2 & -1/2 \\ \hline 1/2 & 1/2 & \cdots\cdots & \cdots\cdots & \cdots\cdots & \cdots\cdots \\ 1/2 & -1/2 & & & & \\ -1/2 & 1/2 & & & & \\ -1/2 & -1/2 & & & & \end{array}$$

Strategy: The integrals $T_{ij} = \langle sm_{ai}sm_{bi}|\mu|sm_{aj}sm_{bj}\rangle$ have to be calculated using the terms in the table for each of the combinations of s and m, of which there are 16, but not as many as this need be calculated because the matrix is symmetrical. The subscripts i and j give the position in the matrix; i is the row, j the column. The vital point is that radiation has one unit of angular momentum. This means that only changes of the m quantum number by ± 1 on one spin of the pair can be coupled by the radiation at any time and hence produce the spectrum. Recall that the wavefunctions of each individual spin are normalized and orthogonal; $\langle m_a|m_a\rangle = 1$ and $\langle m_a|m_b\rangle = 0$. Because the nuclear spin quantum number s is the same for both nuclei the subscripts can be dropped for clarity.

Q8.2 In the ground state of a hydrogen atom, there is a hyperfine coupling which is caused by the coupling of the proton's spin with that of the electron and this causes a small change in the energy levels, the Hamiltonian has a term $aS{\cdot}I$ to account for this and $a = 1.42$ GHz. The energy levels in a hydrogen atom are also split in a magnetic field by the electron and nuclear Zeeman effect. If the field is in the z direction only, the Hamiltonian is

$$H = g\beta S_z B - g_N\beta_N I_z B + aS{\cdot}I,$$

where g is the g factor for the electron and β the Bohr magneton, and g_n and β_n are the proton g value and nuclear magneton respectively. S_z and I_z are the electron and nuclear (proton) spin quantum numbers respectively.

Calculate the energy levels and plot a labelled graph vs B in tesla, (B is often called the magnetic field strength; however, it is the magnetic flux density and has the units $\text{T} = \text{kg A}^{-1}\,\text{s}^{-2}$.)

The constants are $g = 2.002$, $\beta = 9.274 \times 10^{-24}\,\text{JT}^{-1}$, $g_N = 5.586$, $\beta_N = 5.051 \times 10^{-27}\,\text{JT}^{-1}$. See Foote (2005), Cohen-Tannoudji et al. (1977), or Carrington & McLachlan (1969) for a thorough discussion of this problem.

Strategy: Modify the method used to calculate the NMR spectrum shown in the text. The matrix of expectation values can then be written down directly by analogy with equation (8.25). To convert to Hz and tesla, divide the constants by Planck's constant.

Q8.3 Write down the Hamiltonian for the interaction of three nuclei such as in an NMR experiment. Construct the matrix using the method outlined in the text or using Maple by adapting the code given in Algorithm 8.4. It is possible to decide which entries are zero by examining the basis set because the matrix only has entries $H_{i,j}$ when *both* m values change by $+1$ and -1 or vice versa, or s and m are the same see equations (8.19) to (8.21). The basis set has eight terms each with three spin states, (α, α, α), (α, α, β), and so forth. The matrix can be blocked into two 1×1 matrices and two 3×3 matrices if the ordering is so arranged. However, finding the eigenvalues (using Maple) for the 3×3 matrices produces an exceptionally complex set of equations, and if you want to look at the values, it is best to do so numerically.

Strategy: The three spins produce a matrix of order $2^3 = 8$. The three nuclei will have three chemical shifts but interact pair-wise with J coupling terms, J_{12}, J_{13}, J_{23}. The wavefunction is the product of those for three spins but the operator for nucleus a can only work on a spin, not b or c, thus the generic form is

$$\langle sm_{a1}m_{b1}m_{c1}|op_a|sm_{a2}m_{b2}m_{c2}\rangle = \langle sm_{a1}|op_a|sm_{a2}\rangle\langle m_{b1}|m_{b2}\rangle\langle m_{c1}|m_{c2}\rangle$$

The non-zero terms in the matrix can be written down once the basis set ordering is decided on, only those two entries where m changes by +1 and −1 are non-zero; for example, the element between two pairs of basis set values (β, α, β) to (β, β, α) is non-zero.

Q8.4 In a molecule such as ethane the two methyl groups do not undergo free rotation but are hindered by one another. The free-rotor model of methyl group rotation is therefore not very accurate, and instead, the molecule moves in a sinusoidal, threefold potential, with eclipsed configurations having more energy than the staggered ones. Consider that the wavefunction for the rotation of the methyl groups is that of a free rotor but modified by a sinusoidal potential of the form

$$V = \frac{V_3}{2}[1 - \cos(3\theta)]$$

where V_3 is a constant and θ the rotation angle. This extra potential is going to change the energy levels. The normalized free-rotor wavefunction is

$$\varphi_m = \frac{1}{\sqrt{2\pi}} e^{im\theta}$$

with quantum number m restricted to values $0, \pm 1, \pm 2, \cdots$ with Hamiltonian

$$H^0 = -\frac{\hbar^2}{2I}\frac{d^2}{d\theta^2}.$$

The Schrödinger equation is therefore $-\dfrac{\hbar^2}{2I}\dfrac{d^2\varphi_m}{d\theta^2} = E_m\varphi_m$ where I is the moment of inertia of the methyl group.

(a) Show that φ_m is a solution of this last equation with quantum numbers m. Find the energy E_m and show that the diagonal matrix elements are $m^2\hbar^2/2I$ and that the off-diagonal values are zero.

(b) If the total Hamiltonian is changed to $H^0 + H^1$, show that the matrix elements of H^1, using the potential V and the wavefunctions φ are

$$(H^1)_{m,m'} = \frac{V_3}{2}\delta_{m,m'} - \frac{V_3}{4}\delta_{m',m\pm 3}.$$

(c) Calculate the matrix of expectation values in the basis $m = 0, 1, -1, 2, -2, 3, -3, \cdots$ which means that the wavefunction in the presence of the potential has the form of equation (8.11) *viz*; $\psi = v_0\varphi_0 + v_1\varphi_1 + v_2\varphi_{-1} + v_3\varphi_{-2} + v_4\varphi_4 + \cdots$.

(d) Solve this problem using Maple, and calculate the new energy levels in the presence of the potential V. Check that the levels are sorted in order of their energy. Assume that $m = 3$ and, using the algebraic expression, compare the energies as m increases *versus* those for the free rotor. Use the value $A = \hbar^2/2I = 17\ \text{cm}^{-1}$, which is the value for protons in methanol rotating about the CO bond and, assume $V_3 = 200\ \text{cm}^{-1}$ for the restricted rotor and zero for the free rotor.

Strategy: In (a) you are asked to *show* that the wavefunction is a solution, not prove it; therefore substitute and simplify to get the answer and do the same for part (b). The solution can be found using methods described in Chapter 10.6. Having these answers, construct the basis set and matrix with the same ordering, any basis set ordering can be chosen as long as it is used throughout. Finally, diagonalize the matrix using Maple. The matrix element for any Hamiltonian H and quantum numbers m and m' is $\int \psi_m^* H \psi_{m'} d\theta \equiv \langle m|H|m'\rangle$. The wavefunction is complex; therefore, remember to take the complex conjugate where necessary.

The expectation values of operator H^0 produce only diagonal terms in the matrix, because, as the saying goes, 'H^0 is diagonal in its own eigenstates'. This means that H^0 exactly solves the Schrödinger equation, $H^0\varphi = E_0\varphi$, and E_0 is the energy for each quantum level. In this example, H^0 would correspond to that for the rigid rotor with no internal rotation. The other operator H^1, has the effect of coupling energy levels and therefore, in calculating the expectation values, the two numbers m and m' are different.

Q8.5 The first three levels observed in the hindered rotational spectrum of CH_3CH_2F, are at 242.7, 225.5, 208.4 and 177.0 cm^{-1} (Sage & Klemperer 1963). Using the method of the previous question, and by fitting the observed energy difference to trial values, find the value of $A = \hbar^2/2I$ in cm^{-1}. The potential's magnitude is $V_3 = 1158\ \text{cm}^{-1}$.

By guessing some values and repeating the calculation with a sufficiently large basis set a value of $A = 6.3\ \text{cm}^{-1}$ provides a fairly good fit to the data. Do you agree?

Q8.6 When there are trans and gauche as well as eclipsed forms, the potential has a more complicated shape with terms in θ as well as 3θ and 6θ.

(a) In the free rotor basis used in previous questions, work out the expectation values for the potential

$$V=\frac{V_1}{2}[1-\cos(\theta)]+\frac{V_3}{2}[1-\cos(3\theta)]+\frac{V_6}{2}[1-\cos(6\theta)]$$

(b) Construct the matrix of expectation values.

(c) If you feel confident, write a Maple procedure to work out the eigenvalues.

Q8.7 An ion with an electron in one of the 2p orbitals is placed in a field with orthorhombic symmetry with a potential $V=Ax^2+By^2+Cz^2$. Except at the origin, this potential must obey Laplace's equation $\dfrac{\partial^2 V}{\partial x^2}+\dfrac{\partial^2 V}{\partial y^2}+\dfrac{\partial^2 V}{\partial z^2}=0$, therefore $A+B+C=0$ or $C=-A-B$ making the potential $V=A(x^2-z^2)+B(y^2-z^2)$. (An orthorhombic crystal has 90° angles but different lengths along the x, y, and z-axis.)

(a) Calculate the orbitals' energies by setting up the secular equation using the spherical harmonic functions given below to describe the wavefunctions. Examine the matrix element integrals of the form $\langle Y_{0,\pm1}|V|Y_{0,\pm1}\rangle$ and decide which ones are zero without calculating them all. The radial part of any wavefunction can be ignored as it does not depend on A and B and is the same in all directions. The angular parts of the orbitals are

$$Y_{10}=n\sqrt{2}\cos(\theta),\ Y_{1\pm1}=\mp n\sin(\theta)e^{\pm i\phi}$$

where the first subscript identifies the angular momentum, which is 1 for a P state, and the second, the m component 0, ±1. Use the m quantum numbers as a basis set, therefore the secular determinant will be 3×3. Simplify the result by separating out A and B as factors and replacing other terms with a constant and this will be easier to do if the integrals are evaluated with Maple.

(b) If the orbitals are made into linear combinations as p_x, p_y, and p_z, without any calculation, explain the form of the matrix where elements are $\langle p_x|V|p_y\rangle$ and so forth.

(This question is based on a similar one by Squires 1995, Chapter 10.)

Strategy: The potential is in x, y, and z, the wavefunctions in r, θ, ϕ, so the conversions from Cartesian to spherical polar coordinates are needed. These are $z=r\cos(\theta)$, $y=r\sin(\theta)\sin(\phi)$, $z=r\sin(\theta)\cos(\phi)$, see Chapter 4.11. The Jacobian for the change from Cartesian to spherical polar coordinates is also needed, this is $dxdydz\rightarrow r^2\sin(\theta)drd\theta d\phi$. Examining the integrals first can determine whether they are zero or not. The integration limits on θ are 0 to π, and are 0 to 2π on ϕ. The following results are useful;

$$\int_0^{2\pi}\cos^2(\phi)d\phi=\int_0^{2\pi}\sin^2(\phi)d\phi=\pi \qquad \int_0^{2\pi}\cos^n(\phi)d\phi=\int_0^{2\pi}\sin^n(\phi)d\phi \begin{matrix}=0,\ n \text{ is odd}\\ \neq 0,\ n \text{ is even}\end{matrix}$$

$$\int_0^{2\pi}\cos^2(\phi)e^{\pm 2i\phi}d\phi=-\int_0^{2\pi}\sin^2(\phi)e^{\pm 2i\phi}d\phi=\pi/2 \qquad \int_0^{\pi}\sin^m(\theta)\cos^n(\theta)d\theta \begin{matrix}=0,\ n \text{ is odd}\\ \neq 0,\ n \text{ is even}\end{matrix}$$

There are three values of the m quantum numbers. Therefore, ordering of the basis set could be 0, 1, −1.

8.4 Basis sets and bra-ket algebra

In this section, bra-ket algebra is described more fully together with basis sets and how they are intimately related to the bra and ket. Some algebra for manipulating the bra and ket is also described.

8.4.1 To each ket $|\,\rangle$ belongs a bra $\langle\,|$

In the Dirac notation, a *ket* is a column and a *bra* a row vector. The elements of the vector are those of the basis set used. Each ket always has a corresponding bra. The bra is formed as the conjugate transpose of the ket, which means converting column into row and then taking the complex conjugate of each term. Formally the ket is the Hermitian adjoint of the bra, see Chapter 7.4.8. For example, if the ket is

$$|v\rangle = \begin{bmatrix} a \\ b \\ c \end{bmatrix} \tag{8.27}$$

the bra is

$$\langle v| = [a^* \quad b^* \quad c^*]. \tag{8.28}$$

where * indicates a complex conjugate. The a, b, and c are the elements of the basis set being used. However, these are themselves only *aliases* or references to the actual properties. By manipulating these symbols, problems can be solved so that only at the end of the calculation do numbers have to be used. The bra and ket can be multiplied together in two ways, as an *inner product* (dot product) to produce a scalar number

$$\langle A|B\rangle = [a_1^* \quad a_2^* \quad a_3^* \quad \cdots\cdots] \begin{bmatrix} b_1 \\ b_2 \\ b_3 \\ \cdot \\ \cdot \end{bmatrix} = \sum_i a_i^* b_i \tag{8.29}$$

and an *outer product* is an *operator* and produces a matrix.

$$|B\rangle\langle A| = \begin{bmatrix} b_1 \\ b_2 \\ b_3 \\ \cdot \\ \cdot \end{bmatrix} [a_1^* \quad a_2^* \quad a_3^* \quad \cdots\cdots] = \begin{bmatrix} b_1a_1^* & b_1a_2^* & b_1a_3^* & \cdots \\ b_2a_1^* & b_2a_2^* & \cdots & \cdots \\ \vdots & \vdots & \ddots & \\ & & & \end{bmatrix}. \tag{8.30}$$

which are the same rules for any matrix met in Chapter 7. The only difference in their usage in quantum mechanics is that the elements of the basis set must form an orthogonal set. If they are also normalised then they form an orthonormal set.

8.4.2 Discrete basis sets

The bra and ket, equations (8.27) and (8.28) have terms a, b, and c and these are the elements of the basis set chosen for the calculation. All quantum problems using bra-ket notation must have a basis set defined before the calculation starts. Unlike many calculations using vectors, the basis set does not usually represent spatial coordinates but is instead some other relevant property. The spin state of the proton, which is described next, is an example of this.

The spin state of an atom or nucleus can be described with a spin quantum number s and magnetic (azimuthal) quantum numbers m, which, when $s = ½$ has values +1/2 or −1/2, or, in general, $m = +s, +s - 1, \cdots -s$ and thus is a range of values from s to $-s$ each separated from one another by unity. Sometimes the magnetic quantum number is labelled m_s or m_z; the latter because it is the projection of angular momentum onto a unique axis normally labelled z when a unidirectional magnetic field is present, Fig. 8.4. The magnitude (expectation value) of the angular momentum is $\hbar\sqrt{s(s+1)}$. Nuclear spin angular momentum is usually labelled with I instead of s. If a magnetic field B is along the z-axis, the energy of the spin depends on its orientation and is $E = -\gamma m_z \hbar B$ where γ is the magnetogyric ratio with units of rad T^{-1} s^{-1}.

The spin basis set for a spin half particle (electron, ^1H, ^{13}C, etc.) can be constructed in terms of the two quantum numbers s and m, labelled with two indices as (s, m), and is

$$\{(1/2, +1/2)\ (1/2, -1/2)\},$$

where (1/2, +1/2), is a *label* describing one spin state, it has no mathematical significance, and the label (1/2, −1/2) describes the other. In the case of spin half particles we conventionally reserve symbols α to represent the state (1/2, +1/2) and β to represent the state (1/2, −1/2) and the basis set is now written as $\{\alpha \quad \beta\}$ or as kets $|\alpha\rangle$ and $|\beta\rangle$ as $\{|\alpha\rangle \quad |\beta\rangle\}$.

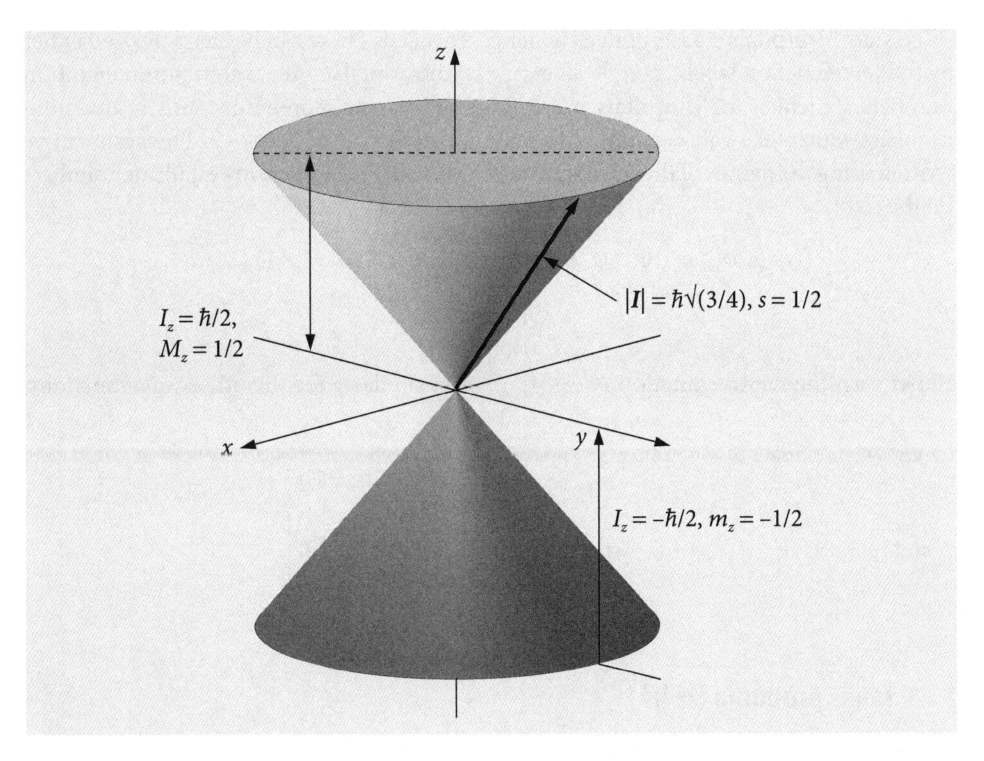

Fig. 8.4 Sketch describing spin quantum numbers and angular momentum for a ^{1}H nucleus. The angular momentum vector I can be at any angle around the cone.

Both notations are common. In this two-dimensional basis, the α spin state ($s = ½$, $m = ½$) is represented as the column vector in the *standard basis set* as $|\alpha\rangle = \begin{bmatrix} 1 \\ 0 \end{bmatrix}$. As α is a basis element (basis ket) of the *standard basis set*, it is zero everywhere except for one element representing its *position* in the basis set and which is given the value of unity. Similarly β ($s = ½$ $m = -½$) is represented as $|\beta\rangle = \begin{bmatrix} 0 \\ 1 \end{bmatrix}$ as it is in the second position. The basis set could be written as

$$\{\alpha, \beta\} \equiv \left\{ \begin{bmatrix} 1 \\ 0 \end{bmatrix}, \begin{bmatrix} 0 \\ 1 \end{bmatrix} \right\}$$

The basis set can be ordered differently; for example, $\{\beta, \alpha\} \equiv \left\{ \begin{bmatrix} 0 \\ 1 \end{bmatrix}, \begin{bmatrix} 1 \\ 0 \end{bmatrix} \right\}$ would be equally valid, it is only necessary to decide in which order the elements are placed.

If there are two electrons (or nuclei) each with spin quantum number +1/2 then we have, depending on spin orientation, a total spin of $S = 1$ or 0, loosely speaking these correspond to spins being 'parallel' or 'anti-parallel' to one another. The complete set of quantum numbers is $S = 0$, $m = 0$, and $S = 1$ with $m = +1, 0, -1$, making four states in all. A minimal basis set therefore has four terms

$$\{(0, 0) \quad (1, -1) \quad (1\ 0) \quad (1\ +1)\}$$

and four basis vectors. The vectors can be labelled with the quantum numbers (s, m) as below, but any four unique symbols could be used, e.g. ξ, ψ, x, y. The basis set order is also arbitrary.

$$|0, 0\rangle = \begin{bmatrix} 1 \\ 0 \\ 0 \\ 0 \end{bmatrix} \quad |1, -1\rangle = \begin{bmatrix} 0 \\ 1 \\ 0 \\ 0 \end{bmatrix} \quad |1, 0\rangle = \begin{bmatrix} 0 \\ 0 \\ 1 \\ 0 \end{bmatrix} \quad |1, 1\rangle = \begin{bmatrix} 0 \\ 0 \\ 0 \\ 1 \end{bmatrix} \qquad (8.31)$$

It is important to realize that in $|0, 0\rangle$ for example, the 0, 0 is just a *label*, it has no other meaning per se but the label tells us that this is state (0, 0); it does not enter into the maths. The basis set above is orthonormal as can be seen by calculating $\langle 0, 0|1, -1\rangle$ or any of the other 11 combinations.

To study d-orbitals a basis of five elements is needed. These can be characterized either by the wavefunction labels, z^2, x^2-y^2, etc. or by the orbital angular momentum quantum numbers, L, which for d-orbitals has 2 units of angular momentum and m quantum numbers which take values from −2 to 2 in unit steps as 2, 1, 0, −1, −2. The basis can be written using the names of the orbitals as shown on the left, or by using quantum numbers on the right,

$$\{\psi_{x^2-y^2}, \psi_{xz}, \psi_{z^2}, \psi_{yz}, \psi_{xy}\} \equiv \{(2,-2)\ (2,-1)\ (2,0)\ (2,+1)\ (2,+2)\}.$$

Either way, the z^2 wavefunction is $\psi_{z^2} = \begin{bmatrix} 0 \\ 0 \\ 1 \\ 0 \\ 0 \end{bmatrix}$, and similarly for the other wavefunctions.

A state such as $\Psi = a\psi_{xy} + b\psi_{xz}$ would be represented as $\Psi = \begin{bmatrix} 0 \\ a \\ 0 \\ 0 \\ b \end{bmatrix}$.

8.4.3 Inner products $\langle \alpha | \beta \rangle$

The *dot* or *inner product* of the two vectors is always a number, i.e. a scalar. Using the normalized and orthogonal (spin) states α and β

$$\langle \alpha|\alpha\rangle = [1 \quad 0]\begin{bmatrix} 1 \\ 0 \end{bmatrix} = 1, \qquad \langle \beta|\beta\rangle = [0 \quad 1]\begin{bmatrix} 0 \\ 1 \end{bmatrix} = 1, \qquad \langle \alpha|\beta\rangle = \langle \beta|\alpha\rangle = 0,$$

There is a somewhat peculiar situation with respect to spin angular momentum as opposed to orbital angular momentum, in that we do not know what the equations describing spin angular momentum are because they are not functions of space. This is, in fact, unimportant, because everything can be derived by simply defining the symbols for these states.

Consider the state $|\varphi\rangle = a|\alpha\rangle + b|\beta\rangle$, which is the linear combination of the α and β basis spin states where a and b are numbers, often complex numbers. This new state can be represented as the column vector $|\varphi\rangle = \begin{bmatrix} a \\ b \end{bmatrix}$. To normalize the function φ, the same procedure as for any vector is followed. The inner or dot product $\langle\varphi|\varphi\rangle$ is calculated

$$\langle\varphi|\varphi\rangle = [a^* \quad b^*]\begin{bmatrix} a \\ b \end{bmatrix} = a^*a + b^*b$$

and the normalization equation defined as $N^2\langle\varphi|\varphi\rangle = 1$ with normalization constant N. Hence $N^2 = a^*a + b^*b$ and *normalized* wavefunction is

$$|\varphi\rangle = \frac{a|\alpha\rangle + b|\beta\rangle}{\sqrt{a^*a + b^*b}}.$$

If the constants are real numbers, the complex conjugates are not needed. If there are equal amounts of a and b in φ, then

$$|\varphi\rangle = \frac{|\alpha\rangle + |\beta\rangle}{\sqrt{2}}.$$

To find out how much of state α is in the linear combination state φ, this is probed with the basis state α. The calculation is

$$\langle\alpha|\varphi\rangle = \frac{1}{\sqrt{a^*a + b^*b}}[1 \quad 0]\begin{bmatrix} a \\ b \end{bmatrix} = \frac{a}{\sqrt{a^*a + b^*b}}$$

If the coefficients a and b are real and normalized to unity when forming φ so that $a^2 + b^2 = 1$, then

$$\langle\alpha|\varphi\rangle = a\langle\alpha|\alpha\rangle + b\langle\alpha|\beta\rangle = a \qquad (8.32)$$

but, in either case, this result is interpreted to be the *probability amplitude* for the state φ to collapse into the state α and the probability of this happening is $\langle\alpha|\varphi\rangle^2$. Similarly, acting on φ with β, gives the probability amplitude $\langle\beta|\varphi\rangle = b$. As the product $\langle\alpha|\varphi\rangle$ is the coefficient a, we can then write the odd looking equation

$$|\varphi\rangle = \langle\alpha|\varphi\rangle|\alpha\rangle + \langle\beta|\varphi\rangle|\beta\rangle.$$

Expanding this out produces

$$|\varphi\rangle = [1 \quad 0]\begin{bmatrix} a \\ b \end{bmatrix}\begin{bmatrix} 1 \\ 0 \end{bmatrix} + [0 \quad 1]\begin{bmatrix} a \\ b \end{bmatrix}\begin{bmatrix} 0 \\ 1 \end{bmatrix} = a|\alpha\rangle + b|\beta\rangle.$$

Two kets in the basis $\{x, y, z, \cdots\}$

$$|A\rangle = a_1|x\rangle + a_2|y\rangle + a_3|z\rangle + \cdots$$
$$|B\rangle = b_1|x\rangle + b_2|y\rangle + b_3|z\rangle + \cdots$$

form the dot or inner product as do any single column or row matrices or vectors

$$\langle A|B\rangle = [a_1^* \quad a_2^* \quad a_3^* \; \cdots]\begin{bmatrix} b_1 \\ b_2 \\ b_3 \\ \vdots \\ \vdots \end{bmatrix} = \sum_i a_i^* b_i.$$

If a continuous basis set is being used with a function such as $R(r)$ or any wavefunction $\psi(x)$ or $\varphi(x)$ then the inner product becomes an integral. The basis set elements are so closely spaced that the summation of the finite basis is replaced by the integral, and then we have the familiar formula

$$\langle\varphi|\psi\rangle = \int \varphi(x)\psi(x)dx.$$

8.4.4 Outer products: $|\alpha\rangle\langle\beta|$

Objects of the form $|\alpha\rangle\langle\alpha|$, $|\alpha\rangle\langle\beta|$ or $|\psi_{z^2}\rangle\langle\psi_{xy}|$ and so forth are always square matrices and correspond to *operators*. This means that when they are placed to the left of a ket they operate on this and a *new* ket is formed. Using the definitions above

$$|\alpha\rangle\langle\beta| = \begin{bmatrix} 1 \\ 0 \end{bmatrix}[0 \quad 1] = \begin{bmatrix} 0 & 1 \\ 0 & 0 \end{bmatrix},$$

$$|\psi_{z^2}\rangle\langle\psi_{xy}| = \begin{bmatrix} 0 \\ 0 \\ 1 \\ 0 \\ 0 \end{bmatrix}[0 \quad 0 \quad 0 \quad 1 \quad 0] = \begin{bmatrix} 0 & 0 & 0 & 0 & 0 \\ 0 & 0 & 0 & 0 & 0 \\ 0 & 0 & 0 & 1 & 0 \\ 0 & 0 & 0 & 0 & 0 \\ 0 & 0 & 0 & 0 & 0 \end{bmatrix}.$$

Operating on the ket $|\varphi\rangle = a|\alpha\rangle + b|\beta\rangle$ with $|\alpha\rangle\langle\beta|$ produces the new ket, $b|\alpha\rangle$,

$$\begin{aligned}(|\alpha\rangle\langle\beta|)|\varphi\rangle &= |\alpha\rangle\langle\beta|\varphi\rangle \\ &= \langle\beta|\varphi\rangle|\alpha\rangle = b|\alpha\rangle\end{aligned}$$

It is not that obvious how this result is obtained. First, recall that the product $\langle\beta|\varphi\rangle = b$ is the inner or dot product of $\langle\beta|$ and $|\varphi\rangle$ and is a number. However, changing to matrix form clarifies the calculation,

$$(|\alpha\rangle\langle\beta|)|\varphi\rangle = \begin{bmatrix} 0 & 1 \\ 0 & 0 \end{bmatrix}|\varphi\rangle = \begin{bmatrix} 0 & 1 \\ 0 & 0 \end{bmatrix}\begin{bmatrix} a \\ b \end{bmatrix} = \begin{bmatrix} b \\ 0 \end{bmatrix} = b|\alpha\rangle,$$

and the last step is made by identifying with $|\alpha\rangle$. The result of $|\alpha\rangle\langle\beta|$ acting on the ket $|\varphi\rangle$, is to produce another ket which is $b|\alpha\rangle$, and hence, $|\alpha\rangle\langle\beta|$ is an *operator*. The 2×2 matrix produced is zero except for just one place; hence, it will extract just one value. The alternative calculation can be done by multiplying the two right-hand matrices first, operating with $\langle\beta|$ on $|\varphi\rangle$ and then multiplying the answer by $|\alpha\rangle$,

$$|\alpha\rangle(\langle\beta|\varphi\rangle)=\begin{bmatrix}1\\0\end{bmatrix}\left(\begin{bmatrix}0 & 1\end{bmatrix}\begin{bmatrix}a\\b\end{bmatrix}\right)=\begin{bmatrix}1\\0\end{bmatrix}b=b|\alpha\rangle.$$

The operator $|\alpha\rangle\langle\alpha|$ will project $a|\alpha\rangle$ out of the ket $|\varphi\rangle$, and is therefore called a *projection operator*. The calculation is $|\alpha\rangle\langle\alpha||\varphi\rangle=\begin{bmatrix}1 & 0\\0 & 0\end{bmatrix}|\varphi\rangle=\begin{bmatrix}1 & 0\\0 & 0\end{bmatrix}\begin{bmatrix}a\\b\end{bmatrix}=\mathrm{a}|\alpha\rangle$.

8.4.5 Questions

Full solutions are available at **www.oxfordtextbooks.co.uk/orc/beddard**.

Q8.8 Using the bra-ket notation, work out the following matrix multiplications in the orthonormal basis of an s and two p orbitals $\{|s\rangle,|p_x\rangle,|p_y\rangle\}$, for the linear combination of wavefunctions

$$|\psi\rangle=c_s|s\rangle+c_x|p_x\rangle+c_y|p_y\rangle,$$

$$|\varphi\rangle=b_s|s\rangle+b_x|p_x\rangle+b_y|p_y\rangle.$$

Assume that ψ and φ are normalized, which means for example that, $c_s^*c_s+c_x^*c_x+c_y^*c_y=|c_s|^2+|c_x|^2+|c_y|^2=1$ and similarly for coefficients b.

(a) Calculate $\langle s|s\rangle$ and $\langle s|p_x\rangle$ and similarly for the other basis vectors and show that the basis is orthonormal.

(b) Calculate (i) $|s\rangle\langle s|$, (ii) $|s\rangle\langle p_x|$, (iii) $|s\rangle\langle s|\varphi\rangle$, (iv) $|s\rangle\langle p_x|\varphi\rangle$, (v) $\langle\varphi|\psi\rangle$, (vi) $\langle p_y|\psi\rangle$, (vii) $|\varphi\rangle\langle\psi|$, (viii) $|s\rangle\langle p_x|\varphi\rangle$.

Strategy: Check that the matrices are commensurate for each calculation, the column of the left matrix (or vector) must have the same number or rows as the right matrix (or vector), and if so multiply them out. The next important point is to note that in vector form the standard basis sets are orthogonal and normalized, which means they are

$|s\rangle=\begin{bmatrix}1\\0\\0\end{bmatrix}$ and $|p_x\rangle=\begin{bmatrix}0\\1\\0\end{bmatrix}$ and $|p_y\rangle=\begin{bmatrix}0\\0\\1\end{bmatrix}$. The wavefunctions are then the *kets* $|\psi\rangle=\begin{bmatrix}c_s\\c_x\\c_y\end{bmatrix}$ and $|\varphi\rangle=\begin{bmatrix}b_s\\b_x\\b_y\end{bmatrix}$.

Q8.9 Using ψ and ϕ defined in question **Q8.8**, show that $\langle\psi|\varphi\rangle=\langle\varphi|\psi\rangle^*$ is a general statement for any bra-ket pair.

Q8.10 In the NMR spectroscopy of a single ^{1}H, with spin ½ quantum number, a two element basis set can be defined as (α,β) to represent the spin state. The α spin state ($s=½$, $m=½$) is represented as the column vector $\alpha=\begin{bmatrix}1\\0\end{bmatrix}$, and as this is a basis element (basis ket) of the standard basis set it is zero everywhere except for one element which is unity. Similarly, β ($s=½$, $m=-½$) is $\beta=\begin{bmatrix}0\\1\end{bmatrix}$. The angular momentum operator I has components in the x-, y-, and z-directions and because it is an operator it is represented as a matrix. Angular momentum has units of $\hbar$ and the components of the angular momentum operators can be represented as the (Pauli) matrices,

$$I_x=\frac{\hbar}{2}\begin{bmatrix}0 & 1\\1 & 0\end{bmatrix},\qquad I_y=\frac{i\hbar}{2}\begin{bmatrix}0 & -1\\1 & 0\end{bmatrix},\qquad\text{and}\qquad I_z=\frac{\hbar}{2}\begin{bmatrix}1 & 0\\0 & -1\end{bmatrix}.$$

(a) Show that if $I^2=I_x^2+I_y^2+I_z^2$ then $I^2=\dfrac{\hbar^2}{4}\begin{bmatrix}3 & 0\\0 & 3\end{bmatrix}=\dfrac{3}{4}\hbar^2\boldsymbol{1}$ where $\boldsymbol{1}$ is the unit diagonal matrix.

(b) Show that I^2 and I_z commute, $[I^2,I_z]=0$, but that $[I_x,I_y]=i\hbar I_z$ and the other combinations of x, y, and z components do not commute.

(c) Show that when I^2 operates on α, the eigenvalue is $\dfrac{3\hbar^2}{4}$, i.e. show that $I^2\alpha=\dfrac{3\hbar^2}{4}\alpha$ and calculate $I^2\beta$ and $I_z\alpha$ and $I_z\beta$. Comment on the results.

These equations are eigenvalue–eigenvector equations, as is the Schrödinger equation $H\psi=E\psi$ where ψ is the wavefunction, E the energy, and H the operator.

(d) Calculate $I_x\alpha$, $I_x\beta$ and similar terms for the y and z components.

(e) Using the operators $I^+=I_x+iI_y$ and $I^-=I_x-iI_y$, show that these are raising or lowering operators which convert the eigenstate α or β into the other.

Q8.11 Angular momentum raising and lowering (shift) operators L move a system from one state to another. They have the property

$$L^+|L, m\rangle = \sqrt{L(L+1) - m(m+1)}\,|L, m+1\rangle$$

$$L^-|L, m\rangle = \sqrt{L(L+1) - m(m-1)}\,|L, m+1\rangle$$

where the angular momentum quantum number is L and the projection or z quantum number is m. The values m takes run from $-L$ to $+L$ in unit steps.

(a) Show that the Lm basis set used is orthonormal.

(b) Calculate the raising and lowering operators for a state with angular momentum 3/2 in the L, m basis set. Note the representation of the operators will be a matrix.

(c) Using the result from (a) show that the commutator $[L^+, L^-] = 2L_z$ where L_z is given by equation (8.21). (Assume $\hbar$ equals 1 and change s to L.)

(d) Calculate the L^+ operator for spins, 0, 1, 2, 3 and then for half unit spins 1/2, 3/2, 5/2 in the Lm basis set.

Strategy: (b) The basis set must contain all m values therefore the set could be ({3/2, −3/2}, {3/2, −1/2}, {3/2, 1/2}, {3/2, 3/2}) if $L = 3/2$. Define vectors so that they are orthonormal.
(c) The raising and lowering operators form a block diagonal matrix. Individual blocks of which can be calculated as in (a) or a more extensive basis set formed and the whole matrix calculated. It may be useful to define a Maple function to work out terms in the operator according to the L^+, L^- equations.

Q8.12 **(a)** Using the raising and lowering operator in the previous question produce equations (8.19) and (8.20) starting from $L_x = (L^+ + L^-)/2$ and $L_y = (L^+ - L^-)/2i$. (Assume $\hbar = 1$.)

(b) Show that the commutator is $[L^+, L^-] = 2L_z$ where L_z is given by a similar equation to (8.21), $L_z|Lm\rangle = m|Lm\rangle$.

8.4.6 Continuous basis sets

Suppose a wavefunction R represents the radial part of the ground state of a hydrogen atom then a continuous basis set is needed because the wavefunction depends on spatial coordinates, and in particular on the radial distance from the nucleus, r. At this point, R is just the *idea* of the wavefunction as no basis or *representation* has yet been determined. The wavefunction can now be defined by left multiplying with a bra; the result is the value of the function at r,

$$\langle r|R\rangle = R(r).$$

If R were a cosine, the expression would be $\langle r|\cos\rangle = \cos(r)$ where cos is the operator that converts, or maps, r into $\cos(r)$. This is not the way we usually think of functions and it seems to be 'back to front'. It may be useful therefore to think of $\langle r|R\rangle = R(r)$ as extracting the coefficient of R at the point r, which is the value of the function itself, $R(r)$. The equivalent process with a discrete basis set is equation (8.32), which extracts the coefficient a from φ.

If a basis set is defined as the set of an orthogonal polynomial P_n then this basis is $\{P_0, P_1, P_2, \cdots, P_\infty\}$ and the bra is

$$\langle r| = [P_0(r) \quad P_1(r) \quad P_2(r) \quad \cdots \quad P_\infty(r)]$$

where the $P_0(r)$ etc. are *functions* of r. For example, if P are the Legendre polynomials, then $\langle r| = \left[\,1, \dfrac{3x^2-1}{2}, \dfrac{5x^3-3x}{2} \cdots\right]$ and the terms are the polynomials. The corresponding ket is

$$|P_n\rangle = [0 \quad 0 \quad \cdots \quad 1 \quad \cdots]$$

where the 1 is in the nth position for polynomial P_n. For example, the product $\langle r|P_2\rangle = P_2(r) = \dfrac{3x^2-1}{2}$ and is the value of the function at r, and this is consistent with the notion that P_2 operates on r to produce the function $P_2(r)$.

It would have been just as easy to choose a basis in momentum p which is $\{p\}$ rather than r, then the wavefunction would be $\langle p|R\rangle = R(p)$, which is now a function of p the momentum. To represent the whole wavefunction ψ with its radial and angular parts then it is necessary to define a different basis set to encompass all these three coordinates, (r, θ, ϕ) which can be represented as $\{(r, \theta, \phi)\}$ where there are three indices to each basis vector but $|(r, \theta, \phi)\rangle$ is still an infinite length column vector and therefore, $\langle(r, \theta, \phi)|\psi\rangle = \psi(r, \theta, \phi)$ extracts the value at r, θ, ϕ.

At this point, however, it becomes unnecessary and rather complicated to continue with the bra-ket form and it is simpler to revert to normal functions but still holding onto the idea of basis sets. The basis set could comprise almost any set of functions provided that they can be made orthogonal over the range of values needed, alternatively the known orthogonal polynomials, Legendre, Hermite, Chebychev, etc. could be used. A function $f(x)$ can be expanded as a linear combination of orthogonal functions and if these from a set of wavefunctions ψ then

$$f(x) = c_0\psi_0(x) + c_1\psi_1(x) + c_2\psi_2(x) + \cdots \tag{8.33}$$

with coefficients c_n. The wavefunction need not be specified yet, but whatever it is it must be orthogonal given different quantum numbers. This is exactly what is done in the general Fourier series described in Chapter 9. There it is shown that the coefficients are

$$c_n = \int \psi(n, x)^* f(x)dx \tag{8.34}$$

If the function f is represented as the ket $|f\rangle$, suppose that left multiplying by the bra $\langle n|$ will extract the coefficient c_n in the same manner as for a discrete basis set. However, in a continuous basis the bra-ket represents an integral, thus $\langle n|f\rangle \equiv \int \psi(n, x)^* f(x)dx$. Equation (8.34) provides a method of calculating the coefficients of the expansion provided f has the same range as the wavefunction. To illustrate this, particle in a box wavefunctions are used to form the target function $f(x) = 64 - (2 - x)^6$ where the length of the box $L = 4$, or $0 \le x \le 4$. The calculation is in Algorithm 8.5. The series $s(x)$ approximates $f(x)$ in the basis set of the ψ as

$$\begin{aligned} s(x) &= \psi(0, x)\int \psi(0, x)^* f(x)dx \\ &+ \psi(1, x)\int \psi(1, x)^* f(x)dx \\ &+ \psi(2, x)\int \psi(2, x)^* f(x)dx + \cdots \end{aligned}$$

The first five terms in equation (8.33) are shown in Fig. 8.5. To summarize $f(x) = 64 - (2 - x)^6$ is approximated by using a series containing as a basis set the particle in a box wavefunctions $\psi(n, x) = \sqrt{\frac{2}{L}}\sin\left(\frac{n\pi x}{L}\right)$, if more than five terms are used s becomes a progressively better mimic of f.

Algorithm 8.5

```
> restart:
  num:= 5;                    # number of terms in series
  L:= 4:
  psi:=(n,x)-> sqrt(2/L)*sin(n*Pi*x/L);
  f:= x-> 64-(2-x)^6;                          # target equation
# calculate coefficients c eqn (8.34)
  c:= n-> int( psi(n,x)*f(x) ,x=0..L); # coefficients
  s:= x-> add( c(n)*psi(n,x),n=1..num); # n terms, eqn (8.33)
  plot([f(x),s(x)],x= 0..L);
```

Why this method works can be appreciated by realizing that at larger values of n the sine function oscillates more rapidly and so allows for the rapid rising and falling part of

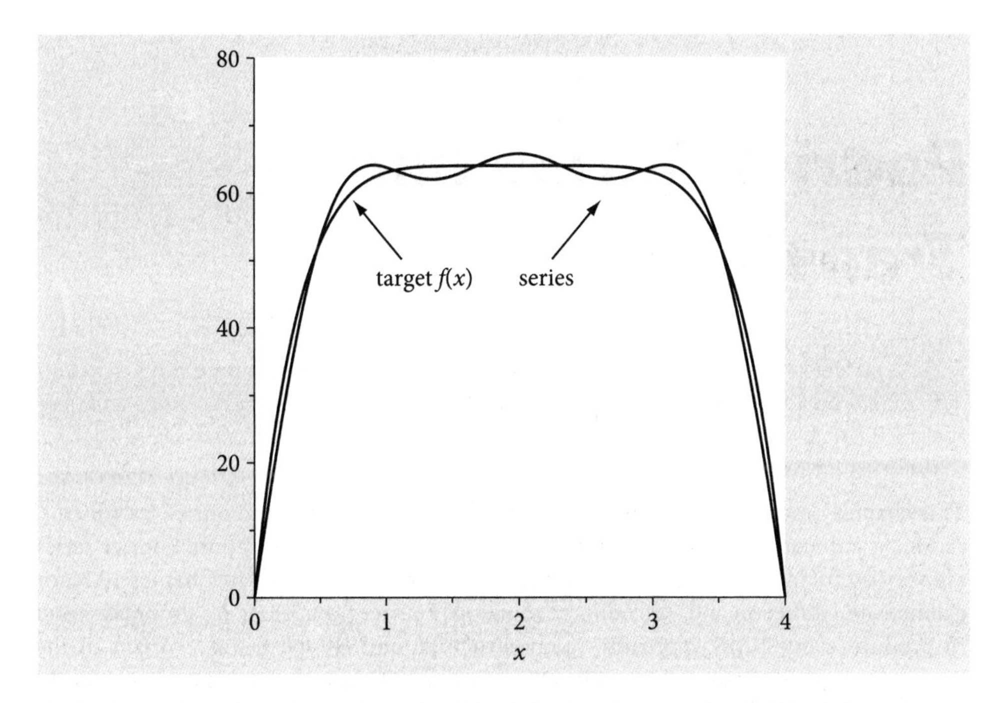

Fig. 8.5 Comparison of a series made up of weighted sine functions equation (8.33) and the target function $64 - (2 - x)^6$. A far better fit is obtained if more than five terms are added together.

the curves. The coefficients automatically adjust the proportion of each sine wave to describe the target equation. Other target functions can be tried quite easily by, changing f, however, the less the function looks like a sine wave, such as $\exp(-x)$, the greater will be the number of terms that are needed to produce a good description of the function, >100 in that case.

9 Fourier Series and Transforms

This chapter describes several related topics: the Fourier series and Fourier transforms, as well as autocorrelations, convolutions, and their numerical calculation. Fourier series are very useful in solving partial differential equations, as explained in Chapter 10.5, for example, the diffusion and Schrödinger equation. Fourier transforms are nowadays used to produce every NMR spectrum, X-ray structure, and IR spectrum recorded in the laboratory; a good grasp of this topic is therefore essential for the molecular scientist. In many experiments the measuring instrument distorts the data and, in this case, what is measured is convoluted with the response of the instrument. Here Fourier transforms can unravel the data to produce the true response. At the end of the chapter a discrete transform, the Hadamard transform is also described.

9.1 Motivation and concept

The Taylor and Maclaurin series expand functions as an infinite series in the powers x^n, and the coefficients needed to do this are the derivatives of the function, see equation (5.15). These series have rather tight restrictions placed upon them; the function must be differentiable n times over and the remainder must approach zero. In a Fourier series, the expansion is performed instead, as trigonometric series in sines and cosines, with two sets of coefficients, a_n and b_n, to describe the nth term, and which are evaluated by integration. The series formally extends to infinity, but in practice, at most only a few tens of terms are needed to replicate most functions to an acceptable level of approximation. The advantage of using a Fourier rather than a Taylor/Maclaurin series is that a wide class of functions can be described by the series, including discontinuous ones. However, by their very nature, Fourier series can only represent *periodic* functions, and this must not be forgotten. Periodic means that the function repeats itself; the repeat interval is normally taken to be $-\pi$ to $+\pi$ but can be extended to the range $-L$ to L and L can even be made infinite.

To illustrate that it is possible to make an arbitrary shaped function by adding a number of sine and cosine waves, four such waves are shown in Fig. 9.1. It is possible to imagine, without any difficulty, that waveforms that are more complex can be formed by using more sine or cosine terms. The lowest and most complicated waveform is simply the sum of the individual waves, and repeats itself with a period of 2π. This waveform could be the signal that is measured on an oscilloscope or spectrometer and recorded on a computer. It might alternatively, be part of an image that is formed from adjacent columns of different waveforms. Periodic components present in a signal represent information in the waveform or image can be retrieved by performing a *Fourier transform*. This unravelling process is discussed later on but now the Fourier series is considered.

9.1.1 The Fourier series

Because the Fourier series extends to infinity, many more sine and cosine terms will be needed to reproduce the waveform exactly as shown in Fig. 9.1 than were used to produce it in the first place; the Fourier series cannot somehow pick out just those sine and cosine waves used to make the original function. If the complicated waveform shown in Fig. 9.1 were truncated so that it was zero outside $x = -\pi$ to $+\pi$ then its Fourier series would have to try

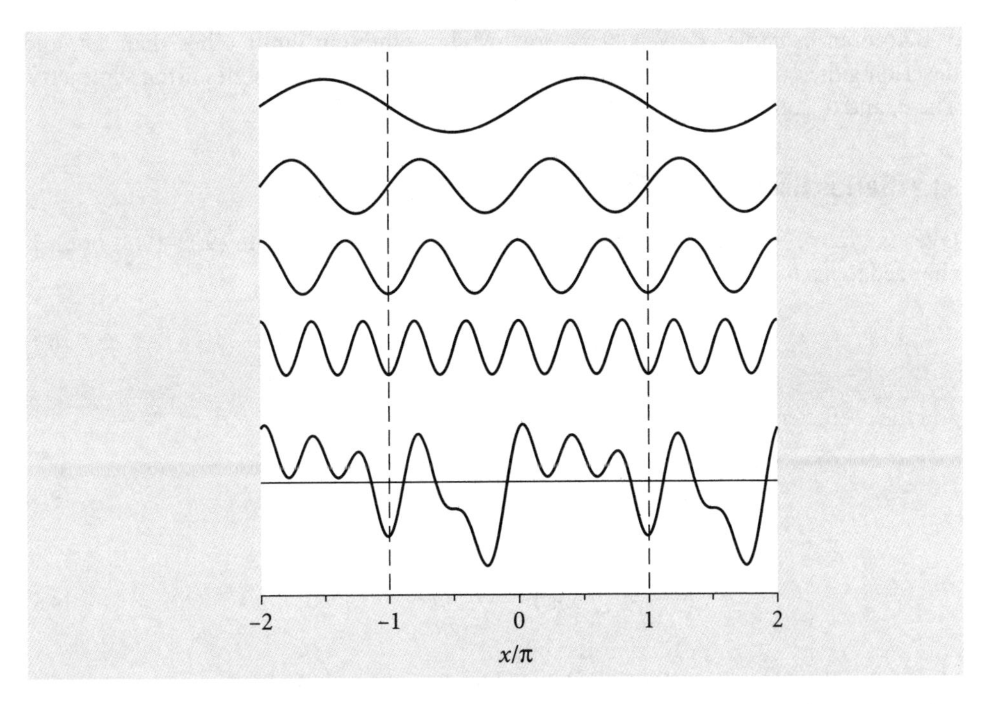

Fig. 9.1 A complex and periodic waveform or function is constructed out of the sum of sine and cosine waves. The complicated waveform, repeats itself with a period of 2π. In the Fourier series, the reconstruction of this waveform will require many more sine and cosine terms to reconstruct its form than are used to generate it, because the Fourier series only represents a function exactly when an infinite number of terms are included in the summation.

to match this new waveform with sufficient terms in the series whose overall effect is to cancel out to zero outside the range $x = -\pi$ to $+\pi$ but reproduce the function inside the range. This produces a poor description of the waveform as illustrated in Q9.2 in which the series for $\sin(x)$ is calculated over the limited range $-L$ to L even when more than 100 terms are added to the series.

Suppose that, over the range $-\pi$ to $+\pi$, a Fourier series $g(x)$ approximates a victim function $f(x)$, which might be x^3 or $(e^{-x} - 1)^2$ and so forth, then the Fourier series can be written down in general and in quite a straightforward manner and is

$$g(x) = \frac{a_0}{2} + \sum_{n=1}^{\infty} a_n \cos(nx) + \sum_{n=1}^{\infty} b_n \sin(nx). \tag{9.1}$$

The summations start at index $n = 1$, and n is a positive integer. The a_n and b_n coefficients are the integrals

$$a_n = \frac{1}{\pi}\int_{-\pi}^{+\pi} f(x)\cos(nx)dx \qquad (n \ge 0) \tag{9.2}$$

$$b_n = \frac{1}{\pi}\int_{-\pi}^{+\pi} f(x)\sin(nx)dx \qquad (n > 0) \tag{9.3}$$

which are normalized by $1/\pi$. When the number of terms in the series is large, then $g(x) \to f(x)$, and when infinite, $g(x) = f(x)$. At each value of x, the Fourier series consists of a constant term, $a_0/2$ plus the sum of an infinite number of oscillating terms in integer multiples of x. Notice, that the target function $f(x)$ appears as part of the expansion coefficients only, and must, therefore, be capable of being integrated. The target function $f(x)$ determines the weighting to be placed on each term in the expansion, and this is how information about the shape of $f(x)$ is included in the expansion.

If $f(x)$ is periodic in time, then the variable x would normally be changed to ωt where $\omega = 2\pi\nu$ is the angular frequency in radians s^{-1}, and ν is the frequency in s^{-1}. If the dimension is spatial, then x is often replaced with $2\pi x/L$ of which $2\pi/L$ can be interpreted as a spatial frequency, with units of radians m^{-1} by analogy with 'normal' frequencies. Often this spatial frequency is called the *wavevector* and given the symbol k.

Before an example is given, it is worthwhile examining limits other than $\pm\pi$, and describing the exponential form of the series and also simplifying series using symmetry. The a_n and b_n constants are also derived.

9.1.2 Series limits from $-L$ to L

Over the range $-L$ to L, the equations to use for $f(x)$ are similar to (9.1)–(9.3) but x is changed to $\pi x/L$. The series is

$$g(x) = \frac{a_0}{2} + \sum_{n=1}^{\infty} a_n \cos\left(\frac{n\pi x}{L}\right) + \sum_{n=1}^{\infty} b_n \sin\left(\frac{n\pi x}{L}\right) \tag{9.4}$$

and the coefficients

$$a_n = \frac{1}{L}\int_{-L}^{+L} f(x)\cos\left(\frac{n\pi x}{L}\right)dx \qquad (n \ge 0) \tag{9.5}$$

$$b_n = \frac{1}{L}\int_{-L}^{+L} f(x)\sin\left(\frac{n\pi x}{L}\right)dx \qquad (n > 0). \tag{9.6}$$

Notice that the arguments, limits, and normalization are each changed compared to those when the range is $-\pi$ to $+\pi$. The integral now has limits $\pm L$ instead of $\pm\pi$, and normalization $1/L$ rather than $1/\pi$.

9.1.3 Exponential representation

Because the sine and cosine functions can be represented as sums and differences of complex exponential terms (see Chapter 2, for example, $\cos(x) = (e^{ix} + e^{-ix})/2$), the most general way of describing the Fourier series is to use the complex exponential form;

$$g(x) = \sum_{n=-\infty}^{\infty} c_n e^{i(n\pi x/L)} \tag{9.7}$$

and the coefficients become

$$c_n = \frac{1}{2L}\int_{-L}^{L} f(x)e^{-i(n\pi x/L)}dx. \tag{9.8}$$

Note the change in sign in this second exponential, and $i = \sqrt{-1}$. The set of coefficients c_n, are sometimes called the *amplitude* spectrum of the transform.

9.1.4 Deriving the integral describing the *a* and *b* coefficients

The integrals describing the coefficients a_n are obtained by multiplying each term in the Fourier series by $\cos(mx)$, where m is an integer, then integrating term by term using the orthogonality of sine and cosine integrals as necessary. The coefficients b_n can be obtained similarly by multiplying the series by $\sin(mx)$ and integrating.

First, the form of the cosine integrals is examined. The integrals containing a product of sine and cosine are all zero for any n and m because the integral has 'odd' symmetry because of the sine, and the limits are symmetrical;

$$\int_{-\pi}^{\pi} \cos(mx)\sin(nx)dx = 0. \tag{9.9}$$

See Chapter 4 for other examples of odd and even functions and integrals. The product of two cosines makes the integral

$$\int_{-\pi}^{\pi} \cos(mx)\cos(nx)dx = \pi\delta_{nm} \tag{9.10}$$

where the (Kronecker) delta function δ_{nm}, is zero only if $n \ne m$, and is 1 if $n = m$.

To calculate the a coefficients, the equation for the Fourier series

$$f(x) = \frac{a_0}{2} + \sum_{n=1} a_n \cos(nx) + \sum_{n=1} b_n \sin(nx) \tag{9.11}$$

is multiplied by $\cos(mx)$ and integrated. The cosine integrals with $m > 0$ are,

$$\begin{aligned}&\int_{-\pi}^{\pi} f(x)\cos(mx)dx \\ &= \frac{a_0}{2}\int_{-\pi}^{\pi} \cos(mx)dx + \int_{-\pi}^{\pi}\left[\sum_{n=1} a_n\cos(nx) + \sum_{n=1} b_n \sin(nx)\right]\cos(mx)dx\end{aligned} \tag{9.12}$$

These integrals can be easily evaluated. The sine is an odd function over the range $-\pi$ to π; therefore, the last integral containing b_n is zero. The first integral is also zero by direct integration,

$$\frac{a_0}{2}\int_{-\pi}^{\pi} \cos(mx)dx = \frac{a_0}{2m}\sin(mx)\Big|_{-\pi}^{\pi} = 0.$$

The integral containing b_n is given by equation (9.10) and is zero when $n \ne m$, therefore, equation (9.12) is reduced to

$$\int_{-\pi}^{\pi} f(x)\cos(mx)dx = \pi \sum_{n=1} a_n \delta_{nm} = \pi a_n.$$

because the delta function picks out just one term from the summation. Rearranging gives

$$a_n = \frac{1}{\pi}\int_{-\pi}^{+\pi} f(x)\cos(nx)dx, \qquad n > 0.$$

The index has been changed to n from m because both n and m are integers and only one index is needed; equation (9.11) was written initially with n, so this is again chosen. When $m = n = 0$, equation (9.12) becomes

$$\int_{-\pi}^{\pi} f(x)dx = \frac{a_0}{2}\int_{-\pi}^{\pi} 1\, dx = a_0\pi$$

making $a_0 = \frac{1}{\pi}\int_{-\pi}^{+\pi} f(x)dx$. Similar arguments lead to the b_n coefficients, see Q9.1.

9.1.5 Odd and even functions

When functions are either even or odd, the Fourier series is simplified to cosine or sine series respectively. An even function has the property $f(-x) = f(x)$ and the sine integration (9.6) producing the Fourier coefficient b_n is odd and therefore $b_n = 0$, because the integration limits are symmetrical. The cosine integral (9.5) is not zero, and can be written as

$$a_n = \frac{2}{\pi}\int_{0}^{+\pi} f(x)\cos(nx)dx. \tag{9.13}$$

When $f(x)$ is odd $f(-x) = -f(x)$, the opposite situation arises; $a_n = 0$ and only the cosine terms remain;

$$b_n = \frac{2}{\pi}\int_{0}^{+\pi} f(x)\sin(nx)dx. \tag{9.14}$$

In cases when $f(x)$ is neither odd nor even, for example $f(x) = \pi/2 - x$, both coefficients have to be evaluated.

As an example, consider calculating the series of $f(x) = x^2$ over the range $-\pi$ to π. Because this is an even function, the sine integral should be zero and $b_n = 0$. The a coefficients are

$$a_n = \frac{1}{\pi}\int_{-\pi}^{+\pi} x^2\cos(nx)dx$$

and when $n = 0$

$$a_0 = \frac{1}{\pi}\int_{-\pi}^{+\pi} x^2 dx = \frac{2}{3}\pi^2.$$

When $n > 0$, a_n can be evaluated by integration by parts. Maple gives the same result;

```
> 1/Pi*Int(x^2*cos(n*x),x): % = value(%);
```

$$\frac{\int x^2\cos(nx)\,dx}{\pi} = \frac{n^2x^2\sin(nx) - 2\sin(nx) + 2\,nx\cos(nx)}{\pi n^3}$$

and adding limits $\pm\pi$, the integral becomes $(-1)^n\dfrac{4}{n^3}$ because the sine terms are zero when n is an integer and the cosine -1 or $+1$ depending on whether n is, respectively, odd or even. Therefore $a_n = (-1)^n\dfrac{4}{n^3}$, when $n > 0$. The b coefficients are calculated similarly and, just to show that they are zero, the integral is

```
> 1/Pi*Int(x^2*sin(n*x),x=-Pi..Pi): % = value(%);
```

$$\frac{\int_{-\pi}^{\pi} x^2\sin(nx)dx}{\pi} = 0$$

However, it was not necessary to do the calculation, because x^2 is an even function and $\sin(x)$ an odd function, making the integral odd overall. The Fourier series for x^2, is therefore,

$$\begin{aligned} x^2 \approx g(x) &= \pi^2/3 - 4\cos(x) + \cos(2x) - 4\cos(3x)/9 + \cdots \\ &= \frac{\pi^2}{3} + 4\sum_{n=1}^{\infty}\frac{(-1)^n}{n^2}\cos(nx) \end{aligned} \tag{9.15}$$

which, as shown in Fig. 9.2, is a good representation of x^2. To plot the result with Maple, the Fourier series is first defined as the function FS in m and x; m is the limit of the summation and so n runs from 1 to m. The Maple add instruction is used to evaluate the summation whenever a numerical result is needed.

```
> FS:=(m,x)->(Pi^2)/3+add((-1)^n*4/n^2*cos(n*x),n=1..m);
  plot( [x^2,FS(3,x)] ,x= -Pi..Pi );
```

The repetitive nature of the Fourier series is clear in the right-hand pane of Fig. 9.2. The fit can be made to a larger x value than π by using equation (9.4) with the correspond coefficients. The size of the first 10 coefficients for the Fourier series of x^2, as shown in Fig. 9.2, are plotted in Fig. 9.3; this is sometimes called the spectrum, and shows why so few terms are needed to make a good approximation to x^2. After 10 terms the amplitude of the coefficients becomes very small, only 1/25. There is clearly no need to sum to infinity.

9.1.6 **Discontinuous functions**

The square wave shown in Fig. 9.4, has a stepwise discontinuity. When $x \ge 0$, $f(x) = 1$, and when $x < 0$, $f(x) = -1$ and the wave repeats itself with a period of 2π. This discontinuous function is sometimes called the odd *signum* function, sgn(x). It can be shown that, if the discontinuity occurs in the mid range $-\pi$ to π, each integral, a_n and b_n, can be split into

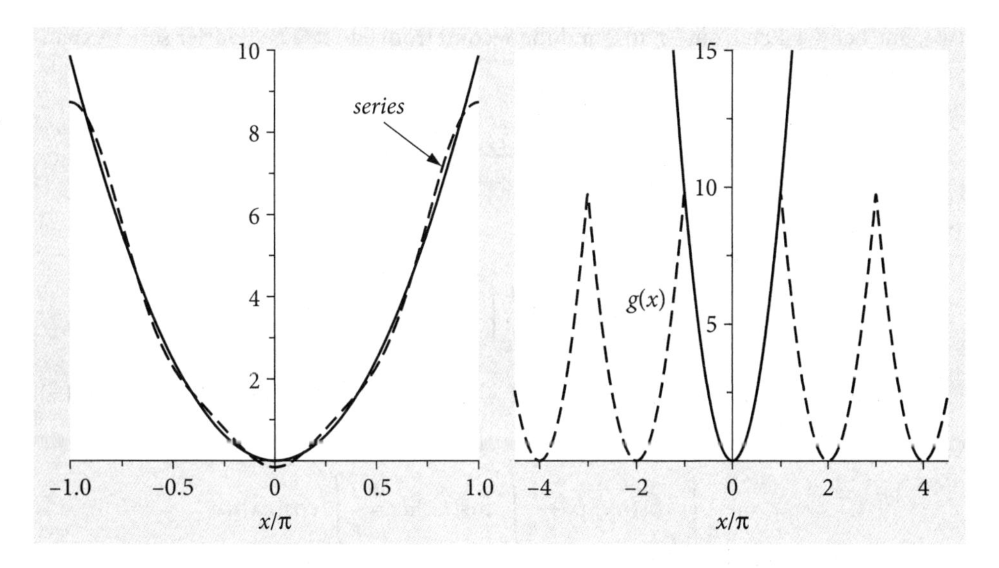

Fig. 9.2 Left: Plot of x^2 and its Fourier series to $m = 3$ showing a fairly good fit to the true function. More terms produce a better fit but only over the range $-\pi$ to $+\pi$ as shown (right) where 30 terms are included in the summation. This plot also shows how the fit is only over the range $\pm\pi$ and then it repeats itself.

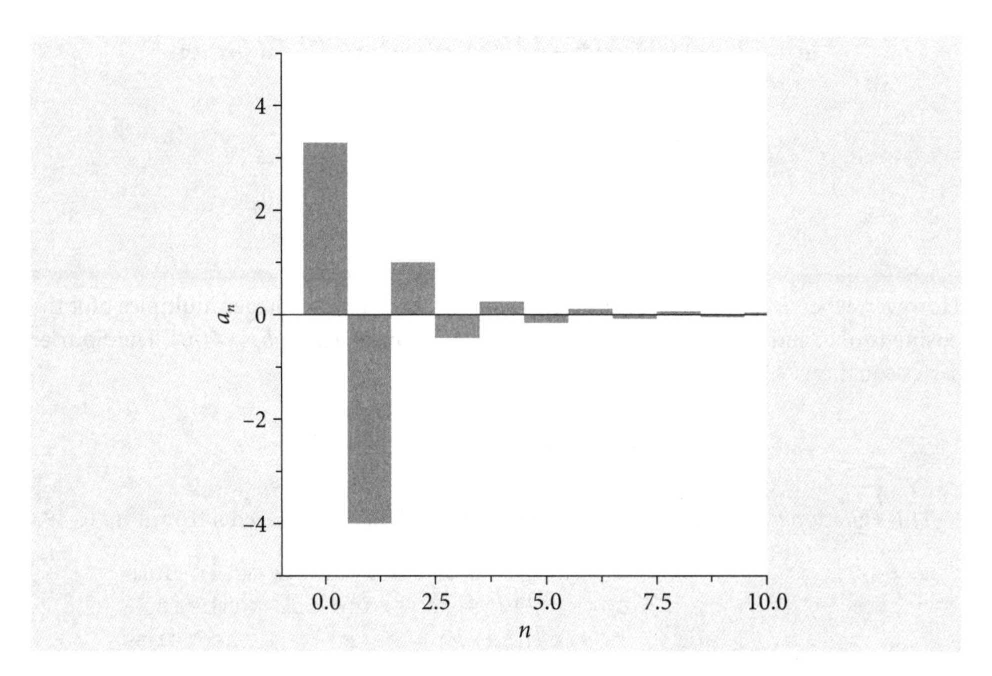

Fig. 9.3 Amplitude of coefficients in equation (9.15) showing the rapid decrease in their value.

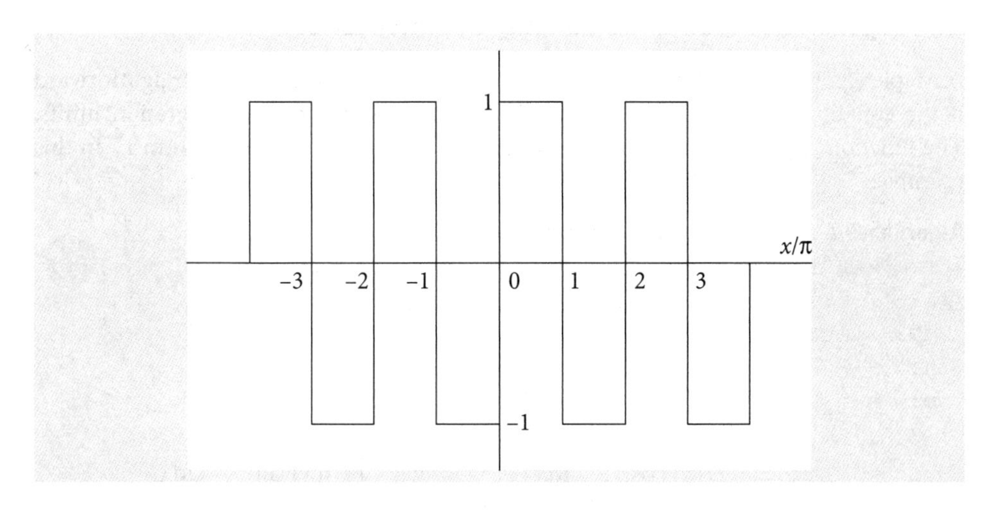

Fig. 9.4 A few cycles of a square wave, which is a discontinuous function.

two, one being taken from $-\pi$ to 0 and the second from 0 to π. The Fourier series expansion has coefficients

$$a_n = \frac{1}{\pi}\int_{-\pi}^{0}(-1)\cos(nx)dx + \frac{1}{\pi}\int_{0}^{\pi}(+1)\cos(nx)dx.$$

When $n = 0$,

$$a_0 = \frac{1}{\pi}\int_{-\pi}^{0}(-1)dx + \frac{1}{\pi}\int_{0}^{\pi}(1)dx = -1 + 1 = 0,$$

and because the function has odd symmetry $f(-x) = -f(x)$, the cosine integrals and all other a coefficients are zero

$$a_n = -\frac{1}{\pi}\int_{-\pi}^{0}\cos(nx)dx + \frac{1}{\pi}\int_{0}^{\pi}\cos(nx)dx = \frac{2}{\pi}\int_{0}^{\pi}\cos(nx)dx$$

$$= \frac{2}{\pi}\sin(nx)\Big|_{-\pi}^{0} = 0$$

The b coefficients are

$$b_n = \frac{1}{\pi}\int_{-\pi}^{0}(-1)\sin(nx)dx + \frac{1}{\pi}\int_{0}^{\pi}(+1)\sin(nx)dx = \frac{2}{\pi}\int_{0}^{\pi}\sin(nx)dx$$

$$= -\frac{2}{n\pi}\cos(nx)\Big|_{0}^{\pi}$$

$$= -\frac{2}{n\pi}\cos(n\pi) + \frac{2}{n\pi}$$

However, when n is even, the coefficient b_n is zero, because for integer multiples of π the cosine is one, and when n is odd the cosine is -1, consequently $b_n = 4/n\pi$. The Fourier series equation (9.1) reduces to

$$g(x) = \sum_{n=odd} b_n \sin(nx) = \frac{4}{\pi}\sum_{n=odd}\frac{\sin(nx)}{n}$$

This equation is plotted in Fig. 9.5 using Maple with odd numbered n from 1 up to 49.

```
> m:= 50;                              # number of terms in sum
  ss:= m -> seq( n, n = 1..m, 2); # generate odd numbers
  FS:= x -> 4/Pi*add(1/n*sin(n*x),n = ss(m)); # add terms
  plot(FS(x*Pi, m),x= -3..3,-1.5..1.5);
```

9.1.7 A general method for calculating a Fourier series

A Maple algorithm to calculate the Fourier series of a function is fairly straightforward if the equations (9.5) and (9.6) are used and care taken over the integration limits. The function is placed in the first line as `f:=x-> ...` and the range is from ±7 in this example.

Algorithm 9.1 General method for Fourier series (−*L* to *L*)

```
> f:= x-> exp(-x/2)*cos(x)^2:   # a victim function
  L:= 7.0:                      # range is -L to +L
  a0:= 1/L*int(f(x),x=-L..L);   # coefficients a and b
  a:= n-> 1/L*int(f(x)*cos(n*Pi*x/L),x=-L..L);
  b:= n-> 1/L*int(f(x)*sin(n*Pi*x/L),x=-L..L);
```

$$a := n \to \frac{\int_{-L}^{L} f(x)\cos\left(\frac{n\pi x}{L}\right)dx}{L} \qquad b := n \to \frac{\int_{-L}^{L} f(x)\sin\left(\frac{n\pi x}{L}\right)dx}{L}$$

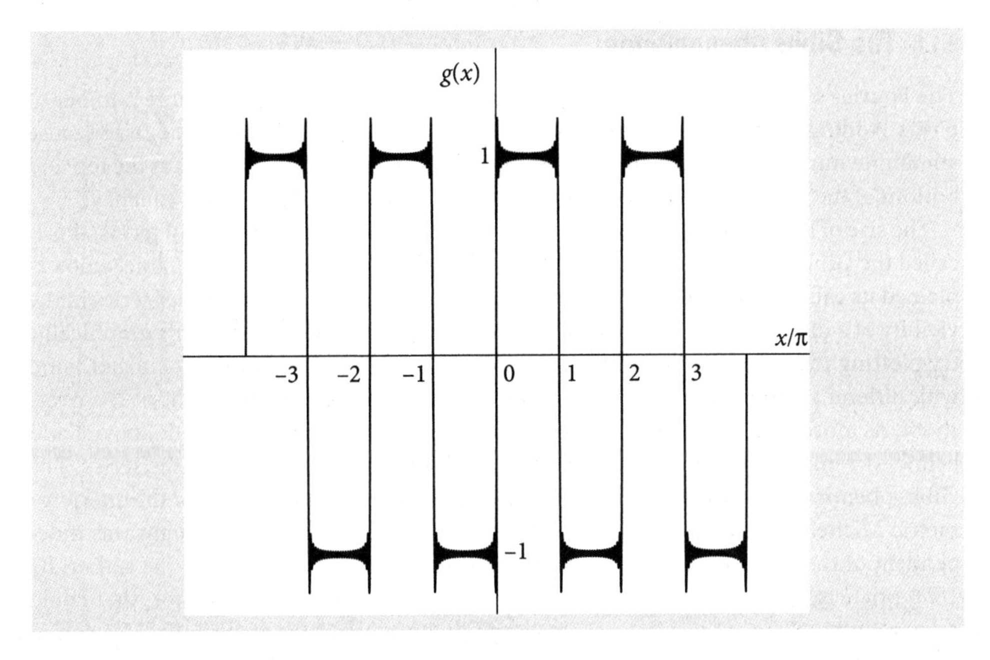

Fig. 9.5 The Fourier series for the square wave, signum or sgn(x) calculated to $n = 49$.

```
> # next calculate series make function FS
  FS:= (x,maxn)-> a0/2
 +add(a(n)*cos(n*Pi*x/L) + b(n)*sin(n*Pi*x/L),n=1..maxn);
```

$$FS := (x, maxn) \rightarrow \frac{1}{2}a0 + add(a(n)\cos(nx) + b(n)\sin(nx), n = 1..maxn)$$

```
  plot([ f(x),FS(x,60) ],x=-L..L);
```

Because the integrals are not evaluated before plotting, the calculation can be slow because Maple is first seeking an algebraic solution. To speed it up, place the integration inside the `evalf(..)` function. The instruction is, for example,

```
> a:= n-> evalf(1/L*int(f(x)*cos(n*x),x=-L..L) ):
```

The result of the calculation is shown in Fig. 9.6 and is compared with the original function, and for most of the curve the fit is tolerably good but not excellent. The main discrepancy is the overshoot, which is the Gibbs phenomenon and is described next. The generalized transform using orthogonal polynomials may make a better job with fewer functions, see Section 9.4 and question Q9.4.

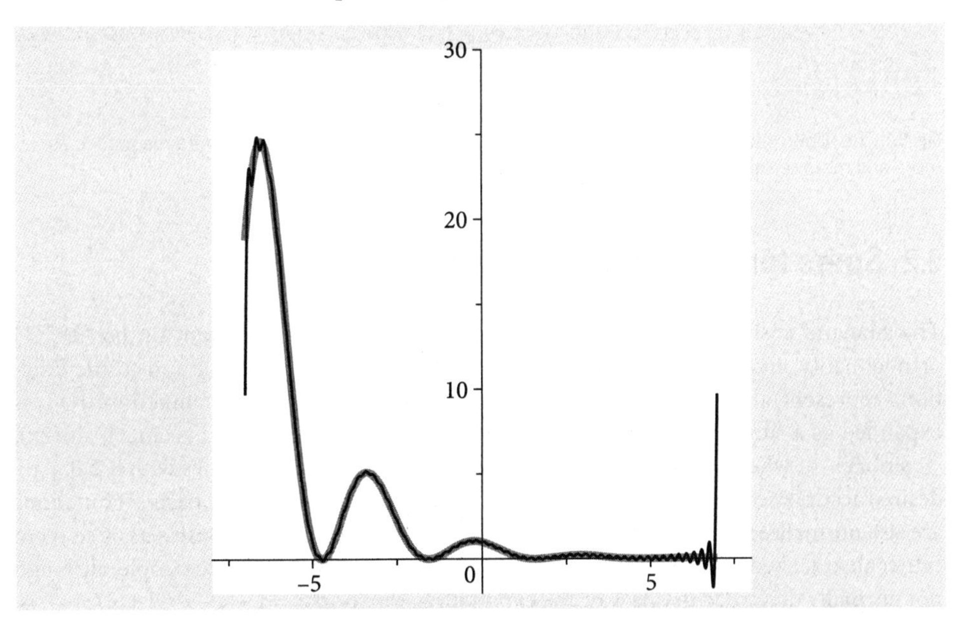

Fig. 9.6 The function $\exp(-x/2)\cos(x)^2$ as a thick grey line and its Fourier series representation.

9.1.8 The Gibbs phenomenon

The Fourier series of a square wave is shown in Fig. 9.5 with a relatively large number of terms. Adding more will improve the fit to the function, but it will never be exact because an infinite number of terms will be needed to follow the right-angled bends at the top and bottom of the wave. This angle effectively corresponds to an infinite sine frequency.

The size of the overshoot remains the same independent of the number of terms; this is called the Gibbs' phenomenon, after J. Willard Gibbs, of thermodynamics fame, who explained its cause as being due to the non-uniform convergence of the Fourier series in the vicinity of a discontinuity. It is difficult to prove, but may be seen quite easily graphically; try plotting the Fourier series approximation to the square wave from −0.2 to +0.2 and with different numbers of terms in the series, and then observe the height of the overshoot. As more terms are added, this becomes closer to zero, but its amplitude above 1 and below −1 remains the same, as shown in Fig. 9.7 for 10 and 100 terms in the series. The Gibbs phenomenon was first observed by Michelson who is more famous for the interferometric Michelson–Morley experiment that determined that the speed of light was independent of the position of the earth and thus disproved the hypothesis of the aether. By 1898 Michelson had constructed a machine, called the harmonic integrator, that could calculate up to 80 terms in a Fourier series and present the results graphically. He noticed the overshoot and wrote to Gibbs for an explanation thinking his machine was in error. You can see a photograph of this machine in *A Student's Guide to Fourier Transforms* (James 1995). This book also contains a clear introduction to Fourier transforms.

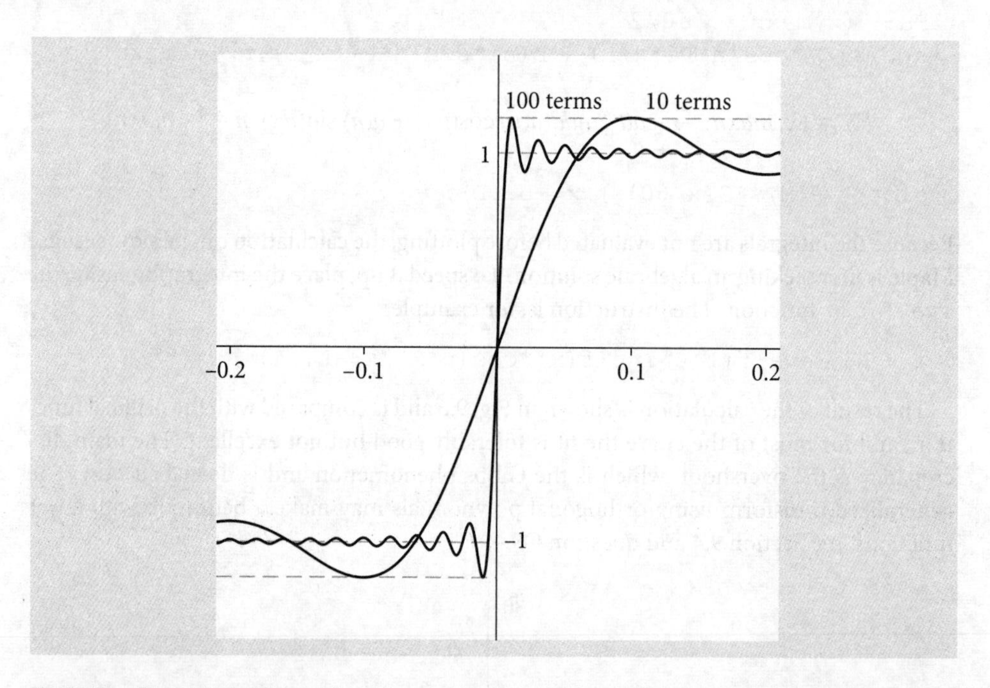

Fig. 9.7 The Gibbs phenomenon where the over- and undershoot remain at the same magnitude as more and more terms are added to the Fourier series.

9.2 Some formal points about the Fourier series

The sine and cosines making up the Fourier series have two important properties: (1) orthogonality, and (2) completeness. In the language of vectors, the sine and cosine functions represent a *complete orthogonal* basis set on which the target function, $f(x)$, is expanded as a sum of N terms. This basis set is of infinite length and is $\sin(x)$, $\sin(2x)$, $\cdots \sin(nx) \cdots$, where $n = 1, 2, \cdots$, and there is a similar set for the cosines with $n \ge 0$. Any desired accuracy can be achieved, provided that N is large enough. If the basis set functions are also normalized, the set is *orthonormal* rather than just orthogonal. Basis sets often seem rather abstract because we do not often need to use them explicitly. For example, although not normally described in this way, the exponential function $e^x = 1 + x + x^2/2! + x^3/3! \cdots$, is, by contrast, an expansion in basis set of x^n and the coefficients are $1/0!$, $1/1!$, $1/2!$, $1/3!$, $\cdots 1/n! \cdots$. This basis set is not orthogonal because the condition for this to apply is that the product of any two elements is zero when taken over the whole range of the set, $\pm\pi$ for sine and cosine. The dot product of any two orthogonal vectors is zero, see Chapter 6.2. Similarly for the sine and cosine basis set even though it is continuous, the condition is $\int_{-\pi}^{\pi}\sin(mx)\sin(nx)dx = \pi\delta_{nm}$, which is zero if $m \neq n$. This is not true of the coefficient of the x^n basis set of the exponential expansion, thus we cannot form a Fourier series based on this.

The great importance and usefulness of the Fourier series is that it represents the best fit, in a least-squares way, to any function $f(x)$ because $\int_{-L}^{L}[f(x) - g(x)]^2dx$ is minimized when $g(x)$ is the series expansion of $f(x)$.

A given set of functions, $S_n(x)$ is said to be *complete* if some other arbitrary function $f(x)$ can be expanded in the set of the S functions. If they have the same boundary conditions, this makes the functions complete and orthogonal in a given range, and in the case of the sine and cosine Fourier series, this range is $-\pi$ to π. The general series describing $f(x)$, is

$$g(x) = \sum_{n=1}^{\infty} q_n S_n(x), \tag{9.16}$$

q_n being constants that are some functions of the target function $f(x)$. If the S functions are orthonormal, then, when n and m are integers, $\int S_n^*(x)S_m(x)dx = \delta_{nm}$.

In modern mathematics, the term 'Fourier series' does not refer just to the original sine and cosine series, or their complex exponential representation, but to a series formed by other functions that form a complete orthogonal basis set. Often the term *generalized Fourier series* is used to describe these, but this is not universal. The sine or cosine functions are not unique in forming series and many other functions could be used provided that they can form an orthogonal set. Other such functions include the Hermite polynomials, used to describe the harmonic oscillator wavefunctions, and the Legendre and Chebychev polynomials. In Section 9.4 it is shown how these can also be used to form series that describe arbitrary target functions $f(x)$.

9.3 Integrating series

The series for x^4 can be obtained from the Fourier series for x^3 by integrating term by term; in addition, integration can lead to a better representation of a function with the same number of terms in the summation. The algebraic result will be different from that of a direct series for x^4, but should be just as good a representation. The series for x^3 is calculated first and as this is an odd function, all the even cosine terms are zero; therefore, all the a coefficients are zero and the series only contains sine or b terms of the form

$$b_n = \frac{2}{\pi}\int_0^{+\pi} x^3 \sin(nx)dx.$$

This and similar integrals can be integrated 'by parts' or the sine converted to an exponential and then integrated; in either case the sine or exponential part is integrated first, so that the power of x is reduced in the second term of the 'by parts' integration. Using Maple produces the result for b_n of

```
> assume(n, integer):
> 1/Pi*Int(x^3*sin(n*x),x=-Pi..Pi): %= value(%);
```

$$\frac{\int_{-\pi}^{\pi} x^3 \sin(nx)dx}{\pi} = \frac{2\,(-1)^{1+n}(-6+\pi^2 n^2)}{n^3}$$

and the `assume(n,integer)` is needed to obtain a simple result. The expansion of x^3 over the range $-\pi < x < \pi$ is, with $a_0 = 0$ and $a_n = 0$,

$$x^3 \approx g(x^3) = \sum_{n=1} b_n \sin(nx) = 2\sum_{n=1}\frac{(-1)^{n+1}}{n^3}(n^2\pi^2 - 6)\sin(nx)$$

$$= 2(\pi^2 - 6)\sin(x) - \left(\frac{2\pi^2 - 3}{2}\right)\sin(2x) + \cdots$$

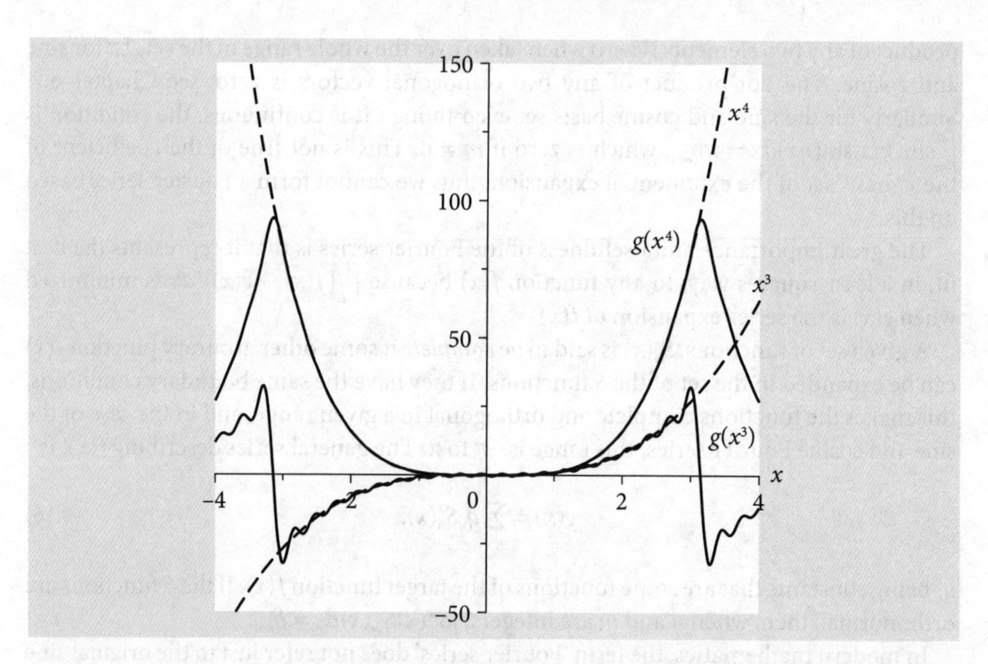

Fig. 9.8 The x^3 and x^4 are shown as dashed lines. The wiggly solid curve shows the result of having only 20 terms in the x^3 summation. The underlying oscillation of the trigonometric terms is obvious, even though the series clearly approximates x^3. The $g(x^4)$ series, after integrating the x^3 Fourier series, is a considerably better fit to x^4 that is than the fit to x^3.

The series expansion is plotted in Fig. 9.8; notice how the difference in the value of the series for x^3 is greatest at the limits $-\pi$ and π, and how the series repeats itself. The Fourier series is zero at both $-\pi$ and π because it assumes that the function x^3 is periodic; thus just before π its value is π^3, and just after $-\pi^3$, and this leads to an overshoot.

Integrating term by term produces a new series, which is

$$x^4 \approx g(x^4) = 2\int_0^x \sum_{n=1} \frac{(-1)^{n+1}}{n^3}(n^2\pi^2 - 6)\sin(nx)dx$$

$$= 4 \times 2\sum_{n=1} \frac{(-1)^{n+1}}{n^4}(n^2\pi^2 - 6)[1 - \cos(nx)],$$

and the initial 4 arises because the integration of x^3 is $x^4/4$. The curve for x^4 and its Fourier series is shown in Fig. 9.8. The fit is good up to the limit $\pm\pi$, the limit of the initial Fourier series. The x^4 series $g(x^4)$ is smoother than the x^3 and also a better fit to the curve, and this may be attributed to the integration, which measures the area under a curve and has the effect of smoothing the function.

The Fourier series can also be used to evaluate summations; for example the expansion of x^2 in the range $-\pi$ to $+\pi$, equation (9.15), gives

$$x^2 = \frac{\pi^2}{3} - 4\left(\cos(x) - \frac{\cos(2x)}{2^2} + \frac{\cos(3x)}{3^2} \cdots\right).$$

When $x = \pi/2$, the series produces the result

$$\frac{\pi^2}{12} = 1 - \frac{1}{4} + \frac{1}{9} - \frac{1}{16} + \cdots \frac{(-1)^n}{n^2} + \cdots$$

provided that the series limit is taken to infinity. In fact, this series converges very slowly; the alternating sign in $\pm 1/n^2$ terms ensures this, and about 360 terms are needed to obtain a value for π accurate to five decimal places: not really a good way to calculate π. When $x = \pi$ the series is

$$\frac{\pi^2}{6} = 1 + \frac{1}{4} + \frac{1}{9} + \frac{1}{16} + \cdots \frac{1}{n^2} + \cdots$$

which converges more rapidly as all the terms have the same sign. Many other unusual summations can be achieved using different Fourier series; however, for us they are only curiosities.

9.3.1 Questions

Full solutions are available at **www.oxfordtextbooks.co.uk/orc/beddard.**

Q9.1 Confirm equation (9.3), describing b_n by using a similar calculation to that used to derive coefficients a_n.

Q9.2 Calculate the Fourier series for sin(x) over the range $-L < x < L$.

Q9.3 **(a)** Calculate the Fourier series of $\pi/2 - x$ over the interval $-\pi$ to π. Choose at least 10 terms in the series. If you choose more terms, observe that the overshoot persists but is not of constant height as it is in the square wave.

(b) Using the general Algorithm 9.1, test it with $f(x) = x - x^3$, $|x|$, $x^2e^{-x^2/2}$ and tanh(x), and plot the graphs. Generalize the code to make the limits $\pm L$ and recalculate over the range ± 20.

Strategy: (a) the function $\pi/2 - x$ is neither odd nor even and both a and b coefficients will have to be calculated.

9.4 Generalized Fourier series with orthogonal polynomials

The Fourier series expands a function in sine and cosines. In the language of vectors and matrices, the sine and cosines form a basis set in which the function f is expanded. The essential property that any basis set of functions must have is orthogonality. If φ is such a function, then the basis set is the functions $\varphi_1, \varphi_2, \varphi_3 \cdots$ and the condition for orthogonality is

$$\int_a^b \varphi_m(x)^* \varphi_n(x)dx = c_n\delta_{n,m} \tag{9.17}$$

where c is a constant and δ is the delta function and is zero if $n \neq m$ otherwise it is unity. The asterisk indicates the complex conjugate if the function φ is complex. If the functions are normalized as well as being orthogonal, then $c_n = 1$. The range of the integral, a to b, depends on the type of function which might be ± 1 or $\pm\infty$. Some orthogonal polynomials, such as the Hermite polynomials, are so defined that a weighting function w, is needed to make them orthogonal. In this case, equation (9.17) is redefined as

$$\int_a^b \varphi_m(x)^* \varphi_n(x)w(x)dx = c_n\delta_{n,m}. \tag{9.18}$$

Suppose $f(x)$ is the function we want to expand as a series in the φ's. The generalized Fourier series is defined in a simpler way than the normal sine/cosine series, because only one function is involved; hence

$$f(x) = \sum_{n=0}^{\infty} a_n\varphi_n \tag{9.19}$$

The next task is to find the coefficients a_n, and this is done by multiplying both sides of (9.19) by φ and any weighting w, and integrating.

$$\int_a^b f(x)\varphi_m(x)w(x)dx = \int_a^b \sum_{n=0}^{\infty} a_n\varphi_n(x)\varphi_m(x)w(x)dx = \sum_{n=0}^{\infty} a_n c_n \delta_{n,m} = a_n c_n.$$

The summation is taken outside the integral in the second step, because it depends on n not x, and the orthogonality (9.18) is used to solve the integral. The summation disappears in the last step because of the property of the δ function. The coefficients a_n are therefore

$$a_n = \frac{1}{c_n}\int_a^b f(x)\varphi_n(x)w(x)dx \tag{9.20}$$

Table 9.1 The range of some functions and orthogonal polynomials and their weighting factors.

Functions	Symbol $f(x)$	Range	Weighting* $w(x)$	normalization c_n
Sine/cosine		$\pm\pi$ or $\pm L$	1	π or L
Complex exponential	$e^{in\pi x/L}$	$\pm L$	1	L
Legendre	$P_n(x)$ or $P(n, x)$	± 1	1	$2/(2n+1)$
Hermite	$H_n(x)$ or $H(n, x)$	$\pm\infty$	e^{-x^2}	$2^n n!\sqrt{\pi}$
Laguerre	$L_n(x)$ or $L(n, x)$	$0 \rightarrow \infty$	e^{-x}	1
Chebychev	$T_n(x)$ or $T(n, x)$	± 1	$1/\sqrt{(1-x^2)}$	π when $n = 0$
(1st kind)				$\pi/2$ when $n \neq 0$

*If the variable x is an angle θ then the volume element for the integration has to be changed from dx to $\sin(\theta)d\theta$ and in these cases $\sin(\theta)$ must multiply the weighting function.

and this equation and (9.19) form the equations for the generalized Fourier transform. The only way that information about the function f enters the calculation, is *via* the coefficients a_n.

The similarity of (9.19) to the expansion of a vector in its basis set is clear. In a vector equation in three dimensions, we might write $V = \sum_{m=0}^{3} v_m \boldsymbol{i}$ where $\boldsymbol{i}$ is the unit vector along the x-axis, and then the v_m are the projections of the vector V onto this axis. If there were n dimensions, which clearly cannot be pictured graphically if $n > 3$, then the v_m would be projections of V onto the mth axis. Similarly, in (9.19), the a_n coefficients are the projections of the function f onto the φ, but in this case there are more than three 'axes'.

The Legendre polynomial $P_n(x)$ is used as an example of a basis set function with which to calculate the generalized Fourier series of $e^{-x}\cos^3(x)$ over the range ± 1. These polynomials appear in problems when an electric charge, perhaps on an ion, is close to other charges, such as from a dipole, or when an electron feels the effect of other electrons and nuclei. All the orthogonal polynomials have several different recursion formulae, and one of the most useful for the Legendre is

$$nP_n(x) = (2n-1)xP_{n-1}(x) + (n-1)P_{n-2}(x)$$

for $n = 2, 3, 4, \cdots$ with $P_0(x) = 1$ and $P_1(x) = x$, and the polynomials are defined only over the range $x = \pm 1$. The weighting function is 1, and their normalization constant $c_n = 2/(2n+1)$. Although it is not apparent from the formulae, these polynomials are oscillating functions with an increasing number of nodes as n increases. The Fourier series using Legendre polynomials instead of sine/cosine is sometimes called the Fourier–Legendre series.

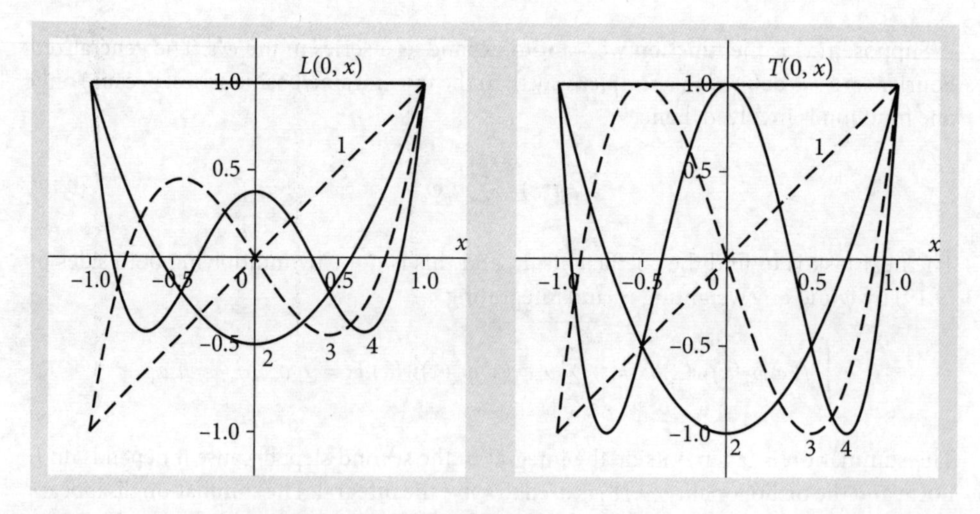

Fig. 9.9 Left: Legendre polynomials. Right: Chebychev polynomials. The numbers indicate the order, n, of the polynomial, $L(n, x)$ or $T(n, x)$. The even numbered ones are drawn with a solid line the odd, dashed. Notice the odd–even nature of the functions corresponds to the index number. Both polynomials are limited to the range $-1 \leq x \leq 1$.

The Fourier calculation is greatly aided by using Maple, especially as it has its own built-in Legendre polynomial function. All the in-built orthogonal polynomials are enabled by using `with(orthopoly)` and the Legendre is called `P(n,x)`. (Maple is unforgiving here: you do need to know beforehand to look for the `orthopoly` instruction in the help manual). The calculation shows that the function is well approximated after 15 terms in the summation Fig. 9.10; 10 terms produce a poor fit, with oscillatory features at x values approaching 1. The integration can be made faster, by surrounding it by an `evalf` statement to force a numerical result; e.g.

```
a:= n-> evalf( int(f(x)*..etc):
```

Algorithm 9.2 Generalized Fourier series using Legendre polynomials

```
> restart: with(orthopoly);
> # Legendre polynoms are P(n,x), range -1 to +1
> R:= 1.0:                         # range of calculation
  f:= x-> cos(x*3)^3*exp(-x);      # function to approximate
  w:= x-> 1.0:                     # weighting function
  c:= n-> 2/(2*n+1);               # normalisation const
  a:= n-> int(f(x)*P(n,x)*w(x),x=-R..R)/c(n): # a coeffs
  ser:= x-> add(a(n)*P(n,x),n=0..15); # series sum
  plot([f(x),ser(x)],x=-R..R, thickness=[5,1]);
```

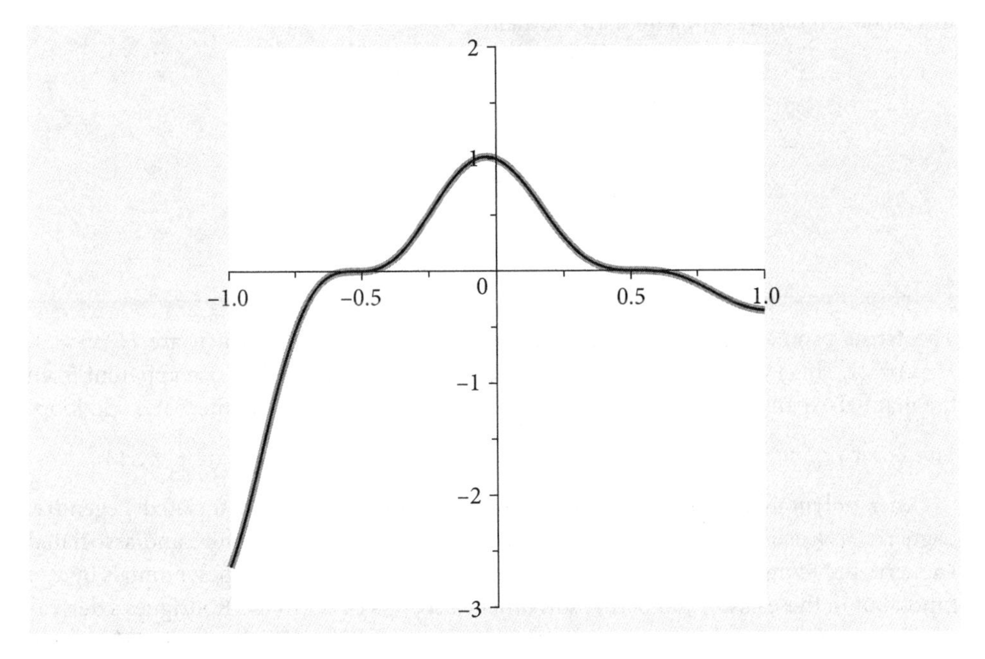

Fig. 9.10 The function $f(x) = e^{-x}\cos^3(3x)$ (thick grey) and its generalized Fourier series (black).

9.4.1 Generating Functions for orthogonal polynomials

The named orthogonal polynomials are the solutions of specific differential equations. They occur widely in physics and chemistry; the Hermite polynomials describe the harmonic oscillator wavefunctions; the Legendre electric charge distributions; and the associated Laguerre the electron distribution in the H atom. The spherical harmonic polynomials are based on the Legendre polynomials, with $x = \cos(\theta)$, and are used to describe the angular momentum in atoms and molecules; they define the shapes of atomic orbitals, the rotational motion of molecules, and the heat flow around spheres.

The orthogonal polynomials, such as the Legendre polynomial used in the previous section, can be produced from a series expansion of a *generating function*. These functions have two variables x and u; expansion is made in the powers of u and the polynomials are the factors belonging to the terms in u^n or $u^n/n!$. A typical series expansion producing polynomial $p_n(x)$ is

$$f(x, u) = p_0(x) + p_1(x)u + p_2(x)\frac{u^2}{2!} + \cdots = \sum_{n=0} p_n(x)\frac{u^n}{n!}.$$

This definition has to be checked, however, as some polynomials do not require u^n to be divided by $1/n!$. As an example, the Hermite polynomials can be calculated as the coefficients of the $u^n/n!$ in the expansion of $f(x, u) = e^{-u^2+2xu}$, and are the coefficients in the series $e^{-u^2+2xu} = \sum_{n=0} H_n(x)\dfrac{u^n}{n!}$. The expansion is

$$\begin{aligned} e^{-u^2+2xu} &= e^{-u^2}e^{+2xu} \\ &= \left(1 - u^2 + \frac{u^4}{2!} - \frac{u^6}{3!} + \frac{u^8}{4!} - \cdots\right)\left(1 + 2xu + 4x^2\frac{u^2}{2!} + 8x^3\frac{u^3}{3!} + \cdots\right) \\ &= 1 + 2xu + (4x^2 - 2)\frac{u^2}{2!} - (8x^3 - 12x)\frac{u^3}{3!} + \cdots \end{aligned}$$

This is clearly going to be an easier calculation if Maple is used to order the terms,

```
> series(exp(-u^2+2*x*u),u,5);
```

$$1 + 2\,x\,u + (-1 + 2\,x^2)u^2 + \left(-2\,x + \frac{4}{3}x^3\right)u^3 + \left(\frac{1}{2} - 2x^2 + \frac{2}{3}x^4\right)u^4 + O(u^5)$$

But now the terms need to be multiplied by $n!$ to make them consistent with the definition. A far better way is to extract the coefficients directly, using the `coeff(...)` instruction, multiply by $n!$ and then list them

```
> f01:= series(exp(-u^2+2*x*u),u,10):                    # up to u^10
  for n from 0 to 4 do coeff(f01,u, n )*n!; end do;
```

$$1$$
$$2\,x$$
$$-2 + 4\,x^2$$
$$-12\,x + 8\,x^3$$
$$12 - 48\,x^2 + 16\,x^4$$

The terms produce the Hermite polynomials, the first three of which are $H_0(x) = 1$, $H_1(x) = 2x$, $H_2(x) = 4x^2 - 2$ as described in chapter 1.7.1 and which is also apparent from the manual expansion. If you want a tidy output, put the following in the `for..do` loop.

```
>printf("%a %s %a\n",Hermite(n,x),"=",coeff(f01,u,n)*n!);
```

Other polynomials that are frequently met are the Legendre, associated Legendre, Laguerre, associated Laguerre, and Chebychev. The associated Legendre and associated Laguerre polynomials are obtained by differentiating their respective polynomials by x, n times, but in these cases, and perhaps in others also, it is easier to use Rodrigues's derivative formula or one of the recursion equations to generate the polynomials. See Margenau & Murphy (1943) and Arkfen (1970) for a full discussion of these polynomials and generating functions.

The generating function definitions, and a derivative formula for integral n and k, are shown in the table. The associated Legendre generating function is omitted because of its complexity.

	Generating function	Derivative formula
Legendre	$\dfrac{1}{\sqrt{1-2xu+u^2}} = \sum_{n=0} P_n(x)\dfrac{u^n}{n!}$	$P_n(x) = \dfrac{1}{2^n n!}\dfrac{d^n}{dx^n}(x^2-1)^n$
Associated Legendre	(too complex to use)	$P_n^k(x) = (1-x^2)^{k/2}\dfrac{d^k}{dx^k}P_n(x)$
Laguerre	$\dfrac{e^{-xu/(1-u)}}{1-u} = \sum_{n=0} L_n(x)u^n$	$L_n(x) = \dfrac{e^x}{n!}\dfrac{d^n}{dx^n}(e^{-x}x^n)$
Associated Laguerre	$(-1)^{k+1}\dfrac{e^{-xu/(1-u)}}{(1-u)^{k+1}} = \sum_{n=0} L_n^k(x)u^n$	$L_n^k(x) = \dfrac{e^x}{x^k n!}\dfrac{d^n}{dx^n}(e^{-x}x^{n+k})$
Hermite	$e^{-u^2+2xu} = \sum_{n=0} H_n(x)\dfrac{u^n}{n!}$	$H_n(x) = (-1)^n e^{x^2}\dfrac{d^n}{dx^n}e^{-x^2}$

It is worth noting in passing that generating functions play a more general role in mathematics than is described here and are used to form various distribution functions and series of numbers. The generating function $1/(1-x)$ produces a square wave $-1, +1$ repeating sequence; the function $1/(1-ax)$ produces the sequence of increasing integer powers of a; and $x/(1-x-x^2)$ generates the Fibonacci sequence. For example,

```
> f01:= series(x/(1-x-x^2),x,20):
  for n from 1 to 18 do coeff(f01,x,n); end do;
 1 1 2 3 5 8 13 21 34 55 89 144 233 377 610 987 1597 2584
```

Other generating functions can be used to work out the number of ways of selecting several items from a list, where the same item can be picked many times.

9.4.2 Questions

Full solutions are available at **www.oxfordtextbooks.co.uk/orc/beddard**.

Q9.4 **(a)** Calculate the series expansion of

(a) $f = \cos^2(x)e^{-x/2}$, using Hermite polynomials over the range ±7, and

(b) $f = x + x^3/10 - 2x^7$ over the range ±1, using Chebychev polynomials.

Use the Maple functions for the polynomials. The polynomial package has to be loaded first using `> with(orthopoly):`. The Hermite polynomials are called `H(n,x)`, the Chebychev `T(n,x)`.

Strategy: Use Algorithm 9.2, taking care to add the correct weighting and normalization terms, these are given in the text also in Section 9.4. The calculation is a little awkward for the Chebychev polynomials, because an exception has to be made for the term $n = 0$. Speed up the calculation by using `evaf(int()..)` when calculating the coefficients.

Q9.5 **(a)** Use the generating function, **(b)** the derivative formula and **(c)** the recursion formula

$$(n+1)L_{n+1}(x) = (2n+1-x)L_n(x) - nL_{n-1}(x), \quad L_0(x) = 1, \quad L_1(x) = 1 - x.$$

to confirm that these give the same results as the first few Laguerre polynomials. Use Maple as necessary.

Q9.6 The *associated Legendre* polynomials are obtained by repeated differentiation of the Legendre polynomials, defined in Section 9.4,

$$P_l^m(x) = (-1)^m(1-x^2)^{m/2}\frac{d^m}{dx^m}P_l(x)$$

where $P_l(x)$ is the Legendre polynomial. The values of m and l are related as $0 \le m \le l$, meaning that m cannot exceed l or be negative. These functions are used to obtain the *spherical harmonics*, which are the functions that describe the shapes of the atomic orbitals, s, p, d, etc. and other angular momentum properties of atoms and molecules. The spherical harmonics are defined as

$$Y_{l\,m}(\theta, \varphi) = \sqrt{\frac{(2l+1)(l-m)!}{4\pi(l+m)!}}P_l^m(\cos(\theta))e^{im\varphi}. \tag{9.21}$$

When m is negative then $Y_{l,-m}(\theta, \varphi) = (-1)^m Y^*_{l,m}(\theta, \varphi)$ where the * indicates the complex conjugate and $x = \cos(\theta)$. If dealing with atomic orbitals, then l would be the overall angular momentum where $l = 0$ for s orbitals, 1 for p and 2 for d. The projection, magnetic, or z-value quantum number is m and defines the orientation in space. In the spherical harmonics, $-l \le m \le l$.

There is a good recursion formula on l for the associated Lagrange polynomial, which is,

$$(l-m)P_l^m = x(2l-1)P_{l-1}^m - (l+m-1)P_{l-2}^m \tag{9.22}$$

and (x) is suppressed for clarity. This formula means that m is chosen and the polynomial with different l values is calculated. Note also that to use this equation $l - 2 \ge 0$. A starting value can be found, when m and l are the same with the relationship (see Prest et al. 1986, pp. 180–2),

$$P_m^m = (-1)^m(2m-1)!!(1-x^2)^{m/2}.$$

A double factorial, in general, $n!!$, reduces the index by 2 each time instead of 1 as in the normal factorial. The series is thus $n!! = n(n-2)(n-4)\cdots(6)(4)(2)$ if n is even and if n is odd, as is $2m-1$, the series is one of odd numbers ending in 1.

A second starting function, is obtained by substituting $l = m+1$ into the recursion equation and letting $P^m_{m-1} = 0$. To form the spherical harmonics, use $x = \cos(\theta)$. The formulae are only valid if $|x| \le 1$.

Using Maple if necessary, calculate the associated polynomials and the spherical harmonics with $l = 0, 1$, and 2 and their associated $\pm m$ values.

Strategy: The first step is to calculate the second starting function. Using the information given, and substituting $l = m+1$ into (9.22) produces

$$P^m_{m+1} = x(2m+1)P^m_m - 2mP^m_{m-1}$$

and, with the last term defined as zero, gives

$$P^m_{m+1} = x(2m+1)P^m_m$$

as the second starting function. Note that in the polynomials $0 \le m \le l$, so if m is zero so is l. Use the Maple built-in function `doublefactorial(..)`. Prest et al. (1986, p. 182) gives a routine that could be used as a basis for your calculation.

9.5 Fourier transforms

9.5.1 Motivation and concept

Fourier transforms are of fundamental importance in the analysis of signals from many types of instruments; these range from MRI and CT scan imaging to seismology. Usually, the data, which might be a string of values taken at many sequential times, is transformed to allow the frequencies present be displayed and analysed. More fundamentally, the measurement of infrared and NMR spectra produce data that is the transform of the spectrum, and similarly, in X-ray crystallography, the image of spots produced on the detector is the Fourier transform of the gaps between the planes of atoms in a crystal. Everyone is familiar with the interference pattern produced by light passing through a pair of slits; this is the (spatial) Fourier transform of the two slits. Even in everyday life, Fourier transforms are important because they are used to produce the images observed in a digital television and in most other forms of digital information processing.

Although we concentrate on Fourier transforms, they are only one in a class of integral transforms. The Abel transform is an integral transform that is used to recover the three-dimensional pattern from its two-dimensional image. It is used in such diverse areas as astronomy and the study of the photo-dissociation pathways of molecules. In photo-dissociation experiments, the fragments (atoms, ions, electrons) are spatially dispersed depending on where the breaking bond is pointing at the instant of dissociation. Their image is captured on a camera and by transforming this, the geometry of the dissociation process can be determined (Whittaker 2007). Other transforms are the Hilbert, used in signal processing, and the Laplace, used to solve differential equations.

Folklore has it that Fourier transforms are formidably difficult and abstruse things. We know that they form the basis of the FTIR and NMR instrument, but secretly hope that nobody asks us how or why! In fact, Fourier transforms are quite straightforward but must be treated with respect. We are used to seeing the NMR or IR spectrum as a set of lines at different fixed frequencies and feel comfortable with this, but the raw data produced is a wiggly signal in which the information needed is almost totally obscured. This makes the process of unravelling this in a Fourier transform seems mysterious: 'I cannot understand the data so where does the spectrum come from?' Contrariwise, we are used to interpreting speech and music that are oscillating signals in time, and would not easily understand either of them if transformed and viewed as a continuously changing spectrum of frequencies.

We shall come back to this shortly but, briefly, a Fourier transform is an integral and therefore it can be evaluated by any of the methods used to solve integrals. The Fourier transform integral is one of several types of *integral transforms* that have the general form

$$g(k) = \int f(x)G(k, x)dx. \tag{9.23}$$

The 'transformed' function is g and the function being transformed is f. The algebraic expression G is called the *kernel* and this changes depending on the type of transform, Fourier, Abel, etc. The exact form of the kernel is also described later. However, whatever form the transform takes it *always* occurs between pairs of conjugate variables, which are x and k in equation (9.23). Often these conjugate pairs are time (seconds) and frequency (1/seconds), or distance and 1/distance. The reciprocal relationship between variables is why the transform converts time into frequency, changing, for example, an oscillating time profile into a spectrum. A second property is that these integral transforms are reversible, also called *invertible*, which means that f can be changed into g and g can be changed into f depending on which one we start with.

Solving the Fourier transform integral both algebraically and numerically will be described starting in Section 9.5.5, but first the role of the Fourier transform in FTIR and NMR experiments, and in X-ray crystallography is outlined.

9.5.2 The FTIR instrument

The Fourier transform infrared (FTIR) spectrometer directly generates the Fourier transform of the spectrum by mechanically moving one mirror of a Michelson interferometer and measuring the signal generated by the interference of the two beams on the detector. Fig. 9.11 shows a (simulated) example of the raw data from the instrument and the IR spectrum produced after this is transformed. After transformation, the displacement from the centre of the interference signal is changed into reciprocal distance or wavenumbers, cm^{-1}, which is proportional to the IR transition frequency.

The FTIR spectrometer is an interferometer, therefore, the waves that have travelled down each of its arms are combined on the detector and this measures the intensity or the square of the wave's amplitude, Fig. 9.12. Constructive interference occurs when the path length in both arms differ by zero or a whole number of wavelengths; destructive interference occurs when they are exactly out of phase and the difference in length is an odd multiple of half a wavelength. If only one wavelength is present, changing the path-length Δ would make the signal on the detector change sinusoidally. The 'coherent' broadband infrared 'light' from the source contains many wavelengths, and at a given path-length, some constructively and some destructively interfere, but the signal is greatest when both paths are the same. The relative path-length of the two arms of the interferometer can be changed by mechanically moving one mirror; the full interference pattern is mapped out as a function of path length and this pattern decreases in an oscillatory manner to some constant, but not zero value, as the difference in path length increases. This is shown in the

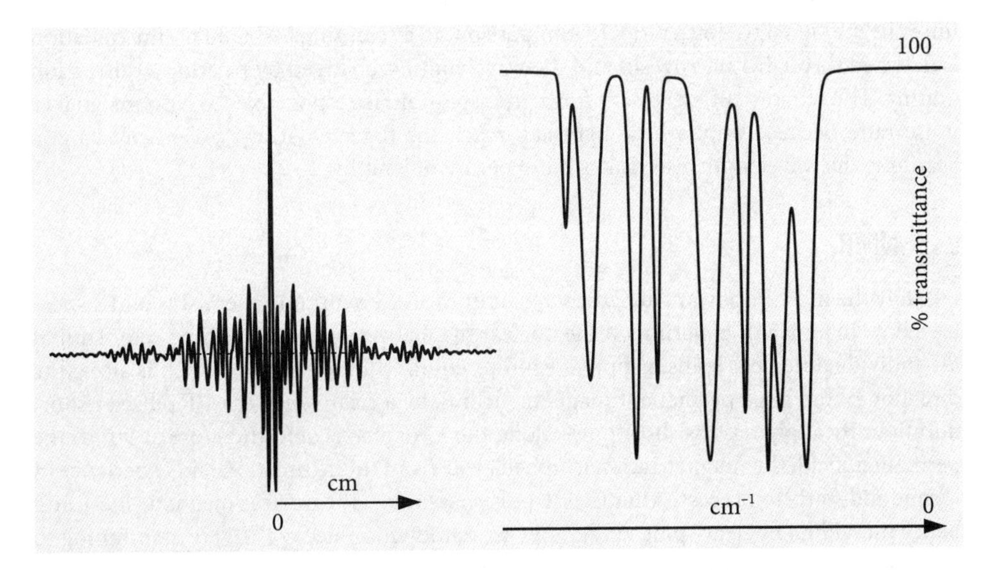

Fig. 9.11 Left: A simulated Fourier transform as might be produced directly by an FTIR spectrometer. Right: the IR spectrum after Fourier transforming and converting into transmittance.

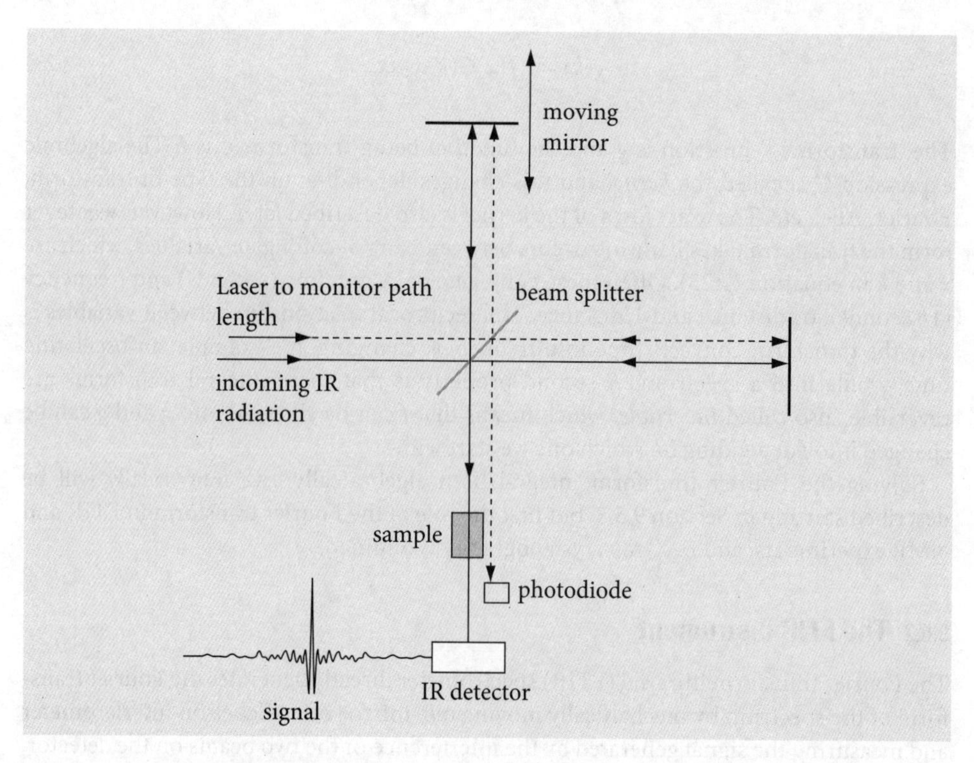

Fig. 9.12 Schematic of an FTIR spectrometer as an interferometer. The laser is used to measure the relative distance of the two arms and need not pass through the sample.

left of Fig. 9.11. Because changing either path's length has the same effect, the signal is symmetrical about zero path difference.

When the sample is placed in the beam, it absorbs only some frequencies depending on the particular nature of the sample, which results in a change in the signal on the detector. When this interference signal is subtracted from that obtained without the sample and is transformed, the infrared absorption spectrum is produced. The distance the mirrors are moved is accurately determined by using a visible laser that follows the same path in the interferometer, but does not pass through the sample. This laser produces an interference pattern on a second (photodiode) detector; the number of fringes passed as the arm of the interferometer moves is counted, and this is used to determine how far one mirror has moved relative to the other.

The FTIR spectrometer has the multiplex (Fellgett) advantage over a wavelength-scanning instrument, because all wavelengths are simultaneously measured on the detector, which also receives a large and virtually noise free signal. Both of these factors improve the signal to noise ratio. In comparison, in a scanning instrument, the radiation is detected through a narrow slit and the wavelength is changed by rotating a diffraction grating. The narrow slit, necessary for high resolution, is responsible for a poor signal to noise ratio because only a little light can reach the detector at any given wavelength. Scanning the wavelength also makes the experiment lengthy.

9.5.3 NMR

Possibly the most important analytical technique for the synthetic chemist is NMR spectroscopy. In an NMR experiment, the nuclear magnetization, which is the vector sum of the individual nuclear spins, is tipped from its equilibrium direction, which is along the direction of the huge permanent magnetic field B, by a relatively weak RF pulse of short duration. By applying this short pulse along the x- or y-axis, and therefore at 90° to the permanent field, the magnetization is tipped away from the z-direction and experiences a torque and starts to precess. After the RF pulse has ended, the nuclear magnetization, and hence individual nuclear spins, undergoes a free induction decay (FID) by continuing to precess about the permanent magnetic field B. The rotating magnetization, Fig. 9.13, is measured by the detecting coil in the x–y plane as an oscillating and decaying signal, which, when Fourier transformed, produces the NMR spectrum.

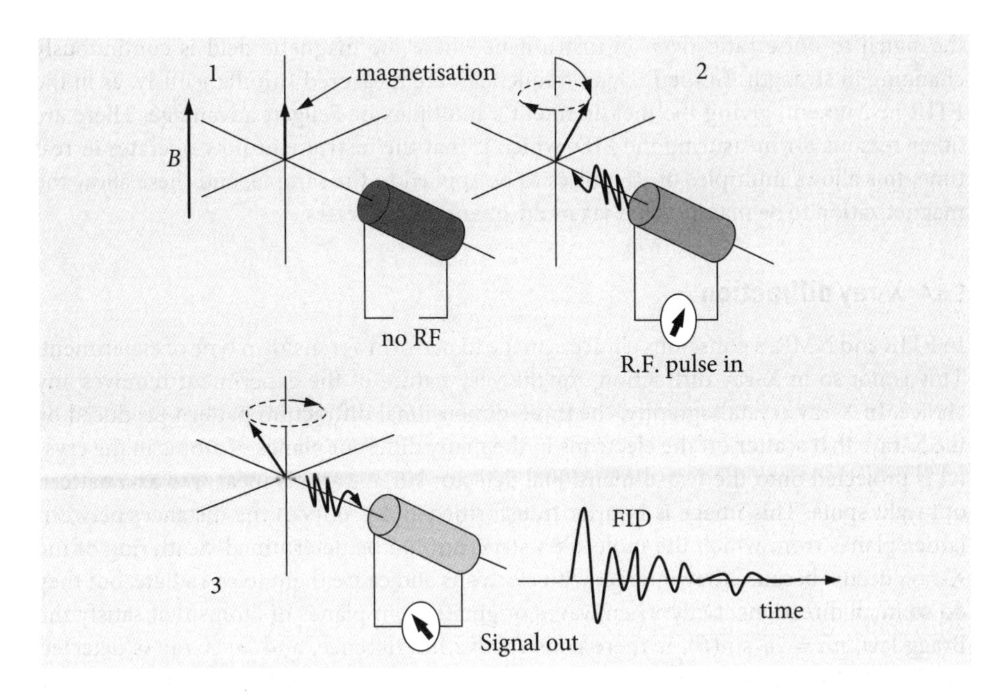

Fig. 9.13 The sequence of the magnetization and the FID produced during a basic NMR experiment.

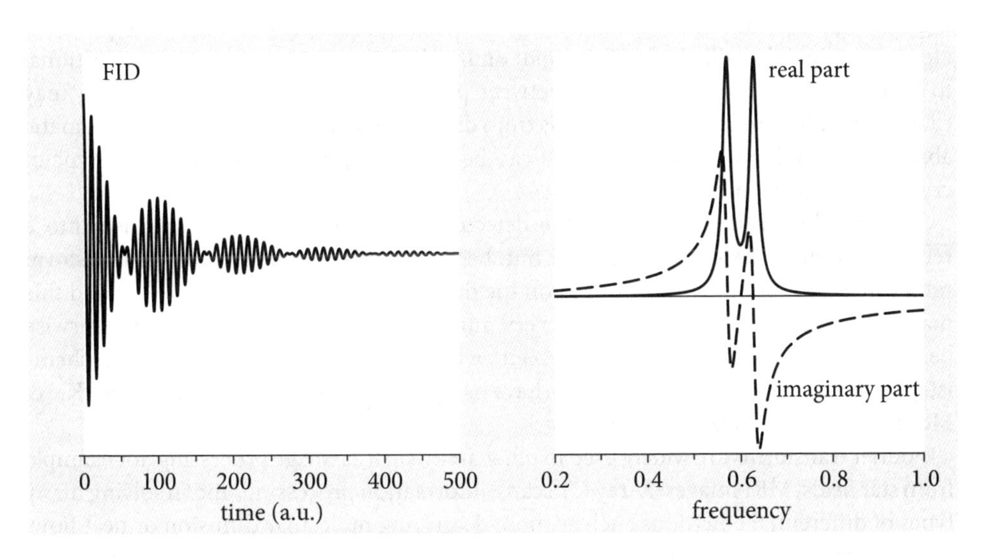

Fig. 9.14 A simulated FID and its real and 'imaginary' parts. The real part is the absorption spectrum or the normal NMR spectrum, the imaginary part (dashed line) the dispersion.

In this experiment, the oscillating and decaying signal is converted into reciprocal time or frequency, which is ultimately displayed as a *frequency shift* δ in ppm from a standard compound, such as tetramethylsilane. In a classical sense, it is possible imagine the rotating nuclear magnetization repeatedly passing in front of the detection coil, thereby inducing a current to flow in it as it does so, and which will cause the output signal to rise and fall. Many such magnetizations from the many groups of nuclear spins in different chemical environments produce many signals, resulting in a complicated oscillating FID. Figure 9.14 shows a synthesized NMR free induction decay of two spins with a frequency of 10 and 11 MHz, and the corresponding real and imaginary parts of the transform, which we will suppose is the NMR spectrum of two lines separated by 1 MHz. The RF pulse used to tip the magnetization contains many frequencies, as may be seen from the Fourier series of a square pulse, and simultaneously excites the nuclear spins in different magnetic environments in the molecule. The analysis of the spectrum provides information about the structure of the molecule, but not bond distances or angles unless sophisticated multiple pulse methods are used (Sanders & Hunter 1987; Levitt 2001).

In an NMR experiment, the data is obtained as an FID rather than directly as a spectrum because this increases the speed of data acquisition and, more importantly, increases the signal to noise ratio over an instrument where the magnetic field is continuously changing in strength. In the FID, all frequencies are measured simultaneously, as in the FTIR instrument, giving the measurement a multiplex or Fellgett advantage. There are other reasons for measuring the FID, which is that the instrument now operates in real time; this allows multiples of RF pulses to be applied to the sample, and these allow the magnetization to be manipulated via multi-quantum processes.

9.5.4 X-ray diffraction

In FTIR and NMR, a conscious choice is made to perform a transform type of experiment. This is not so in X-ray diffraction, for the very nature of the experiment removes any choice. In X-ray crystallography, the three-dimensional diffraction pattern produced by the X-rays that scatter off the electrons in the many different planes of atoms in the crystal is projected onto the two-dimensional detector surface and is measured as a pattern of bright spots. This image is Fourier transformed and produces the distances between lattice planes from which the molecule's structure can be determined. Scattering of the X-rays occurs because they interact with electrons and cause them to re-radiate, but they do so in all directions. Only when waves originate from planes of atoms that satisfy the Bragg law, $n\lambda = 2d\sin(\theta)$, is there constructive interference, and an X-ray is detected on the CCD detector. Everywhere else, there is destructive interference and no waves exist. The CCD detector is similar in nature to the one in a digital camera or mobile phone and the brightness of a spot is proportional to the amplitude squared of the X-rays arriving at that point. The atoms in a crystal form repeating unit cells and each set of planes of atoms, in principle, will produce one spot on the detector and in a position proportional to the reciprocal of the lattice spacing between planes. However, a crystal's symmetry may cause extra interference between X-rays from different planes, which produces systematic absences in the X-ray image and these can be use to distinguish one particular type of crystal lattice from another.

The two-dimensional image on the detector has to be Fourier transformed into a representation of the crystal structure but, because the absolute value of the transform rather than its amplitude is produced on the detector, phase information is lost and this makes the interpretation of the image very much more difficult than it would otherwise be. This is the origin of the 'phase' problem, which is usually outlined in physical chemistry textbooks, and ingenious methods have had to be devised to overcome this (McKie & McKie 1992; Giacovazzo et al. 1992).

Fourier transforms are widely used in other areas, such as image processing, for example from star fields, MRI images, X-ray CT scans, information processing, and in solving many types of differential equations such as those describing molecular diffusion or heat flow. These technologies show that it is essential to be familiar with Fourier transforms whether you are a chemist, physicist, biologist, or a clinician.

9.5.5 Linear transforms

The next few sections describe the Fourier transform in detail, but first some jargon has to be explained. Formally, a Fourier transform is defined as a linear integral transform of one function or set of data into another; see equation (9.23). The transform is *reversible*, or *invertible*, enabling the original function or data to be retrieved after an inverse transform. These last two sentences are in 'math-speak', so what do they really mean?

Integral simply means that the transform involves an integration as shown in equation (9.23). The word 'linear', in 'linear transform', means that the transform T has the property, when operating on two regular functions f_1 and f_2, that $T(f_1 + f_2) = T(f_1) + T(f_2)$. This means that the transform of the sum of f_1 and f_2 is the same as transforming f_1, and then transforming f_2 and adding the result. In addition, the linear transform has the property $T(cf_1) = cT(f_1)$ if c is a constant.

Reversible, or invertible, means that a reverse transform exists that reforms the initial function from the transform; formally this can be written as $f = T^{-1}[T[f]]$ if T^{-1} is the inverse transform. Put another way, if a function f is transformed to form a new function g, as $T[f] = g$, then the inverse transform takes g and reforms f as $f = T^{-1}[g]$. This might seem to be rather abstract, but is, in fact, very common. A straightforward example is the log and exponential functions, as they are convertible into one another as an operator pair: If T is the exponential operator $e^{(\ )}$, and x^2 is the 'function', then $T[x^2] = e^{x^2}$. The inverse operator T^{-1} reproduces the original function: $T^{-1}[T[x^2]] = x^2$ or, by substitution, $T^{-1}[e^{x^2}] = x^2$ if T^{-1} is the logarithmic operator ln( ) because $\ln(e^{x^2}) = x^2$. The Fourier transform is only a more complicated version of an operator than ln( ) or $e^{(\ )}$.

The Fourier transform can be thought of as changing or 'mapping' the initial function f to another function g, but in a systematic way. The new function may not look like the original, but however one might modify the transformed function g, when transformed back to f, it is as if f itself had been modified. Although it is common to use the word 'transform', the word 'operator' could equally well be used although this is not usual in this context. Conversely, a matrix when acting on another matrix or a vector, performs a linear transform, however, a matrix is usually called a linear operator.

9.5.6 The transform

The Fourier transform is used either because a problem is most easily solved in 'transform space', or, because of the way an experiment is performed, the data is produced in transform space and has then to be transformed back into 'real space'. This 'real space' is usually either time or distance; the transform space is then frequency (inverse time) or inverse distance. The time to frequency and the distance to inverse distance are both conjugate pairs of variables between which the Fourier transform operates. In practice, there are two 'flavours' of Fourier transforms. The simpler is the mathematical transformation of a function, such as a sine wave or exponential decay, the other is, effectively, the same process, but performed on real experimental data presented as a list of numbers. The latter is called the discrete Fourier transform (DFT).

The Fourier transform is always between pairs of conjugate variables, time $\leftrightarrow$ frequency, so that $\Delta t \Delta \nu = 1$ or distance $\leftrightarrow$ 1/distance. As the transform changes one variable into its conjugate, it is possible in simple cases to visualize what the spectrum will look like without actually doing the calculation. A long sine wave of one frequency has a Fourier transform that is one line set at the frequency of the wave. Because the transform is in reciprocal space, values near to zero on its horizontal axis correspond either to large values of distance or long times and vice versa. If there are two waves of different frequencies superimposed on one another, two lines will appear after transforming. So far, so good, but the length of the waves is not specified. Are they of finite length and so contain only a finite number of oscillations, or are they of infinite extent? If a sine wave is infinitely long, then only one line is observed in the transform, and will be of infinitesimal width and occur at the frequency of the sine wave. This line is a delta function. If the waves are turned on at some point and off again at another, then there are discontinuities at these points, and some additional frequencies must be associated with turning the signal on and off, which will appear in the transformed spectrum as *new* frequencies. Think of how a waveform is made up of a sum of sine waves of different frequency, Fig. 9.1. If a waveform is to be zero in some regions and not in others, then many waves have to be present to cancel out one another as necessary and these are the new frequencies needed. A broadening of the lines also occurs, because $\Delta t \Delta \nu = 1$ and if Δt, the length of the whole sine wave is finite, then $\Delta \nu$ has a width associated with it. This is observed in FTIR and NMR spectra, but the software provided with many instruments can be set to remove as much of this broadening as possible by *apodizing* the lines (Sanders & Hunter 1987).

The effect of Fourier transforming a short and a long rectangular pulse is shown below. The right-hand plots show the real part of the transform, which is a sinc function, $\text{sinc}(ax) \equiv \sin(ax)/ax$. The result of transforming is mathematically the same for both pulses, of course, but in a fixed frequency range the effect appears to be different. The short pulse has a wide central band set at zero and widely spaced side bands, which decay rapidly at frequencies away from zero and extend to infinity. The longer pulse has a narrower central band, also centred at zero, and higher frequency side bands than in the short pulse case; the results conform to $\Delta t \Delta \nu = 1$, i.e. short Δt with wide $\Delta \nu$ and vice versa.

If a pulse is turned on and off, as shown in Fig. 9.15, the transform must have frequencies associated with these changes. Again, think of the pulse being made of many terms in a Fourier series. Fig. 9.1 shows a few of the terms, but the more of these there are each with a different frequency, the better is a sharp edge or pulse defined. The oscillations in the

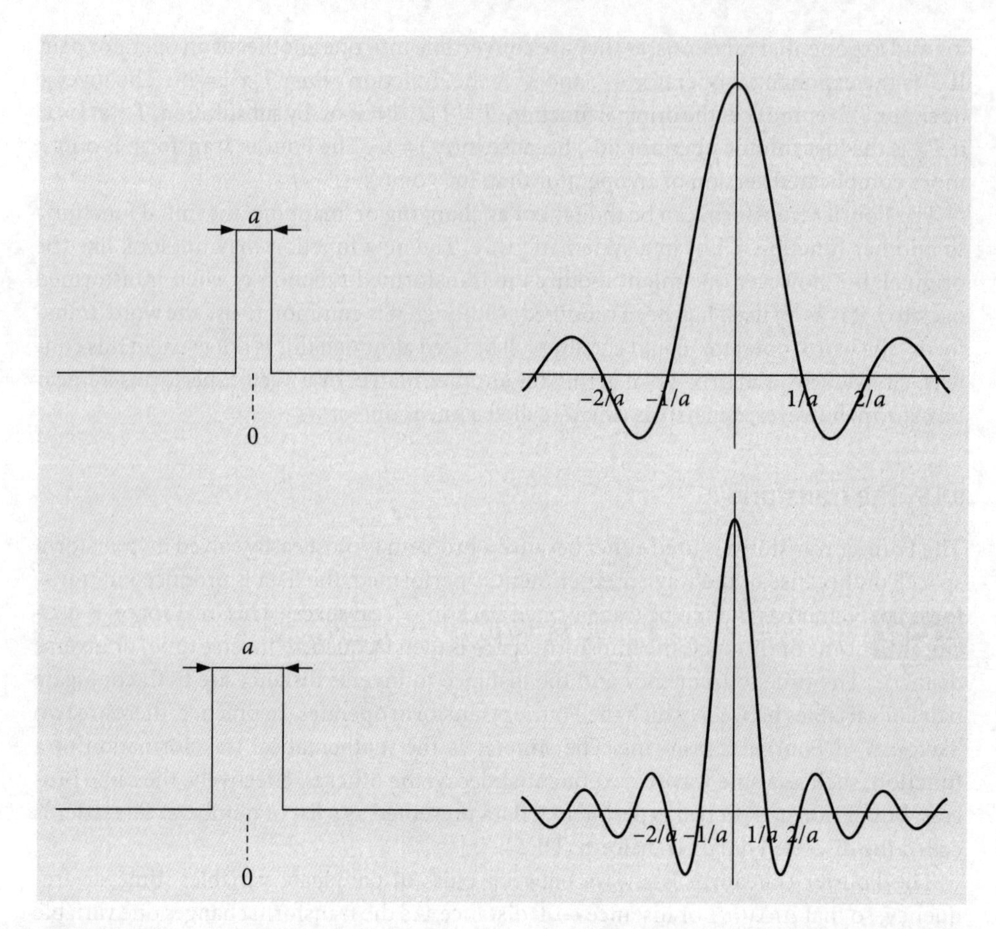

Fig. 9.15 Example of the Fourier transform of a short and a long rectangular pulse each centred about zero. Only the real part of the transform is shown, and is the sinc function, $\sin(ax)/ax$. The transform extends to $\pm\infty$.

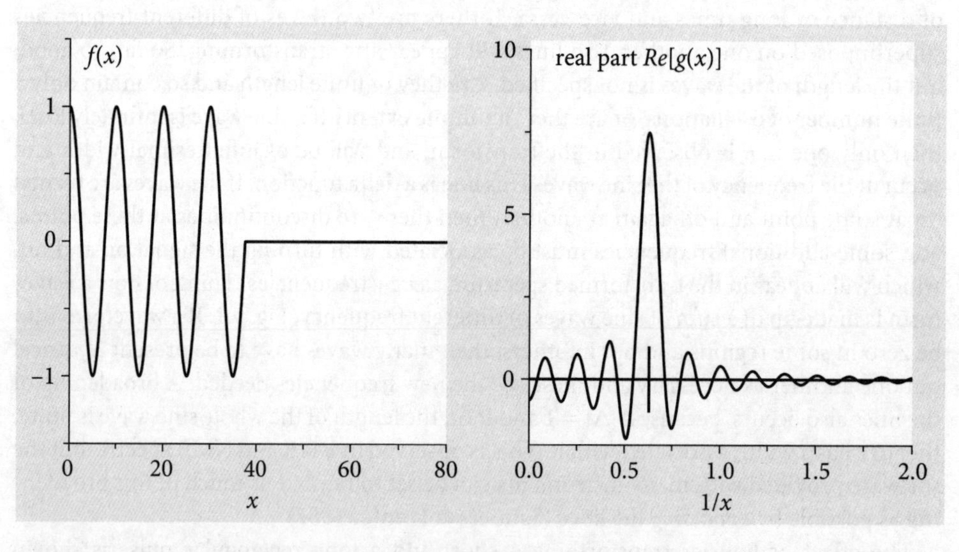

Fig. 9.16 Left: A truncated cosine wave of frequency $2\pi/10$, starting at zero and of length 75/2 or 3.5 cycles. Right: The real, imaginary and absolute parts of its Fourier transform. The value of the wave's frequency is marked with a vertical line.

transform of Fig. 9.15 arise from the many terms needed to describe the rectangular pulse. In fact, to reproduce the original pulse exactly by reverse transforming, an infinite frequency range is needed. If the transform of Fig. 9.15 *exactly* as shown in the right-hand side were reverse transformed, the rectangular pulse shown on the left of the figure would *not* be produced, because on the plot the transform has a limited frequency range.

What is the transform of a cosine wave of finite length? The result is shown in Fig. 9.16 and is somewhat similar to that of the square pulse except that the transform frequency

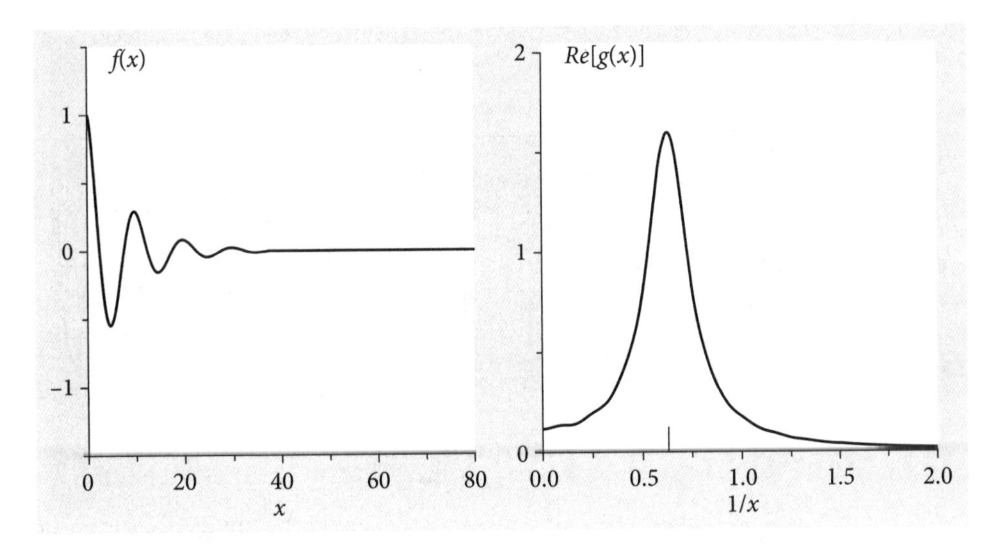

Fig. 9.17 The same cosine wave but now apodized by multiplying by exp(−t/8), which makes the cosine diminish at long times. In the transform (right), one peak is found at the frequency of the wave (small vertical line). All the frequencies associated with suddenly ending the cosine are removed.

cannot be centred at zero because the cosine has a finite frequency. The main peak is almost at the cosine's frequency, and the many other sidebands are needed to account for the fact that the wave is suddenly turned off. Now suppose that the cosine is damped by an exponential function and smoothly decreases in amplitude, then these extra frequencies disappear, because at the end of the cosine wave there is no discontinuity; the exponential makes the cosine gently approach zero. The result is a widening of the feature at the frequency of the cosine wave, Fig. 9.17. The effect of the exponential decay is to *apodize* the transform.

9.5.7 Fourier series and transforms

The connection between the Fourier series and the Fourier transform is important, and it should not be ignored. To produce the Fourier series, which describes a rectangular pulse, infinitely many terms in the Fourier series will be needed, and of ever increasing frequency. The Fourier transform allows us to see these frequencies by transforming to frequency space, so that each frequency in the Fourier series appears as a feature of some greater or lesser width.

In an NMR experiment, a square pulse of RF radiation is used to excite the nuclear spin states in the sample and, as has been seen, the Fourier transform of such a pulse illustrates that it has many frequencies contained within it. In an experiment, the pulse is made of sufficient duration to contain all those frequencies needed to excite the nuclear spins. Of course, these frequencies are not made by the transform, but are there all the time, because to form the pulse in the first place many different sine or cosine waves each of different frequency have to be added together in the electronic circuitry.

To illustrate this further, consider a laser pulse with the duration of a few femtoseconds. Such pulses are made by the process of *mode-locking*. For a laser to work, the light waves in the cavity must fit exactly into its length no matter what the colour of the light, and a node must occur at each of the mirrors; the restriction is that n half wavelengths must equal the cavity length, $n\lambda/2 = L$. If these waves, which have different frequencies, can be forced to be in phase with one another, a pulse results; mode-locking is the process by which this is achieved. Making the phase the same means ensuring that each of the waves has a maximum in the same place, no matter what their frequency is. A pulse results because waves of different frequency must eventually fall out of step with one another away from zero or $\pm n\pi$, where they are in phase. Figure 9.18 shows that a pulse can only result from the addition of many different frequencies, if they are in phase. The pulse is normalized to a maximum of ± 1 in the figure. In a mode-locked laser, $\approx 10^6$ waves may be added together rather than the few shown; consequently, the laser pulse is far better defined.

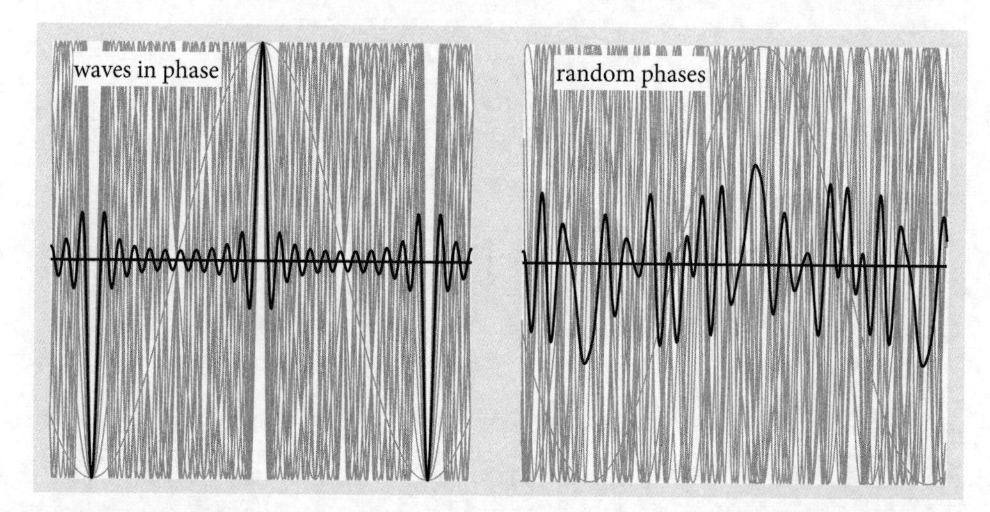

Fig. 9.18 Left: Eleven cosine waves and their sum show that pulses can only be made by adding waves of different frequency together but with the same phase. Right: One possible sum when the waves are added with random phases. The waves are $\cos(nx/2)$ where n is an odd integer. The effect is more pronounced if more waves are used; the pulse becomes shorter and the random noise (right) becomes smaller in amplitude.

To realize mode-locking, a laser must have a broad emission spectrum and nowadays titanium sapphire is often used as the gain medium to produce femtosecond duration pulses, dye-lasers are sometimes still used to produce picosecond pulses. The Ti^{3+} ions have many different sites in the sapphire (Al_2O_3) crystal lattice and therefore have a broad emission spectrum, which is in the far-red part of the visible spectrum and centred around 850 nm. The molecules or ions used to produce the fluorescence/luminescence which give rise to lasing, have a certain wavelength range caused by the nature of their potential energy surfaces and by the inhomogeneity of the host material—a glass or liquid—which shifts energy levels up and down. The coating on the mirrors, and perhaps added optical elements such as gratings, interference or birefringent (Lyot) filters, restrict the wavelengths over which the laser can operate, and this is done to enable the wavelength to be changed. However, if a short pulse is to be produced, the wavelength range has to be so wide that no filters are wanted, quite the opposite, as little restriction as possible on the wavelength range is desirable, as the product $\Delta t \Delta \nu$ is constant this means that a wide frequency (or wavelength) range is necessary if Δt is to be small. This is entirely consistent with the observation that many waves of different frequencies are needed to make a pulse.

9.6 The Fourier transform equations

The derivation of the transform equations is now sketched out by starting with the Fourier series. Butkov (1968) gives the full derivation. The Fourier series, considered in Section 9.1, are all formed from periodic functions, but suppose that the function is thought of as having an infinite period, or to put it another way, if the limits are $-L$ to L then what happens when $L \to \infty$? It is easier here to use the complex exponential form of the series, equations (9.7), and write

$$f(x) = \sum_{n=-\infty}^{\infty} c_n e^{+in\pi x/L} \tag{9.24}$$

with coefficients

$$c_n = \frac{1}{2L}\int_{-L}^{L} f(x)e^{-in\pi x/L}dx \tag{9.25}$$

where n is an *integer* specifying the position in the series, therefore, c_n is one of a series of numbers that could be plotted on a graph c_n vs n. To simplify (9.24), we define $k = n\pi/L$, which gives $\Delta k = (\pi/L)\Delta n$ for a small change in k, and clearly, as L gets larger, k gets smaller. However, there is a problem here, for when $L \to \infty$ it looks as though all values of c_n, equation (9.25), will go to zero, because L is in the denominator. Instead of immediately taking the limit, suppose that the values of n describe adjacent points on a graph of c_n vs n, and because adjacent points are the smallest differences that n can have, then $\Delta n = 1$ and so $\Delta k = (\pi/L)$ or $(L/\pi)\Delta k = 1$. Equation (9.24) can now be multiplied by this factor without difficulty because it is 1, giving

$$f(x) = \sum_{n=-\infty}^{\infty} \left(\frac{L}{\pi} c_n\right) e^{ikx} \Delta k \tag{9.26}$$

and c_n is given by equation (9.25). The limit $L \to \infty$ also means that $\Delta k \to 0$, which makes k into a continuous variable, and the coefficients c_n can now be written as a function of k, i.e. as $c(k)$ instead of the discrete values c_n. Taking this limit also changes $f(x)$ to an integral, because $\Delta k \to 0$,

$$f(x) = \lim_{L\to\infty} \sum_{n=-\infty}^{\infty} \left(\frac{L}{\pi} c_n\right) e^{in\pi x/L} \Delta k$$

$$= \int_{-\infty}^{\infty} c(k) e^{ikx} dk$$

and $c(k) = Lc_n/\pi$, but from equation (9.25) $c(k)$ is

$$c(k) = \frac{1}{2\pi} \int_{-\infty}^{\infty} f(x) e^{-ikx} dx.$$

This equation is conventionally rewritten by defining a new function $g(k)$, where

$$g(k) = c(k)\sqrt{2\pi}.$$

This function is the *forward Fourier transform* and is defined as

$$g(k) = \frac{1}{\sqrt{2\pi}} \int_{-\infty}^{\infty} f(x) e^{-ikx} dx \qquad \text{forward transform} \tag{9.27}$$

and the *reverse Fourier transform* is

$$f(x) = \frac{1}{\sqrt{2\pi}} \int_{-\infty}^{\infty} g(k) e^{+ikx} dk. \qquad \text{reverse transform} \tag{9.28}$$

The two functions form a *Fourier transform pair*; the function $f(x)$ with a positive exponential is the '*reverse*' or '*inverse*' transform, and $g(k)$, equation (9.27), with a negative exponential, the '*forward*' transform because it converts the measured or known function $f(x)$, where x might be distance, into the transformed space k which is inverse distance. Alternatively, if x represents time then k represents frequency.

There are some other points to note.

(i) These equations give the value of the transform at one point only. To obtain the full transform, k has to be varied in principle from $-\infty$ to $+\infty$, but, in practice, a value of k which is far less than infinity can be used because the transform often has an infinitesimal amplitude at large k; see Fig. 9.19 for an example.

(ii) Because the integration involves a complex number, the result might be complex or it might be real; this just depends on what the function is and it might therefore be necessary to plot the real, imaginary, and absolute value of the transform.

(iii) There are different forms of Fourier transform pairs that differ from one another by normalization constants, $1/\sqrt{2\pi}$ in our notation. This can lead to confusion when comparing one calculation with another.

(iv) Finally, note that some authors, engineers in particular, often define the forward transform with a positive sign in the exponential and negative in the reverse, which is a change of phase with respect to our notation. They also often use j instead of i to mean $\sqrt{-1}$.

9.6.1 Plotting transforms

Because the transform is normally a complex quantity, it has a real and imaginary part. In plotting the transform three graphs can be produced; one for each of the real and the imaginary components of the whole transform and one of the square of the absolute value, which is usually called the power or transform *spectrum* and is $g(k)^*g(k) = |g(k)|^2$, the asterisk indicating the complex conjugate.

9.6.2 What functions can be transformed?

To perform the transform, $f(x)$ must be integrable and must converge when the integration limits are infinity; this generally means that $f(x) \to 0$ as $x \to \pm\infty$: a sufficient condition is that $\int_{-\infty}^{\infty} |f(x)|dx$ exists.

9.6.3 How to calculate and plot a Fourier transform

As an illustration, the Fourier transform of a sine wave $f(x) = \sin(\omega x)$ which has an angular frequency $\omega = 2\pi/L$ will be calculated over the range $-L$ to $+L$; this supposes also that the function $f(x)$ is zero everywhere else, Fig. 9.19. Choosing the sine function to have the argument $2\pi x/L$ means that it is zero, i.e. has a node, at $x = \pm L$. (The frequency need not be a multiple of the range of the transform, but the resulting equations are simpler if it is.) Because the function is zero outside $\pm L$, so is the integral, and the integration limits become $\pm L$ rather than $\pm\infty$. The forward transform is

$$g(k) = \frac{1}{\sqrt{2\pi}}\int_{-L}^{L} \sin(2\pi x/L)e^{-ikx}dx.$$

This integral is easily performed by changing the sine to its exponential form first, and can be checked using Maple;

```
> Int(sin(2*Pi*x/L)*exp(-I*k*x),x = -L..L):
   % = simplify( value(%) );
```

$$\int_{-L}^{L} \sin\left(\frac{2\pi x}{L}\right) e^{-Ikx}\,dx = -\frac{4I\pi L\sin(Lk)}{k^2L^2 - 4\pi^2}$$

Note that the capital I in Maple is really the complex number i in mathematical notation. The Fourier transform is, in this particular example, wholly the imaginary part of a complex number. When $k = 0$ and when $Lk = \pm n\pi$, the transform is zero, except when $Lk = \pm 2\pi$ where the maximum or minimum occurs. When $Lk = +2\pi$, the transform has the nominal value of 0/0, which can be evaluated using l'Hôspital's rule (see Chapter 3.8). Remember to stop differentiating when either the top or bottom of the fraction is not zero, the result is

$$\lim_{k\to 2\pi/L}\frac{-4\pi iL\sin(Lk)}{L^2k^2 - 4\pi^2} \to \frac{-4\pi iL^2\cos(Lk)}{2L^2k} = -\frac{2\pi i}{k} = -iL$$

which is the minimum value of the transform. The maximum occurs when $kL = -2\pi$, (see Fig. 9.19), which corresponds to the frequency $k = 2\pi/L \equiv \omega$ in radians, or $1/L$ in Hz, if L measures time. If L is distance, cm, for example, as in an FTIR spectrometer, then $1/L$ is in wavenumbers or cm^{-1}.

To plot the transform, it is necessary to plot either the imaginary part, as above, or its absolute value; there is no real part in this particular example. Notice that there appear to be two frequencies, one at about 0.5 and at -0.5; negative frequencies do not make any sense if the sine wave is a signal from an experiment and for real experimental data, the negative frequencies need to be ignored. If the range $\pm L$ is kept the same, and instead of a sine, a cosine wave of the same frequency transformed, the real frequency part of the Fourier transform would now look like the imaginary part of Fig. 9.19.

A sine wave of infinite extent has a single frequency; the extra frequencies seen in Fig. 9.19 arise due to the fact that this wave exists only between $\pm L$. The sudden change in value of the function at $\pm L$ corresponds to having several different frequencies present, although they are not apparently there. Put another way, if a Fourier series of this truncated sine wave had to be formed very many sines or cosines of a different frequencies

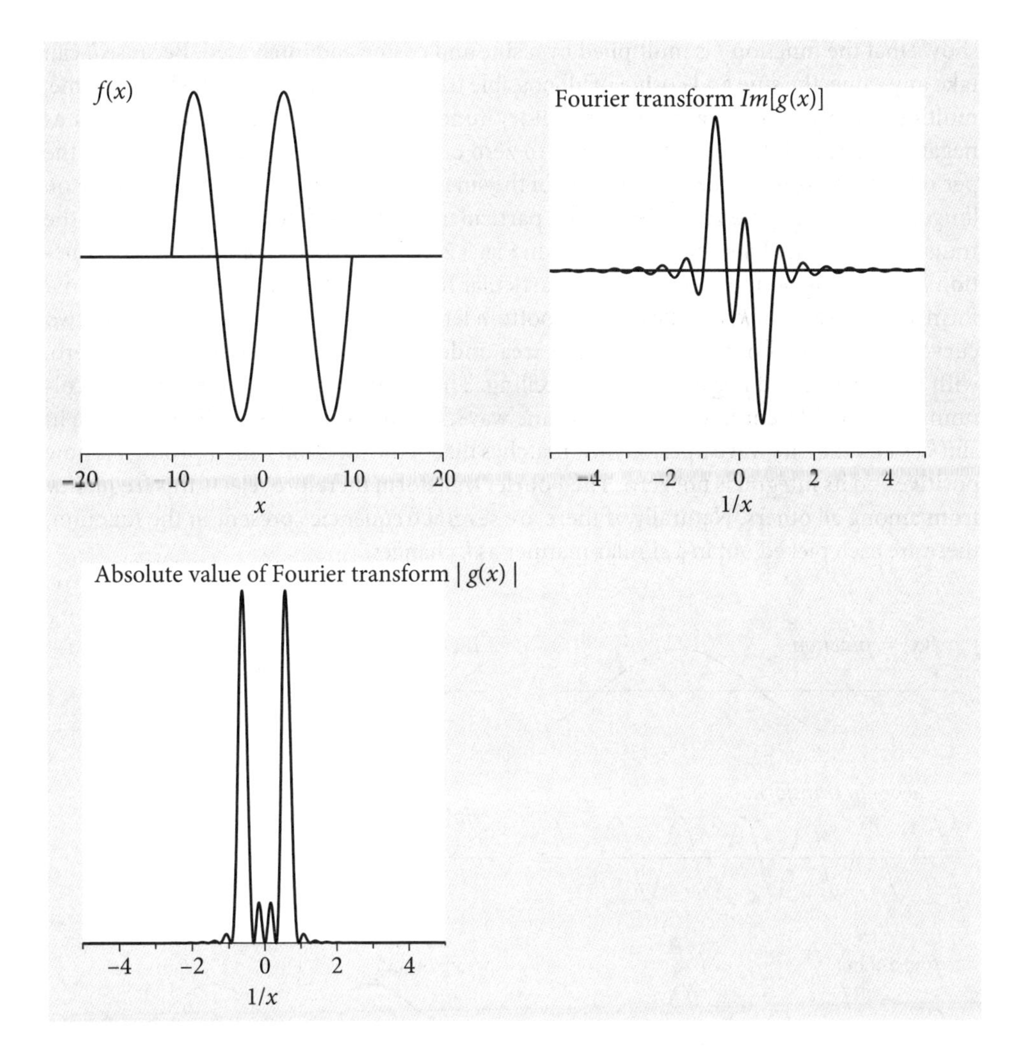

Fig. 9.19 Graphs of sin(x) from $\pm L$ when $L = 10$; its Fourier transform (middle), and its spectrum, the square of its absolute value (right). (The vertical scales are not the same.)

would have to be included. Why several terms? A single sine wave normally extends to infinity, many waves of different frequency are needed to reinforce the values near $k = 0$ and simultaneously to cancel out the part where the amplitude is zero, between $-L$ to $-\infty$ and L and ∞. Although, for practical purposes, these regions in the integration were ignored, this was only because the sine wave is zero here, but this does not mean that waves do not exist to make the amplitude zero. These terms produce the extra frequencies seen in the transform. Put another way, to understand the transform it is necessary to consider all the terms needed to describe the initial truncated function $f(x)$ as a Fourier *series* because it is exactly these terms that appear as frequencies in the transform.

9.6.4 How the Fourier transform works

The transform appears to have the effect of seeking out any repetitive features in a signal. This is true whether it is the discrete transform acting on real data, or the mathematical transform of a sine wave or other function. To understand what the transform does, we must look at the equation,

$$g(k) = \frac{1}{\sqrt{2\pi}}\int_{-\infty}^{\infty} f(x)e^{-ikx}dx.$$

and remember that this only gives the value at one point k. To obtain the transform, k has to vary from $-\infty$ to ∞ although in practice only a limited range is needed to observe the major features of the transform. In this exponential form, the oscillating nature of the argument is not so apparent, but writing it as

$$g(k) = \frac{1}{\sqrt{2\pi}}\int_{-\infty}^{\infty} f(x)[\cos(kx) - i\sin(kx)]dx$$

shows that the function f is multiplied by a sine and cosine and integrated. Because k can take any value, the sine and cosine of all possible frequencies multiply f. Most of the time, multiplication results in a highly oscillatory function, with as many positive parts as negative ones, and the integral evaluates to zero or something very close to it. When the period of f is close to, or the same as, that of the sine or cosine, then when squared these no longer integrate to give zero. Hence, the particular frequency determined by k gives the transform its value. This effect is pictured in Fig. 9.20. The left-hand column shows a function f with a long period compared to a particular frequency of the sine wave in the transform at some value of k, middle left. The bottom left curve shows the product of these two curves. The integral of this product, the area under the curve, almost evaluates to zero, with the positive and negative parts cancelling. The middle graph of the right-hand column shows a different frequency of the sine wave, because k now has a different value in $\sin(kx)$, and the sine wave's period now matches that of the function f; their product is now positive and its integral is not zero. The Fourier transform therefore selects this frequency from among all others. Naturally, if there are several frequencies present in the function, these are each picked out in a similar manner as k changes.

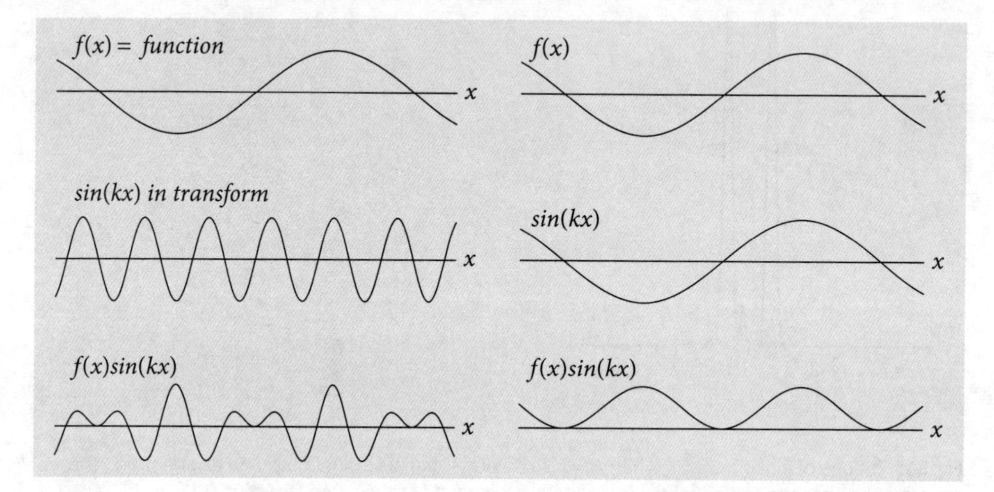

Fig. 9.20 Left column: The function $f(x)$ has a period that is very different from that of the sine wave $\sin(kx)$ middle curve, with some value k. Their product, lowest left curve, oscillates about zero and integrates to zero or very close to it. Right column: The period of the sine wave, which is determined by k, is now changed compared to the left-hand figure and now matches the period of the function. Their product is only positive, and integrates to a finite number and so appears as a peak in the transform.

9.6.5 Phase sensitive detection

In measuring signals buried in noise, the technique of phase sensitive detection is a very effective way of extracting the data and removing noise. In this method, the input to an experiment is modulated at a fixed frequency and the signal produced by the experiment is measured at this same frequency by a device known as a lock-in amplifier. This device illustrates the principle underlying the Fourier transform, although it is not a transform method.

In measuring fluorescence, for example, the light used to excite the molecules, and so stimulate the fluorescence, is modulated by rotating a slotted disc (chopper) in the exciting light's path. The photomultiplier or photodiode detects the modulated (on–off) fluorescence signal together with any noise. The lock-in receives also the reference signal directly from the chopper and it electronically multiplies this with the fluorescence signal, rather as done in Fig. 9.20. Multiples (higher harmonics) of the fundamental reference frequency are filtered away, the resulting signal is integrated over many periods of the fundamental frequency, and a DC output signal is produced. As shown in Fig. 9.20, when the product of reference and signal is integrated, frequencies dissimilar to the reference, $f(x)$ in the figure, will average to something approaching zero.

If the reference signal is $r = r_0 \sin(\omega t)$ and the noise free signal $s = s_0 \sin(\omega t + \varphi)$ then the output of the lock-in is

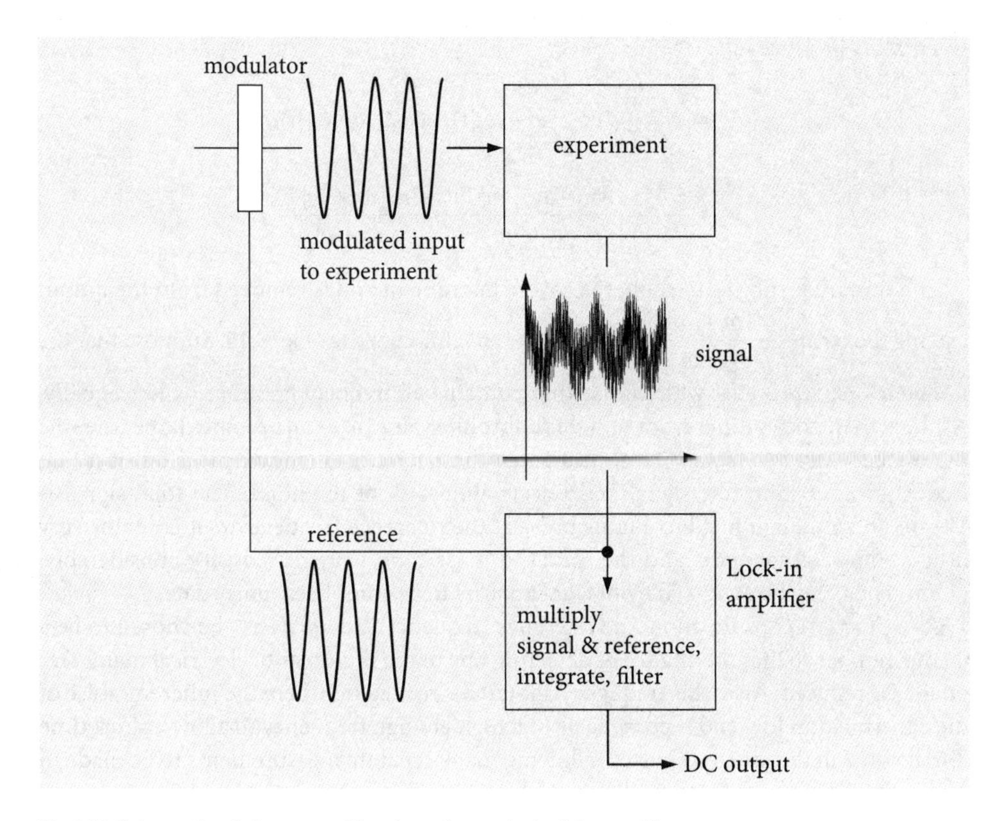

Fig. 9.21 Schematic of phase sensitive detection and a lock-in amplifier.

$$V_s = r_0 s_0 \frac{1}{T}\int_0^T \sin(\omega t + \varphi)\sin(\omega t)dt$$

where $T = 2\pi n/\omega$ and $n \gg 1$ is an integer and φ is the phase (time) delay between the reference and the signal and is due to detectors and the amplifiers and other components in the experiment, but can be changed by the user. Expanding the sines and integrating gives

$$V_s = \frac{r_0 s_0}{2}\frac{1}{T}\int_0^T \cos(\varphi) - \cos(2\omega t + \varphi)dt$$

$$= \frac{r_0 s_0}{4}\frac{1}{\omega T}[\sin(\varphi) + 2T\omega\cos(\varphi) - \sin(2\omega T + \varphi)]$$

The sine at twice the reference frequency is electronically filtered away leaving a signal that is constant because T is the integration time set by the experimentalist and normally ranges from a few milliseconds to a few seconds. The measured signal is

$$V_s = \frac{r_0 s_0}{4}\frac{1}{\omega T}[\sin(\varphi) + 2T\omega\cos(\varphi)] \tag{9.29}$$

and as the phase φ can be adjusted by the user, this signal can be maximized.

Now consider the situation when noise is present and assume that this has a wide range of frequencies. The signal from the instrument is noisy and is represented as

$$s_0 \sin(\omega t + \varphi) + n_1 \sin(\omega_1 t + \varphi_1) + n_2 \sin(\omega_2 t + \varphi_2) + \cdots$$

where s_0, ω, and φ are respectively the amplitude, frequency, and phase (relative to the reference) of the data and $n_{1,2\cdots}$ $\omega_{1,2\cdots}$ $\varphi_{1,2\cdots}$ are the same for the noise. The first term due to the signal produces V_s equation (9.29). We need only consider one noise term, for all the others behave similarly. Multiplying by the reference at frequency ω but ignoring the phase φ, as this adds nothing fundamental but makes the equations more complicated, gives the term,

$$\sin(\omega_1 t)\sin(\omega t) = [\cos([\omega - \omega_1]t) - \cos([\omega + \omega_1]t)]/2$$

Integrating produces

$$V_n = \frac{r_0 n_1}{2}\frac{1}{T}\int_0^T \cos([\omega-\omega_1]t) - \cos([\omega+\omega_1]t)dt$$

$$= \frac{r_0 n_1}{4}\frac{1}{T}\left[\frac{\sin([\omega_1-\omega]T)}{\omega_1-\omega} - \frac{\sin([\omega_1+\omega]T)}{\omega_1+\omega}\right]$$

The sum frequency term is filtered by the instrument and is removed from the output leaving the term $\dfrac{\sin([\omega_1-\omega]T)}{\omega_1-\omega}$ which is the sinc function, see Fig. 9.15. Suppose that the frequency ω_1 represents white noise that contains all frequencies more or less equally. As these frequencies differ from ω and the absolute value $|\omega_1-\omega|$ becomes larger the sinc rapidly becomes very small. This means that the reference sine wave picks out just that frequency containing the signal and rejects almost all of the noise. The total signal is $V + V_n$ and although it still contains noise at the reference frequency ω it contains very little at other frequencies, and the signal to noise ratio is increased very considerably. Often signals can be extracted from what appears to be completely noisy data.

As a practical consideration, the reference frequency should always be chosen to be a prime number so that the chance of detecting one of the multiples of electrical mains frequency is reduced. Also, this frequency should be in a region where the inherent noise of the experiment is low and if possible be of a high enough frequency to allow a short time T to be used in the integration step, allowing many separate measurements to be made in a reasonable time.

9.6.6 Parseval's theorem

This theorem is important because it proves that there is no loss of information when transforming between Fourier transform pairs. This is rather important because otherwise how would it be possible to tell what information has been lost or added? Fortunately, it can be shown that

$$\int_{-\infty}^{\infty} g^*(k)g(k)dk = \int_{-\infty}^{\infty} f^*(x)f(x)dx, \qquad (9.30)$$

where the asterisk denotes the complex conjugate. The Fourier transform of $f(x)$ is $g(k)$, which is integrated over its variable k, and similarly $f(x)$ is integrated over its variable x. As the total integral taken over all coordinate space x and that over its conjugate variable k is the same, all the information in the original function is retained in the transformation, i.e. it looks as if the transform is a different beast but this is only a disguise as it contains exactly the same information. This, of course, means that if something is done to the transform, then, in effect, the same is done to the function.

This theorem is very important in quantum mechanics. Should $f(x)$ represent a wavefunction that varies as a function of distance x, which could be the displacement from equilibrium of an harmonic oscillator, then variable k can be interpreted as the momentum (usually given the letter p) making $g(k)$ the wavefunction in 'momentum space'. This means that calculations can be formed either in spatial coordinates, i.e. distance or in 'momentum space' depending upon which is the most convenient mathematically. The change in displacement, δx, and change in momentum, δp, are conjugate pairs of variables and are linked by the Heisenberg uncertainty principle $\delta x \delta p \ge \hbar/2$.

9.6.7 Some Fourier transform properties

The transform pair is $f(x) \Leftrightarrow g(k)$

Shift	$f(x-x_0) \Leftrightarrow e^{-ikt_0}g(k)$
Modulation	$f(x)e^{-ik_0t} \Leftrightarrow g(k-k_0)$
Reversal	$f(-x) \Leftrightarrow g(-k)$
Complex Conjugate	$f(x)^* \Leftrightarrow g(k)^*$
Scaling	$f(ax) \Leftrightarrow \dfrac{1}{\lvert a \rvert}g(k/a)$

9.6.8 Questions

Full solutions are available at **www.oxfordtextbooks.co.uk/orc/beddard.**

Q9.7 The excited states of isolated atoms and molecules normally decay exponentially in time. When a spectral line is measured, the spectrometer performs a Fourier transform, converting the time profile into a spectral profile that has a Lorentzian shape, and looks something like a Gaussian but is wider in the wings. (We assume that the line measured is isolated from others so that no overlap occurs.) The detector, a photodiode, CCD camera, or photomultiplier are all 'square law' detectors and measure the square of the absolute value of the electric field of the radiation; $E^*E = |E|^2$. Suppose that the emission starts at time zero and decays with a lifetime of τ seconds, with a rate $E(t) = e^{-t/\tau}/\tau$ photons/second, calculate the Fourier transform, and plot this with the decay, assuming that $\tau = 10$. Calculate the half-width of the spectral line produced and comment on its value.

Strategy: The variable is time t rather than x, as in equation (9.27). The limits are from 0 to ∞. The half-width is calculated by evaluating half the value of the height of the transformation at $k = 0$, which represents the spectral line, then solving for k.

Q9.8 If a rectangular light pulse of duration t_p is used to excite a molecule, its electric field can be represented as $E(t) = e^{i\omega_0 t}$, $0 < t < t_p$.

(a) Calculate the real, the imaginary, and the square of the absolute value of the Fourier transform, and thereby convince yourself that the pulse is comprised of many frequencies. Find its principal frequency when the transform is at a maximum. Change the notation in equation (9.27) from k to ω explicitly to acknowledge that k is a frequency. Plot the real and imaginary parts of the transform and $|g|^2$, the absolute square of the transform, using $\omega_0 = 20\,000$ cm^{-1} and a pulse duration of 0.1 ps.

(b) What frequencies are missing?

Strategy: In (a) use the transform equation (9.27) and change k to ω; the integration is over time so change x to t. In (b) the missing frequencies must occur when the transform is zero.

Q9.9 In an NMR experiment, nuclear spins are excited by a rectangular shaped electrical pulse of length τ and amplitude a.

(a) Show that the Fourier transform has the same mathematical form as that calculated with Q9.8, except for a constant frequency offset.

(b) Show that the real part of the transform is proportional to the sinc function $\sin(\omega\tau)/\omega\tau$.

(c) What is the *fwhm* frequency spread of the real part of the pulse if $\tau = 1$ μs?

(d) What is the frequency spread in ppm in a 400 MHz spectrometer? Would this pulse duration be suitable for proton NMR spectroscopy?

Q9.10 In an NMR experiment, two lines are measured with frequencies ν_1 and ν_2. The free induction decay has a decay time (lifetime) of $\tau = 100$ which multiplies the sum of two cosines having frequencies $\nu_1 = 1/10$ and $\nu_2 = 1/11$ making the FID $e^{-t/\tau}[\cos(2\pi\nu_1 t) + \cos(2\pi\nu_2 t)]$. In this and similar equations, the frequencies in Hz (s^{-1}) are multiplied by 2π to make them into radians s^{-1} because all trigonometric functions have these units as their arguments.

(a) Explain what physical processes generate the FID. Why does it decay away?

(b) Fourier transform the FID and reproduce Fig. 9.14.

Use Maple and take limits of the transform integral from zero to $t = 1000$.

Strategy: (b) Having set up the FID, transform it using equations (9.27)–(9.28), integrate from $t = 0$ to some large value, say 1000, but you may have to change this to reproduce the figure. Too short an integration time produces wiggles on the transform, why is this?

Q9.11 **(a)** Show that, if the phase φ of $\sin(2\pi t/t_0 + \varphi)$ is chosen *not* to be $n\pi$, where n is an integer including zero, then both real and imaginary parts of its Fourier transform exist.

(b) When the phase is $n\pi$ show that the transform is imaginary.

Q9.12 Suppose that the electron density of an atom can be represented in one dimension as a delta function at a distance a from the origin; $\delta(x - a)$. The scattering of electromagnetic waves, such as X-rays off an atom's electrons is represented by the Fourier transform with such a delta function.

(a) Calculate this and its probability, or power density, and show that waves scatter equally in all directions. The width of the delta function is infinitesimally narrow, so that a is always greater than this width.

(b) Repeat the calculation with a Gaussian electron density of width $2d$ centred at position a; $e^{-(x-a)^2/d^2}/d$. Comment on the scattering.

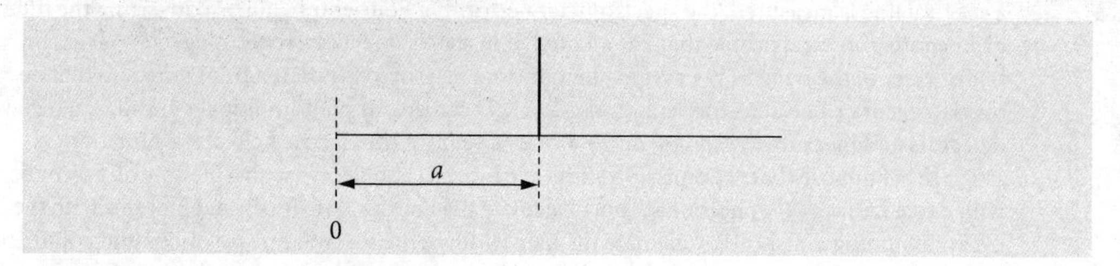

Fig. 9.22 Schematic of one atom's electron density as a δ function.

Strategy: General properties of δ-functions are given in Chapter 1.10. In this question, the useful property is that it extracts the function $f(a)$ out of the integral $\int_{-\infty}^{\infty} f(x)\delta(x-a)dx = f(a)$ and no calculation is necessary. If P is the Fourier transform, then the 'power density' is P^*P.

Q9.13 Following on from the previous question, suppose now that there are two (atom) delta functions separated by a distance $2a$. Show that the Fourier transform is a cosine and calculate the probability density.

Strategy: The delta function at position $\pm a$ is $\delta(x \pm a)$.

Q9.14 Again, following on from Q9.13, consider a comb of atoms represented as δ-function scattering centres that might represent a one-dimensional crystal that can scatter X-rays. Each atom is separated from the next by a distance a. Calculate the transform for a row of atoms extending from $-\infty$ to ∞, and show that it is the same as in the figure. The total transform is the sum of this infinite number of individual transforms. Consider two cases; one where the waves have arbitrary wavelengths, and the other when they have wavelength a.

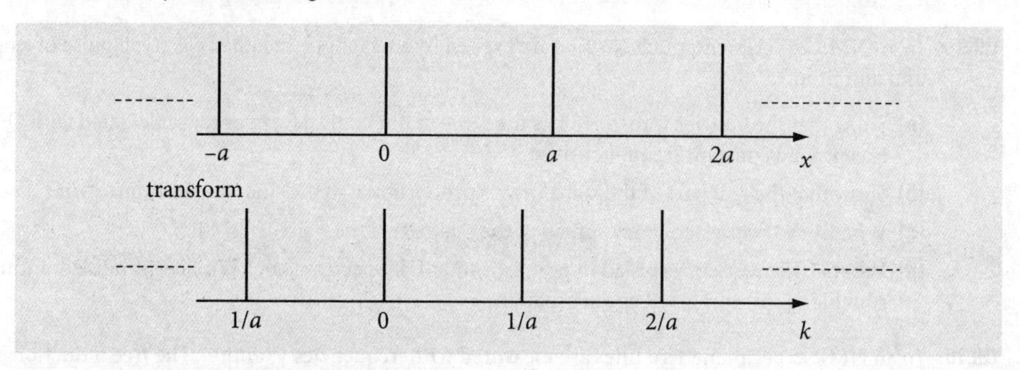

Fig. 9.23 A comb of atoms and its transform

Strategy: The initial task here is to define the δ-function for an atom at some position na. The total transform is the sum of all the individual atoms, in which case, n ranges from $-\infty$ to ∞. The physics involved determines how to solve the infinite sum produced. If the X-rays (or photons/waves in general) scattering off the atoms are in phase, then a different sum is obtained compared to when they are not in phase. Being 'in phase' means that the X-rays scatter, so that the scattering from each atom reinforces the total at the scattering angle; being out of phase means that the scattering is overall zero. If the waves have different wavelengths, in general the overall scattering must be zero, because the scattering from adjacent atoms cannot reinforce one another.

Q9.15 A laser produces an exponentially shaped pulse $p = e^{-|ax|}$, with a duration (fwhm) of $\tau = 10$ fs at 800 nm. **(a)** Calculate a, then calculate the frequency spread of the pulse and **(b)** calculate the number of modes or waves to be added together in the laser to form this mode-locked pulse if the laser cavity is 1.5 m long. The photons fit into the cavity in integer numbers of half wavelengths.

Strategy: Take the wavelength spread to be the full width at half maximum (fwhm) of the frequency spread, which is calculated with a Fourier transform. The scaling has to be correct in this calculation. It will be useful to work in units of fs and wavenumbers.

9.7 Convolution

9.7.1 Motivation and concept

Instruments measure everything: mass, energy, number of particles, wavelength of light, voltage, current, images, and so forth. However, every instrument distorts the data to a greater or lesser extent, and obviously we try to make these distortions insignificant but this is not always possible. In cases when a detector may not respond quickly enough to an event, when very wide slits have to be used in a spectrometer to detect a weak signal, or an electronic circuit does not respond in a linear manner to the input voltage, a distortion to the data is unavoidable. The effect is to *convolute* the ideal response, as defined by the physics behind the experiment, with the instrumental response. Fortunately Fourier transforms can be used to unravel the effect of convolution but this is not always possible.

(i) To be specific, suppose that the lifetime of electronically excited atoms or molecules is to be measured by exciting them with a pulse of light and their fluorescence measured as it decays with time. This fluorescence could be observed with a photodiode or photomultiplier, whose output voltage is measured with an oscilloscope. Before doing this experiment, two questions have to be answered; (i) Is the laser used to excite the molecules of short enough duration that the molecules or atoms can be excited quickly enough before any significant number can decay back to the ground state? (ii) Is the detection equipment (photodiode, oscilloscope) used able to respond quickly enough to measure the decaying fluorescence properly? If either one or both of these conditions cannot be met, then the data will be distorted by the relatively slow response of the instrument. The convolution shows how this distortion affects the data, Fig. 9.24. In this figure, the top curve is the ideal decay of the excited state, but it could represent any ideal response. This behaviour would be observed if the molecules could be excited with an infinitesimally narrow laser pulse and measured with a photo-detector with an infinitely fast time response. The second curve is the actual shape of the laser pulse, or detector response, and is the instrument response drawn on the same timescale. Clearly, this has a width and a rise and decay time that is not so very different to that of the ideal response. The lower curve is the convolution of the ideal response with the instrument response, and is what would be measured experimentally and clearly has characteristics of both curves. A log plot of the data would show that only at long times does the convoluted response have the same slope as the ideal

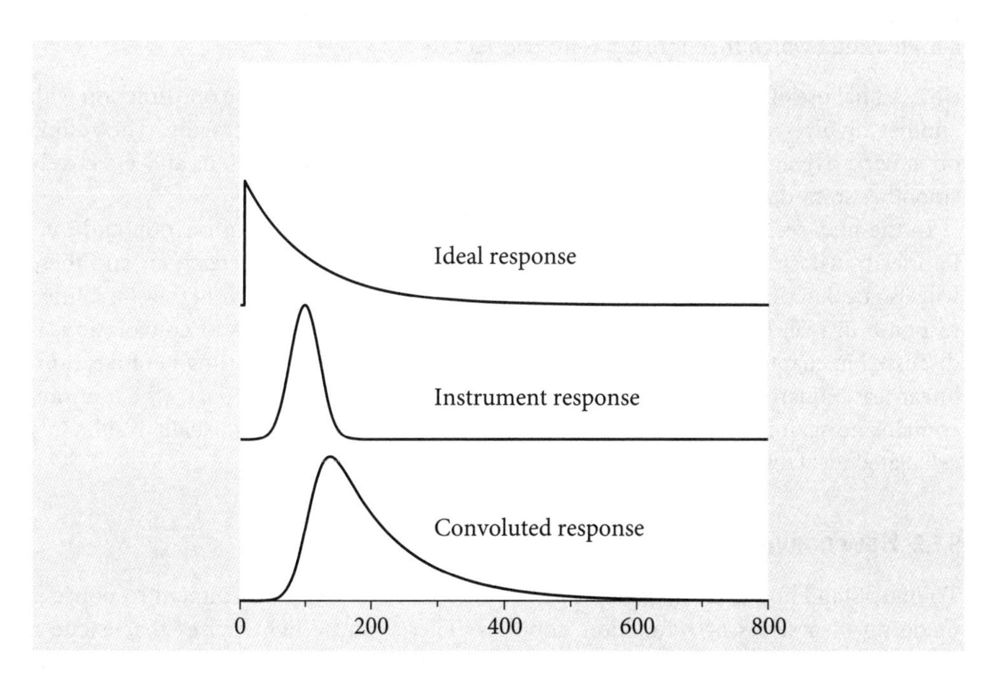

Fig. 9.24 Top: A signal representing the ideal response of an experiment to a sudden impulse. Middle: The actual stimulation used in the experiment represented as the instrument response. Bottom: The measured signal, the convolution of the two upper curves.

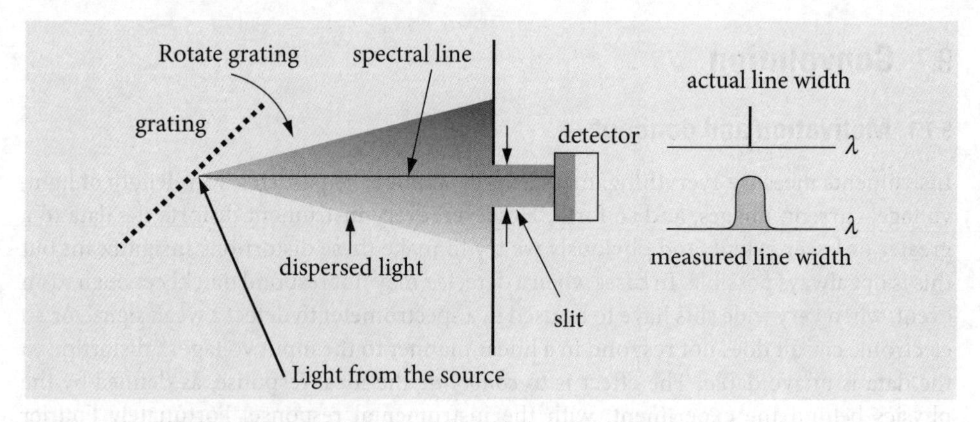

Fig. 9.25 The convolution of a narrow spectral line with a wide slit in a spectrometer.

one. It makes no difference if the instrument response consists of a slow 'driving force' for the experiment, in this case a long-lived light-pulse, or a slowly responding detector or both, because the effect producing the convolution is the same. Fortunately, convolution can be calculated easily and rapidly using Fourier transforms.

(ii) As a second example, consider measuring the width or position of one particular spectral line, such as from a star or a sample of molecules in the lab. The spectrometer has slits on its entrance and exit and these, with the number of grooves in the grating, control the resolution of the spectrometer. Typically, this is 0.1 nm/mm of slit width for a moderately good spectrometer and 1 nm/mm for a general purpose one. If the slits cannot be closed to more than 0.1 mm, then the resolution of the general purpose instrument will be approximately 0.1 nm and a narrow spectral line will appear to have this value even it is many times narrower. This is because the grating is rotated while measuring the spectrum and the spectral line is swept across the slits. The effect is to sequentially place a spectral line at all possible points, and hence wavelengths, across the slit. A signal is recorded at all these wavelengths rather than being measured only at its proper one, and the response measured is the convolution of the ideal width of the spectral line with the instrument response, which is the finite width of the slit. In many instruments, a CCD camera measures all wavelengths simultaneously, and a slit is not needed nor is the grating scanned. However, the same reasoning applies because the individual elements of the camera have a finite width, which therefore act as individual slits.

(iii) A final use of convolution is to smooth data. Because convoluting one function with another involves integration, this has the effect of summing or averaging. The rolling or moving average method (Section 9.10.4) is in effect a convolution, and effectively smoothes spiky data.

In the next sections, a convolution will be calculated by direct summation and by a Fourier transform. Convolution is related to the *auto-* and *cross-correlations* and these will also be described. How to go about estimating the true response from the convoluted response in real, that is experimental data, i.e. reversing the effects of convolution, is discussed in chapter 13 on numerical methods. This is usually done using iterative, non-linear least squares methods, (See 13.6.7), because when using real data, which always contains noise, it is found that reverse transforming the convolution usually results in a calculated ideal response that is so noisy as to be useless.

9.7.2 How convolution works

To understand how convolution works, suppose that the overall instrument response is made up of a series of δ-function impulses. These can be light pulses that excite a molecule, or, in a mechanical analogy, sudden 'blows' to an object we want to move. Suppose these impulses are made at ever shorter time intervals, then the effect is that of smoothly pushing the object. After a while the pushing stops. Each of the impulses elicits an ideal response but because there are many of them, their responses must be added

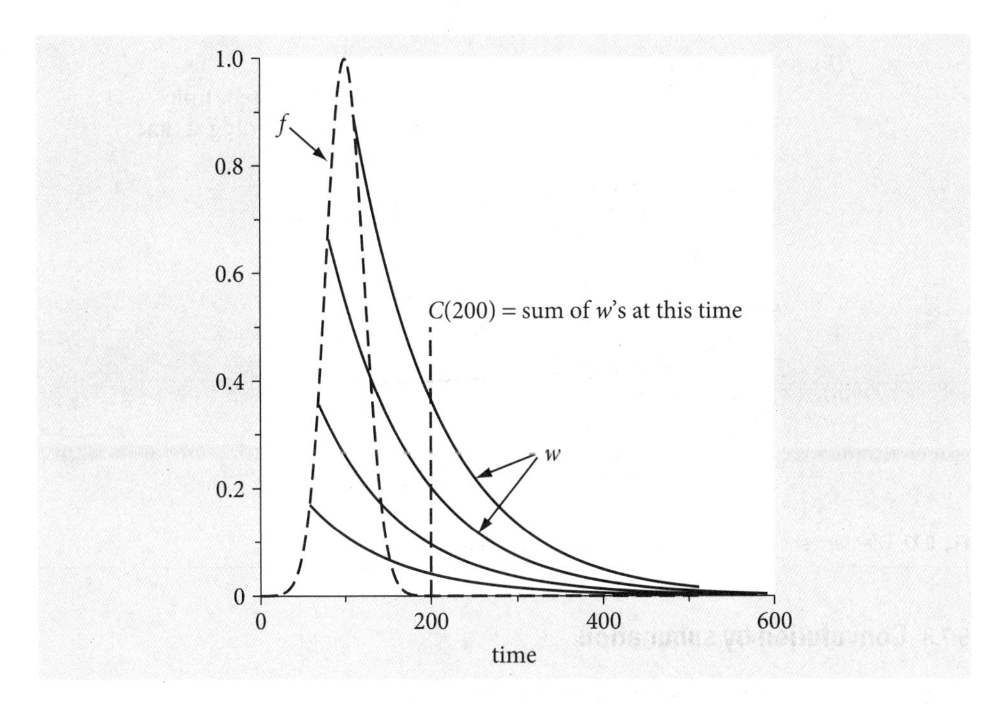

Fig. 9.26 Curves show the instrument response, as a series of impulses (dashed), which produce a response at each point on its profile not all of which are shown. These are then added together in this time delayed manner, to produce the convoluted response.

together. The result is the convolution; the effect is shown in Fig. 9.26. It is always assumed in the convolution that the response is linear with the impulse, which simply means that doubling the impulse doubles the response and so forth.

The blows (or light pulses) occur at each point in the dashed curve, Fig. 9.26. The response from each impulse is the decaying solid curve. To calculate the overall response at any given point along the x-axis, the effect of all previous blows must be added into the calculation. Suppose that the pulse exciting the sample has a shape given by some function f, the ideal experimental response w, and the convolution C. The terms can be written down at each time if it is assumed, for the present, that the impulses are discrete and the data is represented as a series of points at times 1, 2, 3, and so forth; $f(6)$, for example, represents the value of f at the sixth time position. The first point of the impulse is $f(1)$ and this produces the response

$$f(1)[w(1) + w(2) + w(3) + \cdots].$$

The second and third impulses produce

$$f(2)[w(1) + w(2) + w(3) + \cdots] \quad \text{and} \quad f(3)[w(1) + w(2) + w(3) + \cdots].$$

The convolution is the sum of these terms at times 1, 2, 3, and so on;

$$\begin{aligned} C(1) &= f(1)w(1) \\ C(2) &= f(1)w(2) + f(2)w(1) \\ C(3) &= f(1)w(3) + f(2)w(2) + f(3)w(1) \\ C(4) &= f(1)w(4) + f(2)w(3) + f(3)w(2) + f(4)w(1) \end{aligned} \tag{9.31}$$

These sums are shown in Fig. 9.27 by adding the products of f and w vertically. Clearly, only where both f and w are not zero, will this product have a value.

The symmetry in these sums soon becomes apparent, each being the product of one series running to the right, and the other to the left; for instance, look at C(4). The name convolution arises from just this effect; the word also means 'folded' and this is shown in the form of the series where each function is folded back onto the other. Convolution is also the *distribution of one function in accordance with a law specified by another function* (Steward 1987) because the whole of one function w, is multiplied with each ordinate of the other f, and the results added. The ideal response (the 'one function') is distributed, i.e. spread out according to the shape (law) of the driving function f.

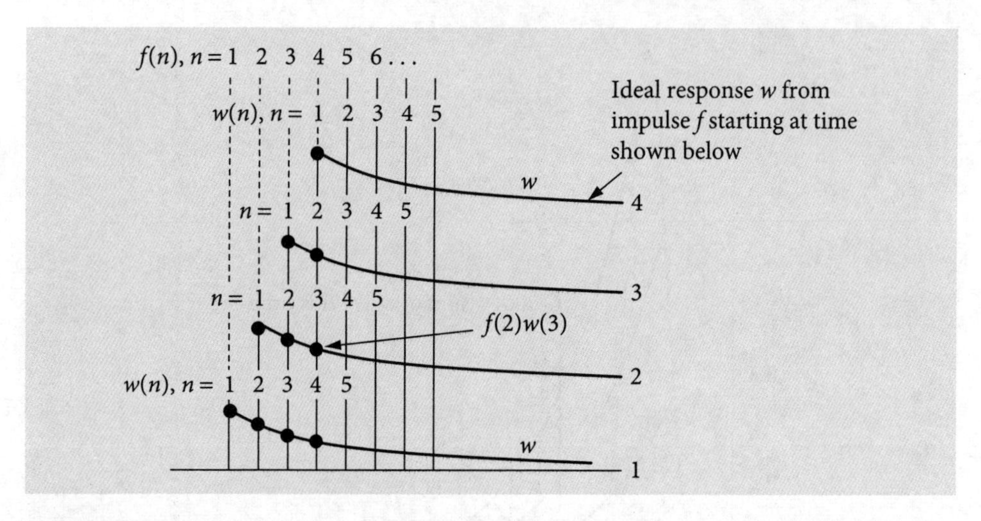

Fig. 9.27 Diagram showing the notation used to calculate a convolution.

9.7.3 Convolution by summation

Written as a summation, the convolution at point k is

$$C(k) = \sum_{i=1}^{k} f(i)w(k-i+1). \tag{9.32}$$

This sum evaluates just one point; to calculate the whole convolution, the index k must now be varied from 1 to n, which is the number of data points, making a double summation. One reason Fourier transforms are used to calculate convolutions is that the fast Fourier transform algorithm, FFT, is far quicker on the computer than calculating the convolution as a double summation, particularly for a large number of data points.

The Maple algorithm to calculate the summation has a double loop to calculate all values of k and to perform the summation in (9.32). The two functions used are those that produced Fig. 9.24, which are $f(t) = e^{-t/100}$ and $w(t) = e^{-(t-100)^2/1000}$, and 2^{10} points will be also be used to mimic the data produced by an instrument. First, because the data is discrete, arrays f and w are made; to hold the data points. Then two `for..do` loops are made, one changes k from 1 to n the and inside one calculates $C(k)$. The indices are arranged as in equation (9.32). The variable s accumulates the sum as the inner do loop progresses. This is a slow calculation because of the double loop.

Algorithm 9.3 Convolution by summation

```
> n:= 2^10;                                    # 1024 points
  f:= Vector(n): w:= Vector(n): C:= Vector(n):
  for i from 1 to n do                          # define data
    f[i]:= exp(-i/100.0):
    w[i]:= exp(-(i-100)^2/1000.0);
  end do:
  for k from 1 to n do                          # loop for n points
    s:= 0.0:                                    # start for each k
    for i from 1 to k do                        # sum C(k) eqn (9.32)
      s:= s + f[i]*w[k-i+1]:
    end do;
    C[k]:= s;
  end do:
> ss:= seq([i,C[i]/s],i=1..n):                  # make list & plot
  plot([ss],axes=framed);
```

9.7.4 Convolution by Fourier transform

The convolution can also become an integral, by supposing that the points are separated by an infinitesimal amount, and therefore, the change $\sum \rightarrow \int$ is allowable. The integral form of the convolution at time u, is

$$C(u) = \int_0^{\infty} f(t)w(u-t)dt, \tag{9.33}$$

which represents the response at time u to an impulse delivered at time t. The limits to the integral are often represented as $-\infty$ and ∞. If the signal is zero at times less than zero, then the lower limit can be made zero as illustrated. The convolution integral is frequently written as,[1]

$$C(t) = f(t) \otimes w(t) \quad \text{or} \quad C = f \otimes w. \tag{9.34}$$

The convolution is performed by Fourier transforming functions f and w separately, multiplying the transforms together and then inverse transforming. The symbol $\otimes$ represents *all* these calculations because the result is returned in the time domain. Sometimes, the convolution is written only as a conversion into the frequency domain as

$$f(t)^* w(t) = \sqrt{2\pi} F(\omega) W(\omega)$$

where F and W are the respective transforms of f and w, ω being angular frequency.

If T represents the Fourier transform and $T[\cdots]^{-1}$ the inverse transform the convolution is formally written as

$$C = T[T(f)T(w)]^{-1} \tag{9.35}$$

which is the same as equation (9.34). If the functions f and w are known, an exponential and a Gaussian for example, then the Fourier transform integral of each can be calculated as described in Section 9.6, the product of these multiplied and the inverse transform integral then calculated. The result is the convolution of the two functions. As an example, consider convoluting a square pulse with two delta functions. Their convolution will produce two square pulses centred on the two delta functions, because, as the pulse is swept past the two deltas, only at their overlap will their product have a finite value. Three stages of the convolution are shown at the top of Fig. 9.28, and the result is shown below this.

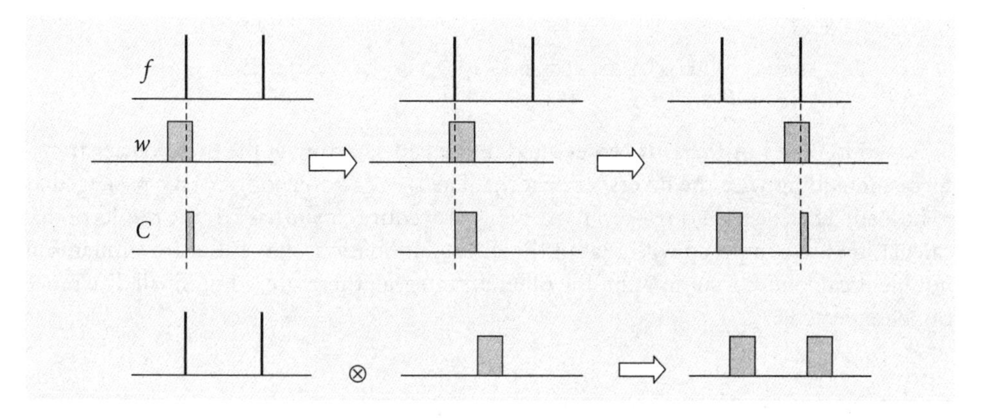

Fig. 9.28 Convolution as Fourier transforms.

Next, the convolution is evaluated using Fourier transforms. The transforms of the two delta functions and the pulse have already been calculated, and are shown in Fig. 9.29. This product of the two transforms is then reverse transformed and two square pulses are produced.

This last convolution is, incidentally, another way of describing the interference due to a double slit, and if many delta functions are used then this describes the effect of a diffraction grating on light waves.

The data needed in a convolution is frequently a list of numbers because it comes from an experiment and in this case a numerical method has to be used to do the transform, which is then called a *discrete Fourier transform*. This is described further in Section 9.9, but here is an example of Maple code to illustrate convolution using discrete Fourier transforms.

[1] The symbols $\otimes$ and $\odot$ are used by different authors to represent either convolution or autocorrelation, or vice versa, rather than 'x' used in multiplication. There is no fixed convention.

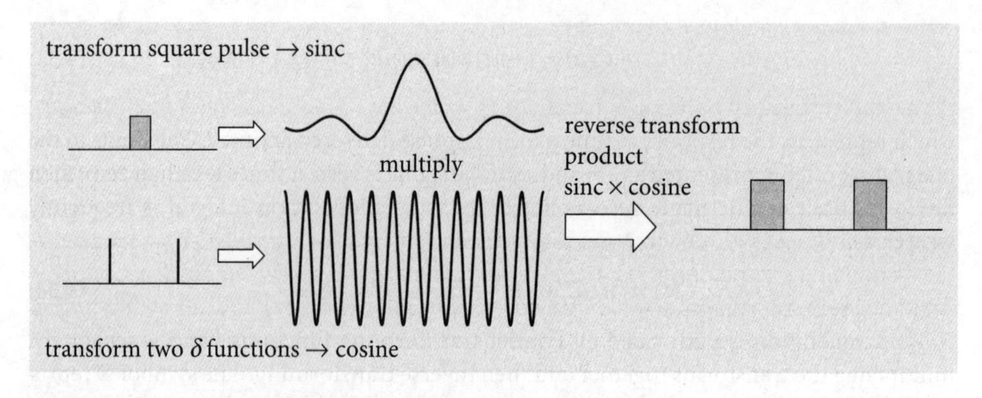

Fig. 9.29 Left: The two waveforms are the Fourier transform of a square pulse (top) and two delta functions (lower). When these are multiplied together and reverse transformed two pulses are produced which is the convolution of the delta functions and the single square pulse. The same method has been used to make Fig. 9.24, even though the functions differ.

Algorithm 9.4 Convolution by Fourier transform

```
> restart: with(DiscreteTransforms): with(plots):
> n:= 2^10:                              # 1024 data points
  f:= Vector(n): w:= Vector(n):          # vectors for data
  P:= Vector(n): C:= Vector(n):
  for i from 1 to n do                   # make data
    f[i]:= exp(-i/100.0):
    w[i]:= exp(-(i-100)^2/1000.0);
  end do:
  FTf:= FourierTransform(f):             # discrete transform
  FTw:= FourierTransform(w):
  for i from 1 to n do                   # multiply transforms
    P[i]:= FTf[i]*FTw[i];
  end do:
  C:= InverseFourierTransform(P):        # convolution result
  listplot(abs(C));
```

Equation (9.35) or (9.34) is represented in the code starting at the first Fourier transform and ending with the inverse transform. The `DiscreteTransforms` package has to be loaded before the `FourierTransform` procedure can be used. If the results of this calculation are compared with that of the direct summation, the results are similar, but not identical; one is a summation the other an integral, therefore, some small difference can be expected.

9.8 Autocorrelation and cross-correlation

A correlation is a function that measures the similarity of one set of data to another or to itself. The data might be a voltage from a detector, or it might be an image or residuals from fitting a set of data. A cross-correlation compares two different sets of data and tells us about the similarity of one signal with another; an autocorrelation compares a set of data with itself. In Fig. 9.30 part of a noisy sinusoidal curve is shown. The second curve is displaced only a little from the first and is clearly only slightly different; the third which is displaced by more is clearly different from the first as it is negative at large x when the first curve is positive. The right-hand figure shows the autocorrelation of curve 1 shown on the left, and as this is an oscillating curve, the autocorrelation also oscillates but eventually reaches zero. The oscillation is a result of the fact that a sinusoidal curve is similar to itself after each period, and the autocorrelation measures this similarity by increasing and decreasing. The autocorrelation is less noisy that the data because it involves summing or integrating over many data points. A random signal with an average of zero will have an autocorrelation that averages to zero at all points except the first, whereas the autocorrelation of an exponential and similar functions will be not be zero, but decay away in some manner. The autocorrelation is a measure of the 'memory' a function has, that is, how

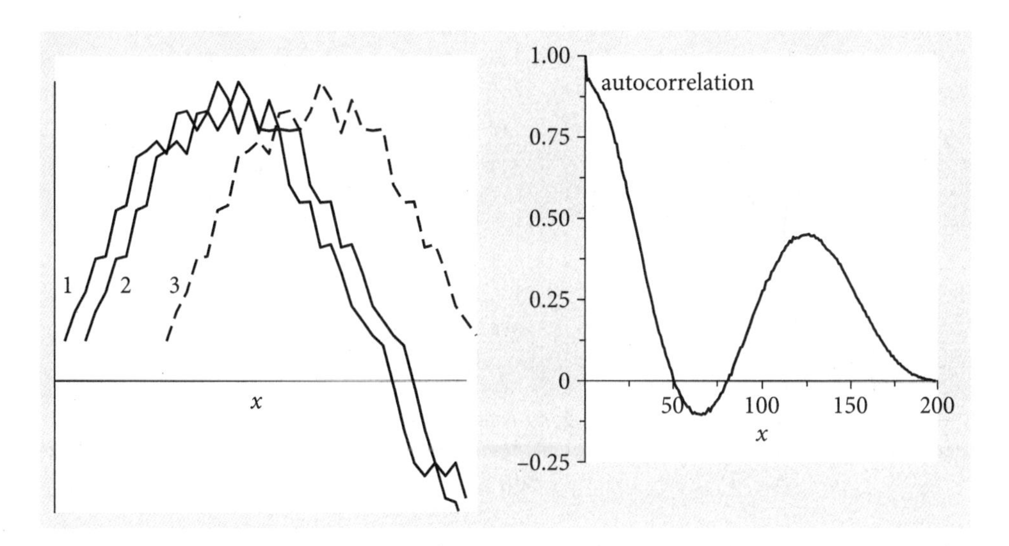

Fig. 9.30 A sketch showing the first part of a set of data. The data is still somewhat similar to itself when displaced by only a few points but much less so, when displaced by many, curve 3. The autocorrelation of all the data is shown on the right.

similar one part of the data is with an earlier or later part. A zero average random signal has no memory because it is random, and each point is independent of its predecessor; this is not true of any other signal. The correlation is therefore a process by which we can compare patterns in data. In data analysis, the residuals, which are the difference between the data and a fitted function, should be random if the fit is correct; the shape of the autocorrelation is therefore a way of testing this.

In ultra-fast laser spectroscopy, autocorrelations are used to measure the length of the laser pulse because no electronic device is fast enough to do this, as they are limited to a time resolution of a few tens of picoseconds at best, but laser pulses can be less than 10 fs in duration. In single molecule spectroscopy, the correlation of the number of fluorescent photons detected in a given time interval is used to determine the diffusion coefficient of the molecules. In the study of the electronically excited states of molecules, the correlation of time resolved spectra, recorded as the molecule moves on its potential energy surface, is a measure of excited state and solvent dynamics.

The correlation function is similar to, but different from, convolution. The autocorrelation is always symmetrical about zero displacement or lag, the cross-correlation is not. In the convolution the two functions f and w are folded on one another, the first point of f multiplying the last of w and so on, until the last point of f multiplies the first of w, equation (9.31). In the auto- and cross-correlation, one function is also moved past the other and the sum of the product of each term is made but with the indices running in the *same* direction, both increasing. A cross-correlation is shown in Fig. 9.31 using a triangle and a rectangle, each with a base line, and for clarity, defined with only six points.

The first term in A occurs when point $f(6)$ overlaps with $w(1)$, when f is to the far left of w. The position at -5 to the left is shown in the figure as $A(-5)$. The middle term in the correlation is at zero displacement, or lag, and there is total overlap of the two shapes and the correlation is at a maximum. The figure on the right shows the last overlap, consisting of just one point in common between the two shapes. There are six terms in the summation of $A(0)$ down to one in each of $A(-5)$ and $A(5)$. The zero lag term is

$$A(0) = f(1)w(1) + f(2)w(2) + \cdots + f(6)w(6)$$

The next term has one point displacement between f and w and five terms are summed,

$$A(1) = f(1)w(2) + f(2)w(3) + f(3)w(4) + f(4)w(5) + f(5)w(6).$$

With two points displaced, there are four terms

$$A(2) = f(1)w(3) + f(2)w(4) + f(3)w(5) + f(4)w(6)$$

and so forth for the other terms. The last overlap is

$$A(5) = f(1)w(6) \qquad (9.36)$$

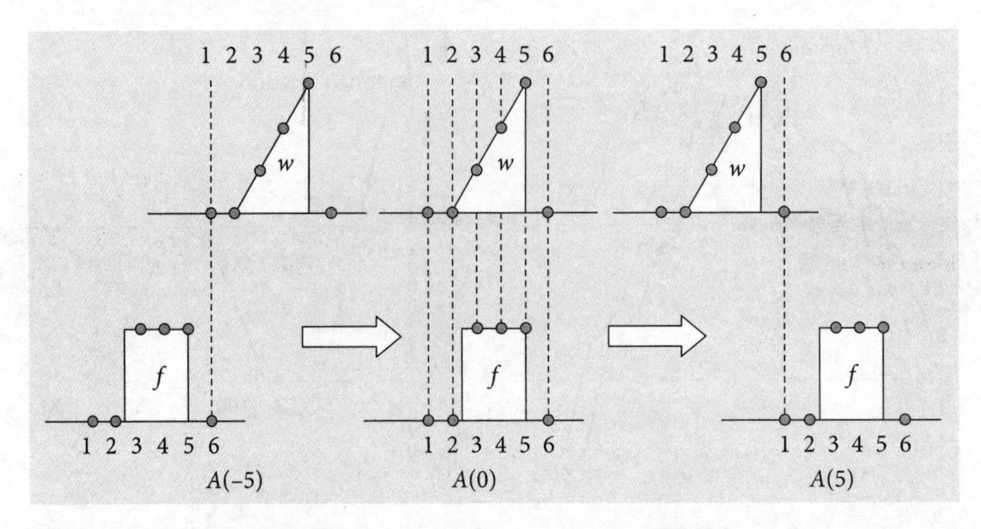

Fig. 9.31 A pictorial description of cross-correlation of the signals (functions) *w* and *f*.

On the negative side, the indices are interchanged, *f* for *w* and vice versa, and the first (far left) term is

$$A(-5) = f(6)w(1)$$

and similarly for the other terms. There are 11 terms in all or, in general $2n - 1$, for data of *n* points. In an autocorrelation, *f* and *w* are the same function and therefore the autocorrelation must be symmetrical and only terms from zero to five are needed, the others being known by symmetry.

The formula for the autocorrelation for *n* data points is

$$A_a(k) = \sum_{i=1}^{n-k} f(i)f(k+i) \qquad k = 0, 1 \cdots \to \cdots n-1 \tag{9.37}$$

where the first value of the displacement *k* is zero, and the last $n - 1$, and both functions are now labelled *f*. Very often the autocorrelation is *normalized*; this means dividing by $\sum f(i)^2$ and the equation becomes

$$A_a^{norm}(k) = \frac{\sum_{i=1}^{n-k} f(i)f(k+i)}{\sum_{i=1}^{n} f(i)^2}. \tag{9.38}$$

These last two formulae produce just half of the autocorrelation. To produce the full correlation, symmetrical about zero lag, the mirror image of equation (9.37) must be added as points $-n + 1$ to -1 to the left-hand part of the data.

The cross-correlation uses a similar formula

$$A_c(k) = \sum_{i=1}^{n-k} f(i)w(k+i) \qquad k = -n+1 \cdots \to \cdots 0 \cdots \to \cdots n-1 \tag{9.39}$$

but now *k* always ranges from $-n + 1$ to $n - 1$. This distinction is crucial, otherwise the whole of the cross-correlation is not calculated.

In calculating a correlation as a summation with a computer, as with a convolution, each term in the correlation is a sum, so this means that two nested 'loops' are needed to calculate the whole function; one loop sums each individual term, the other calculates the sum, $A(k)$.

Some authors define the correlation up to a maximum of *n* in the summation, not $n - k$. There is, however, a pitfall in doing this because, if the correlation is not zero above half the length of the data, then this folds round and what is calculated is the sum of the correlation plus its mirror image. The way to avoid this is to add *n* zeros to the data and the summation continued until $2n$. This should be done routinely if Fourier transforms are used to calculate the correlation.

Correlations and convolution are not restricted to digitized data but apply also to normal functions As an integral, the cross-correlation of a real, i.e. not complex, function is

$$A_c(u)=\int_{-\infty}^{\infty} f(t)w(u+t)dt \tag{9.40}$$

and the autocorrelation of function f,

$$A_a(u)=\int_{-\infty}^{\infty} f(t)f(u+t)dt \tag{9.41}$$

Notice that the sign in the second term is positive in the correlation but negative in a convolution, equation (9.33). If the function contains a complex number, then the conjugate is always placed on the left,

$$A_a(u)=\int_{-\infty}^{\infty} f^*(t)f(u+t)dt.$$

The normalized autocorrelation is

$$G(u)=\frac{\displaystyle\int_{-\infty}^{\infty} f(t)f(u+t)dt}{\displaystyle\int_{-\infty}^{\infty} f(t)^2dt}=\frac{\langle f(t)f(u+t)\rangle}{\langle f(t)^2\rangle} \tag{9.42}$$

and the bracket notation indicates that these are average values. The denominator is the normalization term and is also the value of the numerator with $u=0$.

9.8.1 Calculating an autocorrelation

(i) If the function is periodic then the integration limits should cover one period. The normalized autocorrelation of a cosine $A\cos(2\pi\nu t+\varphi)$, where the period is $T=1/\nu$ and ϕ is the phase, is calculated as

$$G(u)=\frac{\displaystyle\int_0^T \cos(2\pi\nu t+\varphi)\cos(2\pi\nu(u+t)+\varphi)dt}{\displaystyle\int_0^T \cos(2\pi\nu t+\varphi)^2dt}$$

and the result will be independent of the phase. The normalization integral is a standard one and can be looked up or converted to an exponential form to simplify integration. The result is $\int_0^T \cos(2\pi\nu t+\varphi)^2dt=\frac{T}{2}$.

The other integral can similarly be calculated. Using Maple, this is

```
> Int(cos(2*Pi/T*t+phi)*cos(2*Pi/T*(u+t)+phi),t=0..T):
                    %= combine( value(%) );
```

$$\int_0^T \cos\left(\frac{2\pi t}{T}+\phi\right)\cos\left(\frac{2\pi(u+t)}{T}+\phi\right)dt=\frac{1}{2}T\cos\left(\frac{2\pi u}{T}\right)$$

from which it is seen that the autocorrelation is also a cosine $G(u)=\cos(2\pi\nu u)$. If the initial cosine is written as $\cos(\omega t+\varphi)$ then the period $T=2\pi/\omega$.

If the trigonometric function is a complex exponential $Ae^{-i(\omega t+\varphi)}$ rather than a sine or cosine then the complex conjugate of the function is taken in both of the autocorrelation integrals. The normalization could not be simpler $\int_0^T dt=T$. The correlation is also a very straightforward integral

$$G(u)=\frac{1}{T}\int_0^T e^{-i(\omega t+\varphi)}e^{i(\omega(u+t)+\varphi)}dt=\frac{1}{T}\int_0^T e^{i\omega u}dt=e^{i\omega u}.$$

Using the Euler relationship, $e^{-i\theta} = \cos(\theta) + i\sin(\theta)$, the real or imaginary parts of the function give the cosine or sine result respectively.

(ii) If the function is not periodic, then the limits must be determined by the function being used. The normalized autocorrelation $A(u)$ of $f(t) = e^{-at}$, when $t \ge 0$ and $f(t) = 0$ when $t < 0$, will be calculated, and also its full width at half-maximum, fwhm. The integration limits can be changed from those in equation (9.42) because the function is zero for $t < 0$ and the lower limit can be zero. The normalization, using equation (9.42), is

$$\int_{-\infty}^{\infty} f(t)^2 dt = \int_0^{\infty} e^{-2at} dt = \frac{1}{2a}$$

and the autocorrelation

$$\int_{-\infty}^{\infty} f(t)f(u+t)dt = \int_0^{\infty} e^{-at}e^{-a(u+t)}dt$$

$$= e^{-au}\int_0^{\infty} e^{-2at}dt = \frac{e^{-au}}{2a}.$$

Importantly, the autocorrelation must be an *even* function because it is symmetrical; therefore, the value of u must always be positive, hence the autocorrelation is

$$A(u) = \frac{e^{-a|u|}}{2a},$$

and the normalized autocorrelation is $A(u) = e^{-a|u|}$. The $|u|$ does not follow from the mathematics; it is imposed by our knowledge of symmetry of the function.

As a check, at $u = 0$, $A(0) = 1$, which is correct and the function is even or symmetrical about its y-axis, or, $u = 0$. The fwhm is calculated when $A(u_h) = \frac{1}{2} = e^{-a|u_h|}$ or $|u_h| = \frac{\ln(2)}{a}$, and therefore is $2\ln(2)/a$. The initial function has a fwhm of $\ln(2)/a$ and in this instance the autocorrelation fwhm is twice as wide as the function.

(iii) The duration of a short laser pulse is often measured as an autocorrelation with an optical correlator. If the intensity profile I of the short laser pulse is a Gaussian centred at zero $I = e^{-2(t/a)^2}$, it is possible to calculate the width of its normalized autocorrelation. If the calculated autocorrelation shape is compared with an experimentally measured one, an estimation of the laser pulse's duration can be made. The optical correlator to do this measurement is a Michelson interferometer; the path length in one arm is changed relative to the other so that one pulse is moved past the other in time. The pulses are combined in a frequency doubling crystal, and a signal is detected only when the pulses overlap. To achieve this, the doubled frequency, which is in the ultraviolet part of the spectrum, is separated from the fundamental wavelength by a filter. The size of the signal vs the distance the mirror moves, which is proportional to time, is the autocorrelation see Fig. 9.32.

The pulse is centred at zero delay and (theoretically) extends from $-\infty$ to ∞, which are the integration limits of the autocorrelation, equation (9.42). The autocorrelation integral is

$$A(u) = \int_{-\infty}^{\infty} e^{-(t/a)^2}e^{-(u+t/a)^2}dt = a\sqrt{\frac{\pi}{2}}e^{-u^2/2a^2}.$$

The calculation in Maple is

```
> assume( a > 0 );
  Int(exp(-(t/a)^2)*exp(-((u+t)/a)^2),
                    t= -infinity..infinity); %= value(%);
```

$$\int_{-\infty}^{\infty} e^{-(t^2/a^2)}e^{-[(u+t)^2/a^2]}dt = \frac{1}{2}e^{-(1/2)(u^2/a^2)}\sqrt{2}a\sqrt{\pi}$$

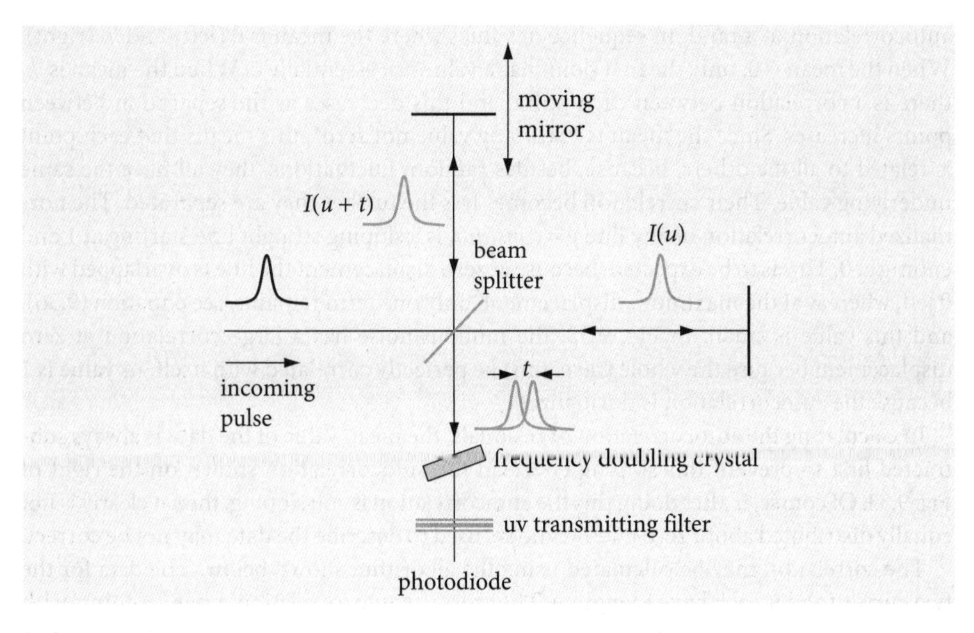

Fig. 9.32 Schematic of an optical autocorrelator used to measure the duration of pico- and femtosecond laser pulses.

The normalization integration can be looked up (Chapter 4.2.13) but need not be worked out because it is the value of autocorrelation when $u = 0$. The normalization equation is therefore

$$\int_{-\infty}^{\infty} e^{-2(t/a)^2} dt = a\sqrt{\frac{\pi}{2}}.$$

The normalized autocorrelation $G(u)$ is also a Gaussian, with a value

$$G(u) = e^{-u^2/2a^2}.$$

The fwhm of this function is calculated when $G(u) = 1/2$ and is $a\sqrt{2\ln(2)}$ and that of the original pulse is $a\sqrt{\ln(2)}$ therefore, the autocorrelation is $\sqrt{2} \approx 1.414$ times wider than the pulse. Knowing this factor provides a convenient way of measuring the duration of a short laser pulse assuming it has a Gaussian profile.

(iv) The randomness or otherwise of the autocorrelation of the residuals obtained from fitting real data to a model (theory) is important when determining the 'goodness of fit'. The function is now a set of data points not an equation. The data in Fig. 9.33 shows the

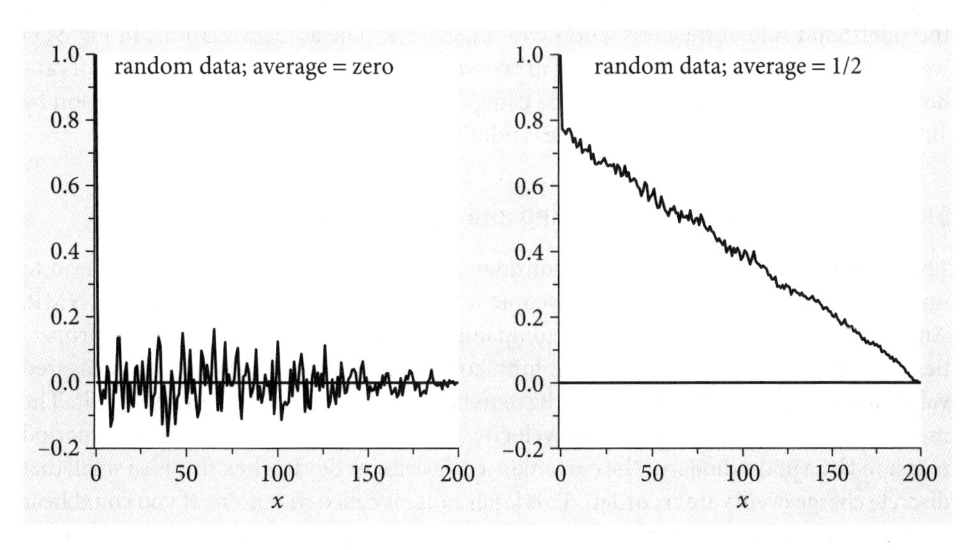

Fig. 9.33 Normalized autocorrelations of 200 random numbers with an average of 0 (left) and an average of ½ (right). Only the right-hand half of the autocorrelation is shown. The left-hand part is the exact mirror image.

autocorrelation of a random sequence of values where the mean is 0 (left) and ½ (right). When the mean is 0, only the first point has a value not essentially 0. When the mean is ½, there is a correlation between each point, and this decreases as the separation between points increases. Since the mean is ½ (or any value not zero), this means that each point is related to all the others, because, besides random fluctuations, they all have the same underlying value. Their correlation becomes less the further they are separated. The normalized autocorrelation of any line $y = constant$, is a sloping straight line starting at 1 and ending at 0. This is to be expected, because at zero displacement the line is overlapped with itself, whereas at the maximum displacement, only one term remains, see equation (9.36), and this value is small. In Fig. 9.33, the random noise has a large correlation at zero displacement because the whole trace must be perfectly correlated with itself; its value is 1 because the autocorrelation is normalized.

In calculating the autocorrelation of residuals, the mean value of the data is always subtracted first to prevent this sloping effect on the autocorrelation shown on the right of Fig. 9.33. Of course, if after doing this the autocorrelation is still sloping, then it clearly is not equally distributed about zero and the model used to describe the data may not be correct.

The correlation may be calculated using the algorithm shown below. The data for the two curves f and w each have a length n. The cross- or autocorrelation A has $2n$ points with indices ranging from 1 to $2n$ with the zero in the centre of the array. The constant term `normL` normalizes the correlation to a maximum of 1.

Algorithm 9.5 Correlation

```
> n:= 200:                    # number of data points
  # correlation into array A; data into arrays f and w.
  f:= Array(1..n): w:= Array(1..n): A:= Array(1..2*n):
  for i from 1 to n do
    f[i]:= ..etc    :         # define equation
    w[i]:= ..etc    :         # define equation
  end do:
  # normalise A. Loop 'for k' etc calculates A
  normL:= sqrt( add(f[i]^2,i=1..n)*add(w[i]^2,i=1..n) ):
  for k from 0 to n-1 do
    A[n-k]:= add(f[i]*w[i+k],i=1..n-k)/normL:
    A[n+k]:= add(w[i]*f[i+k],i=1..n-k)/normL:
  end do:
  s1:= seq([n-i,A[i]],i=1..n):             # plot just rhs
  plot([s1], view=[-1..n,-0.2..1]);
```

The `for k from ..do` loop runs over the displacement of f with respect to w; two calculations are performed both using Maple's built-in add function over the range 1 to `n-k`, as in equation (9.38). One addition calculates the left-hand side with index `n-k`; the other, the right-hand side of the correlation with index `n+k`. The autocorrelations in Fig. 9.33 were calculated in this way by making array `w` the same as `f`, and with `f`, an array of random numbers in the range 0 to 1 made using `f[i]:= rand()/1e12`. The division by 10^{12} is needed because of the size of the random numbers that Maple generates.

9.8.2 Autocorrelation of fluctuating and noisy signals

The autocorrelation of noise is now considered, and in the next section this will lead to understanding the shape of a spectroscopic transition and this is illustrated with NMR. Any experimental measurement is accompanied by noise. When measuring the properties of single atoms, molecules, or photons, considerable fluctuations in their measured values are expected and many events have to be averaged to obtain a precise result. The measured property might be energy, velocity, the number of photons in a given period measured by a photodiode, or the current in a transistor or diode when this is so small that discrete charge events are recorded. This latter noise is called *shot noise*. If you could hear shot noise, the effect would be rather similar to the sound of heavy rain falling on a car's roof. There is thermal noise in all resistors in electrical circuits that causes fluctuations in the current. These fluctuations are caused by the thermal motion of the many electrons as they pass through the inhomogeneous material forming the body of resistor. The frequency of the noise measured on an oscilloscope is determined by the frequency with which the circuitry responds and therefore depends on the capacitance, resistance, and inductance. This generally produces noise with a spread of frequencies of about equal amplitude, except for multiples of mains frequency and those of switched-mode power supplies, and is called *white noise*. At low frequencies, the amplitude of the noise increases in direct proportion to $1/f$ where f is frequency and is therefore called '$1/f$' noise. The origin of $1/f$ noise is not fully understood.

On the macroscopic scale, noise also accompanies experimental measurements. Measuring the amount of any of the many trace gases, such as CO_2, IO_2, and NO_x, in the atmosphere using optical techniques is an inherently noisy process. This is due to the continuous and erratic motion of air packets along the line of sight during the measurement and from one measurement to another. The frequency of the noise is, however, mostly limited to the speed at which the air changes. In the laboratory, all sorts of noise sources can affect an experiment; mostly these are due to voltage or current ripple in DC power supplies. In sensitive laser experiments, noise can be caused by dust particles in the air, vibrations of the building and from the air flow coming from air conditioning units. Atomic force microscopes often have to be suspended inside a sound proof box by elastic bungee ropes, to avoid adding noise to the measurements from the vibrations of the building and from nearby traffic. In an attempt to reduce noise a Fourier transform and an autocorrelation of the signal will provide information about the frequencies present and hence the possible source. The transform can also be used to remove noise as illustrated in Section 9.10.

Suppose that the noise on a measurement is represented by some fluctuating signal $f(t)$, the frequency of which is determined by the nature of the experiment and by the measuring apparatus. This signal will be represented by a general Fourier series similar to that in Section 9.1.1 but where T is the period over which a measurement is made and the summation starts from zero as this makes the resulting equations simpler,

$$f(t) = \sum_{n=0}^{\infty} a_n \cos\left(\frac{2\pi nt}{T}\right) + \sum_{m=0}^{\infty} b_m \sin\left(\frac{2\pi mt}{T}\right).$$

Following Davidson (1962, chapter 14), the time average of f and f^2 is the respective integral divided by the time interval T. The average $\langle f \rangle$ is zero because the noise is random, but the average of f^2 is not; the integral is

$$\langle f^2 \rangle = \frac{1}{T}\int_0^T \left[\sum_{n=0}^{\infty} a_n \cos\left(\frac{2\pi nt}{T}\right) + \sum_{m=0}^{\infty} b_m \sin\left(\frac{2\pi mt}{T}\right)\right]^2 dt,$$

which simplifies considerably because of the orthogonality of the cosine integrals such as $\int \cos\left(\frac{2\pi nt}{T}\right)\sin\left(\frac{2\pi mt}{T}\right)dt = 0$. The result is

$$\langle f^2 \rangle = \frac{1}{2}\sum_n (a_n^2 + b_n^2).$$

This expression can also represent the average of many measurements if the coefficients a and b themselves represent average values. The variance (the square of the standard deviation) on the signal is $\sigma^2 = \langle f^2 \rangle - \langle f \rangle^2$ and in this case the standard deviation is $\sqrt{\langle f^2 \rangle}$ and is the determined only by the amplitudes a, b of the noise and $a^2 + b^2$ represents the energy in the noise.

The autocorrelation of $f(x)$ is

$$A(u) = \langle f(t)f(t+u) \rangle = \frac{1}{T}\int_0^T f(t)f(t+u)dt$$

which looks quite complicated when the substitution for f is made. However, using the formulas for $\sin(A + B)$, $\cos(A + B)$ and the orthogonality rules, a remarkably simple result is produced:

$$A(u) = \frac{1}{2}\sum_n (a_n^2 + b_n^2)\cos\left(\frac{2\pi nu}{T}\right). \quad (9.43)$$

9.8.3 Wiener–Khinchin relations

The autocorrelation (equation (9.43)) is related to the energy or power in a given signal. For example, with electromagnetic radiation the energy is the square of the amplitude of the electric field, the field is given by the constants a and b thus $a^2 + b^2$ represents the energy. This is also true of a sound wave. The power dissipated in a resistor is proportional to the current squared and the kinetic energy of a molecule is proportional to the square of the velocity. Thus, in general if the signal is f, $\langle f^2 \rangle$ represents the average energy or power. The period T (equation (9.43)) is somewhat arbitrary and can reasonably take on any value; therefore, it is possible to define $n/T \equiv \nu_n$ as a frequency. The amount of power P in a small frequency interval from ν to $\nu + d\nu$ is therefore $P(\nu)d\nu = \dfrac{1}{2}(a_\nu^2 + b_\nu^2)$ and the autocorrelation can be written as an integral over frequencies rather than a summation over index n. This effectively means that there are so many terms in the sum that it can be changed into an integral without any significant error, and doing this produces the autocorrelation;

$$A(u) = \int_{\nu=0}^{\infty} P(\nu)\cos(2\pi\nu u)d\nu$$

Comparing this equation with a Fourier transform equation, the power spectrum is

$$P(\nu) = 4\int_{u=0}^{\infty} A(u)\cos(2\pi\nu u)du$$

and these equations are known as the Wiener–Khinchin relationships: the power spectrum and autocorrelation form a Fourier transform pair. The power spectrum is proportional to what we would normally observe in a spectroscopic experiment, as the change in the signal vs frequency. The width of the signal is determined by the autocorrelation and this is determined by the noise. If the noise is due to a random process then it is often found that the autocorrelation decays exponentially as $e^{-t/\tau}$ with rate constant k. In this case the power spectrum $P(\nu)$ is

$$P(\nu) = 4\int_{u=0}^{\infty} e^{-u/\tau}\cos(2\pi\nu u)du = \frac{4\tau}{1 + (2\pi\nu\tau)^2}$$

and the integral is most easily evaluated by converting the cosine to its exponential form. The nature of the random processes contributing to the power spectrum is now considered using NMR as an example. In NMR the function $P(\nu)$ is called the spectral density and usually given the symbol $J(\nu)$.

The nuclear spin angular momentum in a molecule remains in fixed precessing motion governed by the external magnetic field, but the molecules themselves rotate and this causes the nuclear spin to experience a fluctuating magnetic field in addition to the applied external field. Therefore, nuclei undergoing NMR transitions experience this fluctuating field and the effect is to return the nuclear spin population to equilibrium with a lifetime called T1 (Sanders & Hunter 1987; Levitt 2001). The timescale of the fluctuations is of the order of tens of picoseconds because this is the timescale of molecular rotation. It is also similar in frequency to that of the NMR transition (Larmor frequency) and therefore rotational diffusion can greatly influence the return to equilibrium of the nuclear spins and can dominate the T1 and T2 decay processes. Loss of spin coherence is characterized by the lifetime T2. Molecular translational diffusion is far slower than rotation and so causes magnetic field fluctuations at a far lower frequency than the NMR transition and is therefore less important for T1 processes. Similarly, vibrational motion is too high to influence the NMR transition. Large molecules in a viscous solvent have a sluggish response and a small rotational diffusion coefficient, and long rotational relaxation times, and vice versa. However, while different solvents and molecules of different sizes will change the frequency of the random magnetic field fluctuations, the timescale remains comparable to that of the NMR transition. In proteins, while overall rotation can be slow, ≈ tens of nanoseconds, faster local motion of residues called 'wobbling in a cone' motion still occurs. The autocorrelation of rotational diffusion can be shown to be an exponentially decaying function with a lifetime τ proportional to the reciprocal of the rotational diffusion coefficient. Fig. 9.34 shows the spectral density for different rotational relaxation times. The coupling of the magnetic field fluctuations is most effective when $1/2\pi\tau$ is close to the Larmor frequency and therefore molecules of different sizes will be affected differently.

When plotted on a linear scale the spectral density of a slowly decaying exponential autocorrelation is a narrow function centred at zero frequency, whereas the rapidly decaying autocorrelation has the same shape but is wide. Zero frequency here means the transition frequency. The line width is a consequence of the time-energy or time-frequency uncertainty, causing a wide spectral line when processes are rapid and vice versa. When plotted on a linear–log scale the power spectrum is constant over a wide range of low frequencies, and this is called 'white noise'. It rapidly decreases, centred about the frequency $1/2\pi\tau$ as is shown in the figure. If the noise were completely random, the power spectrum would be constant at all frequencies.

The Weiner–Khinchin theorem also shows that the autocorrelation of the signal f is the squared modulus of its Fourier transform $g(k)$. Apart from a constant of proportionality, this is

$$A(u) = \int_{-\infty}^{+\infty} f^*(t)f(u+t)dt = |g(k)|^2.$$

Because the squared modulus of the Fourier transform is produced, the autocorrelation has lost all phase information so it is not possible to invert or reverse $g(k)$ to produce the original function which is f. Thus, in the NMR case, it is not possible to measure the spectral density, which is proportional the shape of the NMR transition, and work backwards to obtain the function that produced this shape. All that can be done is to generate a model of the interactions, such as rotational diffusion, and fit this theoretical model to the data.

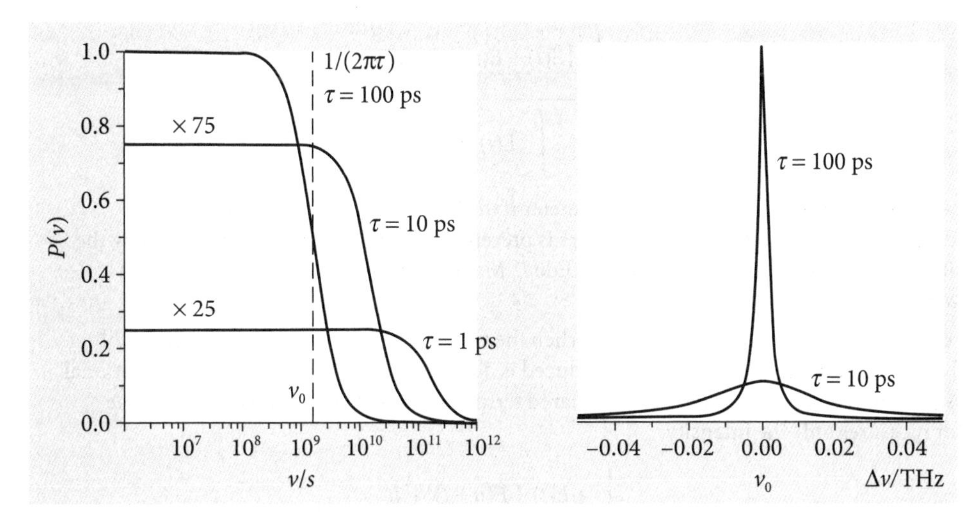

Fig. 9.34 Left: Power spectra (or spectral density) vs. frequency for a signal that has an exponential autocorrelation function, the decay lifetimes of the exponentials are from 1 to 100 ps. The density of the fluctuation in the noise is almost constant at lower frequencies and this is called 'white noise'. Right: Two of the same functions plotted on a linear scale showing more clearly the spectral width, which is proportional to the shape of an NMR transition. (The abscissa is in THz ≡ 10^{12} Hz.)

9.8.4 Questions

Full solutions are available at **www.oxfordtextbooks.co.uk/orc/beddard.**

Q9.16 In example (iii) of Section 9.8.1 describing the optical correlator, Fig. 9.32, a special case has been examined because the laser intensity was given as $e^{-(t/a)^2}$ which represents the envelope of the pulse and ignores the electric field itself. However, it is the *amplitude* of the laser's electric field E that is important, the square of which is the intensity I. A *fringe-resolved* autocorrelation is recorded if a fast responding photodiode is used to detect the signal; this has the shape of a pulse but with sinusoidal type oscillations within the pulse envelope. The amplitude of E describes how the electric field of the light varies throughout the pulse. In a transform-limited pulse, the frequency is constant across the pulse duration. In a *chirped* pulse, the frequency varies, either increasing or decreasing as time progresses. If this could be heard, it would, as the name suggests, sound like a bird's chirp.

The *amplitude* profile of a transform limited pulse, can be written as $E(t) = \cos(\omega t/a)e^{-(t/a)^2/2}$ where $a = \tau/\sqrt{2\ln(2)}$ and τ is the pulse fwhm and ω its central wavelength; Fig. 9.35 shows the pulse shape.

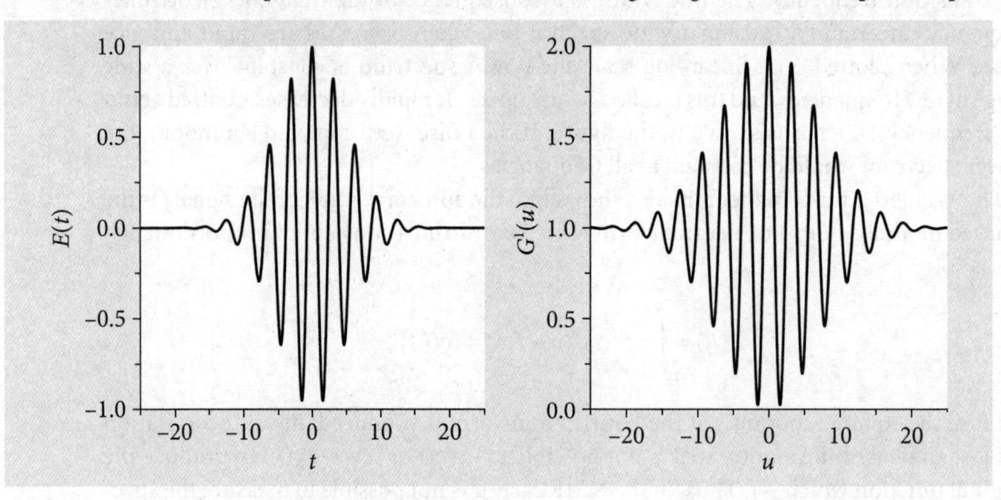

Fig. 9.35 Electric field $E(t)$ of a laser pulse, left, and right, first order autocorrelation $G^1(u)$ normalized to 1 in the wings. The constants are $a = 5$ and $\omega = 10$.

(a) Start by using algebra then Maple to calculate the (first-order) amplitude correlation of a laser pulse

$$G^1(u) = \frac{\int_{-\infty}^{\infty} |E(t) + E(u+t)|^2 dt}{\int_{-\infty}^{\infty} E(t)^2 dt}$$

which would be measured from the interferometer if the beam is directly measured on a photodiode at its usual wavelength. The square is present because the detector measures only the intensity of the light $|E^*E|$ not the field amplitude E. Make the result into a function and plot using $a = 5$ and $\omega = 10$.

(b) The second-order correlation $G^2(u)$ is made when the two beams overlap in a frequency doubling crystal and then only the ultraviolet light produced is detected. The amount of this is proportional to the square of the total field. The second squared term in the equation is present because the detector measures only the intensity.

$$G^2(u) = \frac{\int_{-\infty}^{\infty} |[E(t) + E(u+t)]^2|^2 dt}{\int_{-\infty}^{\infty} E(t)^4 dt}$$

The analytical result is complicated, so use Maple to calculate and plot the autocorrelation, again using $a = 5$ and $\omega = 10$.

(c) Ignore the cosine part of the field E and so calculate the upper and lower bounds of the second order autocorrelation and add these to the graph. Calculate the fwhm of the upper curve and determine how much wider this is than the pulse $t_{1/2}$.

(d) The pulse seems to have only one frequency according to the equation for $E(t)$, but the pulse is short and by time-energy uncertainty a short pulse must have a wide spectral width. Calculate this by Fourier transforming the pulse if a is 5 fs and $\omega = 10\ \text{fs}^{-1}$.

Strategy: As the pulse is real, calculating the absolute value has no effect and the squares can be taken directly. To simplify the calculation, expand out the terms first and collect similar terms together. For example, the second order autocorrelation has terms $\int E(t)^4 dt = \int E(u+t)^4 dt$ and $\int E(t)E(u+t)^3 dt = \int E(t)^3 E(u+t) dt$ both of which could be checked with Maple.

Q9.17 Femtosecond laser pulses have been assumed to have different shapes; $\exp(-|x|/a)$ and $\text{sech}^2(x/a)$ where sech is the reciprocal of the hyperbolic cosine which is cosh, and $\cosh(x) = (e^x + e^{-x})/2$. The sech^2 function looks very much like a Gaussian.

Calculate the autocorrelation of these two pulse shapes, and work out the pulse-widths and fwhm from each autocorrelation and how much wider these are than the laser pulse that generated them. You will need to solve an equation numerically and do this using the Newton–Raphson method (Chapter 3.10); using Maple's `fsolve` does not always give all the solutions.

Strategy: Start with equation (9.42), but notice that the absolute value of x is used. Using Maple, the integration is far simpler if it is assumed that a is positive and that x real. There is a discontinuity in the sech^2 integral but this can be overcome with the help of Maple's help pages and using the `continuous` instruction.

Q9.18 Bats produce a range of high frequency acoustic signals that they use for echolocation. In some species, this is a narrow-band constant frequency and in others, it is broadband and chirped in frequency. In yet other species both signals are used; the constant frequency when locating prey and the chirped signal when attacking, see Altringham (1996). A chirped pulse has a frequency that changes during its duration. As bats hunt where there are many other objects around besides insects, tree branches are an example, being able to discriminate one object from another down to at least 3 cm is clearly important because this is the size its prey. Simmons (1971) has devised experiments suggesting that the amplitude of the envelope of the correlation of a bat's acoustic pulse with that scattered off its prey is used to discrimination between objects. He also suggests that the autocorrelation rather than the cross-correlation is sufficient for this purpose. The larger the amplitude of the correlation envelope between the outgoing and returning signal is, the more difficult it is for the bat to discriminate its prey from another nearby object. When the correlation envelope is at its maximum amplitude there is only a 50% chance that the prey has been correctly identified; when the convolution envelope is smaller, discrimination of objects separated by 3 cm rises to almost 100%.

The amplitude of the envelope of the correlation is hard to determine, but its integral with time is not since at any time it is approximated as the sum of the absolute value of the correlation from all previous times. We will suppose that the size of this integral can instead be used as a measure of the success of prey discrimination.

The function

$$S(t) = (e^{-12t} - e^{-13t} + 0.02e^{-10(t-0.5)^2})\cos(150t - a/(t+1)^2),$$

very approximately mimics the acoustic pulse of insectivorous bats (*Eptesicus fuscus*), where time is measured in milliseconds. An object at 12 mm returns a signal after 70 μs, and this allows the calibration in distance instead of time.

(a) Plot $S(t)$ for an un-chirped pulse, $a = 0$, and a chirped pulse, $a = 60$. Next, using Maple, convert it into an array, which will make a set of data points as if they had come from an instrument, and then calculate and plot the autocorrelation.

(b) Calculate the sum of the autocorrelation with time, and compare this with the experimental data below for the percentage of successful discrimination. (Data measured from Simmons 1971, Fig. 2.)

Distance	0	0.5	1	1.5	2	3	4	5	6	7	8	9	10
% correct	52	61	70.5	78	88	96	95	97	94	96	97.5	96	96

(c) Explain why an un-chirped pulse is very poor compared to the chirped one, when discriminating between two objects if the bat uses autocorrelation.

Strategy: Use Algorithm 9.5, to calculate the autocorrelation. If you use $n = 500$ data points to convert to data points to calculate the autocorrelation, change the arrays defined in the algorithm to

```
> f[i]:= S(i/n):
  w[i]:= S(i/n):
```

This makes the maximum value of t equal to 1, and therefore the maximum time equal to 1 ms which is 2 μs per point. To plot the summed signal up to a given time, convert time to distance with 12 mm = 70 μs.

9.9 Discrete Fourier transforms (DFT) and fast Fourier transforms (FFT)

9.9.1 Concept

Data is normally produced by an instrument as a series of numbers measured at equal intervals, and not as an equation; therefore, any Fourier transform has to be calculated numerically. The sampling intervals could be time, distance, or some other quantity depending on the experiment. Time will be used in the following examples. The numerical transform is usually called a discrete Fourier Transform (DFT) and the algorithm used is called the fast Fourier Transform FFT. A frequently used algorithm was devised by Cooley and Tukey (1965) (Bracewell 1986, p. 370). It is not necessary to devise one of these transforms, because the method is very well established and fast operating code has been written and checked. Explanations of its working can be found in texts such as *Numerical Recipes* (Prest et al. 1986). Both Maple and Mathematica have built in discrete Fourier transform routines as do numerical packages such as MathCad, Matlab, IDL, IGOR and Origin. These routines can be used as black boxes, nevertheless a clear understanding of the discrete transform is essential to avoid making mistakes. Some examples of Fourier transforms are shown schematically in Fig. 9.36.

The cosine transform produces two delta functions at frequencies $\pm 1/p$ in the top figure, and $\pm 2/p$ in the second figure. Because the cosine period is halved, the frequency is doubled. The 'typical data', row (C), is supposed to represent some experimental data and if ideally transformed, with an infinite number of sampled points, will give the schematic spectrum on the right, which is shown as its real part because the transform may be complex. The imaginary part would look something like the derivative of the real part. The data has a large number of low frequencies in it but few high frequencies, which is why the transform tails off as the frequency increases. The maximum frequency present in the data is found to be ν_m. In row (D), the function is sampled at times starting at zero and separated by Δ, and a discrete transform is calculated. The highest frequency present in the transformed data can only be $1/\Delta$; the lowest frequency is $1/T$ where T is the length of the data. To correctly sample the data $1/(2\Delta) \equiv \nu_c > \nu_m$; this is known as the Nyquist condition and is explained further below. The discrete transform is usually plotted in an unfolded form as shown in row (D), Fig. 9.36. Row (E) shows it in the same form as the ideal or mathematical transforms. Notice also that the frequency axis is ν/N, where ν, the frequency integer, is the index number of each point sampled and N is the number of sampled points. The ratio ν/N is not the same as a conventional frequency, because both ν and N are integers.

There are a number of important differences between discrete transforms that relate to real experimental data and the ideal transforms of mathematical functions. These all relate to how the often the data has been sampled, which means how many data points are measured compared to how rapidly the data changes with time. Typically, $2^{13} = 8192$ to $2^{16} = 65\,536$ data points will be needed with actual data but for speed of calculation, smaller numbers are used in the examples below.

The discrete transform $g(\nu)$ of a signal $f(t)$ sampled at discrete points t, is defined as

$$g(\nu) = \frac{1}{\sqrt{N}} \sum_{t=0}^{N-1} f(t) e^{-i2\pi(\nu/N)t}. \tag{9.44}$$

where N is the number of data points and ν and t are integers. The ratio ν/N, which can only take on certain values because both ν and N are integers, is unlike a true frequency, $\upsilon \equiv \omega/2\pi$ which can take any value. The values of ν range from 0 to $N-1$, and the ratio takes only N values which are $\nu/N \rightarrow 0, 1/N, \cdots (N-1)/N$. Transforms with ν/N above 1/2 do not contain any new data, but a copy of the earlier part of the spectrum. Therefore the transform must be treated with some care to avoid interpreting data in this second half of the spectrum as something different to that in the first. This means not interpreting a feature at an apparently high frequency as something different from its twin at low frequency. The symmetry of the transformed data is seen in Fig. 9.36. In terms of frequency, a point ν has a frequency $f = \nu/(N\Delta)$, the maximum frequency is $f = (N-1)/(N\Delta)$ because the last point in the transform is numbered $N-1$ and as Δ is the separation of any two data

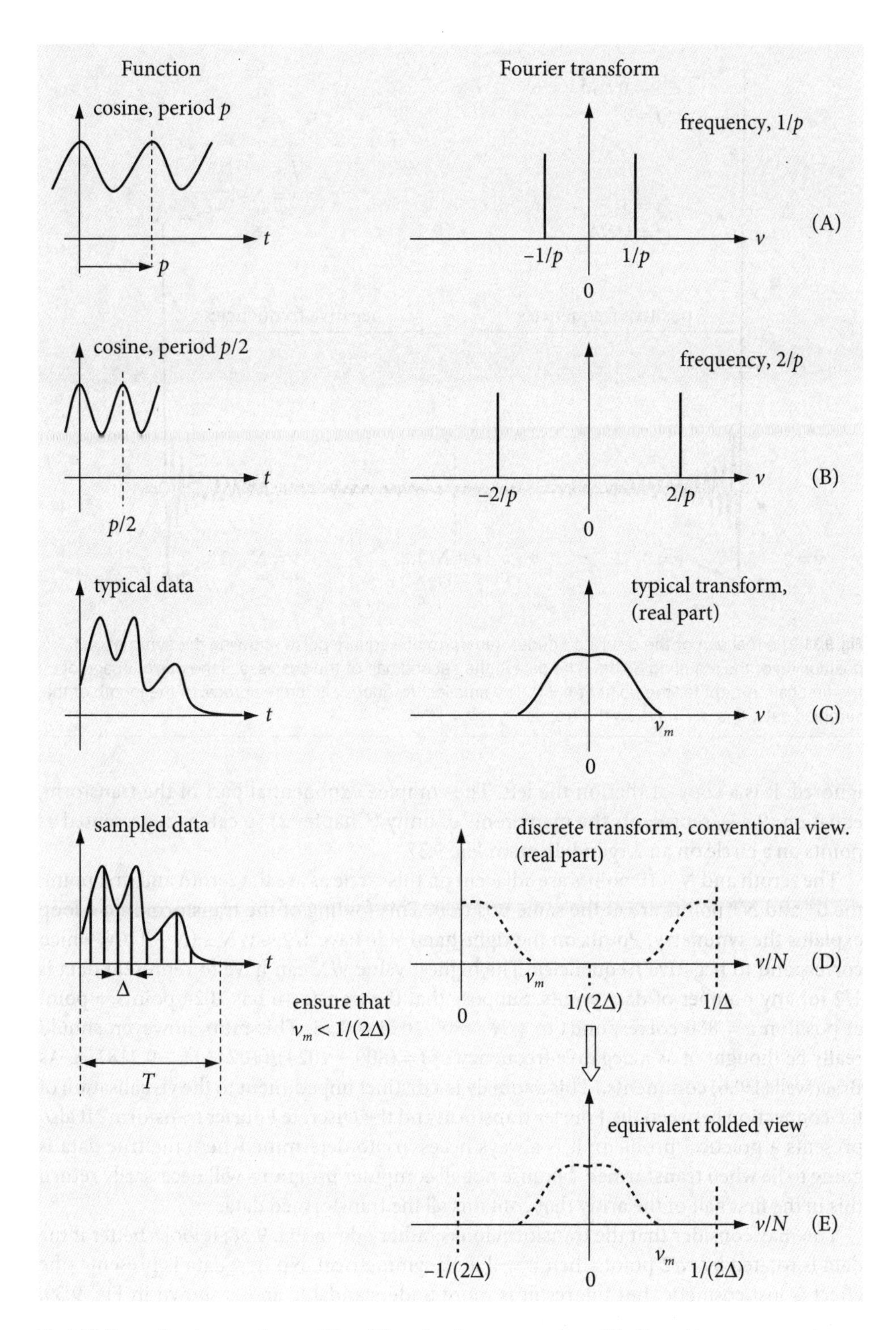

Fig. 9.36 Some Fourier transforms, (A) to (C), and a discrete transform (D). (Based in part on a web diagram of M Levoy, Stanford University.)

points, so $1/\Delta$ is a frequency. See Bracewell (1986) and Prest et al. (1986) for a more detailed discussion.

The discrete transform algorithms require 2^N sampled data points, where N is an integer. If the data does not contain exactly 2^N points, then zeros can be added to the end of the data to make up the number of points required. An example of a discrete transform is shown in Fig. 9.38; notice in particular where the data appears in the transform. The data being transformed is a set of points representing a square pulse and the transform in the figure is symmetrical about its centre except for the point at $\nu = 0$. Any data in the right-hand half of the transform can be

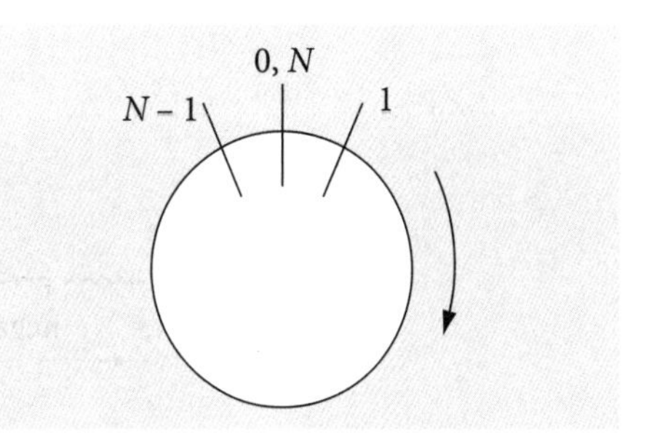

Fig. 9.37 Roots of a complex exponential represented as points on a circle.

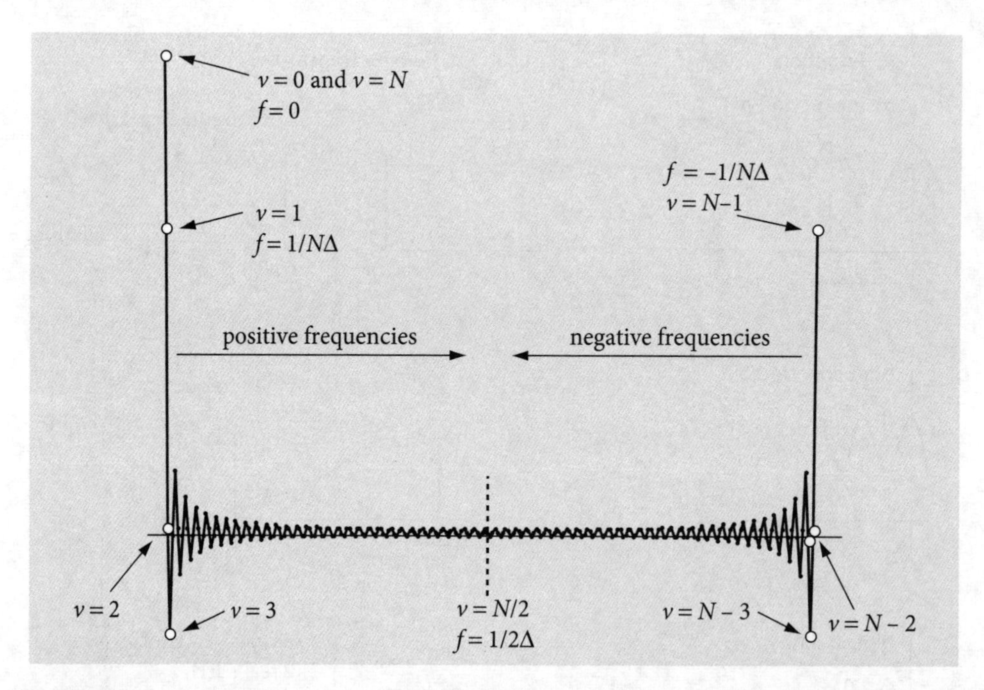

Fig. 9.38 The real part of the discrete Fourier transform of a square pulse showing the symmetrical positioning of the resulting points. The data in the second half of the abscissa is the mirror image of the first half, except for the point at $v = 0$. The smallest frequency is the reciprocal of the length of the sampled data, this is $T = N\Delta$ so the frequency is $f = 1/N\Delta$.

ignored; it is a copy of that on the left. The complex exponential part of the transform, equation (9.44), represents the many roots of unity (Chapter 2) so can be represented as points on a circle on an Argand diagram Fig. 9.37.

The zeroth and $N - 1^{\text{th}}$ points are adjacent on this circle as are the zeroth and first point; the 0th and N^{th} points are at the same position. This folding of the transform into a loop explains the symmetry. Points on the right-hand side have $1/2 < v/N \le (N - 1)/N$, which correspond to negative frequencies. The highest value v/N can have to represent data is 1/2 for any number of data points. Suppose that the transform has 1024 points; a point at position $v = 800$ corresponds to $v/N = 800/1024 \approx 0.78$. This ratio, however, should really be thought of as a negative frequency of $f = (800 - 1024)/(1024\Delta) = -0.2187/\Delta$. As Bracewell (1986) comments, 'This anomaly is a distinct impediment to the visualisation of the connection between the Fourier transform and the Discrete Fourier transform'. It also presents a practical problem; it is always necessary to determine where the true data is going to lie when transformed, because not all computer programs will necessarily return this in the first half of the array that contains all the transformed data.

You may consider that the transform looks rather odd in Fig. 9.38; it looks better if the data is rotated by $N/2$ points then it will look symmetrical. No new data is present—the effect is just cosmetic, but the result is more understandable and is shown in Fig. 9.39. The transform now looks like the mathematical transform of a square pulse; see the sinc

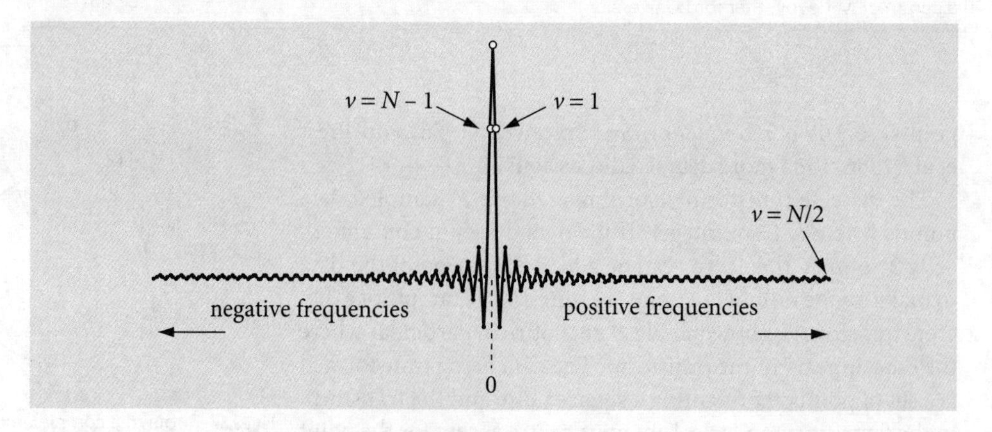

Fig. 9.39 The transform of Fig. 9.38 rotated by $N/2$ points, and clearly showing its symmetrical nature.

function, Fig. 9.15. If the centre frequency is zero, the idea of positive and negative ν/N now seems more intuitive.

9.9.2 Nyquist critical frequency and Shannon sampling theorem

The Nyquist critical frequency has been mentioned in passing, but now it is explained more fully. An experimental set of data is band-limited (bandwidth limited), if its maximum frequency has essentially zero amplitude, as sketch (C) shows in Fig. 9.36. Because the signal's frequency has a maximum, this means that any feature in the signal cannot change faster than the reciprocal of this frequency. If the signal were conveying information, then anything changing faster than this frequency would be lost. The maximum sinusoidal wave that can be extracted or recovered from data by a discrete (Fourier) transform is given by the Nyquist critical frequency $\nu_c = 1/2\Delta$, where Δ is the (time) interval between any two consecutive data points. Consider a set of experimental data points represented by a series of equally spaced points on the time axis separated by Δ. The highest frequency in the data, is clearly when consecutive points have the opposite sign,

$$\begin{aligned} data &= [+1 \quad -1 \quad +1 \quad -1 \quad +1 \quad \cdots] \\ index\ t &= [\ 0 \quad\ \ 1 \quad\ \ 2 \quad\ \ 3 \quad\ \ 4 \quad \cdots], \end{aligned}$$

and this series is equivalent to the function $e^{i\pi t/\Delta}$ if $\Delta = 1$, $i = \sqrt{-1}$ and t the index number of a data point, 0, 1, 2 $\cdots$. The exponential series with integer t is

$$e^0 = 1,\ e^{i\pi} = -1,\ e^{2i\pi} = 1,\ e^{3i\pi} = -1,\ \cdots.$$

The highest frequency in the data $\omega_{\max}$ occurs when $e^{it\omega_{\max}} = e^{i\pi t/\Delta}$ or $\omega_{\max} = \pi/\Delta$ and as $\omega = 2\pi\nu$ the

$$Nyquist\ critical\ frequency\ (in\ \text{Hz})\ \text{is}\ \nu_c = 1/2\Delta \tag{9.45}$$

or two points per cycle. The lowest frequency is zero when all the points have the same value and sign; [+1 +1 +1 +1 +1 $\cdots$]. A general digitized waveform therefore has frequencies that can range from 0 to the Nyquist critical frequency; the data may contain higher frequencies than this, but these cannot be measured if they are higher than the critical frequency. Therefore, if twice as many data points than before are sampled in a given time period, the Nyquist frequency increases, because Δ is then smaller and higher frequencies in the data can be transformed properly. Conversely, if Δ is too large, not enough sampling occurs and the data is not properly described, but this begs the question; how many points are too many and how many too few? The Shannon sampling theorem dictates exactly the minimum number of data points needed to ensure that data of a certain frequency is properly sampled. This theorem states that, for reliable replication of any waveform, two or more data points per period are required to define each sinusoid in the data, which means that the sampling frequency must be greater than twice the highest frequency in the signal

$$\nu_{sample} \ge 2\nu_{max}. \tag{9.46}$$

Whether the Nyquist or Shannon condition applies depends on your particular situation. If you have some data, and for an instrumental reason can sample no higher than a certain value, any frequencies higher than the Nyquist will be under-sampled and not be measured properly. If on the other hand you can sample as often as you wish, no more samples than those predicted by the Shannon limit are needed. However, this sampling rate must apply to *all* the data even though its value is dictated only by the highest frequency present. If the data is not sampled frequently enough, then *aliasing* occurs and extra false signals are observed in the Fourier transform. This effect is described after a discrete transform has been calculated.

The Nyquist frequency has great implications for NMR and effectively dictates how the experiment is performed. A typical spectrometer operates at 400 MHz so that sampling should be done at 800 MHz or every 1.25 ns, which is a very tall order even with today's fast electronics. The amount of data collected would also be vast because the FID can last for several seconds, therefore, $\approx 8 \times 10^9$ points over 10 seconds and perhaps four times as many bytes depending on the word size of the computer. However, the small frequency shift in the NMR signal from the RF frequency used to excite the nuclei comes to the

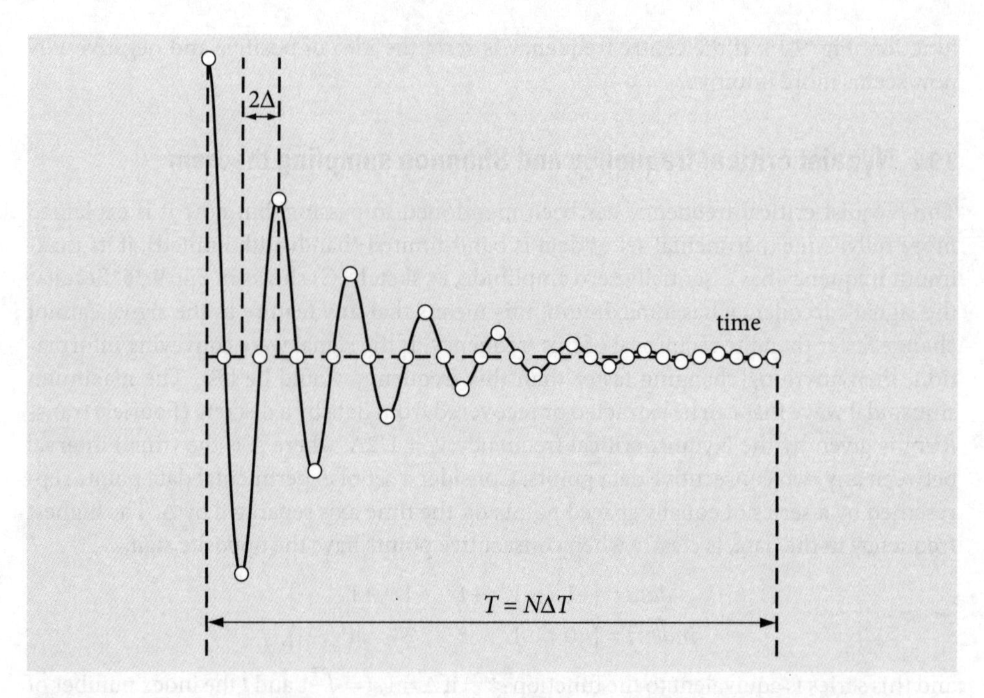

Fig. 9.40 A sine wave of decreasing amplitude and its sampling at 2 points per cycle; no more points are needed for such a sine wave. If the frequency changes with time, for example if the signal is chirped, then more data points are needed across the whole curve to describe the highest frequency correctly.

rescue. This shift is only a few kHz, and if the NMR signal from the detector is mixed (heterodyned), with the frequency of the RF source beating will occur and the KHz frequency shift is extracted. (An analogue radio extracts the audio signal from the MHz broadcast frequency by a similar process.) After heterodyning, it is now relatively easy to sample the low frequency signal at a sufficiently high time resolution, in milliseconds not nanoseconds, to satisfy the Nyquist condition and accurately reproduce all frequencies in the signal. After a discrete Fourier transform, this signal becomes the NMR spectrum.

9.9.3 Calculating a discrete Fourier transform (DFT)

As an illustration of numerically obtaining a discrete transform using the fast Fourier transform algorithm, some data will be synthesized and then transformed. Each data point is in principle a complex number, and therefore can be displayed in different forms: as the real and imaginary parts, as the absolute value, and as the phase. If Z is a complex number, recall that the absolute value, which is also called the modulus or magnitude, is $|Z^*Z|^{1/2}$. The square of this, when plotted against frequency, is called the power spectrum. The phase of the signal, in radians, is defined as $\theta = \tan^{-1}\left(\dfrac{\mathrm{Im}(Z)}{\mathrm{Re}(Z)}\right)$.

The data shown in Fig. 9.41, simulates the FID from a mock NMR experiment, with two spins producing NMR transition frequencies of 1/35 and 1/20 Hz. The FID comprises two cosines multiplied by an exponential decay with a 200 s decay time, and extends to 1000 seconds. In a real NMR experiment, 16384 (2^{14}) points would be typically used and the total time range would be a fraction of a second to a few seconds duration. The transition frequencies are typically in the 100 to 500 MHz, region but the actual frequency depends on the strength of the permanent magnetic field.

The time between any two data points is $\Delta = T/N$ where T is the total time range of the data. In this example, this is 1000 seconds, and the number of data points $N = 2048$. This time increment per point Δ is called `tp` in the calculation, and is $1000/2048 \approx 0.488$. The maximum, or Nyquist frequency that can be calculated using this data is $1/(2\Delta)$ or 1.024 Hz and both frequencies are well below this. The real, imaginary, and phase parts of the transform are calculated and plotted against frequency in Hz, which is $(\nu - 1)/(tp \times N)$ where ν is the index number. We could also plot against the index number itself, or against angular frequency $2\pi(\nu - 1)/(tp \times N)$. It is important to decide what units are to be plotted. Note that the Nyquist frequency is measured in Hz, not angular frequency.

Algorithm 9.6 Discrete Fourier transform

```
> restart: with(DiscreteTransforms):
> nu1:=1/35.0: nu2:=1/20.0:     # define frequencies in Hz
  # define FID as a function called data
  data:= x-> exp(-x/200.0)*(cos(2*Pi*nu1*x)+cos(2*Pi*nu2*x));
  T:= 1000:                                  # maximum time
  plot( data(x),x= 0..T,title= "FID");
  N:= 2^11;                                  # number points
  tp:= evalf(T/N);                           # time per point;
  FID:= Array(1..N):                         # FID array
  # sample data integer index is t
  for t from 0 to N-1 do
    FID[t+1]:= data(tp*t):                   # FID data
  end do:
  FT:= FourierTransform(FID):                # perform FFT
  # next make real part of data and x axis values in frequency
  FT_data:= seq( [ (v-1)/(tp*N), Re(FT[v]) ], v=1..N ):
  plot([FT_data],view=[0..0.1,0..5],title="spectrum");
```

The imaginary part of the transform and the phase are shown in Fig. 9.42. A new sequence is defined in each case and then plotted as shown above. For instance;

```
> phase_data:=
seq([ (v-1)/(tp*N),arctan(Im(FT[v])/Re(FT[v])) ],v=1..N):
```

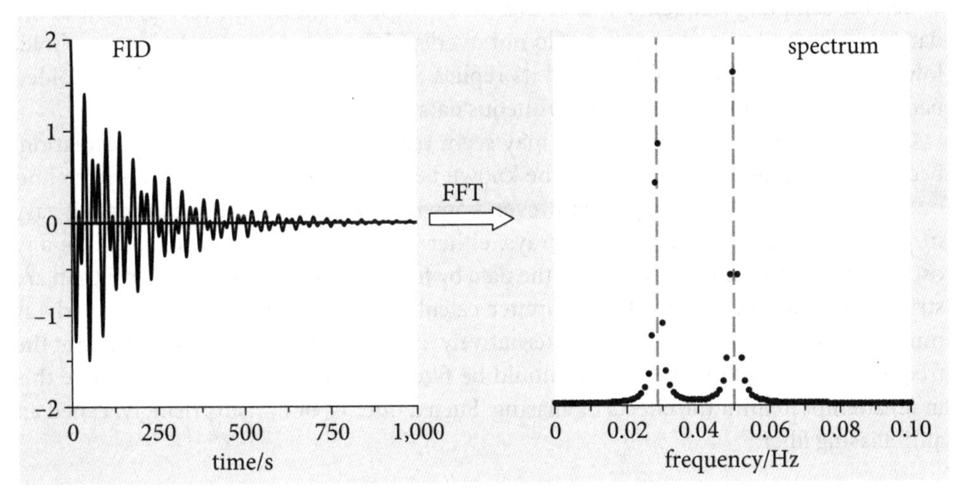

Fig. 9.41 The time decaying FID, and (right) the real part of its discrete Fourier transform with frequencies of 1/35 = 0.0286 and 1/20 Hz. Only the first part of the spectrum is plotted; its mirror image at higher frequencies than the Nyquist limit of 1.024 Hz is not shown.

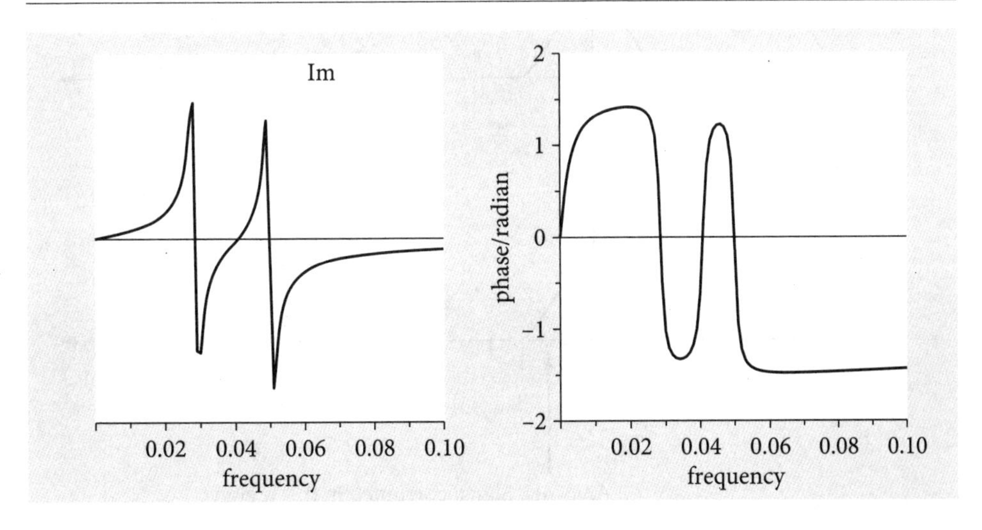

Fig. 9.42 Imaginary and phase plots of the transformed data. The phase is in radians, 2π of which equal 180°. At larger frequencies, the phase approaches zero, reaching it at point $N/2$ or 1.024 Hz, which is far to the right of the plotted range.

If the real and imaginary parts of the transform are plotted against one another, rather than against frequency, a curve resembling a figure of eight is produced.

9.9.4 Aliasing

Under-sampling means sampling at less than twice the highest frequency component in the data. This means that adjacent sampling points do not have a small enough separation to represent the highest frequency feature. The consequence of under-sampling a signal is erroneously to fold high frequency parts into the lower frequency parts of the Fourier transform; the effect is called *aliasing*.

A sufficient condition for the proper sampling of a sinusoidal signal is that the sampling frequency ν_s obeys $\nu_s/2 \ge \nu_{\max}$. Suppose that for some reason, this condition cannot be met, then as the frequencies of sinusoids in the data increase from $\nu_s/2$, which is the maximum frequency for correct sampling, to ν_s, these sinusoids are under-sampled and their apparent frequency moves down from $\nu_s/2$ to zero. To emphasize this, another name for the Nyquist frequency $\nu_s/2$ is the folding frequency, and the aliased data is sometimes called an 'image' or 'replica'. When a sinusoid of frequency ν_1 is sampled with frequency ν_s, the resulting samples are indistinguishable from those of replicate sinusoids at frequency $|\nu_1 - n\nu_s|$ where n is any integer. The replica signals occur because a cosine with frequency $2\pi\nu$ is indistinguishable from one at frequency $2\pi\nu n$, because the cosine is periodic. When aliasing occurs, data from the adjacent higher frequency replica leak into the lowest primary one. The sketch, Fig. 9.43, attempts to show this. In the top sketch, the primary (normal) Fourier transform is shown as a solid line and replicates as dotted. When the sampling frequency ν_s is at least twice the maximum frequency present in the data, $\nu_{\max}$ the replica and transform do not overlap. When this condition does not hold, lower sketch, then the transform and its replica do overlap, and frequencies are folded back into the transform, leading to erroneous data.

Choosing the sampling frequency may seem rather like a chicken and egg situation, because the highest frequency has to be known before performing the transform to find out what the highest frequency is. However, knowing how often to sample the data is easy in practice and can be done in two ways: either ensure that several points sample any oscillatory feature, or frequency limit the data by filtering. More sampling points than are strictly necessary only make the computer calculation slower, but this is not nearly as much of a problem as it used to be. Alternatively, if your instrument cannot sample at the frequency you would like, the data should be filtered to remove frequencies above this in an attempt to limit the effects of aliasing. Such a filter is, not unsurprisingly, called an anti-aliasing filter.

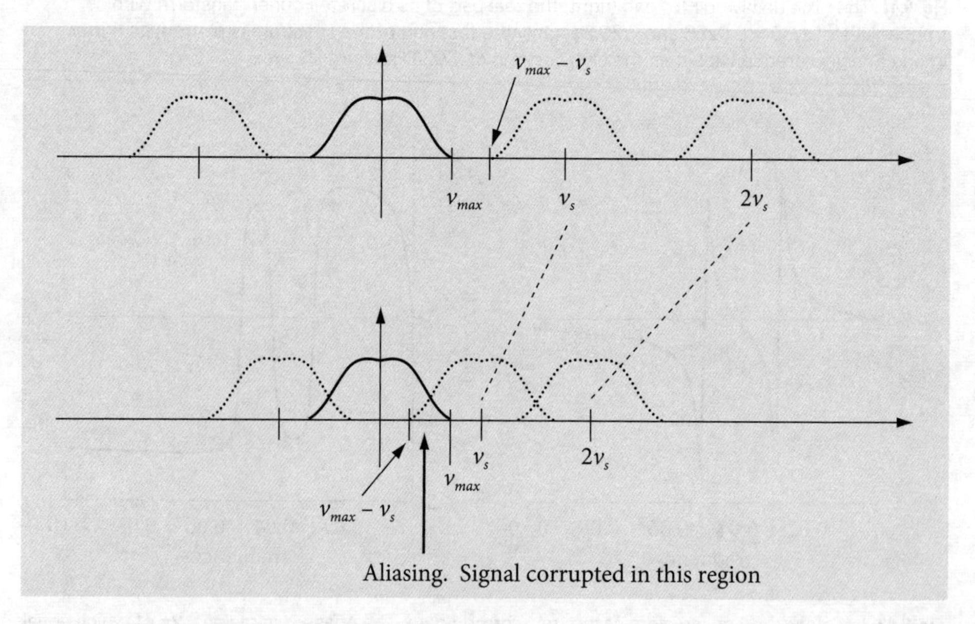

Fig. 9.43 Correct sampling (top) and under-sampling (bottom) leading to aliasing.

An everyday example of aliasing may be seen on your television. You may have noticed how the spokes of a car or wheel, or markings on a tyre appear to move slowly backwards or forwards at a rate far smaller than is expected to be the true rate of rotation. This is very apparent in F1 motor racing and also in old Westerns with horse-drawn carriages. The effect is due to the camera sampling the scene at too slow a rate to measure rapidly moving objects properly. Suppose spoke 1 is observed when the image is sampled. A short time later, the image is updated on the screen, but the wheel has rotated by a large angle and at the next observation time spoke 2 or even 3 is in the same, or almost the same position, as was spoke 1. As all the spokes look alike, spoke 1 appears to have moved only a small amount forwards or even backwards and the effect is very odd because the rapidly moving vehicle is seen with its wheel apparently moving backwards. The same effect may also be observed directly from the spokes of a car wheel, when viewed through vertical railings.

In the Fourier transform, aliasing has to be taken very seriously or data will be misinterpreted. Consider the curves shown below Fig. 9.44; the one on the left is an under-sampled, damped sine wave. The data, which would be recorded as the set of points shown, would appear as an exponential decay. Alternatively, if the samples were taken at a different time a straight line along the axis would be reported or perhaps as a rising exponential. This shows that the phase of the sampling is important; phase here means only where sampling is started. On the right, the data is over-sampled; more data points are present than are strictly necessary, but this is not a problem.

The calculation shown in Fig. 9.45, illustrates the effect of aliasing on an oscillating signal with a constantly changing frequency. The data, accurately sampled with 4096 points, is shown on the left and that with 256 sampled points, on the right. The abscissa on each transform is ν/N and ranges from 0 to $(N-1)/N \approx 1$. The failure of the sampling to reproduce the high frequency part of the data is clearly seen (B), where some data is missing. The real part of the transform (D), apparently has much higher frequency data present, but this is the effect of aliasing as seen by comparison with the accurate real part of the transform (C). The inset in (C) shows details up to $\nu/N = 0.06$. The transform (D) is still symmetrical about $\nu/N = 1/2$, and does also contain some correctly transformed data but only up to the Nyquist critical frequency of the 256 data points. You can see how the inset in (C) has a similar form to that in (D) at low frequency, but only up to about five periods of the wave. The amplitude is also smaller than (C), because the total amount of transformed data is the same but spread out over more frequencies. Reconstructing the data from (D) by reverse transforming produces the curve in (B) which is a very different waveform to that shown in (A).

To prevent aliasing, either the sampling rate has to be increased, or if this is not possible, the original data has to be filtered to remove higher frequencies. If the latter, the remaining transform will then be accurate, but only up to a certain frequency depending upon how filtering is performed.

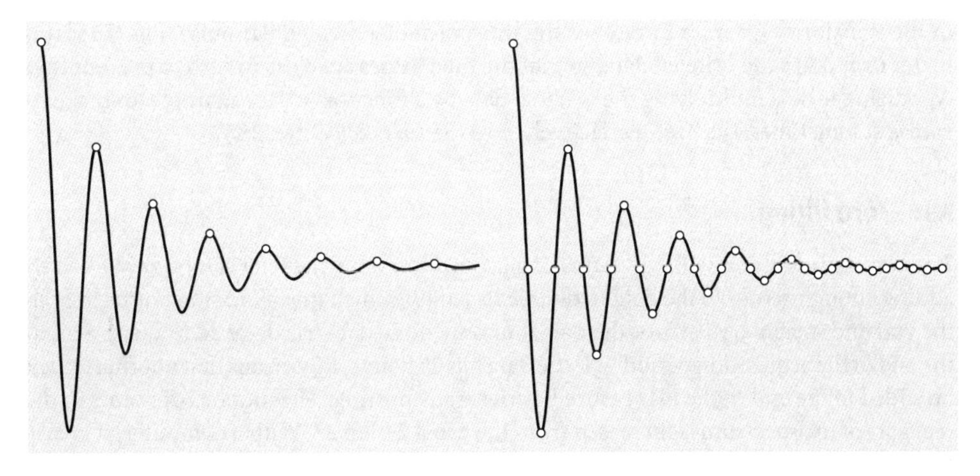

Fig. 9.44 The figure on the left shows the under-sampling of a sine wave. The data, which is recorded as the set of points, would erroneously appear as an exponential decay not as a damped sine wave. On the right, the data is over-sampled; more data points are present than strictly needed.

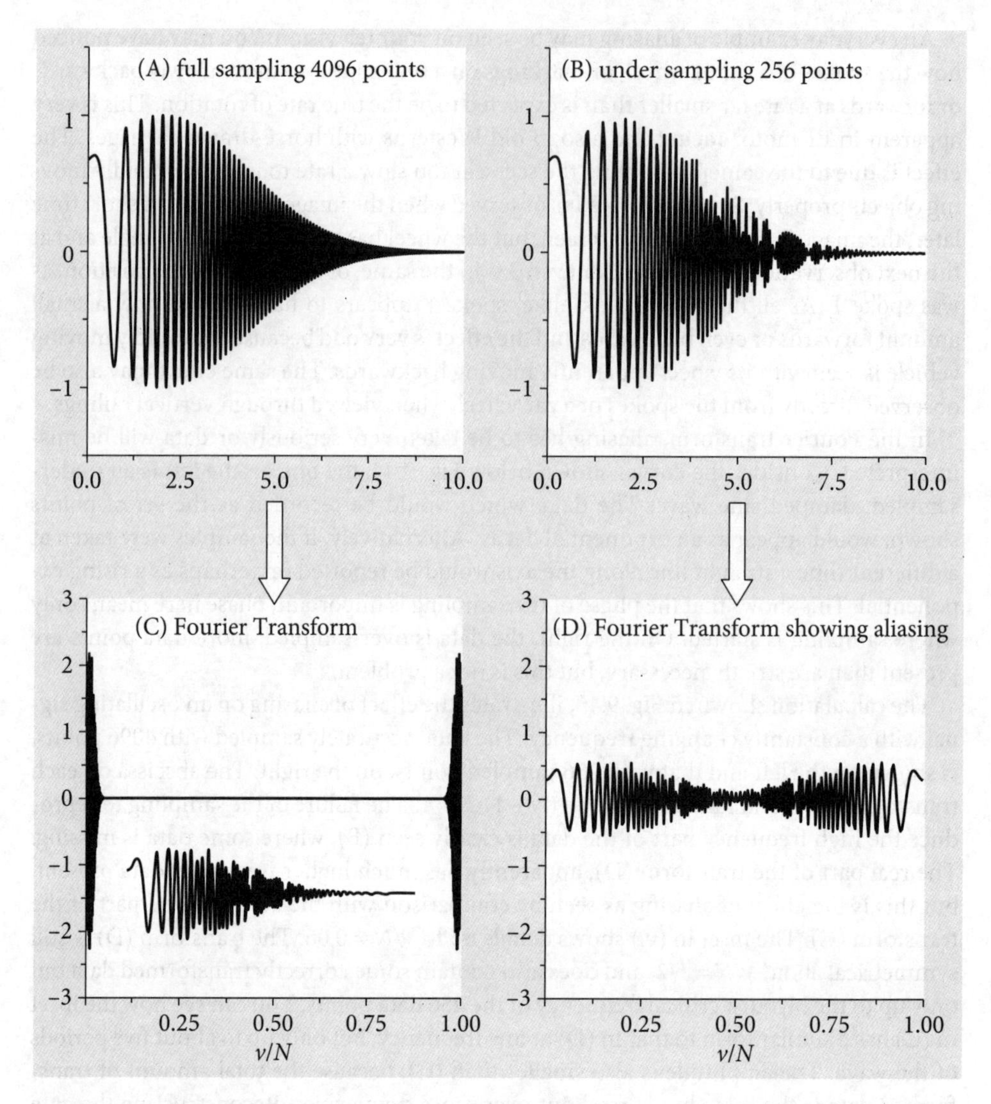

Fig. 9.45 Calculations showing proper sampling, left-hand column, and under-sampling, (right), that produces the effect of aliasing.

Finally note that aliasing is not the only way that erroneous transforms can be produced. If the data suddenly ends, the transform will 'assume' that there is a sudden change in frequency, in effect a step function, and this will generate many unwanted frequencies in the transform. As an example, see the transform of a rectangular pulse, Fig. 9.15. It is better to apodize the original data, so that the final values are close to zero to prevent this. Apodizing means multiplying the original data by a function, e^{-x} for example, to make its values at long times, (its 'feet'), effectively zero (Sanders & Hunter 1987).

9.9.5 Zero filling

To increase resolution without increasing acquisition time, the data can be zero filled by adding enough zeroes to the data to double its length, which means adding points to both the real and imaginary parts of the data, but only if the data has decayed to quasi-zero at the end of the acquisition period T. If the data has 2^N points, this means that another 2^N can be added to the end of the FID before Fourier transforming. The number of points in the real part of the spectrum is increased from the usual 2^{N-1} to 2^N. With a computer, it is easy to add sets of 2^N points repeatedly to the end of the data. However, this does not increase resolution but in effect interpolates between data points. The effect is superficial; no extra information is obtained by adding a second and subsequent zero fills, but it does give nice smooth graphs!

9.10 Using Fourier transforms for filtering, smoothing, and noise reduction on data

9.10.1 Motivation

To the experimentalist, noisy data always presents a problem. The best option is always to go and do more experiments, but this is not always possible for any number of reasons. One option to improve the data is smoothing. Although it is rarely thought of as smoothing, one way of doing this on data showing a trend, is to perform a least squares analysis. The parameters obtained, for example, the slope and gradient of a straight line with their associated standard deviations, provide a description of the data. If the data is not a straight line, then a polynomial fit may be used, and, if an exponential or other complex function, then a non-linear least squares may be necessary. However, smoothing data is often done independently and before any fitting is performed in an attempt to reduce the noise. As a rule of thumb, the noise and data can be split into three approximate classes: (i) The data is a slowly changing function and the noise is spiky, i.e. of high frequency; (ii) the converse, the noise is slowly varying and the data is of high frequency; and (iii) noise and data are of a similar frequency. In cases (i) and (ii) it is generally possible to separate noise and data; in (iii) this may not be possible.

The Savitsky–Golay method is an optimal polynomial fitting method that spans several data points at a time, replacing each with a mean value. It is available on most data analysis and graphics packages, the details of the calculation are complex and will not be examined here, but is a type of convolution and is good at removing noise spikes on data (Gorry 1990). If the data had an underlying slow trend on it, the rolling average method is also good if the noise is random and spiky on top of the data, see Section 9.10.4.

However, smoothing data is always a risky procedure because some of the data may inadvertently be removed with some of the noise, without you realizing it. Thus, the advice is always: do not smooth the data if it can be avoided. Having said this, a noisy signal must sometimes be 'cleaned up' to extract the signal from the noise, and a Fourier transform is a good way to do this. This approach is a little different to other smoothing methods, because by transforming the data it is possible to be very specific about what noise is removed. If the data to be improved is a one-dimensional time profile or a noisy image, the method is essentially the same: make a Fourier transform, look for the signal, remove unwanted frequencies and transform back again. If the noise is 'white' that is to say equally spread over all frequencies, and the signal wanted is a slowly varying periodic function or the converse then a Fourier transform may be what is needed.

9.10.2 Filtering and apodizing

To filter the data first transform it, then remove some of the frequencies with a filter function. This filter will probably have to be chosen by hand, and will involve some trial and error. Finally, reverse the transform to obtain the smoothed data. There are a number of functions that can be used as filters; they are designed to extract the data and not to add unwanted frequencies back into it after the calculation. Filtering is similar to apodizing the data. Apodizing means multiplying the signal by an exponential, Gaussian, or other function, before transforming. The apodizing filters are designed to remove extra oscillations at the foot of the signal, hence their name, but they do have the effect of broadening any features in the data. The Connes filter is commonly used in FTIR spectroscopy; it has the form $(1 - (x - x_0)^2/a^2)^2$ where $x = \nu/N$ which ranges from 0 to 1, and a is a parameter that the user can vary to optimize the signal, and x_0 is the position of the signal to be enhanced. This function looks very like the Gaussian bell-shaped curve. See Bracewell (1986) for several other filters.

9.10.3 Filtering a transform to remove noise

The example of filtering now worked through uses a square filter function, but the method is quite general. In filtering, the data is transformed, this is then filtered, and finally this is reverse transformed to recover the filtered data. The target signal is a simple sinusoid buried in noise, and is $\cos(2\pi\omega j) + \text{rand}(\,)$, with $\omega = 50/n$ and where rand( ) is a uniformly distributed random number between −5 and 5, making the 'noise' from the random numbers rather large compared to the signal; j is the index number ranging from 1 to $n = 2^{10} = 1024$. The signal is so noisy that the casual observer might question whether it is anything but noise; Fig. 9.46. The Fourier transform immediately shows a single frequency and indicates that there is a single sine wave present; its frequency could be measured easily, but in most real experiments, the signal could be more complex so the data is filtered, then inverse transformed, to get at the true signal.

Algorithm 9.7 Fourier Transform of noisy data. Part 1

```
> # make noisy data then Fourier Transform
> restart:
> with(DiscreteTransforms): with(plots):
> n:= 2^10;                                    # 1024 data points
  Seed:= randomize():                          # start random number
  sig:= Vector(1..n):                          # vector to hold data
  for j from 0 to n-1 do                       # make noisy data
    sig[j+1]:= cos(2*Pi/n*j*50) + 10*(rand()/1e12-1/2):
  end do:
  listplot(sig, view=[0..200,-10..10], axes=framed);
> FT:= Vector(1..n,datatype=complex[8]):# define vector
  FT:= FourierTransform( sig ):                # do FT
  # make RE value of FT and convert to list to plot
  pFT:= seq([(i-1)/n, Re(FT[i])],i=1..n): # get real data
  plot([pFT], view =[0..0.2,-1..20]);
```

The signal and its Fourier transform are shown in Fig. 9.46; modify the `plot` with parameters in `view` to view any part of the transform you want. The number of points is 2^{10}; a multiple of 2 makes the transform faster. The `for..do` loop calculates the noisy data that is stored in the vector `sig`. The 2^{12} dividing `rand()` is there because `rand()` generates a 12 bit integer and we want a number between 0 and 1. The random numbers vary from −5 to 5 in `sig`. The instruction `seq([(i-1)/n,Re(FT[i])],i=1..n):` makes pairs of data points with the abscissa as frequency, so that it can be plotted, and also extracts the real part of the FT data.

The spike in the transformed data at a frequency of 50/1024 is the single frequency of the sine wave. The signal (the cosine wave) is extracted, by multiplying all those values at a lower and higher frequency than the spike by zero, and then inverse transforming. The result is shown below; clearly, the filtering works well in this case where there is a single well-defined frequency. The filter is very simple; a new vector `Z3` is made zero, everywhere except where the data needs to be extracted, in this case from channels 48 to 52; see Fig. 9.46.

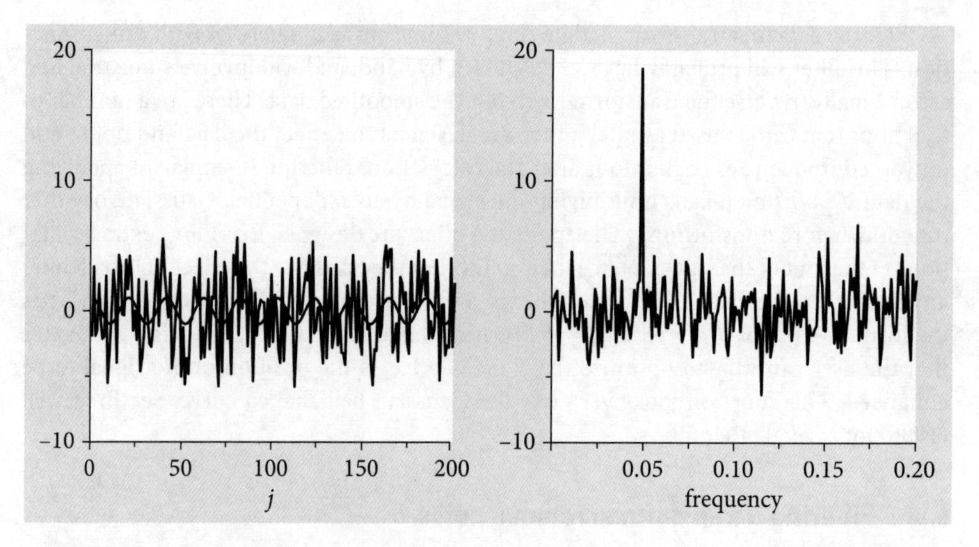

Fig. 9.46 Noisy data (left) (with the underlying cosine superimposed) and its Fourier transform (right) illustrating that a single frequency hidden in noise can be identified. The left plot has an abscissa in index numbers; only the first part of each data set is shown.

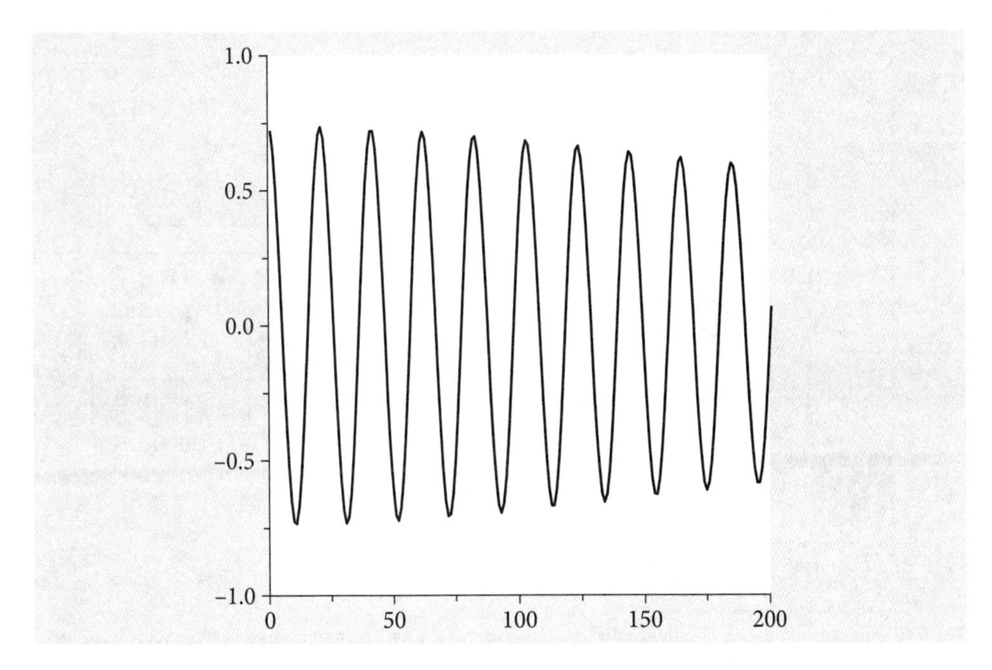

Fig. 9.47 Reconstructed data obtained by filtering the Fourier transform and reverse transforming. Because of the Shannon theorem, very few data points are needed to reconstruct this signal. The original data is shown in Fig. 9.46.

Algorithm 9.8 Fourier Transform filter, part 2

```
> # filter transformed data, then transform back.
  aFT:= Vector(1..n, datatype = complex[8]):
  for i from 1 to n do                    # simple square filter
      aFT[i]:= 0.0 + 0.0*I:               # make data zero
      if i > 48 and i < 52 then           # select some values
        aFT[i]:= FT[i]
      end if
    end do:
  iFT:= InverseFourierTransform(aFT):
  amp:= seq([i,Re(iFT[i])],i=1..n):# get real part of data
  plot([amp],view=[0..200,-1..1],axes=framed);
```

The filter function above removes all but three non-zero data points, which is a very small number, compared to the 1024 in the data. This is only possible because the spike in the transform is very sharp because it is a single frequency. If the signal were an exponential decay instead of a sine wave, then this Fourier method would perform very badly. The exponential is not periodic and is not described by single frequency but by a range of them in the form of a Lorentzian; in this and many other circumstances it is not easy to isolate the data from the noise. If Fourier and other smoothing methods fail then least squares fitting to a line or polynomial may be acceptable as a last resort.

9.10.4 Rolling or moving average

This method is widely used in the financial sector to smooth out fluctuations in the price of stocks and shares in an attempt to predict trends. It is also a particularly good method of removing unwanted noise spikes in experimental data. These are not part of the noise normally expected on data but occur nevertheless because someone switches on an instrument elsewhere in the lab or there is a glitch in the mains power and so forth. Such spikes can completely ruin data so some way of removing them is necessary. The moving average method starts at the beginning of the data, and averages over a small odd number of the next few data points and this average is recorded. This averaging 'window' is moved by one point along the data and the new average recorded and so on. This process weights in some future and past data into any point and so if a spike is preset it is averaged away by the normal data on either side of it, see Fig. 9.48.

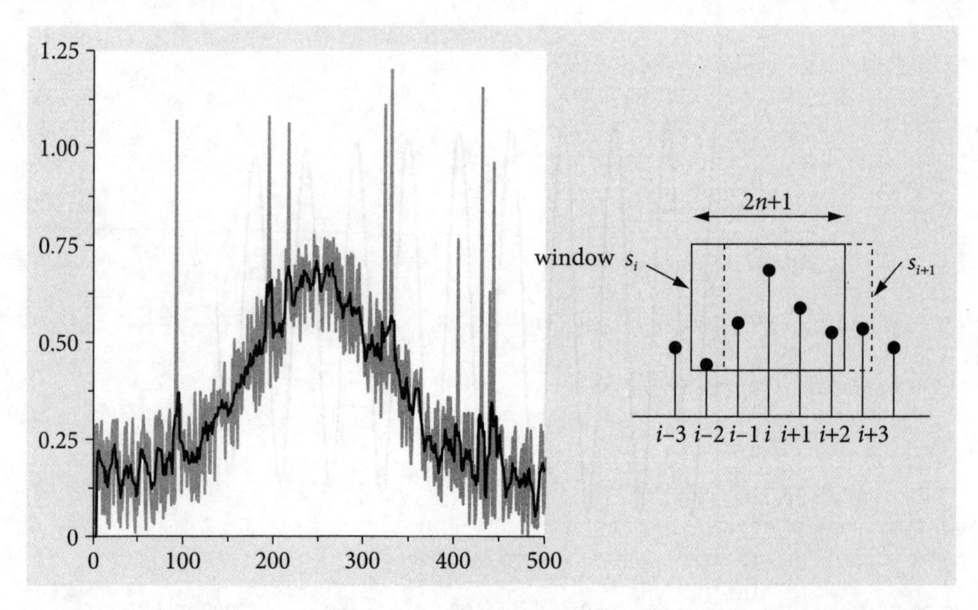

Fig. 9.48 Illustrating the moving averaging smoothing to remove spikes in noise. A filter with $n = 3$ was used which is a window of 7 points. The window that calculates point s_i is moved one point to the right at each step of the calculation. The original data is grey, the smoothed data, black.

The mid-point moving average s_i is defined as

$$s_i = \frac{1}{2n+1}\sum_{j=i-n}^{i+n} y_j$$

where $2n + 1$ is the width of the data window and i runs along the data starting at point n and ending $n - 1$ points from the end. The method can be made into a recursive algorithm where the updating equation is

$$s_i = s_{i-1} + [y_{(i+n)} - y_{(i-n-1)}]/(2n+1)$$

where the first point has to be explicitly calculated. The recursion works by considering the window at adjacent positions, subtracting the value that has just left the window and adding the one just entered. Although this method is not a Fourier method, it can be considered as a convolution of the data with a kernel function, which is a rectangular pulse with each of its m points of size $1/m$.

9.10.5 Questions

Full solutions are available at **www.oxfordtextbooks.co.uk/orc/beddard.**

Q9.19 **(a)** Use Maple to create an FID, such as might be observed from an NMR experiment on two spin ½ nuclei, and then calculate the absolute value of the spectrum using an exponential decay, with a lifetime of 300 channels and frequencies of 10/48 and 10/45; assume the FID has the form

$$e^{-x/300}[\sin(2\pi x/4.8) + \sin(2\pi x/4.5)]$$

where x is the time, which is in units of the channel number of the data, i.e. x ranges from 1 to 2^{12} which is the maximum number of points.

(a) Calculate the FFT and plot its absolute value. This is the spectrum.

(b) Add noise to the FID of magnitude ±1.5; recalculate the FFT.

(c) Apodize the FID by multiplying by $e^{-x/200}$ and recalculate the spectrum; comment on its appearance.

(d) Explain why the apodization improves the signal to noise of the spectrum.

Q9.20 The shape of the electric field from a femtosecond laser is measured and found to be chirped but is noisy. The pulse has the equation

$$f(t) = e^{-(t-1000/200)^2} \sin(t^2/400^2) + noise$$

The pulse is chirped because its frequency changes throughout the duration of the pulse. The noise is uniformly distributed about zero with a maximum value of 2. Time t is in femtoseconds. Generate a set of noisy data, then use a Fourier transform method to clean up the data and extract the original signal.

Strategy: Use the method described in Section 9.10.3. Decide how many data points to use, which will depend on the highest frequency of the pulse, and is determined by the Nyquist frequency, Section 9.9.2.

Q9.21 **(a)** Use the moving average method to write a recursive algorithm in Maple to effect smoothing on a set of data of your choice as shown in Fig. 9.48. **(b)** Use the Maple convolution algorithm given in the text to perform the same calculation. Generate the data by choosing a function and adding noise or use some experimental data.

9.11 Hadamard transform: encoding and decoding

9.11.1 Concept and motivation

To obtain the average value of a quantity, such as weight, individual measurements are usually made, then added together and divided by the number of measurements made. However, this is not the only way of obtaining the average. You may be familiar with the method of weighing several objects at a time and know that doing this will reduce the error in the average value. This multiple weighing method is an example of using a Hadamard transform and anything that can be measured in groups can be treated in the same way, for example, a spectrum or an image, thus the method is quite general. The reason that this multiple measuring method works is that the error is introduced by the balance not by the objects being weighed. A large weight therefore has the same error associated with its measurement as a smaller one does.

The reason for doing any transform experiment is always the same and either this is to achieve an improvement in signal to noise, or, a reduction in the time taken to do an experiment at a fixed signal to noise, which is effectively the same thing. Normally for n measurements, the signal to noise increases only as $\sqrt{n}$ but in the Hadamard approach, the signal to noise achievable increases at least as $n/2$, which is a huge improvement if n is 100, for example.

9.11.2 The Hadamard transform

The Hadamard transform is a purely discrete transform and instead of forward and back transforming, as in the Fourier transform, the equivalent steps are encoding and decoding. The encoding is done by adding several measurements together according to a set of rules or algorithm. The rule is always written down as a matrix, two forms of which H and S can be used; in the first H is a matrix of 1's and −1's, the other S, is a matrix of 0's and 1's. We shall concentrate of the S matrix form; it is the most useful one to use experimentally because it involves making only one measurement at a time; the H matrix method involves making two measurements. Harwit & Sloane (1979) describe the Hadamard transform method in detail, but see Marshall (1978) for a brief description.

9.11.3 Encoding and decoding with *S* matrices

Suppose that there are three samples to be weighed, they could be grouped as $x_1 + x_2$, $x_1 + x_3$ and $x_2 + x_3$ and weighed two at a time on a single pan balance. Written as equations where the z's are the measured values then

$$
\begin{aligned}
z_1 &= x_1 + x_2 + 0\\
z_2 &= x_1 + 0 + x_3\\
z_3 &= 0 + x_2 + x_3
\end{aligned}
\tag{9.47}
$$

These equations can be solved simultaneously to find the *ms*, for example, $2x_1 = z_1 + z_2 - z_3$ and so forth. Instead of doing this, which would be hard if there were 100 equations, the coefficients in equation (9.47) can be put into a matrix called an *S* matrix, (also called a Simplex matrix) and for this example this is

$$S = \begin{bmatrix} 1 & 1 & 0\\ 1 & 0 & 1\\ 0 & 1 & 1 \end{bmatrix}.$$

Note how the pattern is the same as that of the coefficients in the equations and that it cycles around so that each column is related to the next by one position of cyclic rotation. As a matrix equation, equation (9.47) is

$$z = Sx$$

and to solve for (column) vector *x*, (column) vector *z* is multiplied by the inverse of matrix *S* and the result is

$$x = S^{-1}z. \tag{9.48}$$

To show that this works, suppose that the *x* values are 12, 5, and 2, then the individual *z* values are $x_1 + x_2 = 17$, $z_2 = 14$, and $z_3 = 7$ and the calculation $S^{-1}z$ is

$$\begin{bmatrix} 1/2 & 1/2 & -1/2\\ 1/2 & -1/2 & 1/2\\ -1/2 & 1/2 & 1/2 \end{bmatrix}\begin{bmatrix} 17\\ 14\\ 7 \end{bmatrix} = \begin{bmatrix} 12\\ 5\\ 2 \end{bmatrix}$$

Using Maple as a check

```
> S := < < 1|1|0 >,< 1|0|1 >,< 0|1|1 > > ;
  z:= < 17 , 14 , 7 >;
  x:= S^(-1).z;                    # S^(-1) is the inverse of S
```

$$x := \begin{bmatrix} 12\\ 5\\ 2 \end{bmatrix}$$

If there are more than three values then a different *S* matrix will be used. The rules are that each column must be orthogonal to every other one and that each column must be related to the next by cyclic rotation, and thus they contain the same number of 0s and 1s. The rules for producing the *S* matrix are described in Section 9.11.6.

9.11.4 Signal to noise improvement

To see why this method works to improve signal to noise some error has to be added to each measurement. If each measurement has a standard deviation determined by the scales used then a measurement of *x* if done individually also has this error. When weighed in pairs, each pair of *x*'s has the same standard deviation σ because this is a property of the scales not the weights. The mean square error between the true and estimated values is $mse \equiv \langle m^2\rangle = \sum_i (\psi_i - m_i)^2$ and is used to estimate the overall error. If `z_est` is the estimated value, the calculation for weighing in groups is,

```
> z_est:= < m[1]+m[2]+sigma ,
            m[1]+m[3]+sigma ,
            m[2]+m[3]+sigma >;
  psi:= S^(-1).z_est;
```

$$\psi := \begin{bmatrix} m_1 + \frac{1}{2}\sigma\\ m_2 + \frac{1}{2}\sigma\\ \frac{1}{2}\sigma + m_3 \end{bmatrix}$$

```
> mse:= add( (psi[i]-m[i])^2, i= 1..3);
```

$$mse := \frac{3}{4}\sigma^2$$

The mean square error of any single measurement is σ^2 so that an improvement is obtained by measuring in groups, although it is small in this case. If n measurements are made then mean square error is reduced by $(n+1)^2/4n$ and the signal to noise improved by $(n+1)/\sqrt{4n} \approx \sqrt{n}/2$.

9.11.5 Implementation

Instead of weights, suppose that a spectrum is to be measured. To do the experiment, the detector is placed at the focusing plane of a spectrometer, the exit slits of which are removed and then a mask consisting of strips of opaque (0) and transparent (1) regions is placed there instead, see Fig. 9.49. At the first position the total amount of light falling on the detector is measured, this is z_1 and corresponds to measuring at positions $1+2+3+5$. Next, the mask is moved by one position, z_2 is measured which corresponds to light transmitted by $2+3+4+6$, and so on until all measurements are taken, 7 in this example. Each measurement corresponds to moving from one column to the next in the S matrix and the total light measured forms the z matrix. Once this encoded z matrix is established it is multiplied by S^{-1} and the signal x values recovered. Experiments have been performed by physically moving a mask etched in glass but a programmable liquid crystal mask would be easier to use. If there are n elements in the mask, then n different wavelengths are measured at the end of the n experiments. The resolution is determined by the width of the mask compared to the wavelength spread it covers.

In Fig. 9.49 is shown a simulated comparison of data taken in the normal way and with that taken using the Hadamard encoding method. The improvement in signal to noise is clear. In the calculation normally distributed noise with a standard deviation of 0.5 was added to an exponential $\exp(-t/50)$. The data is the average of n decays of n data points each and the Hadamard average is of n experiments also of n data points each. The expected improvement in signal to noise is the ratio of the mean square error of the two sets of data. The expected value for the increase in signal to noise ratio is $(n+1)^2/4n = 26.25$ for the 103 data points used and the measured value for this particular calculation is 25.4.

The Hadamard technique does not appear to have been applied to time-resolved measurements. However, this seems feasible and its use in such experiments is now proposed.

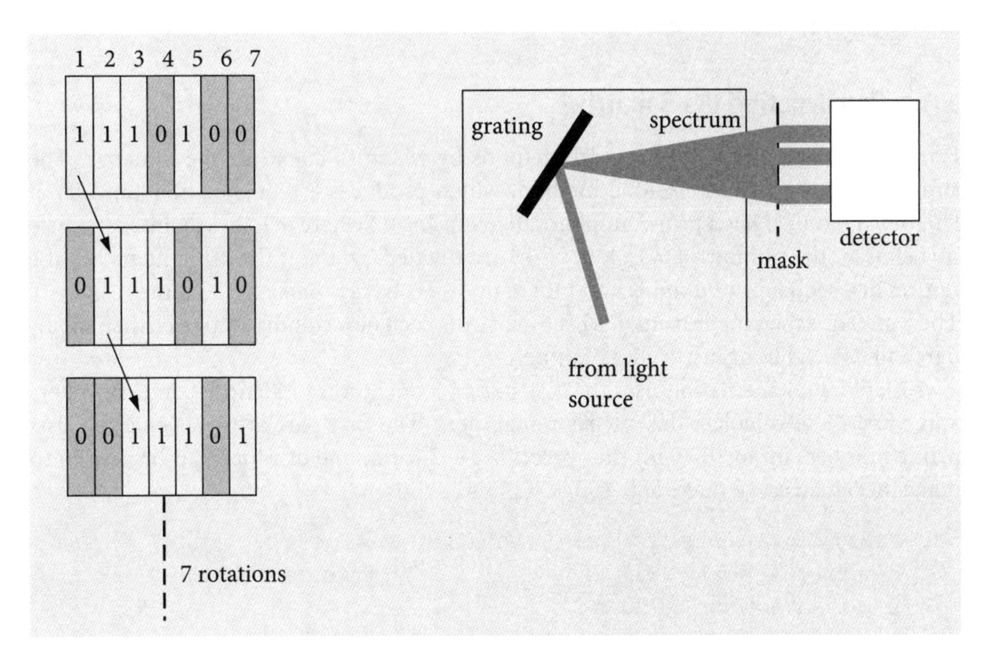

Fig. 9.49 The pattern of the mask replaces the slits of the spectrometer. The detector measures all the light transmitted by the mask at each position. Each mask is rotated by one element from the previous one. All the mask elements when placed together form the ***S*** matrix. A possible experimental set up is shown on the right.

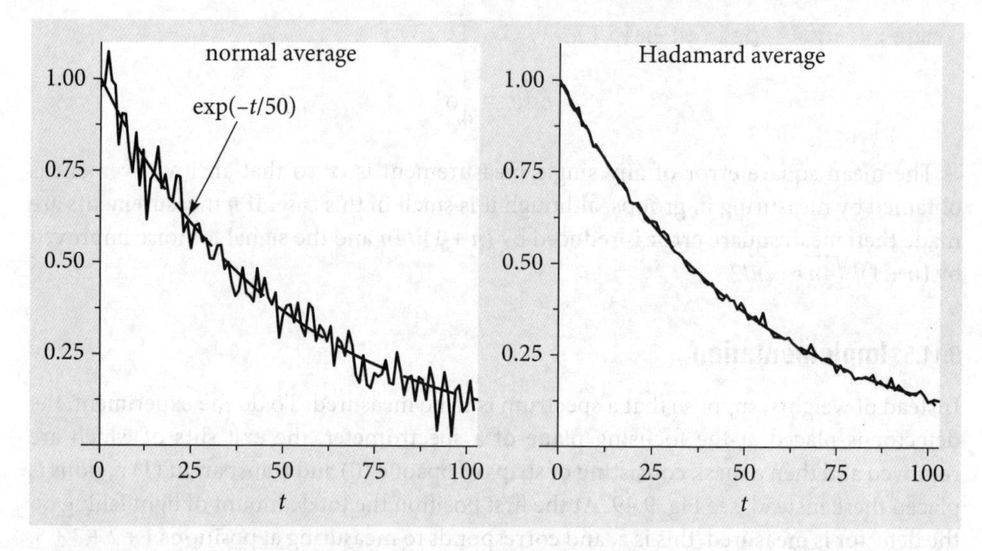

Fig. 9.50 Comparison of normal averaging and Hadamard averaging on the same set of data.

To perform a time-resolved absorption measurement, a train of pulses whose intensity is in the 0 and 1 pattern of one column of the S matrix is passed through the sample each time the reaction is started. Starting the reaction can be achieved by mixing reactants in a stopped flow reactor or by a flash of light in a flash photolysis or pump probe experiment. The *total* light transmitted by the sample is recorded after each pulse train has finished. The first train of pulses produces point z_1 as described above. The experiment is repeated by exciting the sample again, rotating the sequence of pulses by one element and the total light transmitted is recorded again, producing point z_2, and this is repeated for n experiments. The z data, a column matrix of n elements, is the Hadamard encoded data and is decoded as in equation (9.48) to produce the x column matrix which is the intensity transmitted by the sample at each time point. The timescale of the measurement is set by the spacing between any two pulses in the sequence and the total measurement time is n times this. The train of pulses can conveniently be produced by an electro or acousto-optic modulator, either by extracting pulses from a train of mode-locked laser pulses or from a continuous laser or other light source. If pulses could be produced close enough together in time, this method would remove the need for delay lines in picosecond and femtosecond pump-probe experiments.

9.11.6 Constructing the S matrix

Harwit & Sloane (1979) give several methods by which to construct the S matrix. The simplest is the quadratic residue method, which produces a sequence of 1s and 0s of length n but only if n is a prime number satisfying $4m + 3$ where m is also an integer. Once n is chosen, the numbers $i = 1, 4, 9, \cdots n^2$ are divided by n and the remainders are the *indices* in a sequence of numbers and these numbers have a value of 1 and the rest are 0. The S matrix is then made from this list by rotating each new column by one element compared to its neighbour in a cyclical manner.

A Maple implementation uses the `isprime()` operation to identify prime numbers and `irem()` to calculate the integer remainders. The first part of the calculation lists prime numbers up to 107 with the correct $4m + 3$ form, one of which can be chosen to make the S matrix. To make Fig. 9.50, $n = 103$ was chosen.

```
> restart: with(ListTools): with(LinearAlgebra):
  for n from 1 to 110 do                        # produce numbers
    for m from 1 to 1000 do
      if isprime(n) and n = 4*m+3 then
        print('number', n);
      end if
    end do;
  end do:
```

The sequence and S matrix is made as follows,

```
> Srow:= Vector(n, datatype = integer):
  A:= Vector(n, datatype = integer):
  for i from 1 to n do
    A[i]:= irem( (i^2),n);          # integer remainder
    Srow[A[i]+1]:= 1;
  end do:
  S:= Matrix(n, n):
  SL:= convert(Srow, list):         # make list to rotate
  for i from 1 to n do
    for j from 1 to n do
      S[I,j]:= SL[j];
    end do;
  SL:= Rotate(SL, 1):               # rotate 1 element
  end do:
```

The result is the S matrix. This can be inverted in Maple as `S^(-1)` but for the particular form of this matrix, inversion can be obtained more quickly using

$$S^{-1} = \frac{2}{n+1}(2S^T - J_n)$$

where the superscript T means transpose and J_n is an $n \times n$ matrix where each element is 1 (Harwit & Sloane 1979). The array `data[i]` is the signal to be measured and `Hseq` the sequence of light pulses or open and closed parts of a mask if a spectrum is being recorded. The coding of some test data, in array `data[i]` and its decoding can be done as follows:

```
Sinv:= s^(-1):                        # invert S once only
Hseq:= convert(S[1..n,1],list);       # get mask from S matrix
for k from 1 to n do
  HS:= Hseq:
  for j from 1 to n do                # coding
    signal[j]:= add( data[i]*HS[i], i=1..n) + noise();
    hh:= Rotate(hs,j);                # rotate mask
  end do:
  HT:= Sinv. signal:                  #decode: matrix multiply
  for j from 1 to n do                # add for average
    ss[j]:= ss[j] + HT[j]:
  end do;
  end do:
> pp:= seq([i-1,ss[i]/reps],i=1..n): # plot data
  plot([pp]);
```

In the coding step, the noise is added after coding which means that it is noise in the detector not in the light source. For the purposes of calculation the noise could be sampled from a uniform distribution then `noise()` could be replaced by `rand()/1e12`, averaged data similar to that on the right of Fig. 9.50 should be produced.

The same experiment done in the conventional way produces the data

```
> for k from 1 to reps do
    for j from 1 to n do
      Ndata[k,j]:= data[j] + noise():
    end do
  end do:
  for j from 1 to n do      # averaged data
    av_data[j]:= add( Ndata[k,j], k= 1..n)/n:
  end do:
```

which is plotted in the left-hand part of Fig. 9.50.

10 Differential Equations

10.1 Motivation and Concept

10.1.1 Background

To understand many aspects of chemistry, physics, and biology it is necessary to construct mathematical models that enable us to take a dynamic view of phenomena. It is usually possible to write down the *rate of change* of some quantity, but impossible to write down how the quantity itself changes, and so these models involve differential equations. In chemistry, differential equations are used in chemical kinetics, quantum mechanics, and transport phenomena such as diffusion. In biology, the diffusion equation is also important and, as in chemistry, rate equations are used to describe oscillating reactions, for example, glucose oxidation. Differential equations also describe nerve impulses (the Fitzhugh–Nagumo equations) and the spread of diseases in a population or of animals across a continent. Often these equations are complicated and need numerical solutions, some of which are described in Chapter 11. In physics, there are additionally the equations of dynamics, the harmonic oscillator describing molecular vibrations, the motion of pendulums or the orbits of planets, and the equations of magnetic and electric fields, as well as a host of non-linear phenomena, for example, that describe how the laser works, or the stop-start nature of heavy traffic flow on a motorway.

Differential equations are characterized into ordinary and partial ones. Ordinary differential equations only involve one independent variable and so have ordinary (total) derivatives; for example, the equation for a damped harmonic oscillator is $m\dfrac{d^2y}{dt^2}+a\dfrac{dy}{dt}+bx=0$ with time t as the independent variable, m, a, and b are constants, and y is the displacement at time t. Partial differential equations, such as the diffusion equation $D\dfrac{\partial^2 y}{\partial x^2}=\dfrac{\partial y}{\partial t}$, have two or more independent derivatives, t and x.

Differential equations are solved by integration and as with any integration, constants of integration are produced. These constants determine the exact solution of a differential equation and are determined by the initial values and/or the boundary values y is given at some x and t values.

10.1.2 Initial value problems (IVP)

In these problems, the starting conditions only are specified. For example, with the equation $dy/dt=ay$, the initial condition (or initial value) could be $y(t_0)=2$ where t_0 is the initial time, which is often zero. There is one initial condition because only a single integration step is necessary to solve the equation; the constant of integration is satisfied if y is specified at one value of t. The initial value of y and t_0 depends entirely on the problem being studied and is the equivalent of specifying the constant of integration in a normal integral.

If the equation is of second or higher order, two, three, or more initial conditions apply. One condition is needed for each stage of integration. Suppose the equation $\dfrac{d^2y}{dt^2}=t\dfrac{dy}{dt}+t+1$ has the particular set of initial conditions $y(t_0)=0$, $\left.\dfrac{dy}{dt}\right|_{t_0}\equiv y'(t_0)=1$ at $t_0=0$. The second condition is the gradient evaluated at time t_0 and what this gradient is depends on the problem being analysed. You can appreciate these initial conditions if you play football, rugby, or golf; to get the ball to where you want it go to, the initial direction and speed (dy/dt) with which it is hit clearly matter. The first stage of integration leaves us with one constant of integration and an equation in dy/dt and therefore dy/dt is needed at t_0 to find this constant. Integrating this equation produces a second constant; hence, two initial conditions are needed. Very often in differential equations, t is time, in which case the notation y' or $\dot{y}$ is often used to represent velocity, and y'' or $\ddot{y}$, acceleration.

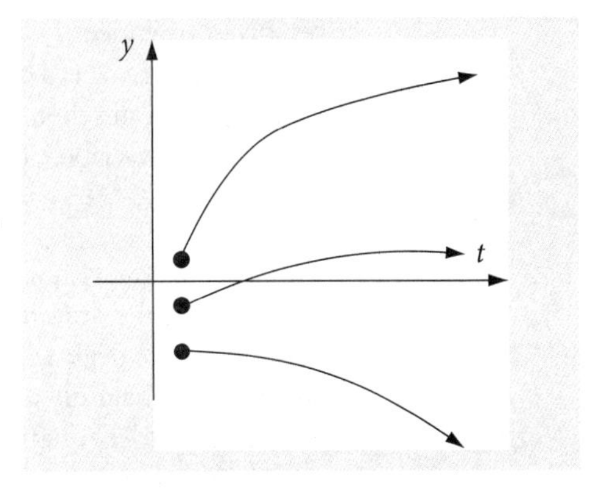

Fig. 10.1 Different initial conditions lead to different trajectories from the same equation.

In any initial value problem once the calculation is started, there is no telling what value y will have since only the initial value has been fixed. Because many potential initial conditions could apply, all trajectories could start at t_0, and have different $y(t_0)$ and gradients or start at different t_0 with the same gradient and so forth. These different starting conditions generate a *field or swarm* of solutions, or trajectories, Fig. 10.1. Solutions are sometimes represented on a graph by sets of arrows to indicate trends at different places and these are drawn as well as a particular solution.

10.1.3 **Boundary value problems (BVP)**

The boundary value problem defines conditions at two places that have to be satisfied throughout the calculation, thereby constraining the solutions. The Schrödinger equation describing the quantum mechanical particle in a box is solved with boundary conditions, because the wavefunction must always be zero at the edges of the box. The equation equates the kinetic energy with the total energy, $-\dfrac{\hbar^2}{2m}\dfrac{d^2\psi}{dx^2} = E\psi$, because the potential energy is zero, and the boundary conditions are $\psi(x_0) = 0$ and $\psi(x_L) = 0$ where x_0 and x_L define the positions of the ends of the box.

10.2 **Separable variables**

The rate of a chemical reaction can always be written down because this is proportional to the rate of loss or gain of a molecule's population. For example, in a *first-order* process $A \rightarrow B$, such as a cis-trans isomerization or decay of a radioactive nucleus, the rate of change of species A is proportional to the amount of A left unreacted or

$$d[A]/dt = -k_1[A]. \qquad [A]_0 = A_0 \tag{10.1}$$

where A_0 is the amount initially present. The constant of proportionality k_1 is the rate constant; more properly called the rate coefficient because in a chemical reaction, it depends on temperature. The negative sign is present because A decays into something else, and $[A]$ is the amount of A unreacted at time t. It is important to remember that $[A]$ is changing with time, although this is never explicitly stated in the equations; i.e. we do not normally write $[A]_t$. If the equation were written $dA/dt = +k_1A$, with a positive sign, then it could, for instance, describe the rate of growth of the number of bacteria A in the presence of an unlimited food supply. Note that, as the notation $[A]$ or $[B]$ can be rather cumbersome, A and B will normally be used instead to represent concentrations.

First-order equations are derived by considering the difference in the number of molecules N present at a time t and the number reacted during a small time interval δt. This can be expressed as, number unreacted at time $t + \delta t$ = number at time t – number reacted in time δt, which can be written as $N(t + \delta t) = N(t) - k_1N(t)\delta t$ from which

$$\frac{N(t+\delta t) - N(t)}{\delta t} = -k_1N(t)$$

and in the limit when dt becomes small, this equation becomes $\dfrac{dN}{dt} = -k_1N$. This becomes equation (10.1) if the number of molecules is converted into a concentration and can be solved provided the amount of $[A]$ present initially is defined; the initial condition is $[A]_0 = A_0$ where A_0 is a constant.

If the reaction scheme is more complex, $A \rightarrow B \rightarrow C$, then the rate of decay of A is the same as just described, but B changes as

$$d[B]/dt = k_1[A] - k_2[B]$$

which is the rate with which A converts to B, less the rate that B reacts to form C. To solve this scheme, the amount of $[A]$ at time t is inserted into this equation. Reactions with many steps are dealt with in a similar manner, but eventually the scheme can become so complicated that only a numerical solution to the equations is possible.

The first-order equation is found to describe many physical phenomena and is also representative of many types of equations where the variables are separable. This means that the equation can be written with y on one side and x on the other, or, for a first-order chemical reaction A on one side and t on the other. The general form of the separable equation is

$$Xdx + Ydy = 0, \tag{10.2}$$

which integrates to

$$\int Xdx + \int Ydy = c, \tag{10.3}$$

which is the *general* solution since it contains an arbitrary constant c. The *particular* solution is that obtained when either the initial or the boundary conditions are used. In the first-order reaction of species A, the steps in the integration are

$$\frac{dA}{dt} = -k_1A, \qquad \frac{dA}{A} = -k_1dt, \qquad \int \frac{dA}{A} = -k_1\int dt, \qquad \ln(A) = -k_1t + c,$$

and the result should be familiar as the integration is a standard one. If the initial condition is that $[A]_0 = A_0$ at $t = 0$ then the solution is $\ln(A/A_0) = -k_1t$ which is usually written as $A = A_0e^{-k_1t}$ and indicates more clearly how the concentration varies with time. If the initial amount is A_0 at time $t = t_0$, then the equation is $A = A_0e^{-k_1(t-t_0)}$.

As species A decays it must form another B, but if this does not decay then B can be obtained from the initial conditions because the total number of molecules must be constant; therefore $B = A_0(1 - e^{-k_1t})$ and B rises at the same rate as A falls. If, however, B decays to C, the scheme being $A \rightarrow B \rightarrow C$, then there is another rate equation $dB/dt = k_1A - k_2B$ and we can substitute A into this to obtain

$$\frac{dB}{dt} = k_1A_0e^{-k_1t} - k_2B$$

but now the terms are not separable, and another method is needed, which is described in Section 10.4. Sets of equations for sequential and parallel reactions such as $dA/dt = \cdots$, and $dB/dt = \cdots$ can alternatively be solved by eigenvalue methods described in Chapter 7.13 provided no terms with a product or ratio of concentrations is present, i.e. no $A \times B$ terms, unless pseudo first-order conditions apply.

10.2.1 Steady state

In many chemical schemes, after the reaction has started it enters a period where intermediate or transient species can be identified, and their rate of change is effectively zero. This does not mean, however, that their concentration has to be small but if the rate of change is zero, then most rate equations can be solved relatively easily. This is of great utility because the complete solution can be very complex and often only numerical solutions are available.

(i) The scheme

$$I + M \underset{k_{-1}}{\overset{k_1}{\rightleftharpoons}} IM$$

$$I + IM \xrightarrow{k_2} I_2 + M$$

describes how iodine atoms are recombined to form I_2 in the presence of an inert buffer gas M. The transient collision complex IM can be supposed to exist at steady state once the reaction has started then,

$$\frac{d[IM]}{dt} = k_1[I][M] - (k_{-1} + [I])[IM] = 0$$

making $[IM] = \dfrac{k_1[I][M]}{(k_{-1} + [I])}$. At steady state the rate of formation of I_2 molecules is

$$k_2[I][IM] = \frac{k_1 k_2 [I]^2[M]}{(k_{-1} + [I])}$$

and this could be tested experimentally.

(ii) Another example is the calculation of the fluorescence yield of a molecule. This yield φ is the fraction of molecules excited to those that emit a photon and is defined as

$$\varphi = \frac{\text{rate of emission}}{\text{rate of absorption}}$$

If G is the ground state, S_1 the excited singlet and T the triplet state, the scheme is

$$\begin{aligned} G &\xrightarrow{k_a} S_1 \\ S_1 &\xrightarrow{k_f} G + h\nu \\ S_1 &\xrightarrow{k_s} T \end{aligned} \tag{10.4}$$

In this scheme k_a is the rate constant for absorption of a photon by the ground state, k_f the rate constant for fluorescence and k_s the intersystem crossing rate constant forming the triplet state. The S_1 rate equation is

$$\frac{dS_1}{dt} = k_a G - (k_f + k_s)S_1$$

where $k_a G$ is the rate of absorption. At steady state $dS_1/dt = 0$, making $S_1 = \dfrac{k_a G}{(k_f + k_s)}$. The fluorescence yield is therefore

$$\varphi = \frac{\text{rate of emission}}{\text{rate of absorption}} = \frac{k_f S_1}{k_a G} = \frac{k_f}{k_f + k_s} \tag{10.5}$$

and is the fraction of molecules that fluoresce. A molecule fluoresces with a rate constant that is the sum of all process destroying the excited state; therefore, if $k = k_f + h_s$ and as the fluorescence lifetime is the reciprocal k or $\tau = 1/k$, the yield becomes $\varphi = k_f \tau$.

10.2.2 The phase portrait

The steady state is found when $dy/dx = 0$. In non-linear differential equations which have higher powers of x, such as $dy/dx = x + ax^2 + bx^3$, there are many 'steady states', which are also called *critical*, *fixed*, or *equilibrium* points, and it is now useful to see where these are by plotting dy/dx vs x. This graph is called the *phase portrait*. The new feature is that some equilibrium points are stable and others not. If an initial value of x is chosen close to a *stable* equilibrium point, conditions will change so that this point will be reached; if it is an unstable point, conditions now change so that this point is avoided. This may or may not mean that another stable point will be found. The graph, Fig. 10.2, shows a schematic phase portrait. Some steady state points therefore can also be called equilibrium points because once the system reaches one of these it will remain there. In Fig. 10.2 the arrows indicate the direction x will follow depending on its initial value; for example if $3 \le x < 2$, then x will move towards, and end at, the stable point $x = 3$. The rule of thumb is that if the curve is above the x-axis the arrows move to the right, and they move to the left if the curve is below the line. A point starting in the range 0–1 or 1–2 moves to the stable point 1, and point 2 is unstable. Starting in the range > 2, moves to stable point 3, but starting at zero or any negative value is unstable. A negative value of d^2y/dx^2 at a steady state point also indicates stability.

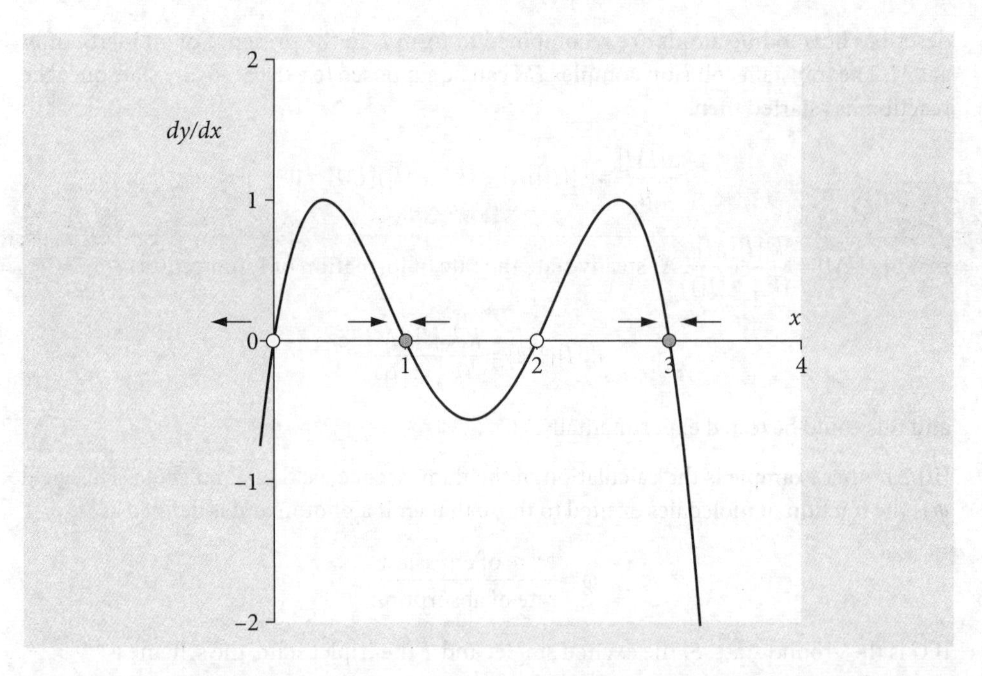

Fig. 10.2 The stable (filled circles) and unstable equilibrium points in a schematic phase portrait.

If a problem is described by two coupled non-linear differential equations $dy/dx = f(x,y)$ and $dz/dx = g(x,y)$ then the phase portrait is not usually plotted but instead the *phase plane* in which is y vs z is plotted. Chapter 11 gives examples of using the phase plane. Strogatz (1994) discusses phase portraits and phase planes in detail and illustrates these with many interesting examples.

10.3 Phase planes and solving equations by separating variables

(i) A laser has two basic forms, an amplifier and an oscillator. A laser amplifier is usually a single or double pass device in which an input laser is amplified by spontaneous emission in the gain material in which there is a population inversion. If the gain material contains atoms, ions, or molecules this inversion contains many more excited states than ground states and an incoming photon of the correct energy will stimulate an excited state to enter a lower state by releasing its energy as a new photon. The population of this lower state has to be kept close to zero if the laser is to work well, because the transition rate is proportional to the population difference. The initial inversion is produced by an external source, for example, by using another laser, flash lamps, or an electric current. In a laser oscillator, the gain material is placed between mirrors so that feedback of photons between these is possible. The lasing builds up to a steady state because a little of the spontaneous emission from the population inversion is captured by the mirrors and repeatedly passes back and forth through the gain material causing each photon to stimulate another every time it does so. The process is therefore non-linear. If you have tried to align a laser, you will have noticed that as soon as the alignment is correct, and the losses are therefore drastically reduced, the laser instantly lights up. A steady state is reached because there are losses in the cavity as well as gain and these balance one another.

A basic model of a laser measures the rate of change of the number of photons dn/dt as the difference between the number due to gain and loss and, clearly, if the laser is to work the gain must initially exceed the loss. The threshold to lasing occurs when gain equals loss, (see Haken 1978, p. 127; and Svelto 1982). The increase or gain in the number of photons is due to stimulated emission. The losses are caused by reflections from surfaces in the laser cavity, such as those on the laser rod or the quartz of a dye cell, and from transmission through the output-coupling mirror. Some gain materials also absorb at the same wavelength as the laser operates, and this is a cause of loss. The gain minus loss in the number of photons is therefore

$$dn/dt = gNn - kn,$$

where g represents the gain coefficient, N the number of excited states in the gain medium, which assumes that the lower state has zero population. The rate constant for decay of photons out of the cavity k accounts for all the losses. This equation is non-linear although it appears not to be. The non-linearity is introduced because the number of excited states N is not expected to be constant since photons are stimulated out of it. Suppose, therefore, that by continuous external excitation, the number of excited states is kept constant at N_0, then $N = N_0 - \gamma n$ where γ is a constant. The rate equation now becomes non-linear; $dn/dt = g(N_0 - \gamma n)n - kn$ which can be rewritten as

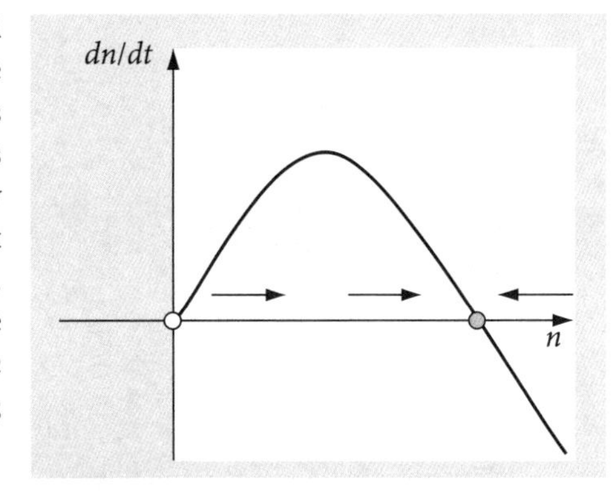

Fig. 10.3 Gain threshold and stability in a basic model of a laser when $\alpha > 0$.

$$dn/dt = \alpha n - \beta n^2,$$

with the constant $\alpha = gN_0 - k$ and $\beta = g\gamma$. If the excitation producing the excited states is weak, then α is negative and the number of photons does not increase. The lasing threshold condition occurs when $\alpha = 0$ giving the minimum N_0 as k/g. Thus, if the laser gain coefficient g is high, the lasing threshold is small. If the rate of loss of photons in the cavity k is small, which corresponds to a long cavity lifetime, the threshold is also small. If excitation is strong, making N_0 large, then α is positive and the laser operates.

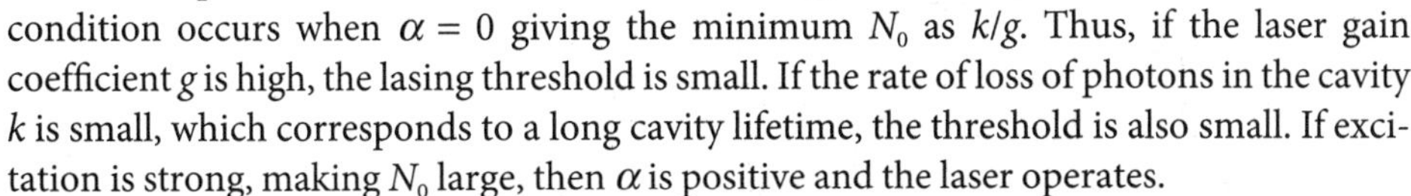

The phase portrait dn/dt vs n is now an inverted curve as $\alpha > 0$ and has steady states at $n_{ss} = 0$ and at $n_{ss} = \alpha/\beta$. Since the rate of change of the number of photons has to be positive for the laser to work, the gradient at threshold must be negative; see Fig. 10.3. The second derivative at threshold is $-\alpha$, which is negative, and confirms the stability as α is positive. The presence of the steady state means that the laser intensity will rise to a constant value, given by the pump intensity. When this is achieved, the gain must balance the loss because the rate of change of the number of photons is zero. Note that loss includes the number of photons that form the laser beam itself.

The rate equation is solvable by separating variables and integrating

$$\int \frac{dn}{\alpha n - \beta n^2} = t + c.$$

The integral can be separated into simpler terms using partial fractions, then looked up in Chapter 4.2.13, or Maple used. The result is

$$\int \frac{dn}{\alpha n - \beta n^2} = \frac{1}{\alpha}\ln\left(\frac{n}{\beta n - \alpha}\right)$$

Hence $\dfrac{1}{\alpha}\ln\left(\dfrac{n}{\beta n - \alpha}\right) = t + c$ where c is a number determined by the initial conditions. Using Maple `dsolve()` solves the differential equation directly with the initial condition that the population is n_0 at time zero,

```
> eqn:= diff(n(t),t) = alpha*n(t)-beta*n(t)^2;
  inits:= n(0) = N0:                    # initial condition
  dsolve( [eqn, inits] );               # include initial cndtns
```

$$n(t) = -\frac{\alpha N0}{(-\alpha + \beta N0)e^{-\alpha t} - \beta N0}$$

Note that n is written as `n(t)` to tell Maple what the variable is. In `dsolve()`, square or curly brackets must be used to include both the equation and initial conditions. If n is plotted vs time, the curve rises or falls to reach a steady state number of photons, $n_{ss} = \alpha/\beta$. This is reached because the exponential terms tend to zero as t increases. This steady state can be greater or smaller than N_0 because $n_{ss} = (gN_0 - k)/g\gamma$ and when the gain g is large and k the rate of loss of photons is small, then $n_{ss} > N_0$ and the number of photons increases from N_0 to the steady state value. The threshold condition is k/g and this needs to be small for lasing to occur. The expression for the steady state also shows that the number of photons increases linearly with pump power, which is proportional to N_0. Thus, by examining the phase portrait, the same conclusions are arrived at as by solving the equations and then

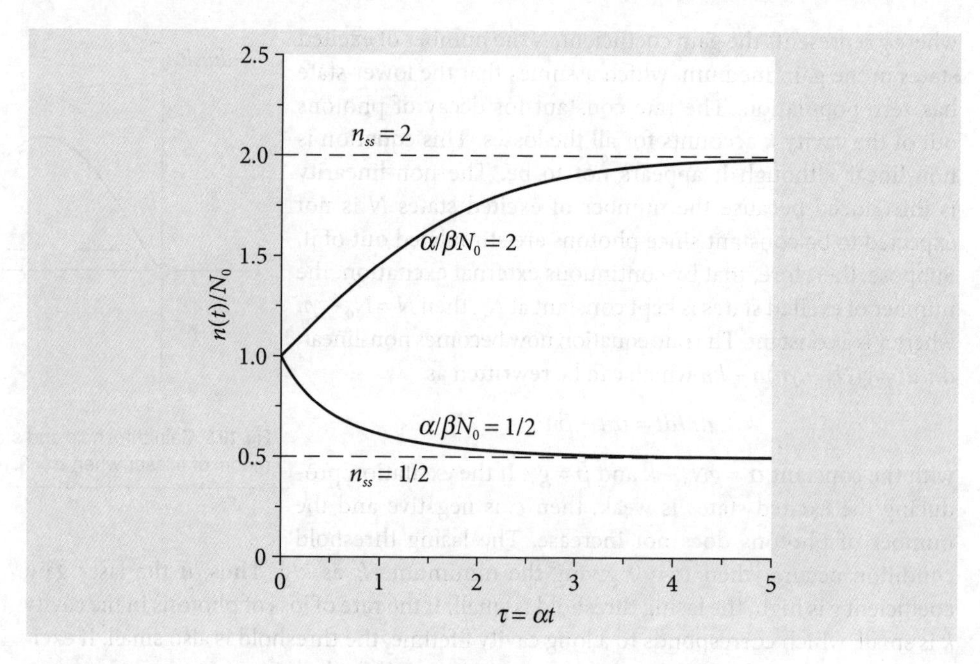

Fig. 10.4 Relative population $n(t)/N_0$ vs αt for a laser with $\alpha' = \alpha/\beta N_0 = 2$ and 1/2.

looking for the long time limit. The growth or decrease in relative population $n(t)/N_0$ is shown in Fig. 10.4 plotted vs reduced time $\tau = \alpha t$ and with dimensionless parameter $\alpha' = \alpha/\beta N_0$. Making these changes reduced the equation to a simpler form

$$\frac{n(t)}{N_0} = \frac{\alpha'}{(\alpha' - 1)e^{-\tau} + 1}$$

with only one dimensionless parameter, α'.

The form of the rate equations describing the laser is mathematically the same as those for an autocatalytic reaction; this is not so very surprising because in a laser one photon stimulates another. In an autocatalytic reaction, one molecule produces a copy of itself: $A + B \to P + 2B$. An autocatalytic reaction is described in question Q9.18.

(ii) Simple harmonic motion is described by the generic equation $\dfrac{d^2x}{dt^2} + \omega^2 x = 0$ where ω is the oscillator frequency, x position, and t time. As the phase portrait is a plot of dx/dt vs x, the equation has to be integrated once to get it into this form. This can be done by defining $v = dx/dt$, then $\dfrac{d^2x}{dt^2} = \dfrac{dv}{dt}$. Next, using the chain rule, $\dfrac{dv}{dx} = \dfrac{dv}{dx}\dfrac{dx}{dt} = v\dfrac{dv}{dx}$ and the equation becomes

$$v\frac{dv}{dx} + \omega^2 x = 0.$$

Integrating by separating variables gives

$$v^2 + \omega^2 x^2 = c$$

where c is a constant. The trajectories in the phase plane are ellipses whose exact values are determined by their initial conditions. However, all of the trajectories take the same time to complete, as only one frequency is associated with the oscillator. The origin of the phase portrait is a position of stable equilibrium, and corresponds to the oscillator being stationary. If a pendulum is observed, with x as the initially positive angle and with an initial velocity of zero, then the starting point is on the positive x-axis. The phase plane is followed clockwise, as the pendulum increases in (negative) angular velocity to the left and the angle decreases towards zero, at which point the pendulum is vertically downwards and the angular velocity has its maximum negative value. The pendulum continues to move until the velocity is again zero at the opposite angle to the starting point. It now changes direction, the velocity reversing (becoming positive), and returns to the starting point, travelling over the top of the phase plane.

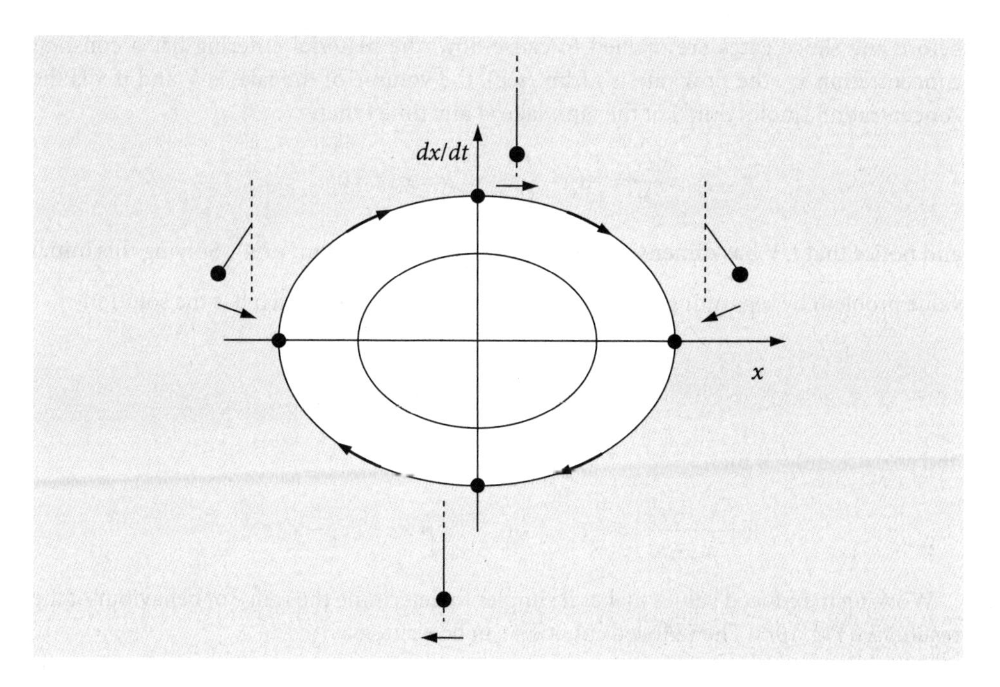

Fig. 10.5 Phase portrait for the simple harmonic oscillator. The position of a pendulum at various points is shown. Motion to the left is a negative velocity.

10.3.1 Chemical kinetics

(i) In a bimolecular reaction, such as $A + B \rightarrow C$, the rate of reaction of A would normally be written as

$$\frac{dA}{dt} = -k_2AB,$$

and is of little help in solving the equation. If the initial amounts of A and B are $[A]_0 = a$, $[B]_0 = b$, and if x moles have reacted at time t, then the rate equation can be solved as

$$\frac{d(a-x)}{dt} = -\frac{dx}{dt} = -k_2(a-x)(b-x), \qquad x = 0, t = 0.$$

The variables are separable, and to integrate the right-hand side this can be split into partial fractions,

$$\int \frac{dx}{(a-x)(b-x)} = k_2 \int dt \qquad \text{or} \qquad \frac{1}{a-b}\int \frac{1}{b-x} - \frac{1}{a-x}dx = k_2 \int dt.$$

Integrating and combining terms produces

$$\frac{1}{a-b}\ln\left(\frac{a-x}{b-x}\right) = k_2t + c$$

and the constant can be evaluated because $x = 0$ at $t = 0$ giving,

$$\frac{b(a-x)}{a(b-x)} = e^{(a-b)k_2t}$$

which can be solved for x the amount reacted time t.

(ii) Water entering and leaving a tank, reaction vessel, or a lake can be modelled by calculating the difference of material flowing in and out, viz.,

Rate of change of material = rate in − rate out

in an analogous way to a chemical reaction. It has to be assumed not only that the rate of inflow is the same as outflow, otherwise the lake will fill or empty, but also that material entering is fully mixed before it leaves (a well-stirred reactor), otherwise a concentration gradient would exist and this will complicate the calculation. If the concentration is x_0

before any sluice gates are opened to cause flow, the material entering has a constant concentration x_{in}, the flow rate is f (dm^3/sec), the volume of the lake is V and if x is the concentration (moles dm^{-3}) of the tank/lake at any time t then

$$\frac{dx}{dt}=\frac{f}{V}x_{in}-\frac{f}{V}x \qquad x=x_0,\, t=0,$$

and notice that f/V has dimensions of a first-order rate constant or s^{-1}. Solving this initial value problem by separating variables gives $\displaystyle\int\frac{dx}{x_{in}-x}=\frac{f}{V}\int dt$, which has the solution

$$-\ln(x_{in}-x)=\frac{f}{V}t+c$$

and as $x=x_0$ at $t=0$ then

$$\ln\left(\frac{x_{in}-x_0}{x_{in}-x}\right)=\frac{f}{V}t \qquad \text{or} \qquad x=x_{in}-(x_{in}-x_0)e^{-\frac{f}{V}t}$$

Working in reduced values makes it simpler to determine the range of behaviour of the results, see Fig. 10.6. The reduced equation can be written as

$$r=r_{in}-(r_{in}-1)e^{-\tau}$$

where $r=x/x_0$, $r_{in}=x_{in}/x_0$ and $\tau=ft/V$ so that now only the ratios need be considered. If $r_{in}>1$, the concentration ratio increases with time and the lake becomes polluted. At values of $r_{in}<1$, which means that $x_0>x_{in}$, the lake becomes more dilute and if pure water is added, $r_{in}=0$, the polluted lake would be cleaned after a time larger than approximately $t=5V/f$. Barnes & Fulford (2002) discuss several similar models.

(iii) When a solid solute is dissolved in a solvent, the rate equation is found by considering the change in the amount dissolved in solution during a short time period. In dissolving a solid, a saturated solution will eventually be formed and this limits how much solid will dissolve. If x_0 is the initial amount of solid to be dissolved in m grams of solvent, s_x/m the saturated solution concentration, k the rate of dissolution (mass s^{-1}) and x the number of grams of solid remaining at time t, then

Amount of x dissolved in time $t+\delta t$ = amount x at t − amount dissolved in δt

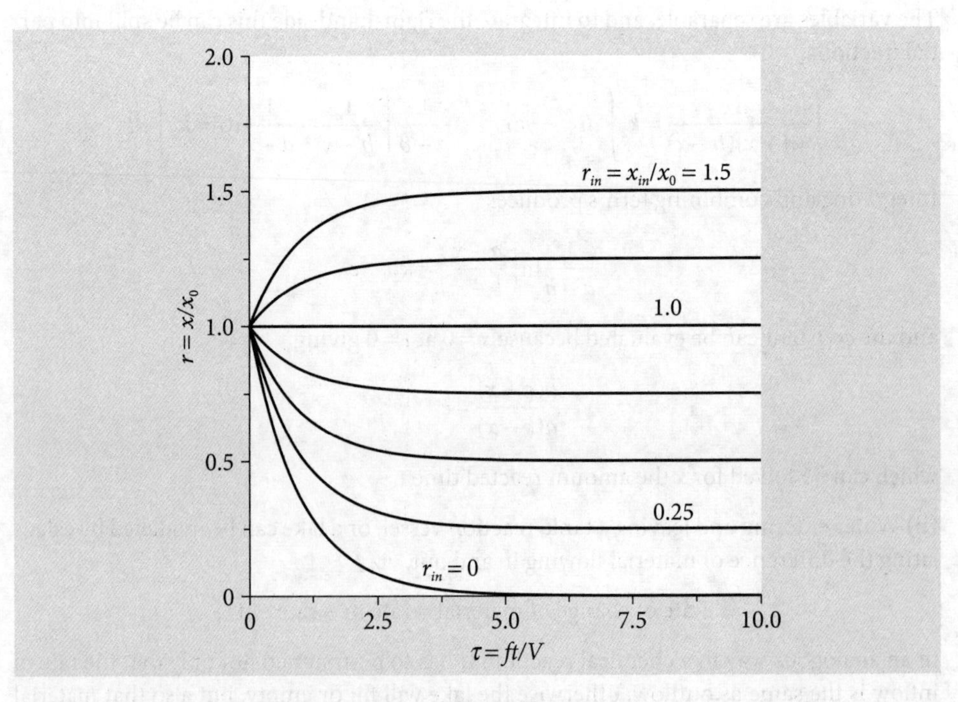

Fig. 10.6 The change in the ratio x/x_0 vs reduced flow ft/V for different $r_{in}=x_{in}/x_0$ ratios, showing the range of curves obtainable with different starting conditions.

The amount dissolved in time δt is proportional to the amount of solid undissolved x, multiplied by the difference in concentration compared to that of a saturated solution, and this product is $kx\left(\frac{x_s}{m}-\frac{x_0-x}{m}\right)\delta t$. The term $(x_0-x)/m$ is the concentration dissolved. The rate equation is

$$\frac{dx}{dt}=-kx\left(\frac{x_s}{m}-\frac{x_0-x}{m}\right) \tag{10.6}$$

To solve, separate variables and with the abbreviation, $a=x_s-x_0$,

$$\int\frac{dx}{ax+x^2}=-\frac{k}{m}\int dt$$

which is a standard integral as a is a constant. If this is not recognized, then converting to partial reactions gives,

$$\frac{1}{a}\int\frac{1}{x}-\frac{1}{a+x}dx=-\frac{k}{m}\int dt$$

and the integral is

$$\ln\left(\frac{x}{a+x}\right)=-\frac{akt}{m}+c.$$

If $x=x_0$ at $t=0$ then $c=\ln\left(\frac{x_0}{a+x_0}\right)$ and therefore

$$\ln\left(\frac{(a+x)}{x}\frac{x_0}{(a+x_0)}\right)=\frac{akt}{m},\qquad a=x_s-x_0$$

which can be solved for x if the concentration profile with time is needed. This result shows that the amount of solid x decreases exponentially to a constant value determined by x_s-x_0 from an initial value x_0 provided $x_s>x_0$, otherwise the amount of solid material becomes zero.

A rate equation with the form of equation (10.6) also appears in a completely different context, which is that of spreading a disease among N individuals if the rate of spread is proportional both to the number infected x and the number who are not infected. The equation then has the form

$$\frac{dx}{dt}=kx(N-x), \tag{10.7}$$

where k is the rate of spreading the infection in units of number of individuals time^{-1}.

10.3.2 Diffusion of heat and molecules

At steady state, the constant quantity of heat, or the heat flux, Q in watts (J s^{-1}) flowing through an area A, is proportional to the *spatial* rate of heat loss or

$$Q=-kA\frac{d\theta}{dx} \tag{10.8}$$

where θ is the temperature, x the thickness, and k the thermal conductivity coefficient. This equation is called Fourier's law of heat conduction. The heat flux Q is also $dH/dt=$ constant, if dH is the constant amount of heat transferred in time dt. In general, the rate of heat transferred from one body to another will depend on the shape of the body and its composition, as well as the temperature difference and whether these are held constant. Therefore, the flux and $d\theta/dx$ may change across an object. This detail, however, adds nothing new, *per se*, to the problem. The corresponding steady state equation for the mass diffusion of fluids such as gases and liquids, is Fick's first law, which is

$$J=-D\frac{dc}{dx} \tag{10.9}$$

where J is the *flux* which is the amount of material diffused/unit area/unit time and this is a constant quantity. The diffusion coefficient is D ($m^2\ s^{-1}$), c is the concentration, and x the distance over which diffusion occurs. If these last two 'diffusion' equations were to be rearranged, they would have the mathematical form of *zero-order* rate equations because the rate of change does not depend on θ or c but is a constant, for example $\dfrac{dc}{dx}=-\dfrac{J}{D}$. An example from chemical kinetics of a zero-order rate equation is $\dfrac{dc}{dt}=-k_0$.

(i) The heat and diffusion equations can be used to solve a variety of problems. For example, a fridge has a wall that is 5 cm thick, and is kept 20° cooler than the room. The thermal conductivity coefficient $k = 0.1\ J^{-1}\ s^{-1}\ m^{-1}\ K^{-1}$ (or 0.1 watt/metre kelvin), which is typical of good insulating materials, then the steady heat flow into the fridge is $Q=-kA\Delta\theta/\Delta x=-40$ W for each m^2 of surface area. Because the temperature change is fixed as is the wall thickness, then $d\theta/dx=\Delta\theta/\Delta x$.

(ii) The following example concerns ice forming on a still lake. The ice layer increases with the square root of time when the water temperature is 0 °C and the air temperature is lower and constant. The ice acts to insulate the water from the colder air above. A quantity of heat, dH, is taken from the water to freeze it in time dt, and the rate of heat transfer is dH/dt. By Fourier's law, equation (10.8) this is proportional to the temperature gradient $\Delta\theta/x$ across the ice, therefore $dH/dt \propto \Delta\theta/x$. As the ice thickens, the temperature gradient decreases and so heat transfer is reduced per unit time, which must be proportional to the thickness $dH \propto dx$, hence $dx/dt = \alpha\Delta\theta/x$ where α is a constant of proportionality. When integrated, this equation shows that the ice thickness increases as the square root of time; $x \propto \sqrt{t}$. This, and the fact that ice is less dense that water, means that lakes do not generally freeze solid; there are usually not enough cold days.

An interesting consequence concerns the freezing of water droplets in the atmosphere. Non-linear freezing means that the outside of the liquid droplet freezes rapidly, forming a shell. This may become strained as it thickens, causing the ice to crack thus releasing a spurt of liquid water from the interior in the form of micro-droplets, which rapidly freeze in the cold air. Such a process may be important in cloud formation.

(iii) A third example concerns diffusion. A thin porous pipe of outer diameter b has gas at pressure p_0 flowing inside it. The inside of the pipe has diameter of a and the concentration of gas outside the pipe is zero. If the diffusion coefficient through the pipe to the outside is D, the rate of gas loss can be calculated using Fick's first law. The 'pipe' could be, for example, a capillary in the lungs containing O_2 and CO_2.

The flux per unit area/second through the pipe is a fixed quantity, which is given by $J=-Ddc/dr$, where r is the radial distance from the centre of the pipe and c the concentration of the gas. The thickness of the pipe's wall is $b-a$ and the surface area is $2\pi r$ per unit length, hence $J=-2\pi rDdc/dr$ per unit length. Integrating through the wall of the pipe, from a to b gives $\displaystyle\int dc=\frac{J}{2\pi D}\int\frac{dr}{r}$ with the result

$$c=\frac{J}{2\pi D}\ln(r)+q$$

where q is the integration constant. With the initial conditions $c(b)=0$, $c(a)=c_0$, the constants J and q can be found and eliminated using $0=\dfrac{J}{2\pi D}\ln(b)+q$ and $c_0=\dfrac{J}{2\pi D}\ln(a)+q$ which results in

$$c=\frac{c_0}{\ln(b/a)}\ln(b/r)$$

To find the mass flow J, differentiate this equation with respect to r and substitute into Fick's law. The flow rate is

$$J=\frac{2\pi D}{\ln(b/a)}c_0.$$

Does this make sense? Common sense would suggest that the flux of material is going to be large if the diffusion coefficient and initial concentration is large, and if the wall is thin.

If the wall is very thin then $a \to b$ and the log becomes small ($\ln(1) = 0$) and the flux increases rapidly as the wall becomes thinner. The effect is quite dramatic; if initially the pipe has $b/a = 2$ and then this ratio is reduced to 1.25, a reduction by a factor of 1.6 times, the flux is ≈6.5 times greater. If two gases are in the pipe, the ratio of the amount of each transported outside is in the ratio of their diffusion coefficients and their concentrations. This is important for O_2/CO_2 diffusion in the lungs.

(iv) In the presence of a chemical reaction where the molecule is also diffusing, the diffusion equation is

$$\frac{\partial c}{\partial t} = D\frac{\partial^2 c}{\partial x^2} - R$$

where R is the rate of reaction, which, for simplicity, is assumed to be a constant, independent of c. This can be the case for O_2 diffusing into a cell, or the reaction of ATP molecules when their concentration is high enough. ATP is used by the molecular motor protein ATPase. In this protein, proton flow across a membrane containing the ATPase drives a rotor that then forces closed one part of the protein and thereby splits ATP to ADP and releases phosphate. The reverse reaction can occur depending on the prevailing conditions. ATP is also used to drive similar molecular motors to ATPase that operate the flagellum of a micro-organism and so enable it to move. In this case, the ATP has to diffuse down the flagellum from its base where it has a constant concentration. The possibility exists that diffusion cannot supply enough ATP molecules, and hence energy, to move the flagellum and using the diffusion equation this condition is sought. The general model is therefore of the fixed concentration c_0 of a species outside a boundary, which diffuses (in one dimension in this model) into the bulk where it also reacts and we want to know to how far into the bulk reaction can occur.

Integrating at steady state when $\partial c/\partial t = 0$ produces $\dfrac{\partial c}{\partial x} = \dfrac{R}{D}x + a$, and if this is zero at the end of the flagellum where $x = L$, the integration constant a can be found. Integrating again produces

$$c = \frac{R}{D}\left(\frac{x^2}{2} - Lx\right) + c_0$$

if at $x = 0$, $c = c_0$. At the end of the flagellum, the concentration is $c = c_0 - \dfrac{R}{D}\dfrac{L^2}{2}$ which obviously cannot be negative; the limiting condition is, therefore, that $c_0 \geq \dfrac{R}{D}\dfrac{L^2}{2}$. As a figure of merit, it is found that if $\dfrac{R\,L^2}{D\,C_0} < 1$, diffusion is sufficient to supply enough ATP to the molecular motors or to supply O_2 into cells. Using typical values, reaction can only take place within 10 microns at most from the boundary. More sophisticated reactions with, say, spherical geometry lead to similar conclusions. Thus, it is understandable why small insects, for example, have tracheal tubes to increase the O_2 available to diffuse into their muscles and so maintain the high metabolic rate necessary for flight.

(v) The time taken to cool a cup of coffee or tea follows Newton's law of cooling. This law states that the rate of cooling depends on the *difference* in temperature compared to the surroundings and the form that the rate of heat transferred from one body is the same as that gained by another. This means that the rate is proportional to the temperature difference or

$$\frac{d\theta}{dt} = k(\theta_s - \theta) \tag{10.10}$$

where θ_s is the temperature of the surrounding air, and k is the constant, measuring the convective transfer of the heat from one medium into the other. When divided by the surface area, this is sometimes called Newton's cooling coefficient with units of watts $m^{-2}\,K^{-1}$. The cooling equation assumes that the cooling is convective and the object is in a slight draught of air so that the air temperature next to the object θ_s is constant. If the draught becomes a forced flow of air, then the cooling coefficient becomes dependent on the air speed and a 'wind chill' effect occurs. This is soon felt if you are outside on a cool day and move into a strong wind from a sheltered spot. Note that Newton's law of cooling has the same form as a first-order rate equation. The rate of change of temperature is proportional to the temperature of the body.

Mammals cool themselves by evaporating water from their bodies, by sweating or panting. The heat balance equation must, therefore, have extra positive terms due to the chemical reactions keeping them alive (their metabolism) and negative terms due to evaporation. This last term could be incorporated into the constant k. Heat flow is also important in simple chemical reactions and exothermic reactions are usually cooled. If the heat loss from the reaction vessel is not sufficient, the heat generated by an exothermic reaction, such as fermentation, can result in spontaneous combustion as used to happen with hayricks.

In radical polymerization reactions, as the reaction proceeds towards completion, the mixture must become very viscous. This means that the termination steps cannot occur and the rate of reaction increases because this is inversely proportional to the rate of termination. Secondly, the increased viscosity means that stirring can become ineffective and the only cooling mechanism is thermal diffusion to the vessel walls, which is slow. Consequently, if heat gain is too great, any gases or vapours trapped in the polymer may cause it to explode. This is called the Trommsdorff–Norrish effect and is an example of an auto-acceleration process or one with positive feedback. The rate of temperature change has an extra term for the heat generated and should have the generic form $\dfrac{d\theta}{dt} = k(\theta_s - \theta) + k_r e^{-E_a/R\theta}$ at temperature θ. The activation energy is E_a, and k_r is a constant proportional to the pre-exponential term from the Arrhenius equation and the heat capacity of the reaction mixture.

10.3.3 The centrifuge: forced separation

If a cylinder of radius r and height h is filled with a fluid whose components need to be separated and is spun about its axis at ω rad s^{-1}, the heavier components are forced to come to equilibrium further towards the outside of the cylinder than the lighter ones. The balance is between centripetal forces acting towards the axis of rotation and the centrifugal forces acting in the opposite direction. A small cylindrical shell of fluid in the region r to $r + dr$ from the rotation axis, and of height h, experiences a net force inwards of $2\pi hr \times dP$ where P is pressure (Margenau & Murphy 1943). This is equal to the inwards centripetal force due to the rotational motion which is $m\omega^2 r$. As the mass of the cylindrical shell is $2\pi rhdr\,\rho$, with ρ being the density, equating produces $2\pi rhdP = 2\pi rh\rho\omega^2 rdr$ or

$$\frac{dP}{dr} = \rho\omega^2 r.$$

In a liquid, the density ρ is a constant, and integrating produces the pressure $P = \rho\omega^2 r^2/2 + P_0$. This also explains why, if a liquid is spun in a beaker about its vertical axis, the surface obtains a parabolic profile. This can be seen clearly if two immiscible liquids are spun and one of them is coloured with some dye.

In a centrifuge, imagine strips taken out of the rotating cylinder and each replaced with a tube. Several of these are spun about the axis to balance the rotor. The distance r is therefore the radial distance from the axis, and h the height of the tube when it is spinning in a horizontal plane, which is normally its width.

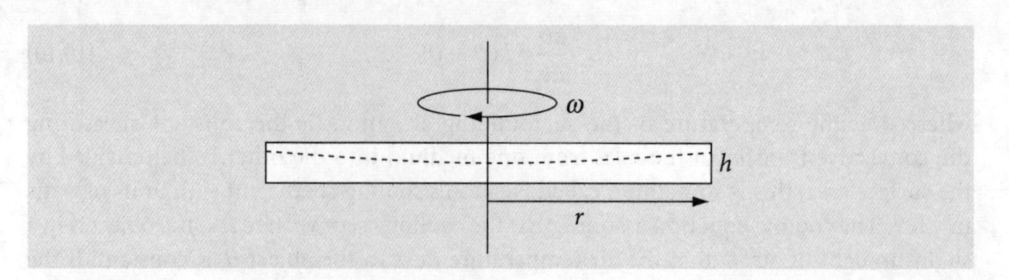

Fig. 10.7 Spinning disc; the fluid level is shown by the dashed line.

In an ideal gas, the pressure is proportional to the density since $P = nRT/V = RT\rho$. Solving the equation $\displaystyle\int\frac{dP}{P}=\frac{\omega^2}{RT}\int rdr$ produces $P=P_0e^{\frac{\omega^2r^2}{2RT}}$ if the pressure is P_0 when $r=0$.

10.3.4 Absorption of photons and scattering of X-rays and electrons

Photons in the ultraviolet and visible part of the spectrum excite electrons from the HOMO to LUMO orbitals of molecules giving rise to the colours we see around us. Vibrational and rotational transitions are excited by infrared light. When photons are absorbed, if the cross section for absorption at a certain wavelength is σ, then σndL photons are absorbed by n molecules per second, in a layer of thickness dL. The fractional decrease in intensity of the photons is then

$$\frac{dI}{I}=-\sigma ndL.$$

If the initial intensity is I_0 then the transmitted intensity is $I = I_0e^{-\sigma nL}$, which is called the Beer–Lambert law. The absorbed light intensity is $I_{abs} = I_0(1 - e^{-\sigma nL})$ because the total number of photons is those absorbed plus those transmitted. In many instances, the quantity σnL, the optical density, is replaced with $\varepsilon[c]L$ where ε is the extinction coefficient at a certain wavelength with units of $dm^3\ mol^{-1}\ cm^{-1}$, [c] is the solution concentration and L the path length.

A beam of X-rays can be scattered by the electrons in a molecule's atoms, rather than being absorbed. In a single crystal, parallel planes of atoms occur at regular intervals. The scattered intensity from the many planes similarly oriented to the X-ray beam can add in phase or cancel out, and this forms the basis of X-ray crystallography. The beam of X-rays has intensity I, and if a sample contains n electrons per unit volume, the first layer of thickness dL intercepts ndL photons, each electron presenting an area of $4\pi d^2$ where $d= e^2/mc^2$ and is approximately the closest approach distance of two electrons. An equation similar to the Beer–Lambert equation describes scattering but if this is small, and as nL is the number of scattering electrons in the sample, the fractional scattering of a single electron becomes $I_s/I_0 = 4\pi d^2 = 4\pi\left(\dfrac{e^2}{mc^2}\right)^2$, see Q10.5.

10.3.5 Velocity and acceleration

A number of problems involving the motion of bodies can be solved by using Newton's laws to equate forces, and separating the variables of the equation produced. Usually, the force acting downwards is gravity, and it is often convenient to choose the positive direction as downwards. By Newton's second law of motion, the net force f acting on a body, is equal to the rate of change of momentum, or

$$f=m\frac{dv}{dt}$$

where m is the (constant) mass and v the velocity at time t. This equation can also be written as, force equals mass times acceleration or

$$f=m\frac{d^2x}{dt^2},$$

x being the position at time t. Both of these equations are of fundamental importance in solving problems in dynamics.

Consider a skydiver who jumps from an aeroplane and falls under gravity, but with air resistance that is linearly proportional to velocity. Naturally, we shall want to work out how velocity increases with time, and if it will reach a constant, limiting value. After a certain time, the parachute must open, and the resistance to falling will increase so that it is now proportional to the square of velocity. At any time the forces due to gravity and air resistance are balanced, and therefore act in opposite directions and this is used to solve the problem. The force of gravity is mg, g being the acceleration due to gravity and m the mass. The air resistance is av where a is a constant with units of $kg\ s^{-1}$ and v_0 is the initial downwards velocity which could be zero. The equation of motion in terms of force is

$$m\frac{dv}{dt} = mg - av \qquad v(t_0) = v_0, \quad t_0 = 0,$$

where the initial condition of velocity being v_0 at zero time is included. Note that v is a function of time, although this is not shown explicitly in the equation. To calculate the result it is easier to change the force equation to

$$\frac{dv}{dt} = g(1 - hv)$$

where $h = a/gm$. Separating variables and integrating gives

$$\int \frac{dv}{1 - hv} = g\int dt, \qquad -\frac{1}{h}\ln(1 - hv) = gt + c$$

and with the initial values, $c = -\frac{1}{h}\ln(1 - hv_0)$ this produces $\frac{1}{a}\ln\left(\frac{1 - hv}{1 - hv_0}\right) = -gt$. After substituting for h and rearranging, the velocity at any time t is,

$$v = \frac{mg}{a} - \left(\frac{mg}{a} - v_0\right)e^{-at/m}.$$

To find the terminal velocity, the limit at long time is taken, which produces $v_{term} = \frac{mg}{a}$. The distance from the ground at any time is found by integrating the velocity because $v = dx/dt$,

$$\frac{dx}{dt} = \frac{mg}{a} - \left(\frac{mg}{a} - v_0\right)e^{-at/m} \qquad x(t_0) = x_0, \quad t_0 = 0$$

if the initial height is x_0 the result is

$$x = \frac{mg}{a}t + \frac{m}{a^2}(mg - av_0)(e^{-at/m} - 1) + x_0.$$

The time to reach the ground is found when $x = -x_0$ and then the resulting transcendental expression has to be solved for t.

10.3.6 Questions

Full solutions are available at www.oxfordtextbooks.co.uk/orc/beddard.

Q10.1 Calculate the phase portrait for $\frac{d^2x}{dt^2} - \omega^2 x = 0$. Show by plotting some trajectories that the centre is a saddle point which means that the curvature along the x-axis is the opposite to that along $v = dx/dt$.

Q10.2 Solve **(a)** $\frac{dy}{dx} = \frac{xy}{1 - x^2}$, **(b)** $\frac{dy}{dx} = e^x\tan(y)$, **(c)** $\frac{dy}{dx} = e^{x-y}\sin(x)$ with the initial condition that $y_0 = c$ when $x = x_0$.

Q10.3 Find A vs t for the reaction $2A \rightarrow$ product.

Q10.4 **(a)** Solve equation (10.7).

(b) Find the time for half the population to be infected if initially, only one individual is infected.

Strategy: (a) To integrate separate variables. The initial condition is $x = 1$ at $t = 0$ and the total population is N. (b) The time when $x = N/2$ is the half life.

Q10.5 Show that the fractional scattering of X-rays off a single electron is $I_s/I_0 = 4\pi\left(\frac{e^2}{mc^2}\right)^2$

Q10.6 When the skydiver's parachute opens we assume that the drag instantly becomes proportional to the velocity squared, i.e. kv^2 where k is a constant. Calculate the terminal velocity.

Q10.7 If x moles of a substance A reacts in time t with another B, it is found that $\dfrac{x}{a-x}=akt$, where a is the number of moles of A or B initially preset, k is the rate constant and at $t=0$, $x=0$. Show that the rate equation for this reaction varies as $(a-x)^2$. What are the units of k?

Q10.8 A jar of water at 15 °C is placed in a temperature of −12 °C and its temperature falls by 5° in 8 minutes. How many minutes will it be before ice could form?

Q10.9 A certain amount of crystalline iodine is dropped into a heated chamber at constant temperature and sublimes to form I_2 vapour. Suppose that the sublimation occurs at a rate that is proportional to the amount of crystalline iodine present at that time. Write down the rate equation for the disappearance of the solid and also the change in pressure p caused by sublimation. Let p_∞ be the final pressure when the entire solid has sublimed.

Q10.10 The Stern–Volmer equation describes the ratio of fluorescence in the presence and absence of a quencher at concentration Q. The excited states behave according to the scheme (10.4). The additional quenching step is $S_1+Q\xrightarrow{k_q}G+Q$ and the Stern–Volmer equation is $\dfrac{\varphi}{\varphi_Q}=1+k_{SV}[Q]$ where k_{SV} is a constant and φ is the fluorescence yield in the absence if quencher and φ_Q that in its presence.

(a) Confirm the Stern–Volmer equation and work out what the rate constant k_q is.

(b) Work out the equivalent equation in terms of the ratio excited state lifetimes not yields, i.e. τ/τ_q.

Strategy: Use steady state conditions to solve the equations.

Q10.11 In the reaction scheme, $A\xrightarrow{k_1}B\xrightarrow{k_0}C$ where $[A]_0=a$, $[B]_0=[C]_0=0$ suppose that species A decays to B in a first-order process, but B decays by *zero* order.

(a) Solve the scheme to find the A and B concentration profiles.

(b) In the body, alcohol like many other drugs is ingested rapidly. It is then enzymatically catalysed in the liver, initially to acetaldehyde in a zero-order reaction with rate constant $k_0=0.19$ g dm^{-3} hr^{-1}. If you rapidly drink 70 g alcohol, or approximately 3 pints of 5% beer or lager, what is the maximum blood alcohol if the relevant body volume is 60 litres, and how long will it take to reach this level?

(c) The maximum level of blood alcohol with which it is legal to drive in the UK is 80 mg/100 ml. How long will it be before this level is reached and how long before no significant alcohol remains? The body volume for a man is about 0.82 times his weight and 0.67 times that for a woman. Take the rate constant $k_1=5$ hr^{-1}. See Marshall (1978) for a short discussion of this topic and others in pharmacokinetics.

Strategy: The rate of change of a zero-order reaction is a constant and therefore does not depend on the change in concentration of any species. In the body when the alcohol level is high, it reacts at a rate that is independent of the alcohol or enzyme concentration and is therefore zero order. At low concentrations, and hence at long times, this is not true and the reaction scheme breaks down. (At low alcohol concentration, k_0 can empirically be replaced by $k_0B/(k_0B+5\times10^{-5})$, which better represents its behaviour; see Barnes & Fulford 2002).

Q10.12 A catalytic pellet, made of Pt on an Al_2O_3 support, is poisoned by one of several sulphur compounds after this has diffused only a small way into its surface. The spherical pellet has a radius r_0, the diffusion coefficient in the pellet is D and the poison has penetrated to a radius r_p. If the poison has a fixed concentration in solution of c_0, what is its concentration c_p in the catalyst at radius r_p if the flux into the pellet is equal to the rate of reaction kc_p?

Q10.13 The rate of reaction of hydroxyl and of ethyl radicals has been measured in solution, as have those of many other species, and found to be diffusion controlled. The quenching of a molecule M by a species Q had been found to have a rate constant equal to the diffusion controlled value k_d. The Smolukowski model (Eyring, Lin & Lin 1980) supposes that reaction occurs at a radius R_A around A, see Fig. 10.8. Inside this spherical surface, the concentration of Q is assumed to be zero. Show, assuming that steady state conditions apply and using Fick's first law, that the diffusion controlled rate constant is $4\pi DR$ where D is the sum of the diffusion coefficients of the two species.

Strategy: If M and Q are the concentrations of the species, then the rate of reaction is k_2MQ. This is equal to the total flux entering the spherical surface of radius r, which is $J = +4\pi r^2 D\dfrac{dQ}{dr}$, multiplied by the concentration M. The initial condition is that $c_0 = 0$ at $r = R$. Calculate J first then k_2.

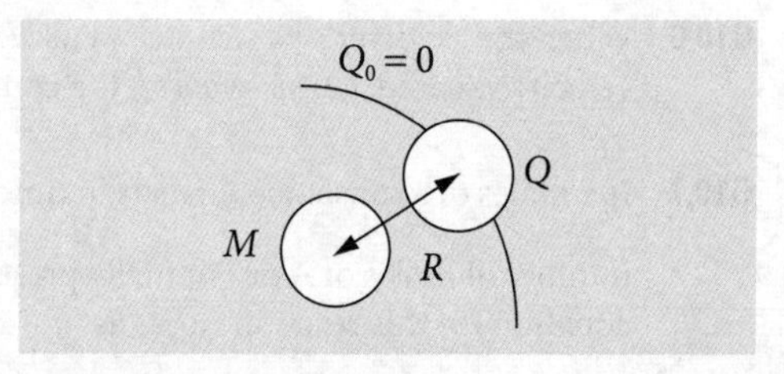

Fig. 10.8 Quenching at distance R.

Q10.14 A gas phase chain-branching reaction is stable under some conditions, but can lead to an explosion under others. Chain branching is the cause of the first explosion limit in the H_2/O_2 reaction. The general radical chain reaction scheme can be drawn pictorially, as shown in Fig. 10.9. The initial species M reacts to produce some chain carriers R, which can produce more of themselves, αR and also the reaction product P. A chain carrier can also be deactivated by reacting with other species or with the walls of the reaction vessel, both of which lead to termination. The branching number α, must be greater than 1 if branching is to occur, because by definition the propagation step produces one radical as well as product, e.g. $R + M \rightarrow R + product$. In the H_2/O_2 reaction the branching steps are,

$$H + O_2 \rightarrow OH + O$$
$$O + H_2 \rightarrow OH + H$$

so that one radical H or O produces an extra one making $\alpha = 2$. The rate of reaction is the radical concentration multiplied by the propagation rate constant, k_pR.

Using Fig. 10.9, write down dR/dt and integrate this equation to find the radial population R vs time, assuming that no radicals are present at $t = 0$. Show that explosion occurs when the rate of branching is greater than the rate of termination, as might be anticipated, and, if not, that a steady state is reached.

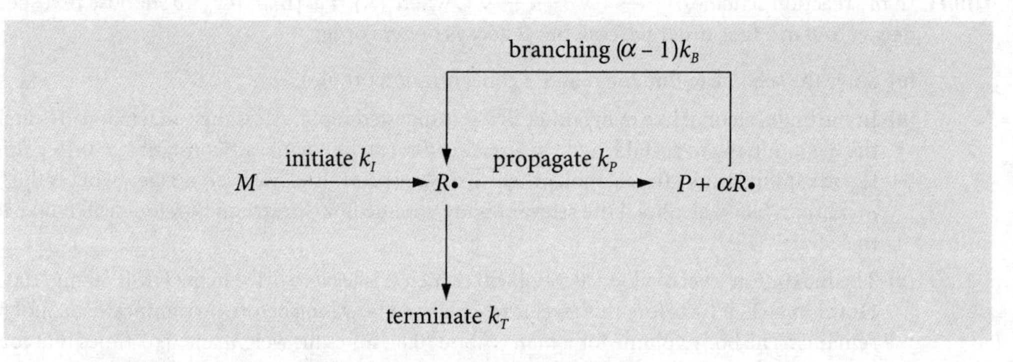

Fig. 10.9 Chain branching; P is the product; R is the radical chain carrier.

Q10.15 The recombination of iodine atoms after the photo-dissociation of molecular iodine in the presence of inert buffer gas, was one of the first reactions followed by Porter using the technique of flash photolysis he had invented and for which he won a share of the Nobel Prize in Chemistry (Porter 1967; Christie et al. 1952; Porter & Smith 1961; Phillips & Barber 2006). This technique uses a flash of light to initiate a reaction followed by a second weaker flash at a predetermined time to probe the species produced.

The first observation made was that the reciprocal pressure of iodine atoms increased linearly with time as they recombined at a constant buffer gas pressure and the second was that as the temperature was raised that the recombination rate constant decreased, i.e. the reaction had negative temperature dependence and hence the experimentally measured activation energy was *negative*. The scheme proposed was

$$I + M \underset{k_{-1}}{\overset{k_1}{\rightleftharpoons}} IM$$
$$I + IM \xrightarrow{k_2} I_2 + M$$

where the species IM is a complex formed between the iodine atom and the buffer gas due to van der Waals intermolecular forces. The stronger the forces are the greater the chance of forming a complex. The photolysis flash is far shorter than any reaction time and so the production of I atoms from I_2 was considered as instantaneous and was ignored in the reaction scheme.

(a) Solve this scheme and find the rate of change of iodine molecules vs time by assuming that the equilibrium with complex *IM* is rapid. Rearrange the result so that a straight-line plot can be made.

(b) Explain the negative activation energy by assuming that each rate constant can be described by an Arrhenius equation.

Strategy: Assume that *IM* is at steady state and use this to simplify and then integrate the rate equation.

Q10.16 The mutarotation of glucose in water or the racemization of biphenyls are reversible unimolecular reactions with the scheme $A \underset{k_{-1}}{\overset{k_1}{\rightleftarrows}} B$. Transitions in membrane proteins also follow this scheme if the transitions are induced by a voltage, and then *A* and *B* indicate states when the protein is open or closed to the passage of ions. If *a* and *b* are the initial amounts of *A* and *B* respectively and *x* the amount of *A* reacted at time t,

(a) Calculate the change in *A* as equilibrium is reached.

(b) Express your result is terms of the equilibrium amount of *x*, which is x_e.

(c) If the optical rotation α is measured with time the final value being $\alpha_\infty - \alpha_0$, express the answer in terms of the difference in optical activity $\alpha_t - \alpha_0$ if α_0 is that initially measured. Show that the units of the measuring parameter α do not matter. This is always the case for a first-order reaction.

(d) Instead of starting with an excess of *A*, suppose that equilibrium has been reached and this is perturbed by a small amount *x* caused by heating the sample by a few degrees with a laser or electrical pulse whose duration is shorter than the relaxation time back to equilibrium. Use the same rate expression as in (a) but substitute for $\Delta = x - x_e$ to obtain $d\Delta x/dt = \ldots$ then integrate to find Δx.

(e) If a voltage *V* is applied across a membrane such that it is zero at the centre, show that the first-order rate constant for a protein opening and closing is a minimum when $V = 0$. The barrier height is changed by the application of a voltage, and then the rate constants have the form $k_{1,-1} = k_0 \exp(-(E_0 \pm V/2)/RT)$ where E_0 is the barrier in the absence of the potential. A typical voltage is 100 mV across a membrane of 50 nm width, which is equivalent to 2 MV m^{-1}. In converting to energy, this becomes a barrier of 0.1eV $\times$ 96.4 kJ mol^{-1} eV^{-1} $\approx$ 9.6 kJ mol^{-1} across the membrane. This is comparable to the barrier (activation energy) for a hydrogen bond, but larger than thermal energy at room temperature which is $RT \approx 2.5$ kJ mol^{-1}.

Strategy: Let *x* be the amount of *A* consumed at time *t* and use this to calculate the loss in *A* and the increase in *B*.

Q10.17 In the relaxation methods of studying chemical kinetics, the reaction at equilibrium is perturbed by a small amount and the return to equilibrium under the new conditions followed. In one method, water is heated up to a few degrees above room temperature by an infrared laser pulse of nanosecond duration, and the equilibrium

$$H_2O \underset{k_2}{\overset{k_1}{\rightleftarrows}} H^+ + OH^-$$

is thereby perturbed by a *small* amount. Let *a* be the initial concentration of water and *x* that of the ions produced.

Show that the relaxation back to equilibrium is first-order.

Strategy: Let *x* be the amount of water dissociated or ions produced, and write down dx/dt. Then to allow for a small change let $\Delta x = x - x_e$ where x_e is the equilibrium concentration of ions, and substitute this into the rate equation to find $d\Delta x/dt$. Assume Δx^2 is small compared to Δx and x_e^2.

Q10.18 There are many different classes of catalytic reactions. An important set of reactions is the Michaelis–Menten scheme $S + E \rightleftarrows SE \rightarrow P + E$. The enzyme *E* catalyses the substrate *S* to form product *P* and is returned to its active state. The classic example is trypsin produced from trysinogen. This scheme cannot be solved analytically, except at steady state, but is easily solved numerically, see Chapter 11.

Simpler catalytic reaction schemes are found between acetone and iodine in an acidic solution, or Ru^{2+} oxidized by acidic bromate ions containing traces of BrO_2. The stoicheiometry for the reaction of acetone is

$$CH_3COCH_3 + I_2 + H^+ \rightarrow CH_3COCH_2I + I^- + 2H^+$$

and the reaction proceeds by rapidly protonating the oxygen atom. This species then rearranges, losing a proton and reacting with iodine. The experimental rate expression is $d[acetone]/dt = -k[acetone][H^+]$. As both the ruthenium and acetone reactions are autocatalytic, the product acts to accelerate the reaction.

The general *autocatalytic* reaction of species A with catalyst B is

$$A + B \rightarrow P + 2B.$$

and the rate equation is deceptively simple and is

$$da/dt = kab$$

if a and b are the concentration A and B.

(a) Write down and solve the general autocatalytic scheme if the concentration of A is a_0 and of B is b_0 initially. Plot the graphs of A and B concentrations if $a_0 = 1$, $b_0 = 0.005$, and $k = 0.05$.

(b) Sketch the phase portrait and characterize the equilibrium points as stable or unstable.

Strategy: Let x be the concentration of acetone and b that of acid. As acetone reacts more acid is formed, therefore, $a_0 - x = b - b_0$. Use this to solve the scheme. (The form of this equation is similar to the laser model, see Section 10.1.7.)

Q10.19 The Gompertz population growth law appears to be good at describing the growth of tumours (Strogatz 1994). The rate of change of the size n of a tumour (as number of cells) is represented as

$$dn/dt = -kn\ln(n/a), \qquad n(0) = n_0, \qquad k > 0, a > 0.$$

(a) Solve this equation, show that the tumour size is self-limiting, and suggest what the constants represent.

(b) Plot the graph in dimensionless units; n/n_0 vs kt with two initial populations above and below the long-time value.

(c) Two interacting populations (or chemical species) A and B, change according to the rate equations

$$dA/dt = k_2AB, \qquad A(0) = A_0,$$
$$dB/dt = -k_1B, \qquad B(0) = B_0,$$

where A and B represent the populations at time t. Show that these equations will produce the Gompertz equation.

(d) If, instead of a population, B is the tumour growth rate (1/time), then k_1 is the constant (1/time) describing its retardation and B_0 the initial rate. (Note that B is a rate 'constant' that changes in time. This may seem strange but time-dependent rate constants also appear naturally in diffusional quenching of molecules in solution.) Variable A now represents the number of cells in the tumour at time t and the constant k_2 is irrelevant and can be set to unity. Using your answer to (c), conclude what the condition is, in terms of k_1 and B_0, for the tumour to shrink in size.

Strategy: (a) try the substitution $n/a = e^u$.

Q10.20 Rate equations can be used to study populations of all sorts of animals, fish, or bacteria. Suppose that a number of creatures N are born with a rate constant k_1 (time^{-1}) but die with a rate constant that has a value that is not only dependent on the rate of natural deaths k_2 but also on the number of animals k_3N. The units of k_2 are the same as k_1, and k_3 are number^{-1} time^{-1}. The factor k_3N represents death caused by 'crowding' which is not meant literally, but is due to competition for resources or shelter and so forth. There are N_0 animals are present initially.

(a) Write down the rate equation and work out the steady state conditions. (The rate equation produced is called the logistic equation and should not be taken too literally when describing animal populations, but used as a starting point; see Strogatz 1994).

(b) Solve the rate equation for the population at time t. Plot several curves of the solution with the rate constants, such as $k_1 = 2$, $k_2 = 1$ (unit: time^{-1}), $k_3 = 0.001$ (unit: number^{-1} time^{-1}) and N_0 from 200 to 2000 to investigate the properties of this equation. Why does k_3 have to be far smaller than the other rate constants?

(c) When the animals are harvested with a rate constant k_h, (number / time), which is independent of the number present, the population may collapse or recover, depending on how aggressive the harvesting is relative to the initial number present. Write down the rate equation and plot the phase portrait and also that for part (a). Use the parameters above with $k_h = 200$. Calculate the steady state populations, and in both cases locate them on the phase portrait and also add them to a plot of population vs time. This model will produce negative populations under some conditions, which is of course wrong. Strogatz (1994, Chapter 3) describes a modification that overcomes this.

(d) Solve the rate equation in the presence of harvesting using Maple to perform the integration as it is complicated. The algebraic result can be simplified although this is not necessary to be able to plot graphs of the population vs time. Use the parameters above and initial populations starting at 100 and increasing to 1000.

Barnes & Fulford (2002) give several more examples describing the change of animal populations and a more detailed analysis is given by Strogatz (1994).

Strategy: The rate of population change is the difference in that due to an increase and decrease in the population. The steady state conditions are found when the rate of change is zero. The phase portrait is the plot of dN/dt vs N; see Section 10.1.6.

Q10.21 A solution of an aromatic molecule in its electronic ground state G is excited by a pulse of light, and its first excited singlet state S_1 is produced. Some of these molecules fluoresce while others form triplet states via intersystem crossing. The triplet excited state T, is far longer lived than the singlet, typically by 1000 times. In solution, the triplets can annihilate one another after diffusing together and this can make measuring the correct triplet lifetime difficult. The scheme is

$$G \to S_1$$
$$S_1 \xrightarrow{k_f} G + h\nu$$
$$S_1 \xrightarrow{k_s} T$$
$$T \xrightarrow{k_T} G$$
$$T + T \xrightarrow{k_a} G + G \text{ or } (S_1 + G)$$

where k_a is the second order annihilation rate constant, k_f is the first-order rate of fluorescence, and k_s that of forming the triplets T via intersystem crossing. The triplets also decay with a rate constant k_T, which is the sum of phosphorescence and non-radiative decay rate constants back to the ground state G.

In the antennas of photosynthetic organisms, intense laser excitation can cause singlet–singlet annihilation $S_1 + S_1 \to S_1 + G$ and this is due to energy migration among the (chlorophyll) pigments. This process can be used to estimate the rate of energy migration that is otherwise hard to obtain. We will assume, however, that the S_1 states are too short lived and the excitation intensity too low for this to be a significant process and it can be ignored.

Write down the full rate equations for S_1 and T_1 in this scheme, but solve only for the triplet population by assuming that S_1 has decayed to zero before any significant T-T annihilation occurs. Take the initial concentration of excited singlet states as S_0 and triplets T_0 and show that the triplet population is $\dfrac{1}{T} = \left(\dfrac{1}{T_0} + \dfrac{k_a}{k_T}\right)e^{k_T t} - \dfrac{k_a}{k_T}$. Plot T vs time if $k_a = 10^9$ dm^3 mol s^{-1}, and $k_T = 10^2$ to 10^4 s^{-1} and $T_0 = 10^{-4}$ mol dm^{-3}. Show that at long times the decay is first-order and k_T can be measured, and that at short times the decay is second order.

10.4 Integrating factors

10.4.1 Homogeneous equations

A homogeneous differential equation has the form

$$Mdx + Ndy = 0$$

if M and N are functions of the *same degree* in x and y. This constraint means that M/N is a function of y/x and the homogeneous equation has the form

$$\frac{dy}{dx} = f\left(\frac{y}{x}\right),$$

where f is the function of y/x. To solve this equation the substitution $y = ux$ is used. The derivative is $dy/dx = u + xdu/dx$ then

$$u + xdu/dx = f(u)$$

which can have its variables separated to produce

$$\frac{du}{f(u) - u} = \frac{dx}{x} \tag{10.11}$$

which is integrated to give the solution. The equation $\frac{dy}{dx} = (2x^2 + 3y^2)/xy$ satisfies the condition $Mdx + Ndy = 0$ because the degree of xy and x^2 and y^2 is the same, and is 2, on both on the top and bottom of $(2x^2 + 3y^2)/xy$. The function $f(u)$ then becomes

$$f(u) = (2x^2 + 3u^2x^2)/ux^2 = (2 + 3u^2)/u$$

and then the solution is found using equation (10.11) as

$$\int \frac{u}{2(1+u^2)} du = \int \frac{dx}{x}$$

and which is, with $\ln(c)$ as the integration constant,

$$\frac{1}{4}\ln(1 + u^2) = \ln(x) + \ln(c).$$

and after substituting and rearranging, the general solution is $y = x\sqrt{c^4x^4 - 1}$.

10.4.2 Exact equations

The exact equation also has the form $Mdx + Ndy = 0$ with the additional constraint that

$$\frac{\partial M}{\partial y} = \frac{\partial N}{\partial x}.$$

and this type of equation is frequently met in thermodynamics, see Chapter 3.12.8.

If a differential equation

$$Mdx + Ndy = 0,$$

is multiplied throughout by a suitable term it can become exact; i.e. the equation is now

$$(Mdx + Ndy)G(x, y) = 0.$$

The term $G(x, y)$ is called an integrating factor which in general can be difficult to uncover except in the case of linear first-order equations described next.

10.4.3 Linear first-order equations

Linear first-order equations have the form,

$$\frac{dy}{dx} + Py = Q$$

where P and Q are functions of x only. These equations can be integrated using integrating factors. First we consider the particular case when $Q = 0$ or

$$\frac{dy}{dx} + Py = 0. \tag{10.12}$$

Writing the equation as $\frac{1}{y}dy + Pdx = 0$ and integrating gives

$$\ln(y) + P\int dx = \ln(c),$$

where the constant is $\ln(c)$. Simplifying this result gives $ye^{\int Pdx} = c$. If this equation is differentiated, it produces $\frac{d}{dx}(ye^{\int Pdx}) = 0$. From this result, it can be shown that

$$\frac{d}{dx}ye^{\int Pdx} = e^{\int Pdx}\left(\frac{dy}{dx} + Py\right)$$

and the right-hand side is the starting equation (10.12), multiplied by the *integrating factor* $G(x) = e^{\int Pdx}$. It follows that the solution of the differential equation becomes the two integrals,

$$ye^{\int Pdx} = \int Qe^{\int Pdx} + c.$$

As an example of this method, the equation, $dy/dx + y/x = 3\sin(x)$ is solved. In this equation $P = 1/x$ and $Q = 3\sin(x)$. The integrating factor is simple in this case and is

$$\int Pdx = \ln(x)$$

and as $e^{\ln(x)} = x$ then $yx = 3\int x\sin(x)dx + c$.
Integrating this by parts and rearranging, produces

$$y = 3\sin(x)/x - 3\cos(x) + c,$$

and the constant is determined once the initial conditions are specified.

While this method is very useful, it does have its limitations particularly when $Q \neq 0$; for instance the similar equation $dy/dx + xy = 3\sin(x)$ has the integrating factor $e^{\int xdx} = e^{x^2/2}$. The solution is

$$ye^{x^2/2} = 3\int e^{x^2/2}\sin(x)dx + c,$$

and the remaining integral is now a difficult one. Using Maple to do this produces

```
> Int(exp(x^2/2)*sin(x),x):%= simplify(value(%),symbolic);
```

$$\int e^{\frac{1}{2}x^2}\sin(x)\,dx = -\frac{1}{4}\sqrt{\pi}e^{\frac{1}{2}}\sqrt{2}\left(\text{erf}\left(\frac{1}{2}I\sqrt{2}x - \frac{1}{2}\sqrt{2}\right) - \text{erf}\left(\frac{1}{2}I\sqrt{2}x + \frac{1}{2}\sqrt{2}\right)\right)$$

and the final solution is $y = -\dfrac{3\sqrt{2\pi}}{4}e^{(1-x^2)/2}\left[erf\left(\dfrac{ix-1}{\sqrt{2}}\right) - erf\left(\dfrac{ix+1}{\sqrt{2}}\right)\right] + c.$

(i) One particularly useful application of this method is to solve coupled kinetic equations. If a molecule reacts in a scheme $A \xrightarrow{k_1} B \xrightarrow{k_2} C$ the rate equations are readily written down and are

$$\frac{dA}{dt} = -k_1A$$

$$\frac{dB}{dt} = k_1A - k_2B$$

If the initial amount of A is A_0 and $B_0 = 0$, then integrating the first equation gives $A = A_0e^{-k_1t}$. Substituting this into the second and rearranging gives

$$\frac{dB}{dt} + k_2B = k_1A_0e^{-k_1t}$$

which has the form $\dfrac{dB}{dt} + PB = Q$ and can be solved using the integrating factor $e^{\int Pdt} = e^{k_2t}$ with the solution,

$$Be^{k_2t} = k_1A_0\int e^{k_2t-k_1t}dt + c,$$

or

$$Be^{k_2t} = \frac{k_1A_0}{k_2 - k_1}e^{k_2t-k_1t} + c.$$

The initial conditions are $B = 0$ when $t = 0$ then

$$B = \frac{k_1A_0}{k_2 - k_1}(e^{-k_1t} - e^{-k_2t}),$$

and this has the expected form; it is zero when $t = 0$ and again when $t = \infty$ and passes through a maximum when $dB/dt = 0$.

10.4.4 The Bernoulli equation

The equation

$$\frac{dy}{dx} + Py = Qy^n$$

where P and Q are functions of x alone, is called the Bernoulli equation. If integer $n = 1$ the equation is solved by separating variables in the usual way. If $n \neq 1$ the equation can be solved with an integrating factor with the substitution

$$y^{1-n} = u.$$

Differentiating this expression gives $\dfrac{(1-n)}{y^n}\dfrac{dy}{dx} = \dfrac{du}{dx}$ and the equation becomes

$$\frac{du}{dx} + (1-n)Pu = (1-n)Q.$$

The equation $x\dfrac{dy}{dx} + y = x^2e^xy^2$ is seen to be of the Bernoulli form because if divided through by x it becomes

$$\frac{dy}{dx} + \frac{y}{x} = xe^xy^2.$$

With $n = 2$ the substitution is $u = y^{-1}$ and then $\dfrac{du}{dx} - \dfrac{u}{x} = -xe^x$. The integrating factor is $e^{-\int dx/x} = 1/x$ then $\dfrac{u}{x} = -\int e^x dx$ or $u = -x(e^x + c)$. Substituting back gives $y = -\dfrac{1}{x(e^x + c)}$.

10.4.5 Questions

Full solutions are available at www.oxfordtextbooks.co.uk/orc/beddard.

Q10.22 A molecule is continuously excited by visible light at a rate R and its excited singlet state S_1 is produced. $R = I_{abs}\sigma[G]$ where I_{abs} is the light intensity, σ the absorption cross section, and $[G]$ the ground state concentration. The state S_1 decays to a triplet state T in a process called intersystem crossing with rate constant k_s, and fluoresces with rate constant k_f. The triplet decays back to the ground state with rate constant k_T.

Solve the scheme for singlet S_1 and triplet T populations. The scheme is shown in the sketch, Fig. 10.10. The initial populations at $t = 0$ are $S_1(0) = S_0$ and $T(0) = 0$.

Strategy: Write the rate equations in the usual way accounting for loss and gain for each species. Solve the equation for S, substitute the result into that for T and use the integrating factor method.

Q10.23 If a laser pulse of duration $I(t)$ initiates a reaction in a molecule A with rate constant k_s, then the rate of change of A can be written as

$$\frac{dA}{dx} = k_aI(t) - k_sA.$$

If the initial amount of A is A_0, solve this equation,

(a) when the pulse is so short it can be considered as a delta function $I(t) = I_0\delta(t - t_0)$ where I_0 is the instantaneous intensity at time t_0, and

(b) with an exponential pulse, $I(t) = I_0e^{-at}$. Discuss the effect of a short and long duration pulse.

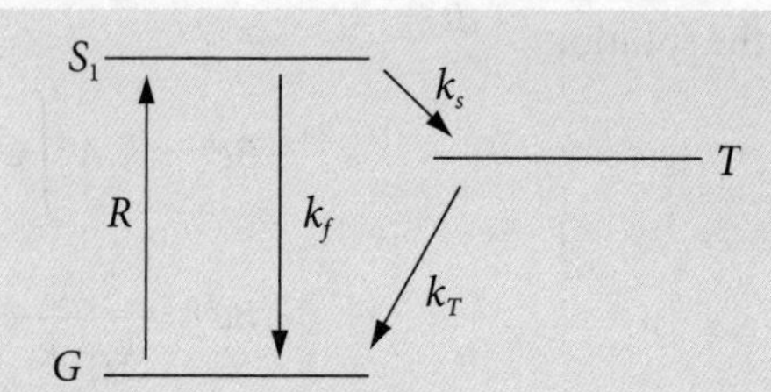

Fig. 10.10 Singlet–triplet scheme.

10.5 Second-order differential equations

10.5.1 Newton's laws and differential equations describing motion

Differential equations dominate the study of the motion of the planets and of molecules. In single molecules or in ensembles of them, molecular dynamics calculations rely on solving many simultaneous equations according to Newton's laws of motion, linked to potentials describing intermolecular interactions. Newton's laws are

(i) Every particle remains either in a state of rest or of constant speed in a straight line unless acted on by a force to change that state.

(ii) The rate of change of the momentum of a particle is proportional to the force acting on it and is in the direction of the force.

(iii) Action and reaction are equal and opposite.

The first law states that any acceleration experienced by a particle is caused by the action of an external force. The second law proposes that force f is equal to the product of mass and acceleration since momentum is the product of mass and velocity;

$$f = m\frac{d^2x}{dt^2} = m\frac{dv}{dt}$$

where x is position, v velocity, and t time. The second law includes the first, for if the force is zero then so is the acceleration and the body must remain unchanged.

The third law asserts that if two particles exert forces on one another, the force exerted by the first on the second is equal to that exerted by the second on the first. This law can be used to define the mass of a particle.

Newton's law of gravity was formulated to understand the motion of the planets. It states that the force of attraction of two bodies is proportional to the product of their masses and the inverse square of their separation.

$$f = \frac{Gm_1m_2}{r^2}$$

and G is the gravitational constant 6.673×10^{-11} N m^2 kg^{-2}. This law of attraction takes a simpler form in the case of a small body falling to earth from heights that are small compared to its radius. In this case, r is sensibly constant and the law becomes

$$f = mg$$

where g is the acceleration due to gravity, $g = 9.81$ m s^{-2}. Clearly, g will be different on the moon to that on the earth. The differential equation for a body falling from a height x under gravity, is therefore

$$\frac{d^2x}{dt^2} = g$$

and is independent of the mass. The distance x is positive in the downwards direction. Positive upwards would change g into $-g$. If this equation is integrated once then

$$\frac{dx}{dt} = gt + v_0$$

where the integration constant v_0 is the initial velocity at $t = 0$, and $v = dx/dt$ is the velocity at time t producing the familiar equation $v = v_0 + gt$. Integrating again produces the distance travelled from the starting or reference point x_0,

$$x = \frac{gt^2}{2} + v_0t + x_0.$$

This equation contains all the information about a freely falling body.

10.5.2 Simple harmonic motion

Simple harmonic motion is the most important form of periodic motion. It describes the small angular oscillations of a pendulum as well as the vibrations of molecules, and this is called the harmonic oscillator model. A spring that hangs vertically with a mass attached to its end, a taut wire with a mass attached at it centre, or the oscillations of a ship or of a hydrometer, all follow simple harmonic motion, if the inertia of the spring, wire, or liquid is ignored. In this motion, the acceleration is proportional to the displacement of the particle from its central position.

The equation of motion is formed by equating force, as mass multiplied by acceleration, to the force $f(x)$ on the particle, such as given by Hooke's or some other law,

$$m\frac{d^2x}{dt^2}=f(x). \tag{10.13}$$

If the force is due to gravity, then $f(x)=mg$; if describing simple harmonic motion based on a *small* extension obeying Hooke's law, then $f(x)=-kx$ where k is the force constant and in molecules this has values of a few hundred newton metre^{-1}. This type of differential equation can always be solved by multiplying both sides by dx/dt which produces

$$m\frac{dx}{dt}\frac{d^2x}{dt^2}=f(x)\frac{dx}{dt}$$

and, although this does not look too promising, the left-hand terms are the *derivative* of $(dx/dt)^2/2$; therefore, integrating both sides gives

$$\begin{aligned}\frac{m}{2}\left(\frac{dx}{dt}\right)^2 &= \int f(x)\frac{dx}{dt}dt+c\\ &= \int f(x)dx+c \qquad (10.14)\\ &= F(x)\end{aligned}$$

where c is the constant of integration, and F represents the result of the integration. This is the 'equation of energy'; the left-hand side is the kinetic energy since dx/dt is velocity and the right hand side is the potential energy.

The time to reach position x can also be found. First, rearrange the energy equation to

$$\frac{dt}{dx}=\pm\sqrt{\frac{2}{m}}\frac{1}{\sqrt{F(x)}}$$

and then integrate

$$t=\pm\sqrt{\frac{2}{m}}\int\frac{dx}{\sqrt{F(x)}}+c_1 \tag{10.15}$$

This equation contains two arbitrary constants, c_1 and c, of which c is already included in $F(x)$. Two constants are needed because the acceleration has to be integrated twice.

The equation of energy can be obtained in a slightly different way if the velocity v is required. Taking x as the dependent variable and using $\dfrac{dv}{dx}=\dfrac{dv}{dt}\dfrac{dt}{dx}$, equation (10.13) can be written as

$$mv\frac{dv}{dx}=f(x) \tag{10.16}$$

Integrating this equation produces

$$\frac{mv^2}{2}=\int f(x)dx+c,$$

which is the same as equation (10.14).

The calculation can be completed when different types of forces are assigned to $f(x)$. In the case of a force in a line towards the origin that is a fixed point, simple harmonic motion ensues if the force is described by Hooke's law $f(x)=-kx$ where k is the force constant and x the *displacement* from the origin. The minus sign indicates that the force is towards the origin; when d^2x/dt^2 is negative x is positive and vice versa. The equation of motion is now

$$\frac{d^2x}{dt^2} = -\frac{k}{m}x.$$

The frequency of vibration or oscillation is defined as $\nu = \frac{1}{2\pi}\sqrt{\frac{k}{m}}$ s^{-1} and as $\omega = 2\pi\nu$ (radian s^{-1}), the vibration's *angular* frequency is $\omega = \sqrt{k/m}$ and for clarity this is now substituted making the equation of motion,

$$\frac{d^2x}{dt^2} = -\omega^2 x \qquad (10.17)$$

which when integrated gives

$$\left(\frac{dx}{dt}\right)^2 = -\omega^2x^2 + c^2,$$

see equation (10.14), and the constant is made c^2 as this has to be positive because the velocity dx/dt must be a real, not complex, number. Rearranging this equation to find the time gives

$$t = \int \frac{dx}{\sqrt{c^2 - \omega^2x^2}} + c_1$$

which is a standard integral[1] giving $t = \frac{1}{\omega}\sin^{-1}\left(\frac{\omega x}{c}\right) + c_1$. Thus

$$x = \frac{c}{\omega}\sin(\omega t - \omega c_1).$$

and the constants c and c_1 are determined by the initial conditions. As c and c_1 are constants this equation can just as correctly be written as

$$x = A\sin(\omega t + B) \qquad (10.18)$$

by defining A and B as constants also fixed by the initial conditions. B is called the *phase angle* of the sine wave and A is the *amplitude*. Some authors give a cosine solution but this differs only by the phase from the sine; $\sin(x) = \cos(x \pm \pi/2)$. Furthermore, by the properties of sine and cosine functions, we can also write

$$x = \alpha\sin(\omega t) + \beta\cos(\omega t). \qquad (10.19)$$

where α and β are new constants but still determined by the initial conditions.

If the initial velocity is v_0 and the position x_0 at $t = 0$, then from (10.19) $x_0 = \beta$ and

$$v = \frac{dx}{dt} = \alpha\omega\cos(\omega t) - \beta\omega\sin(\omega t)$$

which produces $v_0 = \alpha\omega$, thus

$$x = \frac{v_0}{\omega}\sin(\omega t) + x_0\cos(\omega t). \qquad (10.20)$$

and is dimensionally correct since v_0/ω has dimensions of distance. This equation completely describes simple harmonic motion.

(i) Problems with springs often state that the spring has a weight attached, which causes the spring to extend by a certain amount. It is then extended or compressed by a further amount and let go. The equation of motion is then sought.

As a specific example, suppose that a helical spring has a mass of 2 kg attached to it and this extends it by 5 cm. It is then displaced by 3 cm above its new equilibrium position and let go. At equilibrium, before the mass is moved, the tension in the spring due to Hooke's law is balanced by the force of gravity; $mg = ks$ where s is the displacement from equilibrium before the mass is moved. If the spring is now extended by x the tension T increases to $k(s + x)$. As the spring has no forces acting on it other than gravity, the force equation is

[1] The integration may be found as a $\tan^{-1}$ but this can be converted into $\sin^{-1}$ using a right-angled triangle; see Chapter 1.5 and 1.6.

$$m\frac{d^2x}{dt^2} = mg - T.$$

The tension is $T = k(x + s)$ and as $mg = ks$ then $m\frac{d^2x}{dt^2} = -kx$, which is the same as equation (10.17) and has the solution (10.20) when the frequency squared is substituted for k/m. To evaluate this equation, the oscillation frequency must be found, as must x_0 and v_0, which are set by the initial conditions.

As $v_0 = 0$, when the mass is let go (it is not pushed), then the solution is immediately given by

$$x = x_0 \cos(\omega t).$$

As the initial displacement is above the equilibrium position $x_0 = -3$, therefore $x = -3\cos(\omega t)$. Finally, the frequency is found from Hooke's law. At equilibrium, $mg = ks$ and $s = 5$ is the displacement from equilibrium before the mass is moved. Therefore, the force constant $k = 2 \times 9.81/5$ N m^{-1} and the frequency squared is $\omega^2 = k/m$ or $\omega = 1.4$ rad s^{-1}. This corresponds to a natural frequency of $\nu = \omega/2\pi$ or 0.223 s^{-1} and period $T = 1/\nu = 4.49$ seconds. The position of the mass at any time t is found to be $x = -3\cos(1.4t)$. Note that if the question stated that the *weight* was 2 kg then this would correspond to mg not m and the mass to use would therefore be 2/9.81 kg.

10.5.3 Total energy

The total energy is the sum of the kinetic and potential energy and is a constant as no external force acts on the system. The kinetic energy is that due to motion $mv^2/2$, and the potential energy can take several forms depending on how the motion is achieved. On extending a spring, this is force multiplied by distance displaced and the force is given by Hooke's law. If the potential energy is that due to displacing a mass in a gravitational field, as occurs with a pendulum, then the potential is mgh where h is the height change.

If the displacement at time t is $x = A\sin(\omega t + B)$, then the kinetic energy is

$$\frac{m}{2}\left(\frac{dx}{dt}\right)^2 = \frac{mA^2\omega^2}{2}\cos^2(\omega t + B).$$

The potential energy determined by Hooke's law force is linearly proportional to extension and is

$$V = k\int_0^x x\,dx = \frac{1}{2}kx^2 = \frac{k}{2}A^2\sin^2(\omega t + B).$$

As $\omega^2 = k/m$ the total energy, the sum of the potential and kinetic terms is

$$E = \frac{k}{2}A^2$$

which is a constant.

10.5.4 The simple pendulum

The pendulum will consist of a light rigid rod with a mass m at its end; the pivot holding the pendulum is frictionless and no air or other resistance hinders the motion. The variable φ is the angle in radians away from the vertical; Fig. 10.11. It is found that the pendulum's angular frequency is $\omega = \sqrt{g/L}$ rad s^{-1} if g is the acceleration due to gravity and L its length. The mass of the pendulum m is used to calculate forces but cancels out in the result.

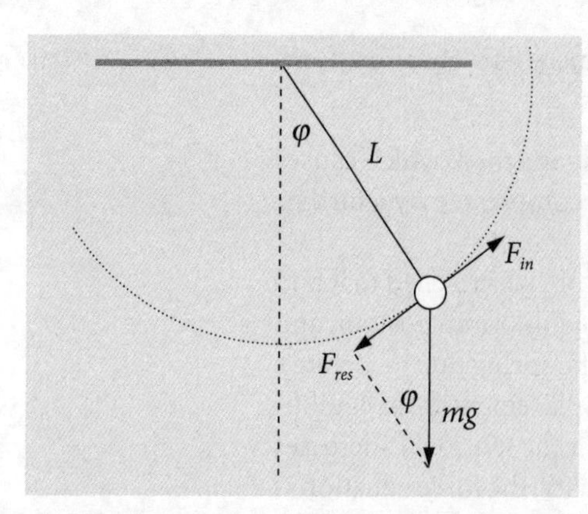

Fig. 10.11 The geometry used to calculate the forces acting on the pendulum.

The equation of motion can be derived in a number of different ways; here we make the inertial force of the moving pendulum equal to the applied force due to gravity, $F_{in} = F_{res}$.

The inertial force F_{in} is found quite easily in terms of the angle φ as follows. The pendulum travels a distance

$(2\pi L)(\varphi/2\pi) = L\varphi$ around the circumference to reach an angle φ; its acceleration is therefore $Ld^2\varphi/dt^2$ and the inertial force F_{in}, is m times this since force is mass × acceleration. The restoring force F_{res}, is the projection of the gravitational force, acting vertically down on the line of travel of the pendulum and is $-mg\sin(\varphi)$. The normal component of the force acts along the pendulum's arm. Equating forces give

$$mL\frac{d^2\varphi}{dt^2} = -mg\sin(\varphi)$$

which, by cancelling the mass and substituting for ω, produces the equation of motion

$$\frac{d^2\varphi}{dt^2} = -\omega^2\sin(\varphi).$$

It is worth noting in passing that mass, inertial and gravitational, cancels in this equation so that the period depends only on the length of the pendulum and the force of gravity. There is no reason why the inertial mass has to be the same as gravitational mass, but it is. Galileo, it is thought, first realized this by dropping different masses of the same size from the leaning tower of Pisa and observing that they struck the ground at the same time. Einstein did not ignore the equality of the inertial and gravitational mass, but developed this into a theory of gravity known as the general theory of relativity.

When the angular displacement is small, $\sin(\varphi) \to \varphi$, which means that $\varphi < 0.15$ rad or $\approx 8°$, the pendulum's motion becomes a simple harmonic oscillator of frequency ω. The equation of motion is $\dfrac{d^2\varphi}{dt^2} = -\omega^2\varphi$ which has the same general solution

$$\varphi(t) = \varphi_0\cos(\omega t) + \frac{v_0}{\omega}\sin(\omega t) \tag{10.21}$$

as calculated in equation (10.20) but with φ (in radians) instead of x as displacement. The initial angle is φ_0 from the vertical and initial velocity v_0 rad s^{-1}.

If the angular displacement is not small then the problem becomes rather more difficult, and after some effort (Arkfen 1970) an elliptical integral is produced with which to calculate its period but which has to be calculated numerically. The numerical integration of the pendulum's equation of motion for any displacement is somewhat easier and is examined in Chapter 11.6.3.

10.5.5 The compound pendulum and torsional oscillations

The compound pendulum is a rigid body freely suspended from a point that is not its centre of gravity and allowed to oscillate. The equation of motion is

$$I\frac{d^2\theta}{d\theta^2} = -mgh\sin(\theta)$$

where I is the moment of inertia of the body about the axis of suspension, m its mass, and h the distance from the centre of gravity to the point of suspension. This is the same equation as for a simple pendulum of length $L \equiv I/(mh)$.

If a disc is suspended by a wire attached to its centre and twisted in the horizontal plane, torsional oscillations occur. The elasticity of the wire produces resistance to twisting and provides the restoring force. An angular displacement of θ radians produces a force of $k\theta/L$ if the wire is of length L. The constant k is the torsional modulus of the wire. This can be related to the rigidity of the material comprising the wire and its geometry. The equation of motion is $I\dfrac{d^2\theta}{d\theta^2} = -\dfrac{k\theta}{L}$ where I is the moment of inertia of the mass about its point of suspension.

10.5.6 Inverted pendulum: unstable equilibrium

In the normal pendulum, acceleration is towards the vertical or origin of the motion. In the inverted pendulum with a rigid rod instead of a wire, acceleration is away from the vertical and the motion is unstable. The equation of motion for *small* displacements from the vertical, is similar to that of the pendulum but with a positive sign on the right-hand side of the equation,

$$\frac{d^2x}{dt^2} = n^2x \tag{10.22}$$

and n is a constant. The horizontal component of the thrust outwards is mgx/L, provided x is small; the constant is therefore $n^2 = g/L$. The solution for angular position is

$$\varphi = Ae^{nt} + Be^{-nt} \tag{10.23}$$

where A and B are constants determined by the initial conditions. The general method of solution is given in Section 10.4.3. If the initial position is x_0 and velocity v_0, then $\varphi_0 = A + B$ and $v_0 = n(A - B)$. Changing the exponentials into hyperbolic sinh and cosh and substituting for the constants gives

$$\varphi = \varphi_0 \cosh(nt) + \frac{v_0}{n}\sinh(nt),$$

which may be compared with (10.21).

10.5.7 Simple harmonic motion with a constant force

If a particle is attracted to the origin, and subject to a further constant force W, the equation of motion takes the form based on equation (10.13)

$$m\frac{d^2x}{dt^2} = f(x) + W$$

If a spring is displaced, then the force is $f(x) = -kx$ at extension x, and the equation of motion is

$$m\frac{d^2x}{dt^2} = -kx + W$$

If this is rewritten as $\dfrac{dx^2}{dt^2} = -\omega^2x + \dfrac{W}{m}$ and again as

$$\frac{d^2}{dt^2}\left(x - \frac{W}{m\omega^2}\right) + \omega^2\left(x - \frac{W}{m\omega^2}\right) = 0$$

then it has the form of (10.13) with a displacement about a new equilibrium position $W/(m\omega^2)$ rather than zero. This is understandable since a constant force is applied to the mass which must displace it by a *constant average* amount even though it is oscillating. The position at time t is therefore obtained directly from equation (10.20) as

$$x - \frac{W}{m\omega^2} = \frac{v_0}{\omega}\sin(\omega t) + x_0\cos(\omega t),$$

if x_0 and v_0 are the initial position and velocity. The frequency ω is the same whether the constant force is present or not.

10.5.8 Particle in a one-dimensional box

Two quantum mechanical problems that can easily be solved using the methods just described are a particle (electron, proton, C_{60}) in a one-dimensional box and a particle on a ring. Both involve integrating the Schrödinger equation, as must always be the case in quantum mechanics, but because the potential energy is zero the equations are considerably simpler than, say, that for the hydrogen atom or the harmonic oscillator. The one-dimensional Schrödinger equation in its general form is

$$-\frac{\hbar^2}{2m}\frac{d^2\psi}{dx^2} + V(x)\psi = E\psi$$

where $V(x)$ is the potential energy, m the mass of the particle and E the total energy. The length of the box is L. In the box, $V(x) = 0$ and additionally the boundary conditions are that the wavefunction ψ, which is a function of position x, is zero, i.e. has a node at the ends of the box; the sides of the box are vertical and assumed to be infinitely high so the particle cannot tunnel into the walls, see Fig. 3.54 (of the online solutions). The equation to solve is

$$\frac{d^2\psi}{dx^2}+\omega^2\psi=0 \qquad \text{with} \qquad \omega^2=2mE/\hbar^2,$$

and the solution can be written down directly from equation (10.19) and is

$$\psi=\alpha\sin(\omega x)+\beta\cos(\omega x).$$

where α and β are constants. The boundary conditions are that $\psi=0$ at both $x=0$ and at $x=L$, but these are not sufficient to enable both arbitrary constants in the problem to be calculated. To obtain both, the wavefunction has also to be normalized. When $x=0$ then we must make $\beta=0$ and therefore $\psi=\alpha\sin(\omega x)$ because the cosine would not have a node at the wall so is physically unacceptable. When $x=L$, then $\alpha\sin(\omega L)=0$ and as α cannot be zero as well as $\beta=0$ (the wavefunctions would then always be zero), then it follows that $\omega L=n\pi$ where n is an integer. The wavefunctions are therefore

$$\psi_n=\alpha\sin(n\pi x/L).$$

When $n=0$ then $\psi=0$, which is not a physically acceptable solution. Positive and negative integer n each produce acceptable solutions, but they are not independent of one another; therefore $n=1, 2, 3, \cdots$ are used to give unique solutions. As the sine function is periodic, increasing n means that waves with smaller periods can fit into the same box; the nth wavefunction has $n-1$ nodes between the walls, the lowest wavefunction having no nodes is just half a sine wave.

The constant α can be obtained by using the normalization condition $N^2\int_0^L \psi_n^*\psi_n dx=1$ and then $N=\sqrt{2/L}$. The normalized wavefunctions are

$$\psi_n=\sqrt{\frac{2}{L}}\sin(n\pi x/L) \qquad n=1, 2, 3, \cdots \tag{10.24}$$

The energy is found by putting ψ_n back into the Schrödinger and is

$$E_n=\frac{\hbar^2}{2m}\left(\frac{n\pi}{L}\right)^2 \qquad n=1, 2, 3, \cdots, \tag{10.25}$$

which has units of joules. This means that there are n energy levels, one for each wavefunction, and so E is labelled with this quantum number as E_n. Notice that the energy is never zero because there is a zero-point energy in accord with the Heisenberg uncertainty principle. The box is of finite length and the particle is thus in a restricted region of space. However, we cannot determine exactly where it is in this region and at the *same time* know its momentum; the condition is $\Delta x\Delta p \ge \hbar/2$ where Δx is the uncertainty in position and Δp that in momentum. The uncertainty in position is given by its standard deviation $\Delta x=\sqrt{\langle x^2\rangle-\langle x\rangle^2}$, and that in momentum is $\Delta p=\sqrt{\langle p^2\rangle-\langle p\rangle^2}$. The value $\langle x\rangle$ is the expectation of operator x, and $\langle x^2\rangle$ the expectation value for operator x^2. The operator for momentum is $-i\hbar d/dx$. A general operator Q is calculated as

$$\langle Q\rangle=\frac{\displaystyle\int \psi^* Q\psi dx}{\displaystyle\int \psi^*\psi dx},$$

see Chapter 4.8. Evaluating the integrals gives $\Delta x\Delta p=\left(\frac{n\pi}{\sqrt{3}}\right)\frac{\hbar}{2}$ which is greater than $\hbar/2$ for $n>0$.

10.5.9 The rigid rotor and a particle on a ring

If the distance between two atoms is fixed, or if a particle is constrained to move on a circle, then the Schrödinger equation takes a simple form and the equation is essentially the same as for a harmonic oscillator. To describe motion in more than one dimension, the Schrödinger equation has coordinates in x and y if two dimensional, or x, y, and z if the motion is three dimensional. The equation for two-dimensional motion is

$$-\frac{\hbar^2}{2m}\left(\frac{\partial^2\psi}{\partial x^2}+\frac{\partial^2\psi}{\partial y^2}\right)+V(x,y)\psi=E\psi$$

which is a partial differential equation and is clearly rather complicated. First, to simplify the calculation, the rigid rotor or particle on a ring have zero potential energy so $V = 0$. It only remains to simplify the derivatives and this is done with a change to plane polar coordinates using (see Chapter 1.6.1)

$$x=r\cos(\theta),\qquad y=r\sin(\theta),\qquad r^2=x^2+y^2$$

The differential operator $\nabla^2=\dfrac{\partial^2}{\partial x^2}+\dfrac{\partial^2}{\partial y^2}$ is changed into r, θ coordinates which gives $\nabla^2=\dfrac{\partial^2}{\partial r^2}+\dfrac{1}{r}\dfrac{\partial}{\partial r}+\dfrac{1}{r^2}\dfrac{\partial^2}{\partial\theta^2}$. (See question Q3.114.) Substituting produces

$$-\frac{\hbar^2}{2m}\left(\frac{\partial^2}{\partial r^2}+\frac{1}{r}\frac{\partial}{\partial r}+\frac{1}{r^2}\frac{\partial^2}{\partial\theta^2}\right)\psi+V(r)\psi=E\psi. \tag{10.26}$$

The term in θ describes the angular motion, and the other terms in r, the radial motion. In a rigid rotor or a particle on a ring, the radial part is constant which means that this part of the kinetic energy can be omitted, and as the potential energy V is only a function of r it is also a constant and can be ignored. Only the angular parts remain. There are two boundary conditions because the equation has second derivatives. Because the motion is circular, these conditions are (i) any wavefunction must have the same value after 2π (360°) rotation, or multiples of this, and (ii) that it has the same slope at this point. This means that the wavefunction repeats itself without a discontinuity see Fig. 10.26. The equation is now simplified to

$$-\frac{\hbar^2}{2m}\frac{1}{r^2}\frac{\partial^2\varphi}{\partial\theta^2}=E\varphi\qquad\text{or}\qquad\frac{\partial^2\varphi}{\partial\theta^2}=-k^2\varphi.$$

where φ is used to represent the angular part of the wavefunction and $k^2 = 2mr^2E/\hbar^2$. The distance r is a constant, by definition for the rigid rotor, and mr^2 is the moment of inertia I making $k^2 = 2IE/\hbar^2$ and the energy is

$$E=\frac{\hbar^2k^2}{2I}.$$

The solution to the differential equation can be written down either as the sum of a sine and cosine or as exponentials. The latter is conventionally chosen making the solution

$$\varphi=Ae^{ik\theta}+Be^{-ik\theta}.$$

The boundary conditions are that the wavefunction must reproduce itself exactly after 2π radians, or $\varphi(\theta) = \varphi(2\pi + \theta)$. This condition means that

$$Ae^{ik\theta}+Be^{-ik\theta}=Ae^{ik(\theta+2\pi)}+Be^{-ik(\theta+2\pi)},$$

which will be true for any A and B if $e^{ik\theta} = e^{-ik\theta} = 1$. This condition means that k must be an integer with values $k = 0, \pm1, \pm2, \cdots$. Conventionally the solution

$$\varphi=Ae^{ik\theta}$$

is chosen and using the normalizing condition $N^2\int_0^{2\pi}\varphi^*\varphi d\theta=1$, the wavefunction is

$$\varphi=\frac{1}{\sqrt{2\pi}}e^{ik\theta} \tag{10.27}$$

The quantum numbers are $k = 0, \pm1, \pm2, \cdots$ and as the lowest value is zero, this means that the minimum energy is zero and the rotor is stationary. As $k \ne 0$ is positive and negative it indicates that the rotor moves to the right or left and that these levels are each doubly degenerate. As the lowest energy is zero it initially suggests that the Heisenberg uncertainty principle $\Delta\theta\Delta p \ge \hbar/2$ is not obeyed. However, when $J = 0$ the wavefunction is a constant, $1/\sqrt{(2\pi)}$, so we cannot know what angle the rotor has and this means that

$\Delta\theta\Delta p \neq 0$. Ratner & Schatz (2001) give a clear description of this problem and others in quantum mechanics relevant to chemistry.

10.6 The '*D*' operator. Solving linear differential equations with constant coefficients

To be able to analyse more complex problems, linear second-order differential equations of the form $\dfrac{d^2y}{dx^2}+a\dfrac{dy}{dx}+by=f(x)$ are needed. These equations occur in a number of different situations from chemical kinetics, to the motion of molecules damped by the viscosity of a fluid, or the forced oscillations of the harmonic oscillator. A straightforward mathematical approach is available, which is the '*D*' operator method. Note that this only applies when the *coefficients are constant*; for example, the equation $\dfrac{d^2y}{dx^2}+x\dfrac{dy}{dx}+by=f(x)$ does not have constant coefficients nor does the Schrödinger equation.

The solution of linear differential equations with constant coefficients will always be a sum of exponential terms. If the exponentials are complex, and so contain i, then the solution can usually be expressed as sine and cosine functions. First, the operator rules are described,[2] which we are familiar with from 'normal' differentiation. The equations with $f(x)=0$ are solved next and then those when this is not the case.

10.6.1 Rules of differentiation

It turns out that if the differential operator dy/dx is replaced by D and d^2y/dx^2 by D^2 and so forth, the rules of differentiation can be worked out by allowing D to operate on the functions that follow it. For equations with *constant* coefficients, the D operators formally follow the laws that are valid for polynomials. Some of these results are worked out here, as they will be needed later on.

$$D \to dy/dx$$
$$D^2 \to d^2y/dx^2$$

(i) If u and v are functions of x, then

$$D(u+v)=Du+Dv.$$

(ii) The operator can also act on itself just as in repeated differentiation,

$$D(Dy)=D^2y.$$

It can also act as a normal variable; thus, it can be multiplied but division is not possible;

$$(D-1)(1+D^2)=D^3-D^2+D-1$$

Operating on a function gives

$$\begin{aligned}(D-1)(1+D^2)\sin(3x) &= (D^3-D^2+D-1)\sin(3x)\\ &= -24\cos(3x)+8\sin(3x),\end{aligned}$$

and the same result is obtained without expanding the Ds first, but operating with $1+D^2$ and then $D-1$ or vice versa.

(iii) When operating on exponentials the result is

$$De^{kx}=ke^{kx} \qquad \text{and} \qquad D^ne^{kx}=k^ne^{kx},$$

if n is a positive integer and k a constant. In conventional notation $\dfrac{d^n}{dx^n}e^{kx}=k^ne^{kx}$.

(iv) If $\varphi(D)$ represents a polynomial in the D operator, such as D^2+D-3, the effect it has on an exponential function is

[2] Parts of the following sections using D operators were adapted from Gow (1964).

$$\varphi(D)e^{kx} = \varphi(k)e^{kx} \quad (10.28)$$

where $\varphi(k)$ is a polynomial in k. For example, $(D^2 + D - 3)e^{kx} = (k^2 + k - 3)e^{kx}$.

(v) If u is a function of x then

$$D(e^{kx}u) = ke^{kx}u + e^{kx}Du$$
$$= e^{kx}(D + k)u$$

and

$$D^n(e^{kx}u) = e^{kx}(D + k)^n u.$$

(vi) Generally, if φ is a polynomial function in D,

$$\varphi(D)(e^{kx}u) = e^{kx}\varphi(D + k)u. \quad (10.29)$$

For example,

$$(D - 2)e^{2x}\cos(x) = e^{2x}(D)\cos(x)$$
$$= -\sin(x)e^{2x}.$$

Notice, that in $\varphi(D)$, each D is replaced by $\varphi(D + k)$, and therefore $D - 2$ becomes D as $k = 2$, and similarly,

$$(D^2 - 1)e^{kx}\sin(x) = e^{kx}((D + k)^2 - 1)\sin(x)$$
$$= e^{kx}(D^2 + 2kD + k^2 - 1)\sin(x)$$
$$= e^{kx}(k^2\sin(x) + 2k\cos(x) - 2\sin(x)).$$

10.6.2 There are many solutions to a linear differential equation

If the differential equation is expressed as $\varphi(D)y = 0$ where $\varphi(D)$ is a polynomial in the operator D, such as $(D - 1)(D + 2)$, then there are two solutions if the equation is second order, and three if third order, and so forth. These solutions are y_1 and y_2, but the sum of these two solutions is also a solution when y_1 and y_2 are multiplied by constants a_1 and a_2. Thus

$$y = a_1y_1 + a_2y_2$$

is the general form of the solution. The functions y_1 and y_2 are functions of x (they are often but not always exponentials) and are found systematically as shown next, and a_1 and a_2 are determined by the initial or boundary conditions. The last equation can be shown to be true by operating on it with φ(D); for example,

$$\varphi(D)(a_1y_1 + a_2y_2) = a_1\varphi(D)y_1 + a_2\varphi(D)y_2 = 0,$$

and y_n $(n = 1, 2)$ by definition solves this equation, hence $\varphi(D)y_n = 0$.

The two solutions y_1 and y_2 *must* be linearly independent of one another, which they are, if one is not a constant multiple of the other. This means that if c_1 and c_2 are constants and both are not zero, then $c_1y_1 + c_2y_2 = 0$ over the range of x values defined for the particular equation. A determinant of derivatives of the possible solutions, called a Wronskian (Arkfen 1970; Jeffery 1990), can be formed to check if the solutions are independent, see Section 10.4.3 case (iii).

10.6.3 Using auxiliary equations

The first-order equation

$$(D - a)y = 0$$

is the same as $\dfrac{dy}{dx} - ay = 0$ and has the solution $y = Ae^{ax}$ where the constant A is determined by the initial conditions. If the equation is written as $(D - a)e^{ax} = 0$, the solution is proved by differentiating $(D - a)e^{ax} = ae^{ax} - ae^{ax} = 0$.

In a second-order equation,

$$(D - a)(D - b)y = 0$$

we hypothesize that the equation is satisfied by $y = e^{kx}$ where k is a constant to be determined. The way to find k is to substitute the answer into the equation $(D - a)(D - b)e^{kx} = 0$ and expand, viz.,

$$(D^2 - (a+b)D + ab)e^{kx} = (k^2 - k(a+b) + ab)e^{kx} = 0.$$

The *auxiliary equation* is

$$k^2 - k(a+b) + ab = 0$$

which has roots of $k_1 = a$ and $k_2 = b$. The solution is therefore the *sum of two exponentials*

$$y = Ae^{ax} + Be^{bx}$$

where A and B are arbitrary constants determined by the initial conditions.

There are three different types of roots to the auxiliary equation.

Case (i) The roots are real: the result y is the sum of exponentials.

Case (ii) The roots are complex: the result y is the sum of complex exponentials, which can be converted into sine and cosines.

Case (iii) The roots are the same. In this case, one solution is missing and a test solution has to be tried.

Case (i) Suppose that the equation is $\dfrac{d^2y}{dx^2} - \dfrac{dy}{dx} - 12 = 0$ or $D^2 - D - 12 = 0$ and the initial conditions $y = 0$ and $dy/dx = 2$ at $x = 0$.
The general solution can be written down immediately as $y = Ae^{k_1x} + Be^{k_2x}$.
Next, $(D^2 - D - 12)e^{kx} = 0$ produces $(k^2 - k - 12)e^{kx} = 0$. The roots of the auxiliary equation

$$k^2 - k - 12 = 0$$

are $k_1 = -3$ and $k_2 = 4$. The solution is therefore

$$y = Ae^{-3x} + Be^{4x}$$

The initial condition, $y = 0$ at $x = 0$, gives $A = -B$ and from the gradient found by differentiating the solution, $2 = -3A + 4B$. Then $A = -2/7$ and $B = 2/7$ and the final result is

$$y = \frac{2}{7}(e^{4x} - e^{-3x}).$$

Case (ii) The equation $\dfrac{d^2y}{dx^2} + 16 = 0$ or $(D^2 + 16) = 0$ has the auxiliary equation $k^2 + 16 = 0$ with roots, $k = \pm 4i$. The solution is

$$y = Ae^{4ix} + Be^{-4ix}$$

The complex numbers can be represented as

$$e^{4ix} = \cos(4x) + i\sin(4x) \qquad \text{and} \qquad e^{-4ix} = \cos(4x) - i\sin(4x)$$

therefore y can be expressed as

$$y = (A+B)\cos(4x) + i(A-B)\sin(4x),$$

which could be written with new constants a and b as

$$y = a\cos(4x) + b\sin(4x).$$

This is the most compact form of the general solution except perhaps the equivalent form $y = R\cos(4x + \varphi)$ where R is the amplitude and φ the phase.

Case (iii) The equation $\dfrac{d^2y}{dx^2} - 4\dfrac{dy}{dx} + 4y = 0$ or $(D^2 - 4D + 4)y = 0$ has the auxiliary equation $k^2 - 4k + 4 = 0$ with both roots $k = 2$. This means that one solution is missing. We can, so far, write only $y = Ae^{2x} + y_2$ and have to find y_2.

To find the missing solution, try $y_2 = xe^{2x}$. To test this, it has to be differentiated and put into the starting equation,

$$\frac{dy_2}{dt} = e^{2x} + 2xe^{2x}; \qquad \frac{d^2y_2}{dt^2} = 4e^{2x} + 4xe^{2x}.$$

Putting these into the starting equation gives

$$4e^{2x} + 4xe^{2x} - 4(e^{2x} + 2xe^{2x}) + 4xe^{2x} = 0,$$

Hence y_2 is a solution and is clearly linearly independent of y_1 which is e^{2x}. The final solution is therefore

$$y = Ae^{2x} + Bxe^{2x}$$

where B is the second constant determined by the starting conditions.

The Wronskian is the determinant of the solutions to a differential equation and their derivatives. If the determinant is zero, the solutions are not independent. In this example, the determinant is two dimensional as there are only two solutions,

$$W = \begin{vmatrix} y_1 & y_2 \\ \dfrac{dy_1}{dx} & \dfrac{dy_2}{dx} \end{vmatrix} = \begin{vmatrix} e^{2x} & xe^{2x} \\ 2e^{2x} & e^{2x} + 2xe^{2x} \end{vmatrix} = e^{4x} \neq 0$$

and as the determinant is not zero, the solutions are confirmed to be independent.

In case (iii), we guessed a form of the solution. In general, when the ks are the same, the complete solution can have the form

$$y = (A_1 + A_2x + A_3x^2 \cdots)e^{k_1x} + Be^{k_2x} + Ce^{k_3x} + \cdots$$

10.6.4 Solving the equation $\varphi(D)y = f(x)$ when $f(x) \neq 0$. Inverse operators

The equation $\varphi(D)y = f(x)$ is described as being *non-homogeneous*. To solve it, the calculation is split into two parts. The first part takes $f(x) = 0$, as in the previous sections, and solves the *complementary function* and finds the *homogeneous* solution y_h which includes the arbitrary constants. This is the solution to $\varphi(D)y = 0$ and is a sum of exponential terms. Next, the *particular integral* is solved and this involves $f(x)$. The final solution is the sum of these two solutions $y = y_h + y_p$.

If $y = u$ is a general solution to the equation $\varphi(D)y = 0$, the solutions to $\varphi(D)y = f(x)$ can be seen to be separable if the substitution $y = u + v$ is made where u and v are, like y, each functions of x.

The equation becomes

$$\varphi(D)(u + v) = f(x).$$

By the rules in Section 10.4.1 this is $\varphi(D)u + \varphi(D)v = f(x)$ but as u is a solution, $\varphi(D)u = 0$ hence

$$\varphi(D)v = f(x). \tag{10.30}$$

The equation $\varphi(D)u = 0$ was solved in Section 10.4.2–3 and forms the complementary function and homogeneous solution. To solve (10.30) it is symbolically written as

$$v = \frac{1}{\varphi(D)} f(x)$$

and what this *inverse operator* means has to be worked out. The $\varphi(D)$ is usually a polynomial so if the equation is $\dfrac{d^2y}{dx^2} - \dfrac{dy}{dx} - 12y = f(x)$ or $(D + 3)(D - 4)y = f(x)$ the operator has the form

$$\frac{1}{(D + 3)(D - 4)} f(x)$$

To work out what $1/D$ is, let $\varphi(D) = D$ and then $D\left(\dfrac{1}{D}y\right) = y$. However,

$$D\left(\int ydy\right) = y$$

which indicates that the inverse operator, $1/D$ or D^{-1}, represents integration. Notice, that in this integration, no constants of integration are needed because they are always included in the complementary function.

10.6.5 Inverse operators applied to exponentials $f(x) = ce^{mx}$

If the function on the right of the equation is an exponential, $f(x) = ce^{mx}$, where c and m are constants, equation (10.28) shows that

$$\varphi(D)e^{mx} = \varphi(m)e^{mx}$$

and by repeated differentiation

$$\frac{1}{D^n}e^{mx} = \frac{1}{m^n}e^{mx}.$$

The most general form is

$$\frac{1}{\varphi(D)}e^{mx} = \frac{1}{\varphi(m)}e^{mx} \qquad \text{if} \qquad \varphi(m) \ne 0 \tag{10.31}$$

which means that the particular integral can be found if it is an exponential. To calculate $\varphi(m)$, let m replace each D in the function φ. For example, if the exponential is e^{3x}, then $m = 3$ and if the function is $\varphi(D) = (D - 4)$ then $\varphi(m) = 3 - 4 = -1$, thus $\dfrac{1}{(D-4)}e^{3x} = -e^{3x}$.

When $\varphi(m)$ is zero, this method will fail and that of section 10.6.7 has to be used.

(i) As an example, the equation $\dfrac{d^2y}{dx^2} + \dfrac{dy}{dx} - 6y = 10 + e^{4x}$, which is also $(D^2 + D - 6)y = 10 + e^{4x}$, can now be solved.

If the constant 10 is represented as $10e^{0x}$, then the right-hand side is the sum of two exponential terms. The complementary function is the solution to the homogeneous equation $(D - 2)(D + 3)y = 0$, the solution to which can be written down immediately as,

$$y_h = Ae^{2x} + Be^{-3x},$$

because the roots of $(k - 2)(k + 3) = 0$ are $k = +2$ and -3. The particular integral is

$$y_p = \frac{1}{D^2 + D - 6}(10e^{0x} + e^{4x})$$

and is found using (10.31). The first term with $m = 0$ gives

$$\frac{1}{D^2 + D - 6}10e^{0x} = -\frac{1}{6}10e^{0x} = -\frac{5}{3}.$$

The second term with $m = 4$ produces $\dfrac{1}{D^2 + D - 6}e^{4x} = \dfrac{1}{16 + 4 - 6}e^{4x} = \dfrac{1}{14}e^{4x}$ and the full, or general solution is

$$y = Ae^{2x} + Be^{-3x} + \frac{e^{4x}}{14} - \frac{5}{3}.$$

Maple does this rather more easily, but you can now appreciate how it is done,

```
> eqn:= diff(y(x),x,x)+diff(y(x),x)-6*y(x)=10+exp(4*x);
  dsolve(eqn);
```

$$y(x) = e^{-3x}\_C2 + e^{2x}\_C1 - \frac{5}{3} + \frac{1}{14}e^{4x}$$

10.6.6 Inverse operators applied to sine and cosine functions; $f(x) = \sin(ax)$ and $f(x) = \cos(ax)$

Equations of the form $\varphi(D)y = b\sin(mx)$ and $\varphi(D)y = b\cos(mx)$ often describe forces and simple harmonic motion where the driving force is a sine or cosine. To solve this type of equation, use e^{imx} to replace the sine or cosine term. The real part is then extracted if $f(x)$ is a cosine, and the imaginary part if a sine, because $e^{\pm iax} = \cos(ax) \pm i\sin(ax)$. This method is therefore effectively the same as the previous section.

If the equation is $2\frac{d^2y}{dx^2} - 5\frac{dy}{dx} - 3y = 7\sin(4x)$ the complementary function is found with $(2D+1)(D-3)y = 0$ which has roots $-1/2$ and 3 and is

$$y_h = Ae^{-x/2} + Be^{3x}.$$

Replacing the sine by an exponential makes a particular integral

$$y_p = 7\frac{1}{2D^2 - 5D - 3}e^{4ix}.$$

Using $m = 4i$ gives the solution $y_p = 7\frac{1}{32i^2 - 20i - 3}e^{4ix}$. This can be simplified by multiplying top and bottom by the conjugate of the complex number,

$$y_p = 7\left(\frac{1}{-35-20i}\right)\left(\frac{-35+20i}{-35+20i}\right)e^{4ix} = \left(-\frac{49}{325} + \frac{28}{350}i\right)e^{4ix}.$$

Now replacing the exponential gives

$$y_p = \left(-\frac{49}{325} + \frac{28}{325}i\right)(\cos(4x) + i\sin(4x))$$

$$= \left(-\frac{49}{325} + \frac{28}{325}i\right)\cos(4x) + \left(-\frac{49}{325}i - \frac{28}{325}\right)\sin(4x)$$

and the imaginary part chosen as $f(x)$ is a sine function giving

$$y_p = \frac{28}{325}\cos(4x) - \frac{49}{325}\sin(4x)$$

and the general solution is $y = Ae^{-x/2} + Be^{3x} + \frac{28}{325}\cos(4x) - \frac{49}{325}\sin(4x)$.

10.6.7 Inverse operators applied to $f(x) = e^{mx}G(x)$

When the function on the right of the differential equation is more complex, the particular integral can still be found. In this case, the relationship

$$\frac{1}{\varphi(D)}e^{mx}G(x) = e^{mx}\frac{1}{\varphi(D+m)}G(x) \tag{10.32}$$

is used. This is done by replacing each D by $D+m$ with the m coming from the power in the exponential. If, for example, $\varphi(D) = D+3$ then $\varphi(D+m) = D+3+m$. If the function $G(x) = 1$, then the $f(x)$ reverts to a pure exponential. The formula (10.32) allows a solution in cases where $\varphi(m) = 0$.

(i) The equation $(D^2 + D - 2)y = 2\sinh(2x)$ has the complementary function

$$y_h = Ae^{-2x} + Be^{x}.$$

The sinh is expanded into exponentials making the particular integral

$$y_p = \frac{1}{(D-1)(D+2)}(e^{2x} - e^{-2x}).$$

Using the method from the previous section, the first term is

$$\frac{1}{(D-1)(D+2)}e^{2x} = \frac{1}{4}e^{2x}$$

The second exponential term is $\frac{-1}{(D-1)(D+2)}e^{-2x} = \frac{-1}{0}e^{-2x}$, which cannot be solved in this way as $\varphi(m) = 0$. Instead, the solution is found in two steps; the first step in the usual way, substituting into $D-1$, and the second using equation (10.32). The first step is

$$\frac{-1}{(D-1)(D+2)}e^{-2x} = \frac{1}{(D+2)}\frac{1}{3}e^{-2x}.$$

Then letting $1 = G(x)$,

$$\frac{1}{3}e^{-2x}\frac{1}{(D+2)}(G(x)) = \frac{1}{3}e^{-2x}\frac{1}{(D+2)}(1)$$

Next using (10.32), $\varphi(D)$ in this equation is replaced by the actual value in the equation, which is $D+2$ or $\varphi(D) \rightarrow (D+2+m)$ so that as $m=-2$, $(D+2+m) \rightarrow D$ and

$$\frac{1}{3}e^{-2x}\frac{1}{(D+2)}(1) = \frac{1}{3}e^{-2x}\frac{1}{D}(1)$$
$$= \frac{1}{3}xe^{-2x}$$

and x is produced because 1 is integrated to x or $D(x) = 1$. The final solution is $y_h + y_p$ and is

$$y = Ae^{-2x} + Be^{x} + \frac{e^{2x}}{4} + \frac{x}{3}e^{-2x}.$$

The solution can be confirmed by differentiation and substitution into the original equation and should produce zero.

10.6.8 Inverse operators applied to polynomials

If $f(x)$ is a polynomial, then *each* reciprocal D term produced by the original equation can be expanded out as a series.

$$\frac{1}{D+a}f(x) = \frac{1}{a}\frac{1}{(1+D/a)}f(x)$$
$$= \frac{1}{a}\left(1 - \frac{D}{a} + \frac{D^2}{a^2} - \frac{D^3}{a^3} + \cdots\right)f(x)$$

and the polynomial $f(x)$ is differentiated by the series until no terms are left. Often a term in $1/D$ is present after differentiating. This term cannot be expanded as a series and instead it is used to integrate each x term remaining and this is illustrated in the next example.

(i) If the equation to solve is $(D^2 + D - 2)y = x^3e^{-2x}$ then the complementary function is the same as that in the last example and is $y_h = Ae^{x} + Be^{-2x}$.

In the particular integral, the exponential part is solved first using (10.32) with $m=-2$, producing

$$y_p = \frac{1}{(D-1)(D+2)}x^3e^{-2x} = e^{-2x}\frac{1}{D(D-3)}x^3.$$

Now the x^3 part is solved. Changing $D - 3$ into $-3(1 - D/3)$ and expanding the series produces

$$y_p = -\frac{1}{3}e^{-2x}\frac{1}{D}\left(1 + \frac{D}{3} + \frac{D^2}{9} + \frac{D^3}{27} + \cdots\right)x^3$$
$$= -\frac{1}{3}e^{-2x}\frac{1}{D}\left(x^3 + x^2 + \frac{6x}{9} + \frac{2}{9}\right) \qquad (10.33)$$
$$= -\frac{1}{3}e^{-2x}\left(\frac{x^4}{4} + \frac{x^3}{3} + \frac{x^2}{3} + \frac{2}{9}x\right)$$

The general solution is

$$y = Ae^{x} + Be^{-2x} - \frac{1}{3}e^{-2x}\left(\frac{x^4}{4} + \frac{x^3}{3} + \frac{x^2}{3} + \frac{2}{9}x\right).$$

(ii) To solve $(D^2 + 3D - 4)y = x$ the roots of $k^2 + 3k - 4 = 0$ produce $k = 1$ and -4 and therefore the complementary function is $y_h = Ae^{-4x} + Be^{x}$.

The particular integral is,

$$y_p = \frac{1}{(D-1)(D+4)}x.$$

Expanding this expression as *two* series produces

$$y_p = -\frac{1}{4}(1 + D + D^2 + \cdots)\left(1 - \frac{D}{4} + \frac{D^2}{16} - \cdots\right)x$$

$$= -\frac{1}{4}\left(1 + D - \frac{D}{4}\right)x = -\frac{1}{4}\left(x + \frac{3}{4}\right)$$

and therefore the general result is $y_h = Ae^{-4x} + Be^x - x/4 - 3/16$.

10.6.9 Alternative evaluation method for the particular integral

Although the inverse operator method can be used, this can be formed into a formula and used to determine the particular integral. The integral produced can sometimes be complicated. Suppose that the differential equation is $(D^2 + aD + b)y = f(x)$ and has roots k_1 and k_2 then

$$(D - k_1)(D - k_2)y = f(x).$$

The particular integral is then calculated as the nested integral

$$y_P = e^{k_1 x}\int e^{(k_2 - k_1)x}\left[\int e^{-k_2 x} f(x)dx\right]dx \tag{10.34}$$

If we consider again $(D^2 + D - 2)y = x^3e^{-2x}$, the roots are $k_1 = 1$ and $k_2 = -2$ then

$$y_P = e^x \int e^{-3x}\left(\int e^{2x}x^3e^{-3x}dx\right)dx.$$

This is not a difficult integration and each one can be done by parts as it has the form $x^n e^{-mx}$ with n and m as integers. Alternatively, the general form can be looked up in a table. A Maple function provides a convenient way of doing the calculation,

```
> P_int:=(m, n, f, x)->
       exp(n*x)*int(exp((m-n)*x)*int(exp(-m*x)*f, x),x);
```

$$P\_int := (m, n, f, x) \to e^{nx}\left(\int e^{(m-n)x}\left(\int e^{-mx} f dx\right)dx\right)$$

The function to integrate is f and is passed *symbolically*. The particular integral is therefore

```
> f:= x^3*exp(-2*x):
  k1:= 1:   k2:= -2:
  P_int(k1,k2,f,x);
```

$$-\frac{1}{108}e^{-2x}x(8 + 12x + 12x^2 + 9x^3)$$

which gives the same result as (10.33) if simplified a little. It does not ultimately matter which root, k_1 or k_2, is placed first when doing the calculation. However, the order does make a difference for this part, but when combined with the homogeneous solution and the limits are added, the result is the same.

Finally, note that if the differential equation has three roots then the particular integral has one more level of nesting, viz.;

$$y_P = e^{k_1 x}\int e^{(k_2 - k_1)x}\left(\int e^{(k_3 - k_2)x}\left[\int e^{-k_3 x} f(x)dx\right]dx\right)dx.$$

10.6.10 Damped simple harmonic motion

The equations we have been solving are of the type found when oscillators have, in addition to acceleration, a damping term due to viscous effects, and another term if an external force drives the oscillator. The effect of damping is to dissipate the energy from the moving body into the surrounding medium where it ends up as heat. Consequently, if an external force does not drive the body, the damping eventually brings the body to rest. In the presence of a constant force such as gravity the damping due to air resistance will cause a falling object to reach a constant terminal velocity. On the molecular scale, a charged protein moving in the presence of an external electric field will be accelerated and driven along by the field but this motion will be constrained by the viscous effects due to interaction with the solvent. The equations used to describe the motion of bodies in fluids can be 'mapped' onto those describing electronic circuits consisting of resistors, capacitors, and inductors.

If the oscillator is not driven but is in a viscous medium, a pendulum in air for example, the equation of motion is modified from that given in Section 10.3.2, and the motion is described as exhibiting *free damped oscillations.* The damping naturally resists motion, and it is often assumed to be linearly proportional to the speed. The force it produces is $-c dx/dt$ where c is a constant of proportionality and is the damping coefficient and has units kg s^{-1}. The negative sign ensures that damping resists motion.

The equation of motion (10.13) becomes

$$m\frac{d^2x}{dt^2}=f(x)-c\frac{dx}{dt}$$

where $f(x)$ is the expression describing the external force on the oscillator. If this is caused by Hooke's law, then it is $f(x)=-k_f x$ if x is the displacement from equilibrium and k_f the force constant. The equation to solve is

$$\frac{d^2x}{dt^2}+2h\frac{dx}{dt}+\omega^2 x=0 \qquad (10.35)$$

where for clarity the abbreviations $2h=c/m$ and $\omega^2=k_f/m$ are made. In D operator notation, this equation is $D^2+2hD+\omega^2 x=0$. Using the solutions developed in Section 10.4.4–9 the characteristic function can be written down directly after finding the roots of the equation $k^2+2hk+\omega^2=0$. These are $k=-h\pm\sqrt{h^2-\omega^2}$, therefore, the homogeneous solution for the displacement at time t is

$$x=Ae^{(-h+\sqrt{h^2-\omega^2})t}+Be^{(-h-\sqrt{h^2-\omega^2})t} \qquad (10.36)$$

where A and B are constants determined by the initial conditions. This equation is more conveniently written as $x=Ae^{m_1 t}+Be^{m_2 t}$ using the abbreviations $m_1=-h+\sqrt{h^2-\omega^2}$ and $m_2=-h-\sqrt{h^2-\omega^2}$.

Three cases can now be easily identified depending on the terms in the square root.

(i) When $h^2-\omega^2>0$ the values of m_1 and m_2 in the exponentials are both negative because $\sqrt{h^2-\omega^2}<h$ and this is called the *over-damped* case. In this situation, the oscillator drifts to a standstill in a gradual and uniform manner without oscillating. This situation would correspond to that describing the motion of a light pendulum moving in thick oil. If the initial displacement is d and the motion starts with an initial velocity of zero, the initial conditions are $x(0)=d$ and $dx/dt=0$ when $t=0$. With these conditions $b=A+B$ and $\dfrac{dx}{dt}=0=m_1A+m_2B$ and therefore $A=-m_2d/(m_1-m_2)$ and $B=m_1d/(m_1-m_2)$. The equation of motion is

$$x=\frac{m_2 d}{m_1-m_2}(e^{m_2 t}-e^{m_1 t})$$

and will decay bi-exponentially to zero, as both m are negative numbers.

(ii) When $h^2-\omega^2<0$, the motion is *under-damped* and the m are complex numbers, which means that exponentials can be represented as sines and cosines. If $h^2-\omega^2=-\alpha^2$ the displacement is

$$x=Ae^{(-h+i\alpha)t}+Be^{(-h-i\alpha)t}.$$

With the initial conditions given in (i) and after some rearranging, this equation can be rewritten as

$$x=de^{-ht}\left(\cos(\alpha t)+\frac{b}{\alpha}\sin(\alpha t)\right).$$

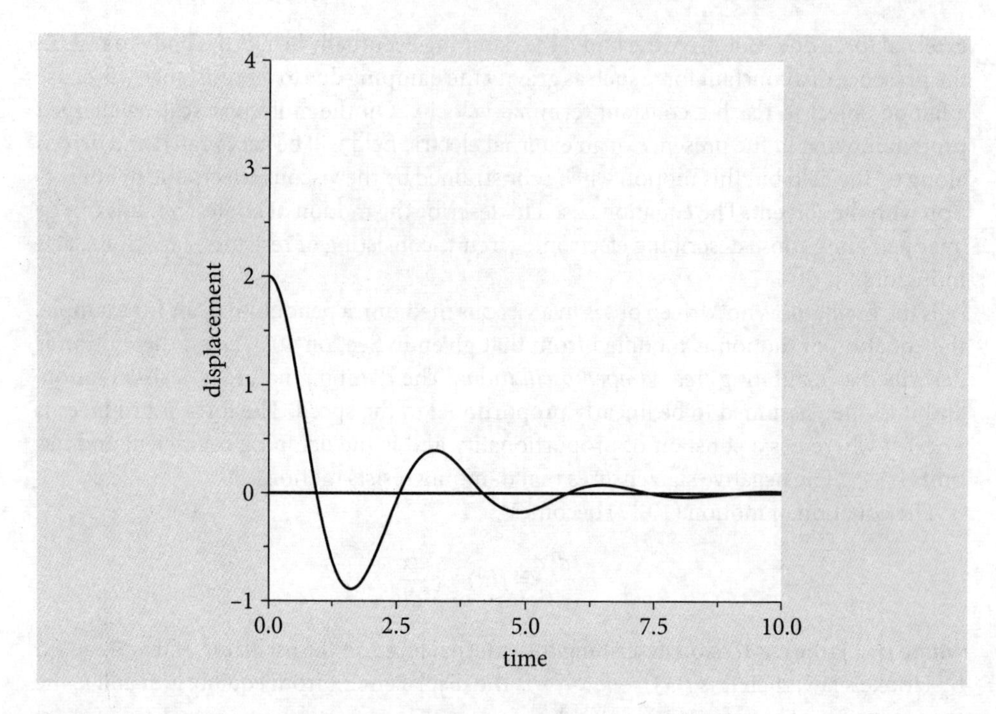

Fig. 10.12 An under-damped oscillator, $\omega = 2$ and $h = 0.5$.

The motion is still oscillatory but the amplitude of the motion decreases exponentially with a rate determined solely by the damping coefficient h, Fig. 10.12. This means that if the exponential $\pm de^{-bt}$ were drawn, the curve would touch either all the crests or all the troughs of the decay. The oscillation has a frequency $\alpha = \sqrt{\omega^2 - h^2}$ radians s^{-1}, which is lower than that of the undamped case, ω, and the period is longer $T = 2\pi/\alpha = 2\pi/\sqrt{\omega^2 - h^2}$. A mechanical analogy is the motion of weight suspended by a (real) spring or elastic band. It is possible to characterize the number of oscillations by defining a quality factor, $Q = \dfrac{\omega}{2h}$. The amplitude of the motion has decayed to 1/e when $ht = 1$; therefore, substituting for h and the period T produces $Q = \pi\dfrac{t}{T}$. The ratio t/T is the number of periods for the oscillator to lose 1/e of its amplitude. If the oscillator is lightly damped, then the Q is large and vice versa. The idea of Q factors is, however, more important in the driven damped oscillator; the closer the driven frequency is to the free or natural oscillator frequency, the greater the Q factor is. The (under) damped oscillator in Fig. 10.12 has a Q of $\approx \pi$, because $t/T \approx 1$, and is very low; it is clear that very few oscillations occur before the motion effectively ceases.

(iii) When $\omega^2 = h^2$ and $\alpha = 0$, the motion is *critically damped*. The characteristic equation is now $k^2 + 2hk + h^2 = 0$ with double roots $-h, -h$. The solution is sought from the method outlined in Section 10.4.3(iii). One solution is $x_1 = e^{-ht}$ and a test solution tried as $x_2 = te^{-ht}$, which is a suitable function. This is tested by differentiating the result,

$$x = Ae^{-ht}(1 + Bt).$$

With the initial conditions as before, $x = de^{-ht}(1 + ht)$, which decreases uniformly to zero. If the initial velocity is not zero then x may rise before falling or fall and undershoot, before reaching zero; Fig. 10.13.

10.6.11 Driven simple harmonic motion

When a mass is driven by an external power source, it can be made to move continuously and therefore has behaviour not seen in a free oscillator. If the oscillator is also damped some of the energy is dissipated as heat into the surroundings. However, if damping is essentially absent and if the driving force is at the same period and phase as the natural period of the oscillator, resonance occurs. In this situation, the energy supplied cannot

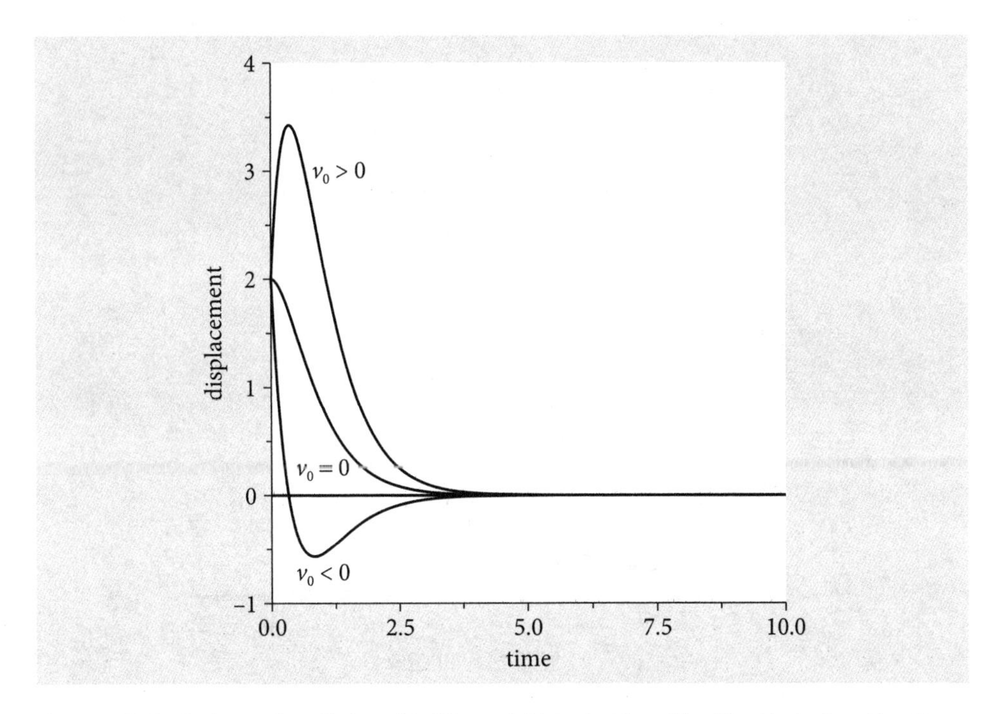

Fig. 10.13 Critically damped oscillation with different initial velocities of 0, ±10 with $d = 2$ and $h = 2$.

be dissipated and the oscillator can be destroyed by being forced to undergo ever-larger displacements from its equilibrium position.

If the external force is constant then the equation of motion is

$$\frac{d^2x}{dt^2} + 2h\frac{dx}{dt} + \omega^2 x = W$$

and, as in Section 10.4.10, this can be rewritten using equation (10.35) as

$$\frac{d^2}{dt^2}\left(x - \frac{W}{\omega^2}\right) + 2h\frac{d}{dt}\left(x - \frac{W}{\omega^2}\right) + \omega^2\left(x - \frac{W}{\omega^2}\right) = 0,$$

for which solutions are already given in Section 10.4.10 provided x is replaced by $x' = x - W/\omega^2$. In D notation the equation is $D^2 + 2hD + \omega^2 x' = 0$.

When the driving force is not constant then its form is clearly going to determine the nature of the solution. The general equation is

$$\frac{d^2x}{dt^2} + 2h\frac{dx}{dt} + \omega^2 x = f(t)$$

where $f(t)$ is the driving force. As before, h is the damping coefficient and ω is the constant frequency of the free oscillator. This equation, when written in D operator form is $D^2 + 2hD + \omega^2 x = f(t)$ and some solutions for different $f(t)$ have been developed in Section 10.4. If the oscillator is driven by a force $f(t) = a\sin(\omega_f t)$ then

$$\frac{d^2x}{dt^2} + 2h\frac{dx}{dt} + \omega^2 x = a\sin(\omega_f t).$$

The homogeneous equation is $y_h = Ae^{m_1 t} + Be^{m_2 t}$ as found in equation (10.36) and $m_{1,2} = -h \pm \sqrt{h^2 - \omega^2}$. The particular integral is obtained by converting the sine to an exponential, integrating and then extracting the imaginary part,

$$y_p = a\frac{1}{(D + m_1)(D + m_2)}e^{i\omega_f t}.$$

Substituting for $i\omega_f$ and simplifying gives

$$y_p = a\frac{m_1 m_2 + \omega_f^2 + i\omega_f(m_1 + m_2)}{(m_1^2 + \omega_f^2)(m_2^2 + \omega_f^2)}e^{i\omega_f t}$$

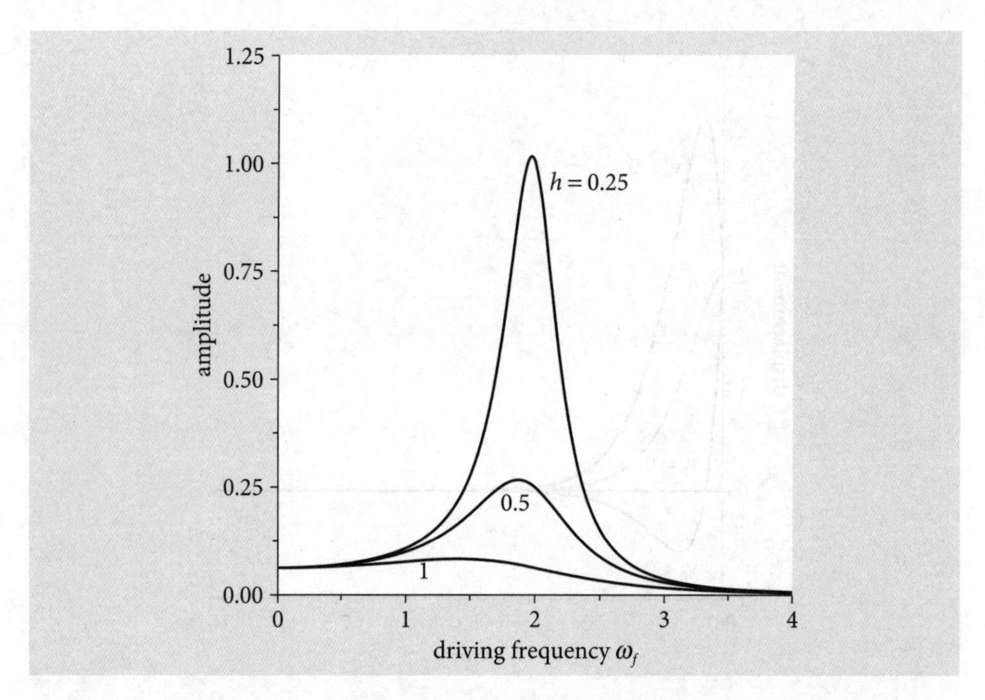

Fig. 10.14 The amplitude of a forced or driven oscillator vs its driving frequency ω_f with different damping coefficients h. The resonant frequency $\omega = 2$.

Replacing the exponential by $e^{i\omega_f t} = \cos(\omega_f t) + i\sin(\omega_f t)$, substituting for m_1 and m_2, and collecting just the imaginary part because the driving force is a sine function, gives

$$y_p = \frac{a}{(\omega_f^2 - \omega^2)^2 + 4h^2\omega_f^2}[(\omega^2 - \omega_f^2)\sin(\omega_f t) - 2\omega_f h\cos(\omega_f t)]$$

which is the long time solution after the exponential terms in y_h, which are of little interest in many physical situations have been damped out. The amplitude is the term outside the bracket containing the sine and cosine and this is drawn vs the driving frequency ω_f in Fig. 10.14. As the resonance is approached the oscillation amplitude increases significantly, and the frequency where it peaks approaches that of the free oscillator. The frequency at maximum amplitude is found in the usual way by differentiating and setting the result to zero. The maximum value is the frequency

$$\omega_f = \sqrt{\omega^2 - 2h^2},$$

which is always at a lower value than the free oscillator.

The Q factor (p 560) can be redefined as the ratio of the resonant frequency to the full width at half maximum of the curve of power absorbed vs driving frequency or $Q = \omega/\Delta\omega$. This can be shown to be $Q = \omega m/h$. The characteristic behaviour of an oscillator with a high Q factor (low damping) is that it responds by peaking with large displacements near to resonance, and is relatively unperturbed by other driving frequencies. The displacements are large, because the denominator in the amplitude becomes small, both as a result of h being small and $\omega_f \rightarrow \omega$. If the damping should be zero ($h = 0$), then a 'resonance catastrophe' can occur as the amplitude becomes infinity when $\omega_f = \omega$. When the Q is low, which means that the damping factor is large, the oscillator amplitude remains small over a wide range of driving frequencies. A detailed account of damped and forced oscillators is given by Pain (1993).

Although the damped, forced oscillator is described in terms of a mechanical oscillator, such as a pendulum or a spring, it also forms a classical model of how an atom or molecule interacts non-resonantly with a light wave. The electric field of the radiation produces a sinusoidal force on each electron. The displacement of the electron δx produces a dipole moment $e\delta x$ if e is the electronic charge. The polarization or dipole moment/unit volume is obtained from this model by summing over all the electrons in a given volume. The refractive index and conductivity can be derived from the polarization. However, the classical model of absorbing a photon leads to very unusual behaviour because energy would be absorbed over a wide range of frequencies and in increasing amounts as resonance is approached, which is clearly not what is observed experimentally. Only a quantum model properly describes absorption.

Q-switching is used to produce intense nanosecond duration laser pulses, which are used to measure rates of chemical reactions; they are also used in cutting and welding materials and in laser ranging or 'lidar'. Siegman (1986) describes Q-switching in detail but the basic idea is as follows. The gain material is excited by flash lamps but the laser cavity is initially held in a condition of low gain, so it cannot lase even though the gain medium, such as Nd^{3+} ions in YAG, is being excited. The gain is forced to be low by an electro-optic polarizer inserted inside the laser cavity. This is aligned so that those photons that reach the cavity mirrors are not fed back into the gain medium with the correct polarization and are therefore not amplified. As the gain material is excited, a large population inversion soon results and consequently it contains a large amount of stored energy. A laser pulse is now made by suddenly increasing the Q of the laser cavity by increasing the gain. This can be done by switching the state of the electro-optic polarizer so that light can now travel back and forth between the mirrors. The effect that this has is to cause a very rapid rise in the gain of the cavity because, by stimulated emission, the stored energy is converted into photons, and a laser pulse of a few nanoseconds duration is produced. The photons deplete the gain far more rapidly than it can be replenished by excitation from the flash lamps, and the gain drops, terminating the laser pulse. The Q of an optical cavity is given by $Q = \omega E/(-dE/dt)$, where ω is the laser frequency, E the energy, and dE/dt the power or rate of change of energy. The Nd YAG laser operates at 1064 nm and can typically produce a 1 J, 10 ns duration Q-switched laser pulse. If the cavity gain is switched in 10 ns, then the Q of the cavity is, for a short period, extremely high at $\approx$3 million.

The Q of mechanical systems is also important. The scanning atomic force microscope, AFM, has a microscopic cantilever with an even smaller tip at its base. The first type of instruments worked in contact mode where the tip was dragged over a surface using piezoelectric actuators to control its position. The distance from the surface was measured by how much the cantilever was bent by the surface forces. In this mode, the surface can be damaged by the tip. To prevent this and to increase the quality of the images, the tip can be made to oscillate up and down when driven by a piezoelectric crystal and it only 'taps' the surface once per period. The electrons in an atom on the surface interact with the tip as it approaches and it does not have to touch a molecule for its presence to be felt. The oscillation's amplitude ($\approx$20 nm) is maintained by a feedback loop and, as it approaches a molecule on the surface, the oscillation frequency changes and so the feedback changes also thus providing a measure of the surface shape. If the Q of the cantilever is high, then the amplitude of the cantilever's response is large when the driving frequency is close to that of its natural frequency and is small at other frequencies. When a molecule is encountered, the force on the cantilever changes its frequency by a small amount, which because the Q is high, causes a large change in its amplitude; see Fig. 10.14 thus increasing sensitivity. The Q of the cantilever can be as high as 20 000 when operated at 250 kHz and images with resolution of a few angstrom are possible.

10.6.12 **Brownian motion**

In 1828, the naturalist Robert Brown (Brown, 1828) reported that the motion of pollen seen under the microscope never ceases. The same was soon shown to be true of microscopic particles of glass, minerals, petrified wood, and even stone from the Egyptian Sphinx. Various causes were investigated—convection currents, uneven evaporation, capillary action, and so forth—but none proved to be responsible. We now understand that Brownian motion is due to the ceaseless motion of all molecules colliding with one another and with larger particles and in doing so moving them. Because the number of collisions with a particle is not the same in all directions at each instant in time, the forces will not always cancel and the particle will be knocked about in a random manner. In 1910, Perrin observed and recorded the random motion of a single particle under the microscope and measured the average of the square of the distance moved $\langle x^2 \rangle$ in periods of 30 seconds, the scale being of a few tens of micrometres. Einstein had previously calculated what the average displacement should be in a given time period.

The total energy of a macroscopic particle in a liquid or gas is the sum of its kinetic and potential energy. Its average energy can be found by statistical mechanics and is $\langle mv_x^2/2\rangle = k_BT/2$. The equipartition theorem states that each *squared term* in the equation for the displacement of a particle contributes $k_BT/2$ to the average energy. In three dimensions the average is therefore $3k_BT/2$. A particle moving in a harmonic potential has a contribution of $k_BT/2$ from its kinetic energy and $k_BT/2$ from its potential energy, since this energy is $kx^2/2$ for a displacement x and force constant k. The total average energy is therefore k_BT.

To find an expression for $\langle x^2\rangle$, an equation of motion has to be solved, and for the driven and damped motion of a particle in a fluid, this equation has the form

$$m\frac{d^2x}{dt^2}+\gamma\frac{dx}{dt}=f(t)$$

where γ is a damping coefficient and f is a fluctuating random force. This equation cannot be usefully solved because f is a random quantity, but, in any case, the experimental measurement is the average of the square of the displacement $\langle x^2\rangle$, not x the displacement from equilibrium, so this average quantity is sought. There are two steps to doing this. First, a differential equation for x^2 is sought and, secondly, its average over time is found and the equipartition theorem is used to do this. Although the averaging can be done before integrating, as this greatly simplifies the calculation, it is not essential because integration is a sum and the order of integrating and averaging can be interchanged. In physical terms, averaging before integrating is valid because we can repeatedly observe the displacement of one particle at each of several time steps and average the results, or we can observe many particles simultaneously, averaging after each time step. The ergodic hypothesis states that these approaches are equivalent.

The first step is to find an expression for x^2; the 'trick' is using the relationship

$$\frac{1}{2}\frac{d^2}{dt^2}(x^2)=\left(\frac{dx}{dt}\right)^2+x\frac{d^2x}{dt^2};$$

(treat x as a function of t), and as $dx^2/dt = 2x dx/dt$ this gives, after substituting for d^2x/dt^2 and with some rearranging,

$$m\frac{d^2}{dt^2}(x^2)+\gamma\frac{d}{dt}(x^2)=2m\left(\frac{dx}{dt}\right)^2+2xf(t).$$

Next, the average of x^2 has to be obtained. Because f is a random force, x and f are uncorrelated in time, therefore $\langle xf\rangle = 0$. The average kinetic energy can also be replaced with $m\left\langle\left(\frac{dx}{dt}\right)^2\right\rangle = k_BT$ and the equation of motion simplifies to

$$m\frac{d^2}{dt^2}\langle x^2\rangle+\gamma\frac{d}{dt}\langle x^2\rangle=2k_BT.$$

This equation is solved with the initial conditions that at time zero $\langle x^2\rangle = 0$, and its first derivative is also zero at $t = 0$. The methods of previous sections could be used but using Maple with the abbreviation $s \equiv \langle x^2\rangle$ and $g \equiv \gamma$ produces

```
> eqn:= m*diff(s(t),t,t )+ g*diff(s(t),t)-2*k*T = 0;
  init:= s(0)=0, D(s)(0)=0;
  simplify( dsolve( [eqn, init] ) );
```

$$s(t)=\frac{2kBT(me^{-\frac{gt}{m}}+gt-m)}{g^2}$$

and at long times, $t \gg \gamma/m$, when the transient (exponential) behaviour is finished, the result is

$$\langle x^2\rangle=\frac{2k_BT}{\gamma}t,$$

which shows that the average displacement increases linearly with time. Einstein further derived the famous formula

$$\langle x^2\rangle = 2Dt,$$

where D is the diffusion coefficient. This can also be obtained from an analysis of a random walk and is found by solving the diffusion equation. Therefore combining the results gives

$$D = k_BT/\gamma$$

and by Stokes's law

$$\gamma = 6\pi\eta r$$

where η is the solvent viscosity and r the radius of the particle. Together they produce the Stokes–Einstein equation

$$D = \frac{k_BT}{6\pi\eta r}$$

This equation has proved very successful in predicting the diffusional properties of molecules and properties depending on diffusion, such as the rate constants of diffusion controlled chemical reactions, even though the equations strictly relate only to macroscopic particles. As a rule of thumb, we can feel confident in using the equation if the solvent is smaller than the molecule diffusing in it, because the derivation assumes a continuous structureless solvent, and not a molecular one. When you use these equations, note that the SI unit of viscosity is Pa s (kg m^{-1} s^{-1}) but the common unit is centipoise and 1 Pa s = 10^{-3} cP. The diffusion coefficient D has units of m^2 s^{-1}.

10.7 **Simultaneous equations**

(i) Sequential equations

Complex chemical reactions can often be represented as a set of simultaneous reactions. The sequential scheme $A \xrightarrow{k_1} B \xrightarrow{k_2} C$ has already been solved with the integrating factor method in Section 10.2, and as an eigenvalue–eigenvector equation in Chapter 7.12.3. Here it is converted into a second-order equation and solved using the D operator method. The rate equations are,

$$\frac{dA}{dt} = -k_1A$$

$$\frac{dB}{dt} = k_1A - k_2B$$

Differentiating the second equation and substituting dA/dt produces $\dfrac{d^2B}{dt^2} = -k_1^2A - k_2\dfrac{dB}{dt}$.

Substituting for A and rearranging gives

$$\frac{d^2B}{dt^2} + (k_1 + k_2)\frac{dB}{dt} + k_1k_2B = 0 \qquad \text{or} \qquad (D^2 + (k_1 + k_2)D + k_1k_2)B = 0$$

The solution can be written down after solving the characteristic equation $m^2 + (k_1 + k_2)m + k_1k_2 = 0$, which has roots $m = -k_1, -k_2$. The homogeneous equation is

$$B = c_1e^{-k_1t} + c_2e^{-k_2t}$$

and the constants c_1 and c_2 are, as usual, determined by the initial conditions. If for instance, A_0 is the initial concentration of A, and B and C are initially zero, then since C does not decay, $C = A_0 - A - B$. Because B is initially zero, then

$$B = c_1(e^{-k_1t} - e^{-k_2t})$$

and the final step is to find c_1 when $A = A_0e^{-kt}$. This is done by differentiating B and equating this with the rate equation $dB/dt = k_1A - k_2B$ after substituting for A and B. The resulting equation produces $c_1 = k_1A_0/(k_2 - k_1)$ making

$$B = \frac{k_1A_0}{(k_2 - k_1)}(e^{-k_1t} - e^{-k_2t})$$

which is the same result as that from the integrating factor method.

(ii) Equal rate constants

Consider now the case when the rate constants are all equal at k in the scheme $A \to B \to C$. This does not mean that the concentration of B is zero at all times, even though it is formed at the same rate as it decays. Taking the last result, it is not possible to make $k_1 = k_2$ because the concentration of B becomes undefined as 0/0, however, l'Hôspital's rule (Chapter 3.8) could, in this case, be used to find the limit $k_2 - k_1 \to 0$. Instead, we start with the rate equations,

$$\frac{d^2B}{dt^2} + 2k\frac{dB}{dt} + k^2B = 0 \qquad \text{or} \qquad (D^2 + 2kD + k^2)B = 0. \tag{10.37}$$

The roots of the auxiliary equation are $m = -k, -k$, and, as they are the same, the method of Section 10.4.3(iii) has to be used. The first part of the solution is $B = c_1e^{-kt} + B_1$ where B_1 is a new function of t, which has to be found. This solution is normally guessed, and then tried to see if it fits the equation. As a test, let $B_1 = c_2te^{-kt}$ making the solution

$$B = c_1e^{-kt} + c_2te^{-kt}$$

If B is differentiated and put back into (10.37), the result is zero showing that this is indeed a solution. To evaluate the constants, this result should be differentiated and A and B substituted into the rate equation for B, as was done in the previous calculation. The initial conditions are normally $A = A_0$ and $B = 0$ at $t = 0$. As $B = 0$, then $c_1 = 0$ and, after substituting into the rate equation $c_2 = kA_0$, this makes

$$B = A_0kte^{-kt}.$$

Plotting the A and B concentrations shows that as A falls exponentially, B rises and falls as intuition dictates. If there are a series of reactions, $A \to B \to C \to D \to$ etc. and only A is present initially and k is the rate constant for each step, then the concentration of the nth species C_n is found to be $C_n = A_0k\dfrac{t^n}{n!}e^{-kt}$.

Although rate constants for the reaction of different molecules can be the same, they are generally different. One situation where this is not true is the unfolding of concatenated proteins by mechanical force. A concatamer is a macroscopic molecule made by attaching several identical proteins in tandem. When one end of it is anchored to a substrate and the other to the tip of an atomic force microscope (AFM), the concatamer can be stretched by the AFM and each protein will unfold but in a random order with respect to its position in the concatamer (Reif et al. 1977; Brockwell et al. 2003). See Fig. 3.21 for a sketch of the experiment. The unfolding is registered by measuring the force and, to a good approximation, the unfolding rate constants for each protein are all the same. However, in this experiment, unlike normal chemical kinetics, the probability or chance S that a protein is still folded at any given force is measured. If there are four proteins the reaction sequence is $S_4 \to S_3 \to S_2 \to S_1$ where S_4 is the chance that all proteins are folded, S_3 that three remain folded etc. at a given force.

The rate constants k_0 of thermally induced chemical reactions are constant at fixed temperature and are the same for each protein because they are identical. In the mechanical pulling experiment, the rate constant depends on the force f, because the barrier to unfolding is lowered by the applied force (Bell 1978; Evans & Richie 1997). The rate constant also depends on how fast the force is applied, and the protein becomes stiffer, i.e. more force is needed to unfold it the faster the force is applied. This occurs because the barrier can be lowered more quickly than the average time between thermally induced protein fluctuations that lead to barrier crossing.

The unfolding rate constant is given by $(k_0/L)\exp(f/f_0)$ where $f_0 = kT/x_u$ and L is the load rate in pN s^{-1}. The constant k_0 is the (thermal) unfolding rate constant at zero force and x_u is the distance from the minimum of the potential well to the top of the barrier separating the folded and unfolded protein. It is a measure of the distance the protein conformation has to change to reach the unfolding transition state and is typically 0.25 nm. The unfolding forces are 100 to 300 piconewton and $f_0 \approx 15$ pN. Since all proteins are identical, and we assume that each protein experiences the same force irrespective of how many are folded or unfolded, the unfolding scheme for N proteins can be written as

$$\frac{dS_N}{df} = -(k_0/L)\exp(f/f_0)S_N$$

$$\frac{dS_{N-1}}{df} = -(k_0/L)\exp(f/f_0)S_{N-1} + (k_0/L)\exp(f/f_0)S_N$$

$$\vdots$$

$$\vdots$$

$$\frac{dS_1}{df} = -(k_0/L)\exp(f/f_0)S_1 + (k_0/L)\exp(f/f_0)S_2$$

$$\frac{dS_{unfold}}{df} = +(k_0/L)\exp(f/f_0)S_1$$

where the expressions are derivatives with force not time. This set of equations also applies if a set of parallel bonds is unzipped (or unpeeled); they could be the hydrogen bonds holding a protein β-sheet or double stranded DNA together when an opposite force is applied to the top of each strand Fig. 10.15. The set of rather formidable looking simultaneous equations is not linear in f; however, the substitution $u = (f_0k_0/L)(\exp(f/f_0) - 1)$ greatly simplifies them. The derivative is $du/df = (k_0/L)e^{f/f_0}$ and substituting using $\dfrac{dS}{df}\dfrac{df}{du}$ produces,

$$\frac{dS_N}{du} = -S_N$$

$$\frac{dS_{N-1}}{du} = -S_{N-1} + S_N$$

$$\vdots$$

$$\vdots$$

$$\frac{dS_1}{du} = -S_1 + S_2$$

$$\frac{dS_{unfold}}{du} = +S_1$$

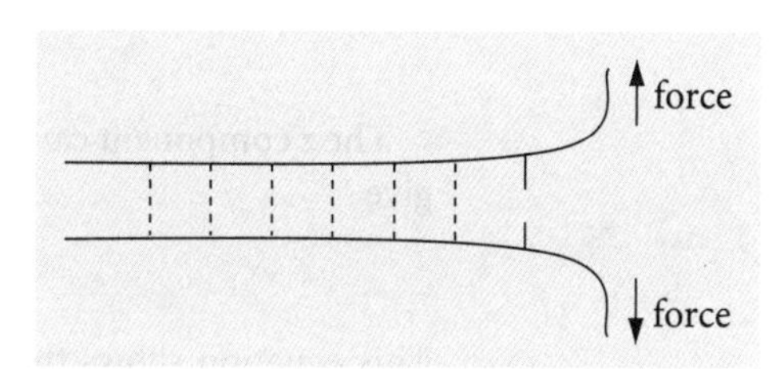

Fig. 10.15 Highly schematic sketch of several parallel hydrogen bonds holding two strands of a protein or two lengths of DNA together and being unzipped by an applied force

This set of equations can be solved by an extension of the method of equation (10.37) and then u substituted for f. If the initial probability of unfolding is S_0 and all other S are zero when $f = 0$, then $S_N = S_0e^{-u}$, $S_{N-1} = S_0ue^{-u}$, $S_{N-2} = S_0\dfrac{u^2}{2}e^{-u}$ and by continuing the calculation it is found that $S_{N-m} = S_0\dfrac{u^m}{m!}e^{-u}$ and S_{unfold} is the integral of this result from 0 to u.

(iii) Bloch equations and NMR

In an NMR experiment, the sample is placed in a large permanent magnetic field aligned along the z-axis. The magnetic field splits the energy of the relevant nuclear spin states and if circularly polarized RF radiation of the correct frequency is applied this will cause transitions between the energy levels. The time and spatial evolution of the magnetization (the vectorial sum of the spin magnetic moments/unit volume) is described by the Bloch equations, which are basic to NMR (Flygare 1978; Günther 1992; Levitt 2001). The magnetization M has components in the x, y, and z-directions and decays by T_1, longitudinal or population relaxation and by T_2 transverse, or spin-spin, relaxation due to loss of spin coherence. See Fig. 9.13 for a sketch of the geometry of an experiment.

The equilibrium magnetization is M_0 and the magnetic field can have components in the x-, y-, and z-directions, $B_{x,y,z}$. The Bloch equations describing the experiment are

$$\frac{dM_x}{dt} = \gamma(M_yB_z - M_zB_y) - \frac{M_x}{T_2}$$

$$\frac{dM_y}{dt} = \gamma(M_zB_x - M_xB_z) - \frac{M_y}{T_2}$$

$$\frac{dM_z}{dt} = \gamma(M_xB_y - M_yB_x) - \frac{M_z - M_0}{T_1}$$

Consider now the case where the external field only has a z component then the equilibrium magnetization vector M_0 points along the z-axis. In a normal NMR experiment, the magnetization is moved from the equilibrium position by the magnetic component of an RF pulse applied in the x–y plane. After the RF pulse has finished (defined as $t = 0$) the spins precess only in the external B_z field starting with the magnetization where it ended up after the RF pulse ended. This will have components along the three axes of $M_x(0)$, $M_y(0)$, and $M_z(0)$. Because only B_z is not zero, the Bloch equations are now simplified to

$$\frac{dM_x}{dt} = \gamma M_yB_z - \frac{M_x}{T_2}$$

$$\frac{dM_y}{dt} = -\gamma M_xB_z - \frac{M_y}{T_2}$$

$$\frac{dM_z}{dt} = -\frac{M_z - M_0}{T_1}$$

The z component can be integrated directly with initial condition $M_z = M_z(0)$ at $t = 0$ to give

$$M_z = M_0 + (M_z(0) - M_0)e^{-t/T_1}.$$

This equation shows that the z component decays only with a T_1 lifetime, which is the decay of the spin population back to equilibrium.

The M_x and M_y equations can be integrated as a pair as in section (i), but in this case it is simpler if they are combined as $M_+ = M_x + iM_y$ where $i = \sqrt{-1}$ and afterwards separated into real and imaginary parts. They become

$$\frac{d(M_+)}{dt} = \left(\gamma iB_z - \frac{1}{T_2}\right)M_+$$

which is a first-order equation and integrates to

$$\begin{aligned} M_+ &= [M_x(0) + iM_y(0)]e^{-i\gamma B_z t}e^{-t/T_2} \\ &= [M_x(0) + iM_y(0)][\cos(\gamma B_z t) + i\sin(\gamma B_z t)]e^{-t/T_2}. \end{aligned}$$

and Euler's relationship was used to convert the exponential. If this equation is expanded and split into real and imaginary parts, M_x and M_y are obtained as

$$M_x = [M_x(0)\cos(\omega t) + M_y(0)\sin(\omega t)]e^{-t/T_2}$$

$$M_y = [-M_x(0)\sin(\omega t) + M_y(0)\cos(\omega t)]e^{-t/T_2}$$

where the substitution $\omega = \gamma B_z$ is also made. This is the NMR transition and Larmor frequency, which is the frequency of precession about the z-axis.

The motion of the magnetization is the vector of the three components. This spirals around the z-axis at the Larmor frequency ω until it reaches the equilibrium magnetization pointing along the z-axis, see Fig. 10.16. The NMR signal is the magnetization's component as it crosses the x- or y-axis, whichever contains the detecting coil, and this produces the free induction decay or FID, which decays away with lifetime T_2.

To analyse the complete NMR experiment, the effect of the weak field from the coil in the x–y plane has to be included. In this case, B_x and B_y vary as a cosine and sine respectively, and the Bloch equations are far harder to solve. It usual to make the transformation into the *rotating frame* which means that the axes are changed so that they rotate at the frequency of the *R.F.* radiation. Flygare (1978), Allen & Eberly (1987), and Günther (1992) discuss this in detail.

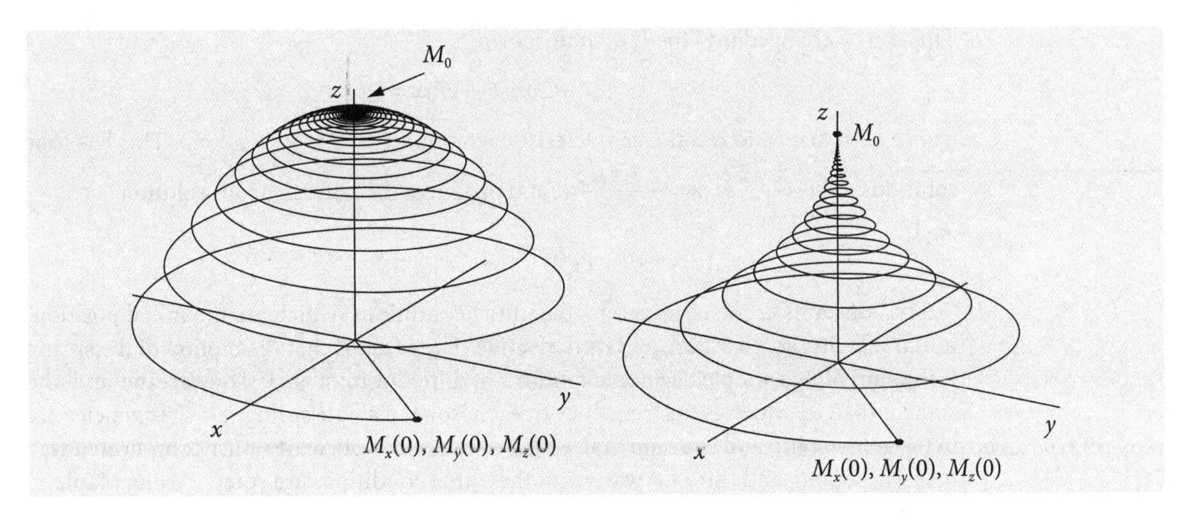

Fig. 10.16 Magnetization vs time when $T_1 = T_2/2$ (left) and when $T_1 = 2T_2$. The magnetization starts at $M_x(0)$, $M_y(0)$, $M_z(0)$ and ends at M_0.

The magnetization in a real NMR experiment, while having the form shown in the figure, has many more rotations than could possibly be shown. The typical resonance frequency for a proton is 400 MHz and T_1 lifetime 10 s. To make the graphs a frequency of 0.1 was used and $T_1 = 200$ and $T_2 = 400$ was used in the left figure.

Using the Maple `spacecurve()` routine enables the curves to be pictured. Define `Mx:=t->..` and similarly for M_y and M_z then use

```
> Mx0:= -1.0: My0:= -1.0: Mz0:= 0.0:
  Mx:= t->..
  .. etc ..
  > spacecurve( [ Mx(t), My(t), Mz(t)], t = 0..800*Pi);
```

The initial values for T_1, T_2 and the initial position of the vector $M_x(0)$ etc. must be defined also. M_0 was calculated as the length of the initial vector. In the figure the initial magnetization vector was at $\{-1, -1, 0\}$. The shape of the FID can be seen by plotting M_x or M_y.

(iv) Second-order equations

Pairs of second-order equations can be solved using the operator method. In this example, the motion of a pair of masses and springs is calculated. This calculation is an alternative to the matrix method of Chapter 7.12.

Consider two identical springs each supporting a mass as shown in Fig. 10.17. This could be a simple model for part of a vehicle's suspension. The displacement from equilibrium of the upper mass m_1 is y, and that of the lower one x. The force constants are k_1 and k_2. If the springs were isolated, they would exert a force equal to $-k_2x$ or $-k_1y$ on their respective masses. When connected together, the upper spring now exerts a force equal to $-k_1(y-x)$ on m_1 as the displacement is changed by the lower spring. The total force on the lower spring is also changed and is $-k_2x + k_1(y - x)$. Together, these produce the force equations,

$$m_1\frac{d^2y}{dt^2} = -k_1(y-x)$$

$$m_2\frac{d^2x}{dt^2} = -k_2x + k_1(y-x)$$

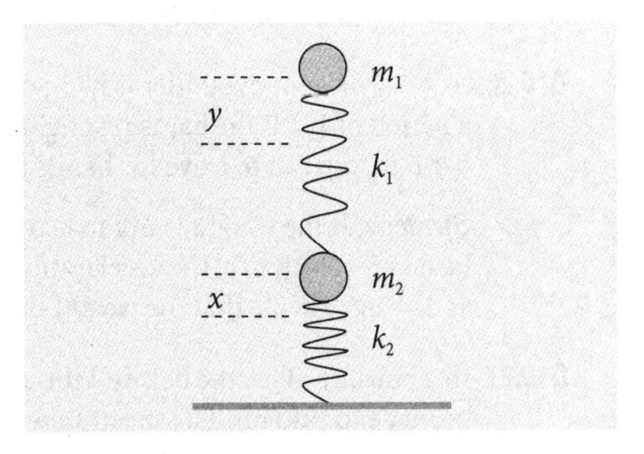

Fig. 10.17 Two coupled springs.

To make the algebra simpler, suppose that the masses are each m and the spring force constants, k. Differentiating twice gives $m\dfrac{d^4x}{dt^4} = -2k\dfrac{d^2x}{dt^2} + k\dfrac{d^2y}{dt^2}$, which reduces to

$$\frac{d^4x}{dt^4} = -3\frac{k}{m}\frac{d^2x}{dt^2} - \left(\frac{k}{m}\right)^2 x.$$

This can be changed into the D operator form

$$(D^4 + 2\omega^2 D^2 + \omega^4)x = 0,$$

where $\omega^2 = k/m$, and has the characteristic equation $\chi^4 + 2\omega^2\chi^2 + \omega^4 = 0$. This has four solutions $\chi = i\dfrac{1 \pm \sqrt{5}}{2}\omega$, $\chi = -i\dfrac{1 \pm \sqrt{5}}{2}\omega$ and produces the homogeneous solution

$$x = Ae^{i(1-\sqrt{5})t/2} + Be^{i(1+\sqrt{5})t/2} + Ce^{i(-1-\sqrt{5})t/2} + De^{i(-1+\sqrt{5})t/2}.$$

The constants are determined by the initial conditions, which are the initial position, initial velocity, acceleration, and its derivative. The result is that the motion of the spring is the sum of two simple harmonic motions of different frequencies, one greater and one smaller than ω, which is the frequency of each isolated single spring. The frequencies are in the golden ratio and its reciprocal, with respect to ω. Notice the rather complicated way the first, second, and third derivatives of the initial conditions are entered using Maple.

```
> eqn:=
     diff(x(t),t$4)+3*omega^2*diff(x(t),t,t)+omega^4*x(t);
  inits:=x(0)=1, D(x)(0)=0, D(D(x))(0)=0, D(D(D(x)))(0)=0;
  dsolve([eqn,inits]):
  x_disp:= collect(simplify(rhs(%),symbolic),cos);
```

$$eqn := \frac{d^4}{dt^4}x(t) + 3\omega^2\left(\frac{d^2}{dt^2}x(t)\right) + \omega^4 x(t)$$

$$inits := x(0) = x0, D(x)(0) = 0, D^{(2)}(x)(0) = 0, D^{(3)}(x)(0) = 0$$

$$x\_disp := -\frac{1}{10}x0(-5 - 3\sqrt{5})\cos\left(\frac{1}{2}(\sqrt{5}-1)\omega t\right) - \frac{1}{10}(3\sqrt{5}-5)x0\cos\left(\frac{1}{2}(\sqrt{5}+1)\omega t\right)$$

To obtain y, which is the displacement of mass m_2, the equations have to be recalculated starting with d^4y/dt^4.

10.7.1 Questions

Full solutions are available at www.oxfordtextbooks.co.uk/orc/beddard.

Q10.24 If $H_n(x)$ represents the Hermite polynomial, show that the *operator* $(2x - d/dx)^n f = H_n(x)$ or equivalently $(2x - D)^n f = H_n(x)$ is true when the function $f = 1$ by working out the first few terms then inferring the rest. Chapter 9.4 and Chapter 1.8.1 describe the Hermite polynomials.

Strategy: As $D \equiv d/dx$ is an operator the order of multiplication is vitally important; for example, if a and b are operators then $(a + b)^2 = aa + ab + ba + bb$ and ab may not expected to be the same as ba. The operator acts on the term to its right. An expression D^2f is the same as $D(Df) \equiv DDf \equiv d^2f/dx^2$.

Q10.25 If a simple harmonic motion is described by $x = A\sin(\omega t + B)$, and if the initial velocity is v_0 and position x_0 at $t = 0$, find the amplitude A and phase B.

Strategy: B is easier to find. Substitute B back into one of the equations to find A. Change the $\tan^{-1}()$ into $\sin^{-1}()$ using a right-angled triangle, and then use the fact that $\sin(\sin^{-1}(x)) = x$.

Q10.26 At the gym, a bodybuilder is lying on his back repeatedly pushing weights on a bar and doing so with a period of 1 s. If the bar is resting on his palms, what is the maximum amplitude displacement of the bar if it does not to leave his hands and assuming that the motion is simple harmonic?

Strategy: If the weight is not to leave his hands, then the maximum acceleration moving the bar must be no greater than g, the acceleration due to gravity. The period T is related to the angular frequency as $T = 2\pi/\omega$. Note that the weight lifted does not enter the final calculation.

Q10.27 In a bowling alley, the ball and the surface of the alley are both smooth. This means that the ball can be made to skid until its speed slows enough that friction prevents this and the ball starts to roll and then does so at constant velocity (Lamb 1947). The ball has a mass m, radius r, and friction coefficient

α between the point of contact of the ball and the surface. The initial linear velocity is v_0 and angular velocity ω_0. When $v_0 > r\omega_0$ then skidding occurs and the frictional force αmg opposes the linear motion, but when $v_0 = r\omega_0$ the ball rolls.

(a) Write down two equations of motion for the linear and angular velocity assuming that the ball is skidding, and integrate them to find the linear and angular velocity. (It is convenient to use the moment of inertia as $I = k^2m$ where k is the radius of gyration and write equations of motion as $mdv/dt = \cdots$ and $Id\omega/dt = \cdots$).

(b) Next, assume that $v = r\omega$ and find the time that the ball starts to roll and its linear speed. Show that v is independent of the friction coefficient but that the time is not. Explain these observations.

(c) Calculate the loss in kinetic energy.

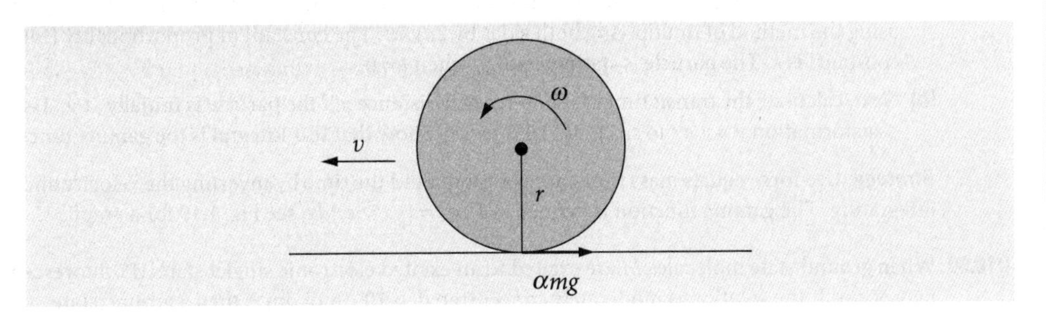

Fig. 10.18 A ball skidding/rolling on a surface with friction.

Strategy: Equate the linear force forwards with the negative value of the frictional force. Angular velocity is always calculated by using the moment of inertia instead of mass. The frictional force for the linear motion acts at the point of contact with the surface but with angular motion the frictional force is referenced to the centre of the ball so is a *couple*, i.e. the force times a distance, which in this case is the radius. The equations for linear and angular motion are independent of one another, and Euler discovered this in 1749 in connection with the motion of ships.

Q10.28 A particle is attracted with a force that varies inversely as the square of its distance from the origin.

(a) If α is its acceleration at *unit* distance, write down and solve the equation describing the velocity at distance x from the origin of the force. Assume that the particle is initially stationary and starts at distance R.

(b) Calculate the time taken to reach distance x from a very distant starting point R after starting from rest. Use the substitution $x = R\cos^2(\theta)$ to solve the integral. If at $t = 0$, $\theta = 0$ and finally $\theta = \pi/2$, calculate the time to reach the origin.

(c) Consider now that the 'particle' is the earth and that the earth's orbital motion has been arrested. Using the last result, how long would it take the earth to fall into the sun given that the time of revolution in a circular orbit is $t_c = 2\pi\sqrt{R^3/\alpha}$?

(d) A meteorite is initially stationary and at a great distance from the earth, show that its velocity at the earth's surface is ≈ 11.2 km s^{-1}. This is numerically equal to the escape velocity.

(Notes: Assume no attenuation by the atmosphere. The constant in the force equation is now $\alpha = gr^2$ where r is the radius of the earth and g the acceleration due to gravity. The acceleration due to gravity

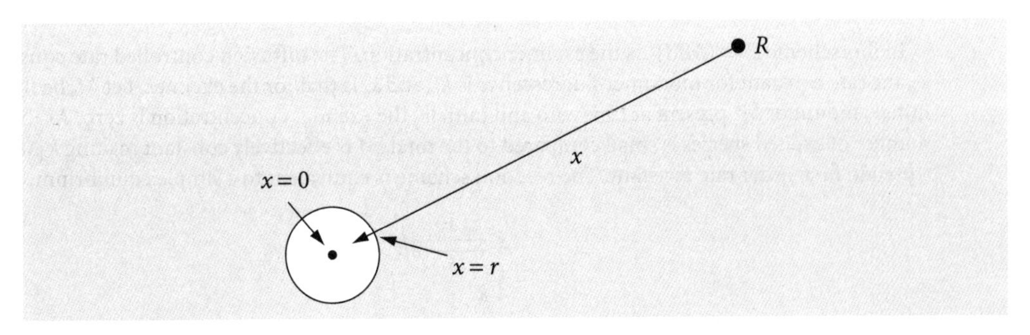

Fig. 10.19 A small particle starts from rest at a distance R from a massive body.

varies as the inverse square of the distance from the earth's centre, where $x = 0$. Assume that the initial distance of the meteorite R is very large making $1/R$ negligible. The average radius of the earth is 6371 km and the earth–sun average distance is 1.49×10^8 km, $g = 9.81$ m s^{-2}.)

Strategy: (a) As the velocity is required, use equation (10.16); the force is $-m\alpha/x^2$ if x is the distance from the origin (centre of earth) and m the mass. (b, c) Use equation (10.15) to calculate the time. The substitution $x = R\cos^2(\theta)$ greatly simplifies the integration. (d) The velocity is found when $x = r$ at the surface of the earth. The starting point is so distant that the constant of integration can be set to zero.

Q10.29 A particle of mass m is moving in a straight line because it is attracted to a point with a force that is inversely proportion to its distance y from the point.

(a) Write down the differential equation for the force and integrate once to find the velocity dy/dt using the method of multiplying both sides by $2dy/dt$. The constant of proportionality (force constant) is k. The particle is stationary at y_0 when $t = 0$.

(b) Next, calculate the transit time taken to reach a distance y if the particle is initially at y_0. Use the transformation $y = y_0 e^{-x}$ to simplify the integral. Show that this integral is the gamma function.

Strategy: Use force equals mass times acceleration. Find the time by inverting the velocity and integrating. The gamma function is defined as $\Gamma(\alpha) = \int_0^\infty x^{\alpha-1}e^{-x}dx$, see Fig. 1.19 for a graph.

Q10.30 When ground state molecules M are excited to an excited electronic singlet state M^*, fluorescence may occur. If the solution is moderately concentrated, $> 10^{-3}$ mol dm^{-3}, then a ground state molecule may diffuse sufficiently to collide with an excited state before the latter has had a chance to fluoresce. In this case, an *excimer* or excited dimer can be formed and this species has an attractive potential well in its excited state $(MM)^*$ but is repulsive in its ground state (MM). The reaction is $M + M^* \rightarrow (MM)^*$ and the excimer $(MM)^*$ can fluoresce but does so at longer wavelength than M^* and its emission spectrum is broad and structureless, because the ground state is repulsive. The interaction between M and M^* is caused by the combination of Coulomb and exchange interactions between the excited and ground state of the molecules. Excimers are formed with several types of planar aromatic molecules, pyrene being the most studied example (Birks 1970).

Excimer emission has found considerable use in determining the lateral diffusion coefficients of molecules in biological membranes. The rate of excimer formation is sensitive to the diffusion coefficient via rate constant k_d (see below). When the emission yield or, preferably, the time profile of the monomer and excimer fluorescence are analysed, k_d can be determined. Phase transitions, where viscosity may change suddenly can also be observed in this way. Excimer and exciplex fluorescence can also be used to measure the proximity of one molecule to another, such as bases on complementary strands of DNA if they are suitably modified. If the two molecules are of different types, then the excited species is called an *exciplex*, short for excited complex. Partial charge transfer occurs between the two molecules in an exciplex therefore the emission is sensitive to the polarity of its environment and can be used as a probe of this. The two species in an excimer or exciplex need not be separate entities but can be linked to one another and, if this is long and flexible enough, intramolecular excimers and exciplexes can be formed.

The intermolecular excimer reaction scheme is

$$M + M^* \underset{k_1}{\overset{k_d}{\rightleftharpoons}} E^*$$

$$M^* \xrightarrow{k_f} M$$

$$E^* \xrightarrow{k_e} M + M$$

In this scheme $E^* \equiv (MM)^*$ is the excimer concentration. The diffusion controlled rate constant is k_d, the rate constant for monomer fluorescence is k_f, and k_e is that for the excimer. Let M_0 be the initial amount of M^* present at time zero and initially the excimer concentration is zero. As the number of excited species is small compared to the total, M is effectively constant making $k_dM \equiv k_2$, a pseudo first-order rate constant. The reaction scheme is equivalent to a simple equilibrium;

$$A \underset{k_1}{\overset{k_2}{\rightleftharpoons}} E^*$$

$$\downarrow k_f \qquad \downarrow k_e$$

(a) Describe what you would expect to see in an experiment to measure the time profiles of excimer E^* and M^* fluorescence.

(b) Solve the differential equations. Using the constants, $k_f = 0.02$, $k_e = 0.01$, $k_2 = 10^{10} \times M_0$, $k_1 = 0.05$, each in units of ns^{-1}, and $M_0 = 3 \times 10^{-3}$ dm^3 mol^{-1} plot the time profiles of E^* and M^* populations to see if your intuition in (a) was correct. Vary the rate constants and concentration to observe how the species respond. Initially $E^* = 0$.

***Strategy*:** Form a second-order rate equation from M* and E*.

Q10.31 The simplest model of a diffuse electrical double layer formed at an interface between an electrolyte in solution and a metal, a bilayer membrane, a protein or a colloid, consists of an excess of ions and electrons at the surface and an equivalent amount of oppositely charged ions randomly distributed in solution. This means that the electric field decreases more rapidly when moving away from the surface than would be the case in a vacuum. The Poisson–Boltzmann equation describes the change in potential with distance and combines the Poisson equation of electrostatics with the Boltzmann distribution. In particular, the equation relates the time average of the space-charge density ρ to the potential φ at the interface. In one dimension, the equation has the form

$$\frac{d^2\varphi}{dx^2} = -\frac{4\pi}{\varepsilon}\rho,$$

and the charge density is

$$\rho = e_0 \sum_{i=1} z_i N_i \exp\left(-\frac{z_i e_0 \varphi}{k_B T}\right)$$

for i different species with charge z_i. The boundary and initial conditions are that at $x = \infty$, $\varphi \to 0$, and $d\varphi/dx \to 0$. At $x = 0$, $\varphi = \varphi_0$. The charge on the electron is e_0 and ε the permittivity or dielectric constant of the solvent.

(a) Solve the equation and show that the electrical potential is $\varphi = \varphi_0 e^{-\kappa x}$ if the solution contains only a binary $A^{z+}B^{z-}$ electrolyte with charges $\pm z$ and $N_i = N$ and simplify the final result of the integration by assuming that $k_BT >> ze_0\varphi$. The constant is $\kappa = \sqrt{\dfrac{8\pi N z^2 e_0^2}{\varepsilon k_B T}}$.

(The double layer and Poisson–Boltzmann equation is described by Koryta et al. (1993), Kuhn & Fursterling (1999) and Jackson (2006).)

***Strategy*:** Simplify the charge density first. Multiply both sides of the equation by $d\varphi/dx$ and use the relationship, $2\dfrac{d\varphi}{dx}\dfrac{d^2\varphi}{dx^2} = \dfrac{d}{dx}\left(\dfrac{d\varphi}{dx}\right)^2$, on the left hand side (see equation (10.14)).

10.8 Linear equations with variable coefficients

A more difficult equation to solve than those met previously is

$$a(x)\frac{d^2y}{dx^2} + b(x)\frac{dy}{dx} + c(x)y = g(x)$$

where the coefficients a, b, c are functions of x. This equation is usually put into the form

$$\frac{d^2y}{dx^2} + P(x)\frac{dy}{dx} + Q(x)y = f(x)$$

by dividing through by $a(x)$. The general method to use is the variation of parameters, because this can be applied to all differential equations. However, it is quite complicated to use and specialized texts should be consulted (Jeffery 1990; Bronson 1994; Aratyn & Rasinariu 2006). There are, however, at least two methods that can still be used: the main one is to solve the equation as a series expansion and the second reduces the equation to a simpler one, but this can be used only in certain cases.

10.8.1 Reduction to simpler forms by change of variable and substitution

(i) Sometimes a simple substitution can produce constant coefficients, but success in this depends very much on the exact form of the equation. The equation

$$\cos(x)\frac{d^2y}{dx^2}+\sin(x)\frac{dy}{dx}+a\cos^3(x)y=0$$

can be simplified with $u=\sin(x)$. The derivatives are $du/dx=\cos(x)$, $d^2u/dx^2=-\sin(x)$ and as $\dfrac{dy}{dx}=\dfrac{dy}{du}\dfrac{du}{dx}$ then $\dfrac{dy}{dx}=\cos(x)\dfrac{dy}{du}$ and

$$\frac{d^2y}{dx^2}=-\sin(x)\frac{dy}{du}+\cos(x)\frac{d^2y}{du^2}\frac{du}{dx}$$

$$=-\sin(x)\frac{dy}{du}+\cos^2(x)\frac{d^2y}{du^2}$$

Substituting into the differential equation produces

$$\frac{d^2y}{du^2}+ay=0,$$

which is solved by standard methods to find y in terms of u. Notice that if the $\cos^3(x)y$ term in the original equation was replaced by $\cos(x)y$, then the equation produced is not as simple and would be $(1-u^2)\dfrac{d^2y}{du^2}+ay=0$.

(ii) Equations of the particular type

$$x^3\frac{d^3y}{dx^3}+ax^2\frac{d^2y}{dx^2}+bx\frac{dy}{dx}+cy=f(x)$$

where terms in x have the same power as the differential are called Euler's or Cauchy's equation and can be simplified with the substitution $x=e^z$. The derivative is $dz/dx=e^{-z}$ and if $D\equiv d/dz$ then

$$\frac{dy}{dx}=\frac{dy}{dz}\frac{dz}{dx}=e^{-z}Dy, \qquad \frac{d^2y}{dx^2}=\frac{d^2y}{dz^2}\left(\frac{dz}{dx}\right)^2+\frac{dy}{dz}\frac{d^2z}{dx^2}=e^{-2z}(D^2-D)y$$

This last derivative is obtained from $\dfrac{d^2y}{dx^2}=\dfrac{d}{dz}\left(\dfrac{dy}{dz}\dfrac{dz}{dx}\right)\dfrac{dz}{dx}$ using the product rule and $\dfrac{\partial}{\partial z}\dfrac{\partial z}{\partial x}\equiv\dfrac{\partial}{\partial x}$. The different derivatives can now be written down as

$$x\frac{dy}{dx}=Dy, \qquad x^2\frac{d^2y}{dx^2}=(D^2-D)y, \qquad x^3\frac{d^2y}{dx^2}=(D^3-3D^2+2D)y$$

and used to simplify the differential equation. Note that the right-hand side has derivatives in dy/dz.

To solve the equation

$$x^2\frac{d^2y}{dx^2}+2x\frac{dy}{dx}-y=0$$

simplify this to $(D^2+D-1)y=0$ using the derivatives above. This characteristic equation has roots $\dfrac{\sqrt{5}-1}{2},\dfrac{-\sqrt{5}-1}{2}$. However, the operator D is a function of z not of x, hence, the solution is a function of z,

$$y=Ae^{(\sqrt{5}-1)z/2}+Be^{(-\sqrt{5}-1)z/2}$$

where A and B are constants determined by the initial conditions. As $x=e^z$ or $z=\ln(x)$, the solution is rewritten as $y=Ae^{(\sqrt{5}-1)\ln(x)/2}+Be^{(-\sqrt{5}-1)\ln(x)/2}$ and finally as

$$y=Ax^{(\sqrt{5}-1)/2}+Bx^{(-\sqrt{5}-1)/2}.$$

(iii) If the equation can be put into the form,

$$\frac{d^2y}{dx^2}+P(x)\frac{dy}{dx}+Q(x)y=0,$$

then it can be reduced by the transformation $y=uv$ where $v=e^{-\frac{1}{2}\int pdx}$. The equation becomes

$$\frac{d^2u}{dx^2} - w(x)u = 0, \qquad \text{with} \qquad w(x) = \frac{1}{2}\frac{dP}{dx} + \left(\frac{P}{2}\right)^2 - Q,$$

and the final solution is $y = uv$.

The equation $\dfrac{d^2y}{dx^2} + x\dfrac{dy}{dx} + (x-1)y = 0$ can be solved by this method with $v = e^{-\frac{1}{2}\int x dx} = e^{-x^2/4}$ and $w = 3/2 + x^2/4 - x$, which produces the equation with which to find u as,

$$\frac{d^2u}{dx^2} + \left(\frac{3}{2} + \frac{x^2}{4} - x\right)u = 0.$$

10.8.2 Series solution of differential equations

While many types of equations can be solved using the methods described so far, there are a number of problems whose equations can only be solved by a series expansion. These are often different forms of the Schrödinger equation, examples of which are the quantum harmonic oscillator and the radial and angular solutions to the hydrogen atom. While the series solution method will be described in general, in quantum mechanics and in some other problems, the equations often have a form whose solution is well known because the equation has a specific name. The harmonic oscillator is solved using Hermite's differential equation and the hydrogen atom requires Legendre's and Laguerre's equations. Other commonly used equations are named after Helmholtz, Laplace, and Bessel.

Many functions can be expanded as a power series that has the form

$$y = a_0 + a_1x + a_2x^2 + a_3x^3 + \cdots + a_nx^n + \cdots \tag{10.38}$$

with constants a_0, a_1, etc. as described in Chapter 5. By taking derivatives of y based on this expansion, the differential equation can be reconstructed in terms of the powers of x and these constants. To find the constants, the powers of x are grouped together and the resulting groups of constants solved assuming that each group of them is zero. What results is a recursion formula with which to calculate the constants and so the equation is solved without any formal integration. It is assumed that the solution can be found with this method, because this is usually the case for problems in chemical physics. There is a simple way to check if a series solution is possible and if not, another related method due to Frobenius can be used to solve the equations in these cases.

(i) The basic series method is illustrated with the equation

$$\frac{d^2y}{dx^2} + x\frac{dy}{dx} + y = 0$$

and the solution is assumed to have the form of equation (10.38). The strategy is

(a) to differentiate the series solution, and substitute the results into the differential equation;

(b) collect together coefficient with the same power of x and set each result to zero;

(c) find a recursion equation in the coefficients $a_0, a_1, a_2, \cdots$ using the initial conditions as the starting points.

(a) The differential equation is formed out of the series solution by taking the derivatives. These are

$$\frac{dy}{dx} = a_1 + 2a_2x + 3a_3x^2 + 4a_4x^3 + \cdots + (n-1)a_{n-1}x^{n-2} + na_nx^{n-1} + (n+1)a_{n+1}x^n \cdots,$$

$$\frac{d^2y}{dx^2} = 2a_2 + 6a_3x + 12a_4x^2 + \cdots +$$

$$n(n-1)a_nx^{n-2} + n(n+1)a_{n+1}x^{n-1} + (n+1)(n+2)a_{n+2}x^n + \cdots$$

and the $n-2$, $n-1$ and n^{th} terms are tabulated below as these are needed later on. Notice that these derivatives are the same for *all* differential equations solved by the expansion in (10.38). These coefficients are shown in the table and terms that are useful in other equations are also shown.

	x^{n-2}	x^{n-1}	x^n
y	a_{n-2}	a_{n-1}	a_n
y'	$(n-1)a_{n-1}$	na_n	$(n+1)a_{n+1}$
y''	$n(n-1)a_n$	$n(n+1)a_{n+1}$	$(n+1)(n+2)a_{n+2}$
xy	a_{n-3}	a_{n-2}	a_{n-1}
xy'	$(n-2)a_{n-2}$	$(n-1)a_{n-1}$	na_n

(b) The next step is to substitute these derivatives into the differential equation and then to group all the coefficients with $x^0, x^1, x^2, x^3 \cdots$ together and make each group zero. The groups of coefficients for each power of x must be zero, as the differential equation is itself equal to zero. For this particular equation, substituting produces

$$\begin{aligned} &2a_2 + 6a_3x + 12a_4x^2 + \cdots &&+(n+2)(n+1)a_{n+2}x^n + \cdots \\ &+\ a_1x + 2a_2x^2 + 3a_3x^3 + \cdots &&+ na_nx^n + \cdots \\ &+a_0 + a_1x + a_2x^2 + a_3x^3 + \cdots &&+ a_nx^n + \cdots = 0 \end{aligned}$$

Grouping the coefficients gives

$$\begin{aligned} &a_0 + 2a_2 = 0 && \text{coefficients of } x^0 \\ &6a_3 + 2a_1 = 0 && \text{coefficients of } x^1 \\ &12a_4 + 2a_2 + a_2 = 0 && \text{coefficients of } x^2 \\ &\cdots\cdots\cdots \\ &(n+2)(n+1)a_{n+2} + na_n + a_n = 0 && \text{coefficients of } x^n. \end{aligned}$$

(c) The recursion formula from the x^n term is

$$a_{n+2} = -\frac{(n+1)a_n}{(n+1)(n+2)}.$$

The first two coefficients are a_0 and a_1, and the other coefficients are expressed in terms of these as

$$n = 0,\ a_2 = \frac{1}{2}a_0, \qquad n = 1,\ a_3 = \frac{1}{3}a_1.$$

$$n = 2,\ a_4 = \frac{1}{4}a_2 = \frac{1}{8}a_0 \qquad n = 3,\ a_5 = \frac{1}{15}a_1.$$

The solution is therefore

$$y = a_0\left(1 - \frac{x^2}{2} + \frac{x^4}{8} - \frac{x^6}{48} + \cdots\right) + a_1\left(-\frac{x}{3} + \frac{x^3}{15} - \frac{x^5}{105} + \cdots\right).$$

The two solutions are independent of one another, and the even powered series is the expansion of $e^{-x^2/2}$, while the odd series is not so easily identified. The coefficients are determined by the initial conditions and, if $y(0) = a_0$ and $dy/dx = 0$ at $x = 0$, then the solution is $y = a_0e^{-x^2/2}$.

(ii) In this example, the equation is

$$\frac{d^2y}{dx^2} + (\alpha - \beta x^2)y = 0,$$

and, following the previous calculation, the recursion equation is

$$a_{n+2} = \frac{\beta a_{n-2}}{(n+1)(n+2)} - \frac{\alpha a_n}{(n+1)(n+2)},$$

and is not valid when n is 0 or 1 because a_{-2} and a_{-1} are not defined. The coefficients of the first two terms in the expansion must then be examined. These are

$$2a_2 + \alpha a_0 = 0 \qquad \text{and} \qquad 6a_3 + \alpha a_1 = 0$$

from which a_2 and a_3 can be found in terms of a_0 and a_1 and these then used as the starting points for the recursion formula. The series solution can be obtained using Maple to do the recursion and form the solution. Only the first few terms are shown;

```
> a[2]:=-alpha*a[0]/2:
  a[3]:=-alpha*a[1]/6:
  for n from 2 to 4 do
    a[n+2]:=(beta*a[n-2]-alpha*a[n])/((n+1)*(n+2));
  end do;
```

$$a_4 := \frac{1}{24}\alpha^2 a_0 + \frac{1}{12}\beta a_0$$

$$a_5 := \frac{1}{120}\alpha^2 a_1 + \frac{1}{20}\beta a_1$$

$$a_6 := -\frac{1}{30}\alpha\left(\frac{1}{24}\alpha^2 a_0 + \frac{1}{12}\beta a_0\right) - \frac{1}{60}\beta\alpha a_0$$

$$a_7 := -\frac{1}{42}\alpha\left(\frac{1}{120}\alpha^2 a_1 + \frac{1}{20}\beta a_1\right) - \frac{1}{252}\beta\alpha a_1$$

```
> s:=0:
  for n from 0 to 4 do
    s:= s + a[n]*x^n;
  end do:
  y:= s;
```

$$y := a_0 + a_1 x - \frac{1}{2}\alpha a_0 x^2 - \frac{1}{6}\alpha a_1 x^3 + \left(\frac{1}{24}\alpha^2 a_0 + \frac{1}{12}\beta a_0\right)x^4 + \left(\frac{1}{120}\alpha^2 a_1 + \frac{1}{20}\beta a_1\right)x^5$$

There are two series in the result; one with terms in a_0 and the other in a_1. These are independent solutions. Maple can produce the series solution directly and this is done with the instructions,

```
> eq:= diff(y(x),x,x)+(alpha-beta*x^2)*y(x);
  init:= y(0)= a0, D(y)(0)= a1;           # initial conditions
  dsolve([eq, init],y(x), type=series );
```

(iii) The Hermite equation is important because it leads to the solution of the Schrödinger equation for the quantum mechanical harmonic oscillator. The equation has the form

$$\frac{d^2y}{dx^2} - 2x\frac{dy}{dx} + 2\gamma y = 0 \tag{10.39}$$

where γ is a real number. Solving this equation as a series leads to the coefficients

$$\begin{aligned}
&2a_2 + \ 6a_3 x + 12a_4 x^2 + \cdots && + (n+2)(n+1)a_{n+2}x^n + \cdots \\
&-2a_1 x - \ 4a_2 x^2 - \cdots && - 2na_n x^n + \cdots \\
&+2\gamma a_0 + 2\gamma a_1 x + 2\gamma a_2 x^2 + \cdots && + 2\gamma a_n x^n + \cdots = 0
\end{aligned}$$

The recursion equation is therefore

$$(n+2)(n+1)a_{n+2} - 2na_n + 2\gamma a_n = 0 \quad \text{or} \quad a_{n+2} = 2\frac{(n-\gamma)}{(n+2)(n+1)}a_n$$

and evaluating the coefficients gives the series

$$y_\gamma = a_0\left(1 - \gamma x^2 - \frac{(2-\gamma)\gamma}{6}x^4 - \frac{(4-\gamma)(2-\gamma)\gamma}{90}x^6 - \cdots\right)$$

$$+ a_1\left(x + \frac{(1-\gamma)}{3}x^3 + \frac{(3-\gamma)(1-\gamma)}{30}x^5 + \cdots\right)$$

These series form the Hermite polynomials when γ is a positive integer. For example, when $\gamma = 3$, the odd power series terminates to give

$$y_3 = a_1\left(x - \frac{2}{3}x^3\right)$$

and when $\gamma = 4$, the even power series terminates as

$$y_4 = a_0\left(1 - 4x^2 + \frac{4}{3}x^4\right).$$

Each series is limited to a few terms because of the way the coefficients are formed. The first few Hermite polynomials are

$$H_0(x) = 1 \qquad H_1(x) = x$$
$$H_2(x) = 4x^2 - 2 \qquad H_3(x) = 8x^3 + 12x$$
$$H_4(x) = 16x^4 - 48x^2 + 12 \qquad H_5(x) = 32x^5 - 160x^3 + 120x$$

and choosing the constants as

$$a_0 = (-1)^{\gamma/2}\frac{\gamma!}{(\gamma/2)!},$$

with γ as an even integer and

$$a_1 = (-1)^{(\gamma+1)/2}\frac{(\gamma+1)!}{((\gamma+1)/2)!}$$

with γ as an odd integer, converts the series solution to the differential equation into the sum of two Hermite polynomials. The general solution can then be written as

$$y_\gamma = c_1 H_\gamma(x) + c_2 H_{\gamma+1}(x)$$

where c_1 and c_2 are two new constants that are determined by the initial conditions.

The equation,

$$\frac{d^2y}{dx^2} + (2\gamma + 1 - x^2)y = 0, \tag{10.40}$$

with γ as a constant, is related to the Hermite equation as can be seen if $y = e^{-x^2/2}v$ then, as v is a function of x,

$$\frac{d^2y}{dx^2} = \left(\frac{d^2v}{dx^2} - 2x\frac{dv}{dx} + (x^2 - 1)v\right)e^{-x^2/2}$$

and (10.40) becomes $\dfrac{d^2v}{dx^2} - 2x\dfrac{dv}{dx} + 2\gamma v = 0$, which is identical to (10.39). The solutions of (10.40) are therefore Hermite polynomials multiplied by the Gaussian $\exp(-x^2/2)$.

The Schrödinger equation for the harmonic oscillator can be written as

$$\frac{d^2\psi}{dx^2} + \frac{2m}{\hbar^2}\left(E - \frac{k}{2}x^2\right)\psi = 0,$$

where k is the force constant related to the frequency as $\omega = \sqrt{k/m}$. The equation can be simplified to $\dfrac{d^2\psi}{dx^2} + (a - b^2x^2)\psi = 0$ with the abbreviations $a = 2mE/\hbar^2$ and $b^2 = mk/\hbar^2$. Next, the substitution $z = \sqrt{b}x$ is used to remove b from the x^2 term and make the equation similar to (10.40). This change produces

$$\frac{d^2\psi}{dz^2} + \left(\frac{a}{b} - 1 + 1 - z^2\right)\psi = 0$$

which has the correct form with the abbreviation $a/b - 1 = 2\gamma$. The solutions are Hermite polynomials multiplied by $\exp(-z^2/2)$ and these have the characteristic form of the harmonic oscillator wavefunctions. These decay exponentially to zero when z has a large positive or negative value but oscillate near to zero. The different wavefunctions are found by substituting $E = \hbar\omega(n + 1/2)$ for the quantum number n.

10.8.3 Checking whether a series solution is possible

A solution is usually expanded as a series about $x = 0$; therefore, the first thing to check is whether the series method is applicable. This means examining the differential equation to ensure that $x = 0$ is analytic, hence not singular, which in turn means that $x = 0$ does produce infinity when substituted into the equation.

The general equation is

$$\frac{d^2y}{dx^2} + P(x)\frac{dy}{dx} + Q(x)y = f(x).$$

If the equation is $\frac{d^2y}{dx^2} + \frac{dy}{dx} + y = 0$, then this is analytic at $x = 0$, as P and Q are both 1.

The equation

$$\frac{d^2y}{dx^2} + \frac{x}{(x-1)}\frac{dy}{dx} + \frac{1}{(x-1)}y = 0$$

is also analytic at $x = 0$, and an expansion about 0 is possible. The point $x = 1$ is *singular*, and the function $1/(x-1)$ is not defined, so the series expansion will not be possible here and is valid only in the range $-1 < x < 1$. The equation $\frac{d^2y}{dx^2} + x(x-1)\frac{dy}{dx} + (x-1)y = 0$ can be expanded, because the $x(x-1)$ and $(x-1)$ are polynomials and every point is normal.

Bessel's equation is $x^2\frac{d^2y}{dx^2} + x\frac{dy}{dx} + (x^2 - n^2)y = 0$ where n is a constant and when rearranged is $\frac{d^2y}{dx^2} + \frac{1}{x}\frac{dy}{dx} + \left(\frac{x^2 - n^2}{x^2}\right)y = 0$, so has a singular point at $x = 0$. In this case, the equation can be shown to have solutions

$$y = x^m \sum_{k=0}^{\infty} a_k x^k$$

and where $a_0 \neq 0$. Using this series is often called Frobenius' method and both the index m and coefficients a_k have to be determined. The method is very similar to that already described and is given in more advanced textbooks (Boas 1983; Jeffery 1990; Bronson 1994; Aratyn & Rasinariu 2006).

10.8.4 **Questions**

Full solutions are available at **www.oxfordtextbooks.co.uk/orc/beddard**.

Q10.32 Solve $(1 + x^2)\frac{d^2y}{dx^2} - 2y = 0$ by the series method.

Strategy: Use the nth terms of the derivatives given in the text to work out the recursion equation.

Q10.33 Particles with energy E, such as electrons, interact with a one-dimensional energy potential barrier which extends from 0 to $+a$, and has an energy height V_0 and is zero elsewhere. The particles have a certain chance of being transmitted through or reflected off the barrier, the sum of these being unity.

(a) Write down the Schrödinger equation for the region inside the potential barrier.

(b) When $E > V_0$ calculate the *two* wavefunctions $\varphi_{1,2}$ which have an odd and even nature and are found with the initial values $\varphi(0) = 0$, $d\varphi/dx|_0 = 1$ and vice versa.

(c) Show that the transmission is

$$T = \frac{4(E - V_0)E}{4(E - V_0)E + V_0^2 \sin^2(\sqrt{2m(E - V_0)}a/\hbar)}$$

and plot this vs relative energy E/V_0, with $\sqrt{2mV_0}a/\hbar = 5$. To do this calculation use

$$A_T = -\frac{e^{-iak}}{2}\left(\frac{L_1 + iak}{L_1 - iak} - \frac{L_2 + iak}{L_2 - iak}\right) \quad \text{with} \quad L_{1,2} = \frac{a}{\varphi_{1,2}(a)}\frac{d\varphi_{1,2}(a)}{da}.$$

The transmission is $A_T^*A_T$; use Maple to do the algebra, and then use $\sin^2(2x) = 4\cos^2(\theta) - 4\cos^4(x)$ to simplify the result.

(d) Comment on the nature of the transmission vs energy. What causes the oscillations?

Strategy: Write down the standard Schrödinger equation with potential V_0. It is easier to solve if you simplify using $c^2 = \frac{2m}{\hbar^2}(E - V_0)$. Use the two conditions given to remove the arbitrary constants in the solution. To plot T vs reduced energy rearrange the transmission equation into terms in E/V_0 then plot using this ratio as abscissa.

10.9 Partial differential equations

10.9.1 Background

Many important phenomena in chemistry, physics, and biology involve partial differential equations; for example, the Schrödinger, wave, and diffusion equations. Although the general theory of partial differential equations is well beyond the scope of this book, the equations mentioned are among those whose solutions can be obtained by a powerful method known as the *separation of variables*.

The simplest type of partial differential equation has two independent variables, x and y, and its solution must represent a surface rather than a curve as does that of an ordinary differential equation. The ordinary differential equation has arbitrary constants as a result of integration, which are determined by the initial or boundary conditions; the partial differential equation has instead arbitrary *functions of integration* and these have to be eliminated to obtain a particular solution. The problem is not that there are too few arbitrary functions to solve the equation but that there are too many. An example given by Stephenson (1996) illustrates this point. Consider the equation formed by differentiating $w = yf(x)$ with respect to y, and where $f(x)$ is a general, unspecified function of x. The differential is $\frac{\partial w}{\partial y} = f(x)$ and by substitution to eliminate $f(x)$ the partial differential equation is

$$y\frac{\partial w}{\partial y} - w = 0$$

which is a first-order equation. The solution to this is $f(x)$ but as this function was not specified, it is therefore *any* arbitrary function. Exactly what this function might be has to be determined by using the initial or boundary conditions imposed on the problem.

As a second example, the one-dimensional wave equation will be 'solved'. This equation has the form $\frac{\partial^2 u}{\partial t^2} = a^2\frac{\partial^2 u}{\partial x^2}$. Consider now the equation,

$$u = f_1(x + at) + f_2(x - at)$$

where the functions f are arbitrary and are not defined in any real sense. To find a solution we differentiate twice by x then twice by t and add the results. Let $x + at \equiv r$ and $x - at \equiv s$, then

$$\frac{\partial u}{\partial x} = \frac{\partial(f_1 + f_2)}{\partial r}\frac{\partial r}{\partial x} + \frac{\partial(f_1 + f_2)}{\partial s}\frac{\partial s}{\partial x}$$
$$= f_1'(x + at) + f_2'(x - at)$$

and similarly,

$$\frac{\partial^2 u}{\partial x^2} = f_1''(x + at) + f_2''(x - at); \qquad \frac{\partial^2 u}{\partial t^2} = f_1''(x + at)a^2 + f_2''(x - at)a^2.$$

Adding these two equations and eliminating the functions produces the wave equation $\frac{\partial^2 u}{\partial t^2} = a^2\frac{\partial^2 u}{\partial x^2}$. To convince yourself of the arbitrary, and hence not very useful nature of the functions f_1 and f_2 that solve the wave equation, use Maple to do the algebra on a few functions with arguments $x \pm at$;

```
> f:=sin(exp(x+a*t))+ exp(sin(x-a*t)):       # is a solution
  Diff(f,t,t)- a^2*Diff(f,x,x):%=simplify(value(%));> ;
```

$$\frac{\partial^2}{\partial t^2}(\sin(e^{x+at}) + e^{-\sin(-x+at)}) - a^2\left(\frac{\partial^2}{\partial x^2}(\sin(e^{x+at}) + e^{-\sin(-x+at)})\right) = 0$$

```
> f:=(x+a*t)^3+1/(x-a*t):                    # also a solution
  Diff(f,t,t)- a^2*Diff(f,x,x): %= simplify(value(%));
```

$$\frac{\partial^2}{\partial t^2}\left((x+at)^3 + \frac{1}{x-at}\right) - a^2\left(\frac{\partial^2}{\partial x^2}\left((x+at)^3 + \frac{1}{x-at}\right)\right) = 0$$

This method is therefore a way of solving the equation, but in many cases, the method of separating variables is more satisfactory and is described next.

10.9.2 Separation of variables

This method assumes that the solution to the partial differential equation is based on splitting the solution into two parts. In the diffusion equation the quantity required is usually the concentration of a species as a function of time and of position. The solution is therefore found using $c(x,t) = c(x)c(t)$ and splitting the partial differential equation into two parts, one, which depends on time alone and the other on position alone. Both these equations are then made equal to a constant and each can be solved. To find the constants introduced by integration, and thus the exact form of the solution, the initial and boundary conditions are now used. Finding these constants often, but not always, involves expanding the solution as a Fourier series. Solving the differential equations is often the simpler part of the calculation; handing the boundary conditions can involve more work. Some examples are now given.

(i) Particle in two-dimensional box

The energy and wavefunctions of a particle in a two-dimensional box with sides of length a and b are calculated from the Schrödinger equation

$$-\frac{\hbar^2}{2m}\left(\frac{\partial^2\varphi}{\partial x^2} + \frac{\partial^2\varphi}{\partial y^2}\right) = E\varphi, \tag{10.41}$$

which is a partial differential equation in x and y. The separation of variables assumes that the wavefunction φ which is a function of x and y, is also the product of two wavefunctions, one in x and the other in y, thus

$$\varphi(x,y) = \psi(x)\psi(y) \equiv \psi_x\psi_y.$$

Substituting gives

$$-\frac{\hbar^2}{2m}\left(\psi_y\frac{\partial^2\psi_x}{\partial x^2} + \psi_x\frac{\partial^2\psi_y}{\partial y^2}\right) = E\psi_x\psi_y$$

and rearranging

$$-\frac{\hbar^2}{2m}\left(\frac{1}{\psi_x}\frac{\partial^2\psi_x}{\partial x^2} + \frac{1}{\psi_y}\frac{\partial^2\psi_y}{\partial y^2}\right) = E$$

The next step is essential to the method of separating variables and this is that each of the terms on the left of the equation must be equal to a constant. This is true because the two terms are variables of either x or y. If x is varied, only the first term of the equation changes, the other is a constant as is the energy E. However, for the whole equation to be satisfied, the derivative in x must be equal to a constant, and this is labelled as E_x, and is called a *separation constant*;

$$-\frac{\hbar^2}{2m}\left(\frac{1}{\psi_x}\frac{\partial^2\psi_x}{\partial x^2}\right) = E_x. \tag{10.42}$$

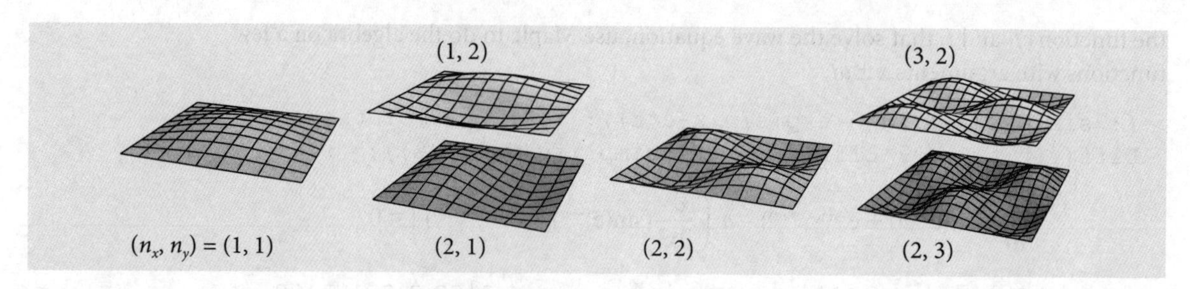

Fig. 10.20 Wavefunction shapes in the square well.

Similarly for changes in y, with x constant,

$$-\frac{\hbar^2}{2m}\left(\frac{1}{\psi_y}\frac{\partial^2\psi_y}{\partial x^2}\right)=E_y \tag{10.43}$$

and then

$$E_x+E_y=E. \tag{10.44}$$

The two equations in x and y can now be solved separately in the normal way. For example, $\frac{\partial^2\psi_x}{\partial x^2}+k^2\psi_x=0$ with $k^2=2mE/\hbar^2$ which are the solutions to the particle in a one-dimensional box, see Section 10.3.8. Combining the equations gives

$$\varphi=\sqrt{\frac{4}{ab}}\sin\left(\frac{n_x\pi x}{a}\right)\sin\left(\frac{n_y\pi x}{b}\right)$$

and the energies are $E=\frac{\hbar^2}{2m}\pi^2\left(\frac{n_x^2}{a^2}+\frac{n_y^2}{b^2}\right)$, where n_x and n_y are positive integer quantum numbers which must be greater than zero. Notice that in a square box some levels will be degenerate depending upon how the quantum numbers add, for example, $n_x=2$ and $n_x=1$ and vice versa. The first few wavefunctions are shown in Fig. 10.20.

(ii) Particle in a circular well

The pattern of nodes of the radial part of a wavefunction of a particle in a deep circular well is described by a Bessel function. The properties of these functions are well established and look like damped sine and cosine waves. The pattern of vibrations on a circular drum head is similar.

There are two boundary conditions. One ensures that the angular solutions repeat themselves around the circumference and the gradients match at the same point, just as for a particle on a ring. The other ensures that the radial part of the wavefunction is zero at the edge of the disc, just as is the case for a particle in a box at its edges.

The wave equation starts out as (10.41) but for a disc has to be written in plane polar coordinates to be solvable and is

$$-\frac{\hbar^2}{2m}\left(\frac{\partial^2}{\partial r^2}+\frac{1}{r}\frac{\partial}{\partial r}+\frac{1}{r^2}\frac{\partial^2}{\partial\theta^2}\right)\psi+V(r)\psi=E\psi \tag{10.45}$$

and the potential energy V is zero. The disc has a radius of a. (See Section 10.5.9 for the conversion equations to polar coordinates.) With the abbreviation $k^2=2mE/\hbar^2$ (units of k are m^{-1}) the equation can be rewritten as

$$\frac{\partial^2\psi}{\partial r^2}+\frac{1}{r}\frac{\partial\psi}{\partial r}+\frac{1}{r^2}\frac{\partial^2\psi}{\partial\theta^2}+k^2\psi=0 \tag{10.46}$$

Separating variables into r and θ assumes that the solution has the form $\psi=R(r)\varphi(\theta)$. Substituting for ψ, then dividing by $R\varphi$ and multiplying by r^2 gives

$$\frac{r^2}{R}\frac{\partial^2R}{\partial r^2}+\frac{r}{R}\frac{\partial R}{\partial r}+\frac{1}{\varphi}\frac{\partial^2\varphi}{\partial\theta^2}+r^2k^2=0$$

and *each* term must be equal to a constant and is therefore independent of r and θ. The right-hand side of this equation is zero hence if we choose the separation constant to be n^2 this is $-n^2$ for one equation and $+n^2$ for the other. The θ equation is $\dfrac{1}{\varphi}\dfrac{\partial^2\varphi}{\partial\theta^2} = const$ and using the separation constant $-n^2$ this can be written as

$$\frac{\partial^2\varphi}{\partial\theta^2} + n^2\varphi = 0, \tag{10.47}$$

and its solution is given by equation (10.27) and is

$$\varphi = \frac{1}{\sqrt{2\pi}}e^{in\theta}.$$

The angular boundary conditions make the wavefunction φ repeat itself after each $2\pi n$ radians where n has values $n = 0, \pm1, \pm2$, etc.

The radial equation is

$$\frac{r^2}{R}\frac{\partial^2 R}{\partial r^2} + \frac{r}{R}\frac{\partial R}{\partial r} + r^2k^2 = n^2 \tag{10.48}$$

and is more difficult to solve but, with the substitution $x = kr$, has the form of Bessel's equation

$$x^2\frac{d^2R}{dx^2} + x\frac{dR}{dx} + (x^2 - n^2)R = 0$$

which has the solution $J_n(x)$ where J is Bessel's function of the first kind of order n. All Bessel's functions are described by an infinite series in x just as sine and cosine are. See Margenau & Murphy (1943), Abramowicz & Stegun (1965), or Arkfen (1970), for the series. Using Maple to solve the equation produces

```
> eq:= x^2*diff(R(x),x,x)+x*diff(R(x),x)+(x^2-n^2)*R(x);
```

$$eq := x^2\left(\frac{d^2}{dx^2}R(x)\right) + x\left(\frac{d}{dx}R(x)\right) + (x^2 - n^2)R(x)$$

```
> dsolve(eq);
```

$$R(x) = \_C1\ \text{BesselJ}(n, x) + \_C2\ \text{BesselY}(n, x)$$

and the constants will be determined by the boundary conditions. The radial boundary condition is that the wavefunction is zero at the edge of the disc, which means that the Bessel function has to be zero here. Additionally the wavefunction has to be normalized. This latter requirement means that the second constant `_C2` has to be zero because the second Bessel function `BesselY` is $-\infty$ at $r = 0$ and a wavefunction based on this function could not be normalised. We can arbitrarily set `_C1` to N, which we will assume normalizes the wavefunction, and the radial solution is therefore

$$R(r) = NJ_n(kr).$$

The radial boundary condition means that $J_n(ka) = 0$. The Bessel function is repeatedly zero, as are sine and cosine, which it resembles, and $\rho_{n,l} = k_{n,l}a$ is the number where the nth Bessel function $J_n(k_{nl}r)$ crosses zero for the lth time. The wavefunction is

$$R(r) = NJ_n(\rho_{nl}r/a).$$

The values of r_{nl} are tabulated and can also be obtained using the Maple `BesselJZeros(k,n)` instruction. The first few Bessel functions are shown in Fig. 10.21. Except for the first one J_0, they all start at zero as do sine functions.

The shape of the wavefunction at a fixed angle θ has the profile of the Bessel function with the radius set at each of the zero crossings, the lowest energy is found at the first crossing, $\rho_{0,1}$ the next at crossing $\rho_{1,1}$ and so on with increasing r, see Fig. 10.21. The wavefunction for the lowest state ($k = 0, l = 1$) does not have a node before reaching the perimeter, the second ($k = 1, l = 1$) has one node as so on, just as for the particle in a box. Using the definition of k^2 the energy is

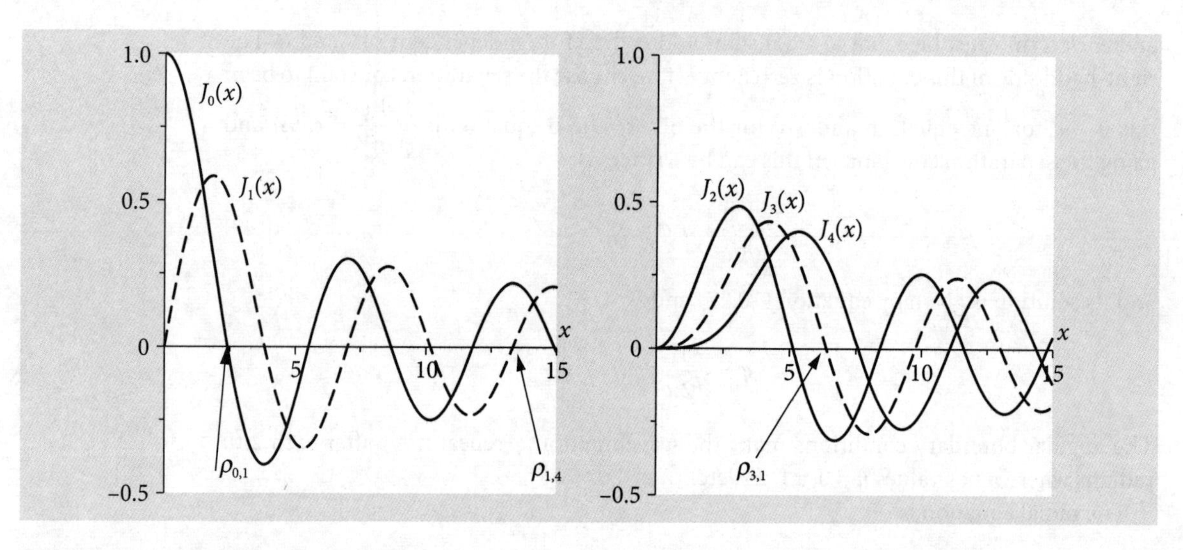

Fig. 10.21 Bessel's functions as a function of $x = kr$.

$$E_{nl} = \frac{\hbar^2}{2m}\left(\frac{\rho_{nl}}{a}\right)^2$$

and the order of the first few energy levels is that of the value of $\rho_{k,n}$. Notice how similar this is to the energy of a particle in a box. The energy levels are ordered with the quantum numbers, n and l, but in this case act through the value ρ, and this is a little different to that of the square well or box where the quantum numbers are included directly in the energy.

n	0	1	2	0	3	1
l	1	1	1	2	1	2
ρ_{kl}	2.40	3.83	5.13	5.52	6.38	7.01

The total wavefunction is

$$\psi = \frac{N}{\sqrt{2\pi}} J_n(\rho_{nl} r/a) e^{in\theta}$$

and as n can take values 0, ±1, ±2, etc. all levels except the first are doubly degenerate. The quantum number n has positive integer values > 0.

Equation (10.46), if slightly modified, describes the shape of the normal modes of a circular drum, and the corresponding equations for a square box those of a square drum. The change to make is $k^2 = \omega^2/c^2$ where $c^2 = \sqrt{T/\sigma}$ and ω is the normal mode's vibrational frequency in rad s^{-1}. If the drum skin were slit a tension of T newton per metre would be needed to keep it closed. The density of the drum skin ρ units of km m^{-2} giving c units of velocity. The normal mode frequencies are $\omega_{nl} = \dfrac{c\rho_{nl}}{a}$ and because the overtones are never an integer multiple of the fundamental, because of the values ρ takes, a drum makes a noise rather than a pure sound.

(iii) Steady state temperature profile in two dimensions

The Laplace equation $\dfrac{\partial^2 V}{\partial x^2} + \dfrac{\partial^2 V}{\partial y^2} + \dfrac{\partial^2 V}{\partial z^2} = 0$ finds application in areas as diverse as steady state heat flow and electrostatics. This equation is so important that it is often abbreviated to $\nabla^2 V = 0$ with the symbol ∇^2 which is called 'del' squared. In two dimensions, $\dfrac{\partial^2 V}{\partial x^2} + \dfrac{\partial^2 V}{\partial y^2} = 0$ and the solution can be obtained in a similar manner to the last example. Separating variables produces

$$\frac{1}{V_x}\frac{\partial^2 V_x}{\partial x^2} + \frac{1}{V_y}\frac{\partial^2 V_y}{\partial y^2} = 0$$

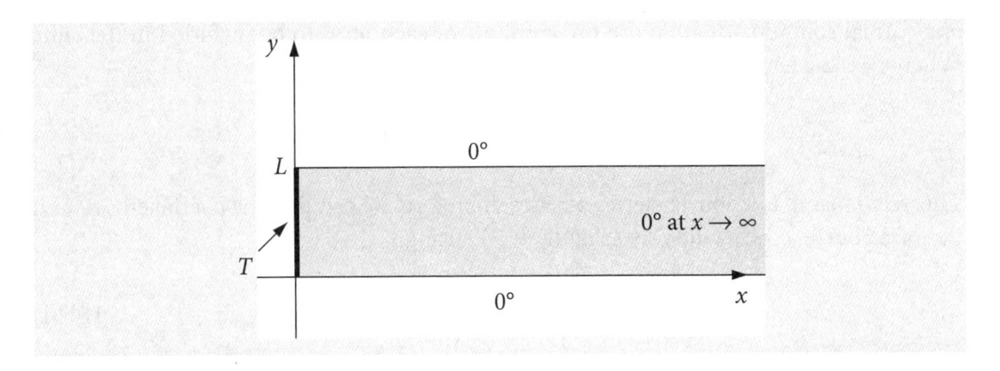

Fig. 10.22 Bar with one end held at temperature T. The sides are held at 0°.

but each derivative must now be equal to the same constant as their sum is zero. Thus,

$$\frac{1}{V_x}\frac{\partial^2 V_x}{\partial x^2} = -\frac{1}{V_y}\frac{\partial^2 V_y}{\partial y^2} = k^2$$

and k^2, where $k \ge 0$, is chosen to make the solutions simpler. The general solutions to these equations are

$$V_x = Ae^{+kx} + Be^{-kx} \qquad \text{and} \qquad V_y = C\sin(ky) + D\cos(ky)$$

and the final result is the product V_xV_y with linear combinations of these solutions because the differential equation is linear and as such any linear combinations of solutions is also a solution. The next step is to find solutions that satisfy the initial or boundary conditions and these are defined by the problem being solved.

The temperature profile of a bar will be calculated. This has a width is L and a length that is so long as to be effectively infinite. The sides of the bar are kept at zero degrees and the end constantly heated to T degrees; for generality, this temperature profile is called $f(y)$. The sketch, Fig. 10.22 shows the situation.

The boundary conditions impose the constraints

$$\begin{aligned} y&=0, \quad V=0, \qquad & y&=L, \quad V=0, \\ x&=0, \quad V=f(y), \qquad & x&=\infty, \quad V=0 \end{aligned}$$

The value $f(y)$ is some normal function that specifies only how V varies with position y. The four solutions are written without the integration constants because these are realized when the boundary conditions are applied and are

$$\begin{aligned} V &= e^{+kx}\sin(ky) \\ &= e^{-kx}\sin(ky) \\ &= e^{+kx}\cos(ky) \\ &= e^{-kx}\cos(ky) \end{aligned} \tag{10.49}$$

When $x = \infty$, the solution cannot contain e^{+kx} because the temperature is zero not infinity, so $e^{+kx}\sin(ky)$ and $e^{+kx}\cos(ky)$ are not solutions. At $y = 0$, the temperature is zero and so the solution cannot be $e^{-kx}\cos(ky)$ because $\cos(0)$ is not zero. This leaves the solution as

$$V = e^{-kx}\sin(ky)$$

and k is to be determined. When $y = L$ the temperature is zero and $\sin(kL) = 0$, which it will be for all $k = n\pi/L$ if n is an integer. Thus, the general solution is

$$V = e^{-n\pi x/L}\sin(n\pi y/L).$$

However, this does not satisfy the condition that at $x = 0$ the temperature is $f(y)$ which could be some constant value, say T. To obtain a solution, a sum of all possible solutions is tried and this produces

$$V = \sum_{n=1}^{\infty} b_n e^{-n\pi x/L}\sin(n\pi y/L), \tag{10.50}$$

as a formal solution where b_n are the amounts of each term to be included in the sum. When $x = 0$ the temperature is $f(y)$, thus,

$$f(y) = \sum_{n=1}^{\infty} b_n \sin(n\pi y/L)$$

This relationship is a Fourier series of sines, therefore, we can find the coefficients a_n as if they are Fourier coefficients (see Chapter 9.1.7) using

$$b_n = \frac{2}{L}\int_0^L f(y)\sin\left(\frac{n\pi y}{L}\right)dy \tag{10.51}$$

If, for example, the temperature function $f(y) = T$ and $L = \pi$, then the coefficients are

$$b_n = \frac{2T}{\pi}\int_0^{\pi} \sin(n\pi y)dy,$$

which evaluate to $\frac{4T}{\pi}, \frac{4}{\pi}\frac{T}{3}, \frac{4}{\pi}\frac{T}{5}, \cdots$ and the temperature profile is

$$V = \frac{4T}{\pi}\left(e^{-x}\sin(y) + e^{-3x}\frac{\sin(3y)}{3} + e^{-5x}\frac{\sin(5y)}{5} + \cdots\right)$$

The graph shows the temperature profile plotted as contours and the Maple used in the calculation was modified from Algorithm 9.1. The left-hand figure shows the calculation for a constant temperature $T = 10$ along the y-axis. In the right hand figure, the temperature varies as $10e^{-y}$ so it is greater near to the origin. The way the heat spreads out is consistent with this.

Algorithm 10.1 Steady state diffusion

```
> f:= y-> 10;                                    # a victim function
  L:= Pi:                                        # range
  b:= n-> 2/L*int(f(y)*sin(n*Pi*y/L),y=0..L);
# next calculate series make function V
  V:= (x,y, maxn )->
     add( b(n)*exp(-n*Pi*x/L)*sin(n*Pi*y/L),n= 1..maxn);
  plot3d(10-V(x,y,50),x=0..5,y=0..L,style=surfacecontour);
```

Finally, note that the flow of heat has, mathematically, the same form as molecular diffusion when the appropriate changes in variables are made. Thus, the contours in Fig. 10.23 (left figure) could be those of concentration at a steady state if the concentration along the top and bottom is held at zero and held constant at $x = 0$.

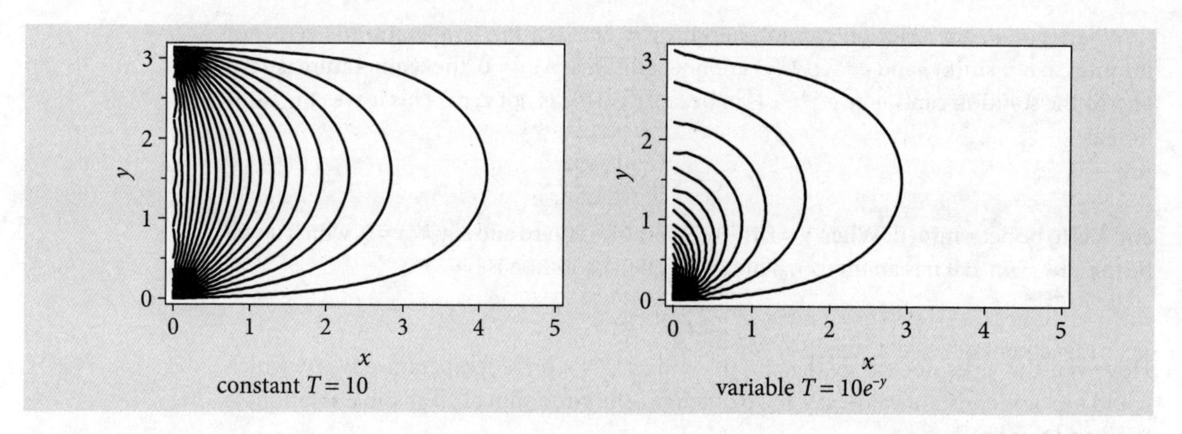

Fig. 10.23 Steady state temperature contours with a temperature of zero along the top and bottom of a plate and 10 and $10e^{-y}$ held along the y-axis. The plate has a height of π for simplicity of the calculation and the temperatures decrease to the right as the temperature is held at zero at large x.

(iv) One-dimensional diffusion

To find the time profile of the concentration of a solute in solution, the diffusion equation, or Fick's second law, is required. Imagine that a partition separates a solution from pure solvent. When the partition is removed, diffusion of solute into the solvent occurs and vice versa. At any given place the concentration is changing as time progresses. At any given time, there is a concentration profile along the whole sample. Fick's first law states that the flux is proportional to the negative gradient of concentration with distance, multiplied by the diffusion constant, or,

$$J = -D\frac{\partial c}{\partial x}.$$

The minus sign ensures that the flux of molecules is from a high concentration to a lower one. However, the flux is also the gradient of the concentration with time, and combining this with Fick's first law produces the second law, which relates the time profile of the concentration to the distance diffused, and is

$$\frac{\partial c}{\partial t} = D\frac{\partial^2 c}{\partial x^2},$$

where t is time, x distance, and c concentration. D is the diffusion coefficient in units of m^2s^{-1}. Typical values are $\approx 10^{-9}$ for a small molecule in water, to $\approx 10^{-11}$ m^2s^{-1} or less in a lipid membrane.

The equation is solved by separating variables, assuming $c = c_x c_t$ and then

$$\frac{1}{Dc_t}\frac{\partial c_t}{\partial t} = \frac{1}{c_x}\frac{\partial^2 c_x}{\partial x^2}.$$

As each term must be constant, such as $-k^2$, two equations are produced

$$\frac{\partial c_t}{\partial t} = -Dk^2 c_t \quad \text{and} \quad \frac{\partial^2 c_x}{\partial x^2} = -c_x k^2. \tag{10.52}$$

The time dependence has the solution $c_t = e^{-Dk^2t}$ and the spatial dependence is $c_x = \sin(kx)$ and $c_x = \cos(kx)$; by hypothesis the solutions are either

$$c = e^{-Dk^2t}\sin(kx) \quad \text{and} \quad c = e^{-Dk^2t}\cos(kx). \tag{10.53}$$

or the sum of both terms.

$$c = e^{-Dk^2t}[A\sin(kx) + B\cos(kx)] \tag{10.54}$$

with A and B as arbitrary constants that are found from the initial and boundary conditions. The constant is chosen to be $-k^2$, and is negative because the concentration might fall, but could not increase to infinity, which it would do if $+k^2$ were used. The constant is squared only to simplify the resulting equations.

The actual solution is determined by the initial and boundary conditions for the problem at hand. Consider, for example, the diffusion out of a slab of thickness L with the faces being kept at a concentration of zero for $t > 0$ and initially the concentration in the slab is $f(x)$ but which would normally be constant. The boundary conditions are:

$$t = 0;\ c = f(x) \text{ and } 0 < x < L.$$
$$t > 0;\ c = 0 \text{ at } x = 0 \text{ and } c = 0 \text{ at } x = L.$$

Considering the boundary conditions, as for the previous example, the solution is based on $c = e^{-Dk^2t}\sin(kx)$ with $k = n\pi/L$, and the general form of the solution is based on equation (10.50) and is

$$c = \sum_{n=1}^{\infty} b_n e^{-Dn^2\pi^2t/L^2}\sin(n\pi x/L); \qquad b_n = \frac{2}{L}\int_0^L f(x)\sin\left(\frac{n\pi x}{L}\right)dx \tag{10.55}$$

If the initial concentration in the slab is a constant, then $f(x) = c_0$ and the solution is

$$b_n = \frac{2c_0}{L}\int_0^L \sin\left(\frac{n\pi x}{L}\right)dx = \frac{2c_0}{n\pi}(1-(-1)^n)$$

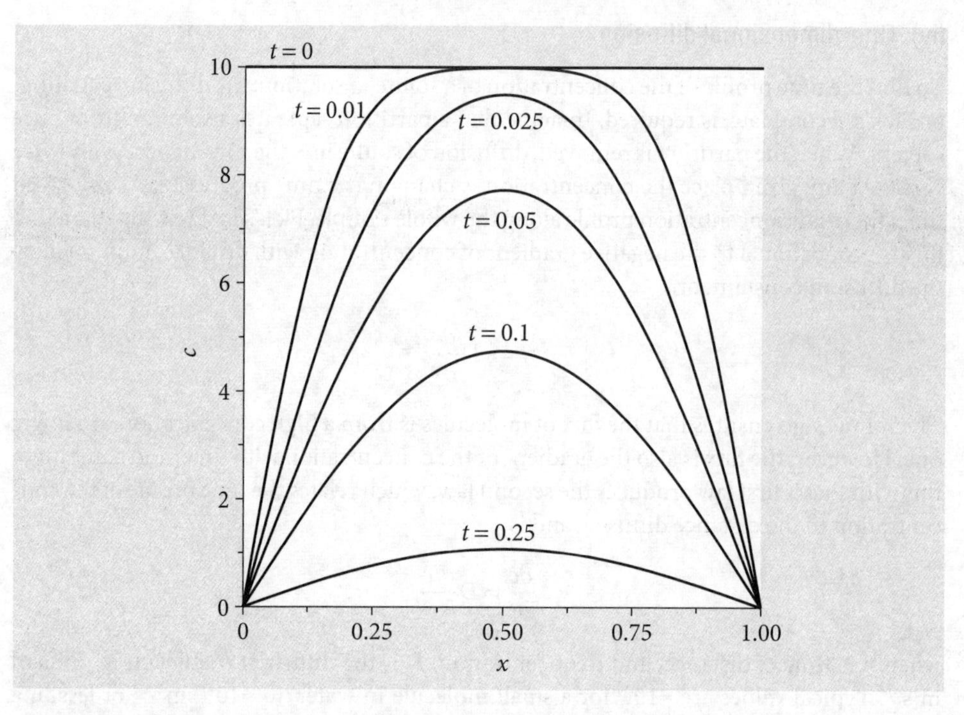

Fig. 10.24 Concentration profiles vs position at different times inside the slab. Molecules are diffusing out and the edges are kept at zero concentration for example, by washing material away. The diffusion coefficient is 1, the initial concentration 10.

which means that when n is even, b_n is zero. The concentration profile becomes

$$c = \frac{2c_0}{\pi}\sum_{n=1}^{\infty}(1-(-1)^n)\frac{e^{-Dn^2\pi^2 t/L^2}}{n}\sin(n\pi x/L) \tag{10.56}$$

(The change $n = 2\upsilon + 1$ will simplify this series which should then start at $\upsilon = 0$). Modifying the equations in Algorithm 10.1, the concentration profile at different times is produced and plotted in Fig. 10.24.

(v) Diffusion in a closed tube or isolated bar

If the ends of a long tube are filled with solvent and then closed and some solute injected, diffusion will ensure that equilibrium will eventually be reached. Similarly, if the end of an otherwise insulated bar is heated for a short while, as heat cannot escape, a uniform temperature will be reached. From these results, we know that, because the initial concentration or temperature profile levels out to a constant value, the solution must have both a time-dependent and a constant part. At long times the temperature or concentration will become uniform and, if $f(x)$ is the amount initially added and $0 < x < L$, the long time value is $\displaystyle\frac{1}{L}\int_0^L f(x)dx$. The initial condition is given by the shape of the concentration profile $f(x)$.

If $c(x, t)$ is the concentration at position x and time t, then, because no heat or material leaves, the concentration gradients at the ends of the tube are zero at all times, or

$$\frac{c(0,t)}{dx} = \frac{c(L,t)}{dx} = 0, \qquad t > 0.$$

Starting with the diffusion equation $\dfrac{\partial c}{\partial t} = D\dfrac{\partial^2 c}{\partial x^2}$ and separating variables as before gives $\dfrac{\partial c_t}{\partial t} = -Dk^2 c_t$ and $\dfrac{\partial^2 c_x}{\partial x^2} = -c_x k^2$. The solution is $c = c_t c_x$.

The time-dependent equation integrates to $c_t = e^{-Dk^2 t}$, but the spatial part needs special attention to incorporate the boundary conditions. The separation parameter k can have values of 0, or it can be positive or positive and imaginary, i.e. ik. In this case $-(ik)^2 \to k^2$

and integrating produces the exponential solution $c_x = ae^{kx} + be^{-kx}$, however the boundary conditions are then only met when $a = b = 0$ which is not a useful result.

When $k = 0$ the equation to integrate is $\dfrac{\partial^2 c_x}{\partial x^2} = 0$ and the solution is $c_x = ax + b$ and the final solution $c_t c_x = ax + b$, because $k = 0$ and the exponential term is unity. The boundary conditions are only met when $a = 0$ but b is undefined so this solution cannot be complete.

Finally, when k is positive the solutions are $c_x = a\sin(kx) + b\cos(kx)$. The gradient boundary condition ensures that the solution is the cosine term because the derivative, $kb\sin(kx)$, is zero at $x = 0$ and at L provided that $k = n\pi/L$ where n is an integer. To make the next equations clearer k is used as if it were an integer and its value only substituted at the end. The solution is

$$c_k = e^{-k^2Dt}\cos(kx), \qquad k = 0, 1, 2, \cdots$$

and the final solution is a linear combination of the k terms including $k = 0$ and this will give a term that is independent of time. The initial condition ensures that

$$f(x) = \sum_{n=0}^{\infty} b_k \cos(kx),$$

giving

$$b_k = \frac{2}{L}\int_0^L f(x)\cos(kx)dx,$$

where $k = n\pi/L$. The complete solution is

$$c = \frac{b_0}{2} + \sum_{n=1}^{\infty} b_k e^{-(n\pi/L)^2Dt}\cos(n\pi x/L)$$

The first term $b_0/2$ is divided by 2 to make it equal to the average concentration at long times, $\dfrac{1}{L}\displaystyle\int_0^L f(x)dx$, because by its defining equation b_0 is otherwise 2 times too large. The results show how the initial profile eventually reaches a constant, time-independent value, the total amount of material or heat being conserved.

If the time-dependent Schrödinger equation is being solved, the diffusion coefficient is replaced by $D \to i\hbar/2m$ and the wavefunction is calculated which may be a complex quantity, but whether it is or not, $\psi^*\psi$ would normally be plotted as this is the probability.

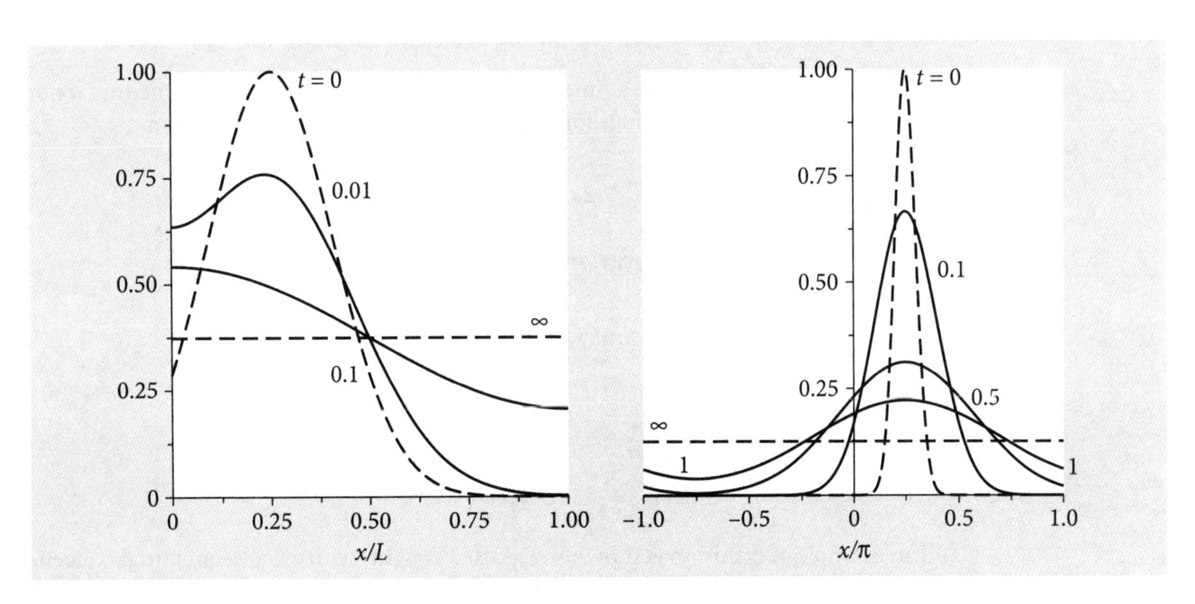

Fig. 10.25 Left: One-dimensional diffusion in a closed tube. Right: Diffusion on a ring or circle. The initial concentration profile (dashed) is the Gaussian $e^{-20(x-L/4)^2}$, the diffusion coefficient $D = 1$. The profile at different times is shown. A uniform profile is reached when $t > 4$, times which are effectively infinity.

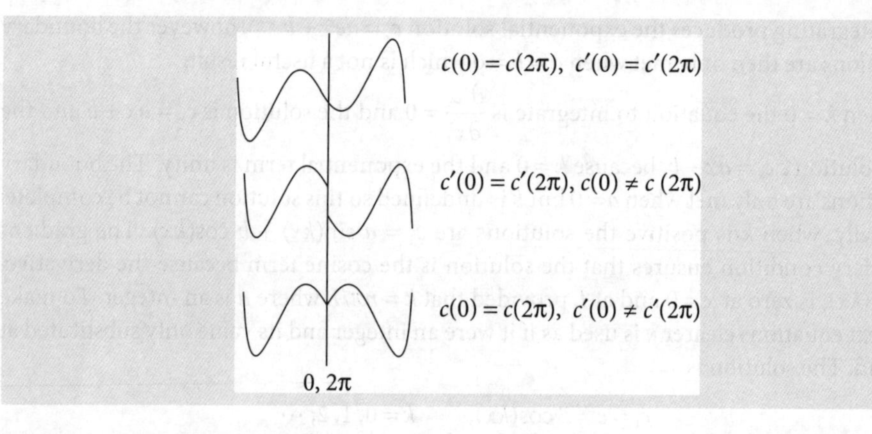

Fig. 10.26 Periodic boundary conditions. Only in the top curve are both conditions simultaneously satisfied.

(vi) Periodic boundary conditions

When diffusion is on a circle, the boundary conditions are periodic. The effect is similar to the insulated bar in that an initial concentration profile becomes constant at long times. If the concentration is initially $f(\theta)$, $-\pi < \theta < \pi$, the amount initially present at long times is $\dfrac{1}{\pi}\displaystyle\int_{-\pi}^{\pi} f(\theta)d\theta$. The boundary conditions ensure that the concentration and its gradient are the same after a period,

$$c(0, t) = c(2\pi,t) \qquad c'(0, t) = c'(2\pi, t) \qquad t > 0.$$

These conditions are most easily illustrated with a figure (Fig. 10.26). Only when both conditions are met is the solution continuous at the boundary.

Following the derivation (10.52) or that for the insulated bar, the general solution is found when k is zero or positive. Rather than changing the notation x is used to represent the angle that varies from $-\pi$ to π. The general solution is

$$c_x = a\sin(kx) + b\cos(kx)$$

and the boundary conditions are now applied. To make the concentration the same when $x = -\pi$ and π, and with $k = 0, 1, 2, \cdots$

$$a\sin(-\pi k) + b\cos(-\pi k) = a\sin(\pi k) + b\cos(\pi k)$$

and the derivatives

$$-a\cos(-\pi k) + b\sin(-\pi k) = a\cos(\pi k) - b\sin(\pi k).$$

Because the sine terms are zero at integer values of π radians, it follows that neither a nor b is zero. Therefore, the solution is found by expanding the complete solution

$$c(x) = \sum_{k=0}^{\infty}[a_k\sin(kx) + b_k\cos(kx)]$$

with the initial condition $f(x)$ providing the coefficients as

$$a_k = \frac{1}{\pi}\int_{-\pi}^{\pi} f(x)\sin(kx)dx \qquad b_k = \frac{1}{\pi}\int_{-\pi}^{\pi} f(x)\cos(kx)dx.$$

The complete solution is

$$c = \frac{b_0}{2} + \sum_{n=1}^{\infty} e^{-n^2Dt}[a_k\sin(nx) + b_k\cos(nx)].$$

and an example of diffusion is shown in Fig. 10.25 (right) with a Gaussian initial concentration profile. Notice how the concentration matches both in value and slope at $\pm\pi$ as required by the boundary conditions.

The X-ray structure of the chlorophylls in the photosynthetic antenna protein LHCII shows that they are arranged in the form of a ring. These chlorophylls are all coupled to one another and when electronically excited by a femtosecond laser pulse a localized wavepacket is produced, which, because of interactions with the surrounding solvent molecules and protein residues, will then spread its energy among the pigments' excitonic energy levels. In this case, the time-dependent Schrödinger equation, rather than the diffusion equation, is used. If we assume that the energy levels are close enough to be considered continuous, the wavepacket's motion as it relaxes around the ring of molecules would have behaviour reminiscent of diffusion as shown in Fig. 10.25.

(vii) Diffusion across an interface and from a point

Imagine a very long, narrow tube is half filled with a dye solution and the other half contains pure solvent. A partition between the two halves is removed and diffusion from either half into the other begins. The tube is so long that its ends do not affect the behaviour on an experimental time scale. Immediately after opening the partition the concentration at the interface is $c(x) = c_0$ for $x \ge 0$ and $c(x) = 0$ for $x < 0$ both at $t = 0$. This initial condition is quite specific, and practical, but instead of keeping the value c_0 when $x \ge 0$ and $t = 0$ it can generally be a function of position, $f(x)$ over the whole range of x.

To obtain a solution the equation (10.52), $\dfrac{\partial^2 c_x}{\partial x^2} = -c_x k^2$, is solved in a general way using exponentials as

$$c_x = \alpha_k e^{ikx} + \beta_k e^{-ikx} \tag{10.57}$$

where α_k and β_k are constants determined by the initial and boundary conditions. This solution may generally also be taken as

$$c_x = b_k e^{-ikx},$$

if k is allowed to be positive or negative, the general solution can be written as the product of the time and spatial solutions and integrated,

$$c = \int_{-\infty}^{\infty} b_k e^{-Dk^2t - ikx} dk. \tag{10.58}$$

The change into an integral, compared with the sum previously used (10.55), is made because it is assumed that so many terms are needed in the summation it is equivalent to an integral. Using the initial condition that at $t = 0$ the concentration is $f(x)$, equation (10.58) produces

$$f(x) = \int_{-\infty}^{\infty} b_k e^{-ikx} dk,$$

and the constants d_k are found with the Fourier transform

$$b_k = \frac{1}{2\pi}\int_{-\infty}^{\infty} f(x')e^{ikx'} dx'.$$

The complete solution becomes

$$c = \frac{1}{2\pi}\int_{-\infty}^{\infty}\int_{-\infty}^{\infty} f(s)e^{-Dk^2t - ik(s-x)} ds dk$$

and s is used as a dummy variable to perform the integration. Solving the integral in k, keeping t constant, gives the result for the concentration of the dye at position x and time t,

$$c(x, t) = \frac{1}{\sqrt{4\pi Dt}}\int_{-\infty}^{\infty} f(s)e^{-\frac{(s-x)^2}{4Dt}} ds \tag{10.59}$$

where $f(s)$ is the initial concentration profile. At zero time, this equation reduces to $f(x)$, the initial profile.

We started with a step function in concentration as the initial condition. To use this, $f(x) = c_0$ when $x \ge 0$ and zero otherwise, and the last equation becomes

$$c(x,t) = \frac{c_0}{\sqrt{4\pi Dt}}\int_{s=0}^{\infty} e^{-\frac{(s-x)^2}{4Dt}}\,ds, \tag{10.60}$$

because the function $f(x)$ is zero at $x < 0$ but no other conditions apply when $t > 0$. This integral cannot be evaluated in terms of a finite number of functions and must be calculated numerically as the error function. The error function is defined as $erf(x) = \dfrac{2}{\sqrt{\pi}}\displaystyle\int_0^x e^{-s^2}ds$ and has values only between 0 and 1; it is the area underneath fractions of the normalized Gaussian or error curve. The complementary error function is

$$erfc(x) = 1 - erf(x) = \frac{2}{\sqrt{\pi}}\int_x^{\infty} e^{-s^2}ds$$

which with a transformation of variables has the form of our result. Notice that x appears as the argument of the function and as a *limit* in the integration. Letting $z = (s - x)/\sqrt{4Dt}$ then $dz = ds/\sqrt{4Dt}$

$$\begin{aligned} c(x,t) &= \frac{c_0}{\sqrt{4\pi Dt}}\int_{s=0}^{\infty} e^{-\frac{(s-x)^2}{4Dt}}\,ds \\ &= \frac{c_0}{\sqrt{\pi}}\int_{-x/\sqrt{4Dt}}^{\infty} e^{-z^2}dz \\ &= \frac{c_0}{2}erfc\left(-\frac{x}{\sqrt{4Dt}}\right) \end{aligned}$$

Some concentration profiles are shown in the Fig. 10.27 at different values of the product $\sqrt{(4Dt)}$, which has units of distance. The mean distance diffused in one dimension in time t is $\sqrt{(2Dt)}$. A typical diffusion coefficient in a normal solvent is $\approx 10^{-9}$ m^2 s^{-1}; if $\sqrt{(4Dt)} = 25$ cm then $t \approx 181$ days. Diffusion is a very slow process indeed.

If the 'source term' $f(x)$ is a delta function $\delta(x)$, then all the material is piled up at $x = 0$ in the tube and the integral reduces to

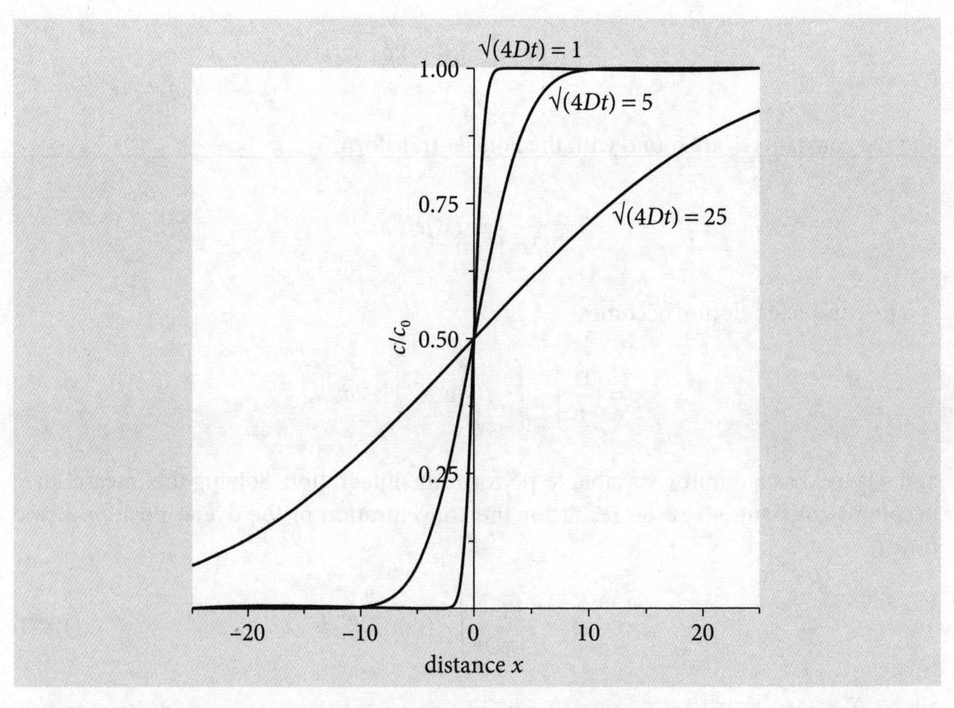

Fig. 10.27 Diffusion from after a step change in concentration. The shape of each curve is the same.

$$c(x,t)=\frac{1}{2\sqrt{\pi Dt}}\int_{-\infty}^{\infty}\delta(s)e^{-\frac{(s-x)^2}{4Dt}}ds=\frac{1}{2\sqrt{\pi Dt}}e^{-\frac{x^2}{4Dt}}$$

because $\delta(0)=1$ and is otherwise zero. This function is a Gaussian or bell-shaped curve as material spreads out from the initial point on the line; see Chapter 12.3.3.

The general solution (10.59) can now be thought of as the convolution of a function $f(x)$ with a impulse δ function in much the same way as described in Chapter 9.7 where fluorescence, equivalent to $f(x)$, is stimulated by a laser pulse which is equivalent to the exponential term in (10.59).

One important application of equations similar to, but more complex than those described here is in the technique known as FRAP, which stands for fluorescence recovery after photo-bleaching. In this technique, a dye molecule is introduced into a cell and then an intense laser is used to bleach the dye rapidly in the small circular spot of the focused laser. A second weaker laser stimulates the dye's fluorescence and the intensity of this is measured as the dye molecules diffuse into the bleached area. The rate at which the fluorescence recovers can be used to determine the diffusion coefficient of the dye molecules or proteins to which the dye may be attached. Should the protein contains an intrinsic chromophore, as does the green fluorescent protein (GFP), this can be used directly.

(viii) Asymmetric boundary conditions: reaction at a plane

In some problems, asymmetric boundary conditions apply. One such case is a (semi) infinite tube that is closed at one end and filled with a solution at a concentration of c_0. Diffusion is caused by the fast removal of material at the closed end, for instance by an electrode reaction that causes precipitation, see Fig. 10.28. The concentration at t and x is found using the initial conditions

$$t=0,\ x>0,\ c=c_0,$$

which means that initially the concentration is c_0 everywhere, and the boundary conditions are

$$t=0,\ x\rightarrow\infty,\ c=c_0,\qquad\text{and}\qquad t>0,\ x=0,\ c=0.$$

Starting with the diffusion equation $\dfrac{\partial c}{\partial t}=D\dfrac{\partial^2 c}{\partial x^2}$, the variables are separated and the general solution obtained as equation (10.57)

$$c(x,t)=\int_{-\infty}^{\infty}\alpha_k e^{-Dk^2t-ikx}+\beta_k e^{-Dk^2t+ikx}dk.$$

The general solution has this form because there are both initial and boundary conditions to satisfy. The initial condition is that at $t=0$ the concentration is c_0 when $x>0$ and this gives $c_0=\int_{-\infty}^{\infty}\alpha_k e^{-ikx}+\beta_k e^{+ikx}dk$. The boundary condition, $c=0$ at all times at $x=0$ is used to find the constants α_k and β_k. These conditions produce $0=\int_{-\infty}^{\infty}\alpha_k e^{-Dk^2t}+\beta_k e^{-Dk^2t}dk$ and since t is a constant in this expression, the exponentials cancel. It then follows that $\alpha_k=-\beta_k$ and $c_0=\alpha_k\int_{-\infty}^{\infty}e^{-ikx}-e^{+ikx}dk$. The constants α_k can be found by Fourier transforming this equation to give

$$\alpha_k=\frac{c_0}{2\pi}\int_{-\infty}^{\infty}e^{+ikx}-e^{-ikx}dx.$$

Replacing α_k into the general solution and changing the integration variable in distance from x to s for clarity gives

$$c(x,t)=\frac{c_0}{2\pi}\int_{-\infty}^{\infty}\int_{-\infty}^{\infty}e^{-Dk^2t+ik(s-x)}-e^{-Dk^2t-ik(s+x)}dsdk$$

Using Maple to do this integration produces

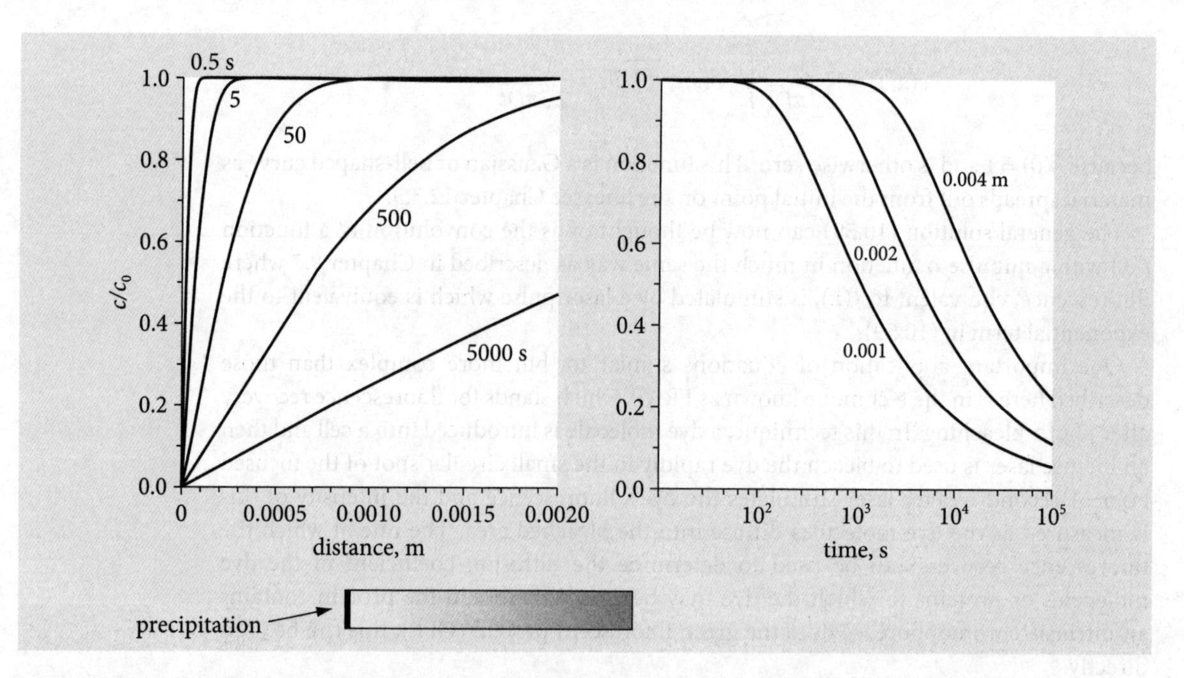

Fig. 10.28 Diffusion profiles where the concentration is zero at $x = 0$, such as may pertain on an electrode. Left: Concentration vs distance at different times. Right: Concentration vs time at different distances. Note the logarithmic time scale. The diffusion coefficient used is $D = 10^{-9}$ m² s.

```
> assume(Dt > 0):
  c0/(2*Pi)* Int( Int(
          exp(-Dt*k^2-I*k*(s-x)) - exp(-Dt*k^2+I*k*(s+x) ),
                   k= -infinity..infinity), s = 0..infinity):
  %= simplify(value(%));
```

$$\frac{1}{2}\frac{c0\left(\int_0^\infty\int_{-\infty}^\infty (e^{-Dtk^2-Ik(s-x)} - e^{-Dtk^2+Ik(s+x)})dkds\right)}{\pi} = c0\ \mathrm{erf}\left(\frac{1}{2}\frac{x}{\sqrt{Dt}}\right)$$

or

$$c(x, t)/c_0 = erf(x/\sqrt{4Dt}).$$

The concentration vs x is plotted in Fig. 10.28 at several times with $D = 10^{-9}$ m² s^{-1} and vs time at several distances. The flux, which is the concentration gradient at any point x at time t, is $\dfrac{\partial c}{\partial x} = \dfrac{c_0}{\sqrt{\pi Dt}}e^{-x^2/(4Dt)}$ and this will tend to zero at $t \to \infty$.

Diffusion to a plane can occur in an electrochemical cell. Initially, suppose that the concentration of oxidant $[Ox]$ is c_o everywhere, and reductant $[R]$ is zero. The potential is then suddenly changed and the reaction starts. The oxidant begins to be reduced at the electrode. The concentration vs distance is shown in Fig. 10.28 (left) and as time increases the amount of oxidant decreases away from the electrode. For example, after 5000 s it is only about 50% of its initial value, 2 mm from the electrode. The reductant concentration shows the opposite trend since at all times oxidant + reductant = c_0. This argument assumes that the oxidant and reductant have the same diffusion coefficient. If the reaction $Ox + ne \rightleftarrows R$ is reversible, then Nernst's equation applies at the electrode surface and is

$$E = E^{\circ} - \frac{RT}{nF}\ln\left(\frac{R}{Ox}\right)$$

where F is the Faraday constant and n the number of electrons transferred in the reaction. For different values of $\Delta E = E - E^{\circ}$ the ratio $s = Ox/R$ can be plotted vs distance from the electrodes. The expression $f_R = (1 - c(x, t))/(1 + s)$ plots the fraction as reductant and $f_{Ox} = (s + c(x, t))/(1 + s)$ the fraction oxidant, their sum being one. At oxidative, or positive, potentials the ratio Ox/R is large, and the electrode remains almost in equilibrium with the Ox concentration, being almost c_0. If the potential is negative, $Ox/R < 1$, and almost all of

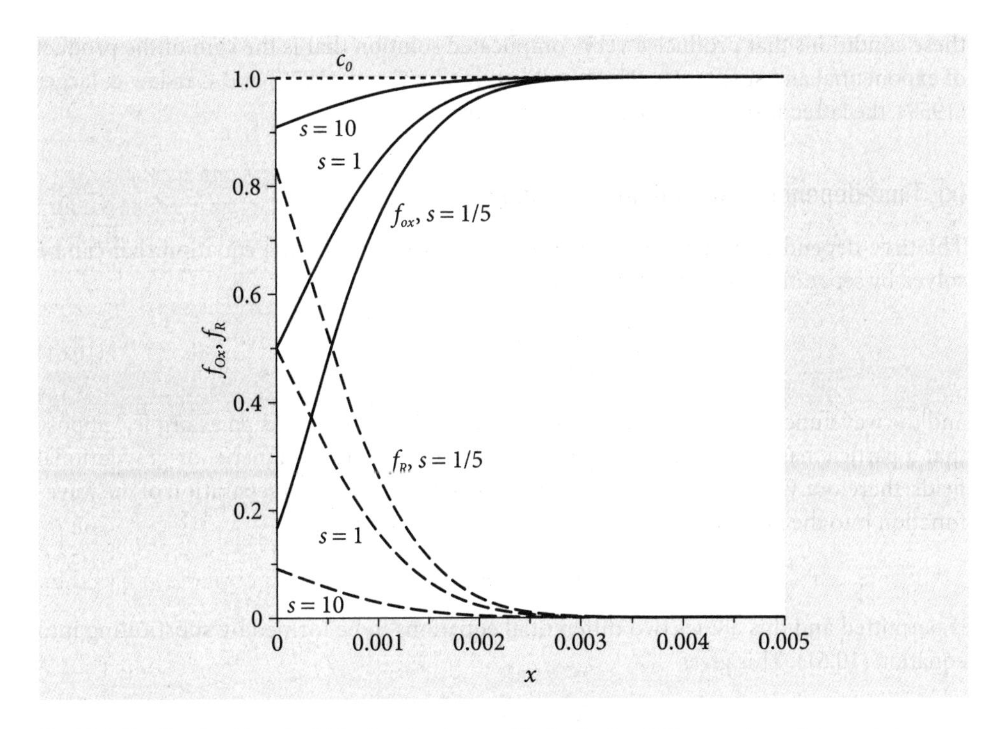

Fig. 10.29 Profile of oxidant (solid line) and reductant (dashed) after 500 seconds vs distance (m) away from an electrode at different ratios of $s = Ox/R$, hence different potentials. The potentials are $\Delta E = 0$, $s = 1$, $\Delta E = 59.5$ mV, $s = 10$, and $\Delta E = -41.6$ mV, $s = 1/5$, for $n = 1$ and at 300 K. At longer times, the curves reach zero or c_0 at larger x.

the oxidant in contact with the electrode reacts, eventually the region of reduced material expands away from the electrode, Fig. 10.29. The current measured in the electrochemical cell is initially large because the slope of concentration with distance is large, this is the flux, and the current is proportional to the flux. At longer times, the slope is smaller at the electrode, hence the flux is reduced and so is the current.

(ix) Diffusion into or out of a sphere

When a substance diffuses into or out of a uniform sphere, the angular coordinates can normally be ignored as diffusion can occur equally well from any place on the sphere's surface. Only the radial coordinate is important and then the diffusion equation is

$$\frac{\partial c}{\partial t} = D\left(\frac{\partial^2 c}{\partial r^2} + \frac{2}{r}\frac{\partial c}{\partial r}\right)$$

where c is concentration and r is the radial distance from the centre. Sometimes this equation is written as $\dfrac{\partial c}{\partial t} = \dfrac{1}{r^2}\dfrac{\partial}{\partial r}\left(Dr^2\dfrac{\partial c}{\partial r}\right)$. The equation appears to be one of great complexity but can be solved with the substitution $u = cr$, which produces the familiar diffusion equation

$$\frac{\partial u}{\partial r} = D\frac{\partial^2 u}{\partial r^2},$$

which is solved by separating variables and applying the boundary and initial conditions in the normal way. If the sphere has a radius R, then $0 < r \le R$, and we will assume that the concentration at the sphere's surface is always held constant at c_s. The boundary conditions are

$$t > 0, \quad u = 0 \text{ at } r = 0, \quad \text{and} \quad u = Rc_s \text{ at } r = R.$$

and the initial condition is $t = 0$, $u = rf(r)$, where $f(r)$ is a function describing the initial concentration profile inside the sphere. Often this is taken to be constant. This problem is now very similar to that of equation (10.54), but the boundary conditions are slightly different, one condition is that the concentration is zero the other that it is a constant. It is these conditions that produce a very complicated solution that is the sum of the product of exponential and sine terms. The result is given by Crank (1979) and Carslaw & Jaeger (1959), the latter also giving a detailed derivation.

(x) Time-dependent Schrödinger equation

The time-dependent Schrödinger equation is a partial differential equation that can be solved by separating the variables. The equation is

$$-\frac{\hbar^2}{2m}\frac{\partial^2 \psi}{\partial x^2} + V(x)\psi = i\hbar\frac{\partial \psi}{\partial t} \tag{10.61}$$

and the wavefunction is a function of both time t and position x. As an example, suppose that a particle has kinetic energy E but is free of any electric, magnetic, or gravitational fields; therefore $V(x) = 0$. There is no expression in both x and t, so separation of the wavefunction into the product

$$\psi(x, t) = \varphi(x)u(t)$$

is permitted and this allows two differential equations to be formed by substituting into equation (10.61). This gives

$$-\frac{\hbar^2}{2m}\frac{1}{\varphi}\frac{\partial^2 \varphi}{\partial x^2} = i\hbar\frac{1}{u}\frac{\partial u}{\partial t}$$

and both terms must be equal to the same constant. The left-hand side of this equation is the Hamiltonian operator for the kinetic energy only, as $V = 0$, and this is equal the energy $E = \hbar\omega$. The energy equation becomes

$$\frac{\partial^2 \varphi}{\partial x^2} + \frac{2m\omega}{\hbar}\varphi = 0, \qquad \text{or} \qquad \frac{\partial^2 \varphi}{\partial x^2} + k^2\varphi = 0$$

where k is a positive constant, $k^2 = \dfrac{2m\omega}{\hbar}$. The solution to this equation is the standard result

$$\varphi = Ae^{ikx} + Be^{-ikx}$$

where A and B are constants depending on the initial or boundary conditions.

The time-dependent part is solved with

$$\frac{i\hbar}{u}\frac{\partial u}{\partial t} = \hbar\omega \qquad \text{or} \qquad \frac{\partial u}{\partial t} + i\omega u = 0,$$

from which $u(t) = e^{-i\omega t}$. The constants of integration need not be added because these have been included in φ. The complete wavefunction is therefore

$$\psi = Ae^{i(kx-\omega t)} + Be^{-i(kx+\omega t)}$$

which is the equation of two plane waves travelling in opposite directions. The wave's phase velocity is $v = \omega/k$. The probability of being position x is

$$\psi^*\psi = |A|^2 + |B|^2 + AB^*e^{2ikx} + BA^*e^{-2ikx}$$

which demonstrates that the particle shows interference and is thus localized periodically. If the particle does not have any boundary condition imposed on it either A or B be can arbitrarily set to zero and then the particle travels either to the right or left and the wavefunction is $\psi = Ce^{i(kx-\omega t)}$, where $\psi^*\psi = |C|^2$ and $\sqrt{C}$ is the normalization.

The particle's energy is $E = \hbar\omega$ or $E = \dfrac{(\hbar k)^2}{2m}$, because momentum is $p = \hbar k$ and we have defined $k^2 = 2m\omega/\hbar$. The classical velocity of the particle is $v = p/m$ or $v = \hbar k/m$ and this is identical to the group velocity, $v_g = d\omega/dk$. The phase velocity is, by definition, $v_p = \omega/k$, and this velocity is also E/p because kinetic energy $E = p^2/2m$. The phase velocity is therefore $v/2$ and half the classical and group velocity. See Flugge (1999) for more details of this problem and many other interesting one-dimensional problems.

10.9.3 **The wave equation**

A travelling wave can be imagined as a disturbance that moves both in space, with a speed $|c|$, and in time. Its general form is $u(x, t) = f(x - ct)$ where f is a function of one variable (Knobel 2000). The profile f of the travelling wave remains the same as time passes; one of many travelling waves is $y = e^{-(x-ct)^2}$, which will move with a speed c.

A standing wave varies in time and, although it appears to be fixed in space, it is the sum of two travelling waves moving in opposite directions. A taut wire when plucked will produce waves with the displacement depending upon its position x along the wire and on time t. The vibrational frequency of such a wire is $\nu_n = \dfrac{n}{2L}\sqrt{\dfrac{T}{\sigma}}$ where $n > 1$ defines the normal mode and is an integer. The wave velocity is $\sqrt{T/\sigma}$. In a 0.5 m long steel guitar string of density $\sigma = 0.5$ g m^{-1}, applying 8 N m tension, T, produces sine waves with frequencies of $n \times 400$ Hz. Sine waves are produced because the displacement of the guitar string is zero at both ends. Tuning the string to the required frequency is relatively easy as this is proportional to the square root of the tension. It is possible to appreciate why strings of different densities and lengths are used to cover the musical scales.

The general wave equation is

$$\frac{\partial^2 u}{\partial t^2} = c^2 \frac{\partial^2 u}{\partial x^2}$$

and one general solution of this equation was indicated in the introduction. Here we solve this equation by separating variables in exactly the same manner as for the diffusion and other partial differential equations. The separation has the form $u(x, t) = w(t)v(x)$ and dividing both sides by u gives

$$\frac{1}{w}\frac{\partial^2 w}{\partial t^2} = \frac{c^2}{v}\frac{\partial^2 v}{\partial x^2} = k^2$$

where k^2 is the separation constant and different solutions are obtained depending on whether k^2 is zero, positive or negative. The solutions are readily found; when $k = 0$, then $\dfrac{\partial^2 w}{\partial t^2} = 0, \dfrac{\partial^2 v}{\partial x^2} = 0$. Integrating twice produces $w = A + Bt$, $v = C + Dx$ where A, B, C, D are constants determined by the initial and boundary conditions. The standing wave solution is

$$u = (A + Bt)(C + Dx).$$

If $k^2 > 0$ then the solutions are found from $\dfrac{\partial^2 w}{\partial t^2} = k^2 w, \dfrac{\partial^2 v}{\partial x^2} = \left(\dfrac{k}{c}\right)^2 v$ and which produce

$$u = (Ae^{-kt} + Be^{kt})(Ce^{-kx/c} + De^{kx/c})$$

When k^2 is changed to $-k^2$, the solutions are

$$u = [A\sin(kt) + B\cos(kt)][C\sin(kx/c) + D\cos(kx/c)]. \tag{10.62}$$

Which of these general solutions should be used depends on the initial and boundary conditions, and just as with other partial differential equations, linear combinations of these solutions may form the standing wave solution to a particular problem.

Returning to the guitar string, suppose that it has length L and as its ends are fixed in place; this determines two boundary conditions, which are

$$u(0, t) = 0; \qquad u(L, t) = 0.$$

However, although the initial displacement or velocity of the string is not yet defined these partial boundary conditions will only permit some solutions. At one end of the string, the condition is $u(0, t) = 0$; therefore, $u(0, t) = w(t)v(0) = 0$, which means that either $v(0)$ or $w(t)$ must be zero for all t. A wave that does not vary in time is not considered to be a standing wave, and this means that $v(0) = 0$ is a boundary condition. Similar reasoning with $u(L, t) = 0$ indicates that the standing wave must satisfy

$$v(0) = 0; \qquad v(L) = 0.$$

The only solution satisfying these two conditions is the sine/cosine solution (10.62). In this solution, only $\sin(kx/c)$ is zero when $x = 0$, and $\sin(kL/c)$ can be zero when $x = L$ depending on the value of k. Picking a value $k = n\pi c/L$ where n is an integer satisfies this condition making the general solution

$$u = C[A\sin(n\pi ct/L) + B\cos(n\pi ct/L)]\sin(n\pi x/L) \qquad (10.63)$$

This has the form of a standing wave as the amplitude of the x wave is changed in time by the term in square brackets. The different integer values of $n > 0$ produce different normal modes of the vibrating string. The fundamental, or first harmonic, has $n = 1$; the second harmonic or first overtone $n = 2$, and so forth. Any particularly complicated vibration that is produced when a string is struck or plucked is always a linear combination of these normal modes, because these modes are the only elementary vibrations present.

Suppose that the initial amplitude and velocity of the guitar string are defined by some more initial conditions. If the string is plucked, just before it is released it has some particular shape; this is defined as $f(x)$, but can be defined more precisely later on. Also, supposing that all points on the string are stationary at $t = 0$, the new conditions will therefore be

$$u(x, 0) = f(x); \qquad \partial u(x, t)/\partial t = 0.$$

As this time derivative is zero at time zero, the sine term in t in the general solution must be zero, because its time derivative is a cosine which is not zero at $t = 0$. The general solution is

$$u = \sum_{n=1}^{\infty} b_n \cos(n\pi ct/L)\sin(n\pi x/L)$$

where it is assumed that it will be necessary to form a linear combination of solutions. The coefficient b_n are found as before by a Fourier series, since initially

$$f(x) = \sum_{n=1}^{\infty} b_n \sin(n\pi x/L)$$

and therefore,

$$b_n = \frac{2}{L}\int_0^L f(x)\sin(n\pi x/L)dx.$$

If the initial displacement has the form $f(x) = e^{-a|2x-L|} - e^{-aL}$, which is peaked at the centre of the string, the solution is found by separating the integration into two parts to overcome the absolute value in the integral. The first part is from 0 to $L/2$, and the second from $L/2$ to L. If the guitar string's length $L = 1/2$ m, which has been calculated to have a fundamental frequency of 400 Hz, the wave velocity is $c = \lambda\nu$ which is 100 m s^{-1} because the wavelength is $L/2$. The Maple used to do the calculation is based on Algorithm 10.1 and if L is 1/2 then

```
> L:= 0.5: c:=100.0: T:= 2*L/c:                # T = period
  f:= x-> exp(-abs(2*x-L)*5)-exp(-L*5);        # initial function
  b:= n-> 2/L*int(f(x)*sin(n*Pi*x/L),x=0..L);
  u:= (x,t, maxn )->
      add( b(n)*cos(n*Pi*c*t/L)*sin(n*Pi*x/L),n= 1..maxn);
```

The figure shows the profile of the string at different times with 20 terms added in the series; `maxn: = 20:`. The displacement is clearly greatly exaggerated compared to the true value. It is clear that the wave is composed of waves moving in opposite directions as can be seen by the way that the peaks of the wave move with time.

In the case when the initial velocity is not zero but the displacement is zero, the initial conditions will be

$$u(x, 0) = 0, \qquad \partial u(x, t)/\partial t = f(x).$$

Notice that the initial velocity is a function of *position*; i.e. this describes the initial velocity at each point along the wire. As the velocity is not zero, the sine term in time in the

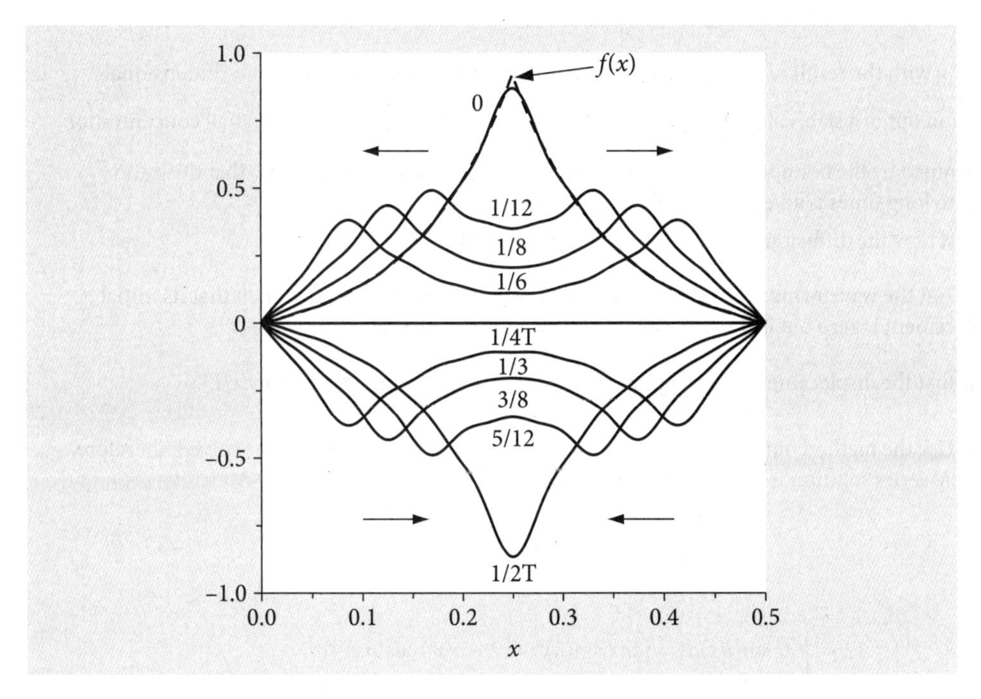

Fig. 10.30 Wave formed by plucking a wire in its centre showing displacement (greatly exaggerated in amplitude) at different times as fractions of the period T. The arrows show how the wave appears as two travelling waves moving in opposite directions.

general equation (10.63) is now used, as its derivative is a cosine which is not zero at $t = 0$. The general term is found by differentiating the general solution giving

$$\frac{\partial u}{\partial t} = \frac{n\pi c}{L}\sin(n\pi ct/L)\sin(n\pi x/L) = f(x).$$

The coefficients are now $b_n = \frac{2}{L}\frac{L}{n\pi c}\int_0^L f(x)\sin(n\pi x/L)dx$ and the general solution is

$$u = \sum_{n=1}^{\infty} b_n \sin(n\pi ct/L)\sin(n\pi x/L). \qquad (10.64)$$

As the initial displacement is zero, but the velocity is not zero, plotting the displacement with time shows that it initially increases; which is the opposite behaviour to that observed in Fig. 10.30.

If both the displacement and velocity are some function of x, then both the sine and cosine terms in time appear in the general solution.

10.9.4 **Questions**

Full solutions are available at **www.oxfordtextbooks.co.uk/orc/beddard**.

Q10.34 In example (iv) of Section 10.6.2 the diffusion out of a slab was calculated. Repeat this calculation, but now suppose that a reaction takes place so that the term $-k_1c$ is added to the diffusion equation making it $\frac{\partial c}{\partial t} = D\frac{\partial^2 c}{\partial x^2} - k_1c$. The boundary conditions are that the concentration is C_0 everywhere at $t = 0$ and zero at the plate edges at $t > 0$.

Strategy: Use the separation of variables method and base the solution on that of example (iv),

Q10.35 While it is possible to calculate the concentration profile of diffusing molecules, this is not usually measured, but the amount of material diffused out of a region as a function of time is measured instead. This quantity is $c_{av}(t) = \frac{1}{L}\int_0^L c(x, t)dx$.

(a) Starting with the result of example (iv) shown in Fig. 10.24, which describes one-dimensional diffusion out of a slab, calculate c_{av} and then the ratio $r = \dfrac{c_{av} - c_f}{c_i - c_f}$ where c_f is the final concentration determined by the boundary conditions, and c_i the initial concentration. Show that this ratio taken to long times is an exponentially decaying function.

(b) Suggest how the diffusion coefficient might then be measured.

Q10.36 **(a)** Work out the waveforms produced if a taut piano wire of length L is struck such that its initial displacement is zero but its initial velocity everywhere has the shape $f(x) = xL - x^2$.

(b) Show that the displacement is well approximated by $u = 4\left(\dfrac{L}{\pi}\right)^3 \sin(\pi ct/L)\sin(\pi x/L)$.

Strategy: Use the method outlined in the text. The integral for coefficients b can be solved; therefore, an algebraic series solution is possible. The integrals are

$$\int x\sin(ax)dx = [\sin(ax) - ax\cos(ax)]/a^2,$$

$$\int x^2\sin(ax)dx = [2ax\sin(ax) + (2 - a^2x^2)\cos(ax)]/a^3,$$

and can also be worked out by converting to exponentials.

Q10.37 A wave that retains its shape is not dispersive, this means that its wavelength does not change with time or distance, and is called a *soliton*. In 1832, Russell published an account of his observation of a solitary wave travelling unaided along the Union Canal near Edinburgh. It was produced after a barge suddenly stopped but the mass of water it was moving did not. The same type of soliton wave is more famously observed as the Severn Bore. Solitons are also formed in some types of lasers and are important for communications, as the wave does not lengthen and one soliton can cross another unchanged. A soliton can also be formed in all-trans polyacetylene. In this case, it is an electron in a non-bonding orbital situated between regions of opposite double bond alternation. This electron can migrate up and down the polyacetylene without spreading.

The Korteweg–de Vries (KdV) equation $\dfrac{\partial u}{\partial t} + u\dfrac{\partial u}{\partial x} + \dfrac{\partial^3 u}{\partial x^3} = 0$ describes the motion of the classical soliton wave. The wave displacement at x and t is u and the shape of the wave is not unlike that of a Gaussian.

Show that $u(x, t) = 2c\,\mathrm{sech}^2\left(\dfrac{\sqrt{c}}{2}(x - ct)\right)$ is a solution of the KdV equation where c is the wave speed.

Numerical Methods

11

This chapter outlines several numerical techniques and associated algorithms which are needed to perform integrations and solve differential equations. These are written generically and with Maple. Accurate and efficient ways of calculating results from theory is an essential tool for any scientist. Using computer algebra programs such as Maple is a great boon because calculations can be done to great numerical precision and accuracy. However, in line with our objective of understanding the principles of how to solve problems, several techniques, which contain the essence of the method to be used, are outlined. Once these are understood, you will have the knowledge to understand the more complex 'black-box' methods that Maple, Mathematica, or other software use.

To start this chapter, some details on numerical accuracy are reviewed; next simple integration methods are described with a few examples and problems. More difficult, and interesting, is the numerical integration of differential equations, some of which only have numerical solutions. The various methods used are explained. A different numerical approach is taken in the next chapter on Monte Carlo methods. Simple integration is rather slow by this method but integrating differential equations is very robust. Monte Carlo methods are very versatile at *simulating* physical processes, such as diffusion or fluorescence, and occasionally this is the only way to do the calculation. Simulation also provides a good way to understand the processes involved, some of which can be very complex and not always easily translated into differential equations.

11.1 Numerical accuracy

11.1.1 Rounding errors

In many calculations, the errors introduced by the computer due to the way its integers and real numbers are manipulated do not matter. Sometimes, the answer the computer returns may be precise and will be printed to many decimal places; however, it may not be accurate. To understand this, the expression $(1-x)^6$ is expanded as the series $1-6x+15x^2-20x^3+15x^4-6x^5+x^6$ and this is plotted close to its minimum at $x\approx 1$ in Fig. 11.1. While the scale is small, $\approx 10^{-13}$, determining the exact root of the equation numerically is difficult. Of course, we know it is at 1, but the computer will not give this answer even using a Newton–Raphson iterative scheme. The reason for the noisy plot is because the alternating positive and negative terms in the summation add up to almost zero, and the rounding error in the precision with which the numbers are calculated becomes important, and this limits the accuracy of the calculation.

A real number is converted in the computer into a floating-point number represented as $x=\pm 0.d_1d_2d_3d_4\times 10^y$ which only approximately represents the number; i.e. it is only accurate to a certain number of decimal places. The mantissa, is the fraction $0.d_1d_2d_3d_4$, where d_1 etc. are the digits $0\cdots 9$. A floating-point number is usually described as a single or double precision and the number of decimal places is typically 7 and 14 respectively; this depends on the type of computer being used. In Maple and Mathematica, the precision can be determined by the user. In Maple, the default is 10 decimal places but when plotting, the hardware floating-point arithmetic is used instead and this depends on the type of computer used. On my computer, this is 17 decimal places, but if the `Digits` setting of

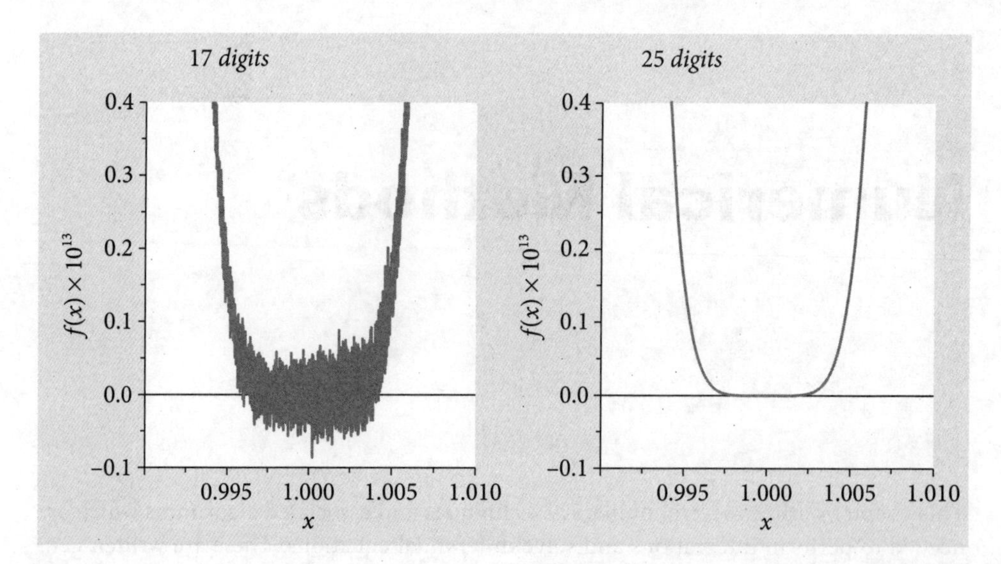

Fig. 11.1 Plots of the series expansion of $(1 - x)^6$ with two different values of floating point precision. (Notice the axes range.)

Maple is greater than 17, this is used instead. However, as seen in Fig. 11.1, 17 digits are not good enough to find the root; 25 decimal places give a far more accurate result. The instruction `Digits:= 25`, sets 25 decimal place precision for floating point numbers.

Using the Newton–Raphson method, the root cannot be found accurately: the method is (see Chapter 3.10)

```
> f:= x-> 1-6*x+15*x^2-20*x^3+15*x^4-6*x^5+x^6;
> df:= x-> -6+30*x-60*x^2+60*x^3-30*x^4+6*x^5; # df/dx
> s:= 0.95; for i from 1 to 30 do s:=s-f(s)/df(s); end do;
```

With 10 decimal places, the root found is 1.03387596, which is not even on the graph shown above. Using 25 decimal places, the result is better but only becomes constant at 1.00013639 after about 80 iterations.

If the function $(1 - x)^6$ is plotted directly without expanding, and hence without accumulating rounding errors, then the curve looks like that obtained with 25 digits precision. This illustrates the important point that in many cases a small amount of algebra can significantly improve numerical calculations; errors rapidly accumulate with the repeated adding and subtracting of similarly sized numbers, and once an error is present it contaminates all subsequent results.

11.1.2 Truncation errors

In summing a series, the number of terms to be added has to be decided in advance. This automatically leads to an error that is nothing to do with arithmetical precision. The calculation should be repeated with more terms until the number of decimal places you require is continuously obtained. Rounding errors must additionally be considered, especially if the series contains alternating positive and negative terms; the errors can accumulate as the summation continues but will not be the same for all values of x.

Consider the series

$$\sin(x) = 1 - \frac{x^3}{3!} + \frac{x^5}{5!} - \cdots$$

which is valid for all x. If the series is summed up to the term $x^n/n!$ with $x > 0$, the error is less than $\left|\dfrac{x^{n+2}}{(n+2)!}\right|$. The values of sin(0.5) is 0.47942553 whether calculated directly using Maple, or as the summation up to the x^7 term. If, however, x is large, for instance $x = 5.5\pi$, then $\sin(x) = -1$, but by summation up to x^7 the result is −79242.06: clearly is ridiculous because $\sin(x) \le 1$ for all x. Only if the number of terms in the summation is increased to at least terms $> x^{59}$, and the numerical precision increased to 20 decimal places can an accurate result be obtained from the series approach. In this instance, we know that the answer from the small summation is incorrect because the properties of sin(x) are well known, but if you are summing a series whose result is unknown, then great care is necessary.

11.1.3 Unstable recursion: magnification of errors

In solving integrals, a recursion formula can often be found, particularly when the integration is done by parts. An example met before (Q4.22) is

$$I_n = \int_0^1 x^n e^{x-1}dx = x^n e^{x-1}\Big|_0^1 - n\int_0^1 x^{n-1}e^{x-1}dx$$
$$= 1 - nI_{n-1}$$

To evaluate the integral with different n, the starting value with $n = 0$ is required, which is $I_0 = 1 - 1/e$. In the recursion calculation, `s` has, arbitrarily, been chosen as the name for the integral because `I` and `int` are reserved names. Recall that `s:= 1 - n*s;` is an assignment, meaning that the new value of s is $1 - n \times$ (current value).

```
> restart: Digits:= 10;                         # digits precision
> f:= n-> int( x^n*exp(x-1.0),x = 0..1):        # algebraic integral
  s:= 1-1/exp(1.0):                             # first value I0
  for n from 1 to 20 do
    s:= 1 - n*s;                                # recursion
    f(n);                                       # direct integration
  end do:
```

Some of the results are tabulated below; $f(n)$ is the result of using Maple's algebraic integration, s the recursion.[1]

The failure of both the Maple integration and the recursion formula indicates how careful one needs to be with some apparently innocent numerical calculations. After about $n = 10$, it is difficult to judge which calculation gives the most accurate result, and by $n = 15$, both are clearly wrong. In this example, 10 digits precision only was used. To obtain results to 7 decimal places with $n = 20$, the precision needs to be increased to 25 digits. Above $n = 20$, the results again become incorrect; to get to $n = 30$, at least 40 digits precision are needed!

While this calculation is a little unusual, it is not uncommon. When calculating wavefunctions using Hermite or other polynomials, large powers of x, exponentials and recursion formulae are commonly used. Recursion and difference formulae are often used in the numerical integration of differential equations, see Section 11.3, consequently, their numerical integration is akin to an art rather than a science. However, you should not be

Table 11.1 Result of recursion and direct integration

n	s	f(n)
1	0.367879	0.367879
5	0.145533	0.145533
10	0.0837734	0.083877
11	0.0784922	0.07735
12	0.0580941	0.0718
13	0.244777	0.067
14	-2.42688	0.07
15	37.4032	0
16	-597.451	1
17	10157.7	-100
18	-182837	0
19	3.47390e+06	-20000

[1] To print the table, the statement, `printf("%3g %13g %13g\n",n, s, f(n));` was placed in the `for..do` loop

too despondent or sceptical about numerical methods because with care, and sometimes cunning, accurate results can be obtained.

11.1.4 Organized chaos

You need to be aware that recursion can sometimes lead to chaotic behaviour as a normal outcome of a calculation. This is a different sort of chaos to that just calculated; it is very much like 'organized chaos' and is not due to rounding or precision errors. Often this chaos is termed 'deterministic'. One well-known example, first described in detail by May (1974, 1987) and May & Oster (1976), arises from the recursive equation $x_{n+1} = cx_n(1 - x_n)$ where c is a constant. This equation is sometimes called the *logistic map*. If c takes values in the range from 2.5 to about 4, and $0 \le x \le 1$, a *bifurcation* of x can be observed at $x = 3$; this means that at the same value of c, two different but constant values of x, are produced after several hundred iterations of the equation. As c increases the doubling multiples to 4, 8, 16 and more numerous sets of numbers produce a random looking but regular pattern. When these values of x are plotted vs c, as abscissa, apparently random behaviour from otherwise regular behaviour is observed, Fig. 11.2. Pain (1993, chapter 12) describes non-linear and chaotic effects that have been observed experimentally, as well as a description of the logistic map.

To calculate the map the equation was iterated 4000 times, and the last 400 numbers stored and plotted as points x (as ordinate) vs each value of c, as abscissa, from 2.5 to 4 in 400 steps in both x and c.

Algorithm 11.1 Bifurcation caused by repeated iterations

```
> n:= 400: m:= 4000:                    # 4000 iterations, save 400
  pop:= Matrix(1..n,1..n):              # store data
  c:= 2.5:                              # initial c
  delta_c:= (4.0-c)/n;                  # increment
  x:= 0:                                # initial x
  for j from 1 to n do                  # loop over n, c values
     for i from 1 to m do               # iterate m times
       x:= c*x*(1-x):                   # calculate new x
       if i > m-n then pop[j,i-m+n]:= x end if; # save last n
     end do:
c:= c + delta_c:                        # increment c
x:= x + 1.0/n;                          # increment x
end do:
```

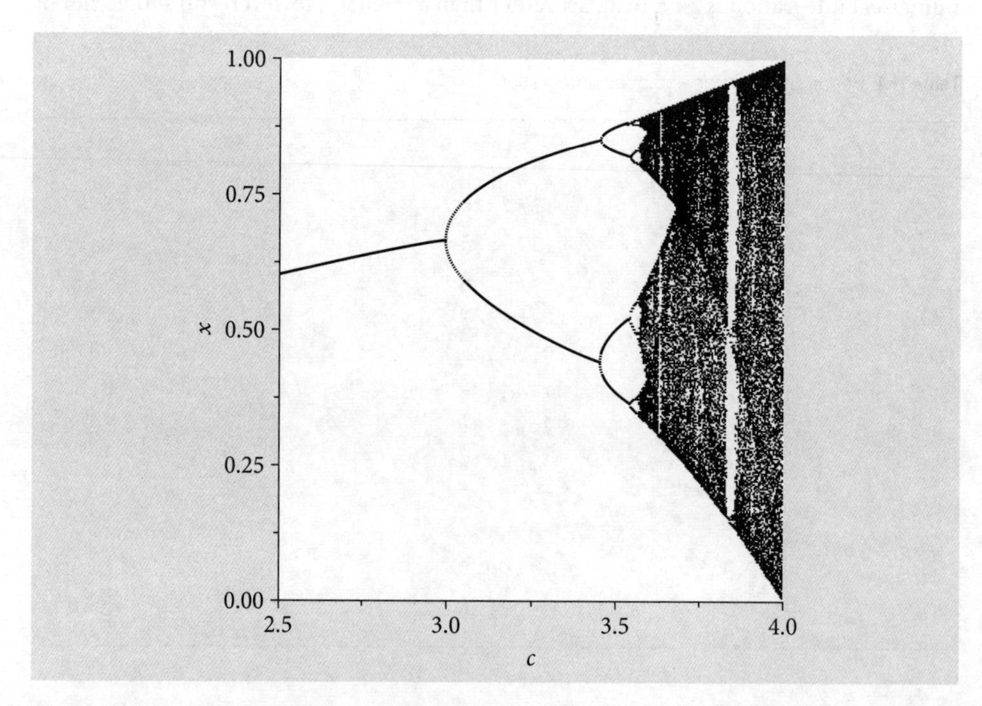

Fig. 11.2 Bifurcations and ordered chaos generated by iterating $x_{n+1} = cx_n(1 - x_n)$.

The data in matrix pop is plotted in Fig. 11.2. The bifurcations are clear at about $c = 3$, 3.44, and 3.54 and after that become too numerous to see on this plot. The data are plotted as points without regard for the size of each datum.

In classical and quantum mechanics, for example in molecules, it is also observed that regular motion can suddenly become chaotic even though the underlying equations of motion do not indicate that this will necessarily happen. By 'chaotic' is meant that two starting conditions placed arbitrary close to one another in value, will lead to entirely different but repeatable outcomes at some later time. If the cause of the chaos is due to rounding errors, the outcome is not repeatable but unpredictable.

11.2 **Numerical methods to find the roots of an equation**

Sometimes in integration or in solving differential equations, the root of some function is required; three related numerical methods, the secant, regula falsi, and bisection, are now described. The Newton–Raphson method has already been met in Chapter 3.10 and is used when the equation can be differentiated.

The secant method is similar in idea to the Newton–Raphson method except that the derivative is expressed as a difference equation. If $f(x) = 0$ is the equation whose roots or x values we want to find, then, in a recursive scheme, the next x value is given by Newton–Raphson as $x_{n+1} = x_n - f(x_n)/f'(x_n)$ with $n = 0, 1, 2\ldots$ and f' is the derivative with respect to x. The x_0 value to start the calculation must be somewhere near to the root, and can be estimated by plotting the function. At the end of each step, the new value x_{n+1} is substituted into the right-hand side of the equation, and the process is repeated until the absolute value of the function becomes less than some preset tolerance or the number of iterations is exceeded.

In the secant method, the derivative is approximated as

$$f'(x_n) \approx \frac{\Delta f}{\Delta x} = \frac{f(x_n) - f(x_{n-1})}{x_n - x_{n-1}}$$

then

$$x_{n+1} = x_n - f(x_n)\frac{x_n - x_{n-1}}{f(x_n) - f(x_{n-1})} \qquad (11.1)$$

which is the working equation for the secant method. Two initial guesses of the root are required, because the equation has to have points at x_1 and x_2 to begin the iterative process and these are usually to one side of the root. Once the method is close to a root, the number of digits of accuracy doubles for every two iterations. However, the method can sometimes fail because the particular curvature of the function may produce $f(x_n) = f(x_{n-1})$. In each of the following three algorithms the new value x_{n+1} is called xm and is given a dummy value greater than the tolerance to start with. The root and the value of the function at the root, which should be zero or very close to it, are printed. The equation is $x^5 + x = 1$.

Algorithm 11.2 Secant method

```
> f:= x-> 1-x-x^5;                          # a root is at ≈ 0.75
> x1:= 0.0: x2:= 0.5; xm:= 2*x2:            # set initial values
  xtol:= 0.001:                             # set tolerance in f(x)
  while abs(f(xm)) > xtol do
    xm:= x1-f(x1)*(x2-x1)/(f(x2)-f(x1)); # new value xn+1
    x1:= x2: x2:= xm:                       # update
  end do:
  print('root', xm, 'function value' = f(xm));
```

If the starting points span the root, then the false position or *regula falsi* method is used. This method uses the same equation as the secant method but a slightly different algorithm is used to generate the next point. This ensures that the root is always straddled. The algorithm is

Algorithm 11.3 Regula falsi

```
> x1:= 0.0: x2:= 1.0;      xm:= 2*x2: # set initial values
  xtol:= 0.001:                        # set tolerance in f(x)
  while abs(f(xm)) > xtol do
    xm:= x1 - f(x1)*(x2-x1)/(f(x2)-f(x1));
    if f(x1)*f(xm) > 0 then x1:= xm else x2:= xm: end if;
  end do:
  print('root', xm, 'function value' = f(xm));
```

The bisection method is the simplest and most robust method. The initial estimates must span the root; therefore, the function is calculated from some point until its sign is changed, if the whereabouts of the root is unknown. At this point, the bisection method is repeatedly used to divide the region into finer parts.

Algorithm 11.4 Bisection method

```
> x1:= 0; x2:= 1;                # set initial values
  xs:= 10*x1: xtol:= 0.001:      # set tolerance in f(x)
  while abs(f(xm)) > xtol do
    xm:= (x2 + x1)/2.0;          # mean value
    if f(x1) * f(xm) > 0 then x1:= xm else x2:= xm end if;
  end do:
  print('root', xm, 'function value' = f(xm));
```

All three methods give the root as 0.754 but differ in the next and subsequent decimal places. Confirm these results for yourself. It is generally advisable to set a limit on the number of iterations in the `while..do` loop just in case the root is not in the region specified or some other error occurs. This is easily done by replacing the while statement with

```
...
m:= 0:
while abs(f(xm)) > xtol or m < 20 do
m:= m+1;
...
```

and the loop finishes when m is 20. Hopefully the root has been found first, but a check on m can always be made. One 'trick' that can sometimes be used in the secant and regula falsi method, should they fail, is to add a small number, $<10^{-6}$, to the denominator, to prevent it accidentally becoming zero.

11.3 Numerical integration

Two methods of numerical integration have been briefly described in other chapters, one is the series expansion of a function and then a numerical evaluation of the terms; the other is the Euler–Maclaurin formula Chapter 5.7 and care needs to be exercised with both these methods. Three general and simpler methods are now outlined; these are the mid-point, the trapezoid, and Simpson's rule. More sophisticated numerical methods than these are described in *Numerical Recipes* (Prest et al. 1986) and similar specialized texts.

Each of these methods divides the function to be integrated into small strips at predetermined abscissas, which are then multiplied by a certain constant number depending on the method, and added together. The integration becomes a summation, Fig. 11.3, and the general name for such a numerical integration method is *quadrature*. These methods are similar to the basic way we think about integration as the limiting area of a summation of rectangular strips as the width of each strip tends to zero. If the strips are fine enough, the result is a good approximation to the true value. In general, however, we do not initially know how many strips will be needed and some experimentation with this number is always necessary.

It should be remembered that whatever numerical method is used, a polynomial is being integrated and not the original equation, and thus the ability of this to describe the curve will affect the results. The more sophisticated the method becomes, the higher the order of the polynomial used.

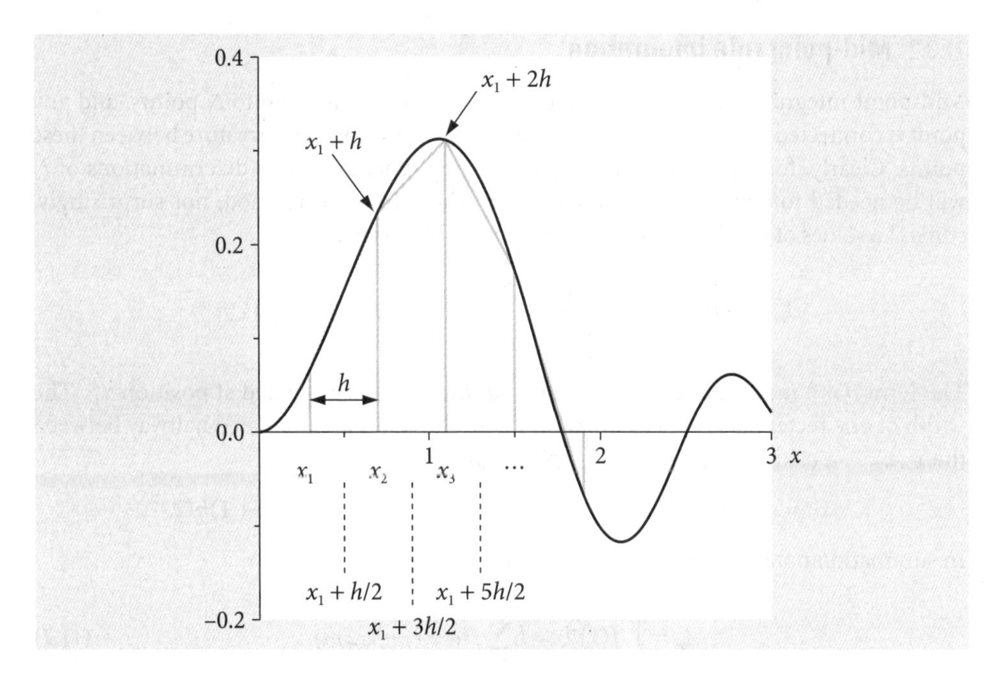

Fig. 11.3 The trapezoidal rule approximates an integral with straight lines between points $x_1, x_2, \cdots$, although not very well in this example, where the strips are too far apart to represent the function accurately. The mid-point method evaluates the function half way between each point, three of which are shown by the vertical dotted lines.

11.3.1 **Always try some algebra first**

There is always a tendency when using numerical methods to submit the equation directly to the computer. In many cases, however, performing some algebra first will help. This might mean simplifying the equation, removing constants to outside the integration, and, in particular, substitution of variables, which is done in standard integration. Consider the integral $\int_0^\infty (1 + x^2)^{-1}dx$. This converges very slowly and makes the upper limit very large, which results in a slow calculation and potentially inaccurate numerical result. In fact, the result will get worse as the numerical integration proceeds, due to rounding errors. Some values of $1/(1 + x^2)$ are given in Table 11.2 and show that the maximum x in the integration has to be at least 100 for four-figure accuracy. This means that at least 1000 points have to be used in the integration and even possibly 10 000.

What is needed is to transform the function into something less extended. The substitution $x = e^{uz} - 1$ where u is a small constant will do this. Typically, $u = 5$, although this value is not critical. Differentiating $x = e^{uz} - 1$ gives $dx = ue^{uz}dz$, but the limits are still the same; $z = 0$ when $x = 0$ and $z = \infty$ when $x = \infty$. Making substitutions and simplifying a little makes the integral,

$$\int_0^\infty \frac{u}{2e^{-uz} + e^{uz} - 2}dz.$$

The upper limit is still ∞, but now this is not really important. If $z = 3$, the function in the integral has the value $\dfrac{5}{2e^{-15} + e^{15} - 2} \approx \dfrac{5}{e^{15}} \approx 1.5 \times 10^{-6}$ making the whole expression virtually zero beyond this value, and so 3 is the effective upper limit; quite an improvement on infinity. Additionally, fewer points in the integration are needed for the same precision.

The individual integration methods are now described.

Table 11.2 Showing slow convergence and range of x.

x	0.01	0.1	1	10	100
$1/(1+x^2)$	0.999900	0.990099	0.50	0.009900	0.000099

11.3.2 Mid-point rule integration

Mid-point integration assumes that the function is split equally into N points, and any point is connected by a straight line to the next, thus ignoring any curvature between these points. Clearly, for a rapidly oscillating or varying function, many determinations of N will be needed to obtain an accurate answer. The mid-point method, not surprisingly, sums the values of the function at the *mid-point* of each strip,

$$I_m = \int_a^b f(x)dx \approx \frac{(b-a)}{N}[f(x_1^m) + f(x_2^m) + f(x_3^m) + \cdots f(x_N^m)]$$

The term $f(x_1^m)$ mean that the function being integrated is evaluated at position x_1^m. The width of one rectangle is $h = (b - a)/N$ and then the terms are evaluated halfway between the $x_1, x_2, \cdots$ positions that define the points, giving

$$x_1^m = a + h/2, \qquad x_2^m = a + 3h/2 \qquad \cdots \qquad x_N^m = a + (N+1)h/2.$$

In summation notation the equation is

$$I_m = \int_a^b f(x)dx \approx h\sum_{j=1}^{N-1} f(a + (j - 1/2)h). \tag{11.2}$$

One advantage of this rule is that it enables functions with singularities in the integration range to be calculated. For instance, $1/(1 - x)$ is infinite at $x = 1$, but an integration spanning this point need not fail with the midpoint method but will do so with all other methods that calculate the function at integer increments of h.

The error in the integral calculated with the mid-point method is $\varepsilon = \pm\dfrac{(b-a)^3 f''(c)}{24N^2}$ and this will be greatly reduced by increasing the number of points N. The size of this error is found by calculating two points c where the second derivative has its maximum and minimum in the range $a \le c \le b$, and choosing the largest absolute value. This can be the most difficult part of the calculation and is usually not worth the effort. It is easier to calculate the integral by doubling the number of points each time until a sufficient number of decimal places are unchanging in the result. The mid-point method is the simplest to calculate and gives accurate answers, yet it is hardly mentioned in textbooks compared to the two methods described next. Notice that, because the mid-points are used, there is one less point in the summation using the mid-point method than in the trapezoid or Simpson's method.

11.3.3 Trapezoidal rule integration

Trapezoidal integration is also straightforward. The equation to approximate the integral is

$$\int_a^b f(x)dx \approx \frac{b-a}{2N}[f(x_1) + 2f(x_2) + 2f(x_3) \cdots + f(x_N)] + \varepsilon \tag{11.3}$$

whereby the integration is again split into N strips, Fig. 11.3, and N must be an *even* number. Calculating points $x_1, x_2, \cdots$ is easily done because they are related to the gap between the points; $x_2 = x_1 + h$, $x_3 = x_1 + 2h$, and so forth; see Fig. 11.3, and $h = (b - a)/N$ is the width of one strip.

The integral can be calculated either, as in equation (11.3), by adding up the first and last value of f and twice all the rest and then multiplying by $h/2$, or by adding up all the values of f then subtracting half the value of the first and last points and finally multiplying the result by h. Usually, the result is calculated with several increasing values of N to see if convergence is achieved. The error ε indicates how many decimal places the result is accurate to and is reduced as $1/N^2$, the same as for the mid-point rule.

To illustrate the method, the function e^{-x^2} will be integrated by the trapezoidal rule and then again by Simpson's rule (see next section) with 10 points over the range 0 to 1. The summation will be done in a `for..do` loop. The equation to be integrated is defined first as a function so that it can be calculated at each value of x. When we want the value of a function this is written as `f(3.0)` or `f(a)` and so forth. The expression `f(a+(i-1)*h)` calculates the function at each x value starting at a.

Algorithm 11.5 Trapezoidal integration

```
> f:= x-> exp(-x^2);                # equation to integrate
> n:= 10:                           # number of points
  a:= 0.0: b:= 1.0:                 # integration limits
  h:= (b - a)/n:
  s:= 0.0:                          # initial value of sum
  for i from 1 to n+1 do            # sum terms into s
    s:= s + f(a+(i-1)*h):
  end do:
  s:= h*( s -( f(a)+f(b) )/2):   # -1/2*(first + last point)
  printf("%s %g ",`Trapezoidal integral =`,s);

                Trapezoidal integral = 0.74621

# compare with Maple's own numerical integrator.
> Int(f(x),x= a..b ): %= evalf(value(%));
```

$$\int_0^1 e^{-x^2}dx = 0.74682413$$

The small number of rectangles in the trapezoidal integration, $n = 10$, give quite a good result, accurate to three decimal places.

11.3.4 Simpson's rule integration

Simpson's rule is slightly more sophisticated than the trapezoidal method, as it fits a quadratic function to data points rather than straight lines; the formula is

$$\int_a^b f(x)dx \approx \frac{b-a}{3N}[f(x_1) + 4\sum f(x_{odd}) + 2\sum f(x_{even}) \cdots + f(x_N)] + \varepsilon. \tag{11.4}$$

The sum is made by adding the first and last point to four times the sum of the odd indexed points, and to twice the sum of the even numbered points. The number of data points must be an even number. The error is reduced as $1/N^4$ so decreases rapidly as the number of data points increase.

Algorithm 11.6 Simpson's rule integration

```
> f:= x-> exp(-x^2);
  n:= 10:                             # num points
  a:= 0.0: b:= 1.0:                   # limits
  h:= (b-a)/n:
  s:= -f(a)-f(b):                     # 1st + last points in sum
  j:= 1:                              # switch odd/even
  for i from 1 to n + 1 do            # final point n+1
    if j = 1 then
      s:= s + 2*f(a+(i-1)*h):         # odd
    else
      s:= s + 4*f(a+(i-1)*h):         # even
    end if;
    j:= -j:                           # switch back
  end do:
  s:= h*s/3:
  printf(" %s %g ",`Simpson's method: integral =`,s);

            Simpson's method: integral = 0.74683
```

This result is accurate to seven figures, far higher than the trapezoidal method with the same number of points. Notice that in the algorithm, `j` is used to switch between odd and even terms by simply negating it each time in the loop.

In Bjerrum's theory of ionic association in solution, the concentration of ion pairs is

$$n_{12} = 3n_1\left(\frac{e^2}{4\pi\varepsilon_0 k_B T\varepsilon a_0}\right)^3 \int_2^b e^x x^{-4} dx$$

where $a_0 = |z_1 z_2| e^2/(2\varepsilon 4\pi\varepsilon_0 k_B T)$ and $b = 2a_0/\sigma$ is a number greater than 2. The ionic charges are z_1, z_2, the solvent dielectric constant ε, the minimum possible separation of the ions, when pressed together, is σ. The limit b has, in practice, a maximum value of ≈ 50. The integral does not have an analytical solution and is normally tabulated. To evaluate this integral, a procedure will be made of the trapezoidal method so that it can then be used repeatedly. A similar process could be used to make procedures of the other integration methods.

The Maple procedure `Trap`, is made by defining `Trap:= proc(a,b,n,f)` which ends with an `end proc:` statement. The values for `a, b, n`, and `f`, which are, respectively, the integration limits, the number of points, and the function itself, which are used inside the procedure, have to be defined before it is used. The procedure is activated by writing `Trap(a,b,n,f);`.

Before doing the calculation, the function $e^x x^{-4}$ should be examined. It changes slowly with x, as may be seen by plotting, and has a minimum at $x = 4$, where the ions are at separation a_0. At larger values of x, the function rises rapidly as the exponential term dominates the x^{-4} term. The value of a_0 for a 1-1 electrolyte is 0.350 nm in water at 25 °C taking the dielectric constant to be 80. The calculation is

$$a_0 = (1.602 \times 10^{-19}\text{C})^2/(8\pi \times 8.854 \times 10^{-12}\ \text{Fm}^{-1} \times 1.381 \times 10^{-23}\ \text{JK}^{-1} 298\ \text{K} \times 80))$$

The unit is $\text{C}^2/(\text{Fm}^{-1}\text{JK}^{-1}\text{K})$ and it is not immediately obviously that this is metres. A farad (F) is a coulomb/volt and a volt is joule/coulomb, therefore a farad is $\text{F} = \text{C}^2\text{J}^{-1}$ giving a_0 units of distance. Returning now to the numerical calculation, the integral at $b = 2$ is clearly zero, starting at $b = 3$ and incrementing by 3 to 30 in a loop, gives a reasonable number of different values.

Algorithm 11.7 Trapezoidal rule procedure

```
> f:= x-> exp(x)/x^4;                          # define function
  Trap:= proc(a, b ,n, f)                      # define procedure
  local h, s, i:                               # local variables
    h:= (b-a)/n:
    s:= 0.0:
    for i from 1 to n+1 do
      s:= s + f(a+(i-1)*h):
    end do:
    s:= h*( s- ( f(a)+f(b) )/2 ):              # s is integral
    printf("%3g %5g %13.5f\n",a, b, s); # formatted output
  end proc:                                    # end procedure
# now call procedure with different limits
  printf("%s\n"," limits and integral"); # title
  n:= 200:                                     # points in integration
  for k from 3.0 by 3 to 30 do                 # k is limit b
    a:= 2.0: b:= k:                            # set limits
    Trap(a, b, n, f);                          # 'call' procedure
  end do:
```

Some of the results using the trapezoidal rule are

```
limits  and  integral
2   3        0.32566
2   6        1.04079
..
..
2  30        15341782.64000
```

and which are accurate to three or four significant figures with 200 points in the integration. Six-figure accuracy would be obtained by using Simpson's rule.

11.3.5 Integrating data from an experiment

Integrating a curve formed from data points obtained from an experiment has to be done numerically. The safest method is to draw straight lines between the data points and calculate the area as rectangles and triangles, which is what the trapezoidal method does, and this works even if the data is not evenly spaced. You might in addition want to fit the data by a least squares method, which has the effect of smoothing the data, integrating the resulting equation and comparing the results. Alternatively, fitting a polynomial to the data and then integrating may give the required solution.

11.3.6 Atomic and molecular beam scattering

In this next extended example, more of a 'real world' rather than a purely mathematical calculation is described; it is the classical model of the elastic scattering of one particle, an atom or molecule, off another. The formulae are developed first, and then integrated using the Coulomb and Lennard-Jones 6-12 potentials, the latter being done numerically. In this example, the scattering angle is calculated. In a later example, the actual trajectory followed by a particle during elastic scattering is calculated and drawn.

11.3.7 Background and derivation of formulae

Geiger and Marsden performed one of the most famous scattering experiments during the first decade of the twentieth century. Alpha particles were passed through a very thin Au foil and the positions of these particles detected by scintillation counting after scattering off the Au atoms. In 1910, while at Manchester University, Rutherford deduced that the mass of the atom was not distributed like that in a raisin pudding throughout its volume, but concentrated at its centre in a very small dense nucleus, thus confirming a 'planetary' model of the atom. This laid the foundation for the old quantum (Bohr) model of the atom.

In the modern version of this experiment, the nature of the force between two particles, such as atoms, molecules, or ions, is sought. A cold beam of atoms or molecules (particles) can be made by passing its gas through a small conical shaped aperture (a skimmer) into a region of high vacuum, typically $<10^{-6}$ torr. If the particles in one beam strike those in a second beam made in the same way, but at right angles, the particles can scatter off one another elastically. This scattering is not exactly like that of one snooker or billiard ball bouncing off another, but depends on the intermolecular potential energy between the particles. This may be mainly Coulombic if the particles are charged, but if neutral, then a Lennard-Jones potential better describes the interaction; see Fig. 5.30 (of the online solutions) for a graph of the potential. The angular position of the scattered particles can be used to calculate the form of the interaction potential. Nowadays detection of the particles' position is measured by a CCD camera with a scintillation plate in front of it. This converts a particle's kinetic energy into a photon. Alternatively, if this is not sensitive enough a specially designed particle detector has to be made and this moved from place to place to record the complete scattering pattern.

The geometry of the experiment is sketched in Fig. 11.4. The incident beam of particles, which has the shape of a cylindrical rod, is aimed towards the target atom. The potential energy between particles depends only on the separation of the particles; this is called a central force and the motion of the two particles about their centre of mass, can always be reduced to an equivalent one-particle problem, which means that the target can be considered to have infinite mass and therefore cannot move. If the incoming particles are initially off-axis they are deflected by the target and generally continue to travel away from their source but if they are more or less on axis they can be backscattered, orbit or slingshot around the target.

The incident particle starts at $-\infty$ and a distance b off the axis, and has a total energy of E_0. The distance b is called the *impact parameter*, Fig. 11.4, and is the distance by which the particles would miss one another if there were no interaction between them. The distance between the particles is the vector $\boldsymbol{r}$, and the closest that they approach to one another is r_0. This minimum distance occurs in a repulsive potential at a distance greater than b, because the initial energy causes the incoming particle to rise up the potential 'hill' to an extent depending on its total energy. A more realistic potential, such as a Lennard-Jones

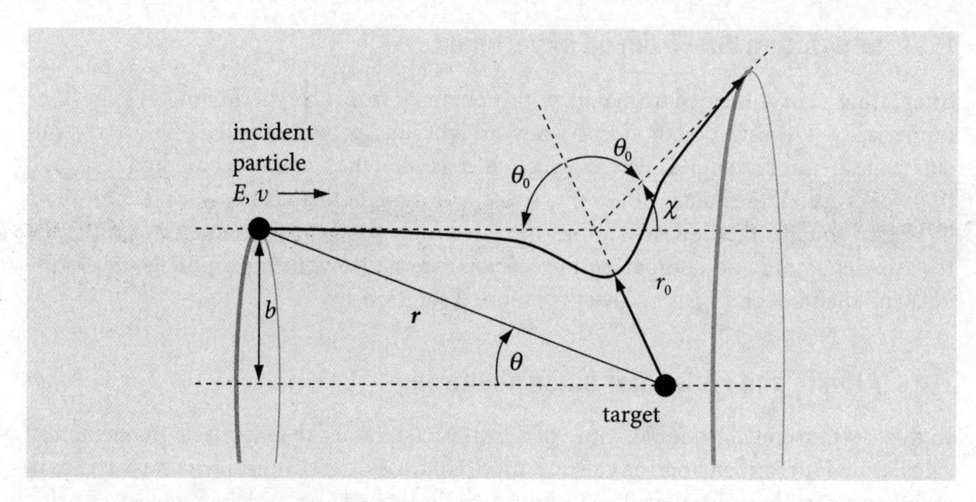

Fig. 11.4 Geometry and definition of angle in particle–particle scattering. θ is the angle at any position $\boldsymbol{r}$, θ_0 the angle when the particles are closest at r_0 and χ is the scattering angle. The incoming beam has cylindrical symmetry.

6-12, has both attractive and repulsive parts. The attractive part is usually more extended than the repulsive, so that the particles are attracted to one another at long range and repel at short range, consequently, the distance of closest approach can now be smaller than b. The particle has a final asymptotic angle χ and the geometry shown in the figure gives this angle as

$$\chi = \pi - 2\theta_0, \tag{11.5}$$

where θ_0 is the angle between the particles when they are closest. The aim of the calculation is to find θ_0 as a function of b and E_0, and hence χ, which can be measured experimentally.

During a collision, the total energy and angular momentum are both conserved and are therefore constant, because no external forces act on the particles. The initial energy is $E_0 = \mu v_0^2/2$, where v_0 is the asymptotic relative radial velocity and the reduced mass of the two particles is μ. The initial energy is equal to the total energy, which is the sum of the kinetic and potential energy $U(r)$ giving

$$E_0 = \frac{\mu v^2}{2} + \frac{L^2}{2\mu r^2} + U(r). \tag{11.6}$$

The kinetic energy is $\mu v^2/2$, where $v = dr/dt$ is the radial velocity at time t during the collision, and L the angular momentum. The centrifugal energy is $L^2/2\mu r^2$, which is positive, and when this reaches its maximum, the radial velocity v is zero, and the particles are at their closest point, r_0. Since the initial velocity v_0 is perpendicular to b, the angular momentum is

$$L = \mu b v_0 = b\sqrt{2E_0\mu},$$

making

$$E_0 = \mu v^2/2 + E_0(b/r)^2 + U(r). \tag{11.7}$$

In this form the centrifugal energy is a function of r and the total potential energy can be considered to be the effective potential $U_{eff}(r) = E_0(b/r)^2 + U(r)$ instead of simply $U(r)$.

By definition, angular momentum is $L = \mu r^2 d\theta/dt$ and, as it is also $L = \mu b v_0$, therefore

$$d\theta = b v_0 r^{-2} dt.$$

Our aim is to find θ the scattering angle and, because the last equation involves $d\theta$ and dt, an integration will be needed, and a change of variable to r from t. This change is found using the energy which, by definition, is $E = \mu(dr/dt)^2/2$ giving $dt = \sqrt{\mu/2E}dr$. Combining this with $d\theta = bv_0 r^{-2}dt$, substituting $E = \mu v^2/2$, using equation (11.7), and simplifying a little gives the differential equation;

$$\frac{d\theta}{dr} = -\frac{b}{r^2}\left(1 - \frac{U(r)}{E_0} - \frac{b^2}{r^2}\right)^{-1/2}. \tag{11.8}$$

The leading negative sign is introduced because $d\theta/dr$ describes the incoming particle. The angle θ can be calculated by integrating from its minimum value at $r = r_0$ to infinity, and takes the form,

$$\theta_0 = -\int_\infty^{r_0} \frac{d\theta}{dr} dr = +b\int_{r_0}^{\infty} \frac{dr}{r^2[1 - U(r)/E_0 - b^2/r^2]^{1/2}}. \tag{11.9}$$

This equation is also conveniently rewritten as $\theta_0 = \int_{r_0}^{\infty} g(r)dr$, where

$$g(r) = br^{-2}[1 - U(r)/E_0 - b^2/r^2]^{-1/2}, \tag{11.10}$$

and g will be used in the numerical calculations. The angle θ_0 depends on the impact parameter b, the initial energy of the particle E_0, and the potential U, each of which must be known, as must r_0, before the calculation can be completed.

11.3.8 Calculations

The scattering by two different potentials, the Coulomb and the Lennard-Jones 6-12 will be illustrated. The first step in the calculation is to find the equation for the separation r_0. This is done by using conservation of energy and then knowing that the *radial* velocity v is zero at the turning point r_0. From (11.6) with $v = 0$ and $r = r_0$, the energy is

$$E_0 = L^2/2\mu r_0^2 + U(r_0).$$

Substituting for $L = b\sqrt{2E_0\mu}$ and rearranging gives

$$1 - b^2/r_0^2 - U(r_0)/E_0 = 0. \tag{11.11}$$

This is solved for r_0 and the largest root used, if there is more than one. Depending on the potential U, r_0 may have to be found numerically, for example, by using the bisection method, see 11.2 and algorithm 11.4.

11.3.9 Coulomb potential

The next step in the calculation depends on the potential. The Coulomb potential energy between charges q_1 and q_2 at separation r is $U(r) = \alpha/r$ where $\alpha = q_1q_2/4\pi\varepsilon_0$ in SI units. Depending on the charges, the Coulomb potential can be repulsive or attractive. This $1/r$ form of potential also describes that due to gravity, where $\alpha = -Gm_1m_2$. For example, it will describe a comet being attracted to and flung around the sun or a planet. G is the gravitational constant and m_1 and m_2 the masses.

Considering now the Coulomb potential, the smallest separation of the particles is found from equation (11.11), which is a quadratic in r_0. The result is

$$r_0 = \frac{\alpha \pm \sqrt{\alpha^2 + 4b^2E_0^2}}{2E_0}$$

and the larger root should be taken. The smallest value r_0 can take occurs when $b = 0$, and is α/E_0, which is the minimum separation because it is limited by the initial energy. The angle θ_0 is given by equation (11.9),

$$\theta_0 = +b\int_{r_0}^{\infty} r^{-2}\left[1 - \frac{\alpha}{rE_0} - \frac{b^2}{r^2}\right]^{-1/2} dr$$

which can, conveniently, be solved using Maple:

```
> assume(b, positive):
> b*Int(1/r^2*1/(sqrt(1-b^2/r^2-U(r)/E0)),r):
    %= simplify(value(%),symbolic);
```

$$b\int \frac{1}{r^2\sqrt{1 - \frac{b^2}{r^2} - \frac{\alpha}{rE0}}} dr = -\arctan\left(\frac{1}{2}\frac{2b^2E0 + \alpha r}{\sqrt{r^2E0 - b^2E0 - \alpha r}\sqrt{E0}b}\right)$$

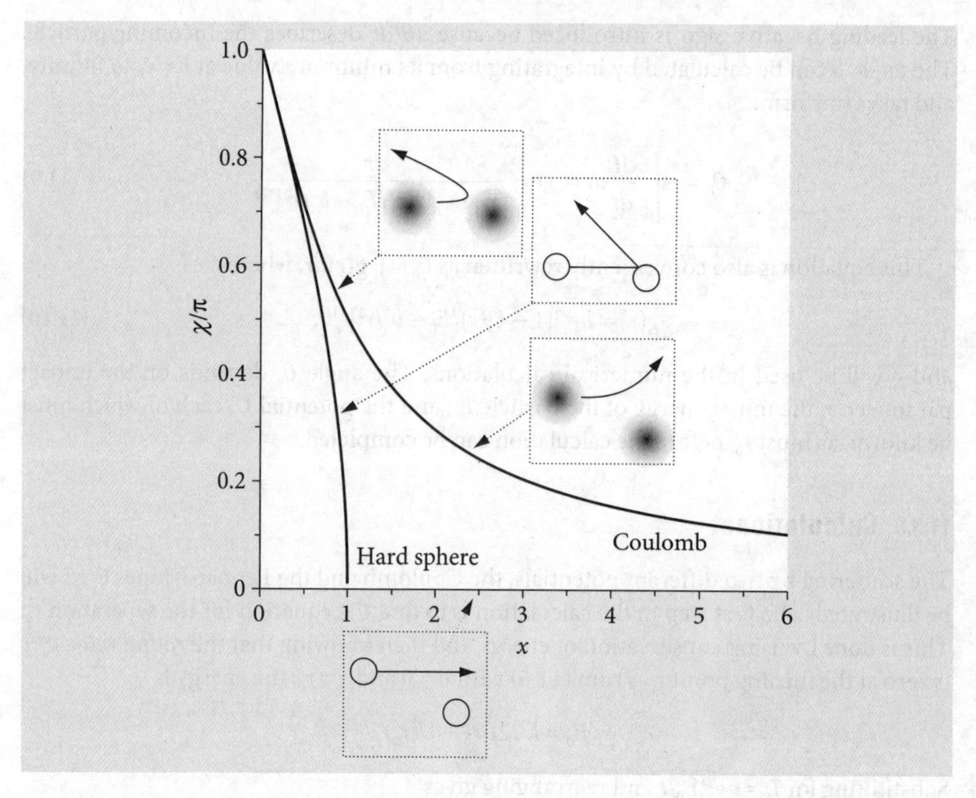

Fig. 11.5 Scattering angle as fraction of π radians or χ/π vs dimensionless parameter x for hard sphere and repulsive Coulomb potentials (see text). The inset diagrams show, schematically, the hard sphere (white circles) and Coulomb deflections (diffuse filled circles), each pair at the same value of x.

To obtain the limit when $r = \infty$, make r large so that $r^2 \gg r$ and $r > b^2E_0$, and then cancel the terms. The other limit, $r = r_0$, is $\tan^{-1}(\infty) = \pi/2$ and together with $2\theta_0 + \chi = \pi$, these limits make the scattering angle

$$\chi = 2\tan^{-1}\left(\frac{\alpha}{2E_0 b}\right), \tag{11.12}$$

which lies between 0 and π radians, because the Coulomb potential is wholly repulsive. At $b = 0$, $\chi = \pi$ and a head-on collision occurs; at larger impact parameters the scattering angle gradually decays to zero. The particle never really misses the target, on an atomic scale, because the Coulomb interaction spreads out into space to infinity. Fig. 11.5 shows the scattering angles for the Coulomb collision, plotted as χ/π vs $x = \alpha/2E_0b$ which is dimensionless and for a hard sphere collision vs $x = d/b$ (see Q11.7). The quantity $\alpha/2E_0$ has dimensions of distance because $\alpha = q_1q_2/4\pi\varepsilon_0$ and has units $C^2/C^2J^{-1}m^{-1}$ where C is the Coulomb, and is therefore the effective diameter of the target. In the Coulomb case, the potential spreads out as r^{-1} and one particle influences the other and is the cause of the small angular deflection at large x values. At small x, which is close to a 'head on' collision, Coulomb scattering is similar to that of a hard sphere. With the hard sphere, because there is no interaction between particles, the scattering angle χ is zero when the separation of the particles is greater than the sum of their diameters.

11.3.10 **The Lennard-Jones potential**

The LJ potential $U(r) = 4\varepsilon[(\sigma/r)^{12} - (\sigma/r)^6]$ has both attractive and repulsive parts to it; see Fig. 5.30 (of the online solutions). The well depth is $-\varepsilon$ and σ is a measure of the extent of the potential. The minimum energy occurs at $2^{1/6}\sigma$. Both the integration to find θ_0 and the initial value r_0, have to be calculated numerically. The interaction between two heavy noble gas atoms is described well with this type of potential, but molecules more approximately so. Substituting for the potential into (11.9) gives

$$\theta_0 = +b\int_{r_0}^{\infty} r^{-2}\left[1 - 4\frac{\varepsilon}{E_0}\left[\left(\frac{\sigma}{r}\right)^{12} - \left(\frac{\sigma}{r}\right)^{6}\right] - \frac{b^2}{r^2}\right]^{-1/2} dr. \tag{11.13}$$

This integral has to be solved numerically; it also has a difficult limit and the function tends to infinity when $r = r_0$ and decays slowly to zero as r increases, a value of at least $r = 1000$ is typically needed to get an accurate integration when σ, b, ε, $E_0 \approx 1$. Using the substitution $r = r_0/(1-w^2)$ simplifies the limits and also makes the calculation more accurate. When $r = r_0$ then $w = 0$ and when $r = \infty$, $w = 1$ thus the new limits are $w = 0$ and 1. Differentiating r gives $dr = 2r_0w/(1-w^2)^{-2}dw$ and this also has to be substituted into the integral giving

$$\theta_0 = +\frac{2b}{r_0}\int_0^1 w\left[1 - \frac{U(r_0/(1-w^2))}{E_0} - \frac{b^2(1-w^2)^2}{r_0^2}\right]^{-1/2} dw. \tag{11.14}$$

The distance r_0 still has to be found and using (11.11) it is the largest root of

$$r = \frac{b}{\sqrt{1 - 4\frac{\varepsilon}{E_0}\left[\left(\frac{\sigma}{r}\right)^{12} - \left(\frac{\sigma}{r}\right)^{6}\right]}} \tag{11.15}$$

for each value of b required. The bisection method is easy to use for this calculation. However, as it is important to take the largest positive root, this can be found by starting at a large value of r and decreasing this in steps until a change of sign in (11.15) occurs. Then the bisection method can be used to find r_0; see Algorithm 11.4.

The schematic of the algorithm for the whole calculation over a range of b values is shown below. The calculation only requires the ratio E_0/ε as input and to plot universal curves b is reduced to b/σ so that $\sigma = 1$ and $\varepsilon = 1$ is chosen. For simplicity, Maple's internal integration routine will be used in step 2.(iii).

(1) Define the constants, E_0, ε, etc., range, increment in b, and number of points.
(2) Loop over several b values in the range 0 to 3
 (i) Search for the root, equation (11.15), starting at maximum r which is bm.
 (ii) Find an accurate root r_0 using the bisection method.
 (iii) Calculate the integral equation (11.9) or (11.10).
 (iv) Calculate angle χ
 (v) Store results
 (vi) Increment b
 End Loop
(3) Plot graph.

One possible Maple code to do this is:

Algorithm 11.8 Elastic collision with Lennard-Jones potential χ vs b

```
> E0:= 0.2;                          # energy
  eps:= 1.0; sigma:= 1.0;            # ε,σ for LJ potential
  n:= 150:                           # number of b values
  bm:= 3.0;                          # max b value
  delta_b:= bm/n;                    # change in b
  b:= delta_b:                       # initial b value
  angle:= Array(1..n):
  bval:= Array(1..n):                # arrays for results
  U:= r-> 4*eps*((sigma/r)^(12)-(sigma/r)^6);       # LJ 6-12
  g:= (r,b)-> b/(r^2*sqrt(1.0-U(r)/E0-(b/r)^2)); # eqn (11.10)
  angle[1]:= Pi:
  bval[1]:= 0.0:                     # first result angle=Pi, b=0
  for j from 2 to n do               # step (2) changing b
    f:= r0-> r0^2*(1 - U(r0)/E0) - b^2: #eq (11.11) solve for r0
    r0:= bm;                         # start at max value bm
    while f(r0) > 0.0 do             # step (i)
      r0:= r0 - delta_b;             # look for largest root
    end do:
# start bisection method, root close to x0, step (ii)
    x1:= r0 - 0.2: x2:= r0 + 0.2:   # set initial values
    xm:= x1:
```

```
    xtol:= 1.0*10^(-5):               # tolerance on root
    m:= 0;                            # m is limit to stop calc
    while abs(f(xm)) > xtol and m < 50 do
      m:= m + 1;
      xm:= (x2 + x1)/2.0;
      if f(x1) * f(xm) > 0.0 then x1:= xm
      else x2:= xm
      end if;
     end do:
    r0:= xm + 1e-3;                   # largest positive root
# end bisection,                      # end step (ii)
# Integrate with substitution included
    ff:= g( r0/(1.0-w^2), b)*2*r0*w/(1.0-w^2)^2;
    theta:= int(ff, w= 0.0..1.0); # integrate (11.10) step (iii)
    chi:= Pi - 2*theta;               # angle χ in radians
    angle[j]:= chi:                   # store results, step (v)
    bsig[j]:= b/sigma;
    b:= b + delta_b;                  # increment b step (vi)
   end do:                            # end step (2)
```

A few points should be made about the calculation before considering the results. The integration is very tricky when the deflection angle is large and negative, because this angle can become $-\infty$ and occasionally the integration will fail because here the square root becomes zero or imaginary. This can usually be avoided by adding a small amount to r_0; 10^{-3} was found to be sufficient to prevent this and not significantly affect the results. The largest positive root is sought by calculating from the largest r, which can be taken to be the same as the largest b, to the smallest r, finding a change in sign and then starting the bisection method. This is necessary because the r_0 equation can have three roots when b is large. The function g, equation (11.10) is a function of r and b, and when the substitution is made, r is replaced. The differential dw also has to be substituted for dr, and this can be seen in the expression `ff:=...`

The angle χ vs b/σ is plotted below for three values of the ratio E/ε. The plot is made by making a list out of the `angle` and `bsig` and plotting this. This can be done as

```
ss:= seq( [ bsig[i],angle[i]],i=1..n): plot([ss]);
```

and you can modify the plot using different options.

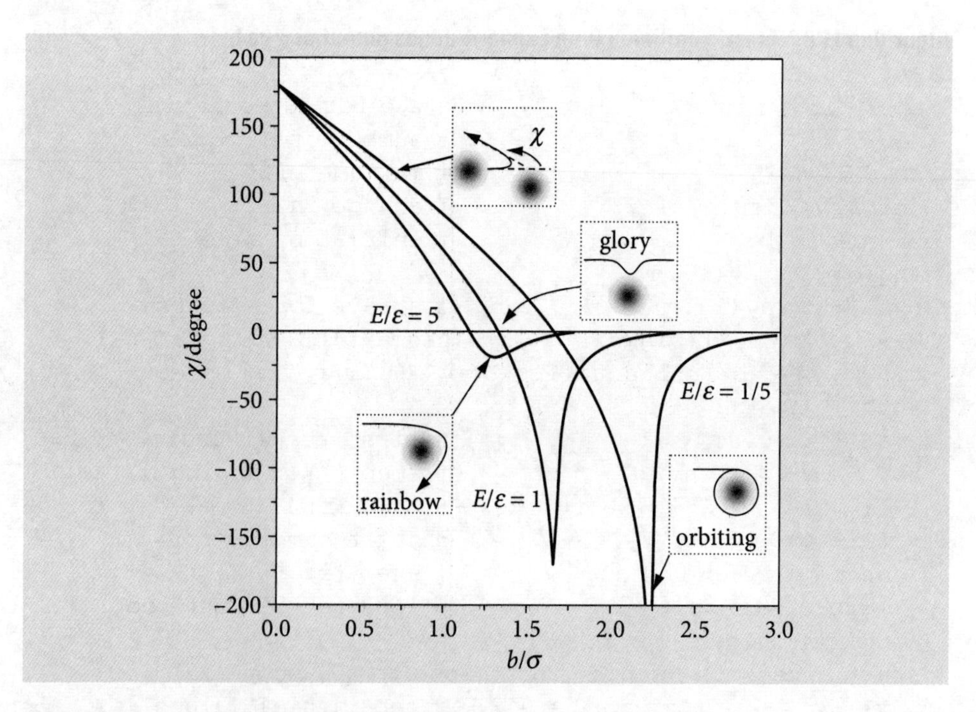

Fig. 11.6 Scattering angles (degrees) with the Lennard-Jones potential.

11.3.11 Analysis of results

When the impact parameter b is small ($b/\sigma < 1$), the recoil angle χ is large and positive and the repulsive part of the potential is dominant. When b increases, the overall deflection can become zero. This is called a 'glory' and the initial energy and values of b mean that the particles are initially attracted to one another, but the impact angle is glancing and, as the repulsive part of the potential is reached, just enough repulsion is present to return the particle on to its original path; the attractive and repulsive interactions exactly cancel. Rainbow scattering occurs when $d\chi/db = 0$ at the minimum of the curve shown in Fig. 11.6. At some b values, the approaching particle can become captured, the angle increases past $-\pi$ to $-\infty$ and the particles orbit one another. This is seen with $E/\varepsilon = 1/5$ at $b/\sigma \approx 2.25$. The trajectories are shown in Fig. 11.10, which follows an algorithm with which to calculate them.

In actual experiments and in quantum calculations, interferences between the different paths the same particle can take are observed. These must be absent in any classical calculation. Scattering is observed experimentally by performing molecular beam experiments and the scattering is, fortunately, not always elastic and chemical reactions can occur and is called reactive scattering. These experiments have provided lots of fundamental information about chemical reactions and there are a number of clear descriptions of this, see, for example, Rigby et al. (1986), Steinfeld et al. (1999), Bernstein (1982), Eyring et al. (1980), and Hirst (1990).

11.3.12 Questions

Full solutions are available at **www.oxfordtextbooks.co.uk/orc/beddard**.

Q11.1 The equation $x_{n+1} = cx_n(1 - x_n)$ *maps* $cx_n(1 - x_n)$ onto x, see Fig. 11.2, but many other expressions can do this. Try $x_{n+1} = c\sin(-x_n)$ and $x_{n+1} = ce^{(-x_n)}$ and see what happens. You may need to change the range of c.

Q11.2 Calculate the integral of $\sin(x^2)$ over the range 0 to 10, by the trapezoidal rule and by Simpson's rule. Choose the number of points sufficient to guarantee accuracy to three decimal places in each case. This is a highly oscillating function, varying between ± 1, so many terms will be needed for an accurate calculation.

Q11.3 Numerically calculate the integral $\displaystyle\int_0^3 \frac{x^6}{1-x}dx$.

Strategy: As this function will become infinite at $x = 1$, the mid-point rule should be used. Choose 1000 points to begin with and then alter this to see how the accuracy varies. The reason for the large number of points is that towards the asymptote $y \to \infty$ at $x = 1$ the function has to be accurately evaluated and values either side of this point have opposite sign and similar value and so almost cancel each other out.

Q11.4 The equation $E = \dfrac{9N_0k_BT}{x_m^3}\displaystyle\int_0^{x_m} \frac{x^3}{e^{+x} - 1}dx$ describes the total energy per mole of a crystalline solid and is used in the Debye theory of the heat capacity of solids. This theory assumes that a solid has a range of vibrational frequencies from zero up to a maximum υ_m, each of which contributes to the total energy. The integral cannot be evaluated algebraically other than at limits of large and small x. The variable x is defined as $x = h\upsilon/k_BT$ and the upper limit x_m is defined similarly but with a maximum frequency υ_m which, for example, in solid Li is $8.06 \times 10^{12}\ \text{s}^{-1}$.

(a) Evaluate the integral by three numerical methods at 1000 K using a 50 point integration and quote the answer to six figures.

(b) Calculate the low and high temperature limits to the integral algebraically, and calculate the heat capacity C_V and compare with the numerical results. Show that at low temperature the heat capacity varies as the cube of the temperature.

Strategy: The trapezoidal and Simpson's rule will fail because at exactly $x = 0$ the denominator is infinity, and the limit making the function zero at $x = 0$ cannot be evaluated by the algorithm. This can be overcome by making the lower limit slightly larger than zero, for example 10^{-6}, which can be done in this instance only because the function's value is almost zero when x is close to zero, and will not appreciably affect the integral.

Q11.5 As discussed in the text, the integral $\int_0^\infty \frac{1}{1+x^2}dx$ converges slowly as x increases and when integrated numerically, the result gets worse as the integration upper limit increases. Use the transformation $x = e^{uz} - 1$ to evaluate the integral numerically.

Strategy: The algebra to do the calculation is shown in the text. The choice of u is not crucial, it could be 1 or 10, but the numerical limits have to be inspected to ensure the entire curve is included in the summation.

Q11.6 Quantum mechanical tunnelling is a commonly observed phenomenon. It happens when alpha particles leave the nuclei of heavy radioactive nuclei, in the scanning tunnelling microscope, where electrons tunnel from the tip to the substrate, and in molecules it is observed when H and D atoms pass through a potential barrier. The probability of tunnelling through a finite width barrier at energy E is given by G, where $\ln(G_E) = -\frac{\sqrt{2m}}{\hbar}\int_{x_1}^{x_2}(V(x) - E)^{1/2}dx$, and $V(x)$ is the barrier shape and E the energy. The integral is the area of the barrier above energy E. The probability of crossing the barrier decreases exponentially with this area.

Numerically integrate and obtain the permeability at several values of the energy $0 \le E \le E_0$ for the Eckardt barrier $V(x) = E_0\,\text{sech}^2(ax)$ which is symmetrical and has a width L. For simplicity, assume $\sqrt{m/h}$ and E_0 are unity and $a = 0.5$. Repeat the calculation by doubling $\sqrt{m/h}$ and a and observe what happens.

Strategy: Before the integration can be performed, the limits have to be determined. As the barrier is symmetrical, the limits are also. At an energy $V(x) = E$, the value of x is $x_E = \frac{1}{a}\text{sech}^{-1}\sqrt{E/E_0}$ and the limits are therefore $\pm x_E$.

Q11.7 Calculate the scattering angle χ vs impact parameter b for a hard sphere potential of radius d with initial energy E_0. The hard sphere has a potential of zero at distances $r > d$ and is otherwise infinite. The result is shown in Fig. 11.5.

11.4 Numerical solution of differential equations

Although many differential equations can be explicitly integrated, as illustrated in Chapter 10, there are many others of interest, particularly in quantum mechanics, that have no such solution other than by a series expansion and even this may only be approximate. In these cases a numerical solution must be sought; however, this is always preceded by as much algebraic analysis of the problem as is possible.

As with any integration, a constant is produced for each integration step; two constants are necessary for an equation with derivative d^2y/dx^2 because two integrations are necessary to produce y. In differential equations, integration constants are found by using the initial values and/or the boundary values to y and its derivatives, and these are determined by the particular problem being solved. These are described in Chapter 10.

11.4.1 Euler's and other methods of numerical approximation

The analytic or algebraic solution to an equation produces a function whose values are continuous in x, y, or t, and that allows us to calculate its value at any point we might choose, there being, effectively, an infinite number to choose from. A general feature of the numerical solution of a differential equation is that it is converted into *recursive finite-difference equation* and this is because a numerical solution can be obtained only at relatively few discrete values. The general first-order differential equation is

$$dy/dt = f(t, y) \tag{11.16}$$

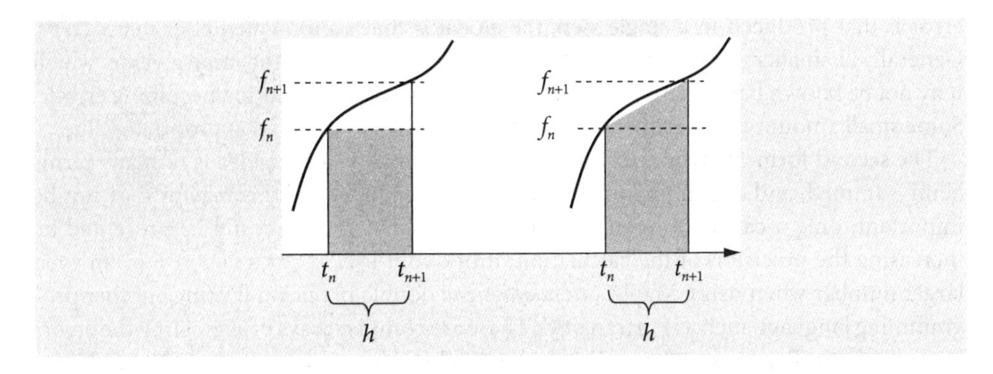

Fig. 11.7 The grey area calculated in the Euler method (left) measures only approximately that under the curve. The second diagram (right) shows the area calculated in the improved Euler method equation (11.23) and is clearly a better approximation. For clarity, the abbreviation $f_n \equiv f(t_n, y_n)$ is used.

which means that $f(t, y)$ is a function of y and t such as $te^{-t} + y + 3$. The initial condition is $y(t_0) = y_0$. The differential has to be approximated in some way, because we can only evaluate at discrete points, and the simplest way to do this is to write

$$\frac{\Delta y}{\Delta t} \approx \frac{dy}{dt} = f(t, y). \tag{11.17}$$

If $\Delta t = t_{n+1} - t_n \equiv h$ is small enough, then this first-order approximation is a good one. The change in y is calculated similarly,

$$\frac{\Delta y}{\Delta t} = \frac{y_{n+1} - y_n}{h} \approx f(t_n, y_n). \tag{11.18}$$

Rearranging gives the Euler method equations

$$\begin{aligned} &y_{n+1} = y_n + hf(t_n, y_n), && \text{(a)} \\ y_0 = y(t_0), \quad &t_n = t_0 + nh, \quad n = 0, 1, 2 \cdots N-1. && \text{(b)} \end{aligned} \tag{11.19}$$

The nth value of t is $t_n = t_0 + nh$, t_0 being the initial time when $n = 0$, and the initial condition is $y(t_0) = y_0$. The time values are $t_0, t_1 = t_0 + h, t_2 = t_0 + 2h$, and so forth. The grey area in Fig. 11.7 (left) is $hf(t_n, y_n)$ which approximates the whole area under the curved line in the Euler method.

The calculation is easy to implement, but several terms have to be defined before starting. These are the initial condition, which means t_0 and $y(t_0) = y_0$, the final time to stop the calculation t_{max}, and the number of points N. These will define h as $h = \dfrac{t_{\max} - t_0}{N}$. Choosing N is not difficult; as a rule-of-thumb start with 100 points and see what happens, then repeatedly double or treble this value. Increasing N reduces h and this will often have to be $\ll 0.01$, but if it is too small, the calculation will take a considerable time to complete and therefore some compromise is always necessary.

Sometimes, the solution of the equation is sought to smaller values than t_0, effectively running the calculation backwards, which means making h negative. In this case, the time has to be reduced as

$$t_n = t_0 - nh$$

and the consecutive values of y calculated as

$$y_{n+1} = y_n - hf(t_n, y_n). \tag{11.20}$$

11.4.2 **Errors**

Before carrying out the calculation, the error implicit in this method should be considered. There will always be an error associated with a numerical method because an approximation to the derivative has always to be made. The error takes two forms; the approximation to dy/dx produces a truncation error; a better approximation could be made by expanding this as a series. The truncation errors can be *local* or *global*, the local error is that produced in a single step, the global is that accumulated after many steps. Generally, a smaller value of h reduces these errors; conversely, too large a value, which may not be known beforehand, can lead to instability with the solution becoming erratic. Some small amount of experimenting is often required to fix h to an appropriate value.

The second form of error is due to rounding. A small value of h leads to many terms being summed, and depending on the particular form of $f(y, t)$ this may or may not be important; only a calculation will reveal this. However, it can mostly be prevented by increasing the precision of the calculation; if in doubt use `Digits:= 25:` or an even larger number when using Maple, and *always* use double precision if using another programming language such as Fortran or C. There are complex ways of estimating the errors but, pragmatically, it is simpler to double the number of points repeatedly until no change in the calculated data is observed.

11.4.3 Using Euler's method

To illustrate the method, the equation $dy/dt = -2e^{3-t} - 3y$ will be integrated from $t = 2$ to 7 with initial condition $y(t_0) = 1$, hence $t_0 = 2$. This equation can be integrated analytically and this allows a check on the algorithm. The solution looks something like a Lennard-Jones potential between two molecules, with a single minimum and tending to infinity at $t = 0$ and to zero at large t. The calculation starts by defining the initial values and other constants such as the number of points for the grid over which to do the calculation. Two arrays, `Eulery` and `dtime`, are used to store the results and the initial values are placed into the first element of each array. Notice that the values of y and t are both incremented in the `for..do` loop which changes y and t for each new step. The function to be integrated is always the right-hand side of the differential equation when written as $dy/dt = \cdots$.

The outline algorithm is

(1) Define equation.

(2) Set initial parameters, y_0, t_0, number of integration points, N.
Calculate h.
Save initial values.

(3) Start loop on N
  (i) Increment y equation (11.19).
  (ii) Increment t.
  (iii) Save y and t.
End loop.

Algorithm 11.9 Euler's method

```
> dydt:= (t,y)-> -2*exp(3-t)-3*y;        # RHS of eqn
  t0:= 2.0:                              # initial t
  y0:=1.0:                               # initial y
  maxt:= 7.0:                            # max time
  N:= 50:                                # numbr points
  Eulery:= Array(1..N+1):                # results
  dtime:= Array(1..N+1):
  h:= (maxt-t0)/N:                       # time incr'mt
  print('step size =',h);
  y:= y0:          t:= t0:               # start Euler
  Eulery[1]:= y0:  dtime[1]:= t0:        # init values
  for i from 2 to N+1 do                 # loop step(3)
    y:= y + h*dydt(t, y):                # inc y step(i)
    t:= t + h:                           # incr'mt time
    Eulery[i]:= y;                       #save y step(iii)
    dtime[i]:= t;                        # save t
  end do:                                # end loop
```

The step `y:= y + h*dydt(t,y):` increments the value of y making the new value equal to the old value plus `h*dydt(t, y)`. The numerical data stored as `Eulery` for two values of the number of iterations $N = 50$ and 500, and the analytic solution is plotted

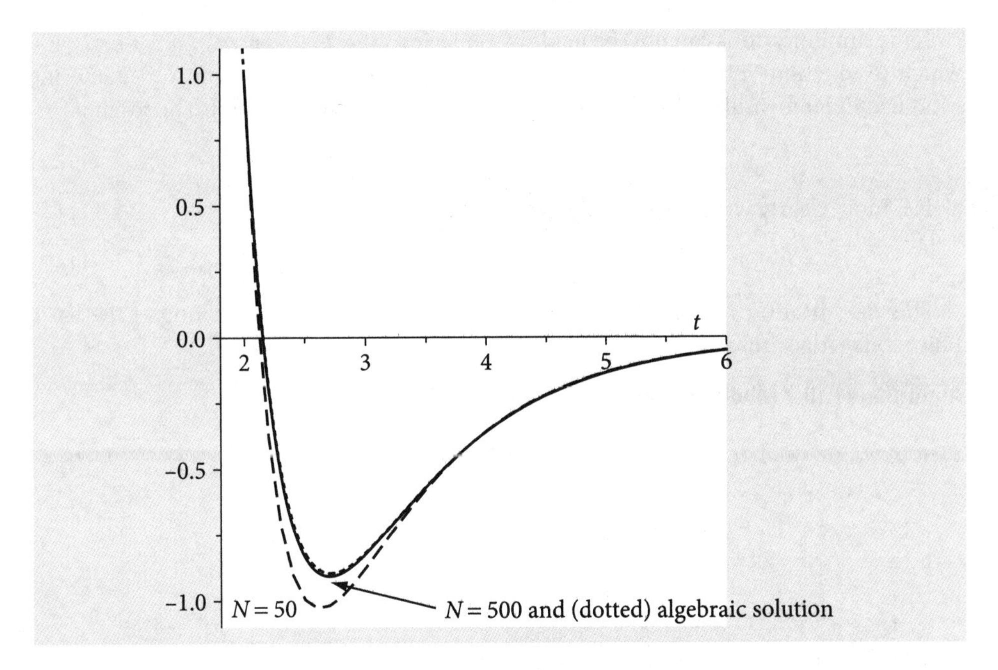

Fig. 11.8 The algebraic solution to $dy/dt = -2e^{3-t} - 3y$ (dotted line) and two Euler method numerical solutions with $N = 50$ and 500. The initial condition is $y(t_0) = 1$ and $t_0 = 2$. The improved (modified) Euler method and the Runge–Kutta methods, Section 11.7, produce curves essentially identical to the algebraic solution.

in Fig. 11.8. With $N = 500$, only a slight difference between the analytic solution and numerical is observed and this becomes better if more points are used. The improved (modified) Euler method and the Runge–Kutta methods, see Section 11.7, produce curves essentially identical to the algebraic solution.

The basic plot is made using

```
> ss:= seq([dtime[i], Euler[i]], i= 1..N ):
  p1:= plot([ss]):
```

and different plots on the same graph made using `display(p1,p2,p3)`, where `p2` and `p3` are made in the same way as `p1`. The plots can be customized when the calculation is complete.

11.4.4 Improving on the Euler method

Many of the examples used are not particularly sensitive to the integration method; however, some equations are generally exquisitely sensitive, while others are only so for some sets of initial conditions and not for others. It is not obvious beforehand whether this is going to be the case or not and because of this more sophisticated integration schemes are necessary. The two described here are an improvement on the Euler algorithm and the Runge–Kutta method.

Modified Euler

The Euler method is rather crude but is easy to implement and is easily improved. Consider again equation (11.19). The updated y value was calculated using $y_{n+1} = y_n + hf(t_n, y_n)$, but a better approximation to the area under the curve between y_{n+1} and y_n can be made than $hf(t_n, y_n)$ which is a rectangle, Fig. 11.7. This better approximation makes the area into a trapezoid and this is shown as shaded in the right-hand panel of Fig. 11.7. Its value is

$$(h/2)[f(t_n, y_n) + f(t_{n+1}, y_{n+1})],$$

The better estimate to update y is

$$y_{n+1} = y_n + (h/2)[f(t_n, y_n) + f(t_{n+1}, y_{n+1})]. \tag{11.21}$$

Some simplifications can now be made by replacing $t_{n+1} = t_n + h$ and $y_{n+1} = y_n + hf(t_n, y_n)$, which is equation (11.19), into the second occurrence of the function to become the Modified Euler formula, which is sometimes also called Heun's method. The result is

$$y_{n+1} = y_n + (h/2)[f(t_n, y_n) + f(t_n + h, y_n + hf(t_n, y_n))]. \tag{11.22}$$

This is more clearly written in three steps;

$$k_1 = f(t_n, y_n), \qquad k_2 = f(t_n + h, y_n + hk_1), \qquad y_{n+1} = y_n + h(k_1 + k_2)/2 \tag{11.23}$$

This modification replaces the simpler expression in the `for..do` loop of the basic Euler code (Algorithm 11.9).

Algorithm 11.10 Modified Euler's method or Heun's method

```
.. ..
for i from 2 to N+1 do        # modified Euler eqn (11.23)
  k1:= dydt(t,y):
  k2:= dydt(t+h, y+h*k1):
  y:= y + h*(k1+k2)/2:        # increment y
  t:= t + h:                  # increment time
  Eulery[i]:= y;              # save y
  dtime[i]:= t;               # save t
end do:                       # end loop
```

When the equation is second order and the equations are coupled then the method changes slightly and is given in Section 11.5.

Runge–Kutta

There are other more sophisticated methods than the modified Euler of which the Runge–Kutta method is often the method of choice and this is what Maple will usually use if you ask it to do a numerical calculation using `dsolve(.., numerical);`. The Runge–Kutta method uses the average of the gradient at the mid- and end-points of an interval to calculate the next y value. This is an example of a predictor–corrector method, which as the name suggests, predicts the next value then makes a weighted correction to this (Prest et al. 1986). In this method, four quantities are required for each equation being solved. The equation being solved is $dy/dt = f(t, y)$ and h is the increment in t.

$$k_1 = f(t_n, y_n)h \qquad k_2 = f(t_n + h/2, y_n + k_1/2)h$$
$$k_3 = f(t_n + h/2, y_n + k_2/2)h \qquad k_4 = f(t_n + h, y_n + k_3)h$$
$$y_{n+1} = y_n + (k_1 + 2k_2 + 2k_3 + k_4)/6 \qquad t_{n+1} = t_0 + h$$

Note that this method would be Simpson's rule if $f(t, y)$ did not depend on y. If the equation to integrate is written as `dydt:=(t,y)-> ...` as in Algorithm 11.9 then the Runge–Kutta code could be

Algorithm 11.11 Runge–Kutta method

```
.. ..
for i from 2 to N+1 do                          # Runge Kutta
  k1:= dydt(t,y):
  k2:= dydt(t + h, y + h*k1/2):
  k3:= dydt(t + h, y + h*k2/2):
  k4:= dydt(t + h, y + h*k3/2):
  y:= y + h*(k1 + 2*k2 + 2*k3 + k4)/6:  # increment y
  t:= t + h:                            # increment time
  Eulery[i]:= y;                        # save y
  dtime[i]:= t;                         # save t
end do:                                 # end loop
```

The equations, k, etc, in this algorithm are different from but equivalent to those in the text. You can use this method for any calculation for which the Euler or modified Euler method is appropriate; it will take slightly longer to calculate but should produce results that are more accurate. The method for coupled equations is described in Section 11.5.

The relative errors in the Euler, modified Euler, and Runge–Kutta vary as $O(h^2)$, $O(h^3)$, $O(h^5)$ respectively, where the local error is estimated as $|f(x) - f_{numeric}(x)|$ with f as the true algebraic solution at some point x.

11.4.5 Other Euler type methods

There are also some methods that have been developed for particular types of problems. One of them, the Euler–Cromer method, is used to improve the calculation of periodic systems such as the pendulum and is described in Section 11.6.4. Another, and very widely used method, is the *Verlet algorithm*, which is described next, and is used to integrate equations of motion in molecular dynamics (MD) simulations; in dynamical systems in general, e.g. bouncing balls, motion of planets, and to make animations in computer games look realistic; see Frenkel & Smit (1996).

11.4.6 Verlet algorithm: calculating dynamics and performing MD simulations

An MD simulation is a technique to calculate equilibrium and transport properties of many body systems, such as proteins, DNAs, polymers, or simple liquids. The calculation is normally classical, and integrates Newton's equation of motion. As classical mechanics is used, quantum effects such as tunnelling are ignored. The speed and capacity of modern computers make MD a common tool for the study of bio-molecules in particular, and a vast literature exists.

The MD simulation proceeds by calculating the forces between pairs of particles and integrating them at some time t; this time is now incremented by a small amount Δt and the process repeated. In a protein or DNA, the forces between the atoms will comprise those of individual chemical bonds, with their own particular force constants for bond stretching, bending, and rotating. Between non-bonded atoms, Lennard-Jones type potentials could be used and the Coulomb potential between charged groups. These interactions can become rather complicated and are usually parameterized and standard sets of values used.

In the Verlet algorithm, the equations of motion are not solved as such, which means that the starting point is not to simplify the differential equation and then approximate them as done in the Euler and Runge–Kutta methods. Instead, a Taylor expansion in time about the initial position is made. To use the Verlet algorithm, the force or acceleration of each particle must be known; this is always the case once the potential is known because force is the negative derivative of the potential with distance; $f_x = -dU(x)/dx$. As the equations of motion are second order, two initial conditions are necessary for each atom or molecule and these are the starting point and the initial velocity. From these, the next position, a very small distance away, is calculated, perhaps by using Newton's laws, and from these two positions all the other positions, and the corresponding velocities, can be found from the algorithm.

The Verlet calculation starts by making a Taylor expansion (see Chapter 5) about the position at time $t+\Delta t$ and $t-\Delta t$ and the sum and difference between these two positions calculated. The formula for a Taylor series of any function $f(x)$ expanded about a point x_0, is

$$f(x) = f(x_0) + (x - x_0)\left(\frac{df}{dx}\right)_{x_0} + \frac{(x-x_0)^2}{2!}\left(\frac{d^2f}{dx^2}\right)_{x_0} + \cdots.$$

Changing notation, from the general to the specific, the function f represents position r which is a function of time; the notational change is $f(x) \to r(t)$. Expanding r about time t_0 to $t = t_0 + \Delta t$, where Δt is a small time step, gives

$$r(t_0 + \Delta t) = r(t_0) + v(t_0)\Delta t + \left(\frac{d^2r}{dt^2}\right)_{t_0}\frac{(\Delta t)^2}{2!} + \left(\frac{d^3r}{dt^3}\right)_{t_0}\frac{(\Delta t)^3}{3!}t\cdots$$

Similarly for a small decrease in time

$$r(t_0 - \Delta t) = r(t_0) - v(t_0)\Delta t + \left(\frac{d^2r}{dt^2}\right)_{t_0}\frac{(\Delta t)^2}{2!} - \left(\frac{d^3r}{dt^3}\right)_{t_0}\frac{(\Delta t)^3}{3!} + \cdots$$

where the velocity is $v(t_0) = (dx/dt)_{t_0}$ and the acceleration is given by

$$(d^2r/dt^2)_{t_0} = f(t_0)/m,$$

where m is the mass and f is the force, which is calculated from the potential at the position corresponding to time t_0. The sum and the difference of increments in r are

$$r(t_0+\Delta t) + r(t_0-\Delta t) = 2r(t_0) + \left(\frac{d^2r}{dt^2}\right)_{t_0}(\Delta t)^2 + \cdots \tag{11.24}$$

$$r(t_0+\Delta t) - r(t_0-\Delta t) = 2v(t_0)\Delta t + 2\left(\frac{d^3r}{dt^3}\right)_{t_0}\frac{(\Delta t)^3}{3!} + \cdots \tag{11.25}$$

If the time increment Δt is small, the third and higher derivatives can be ignored giving from equation (11.24)

$$r(t_0+\Delta t) \approx 2r(t_0) - r(t_0-\Delta t) + \frac{f(t_0)}{m}(\Delta t)^2 \tag{11.26}$$

which will be the new position of the particle after time step Δt. The velocity is obtained from equation (11.25) as the difference

$$v(t_0) = \frac{r(t_0+\Delta t) - r(t_0-\Delta t)}{2\Delta t}, \tag{11.27}$$

and not directly. These two equations give the new position and velocity and form the basis of the Verlet algorithm.

To begin the calculation, two positions have to be known; equation (11.26) indicates that r is needed at times t_0 and $t_0 - \Delta t$. If Δt is small, the latter can be approximated as $r(t_0) - v_0\Delta t$ with initial velocity v_0. It seems strange to have to calculate values before the initial time, which could be negative, but this is what is required. The next step of the algorithm updates positions and forces and increments the time. The positions at the old time $t_0 - \Delta t$ can be discarded; the current positions become the old ones and the new positions become the current ones. The force is also re-evaluated at the new position. This process is repeated in a loop with as many steps as are required. The forces are initially calculated outside the loop, and are calculated again at the end of the loop and are therefore ready to use when the loop restarts.

The elementary Verlet algorithm has the form:

(1) Define potential energy or force & number of time steps.
(2) Define initial position $r(t_0)$, velocity v, and force.
    Calculate second position at $t_0 - \Delta t$.
(3) Loop over time steps
    (i) Calculate new positions and velocities if needed.
    (ii) Calculate and save quantities to be measured; energy, distance, etc.
    (iii) Increment time.
    (iv) Recalculate forces.
End loop. Continue until total time is reached.

Before an example is worked through, a couple of topics need a brief mention.

11.4.7 Molecular dynamics simulations

The Verlet is widely used in MD simulations because it is time symmetric and so conserves energy well. As it is less complicated, it works faster than other integration schemes. In an MD simulation, time steps are typically 10^{-15} s, when put into real units, because chemical bond vibrational periods are $\approx 10^{-14}$ s and several samples will be needed in each period. As with any numerical integration, the time step has to be short enough to allow an accurate calculation, but not so short that the calculation becomes unnecessarily long. The implementation of MD schemes is too long to discuss here; obviously many hundreds of atoms are present, the velocity and position of each has to be calculated in turn, subject to the interaction potential a molecules has with each of its neighbours. The way that coupled equations are dealt with is described in Section 11.5, the MD calculation takes this to extreme with thousands of equations but the basic idea is the same. Frenkel & Smit (1996) give basic algorithms, and detailed ones can be found in Allen & Tilldesley (1987).

11.4.8 Other Verlet algorithms

There are other widely used versions of this algorithm, examples are the velocity Verlet and the Beeman algorithm, both of which give a better estimation of the velocities (see Frenkel & Smit 1996). This is because there is no subtraction of two large numbers as in equation (11.27), which is the main disadvantage of the basic Verlet algorithm.

The equations for the velocity Verlet method can be shown to be the same as for the normal Verlet. The position and velocity equations are

$$r(t_0+\Delta t)=r(t_0)+v(t_0)\Delta t+\frac{f(t_0)}{2m}\Delta t^2 \tag{11.28}$$

$$v(t_0+\Delta t)=v(t_0)+\frac{f(t_0)+f(t_0+\Delta t)}{2m}$$

and because the force is needed in the velocity equation at $t_0+\Delta t$, the calculation has to be done in two steps. The position is calculated using (11.28), then the velocity at half a step calculated using

$$v(t_0+\Delta t/2)=v(t_0)+f(t_0)/2m, \tag{11.29}$$

and the force is calculated at $t_0+\Delta t$, using the position $r(t_0+\Delta t)$, and the velocity finally obtained is

$$v(t_0+\Delta t)=v(t_0+\Delta t/2)+f(t_0+\Delta t)/2m. \tag{11.30}$$

11.4.9 Basic Verlet Algorithm

Consider calculating the position and velocity of a ball dropped from a height of $h=30$ metres until it reaches the ground. The equation of motion is obtained by balancing forces and is $md^2y/dt^2+mg=0$ where m is the mass of the ball, which cancels out, and g the acceleration due to gravity 9.81 ms^{-2}. The time steps have to be determined by trial and error but the calculation is quite accurate with time steps of 0.005 s.

To start the calculation, two positions are needed. If the initial position is y_0 and velocity v_0, the other can be calculated from Newton's laws using $v_{old}=v_0+(-g)(-\Delta t)$ and $y_{old}=y_0-v_{old}\Delta t$. The acceleration is constant at all times and is $-g$. The Verlet algorithm is as follows:

Algorithm 11.12 Basic Verlet algorithm: Falling under gravity

```
> y0:= 30.0: v0:= 0.0:                   #init values
  t:= 0.0: dt:= 0.005: g:= 9.8:          # accel'n is g
  n:= 2000:                              # numbr points
  height:= Array(1..n): atime:= Array(1..n):
  velo:= Array(1..n):                    # arrays for data
  height[1]:= y0: atime[1]:= 0.0: velo[1]:= v0:
  accln:= -g:                            # accel'n
  v1:= v0 + accln*dt;                    # old velocity
  y:= y0;
  yold:= y0 - v0*dt:                     # 1st two y values
  for i from 2 to n do
    ynew:= 2*y - yold + accln*dt^2:      # eqn (11.26)
    v:= (ynew - yold)/(2*dt);            # veloc eqn (11.27)
    yold:= y:            y:= ynew:       # update
    if y < 0 then y:= 0; v:= 0; end if;  # check if -ve
    height[i]:= ynew;                    # save data
    velo[i]:= v:                         # save data
    atime[i]:= t:
    t:= t + dt:                          # new time
    accln:= -g:                          # accel'n const
  end do:
```

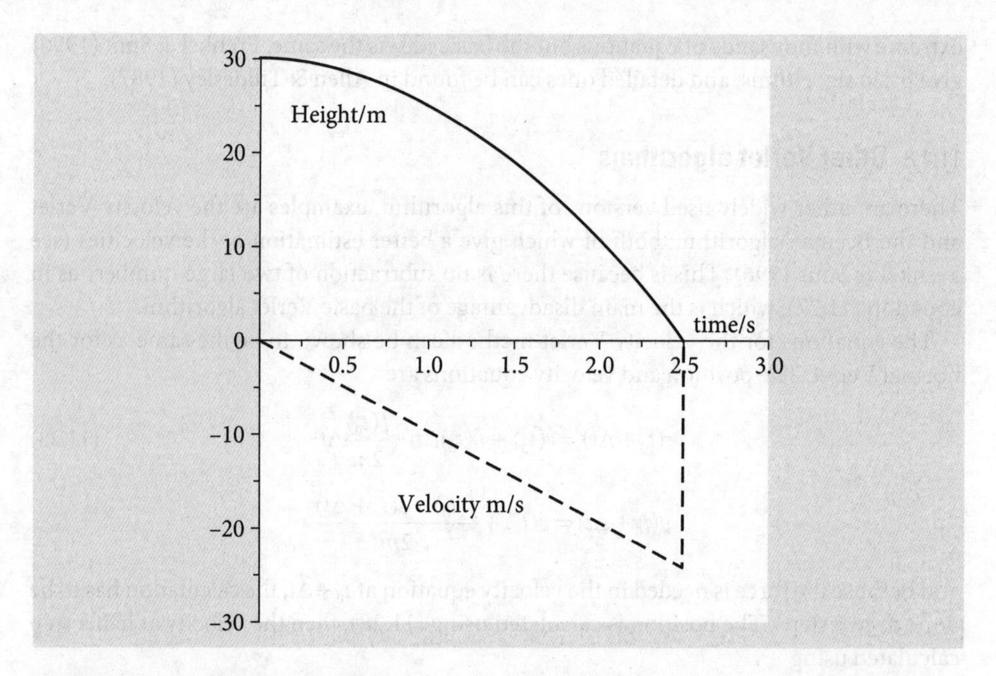

Fig. 11.9 Verlet calculation showing height and velocity vs time for a ball dropped under gravity. The velocity instantaneously becomes zero when the ball hits the ground.

The calculation is quite accurate with the small time step used, but more accurate if this is made smaller. Note that a check is made to prevent y becoming negative because the ball cannot go into the ground. The rate of change of velocity with time, the acceleration, is constant as expected, because the force of gravity is constant.

11.4.10 The Verlet algorithm used to calculate atom scattering trajectories

The trajectory, followed by the scattering by one particle off another, described in Section 11.3.6, will be calculated with a Lennard-Jones 6-12 (LJ6-12) potential. Fig. 11.4 shows the geometry, and the accompanying text describes the background to this problem. The LJ potential has a wide but shallow attractive well and a narrow and very steep repulsive wall at short range; its equation is $U(r) = 4\varepsilon[(\sigma/r)^{12} - (\sigma/r)^6]$ and for simplicity $\varepsilon = \sigma = 1$ is used. The initial energy of the incoming particle is $E = 1/5$. The reduced mass of the particles is assumed to be 1. The impact parameter b, is in the y-direction; the starting x value is $x_0 = -10$, which is sufficiently far away to only weakly feel the potential. The particle initially approaches parallel to the x-axis; the initial velocity in the x-direction is `vx` (Algorithm 11.13) and the velocity in the y direction is zero. This initial distance y_0, is the impact parameter b. The potential U and force $-dU(r)/dr$, is calculated from the radial distance r between the two particles. However, our calculation uses x and y coordinates not r. The radial distance is related to x and y via $r = \sqrt{x^2 + y^2}$. The algorithm also requires the forces along x and y and these components are calculated as

$$-\frac{dU(r)}{dx} = -\left(\frac{dU(r)}{dr}\right)\left(\frac{dr}{dx}\right)$$

and similarly, for y. Unsurprisingly, these forces are called `fx` and `fy` in the calculation. The time step is 0.02 and 2000 steps are taken in total. The initial radial velocity is calculated from the energy as $v_r = \sqrt{2E_0}$, as we are assuming the reduced mass is unity. To start the calculation, the position at two times t_0 and $t_0 - \Delta t$ has to be known, see (Equation 11.26). The coordinate at t_0 is at $[x_0, y_0]$ and that at $t_0 - \Delta t$ has to be found. This is crucial because the velocity of the particle is not used elsewhere in this calculation. If this is not approximated accurately, the consequence is that a different initial energy rather than E_0 will have been used. The calculation is started at a large negative x value, where the particle is approaching almost parallel to the x-axis at a height y_0. The positions at $t_0 - \Delta t$ can be approximated by taking fractions of the velocity in the x and y directions to give $[x_0 + v_r\Delta t x_0/r \;\; y_0 - v_r\Delta t y_0/r]$. Although these work reasonably well, they are not quite accurate enough as they give the approaching particle too much energy. This can be observed by calculating the minimum approach distance r_0 and comparing that with a trajectory. It is found that, given its initial energy, the particle passes closer to the target particle than it should; it crosses the boundary defined by a circle of radius r_0, which is given by equation (11.15). A better approximation is to use equation (11.7) to calculate v. This gives the particle at $t_0 - \Delta t$ the coordinates

$$\left[x_0 - \frac{x_0 \Delta t}{r}\sqrt{2}\sqrt{E_0 - E_0(y_0/r)^2 - U(r)}, y_0\right] \tag{11.31}$$

where only the x value is changed because equation (11.7) assumes that the particle is parallel to the x-axis. This gives the trajectory a good approximation to r_0, the distance of closest approach.

The initial velocity only is used in the initial step; the velocity of the particle is not needed elsewhere in this particular calculation but if needed then equation (11.27) can be used to find v_x and v_y. The radial velocity v_r at any given time, at point $[x, y]$, is the sum of the x and y components calculated as

$$v_r = \frac{dr}{dx}\frac{dx}{dt} + \frac{dr}{dy}\frac{dy}{dt} = \frac{x}{r}v_x + \frac{y}{r}v_y.$$

The radial velocity v_r is zero when the two particles are at their closest point although the individual components may not be zero.

Algorithm 11.13 Scattering trajectory using the Verlet algorithm

```
> eps:= 1.0: sig:= 1.0: E0:= 0.2:           # LJ potential params
  U:= r-> eps/4.0*((sig/r)^(12)-(sig/r)^(6));# LJ potential
  n:= 2000:                                 # number of points
  atime:= Array(1..n):                      # arrays x, y, t
  posx:= Array(1..n):
  posy:= Array(1..n):
  x0:= -10.0: y0:= 1.65:                    # initial x and y
  t:= 0.0: dt:= 0.02:                       # initial t and dt
  posx[1]:= x0: posy[1]:= y0: atime[1]:= 0.0: # init'l values
  r:= sqrt(x0^2 + y0^2):                    # init'l separation
  x:= x0: y:= y0:
  ff:= 24*eps*(2*(sig/r)^(13)-(sig/r)^(7)); # L-J force
  fx:= ff*x/r:                              # force x step(2)
  fy:= ff*y/r:                              # force y step(2)
  xold:= x0+sqrt(2*E0-2*E0*(y0/r)^2-2*U(r))*dt*x/r;
  yold:= y0;                                #initl vals eq (11.31)
  for i from 2 to n do                      # start Verlet loop
  xnew:= 2*x - xold + fx*dt^2:              # x step(i)
  ynew:= 2*y - yold + fy*dt^2:              # y step(i)
  xold:= x: yold:= y:                       #replace old←current
  x:= xnew: y:= ynew:                       #replace current←new
  posx[i]:= x;                              # save step (ii)
  posy[i]:= y:                              # save (ii)
  atime[i]:= t:                             # save (ii)
  t:= t + dt:                               # step(iii)
  r:= sqrt(x^2+y^2):                        # new separation
  ff:= 24*eps*(2*(sig/r)^(13)-(sig/r)^(7));# new force (iv)
  fx:= ff*x/r:                              # new force along x
  fy:= ff*y/r:                              # new force along y
  end do:                                   # end Verlet loop
```

Several trajectories, x vs y, are plotted in Fig. 11.10 with increasing impact parameter b starting at zero. Plotting the final angles χ relative to the horizontal, except that at $b \approx 2.25$, produces a curve similar to that shown in Fig. 11.6 for the same initial energy. The inner grey circle is the potential at $E = 1/5$, and the wide grey line follows the bottom of the potential well. The middle circle is the average value of r_0, which ≈ 0.99 because the repulsive potential is so steep.

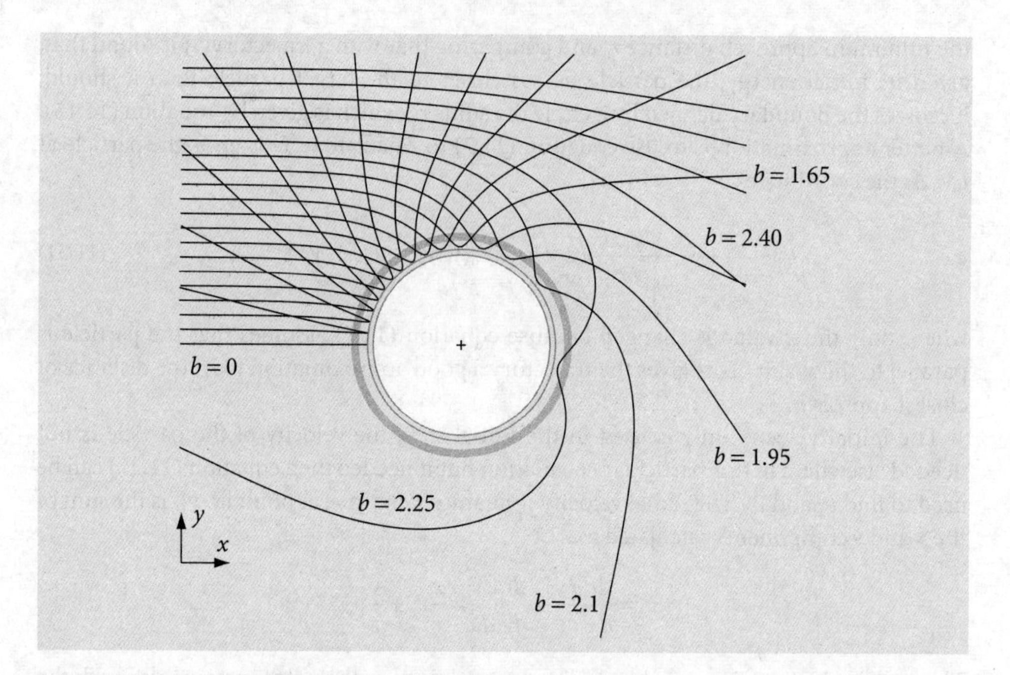

Fig. 11.10 Trajectories calculated with a Lennard-Jones potential ($\varepsilon = \sigma = 1$) with an initial energy of $E = 1/5$. The impact parameter b starts at zero and is incremented by 0.15 in the y-direction. The inner grey circle is the potential at energy E, the outer and wide grey circle is the potential minimum and the middle circle has a radius 0.99, which is a typical value for r_0. The centre of the circle is at coordinate (0, 0). The value at $b = 1.65$ is a Glory scattering as the overall beam is undeflected.

The effect that the attractive and repulsive parts of the potential have is clear. At small impact parameter, the repulsive part of the potential acts not unlike a hard sphere and the approaching particle bounces almost straight off. As the impact parameter is increased, more of the attractive and less of the repulsive part of the potential is felt. Eventually the attractive potential only gradually draws the particle in only to force it to be violently repelled at a smaller separation. Then at an impact of about 2.25 to 2.295 the attraction almost exactly balances the kinetic energy and the incoming particle is trapped for a short period and performs a complete orbit or two before leaving. At still greater impact parameter, all that the potential can now do is to bend the incoming particle from its initial course. The trajectory at $b = 1.65$ has almost zero displacement and is called a 'Glory', see also Fig. 11.6 which shows the scattering angle vs impact parameter.

11.4.11 **Questions**

Full solutions are available at **www.oxfordtextbooks.co.uk/orc/beddard.**

Q11.8 In this problem, a differential equation is solved in two different ways but by three different numerical methods.

(a) Solve $dy/dt = y\sin(t^2)$ by separating variables and show that an integral is formed that can only be solved numerically.

(b) Solve the differential equation by a numerical method with $t_0 = 0$, $y(t_0) = 3$ over the interval −7 to 7.

(c) Confirm that Maple's `dsolve` procedure agrees with your calculation. (To do this second part you will have to look up Maple's help pages, use `dsolve` and `odeplot`, or see appendix 1.)

***Strategy*:** (a) Separating variables means rearranging to put terms only in t and only with y on different sides of the equation. (b) Because the starting time is less than t_0, equation (11.20) should be used up to the value t_0. Alternatively, a negative h should be used. The time values start at t_0 and decrease or increase depending on the sign of h. The simplest, although not the most elegant way to do this is to define two arrays to hold the y values calculated to smaller and bigger values of t_0 and two arrays for the time. An alternative way would be to use a procedure to work out the values and call this with h positive, and then negative. In this example, $t_0 = 0$ so that the range is symmetrical about zero; if t_0 was not zero then more care would have to be taken to produce the range required. Algorithm 11.9 can be modified to do this calculation.

Q11.9 Calculate the scattering trajectories produced by a Coulomb potential and by a screened Coulomb or Yukawa potential, $U(r) = \alpha e^{-\beta r}/r$. Before doing the calculation describe what the scattering trajectories should look like.

Q11.10 In a scattering experiment, the incoming atomic or molecular beam has a cylindrical symmetry and the elastically scattered beam retains this symmetry. The *differential cross section* $I(E_0, \chi)$ for scattering is defined as the number of particles entering a unit solid angle, in unit time, divided by the incident flux density. A solid angle, measured in steradians, is really an area in the sense that it is the angle subtended by circle drawn on the surface of a sphere. A sphere subtends a solid angle of 4π sterads, a hemisphere of 2π and so forth. The differential cross section is given by $I(E_0, \chi) = \dfrac{b}{\sin(\chi)}\left|\dfrac{db}{d\chi}\right|$.

Calculate this for the hard sphere and Coulomb potentials.

Q11.11 A football is thrown upwards from a height h with initial velocity v_0; the friction slowing the ball is proportional to its velocity. The ball bounces elastically, when it reaches the ground. The equation of motion, found by balancing forces, is

$$md^2y/dt^2 + mcdy/dt + mg = 0$$

where m is the mass of the ball, c a damping constant, and g the acceleration due to gravity 9.8 ms^{-2}.

Calculate how the height and velocity of the ball changes with time. Choose your own values for the initial height, mass, and damping constant, which can be zero, but values of 30 m and c in the range 0 to 1 kg s^{-1} will work with time steps of 0.005 s. Although this problem concerns a ball, the form of the equation is general and a molecule, such as a protein or DNA experiences acceleration and friction when in an electric field, in a centrifuge or in a flowing fluid. However, the friction imposed by the solvent is usually proportionally far greater compared to that of air on the ball.

Strategy: Use equations (11.26) and (11.27) for the velocity and follow the method in Algorithm 11.13. To make the ball bounce, reverse the coordinates when the value of y becomes less than zero. Plot y vs time and v vs time. When the damping is zero, the ball should always return to the same height and this can be used to check that a sufficiently small time step has been chosen. From the equation of motion, the acceleration is no longer constant and is $d^2y/dt^2 = -cv - g$ where $v = dy/dt$.

Q11.12 Re-calculate the last problem using the velocity Verlet algorithm.

11.5 Coupled equations

Similar numerical methods to those used for single equations can be used to solve coupled or simultaneous ones. Coupled reactions are commonly found in chemical kinetics, MD simulations, enzyme kinetics, and catalysis; also in the decay of radioactive atoms as well as in the spread of diseases and the predator–prey type behaviour of animals. In each case, species interact or convert into one another and at least two differential equations are needed to describe this. In many cases, these sets of equations cannot be solved algebraically and a numerical method has therefore to be used. As an example, we deliberately start with a pair of equations that can be solved analytically, and this will allow comparison of the numerical solution with the algebraic one. The equations are

$$\begin{aligned} dy/dt &= \cos(t) - x \\ dx/dt &= \sin(t) - 1 - y \end{aligned} \qquad (11.32)$$

subject to the initial condition $x(t_0) = 1, y(t_0) = -3$ when $t_0 = 0$. Notice that the first equation $dy/dt = \cdots$ is a function of x and t and the second a function of y and t but in general dy/dt and dx/dt could be functions of x, y, and t. These equations can be solved algebraically using the method of operators (Chapter 10.4) by making a second-order equation out of the pair and the solutions are $y = \sin(t) - \sinh(t) - 1 - 2\cosh(t)$ and $x = \cosh(t) + 2\sinh(t)$. These functions are plotted in Fig. 11.11 as dashed lines. The result of a modified Euler integration using Algorithm 11.15 from $t = 0 \cdots 3$ is also shown as the solid lines with $N = 200$ steps.

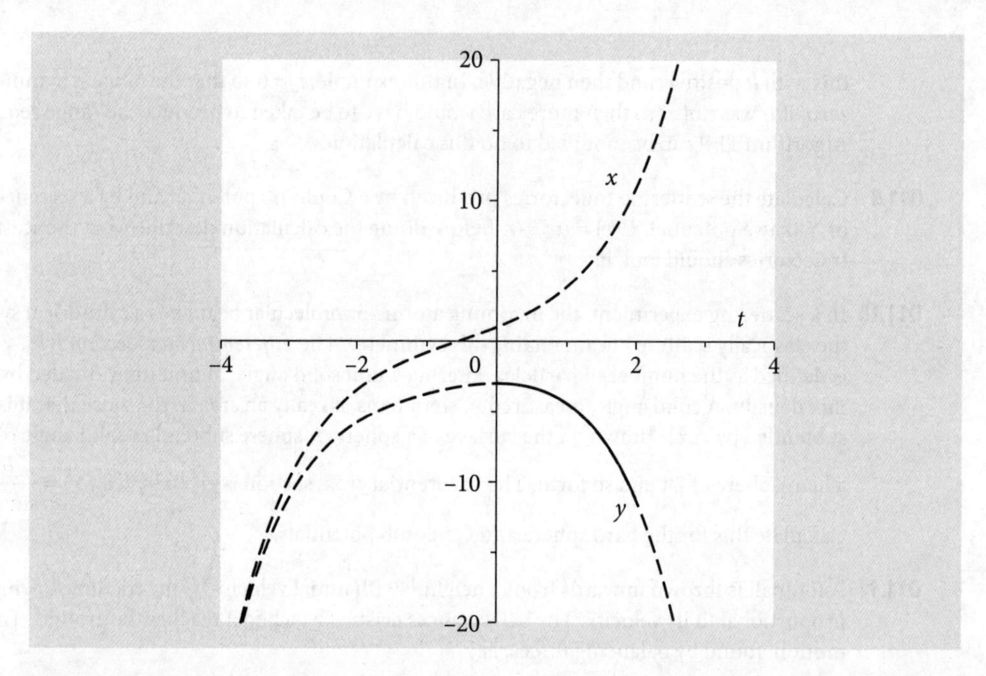

Fig. 11.11 The algebraic solution to the two equations (11.32) (dashed lines) over the range −4 to 4 and the modified Euler numerical solution (solid line) with 200 steps over the range $t = 0$ to 3.

The code for the calculation is similar to that in Algorithm 11.9, except that two functions are defined and *both* x and y are incremented. The modified Euler method can be used just as easily as with a single equation, but for clarity, the simpler method is illustrated.

Algorithm 11.14 The Euler method with coupled equations

```
> dydt:=(t, x)-> cos(t)-x;                    # dy/dt
dxdt:=(t, y)-> sin(t)-1-y;                    # dx/dt
t0:= 0.0:      maxt:= 2.0:                    # initial values
x0:= 1.0:      y0:= -3.0:                     # initial values
N:= 200:                        # number of integration points
Eulery:= Array(1..N+1):     Eulerx:= Array(1..N+1):
dtime:=  Array(1..N+1):
h:= (maxt-t0)/N:                              # step size
x:= x0:  y:= y0:  t:= t0:                     # set initial values
Eulery[1]:= y0: Eulerx[1]:= x0: dtime[1]:= t0:
for i from 2 to N+1 do                        # loop N times
  y:= y + h*dydt(t,x):                        # increment y
  x:= x + h*dxdt(t,y):                        # increment x
  t:= t + h:
  Eulery[i]:= y;                              # save y and x
  Eulerx[i]:=x;
  dtime[i]:= t;                               # save times
end do:
```

11.5.1 Modified Euler and Runge–Kutta equations

The Euler method for coupled equations can be improved in the same way as for equation (11.23). The resulting equations with which to calculate $dy/dt = f(t, x, y)$ and $dx/dt = g(t, x, y)$ are changed from

$$y_{n+1} = y_n + hf(t_n, x_n, y_n) \qquad \text{and} \qquad x_{n+1} = x_n + hg(t_n, x_n, y_n)$$

to

$$y_{n+1} = y_n + (h/2)[f(t_n, y_n) + f(t_n + h, x_n + hg(t_n, y_n, x_n), y_n + hf(t_n, y_n))]$$

and

$$x_{n+1} = x_n + (g/2)[f(t_n, y_n) + f(t_n + h, x_n + hg(t_n, y_n, x_n), y_n + hf(t_n, y_n))]$$

which is more clearly written as

$$\begin{aligned} k_1 &= f(t_n, x_n, y_n), & k_2 &= f(t_n + h, x_n + hL_1, y_n + hk_1), \\ L_1 &= g(t_n, x_n, y_n), & L_2 &= g(t_n + h, x_n + hL_1, y_n + hk_1), \\ x_{n+1} &= x_n + h(L_1 + L_2)/2 & y_{n+1} &= y_n + h(k_1 + k_2)/2. \end{aligned} \tag{11.33}$$

Translating this into Maple, places the following lines into the do loop of Algorithm 11.14 instead of the code for the simpler method.

Algorithm 11.15 The modified Euler method for coupled equations

```
> for i from 2 to N+1 do
    k1:= dydt(t, x, y):
    L1:= dxdt(t, x, y):
    k2:= dydt(t + h, x + L1*h, y + k1*h):
    L2:= dxdt(t + h, x + L1*h, y + k1*h):
    y:= y + h*(k1 + k2)/2:
    x:= x + h*(L1 + L2)/2:
    t:= t + h:
    ..
    ..
  end do:
```

The Runge–Kutta method can also be used with coupled equations in a similar way as in the last algorithm; notice how the k's are defined in terms of *dydt* and add to terms in y and L's to those in x. Based on Algorithm 11.11 the modified equations are

Algorithm 11.16 The Runge–Kutta method for coupled equations

```
.. ..
for i from 2 to N+1 do                          # Runge-Kutta
  k1:= dydt(t,x,y):
  L1:= dxdt(t,x,y):
  k2:= dydt(t + h, x + h*L1/2, y + h*k1/2):
  L2:= dxdt(t + h, x + h*L1/2, y + h*k1/2):
  K3:= dydt(t + h, x + h*L2/2, y + h*k2/2):
  L3:= dxdt(t + h, x + h*L2/2, y + h*k2/2):
  K4:= dydt(t + h, x + h*L3,   y + h*k3):
  L4:= dxdt(t + h, x + h*L3,   y + h*k3):
  x:= x + h*(L1 + 2*L2 + 2*L3 + L4)/6:   # increment x
  y:= y + h*(k1 + 2*k2 + 2*k3 + k4)/6:   # increment y
  t:= t + h:                             # increment time
  Eulery[i]:= y;                         # save y
  dtime[i]:= t;                          # save t
end do:                                  # end loop
```

11.5.2 Second and higher order differential equations

Second and higher order equations are placed here only because they can be easily reduced to a set of first-order equations and solved as coupled equations. Second-order equations can also be converted directly into difference equations for numerical evaluation and, in some cases, such as with the diffusion equation, this is the most convenient way to solve them.

The equation

$$\frac{d^2y}{dx^2} + 3\frac{dy}{dx} + 5y = 0 \tag{11.34}$$

is second order and can be solved algebraically making possible a comparison with a numerical method. Suppose that the initial conditions are $y_0 = 1$ and $dy/dx|_0 = 0$. These mean that the value of y at $x = 0$ is 1, and that the gradient, also at $x = 0$, is zero. The strategy is to take the lowest derivative dy/dx, even if it is not explicitly present in the equation, and make this equal to a new variable z; for instance, $z = dy/dx$. The equation now becomes

$$dy/dx = z, \qquad dz/dx = -3z - 5y. \tag{11.35}$$

As a check, by substitution the original equation is reformed. The equations, initial values, and the Euler method based on that in Algorithm 11.14 are written as:

```
> dydx:= (z) -> z;                 # dy/dx
  dzdx:= (y,z)-> -3*z - 5*y;       # dz/dx
  y0:= 0:                          # initial y
  z0:= 1:                          # initial slope
  x0:= 0:                          # initial x
```

and in the `for..` do loop the increments for x any y become

```
> z:= z + h*dzdx(y,z):
  y:= y + h*dydx(z):
  x:= x + h:
   ...
```

The numerical result is shown in Fig. 11.12 with 800 points used in the numerical integration, together with the algebraic solution shown as the dashed line. A fairly good fit to the algebraic line is obtained, but to improve this, more points could be used in the integration or a better integration method used, such as illustrated in Algorithm 11.15. The algebraic solution can be found using Maple, and has the form of an exponentially damped sine wave.

```
> eqn:= diff(y(x),x,x) + 3*diff(y(x),x) + 5*y(x);
> ivs:= y(0)=0, D(y)(0)=1;         # initial values
> dsolve([eqn,ivs],y(x));
```

$$y(x) = \frac{2}{11}\sqrt{11}\,e^{-3/2x}\sin\left(\frac{1}{2}\sqrt{11}\,x\right)$$

The higher order equation

$$\frac{d^3y}{dx^3} + a\frac{d^2y}{dx^2} - y = 0$$

can similarly be reduced to a number of first-order equations by making substitutions. Two substitutions are needed here because of the third-order derivative. These can be

$$dy/dx = z, \qquad dz/dx = w, \qquad dw/dx = -aw + y,$$

and the substitutions have the following effect, $dw/dx = d^2z/dx^2 = d^3y/dx^3$ and $w = dz/dx = d^2y/dx^2$.

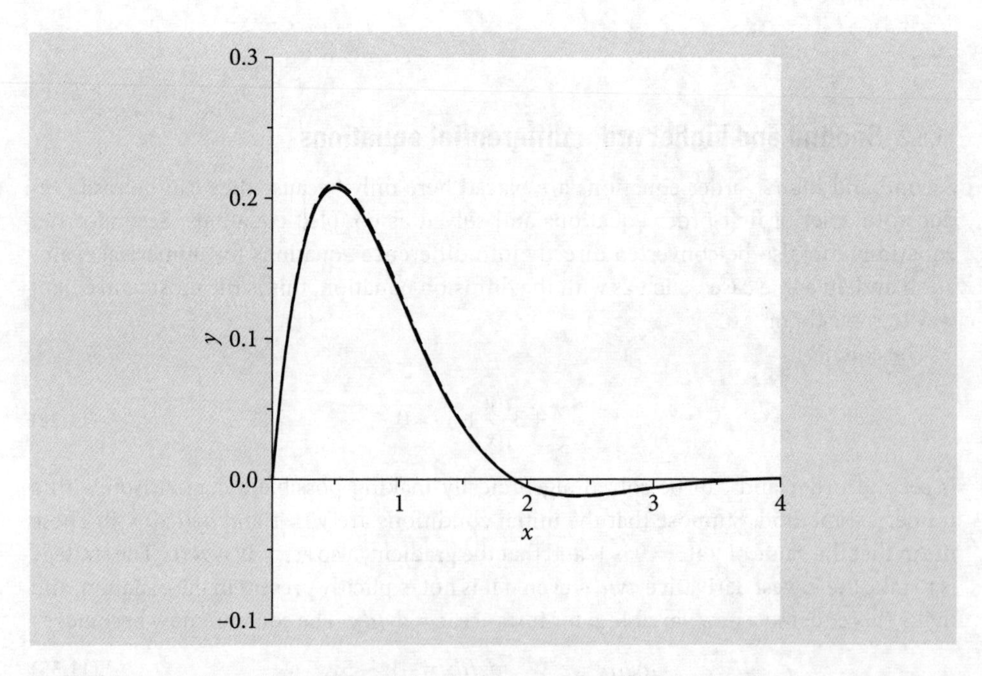

Fig. 11.12 Numerical (solid line) and algebraic solution to equation (11.34).

The initial conditions must also be specified before a solution can be found. In this case, there are, in effect, three integration steps because of the third power of the derivative. The three initial conditions must specify, at the initial x chosen, a value of y, dy/dt and d^2y/dt^2.

11.5.3 **Questions**

Full solutions are available at **www.oxfordtextbooks.co.uk/orc/beddard**.

Q11.13 This question illustrates the importance of using accurate numerical algorithms. Compare the Euler, modified Euler, and Runge–Kutta method to integrate the coupled equations (11.32) from 0 to 20 with initial values, $x_0 = 2$, $y_0 = 1$ at $t = 0$ and integration 500 points. Plot y vs time for the three methods. The exact solution for y to the equation, which can be obtained by the methods of Chapter 10, is $y = \sin(t) + 2e^{-t} - 1$ and thus at small times the function should decay exponentially then become sinusoidal.

Q11.14 Change Algorithm 11.14 to include the modified Euler or Runge–Kutta equations and then solve the system of equations

$$dy/dt = -(k_f + k_2)y + k_1x, \qquad dx/dt = -(k_1 + k_e)x + k_2y$$

over the range $0 \le t \le 1$with initial conditions $y_0 = 1$, $x_0 = 0$ and with constants $k_f = 10$, $k_1 = 2.5$, $k_2 = 2$, $k_e = 5$. These equations describe two coupled species as shown in the reaction scheme and are similar to the kinetics for excimers described in Q10.30.

$$\begin{array}{ccc} y & \underset{k_1}{\overset{k_2}{\rightleftharpoons}} & x \\ \downarrow k_f & & \downarrow k_e \end{array}$$

Q11.15 The Michaelis–Menten scheme is the simplest description of an enzyme catalysed reaction. The enzyme E and substrate S come into equilibrium with an intermediate complex ES that breaks up into reactants or produces product P and the enzyme is returned to its functioning state having acted as a catalyst by converting S into P. The rate equations are derived from the scheme

$$E + S \underset{k_{-1}}{\overset{k_1}{\rightleftharpoons}} ES \xrightarrow{k_2} P + E$$

However, these cannot be solved analytically to produce a formula but either have to be solved at steady state by setting the rate of change of the intermediate ES to zero (see Chapter 10) or have to be solved numerically.

(a) Using the Euler method, write down and solve the rate equations and plot each species vs time up to 10 seconds using a time step of milliseconds. The initial concentrations are $S(0) = 5 \times 10^{-3}$ mol dm^{-3}, $E(0) = 1.5 \times 10^{-3}$ mol dm^{-3}, $ES(0) = 0$ mol dm^{-3} and the rate constants are $k_1 = 1$ dm^3 mol^{-1} s^{-2}, $k_{-1} = 0.3 \times 10^{-3}$ s^{-1}, $k_2 = 10^{-3}$ s^{-1}.

(b) Explain the shape of the curves produced and identify where the steady state is likely to be valid.

Strategy: The Euler method algorithm has to be changed slightly to add new species instead of the two used in most examples so far; for example, the product is calculated inside the 'do loop' with a term such as

```
> p:= p + h*dpdt(x,y,z):
```

and this will have to be saved in an array such as

```
> Eulerp[i]:= p;
```

for plotting. The rate equations have to defined first, as must the initial concentrations and the arrays to hold the concentrations of the four species. Using a time step of a millisecond means that the time increment is $10^4/N$ for N steps and use $N \approx 2000$.

Q11.16 Solve the equation

$$\frac{d^3y}{dx^3} + x\frac{dy}{dx} + y - x + 1 = 0 \qquad (11.36)$$

from $x = 0$ to 10 with the initial conditions $y_0 = 5$, $dy/dx|_0 = 1$, and $d^2y/dx^2|_0 = 2$. Note that there are three initial conditions, each evaluated at $x = 0$.

Strategy: Define two new variables to represent the derivatives and so split the equation into three and solve numerically. This equation can be solved directly by Maple, but only by producing integrals that have to be solved numerically.

11.6 The phase plane, nullclines, and stable points

In the study of two-dimensional linear and non-linear differential equations, the *phase plane*, *nullclines*, and *fixed points* are very useful tools for analysing the equations before any numerical or algebraic calculation. The phase plane allows fixed points to be found, these are also called steady state, or equilibrium points and occur when $dy/dt = dx/dt = 0$. Some examples are also given in Chapter 10. *Closed orbits* and *limit-cycles* can also be observed in the phase plane and are described shortly.

Written generally, pairs of differential equations are

$$dy/dt = f(x, y, t) \quad \text{and} \quad dx/dt = g(x, y, t)$$

where the functions f and g may contain terms in x, y and perhaps t. For example,

$$dy/dt = xy + x^2 - y^2 \qquad dy/dt = -\sin(x) + (y-1)^2$$

and so forth.

11.6.1 The phase plane[2]

The phase plane is the plot of y vs x and is found by calculating dy/dx by using the chain rule. In practice, this means dividing the equation for y by that for x; therefore the phase plane does not explicitly contain time. If the resulting equation can be integrated, the family of curves produced can be plotted for different values of the integration constant, which means for different initial values of y and x. If the integration cannot be done analytically, then a numerical solution of the two initial equations has to be found and, again, y plotted vs x at various times until the entire phase plane is produced.

For example, if the equations governing the motion of a particle in a particular double-well potential, Fig. 11.13, are

$$dy/dt = x - x^3 \qquad dx/dt = y,$$

calculating the phase plane means integrating $dy/dx = (x - x^3)/y$. By separating y and x then $\int y dy = \int (x - x^3) dx$, which integrates to

$$y = \sqrt{(x^2 - x^4/2) + c} \tag{11.37}$$

and this is the equation describing this phase plane. The values of c depend on the initial energy of the particle. If it has sufficient energy, then the barrier will be surmounted and oscillation will occur between both wells. If not, the motion is restricted to one well only. The *phase portrait* is the collection of curves drawn on the phase plane with different initial conditions, in this case, different values of c.

The stable, steady state, or equilibrium points are found when the rate of change is zero, and are therefore $dy/dt = x - x^3 = 0$ and $dx/dt = y = 0$ making fixed points at $x = 0, y = 0$ and $x = \pm 1, y = 0$. A more detailed analysis of fixed points in general shows that they may be stable, unstable or saddle points, see Jeffrey (1990) and/or Strogatz (1994). A stable steady state has the property of returning to that state after a small perturbation to it is made, and clearly, the points $\{\pm 1, 0\}$ are of this nature, Fig. 11.13 as they are at the bottom of the wells. The origin is a saddle point, and is not fully stable because moving in any direction from the origin the gradient is negative, except moving up or down the y-axis.

11.6.2 Isoclines and nullclines

The equation produced when the rate of change is zero is sometimes called an *isocline* or *nullcline*. Just as on an incline we move up, or on a decline down, an isocline means that the gradient is always the same and must therefore be a constant number, and a contour path is followed in the phase plane. A nullcline occurs when this constant is zero The nullclines are $x - x^3 = 0$ and $y = 0$, which in this case are both straight lines. Isoclines are the lines when $x - x^3 = a$ and $y = a$ where a is some constant. Assuming that y is plotted

[2] The name is historical and apparently was originally used in dynamics as the plane containing position x and momentum mdx/dt of a object as it moved under the influence of a force.

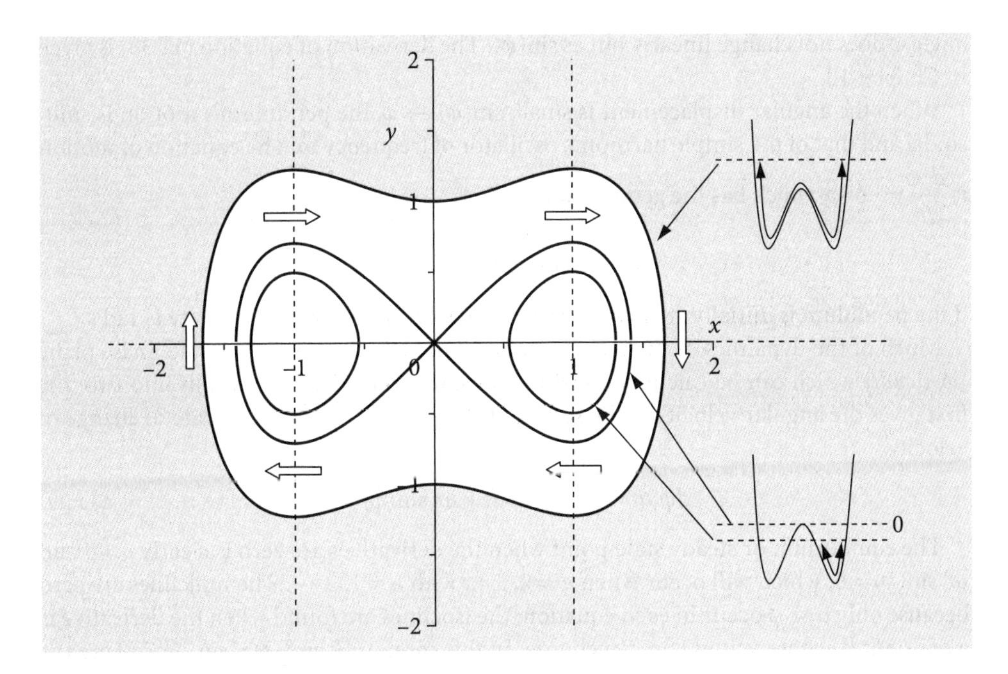

Fig. 11.13 Example of a phase plane with a few contours (isoclines) making up the phase portrait. The potential energy profile has the barrier's top at 0 energy. The motion of a particle starting at different points is shown on the right. The $dy/dt = 0$ nullclines are shown dotted, the other nullcline, $dx/dt = y = 0$ is the x-axis. The arrows show the direction of motion around the phase plane. The separatrix is the 'figure of eight' and passes through the origin.

vertically and x horizontally, the 'flow' or vector showing the direction of change, is always horizontal at each point along the nullcline, $dy/dt = 0$, no matter what its curve is, and vertical on the nullcline $dx/dt = 0$. The nullclines also partition the phase plane into areas where the derivatives have different sign; exactly what these are depends upon the particular equations. Figure 11.13 shows the phase portrait with isoclines at different c values and the nullclines, which are dotted. The figure-of-eight curve is the *separatrix* and, in this example, is the point when the particle has just enough initial energy to cross the barrier separating the region of oscillation in one well from motion over the barrier, and hence, motion between both wells. When the particle is placed in the bottom of either well, it has zero potential and zero kinetic energy. If it is not pushed, it will remain in this stable state at $\{\pm 1, 0\}$.

Finally, in this short introduction to the phase plane, it must be remembered that although the phase plane equation (11.37) does not explicitly contain time, x and y are still functions of time and that time passing on the phase plane is not measured by equal x and y motion, but in a very non-linear manner. This can only really be observed by plotting pairs of x and y coordinates on the phase plane at various times after solving the coupled equations.

11.6.3 Non-linear equations: the pendulum

A rigid pendulum with a heavy bob at its end can move in two ways. When the energy is small, it will oscillate about the vertical in a good approximation to simple harmonic motion, and when the energy is large enough it will rotate continuously in the vertical plane. If the displacement from the vertical is not small, the motion is non-linear and the equation of motion has no exact analytical solution. If it is assumed that the pivot holding the pendulum is frictionless and that no air or other resistance hinders the motion, then the equation of motion is

$$\frac{d^2\varphi}{dt^2} + \omega^2 \sin(\varphi) = 0. \tag{11.38}$$

The variable φ is the angle in radians away from the vertical, and ω is an angular frequency defined as $\omega = \sqrt{g/L}$ rad s^{-1} where g is the acceleration due to gravity and L the length of the pendulum. The frequency ω is the frequency that the pendulum has when it undergoes infinitesimally small oscillations. The mass of the pendulum is m; it is used to calculate forces but cancels out in the result. Equation (11.38) is described as non-linear because the angle φ does not change linearly but as $\sin(\varphi)$. The derivation of equation (11.38) is given in Chapter 10.

When the angular displacement is small, $\sin(\varphi) \to \varphi$, the pendulum's motion is sinusoidal and that of the simple harmonic oscillator of frequency ω. The equation of motion is $\dfrac{d^2\varphi}{dt^2} = -\omega^2\varphi$, which has the general solution

$$\varphi(t) = \varphi_0 \cos(\omega t) + v_0 \sin(\omega t)/\omega$$

if the pendulum is initially at an angle φ_0 rad and starts moving with velocity v_0 rad s^{-1}.

Much of the dynamics of the real pendulum can be understood from the phase plane $\{\varphi, d\varphi/dt\}$ which can be calculated easily if the equation of motion is split into two; the first gives the angular velocity v, the second the angular acceleration or rate of change of velocity,

$$d\varphi/dt = v \qquad dv/dt + \omega^2 \sin(\varphi) = 0. \tag{11.39}$$

The equilibrium or steady state point when the derivatives are zero is clearly $v = 0$ and $\omega^2 \sin(\varphi) = 0$, which will occur when $\varphi = 0, \pm n\pi$ with $n = 1, 2 \cdots$. The nullclines are zero because only v or φ occur in each equation; the isoclines are found when the derivative is a constant k, whose values you can choose. In this case, $v = k$ and, $\varphi = \sin^{-1}(k/\omega^2)$ are the isoclines. The phase plane is obtained by first using the chain rule to give

$$dv/d\varphi = -\omega^2 \sin(\varphi)/v$$

and then variables v and φ can be separated and the equation integrated to give

$$v = \sqrt{2\omega^2 \cos(\varphi) + 2c}$$

where c is a constant of integration. This constant will be determined by the starting conditions; these are the angle that the pendulum is released from and its angular velocity at the point of release. If the initial velocity is v_0 and the release angle φ_0, then

$$v = \sqrt{v_0^2 + 2\omega^2[\cos(\varphi) - \cos(\varphi_0)]}$$

describes the phase plane shown in Fig. 11.14.

The line crossing through $\varphi/\pi = \pm 1$ on the abscissa, is called the *separatrix*; this is produced in this example when the integration constant $c = 1$ if $\omega = 1$. At all points between

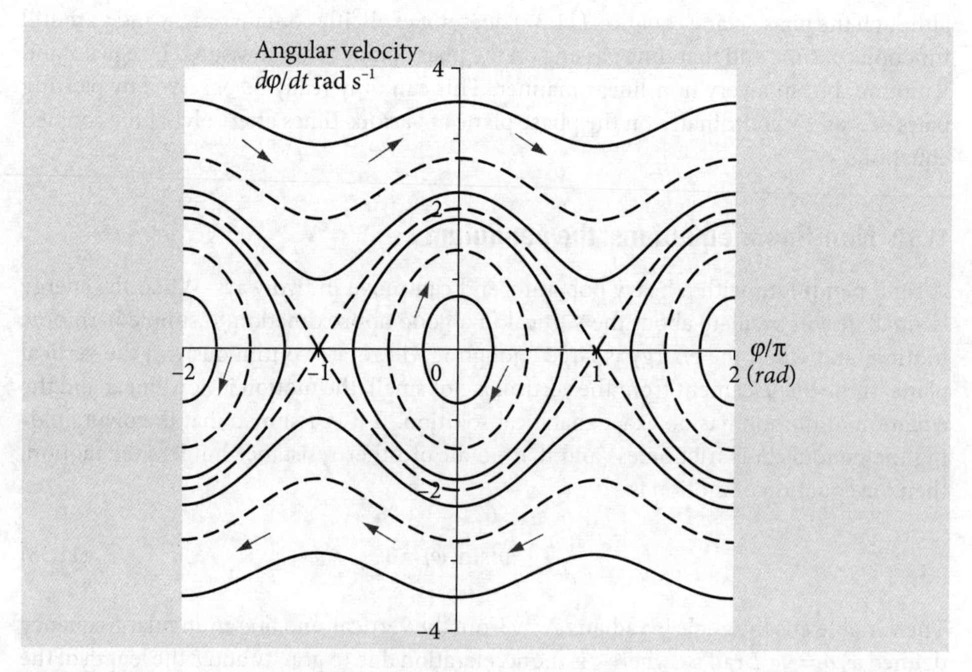

Fig. 11.14 Phase plane, angle vs velocity for the pendulum with $\omega = 1$ and various initial conditions. The separatrix are the lines crossing at $\varphi/\pi = \pm 1$, they separate regions of oscillation, inside the closed curve, from complete rotation of the pendulum.

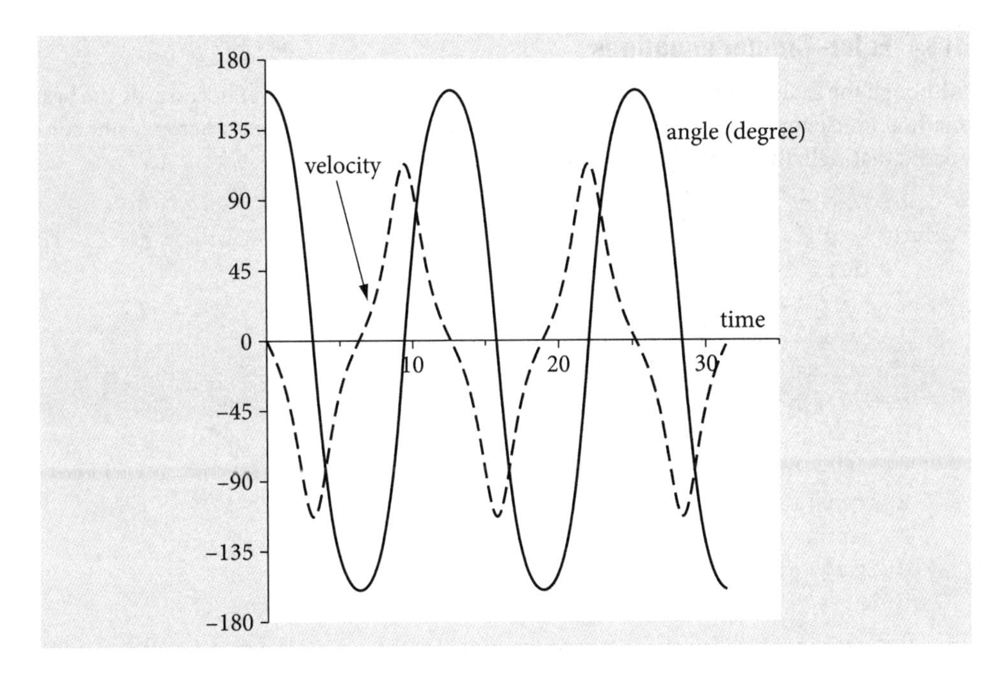

Fig. 11.15 Angle (degrees) and velocity (radians/sec) vs time of the non-linear pendulum, with $\omega = 1$ and an initial angle of $8\pi/9$. Notice that the velocity is zero when the potential energy is a maximum. This is when its angle is greatest or smallest and vice versa.

the separatrixes, the pendulum does not complete more than one revolution, i.e. oscillates back and forth, and the motion appears as closed curves in the figure. If the pendulum starts from a stationary position at any angle except zero, up to a fraction short of π radians, a position almost upside down then, it will swing, ad infinitum, to a similar position on the other side and then back again. If the pendulum starts exactly upside down and also isn't given a push, i.e. initial angular velocity is zero, then it should remain upside down for ever in this metastable state. However, no matter what angle the pendulum is in initially, if it is given a sufficient push and acquires energy in excess of $2mgL$, then it can repeatedly rotate though 360°. This is shown on the phase plane by the lines above the separatrix that do not cross the horizontal φ/π axis.

The direction of the motion can also be determined from the plot, starting at $\varphi/\pi = 1/2$ or 90°, and at zero initial velocity, the pendulum loses potential energy and gains kinetic energy. The angle decreases as the pendulum moves towards its lowest point; this means that the velocity is negative and becomes increasingly so, reaching its largest negative value when the pendulum is pointing vertically down. The motion is therefore clockwise around the closed curves as shown by the arrows on the plot. The motion continues forever, because energy is conserved in this model of the pendulum.

The change in angle and velocity with time can be found by numerically solving equations (11.39) as shown in Fig. 11.15 using the method outlined in Algorithm 11.14 with change in notation from x and y to v and φ,

```
> dphidt:= v -> v;
  dvdt:= phi -> -omega0^2*sin(phi);
```

and changing the steps in the loop of the Euler algorithm to

```
v:= v + h*dvdt(phi):
phi:= phi + h*dphidt(v):
h:= h + t;
```

and 1000 points were used in the calculation. An exact algebraic solution is only possible when the angle is small and $\sin(\varphi) \to \varphi$, which is the harmonic oscillator and has a frequency ω and a period of $2\pi/\omega$ seconds. When the starting angle is not small, the angular motion is not purely sinusoidal, as may be seen in the figure, but spends longer near to the turning points at the top of the swing.

11.6.4 Euler–Cromer equations

Although the Euler method will work well in all our examples, it is not necessarily the best method to use for oscillating systems, such as the pendulum, because energy is not conserved that well. By changing the algorithm 11.14 to

```
> dphidt:= v -> v;
  dvdt:= phi -> -omega0^2*sin(phi);                  # equations 11.39
  ...# define initial values and arrays etc
  ...
  Eulerv[1]:= v0: Eulerphi[1]:= phi0:
  for i from 2 to N+1 do
    v:= v + h*dvdt(phi):
    phi:= phi + h*v:         # Cromer change here
    t:= t + h:
    Eulerv[i]:= v;
    Eulerphi[i]:= phi;
    dtime[i]:= t;
  end do:
```

then the error in the energy becomes proportional to Δt^3 which is a significant improvement over Δt when the step size is small. The angle is calculated as `phi:= phi + h*v` instead of `phi:= phi + h*dphidt` as in the Euler method. See Gould et al. (2007) for more details of this and related methods.

11.7 SIR equations and the spread of diseases

A very interesting, and relatively straightforward example of coupled equations is the spread of an infectious disease, because, besides being intrinsically interesting, it allows a clear illustration of a number of features such as the phase plane and nullclines. An epidemic is defined as the number of infected persons, increasing with time to a number above those initially infected. Kermack & McKendrick (1927), were the first to describe a realistic disease model, which they used to study the spread of a plague on the island of Bombay in 1905/6. In the SIR model, one or more infected persons are introduced into a community where all are equally susceptible to the disease. The model assumes, first, that the disease spreads by contact one to another; each person runs the course of the disease and then cannot be re-infected; and, secondly, that the duration of the infection is short compared to an individual's lifetime, so that the total number of people is constant. Finally, the number of individuals is fixed once the infection has begun; therefore this model only describes infection in a closed community. This is called the S-I-R model, because individuals are either susceptible (S), infected (I), or removed (R). The scheme is

$$S + I \xrightarrow{k_2} 2I \qquad (11.41)$$
$$I \xrightarrow{k_1} R$$

and the aim is to calculate how R, S, and I change with time. The first step describes the transmission of the infection, and the second, the recovery from infection, hence, the number infected I must reach zero at long times as the infection ends. The susceptible persons become infected by reacting with someone who is already infected with a rate constant k_2. In chemical terms, the rate of such a second-order reaction is $k_2[\mathrm{S}][\mathrm{I}]$, supposing that $[S]$ and $[I]$ are concentrations. The second step, infected to removed, has a rate constant k_1, the reciprocal of which is the average time that an individual once infected takes to move into the removed class; the rate for this is $k_1[I]$. Out of the constant total number of individuals, $N = S + I + R$; the number R_0 before the infection starts are those that are immune and clearly at least one infected person has initially to be present. The second equation shows that given time, all individuals will end up in the removed class R, and play no further part in the infection, being immune, isolated, or dead. In this model, the epidemic is assumed to run its course without the intervention of medication, which, given at random times to different individuals, would prematurely cause its end.

If S, I, and R were chemical species, the scheme above would represent a quadratic autocatalytic reaction where R is the product that takes no further part in the reaction. To start the reaction, some initial amount of species I has to be present.

11.7.1 Rate equations

The rate equations of scheme (11.40) are

$$\begin{aligned} dS/dt &= -k_2SI \\ dI/dt &= +k_2SI - k_1I \\ dR/dt &= +k_1I \\ N &= I + S + R \end{aligned} \tag{11.41}$$

where S and I represent the number of individuals (or the concentration of chemical species in an autocatalytic reaction). The initial number infected is I_0, those susceptible S_0, and removed R_0. The total number of individuals is a constant N, and, because of this, the last differential equation is not needed because R can be calculated by subtracting from N the amount of S and I at any time. The rate constants are k_2 the spreading rate constant and k_1 the removal rate constant, and have units of number^{-1} time^{-1} and number time^{-1} respectively. By writing down rate equations, it is implicitly assumed that the number of individuals present is large and can be a continuous variable, not an integer, as is really the case.

Before the equations are numerically integrated, a complete analytical solution not being possible, some analysis of the problem will be carried out. The actual values of the constants are important if a real disease is to be modelled, and before trying to fit the data, it is necessary to know what range of parameters will produce an epidemic and what the expected populations will look like.

Intuitively, the scheme (11.40) suggests that the number infected I, which is initially small (for instance, one person), increases rapidly, passes through a maximum, then slowly decays away. However, this will occur only if $k_2S_0 > k_1$ because when S is large, I is initially formed more rapidly than it is consumed. If the opposite is true, then I is consumed more rapidly and its population cannot become large and no epidemic occurs. To be quantitative, $R_R = k_2S_0/k_1$ is defined as the *reproductive ratio*,[3] which is the number of secondary infections caused by one infected person if all the population is equally susceptible. An epidemic must ensue if the reproductive ratio is greater than one, because more individuals will become infected with time. This can also be appreciated by examining the rate of change of I in the second of equations (11.41) at $t = 0$; $dI/dt|_{t=0} = I_0(k_2S_0 - k_1)$. If $k_2S_0 > k_1$ then $dS/dt > 0$ and an epidemic will occur. Typical values for the reproductive ratio are smallpox = 4; mumps = 5; German measles (rubella) = 6; measles = 12; malaria $\approx$100, (see Britton 2003).

11.7.2 The SIR phase plane

A graph of I vs S is the *phase plane*. The phase plane shows how the number of infectives I and susceptibles S change with time, even though time is only implicit on the graph. The relationship between I and S is found by using the chain rule $\dfrac{dI}{dt} = \dfrac{dI}{dS}\dfrac{dS}{dt}$ and then integrating. The chain rule gives

$$\frac{dI}{dS} = \frac{k_2S - k_2}{-k_2S} = \frac{k_2}{k_2S} - 1$$

and integrating $\displaystyle\int_{I_0}^{I} dI = \int_{S_0}^{S}\left(\frac{k_1}{k_2S} - 1\right)dS$ with initial values I_0 and S_0 produces

$$I = \frac{k_1}{K_2}\ln\left(\frac{S}{S_0}\right) - S + N$$

where the initial values have been substituted with $I_0 + S_0 = N$; at $t = 0$, $R = 0$. Next, dividing by N to make the calculation independent of the number of individuals produces

$$I_N = \frac{k_1}{k_2N}\ln\left(\frac{S_NN}{S_0}\right) - S_N + 1, \tag{11.42}$$

[3] Many texts call the reproductive ratio R_0, which is unfortunately confusing with R_0, the initial number in the removed class.

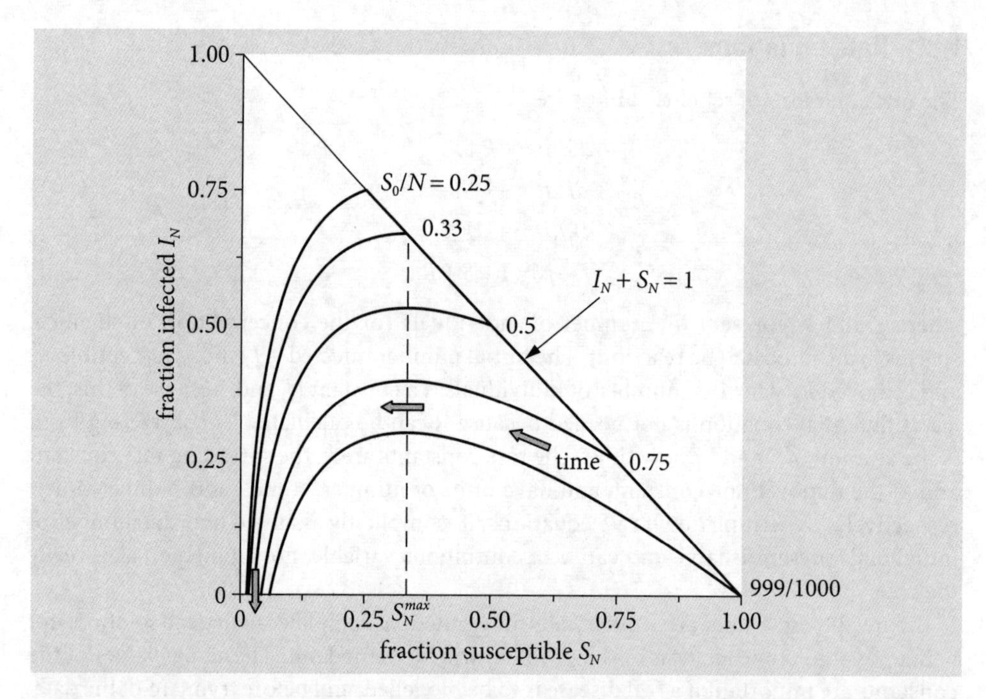

Fig. 11.16 Phase plot of equation (11.42). Different fractions of initially susceptible individuals I_N are shown calculated with $S_n^{max} = 1/3$. An epidemic occurs when a curve starts to the right of S_n^{max}. The arrow shows the direction of change with time and are horizontal on the (vertical) nullcline, $S = k_1/k_2$, and vertical on the $I = 0$ nullcline or horizontal axis.

where the notation is $I_N = I/N$ and similarly for S_N. The graph of I_N vs S_N is shown in Fig. 11.16 at different values of S_0/N, which is the fraction initially susceptible. Time does not explicitly appear in this equation, but this does not mean that the curves are time independent; far from it, because S and I both depend on time. The curves, equation (11.42), must start at the line $S_N + I_N = 1$, or $S + I = N$, which is the diagonal line in Fig. 11.16, because no R (removed class) individuals are present initially, and must move to the left as time progresses. At very long times the fraction infected must become zero.

The definition of an epidemic is that the number of individuals infected increases above those infected initially. In Fig. 11.16, the initial number infected is found where a curve touches the diagonal line, this is 0.25 with $S_0/N = 0.75$, and I_N increases to ≈0.4 at its maximum and therefore, an epidemic may occur. Starting at $S_0/N = 0.25$, the number infected decreases continuously and therefore an epidemic cannot occur. This simple approach indicates the importance of immunization and vaccination. Immunizing or vaccinating a population reduces those susceptible, reducing S_0 and the reproductive ratio R_R, and so making an epidemic less possible. To the right of S_N^{max}, Fig. 11.16, an epidemic occurs, although it may not be severe if the initial value of S_0/N is close to the maximum; to its left the infection dies out. The turnover from epidemic to no epidemic is the point where S_N^{max} touches the diagonal. In the figure this occurs at $I_N^{max} = 1 - 1/3 = 0.6\dot{6}$, or 66% immunization is needed to prevent an epidemic, which is a low value. With an infectious disease such as mumps or German measles, this value has to be ≈0.85, meaning that 85% of the population has to be immunized to prevent an epidemic. Notice that not everyone needs to be immunized to prevent an epidemic; this is called *herd immunity*. A few individuals will by chance, never meet an infected person. An immunization/vaccination level of 85% may be difficult to achieve in a population by voluntary mass vaccination. Should the level of immunization fall by only a small amount, the threshold at S_N^{max} may be crossed and an epidemic could occur. The number of individuals being immunized can suddenly fall, as happened in the UK in the late 1990s and early in this century, due to news media stories about the MMR vaccine for children. Some parents were reluctant to have their children vaccinated even though the risk of damage to health and even death due to measles and mumps, are greater than receiving the vaccine itself.

11.7.3 Steady states, isoclines, and nullclines

In a rate equation, a steady state is produced when the rate of change is zero. In the SIR model $dS/dt = 0$ and $dI/dt = 0$. When molecules, or species in general, interact more than one steady state can be present, and not all of these are necessarily stable. The nullclines on the phase plane of the SIR model are particularly simple and are $I = 0$, or along the S-axis, and $S = k_1/k_2$, which is the vertical line at $S_N^{\max}$ and divides the region where the infected population increases from that where it decreases. The nullclines divide the phase plane into four areas, two areas are below the S-axis in this case, and as negative number of individuals, do not make any sense only two of the four regions have any meaning. Assuming that I is plotted vertically and R horizontally, the 'flow' or vector, arrows Fig. 11.16, showing the direction of change is always vertical at any point of the I nullcline, (horizontal axis), no matter what its curve is, and horizontal on the S nullcline when $dS/dt = 0$. A steady state point is found where the nullclines meet, in the SIR model this is in the S-axis at the point $[k_1/k_2, 0]$ which is the foot of the vertical line S_N^{max}.

11.7.4 Threshold for an epidemic and maximum and total number infected

The maximum fraction of infected individuals is found when $dI_N/dS_N = 0$, and this occurs at the constant value $S_N^{\max} = \dfrac{k_1}{k_2N}$ for any fixed N. When $dI_N/dS_N = 0$ is reached, the infection has peaked and must start to decrease. The maximum *fraction* infected at any one time, is from (11.42)

$$I_N^{\max} = 1 + \frac{k_1}{k_2(S_0 + I_0)}\left[\ln\left(\frac{k_1}{k_2S_0}\right) - 1\right] \tag{11.43}$$

and, when multiplied by the total number $I_0 + S_0$, gives the maximum number of hospital beds necessary to treat the infection. The maximum I_N^{max} may occur mathematically to the right of the diagonal line, Fig. 11.16, but clearly this is not physically possible, because the maximum value I_N can ever take is subject to the condition $S_N^{\max} + I_N^{\max} \le 1$ and this occurs when $R_R \ge 1$.

When the number (or fraction) of infected individuals is zero, there are still some who remain susceptible, see Fig. 11.16, who did not catch the disease even in an epidemic. This fraction is in the range 0.02 to 0.08 in the figure, and is the extent of herd immunity. To be more quantitative, equation (11.42) describes I vs R, and when $t \to \infty$ then I is zero, giving

$$0 = \frac{k_1}{k_2N}\ln\left(\frac{S_N^\infty N}{S_0}\right) - S_N^\infty + 1, \tag{11.44}$$

which is transcendental, and has to be solved numerically for S_N^∞, the fractional amount of S remaining at $t \to \infty$. The Newton–Raphson method (Chapter 3.10) could be used to solve the equation. However, for a strong epidemic the *fractional* amount of S left at the end is very small; $S_N^\infty \ll 1$ hence

$$\frac{k_1}{k_2N}\ln\left(\frac{S_N^\infty N}{S_0}\right) = -1. \tag{11.45}$$

When rearranged,

$$S_N^\infty = \frac{S_0}{N}e^{-k_2N/k_1} \quad \text{or} \quad S_N^\infty = \frac{1}{I_0/S_0 + 1}e^{-k_2S_0/k_1(1+I_0/S_0)} \tag{11.46}$$

by substituting for N. Finally, if $I_0/S_0 \ll 1$, for example, if only one person is infected initially, then

$$S_N^\infty \approx e^{-k_2S_0/k_1}.$$

As a check, using the lowest curve in Fig. 11.16, which has been calculated with the ratio $k_2N/k_1 = 3$, $S_0 = N - I_0$ and $I_0/S_0 = 1/1000$, the fractional amount of S remaining at the end of the epidemic, calculated using the approximate formula, is $S_N^\infty \approx 0.0498$ or 4.98% of the population were never infected. This is close to the exact value of ≈5.94%.

Finally, starting with equation (11.45) the total number infected is approximately

$$I_{tot} \approx N - S_0e^{-k_2N/k_1}. \tag{11.47}$$

or $N(1 - 0.0498)$ and this is the size of the infection, and in effect defines the *total* number of hospital beds needed over the course of the epidemic.

11.7.5 Calculating the time profile of an epidemic

As an illustration of using the SIR model, suppose that you are presented with this specific problem: The following data for the incidence of influenza was recorded at a boys' boarding school. Starting on day zero, the number of boys infected each day was

$$1,\ 3,\ 7,\ 25,\ 72,\ 222,\ 282,\ 256,\ 233,\ 189,\ 123,\ 70,\ 25,\ 11,\ 4.$$

One infected boy started the flu epidemic, and 763 boys were resident. This example is given by Murray (2002, chapter 10, p. 326). The SIR model can be used to estimate the rate constants describing the data, the maximum number infected at any time, and the total number of boys that have been infected at the end of the epidemic.

The strategy is to work out what is already known from the data. First, the timescale is in days; the number of boys susceptible is 763, making $S_0 = 763 - 1$, assuming that only one boy was initially infected. If the infectious period is about 2.5 days, this means that, by definition, $k_1 = 1/2.5$, which can be used as a starting value leaving only k_2 to be estimated. We know that $R_R = k_2S_0/k_1$ has to be greater than 1 and if all the boys are susceptible except one, then $k_2 \times 762 \times 2.5 > 1$ which makes $k_2 > 5 \times 10^{-4}$; the maximum value of R_R, is approximately 20 for common infections, making $k_2 < 0.01$ and this should give a range of rate constants to start the calculation.

Only two quantities need to be calculated; the third, R, is evaluated via $I + S + R = N$. The Euler method code to integrate the differential equations is outlined in Algorithm 11.14 or Algorithm 11.15. Note, that in the calculation, `I` cannot be used because this is a reserved name in Maple; instead we use `In`. The calculation starts as

```
> data:=[ 1, 3, 7, 25, 72, 222, 282, 256, 233, 189, 123,
          70, 25, 11, 4 ];                     # 15 data points
> k2:= 0.00218;          k1:= 0.452;           # initial k's
  num:= 763;                                   # num individual
  dSdt:= (S,In)-> -k2*S*In;                    # dS/dt
  dIdt:= (S,In)-> +k2*S*In - k1*In;            # dI/dt
  t0:= 0.0:    maxt:= 25:
  S0:= 763-1: In0:= 1:   R0:=0                 # initial values
  Np:= 100:                  # number of points in integration
  ...
```

and then follows that in Algorithm 11.14, with the necessary changes in names of the variables. Note, that the functions `dSdt` and `dIdt`, depend on both S, and I but not time directly. The result of numerical integration is shown in Fig. 11.17, with $k_2 = 0.00218$, day^{-1},

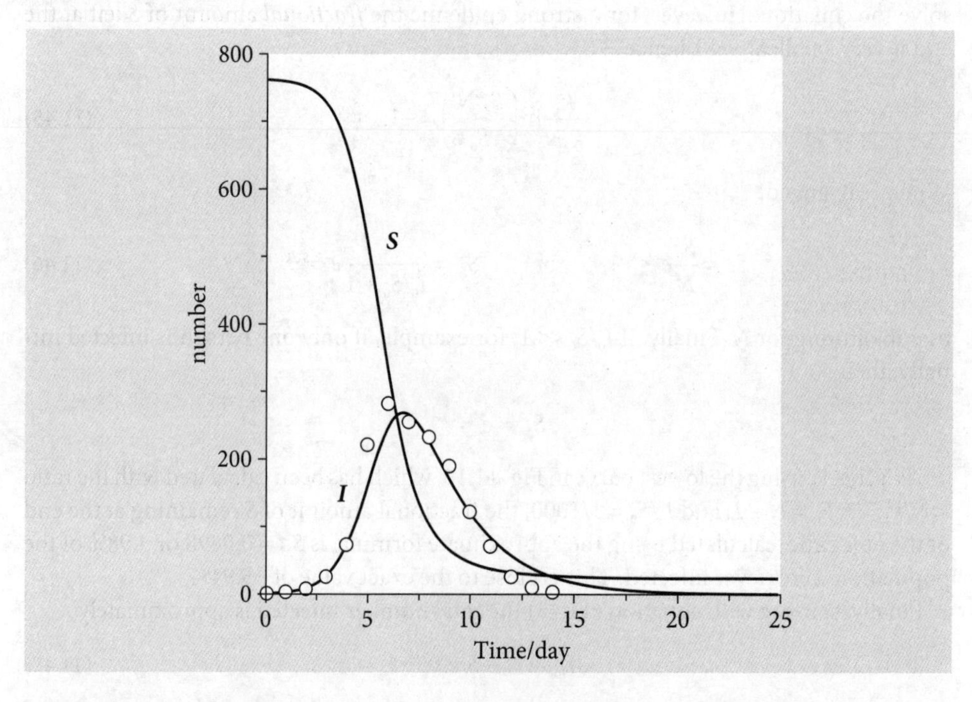

Fig. 11.17 SIR data calculated for boys infected with influenza with $k_2 = 0.00218$, day^{-1} and $k_1 = 0.452$ day^{-1} with 763 boys one of whom was initially infected.

and $k_1 = 0.452$ day^{-1}; the data were fitted as outlined in Q11.18. The number of susceptible individuals initially changes slowly because the product SI in the first rate equation (11.41) is small, I being small. As I increases, S also decreases, but, their product is larger (1×763 is smaller than 2×762 and so forth) and therefore the population of S starts to decrease rapidly as that of I increases; see the first of equations (11.41). The population of I goes through a maximum because in the second equation, the term $k_{IR}I$ eventually starts to dominate and removes I.

The maximum number infected at any time is calculated with equation (11.43); substituting in the numbers gives 286 boys infected, which is seen on the graph where it peaks between days 6 and 7. The number remaining susceptible is not zero at the end of the calculation indicating that not all the boys became infected even though they were all susceptible. The total number who were not infected, using the approximation in equation (11.45) is 19, making 744 who contracted the disease. The numerical calculation gives 24 that were not infected, showing that the approximation, equation (11.45), is quite good.

11.7.6 Questions

Full solutions are available at **www.oxfordtextbooks.co.uk/orc/beddard**.

Q11.17 Calculate Fig. 11.17, and confirm the numbers in the calculation.

Strategy: Use the code and parameters given in the text. Look at the Maple appendix if you are not sure how to plot the data. Vary the rate constants to get an idea of the sensitivity of the fit to these values.

Q11.18 Use Maple's `dsolve(..)` routine to integrate the SIR equations using the data in Fig. 11.17, but automate the fitting by using a grid search to vary k_2 and k_1 with 10 values each, incrementing each value and so find the optimum values. Use two do loops to do this, setting back to the initial values one of the rate constants after the inner loop is completed but allowing the other to vary. Estimate the 'goodness' of each fit by calculating the residual which is the sum of the square of the difference between each of the calculated values and the corresponding data point; $\sum_i (calculated_i - data_i)^2$. Look for the smallest value. The square is taken because some differences may be negative. As smaller values of the residual are found, narrow the search area to 'home in' on the best estimates. The grid search is a slow but effective way of searching.

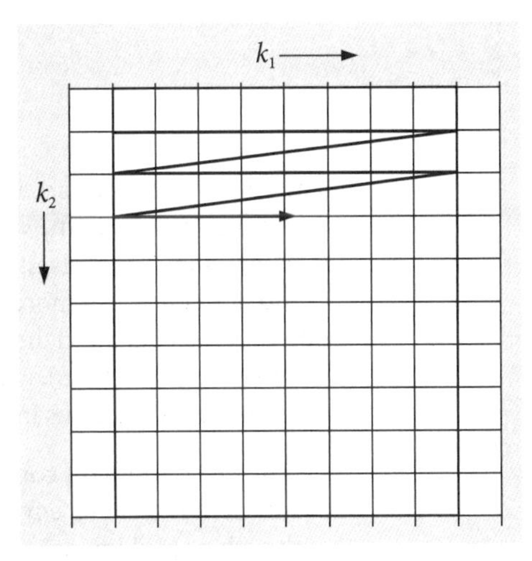

Fig. 11.18 Grid search

Using the SIR model, confirm that k_2 is in the range 0.001 to 0.004 day^{-1} and k_1 is in range 0.2 to 0.6 day^{-1}.

Q11.19 Recalculate equation (11.32) but with initial values $x_0 = 2$, $y_0 = 1$, and $t_0 = 0$ over the range 0 to 10 and then with the modified Euler method. The result is unexpected, increase N and the precision of the calculation. Try to explain why the calculation fails. Repeat with Maple's `dsolve()` routine.

Q11.20 A group of soldiers under 21 years old who were all camping in the same field and doing the same work contracted influenza. The number infected was recorded on consecutive days, starting at day zero, as follows:

1, 5, 25, 28, 10, 18, 15, 7, 8, 3, 2, 2, 1, 0, 0.

The infectious period was reported as 3.5 days. Calculate whether the epidemic follows the SIR model, and if so, estimate the number of susceptible and infected soldiers over a period of 15 days, which means fitting the data with rate constants k_1 and k_2. The grid search method could be used; see question Q11.18 or values guessed, based on initial estimates. The data is somewhat noisy, so some judgement is needed as to what range of parameters will fit the data.

11.8 Reaction schemes with feedback

Feedback in a chemical reaction implies that there are at least two reactions for which the product of one is the reactant for the other, and vice versa; for example, $X + Y \rightarrow 2X$ and $X \rightarrow Y$. Interesting dynamics can be observed in reactions with feedback, because a product that is also a reactant catalyses its own production. There are many such reactions that are now known for example the Belousov–Zhabotinsky (bromomalonic acid/Ru^{2+}/BrO_3^- and traces of BrO_2) and chlorine dioxide/I_2/malonic acid reactions, which involve a complex series of coupled chemical reactions inhibiting and also feeding back on one another, and so catalysing the reaction. Feedback is also common in biology, in the interactions of animals with one another, such as the synchronizing of the flashing of fireflies and of the behaviour of predators and prey. In physiology, the electrical response of nerve and cardiac muscle cells show feedback and the physio-chemical system is termed *excitable*, meaning that under certain specified conditions, far from equilibrium, oscillations in the concentration of different species, or of electrical impulses can occur. See Scott (1995) for a detailed description of oscillating chemical reactions, and Strogatz (1994) for non-linear processes in general. Examples of a few of these processes are now presented.

11.8.1 Predator–prey equations

One of the simplest set of reactions was first studied by Lotka (1925), it involves two species X and Y and some amount of each is present initially.

$$\begin{aligned} Y &\xrightarrow{k_1} Y+Y \\ Y+X &\xrightarrow{k_2} X+X \\ X &\xrightarrow{k_3} D \end{aligned} \tag{11.48}$$

The amount of species Y is doubled in the first step and is then lost by reaction with X to produce more X in the second. Initially, Y increases rapidly, but as it does so, the rate of reaction with X increases. This makes more X which accelerates the reaction, and Y is eventually consumed more rapidly than it is formed and its population falls. The population of X then falls, and the process repeats itself.

Later, Volterra (1926), who was studying the variation of animal populations, described the same set of reactions in this way.

> 'The first case I have considered is that of two associated species, of which one, finding sufficient food in its environment, would multiply indefinitely when left to itself, while the other would perish for lack of nourishment if left alone; but the second feeds upon the first, and so the two species can coexist together. The proportional rate of increase of the prey diminishes as the number of individuals of the predator increases, while the augmentation of the predator increases with the increase of the number of the prey.'

The longest time series of this oscillatory behaviour appears to be the record of the number of lynx and hare pelts sold by trappers to the Hudson Bay Company in Canada over the period 1848 to 1907. The actual data, while oscillatory, is also rather chaotic, illustrating that the actual behaviour of animals is always going to be rather difficult to model due to the uncontrolled nature of the experiment. Nevertheless, some insight into predator–prey behaviour can be obtained with the rate equations;

$$\begin{aligned} \frac{dY}{dt} &= k_1Y - k_2YX \qquad \text{prey} \\ \frac{dX}{dt} &= k_2YX - k_3X \qquad \text{predator} \end{aligned} \tag{11.49}$$

where Y represents a population of prey and X that of predator. Besides hares and lynx, the creatures could be two types of fish or aphids predated on by ladybirds; you can imagine many other examples. This model is a great simplification of actual predator–prey interactions; mathematically, the animals are assumed to be so numerous that they can be treated as if they were molecules in a chemical reaction, but more fundamentally, due to the simplicity of their interactions one with another. A more detailed, and necessarily more complex, description is to be found in Britton (2003) and in Murray (2002). Nonetheless, treating the problem as if it were a chemical one, the prey are breeding at a rate k_1Y, the mother producing one offspring in each unit of time. The fuel to do this and driving the whole predator–prey scheme is grass, or similar vegetation, assumed to be in unlimited supply. The rate constant k_1 is therefore really a pseudo, first-order rate constant and effectively contains a term allowing for the quantity of vegetation available to be eaten. In the second equation the prey is killed at rate k_2YX, which allows the predator to breed. The predators die through natural causes at rate k_3X; however, all the prey are killed by predators and their population does not die off naturally: there is no term $Y \xrightarrow{k_4} D$, where k_4 is the rate of natural deaths. It is assumed in our model for simplicity only that the rate constant for encounter of prey and predator, k_2, is the same as the rate of birth of predators, which it may not be. All the rate constants are positive.

In a molecular example, the equations have to be changed slightly; the first step becomes $C + Y \xrightarrow{k_1'} Y + Y$ where C is some compound whose concentration is unchanged during reaction and provides the material and fuel or free energy to form another Y; it is therefore always at vast excess over Y. To make these equations the same as in (11.48), the substitution $k_1 = k_1'C$ is made and so k_1 is a pseudo, first-order rate constant.

The first step in analysing these equations is to calculate the nullclines and steady state conditions that are also called the equilibrium points. Then the phase plane will be calculated, which plots the number density between predator and prey and finally the time profiles of species Y and X. The initial values to do this will be $X_0 = 60$, $Y_0 = 100.0$, $k_1 = 1.0$, $k_2 = 0.01$, $k_3 = 0.5$, and time from 0 to 100 units. The unit of time could be in seconds or years; this depends on the situation. We need not specify it here, but clearly, the time would be something of the order of a year for hares and lynx. The numerical calculations used to produce the data in Fig. 11.19, was based on Algorithm 11.14 with equations

```
> dYdt:= (Y,X)-> k1*Y - k2*Y*X;   # prey Y
  dXdt:= (Y,X)-> k2*Y*X - k3*X;   # predator X
```

and 1000 points in the integration and a maximum time of 100. In the `for..do` loop, the equations are changed to

```
y:= y + h*dYdt(Y,X):   # increment y
```

and similarly for x. A more sophisticated integration method, such as the modified Euler or Runge–Kutta, could also be used.

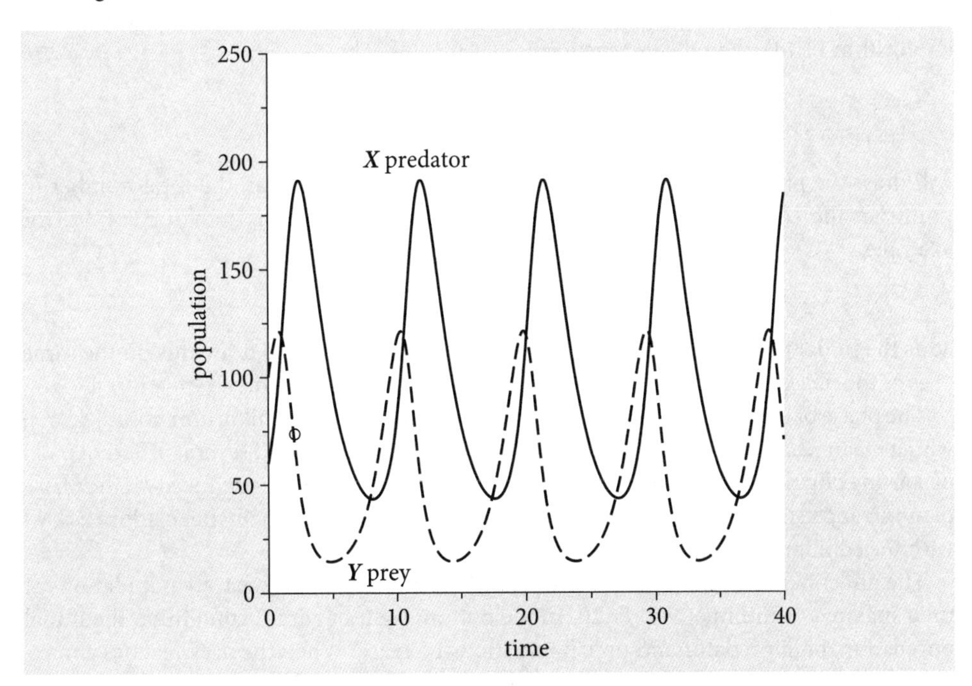

Fig. 11.19 The time profiles of the predator X and prey Y. The prey's population rises first; the predator population always lags behind. The small circle is at $t = 2$ and is shown in Fig. 11.20. $X_0 = 60$, $Y_0 = 100$, $k_1 = 1$, $k_2 = 0.01$, $k_3 = 0.5$.

The steady state conditions are found when the rate of change of each population is zero;

$$\frac{dY}{dt}=k_1Y-k_2YX=0 \qquad \frac{dX}{dt}=k_2YX-k_3X=0$$

giving $Y=X=0$ as one solution and $Y_e=k_3/k_2$, $X_e=k_1/k_2$ as the other. The nullclines are the equations produced when the derivatives are zero, and in this instance, they are the horizontal and vertical straight lines crossing the axes at the equilibrium values, Y_e and X_e. In more complex sets of equations the nullclines could be curves. In Fig. 11.20, the nullclines are plotted as dashed lines. The first equilibrium point at the origin is obvious; no rabbits and no foxes. The second at, $X_e=50$, $Y_e=100$ means that the populations can remain stable given these ratios of rate constants. If the populations are initially different, then they will oscillate in value, ad infinitum, to a greater or lesser extent.

To obtain the phase plane equation, the rate of change of the predator with prey is needed. This is

$$dX/dY=(dX/dt)(dt/dY) \quad \text{or} \quad \frac{dX}{dY}=\frac{(k_2Y-k_3)X}{(k_1-k_2X)Y}.$$

Next, separating variables gives

$$\int\frac{(k_2Y-k_3)}{Y}dY=\int\frac{(k_1-k_2X)}{X}dX,$$

and integrating produces

$$k_2Y-k_3\ln(Y)=k_1\ln(X)-k_2X+c,$$

with an arbitrary constant c. Although this equation does not explicitly contain time, X and Y do change with time. The constant c is evaluated from the initial conditions and the curve produced moves around the non-zero stable point in an anticlockwise manner. You can see that the last equation describing the phase plane is a rather hard one because it is transcendental. We want to plot Y vs X, but the equation cannot easily be put in the form $Y=\cdots$. Numerically solving the equation $k_2Y-k_3\ln(Y)=Q$ for each value of X where Q is the value of the right-hand side is very tedious, and is made difficult because the equation produces real as well as complex solutions. The simplest way to draw the phase plane curve is to plot the populations for a particular set of parameters. If `EulerX` and `EulerY` are the arrays containing the X and Y populations calculated with the Euler method (Algorithm 11.14), then the instructions

```
> se:= seq([ EulerY[i],EulerX[i] ],i=1..N):
> plot([se],view=[0..200,0..200]);
```

will draw the phase plane curve. You can choose N to be less than the total number of points in the calculation, because the curve repeats itself. Using `pointplot()`, for example,

```
> pointplot([[Y0,X0],[EulerY[50],EulerX[50]]],symbol=CIRCLE);
```

adds the initial point and point 50 of the calculation. However, to draw this on the same axes as the phase plane, the `display()` instruction will be needed, see appendix 1.

The phase plane curve is closed and orbits the steady state or equilibrium point $[X_e, Y_e]$, which means that the oscillations in population continue for ever. If several different pairs of starting population are used, a series of separate closed curves are produced on the phase plane arranged one inside the other, as if they were contours drawn on the inside of a bowl with the equilibrium point at the centre.

The nullclines clearly split the phase plane into four areas and separate populations at their maxima or minima, Fig. 11.20. In the bottom right quadrant containing the initial population, both predator and prey populations increase. When the prey reaches a maximum, the $X=100$ nullcline is crossed vertically, and the predator population increases while the prey decreases. This quadrant contains the point that is also shown in Fig. 11.19. At the $Y=50$ vertical nullcline, the predator population peaks. In the next quadrant (top left), both predator and prey population decrease, and in the final quadrant the predator

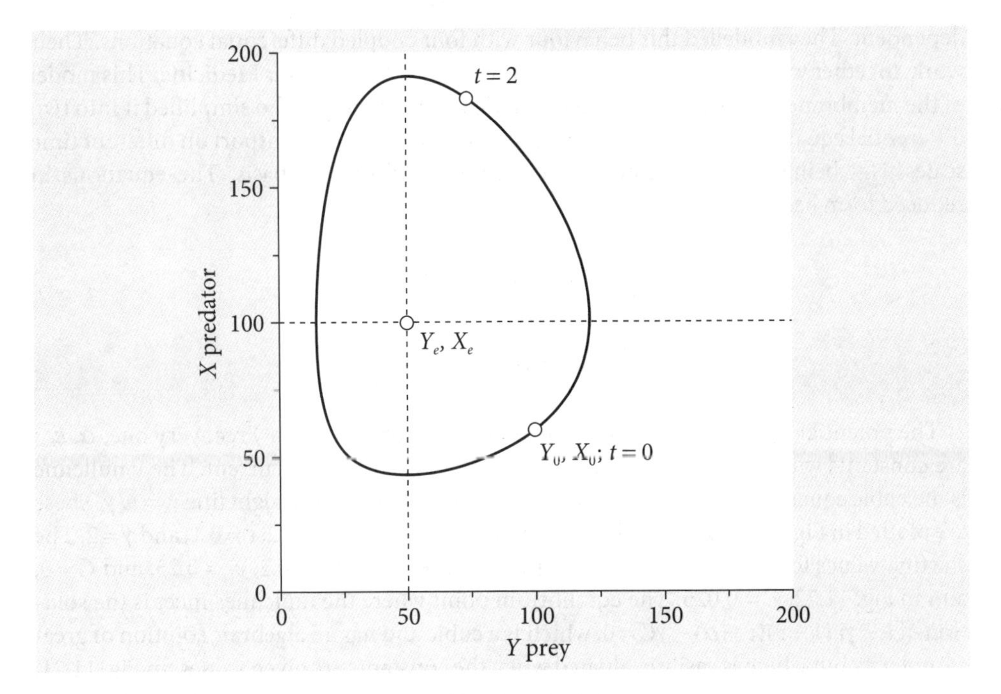

Fig. 11.20 Phase plane plot with the parameters $X_0 = 60$, $Y_0 = 100$, $k_1 = 1$, $k_2 = 0.01$, $k_3 = 0.5$. X represents the predator, Y the prey. The equilibrium and initial values are labelled. The unmarked circle is at $t = 2$ and shows that time moves anticlockwise on this plot. (Note that X is plotted on the vertical axis.)

continues to decrease while the prey recovers. It is interesting to note that the average predator lifetime, in the absence of food from prey, is very short at 2 units of time, which is shown on Fig. 11.19.

The closed form of the phase plane curve shows that the populations oscillate periodically, and will continue to do so for ever. If different initial conditions apply, then a similar plot is produced but with larger or smaller amplitudes. In fact there are an infinite number of these all circling the stable point, X_e, Y_e. This behaviour is in contrast to the *limit-cycle* behaviour of some oscillating reactions, as observed with the Fitzhugh–Nagumo equations, Section 11.8.2. The time profiles, Fig. 11.19, show that the prey population rises before that of the prey. Initially, the prey (hares) breeds and its population increases exponentially. As the predator (lynx) kills the prey, its population growth is limited, reaches a maximum, and starts to fall. However, the predator population has grown too much, and as there is now less prey, the predator population falls as these die off by natural causes determined by rate constant k_3. Since the predator population has fallen, the prey population, fed on an everlasting amount of grass, can now recover and their population increases and the sequence repeats itself.

This simple model only gives an indication of what may happen between predator and prey. It is a starting point from which a number of interesting questions can be asked about how animals interact in a more realistic way or even as to how ecosystems behave. One simple change to the model is to limit the amount of grass available to the prey, and hence to their total population in the absence of predators.

11.8.2 Nerve impulses and the Fitzhugh–Nagumo equations

The biological cell membrane has a potential difference between its inner and outer surfaces. This potential, along with a pH difference, is used by the molecular motor protein ATPase, either to phosphorylate ADP to ATP, or to hydrolyse ATP to ADP. The membrane in its simplest form can be described as a capacitor and resistor in parallel. However, the membrane's electrical properties are not passive but excitable, which means that if a current impulse above a certain limit is applied, the membrane potential subsequently oscillates continuously. From 1948 to 1952, Hodgkin and Huxley conducted experiments on the axon of the giant squid. These 'patch-clamp' experiments were analysed by assuming that channels for Na^+ and K^+ ions existed and that the resistance of the axon was voltage dependent. They modelled this behaviour with four coupled differential equations. Their work, together with that of Eccles received the 1963 Nobel Prize for Medicine. This model of the membrane was itself subsequently modelled by Fitzhugh who simplified it into two differential equations. This was possible because different ions transport on different time scales, Na^+ being slow, and rates could be separated on this basis. The equations in reduced form become

$$\frac{dv}{dt} = v(1-v)(v-\alpha) - w + C$$

$$\frac{dw}{dt} = \varepsilon(v - \gamma w)$$

The potential v is the fast responding voltage; w is the slow (Na^+) recovery one; α, ε, γ are constants with $0 < \alpha < 1$, $\varepsilon \ll 1$; and C is an optional applied current. The v nullcline is the cubic equation $w = v(1-v)(v-\alpha) + C$; the w nullcline is a straight line $w = v/\gamma$. These are plotted in Fig. 11.21 as dotted lines with the parameters, $\alpha = 0.02$, $\varepsilon = 0.1$, and $\gamma = 2$. The starting values for the calculation of the phase plane were $w_0 = -0.2$, $v_0 = 0.25$, and $C = 0$, and in Fig. 11.22, $C = 0.025$. The equilibrium point where the nullclines meet is the solution of $v - \gamma v(1-v)(v-\alpha) - \gamma C = 0$, which is a cubic and has an algebraic solution of great complexity but which is easily evaluated when the constants are given values. In Fig. 11.21, the applied current C is zero and the response of the nerve is to produce one spike and then a highly damped oscillation in v and w signals. The v and w signals are calculated numerically using the Euler method, Algorithm 11.14 and more accurately, with Algorithm 11.15.

The phase plane shows that the v signal initially increases far more rapidly than does w, then v decreases slowly as w increases following the nullcline, but the response breaks close to the maximum in the v nullcline. The response then jumps to the other branch of the nullcline that is then followed back to zero about which a few oscillations occur before reaching the stable point. The oscillations are small and difficult to see on the scale of the plot.

When C is not zero, the v nullcline is raised and now instead of oscillating about the stable point where the nullclines meet, a *limit-cycle* is produced and the phase plane continuously cycles the stable point. A limit-cycle means that the parameters controlling the rate equations are such that the *same* closed curve is produced whatever the initial v and w values are. This is an important result because the oscillation frequency becomes a well-defined function of the physio-chemical state of the system, whereas in the Lotka–Volterra case the oscillation frequency is arbitrary and changes with the initial conditions given to the differential equations.

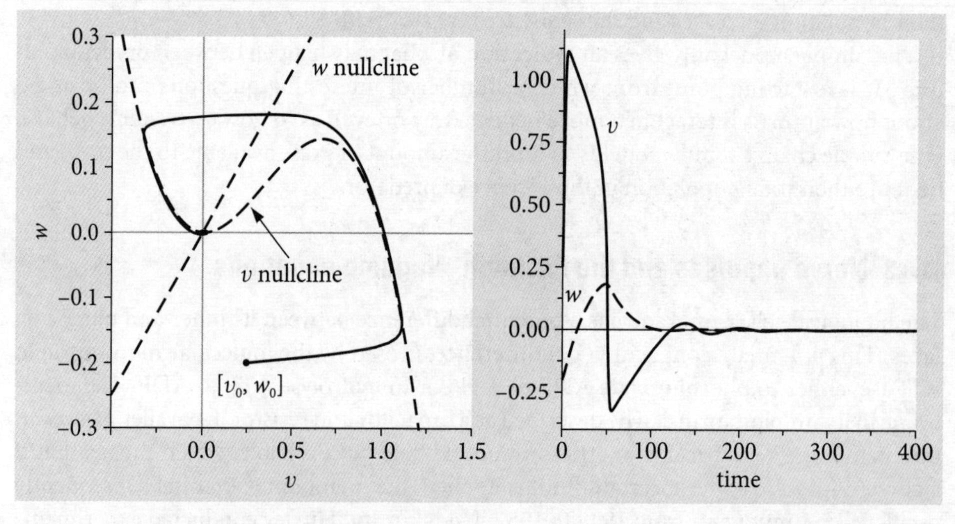

Fig. 11.21 FitzHugh–Nagumo equation's phase plane plot and time profile with $C = 0$ and other values as in the text. The trajectory focuses onto the equilibrium point.

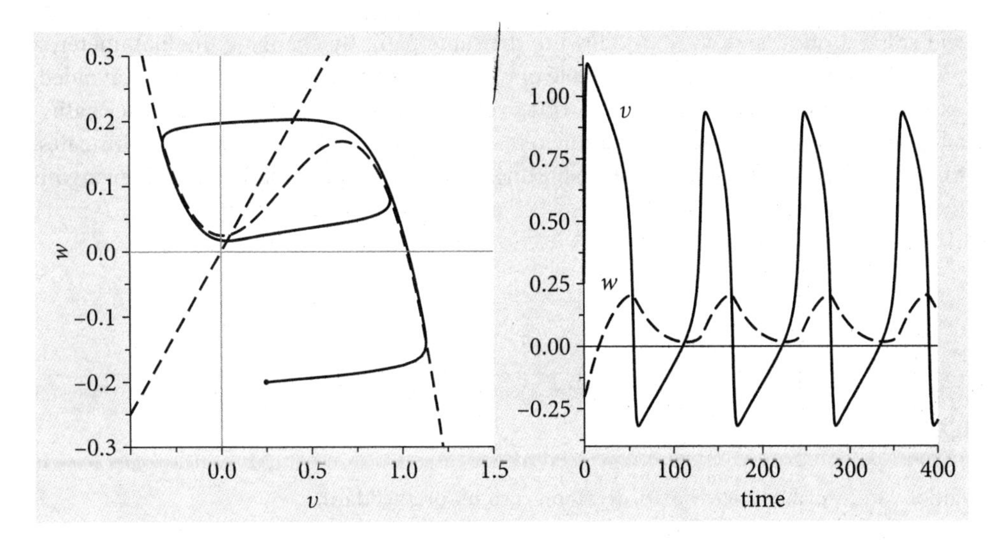

Fig. 11.22 FitzHugh–Nagumo equations phase plane plot and time profile with $C = 0.025$ and other values as in the text. The excitable medium in this case produces a limit-cycle. An equilibrium point occurs where the nullclines cross and is not reached by the trajectory following the limit-cycle.

When the v nullcline is raised up sufficiently by a large value of C, the w nullcline crosses it past its maximum and now the new stable point is formed here rather than near to the lowest point and no oscillations occur; this is not shown and is left as a problem. The limit-cycle is formed when the (stable) point formed, which is where the nullclines cross, is between the maximum and minimum of the v nullcline.

The current C was constant in these examples, but it can be pulsed or made into two pulses separated in time, in which case, chaotic behaviour can be produced. A full discussion can be found in Murray (2002).

11.8.3 Limit-cycles

In Fig. 11.21 and Fig. 11.22, depending on the starting conditions, the trajectory focuses on the equilibrium point or studiously avoids it. The latter produces a limit-cycle that can most simply be understood by considering the mechanical analogy of a harmonic vs a double well potential. In one dimension they can both be represented by $V(x) = ax^2 + bx^4$. The harmonic potential has $a \ge 0$; the double well $a < 0$, Fig. 11.23. Now suppose that these represent a profile cut through a cylindrical surface formed by rotating about the vertical. The trajectory of a ball released anywhere and at any angle on the harmonic surface, $a \ge 0$, will always reach the minimum. In the double well potential, which has the same equation, but with $a < 0$, has a minimum that forms a valley running around the bottom of the potential and any trajectory will end up here. The path around the minimum is the equivalent to the path followed in the limit-cycle; no matter where the trajectory starts from it

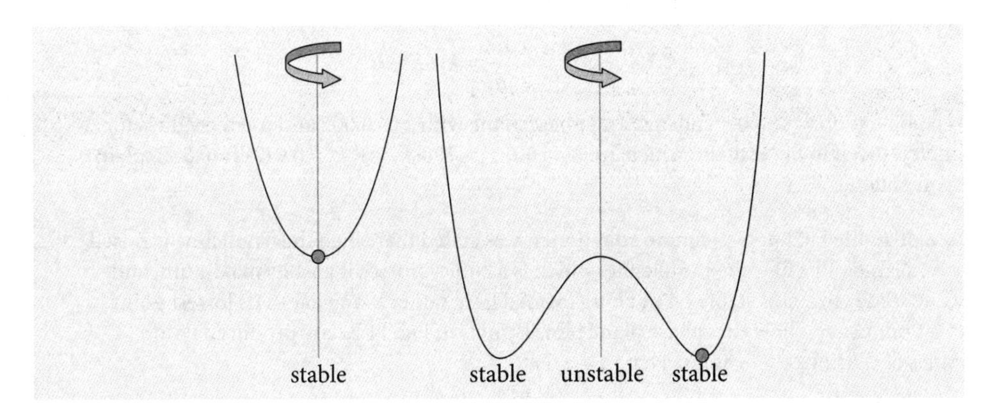

Fig. 11.23 Two potentials with the form $V(x) = ax^2 + bx^4$ with the same value of b but with a either positive or negative. The double well has $a < 0$. The double well has both unstable and stable points.

will spiral around and end up following the same path. By changing one parameter, a stable equilibrium becomes an unstable one, the former equilibrium position is avoided, and completely different behaviour is observed. The idea of *bifurcation* now arises naturally, because, by changing one parameter a stable point splits into two, or bifurcates, forming an unstable point and two stable ones. This was met in a mathematical sense with the logistic equation, see Section 11.1.4.

11.8.4 **Questions**

Full solutions are available at **www.oxfordtextbooks.co.uk/orc/beddard.**

Q11.21 The Lotka–Volterra equations show that the period varies with initial values of X_0, Y_0. Plot several population curves vs time and the corresponding phase planes to convince you of this. Use the Euler method to calculate the populations and values given in the text as starting points. Use 100 points in the calculation.

Q11.22 Using the Lotka–Volterra equations (11.49),

(a) Sketch what happens if the predators do not die by natural causes $k_3 = 0$ and confirm this by calculation.

(b) What happens if an extra term is introduced into each equation; $-k_{11}Y$ into dY/dt and $-k_{22}X$ into the other equation, which removes by poisoning both prey (pest) and predator in proportion to their populations? This has been used as a way of controlling pests; does it work? Assume that the predators do die off naturally so that $k_3 \neq 0$. Use values given in the text and $k_{11} = 0.9$ and $k_{22} = 0.2$.

Strategy: Consider the steady state or equilibrium populations. Plot the phase plane and nullclines, and compare these with the case when k_{11} and k_{22} are zero. Solve the equations using the Euler method with 1000 points in the calculation. The text gives values $X_0 = 60$, $Y_0 = 100$, $k_1 = 1$, $k_2 = 0.01$, $k_3 = 0.5$.

Q11.23 Rewrite the predator–prey equations in reduced or dimensionless form by defining new parameters $x = X/X_e$ and $y = Y/Y_e$, thereby making the equilibrium point [1,1] on the x, y phase plane. If the time is also redefined to be dimensionless as $\tau = k_1 t$, then the equations can be written with only one parameter, which is the ratio of k_3/k_1. Recalculate the time profiles and phase plane for different ratios k_3/k_1. A series of closed curves should be seen one inside the other as if contours on a bowl.

Q11.24 A more realistic model for predator–prey interactions than equation (11.48) is to modify the equations to limit the prey population so that its numbers do not grow exponentially in the absence of predators. The numbers will be limited by the amount of food, irrespective of predation. One way to do this, starting with the basic predator–prey equations is to add a term to the prey equation that limits its population, depending on the *square* of its population: this is a term dependent on its 'density' and is called density dependent growth. This in chemical terms is a bimolecular process $-k_dY^2$ and the squared term represents two Y species recombining to produce an inert product that takes no further part in the reaction. The equations are

$$\frac{dY}{dt} = k_1Y - k_2YX - k_dY^2 \qquad \frac{dX}{dt} = k_2YX - k_3X.$$

Calculate the populations' time dependence and phase plane with $k_d = 0.002$ and use the values for other parameters given in the text and which are $X_0 = 60$, $Y_0 = 100$, $k_1 = 1$, $k_2 = 0.01$, $k_3 = 0.5$. Explain the results you obtain.

Q11.25 In the discussion of the Fitzhugh–Nagumo equation, it was stated that when the v nullcline is raised sufficiently by a large value of C the w nullcline, which is a cubic, crosses it past its maximum, and this large value results in a new stable point being formed here rather than near to its lowest point and no oscillations occur. Show that phase plane plots similar to Fig. 11.22 are produced with different values of C using other values given in the text.

Q11.26 A molecule A reacts with another C to reproduce itself under conditions where C is kept at a constant concentration, i.e. is in vast excess or is continuously supplied. A also produces B which subsequently catalyses A's destruction to an inert species X. B decomposes by a first-order process also to an inert species X. The reaction scheme is

$$\begin{aligned} A + C &\xrightarrow{k_1'} 2A \\ A &\xrightarrow{k_2} B \\ A + B &\xrightarrow{k_3} B + X \\ B &\xrightarrow{k_4} X \end{aligned} \tag{11.50}$$

Because C is held constant, the first reaction can be rewritten with a pseudo, first-order rate constant, $k_1 = k_1'C$, equivalent to $A \xrightarrow{k_1} A + A$. The rate of change of B is zero in the third step because it catalyses this reaction. This scheme also has a biological interpretation; A, is a micro-organism that produces a chemical toxin B that will ultimately poison it.

(a) Write down the rate equations.

(b) Describe the equilibrium (steady state) points and nullclines.

(c) Calculate the concentration of species A and B up to time $t = 600$, and also plot the phase plane with $k_1 = 0.2$, $k_2 = 0.1$, $k_3 = 0.01$, $k_4 = 0.02$ and initial concentrations of $A_0 = 50$, $B_0 = 0$.

(d) Comment on the results obtained.

Strategy: The rate equations are calculated by working out the change in the number of species in the usual way. A is gained in the first step and lost in the other two, while B is produced in the second step, unchanged in the third and lost in the last. The equilibrium points are calculated at steady state, $dA/dt = dB/dt = 0$ and the numerical solution is calculated using Euler's method.

Q11.27 In this question, the motion of a rigid pendulum is examined, see Section 11.6.3 for the equations to use and use $\omega = 1$.

(a) Set the pendulum initially to $\varphi_0 = \pi$ but with zero velocity and observe what happens with the Euler, modified Euler, and Euler–Cromer methods. This is a test of how good the numerical algorithm is, because the pendulum should remain vertical forever. Repeat the calculation with the initial angle at 3.0 radians.

(b) Calculate the angle, velocity, and phase plane at several initial starting angles. Plot a graph of velocity vs time to illustrate how the non-linear motion increases with increase in the initial angle.

Strategy: (a) Use 5000 points in the calculation and use `pi:= evalf(Pi,50)` for an accurate value of π. (b) 500 points will suffice here and initial angles between π and zero radians; the values 3, 1.5, 0.5 and 0.25 radians should work.

Q11.28 A damped pendulum can be made by having a pivot that has some resistance to motion or a bob that experiences air resistance. The equation of motion becomes

$$\frac{d^2\varphi}{dt^2} + f\frac{d\varphi}{dt} + \omega^2 \sin(\varphi) = 0$$

where f is the friction coefficient with units of rad s^{-1}. Because of friction, the pendulum loses energy and will eventually become stationary. If $\omega = 1$ and $f = 0.25$, calculate the profile, angle vs time, with initial angle $8\pi/9$ and a zero initial velocity. Calculate the corresponding phase plane, which cannot easily be done algebraically.

Q11.29 Suppose that there are two coupled oscillators. In the limit that one of them has motion with a period far greater or smaller than the other, then the oscillators hardly affect one other. This decoupling always occurs when vastly different frequencies are involved is a universal types behaviour that is just as common in simple dynamical systems as it is in the complicated vibrational, rotational, and electronic behaviour of molecules.

This question shows how two different oscillators, a pendulum attached to a rotating shaft, can produce a complicated but understandable motion. The figure shows a rigid pendulum freely suspended beneath a rotating rod. The equation of motion is

$$\frac{d^2\varphi}{dt^2} + \left[\frac{\alpha}{L} - \omega^2\cos(\varphi)\right]\sin(\varphi) = 0.$$

(a) Calculate the equilibrium points, nullclines, phase plane, and the angle and the change of the velocity of the pendulum with time.

(b) Consider the cases when L is very long so that $a/L \ll \omega^2$ and when very short.

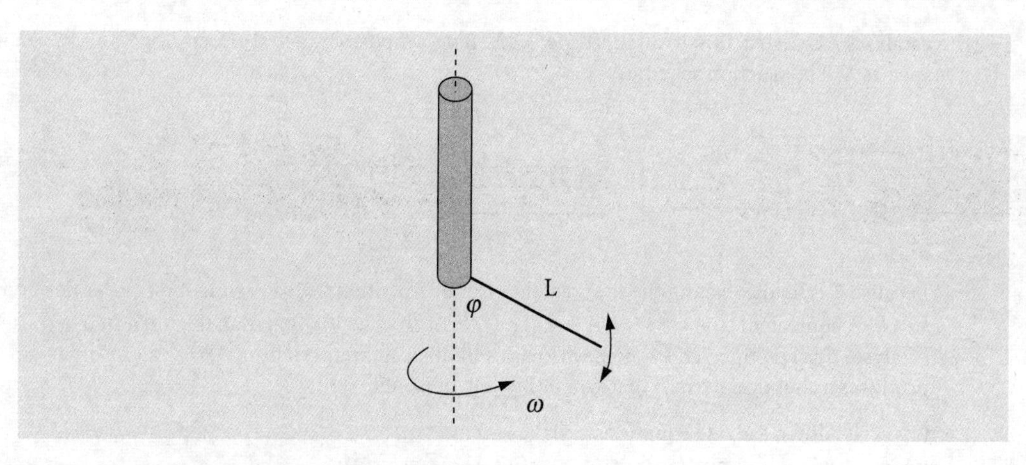

Fig. 11.24 A rigid pendulum attached to the base of a rotating rod

Q11.30 Starting with Q11.29 in which a rigid rod freely hangs from a rotating shaft and has the equation of motion $\dfrac{d^2\varphi}{dt^2}+\left[\dfrac{a}{L}-\omega^2\cos(\varphi)\right]\sin(\varphi)=0$, calculate the period of the pendulum when **(a)** $\varphi\to 0$ and **(b)** $\varphi=\varphi_0$ where φ_0 is the initial angle given by $\varphi_0=\cos^{-1}\left(\dfrac{a}{L\omega^2}\right)$. **(c)** Assume that $\varphi-\varphi_0$ is not small show that $\dfrac{d^2\varphi}{dt^2}=\omega^2\sin^2(\varphi_0)\varphi$ by ignoring a φ^2 term. What does this equation represent?

Strategy: In (a) approximate the series expansion to small angles. In (b) Use a Taylor series expansion about φ_0, and take only the first two or three terms to approximate to small values. In (c) It will be useful to change the angle to $\varphi\to\varphi-\varphi_0$ in the equation of motion.

Q11.31 Suppose that a light, rigid pendulum of length L has a heavy bob of mass m at its end and its pivot is driven up and down at frequency ω by a piston which has a displacement $z=a\sin(\omega t)$. The pendulum's equation of motion is

$$\frac{d^2\varphi}{dt^2}+\sin(\varphi)(\omega_0^2+h\omega^2\cos(\omega t))=0$$

where $h=a/L$ and $\omega_0=\sqrt{g/L}$ is the natural frequency of the pendulum. This equation may be arrived at by replacing the acceleration due to gravity in the normal pendulum equation by $g-\dfrac{d^2z}{dt^2}$ where $-d^2z/dt^2$ is the downward acceleration of the pivot.

(a) Calculate the time profiles and plot the phase plane when the piston's frequency is greater than the pendulum's natural frequency. The fact that the pendulum is driven means that the two motions are coupled and this can lead to rather complicated motion, but with the following parameters the overall motion is clear: $\omega=10$, $h=0.3$, $\omega_0=1$, and $\varphi_0=3\pi/4$. The pendulum is initially 45° from the vertical. Explain the motion produced.

This calculation is very sensitive to the parameters used, using either the modified Euler method or Maple's own numerical algorithm `dsolve(.. type=numeric)` with `odeplot(..)`, the values here will work, but if h and φ are increased by only a small amount, the calculation appears to become unstable. This may be partly overcome by adding a damping term $+cdy/dt$ to the left-hand side of the equation and where c is a small constant ≈ 0.5 or smaller. This interesting system and its regions of stability are described by Acheson (1997). See also Gould et al. (2007) for suggestions for parameters with which to study this system.

Q11.32 The Mathieu equation is found in a number of physical situations one of which is describing the motion of ions through a quadrupole mass spectrometer and in the related Paul ion trap. The equation has the form $\dfrac{d^2\varphi}{dt^2}+\left[\left(\dfrac{\omega_0}{\omega}\right)^2+h\cos(t)\right]\varphi=0$ and is therefore similar to that of a pendulum handing beneath an oscillating mount, which was described in Q11.31, provided that $\sin(\varphi)\to\varphi$. This means that the initial angle measured away from the vertical is small. The natural period of the pendulum is ω_0, and the mount moves with frequency ω.

If the initial angle is exactly zero and so is the initial velocity, then the motion of the pendulum is purely vertical and no oscillation should occur, there being no forces other than purely vertical ones. If, however, a small perturbation occurs so that the initial angle in not zero or initial velocity not zero the pendulum starts to oscillate and its amplitude may grow indefinitely or may be stable.

Show that with an initial angle of 0.1 and zero initial velocity that

(a) the motion is unstable if $h = 0.25$ or 1 and $(\omega/\omega_0)^2 = 0.25$ because the pendulum's amplitude continuously grows.

(b) when $h = 1$ and $(\omega/\omega_0)^2 = 1.75$ the motion is stable and also when $h = 0.1$ and $(\omega/\omega_0)^2 = 0.2$

Strategy: Preferably, use the fourth-order Runge–Kutta method to solve the equations, which can be separated as in Q11.31. Use a minimum of 10^3 steps and time up to $t = 200$.

Exercise: You can investigate the stability regions by observing if the phase plane spirals outwards or not for different sets of parameters. Choose h starting at zero and $(\omega/\omega_0)^2$ from zero to 2. A graph of the stability regions are given by Acheson (1997).

Q11.33 The chlorine dioxide–iodine–malonic acid reaction is known to be an oscillating chemical reaction; see Scott (1995). The reaction conditions are made in such a way that, initially, concentrations are far from equilibrium, as they must if any chemistry is going to happen. The concentration of I^- and ClO_2^- ions are observed to oscillate in time, but eventually this ends as the reaction reaches equilibrium. The free energy (ΔG) of the reaction continuously decreases to its minimum value at equilibrium. If fresh reactants are continuously added to a stirred reactor, then the oscillations can be made to persist.

If MA is malonic acid, then the stoicheiometric reactions are

$$MA + I_2 \rightarrow IMA + I^- + H^+$$

$$2ClO_2 + 2I^- \rightarrow 2ClO_2^- + I_2$$

$$ClO_2^- + 4H^+ + 4I^- \rightarrow Cl^- + 2I_2 + 2H_2O$$

The reaction scheme is more complicated than this stoicheiometry suggests, but has be simplified by Lengyel (1990) using a quasi, steady state approximation. This is possible because the malonic acid, iodine, and chlorine dioxide concentrations vary rapidly during the reaction, and far more so than the intermediates I^- and ClO_2^-, which can be replaced by their average values. When this is done and the reaction scheme is put into dimensionless form, two equations result

$$\frac{dx}{dt} = a - x - \frac{4xy}{1+x^2} \qquad \frac{dy}{dt} = bx\left(1 - \frac{y}{1+x^2}\right)$$

where $x \equiv [I^-]$, $y \equiv [ClO_2^-]$ and a and b are positive constant numbers depending on fixed concentrations and rate constants. It is assumed in this scheme that an unending supply of reagents is available.

(a) Find the stable points and the nullclines, and plot on the phase plane together with some trajectories, each calculated numerically, with constant parameters $a = 8$ and 10 and $b = 2$. Choose a few starting values close to the equilibrium point and a few some distance away.

(b) Comment on the different behaviour of the two parameter sets.

Q11.34 Show that in the Fitzhugh–Nagumo equations, the oscillation period is always the same in a limit-cycle by repeating the calculation with different starting conditions v_0, w_0 using the parameters in the text.

11.9 Boundary value problems: shooting method

The previous examples have been treated as initial value problems, but, in many cases, the equation being examined requires that the solution has predetermined value at two places, and a number of engineering and quantum mechanical problems have this restriction. Fig. 11.25 illustrates, in a schematic way, the difference in the initial value and boundary conditions for a second-order equation; $d^2y/dx^2 = f(x,y)$. The top sketch shows the two initial conditions chosen at $x = 0$ to be $y = 1$ and $dy/dx|_0 = 2$; the gradient is shown as an arrow. The lower figure illustrates the situation if the boundary condition on y is 1 when $x = 0$, and 0 when $x = 2$, which is the limit of the calculation; x can only range from 0 to 2.

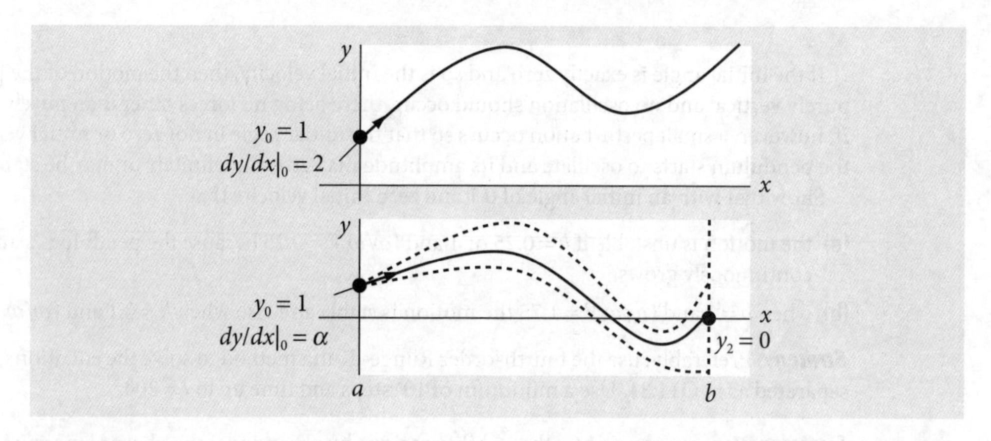

Fig. 11.25 Initial value conditions (top panel), and boundary value conditions (lower panel). This shows three curves following three different initial values α, in addition to $y_0 = 1$, only one of which achieves the boundary condition $y = 0$ when $x = 2$. The shooting method guesses/iterates to the correct initial value α to satisfy the boundary condition.

The only way to solve differential equations is to know what the initial conditions are. In the boundary value problem, the true initial gradient $dy/dx|_0 = \alpha$ and α is unknown, but whose value will produce the required result that at $x = 2$, $y = 0$. Initially α has to be guessed, the calculation performed and then repeated with new guesses until the boundary condition is satisfied. Fortunately, this guessing can be made into an iterative process gradually homing in on the true value.

Suppose the equation we want to solve is the one examined before in Section 11.5.2.

$$\frac{d^2y}{dx^2} + 3\frac{dy}{dx} + 5y = 0 \tag{11.51}$$

but now with boundary conditions $y_0 = 1$ and $y_2 = 0$. This last condition means that the value of y is fixed at zero when $x = 2$. Because this equation can only be solved with initial conditions, the condition $y_2 = 0$ must be replaced with $dy/dx|_0 = \alpha$, and α must be found so that when the equation is solved, $y_2 = 0$. The method to be used is a kind of trial and error techniques called *shooting*.

An outline shooting algorithm, using the particular initial conditions $y_0 = a$ and $y_2 = b$ is;

(1) Set $y_0 = a$. Guess initial α.
  Set precision Q to restrict $|y_{2\alpha} - b| < Q$.
  Set loop limit in case no result is found.
  Solve equation with guessed α to get initial $y_{2\alpha}$.

(2) Loop until $|y_{2\alpha} - b| < Q$ or loop limit exceeded
  (i) Numerically solve the equation in the range $a \le x \le b$.
  (ii) Save $y_{2\alpha}$ the value of y at $x = 2$ found with value α.
  (iii) Update α.
  Loop to (2)

(3) Print result and plot y vs x.

The y value calculated at the last data point ($x = 2$ in this case) is the boundary value $y_{2\alpha}$ for each α, and this has to be stored after each solution and compared with the boundary condition $y_2 = b$. A new value of α is now chosen, and the new $y_{2\alpha}$ compared with the boundary value; when the difference is small enough the calculation ends. Ideally, the difference between the estimated and true value is zero; therefore, this calculation is the same as numerically finding the root of an equation. The new value of α is the root that can be found by linear interpolation (e.g. secant method), using the last two values calculated. Choosing the initial value of α is quite an art. Some equations are 'forgiving' and a value close to y_0 will produce a converging solution at the boundary; in other equations some experimenting with different α values is needed before a solution is reached; the innocuous looking equation $d^2y/dx^2 = 1/(1 + y^2)$ is difficult to solve unless α starts close to zero and the next α value chosen is also very small.

Before working out a specific example, it is worth noting that the boundary conditions and the equation can be written generally, and in numerical methods textbooks may appear as

$$\frac{d^2y}{dx^2}=f(x, y, dy/dx), \qquad a \le x \le b;$$
$$y_a = r \qquad \text{boundary value at } a$$
$$y_b = s \qquad \text{boundary value at } b.$$

The function $f(x, y, dy/dx)$ simply means that the differential equation has terms in one or more of x, y, and dy/dx such as $d^2y/dx^2 + dy/dx + x^2 + y = 0$. The initial condition needed to solve this equation has the form $dy/dx|_a = \alpha$; note that this is defined at a, the initial point, and uses the value α that must be found. The boundary condition $y_b = s$ can be thought of as the non-linear equation $y_b(\alpha) - s = 0$ in the parameter α, and this is the equation that has to be solved iteratively to find its root.

To implement the shooting algorithm, the differential equation has to be solved several times with different initial values. In this case, it makes sense to put this part of the calculation into a small procedure so that the same code does not have to be repeated in several places. A procedure is shown below and is based on Algorithm 11.14. This code is placed inside the procedure and the parameters needed are passed through the header. The parameters used only inside the procedure, are listed as `local`, but those to be used outside, are called `global`. The equation to be solved, $d^2y/dx^2 + 3dy/dx + 5y = 0$, is written into the code by splitting the equation into two coupled equations which produces

$$dy/dx = z, \qquad dz/dx = -3z - 5y.$$

The first equation only involves z, and this means that only the z terms in the modified Euler equation are needed.

Algorithm 11.17 Modified Euler method for coupled equations

```
> solve_diff_eqn:= proc(x0,y0,z0,xN,N)
  local x,y,z,h,i,dydx,dzdx;               # used only in proc
  global Eulery, dtime;                    # used outside proc
  # equation to be solved is d²y/dx² + 3dy/dx + 5y = 0
  dydx:= (y, z)-> z;                       # dy/dx
  dzdx:= (y, z) -> -3*z - 5*y;             # dz/dx
  h:= (xN-x0)/N:                           # step size
  x:= x0: y:= y0: z:= z0:                  # initial values
  Eulery[1]:= y0: dtime[1]:= x0:           # save results
  for i from 2 to N+1 do                   # modified Euler
    k1:= dydx(z):
    L1:= dzdx(y,z):
    k2:= dydx(z+L1*h):
    L2:= dzdx(y+k1*h,z+L1*h):
    y:= y + h*(k1+k2)/2:
    z:= z + h*(L1+L2)/2:                   # increment values
    x:= x + h:                             # increment x
    Eulery[i]:= y:                         # save y and x
    dtime[i]:= x:
  end do:
  end proc:
```

The remaining calculation, defines the arrays `Eulery[]` and `dtime[]` (to hold the x value) and the initial values. The secant rule (Section 11.2) is used to estimate the new value of the parameter α inside the `while ..do` loop. Most of the calculation is taken up with defining the parameters. The `while..do` loop iterates using the secant rule to find new values of α; this finishes when the absolute difference between the last point calculated, `Eulery[N+1]`, at index $N+1$ which is at the boundary, and the y value of the boundary, yxN, (called y_2 above) is less than a small tolerance value Q. You can choose Q to be any value you want, but clearly, it has to be small if an accurate result is to be calculated.

Algorithm 11.18 Shooting method

```
> N:=100:                                        # num points
  Eulery:= Array(1..N+1): dtime:= Array(1..N+1):
  x0:= 0.0;                                      # initial x
  y0:= 1.0;                                      # init y (point a)
  xN:= 2.0;                                      # boundary and max x
  yxN:= 0.0;                                     # boundary value at xN
  alpha0:= 0.0;                                  # init guess dy/dx=α
  alpha1:= 2.0 +alpha0;                          # second guess
  Q:= 0.0001;                                    # stopping value
  solve_diff_eqn(x0,y0,alpha0,xN,N):             # solve with init α
  yN0:= Eulery[N+1];                             # save point at boundary
  k:= 0:
  while k < 5 and abs(yN0-yxN) > Q do            # look for root
    solve_diff_eqn(x0,y0,alpha1,xN,N):           # solve eqn again
    yN1:= Eulery[N+1]:                           # secant method
    alpha1:= alpha1-(yN1-yxN)*(alpha1-alpha0)/(yN1-yN0);
    alpha0:= alpha1:
    yN0:= yN1:                                   # end secant
    k:= k+1;                                     # counter
  end do:
print('iterations=',k-1,'error=',yN0-yxN,'alpha=',alpha1);
```

$$iterations =, 1, error =, 1.369600000\ 10^{-7}, alpha =, -10.9254230100$$

The results are printed out at the end of the calculation; the number of iterations is limited to 5 to keep the calculation short; you can increase this. There is no check on the stability of the calculation; you will have check to see if α is a sensible value by looking at the size of the error. If this is large, $\gg Q$, then it may be necessary to choose another value of α to start with. A good initial value can only be chosen by trial and error. (Occasionally the secant method fails when `yN1 - yN0 = 0` and a very small value 10^{-10} can be added to `alpha1` to prevent this.) The result of the calculation for equation (11.51) is shown in Fig. 11.26 with the boundary conditions $y_0 = 1$ and $y_2 = 0$. The initial conditions satisfying these boundary conditions are $y_0 = 1$ and $\alpha = dy/dx|_0 = -10.92$.

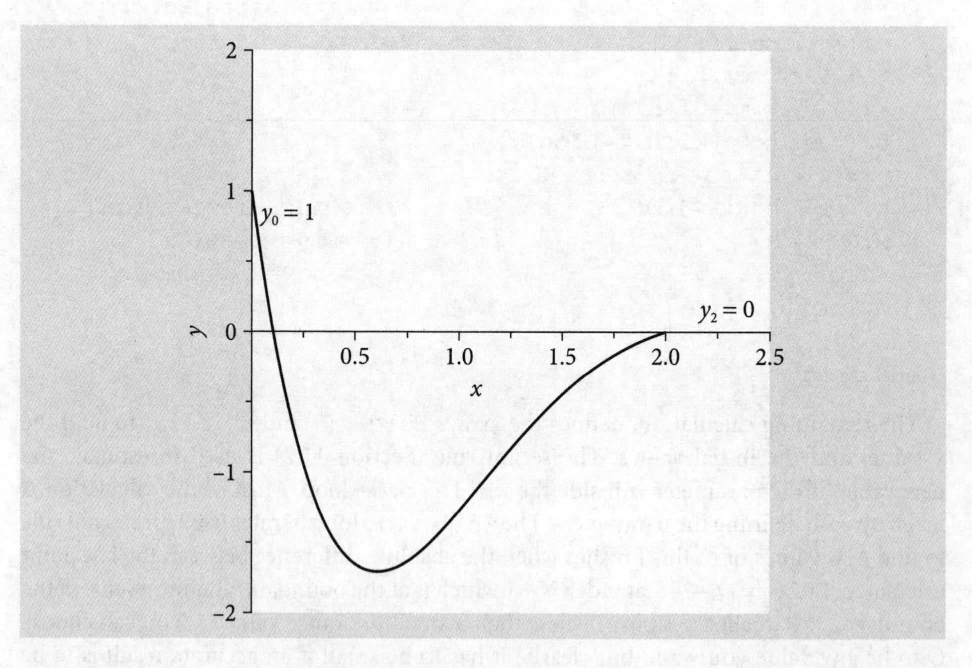

Fig. 11.26 Solution to boundary problem equation (11.51). The initial values $y_0 = 1$ and $dy/dx|_0 = -10.92$ were calculated and are consistent with boundary values $y_0 = 1$ and $y_2 = 0$.

11.10 Numerical integration of the Schrödinger equation

In one dimension, the Schrödinger equation for a particle of mass m with an energy E in a potential $V(x)$ is

$$-\frac{\hbar^2}{2m}\frac{d^2\psi(x)}{dx^2}+V(x)\psi(x)=E\psi(x).$$

The first term represents the kinetic energy, the second the potential, and E is the total energy. At each of certain discrete energies, called the eigenvalues, this equation has a wavefunction that satisfies the condition $\psi(x)\to 0, (x\to R)$. However, at almost any value of E the equation has a solution, which means that it can be integrated to find $\psi(x)$, but only when $\psi(x)\to 0, (x\to R)$ does $\psi(x)$ have a physical interpretation. The range R is taken to be the extent over which the particle can exist; while this is often ± infinity, as in a harmonic oscillator, it is $\pm L/2$ in an infinitely high square well potential of length L and 0 to ∞ for atoms.

Imposing the physics onto the mathematics leads to a huge reduction in the number of possible solutions, but presents us with the interesting task of finding just those solutions that are physically meaningful. Suppose that $\psi(x)$ is the value of a wavefunction of a (quantum) particle at some position x. The word 'particle' is used generically; it might be, for example, an electron, a proton, a rotational, or a vibrational quantum of a molecule. One of the axioms of quantum mechanics is that the probability at time t of finding a particle between coordinate x and $x+dx$ is $\psi(x)^*\psi(x)$ and, since the particle must exist somewhere in the range $-\infty\le x\le\infty$, the total probability must be 1: $\int_{-\infty}^{\infty}\psi(x)^*\psi(x)dx=1$. Because the probability over all space is 1, this ensures that that every wavefunction is zero at $\pm\infty$: $\psi(x)\to 0, (x\to\pm\infty)$. The derivative of ψ at infinity is also zero $d\psi/dx|_\infty\to 0$. The wavefunctions only have this property at certain discrete values of E, and these are called the *eigenvalues*. The shape of the wavefunction changes depending on the energy; for example, 1s, 2s, and 3s atomic orbitals do have different shapes, as do the wavefunctions of the harmonic oscillator or of the 'particle in a box'.

Mathematically, the restriction $\psi(x)\to 0, (x\to R)$ makes the integration of the Schrödinger equation a boundary value problem. In solving for the energies and wavefunctions of the hydrogen or other atoms where the wavefunction's range is $r=0$ to infinity, and the wavefunction is not always zero at $r=0$, the spherically symmetric nature of the problem means that $\psi(r)^*\psi(r)$ has to be multiplied by $4\pi r^2$ to take into account the volume element of spherical coordinates. The probability density is then always zero at $r=0$ and at infinity.

In the numerical calculation, the discrete energies (eigenvalues) E in an attractive potential, such as a harmonic oscillator, can be estimated by several different methods. One of the easiest to use is the shooting method. Once an eigenvalue E is found either the Euler, or a more sophisticated numerical method can be used to find its wavefunction that is also called its *eigenfunction*.

Two boundary conditions $\psi(x)\to 0, (x\to R)$ have to be translated into two initial values because the Schrödinger equation is second order. Suppose that one is β, the value the wavefunction has at some starting point, for example at $x=0$, $\psi_0=\beta$. The other value is the slope of the wavefunction at the same point; $d\psi/dx|_0=\gamma$. The quantities β and γ are derived shortly.

To integrate the equation, a shooting method can be used but in a slightly different manner to that described in Section 11.9. Because of the physics involved, the initial values of ψ and $d\psi/dx$ can be fixed at the outset, and instead the calculation is performed by changing the values of E and the result checked to see if the boundary condition $\psi(x)\to 0$, $(x\to R)$ is obtained at a given E. If it is not, which is generally the case, E is incremented by a small amount and the calculation repeated. When the wavefunction changes sign at the boundary point, it has passed through $\psi(x)=0$ and an approximate value for the eigenvalue is found as the average of the last two values of E, Fig. 11.27. At this point the eigenvalue can be estimated more accurately if required, by bisecting the energy range between the last two calculations into smaller parts. When this first eigenvalue has been found, the energy E is then increased by a small amount and the next eigenvalue is sought, and so the eigenvalues are successively found with increasing energy. It is very important for an accurate determination of E that the precision with which the wavefunction

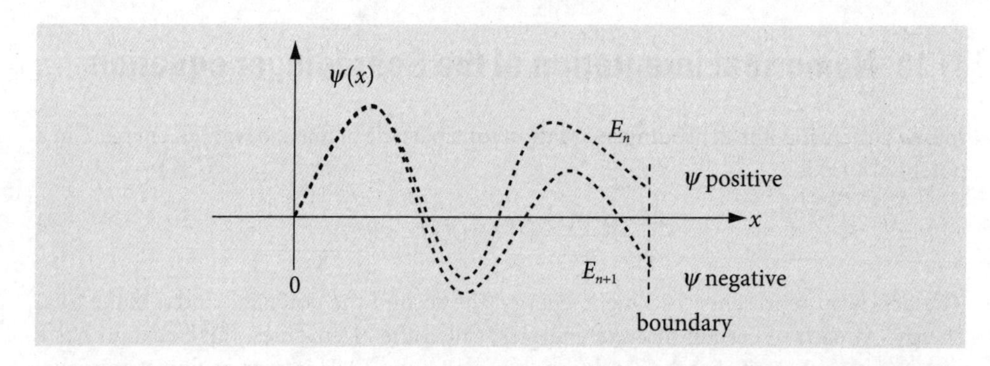

Fig. 11.27 A schematic of two wavefunctions at two energies neither of which is an eigenvalue. A wavefunction is the solution to the Schrödinger equation and if, at the boundary, $\psi(x) \to 0$, then E is an eigenvalue and the wavefunction an eigenfunction. The wavefunction calculated with energy E_n is positive and that with energy E_{n+1} negative; the true energy eigenvalue is between these two values and will produce the true wavefunction with its value of zero at the boundary.

approximates to zero at large x is very high. In the case of a double potential well, such as occurs in the ammonia inversion, if this is not the case energy levels that should be split by a small amount will, erroneously, have the same energy. Decreasing the value at which the bisection is terminated, therefore making the wavefunction closer to zero, will cure this. The stopping value Q may ultimately need to be as small as10^{-8}; see Algorithm 11.18.

The calculation does not have to be carried out to $x = \infty$. Clearly this would be impossible, and an effective infinity is used instead. This is estimated from the shape of the potential; as a rule of thumb, extending the x value in the calculation to twice the width of the potential at the maximum energy to be calculated is usually sufficient. The exception is the particle in the box, where this extends only to the edge of the box.

11.10.1 The shooting method with a quadratic potential; calculating eigenvalues and wavefunctions

As an example, suppose that the potential V is quadratic $V(x) = kx^2/2$, which is symmetric and centred on zero. The value of the force constant k is known for many molecules, 516 N m^{-1} for HCl and 172.1 N m^{-1} for I_2, so that the calculation can be checked with known values of the energy levels, which are spaced as $\hbar\omega(n + 1/2)$ with $\omega = \sqrt{k/\mu}$ if ω is in radians s^{-1} and μ is the reduced mass. The calculation starts by solving the Schrödinger equation for ψ at zero energy and $x = 0$, which happens to be at the lowest point of the potential, and E is incremented by a small amount and at each new value the sign of the wavefunction at the boundary is checked. The energy is incremented until ψ changes sign. When this happen an eigenvalue has been passed and the energy E of the eigenvalue is between these last two values. The bisection method is now used to find an accurate value of the eigenvalue. This is a very stable method but takes several steps to reach the minimum for a given level of precision; the slowness of this method is more than compensated for by its stability because the Newton–Raphson or secant methods are rather subject to instability and can miss eigenvalues. A new energy is now chosen just above the last one found and a new eigenvalue sought and so forth, until the maximum energy required has been reached. The energy increments must be small enough not to miss an eigenvalue but not so small that the calculation takes an inordinate length of time, so some knowledge of the likely energy spacing, based on your knowledge of the chemical physics involved, is going to be useful.

Eigenvalues

The solutions of the Schrödinger equation in a symmetrical potential have either an 'odd' or 'even' parity, meaning that the wavefunction is either symmetrical and has a mirror image about the y-axis, or non-symmetrical and has instead a centre of inversion. If the wavefunction is of even parity then $\psi(x) = \psi(-x)$ and is finite at the origin but has zero slope; the initial condition is therefore

$$\psi(0) = 1, \qquad d\psi/dx|_0 = 0. \tag{11.52}$$

If the wavefunction is odd, then $\psi(x) = -\psi(-x)$ and its value at the origin $x = 0$ must be zero, but its gradient has constant value; therefore

$$\psi(0) = 0, \qquad d\psi/dx|_0 = 1 \tag{11.53}$$

The number 1 is used in these initial values as it only defines the size of the wavefunction. As ψ can be normalized at the end of the calculation these values are not critical. The zeros reflect the odd–even parity and are essential.

For simplicity, Planck's constant and the reduced mass are set equal to unity. If the harmonic potential is $V(x) = 10x^2$, the force constant k is 20 and with quantum numbers $n = 0, 1, 2 \cdots$, the eigenvalues are $\sqrt{20}(n + 1/2)$, which allows an easy comparison with the calculated values. The accuracy of the integration depends not only on the number of steps taken, as with all other integrations, but also on having an accurate integration method. In this example, a Runge–Kutta combined fourth- and fifth-order method is used by using `dsolve`, which is one of Maple's inbuilt routines. This is more a complex algorithm than that given in Algorithm 11.17 but very similar to Algorithm 11.11. Because this is going to be called several times in the calculation, it is placed into its own procedure, `solve_SE`. The potential to be used is written explicitly into the procedure. This will have to be changed for other examples.

Algorithm 11.19 Runge–Kutta method for the Schrödinger eqn (SE)

```
> V:= x-> 10*x^2:                          # potential energy
> solve_SE:= proc(V,x0,y0,z0,xN,E0)  # Maple Runge-Kutta
    local eqn, ivs, ans, yval;
    global yxN_est;
    hbar:= 1.0: mu: = 1.0:
    eqn:= hbar^2/mu*diff(y(x),x,x)-2*(V(x)-E0)*y(x): # SE
    ivs:= y(x0)= y0, D(y)(x0)= z0:     # initial values
    ans := dsolve( {eqn,ivs}, numeric,
           output = listprocedure, method = rkf45 ):
    yval:= subs(ans,y(x)):
    yxN_est:= yval(xN);  # get value at boundary
  end proc:
```

This calculation returns `yxN_est` as the value of the wavefunction at the boundary `xN`, at energy `E0`. These values are passed into the procedure through the header. The use of the instruction `global` ensures that `yxN_est` can be used outside the procedure. The other parameters are `x0` and `y0`, the initial x and y (wavefunction) values, `z0` is the second initial value, and the gradient $d\psi/dx$; `xN` is the x value of the effective infinity.

In the remaining calculation, the first few lines set the initial values of the parameters and constants. The code finds the change of sign of the wavefunction at infinity and calculates the energy by the bisection method. The calculation ends when the difference between two consecutive estimates of the eigenvalue is less that the parameter `Q`, set at 10^{-5}.

Algorithm 11.20 Solution of Schrödinger equation: shooting method

```
> N:= 1000:                         # num points
  x0:= 0.0:                         # initial x
  xN:= 5.0:                         # boundary or infinity x
  yxN:= 0.0:                        # boundary value at maxN
  En0:= 0.0:                        # starting energy
  deltaE:= 1:                       # energy increment
  maxE:= 100.0:                     # max energy
  Q:= 0.00001:                      # stopping value
  Qn:= 0:                           # even quantum number
  y0:= 1.0:                         # y0 even quantum num
  dydx:= 0.0:                       # initial dy/dx guess
  solve_SE(V,x0,y0,dydx,xN,En0):  # get y value at xN
  yN0:= yxN_est:                    # save point at boundary
  En1:= En0 + deltaE:               # second try with energy
  print('quantum number eigenvalue'):
  while En1 < maxE do               # keep on to max energy
```

```
      solve_SE(V, x0, y0, dydx, xN, En1):
      yN1:= yxN_est:                        # save point at boundary
      if yN1*yN0 < 0 then                   # check if zero crossed
  # ** start bisection method **
        while abs(En1-En0) > Q do           # is difference in energy>Q
          Em:=(En1+En0)/2.0;                # mid point
          solve_SE(V,x0, y0, dydx, xN, Em): #recalc at energy Em
          ym:= yxN_est:                     # ψ at boundary
          if yN0*ym > 0 then                # check sign of ψ
            En0:= Em else  En1:= Em
          end if;
        end do:                             # end of while abs()
        En0:= Em; En1:= Em; yN0:= yN1: # replace for next calc
        print( Qn, En1);
        Qn:= Qn + 2:                        # increase quantum num
      end if;
  # ** end bisection **
    En0:= En1: yN0:= yN1:                   # replace for next calc
    En1:= En1 + deltaE;                     # new energy
    end do:                                 # end while En1<maxE
```

Using this code, only even eigenfunctions are calculated. To calculate odd-numbered ones, the initial value y0 will need to be changed 0 and $dy/dx = 1$, as `dydx:= 1.0:`, see equation (11.53). Note also that this code will only work properly if the potential is positive; if it is negative then a constant should be added to make its lowest value zero and then subtracted from the final eigenvalues.

The eigenvalues for two different potentials are shown in Fig. 11.28. The harmonic $V(x) = 10x^2$ and the double well, $V(x) = x^4 - 13x^2 + 169/4$, which has a minimum of zero and a barrier height of 42.25. A few eigenvalues for the harmonic potential are listed below, and the agreement between the numerical and exact eigenvalue energy is good to at least four decimal places.

n	numerical	exact
0,	2.236061095,	2.236067977
1,	6.708198546,	6.708203931
2,	11.18034363,	11.18033988
3,	15.65248107,	15.65247584
4,	20.12461089,	20.12461179
5,	24.59674833,	24.59674775
10,	46.95742789,	46.95742752
15,	69.31811507,	69.31810729

The eigenvalues for the double well show a splitting below the barrier, as expected, because the two wells interact. However, the difference in energy is very small lower down in the well, and the levels are, in this particular potential, effectively accidentally degenerate. Only near the top of the barrier are two closely spaced levels visible in the plot, with quantum numbers $n = 12$ and 13. This is because the barrier is narrow here and there is more interaction between the two halves of the potential. Above the barrier, the potential is suddenly wider and the energy levels now become closer together. Crudely, this can be thought of in a similar way as an infinite square well or 'particle in a box'. A wide well has energy levels closer together than a short one. As the potential energy increases, so does the separation between eigenvalues, just as happens in the square well.

The harmonic potential is unusual because as the energy increases so does the width of the well and it does so in such a way that the energy separation between adjacent levels is constant. In the infinite square well, adjacent levels separate by ever-increasing amounts as the energy increases. In a potential such as $|x^{3/2}|$, the level's separation decreases with an increase in energy as they also do in the hydrogen atom, where the potential is proportional to $1/x$, or in the anharmonic oscillator with the Morse potential. The difference in the spacing between energy levels the harmonic and double well potentials is shown in Fig. 11.28. In the double well potential below an energy of 30 units, the well is narrow, which means it effectively has a large force constant causing the energy levels to be widely

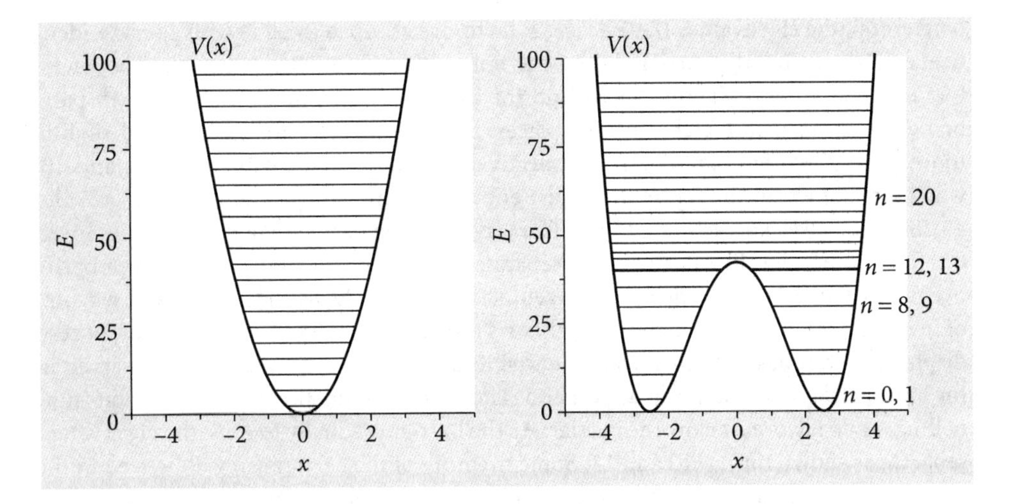

Fig. 11.28 Energy E vs displacement x for two potentials; the harmonic potential $V(x) = 10x^2$ (left), and the double well potential $V(x) = x^4 - 13x^2 + 169/4$ and their associated eigenvalues up to an energy of $E \approx 90$.

spaced, compared to the harmonic potential which is wider, at the same energy, and therefore has a smaller force constant. Well above the barrier, the widths of the double and harmonic potential gradually become similar and the energy spacings are now also more similar to one another.

Wavefunctions

The wavefunctions can also be calculated once the eigenvalues are known. Maple's inbuilt equation solver can be used to do this, although it can also be done by the shooting method described in Algorithm 11.18, which is useful as the Maple boundary value method in `dsolve()` can fail. The wavefunctions are calculated from the centre of the potential to just past the right-hand edge of the potential. If the calculation goes too far past the potential large positive or negative values can be produced here but the remainder of the wavefunction is largely unaffected. When plotted, the symmetry of the wavefunction must be used to produce the whole wavefunction. The left half is either the mirror image or the inverse mirror image of the right half. Figure 11.29 shows some of the wavefunctions, eigenvalues, and the double well potential of Fig. 11.28. At low energy, far below the

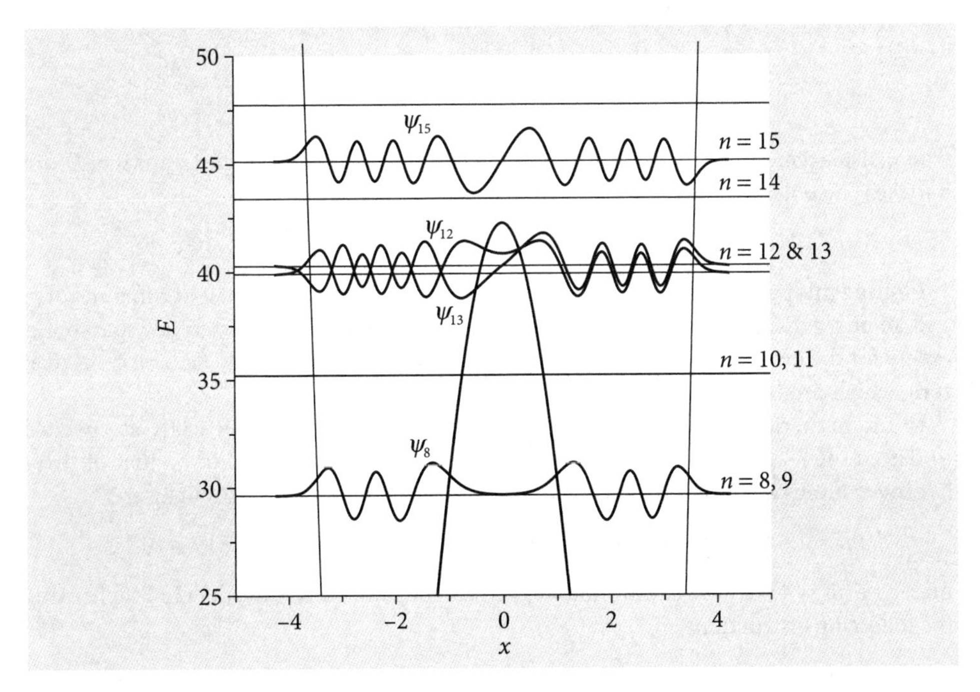

Fig. 11.29 Eigenvalues and wavefunctions vs displacement x for quantum number 8, 12, 13, and 15 superimposed on the double well potential of Fig. 11.28. The eigenvalues with $n = 8$ and 9 are too close to separate on this figure, as are 10 and 11.

barrier top, the eigenvalues $\{E_0, E_1\}$, $\{E_2, E_3\}$, etc. are grouped in pairs with a nearly identical energy because the barrier almost, but not quite, separates the two wells. The lowest two eigenvalues have energy 3.5661 and the next two have energy 10.5325, both pairs being similar to within eight decimal places. Levels 10 and 11 are separated by slightly more energy: 35.1960 and 35.2167 respectively, which is still small. Levels 12 and 13, which are shown in the figure, have just enough of an energy difference (40.2767 and 43.2882) to be seen as separate levels. The wavefunctions with small quantum numbers, $0 \le n \le 9$, are rather like those of two separated wells but still have odd and even parity overall, as does each half. The $n = 13$ wavefunction, Fig. 11.29, has odd parity with a centre of inversion at $x = 0$, and it is clear that there is little chance of finding the molecule at zero displacement. It has to tunnel from one well to the other; an effect famously observed in the 'umbrella' mode of ammonia inversion. The $n = 8$ wavefunction similarly exhibits tunnelling, as do all levels below the barrier. As the barrier gradually lessens, the eigenvalues separate as shown for $n = 12$ and 13. Some memory of the barrier can still be seen in the $n = 15$ level, which above the barrier has an increased amplitude of the wavefunction at a displacement $x \approx \pm 3/4$.

11.10.2 Energy levels for the ammonia inversion normal mode calculated using atomic units

A more realistic example is to calculate the energy levels for ammonia; this also illustrates how to use atomic units to control the huge or tiny numbers present in quantum calculations. The potential energy (Swalen & Ibers 1962), V, of the umbrella or tunnelling normal mode vibration, measured as the displacement x of the N atom from the plane of the H atoms, assuming a C_{3V} point group, is approximated by $V(x) = 30471x^2 + 12478e^{-5.005x^2} - 10335$ cm^{-1}. The displacement x is in Å. The eigenvalues and their spacing will be calculated up to an energy of 4000 cm^{-1}, which is above the barrier.

The normal SI units for Planck's constant and the reduced mass are difficult to use because they generate very large or very small numbers. Converting to atomic units greatly simplifies the calculation. This is done by using an energy scale in hartree, where 1 hartree is 2.1947465225×10^5 cm^{-1} or 27.211396 eV, and the corresponding distance scale, is in units of the Bohr radius a_0, which is 0.529177×10^{-10} m. The hartree is the natural unit of energy and is $e^2/(4\pi\varepsilon_0 a_0)$, where ε_0 the permittivity of free space and e the charge on the electron. The constants in the Schrödinger equation now become $\hbar^2/m_e = 1$, where m_e is the mass in kg of the *electron* and therefore for the reduced mass μ (kg) of the molecular vibration, the ratio $\dfrac{\hbar^2}{m_e}\dfrac{m_e}{\mu} \equiv \dfrac{m_e}{\mu}$ is used. The Schrödinger equation is now

$$\frac{d^2\psi}{dx^2} - \frac{2\mu}{m_e}(V(x) - E)\psi = 0$$

where displacement x is now in units of a_0, and energy in hartree, which means that E is in hartree. The potential has to be changed to achieve this and becomes

$$V(x) = (30471a_0^2x^2 + 12478e^{-5.005a_0^2x^2} - 10335)/2.194746 \times 10^5.$$

Plotting this potential shows that the calculation needs an x range of 0 to 2 in units of a_0 and an energy scale up to 0.02 hartree, Fig. 11.30. The potential is essentially parabolic, except for the low and Gaussian shaped barrier of $\approx$0.0094 hartree (2063 cm^{-1}), centred at zero, which produces two energy minima at ± 0.723 a_0.

In the numerical calculation, energy increments of no more than 0.001 are needed to determine all the levels accurately. The reduced mass is assumed to be that of three hydrogen atoms acting as one mass, and the nitrogen atom. The masses in kg are

$$m_H = 1.673534 \times 10^{-27}, m_N = 2.325267 \times 10^{-26} \text{ and } m_e = 9.1093819 \times 10^{-31},$$

making $\mu/m_e = 4532.769$. The atomic weights can be found using the data in Maple using the following instructions,

```
> with(ScientificConstants):
> evalf(Element( N[14], atomicmass ));
> evalf(Constant(m[e]));      9.109381882 10^-31
```

The initial conditions in the integration have to be changed depending upon whether the energy level has an even or odd parity; by symmetry, the lowest and all even levels have even parity. Using Algorithm 11.20, the Schrödinger equation has to be modified to use atomic units, which means changing

```
eqn:= hbar^2/mu*diff(y(x),x,x)-2*(V(x)-E0)*y(x):
```

to

```
eqn:= diff(y(x),x,x)-2*mu/me*(V(x)-E0)*y(x):
```

in `solve_SE`.

To determine accurately the small splitting between the first two levels, the stopping value `Q` for the bisection method has to be 10^{-8} or smaller; 1000 points are used in the integration. The first eight energy levels (in hartree), with their quantum numbers are

0, 0.00233237	1, 0.00233735
2, 0.00659715	3, 0.00678569
4, 0.00968351	5, 0.01103156
6, 0.01336172	7, 0.01571011

The energy gaps in cm^{-1} compared with the spectroscopically determined values are shown in the table.

Four levels, as two closely spaced pairs, appear in the double well part of the potential. Here the inversion mode tunnels through the barrier. Above the barrier there are single levels, see Fig. 11.30. The results show only a moderately good agreement between theory and experiment. However, the fitting is very sensitive to the parameters of the potential; a change in the potential far too small to see by eye will cause large changes in the calculated

Table 11.3 Calculated and observed transitions in NH_3 between the levels indicated by Δ.

	Calculated/cm^{-1}	Experiment/cm^{-1}
Δ_{10}	1.092	0.7935
Δ_{32}	41.38	35.81
Δ_{21}	934.9	932.5
Δ_{30}	977.4	968.3

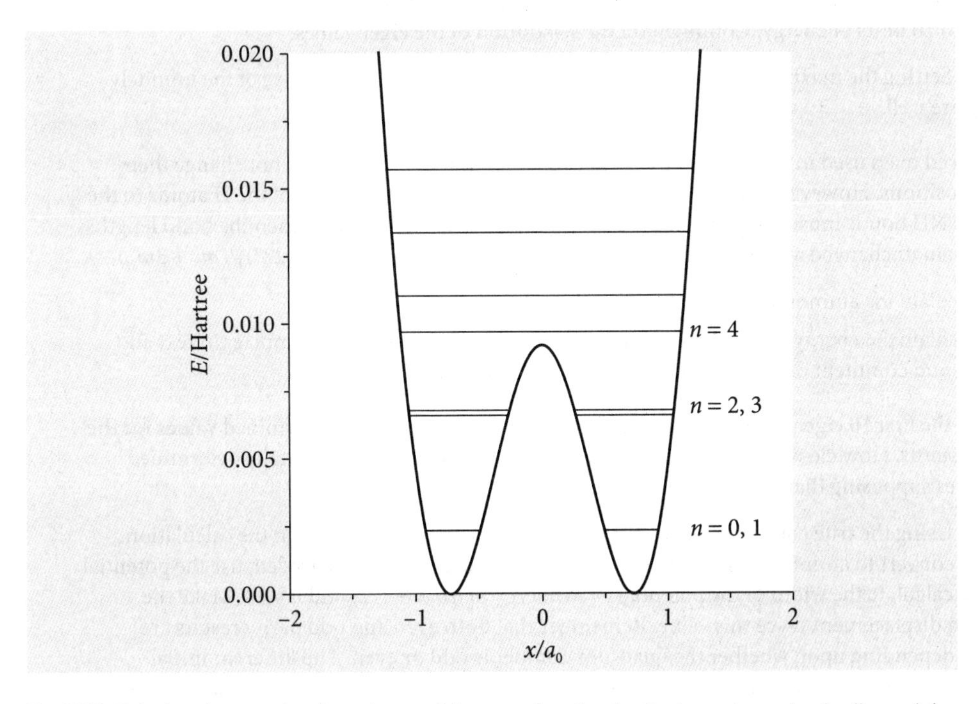

Fig. 11.30 Calculated energy levels and potential energy function for the inversion or 'umbrella mode' vibration in ammonia. The $n = 0$ and 1 levels are split but are too close to be drawn separately.

energy levels. A more sophisticated calculation would vary the parameters in the potential optimally to fit the data. A further sophistication would add terms in x^4 to the potential, but doing this eventually becomes an exercise in fitting the data to an arbitrary potential that may not have any physical reality. Preferably, the potential should be derived from an *ab initio* analysis of the motion of the atoms, rather than assuming the potential shape has a certain form; this is, however, a very difficult calculation.

11.10.3 Questions

Full solutions are available at **www.oxfordtextbooks.co.uk/orc/beddard.**

Q11.35 Solve $d^2y/dx^2 = 1/(1+y^2)+x^2$ with boundary values $y_0 = 10$ and $y_5 = -2$. This equation is sensitive to the initial value of α.

Strategy: Use the shooting method. Try several values of α to find the region where its true value lies, then "home in" on this with better estimates.

11.36 The potential $V(x) = -V_0/\cosh^2(\alpha x)$, sometimes called the Pöschl–Teller potential, describes a potential hole and is used to approximate that produced by the diffusion of material in a quantum well or similar device that is used in opto-electronics. The potential is zero at large $\pm x$ and $-V_0$ at the origin, and the constant α determines the width. Calculate *all* its eigenvalues and wavefunctions if $V_0 = 10$ and $\alpha = 1$. Compare your answer with the analytic eigenvalues

$$E_n = \frac{-\alpha^2}{8}\left[\sqrt{1+8V_0/\alpha^2} - 2n - 1\right]^2.$$

Strategy: Plot out the potential first and estimate the maximum x needed for the integration and also the energy range and increment. Use Algorithm 11.20 to solve the equations. The energy starts at -10 and ends at zero. A step size of 0.1 is a good starting value. It can be made smaller if you suspect that energies have been missed. The Maple code, using the shooting method to calculate the eigenvalues (energy levels) and eigenfunctions (wavefunctions), will only work if the minimum energy is zero. This means that V_0 should be added to the potential energy.

Q11.37 Calculate the eigenvalues and wavefunctions for levels 3, 6, 7, and 8 for the potential hill $V(x) = +V_0/\cosh(\alpha x)$ centred in an infinitely deep square well of width 10 and with $V_0 = 5$ and $\alpha = 1$ up to 10 units of energy. Comment on the separation of the eigenvalues.

Strategy: Setting the maximum and minimum x value effectively defines the size of the infinitely deep square well.

Q11.38 The reduced mass used in the ammonia example assumes that the H atoms do not change their relative positions. However, as the N atom tunnels from one side of the plane of the H atoms to the other, the NH bonds must compress. If the transition only involves bending, then the bond lengths must remain unchanged and the reduced mass is now $\mu = 3m_H[m_N + 3m_H\sin^2(22°)]/(m_N + 3m_H)$.

(a) Recalculate the ammonia energy levels.

(b) Recalculate the energy levels for ND_3, using either of the reduced mass formulae (in text and here), and comment on why the results are different from that for NH_3.

Q11.39 Calculate the first 10 eigenvalues for HCl and I_2 using the experimentally determined values for the force constants. How closely do your calculated results come to the experimentally determined eigenvalues supposing that your potential is harmonic?

Strategy: Using the true constants means having very small or large numbers in the calculation, therefore convert to atomic units. To estimate how large a displacement is needed, use the potential energy to calculate the width of the potential at an energy of $h\upsilon(n+1/2)$ and $n = 10$. Make the maximum displacement twice this value. Remember that both even and odd parity results are required, depending upon whether the quantum number is odd or even. The different initial conditions are given in the text; two sets of calculations should be performed to calculate all the eigenvalues.

Q11.40 Calculate the first four energy levels for the ns orbitals of the hydrogen atom and compare these with the theoretical values.

Strategy: Because this is a real example, work in atomic units. The Schrödinger equation for the H atom is usually written as

$$-\frac{\hbar^2}{2m_e}\frac{d^2\psi}{dr^2}+\left(\frac{\hbar^2}{2m_e}\frac{\ell(\ell+1)}{r^2}-\frac{e^2}{4\pi\varepsilon_0}\frac{1}{r}\right)\psi=E\psi$$

where the term proportional to $\ell(\ell+1)/r^2$ is the kinetic energy associated with the angular motion of the electron. This appears in the equation because it acts as an effective potential, keeping wavefunctions away from the origin when $\ell \neq 0$. In our calculation, $\ell = 0$, because s orbitals are involved, and this term can be ignored. Changing to atomic units gives

$$-\frac{1}{2}\frac{d^2\psi}{dr^2}-\frac{1}{r}\psi=E\psi$$

where the energy is in hartree (1 hartree = 27.211396 eV). The radial separation of electron and nucleus is r and the nucleus is assumed to be of infinite mass, and the electron is in units of the Bohr radius; $a_0 = 0.529177 \times 10^{-10}$ m. Levels with the same principal quantum number n, are degenerate; for example, 2s, 2p are degenerate as are the 3s, 3p, 3d orbitals in this model of the H atom. The term in ℓ does not appear in the energy and the degeneracy of the p and d orbitals can be considered as being accidental. Additional interactions, such as spin-orbit coupling, are needed besides the Coulomb interaction between the proton and electron to split this degeneracy.

12 Monte Carlo Methods

Numerically solving integrals or differential equations using a Monte Carlo method is done by repeatedly guessing the random values of one or more variables—for example, x in $f(x)$—and assessing the result against some criteria. This method differs from other numerical calculations, which proceed by smoothly varying x from start to finish. Although a Monte Carlo method uses random or 'stochastic' values, this is always done according to some algebraic formula or algorithm. However, whatever the exact method used, efficient or not, the final answer is only achieved after averaging many guesses and is therefore only an approximation to the true value.

The Monte Carlo methods fall into two broad areas; the first is the use of random numbers with a formula to perform integrations or solve differential equations; the second is the use of random numbers to *simulate* physical processes at an elementary or molecular level, usually because these processes are too difficult to solve in any other way.

12.1 Integration

To use a Monte Carlo method to numerically integrate a function, the area corresponding to the integral is calculated by repeatedly guessing pairs of x and y values at random and evaluating the function $y=f(x)$ to see if y lies within the area bound by the integral or not, Fig. 12.1. Clearly many guesses will be needed and while many will fall inside the required region, many will not. The ratio of correct guesses to the total number is proportional to the integral. The more guesses that are made, the closer the answer becomes to the true value. The error can then be estimated and the calculation truncated when a satisfactory result is achieved.

There are two considerations. First, the computer generated random numbers must actually be random, or as near random as possible. These are then usually described as being *pseudo-random*. The randomness of numbers is a complicated issue, but if Maple and similar programs are used, then the (pseudo) random number generators can be relied upon; (Prest et al. 1986 give details about random number generators.) In Maple, the repeat sequence length before the same set of numbers is obtained is a really huge number $2^{19937}-1$, which is about 6000 digits long. The second consideration is that numbers should be taken from a *uniform distribution*. A uniform distribution is one in which the chance of obtaining any number in a range, which may be 0 to 1 but is normally a to b, is the same no matter where the number is in that range. In other distributions, this is not true; a Gaussian distribution has more chance of returning a number near to its mean value than elsewhere.

Figure 12.1 explains the Monte Carlo integration method. The integral is $Q=\int_a^b f(x)dx$. Two uniformly distributed random numbers are chosen; one R_x between a and b and another R_y between limits $f(c)$ and $f(d)$, where the points c and d must be chosen to include the minimum and maximum of the function in the range a to b. A large number of pairs of points are chosen, those for which $R_y \le f(R_x)$ are found and then counted up. The integral is approximated as

$$\int_a^b f(x)dx \approx \left(\frac{\text{\# guesses inside/under curve}}{\text{total \# guesses}}\right)\times A, \tag{12.1}$$

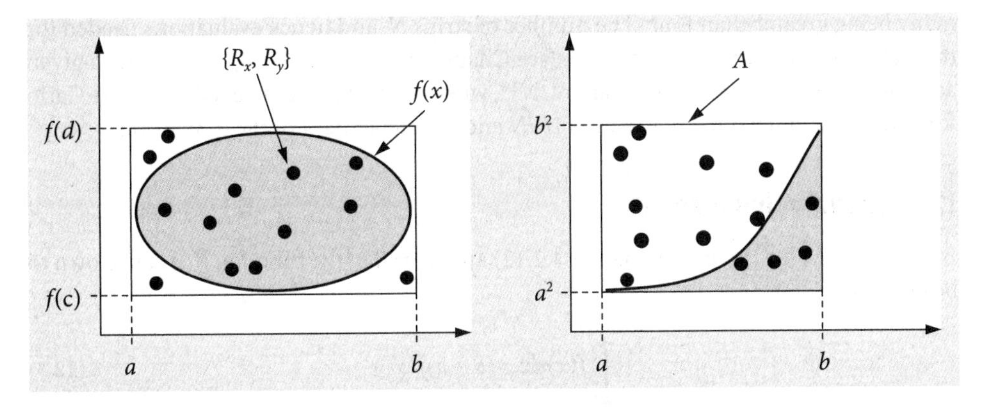

Fig. 12.1 Monte Carlo integration; guess pairs of numbers inside the box and count how many are within the area defined by the function $f(x)$. A is the area in which the integration is performed

where A is the integration area, see Fig. 12.1. If the function is $f(x) = x^2$, and R_x and R_y are random guesses in the range a to b in x, and a^2 to b^2 in y, respectively, only if $R_y \le R_x^2$ is the point counted as being within the function. This approach is clearly rather wasteful; less than half of the points fall within the area for x^2 although this is better for an ellipse, as shown in the figure, because most of the ellipse fills the rectangular area. In practice, this method, while easy to implement would never be used; the mean value method described next works far better and involves only a little more analysis.

The random nature of guessing means that standard statistical analysis methods can be used to estimate the error in the integration just as with a real set of data. If the average (mean) value of the integral calculated is $\langle f \rangle$ with n evaluations, then the error or standard deviation is

$$\sigma = \sqrt{\frac{\langle f^2 \rangle - \langle f \rangle^2}{n}}. \tag{12.2}$$

The function $\sin^2(1/x)$ is symmetrical about $x = 0$ and oscillates ever more rapidly as it approaches zero; see Fig. 12.2. The integral has no algebraic solution but one that involves the sine integral, which has to be evaluated numerically. The integral can, however, be estimated by the Monte Carlo method without using any algebra. The range of y is 0 to 1, because the maximum of the function is 1, and we choose $x = 0$ to 2 so that two random numbers in these ranges are needed. The area of the integration box is $A = 2$ because the y range is 1. The Monte Carlo method using Maple with equation (12.1) is illustrated below. The random number is called using `rand()/1e12` as this generates a number in the range 0 to 1. This is then adjusted to fall in the range of the x and y limits as R_x and R_y respectively. The method works by adding *one* to the total `s` if the function at R_x, `f(Rx)` is greater than R_y. To avoid using the same set of random numbers in the calculation, set `seed:= randomise()` at the start of the calculation once it is working properly.

Algorithm 12.1 Direct Monte Carlo method

```
> f:= x->sin(1/x)^2;                     # define function
  n:= 20000;                             # number of guesses
  xlim1:= 0.0; xlim2:= 2.0:              # x limits
  ymax:= 1.0;                            # max y, min is zero
  s:= 0:                                 # variable for sum
  A:= (xlim2-xlim1)*ymax;                # area A
  for i from 1 to n do
    Rx:= (xlim2-xlim1)*rand()/1e12 + xlim1;    # Rx rand num
    Ry:= ymax*rand()/1e12;                     # Ry rand num
    if Ry <= f(Rx) then s:= s+1: end if:       # make sum
  end do:
  av_f:= A*s/n;                          # estimate of integral
```

A typical result is 1.09, which is quite close to the exact value of 1.084. However, notice that many (20 000) function evaluations are needed to get this result. Even with a minimal box size a Monte Carlo method always involves many evaluations of the function and this method really only comes into its own when the integrations involve many dimensions d, many being greater than four. The number of strips N, and hence evaluations needed for the trapezoidal or Simpson's method, (see Chapter 11.3.3 and 11.3.4), to produce a given accuracy, increases in proportional to $1/N^{2/d}$, whereas the number needed by Monte Carlo for similar accuracy only increases as $1/\sqrt{N}$ and is independent of the dimension.

12.1.1 Mean-value method

By the mean-value theorem (Chapter 4.2.12) an integral of the function $f(x)$ can shown to be given by

$$\int_a^b f(x)dx = (b-a)f(x_m) \tag{12.3}$$

where $f(x_m)$ is the value of the function at x_m. As an integral is equal to an area, then $f(x_m)$ is equivalent to the mean value of the integral if we imagine the area $(b-a)f(x_m)$ as width multiplied by height. Let us see if there really is such a point x_m by trying an integral whose value we know. Suppose the integral is $\int_2^4 e^x dx = e^4 - e^2 = 47.209$, by the mean-value theorem this should be equal to $(4-2)e^{x_m}$ making $x_m = 3.1614$, which is a real number within the limits of the integral. Since $2e^{3.1614} = 47.209$, which is equal to the integral, this proves that the point x_m exists. To calculate this integral by the Monte Carlo method, the mean-value equation is rewritten as,

$$\int_a^b f(x)dx \approx \frac{(b-a)}{n}\sum_{i=1}^{n} f(x_i) = (b-a)\langle f(x)\rangle, \tag{12.4}$$

where $\langle f(x)\rangle$ is the average value of the function calculated over the range a to b. Clearly, for a closed symmetrical function such as an ellipse, Fig. 12.1, the limits a and b must be chosen so that the mean does not turn out to be zero.

A two-dimensional integral is in general $\int_a^b\int_c^d f(x, y)dxdy$. For example, this might be $\int_0^2\int_0^1 \sqrt{4-x^2-y^2}dxdy$ where the limits 0 to 1 refer to the x coordinate, and 0 to 2 to the y. In this case, the Monte Carlo method and the mean-value theorem give

$$\int_a^b\int_c^d f(x, y)dx \approx \frac{(b-a)(d-c)}{n}\sum_{i=1}^{n} f(x_i, y_i) = (b-a)(d-c)\langle f(x, y)\rangle \tag{12.5}$$

Two random numbers are now chosen from a uniform distribution, which are x_i and y_i and the average of the function is calculated as a summation. The extension of this formula to more than two dimensions is done in a similar way.

This Monte Carlo method is now used to calculate the integral of $\sin^2(1/x)$ again; see Fig. 12.2. This is rather a difficult integral because the function oscillates close to zero. The trapezoidal or Simpson methods are generally more efficient (see Chapter 11.3.3 and 11.3.4), but in this case the rapid oscillations require that these methods also have to use a large number of points. The following algorithm will evaluate an integral of the function f, and its standard deviation is calculated as shown in equation (12.2). The average values are multiplied by the x range `xlim2-xlim1` as in (12.3).

Algorithm 12.2 Monte Carlo, mean-value integration

```
> f:= x-> sin(1/x)^2;
  xlim1:= 0.0; xlim2:= 2.0;                          # x limits
  n:= 20000:                                         # num points
  s:= 0.0: s2:= 0.0:                                 # sums
  for i from 1 to n do
    x:= ( xlim2 - xlim1 )*rand()/1e12 + xlim1;       # x range
    s:= s + f(x);
    s2:= s2+ f(x)^2;
  end do:
  av_f:= ( xlim2 - xlim1 )*s/n;                      # <f>
  av_f2:= ( xlim2 - xlim1 )^2*s2/n;                  # <f²>
  sig:= sqrt((av_f2 - av_f^2)/n):                    # std dev
  printf(" %s %f %s %f", "integral =",av_f, "+ -" ,sig );
                  integral = 1.082108 + - 0.004003
```

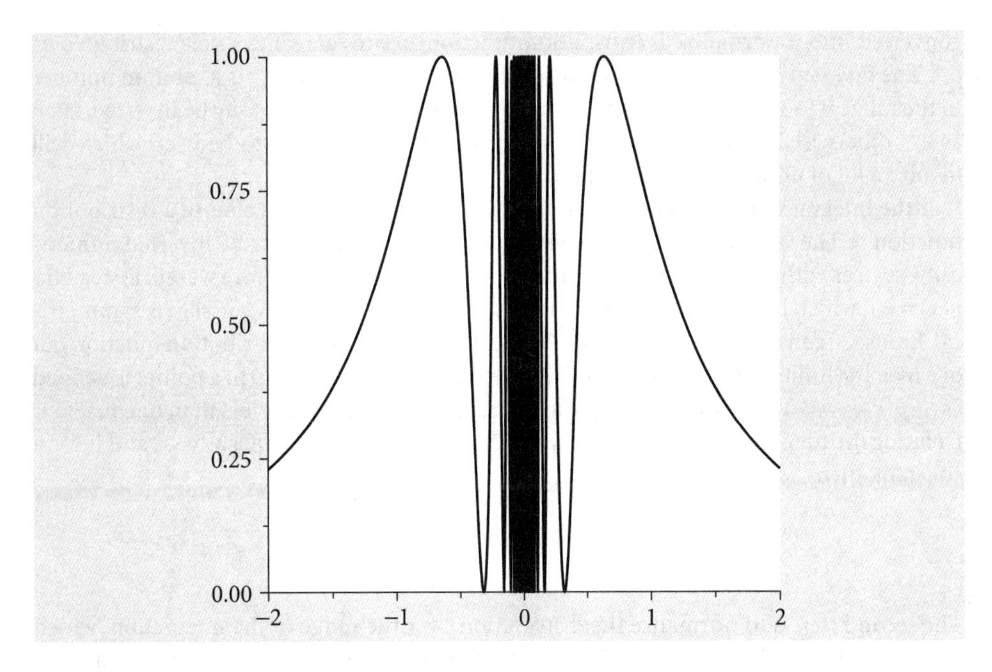

Fig. 12.2 The function $f(x) = \sin^2(1/x)$.

The answer obtained for this integral depends on the actual calculation as for all Monte Carlo methods. Even with 20 000 evaluations, this result has a large standard deviation of approximately ±0.004. As a check, Maple produces the accurate result of 1.0844.

12.1.2 Non-uniform distributions: Importance Sampling

It is easy to imagine functions that are large in only small regions of x and are virtually zero elsewhere; the bell-shaped Gaussian curve being one example. So far a uniform random number distribution has been used to numerically evaluate an integral, but this can waste a lot of effort by repeatedly calculating contributions to the integral that are effectively zero. In some instances, an incorrect result may even be obtained with this approach, see for example Krauth 2006, §1.4. A better method is to use a non-uniform distribution to bias sampling and to add in more of those values that have a large contribution to the integral, and less of those that do not. This is sketched in the figure where many samples are taken where the function is large.

In simulating a number of physical processes, such as the mean energy of an oscillator at a given temperature, or of an array of spins whose energies are distributed by the Boltzmann distribution, and many other examples in statistical mechanics and molecular dynamics, a biased sampling method, called the Metropolis algorithm is used. This is described in Section 12.4.

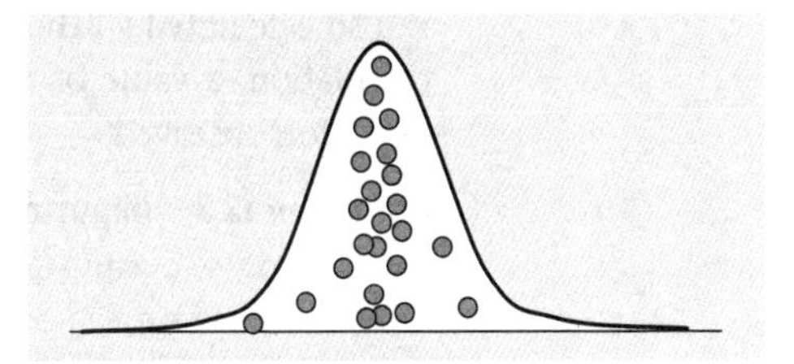

Fig. 12.3 Schematic of importance sampling, more samples are taken where the function is greatest.

In Monte Carlo sampling, the standard deviation is reduced as $\sigma = Q/\sqrt{N}$ for N samples and Q is a constant that depends on the particular problem being solved. Increasing the number of samples N will obviously reduce the standard deviation, but only slowly; additionally N may in practice be limited, because it is proportional to the time the computer takes to do the calculation. However, it turns out that importance sampling can reduce Q and therefore σ by carefully selecting where the samples are taken. This can be done if the function being integrated is changed to another that is effectively 'flatter'. Suppose, therefore, that the integral is multiplied by $p(x)/p(x)$ which is 1, and written as

$$\int_a^b f(x)dx = \int_a^b \left[\frac{f(x)}{p(x)}\right] p(x)dx = \int_a^b \frac{f(p^{-1}(r))}{p(p^{-1}(r))}dx = \int_a^b \frac{f(t)}{p(t)}dt \tag{12.6}$$

where $p(x)$ is a function we need to guess. Preferably, it has the same shape as $f(x)$ so that on division the ratio $[f(x)/p(x)]$ is approximately constant. The remaining $p(x)$ is converted into a normalized distribution function to provide values of x distributed as p^{-1}. The *inverted distribution*, which we shall call t, is $t = p^{-1}(r)$ and r is a random number in the range 0 to 1. The choice of p is limited to functions that can easily be inverted, such as e^{-x}; otherwise, a numerical way of inverting the function has to be used which will involve a lot of unnecessary calculation.

If the integral to be evaluated is $\int_0^b e^{-x^2}dx$, the starting point is to choose a distribution function p. The Gaussian e^{-x^2} is an obvious choice but this can only be inverted numerically, such as with the Box–Muller algorithm (Weisstein), and therefore we will instead try $p(x) = e^{-x}$, which has a similarly decaying form when $x > 0$. (As the Gaussian is symmetrical, limits < 0 can easily be calculated). The next step is to make a distribution function out of p over the limits of the integration. The *cumulative distribution* up to a point t is defined as $\int_{-\infty}^{t} p(x)dx = r$. This is the chance (probability) of choosing a value less than or equal to r. Defining the function p as $p(x) = ke^{-kx}$ when $0 \le x \le \infty$, and zero for negative x, and if k is a constant, then

$$\int_0^t e^{-kx}dx = (1 - e^{-kt})/k.$$

The second step is to normalize the distribution over the range of the integration, which is the integral $\int_0^b e^{-kx}dx = (1 - e^{-kb})k$. Finally, the normalized cumulative distribution is $r = \dfrac{(1 - e^{-kt})}{1 - e^{-kb}}$ and rearranged this gives the inverted distribution

$$t = -\ln[1 - r(1 - e^{-kb})]/k \tag{12.7}$$

when r is a uniform random number between 0 and 1. The same procedure can be followed for any function $p(x)$ provided $p^{-1}(x)$ can be calculated.

The general Importance Sampling algorithm is now straightforward:

(1) Choose a distribution $p(x)$, normalize, and invert.
 $p(x)$ should have a shape similar to $f(x)$.
 Integrate p from $-\infty$ to t to give I_t.
 Integrate p between limits a and b to find the normalisation.
 Invert normalized cumulative distribution to give $t = somefunction(r)$.
(2) Start Monte Carlo
 Loop over the number of trials
  Calculate t from random number r in range 0 to 1
  Calculate the sum of $N \times f(t)/p(t)$ and its square to find std. dev.
 End loop
(3) Calculate the average.

A possible code to do this in Maple is shown below. The square of the sum of the terms is also calculated so that the standard deviation of the result can be determined. In the calculation, a value of $k = 1$ was chosen; other values could be tried and the standard deviation observed.

Algorithm 12.3 Importance sampling using $p(x) = ke^{-kx}$

```
> # derive cumulative distribution and invert
  # first integrate and normalise function, then use result
  # in Monte Carlo.
  p:= x-> exp(-k*x):                      # prob function
  It:= int( p(x),x = 0..t ):              # integrate
  N:= int(p(x),x = a..b):                 # normalisation
  It/N;
```

$$\frac{-1 + e^{-kt}}{e^{-ka} - e^{-kb}}$$

```
# invert normalised cumulative distribution to find t.
> solve(It/N = r, t);                     # eqn (12.7) if a = 0
```

$$-\frac{\ln(1 - r\,e^{-ka} + r\,e^{-kb})}{k}$$

```
# start Monte Carlo integration
> f:= x-> exp(-x^2);                          # function
  k:= 1.0;                                    # choose k in p
  a:= 0.0; b:= 2.0:                           # integrtn limits
  c:= exp(-k*a) - exp(-k*b);                  # constnt in It/N=r
  N:= (exp(-k*a) - exp(-k*b))/k:              # normalisation N
  s:= 0.0: s2:= 0.0:                          # sum and sum²
  n:= 5000:                                   # number of samples
  for i from 1 to n do                        # loop over samples
    t:= -ln( 1.0-rand()/1e12*c)/k:            # eqn (12.7)
    Q:= N*f(t)/p(t);
    s:= s + Q;                                # sum
    s2:= s2 + Q^2;                            # sum²
  end do:                                     # end loop
  av_f:= evalf(s/n):                          # average
  av_f2:= evalf(s2/n):                        # average2
  std_dev:= sqrt((av_f2-av_f^2)/n):           # std deviation
  printf(" %s %f %s %f", `integral =`,av_f,`+ -`,std_dev );

                  integral = 0.884454 + - 0.003744
```

The result is that the integral has the value 0.884 ± 0.004 in this particular run. The same integral, calculated by the mean-value method, had an error ± 0.01, which is a considerably larger error for the same number of samples. The accurate result is 0.88208.

12.1.3 Limits tending to infinity

Integration limits that tend to ± ∞ can present a particular problem for any numerical method. A way of solving this is to try to look at the function at large values of x. Often the function will become negligible and while, technically, the integral may go to infinity the numerical value may have converged to a sensible answer well before this. See question Q12.6 for an example of this.

12.1.4 Questions

Full solutions are available at **www.oxfordtextbooks.co.uk/orc/beddard**.

Q12.1 Calculate the area under the curve $e^{-x}\sin(x^2)$ in the range 0 to 6 using the mean-value method.

Strategy: Use the algorithm in the text modified for the new function and limits.

Q12.2 Integrate $\int_0^2\int_0^1\int_0^{1.5}\sqrt{16-x^2-y^2-z^2}dxdydz$ using the mean value, Monte Carlo method and accurate to two significant figures.

Strategy: Use equation (12.5) modified for three variables. Repeat the calculation until the answer becomes accurate enough for your needs.

Q12.3 Integrate $\int_0^\infty e^{-x}\sin(x^2)dx$ using importance sampling.

Strategy: The distribution function can be chosen as e^{-x} and the importance sampling algorithm followed. This method also enables us to calculate from zero to infinity; however, a cut-off value has to be determined and e^{-10} can be chosen because it is so small compared to 1. The limits are therefore $a = 0$ and $b = 10$.

Q12.4 Calculate the integral $\int_0^2 e^{-x^2}dx$ using importance sampling and the distribution function $p(x) = 1/(x^2 + 1/4)$, which has a similar shape to the function in the integral.

Strategy: Following the text on importance sampling, $p(x)$ has to be made into a distribution. The integral from 0 to t is

```
> Int(1/(x^2+1/4),x=0..t): % = value(%);
```

$$\int_0^t \frac{1}{x^2+\frac{1}{4}}dx = 2\arctan(2t)$$

and the normalization is therefore $N = 2\tan^{-1}(4)$. Working out the cumulative distribution and solving for r produces $r = \tan^{-1}(2t)/\tan^{-1}(4)$. The variable r is sampled from a uniform distribution in the range 0 to 1.

Q12.5 Apparently innocent functions, when calculated by simple Monte Carlo methods, can give incorrect results even when huge numbers of sampling points are taken. One example is the integral

$\Gamma = \int_0^1 x^\gamma dx = \frac{1}{\gamma+1}$ when $\gamma > -1$. (Krauth 2006, section 1.4, p. 66)

(a) Accurate values are obtained when $-0.5 < \gamma < 1$. Show that this is the case using a simple, or mean-value Monte Carlo method with $\gamma = 1/2$ and 2×10^4 samples.

(b) Using the same method show that an incorrect result is produced when $\gamma = -0.8$, and similar values close to -1, even though tens or even hundreds of thousands of samples are taken. (Note that this may take several minutes to complete depending on the computer used and how efficient the algorithm is.)

(c) Next, use the importance sampling method with the distribution $p(x) = x^\lambda$ where $\gamma < \lambda < 0$, and obtain an accurate result.

Q12.6 The virial coefficients are used in the description of real gases. The compression factor $Z = 1$ for an ideal gas, but is expanded as a series for a real gas,

$$Z = \frac{pV}{nRT} = 1 + B_2\left(\frac{n}{V}\right) + B_3\left(\frac{n}{V}\right)^2 + \cdots$$

The constants B_2, B_3, and so forth are the virial coefficients. The second coefficient B_2 can be related to the potential energy of interaction between molecules, which leads to non-ideal behaviour. The constant is (Rigby et al. 1986; Murrell & Jenkins 1994; Stone 1996)

$$B_2 = \int_0^\infty (1 - e^{-U(r)/k_BT})r^2 dr$$

where U is the interaction potential energy at the separation of r between molecules. In the case of a Lennard-Jones 6-12 potential,

$$U(r) = 4\varepsilon\left[\left(\frac{\sigma}{r}\right)^{1/12} - \left(\frac{\sigma}{r}\right)^{1/6}\right];$$

the integral has no analytic (algebraic) solution and has to be calculated numerically.

Using the Monte Carlo method, calculate B_2 with the parameters for CO_2, which are $\varepsilon = 140$ cm^{-1}, $\sigma = 3.943$ Å. Boltzmann's constant is 0.693 cm^{-1} K^{-1}.

Strategy: The limits of the integration need to be addressed, because the limit of infinity is not generally possible with the Monte Carlo or other numerical methods. Additionally, the limit when $r = 0$ needs to be checked because here $U(0) = \infty$. In the latter case, the exponential term rescues the situation because $0(1 - e^{-\infty}) = 0$, therefore the limit $r = 0$ is calculable. When $r = \infty$, U is zero, and the whole expression inside the integral (the integrand) is also zero. This can be seen by plotting $f(r) = (1 - e^{-U(r)/k_BT})r^2$ and $U(r)$. In practice, a maximum value has to be put on r, and using the values given in the question, a plot indicates that $r = 20$ Å $= 2$ nm is quite sufficient. The function $f(r)$ is zero initially, and rises as r^2 for small r; however, this term is overwhelmed by the exponential at larger values of r.

12.2 Solving rate equations

12.2.1 Concept and motivation

A Monte Carlo method of solving differential equations has been developed by Gillespie (1977, 2007) and is just as applicable to chemical reactions as it is to describing S-I-R diseases or predator–prey populations. This method is equivalent to any other numerical method in the sense that it is purely mathematical: no science enters into the method of solution; this is all done when the rate equations are written down. In comparison, in Section 12.3, a heuristic way of solving problems will be described, and in this case, the science behind the processes involved is an essential part of the method.

When considered at the level of individual molecules, any chemical reaction has relatively long time periods during which no reaction is possible followed by small time periods when reaction occurs. The Gillespie method calculates these two times and uses them as the basis of the method to solve the rate equations. Bimolecular chemical reactions rarely occur on first contact of the two reactants because there is usually an energy barrier, the activation energy, between reactants and products. Many millions and perhaps thousands of millions of collisions are needed before reaction occurs if the activation energy is much larger than thermal energy and this time is far longer than the time it takes to cross the barrier. Even if a reaction occurs on first contact, there are periods during which no reaction is possible because the molecules are separated from one another. In a first-order, solution phase reaction, such as a cis-trans isomerization, many collisions with solvent are needed to give the reactant molecule enough energy to surmount the energy barrier between reactant and product. Similar unreactive periods occur even in a molecule isolated from others in the gas phase. This may have enough total energy to react but this is may be spread over all of its $3N-6$ vibrational modes. The energy must flow out of these modes and find its way into the reactive bond and this takes time.

If the reacting molecules are considered individually, the various times that would be observed between reactions have a Poisson distribution, which is a distribution of many events with each having a low probability. The probability of a reaction is therefore also described by a Poisson distribution; see Chapter 13.6.4. Monte Carlo methods mimic the Poisson behaviour of molecules by considering the chance (probability) that no reaction will occur during a certain period of time, which is followed by another smaller time interval during which a reaction does occur. The method, therefore, involves working out these two probabilities and multiplying them together.

12.2.2 The Gillespie method

When two species A and B react, $A+B \rightarrow product$, the rate of reaction is $k[A][B]$ where k is the rate constant and $[A]$ and $[B]$ the concentrations at any time during the reaction, which clearly change with time. To work out what happens in each small time interval means calculating the change in the number of reactant molecules in a given volume. Doing this means calculating two things. The first is the time at which any of possibly several reactions will occur, and the second is to decide which reaction this will be.

In the reaction $A + B \rightarrow product$, there is only one type of reaction but there are very many schemes where there are two or more reactions; for example, the photo-dissociation of halocarbons with UV light and subsequent reactions to destroy ozone.

$$CFCl_3 \xrightarrow{uv} CHCl_2+Cl \qquad Cl+O_3 \rightarrow ClO+O_2 \qquad ClO+O \rightarrow Cl+O_2$$

Given such a reaction scheme, the first step is to decide what reaction will happen next given the current state of the reaction at time t. At the level of considering individual molecules, once the reaction has started, it does not follow sequentially as shown in the scheme. This only happens when a whole ensemble of molecules, typically as large as Avogadro's number is examined, and then normal kinetic equations apply because of the averaging over many events that occurs. In any given time interval when considering molecules one at a time, it is not possible to predict exactly which reaction will happen next. It is quite possible, for instance, that several $Cl+O_3$ reactions may occur before either of the other two. The best that can be done is to guess the outcome, and repeat the calculation many times and this approach is now described.

Suppose that the reactions are labelled with a number R that represents reaction 1 or 2 or 3 etc. The chance of a reaction of type R occurring will be defined as

$$prob(T+dT, R) \equiv a_R dT.$$

Specifically, this is the probability that a reaction of type R occurs in time interval T to $T+dT$ in the reaction volume, given the current state of all reactions at an earlier time t. The values of a_R can be easily worked out, and this is shown later. What is now needed is

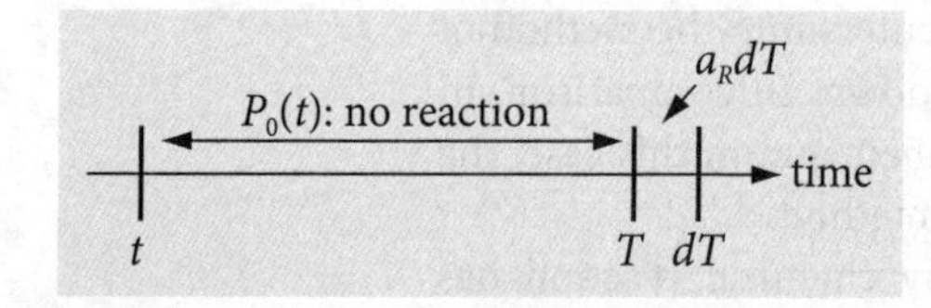

Fig. 12.4 Calculation of reaction probability.

the probability $P_0(T)$ that no reaction of any type occurs up to time T past the present time t, multiplied by the probability that only a reaction of type R occurs in the small time interval T to $T + dT$, which is defined as $a_R dT$. The probability of reaction type R happening is therefore

$$P(T, R) = P_0(T)a_R dT$$

and is shown schematically in Fig. 12.4.

To calculate the probability of *no* reaction occurring, consider any small interval $d\tau$ after some arbitrary time τ, making an interval τ to $\tau + d\tau$. The probability of no reaction is then

$$P_0(\tau + d\tau) = P_0(\tau)\left(1 - \sum_{R=1}^{n} a_R d\tau\right)$$

where the summation is over all n reaction types R. The summation is made because this gives the chance that *any one* of the R reactions will occur, and therefore one minus this number is the chance that no reaction occurs. Rearranged, this equation gives

$$\frac{P_0(\tau + d\tau) - P_0(\tau)}{d\tau} = -\sum_{R=1}^{n} a_R,$$

and in the limit of a small time interval $d\tau$, it becomes the true derivative $\frac{dP_0}{d\tau} = -\sum_{R=1}^{n} a_R$. For clarity only, a notational change is made so that

$$a_0 = \sum_{R=1}^{n} a_R \tag{12.8}$$

therefore, $\frac{dP_0}{d\tau} = -a_0$. The chance of no reaction up to time T is found by integrating, and is

$$P_0(T) = e^{-a_0 T}.$$

The final result for the probability of reaction R occurring at time T is the chance of no reaction up to T multiplied by the chance of reacting at T, which is a_R, giving

$$P(T, R) = a_R e^{-a_0 T} \tag{12.9}$$

an expression that shows that these times are Poisson distributed. Equation (12.9) is a joint probability density function, as it is the chance that reaction occurs in the time interval *and* that it will be reaction R.

Two terms now remain undefined. The first is the individual a_R and their sum a_0, and the second determines which reaction happens next. Calculating a_R is very easy, as each a_R is the product of the number of molecules present in a reaction of type R multiplied with the rate constant per unit volume. If there is just a single reaction $A \xrightarrow{k_1} B$ then $a_1 = k_1 N_A$ where N_A is the number of A molecules present at time T. If the reaction is $A + B \xrightarrow{k_2} C$ then $a_2 = k_2 N_A N_B$ and if both reactions are present in the overall reaction scheme which would then be

$$A \xrightarrow{k_1} B \qquad R_1$$
$$A + B \xrightarrow{k_2} C \qquad R_2$$
$$a_0 = \sum a_R = a_1 + a_2. \tag{12.10}$$

In the special case where a reaction combines two identical molecules $A + A \xrightarrow{k_2}$, the number of indistinguishable combinations of two species must be calculated and this makes $a = k_2 N_A(N_A - 1)/2!$. Normally, in dealing with chemical kinetics, N_A is so vast that $a = k_2 N_A^2/2$, but in the Monte Carlo simulations, N_A is small because individual molecules are dealt with, and the exact formula must be used. Thus, macroscopic, (kinetic), chemical equations are only valid when deviations from the Poisson distribution are negligible.

The final part of the calculation determines which particular reaction step is going to occur and at what time this is done by choosing pairs of numbers that have the distribution $P(T, R)$ equation (12.9). Gillespie (1997, 2007) argues that $P(T, R)$ is equivalent to two distributions, P_1 and P_2;

$$P(T, R) = (P_1)(P_2)$$
$$= (a_0 e^{-a_0 T})\left(\frac{a_R}{a_0}\right)$$

and which can be determined by choosing two uniformly distributed random numbers, r, each in the range 0 to 1. The equation

$$T = -\frac{1}{a_0}\ln(r). \tag{12.11}$$

generates a random number T according to the probability density function $P_1 = a_0 e^{-a_0 T}$. Equation (12.11) produces times that are Poisson distributed, and the time T is added to the current time of the reaction. The second distribution, P_2, is calculated by choosing a uniform random number according to the probability density function a_R/a_0. For example, the chance that reaction 1, (equation (12.10)), occurs is $a_1/(a_1 + a_2)$ and for reaction 2 this chance is $1 - a_1/(a_1 + a_2) = a_2/(a_1 + a_2)$. Which reaction occurs is determined by choosing a random number between 0 and 1 and seeing which of the two regions it falls into. If it falls between 0 and $a_1/(a_1 + a_2)$, reaction 1 occurs; otherwise, reaction 2 occurs. If reaction 1 occurs, $A \rightarrow B$, then 1 is subtracted from the original number of A molecules in each step of the reaction and 1 is added to the number of B molecules. If reaction 2 happens, $A + B \xrightarrow{k_2} C$, then 1 molecule is removed from each of the numbers for A and B. How many molecules are added and subtracted, of course, depends on the stoicheiometry of each particular reaction.

The outline plan for the Gillespie algorithm is

(1) Define the initial amounts of each species.
Define the arrays of data points to hold results.
Set the maximum time.
Set the time to zero; $t = 0$.

(2) Start a loop around the calculation until the time is up or no molecules are left.
Calculate a_1, a_2, a_3 etc. and $a_0 = a_1 + a_2 + a_3 + \cdots$
Calculate time $T = -\dfrac{1}{a_0}\ln(r)$.
Increment time $t = t + T$.
Calculate which reaction occurs: if a random number falls between 0 and a_1/a_0, reaction 1 occurs, and so forth for other reactions.
Add or subtract numbers of molecules from the totals, depending on which reaction has occurred and its stoicheiometry.
Store the results.

(3) Continue the loop until finished.

The first calculation attempted is that of a simple isomerization, or the radioactive decay of an atom, or the decay of an excited state. The scheme is

$$A \xrightarrow{k_1} product.$$

The Gillespie method is worked out in Maple and shown in Algorithm 12.4. Much of the algorithm is involved in checking that something has not gone wrong, such as a negative number of molecules, and in organizing the data to plot it. The method produces a series of times and the number of molecules of each species at each time; however, the times produced are randomly distributed and to plot them a histogram has to be made containing a number of time bins. In the calculation, 100 bins are chosen; the reaction rate constant is 1/20, so the lifetime of the reaction is 20, and a total time of 150 is chosen; `maxt` = 150. You can, of course, change these numbers and a small amount of experimenting is usually necessary to get the calculation as you might want it. No time units are given. This is for you to decide, they could be seconds to picoseconds as appropriate to the reaction. The loop continues until either the time calculated, when converted into an index number for storing the data, has reached its maximum value, or the number of molecules has reached zero, in which case no more reaction is possible.

To make the histogram, the calculated times are made into an integer `indx`, the value of which is to be the *index* number of the array `Acount` that holds the number of A molecules. On each loop round the calculation, a new time is calculated and becomes a bin index, and 1 is subtracted from the content of that bin because another A molecule has reacted. The times are scaled by multiplying by the number of bins/maximum time, and then truncated to make this number into an integer. Times greater than the maximum allowable time maxt will sometimes be calculated; these are excluded by only allowing calculations with bin index less than the maximum. This is done with the instruction

```
if indx < bins-2 then .. end if;
```

If this were not done, an error would occur because the index would exceed the size of the array. When all the molecules have reacted, i.e. no molecules remain unreacted, the calculation has to stop and a check is put in the while..do loop to end it when this occurs. The loop has the form

```
while indx < bins and nA > 0 do
...whole calculation in here..
end do:
```

The first index number in any array has to be 1; in this calculation this is exactly at time zero. The index in the calculation has 2 added because we start at 1, and the next time calculated, may be truncated to zero. If this is the case, the next time point has to be at index 2, since it is actually greater than zero. Finally, so that the decay can be seen clearly the log of the data is required, and any data point that is zero is made into a small number, because $\ln(0) = -\infty$ and this will cause an error in plotting. As the number of molecules cannot be less than zero, this does not affect the plot. To check the result you may want to find its gradient. As a general point, if you calculate a least-squares line through any data you should never alter it, so leave any zero points as zero when doing this. Recall that in Maple we use rand()/1e12 to calculate a uniform random number in the range 0..1.

Algorithm 12.4 Monte-Carlo integration by Gillespie method

```
> restart: with(plots):
> # define parameters
  Seed:= randomize();              # start random numbers
  bins:= 100:                      # number of data points
  maxt:= 150.0;                    # maximum time
  k1:=   1/20.0:                   # decay rate constant
  A0:=   2000:                     # initial number molecs
  Acount:= Array(1..bins):         # define arrays
  dtime:= Array(1..bins):
      # next set initial values into arrays
  for i from 1 to bins do
    Acount[i]:= 0;
    dtime[i]:= evalf( maxt*(i-1)/bins );
  end do:
      # do Monte Carlo and bin results automatically
# make t into the index value of each bin,
# index is 'indx[t+2]' because t can turn out to be < 1 &
# minimum index is 1.
# for clarity t0 is used instead of T as the update time; eqn 12.11
# 'trunc(..)' makes an integer by rounding down.
  nA:= A0:                          # initial number molecs
  t:=0:                             # start time
  indx:= 1:
  Acount[indx]:= nA:                # first data point
  while indx < bins and nA > 0 do   # start loop & check if ok
    a0:= k1*nA:                     # one species; no sum
    t0:= -ln( rand()/1e12 )/a0;     # eqn (12.11) calc time
    t:= t + t0:                     # increment time
    indx:= trunc(t*bins/maxt);      # make into array index
    if indx < bins-2 then           # don't overflow array
      nA:= nA - 1:                  # subtract one A molec
      Acount[indx + 2]:= nA;        # store data
    end if;                         # end of if indx
  end do:                           # end loop
                      # set up to plot data
```

```
# check data for zeros for log plot
for i from 1 to bins do
  if Acount[i] = 0 then Acount[i]:= 1e-10; end if;
end do:
  # make list of times and number of molecules and plot
pt:= seq( [dtime[i], ln(Acount[i]) ], i = 1..bins):
p1:= plot([pt],style=point,
               view=[0..dtime[bins]+5,0..ln(events)+1]):
p4:= plot(ln( Acount[1]*exp(-x*k1)), x=0..dtime[bins]):
display(p1, p4);
```

The result of a calculation is plotted in Fig. 12.5; the line through the data is an exponential with a $\tau = 20$ lifetime; $e^{-t/\tau}$. The 'drop-outs' in the data at long times (> 130) are due to there being too few simulations to fill properly all the time bins and are distinctly unphysical. They are most apparent in a log plot, but in this case only produce a small error that is hardly noticeable on a linear plot, and are an inevitable part of the method; if more time points are used then the drop-outs become more frequent. As they are so noticeable, and clearly unphysical, they can be ignored, if desired. However, they should really be eliminated by starting with more molecules; the only penalty is a longer calculation.

At intermediate times, the limited number of events simulated is also seen as noise on the data, but it is clear that the simulated data is a good match with the theoretical line over about a 1000-fold range of counts. An improved calculation could be made, by starting with a larger initial number or averaging together several single calculations such as this one. The noise on this data is completely different to what would be observed experimentally and to that on Fig. 12.16, which is the result of a simulation of the process.

As a second example, suppose an excited state of a molecule reacts as shown in Fig. 12.6. The excited state decays to the triplet state by intersystem crossing and by fluorescence, to the ground state. The triplet is assumed to have a very long lifetime, so does not convert to the ground state during the time considered in this calculation. The rate constants are k_{isc} to the triplet and k_f to the ground state. The population of all three states involved will be calculated. The decay of the excited state is $\tau = 1/(k_f + k_{isc})$ and this is the lifetime with which the triplet is formed and also that with which the ground state is formed; equivalently the rate constant is $k = k_f + k_{isc}$. In fact, only two states need be calculated as the total number of all molecules is a constant number, but all state populations will be calculated by way of illustration.

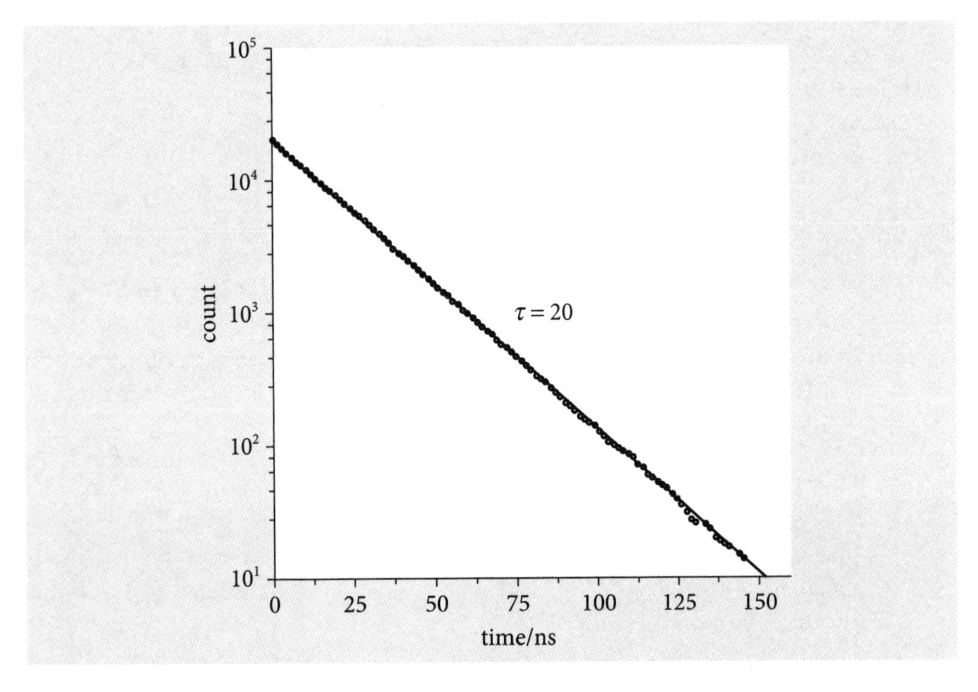

Fig. 12.5 A calculated decay with $k_1 = 1/20$ and a line from the analytical equation, also with the same lifetime to show that the Monte Carlo result is the same, within error. The 'drop-outs' at long times appear as missing data points.

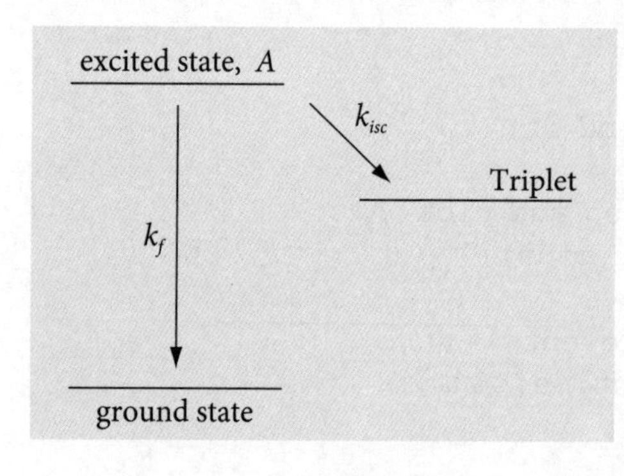

Fig. 12.6 Excited state decay pathways

If the triplet or ground state populations are needed, then one molecule is added each time to these populations; initially these are zero. The chance (probability) of going to the triplet is the fraction or yield $k_{isc}/(k_f + k_{isc})$, and of fluorescing and going to the ground state, is the fluorescence yield $\phi = k_f/(k_f + k_{isc})$. The sum of these yields is 1. Which state the molecule ends up in is determined by casting a die in the form of a uniform random number in the range 0 to 1. If it is in the range 0 to $k_{isc}/(k_f + k_{isc})$, the triplet is produced; if not, the ground state results and the molecule has fluoresced.

The calculation is shown below and is only slightly modified from that of the last example. The fluorescence rate constant is 1/(20 ns), and the intersystem crossing rate constant 1/(10 ns) makes the fluorescence yield 1/3, so that 2/3 goes to the triplet. The prefix t is used for the triplet, such as `tcount` or `nt` for number of triplets, g for ground state, and A the excited state. As before, a check is made in the `while..do` statement to prevent the population of any state becoming negative. The triplet and ground state cannot become zero, because +1 is always added to their populations. The data is placed into 100 bins and is calculated for a total time of 100 ns with 20 000 molecules, initially in the excited state. The units of nanoseconds (ns), need not be explicitly included if we work in these units.

Algorithm 12.5 Excited state decay to ground state and triplet

```
> restart:
  Seed:= randomize():
  bins:= 100:                                    # histogram bins
  maxt:= 100.0:                                  # total time
  kf:= 1/20.0:       kisc:= 1/10.0:              # rate constants
  k1:= kf + kisc;                                # decay rate
  phi:= kf/k1;                                   # fluoresce yield
  A0:= 20000:                                    # initial number A
  Acount:= Array(1..bins):      Tcount:= Array(1..bins):
  Gcount:= Array(1..bins):      dtime:= Array(1..bins):
  for i from 1 to bins do                        # set array values
    Acount[i]:= 0; Tcount[i]:= 0; Gcount[i]:= 0;
    dtime[i]:= evalf( maxt*(i-1)/bins );
  end do:
  nA:= A0: nG:= 0: nT:= 0:                       # initial values
  t:= 0.0:                                       # initial time
  indx:= 1:
  Acount[1]:= nA:  Tcount[1]:= 0: Gcount[1]:= 0:
  while indx < bins and nA > 0 do
    a0:= k1*nA:
    t0:= -ln(rand()/1e12)/a0:                    # equation (12.11)
    t:= t + t0:
    indx:= trunc(t*bins/maxt):                   # index is time
    if indx < bins-2 then                        # don't overflow
    if rand()/1e12 < phi then                 #fluoresce to ground state
        nG:= nG + 1:                            # add 1 to ground state
        Gcount[indx+2]:= nG:
      else                                       # triplet formed
        nT:= nT + 1:                             # add 1 to triplet
        Tcount[indx+2]:= nT:
      end if:
      nA:= nA - 1:                               # A always depleted
     Acount[indx+2]:= nA:
    end if:                                      # end'if indx <..'
  end do:                                        # end while-do loop
```

The results are shown in Fig. 12.7, together with the analytic solution because these equations can be solved analytically. The triplet population increases as $A_0(1-\phi)(1-e^{-k_1 t})$,

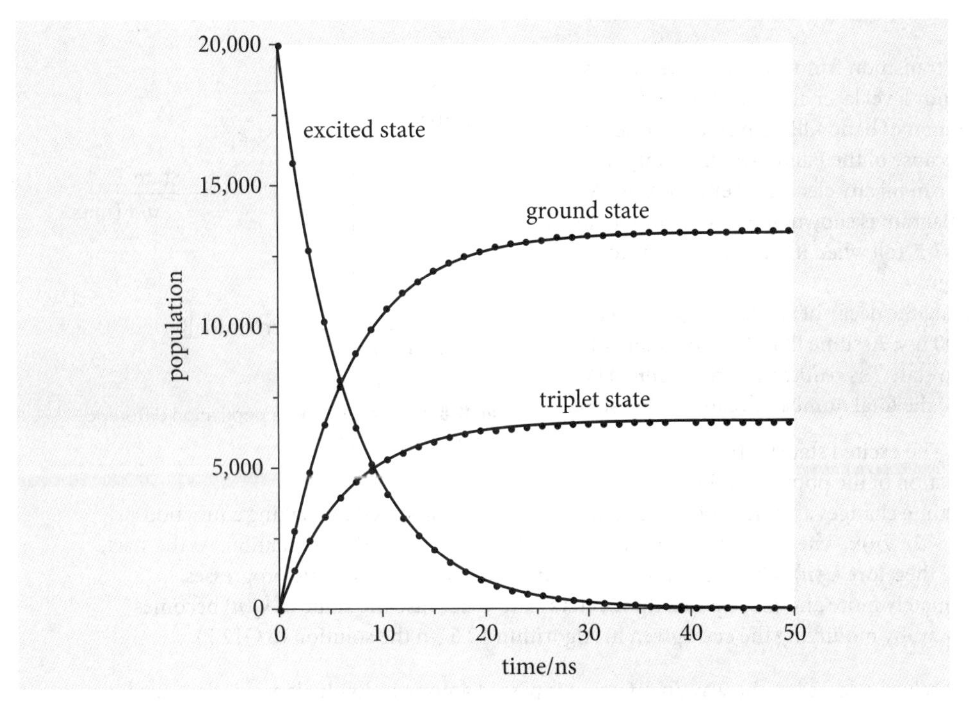

Fig. 12.7 Monte Carlo simulation of the decay of an excited state and the appearance of triplet and ground state (circles) together with theoretical lines (solid). The gaps (drop-outs) in the Monte Carlo data occur because of the limited number of trials; 20 000 in this example.

and that of the ground state similarly increases and matches that of the decay of the singlet $A_0e^{-k_1t}$ but in proportion to the amount entering each state; 2/3 goes to the triplet. The Monte Carlo data and analytic curves are shown in Fig. 12.7, and they match exactly, except for the 'drop-outs' at long times. Fewer drop-outs would be present with a larger initial number of molecules.

12.2.3 **Questions**

 Full solutions are available at **www.oxfordtextbooks.co.uk/orc/beddard**.

Q12.7 A singlet excited state decays to a triplet state by intersystem crossing with a rate constant, $k_{isc} = 2 \times 10^8\ s^{-1}$, as well as fluorescing to the ground state with rate constant $k_f = 1 \times 10^7\ s^{-1}$.

(a) What is the fluorescence lifetime and the fluorescence yield?

(b) Assume that the triplet state has a 20 ns decay time, $k_T = 1/(20\ ns)$. Calculate the fluorescence decay as well as the rise and fall in the triplet population for at least 100 ns, using a Monte Carlo method.

Strategy: The yield and lifetime are very easy to calculate. The fluorescence lifetime is the reciprocal of the sum of the rate constants depleting the excited singlet $\tau = 1/(k_f + k_{isc})$. The fluorescence yield is the fraction of molecules that fluoresce compared to the total number of excited states, which is the same as the rate constant for fluorescence divided by the total of all rate constants causing the state to decay, $\varphi = k_f/(k_f + k_{isc}) = k_f\tau$ (see Turro 1978). All yields must be between 0 and 1. The calculation is similar to that in the examples, but now two reactions are present, the singlet and triplet. The rate equations are

$$S \xrightarrow{k_{isc}} T \qquad S \xrightarrow{k_f} S_0 \qquad T \xrightarrow{k_T} S_0$$

where S is the excited state, T the triplet and S_0 the ground state.

The populations are calculated as in Algorithm 12.5 modified for the decay of the triplet.

Q12.8 We have assumed in previous calculations that the population of the final state does not affect the transition from other states, but, in fact, the rate of a transition is proportional to the *difference* in population between energy levels. Two common situations arise where the final state population can affect the transition from an upper level; one is a three or four level laser and the other is the phenomenon of band filling in semiconductors, where, because of the Pauli principle, only two electrons can fill any electronic energy level. A generic diagram is shown in Fig. 12.8. It is assumed that state T is full when it receives half the total population.

Calculate the decay of state A if $k_1 = 1/(30\text{ ns})$ and $k_2 = 1/(100\text{ ns})$. Assume that the probability of going into state T is *reduced in proportion to the fraction* of the total number already in that state.

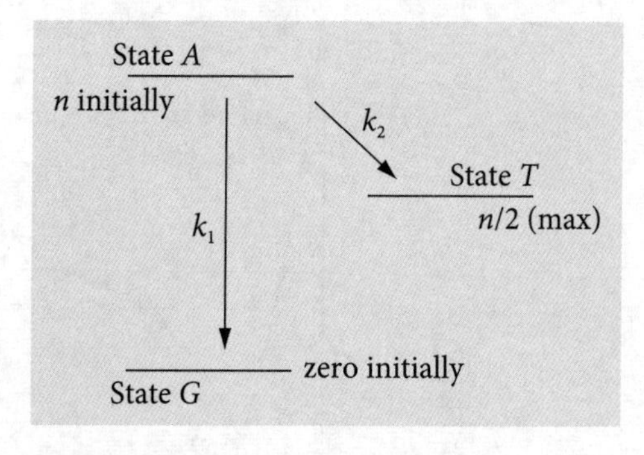

Fig. 12.8 A scheme where population difference becomes important.

Strategy: The excited state lifetime is $1/(k_1 + k_2)$, but k_2 is a function of the population in level T so the lifetime changes as time progresses. This can be accommodated by making a function $k_2(n) = (1 - 2n_T/n)k_2$ where n_T is the number in level T at any state of the calculation. At the start, $n_T = 0$ and therefore $k_2(n) = k_2$, and at the end $n_T = n/2$, and $k_2(n) = 0$. As time progresses, proportionately more and more population enters state G because the value of $k_2(n)$ becomes smaller. Start by modifying the code given in Algorithm 12.5 (in the solution to Q12.7).

Q12.9 The S-I-R scheme describes the way an infection is passed among individuals and is described in detail in Chapter 11.7. The scheme is

$$S + I \xrightarrow{k_{SI}} 2I$$
$$I \xrightarrow{k_{IR}} R$$

and this will be solved using the Gillespie Monte Carlo method. S represents the number of susceptible persons, I the number infected, and R those removed, i.e. those who have had the infection, and are otherwise immune. The rate equations are

$$dS/dt = -k_2SI \qquad dI/dt = +k_2SI - k_1I \qquad dR/dt = +k_1I$$

The data is taken from Chapter 11.7.5 and relates specifically to an infection in a boys' school; $k_{SI} = k_1 = 2.18 \times 10^{-3}$, $k_{IR} = k_2 = 0.452$ hours. The initial S population S_0, is 762 and one person is initially infected, $I_0 = 1$. The equations can of course be solved without knowing the underlying science; the equations could just as easily be (autocatalytic) chemical species where R, once formed, does not react. It will be necessary to repeat the calculation several, i.e. 40 times, to obtain good averaged data.

Strategy: The method to use is similar to that outlined in the examples. The question gives all the information needed except the number of bins to calculate the data over; 100 should be sufficient which is one data point every 5 hours. To improve the calculation, several runs may have to be calculated and averaged.

Q12.10 The Lotka–Volterra, predator–prey scheme is

$$\begin{aligned} Y &\xrightarrow{k_1} Y + Y \\ Y + X &\xrightarrow{k_2} X + X \\ X &\xrightarrow{k_3} Q \end{aligned} \tag{12.12}$$

where Y is the prey and X the predator. The rate equations are

$$\frac{dY}{dt} = k_1Y - k_2YX$$

$$\frac{dX}{dt} = k_2YX - k_3X$$

These equations are described in more detail in Chapter 11.8.1 where they are solved numerically. Solve these equations by using the Gillespie Monte Carlo method to calculate the time profile of species Y and X over 100 time units with initial values, $Y_0 = 200$, $X_0 = 80$, $k_1 = 1.0$, $k_2 = 0.001$, $k_3 = 0.5$. Use 200 000 steps in the reaction and 400 time bins. Comment on the results.

Strategy: Use the model as in the text, making sure that a_0 is calculated correctly. Do not forget that there are three reactions, so three choices have to be made as to which reaction is occurring at any time. If you also plot the phase plane this is, not surprisingly, very random compared to a direct numerical solution.

12.3 Monte Carlo simulations and calculations

12.3.1 Concept and motivation

In numerically integrating differential and other equations, a number of methods are available. Some Monte Carlo methods are described in earlier sections and other methods are described in Chapter 11. In this section *simulations* are described. A simulation uses the underlying physical laws and principles to control the calculation; no differential equations will be used. Usually processes, such as the decay of an excited state, the dispersal of a population of animals, or diffusion of molecules, can be simulated by taking a stochastic approach to the problem, repeating many calculations and slowly building up the results. This approach may seem rather odd; why not just solve the equations and be done with it? Well, sometimes this approach is simply pragmatic; a Monte Carlo approach is sometimes simpler and quicker to use. More often, it is the case that the equations describing some complex phenomenon cannot be solved algebraically or are simply not known because the process being studied is too complex. In these cases, we turn to the underlying physical laws and principles to control a simulation. Other examples of problems that can be studied are reaction kinetics, nuclear spins interacting with one another in a solid, ions moving through membrane pores, animals or people infecting one another, the spreading of forest fires, and oil moving through pores in rocks and so forth. Any computer simulation should be repeated many times to obtain an average value and then repeated again with many different initial values to explore the problem properly and to learn what are the important parameters that control behaviour.

Two classes of Monte Carlo calculations must be clearly distinguished. The class described in the last section uses a Monte Carlo method to solve rate equations or, in general, differential equations. The method now described is to simulate a physical process from first principles without explicitly solving equations. The algorithms to do these two types of calculations are, in many cases, rather similar, but the way they operate is fundamentally different.

12.3.2 Simulating physical processes

As this is a very general approach, and in principle applies to almost any process, there are no fixed rules other than to understand the problem at a fundamental level and then decide how it can be broken down into individual steps. Random numbers are always used to decide what is going to happen next, and the result of many trial calculations summed to remove any bias set up by the random nature of the process.

As an example, consider simulating one-dimensional molecular diffusion. This could describe the diffusion of a molecule inside a long nanotube, along a channel in a zeolite or along a path into an enzyme's active site. At any particular point, the molecule could be buffeted by a solvent molecule and knocked forwards or backwards. To idealize the problem we assume that each displacement or step is of the same length. More colourfully, it has become the convention in textbooks to describe a random walk as the meanderings of the proverbial drunkard.

In the simulation of a random walk, only discrete steps can be taken $1, 2, 3\cdots$ and so on. A random walk of this discrete type is described as a Markov chain of events in which the next event to occur depends only on the current situation or state and has no history of past events because each event has to be independent of all others.

The walk is random and also unbiased, which means that the chance of moving to the right or to the left at any step is equal to $1/2$. Suppose that a walk can have only two steps, then there are four ways of moving and only three places the walk can end up, which is either where it started, or two steps to the left, or two to the right. The chance of two left moving steps, or a left and a right, is $1/4$. Similarly, the chance of a right-hand step and then a left-hand one, or of two right steps, is $1/4$. The chance, therefore, of returning to the starting place is $1/2$, and it is $1/4$ of moving, either two places to the right, or two to the left, of the start. After three steps, the walker can have moved three steps to the right or to the left, or two right and one left, and vice versa, or two left and one right, and vice versa. After an odd number of steps, the walker cannot be at the start of the walk. Clearly, the number of combinations of left and right steps soon becomes complicated but it is, nevertheless, calculable. The probability of ending up at different positions is shown in the table. The sums of each row add to one as these entries represent probability. These probabilities can be calculated by the binomial distribution, and, when fractions are cleared, Pascal's triangle results.

In an unbiased random event if p is the probability of success, the probability of k successes after n attempts is the binomial distribution

$$P(k, n, p) = \frac{n!}{k!(n-k)!}p^k(1-p)^{n-k}.$$

The chance of throwing a die three times so that each time the same number appears is $P(3, 3, 1/6) = 1/6^3$. In terms of the random walk, 'success' means that after a total of n steps, the walker is at position k away from the starting point. To calculate this probability, suppose that in a random walk of length $2n$, there are x steps taken to the right, then $2n - x$ must be taken to the left. Therefore, the walker ends at position $x - (2n - x) = 2(x - n)$. If x steps are taken to the right out of a total of $2n$ then the probability of this is $P(x, 2n, p) = \dfrac{2n!}{x!(2n-x)!}p^x(1-p)^{2n-x}$; however, this can only be true if the walk ends at position $2(x - n) = 2k$. Therefore $x = n + k$, giving

$$P(n+k, 2n, p) = \frac{(2n)!}{(n+k)!(n-k)!}p^{n+k}(1-p)^{n-k}.$$

as the probability of being at k after $2n$ steps. When $p = 1/2$ and four steps are taken ($n = 2$, $k = 2$) this equation gives a probability of 1/16 as shown by the bottom right entry in the table.

steps				–	position x +				
	−4	−3	−2	−1	0	+1	+2	+3	+4
0					1				
1				1/2		1/2			
2			1/4		1/2		1/4		
3		1/8		3/8		3/8		1/8	
4	1/16		4/16		6/16		4/16		1/16

12.3.3 Simulation of one-dimensional diffusion

In diffusion, time, and distance are linked by Fick's second law; Chapter 10.6.2 (iv). However, in a simulation, this equation is not used and the position where the walk ends after a fixed number of total jumps is calculated. This is repeated many times and a histogram made of the final positions. To make the walk last a longer time, a larger total number of jumps is taken because each jump is supposed to take an equal time. In the Monte Carlo simulation, the total number of steps, s, must be decided beforehand. The walk must start at only one place, say x_0, and as there is an equal chance of moving to the right or left, the starting place of the walk will also be the mean value $\langle x_0 \rangle$. The root mean square distance the walker moves in taking s steps, is $\sqrt{s}$,

Using a random number generator in the range 0 to 1 to determine the direction of each step, the position where the walker ends after completing the total number of steps is recorded. This value is stored in an array. It is important to remember to calculate the result only where the walker stops after the *total number* of steps is completed, and to record this value. To prevent the walker leaving either end of the array used to hold the final positions, we could make the array many times $\sqrt{s}$ so the walk cannot reach the end within the number of allotted steps. Alternatively, and more reliably, we can check when the ends are reached, but this will make the calculation a little slower. Finally, the calculation is repeated to improve the statistics.

Start the walk at the centre of the array `prob`, by making its length an odd number n, then the centre is at $(n - 1)/2$. If the walk consists of an even number of steps, then the walker must end up an even number of positions away from the starting point, 0, ±2, ±4,

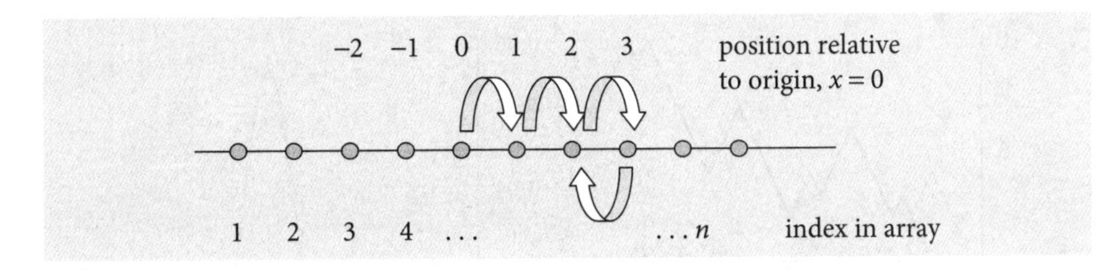

Fig. 12.9 Labelling a random walk.

and so forth. Stored in an array, these steps leave every odd-numbered array index as zero. To avoid having a graph with every other point as zero, the *change* in the number of positions away from the starting point can be divided by two, because it is an even number, and then added to the index of the starting point $(n - 1)/2$. The variable `m` in the code below is the number of jumps *away* from the start. The algorithm below only works, as written, for an even number of steps in a walk, or as $2n$ steps for odd-numbered n. To calculate the distance moved, a second array must contain only even numbers with the value zero at the starting point index $(n-1)/2$; this is array `xdata`. Each jump is taken to be one time step; therefore, the number of jumps is equivalent to time.

In the calculation, the parameters and arrays are defined first. A random number between 0 and 1 is produced and a check is made to see if it is <0.5. If it is, this is taken to be a right-hand jump and 1 is added to the counter `m`, if not, 1 is subtracted from `m`. The loop `for L from 1 to reps do` and the associated `end do` later on, make the calculation repeat `reps` times. The array `xdata` contains the position away from the start of the walk.

Algorithm 12.6 One-dimensional random walk

```
> Seed:= randomize():
  n:= 501:                          # walk array size
  steps:= 50:                       # number of time steps
  reps:= 5000:                      # number of repeats
  prob:=  Array(1..n):              # probability
  xdata:= Array(1..n):              # x position for graph
  c:=( n-1 )/2;                     # start at centre
  for L from 1 to reps do           # repeat whole calc'n
    m:= 0:                          # start; no displacement
    for j from 1 to steps do
      r:= rand()/1e12;              # random number 0 to 1
      if r < 0.50 then              # half left, half right
        m:= m + 1:                  # right move add 1
      else
        m:= m - 1:                  # left move subtract 1
      end if;
    end do :                        # end 'for j ..steps'
    prob[c + m/2]:= prob[c + m/2] + 1:
  end do:                           # end of 'for L ..reps'
  for i from 1 to n do              # x values
    xdata[i]:= (-c + i )*2;         # only even x numbers
  end do:
```

The results are shown Fig. 12.10. On the left, position vs time, in terms of steps taken, is shown for three different walks. On the right, the probability histogram is shown. Before plotting the array, `prob` was normalized by dividing by the number of repetitions and then by two because the distance stepped is always even. The Gaussian or normal distribution $p(x) = \dfrac{1}{\sigma\sqrt{2\pi}}e^{-(x-\langle x\rangle)^2/2\sigma^2}$ is the probability of the walker being found at position x where $\langle x\rangle$ is the mean position, which is zero in this calculation, and σ the standard deviation of the walker's position. This probability is 'continuous', x can take any value, unlike the probability calculated by the discrete random walk; however, the agreement between the two calculations is very good. The graph below shows that if there are s steps in the random walk, then $\sqrt{s} = \sigma$.

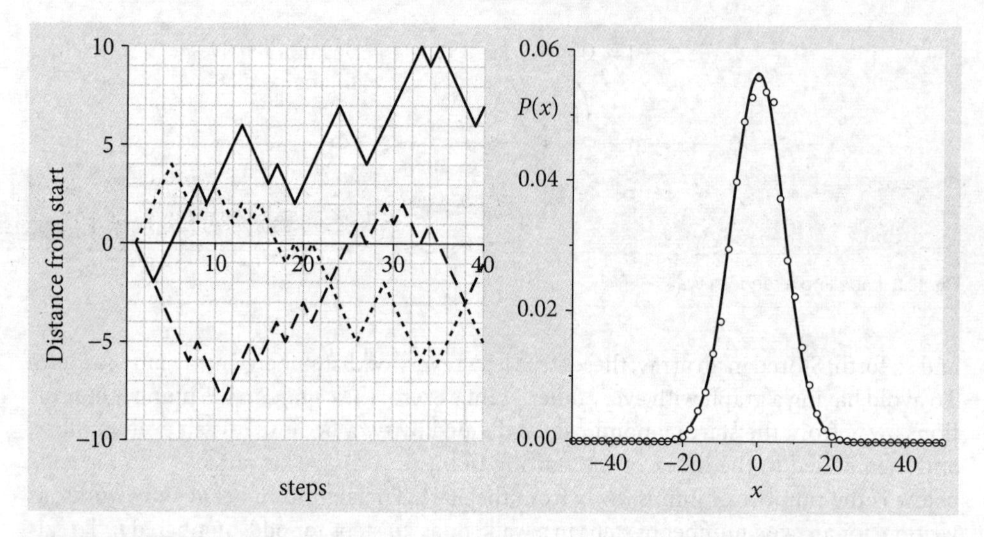

Fig. 12.10 Left: three results of $s = 50$ steps each taken by a random walker starting at point 100. Right: Five thousand simulations of a discrete distribution of 50 steps each and the Gaussian distribution $p(x)$ with $\sigma = \sqrt{50}$ (solid line).

As an exercise, calculate the distribution for other walk lengths. You should find that as the walk gets longer, the peak is less and the wings broader, as common sense would dictate for diffusion.

12.3.4 Reacting molecules

Many molecules react by a first-order process; $A \xrightarrow{k} B$, e.g. $CH_3CN \rightarrow CH_3NC$ or a cis-trans isomerization. A first-order decay was calculated by the Gillespie method in Section 12.2.2, but the method used there is very different. Imagine instantly starting a reaction and then being able to observe each molecule individually and record the instant t that each reacts. Each molecule follows a first-order process, the probability of *not* having reacted up to time t is $p = e^{-kt}$, where k is the rate constant and the lifetime τ for the reaction is $\tau = 1/k$. Observing these times for different molecules allows the probability distribution to be made and the reaction rate constant or lifetime calculated. The lifetimes of excited states and of first-order reactions have a huge range from $\approx 10^{-13}$ to >10 s. The lifetimes of radioactive nuclei range from a few seconds to thousands of years.

When we simulate the population of a species, the distribution of reaction times has to be known. This can be achieved by inverting the exponential equation to produce the time and guessing the probability p, from a uniform distribution between 0 and 1. The list of times produced by repeatedly doing this are then made into a histogram and displayed, and, if enough events are recorded, a decay similar to one that could be observed experimentally will be produced. The time t that a molecule reacts, not having reacted up to that time, is

$$t = -\frac{1}{k}\ln(p) \tag{12.13}$$

and p is guessed from a uniform distribution in the range 0 to 1.

One way of doing the calculation is shown below. A histogram of time intervals is made by making the time into an index of an array and then adding up the number of times this index occurs. The array is called `fcount`; the maximum time is `maxt` and the number of bins in the histogram, `bins`. The method is:

(1) Define parameters; set arrays to initial values;

(2) Repeat the following calculation in a `for..do` loop
Calculate $t = -\ln(p)/k$, where p is a random number 0 to 1.
Convert t to an integer to use as array index with

```
indx:= trunc(t*bins/maxt):
```

Add up the number of events with this index as

```
fcount[indx+1]:= fcount[indx+1] + 1;
```

End of loop.

The Maple code is shown below with 5000 events placed in a histogram of 100 bins. The reaction lifetime τ is 10. In the simulation, the time scale need not be explicitly defined; each time unit could be femtoseconds or years; it all depends on the reaction.

Algorithm 12.7 Simulation of first-order process

```
> restart:
  Seed:= randomize();                         # start random no's
  events:= 5000:                              # total repeats
  bins:= 100:                                 # number of bins
  maxt:= 100.0;                               # maximum time
  tau:= 10.0;                                 # lifetime
  fcount:=Array(1..bins):                     # define arrays
  dtime:= Array(1..bins):
  for i from 1 to bins do                     # set initial values
    fcount[i]:= 0;
    dtime[i]:= evalf(maxt*(i-1)/bins); # time / bin
  end do:
  for i from 1 to events do                   # start calc'n
    p:= rand()/1e12;                          # random number 0..1
    t:= -tau*ln( p );                         # calc time
    indx:= trunc( bins*t/maxt );              # make into integer
    if indx < bins-1 then                     # check index ok
      fcount[indx+1]:= fcount[indx+1] + 1;  # make histogram
    end if;
  end do:                                     # end for i loop
```

Notice how the index is multiplied by `t/maxt` to make it dimensionless and to scale it properly into the array. If the index is too large, a check is made to prevent an error. The array `dtime` has the value `maxt*(i-1)/bins` which starts at zero when the array index is 1; in Maple, arrays cannot have an index of zero. The instruction `trunc`, makes the smallest integer out of a real number, e.g. 3.6 → 3. Some results are shown in Fig. 12.11. On the left, the time a molecule remains excited is plotted vs the event number. Although there are some long times the average time is about 10, which is the lifetime of the ensemble of molecules. The histogram of these times in the left figure produces the right-hand one when plotted on a log scale. This data looks just like that determined experimentally by time-correlated single photon counting experiments used to measure the excited state lifetime of a molecule or atom.

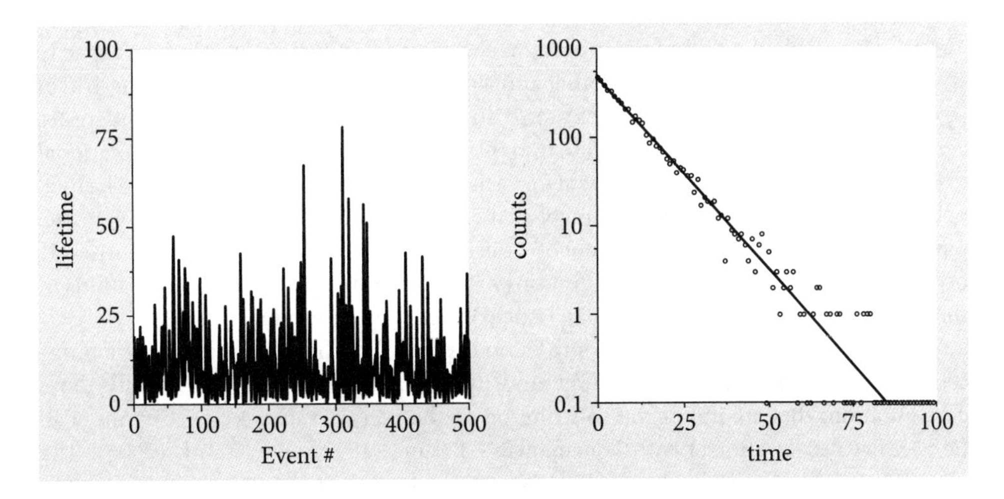

Fig. 12.11 Left: The lifetimes vs event number for 500 events. Right: A histogram of the natural logarithm of the number of counts in each time channel in the simulation of a first-order rate process together with an exponential decay with the same decay time; $\tau = 10$ in this example, and 5000 events have been recorded.

12.4 The Metropolis algorithm

In statistical mechanics, if the partition function or 'sum over states' Z is calculated, other thermodynamic properties such as internal energy or heat capacity, can be evaluated. For example, the internal energy U at constant volume is $U = k_BT^2 \left.\dfrac{\partial \ln Z}{\partial T}\right|_V$. As a result, a great deal of attention is paid in textbooks to evaluating Z for different problems, such as the harmonic oscillator or rigid rotor. The partition function is defined as

$$Z = \sum_n g_n e^{-E_n/k_BT}$$

where the energy of level or state n is E_n, k_B is Boltzmann's constant, and T the temperature. The degeneracy of level n is g_n. The harmonic oscillator's energies are $E_n = h\upsilon(n + 1/2)$, the quantum numbers $n = 1, 2, 3, \cdots$, and levels are singly degenerate; $g_n = 1$. The summation forming the partition function can be evaluated algebraically, but even if no algebraic solution was forthcoming, such as is the case for the anharmonic oscillator, Z could be evaluated numerically because with increasing energy the exponential terms very rapidly become insignificant and contribute negligibly to the total sum. In practice, the summation has a limited, rather than infinite, number of terms. The small value of terms when the energy is large can be appreciated by comparing the $n = 0$ term with that when $n = 100$ where clearly $e^{-h\upsilon/2k_BT} \gg e^{-201h\upsilon/2k_BT}$; the ratio of the two terms is $\approx 10^{88}$.

Now consider a more involved case of calculating the partition function for rotation around bonds in a protein. The purpose of this would be to estimate the lowest energy configuration and so the folded structure of the protein from its amino acid sequence.[1] Suppose that there are 100 amino acids in the protein and the interaction energy between adjacent amino acids, caused by bond rotations, can take only three values, then there are $3^{100} \approx 5 \times 10^{47}$ configurations to calculate! Unlike the harmonic oscillator, there are now only three energies to consider, and these are similar in value. However, there are so many combinations of these that the partition function *cannot* be estimated by summation because the summation does not tend to a limit as more terms are added. This is quite unlike the harmonic oscillator, where terms become progressively and rapidly smaller as more are added. The number of configurations, 5×10^{47}, may not seem so large; large numbers can easily be imagined, Avogadro's number for instance, or even larger a googol 10^{100} or a googolplex $10^{10^{100}}$, for example. To put our number into context, there have only been $\approx 4 \times 10^{17}$ s since the universe was created, and even if it were possible with a supercomputer to calculate 10^{15} configurations each second, this single calculation would still take longer than the age of the universe to complete. Quite a challenge!

A second example is the Ising spin model problem studied by physicists. In this, a two- or three-dimensional grid of paramagnetic transition metal atoms is imagined either with or without an external magnetic field. The unpaired electrons have a spin that is either 'up' or 'down'. The problem is to find the minimum energy, magnetization, or spin heat capacity when the spins interact with one another and with an external magnetic field. If the grid of atoms is small, such as 10×10, as each atom can have only one of two spin states, there are 2^{100} or $\approx 10^{30}$ configurations to search to find the minimum energy; again a computational disaster. Two quantum states are present for each spin, which are similar in energy. The total energy for any configuration is the sum of all these individual but similar terms. Again, the partition function summation does not obviously tend to a limit as it does in the harmonic oscillator case, and hence, all configurations have to be evaluated. Two configurations of the $2^{256} \approx 10^{77}$ possible ones in a 16×16 lattice of spins, are shown in Fig. 2.12.

The Metropolis algorithm is a Monte Carlo method that allows various average properties to be calculated, such as energy and magnetization, by sampling mainly those configurations that are important contributors to the average value; see Metropolis et al. (1953), Frenkel & Smit (1996), Newman & Barkema (1999) and Krauth (2006). The algorithm is general and can be used to perform ray-tracing in computer graphics and to

[1] The problem is more complicated than this because folding is cooperative, but we shall suppose that we do not know this. The allowed and forbidden angles, which are those with large steric repulsion, produce the Ramachandran plot, which shows how much α-helix, β-sheet, and random coil a protein contains. See Chapter 6.20 for a description of how to calculate this from the X-ray structure.

solve the 'travelling salesman' problem, which is to find the shortest route when visiting several cities. The biasing in this case is based on how far the next city is from the one you are presently at, and so gives a very low weighting to routes that zigzag the country compared to those that visit neighbours.

Suppose, that instead of calculating the partition function, the energy or some other quantity could be calculated directly. If this quantity is Q, the equation to evaluate the average is the same as has been met before (Chapter 4.8.2) and is

$$\langle Q\rangle = \frac{\int QP(x)dx}{\int P(x)dx}$$

where $P(x)$ is a probability distribution. In the Metropolis algorithm, the Boltzmann distribution is used: thus sampling is energy biased. If energy levels are discrete, the integral can be replaced with a summation. Alternatively, if only finite changes to a parameter are possible, a spin changing from 1/2 to 1/2 for example, then the integral can also be replaced with a sum. The average, or expectation value of Q becomes

$$\langle Q\rangle = \frac{\sum_n Q_n e^{-E_n/k_BT}}{\sum_n e^{-E_n/k_BT}} = \frac{\sum_n Q_n e^{-E_n/k_BT}}{Z}$$

where Z is the partition function. The Metropolis algorithm allows an estimation of the *ratio* of the two sums to be made in an efficient way. This avoids having to calculate Z, which, as we have seen, may be impossible. The algorithm overcomes the necessity to search all the 'phase space', as configurations are generically called, and biases the random guessing of which configuration to add to the total, by using the Boltzmann distribution of energies. In doing this, the algorithm tries to add to the estimate of Q only those configurations contributing significantly to its value and, in doing so, performs a random walk among the configurations; see Fig. 12.12 for a picture of two configurations. This random walk is also called a Markov process. This is defined as a process whereby a 'system' goes from one state to another in a random fashion, but has no memory of its previous condition.

To sample points in configuration space, according to the Boltzmann distribution, it is sufficient, but not necessary, to impose 'detailed balance' between any two configurations. This means that if w_{12} is the rate to go from configuration (or state) 1 with energy E_1, to a new state 2 with energy E_2, then the reverse transition w_{21} is related by

$$w_{12}e^{-E_1/k_BT} = w_{21}e^{-E_2/k_BT}$$

or more familiarly

$$w_{12} = w_{21}e^{-\Delta E/k_BT}, \qquad \Delta E = E_2 - E_1.$$

The ratio of rates is the same as the ratio of probabilities of going between two configurations. The detailed balance condition ensures that any one configuration can be reached from another in a finite number of steps, and this condition is called *ergodicity*. This condition means that no configurations are systematically missed. Furthermore, if a sufficient number of samples are used in the calculation, all configurations will be sampled in proportion to their importance in contributing to $\langle Q\rangle$. The algorithm can also be viewed as a random walk among the many configurations, preferentially adding in those that contribute most to the average.

All the basics of the algorithm are now assembled. The Metropolis idea is to compare two energies E_1 and E_2 from two guessed configurations 1 and 2. If the new one, 2, has the lower energy, then this configuration is accepted. If not, the chance of the new one, 2, contributing to $\langle Q\rangle$ is guessed by comparing $e^{-(E_2-E_1)/k_BT}$ to a uniform random number from 0 to 1 and accepting configuration 2 if the random number is smaller; otherwise, the old configuration 1 is accepted, and so on.

The Metropolis algorithm to calculate an average quantity $\langle Q\rangle$ is:

(1) Initialize parameters, e.g. $Q_{tot} = 0$
Calculate E_1 for any first configuration or state you choose.
Calculate the initial value Q_1.
Calculate E_2 for another state chosen at random.

(2) Start a loop
   (i) If $E_2 < E_1$, keep the new state and add Q_2 to a total quantity Q_{tot}.
   (ii) If $E_2 > E_1$, then calculate $e^{-(E_2-E_1)/k_BT}$.
   (iii) If $e^{-(E_2-E_1)/k_BT} > r$, where r is a uniform random number between 0 and 1, keep the new state and add Q_2 to Q_{tot}.
   (iv) If $e^{-(E_2-E_1)/k_BT} < r$, retain the old state and add Q_1 to Q_{tot}.
(3) Continue the loop until Q_{tot} is obtained to a sufficient accuracy, or a fixed number of trials has been done.
(4) Average $\langle Q\rangle = Q_{tot}/N$ for N calculations.

Notice that even if a new state is *not* accepted the last used value Q_1 is still added to the total, because each (randomly chosen) configuration has to be added into the total.

As a particular example, the mean potential energy of a harmonic oscillator will be calculated; the quantity Q is the energy and can be calculated exactly and is $k_BT/2$. The mean square energy will also be found which can also be calculated exactly, and is $3(k_BT)^2/4$. The standard deviation of the energy is $\sqrt{\langle E^2\rangle - \langle E\rangle^2} = k_BT/\sqrt{2}$.

12.4.1 Metropolis algorithm. Average energy and displacement of a harmonic oscillator

In this example, the potential energy has the form $E = x^2$ and $k_BT = 0.1$. Random guesses of the values of x, (the configuration or state) are made, and the average energy calculated. The range of x must be large enough to allow e^{-E/k_BT} to be very small, $\approx 10^{-7}$ or less, to allow proper sampling of the distribution. In addition, the range of x must be positive as well as negative. The random number has 1/2 subtracted from it to ensure this. If this is not done, then microscopic reversibility is not achieved because not all the possible configurations, in this case bond extension, can be sampled. The calculation continues for 20 000 samples.

Algorithm 12.8 Metropolis algorithm: <*E*> for the harmonic oscillator

```
> Seed:= randomize():                        # set up random numbers
  N:= 20000:                                 # number of samples
  deltax:= 2.0:                              # max displacement
  kBT:= 0.1:                                 # initial thermal energy
  Etot:= 0.0:                                # initial energy <E>
  E2tot:= 0.0:                               # initial <E²>
  V:= x-> x^2:                               # potential energy PE
  x1:= 0.0:                                  # first guess of x
  E1:= V(x1):                                # first guess of PE
  for i from 1 to N do                       # start loop step (2)
      x2:= x1 + (rand()/1e12-0.5)*deltax:  # new x position
      E2:= V(x2):                            # new PE
      DeltaE:= E2 - E1:                      # energy difference
# next line is Metropolis part(3) and (4)in algorithm
     if DeltaE <= 0.0 or exp(-DeltaE/kBT) > rand()/1e12 then
        x1:= x2:                             # save new configuration
        E1:= E2                              # save new energy
     end if:
     Etot:= Etot + E1:                       # always add to total
     E2tot:= E2tot + E1^2:                   # add to total <E²>
   end do:                                   # end loop
   Eav:= Etot/N;                             # average step (5)
   E2av:= E2tot/N;
   stdev:= sqrt(E2av-Eav^2);
```

$Eav := 0.0497965036 \qquad E2av := 0.0071928019 \qquad stdev := 0.0686520951$

The average energy, and its square and standard deviation, are close to the algebraic values of 0.050, 0.0075, and 0.0707 respectively. Although we have calculated the average energy, the average displacement and its square are just as easily calculated. This is done by adding the lines

```
Xtot:= Xtot + x1:             # add to total <X>
X2tot:= X2tot + x1^2:         # add to total <X²>
```

in the loop and remembering to define and set the new variables to zero at the start of the calculation. The calculated values are $\langle x\rangle \approx -0.00085$ and $\langle x^2\rangle \approx 0.049$, which are close to the theoretical values of zero and 0.05.

12.4.2 **Ising spin model**

A second example is taken from statistical mechanics and is the Ising model of a ferromagnet. In a ferromagnet, the atomic spins are coupled to one another by the quantum mechanical exchange interaction J, and can only have two energies corresponding to being 'up' or 'down'. The coupling ensures that 'up' spins want to be next to other up spins and 'down' spins want to be next to other down spins. At zero temperature, the minimum energy configuration occurs when all the spins align in the same direction, and its energy is $-2JN$ for N spins. The entropy of the spin system is zero, because all spins are aligned in perfect order, and so is the heat capacity C_p. The magnetization, which is the sum of the individual spin magnetic moments, is maximal.

As the temperature rises, the energy absorbed induces thermal motion in the atoms. This disrupts the alignment of the spins and they start to flip over. The heat capacity, which is defined as the rate of change of the internal energy with temperature, increases as more spins gain energy and are flipped. However, destroying the spins' alignment causes the magnetization to fall. At high temperatures, the spins are randomly aligned and because of this the total magnetization approaches zero. The heat capacity also falls, reaching zero at infinite temperature. This happens because the spins only have two energies corresponding to being up or down relative to a neighbour, and cannot absorb any more energy than when they are randomly aligned, which corresponds to infinite temperature. Between these two extremes, a surprising phase transition occurs where the heat capacity becomes infinite as the magnetization drops towards zero. The phase transition is observed if a two-dimensional plot is made of the spin state. Just past the transition, large areas of similar spin alignment dissolve and the spins on the lattice become considerably more disordered; Fig. 12.12.

The phase transition can be understood in thermodynamic terms, by considering the free energy change $\Delta G = \Delta H - T\Delta S$. As only spins are considered there is no pV work; we are not considering a gas, therefore $\Delta H = \Delta U$, where U is the internal energy. This is determined by the exchange interaction energy J. At the temperature of the phase transition, the phases are at equilibrium and $\Delta U = T\Delta S$. At lower temperatures, the entropic term must be relatively small because T is small, and ΔS is also small because the spins are more ordered. The free energy ΔG is therefore overwhelmingly determined by the

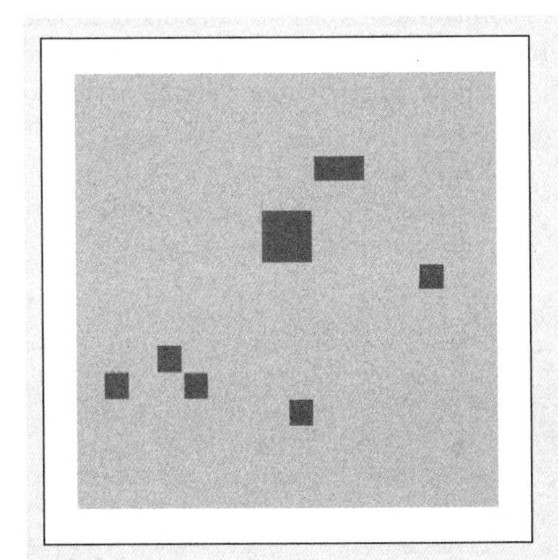

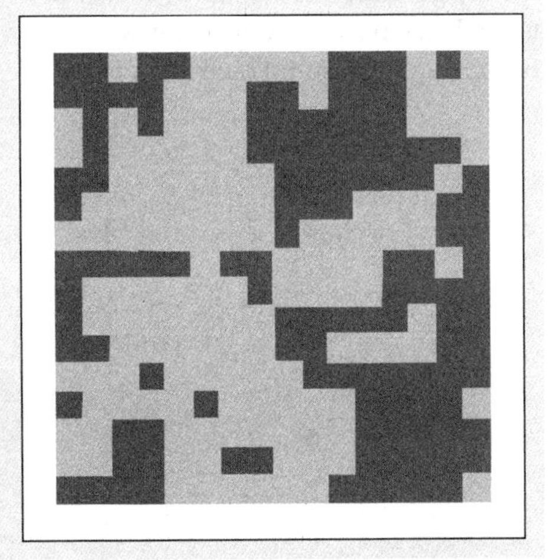

Fig. 12.12 Two examples of 2^{256} or $\approx 10^{77}$ possible configurations of spins in a 16×16 Ising model with temperature $T = 2$ (left) and 3 (right) showing how significantly the disorder increases after the phase transition at $T = 2.269$. The light squares represent spin = 1, and the dark squares, spin = −1. The exchange coupling $J = 1$, and $k_B = 1$.

internal energy. As the temperature increases, the increasing entropic term will make ΔG more negative. Additionally, as the entropy also increases with temperature, the change from internal energy to entropy determining the free energy is sudden in the sense that it occurs over a small temperature range. The heat capacity becomes infinite at the phase transition because the heat supplied does not go into raising the temperature, but in reorganizing the spin states, and that causes an increase in entropy.

The Ising model of a ferromagnet is often studied as a two- or three-dimensional lattice of spins. We shall calculate the properties of two-dimensional lattices for which an exact numerical solution is available and which will allow comparison of our Metropolis Monte Carlo simulation with theory.[2]

Before rushing to the computer, some preliminary calculations should be done which will make the whole calculation run more quickly. Why is this important? To understand this, suppose that the lattice has 5×5 spins, then the number of separate spin configurations is a mere 2^{25} or 33 554 432. If the lattice is 100×100, which is still unimaginably small compared to a real ferromagnet, this number is $\approx 1.27 \times 10^{30}$. It is not unreasonable then, that perhaps an insignificant number of total configurations, perhaps 10^5 to 10^6, may have to be sampled to obtain good statistics, and then the calculation repeated at different temperatures. While 10^6 may be an insignificant number compared to the total number of configurations, this does represent a significant calculation and the need for an efficient code is clear. This is always achieved by an algebraic examination of the problem before any computation begins.

The energy of the Ising spin system is given by

$$E = -J \sum_{\langle \alpha,\beta \rangle}^{n} s_\alpha s_\beta - B \sum_{\alpha}^{n} s_\alpha$$

where J is the exchange coupling energy, and B the external magnetic field, which we will now take to be zero to simplify the calculation. The indices on the summations need some explanation. The exchange interaction operates between two spins and is of short range, decreasing exponentially with the distance apart of the two spins. The decrease is so sharp with separation that only near neighbours have any interaction worth considering. On a square lattice, this means that only spins north, south, east, and west of any one spin interact, as shown in the diagram Fig. 12.13, where the spins are represented as ♦ and ◊.

The double summation term $\sum_{\langle \alpha,\beta \rangle} s_\alpha s_\beta = \sum_{\alpha=1}^{n} s_\alpha \sum_{\beta=1}^{4} s_\beta$, indicates that each spin in turn (index $\alpha = 1 \cdots n$) adds a contribution from each of its four neighbours (north, south, east, and west) to the sum, $\sum_{\beta=1}^{4} s_\beta = s_N + s_S + s_E + s_W$.

The index α in the sum runs over all the spins. If the spins are represented as a two-dimensional array, then α represents the pair of indices covering each row and column. In the Ising model, the change of energy between configurations takes a particularly simple form. It can be obtained directly without having to add up the total energy for each of the two types of states and then subtracting them. The difference in energy, Fig. 12.13, left to

Fig. 12.13 Two patterns of spins in the Ising model. The dotted line shows the near neighbours to the central spin, which is flipped between the two configurations.

[2] The exact calculation is due to Onsager (1944) and was a tour de force in statistical physics.

right is due to flipping one spin. If the spin has a value, either ◇ = +1 or ◆ = −1, then the energy on the left is $E_1 = -J(1+1+1-1) = -2J$ as three of the spin pairs have the same value. The other configuration shown with the central spin flipped is $E_2 = -J(-1-1-1+1) = 2J$; the difference is therefore $E_2 - E_1 = 4J$. Considering all other possible arrangements of the four neighbouring spins, produces only five energy differences. However, because the spin that is flipped is also a neighbour to other spins, the total value of the summations for E always produces an extra factor of 2 compared to the energy differences of only the near neighbours. The values are $\Delta E = -8J, -4J, 0, +4J, +8J$. The energy difference occurring at a spin k can be written in a general way as $\Delta E_k = 2Js_k^0\sum_{\beta=1}^{4} s_\beta$ where β indexes just the four nearest neighbours and $s_k^0 = \pm 1$ is the value of spin k before flipping.

The Metropolis algorithm compares the energy differences between configurations: if ΔE is negative or zero, the difference is accepted directly; if not, $e^{-\Delta E/k_BT}$ is calculated and compared to a random number. This means that only the two positive values of ΔE are ever used to calculate the exponential, and this is very fortunate because calculating exponentials is a relatively slow process on a computer. These values can be calculated once at the start of the calculation and then used repeatedly. The quantities needed at the end of the calculation are usually the energy E, heat capacity C_V, magnetization M, and magnetic susceptibility χ. The total energy can be calculated 'on the fly' by repeatedly adding ΔE to the initial energy which is $-2JN$, if the spins are initially all set at +1. The magnetization is just the sum of the spins $M = \sum_\alpha s_\alpha$, and if this is required at each step, a considerable amount of computer time will be spent just doing this. However, if 10^6 total steps are needed, then only one in every few thousand M's need be calculated to represent how this varies as the calculation proceeds. Alternatively, the last 1000 or so values could be averaged.

The heat capacity is defined as

$$C_V = \frac{1}{Nk_BT^2}(\langle E^2\rangle - \langle E\rangle^2)$$

and magnetic susceptibility as

$$\chi = \frac{1}{Nk_BT}(\langle M^2\rangle - \langle M\rangle^2)$$

for N steps in the calculation. To simplify the calculation, we can make $J = 1$ and $k_B = 1$, and use the temperature to change J/K_BT. This simple change does not affect any calculation, because E can always be scaled with a different value of J at the end of the calculation, but it has the important practical effect of making the calculation wholly that of adding or subtracting integers, which is fast. Finally, any calculation can only use a finite lattice of points and this raises the important question of what to do at the edges. One always uses *periodic boundary conditions* and this jargon means that the lattice is wrapped around on itself, like a snake biting its tail, or like folding a piece of paper onto a cylinder. This is achieved by checking the indices and changing them when they run off an edge and then bringing them back in the same place but on the opposite side.

The periodic boundary is handled with `indx1` to `indx4` in the code below. The energy is added as `Enrg:= Enrg + DeltaE:` and then this result added to the total energy. This ensures that even if a configuration is of higher energy it is counted properly; `Enrg + DeltaE:` is equivalent to $E_1 + (E_2 - E_1)$ where E_1 is the initial energy and E_2 that after a spin flip. Therefore, if a change is made, E_1 becomes the new energy E_2. If not, E_1 is still added into the total energy. This is exactly the same as was done in calculating properties of the harmonic oscillator.

Algorithm 12.9 Two-dimensional Ising Metropolis algorithm

```
>Seed:= randomize():              # set random number generator
  L:= 16:                         # size of lattice side
  n:= L^2;                        # total number of spins
  N:= n*25*10^3;                  # number of samples per spin
  J:=1.0:                         # exchange coupling
  kBT:= 2.0;                      # Boltzmann const x T,
  beta:= J/kBT:
```

```
  R:= 2*rand(0..1)-1:                   # define rand num -1 or +1
  rnd:= rand(1..L):                     # rand num in range 1 to L
  spin:=  Array(1..L,1..L):             # spin lattice
  prob:=  Array(1..8):                  # array for exp(-deltaE/kBT)
 # deltaE has values 2*( -4,-2, 0, 2, 4) only positive used
  prob[4]:= exp(-4*beta):               # define for use later on
  prob[8]:= exp(-8*beta):
  for i from 1 to L do
    for k from 1 to L do
      spin[i,k]:= 1;                    # initial distribution T = 0
    end do:
  end do:
  Enrg:= -2*n:                          # initial energy all spin +1
  tot_Enrg:= 0: tot_Enrg2:= 0:          # initial values zero
  for num from 1 to N do                # start calc
    i:= rnd(): k:= rnd():               # choose pts at random
    indx1:= i+1;
    if indx1 > L then indx1:= 1 end if; # boundary
    indx2:= i-1; if indx2 < 1 then indx2:= L end if;
    indx3:= k+1; if indx3 > L then indx3:= 1 end if;
    indx4:= k-1; if indx4 < 1 then indx4:= L end if;
    sum_nn:= spin[indx1,k] + spin[indx2,k]
             +spin[i,indx3] + spin[i,indx4]:
    DeltaE:= 2*spin[i,k]*sum_nn;
    if DeltaE <= 0 or rand()/1e12 < prob[DeltaE] then
     spin[i,k]:= -spin[i,k];                       # flip spin
     Enrg:= Enrg + DeltaE:                         # increment
    end if;
    tot_Enrg:= tot_Enrg + Enrg;                    # energy
    tot_Enrg2:= tot_Enrg2 + Enrg^2;                # energy²
  end do:
  `<Energy>/spin`:= evalf(tot_Enrg/(N*n));
  `<Energy squared>/spin`:= evalf(tot_Enrg2/(N*n));
  Cv:= beta^2*(tot_Enrg2/N - tot_Enrg^2/N^2)/n;
```

One run of the algorithm with 25 000 samples per spin, or 900 000 samples in total on a 6×6 lattice, produces the result $E = -1.746 \pm 0.28$ and $C_V = 0.686$ with $T = 2$. Figure 12.14 shows C_V and the magnetization per spin on a two-dimensional lattice of 16×16 elements

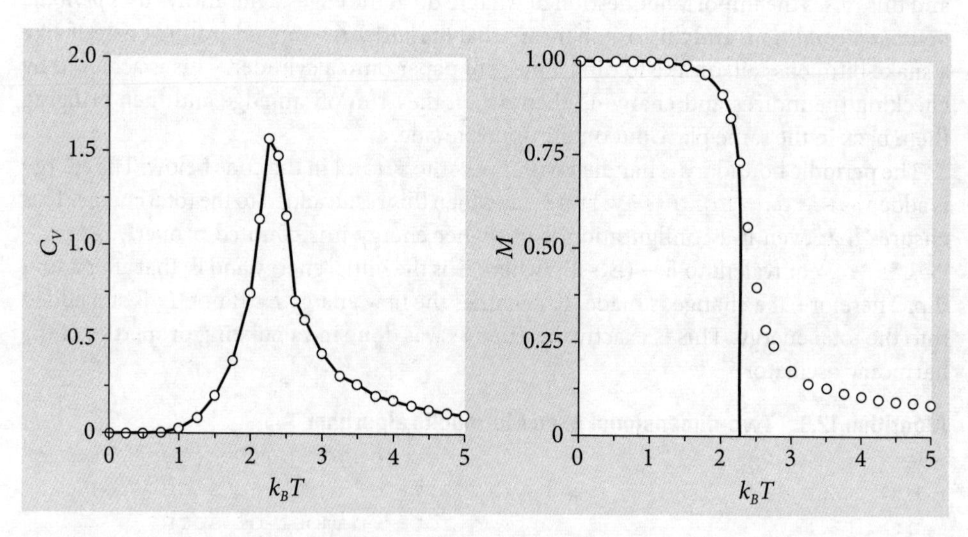

Fig. 12.14 Ising model calculations using the Metropolis method on a 16×16 lattice. Left: Heat capacity vs temperature as k_BT. Right: Theoretical magnetization (solid line) for an infinite lattice and calculated points for the 16×16 lattice.

at 25 000 samples/spin at several values of k_BT/J together with the theoretical result $M_{Th} = (1 - \sinh^{-4}(2J/k_BT))^{1/8}$. Each Metropolis calculation takes an average of ≈1.6 minutes on a 2GHz computer using Maple.

12.5 Forster or dipole–dipole energy transfer

Forster or dipole–dipole energy transfer is a vital process in photosynthesis as it is the mechanism by which energy is transported in the antenna and then to the reaction centre. The same process is used in FRET, which is widely used in the biosciences as a 'spectroscopic ruler' to measure how close one molecule is to another. In this process an excited state of a molecule D^* can, nonradiatively, transfer energy to a nearby acceptor molecule A, provided energy is conserved, which means that the donor's emission spectrum overlaps with the acceptor's absorption;

$$A + D \xrightarrow{h\nu} A + D^* \xrightarrow{k_R} A^* + D.$$

The rate constant varies as the inverse sixth power of the separation between donor and acceptor; therefore, the rate constant is distance dependent. Calculating the decay of the donor's excited state, therefore necessitates summing over acceptors at all distances, not just the donor's nearest neighbours; see Fig. 12.15. In solution the acceptors surround the donor and each other at random, rather like the currants in a cake and to simplify a calculation it is normally assumed that no diffusion of the molecules takes place on the timescale of energy transfer,

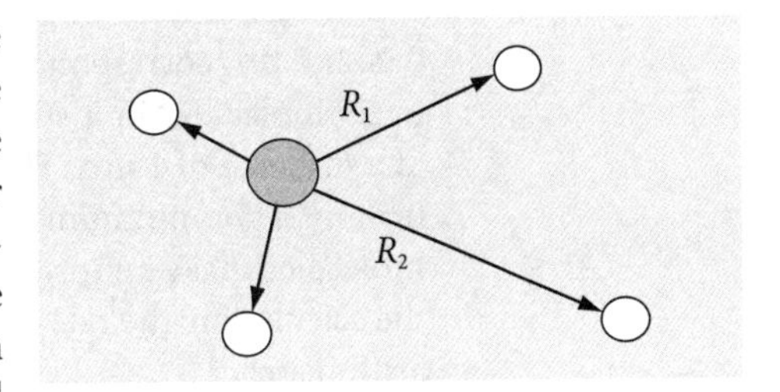

Fig. 12.15 Donor (centre) surrounded by acceptors at different distances. The transfer rate and hence excited state decay of the donor depends upon how far away the acceptors are.

If the donor has a lifetime $\tau = 1/k_f$ when no acceptors are present, the rate constant for Forster, or dipole–dipole, transfer to an acceptor at a distance R nm away from the donor is $k_R = \dfrac{\chi^2}{\tau}\left(\dfrac{R_0}{R}\right)^6$ and consequently the excited state decays with a distance dependent rate constant $k = 1/\tau + k_R$. As some acceptors will be closer than others to the donor, these will interact more rapidly because of the $1/R^6$ distance dependence. This means that the rate constant of energy transfer is not a constant, but depends upon time! This arises because the few nearby acceptors interact faster on average than those far away and, even though there are far more of them farther away, they do not fully counteract this rapid quenching. Overall this leads to faster than expected quenching at short times and therefore to a non-exponential decay. A similar effect occurs in the diffusion quenching of molecules in solution.

The constant R_0 is the critical transfer distance at which the rate constant of fluorescence equals the rate constant of transfer to the donor and is calculated from the overlap of the donor's emission spectrum, fluorescence or phosphorescence, with the acceptor's absorption spectrum; see Turro (1978). This is usually a number that is large compared to the dimensions of a molecule, a typical value of R_0 being 3 nm. The parameter χ^2 describes the orientation of the donor to acceptor transition dipoles, the average value in solution is $\chi^2 = 2/3$.

By calculating the chance of energy transfer at some distance R, the decay of an excited molecule is described by

$$\frac{I(t)}{I_0} = \exp(-k_f t - \alpha\sqrt{k_f t}) \tag{12.14}$$

which means that the energy transfer rate varies as $t^{-1/2}$ and so produces a non-exponential decay at short times. This can be seen by rewriting the intensity as $\dfrac{I(t)}{I_0} = \exp(-tk_{ET}(t))$ where

$$k_{ET}(t) = k_f\left[1 + \alpha\left(\frac{1}{k_f t}\right)^{1/2}\right].$$

This behaviour is shown in Fig. 12.16 and is a direct consequence of the spatial arrangement of several acceptors around a donor.

To calculate the decay of the excited donor using a Monte Carlo method, the donor is placed at the origin. Initially, it might seem to be a good idea to place molecules at random on a grid, record the numbers in an array, and work out the rate constant for transfer for each of them. However, only the distance from the donor is needed, not the coordinates, and each randomly chosen distribution of molecules is only needed once; therefore, the distance can be calculated 'on the fly' by guessing *x*-, *y*-, and *z*-values, but not saving them, and calculating the distance from the origin using Pythagoras' theorem. Only distances up to a certain value are used, so that the enclosed volume is a sphere of radius $6R_0$, this being large enough to give a negligible contribution to the total rate constant. The sum of all these rate constants *plus* that for fluorescence is added up to give the total chance of transferring energy or fluorescing. Because the transfer rate goes to infinity at zero separation, a minimum separation of donor an acceptor has to be determined. This is chosen to be 1 nm, which is only slightly larger than the typical centre to centre separation on contact, of two large aromatic molecules.

The constant $R_0 = 3$ nm, and the cut-off sphere $6R_0$. If the concentration is chosen to be 0.002M, this corresponds to $0.002 \times 6.022 \times 10^{23}/(0.1 \times 10^9)^3 = 1.204\ 10^{-3}$ molecules nm^{-3} or 29 molecules in a sphere of radius $6R_0 = 18$ nm. The maximum rate is set at $\chi^2 k_f R_0^6$ at a distance of 1 nm. This is arbitrary but reasonable for a molecule with a diameter of 0.5 nm. The minimum rate is $1/6^6$ or 46 656 times smaller and is so small that distant molecules make an insignificant contribution to the sum of the rate constants. To speed the calculation, the rate constant calculation is simplified to remove repeated unnecessary arithmetic;

$$k = \frac{\chi^2}{\tau}\left(\frac{R_o}{R}\right)^6$$

$$= \frac{2}{3}\frac{1}{10}\frac{R_0^6}{d^6 R_0^6}\left(\frac{1}{\sqrt{x^2+y^2+z^2}}\right)^6 = s\left(\frac{1}{x^2+y^2+z^2}\right)^3$$

where d is the distance scale chosen, the limit being $6R_0$ with $d = 6$ and where x, y, and z are chosen from a uniform distribution to be in the range −1/2 to 1/2. It may seem rather finicky simplifying k, but this is in the innermost part of the calculation. If there are 20 acceptors and 5000 repeats this means evaluating exactly the same multiplication, division, square root and power, 10^5 times rather than just once. The calculation is run for 50 ns, and the data put into 150 bins.

Algorithm 12.10 Forster energy transfer

```
> seed:= randomize():                          # set rand numbers
  kf:= 1/10.0;                                 # 10 ns lifetime
  c:= 0.002;                                   # conc mol/dm³
  R0:= 3.0:                                    # R₀ in nm
  d:= 6.0:                                     # sphere radius d*R₀
  Avog:= 6.023e23/1e24:                        # num molecs / nm3
  n:= trunc( c*Avog*4/3*Pi*(d*R0)^3); # integer number molecs
  Rmax:= (2/3)*kf*(R0/1)^6;                    # maximum rate
  reps:= 5000:                                 # repeat calc'n
  bins:= 150:                                  # size of histogram
  maxt:= 50.0:                                 # maximum time
  Acount:= Array(1..bins):                     # define arrays
  dtime:= Array(1..bins):
  for i from 1 to bins do                      # initial values
    Acount[i]:= 0;
    dtime[i]:= evalf((i-1)*maxt/bins):
  end do:
  s:= (0.6667*kf/d^6);                         # constants for rate
  for k from 1 to reps do                      # start calc loop
    a0:= kf:                                   # fluorescence rate
    for i from 1 to n do                       # choose acceptors
      x:= rand()/1e12-0.5:                     # x,y,z coordinates
```

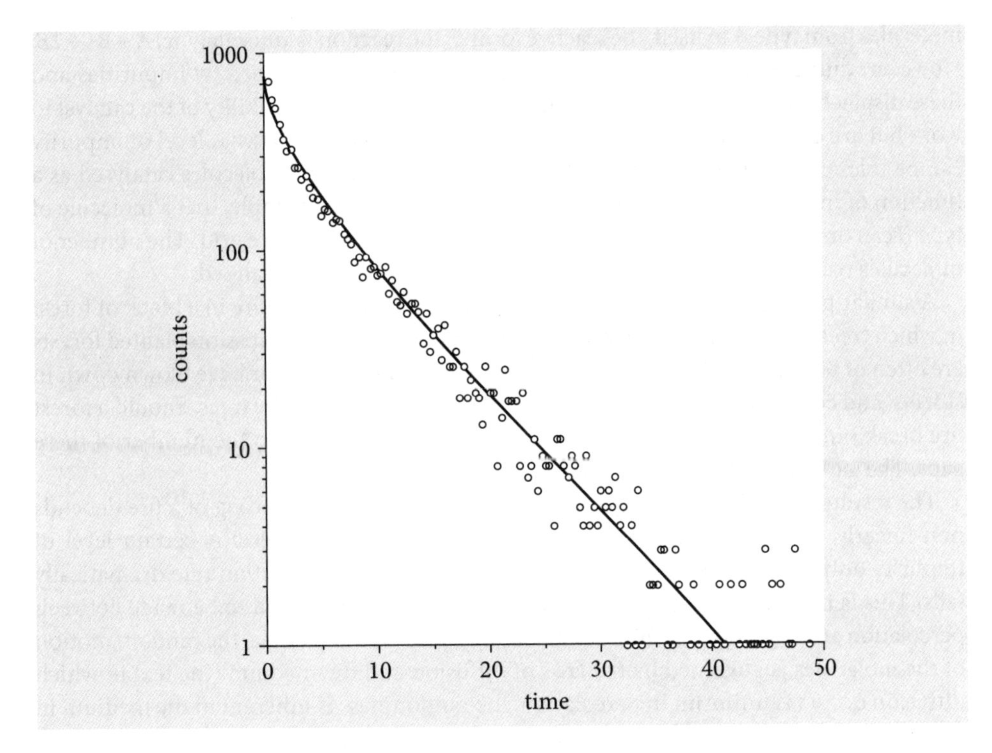

Fig. 12.16 Decay of a donor surrounded by many acceptors and a theoretical fit to equation (12.14) with $\alpha = 1.25$.

```
        y:= rand()/1e12-0.5:
        z:= rand()/1e12-0.5:
        rate:= s/( x^2 + y^2 + z^2 )^3;    # 2/3*kf*(R0/R)^6
        if rate > Rmax then
          rate:= Rmax:
        end if:
        a0:= a0 + rate:                    # sum rate
    end do:
    t:= -ln( rand()/1e12 )/a0:             # calc time eqn (12.13)
    indx:= trunc(t*bins/maxt):             # make index
    if indx < bins-1 then
        Acount[indx+1]:= Acount[indx+1] + 1;  # histogram
      end if:
    end do:                                # end calc loop
```

A log plot of the data fitted to the theoretical equation (12.14) is shown in Fig. 12.16. The fit is good but not exact. At small times, the calculated curve does not match the simulation that well. This may be due to the small number of calculation performed, 5000 only, but could also be due to the fact that the theoretical curve is only an approximation and clearly not exact at very small times because at time zero the rate constant tends to infinity!

It is surprising at first to see that the decay is not exponential. The reason for this is, as explained above, that the few nearby acceptors quench the donor preferentially compared to the more distant ones, which although more numerous, quench with ever decreasing rate constants. Fluorescence and phosphorescence decays measured by many donor–acceptor pairs confirm this type of decay (Birks 1970; Lacowicz 2004).

12.6 Autocatalytic reaction on a surface and the spreading of fires

The spatial arrangement of molecules or other objects can have a special effect on their behaviour. As a molecular example, suppose that a nanoscale device consisting of a small reactive surface of limited area is covered with a monolayer of type *A* molecules. A single reactive molecule *B* is now introduced, which can diffuse around and catalyse other molecules from type A to itself, the reactive form B; the reaction is autocatalytic, $A+B\rightarrow 2B$. However, during the manufacturing process, the surface is poisoned by impurities and these displace type A molecules. These impurities clearly limit the ability of the catalyst to work but are difficult to control, and so it is important to discover what level of impurity can be tolerated. Using a Monte Carlo method, the number of molecules catalysed as a function of impurity level can be calculated by assuming, for example, that a molecule of type B can only catalyse the four molecules adjacent to it on a square grid. The number of molecules reacted vs fraction of impurities x, can be plotted and analysed.

A similar problem, but on an entirely different scale, could be a fire in a block of forest in which trees are spaced regularly on a square grid. Forestry Commission planted forests are often of this type. Over the years, trees have either been felled or have blown down in storms, and consequently spaces exist at random places between the trees. Should a forest fire break out, the extent to which the spreading fire is sensitive to the number of these gaps. The gaps are equivalent to impurities in the chemical model.

The results show that the extent of the reaction and of the spreading of a fire depends non-linearly on the impurity concentration or gaps between trees. A certain level of impurity only can be tolerated; a slight fraction more and the reaction rate dramatically falls. This is behaviour typical of percolation. There is a distinction to be made between percolation and diffusion. In the diffusion of, say, milk in a cup of tea, the random motion of the molecules is governed by the laws of diffusion and the medium (the tea) in which diffusion occurs is uniform. In percolation, the randomness is inherent in the medium in which a fluid finds itself. Thus a solid, such as a sinter glass filter, which is full of randomly interconnected channels exhibits percolation. A fluid will pass through the larger channels but surface tension restricts flow in smaller ones. The manner in which electrons flow through amorphous films can also be described as percolation.

The computational strategy is to represent the surface monolayer as a matrix in which each matrix element is one of three integers to represent A and B molecules and impurity. It does not much matter what these integers are, so zero is chosen for the impurity, 1 for type A and 2 for type B molecules. In the case of the forest, the number 2 can represent the burning trees, 1 those that can catch fire, and 0 the areas with no trees. The initial tree that catches fire or the initial position of molecule B is chosen at random, as are the impurities. The matrix is searched sequentially and the neighbours of each molecule are tested. If a neighbour is of type A, then it is converted into type B. This is repeated until no more can be converted and the number changed is counted. The calculation is then repeated several times with atoms at different random positions. After repeating the calculation at different impurity concentrations, the number of A type molecules converted to B vs the fraction of impurities is plotted.

The algorithm is similar to that used for SIR disease propagation, but now a two-dimensional array (matrix) is used. The algorithm is split into parts: (a) defining constants and initial values, (b) filling the matrix with the different type of molecules, and (c) running the calculation with the restriction that only adjacent molecules can react and are

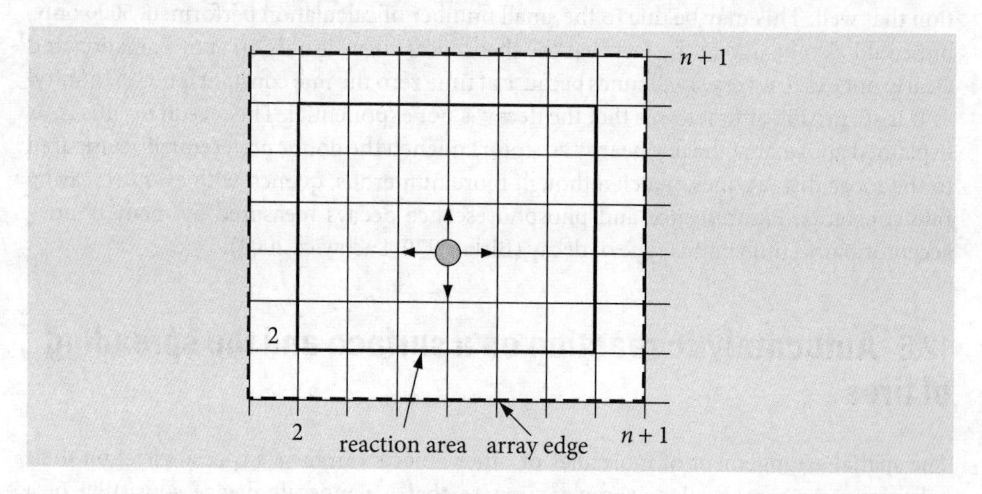

Fig. 12.17 The array dimensions are made larger than the reaction area to avoid edge effects.

those whose index is ±1 in each of the x- and y-directions, making four possibilities. The calculation is continued until the number of changes can go no further, rather than for a number of days, as in the disease propagation problem. This is done in a `while..do` loop; see the code below. A counter c is used and one added to it each time the matrix has a value of 2. The 'while' loop continues to operate until the number of type B molecules are unchanged, and then the loop is left when the condition `b > c` is met. In each iteration, the whole array is searched.

The data arrays are made 2 numbers larger in each direction than the true size, so that when the calculation reaches the edge of the array, the array bounds are not exceeded. These points around the edge of the array are ignored in the calculation; the `for.. do` loops start at 2 and extend to `n+1` for this reason, see sketch Fig. 12.17. The arrows show how a site is connected to another.

The numbers used in the following calculation are only indicative. The grid is relatively small, with 20 units on each side; increasing this will significantly slow down the calcula tion as will increasing the number of repeats, presently set at 10, or the number of fractions to be calculated between 0 and 1.

Algorithm 12.11 Two-dimensional spread of fire, disease, or autocatalysis

```
> # P[,]>:= 1 = A molecs, 2 = B molecs, 0 = impurity.
> restart:
  seed:= randomize():                # start random numbers
  n:= 20:                            # size of grid side
  reps:= 10:                         # repeat calc'n
  fract:= 0.0:                       # initial fraction impurity
  num_fract:= 20:                    # number of fractions calc'd
  delta_fract:= 1/num_fract;         # increment fraction
  data:= Array(1..num_fract):        # array for results
  fdata:= Array(1..num_fract):       # fractions
  for L from 1 to num_fract do       # fraction loop to end
  asum:= 0:
  for k from 1 to reps do            # repeat loop
    P:= Matrix(1..n+2,1..n+2):       # matrix of molecs/ trees
    for i from 2 to n+1 do           # fill matrix
      for j from 2 to n+1 do
        P[i,j]:= 1;
      end do:
    end do;
    fn:= trunc(fract*n*n):           # make into integer
 # part b make fraction of impurities
    num:= 0:
    while num < fn do                # exactly fract impurity
      rr:= rand(2..n+1): ra:= rr():     rb:=rr():
      if P[ra,rb] <> 0 then
        P[ra,rb]:= 0: num:= num + 1:
      end if:
    end do:
    num:= 0:
 # choose one type B molecule
    while num < 1 do
      rr:= rand(2..n+1): ra:= rr():  rb:=rr():
      if P[ra,rb] = 1 then
        P[ra,rb]:= 2:
        num:= 1:
      end if:
    end do:                          # end while num
 ### part c main calculation starts ###
  b:= -1: c:= 0;
  while c > b do                     # check if calc ended
```

```
      b:= c:                          # store old value in b
      c:= 0:                          # reset c
      for i from 2 to n+1 do          # search array
        for j from 2 to n+1 do
          if P[i,j] = 2 then          # if B type then react
            if P[i,j-1]=1 then P[i,j-1]:=2: end if;
            if P[i,j+1]=1 then P[i,j+1]:=2: end if;
            if P[i-1,j]=1 then P[i-1,j]:=2: end if;
            if P[i+1,j]=1 then P[i+1,j]:=2: end if;
            c:= c+ 1;                 # count number changed
          end if;
        end do:                       # end j loop
      end do;                         # end i loop
    end do;                           # end while c > b do
    asum:= asum + c;                  # total number changed
    end do:                           # end number of reps
  # make results
    av:= asum/reps;                   # average number changed
    fdata[L]:= fract*100:             # fraction as %
    data[L]:= av/n^2*100.0:           # number changed as %
    fract:= fract + delta_fract:      # increase fraction
    end do:                           # end loop L on fraction
```

A typical set of data is shown in Fig. 12.18. The right-hand graph shows the relative reaction rate vs percentage of impurity. The sudden fall off in the number of molecules reacted is a sign of percolation. The left-hand graph shows the percentage of A molecules that have reacted vs the percentage of impurities, x. The straight line in this graph is $100 - x$, and shows what would be expected if the number of reacted molecules were linearly proportional to the number of impurities present. Clearly, linearity is not the case. The reason for this behaviour has to do with the connectivity of the array when impurities or gaps are present. In two dimensions, islands of potentially reactive molecules can become isolated by impurities and this limits B type molecules from reaching these molecules. Similarly, gaps between burning trees prevent others from catching fire. The right-hand graph shows that <30% impurity can be tolerated without affecting the reaction rate by very much, but a >45% impurity almost stops the reaction. The maps, Fig. 12.19, show a typical pattern of molecules or trees at different percentages of impurity or gaps. The change from almost complete connectivity to almost none is quite dramatic and typical of percolation; see Sahimi (1994) for a discussion of percolation.

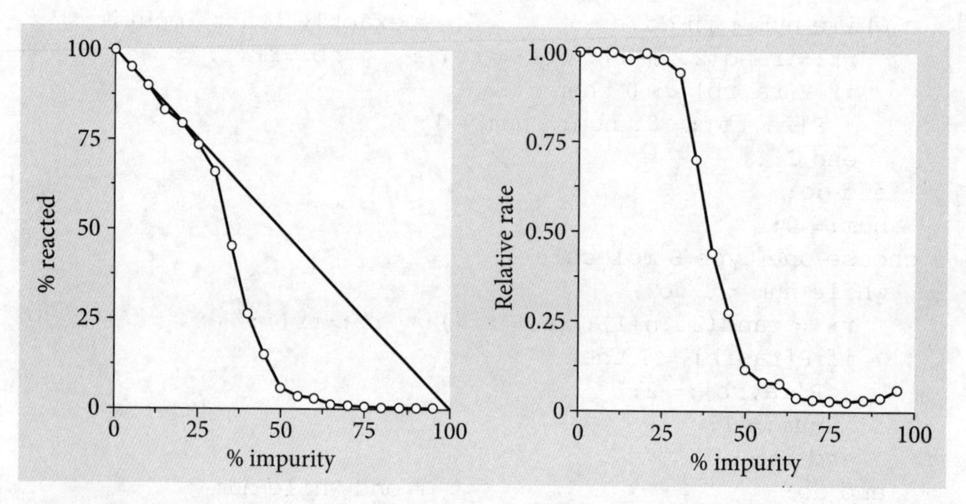

Fig. 12.18 Left: The percentage of molecules reacted (or trees burning) as a percentage of impurities (or gaps between trees). The sudden fall off is a sign of the connectivity ending in the spatial arrangement. The straight line shows what would be expected if the number of reacted molecules were directly proportional to the number of impurities. Right: The relative reaction rate vs % impurity.

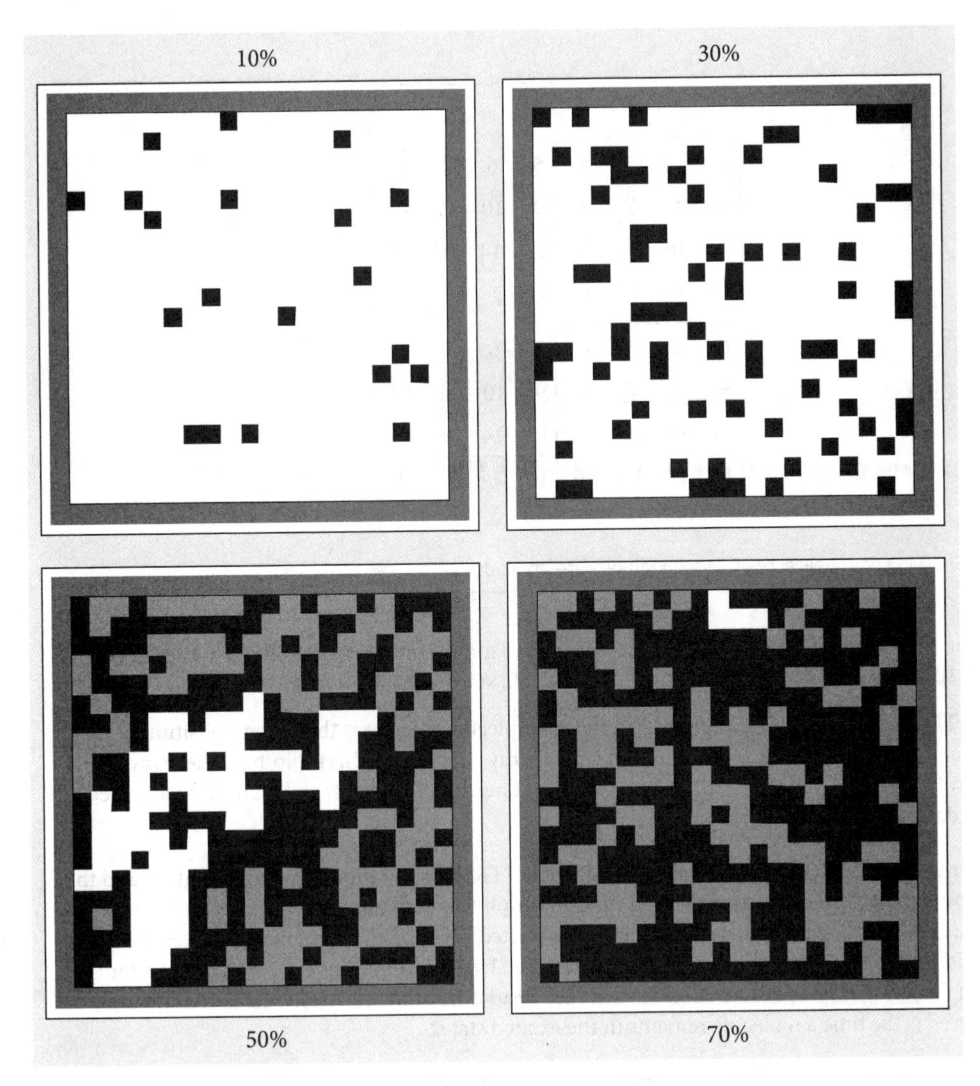

Fig. 12.19 Maps of the amount of *B* molecules (white), *A* type molecules (grey), and impurity (black), with different a percentage of impurity, from top left, top row 10% & 30%, and bottom row 50% & 70%. Alternatively, the picture shows the number of burning trees (white) the number not burning (grey), and the number of gaps (black). The change from 30% to 50% impurity (or gaps) is quite dramatic, the fire being severely limited in the latter. (Note that the active area is surrounded by a dark grey border one square deep.)

12.7 Questions

Full solutions are available at **www.oxfordtextbooks.co.uk/orc/beddard.**

Q12.11 (a) Calculate the heat capacity as in Fig. 12.14. If you choose a bigger lattice, remember that the time for each calculation increases directly with the total number of sites not the size of the lattice side.

(b) Find the change in magnetization per step, and calculate the magnetization by modifying the code. Reproduce the graph. When making the *total* magnetization, add the absolute value of the change. Do this, because at low temperatures the whole spin state can invert back and forth several times during a calculation and a very small average magnetization can sometimes be calculated even though most of the time it is almost +1 or −1.

Q12.12 The two-dimensional lattice can be represented as an array of spins labelled from 1 to n, rather than as a two-dimensional array as in the example. If helical boundary conditions are applied, see Fig. 12.20, then a saving in computer time can be achieved. Rewrite the algorithm to use a linear array to hold the spin state, and use helical boundary conditions.

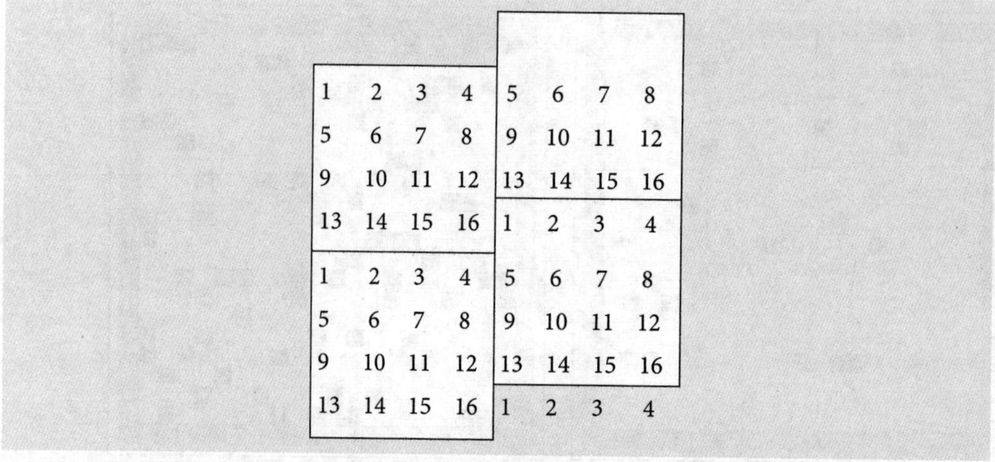

1	2	3	4	5	6	7	8
5	6	7	8	9	10	11	12
9	10	11	12	13	14	15	16
13	14	15	16	1	2	3	4
1	2	3	4	5	6	7	8
5	6	7	8	9	10	11	12
9	10	11	12	13	14	15	16
13	14	15	16	1	2	3	4

Fig. 12.20 Showing how helical boundary conditions are applied.

Q12.13 Modify the random walk diffusion example to produce a graph similar to that shown in the left panel of Fig. 12.10, but only for one walk at a time of length 50, starting at position 0.

***Strategy*:** This is a simpler calculation than in the example; the value of `m` the present position of the walker at a point in the calculation is stored in a new array `d`, and then this is plotted. The array `xdata` and the repeat loop, `for L from ..` are not needed. The graph scale should be adjusted to see the data properly.

Q12.14 In a time-resolved, single-photon counting experiment (TRSPC), the time between excitation and the first photon to be detected, is measured by a time-to-amplitude converter (TAC). This works by charging a capacitor linearly in time; the charging is started before the excitation photon reaches the sample and stops when a fluorescence photon is detected by a photomultiplier or similar detector capable of detecting single photons. Besides the fixed dead time of the electronic circuitry, the time lag measured is the time a molecule remains in the excited state.

Simulate the arrival time of 500 photons, each one from a different excitation of the sample and plot a graph of the number of events on the x-axis and their arrival times on the y-axis. Assume that the molecule has a 10 ns lifetime. A plot similar to that of Fig. 12.11 should result.

***Strategy*:** This question is somewhat 'dressed up' but is quite straightforward. Repeatedly use the equation $t = -\tau\ln(r)$, equation (12.13), to estimate times, and plot t vs the number of the event. The number of molecules is 1 in each case because only 1 photon is detected each time, and the number of events is 500. It is easier to work in units of nanoseconds, so make the lifetime 10. Remember to label the graph as time/ns.

Q12.15 A dye molecule D is intercalated into DNA and, when electronically excited, undergoes electron transfer, which competes with fluorescence and decreases the state's lifetime. Typically, D might be methylene blue whose excited state is oxidized by receiving an electron although only from a guanine base (G) and not from bases A, T, or C. Their redox potential will not allow this to occur at a rate that will compete with fluorescence. The position that the dye intercalates is random, as is the DNA sequence; for example, one of possibly 4^{10} arrangements with 10 base pairs could be

$$\begin{matrix} \cdots & T & A & T & C & G & - & T & G & A & A & C & \cdots \\ & 1 & 2 & 3 & 4 & 5 & Dye & 7 & 8 & 9 & 10 & 11 & \\ \cdots & A & T & A & G & C & - & A & C & T & T & G & \cdots \end{matrix}$$

The electron transfer rate constant depends exponentially on distance as

$$k = k_0 e^{-\beta r} \tag{12.15}$$

and, therefore, the G base does not have to be next to the dye to be quenched; β is a constant that depends on the overlap of the electronic orbitals of the donor and acceptor molecules.

(a) Calculate the rate of electron transfer, and the measured fluorescence decay profile of the dye in the presence of electron transfer. The isolated excited state of the molecule has a 380 ps lifetime, or rate $k_f = 10^{12}/380\ \text{s}^{-1}$. The constant $k_0 = 3.0 \times 10^{12}\ \text{s}^{-1}$ and $\beta = 7.0\ \text{nm}^{-1}$. In DNA, the distance between base pairs is 0.34 nm and this is also the separation of the dye to its nearest base. Assume that all bases are present in equal amounts and make the calculation cover a distance of 5 base pairs either side of the dye. Calculate the decay over 1000 ps and use 1 ps bins to form a histogram, and when the code is working, repeat the calculation 50 000 times.

(b) Show that as β is reduced more quenching of the excited state occurs. Explain why this is.

Strategy: This problem looks as though it could be solved by writing down the rate equation. Only one dye is ever present in each piece of DNA and so the rate equation has the form

$$\frac{d}{dt}[D] = -(k_f + k_3[G_3] + k_6[G_6])[D]$$

if a base G is at position 3 and 6. The rate constant at position 3 away from the dye is, for example, $k = k_0 e^{-3\times0.34\beta}$. However, a little reflection will tell you that there are many different positions for G relative to the dye, and a variable number of G's may also be present. If there are 10 base pairs, then there are $4^{10} = 1\ 048\ 576$ possible ways of arranging the bases, which is a huge number of equations to solve. Instead, the decay rate constant can be estimated by the Monte Carlo method, by placing the G bases at random positions about the dye and repeating the calculation a few thousand times until the result is effectively constant. The decay of the dye is expected to be non-exponential, as previously shown in the problem of donors and acceptors at random, in solution.

Q12.16 One simple model of disease spreading is the S-I-R model, meaning individuals are Susceptible, Infected, or Removed. The scheme is

$$S + I \xrightarrow{k_2} 2I$$
$$I \xrightarrow{k_1} R$$

The differential equations can solved numerically as described in Chapter 11 and also by Monte Carlo integration, see Section 12.2.2. Instead of these approaches, a discrete Monte Carlo simulation will be used to describe how a disease is transmitted and which does not involve integrating the differential equations.

Suppose that just one student becomes infected with flu and then transmits his infection to others; some older students have already had the infection and are immune, while all the others are susceptible to catching it. If there are 1000 students and each day 1 person at random can become infected, provided they are not immune, calculate;

(a) The number infected, and the number susceptible day by day for a period of 40 days. Assume that 1% are initially immune and that each student has a probability of 0.2 of recovering each day after being infected. Calculate also the size of the epidemic, which is the number in the removed category.

(b) Repeat the calculation with different fractions of immunized persons and rationalize your results.

(c) In some diseases, such as colds and influenza, after becoming infected a person is again susceptible; a minor change to the algorithm allows for this. Show that the infection persists.

Strategy: There are three types of students; immunized, susceptible, and infected. Initially, one student is infected out of 1000 students. The calculation can be set up by choosing, at random, 10 students, which is 1% of the total to represent the immune students and then one more of those who are not immune to be the initially infected student. To do this, an array is defined to represent the state of each student and which contains integer values to represent one of the three states; for example, 2 = infected, 1 = susceptible, and 0 = immune. The number infected is found by summing up all those members of the array with a value of 2. The progress of time is represented by a `for..do` loop from 1 to 40. The calculation should be repeated to average a number of initial distributions of infected and immune persons.

The method is

(1) Define constants and parameters, number of students and so forth.

(2) Choose at random those to be immune and one to be infected initially; do not choose anyone that is immune.

(3) Make one `for..do` loop over the number of repeat calculations; inside this, make a second `for..do` loop to range over the number of students. In this latter loop

(i) If an infected student is found, choose two more at random and infect them if they are not already immune. This is the first step $S \rightarrow I$ and depends on both S and I and the rate constant k_2 of 1/day.

(ii) Choose a random number in the range 0 to 1, and if less than the chance of recovery, choose a student at random and if already infected, make immune. This is the removal step with rate constant 0.2/day, and depends only on those infected.

(iv) End the loop over the number of students.

(4) Add up the number of students infected and store the results.

(5) End the repeat loop.

Q12.17 Modify the algorithm in the autocatalytic reaction example (or fire spreading) to include the four nearest diagonal points as well as those along the same row and column. The relative rate should be 50%, at a fraction 0.5694 if a large grid is chosen. Allow trees to burn out during each cycle, so becoming gaps, and calculate the number burning as a fraction. Then do similar calculations on a triangular or hexagonal lattice, which is harder.

Data Analysis

13

13.1 Characterizing experimental data

13.1.1 Experimental accuracy, precision, mean, and standard deviation

The experimentalist always has to contend with noise on data. Some of the causes and methods that can be used to remove it have been discussed in Chapter 9. However, once the mean value of a series of experimental measurements is obtained, some measure of its *precision* and its *accuracy* is always required. This can mean making a comparison with data of the same quantity obtained by others, published in the literature, or with an accepted standard. The type of errors caused by poor experimental method or carelessness, i.e. blunders, can be significant and are not accounted for by any statistical analysis because they are entirely unpredictable. These are not considered further. The errors described in this chapter are predictable, but only in a statistical sense. This means that if the distribution the errors will form is known, then the chance that any given value may occur with can be predicted.

It is to be expected that by repeating a measurement its precision will improve, however, the accuracy may not. This is because accuracy is a measure of how a result differs from the true value of that measurement; the freezing or boiling point of water for example. If an average result is not accurate, then there has to be something systematically wrong in the instrument or experimental method that is causing this. Precision, on the other hand, is a measure of the spread or *dispersion* of the data about its mean value and can be improved by careful experimentation and by repeating the measurements. Clearly, the best result is precise and accurate, and the worst is imprecise and inaccurate. The precision increases if more measurements are made, whereas the accuracy may not.

Because of the random nature of the noise associated with all measurements, it is never possible to measure an exact value, i.e. one with no error. In fact, if someone were to report such a measurement then this would be very suspicious. Similarly, reporting results that are better than could be obtained from the known random nature of experimental results, as described by the normal (Gaussian) or other distributions, should also be a cause for suspicion. Although errors on a measurement are to be expected as the natural consequence of any experiment, this does not mean that they cannot be reduced by good experimental practice and by repeating measurements; but remember that they can never be reduced to zero. Clearly, what is needed is to define a quantity that indicates what an experimental result is and a second quantity that indicates the chance of this being the correct result. The *average* is obviously a good measure of an experimental result, and the *standard deviation* a measure of the dispersion or range that a value can reasonably be expected to have either side of the average. Confidence limits are then constructed from the standard deviation with which it is possible to feel confident to within a certain level, usually 95%, that a measurement will be within that value of the mean. How these calculations are done is described next.

13.1.2 The mean or average value

If several readings have been taken of a quantity x, for example a titration is performed N times with identical solutions in an attempt to be precise, the titration's end-point volume

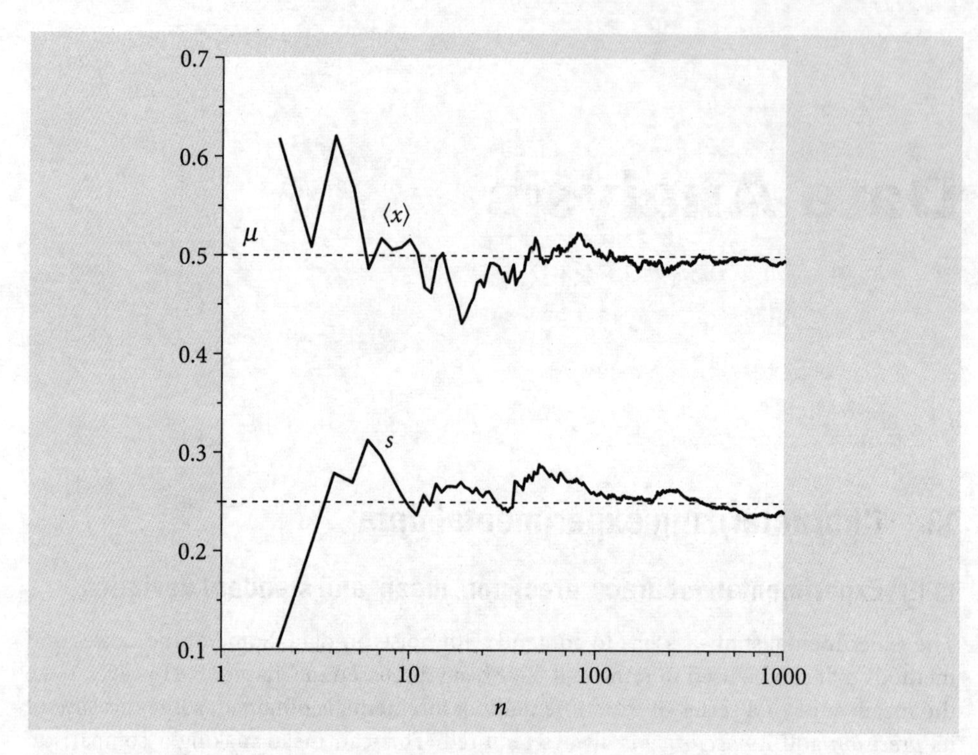

Fig. 13.1 The experimental average value $\langle x\rangle$ tends towards the population or theoretical mean μ as the number of samples increase. The sample standard deviation s is also shown and this tends to a constant value of 1/4 as n increases. (Note the logarithmic abscissa scale). The samples are taken from a Normal distribution with $\mu = 1/2$ and $\sigma = 1/2$.

will be quoted as the *arithmetic sample average* $\langle x\rangle$ of all the n measurements. This is sometimes also labelled $\bar{x}$ (pronounced 'x-bar') and the word 'mean' is often used instead of average. The arithmetic average is

$$\langle x\rangle = \frac{1}{n}\sum_{i=1}^{i=n} x_i = \frac{1}{n}(x_1 + x_2 + x_3 + x_4 + \cdots), \tag{13.1}$$

where $x_1, x_2, \cdots x_i \cdots x_n$ are the measurements. This average value will probably not be the same as any individual value and these will always be spread either side of the mean. The sum of the difference between each point and the average, is always zero; $\sum_i (x_i - \langle x\rangle) = 0$.

One trajectory showing the way the mean value is approached as the number of experiments is increased is shown in Fig. 13.1; the sample standard deviation s (equation (13.2)) is also shown. The samples are selected at random from a normal distribution (Fig. 13.4) with a mean 0.5 and a standard deviation 1/4. In many experiments, the accuracy of each measurement may not be the same and then the mean and standard deviation have to be *weighted* to reflect this. This is described in Section 13.3.8.

13.1.3 Sample standard deviation and parent or population variance

A measure of the spread of the results is the *sample standard deviation s*. The square of the standard deviation is called the *variance*. The sample standard deviation is

$$s = \sqrt{\frac{1}{n-1}\sum_{i=1}^{n} (x_i - \langle x\rangle)^2} \tag{13.2}$$

and is sometimes called the *root mean square* (or *rms*) deviation. This formula produces an unbiased estimate of s, but note that some authors define the standard deviation by dividing by n rather than $n-1$; there is not a single definition of s; see Barlow (1989, p. 11). The standard deviation is one of a class of measures called *dispersion indices*; range, quantile, skew, and kurtosis (peakedness) are others.

In words, the formula for s says 'for each of the n measurements, subtract the average $\langle x\rangle$ from each x value, square the result and then add up all the answers. Next, divide by the total number of measurements less one and finally take the square root'.

The principle of *least squares* is widely used in modelling or analysing data; see Section 13.5.2. The least squares approach minimizes a function such as

$$\sum_{i=1}^{n}(x_i - M)^2$$

with respect to M, where M might represent some 'model' which is an equation or single value describing a set of data. In the definition of the standard deviation s, it appears that $M = \langle x \rangle$, therefore, if s is a least squares estimate, the summation should be at a minimum when $M = \langle x \rangle$. Differentiating the sum of squares with respect to M and setting the result to zero produces

$$\frac{d}{dM}\sum_{i=1}^{n}(x_i - M)^2 = -2\sum_{i=1}^{n}(x_i - M) = -2\sum_{i=1}^{n}x_i + 2nM = 0 \quad \text{or} \quad M = \frac{1}{n}\sum_{i=1}^{n}x_i$$

showing that, indeed, $M = \langle x \rangle$. The mean value makes the sum of squares a minimum, and in this sense it is the best estimate of the deviation.

Suppose that there is an underlying *parent distribution* whose width determines the standard deviation. This has a mean μ, called the *population mean*, and its standard deviation is σ; Greek letters being reserved for parent quantities, then this parent distribution is what an infinite number of ideal experimental results would produce. This ideal distribution is assumed to be the normal (Gaussian) distribution, Figs 13.3 and 13.4. (The other common distribution is the Poisson, Fig. 13.13 which approximates the normal when its mean is ≈ 10 or greater.) The sample mean $\langle x \rangle$ is more likely than not to be different to the population mean μ. If it can be shown that the average of all sample means s equals the population or true mean value μ, then the sample mean is an unbiased estimate of the population mean.

The standard deviation and variance can also be defined with reference to the parent distribution and then this *population* or *parent variance* σ^2 is

$$\sigma^2 = \frac{1}{n}\sum_{i=1}^{n}(x_i - \mu)^2 \tag{13.3}$$

which assumes that μ is known, whereas s, equation (13.2), is obtained only from the data itself. The variance is the single most important parameter when describing the parent population. To calculate σ^2, μ has to be known, but μ can never be exactly known in any set of measurements. The best estimate of this has to be used instead and is obtained from the set of measurements; this is usually taken to be

$$\sigma^2 = \frac{n}{n-1}s^2,$$

which is an unbiased estimator of σ^2. Strictly speaking, the equality symbol = should be replaced by $\approx$ 'approximately' because this is an estimation; however, equality is usually used. The factor $n/(n-1)$ enters because of an argument from statistical theory. The term reflects the degrees of freedom left with which to describe the data. Each parameter that is defined is considered to impose a restraint on the data and is, roughly speaking, equivalent to removing one data point. The more parameters that are measured the fewer data points there are left to describe the data (Parratt 1971).

13.1.4 Standard deviation in the mean

If experimental measurements are repeated, slightly different values of the mean are expected because only a few of the possibly infinite number of values needed to define the true result can be measured. The difference between any two means would be expected to be less than the standard deviation in either set. The standard deviation of the means would then be written in the same way as equation (13.2) as $s_m = \sqrt{\frac{1}{n-1}\sum_{i=1}^{n}(m_i - \langle m \rangle)^2}$ where $\langle m \rangle$ is the *average of the means* of N separate experiments. To evaluate this summation directly would require a huge number of experiments, viz; nN; however, a satisfactory formula is obtained by statistical theory and is

$$s_m = s/\sqrt{N}$$

Experimentally, the square root makes improving precision quite a slow process; 100 measurements are needed to improve the signal to noise by 10 times, which means reducing the s_m relative to the mean by 10 times. The Hadamard transform method, Chapter 9, enables the experimenter to measure in groups and the noise can then be reduced more rapidly than $\sqrt{N}$.

The quantity s_m is sometimes called the *estimated standard error on the mean* or just the *standard error on the mean* or simply the *standard error*. To relate this to the population standard deviation,

$$\sigma_m = \sigma/\sqrt{N},$$

however, this is also called the standard error. Clearly, these names are not fixed so it is always necessary to check the equation being used, which unfortunately, is not always given.

13.2 Central limit theorem

The mean μ and standard deviation σ are used assuming that data is distributed normally. But many quantities measured are not expected to have a normal distribution; for example, when counting photons or particles, a Poisson distribution is produced which is quite unlike a normal distribution at small μ, see Fig. 13.13. The speed of a molecule follows a Maxwell–Boltzmann distribution (see Q3.54), which is clearly skewed or lopsided, again quite unlike a normal distribution. However, remarkably, if several measurements are taken of a given velocity, their distribution will be normal. This is what the central limit theorem predicts and although easy to demonstrate, it is harder to prove. It states that by taking many similar measurements from (almost) any type of continuous distribution, the result always approaches a normal (Gaussian) distribution. This population distribution has the theoretical form

$$p(x) = \frac{1}{\sqrt{2\pi\sigma^2}} e^{-\frac{(x-\mu)^2}{2\sigma^2}} \tag{13.4}$$

where μ is the mean of the distribution and σ the standard deviation. The central limit theorem forms the basis by which the standard deviation of a normal distribution is used to characterize data.

The reasoning put more technically is that the sample mean $\langle x \rangle$, although an unbiased estimate, is unlikely to be exactly equal to the true population mean μ and is itself subject to random variation. By repeating the sampling process, i.e. by repeating the whole set of measurements, a number of different estimates of $\langle x \rangle$ are obtained which are distributed about the true value μ. To simulate this and illustrate the central limit theorem, suppose that there are five measurements A, B, C, D, and E and each is an experimental result with values 1, 2, 3, 4, and 5 respectively. Any other numbers could be used but these are easy to average. The distribution of the numbers is uniform as shown in Fig. 13.2.

Suppose that randomly selected pairs of these values are taken making 25 possible samples. The pairs are shown on the left of the table and the corresponding average (sample mean) is shown on the right. Looking at this table there are five entries with a value of 3, and only 1 entry each with a value of 1 or 5. If plotted as a histogram, this begins to look a little like a normal distribution; Fig. 13.2.

Possible samples					:	Sample means $\langle x \rangle$				
AA	*AB*	*AC*	*AD*	*AE*		1.0	1.5	2.0	2.5	3.0
BA	*BB*	*BC*	*BD*	*BE*		1.5	2.0	2.5	3.0	3.5
CA	*CB*	*CC*	*CD*	*CE*		2.0	2.5	3.0	3.5	4.0
DA	*DB*	*DC*	*DD*	*DE*		2.5	3.0	3.5	4.0	4.5
EA	*EB*	*EC*	*ED*	*EE*		3.0	3.5	4.0	4.5	5.0

Using Maple, the central limit theorem can be demonstrated more convincingly; see Algorithm (13.1). The random numbers are chosen from a uniform, i.e. flat, distribution

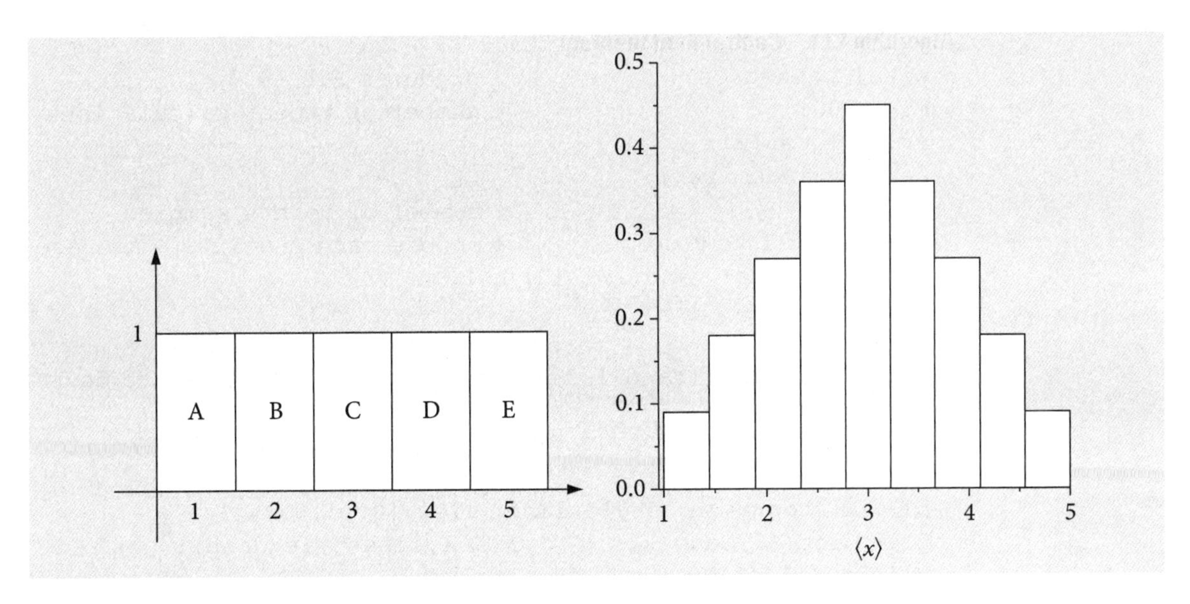

Fig. 13.2 Uniform distribution and the (normalized) histogram formed by sampling pairs of numbers.

between 0 and 1. Five samples are taken each time and the process repeated 5000 times; see Fig. 13.3 for the result. It can be seen that the sample averages that form the histogram are more closely clustered about the population mean and are therefore less variable than the original data. It can be shown that the variation in the mean, which is the uncertainty in the mean, equals the variation in $\langle x \rangle$ divided by the sample size, σ^2/n. With this in mind the normal distribution plotted on top of the histogram of the data is

$$p(x) = \frac{1}{\sqrt{2\pi\sigma^2/n}} e^{-\frac{(x-\mu)^2}{2\sigma^2/n}}.$$

Many other initial distributions can be chosen to start with besides the uniform distribution illustrated here. For (almost) all distributions, a normal distribution will eventually be formed, although this may require a huge number of calculations.

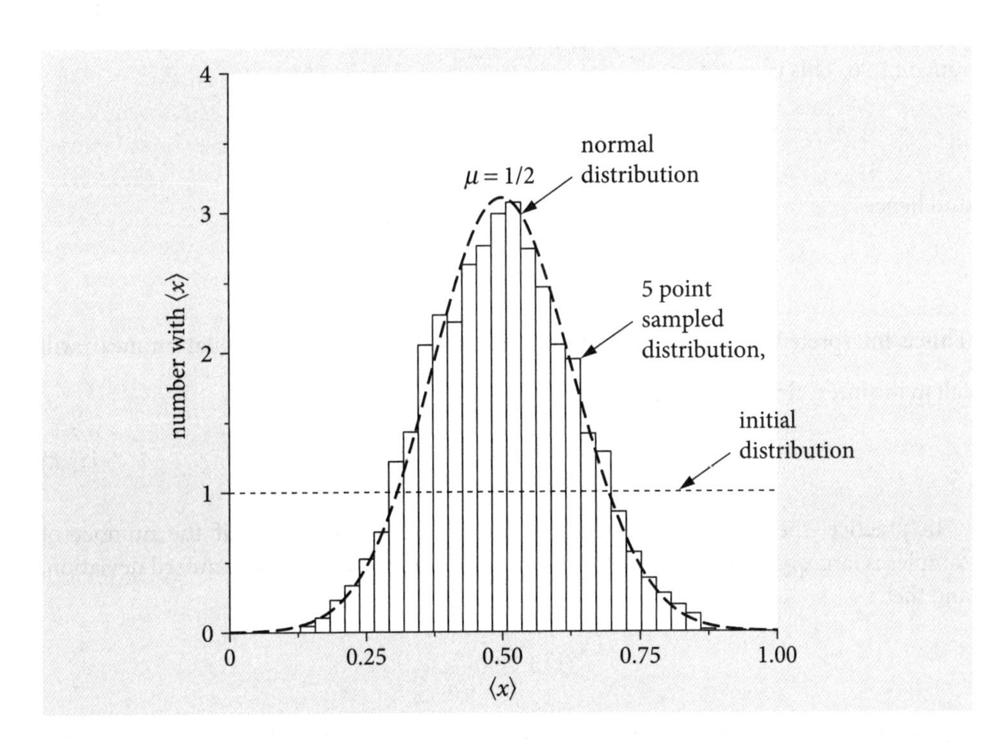

Fig. 13.3 Illustrating the central limit theorem.

Algorithm 13.1 Central limit theorem

```
> with(Statistics):                # produces Fig 13.3
> m:= 5000:                        # number of repeat calculations
  av:= Vector(m):
  seed:= randomize():
  n:= 5:                           # number of points sampled
  for i from 1 to m do             # choose random points & average
    av[i]:= add( rand()/1e12, k=1..n )/n;
  end do:
  av_x:= 0.5:                      # average
  sig:= sqrt(add((rand()/1e12-av_x)^2,i=1..m)/m); # std dev
  Gaus:= (x,sig)->
       exp( -(x-av_x)^2/(2.0*sig^2/n))/sqrt(2*Pi*sig^2/n);
  A:= convert(av,list):
  p1:= Histogram(A, style=line, view=[0..1,0..4]):
  p3:= plot([1,Gaus(x,sig)],x=0..1,linestyle=dash):
  display(p1,p3);
```

13.3 Confidence intervals

The central limit theorem shows that repeated measurements follow a normal distribution (see Section 13.3.1) with a mean μ and variance σ^2/n for n separate measurements. The mean μ may be known from theoretical considerations or from other experimental data. The properties of the normal distribution are used to provide estimates of the probability that, by chance alone, a measurement of the mean will fall inside or outside a certain value.

The standard deviation of the mean is often referred to as the *standard error on the mean* and defined as $\sigma_m = \sigma/\sqrt{n}$. This confirms that measurements get more precise with the square root of the number of measurements. If a new statistic Z is defined as

$$Z = \frac{\langle x\rangle - \mu}{\sigma/\sqrt{n}} \tag{13.5}$$

this is distributed as the standard normal distribution with a mean of 0 and a standard deviation of 1. Because $\langle x\rangle$, the experimental or sample mean value, is unlikely to be equal to μ, there is a need for a measure of the confidence that we have in $\langle x\rangle$. From the properties of the normal distribution (see next section), there is a 95% chance that Z falls within the range ±1.96. This is written as

$$p\left(-1.96 < \frac{\langle x\rangle - \mu}{\sigma/\sqrt{n}} < 1.96\right) = 0.95$$

and hence

$$p\left(\langle x\rangle - 1.96\frac{\sigma}{\sqrt{n}} < \mu < \langle x\rangle + 1.96\frac{\sigma}{\sqrt{n}}\right) = 0.95.$$

This is interpreted to mean that there is 95% confidence that the population mean will fall in the interval $\pm 1.96\frac{\sigma}{\sqrt{n}}$ and is written as

$$\langle x\rangle \pm 1.96\frac{\sigma}{\sqrt{n}}. \tag{13.6}$$

In practice, the standard deviation is almost never known and if the number of samples is large, typically >25, then σ can be replaced by s the *sample* standard deviation, and then

$$\langle x\rangle \pm 1.96\frac{s}{\sqrt{n}}.$$

Confidence limits are not always used when quoting results and it is common to see $\langle x\rangle \pm \sigma$ quoted instead. Unless this is qualified as being 1σ or 2σ etc., what is implied by this is uncertain. If the error is $\pm 1\sigma$ then $\approx$ 68% of all measurements fall in this range. Table 13.1 gives other values.

13.3.1 The normal and standard normal distribution

The normal (Gaussian) distribution with mean μ and variance σ^2 has the form

$$p(x) = \frac{1}{\sigma\sqrt{2\pi}} e^{-(x-\mu)^2/2\sigma^2}$$

which is normalized to 1. $\int_{-\infty}^{\infty} p(x)dx = 1$. The mean, average, or expectation value of x is

$$\int_{-\infty}^{\infty} xp(x)dx = \mu \equiv \langle x \rangle,$$

and the variance is $\int_{-\infty}^{\infty} (x-\mu)^2 p(x)dx = \sigma^2 \equiv \langle x^2 \rangle - \langle x \rangle^2$.

The standard normal distribution describes a Gaussian (bell-shaped) curve with a mean of zero and a standard deviation of one, viz.,

$$p(x) = \frac{1}{\sqrt{2\pi}} e^{-x^2/2}. \tag{13.7}$$

The total area under the curve is 1 and the area between symmetrically placed x values gives the probability of falling within that area. The area $P(x)$ is the probability of an observation being between $\pm x$;

$$P(x) = \frac{1}{\sqrt{2\pi}} \int_{-x}^{x} e^{-x^2/2} dx = erf\left(\frac{x}{\sqrt{2}}\right) \tag{13.8}$$

The area within limits ± 1.96 is $erf(1.96/\sqrt{2}) = 0.950$; hence this is the 95% chance as described by equation (13.6). In Fig. 13.4, the total area in grey adds up to 5% of the total, meaning that a value that differs from the mean should exceed ± 1.96 by pure chance only on 5% of all measurements.

The *probable error* p_e divides the normal distribution area into two with areas placed symmetrically about zero. The areas are 1/4:1/2:1/4; the distribution's x value is $p_e = \pm 0.6745\sigma$. Some values of the area and hence the chance of a value occurring within different standard deviations of the mean is shown in Table 13.1.

Table 13.1

% chance	dispersion index
38.3	$\sigma/2$
50	$0.675\sigma \equiv p_e$
68.3	1σ
82.2	$2p_e$
95	1.96σ
95.45	2σ
99.73	3σ

Specific areas may be calculated using the cumulative distribution function *CDF* function in Maple, or directly by integrating the normal distribution from $-\infty$ to x. For the reverse process—starting with the area to obtain the x value producing that area—the *quantile* function is used. For example, with the normal distribution with a mean of zero and standard deviation of 1 the calculation is

```
> x:- 1.96;                          x := 1.960
  f:= CDF( Normal(0,1),x);           f := 0.975          (13.9)
  Q:= Quantile(Normal(0,1),f);       Q := 1.960
```

The fraction $f = 0.975$ is the total area from $-\infty$ to 1.96σ. This leaves 0.025 not covered by the calculation and is the 2.5% shown as shaded on the right of the figure. The total area $1 - 2 \times 0.025 = 0.95$, gives the chance shown in the table. If the distribution is not normal, for instance the 't' or χ^2 distribution a similar calculation applies but produces different percentages.

The general form for a $(100 - \alpha)$% confidence limit when the population standard deviation σ is known is

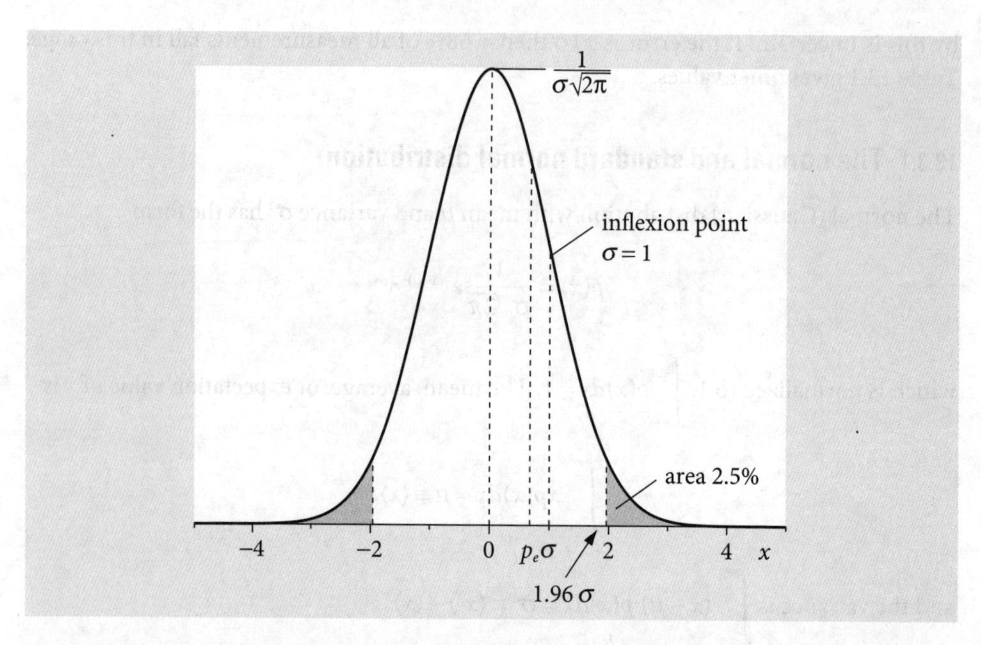

Fig. 13.4 Standard normal distribution with total area in grey adding up to 5%. A measurement that differs from the mean by $\pm 1.96\sigma/\sqrt{n}$ should, by pure chance alone, exceed this value only 5% of the time.

$$\langle x \rangle \pm Z_{\alpha/2}\frac{\sigma}{\sqrt{n}} \tag{13.10}$$

and when σ is unknown, the sample standard deviation is used

$$\langle x \rangle \pm Z_{\alpha/2}\frac{s}{\sqrt{n}}. \tag{13.11}$$

The term $Z_{\alpha/2}$ is the percentage point of the standard normal distribution.

Table 13.2

Confidence Level	$Z_{\alpha/2}$
90%	1.28
95%	1.96
99%	2.58

Example (i) Two hundred samples taken at random were obtained for the K^+ content of the glass used to make Pyrex flasks. The mean value was found to be 136.48 μg with a sample standard deviation of 25.31 μg. The calculation of 95% and 99% confidence limits about the mean mass follows directly from the equations.

$$95\% \text{ limits}: 136.48 \pm 1.96 \times 25.31/\sqrt{(200)} = 136.48 \pm 3.51\ \mu\text{g}.$$

which would normally be rounded to 136.5 ± 3.5, or to 137 ± 4 μg if one were being cautious. This result means that 95% of the samples taken at random should fall between 133 and 141 μg and by chance alone, it could be expected that 5% of results would be outside these limits. The 99% confidence limits produce 136.5 ± 4.62 μg, which will round up to 137 ± 5 μg.

13.3.2 Small sample confidence limits: 'Student's t'

When the number of samples is small, s may not be a very good estimate of σ and, in this case, Student's t test is needed. The distribution is similar in shape to the normal distribution, but is wider in the wings. It is characterized by one parameter $\upsilon = n - 1$ where n is the number of samples being averaged, and υ is called its 'degrees of freedom'. When the sample size increases, the t distribution approaches the normal one. Using a similar argument to that for the normal distribution, the t distribution produces confidence limits,

$$\langle x \rangle \pm t_{\alpha/2}\frac{s}{\sqrt{n}} \tag{13.12}$$

where $t_{\alpha/2}$ is obtained by integrating the distribution, however, only a few values are used regularly, some of which are listed in Table 13.3. (Notice that in the table that υ is one less than the number of data points.)

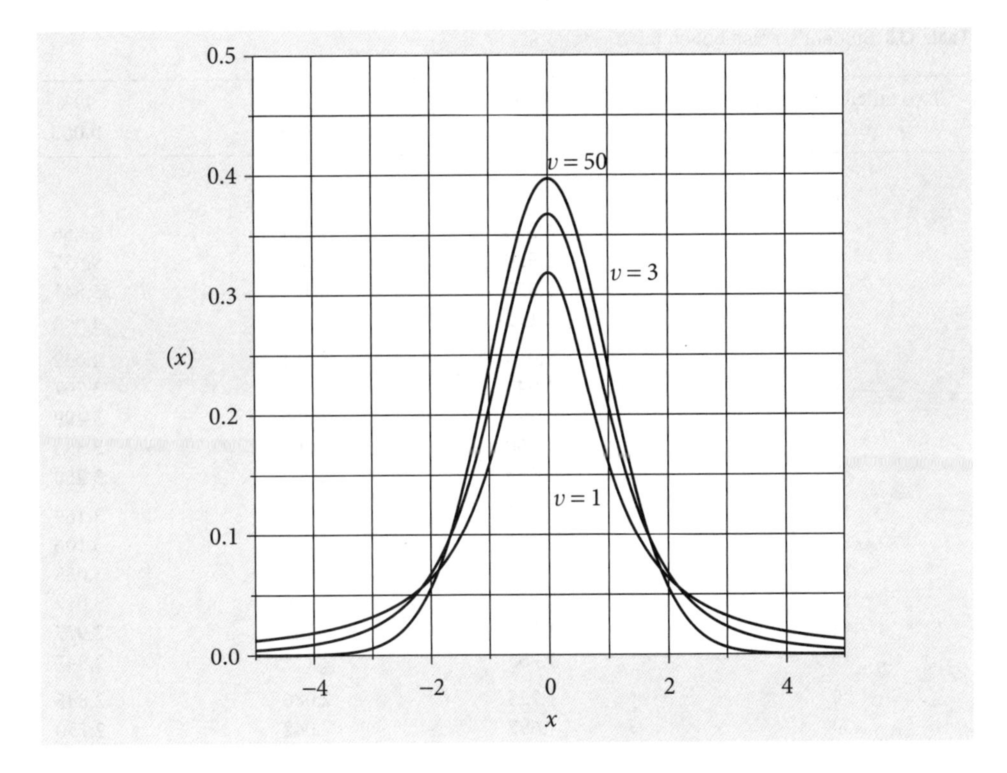

Fig. 13.5 Student's t distribution is wider in the wings than the normal distribution, which it approaches when v is large.

Example (ii) Some typical response times of a certain commercial mass spectrometry service are

$$8.21, 25.15, 11.20, 18.06, 22.55, 16.49 \text{ days},$$

and 95% confidence limits need to be placed on the population mean μ. To calculate how many days on average it will be necessary to wait to be 95% sure of obtaining the results of an analysis, use equation (13.12) with the t value taken from Table 13.3. As σ (population mean) is unknown and the sample is small then

$$\langle x\rangle \pm t_{0.025}\frac{s}{\sqrt{n}}. \tag{13.13}$$

The sample size $n = 6$ so there are $v = n - 1 = 5$ degrees of freedom, and from t distribution tables $t_{0.025} = 2.571$ for 95% confidence. The 0.025 is used because this is the two sided value, the limit being 2.5% to the far right and far left of the t-distribution, in the same way as that shown for the normal distribution in Fig. 13.4. The sample mean $\langle x\rangle = 16.94$ days and the sample standard deviation $s = 6.47$ days. Therefore, a 95% confidence interval for μ is given by

$$16.94 \pm 2.571 \times 6.47/\sqrt{6} = 16.94 \pm 6.79 \text{ days}$$

Rounding the answers gives 17 ± 7 which is between 10 to 24 days and is a very wide variation. It would certainly be worth considering changing your supplier!

13.3.3 Critical values from cumulative Student's *t* distribution

To use Maple to find the t values for n points with $\alpha/2 = 0.025$ use

```
> with(Statistics):
> t:= Quantile( StudentT(n-1), 1-0.025);
```

Note that the 95% confidence level has $\alpha/2 = 0.025$ which is 5/2% because the table is for a 'two-tailed' values on the distribution and $\alpha = 5$. The t value at 95% and for seven points (or six degrees of freedom) is written as $t_{0.025,6}$. Fig. 13.4 shows two-tailed values on the normal distribution.

Table 13.3 Student's *t* distribution table

Two tailed confidence $\alpha/2 \rightarrow$	90% 0.05	95% 0.025	99% 0.005
$v = n - 1$			
1	6.314	12.71	63.66
2	2.920	4.303	9.925
3	2.353	3.182	5.841
4	2.132	2.776	4.604
5	2.015	2.571	4.032
6	1.943	2.447	3.707
7	1.895	2.365	3.499
8	1.860	2.306	3.355
9	1.833	2.262	3.250
10	1.812	2.228	3.169
11	1.796	2.201	3.106
12	1.782	2.179	3.055
13	1.771	2.160	3.012
14	1.761	2.145	2.977
15	1.753	2.131	2.947
20	1.725	2.086	2.845
30	1.697	2.042	2.750
40	1.684	2.021	2.704
50	1.676	2.009	2.678
∞ (Normal distribution)	1.645	1.960	2.576

13.3.4 Hypothesis testing

Suppose that a micro-analytical laboratory has to be certified and one of the tests it has to perform is to determine the ratio of $^{12}C/^{14}N$ on an unknown compound. The examiners know that the ratio should be 50 and a standard deviation of 1/2 is acceptable. The laboratory produces the following set of data,

$$49.8, \quad 50.15, \quad 50.6, \quad 49.9, \quad 50.7, \quad 50.1, \quad 50.9, \quad 49.6$$

and have calculated that they fall within the allowed error bounds. Do you agree?

To solve this problem, a slightly different approach has to be taken and this involves using a common approach to testing data by forming hypotheses. This means testing whether or not there is confidence in a given mean value. Some criterion or *test statistic* is computed and used to make a decision. When using these tests, it is always assumed that the underlying parent distribution is normal and that the samples are independent of one another.

Two related statistics are needed. When the population standard deviation σ is known (it is 1/2 in the problem), the following statistic can be used.

$$z_0 = \frac{\langle x \rangle - \mu_0}{\sigma/\sqrt{n}}. \tag{13.14}$$

This statistic should follow a normal distribution, and if the experimental mean is going to converge on μ_0, if sufficient samples could be taken, then it would be expected that z_0 is 'close' to zero because $\langle x \rangle \rightarrow \mu_0$. The problem is to find critical values with which to test, with a certain confidence, how approximate the statement $\langle x \rangle \approx \mu_0$ actually is. For example, if z_0 *is greater than* ± 1.96 then with 95% certainty $\langle x \rangle \neq \mu_0$; however, we would still expect to observe $\langle x \rangle \approx \mu_0$ on 5% of occasions. Thus, large values of z_0 mean that the experimental mean is probably not the same as the population mean. The data produces

$$\langle x \rangle = 50.22, \text{ and as } \sigma = 1/2 \text{ and } \mu_0 = 50, \text{ then } z_0 = 1.24.$$

This is less than 1.96 which is the value needed for 95% acceptance with a normal distribution, so the conclusion would be that the lab had produced an acceptable set of data. However, the number of samples is small and this last calculation assumes that many measurements have been taken. The statistic would be more discriminating if a second test were used based on the *t* distribution. In this case, the sample standard deviation *s* is used rather than σ. The statistic is

$$t_0 = \frac{\langle x \rangle - \mu_0}{s/\sqrt{n}}. \tag{13.15}$$

Calculating again with $s = 0.466$ gives $t_0 = 1.33$ and from the t distribution, with seven degrees of freedom at the 95% level, $t_{025,7} = 2.365$. As t_0 is less than this and z_0 is less than 1.96, although this is a less critical test, the conclusion reached is that with 95% confidence the data is consistent with a population mean of 50 and standard deviation of 1/2 and the lab is up to standard.

13.3.5 **Comparison of two means**

When a mean value has been obtained, to eliminate systematic errors it is necessary to compare this with another determination, perhaps done on another day, or to compare with a result from different apparatus or with a literature result. The two means are only samples from the true distribution and should not be very different from one another as they are supposedly measuring the same thing. The difference in means should therefore be normally distributed about the true difference $\mu_1 - \mu_2$, which in turn, should be zero. The simplest test is to use the propagation of errors formula, see Section 13.4, to compare the difference in the means $\langle x \rangle_1 - \langle x \rangle_2$ with the difference in standard deviation,

$$s_{1,2} = \sqrt{s_1^2 + s_2^2}$$

If the difference is less than the standard deviation, the two results are probably acceptable.

When there are only a few measurements, two for example, it is natural to try to use the t distribution to quantify their difference which can be done in the following manner after first assuming that the population variances are not different, i.e. $\sigma_1^2 = \sigma_2^2 = \sigma^2$, which means that the data is measuring the same thing. The pooled variance is

$$s_p^2 = \frac{(n_1 - 1)s_1^2 + (n_2 - 1)s_2^2}{(n_1 - 1) + (n_2 - 1)}$$

where data set 1 with mean $\langle x \rangle_1$ is the average of n_1 measurements and similarly for set 2. This variance has $n_1 + n_2 - 2$ degrees of freedom. The t test statistic is

$$t_0 = \frac{\langle x \rangle_1 - \langle x \rangle_2}{s_p}\sqrt{\frac{n_1 n_2}{n_1 + n_2}} \tag{13.16}$$

which is expected to follow a t distribution with $n_1 + n_2 - 2$ degrees of freedom. If this t_0 exceeds the critical value set by the percentage points of the t distribution, then it is clear that the means are *not* the same. Confidence intervals, at 95%, for the difference between population means $(\mu_1 - \mu_2)$ may be obtained using

$$\langle x \rangle_1 - \langle x \rangle_2 \pm t_{0.025} s_p \sqrt{\frac{n_1 + n_2}{n_1 n_2}}. \tag{13.17}$$

Example (iii) The yields of two nominally identical columns used for chromatographic separation have been measured. There are nine and eight experiments on each column under identical experimental conditions and the subsequent yield (mg) for each is as follows:

Column 1:	17.5	21.1	26.6	18.1	23.2	18.4	16.5	21.9	26.8
Column 2:	13.7	12.3	16.3	15.9	21.0	21.9	18.2	14.1	–

To determine if there is evidence of a 'significant' difference in their yield (at the 95% level) the mean and standard deviation for each sample are calculated, then t_0 using equation (13.16), and this compared with the value from the t distribution. The data produces

$$\langle x \rangle_1 = 21.1, s_1 = 3.84 \qquad \langle x \rangle_2 = 16.67, s_2 = 3.46.$$
$$s_p = 3.67 \qquad t_0 = 2.49$$

The t distribution at the 95% level (two-tailed distribution at 0.025, with $v = n_1 + n_2 - 2 = 15$ degrees of freedom) has a value 2.13 (see Table 13.3) and as this is *smaller* than the t_0

calculated from the data, we conclude that the two sets of data are different. The confidence limits (equation (13.17)) are

$$\langle x\rangle_1 - \langle x\rangle_2 \pm 3.8$$

making the lower bound 4.47 − 3.8 = 0.65 mg and the upper bound +8.2 mg. Thus, the difference in mean value at the 95% level is $0.65 \le \langle x\rangle_1 - \langle x\rangle_2 \ge 8.2$ mg which means that it is possible to be 95% confident that column 1 produces between 0.65 to 8.2 mg more on average that column 2.

Example (iv) When repeated measurements are made on the same instrument they are likely to be correlated and therefore not independent of one another. In this example, the *difference* in the experimental values is examined using the t test rather than comparing the two means.

Consider measuring fluorescence from the dye thionine, which is known to intercalate into calf thymus DNA. One sample is treated with protein and measured to see if this has an effect on the amount of fluorescence observed. The data (in arbitrary units) for the fluorescence intensity was as follows:

DNA sample	1	2	3	4	5	6	7
Treated	80.1	64.2	75.4	51.7	71.8	85.9	64.7
Control	88.6	71.3	79.8	60.3	70.2	92.7	65.0

The mean value of the *difference* is −4.78 and the sample standard deviation 4.06. The t_0 statistic (equation (13.15)) is −3.18 with a population mean $\mu_0 = 0$. From the t distribution table with six degrees of freedom $t_{0.025,6} = 2.45$ and as $|-3.18| > 2.45$ the protein does have an effect at the 95% level and the fluorescence is different between the two sets of measurements.

13.3.6 Chebychev's rule and outliers

Sometimes data points seem to be too far from the trend exhibited by all the others and there is then a temptation to remove such points. This must always be resisted. One famous consequence of removing data led to the hole in the Antarctic ozone layer being missed. As the *New Scientist* (31 March 1988) put it, 'So unexpected was the hole, that for several years computers analyzing ozone data had systematically thrown out the readings that should have pointed to its growth.'

When outlying points are found in data, the obvious thing is to check that no numerical or transcriptional error has occurred, then go to the instrument used and check that a simple error has not been made, such as using the wrong solvent or a mistake in the concentration, or amplifier setting and so forth. The instrument could be checked out with a known reference but if everything turns out satisfactorily then it must be assumed that the data point is the result of random chance. Highly unlikely but not impossible. If the experiment cannot be repeated and the data still has to be dealt with, then the Chebychev rule may be useful. This gives a number to the probability that a random variable or the absolute value from a mean $|x - \langle x\rangle|$ exceeds a given number. Suppose that this number is $k\sigma$ where $k > 1$ is an integer and σ is the standard deviation, then the condition is

$$prob(|x - \langle x\rangle| \ge k\sigma) \le \frac{1}{k^2}.$$

This means that the chance that $|x - \langle x\rangle|$ is numerically greater than $k\sigma$ is less than $1/k^2$, or, equivalently, that no more that $1/k^2$ data points should be more than k standard deviations from the mean or, which is the same, that $1 - 1/k^2$ are within k standard deviations. The value of $k > 1$ is for us to choose. The data described above in Section 13.3.4 for the $^{12}C/^{14}N$ ratio has a mean of 50.22 and a standard deviation of $s = 0.466$. If $k = 2$ is chosen then $1 - 1/4 = 0.75$ or 75% of the values should fall in the range $50.22 \pm 2 \times 0.466$. If $k = 3$, then 89% of values fall in the range $50.22 \pm 3 \times 0.466$ and for any points that do not fall in this range there is a sound reason for ignoring them.

Sometimes, a statement may be made along the lines that a measurement has produced a result that is more than '5 sigma from the mean'. This means that the chance of this occurring is $1/25 \equiv 4\%$, which would suggest that it does not belong to the same distribution as other measurements. However, if only a few data points have been taken then there would be less confidence in assuming this, as to opposed to perhaps 100 values in the data set with the mean and standard deviation properly established. Finally, note that the process of smoothing data is akin to removing outliers and should be avoided.

13.3.7 Standard deviation in a single measurement

When only a single measurement has been made, as is often the case in an undergraduate laboratory, the question arises as to what standard deviation it should be given. In such laboratories, many other measurements will undoubtedly have been made so the mean and standard deviation for the experiment will be known. However, in the absence of such knowledge we can appeal to the Poisson distribution (see Section 13.6.4) to determine what value should be given to the standard deviation. It turns out that by calculating the maximum likelihood function that the standard deviation is the square root of the result itself. Thus if the result has a value k then $\sigma_k = \sqrt{k}$.

13.3.8 Weighting

In Chapter 1.9.13 the average energy and average length of trans and gauche butane molecules was calculated by weighting the individual values according to the Boltzmann distribution. Experimental measurements are similarly not always of equal precision and this may be inherent in the nature of the observation. When counting photons, for instance, the precision of each measurement is proportional to the number of counts. For example, the intensity of an emission spectrum varies with wavelength; the precision is therefore different at different wavelengths. Alternatively, it may be that one instrument has twice the resolution of another or simply it may be that one experimentalist is better than another; nevertheless, the average has to be taken. If two measurements of x are made and x_1 is twice as precise as x_2, then the weighted average is made in the proportions, $\langle x \rangle = (2x_1 + x_2)/3$.

In the general case, if w_i are the weights, then

$$\langle x_w \rangle = \frac{\sum_i w_i x_i}{\sum_i w_i}. \tag{13.18}$$

This equation means that the contribution of each measurement to the average is in proportion to w unweighted measurements. This formula has been used in several other guises in other chapters to estimate an expectation value, when w_i was called the distribution of x_i rather than a weighting. The weighted standard deviation is

$$s = \left(\frac{\sum_i w_i (x_i - \langle x_w \rangle)^2}{\sum_i w_i} \right)^{1/2}$$

For experimental measurements, weighting is optimal if it is the reciprocal of the variance,

$$w_i = 1/\sigma_i^2$$

and σ_i must be determined for each of the i observations. In many cases the σ will be equal to one another, then $\langle x \rangle = \frac{1}{N}\sum_i x_i$ which is the unweighted mean of x. The variance of the weighted average is

$$\sigma^2 = \frac{1}{\sum_i w_i} = \frac{1}{\sum_i 1/\sigma_i^2} \tag{13.19}$$

Weightings ensure that more importance is given to the more precise measurement because the smaller the standard deviation is, the larger is the weighting given to it.

Suppose that on one instrument a line in the SO_2 infrared spectrum is measured at 550 cm^{-1} with a standard deviation of $\sigma = 10\ cm^{-1}$; another instrument is then used with $\sigma = 5\ cm^{-1}$ and produces 555 cm^{-1}. The unweighted average is 552.5 cm^{-1}, while the weighted average of the two measurements is 554 cm^{-1}, which is close to that of the higher resolution instrument as might be anticipated. If three measurements are made of a rate constant with values $(3.16 \pm 0.03) \times 10^7$, $(3.21 \pm 0.05) \times 10^7$, and $(3.14 \pm 0.02) \times 10^7\ s^{-1}$, the errors are then taken to be the standard deviations and equations (13.18) and (13.19) are used. The best combined rate constant is $(3.15_2 \pm 0.01_4) \times 10^7\ s^{-1}$.

In many experiments, the standard deviation can be found by looking at an instrument's specification, where a resolution of a certain number of wavenumbers or millivolts, and so forth, is usually given. In other cases this may have to be estimated from the data itself and this can be difficult to do point by point. However, one case in which the standard deviation is known exactly is in a photon counting experiment because the arrival of photons at the detector is Poisson distributed where $\sigma^2 = \mu$ and then $w_i = 1/\mu_i$ where μ_i is the average number of counts in the ith measurement. Fluorimeters often use photon counting to measure fluorescence and phosphorescence spectra, and the standard deviation can then be measure directly from the data.

13.4 Propagation or combination of errors

In many experimental situations, a measurement does not always produce the final result, which will be obtained from further calculations, and may also involve other experimental measurements. For example, the measured value and its associated error may have to be exponentiated and then multiplied by another quantity with its error.

The formula for error propagation (or combination) can be determined by expanding the required function as a Taylor series about its mean value and substituting the result into the variance equation (13.3); see Bevington & Robinson (2003) or Barlow (1989) for the proof. If the functional form is written as $y = f(u, v)$, then the variables that have been measured are u and v and their respective standard deviations σ_v and σ_v. The variance in the final result y is

$$\sigma_y^2 = \left(\frac{\partial y}{\partial u}\right)_v^2 \sigma_u^2 + \left(\frac{\partial y}{\partial v}\right)_u^2 \sigma_v^2 + 2\left(\frac{\partial y}{\partial u}\right)\left(\frac{\partial y}{\partial u}\right)\sigma_{uv}^2$$

where σ_{uv}^2 is the *covariance* between the two variables. This is always assumed to be zero, i.e. the result of one measurement is not influenced by the other; therefore, the result to use is

$$\sigma_y^2 = \left(\frac{\partial y}{\partial u}\right)_v^2 \sigma_u^2 + \left(\frac{\partial y}{\partial v}\right)_u^2 \sigma_v^2. \qquad (13.20)$$

If there are more than two variables, the extra terms are added in the same way;

$$\sigma_y^2 = \left(\frac{\partial y}{\partial u}\right)_{v,w}^2 \sigma_u^2 + \left(\frac{\partial y}{\partial v}\right)_{u,w}^2 \sigma_v^2 + \left(\frac{\partial y}{\partial w}\right)_{u,v}^2 \sigma_w^2 + \cdots. \qquad (13.21)$$

Notice that all the terms are positive because each derivative is squared. The standard deviation σ_y is the square root of the final answer.

Example (v) The gas law states that $p = nRT/V$, and the volume, temperature, and number of moles have been measured, each with their standard deviations. The standard deviation of the pressure is found by taking the partial derivatives of each variable in turn, while holding the others constant and substituting into equation (13.21). The variables are $n \equiv u$, $t \equiv v$, $V \equiv w$,

$$\begin{aligned}\sigma_p^2 &= \left(\frac{\partial p}{\partial n}\right)^2 \sigma_n^2 + \left(\frac{\partial p}{\partial T}\right)^2 \sigma_T^2 + \left(\frac{\partial p}{\partial V}\right)^2 \sigma_V^2 \\ &= \left(\frac{RT}{V}\right)^2 \sigma_n^2 + \left(\frac{nR}{V}\right)^2 \sigma_T^2 + \left(-\frac{nRT}{V^2}\right)^2 \sigma_V^2\end{aligned}$$

which can be simplified by factoring out $(R/V)^2$. The *relative* or fractional uncertainty is found by dividing this result by p^2 producing, in this case, the simpler result,

$$\frac{\sigma_p^2}{p^2}=\left(\frac{1}{n}\right)^2\sigma_n^2+\left(\frac{1}{T}\right)^2\sigma_T^2+\left(-\frac{1}{V}\right)^2\sigma_V^2.$$

Example (vi) The vapour pressure of an organic liquid can be expressed in the form $\ln(p)=\dfrac{m}{T}+c$ and the values of the constants m and c were obtained from a least squares analysis of a plot of log pressure vs $1/T$. The gradient produced $m=-5390\pm 33$ K and the intercept $c=21.89\pm 0.099$. To calculate the liquid's normal boiling temperature, i.e. the boiling temperature when $p=760$ torr or 1 atm pressure, the equation must be rearranged to,

$$T=\frac{m}{\ln(760/1)-c}$$

making $T=353.29$ K. The pressure is written as 760/1 as a reminder that the log must be dimensionless. Using equation (13.21) the error in this determination can be calculated with $\sigma_m=33$ and $\sigma_c=0.099$ and gives

$$\sigma_T^2=\left(\frac{\partial T}{\partial m}\right)^2\sigma_m^2+\left(\frac{\partial T}{\partial c}\right)^2\sigma_c^2$$

$$=\left(\frac{1}{\ln(760)-c}\right)^2\sigma_m^2+\left(\frac{m}{\ln(760)-c)^2}\right)^2\sigma_c^2$$

and evaluating produces $\sigma_T=3.15$ K. The final answer produces a boiling temperature $=353\pm 3$ K. The Maple calculation, which also performs the differentiation, is

```
> T:= m/(ln(p)-c);
  sigTsqrd:= diff(T,m)^2*sigm^2 + diff(T,c)^2*sigc^2;
  sigT:= sqrt(sigTsqrd):
  p:= 760.0;
  m:= -5390;    sigm:= 33;
  c:= 21.89;    sigc:= 0.099;
  'boiling temperature':= T;
  sig:= sigT;
```

$$\textit{boiling temp} := 353.28 \qquad \textit{sig} := 3.1517$$

Example (vii) Continuing with the last example, suppose that the vapour pressure is required and that the temperature has been measured as 353 ± 3 K, and again $m=-5390\pm 33$ K and $c=21.89\pm 0.099$. In this case, $p=e^{\frac{m}{T}+c}$ and the derivatives produce

$$\sigma_p^2=\left(\frac{\partial p}{\partial T}\right)^2\sigma_T^2+\left(\frac{\partial p}{\partial m}\right)^2\sigma_m^2+\left(\frac{\partial p}{\partial c}\right)^2\sigma_c^2$$

$$=\left[\left(-\frac{m}{T^2}\right)^2\sigma_T^2+\left(\frac{1}{T}\right)^2\sigma_m^2+\sigma_c^2\right]p^2$$

and working out the terms produces a standard deviation of 141 torr and a pressure of 750 torr. The resulting standard deviation might seem unusually large, but this is caused by the exponential nature of the pressure equation having a great sensitivity to temperature.

13.4.1 Error propagation formulae

The following table gives some examples of frequently met functions. The variances σ_u^2 and σ_v^2 are assumed to be known. The total variance σ^2 is shown; remember to take the square root before using and note that σ^2 is always positive. The equation to use for many variables is

$$\sigma_y^2=\sum_i\left(\frac{\partial y}{\partial u_i}\right)^2\sigma_i^2 \tag{13.22}$$

Table 13.4 Error propagation formulae

$f(u, v)$		σ_f^2 variance
$u+v$	$u-v$	$\sigma_u^2+\sigma_v^2$
uv		$v^2\sigma_u^2+u^2\sigma_v^2$
$\dfrac{u}{v}$		$\dfrac{v^2\sigma_u^2+u^2\sigma_v^2}{v^4}$
$\dfrac{1}{u}+\dfrac{1}{v}$	$\dfrac{1}{u}-\dfrac{1}{v}$	$\dfrac{v^4\sigma_u^2+u^4\sigma_v^2}{u^4v^4}$
e^{uv}		$(v^2\sigma_u^2+u^2\sigma_v^2)e^{2uv}$
$e^{u/v}$		$\left(\dfrac{v^2\sigma_u^2+u^2\sigma_v^2}{v^4}\right)e^{2u/v}$
$u\,e^{v}$		$(\sigma_u^2+u^2\sigma_v^2)e^{2v}$
$u\ln(v)$		$\left(\dfrac{v^2\sigma_u^2+u^2\sigma_v^2}{v^2}\right)\ln(v)^2$

13.4.2 **Matrix formulation**

Barlow (1989) demonstrates that equation (13.20) can be written in a matrix form, which does not seem to offer any advantage, but this becomes clearer for problems that are more complex. The matrix equation is

$$V = GV_{\sigma^2}G^T \tag{13.23}$$

where V_{σ^2} is a square matrix of variances and G a (Jacobian) matrix of partial derivatives. V is the matrix of the variances for each variable. The total variance is the sum of the terms in V, which may be calculated as

$$\sigma^2 = UV \tag{13.24}$$

where U is a row matrix where each term is 1. In example (vii) the matrix equation is written as

$$V=\begin{bmatrix}\dfrac{\partial T}{\partial m} & \dfrac{\partial T}{\partial c}\end{bmatrix}\begin{bmatrix}\sigma_m^2 & 0\\ 0 & \sigma_c^2\end{bmatrix}\begin{bmatrix}\dfrac{\partial T}{\partial m}\\ \dfrac{\partial T}{\partial c}\end{bmatrix}=\begin{bmatrix}\left(\dfrac{\partial T}{\partial m}\right)^2\sigma_m^2\\ \left(\dfrac{\partial T}{\partial c}\right)^2\sigma_c^2\end{bmatrix}$$

and then

$$\sigma^2=[1\quad 1]\begin{bmatrix}\left(\dfrac{\partial T}{\partial m}\right)^2\sigma_m^2\\ \left(\dfrac{\partial T}{\partial c}\right)^2\sigma_c^2\end{bmatrix}=\left(\dfrac{\partial T}{\partial m}\right)^2\sigma_m^2+\left(\dfrac{\partial T}{\partial c}\right)^2\sigma_c^2.$$

13.4.3 **Questions**

Full solutions are available at **www.oxfordtextbooks.co.uk/orc/beddard.**

Q13.1 In a teaching lab experiment to measure the lattice spacing of graphite, the angle at which the electrons diffract is measured as their accelerating voltage V is changed. By the de Broglie relationship, the electron's wavelength is $\lambda = h/p$ where p is the momentum. The relationship between kinetic energy and accelerating potential is $eV = p^2/(2m)$, hence, $p=\sqrt{2meV}$ where e is the charge on the electron and m its mass. Combining the wavelength and Bragg's law $n\lambda = 2d\sin(\theta)$, produces the equation for the (first-order diffraction) inter-planar spacing d,

$$d = \frac{h}{\sin(\theta)\sqrt{8emV}}$$

In the experiment, the angle θ and the voltage are measured; all the other terms are constants. If the error on the angle θ is σ_θ and on the voltage σ_V, what is the variance on d? Express the result in terms of d^2.

Strategy: Equation (13.20) gives $\sigma_d^2 = \left(\frac{\partial d}{\partial \theta}\right)^2 \sigma_\theta^2 + \left(\frac{\partial d}{\partial V}\right)^2 \sigma_v^2$. After differentiating substitute d into the result to simplify it.

Q13.2 The electrode potential E of a reaction $A_{ox} + ne^- \to A^-_{red}$ is measured to obtain the ratio of activities of oxidized to reduced species, $Q = a_{(red)}/a_{(ox)}$ via the Nernst equation,

$$E = E^{\circ} - \frac{RT}{nF}\ln\left(\frac{a_{red}}{a_{ox}}\right).$$

(a) If the standard deviation of the measured potential is σ_E and σ_T on the temperature, calculate how these affect the measured ratio of activities. Simplify the answer.

(b) At a given temperature, where will the standard deviation be smallest?

Q13.3 Barlow (1989) gives an example of errors in a tracking chamber measured in cylindrical polar coordinates, and how they are related to errors in x, y, and z. This question examines a similar situation. In a molecular beam experiment, molecules are photo-dissociated and the image of the fragments, ions, or electrons is observed on a CCD camera at angle θ and distance r from the origin at the centre of the detector, Fig. 13.6. This is also in the x–y plane. There is an error in both θ and r because the laser source has a certain size, and dissociation occurs from a finite volume in space, and because the molecules are themselves moving.

Calculate the error in x and y on the surface of the CCD camera.

The cylindrical polar coordinates have to be converted to Cartesians and the conversions are $x = r\cos(\theta)$, $y = r\sin(\theta)$, and $z = z$. As the distance to the detector is large and fixed, assume that there is no error in z, and ignore this in the calculation.

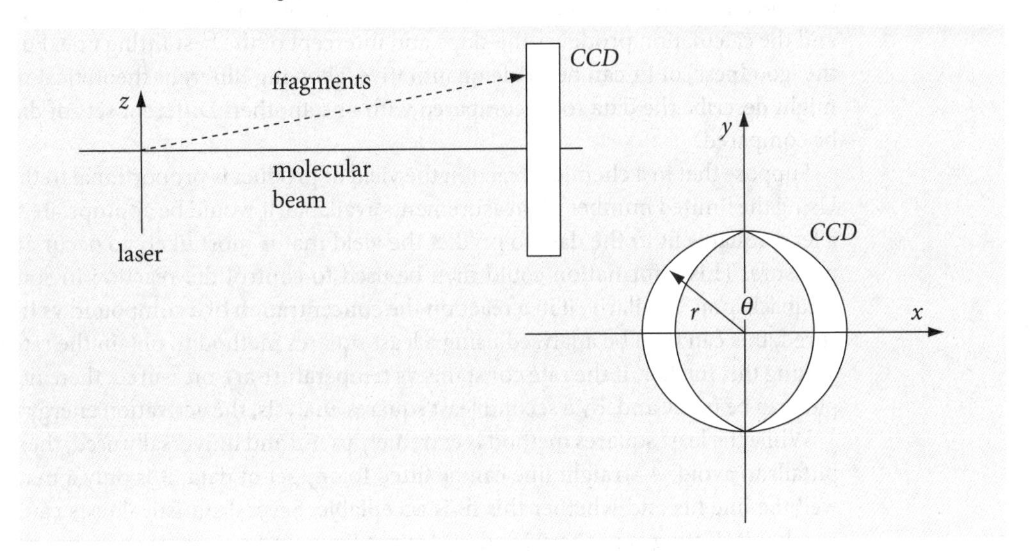

Fig. 13.6 Geometry of a molecular beam photo-dissociation experiment and a sketch of an image as it might appear on a CCD camera. The oval shape is a consequence of the polarization of the absorption and rate of dissociation.

Strategy: Write down the $\boldsymbol{G}$ matrix, which is also the Jacobian transformation between Cartesian and polar coordinates (see Chapter 4.11), and multiply out the matrices.

13.5 Modelling data

13.5.1 Concept

In many situations, the purpose of an investigation is to obtain an equation (regression model) that can be used to predict the value of one variable by knowledge of others.

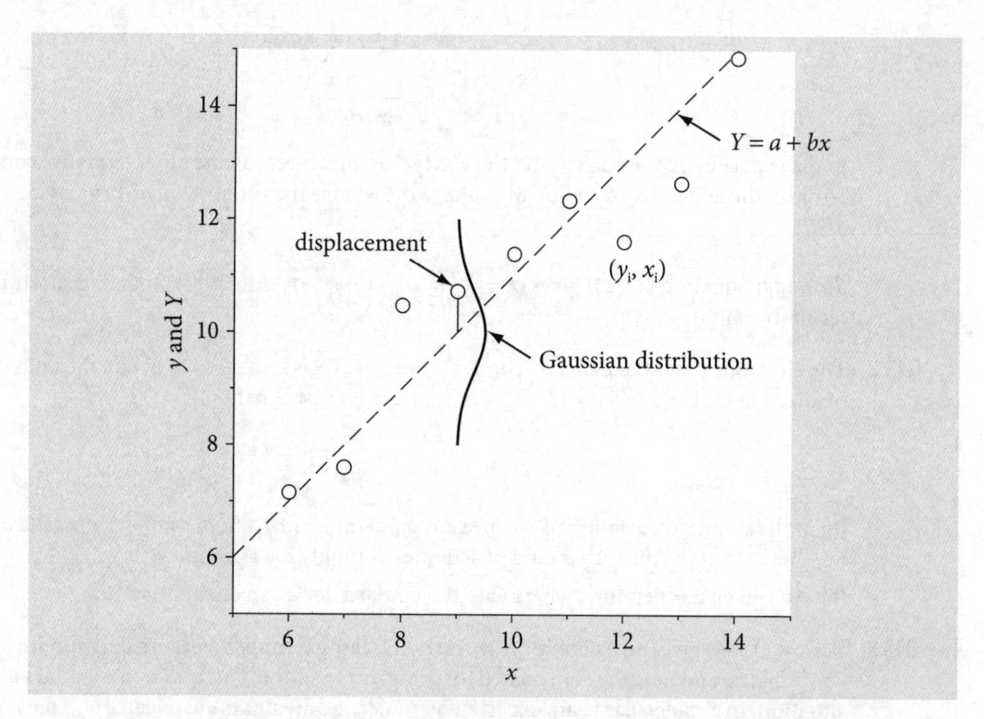

Fig. 13.7 The displacements are assumed to be Gaussian distributed. The line is Y, the experimental points, y_i. The square of all the displacements is minimized.

Probably the most useful and most applied method of doing this is the least squares method. The basic idea is shown in Fig. 13.7, where the square of the y displacement from the line to each data point, is minimized. The minimization is only along y because it is assumed that the x values are known exactly. By minimizing these displacements with an assumed model of the data, a straight line for example, the best fit to the data is obtained, and the calculation produces the slope and intercept of the best fitting line. Furthermore, the 'goodness' of fit can be made quantitative, allowing different theoretical models that might describe the data to be compared with one another. Different sets of data can also be compared.

Suppose that in a chemical reaction the yield of product is proportional to the pressure. Using the limited number of measurements available, it would be appropriate to calculate a least squares fit to the data to predict the yield that is most likely to occur at any given pressure. This information could then be used to control the reaction in some form of feedback loop. Similarly, if in a reaction the concentration of a compound vs time is measured, this can then be analysed using a least squares method to obtain the rate constant. Taking this further, if the rate constants vs temperature are measured, then an Arrhenius plot can be made and, by a second least squares analysis, the activation energy obtained.

While the least squares method is extremely useful and universally used, there are some pitfalls to avoid. A straight line can be fitted to *any* set of data; it is only a matter of how well the line fits and whether this fit is acceptable. Several statistical tests can be used to check this. Fitting is not restricted to straight lines, and quadratic or higher polynomials can be used, as can exponentials or sine and cosines. The function used should always be based on the underlying science. Any set of data could fit equally well to several different functions and the more complicated the function, a polynomial or sum of several exponentials for example, the larger the number of variable parameters will be and the better the fit is going to be. Hence the quip, 'with enough parameters you can fit the shape of an elephant'! However, the fit may describe the data but have no relationship at all to the science, in which case nothing has been achieved because the parameters obtained have no meaning.

Another pitfall is to make false correlation between observables. The reading age of children shows a very good linear relationship to their shoe size, but to suggest that a child with large shoes must be good at reading is nonsense. The obvious correlation is that older children are generally better at reading than younger ones. A less obvious relationship is found in the rate of the rotational diffusion D of molecules in the same solvent. This can be measured by observing a molecule's fluorescence through polarizers. A good linear correlation is found between D and the reciprocal of the molecular mass. This is, however, false. The true correlation is with molecular volume V, which for a limited class of molecules, such as aromatics or dye molecules, has a similar proportionality to mass. The Stoke–Einstein equation $D = k_BT/(6\eta V)$, where η is the viscosity, shows that molecular volume is the important quantity. The D vs reciprocal mass correlation can now easily be tested; an iodo-derivative has a far greater mass but not much greater volume as the protonated molecule. Remember that a strong observed correlation between the variables does not imply a *causal* relationship.

The final thing to look out for when using least squares is outliers. These are data points well away from the trend indicated by other points. It may be argued that these can be ignored as being due to faulty experimentation, but if this is not the case these points have to be dealt with and in Section 13.3.6 such a test was described. The least squares method is inherently very sensitive to these points, because the square of the deviation is used. It is often quite clear that the line does not fit the data and is pulled away from what one would expect to see as the fit; Fig. 13.15. In this case, a more robust method such as least absolute deviation is required, Section 13.6.8, without removing data points.

13.5.2 The least squares calculation for a straight line

Suppose that the straight line

$$Y = a_0 + b_0 x$$

is proposed to describe the data. In a least squares analysis, the test is to determine whether the experimental data y follows the equation

$$y_i = a_0 + b_0 x_i + \varepsilon_i \qquad i = 1, 2, \cdots, n$$

where ε is a random error with a mean of zero and standard deviation of σ. The least squares method produces the best constants a_0 and b_0 that describe the data, in the sense that the Y values calculated are the most probable values of the observations. This is based on the assumption that the data are Gaussian (normally) distributed as is expected to be the case from the central limit theorem.

What the least squares method does is to minimize the square of the displacement between the values calculated from a 'model' function and the experimental data points y. Figure 13.7 shows the displacement for one point and a Gaussian distribution from which that point could have been produced. The statistic used to assess the goodness of fit is called 'chi squared' χ^2 and is defined as

$$\chi^2 = \sum_{i=1}^{n} w_i (y_i - Y_i)^2 \qquad (13.25)$$

where y is the experimental data, Y the model set of estimated data, and w the weighting. The ideal weighting is $w_i = 1/\sigma_i^2$. The χ^2 forms a distribution and the chance that a certain value can be obtained is calculated in a similar way as for the normal or t distributions, see Section 13.5.4.

On the basis that the deviation of each experimental data point from its true mean value is normally distributed, the probability of observing the y_i data points is the product of individual normal distributions, which can be written as

$$P = \left(\frac{h}{\sqrt{\pi}}\right)^n \exp\left(-h^2 \sum_{i=1}^{n} w_i (y_i - Y_i)\right).$$

The most likely values are obtained when this probability is at its maximum, and this is found when $\sum_i (y_i - Y_i)^2 w_i$ is at its minimum. This is the same as minimizing the χ^2 and therefore this is used as a measure of the 'goodness of fit' of the model function to the data. The minimum χ^2 is found by differentiating this with respect to each of the parameters in the model function. This approach is quite general and is called a Maximum Likelihood method.

To fit the straight-line model $Y = a_0 + b_0 x$ to experimental data y_i, the values a and b obtained will be the best estimates of a_0 and b_0 and therefore these are replaced with a and b in the equations. To find the minima, the derivatives $\partial\chi^2/\partial a$ and $\partial\chi^2/\partial b$ are calculated,

$$\frac{\partial}{\partial a}\sum_{i=1}^{n}(y_i - a - bx_i)^2 w_i = -2\sum_{i=1}^{n}(y_i - a - bx_i)w_i = 0 \tag{13.26}$$

$$\frac{\partial}{\partial b}\sum_{i=1}^{n}(y_i - a - bx_i)^2 w_i = -2\sum_{i=1}^{n}x_i(y_i - a - bx_i)w_i = 0 \tag{13.27}$$

which produce two equations and two unknowns; these simultaneous equations are known as the *normal equations*;

$$a\sum_{i=1}^{n} w_i + b\sum_{i=1}^{n} x_i w_i = \sum_{i=1}^{n} y_i w_i, \qquad a\sum_{i=1}^{n} x_i w_i + b\sum_{i=1}^{n} x_i^2 w_i = \sum_{i=1}^{n} y_i x_i w_i \tag{13.28}$$

The best estimate of the slope b is

$$b = \frac{S_{xy}}{S_{xx}} \tag{13.29}$$

where

$$S_{xy} = \sum_i x_i y_i w_i - \frac{\sum_i x_i w_i \sum_i y_i w_i}{\sum_i w_i}, \qquad S_{xx} = \sum_i x_i w_i - \frac{\left(\sum_i x_i w_i\right)^2}{\sum_i w_i} \tag{13.30}$$

$$S_w = \sum_i w_i \tag{13.31}$$

The best estimate of the intercept a is

$$a = \langle y\rangle - b\langle x\rangle \tag{13.32}$$

where the average of x and y are $\langle x\rangle = \dfrac{\sum_i x_i w_i}{\sum_i w_i}$ and $\langle y\rangle = \dfrac{\sum_i y_i w_i}{\sum_i w_i}$. This means that the fitted line goes through the 'centre of gravity' of the data.

The intercept is also given by

$$a = \frac{\sum w_i x_i^2 \sum w_i y_i^2 - \sum w_i x y_i \sum w_i x_i}{S_w S_{xx}} \tag{13.33}$$

Most graphing packages have least squares fitting routines, but the calculation is also made easy by direct calculation and then weighting can be incorporated and confidence curves drawn. Because the differences between two large sums often occur in calculating terms such as S_{xy} and S_{xx}, the possibility of rounding errors can be significant. It is always advisable, therefore, to use a higher precision calculation than would normally be used.

The following data is analysed.

x	200	220	240	260	280	300	320	340	360	380
y	36.2	42.7	44.9	51.8	57.7	60.9	64.4	68.2	76.4	80.1
σ	1.5	1.1	1.8	0.3	2.0	0.9	1.2	1.6	1.9	0.9

The data should be put into three columns, x, y, σ (standard deviation) when read into Maple using Algorithm 13.2. The calculation uses the equations just derived and produces the regression equation $y = -9.27 + 0.234x$ and with 95% confidence limits, $a = -9.27 \pm 3.55$ and $b = 0.234 \pm 0.0128$ which is shown in Fig. 13.8. The residuals are shown in Fig. 13.9. The straight line of the best fit is defined as

```
line:= x-> slope*x + intercept:
```

Equations (13.31)–(13.33) are used to calculate the slope and intercept but the equations for the confidence limits are taken from Hines & Montgomery (1990, chapter 14). These equations are in the algorithm as `C_slope` and `C_intercept` and are given as equations (13.36) and (13.37). The mean square error, `mse` in the calculation, is the reduced χ^2, thus $m_{se} = \dfrac{\chi^2}{n-2}$ and χ^2 is calculated with equation (13.25). The confidence lines are only valid in the range of the data and are extended to zero only to show the large error on the intercept which is caused by the large extrapolation necessary to reach zero.

Algorithm 13.2 Weighted least squares $y = a + bx$

```
> restart: with(plots):with(Statistics):
  Digits:= 30:                     # ensure high precision
> file:="test data.txt":           # 3 columns of data
  filepos(file,0):                 # set to start of file
> data:= readdata(file,3):
  close(file):
  n:= nops(data):                  # get number of data points
  xval:= Vector(n):  yval:= Vector(n):  w:=    Vector(n):
  for i from 1 to n do             # extract data
    xval[i]:=  data[i,1];
    yval[i]:=  data[i,2];
    w[i]:=     1.0/data[i,3]^2;    #make 1 for unweighted data
  end do:                          # end of data input
  Sw:=   add( w[i], i=1..n):                      #  Σw
  Sxw:=  add( xval[i]*w[i], i=1..n):              #  Σx
  Syw:=  add( yval[i]*w[i], i=1..n):              #  Σy
  Sxxw:= add( xval[i]^2*w[i], i=1..n):            #  Σx²
  Syyw:= add( yval[i]^2*w[i], i=1..n):            #  Σy²
  Sxyw:= add(xval[i]*yval[i]*w[i],i=1..n):        #  Σxy
  xbar:= add(xval[i]*w[i],i=1..n)/Sw;             #  <x>
  ybar:= add(yval[i]*w[i],i=1..n)/Sw;             #  <y>
  Sxx:= Sxxw - Sxw^2/Sw:
  Syy:= Syyw - Syw^2/Sw:
  Sxy:= Sxyw - Sxw*Syw/Sw:
  slope:= Sxy/Sxx;                              # slope
  intercept:= ybar-slope*xbar;                  # intercept
  line:= x-> slope*x + intercept:               # calculate fit

# mse = sum of residuals squared /(n-2) this is reduced χ²
  mse:=(Syy - slope*Sxy)/(n-2);
  std_dev_slope:= sqrt(mse/Sxx):
  std_dev_intercept:= sqrt(mse*(1/Sw+xbar^2/Sxx)):
  printf("%s %g \n", "std dev of slope = ", std_dev_slope):
  printf("%s %g \n" ,
              "std dev of intercept = ", std_dev_intercept):
  cov:= -mse*xbar/(Sxx*Sw):
# z is 95 % confidence line
  prec:= 0.975:                          #a/2 95% confidence
  t:= Quantile(StudentT(n-2), prec):
# calculate confidence limit on slope,
# z is equation of confidence line
  C_slope:= t*sqrt(mse/Sxx):
  C_intercept:= t*sqrt(mse*(1/Sw+xbar^2/Sxx)):
  z:=  x-> t*sqrt( mse*(1/Sw  + (x-xbar)^2/Sxx) ):
  printf("%s %g %s %g %s %g %s\n",
         " slope      =",slope , "+- ", t*sqrt(mse/Sxx),
         "at", 100-(1-prec)*200, "%");
  printf("%s %g %s %g %s %g %s", " intercept =",intercept ,
         "+- ",  t*sqrt(mse*(1/Sw+xbar^2/Sxx)),
         "at",100-(1-prec)*200, "%");

                  std dev slope = 0.00557
                 std dev intercept = 1.54
             slope     = 0.234 +- 0.0128 at 95 %
             intercept = -9.27 +- 3.55   at 95 %
```

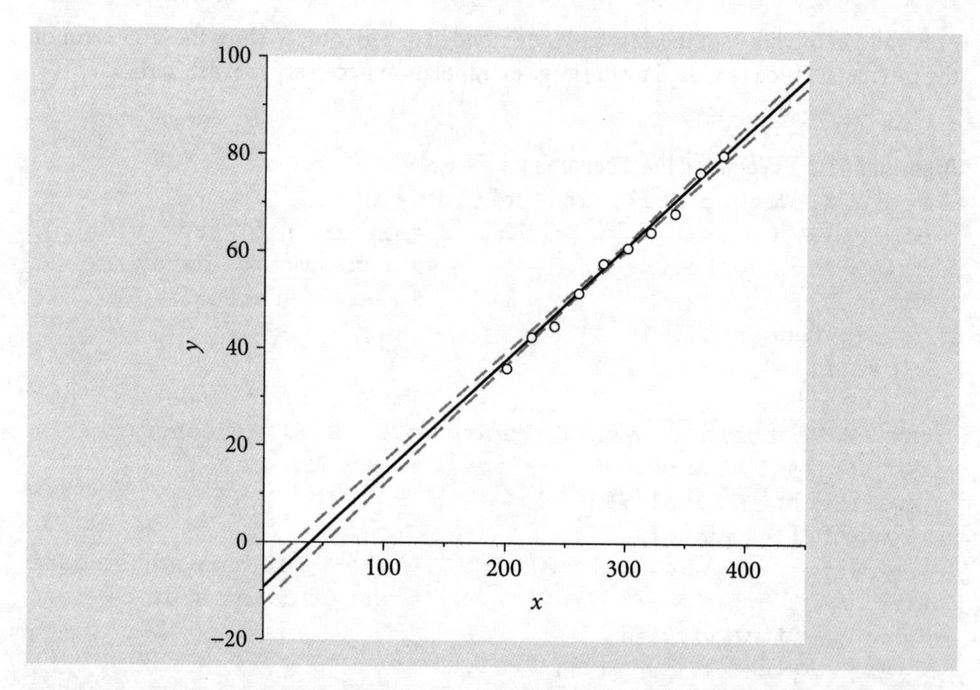

Fig. 13.8 Least squares fit to $y = a + bx$ and 95% confidence lines. These are projected to zero so that the large error on the intercept can be seen but the lines are only valid in the range of the data.

```
# plot data and confidence lines,
# line1 is least squares line
> ss:= seq([xval[i],yval[i]],i=1..n):
  p1:= plot([ss],style=point, symbol=circle,symbolsize=20):
  p2:= plot(
      [line(x)+z(x),line(x)- z(x),line(x)],x= 0..450):
  display(p1,p2,view=[0..450,-20..100]);
```

13.5.3 Residuals

After a fit has been obtained, the emphasis falls not on the plot of the data but on analysing the residuals, a plot of which shows the difference between the data and the fitted line. The residuals are calculated for each data point as

$$\varepsilon_i = (y_i - Y_i)$$

where Y_i is the value of the calculated line at the ith x value. The reduced or normalized residuals

$$r_i = (y_i - Y_i)/Y_i \tag{13.34}$$

are often the best to use, particularly if the data varies in size, as may be the case for exponential data. The reduced residuals between the calculated line and the data are shown in Fig. 13.9 and should be randomly distributed about zero if the fit is good, which it would appear to be.

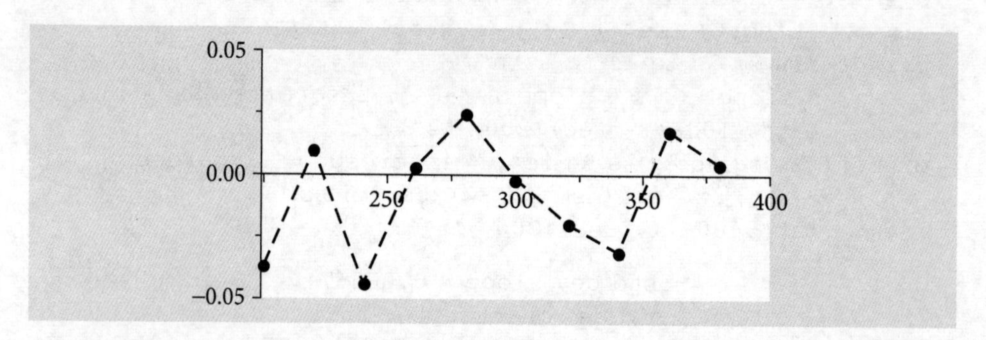

Fig. 13.9 Normalized or reduced residual plot

```
# calculate and plot residuals
> resid:= seq(
  [ xval[i], (yval[i]- line(xval[i]))/line(xval[i]) ]
                                                    ,i=1..n):
  plot([resid], view=[190..400,-0.05..0.05]);
```

13.5.4 Chi squared, χ^2

A good measure of the overall goodness of fit is the χ^2 parameter. This measures the dispersion between the experimental data and the fitted function. If the data is normally distributed then the χ^2 is expected to be equal to the number of degrees of freedom; this is the number of data points less the constraints, two in the case of a linear fit since there are two parameters. (The reduced χ^2 is the same quantity as `mse` used in Algorithm 13.2). The reduced χ^2 should have a value close to 1 if the data is fitted well, and the probability of obtaining this value is 50% since half of the time the χ^2 should exceed the norm. Values are often either very small, which can indicate that the weighting is too small because the standard deviations used are too big, or very large because the model does not fit the data. If the data is Poisson distributed, then the standard deviation is known exactly and the χ^2 can be used quantitatively, otherwise, it can only be used loosely and a probability of 10% or larger is usually acceptable. The Maple instructions to calculate χ^2 are

```
chi_sqr:= add(w[i]*(yval[i]-line(xval[i]))^2,i=1..n)/(n-2);
```

$$\chi sqr := 0.843$$

```
f:= 1 - CDF( ChiSquare(n-2), chi_sqr*(n-2));
```

$$f := 0.564$$

and this means that the probability of obtaining a reduced χ^2 *greater* than 0.84 is 56% for the nine degrees of freedom of the data in Fig. 13.8. The `1- CDF(ChiSquare(..), ..)` is the cumulative distribution function that gives the probability of observing a value greater than the given χ^2. It is the area between χ^2 and infinity on a plot of the distribution. The χ^2 is often tabulated; see Bevington (1969). The χ^2 distribution for n points is

$$f_{\chi^2}(x, n) = \frac{1}{2^{n/2}\Gamma(n/2)} x^{n/2-1} e^{-x/2}$$

where Γ is the gamma function. Because the distribution cannot be less than zero, it is not symmetrical and has a shape skewed towards small values. The chance that the χ^2 is significant can be also calculated using the incomplete gamma function (Prest et al. 1986). In Maple, this calculation is

```
Q:= GAMMA( (n-2)/2, chi_sqr/2 );                    (13.35)
```

$$Q := 5.99$$

If $Q > 0.1$, which it is in this instance, the model definitely describes the data. If $Q > 0.001$ then it probably describes the data but, if less than this, it definitely does not, and a different model may be necessary. This test of goodness of fit is very useful when the weighting for the data is not known, because the χ^2 cannot be used quantitatively.

Other tests can be performed on the residuals to assess the goodness of fit.

(i) The simplest test is to look at the residuals; if they slope, oscillate, or are curved then the model does not fit the data, no matter what the statistics indicate.

(ii) The residuals can be plotted on a normal probability plot and if they are Gaussian (normally) distributed, a straight line is produced.

(iii) The number of positive and negative runs in the residuals can be calculated. A 'run' occurs when consecutive residuals have the same sign, which is unlikely to occur if they are random. Therefore, if the data is random, then there are an equal number of small runs of positive and of negative numbers.

(iv) The autocorrelation of the data should be 1 for the first point, and randomly arranged about zero for the rest, which is expected for a sequence of random numbers.

(v) A scedaticity plot aims to determine if the residuals vary with the size of the data itself, i.e. if the errors are larger when the data value is larger or vice versa. The plot is of residuals vs the experimental y value. The points should be randomly distributed about zero.

A good model will produce a small residual (error) variance and if there are two or more competing models then the model with the smallest error variance should be chosen. The variance of the slope and intercept are calculated by expressing the respective equations as functions of y_i (the experimental data) and using the law of propagation (combination) of errors. The gradient is the ratio $b = S_{xy}/S_{xx}$ and only the numerator depends on y_i. Using the relationships

$$S_{xy} = \sum_{i=1}^{n} w_i(x_i - \langle x \rangle)y_i, \qquad S_{xx} = \sum_{i=1}^{n} w_i(x_i - \langle x \rangle)^2,$$

Hines & Montgomery (1990) give the result

$$\sigma_b^2 = \frac{\sigma^2}{S_{xx}}, \tag{13.36}$$

and a related calculation for the intercept produces

$$\sigma_a^2 = \sigma^2\left(\frac{1}{S_w} + \frac{\langle x \rangle^2}{S_{xx}}\right). \tag{13.37}$$

To use these equations it is necessary to obtain an estimate of the variance σ^2. For the simple linear model an *unbiased* estimate of the variance of the error ε_i is given by the reduced $\sigma_\varepsilon^2 \equiv \chi^2/(n-2)$. This expression can be rewritten in terms of quantities already calculated (Hines & Montgomery 1990) and is called the mean square error mse in Algorithm 13.2.

$$\sigma_\varepsilon^2 = \frac{S_{yy} - bS_{xy}}{n-2}.$$

In this case

$$\sigma_b^2 = \frac{\sigma_\varepsilon^2}{S_{xx}}, \qquad \sigma_a^2 = \sigma_\varepsilon^2\left(\frac{1}{S_w} + \frac{\langle x \rangle^2}{S_{xx}}\right). \tag{13.38}$$

with values 0.00556 and 1.54 respectively for the data in Fig. 13.8.

Alternative estimates of the variance in the coefficients are quoted by Bevington & Robinson (2003).

$$\sigma_a^2 = \frac{\sum_i w_i x_i^2}{S_w S_{xx}} \qquad \sigma_b^2 = \frac{1}{S_{xx}}. \tag{13.39}$$

These variances are calculated by using the equation for error propagation, equation (13.20), differentiating with respect to each y_i. However, these equations should not be used unless the weightings are known, which they are for photon counting experiments. If weightings are unknown and a constant weighting used instead, incorrect results are obtained because both expressions (13.39) now depend only on x.

The covariance between the slope and intercept can also be calculated. If this is large then the slope and intercept are not independent of one another. The covariance for the data used above is -5×10^{-4} which is very small. When several parameters are being estimated such as in a polynomial or nonlinear, least squares calculation then the covariance between pairs of parameters should be examined and is often reported as a matrix of value. Ideally, each values should be small. The covariance for the linear least squares is $\text{cov} = -\sigma_\varepsilon^2\langle x \rangle/(S_{xx}/S_w)$.

13.5.5 **Confidence intervals**

The confidence intervals about any data point, and hence the whole set, can be obtained from the data. The width of these lines at a given confidence level, 95% is typical, is a measure of the overall quality of fit to the data. As the errors ε_i are assumed to be normally distributed and independent of one another, then the slope has the confidence interval (Hines & Montgomery 1990)

$$b_0 = b \pm t_{\alpha/2}\sqrt{\frac{\sigma_\varepsilon^2}{S_{xx}}} \tag{13.40}$$

which gives $b_0 = 0.234 \pm 0.0128$ and for the intercept where $S_w = \Sigma w_i$

$$a_0 = a \pm t_{\alpha/2}\sqrt{\sigma_\varepsilon^2\left(\frac{1}{S_w} + \frac{\langle x\rangle^2}{S_{xx}}\right)} \tag{13.41}$$

which has a value $a_0 = -9.27 \pm 3.55$ or -9 ± 4; see Fig. 13.8.

The confidence for the mean point can be constructed and this is also called the confidence line z for the regression curve. It has the following form (Hines & Montgomery 1990)

$$z = y \pm t_{\alpha/2}\sqrt{\sigma_\varepsilon^2\left(\frac{1}{S_w} + \frac{(x - \langle x\rangle)^2}{S_{xx}}\right)} \tag{13.42}$$

and these lines are shown in Fig. 13.8. The two curves have a minimum width at $\langle x\rangle$ and widen either side of this. Strictly, they are not valid outside the range of the data. Prediction lines can be constructed to project confidence limits past the data and these lines are slightly wider than the confidence lines. The prediction lines are produced by replacing $1/S_w$ by $1 + 1/S_w$ in equation (13.42). This takes into account the error from the model and that associated with future observations. The two sets of curves are almost identical with the particular set of data used in Fig. 13.8.

13.5.6 The analysis of variance (ANOVA) table

When fitting models to data, it is common practice to test the fit to see how well the model equation used fits the data. This is done by partitioning the total variability in the calculated Y into a sum that is 'explained' by the model and a 'residual' or error term unexplained by the regression as shown in the equations,

$$\sum_i (y_i - \langle y\rangle)^2 = \sum_i (Y_i - \langle y\rangle)^2 + \sum_i (y_i - Y_i)^2,$$

$$S_{yy} = SS_R + SS_e.$$

The terms on the right are, respectively, the sum of squares 'explained' by regression SS_R and the residual (error) sum of squares SS_e. Recall that y_i are the data points and Y_i the fitted points. The left-hand summation is

$$S_{yy} = \sum_i y_i w_i - \frac{\left(\sum_i y_i w_i\right)^2}{\sum_i w_i}.$$

and the right-hand one is

$$\sum_i (y_i - Y_i)^2 = \sigma_\varepsilon^2 (n-2)$$

which is $SS_e = S_{yy} - S_{xy}^2/S_{xx}$; the middle term SS_R can be found from the difference. The purpose of the table is to generate a statistic F_0 that can be tested against tables of the f distribution. The ANOVA table is constructed as

Source of variation	Sum of squares *ss*	Degrees of freedom *df*	Mean square *ss*/*df*	F_0
Regression	S_{xy}^2/S_{xx}	1	MS_R (regression)	MS_R/MS_e
Residual error SS_e	$S_{yy} - S_{xy}^2/S_{xx}$	$n-2$	MS_e (error)	
Total	S_{yy}	$n-1$		

The data being considered (Fig. 13.8) produces the following table

Source of variation	Sum of squares *ss*	Degrees of freedom *df*	Mean square *ss*/*df*	F_0
Regression	1499	1	1499	1778
Residual error SS_e	6.7	8	$6.7/8 = 0.84$	
Total	1506	9		

The ANOVA table also leads to a test of 'significance' for the model. It can be shown that if the model has no utility, i.e. if the model linking x to y does not differ 'significantly' from zero, then the statistic

$$F_0 = \frac{\text{Mean Square for Regression}}{\text{Mean Square for Error}}$$

should behave like a random variable from an F-distribution with 1 and $(n-2)$ degrees of freedom. (F-distributions are tabulated in many books on statistics but can be calculated using Maple, as shown below.) Values close to 1.0 support the null hypothesis that the model *does not* fit the data well, while large extreme values, exceeding some critical value from the upper tail of the F-distribution, lead to the conclusion that x has a significant effect upon y.

The null hypothesis, called H_0, is one where model does *not* fit the data, versus H_1 where it does fit the data. To test this, the calculated $F_0 = 1778$ greatly exceeds the critical value 11.26 at the 1% level calculated from the F-distribution $f(0.01, 1, n-2)$. This suggests that x has a significant effect upon y. The value of F_0 is calculated using Maple as

```
>with(Statistics): Quantile( FRatio(1,8),0.99);
```

The result of the ANOVA table is hardly surprising because the data is clearly well described by a straight line.

13.5.7 Correlation coefficients

The correlation coefficient is often listed among the parameters when a spreadsheet or graphing package is used to analyse a straight line fit. It represents the proportion of the variation (information) in y that can be accounted for by its relationship with x. However, in the physical and many of the biological sciences, this is a not a useful quantity and should be avoided as a measure of how well a straight line describes the data. The reason is that what would be considered as a good fit has a correlation coefficient of, for example, $R = 0.99$ and a poor fit perhaps a value of 0.98, which is so similar that it provides very poor discrimination between good and bad fits. The data of Fig. 13.8, produces a $\chi^2 = 0.84$ and $R = 0.997$. This poor discrimination is illustrated if sufficiently large random numbers are added to the data being considered and the least squares fitting repeated so that R decreases slightly to 0.98. However, the χ^2 has now increased to about 10 indicating that the model used is a very poor fit to the data.

If R is the sample correlation coefficient then $0 < R < 1$ and is defined as $R = b\sqrt{\dfrac{S_{xx}}{S_{yy}}}$ so that it is a constant multiplied by the gradient b and this is why it is a poor statistic. The constant is the 'spread' of the x values divided by that of the y.

13.5.8 Questions

Full solutions are available at **www.oxfordtextbooks.co.uk/orc/beddard.**

Q13.4 Calculate the equation for the slope and the variance for a weighted straight line fit through the origin.

Strategy: Generate, then solve, the normal equations for $y = bx$. Use the error propagation equation to calculate the variance and use $\sigma_i^2 = 1/w_i$ to simplify it.

Q13.5 Show that $S_{xx} = \sum_{i=1}^{n} w_i(x_i - \langle x\rangle)^2 = \sum_i w_i x_i - \dfrac{\left(\sum_i w_i x_i\right)^2}{\sum_i w_i}$.

Q13.6 Starting with the error propagation formula calculate

(a) the error in the slope equation (13.39); the slope is $b = S_{xy}/S_{xx}$.

(b) the error in the intercept, $a = \langle y\rangle - b\langle x\rangle$.

It helps to use $S_{xy} = \sum_{i=1}^{n} w_i(x_i - \langle x\rangle)y_i$ and $S_{xx} = \sum_{i=1}^{n} w_i(x_i - \langle x\rangle)^2$.

Strategy: Differentiate with respect to $y_i = 1, 2, 3, \cdots$ which will produce a series of terms. Square these, multiply by the variance for each y_i which is $1/w_i$ and finally sum. The general error propagation equation is $\sigma_b^2 = \sum_i \left(\frac{\partial b}{\partial y_i}\right)^2 \sigma_i^2$.

Q13.7 The following data were obtained in an experiment to determine the lattice spacing d in graphite by electron diffraction. The sine of the diffracted angle vs $1/\sqrt{V}$ is given in the table where V is the acceleration voltage applied to the electrons. The equation describing the lattice spacing is $d = \dfrac{h}{\sin(\theta)\sqrt{8emV}}$; see Q13.1 ($e$ is the charge on the electron and m its mass. h is Planck's constant).
The x data is in volts$^{-1/2} \times 10^4$ and y is $\sin(\theta) \times 10^4$.

x	243	229	218	209	200	192	186	180	174	169	164	160	154	147
y	1140	1110	1060	1010	981	950	899	876	839	820	794	787	766	724

(a) Calculate the slope and the intercept and test whether the data goes through zero or whether there is a systematic error in the measurement of the angle.

(b) Calculate the lattice spacing.

Q13.8 The Stern–Volmer equation is used to analyse the quenching behaviour of electronically excited molecules S^*. The equation is derived from a steady state analysis of the rate equations given in Chapter 10.1.5 and Q10.10 and is

$$\frac{\varphi}{\varphi_Q} = 1 + k_{sv}[Q]$$

where φ is the fluorescence yield in the absence of quenchers and φ_q that with quencher concentration [Q]. The quenching constant is $k_{sv} = k_q\tau$ where k_q is the second-order rate constant for the step $S^* + Q \xrightarrow{k_q} S + Q$ and τ the excited state lifetime in the absence of quencher, $S^* \xrightarrow{1/\tau} S$.

(a) Use the following data, which has been measured from the quenching of quinine sulphate in acidic solution, and determine whether the intercept is 1 within error.

$[Q] \times 10^3$	1	2	3	4	5
φ/φ_Q	1.21	1.44	1.68	1.86	2.10

The standard deviation on each value is 0.02.

(b) Calculate the quenching rate constant if the excited state lifetime is 18 ± 1 ns.

Q13.9 A molecule may undergo simultaneous vibrational and rotational transitions leading to a spectrum with a characteristic shape, Fig. 13.10. A photon has 1 unit of angular momentum and this has to be conserved on absorption or emission. A *pure* vibrational transition therefore does not occur in a molecule because vibrational transitions do not involve angular momentum but instead both the rotational and vibrational energy of the molecule changes. These transitions occur in the infrared part of the spectrum and the rotational transitions occur to smaller (P branch) and to larger frequency (R branch) than the energy of the missing n_{00} vibrational transition. Because the energies of vibrational and rotational transitions are very different, 100 to 1000's cm^{-1} compared to 0.01 to 10 cm^{-1} respectively, to a first approximation, the vibrational and rotational energies of molecules can be regarded as independent, and their energies add.

$$E_{vib,\,rot} = E_{rot} + E_{vib}.$$

While it is clearly illogical to consider a vibrating molecule as a rigid rotator, it is, in fact, a good model because the period of a vibration is far shorter, ≈10 fs, than that of rotation ≈1 ps. Because of this, to a good approximation, the two types of motion are not coupled to one another and their energies can be added.

Normally, it is sufficient to consider a molecule to be a rigid rotator but to be vibrating anharmonically. In units of cm^{-1}, the energy is

$$F(n, J) = BJ(J+1) + (n+1/2)\tilde{\nu}_e - (n+1/2)^2 x_e \tilde{\nu}_e$$

The selection rules are $\Delta n = \pm 1$ for the vibration and $\Delta J = \pm 1$ for rotation, with the proviso that when $J = 0$, $\Delta J = 1$. For a diatomic molecule because $\Delta J = 0$ is not allowed when a photon is absorbed or emitted a change in vibrational energy is always accompanied by a change in rotational energy.

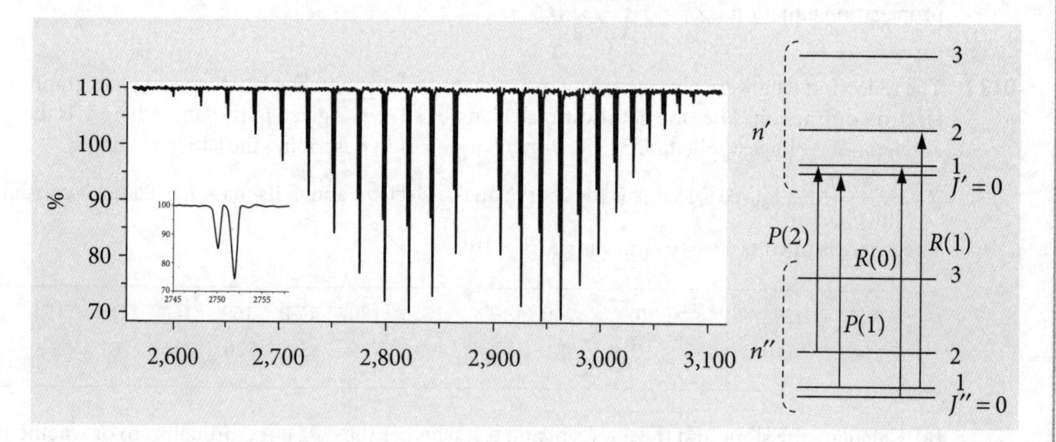

Fig. 13.10 HCL spectrum. The first two *P* and *R* branch transitions are shown on the right. Two vibrational levels are shown, each with a few of their rotational levels but not drawn to scale. The separation between vibrational levels should be far greater than shown. It is conventional to denote *J* in the upper vibrational state, i.e. higher value of *n*, by a single prime i.e. as *J′* and *J* in the *lower* vibrational state by double prime, *J″*.

(a) Split the spectrum into *P* and *R* branches and label with the quantum numbers. Explain the origin of the double lines for each transition. Lines arising from $\Delta J = -1$ are called the *P* branch, the missing line corresponding to $\Delta J = 0$ is called the *Q* branch, and lines arising from $\Delta J = +1$ are called the *R* branch.

(b) Explain why the line spacing is not constant across the spectrum.

(c) Using the equation, calculate the energies, $F(0, J'')$ and $F(1, J')$.

(d) Consider the fundamental band, i.e. the change $n = 0 \rightarrow n = 1$ and calculate the transition frequencies $\tilde{\nu}_{0,J''\rightarrow 1,J'}$ which, because wavenumbers are used, is the same as the difference in $F(n, J)$ values.

(e) Show that the spectrum is a set of equally spaced lines separated by 2B either side of the spectrum's origin at $\tilde{\nu}_0 \equiv \tilde{\nu}_e(1 - 2x_e)$.

(f) Centrifugal stretching of the bond may be neglected to a high degree of approximation but allowance must be made for the effect of vibration on the value of the rotational constant *B* because *B* depends on the moment of inertia of the molecule. This in turn depends on the internuclear distance, which becomes larger in an anharmonic oscillator as quantum number *n* increases. To a good approximation

$$B_n = B_e - \alpha_e(n + 1/2)$$

where B_n is the value of *B* in the nth vibrational level; B_e is the value of *B* at the minimum of potential energy, i.e. when the internuclear separation is r_e, and α_e is a constant which allows for the increased average internuclear separation as *n* increases.

Repeat the calculation in **(c)** with B_n instead of *B*. For the *R* branch let $J' = J'' + 1 \equiv J$, and for the *P* branch, $J'' = J' + 1 \equiv J$. Show that the *P* and *R* transitions depend on J^2.

(g) Let *J* be equal to $m = +1, +2, +3 \cdots$ for the *R* branch, and $-1, -2, -3 \cdots$ for the *P* branch and calculate the difference $\tilde{\nu}_m + \tilde{\nu}_{-m}$ by substituting *m* for $-m$ and show that

$$\tilde{\nu}_m + \tilde{\nu}_{-m} = 2\tilde{\nu}_e(1 - 2x_e) - 2\alpha_e m^2$$

(h) Using the data below for $H^{35}Cl$, plot $\tilde{\nu}_m + \tilde{\nu}_{-m}$ against m^2, which should be a straight line with a slope $-2\alpha_e$, and intercept $2\tilde{\nu}_0 = 2\tilde{\nu}_e(1 - 2x_e)$ and find $\tilde{\nu}_0$ and α_e with their respective standard deviations and 95% confidence limits. Use the value $x_e = 0.0174$.

(i) Repeat the calculation for $H^{37}Cl$ and compare the two force constants using $\tilde{\nu}_e = \dfrac{1}{2\pi c}\sqrt{\dfrac{k}{\mu}}$. Use the t statistic to compare the two mean values obtained and discuss whether the result can be believed. The data is shown in the table. The standard deviation on each value is 0.08 cm^{-1}.

m	$H^{35}Cl/cm^{-1}$	$H^{37}Cl/cm^{-1}$	$-m$	$H^{35}Cl/cm^{-1}$	$H^{37}Cl/cm^{-1}$
1	2906.16	2904.05	−1	2865.04	2862.94
2	2925.83	2923.66	−2	2843.56	2841.5
3	2944.84	2942.66	−3	2821.49	2819.5
4	2963.21	2961.00	−4	2798.86	2796.89
5	2980.92	2978.68	−5	2775.68	2773.74
6	2997.97	2995.72	−6	2751.97	2750.07
7	3014.34	3012.07	−7	2727.71	2725.84
8	3030.01	3027.73	−8	2702.93	2701.12
9	3044.98	3042.68	−9	2677.66	2675.88
10	3059.24	3056.90	−10	2651.92	2650.16
11	3072.77	3070.46	−11	2625.65	2623.99
12	3085.91	3083.24	−12	2598.85	2597.48

13.6 Modelling data with polynomials is simpler using matrices

While the equations for a linear equation are manageable as summations, the polynomial equations become impossibly complicated. However, as the main concern is not in the algebraic form of the equations per se, only in obtaining the solution to a problem, it is then simpler to convert the simultaneous normal equations into a matrix equation and to solve this for the constants.

The equation $y = \alpha_0 + \alpha_1 x$ has already been discussed in some detail; the polynomial $y = \alpha_0 + \alpha_1 x + \alpha_2 x^2$ and exponential $y = \alpha_0 e^{-\alpha_1 x}$ are also commonly found functions that describe data. The polynomial still belongs to the class of *linear* least squares because the normal equations can be solved exactly. The exponential equation will need a *non-linear* least squares method because the normal equations cannot be solved exactly and an iterative method is required. The method of choice is often the gradient expansion method, also called the Levenberg–Marquardt method, and this is discussed in Section 13.8. First, however, a matrix formulation of the *linear* least squares problem is presented and a solution obtained for a polynomial.

13.6.1 General method

A general way of describing a fitting equation is

$$Y = \sum_{i=1}^{m} \alpha_i f_i(x) \tag{13.43}$$

which means that that the model equation proposed to represent the data is the sum of m functions, $f_i(x)$, multiplied by constants a_i whose values we seek, much as was done in a Fourier series. If the equation is a polynomial, then $f_1(x) = 1, f_2(x) = x, f_3(x) = x^2$, and so forth, and, if an exponential then $f_1(x) = e^{-\alpha_2 x}, f_2(x) = e^{-\alpha_4 x}$, etc. It is clear why these last two equations are non-linear because the constants are intimate to the function itself and cannot be separated from it.

The sum of the deviations is minimized just as in the case of the straight line, the equation is

$$S = \sum_k^n w_k \left(y_k - \sum_i^m \alpha_i f_i(x_k) \right)^2$$

and the subscript on x is k because this is the kth x value in the data and k has values $1 \cdots n$. The subscript $i = 1 \cdots m$ identifies the terms in the expansion of the model function equation (13.43). Differentiating the jth term in the ith summation to obtain the minimum, produces

$$\frac{\partial S}{\partial \alpha_j} = -2\sum_k w_k f_j(x_k) \left(y_k - \sum_i \alpha_i f_i(x_k) \right) = 0 \tag{13.44}$$

and the $f(x)$ remains inside the summation because it depends on the x_k. Rearranging the equation gives a (complicated) set of simultaneous equations, because j can take any value in the range 1 to m for m terms in the function expansion.

$$\sum_k w_k f_j(x_k) y_k = \sum_k \left(w_k f_j(x_k) \sum_i \alpha_i f_i(x_k) \right) \tag{13.45}$$

the right-hand side of which is a sum of a sum. For example, the third term is

$$\begin{aligned} \sum_k w_k y_k f_3(x_k) &= \sum_k \sum_i w_k f_3(x_k) \alpha_i f_i(x_k) \\ &= \sum_k w_k f_3(x_k)[\alpha_1 f_1(x_k) + \alpha_2 f_2(x_k) + \alpha_3 f_3(x_k) + \cdots + \alpha_m f_m(x_k)] \end{aligned} \tag{13.46}$$

and all m of these equations can be solved to find the m constants, the α's, if the function is linear. A matrix method can be used to do this, but to convert the simultaneous equations into a matrix equation needs a little cunning. A matrix equation of the form

$$\boldsymbol{B} = \boldsymbol{CA} \tag{13.47}$$

can be made where $\boldsymbol{C}$ contains the coefficients. The left-hand side of this equation is a row matrix with a sum over the k data points for each of m terms and is the left-hand side of equation (13.45),

$$\begin{aligned} \boldsymbol{B} &= \left[\sum_k w_k y_k f_1(x_k) \quad \sum_k w_k y_k f_2(x_k) \quad \sum_k w_k y_k f_3(x_k) \cdots \right] \\ &\equiv [S(wyf_1) \quad S(wyf_2) \quad S(wyf_3) \quad \cdots] \end{aligned}$$

and, in the second line, the abbreviation, $S(\cdots)$, is used to indicate the k summation over the data points. The right-hand side of the normal equations (13.45) can be written as a square matrix $\boldsymbol{A}$ left multiplied by a row matrix $\boldsymbol{C}$ to make another row matrix thus making the right-hand side of the equation equal to the left. To form $\boldsymbol{CA}$, the right-hand side of 13.45 or 13.46 has to be split into a column and square matrix. It is clear that in each term f has indices in i and k so the term in row 1 column 2 in the $\boldsymbol{A}$ matrix can be written using the S abbreviation as $S(wf_1f_2) \equiv \sum_k w_k f_1(x_k) f_2(x_k)$, and contains none of the α's therefore,

$$\boldsymbol{A} = \begin{bmatrix} S(wf_1f_1) & S(wf_1f_2) & S(wf_1f_3) & \cdots \\ S(wf_2f_1) & S(wf_2f_2) & S(wf_2f_3) & \cdots \\ S(wf_3f_1) & \ddots & & \cdots \\ \vdots & \vdots & \vdots & \ddots \end{bmatrix}.$$

If the row matrix $\boldsymbol{C}$ contains only the coefficients, $\boldsymbol{C} = [\alpha_1 \quad \alpha_2 \quad \alpha_3 \quad \cdots]$ the matrix equation is $\boldsymbol{B} = \boldsymbol{CA}$. The right-hand side of equation (13.47) is

$$\boldsymbol{CA} \equiv [a_1 \quad a_2 \quad a_3 \quad \cdots] \begin{bmatrix} S(wf_1f_1) & S(wf_1f_2) & S(wf_1f_3) & \cdots \\ S(wf_2f_1) & S(wf_2f_2) & S(wf_2f_3) & \cdots \\ S(wf_3f_1) & \ddots & & \cdots \\ \vdots & \vdots & \vdots & \ddots \end{bmatrix}.$$

Expanding out the third term produces

$$\alpha_1 S(wf_1f_3) + \alpha_2 S(wf_2f_3) + \alpha_3 S(wf_3f_3) + \cdots + \alpha_m S(wf_mf_3)$$

which is the same as equation (13.46).

The aim is to find the values in the C matrix and this can be achieved by the transformation

$$B=CA \quad \rightarrow \quad BA^{-1}=CAA^{-1} \quad \rightarrow \quad C=BA^{-1}. \tag{13.48}$$

Thus the inverse of A is all that is needed to find the coefficients, furthermore this inverse is also the matrix of the variances and co-variances (Bevington 1969).

13.6.2 **Polynomial fit**

As an example, the functional dependence of the vibrational energy levels of the B $(^3\Sigma_u^-)$ excited state of the oxygen molecule, which has $D_{\infty h}$ symmetry, will be found by analysing some data. The $(^3\Sigma_u^-)$ term symbol means that the molecule is in a triplet spin state and hence has two unpaired electrons, one in each of two orbitals, which together have zero orbital angular momentum, hence the symbol Σ. The orbitals are ungerade (u) or odd, so that they do not have a centre of inversion and the minus sign means that the electronic state is symmetric to C_2 (180°) rotation about the principal axis. The vibrational levels have been measured by observing the absorption spectrum in the ultraviolet starting just above 200 nm and the $n=0$ level has energy of 49 363 cm^{-1} above the $n=0$ of the ground state. The energies of the vibrational levels above the bottom of this potential well and their quantum numbers n, are shown in the table.

G/cm^{-1}	352.3	1039	1705	2340	2961	3553	4113	4652	5153	5613	6040	6433	6780	7083	7335
n	0	1	2	3	4	5	6	7	8	9	10	11	12	13	14

The weights for the data are unknown and hence 1 is used, as this produces an unbiased estimate of the coefficients. An anharmonic potential energy model will be tested. The Morse potential produces vibrational energies (in cm^{-1})

$$G_2=\nu_e(n+1/2)-\nu_e x_e(n+1/2)^2$$

and if this is not sufficient to describe the data, it is common to add another term empirically, producing

$$G_3=\nu_e(n+1/2)-\nu_e x_e(n+1/2)^2+\nu_e y_e(n+1/2)^3,$$

or

$$G_3=\left(\frac{1}{2}-\frac{x_e}{4}+\frac{y_e}{8}\right)\nu_e+\left(1-x_e+\frac{3}{4}y_e\right)\nu_e n+\left(\frac{3}{2}y_e-x_e\right)\nu_e n^2+y_e\nu_e n^3 \tag{13.49}$$

The equilibrium frequency is ν_e (cm^{-1}) and the anharmonicity terms are x_e and y_e. In the Morse potential, $x_e=\nu_e/4D_e$ where D_e is the dissociation energy in cm^{-1}, but y_e is empirical and the aim is to determine x_e, y_e and ν_e. If the quantum number n is represented by x, and the constants by c, then the Morse model G_2 is equivalent to a function $f=c_0+c_1x+c_2x^2$ and the second model $f=c_0+c_1x+c_2x^2+c_3x^3$. The least squares model calculates these coefficients and these can then be related to the spectroscopic parameters ν_e, x_e and y_e. The residuals and χ^2 will be used to test the fit to the data. (Note that different authors us different symbols to represent frequency, and this can be in units of cm^{-1}, s^{-1}, rad cm^{-1} or rad s^{-1}.)

The Maple implementation of the matrix method using $C=BA^{-1}$, is shown below. The inverse matrix A^{-1} is also the error or covariance matrix.

Algorithm 13.3 Least squares polynomial fit

```
> restart: with(LinearAlgebra): with(Statistics):
  Digits:= 30:                          # ensure high precision
# read in data into yval, xval and w
# as in shown in Algorithm (13.2)
  m:= nops(xval)                        # number of data points
  m:= 4:                                # number of coefficients c
```

```
A:= Matrix(1..m,1..m): B:= Vector(1..m):
C:= Vector(1..m):
for i from 1 to m do                              # A & B matrices
  B[i]:= add(w[k]*yval[k]*xval[k]^(i-1), k= 1..n):
  for j from 1 to m do
   A[i,j]:= add( w[k]*xval[k]^(i-1)*xval[k]^(j-1),k= 1..n);
  end do;
end do:
> A;                                                # print matrix>
```

$$\begin{bmatrix} 15.000 & 105.000 & 1015.000 & 11025.000 \\ 105.000 & 1015.000 & 11025.000 & 1.28\ 10^5 \\ 1015.000 & 11025.000 & 1.25\ 10^5 & 1.54\ 10^6 \\ 11025.000 & 1.28\ 10^5 & 1.54\ 10^6 & 1.91\ 10^7 \end{bmatrix}$$

```
> C:= Transpose(B).A^(-1);                          # coefficients
```

$$C := [354.564 \quad 691.758 \quad -8.548 \quad -.374]$$

```
> line:=(x)-> add(C[i]*x^(i-1),i=1..m):        # calculate result
  chi2:=add( (yval[i]-line(xval[i]))^2,i=1..n)/(n-m-1);
```

$$\chi 2 := 8.917$$

```
  Q:= GAMMA((n-2)/2,chi2/2);             # Prest et al. 1986. p. 506
```

$$Q := 224.316$$

```
> resid:= seq([xval[i],(yval[i]-line(xval[i]))/1],i=1..n):
  plot([resid],style= point, symbol= circle):
  data_points:= seq([xval[i],yval[i]],i=1..n):
  plot([dat_points],style= point, symbol= circle);
```

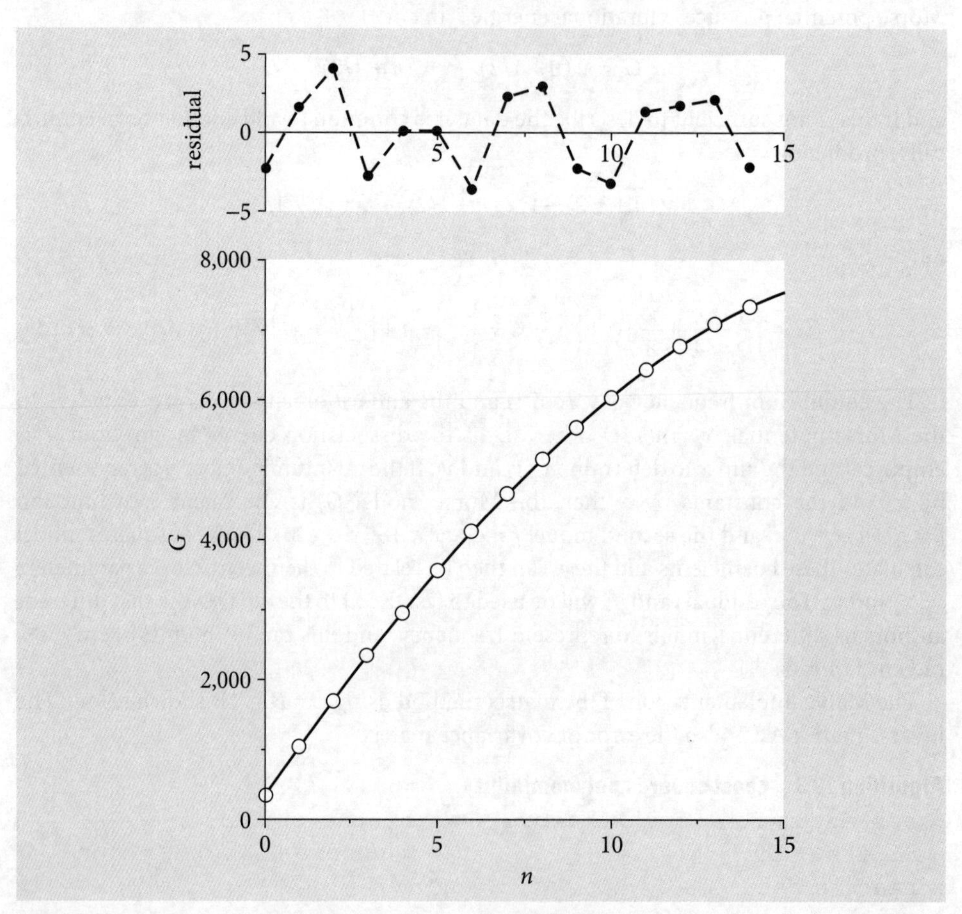

Fig. 13.11 Experimental data points and the function G_3 calculated by least squares. The residuals are shown on the top of the plot. Notice how small the scale is compared to that of the data.

The error matrix is calculated as the inverse of A. The off-diagonal terms are the covariances, notice that the matrix is symmetric about the diagonal. The covariances between c_1 and c_2 are quite large, –0.34, and these are expected to be the least accurate since one value may compensate for the other; the other covariances values are small.

```
> eps:= A^(-1);                                    # error matrix
>
```

$$eps := \begin{bmatrix} 0.673 & -.344 & 0.047 & -0.002 \\ -.344 & 0.276 & -0.045 & 0.002 \\ 0.047 & -0.045 & 0.008 & -0.000 \\ -0.002 & 0.002 & -0.000 & 0.0000 \end{bmatrix}$$

```
for i from 1 to m do
   printf("%a %g %s %G \n", C||i, C[i]," +-", sqrt(eps[i,i]));
end do:
    C1  354.6     +- 0.8
    C2  691.8     +- 0.5
    C3  -8.548    +- 0.09
    C4  -0.3741   +- 0.004
```

To calculate the spectroscopic parameters, the constants have to be compared to the expansion of the G_3 equation (13.49), which is

$$G_3 = \left(\frac{1}{2} - \frac{x_e}{4} + \frac{y_e}{8}\right)\nu_e + \left(1 - x_e + \frac{3}{4}y_e\right)\nu_e n + \left(\frac{3}{2}y_e - x_e\right)\nu_e n^2 + y_e \nu_e n^3.$$

The first term in this equation corresponds to c_1 and is the zero point energy. The constants are $\nu_e y_e = -0.3741$, $\nu_e x_e = 7.98$, and $\nu_e = 692\ \text{cm}^{-1}$, which compare well with the values given by Herzberg (1950, p. 559) which are $\nu_e y_e = -0.3753$, $\nu_e x_e = 8.00$, and $\nu_e = 700.36\ \text{cm}^{-1}$, and the zero point energy is $348.13\ \text{cm}^{-1}$. Because the weighting is unknown, the χ^2 is only relative but the Q factor is large which indicates that the model fits the data, although this is obvious from the residuals anyway.

The calculation with the Morse equation G_2 is achieved by setting `m: = 3` and making no other changes. The calculation produces a fit that looks virtually the same as shown in the figure, but the residuals are about 10 times larger and oscillate noticeably, Fig. 13.12. The $\chi^2 = 737$ and $Q = 10^{-146}$ and the chance of this model being correct is effectively zero. This demonstrates that χ^2 and the residuals are vital in determining the correct fit. If five terms are used in the calculation, the χ^2 is reduced only to 8.6, compared to 8.9 for four terms, which is essentially the same value, therefore, five terms are rejected as being too many to fit the data properly.

13.6.3 Confidence limits

The confidence limits can be calculated in a similar manner to that for the straight line, using the value from the t distribution. The equations look slightly different because the matrix method is used. (See Hines & Montgomery 1990, Chapters 14–15 for the details.) They calculate the mean square error $SS_E/(n - m - 1)$, which is the same as the χ^2 on the whole data set, and give the formula which is equivalent to the one used here;

$$\chi^2 = \frac{\sum_i y_i^2 - CB}{n - m - 1} \tag{13.50}$$

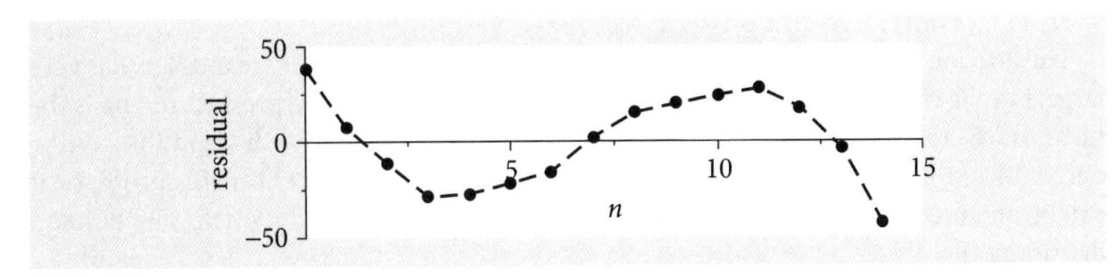

Fig. 13.12 Residuals for the G_2 fitting equation showing oscillations because this model function does not describe the data well.

where CB is a matrix product. The confidence limits at 95% are now

$$c_i \pm t_{0.025,n-m-1}\sqrt{\chi^2(A^{-1})_{ii}}$$

where $(A^{-1})_{ii}$ is the ith diagonal value of the inverted A matrix which is the ith diagonal of matrix `eps` in the Maple calculation and c_i is the ith coefficient.

If the row matrix $x_0 = [1 \quad x \quad x^2 \quad x^3]$ is defined because the highest power of x used in the model function is x^3 then the 95% confidence lines around the fitted curve (`line(x)` the function in the Maple Algorithm 13.3) are calculated as

$$\text{line}(x) \pm t_{0.025,n-m-1}\sqrt{\chi^2 x_0 A^{-1} x_0^T}. \tag{13.51}$$

This curve is not shown in Fig. 13.11 because it is almost identical to the fitted curve, and would not be clear on the graph. Note that the product of the three matrices is a function of x.

13.7 Photon and particle counting and the Poisson distribution

The Poisson distribution is formed by accumulating many events, n, where each have a very small probability, p, of occurring, but the product np, the mean number of events, is moderate. The distribution is asymmetric and skewed towards small numbers, Fig. 13.13, because it is not possible to have a negative number of events. This distribution is observed when photons are counted or particles counted after a radioactive atom disintegrates, provided their decay time is long compared to the observation time. However, the distribution applies to many other types of events, such as the number of faulty CDs produced, the number of misprints on a printed page, or the number of students absent from a class in any week. One of the earliest examples was recorded over a 20-year period during the 1800s, and was the number of deaths of infantrymen in the Prussian army after they were kicked by a horse.

Suppose that a sample of molecules is continuously excited, and their fluorescence viewed through a small aperture or through filters to reduce the intensity so that photons are detected one at a time by a photodiode or photomultiplier. The number of electrical pulses from the detector for, say, 1000 time intervals each of 1 second duration is recorded. If a histogram of the *number* of events recorded in each time interval is made, then this should follow a Poisson distribution. If DNA is exposed to UV light, photo-damage can occur by base pairs forming dimers. If the DNA is spliced into pieces of equal length, the number of dimers in each piece can be recorded. If the pieces are each very small, 10 base pairs for example, most will not contain a dimer, a few will contain one and fewer still will contain two or more and so on. The distribution formed will have a large value for zero dimers and become smaller as the number of dimers increases. In a second experiment, suppose that 100 base pair long segments are examined. Now only a few segments have no dimer, some have one and many have two, three, or four but fewer segments will have five or more. This distribution is now peaked at a value of, say, 2.5. Repeating the experiment with a larger segment will cause the distribution to peak at still larger values until it closely resembles a normal distribution. What has changed between the experiments is only the average number of dimers detected in each segment, thus the Poisson distribution is determined by only one parameter, the average number of 'events'. A similar argument applies to the number of photons or particles detected in progressively longer time intervals.

The Poisson distribution applies when the number of events n that are measured is very large, but the chance p of any one event occurring is very small. The product $\mu = np$ is the mean or average number of events and is of moderate value. The distribution can be derived from first principles, see Hines & Montgomery (1990, §6.8) for this proof, or it can be obtained from the binomial distribution when n is large and p small. The Poisson distribution is

$$P(k, \mu) = \frac{\mu^k e^{-\mu}}{k!},$$

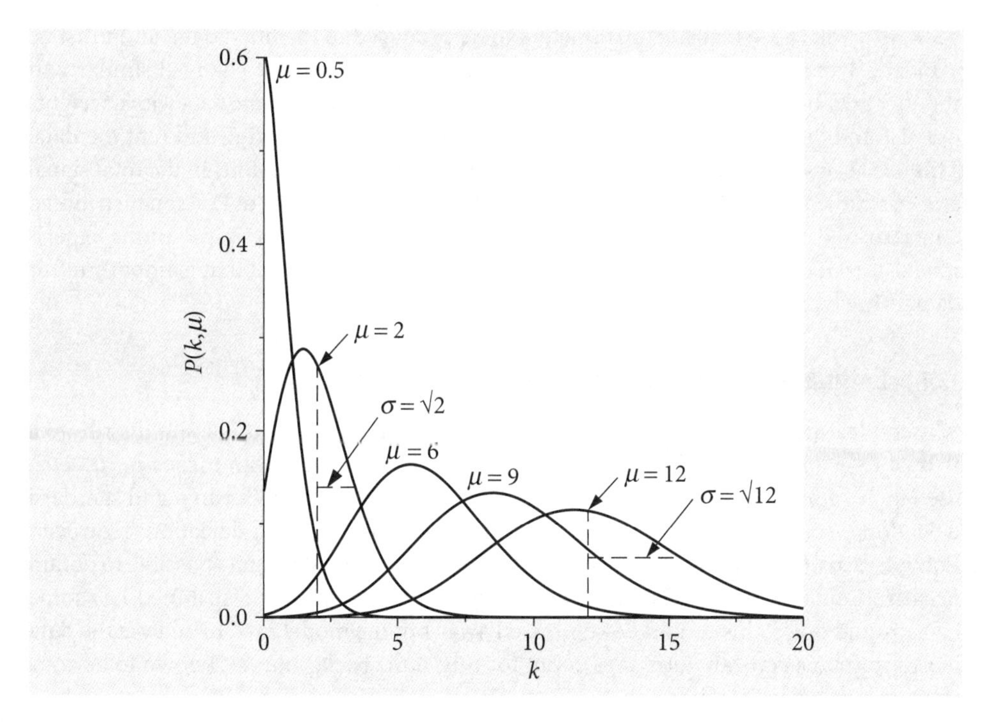

Fig. 13.13 The Poisson distribution for different integer k values and various means values μ. Although k is an integer, the lines are drawn as if it were continuous. The standard deviation $\sigma = \sqrt{\mu}$, is shown for $\mu = 2$ and 12. The distribution becomes more Normal as μ increases.

also called the Poisson frequency distribution or 'density' function. The integer number of events is k and the mean value is μ, which is not an integer. The shape of the distribution is shown in Fig. 13.13 for different mean values μ.

Notice that, when the mean is larger, the distribution resembles a normal distribution and is almost symmetrical. But at small μ, the distribution is very asymmetrical with the mean always to larger k than the peak (the mode or most probable value) of the distribution. The standard deviation of the distribution is also shown and this is

$$\sigma = \sqrt{\mu}$$

where μ is the average number of events detected; for instance, the number of photons detected in a given time period at a given wavelength. This relationship means that the weighting for analysing Poisson distributed data is $w_i = 1/k_i$. The mean and variance are calculated as the average value $\mu \equiv \langle k \rangle = \sum_{k=1}^{n} kP(k, \mu)$ and average of the square $\langle k^2 \rangle = \sum_{k=1}^{n} k^2 P(k, \mu)$ and then in the usual way as $\sigma^2 = \langle k^2 \rangle - \langle k \rangle^2$. Both $\langle k \rangle$ and the variance are equal to μ, see question Q13.10.

The cumulative distribution is found by summing to a given k value

$$CDF = \sum_{k=0}^{n} P(k, \mu) = e^{-\mu}\left(1 + \mu + \frac{\mu^2}{2!} + \frac{\mu^3}{3!} + \cdots + \frac{\mu^n}{n!}\right)$$

The chance of no success, that is of not observing an 'event', is $e^{-\mu}$. The chance of observing one event $k = 1$ is $\mu e^{-\mu}$, of two events $\mu e^{-\mu}/2!$, and of exactly k events is $P(k, \mu)$. The chance of observing more than one event is $1 - e^{-\mu}$. The sum to infinity is unity, and therefore the distribution is normalized.

Photon and particle counting experiments normally follow a Poisson distribution. The weighting to use is $w_i = 1/c_i$ where c_i is the mean number of counts at the ith data point in the data, thus, the weighting is always known. This does lead to a specific problem, viz., what happens when no counts are recorded in a particular channel, $c_i = 0$? This is quite possible at low light levels as occurs in single photon counting fluorescence decay experiments, or, when measuring fluorescence from single molecules through a confocal microscope. The advice to deal with this (O'Connor & Phillips 1984; Lacowicz 2004) is to manipulate the data either by ignoring values below 10 counts, or to add 1 to every data point or to make each zero into 1 and leave all the rest alone. None of this is satisfactory because it will cause incorrect parameters to be produced to the fitted data, and must be avoided. By making these changes, both the data and residuals are changed similarly so the effect can be disguised. The solution (Turton et al. 2003) is to fit the data with a weighting of 1 and then the use values from the fitted curve as the weighting and refit the data. This has been shown to produce unbiased fits down to a very few counts in the total signal and contradicts the widespread assumption that least squares fitting of Poisson distributed data is invalid. One further point worth making is that when conducting counting experiments, it is not the maximum number of counts in any one channel that is important for data fitting but the total number of counts in the whole signal.

13.7.1 Estimating signals buried in noise

A useful feature of Poisson distributed data is that their sum is also Poisson distributed. If $x_1, x_2, x_3 \cdots$ are a set of Poisson distributed random numbers with means μ_1, μ_2, etc., then their sum $\mu = \mu_1 + \mu_2 + \mu_2 + \cdots$ is also Poisson distributed with mean μ and standard deviation $\sqrt{\mu}$. Thus, data can be summed to improve its standard deviation. However, Poisson distributed data cannot be subtracted, because it would be quite possible to obtain negative numbers. This means that when a signal that is Poisson distributed has some background noise, this cannot be subtracted away but the model used to analyse the data has to contain a constant term to account for this, if the background is known to be constant, or a polynomial if it is not.

All instruments produce noise unrelated to the signal and it is often necessary to decide if a signal is present in noise but summing (binning) data does not help because the noise is also summed. For example, a very weak light source may be obscured by thermally induced 'dark' noise from the detector (photomultiplier, photodiode, or CCD) and subsequently amplified in other circuitry. Alternatively, the noise could be from other light close in wavelength to the one being detected or electrical noise added by amplifiers. To decide if a signal is present, data has to be taken with and without what is suspected to be the true signal being there.

Distinguishing between signal and noise depends crucially only on their respective standard deviations. A large and constant background is not a problem; it is the *fluctuation* in the background that limits the signal to noise. To illustrate this, an argument presented by Parratt (1971, Chapter 5.8) based on the combination of errors equation, is described. If there are two sets of counts, c_b for the background and c_{sb} for signal and background, because a signal cannot be detected without background, then the difference in the population *means* is

$$\mu_s - \mu_{sb} = c_b - c_{sb} \pm \Delta s$$

where Δs is the experimental standard deviation of the *difference* in counts. If there are many measurements, the central limit theorem predicts that this difference is normally distributed. Then, using the combination (propagation) of errors equation, $\Delta s = (s_b^2 + s_{sb}^2)^{1/2}\sqrt{2}$ where s_b and s_{sb} are the standard deviations for background and signal plus background respectively. For single measurements of Poisson data the standard deviation is equal to the number of counts giving $\Delta s = (c_b + c_{sb})^{1/2}$. If the two means μ_b and μ_{sb} are for the moment assumed to be the same then $c_b - c_{sb}$ can be tested by comparing with Δs at, say, the 5% probability level of a standard normal distribution to determine the chance whether or not $c_b = c_{sb}$. If the ratio $|c_b - c_{sb}|/\Delta s = 1$, then the probability that $c_b \neq c_{sb}$ is 0.32, (or 32%) which is the area of the normal curve *above* $x/\sigma = 1$, Fig. 13.4 and this is too small to be considered very strong proof that the signal and noise are different. If the ratio $|c_b - c_{sb}|/\Delta s \geq 1.96$ then chance that $c_b \neq c_{sb}$ is measured at the 95% level (see Table 13.1) and the signal would be distinguishable from background.

Consider the case when the counts at a given wavelength in a spectrum are $c_{sb} = 105$ and $c_b = 100$, then $\Delta s = 14.3$ and $c_b - c_{sb} = 5$ and the ratio $|c_b - c_{sb}|/\Delta s = 0.35$. Using the cumulative distribution described in equation (13.9), the chance that $c_b \neq c_{sb}$ is

```
> (2*CDF(Normal(0,1),0.35)-1)*100;                27.36%
```

which is so small that the two signals are indistinguishable. If the counts are $c_{sb} = 150$ and $c_b = 100$ then $\Delta s = 15.8$ and $c_{sb} - c_b = 50$ and percentage chance that $c_b \neq c_{sb}$ is

```
> (2*CDF(Normal(0,1),50/15.8)-1)*100;             99.84%
```

Therefore, the signals can easily be distinguished above the noise at the 99.8% confidence level, which means that the chance that the difference in means will exceed the standard deviation by pure chance alone is only 0.16%.

If the data is not Poisson distributed and/or data sets are added together, then $\Delta s = \left(\frac{s_b^2}{n_b} + \frac{s_{sb}^2}{n_{sb}}\right)^{1/2}$ for n_{sb} sets of signal measurements and n_b background.

13.7.2 **Questions**

Full solutions are available at **www.oxfordtextbooks.co.uk/orc/beddard**.

Q13.10 Show that the mean and variance of the Poisson distribution are both μ the average number of events.

Q13.11 The dissociation energy of a molecule can be calculated by adding up all the differences in vibrational energy until the energy gap is zero. However, it is not always possible to obtain data up to the dissociation limit and the quantum number for this has to be estimated which can be done by plotting the energy gaps vs quantum number and fitting the function to a polynomial. The dissociation energy is the area under the curve.

Using the data for the O_2 molecule's energy levels given in the text Section 13.6.2, calculate the dissociation energy of the ${}^3\Sigma_u^-$ state.

Strategy: Using the best functional form found in the example used in Section 13.6.2, calculate the energy differences $G(n+1) - G(n)$ to find out what polynomial should be used. Plot $\Delta G_{n+1,n}$ vs n, and fit this data. Integrate the resulting equation to find the area but only until the quantum number n that makes $\Delta G_{n+1,n} \approx 0$.

13.8 **Non-linear least squares, gradient expansion, and the Levenberg–Marquardt Method**

In the least squares method, the derivative of the χ^2 is taken with respect to each parameter α_i and minimized. The normal equations of linear least squares analysis (equations (13.28) or (13.44)), have terms only in $a_i x$ or $a_i x^2$ etc., and can be solved exactly. In non-linear least squares, the normal equations cannot be solved exactly. The function $Y = \alpha_1 e^{-\alpha_2 x} + \alpha_3 e^{-\alpha_4 x} + \alpha_5$ has a form that is common to several schemes in chemical kinetics and when analysing the decay of excited states, but in this function the normal equations still contain exponentials in α_i and x, for example $e^{-\alpha_2 x}$ and cannot be solved exactly. An approximate numerical solution has to be sought, usually by an iterative method. Any iterative method follows an algorithm that tries to approach the minimum χ^2 without searching through the whole range of possible values of the parameters (α's). There are several algorithms to choose from, some, such as the simplex, simulated annealing (Prest et al. 1986), and evolutionary (natural selection) algorithms are not least squares methods; however, the best algorithm for non-linear least squares is the Levenberg–Marquardt algorithm (Bevington 1969; Prest et al. 1986; Bevington & Robinson 2003). Additionally it is easy to implement and operates stably, i.e. it hardly ever fails.

Using a numerical method implies (i) that starting values have to be defined for each of the parameters sought; (ii) increments in these have to be determined and changed as the calculation proceeds; and (iii) some ending condition has to be decided upon so that an acceptable fit to the data is found. These requirements mean that this problem is far harder to solve than is linear least squares, partly for the reasons (i) to (iii), but also because there is no guarantee that the *true* minimum will be found even if the calculation appears to end satisfactorily. This last effect has two forms. The first is that close to the minimum solution several related sets of parameters can satisfy the same goodness of fit criterion, so that a unique set is not found, only a range of them. The second effect is that the 'surface' through and about which the minimum χ^2 is sought may be rough, with local minima in which the calculation can become trapped.

The χ^2 is defined as

$$\chi^2 = \sum_{i=1}^{n} w_i(y_i - Y_i(\alpha_j))^2$$

where the subscript i identifies the data points and j the parameters, α_1, α_2, etc. The χ^2 surface would ideally be shaped rather like a bowl so that it can be imagined how starting at one point the minimum can be reached. But it is instead highly complex, and can be represented as a three-dimensional object only if two parameters are used and if there are more, and there usually are, it is almost impossible to have an idea what this object may look like. In the best circumstance, the object is smooth and a minimum is found, but generally, it might be supposed that the objects 'landscape' is 'corrugated' with many local minima separated by 'hills' and 'ridges'. However, not withstanding the complexity of the χ^2 'surface', most equations describing physical and chemical phenomena can be successfully minimized.

Quite often, finding the true minimum starts with initial guesses for the parameters close to the final ones that the calculation will produce. Often, approximate parameters can be guessed by looking at the data; they do not usually have to be that good for the calculation to converge perhaps within a factor of 5 or 10. Alternatively, the literature can be consulted to find values for similar experiments. If this is not possible, as there may be too many variables to find them all, then ranges must be placed on these and perhaps hundreds of sets of parameters can be guessed at random and those with the smallest χ^2 chosen as starting points. Once the calculation is run, if the results are not particularly good it may be because (a) the function is sensitive to the starting values; (b) it ended up in a local not the global minimum; (c) the calculation did not run long enough to converge; or, if it did converge, (d) the model did not describe the data. Starting the calculation with different initial conditions will usually help to sort this out.

The Levenberg–Marquardt Method combines two different approaches to finding the minimum χ^2; Bevington (1969) and Prest et al. (1986) give detailed derivations and equations. The first approach is a gradient-search method that tries to find the steepest descent from any point towards the minimum of the surface that represents the χ^2. This method is good when the calculation is a long way from the minimum, as it allows it to be approached rapidly because the gradient is large. When close to the minimum, this method is poor because the gradient may be almost zero and the calculation will creep along, hardly making any progress. The gradient in the χ^2 is calculated as $\sum_{j=1}^{n} \frac{\partial \chi^2}{\partial \alpha_j}$. The second aspect to the method is to change the equations close to the minimum where the model function Y can be accurately expanded as a Taylor series in the parameters α_i and the χ^2 is minimized, and normal equations calculated. It turns out, fortunately, that the two sets of equations, steepest descent and expansion, can be changed into one another by varying only one parameter conventionally called λ.

It has been demonstrated that the normal equations (13.44) can be written in matrix form (equation (13.48))

$$C = BA^{-1}.$$

In the linear case, these equations can be solved by inverting A and multiplying by B as is done in Algorithm 13.3. The equations in the non-linear case are formally the same as those of (13.48) but now the solution must be obtained iteratively. The iterative part has nothing to do with the Marquardt method; this method changes only the diagonals of matrix A and so moves from a gradient search to a linear expansion of the function. Matrix A is changed so that

$$A_{jj} \rightarrow A_{jj}(1 + \lambda)$$

and λ is changed in the algorithm to go between gradient expansion where λ large, to function linearization where λ is small.

The algorithm (Bevington (1969, Curfit p. 237) and Bevington & Robinson (2003)) is

(i) calculate $\chi^2(\alpha)$

(ii) set $\lambda = 0.001$

(iii) calculate $\chi^2(\alpha + \delta\alpha)$ where $\delta\alpha$ are the increments in parameters α.

(iv) if $\chi^2(\alpha+\delta\alpha) > \chi^2(\alpha)$ then $\lambda \rightarrow 10\lambda$ and go to (iii)

(v) if $\chi^2(\alpha+\delta\alpha) < \chi^2(\alpha)$ then $\lambda \rightarrow \lambda/10$, make $\alpha \rightarrow \alpha+\delta\alpha$ and go to (iii).

(vi) check at (iii) whether a preset minimum difference in consecutive χ^2 is produced or maximum number of iterations reached.

The algorithm is shown below fitted to Poisson distributed data. The data represents a single exponential decay with a constant background and the model function used is $y = c_1e^{-c_2x} + c_3$. The data was simulated, as described in chapter 12.3.4, and the data is

x	1	5	9	13	17	21	25	29	33	37	41	45	49
y	1927	1329	812	568	390	290	171	112	87	48	43	26	26
x	53	57	61	65	69	73	77	81	85	89	93	97	
y	14	13	10	6	9	8	7	3	9	5	4	8	

The true values used to produce the data are $c_2 = 0.1$, $c_3 = 5$, but the initial value of c_1 is unknown as the data has been simulated with 20 000 events and Poisson distributed noise has been added with a mean value of 5. The total number of counts in the data is 5925, which is rather a small number for an experiment, although single molecule fluorescence measured through a confocal microscope may have even fewer than this. This Maple implementation is based on Bevington (1969, programme 11-5 curfit p. 237) but is different in detail because Maple has an `add()` function which is used to replace some 'do loops' and it does not have a 'goto' instruction. In the calculation, matrices ***A*** and ***B*** are divided by `sqrt( AA[j,j]*AA[k,k] )` to improve the precision of the calculation (Bevington 1969). The initial χ^2 (`chiB`) is made extremely large that it exceeds the first calculated χ^2 or (`chiA`) even if the guessed parameters are poor to ensure that the calculation starts properly. The ending condition is that `abs(chiA - chiB) < 0.001` or that `reps` is greater than 40, by repeating the whole calculation in the loop `for L from 1 to reps do .. end do`.

To adapt the calculation for other functions the number of parameters `m:=..`, the line `model:=..`, the initial values `C:=..` and the derivatives must all be changed.

The data is read-in in three columns, x, y, and standard deviation σ. The data is Poisson distributed, therefore the standard deviation is y, and the weight $1/y$.

Algorithm 13.4 Levenberg–Marquardt non-linear least squares

```
> with(LinearAlgebra): with(plots):
> # modified from code Curvfit (Bevington)
> Digits:=30:                          # ensure high precision
# read in data as in Algorithm (2)
  file:="exponential data.txt":        # 3 columns of data, x,y,σ
  filepos(file,0):                     # set to start of file
  data:= readdata(file,3):
  close(file):
  n:= nops(data):                     # get number of data points
  xval:= Vector(n): yval:= Vector(n):  w:= Vector(n):
  for i from 1 to n do                # extract data
    xval[i]:=  data[i,1];
    yval[i]:=  data[i,2];
    w[i]:=     1.0/data[i,3];         #make 1 for unweighted data
  end do:                             # end of data input
# fitting function    y = C1*exp( -C2*x ) + C3
# n is the number of data points defined by reading data
  m:= 3:                                    # number of params α
  lambda:= 0.001:                           # initial value
  reps:= 40:                                # max number calc's
  C:= <11000.0| 0.55| 15>;                  # init params
  model:=(x,C)-> C[1]*exp(-C[2]*x)+C[3]:  # model equation
  nf:= n - m - 1:                           # degrees of freedom
  chiB:= 1e20:                              # large to start calc
```

```
  B:= Vector(1..m):        sig:= Vector(1..m):
  A:= Matrix(1..m,1..m):  deriv:= Matrix(1..m,1..n):
# start main calculation
for L from 1 to reps do                         # start iterations
  beta:= Vector(1..m):                          # define, set to zero
  AA:= Matrix(1..m,1..m):                       # define, set to zero
  for i from 1 to n do                          # calc derivatives
    deriv[1,i]:= exp(-C[2]*xval[i]):            # dy/dC1
    deriv[2,i]:= -C[1]*xval[i]*exp(-C[2]*xval[i]): # dy/dC2
    deriv[3,i]:= 1.0:                           # dy/dC3
  end do:
  chiA:= add( w[i]*(yval[i] - model( xval[i],C) )^2,i=1..n )/nf;
  for j from 1 to m do
    beta[j]:=beta[j]+
      add(w[i]*(yval[i]-model(xval[i],C))*deriv[j,i],i=1..n):
    for k from 1 to j do
     AA[j,k]:= AA[j,k] + add(w[i]*deriv[j,i]*deriv[k,i],i=1..n):
     end do;
  end do;
# difference in chi^2 is one ending condition
  if abs(chiA - chiB) < 0.001 then break end if;
  AA:=(Transpose(AA)+AA)/2.0:                   # symmetrise AA
  while chiB > chiA do                          # step (iii)
    for j from 1 to m do
      for k from 1 to m do
        A[j,k]:= AA[j,k]/sqrt( AA[j,j]*AA[k,k] ):
      end do:
      A[j,j]:= 1.0 + lambda:                    # change diag
    end do:
    A:= A^(-1):                                 # invert matrix
    for j from 1 to m do
  B[j]:= C[j]+add(beta[k]*A[j,k]/sqrt(AA[j,j]*AA[k,k]),k=1..m);
    end do:
    chiB:= add(w[i]*(yval[i] - model(xval[i],B) )^2,i=1..n )/nf;
    lambda:= lambda*10.0;                       # step (iv)
  end do;                                       # end step (iii)
  C:= B;                                        # replace coeffs
  lambda:= lambda/10.0;                         # step(v)
  chiB:= chiA:                                  # replace chi^2
end do:                                         # end iteration for L
for j from 1 to m do
  sig[j]:= sqrt(A[j,j]/AA[j,j]);
  printf(" %s %g %s %g \n",C||j, C[j],"+-",sig[j]);
end do:
                    C1  2131.5        +-  20.0
                    C2  0.10124       +-  0.00076
                    C3  5.0719        +-  0.51

  model(x, C);                                  # print model>
```

$$2131.46\, e^{-0.10x} + 5.07$$

```
  chi_2:= chiA;                                 # reduced χ^2
```

$$\chi 2 := 0.88$$

```
  prob:= (1 - CDF( ChiSquare(nf), chiA*nf))*100;
```

$$prob := 61.33$$

```
> Q:= GAMMA((nf-2)/2,chiA/2);
```

$$Q := 1.2\; 10^5$$

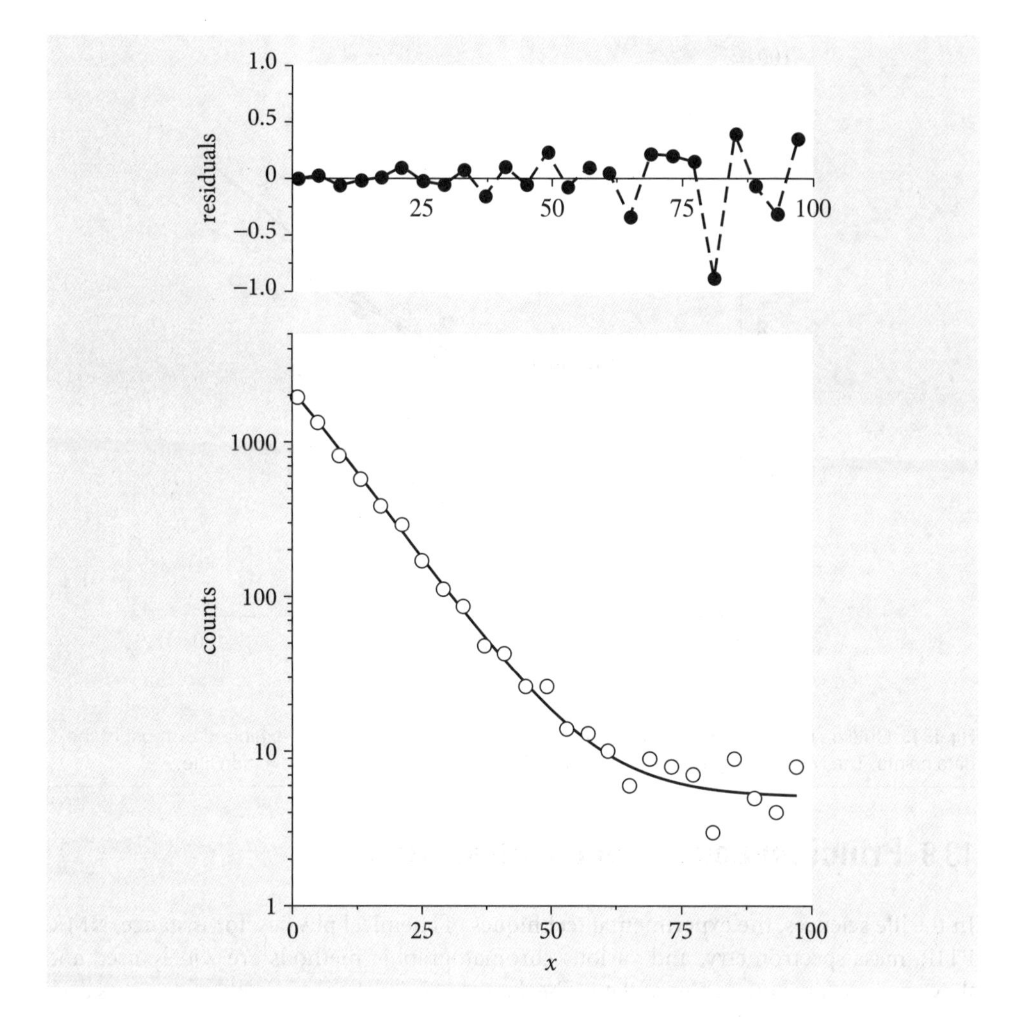

Fig. 13.14 Non-linear least squares fit to $y = c_1 e^{-c_2 x} + c_3$ with $c_1 = 2131$, $c_2 = 0.1$, and $c_3 = 5.07$.

```
total_data:= add(yval[i],i=1..n);            # total counts data
```

$$total\_data := 5925.00$$

```
total_fit:= add(model(xval[i],C),i=1..n); # total counts fit
```

$$total\_fit := 5911.31$$

The calculation produces a χ^2 of 0.9 and that corresponds to a probability of 61% which is near perfect; a perfect value of 1 produces a 50% chance. The model curve clearly fits the data as shown in Fig. 13.14. The total counts predicted by the model are almost the same as in the data itself, also indicating a close fit. The normalized residuals are calculated as $(y_i - \text{model})_i/y_i$.

13.8.1 Least absolute deviation

Any least squares method is very sensitive to outliers in the data; see Fig. 13.15. You can 'eyeball' data and see roughly where a straight line should go through the majority of the data points; however, the least squares line will generally not follow this trend because it is pulled off by the outliers. In such cases, the function to minimize is the absolute value of the deviation rather than the χ^2 and for an unweighted straight-line fit is the sum

$$\sum_i | y_i - a - bx_i |.$$

Using this function now produces a far better fit to the data, but there is no χ^2 with which to estimate how good this fit is. Taking the absolute value of the distance from the best fit line, rather than the square of the distance, means that the effect of large deviations is reduced. Prest et al. (1986) give an algorithm with which to calculate the minimum deviation line and this is shown in Fig. 13.15.

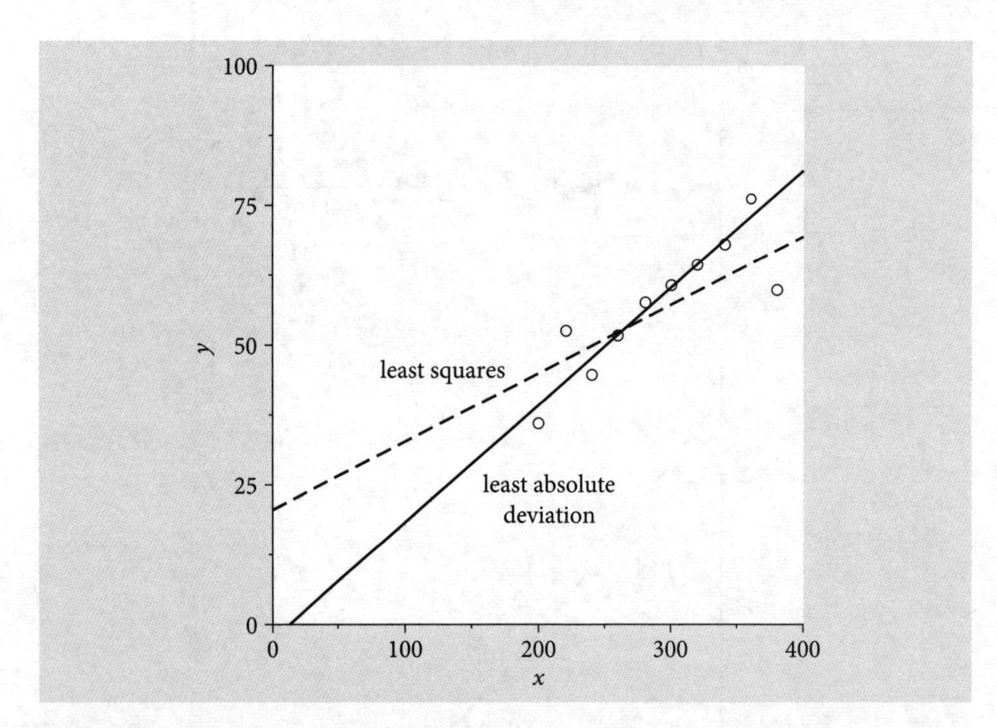

Fig. 13.15 Outliers produce a least squares fit that does not reflect the trend exhibited by most of the data points. The line passing mainly through the data is the least absolute deviation line.

13.9 Principal component analysis (PCA)

In the life sciences, the experimental techniques of chemical physics, for instance, NMR, FTIR, mass spectrometry, and various chromatographic methods are widely used and they produce large quantities of data. The data may consist of hundreds of data sets each containing data from thousands of proteins and metabolites and from which the presence of a few target molecules may be sought. For example, NMR spectra taken on blood plasma will contain signals from thousands of compounds. If samples are taken from a group of patients with a particular disease and a similar group without the disease, the two sets of NMRs can be compared in an attempt to identify markers that may be used to target the disease, either in an attempt either to cure it, or act as an assay to identify its early stages. However, these spectra will be different due to normal variability, will be highly congested and contain numerous overlapping signals from proteins, lipids, sugars, and DNA not involved in the disease. The task is to find the species that are either present or are missing in the diseased group. The problem is that there is an abundance of data but no results.

Data from any instrument is not always in the form that most easily allows the results required to be extracted from it. PCA attempts to rotate the axes with which the data is presented in such a way as to isolate each of its features independently of all of the others. This method produces new variables that are linear combinations of the original data and these are the principal components; mathematically a new orthogonal basis set is produced with which to represent the data (Jolliffe 2002; Miller & Miller 2005). Recall that in changing a basis, the data (a vector) is unaffected only the axes are changed. Some of these new components may contain only background or noise while others contain the data. Sorting out which ones to keep requires some judgement; the process is not automatic. The PCA method is therefore a way of reducing the size of a data set, the penalty is that some data is lost; however, this may not be significant, simply necessary to expose the pertinent data. This type of analysis is therefore always a balance between simplifying the problem to make it understandable and loosing information. Note that principal component analysis is a non-parametric method of analysing data; correlations are sought between data sets without using any model to describe the data. In contrast, the least squares method is a parametric method because a model (equation) is used to test the data.

A set of data can be represented as an expansion of coefficients a in a basis set w of length, or dimension, n,

$$x = a_1 w_1 + a_2 w_3 + a_3 w_3 \cdots a_n w_n$$

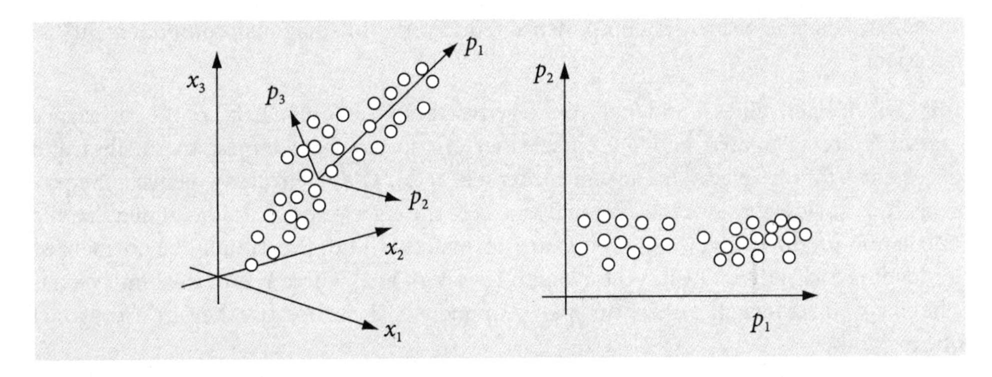

Fig. 13.16 Data on the experimental axes (left), and on the axes of principal components (right).

The same data can also be expressed in another basis set u as

$$x = b_1u_1 + b_2u_3 + b_3u_3 \cdots b_nu_n$$

but with a different set of coefficients, b. This distinction is illustrated in Fig. 13.16 where any data point has approximately equal values of coefficients a, because it is approximately equally positioned between the three experimental x-axes whereas those on the new axes p, have a large initial value b_1, along p_1 and smaller ones along p_2 and p_3. PCA changes the basis set to optimize the variance placing the principal axis along the direction of maximum spread or variance of the data, and other principal components orthogonally to this. Figure 13.16 also shows the new axes p as the principal component axes, two of which are shown on the right and where two regions of data are identified. The missing data in this graph compared to the one on the left is in the direction p_3 and not p_2, and would be seen on a plot of p_1 vs p_3 and in this way, the change of axes (basis set) can identify different groups of data. The next step in PCA is to reduce the size of the basis set used; this means describing the data in the basis set u with a smaller set of terms in the summation; s instead of n. Because the change of axes puts more contributions of each data point along each principal axis, fewer terms are now needed to form an acceptable description of the data, although all the basis set is needed to reproduce the data *exactly*. A reduced data set is often the rationale for using PCA, with the aim of identifying new signals, removing noise and/or isolating known interfering signals.

Matrix methods are used to find the principal components (PC). The data comprises m sets where each set has n data points. The whole data forms an $n \times m$ matrix, $\mathbf{D}$, with each data point in a new row of a given column and these vectors form m columns. The steps to find the principal components are:

(i) Remove the mean value from each of the individual data sets so that the centroid of the data becomes the origin of the new axes. In matrix form subtracting the mean from each data point is

$$\mathbf{X} = \mathbf{D} - \boldsymbol{\mu}$$

where $\mathbf{D}$ is the $n \times m$ matrix of data and $\boldsymbol{\mu}$ a $n \times m$ matrix with the appropriate mean value in each element of its columns. The PC's are also sensitive to the scale of the data, so it may be necessary to divide the data by the variance giving each set a mean of zero and unit standard deviation during the calculation.

(ii) The covariance of the data is then calculated, this is an $m \times m$ symmetrical matrix

$$\mathbf{C} = \frac{1}{n-1}\mathbf{X}^T\mathbf{X} \tag{13.52}$$

The superscript T is the transpose, and the matrix multiplication has dimensions $(m \times n).(n \times m) = m \times m$. The entries in this matrix would be diagonal if the data were random, because each value is then independent of every other. However, this is very unlikely to be the case unless the data contains only random noise; therefore, off-diagonal terms are not zero. The covariance describes how much the data is spread among the axes. If the data were to lie only along each axis the covariance would be diagonal, the rotation of axes accompanying the principal components reduces the off-diagonal component in the covariance.

(iii) The m eigenvalues λ and (column) eigenvectors V ($m\times m$ matrix) of the covariance matrix C are calculated and the eigenvalues are sorted from largest to smallest and the eigenvectors are placed in the same order see 7.12.3. The eigenvalue-eigenvector equation is $CV=\Lambda V$ where Λ is the diagonal matrix of the eigenvalues λ. It is assumed that the calculation produces eigenvectors that are normalized, if not they should be normalized by dividing each column by its vector length $V_i \to V_i/\sqrt{V_i\cdot V_i}$ where V_i is one column vector. The eigenvectors form the new principal component basis set because they are orthogonal one to another.

The eigenvector of the largest eigenvalue forms the principal component of the data, smaller values the other principal components. The largest contains most of the variance or spread of the data and best describes any trend in the data. The smaller eigenvalues do so to lesser extents depending on their size. The largest s eigenvalues are then chosen to approximate the data. The eigenvalues correspond to variances along the principal axes, thus the fractional value of each eigenvalue to the sum of them all is

$$\lambda_i/\sum_{i=1}^{m}\lambda_i$$

and this is the fraction that eigenvalue i contributes to the data. A measure of the error made by approximating the data by s principal components instead of the total number m, is the residual variance,

$$\sigma^2 = 1-\frac{\sum_{i=1}^{s}\lambda_i}{\sum_{i=1}^{m}\lambda_i}.$$

For example, if the first two eigenvalues describe 85% of the data, it may be appropriate to ignore all the others. As a rule of thumb, when selecting data to eliminate, eigenvalues less than ≈ 1 can be ignored.

(iv) To project the data points onto the new axes, the principal components, the dot product of the data X is made with each eigenvector V.

$$Y=V^TX^T \tag{13.53}$$

The matrix dimensions are $(m\times m).(m\times n)=(m\times n)$. Plotting one row vs another, the data in matrix Y produces the data along the various principal component axes as shown on the right of Fig. 13.16. The first column is p_1, the second, p_2, etc. Because the eigenvector (modal) matrix V is square and orthogonal, the relationship $V^{-1}=V^T$ can be used to reform the original data. The transformation steps are to left multiply both sides of the equation with V giving $VV^TX^T=VY$. Simplifying the left-hand side produces $VV^TX^T=VV^{-1}X^T=X^T=VY$ and then VY is transposed again to form X or,

$$X=(VY)^T \qquad \text{and} \qquad D=X+\mu \tag{13.54}$$

and μ restores the mean to the data D.

(v) When only s of the eigenvalues and eigenvectors are being used, the V matrix is reduced by ignoring all but the first s columns and equation (13.54) is used to reconstruct the data.

A Maple algorithm to perform PCA follows. The data is read into a column matrix `data`; the data consists of three sets of 10 data points each. The data is shown in the figure. The `SubMatrix()` command is used to extract part of the eigenvalue matrix rather than doing this element by element.

Algorithm 13.5 Principal component analysis

```
> restart: with(LinearAlgebra):with(plots):
  file:="PCA data.txt":           # 3 columns of data
  filepos(file,0):                #set to start of file
  data:= readdata(file,3):
```

```
  close(file):
  n:= 10;                 # number of rows of data points
  m:= 3;                  # number of columns
  S:= 2;                  # select to calculate first 2 PCs
  data:= convert(data,Matrix);
  pointplot3d(data ,symbolsize=20,axes=boxed);
  means:= Matrix(1..n,1..m):         Sev:= Vector(m):
  VS:= Matrix(m,m):                  X:= Matrix(1..n,1..m):
    for k from 1 to m do
    means[1..n,k]:=add(data[i,k],i=1..n)/n;
  end do:
  X:= data - means;                        # data with mean zero
  C:= Transpose(X) . (X) /(n-1);           # covariance
  (ev,evec):= Eigenvectors(C):             #eigenvalues/eigenvects
  eigval:= map(x-> Re(x), ev):             # make real
  eigvec:= map(x-> Re(x), evec):           # make real
  indx := [ seq( i, i= 1..m ) ];         # make index to sort with
  indx:= sort( indx, (i,j)-> evalb( -ev[i] < -ev[j] ) );
  for i from 1 to m do
    Sev[i]:= eigval[indx[i]];              # sorted eigenvals
    VS[1..m,i]:= eigvec[1..m,indx[i]]; # sorted eigenvects
  end do:
# now select first S eigenvalues
  V:= SubMatrix(VS,1..m,1..S);             # select first S cols
  Y:= Transpose(V).Transpose(X);           # prncpl cmpts eqn 53
  DR:= Transpose(V.Y) + means:             # reform data eqn 54
  pointplot(Y, title="principal axes", axes=boxed);
  pointplot3d(DR, title =" data partly reconstructed",
        axes= boxed, orientation =[120,85], colour= black);
```

The figure shows the data (centred at zero by subtracting the mean values), and the three principal axes. The first eigenvector, (1) the major axis, is drawn in a thick line. The lengths are arbitrary. The eigenvalues are 4.39, 0.103, 0.0750. On the right of the figure, two plots are superimposed. The grey symbols are the data reconstructed from the first two eigenvectors using equations (13.53) and (13.54). The original data is flattened onto the plane of the first two principal axes because the third coordinate is ignored. The second set of data (black symbols) is reconstructed using only the first eigenvalue and is a set of points lying along the principal axis at their relative positions when projected from the other axes. If all three eigenvectors are used the original data is reformed exactly.

Related to PCA is the method of singular value decomposition (SVD). Formally, it is a technique in which one matrix is split into three, one of them is diagonal and the other two are row and column orthogonal as are eigenvector modal matrices (Prest et al. 1986). The

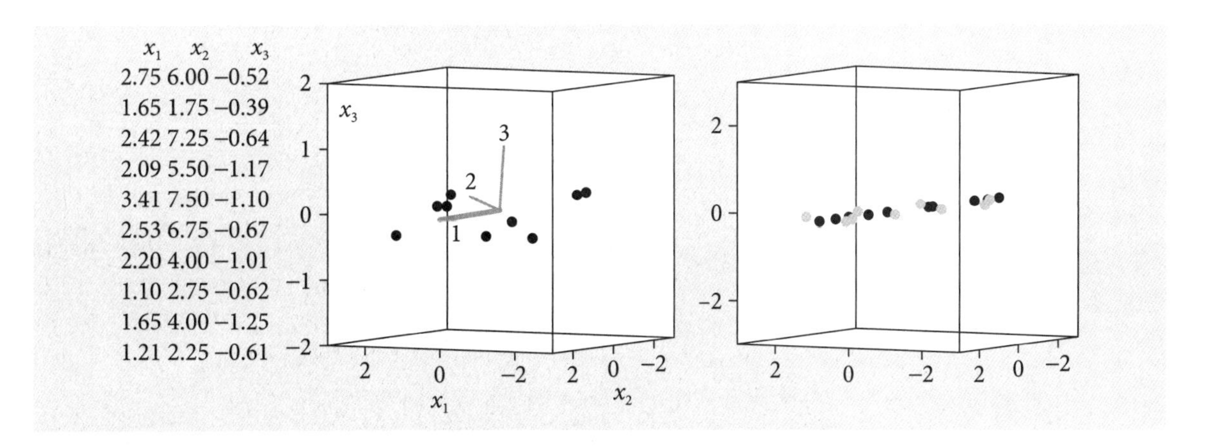

x_1	x_2	x_3
2.75	6.00	−0.52
1.65	1.75	−0.39
2.42	7.25	−0.64
2.09	5.50	−1.17
3.41	7.50	−1.10
2.53	6.75	−0.67
2.20	4.00	−1.01
1.10	2.75	−0.62
1.65	4.00	−1.25
1.21	2.25	−0.61

Fig. 13.17 Left: Data plotted with mean values subtracted together with the principal axes. The major principal axis is (1). Right: Data partly reconstructed (also with mean values subtracted for comparison) with the first two eigenvectors (grey) and also with only the major eigenvalue, black symbols. The data used is shown in the table.

reason for expanding a matrix in this way is that it can now be inverted even though it is close to being singular, i.e. inverting it would normally produce infinity or a number that approximates this, and normal inversion methods fail. This method could be used to invert the *A* matrix in non-linear least squares, in the example given in Algorithm 13.4.

The interest in SVD is however not primarily to do with the mathematics but in unravelling complex information just as in PCA. SVD can be used directly on raw data, not the covariance as in PCA, or as a mathematical method to perform PCA (Jolliffe 2002).

A Maple™ Language Crib

Appendix 1

A.1 Finding your way around

Maple can be used in different modes, the simplest is to use the *worksheet* mode which means that the text is in red preceded by the red symbol >. To force the program to use this mode, click in the [> icon on the toolbar at the top of the screen, then click the *text* icon in the lower tool bar on the left part of the screen. Next, using the *tools* menu, locate and click on *options*, then *interface* and select *default for new worksheets* and using the drop down menu select *worksheet*. Complete the setup by closing the *options* box by clicking on *apply globally*. Maple should start in the worksheet mode next time it is used.

A.2 Some useful points

<table>
<tr><th>Symbol or keystroke</th><th>Meaning</th></tr>
<tr><td>?anything</td><td>Opens help for 'anything'. You can also highlight a word and go to the help menu.</td></tr>
<tr><td>ctrl k</td><td>Inserts a new line above the present line.</td></tr>
<tr><td>shift Enter</td><td>Breaks the current Maple input line.</td></tr>
<tr><td>ctrl Enter</td><td>Makes a page break.</td></tr>
<tr><td>crtl c</td><td>Copies highlighted text.</td></tr>
<tr><td>ctrl v</td><td>Pastes copied text.</td></tr>
<tr><td>crtl x</td><td>Deletes highlighted text.</td></tr>
<tr><td>Return key</td><td>Performs a calculation.</td></tr>
<tr><td>Enter key</td><td>Performs a calculation.</td></tr>
<tr><td>back arrow</td><td>Deletes a character and is the key to the right of += key.</td></tr>
<tr><td>!!!</td><td>On toolbar; recalculates the whole worksheet.</td></tr>
<tr><td>:=</td><td>Assign the right-hand side of the := sign to the name on the left.</td></tr>
<tr><td>=</td><td>is used only in an equation in the mathematical sense
e.g. <code>f01:= x^2 - 3*x = 2;</code></td></tr>
<tr><td>;</td><td>Ends Maple line: performs calculation shows result.</td></tr>
<tr><td>:</td><td>Ends Maple line: performs calculation; DOES NOT show result.</td></tr>
<tr><td>%</td><td>Use previous result; for safety only use this on the same input line as previous expression</td></tr>
<tr><td>#</td><td>On a Maple input (red text) line starts a comment.</td></tr>
<tr><td>^</td><td>Raises to a power. If raising to 1/2 or −20 etc. use brackets, x^(−20)</td></tr>
<tr><td>*</td><td>Multiplies (matrices are multiplied with <code>(mat1 . mat2)</code>)</td></tr>
<tr><td>$</td><td>Used in differentiation, indicates power, <code>x$3</code> is d^3/dx^3</td></tr>
<tr><td><code>Pi</code></td><td>is mathematical π approx 3.14159 ··· note use of capital letter</td></tr>
<tr><td><code>I</code></td><td>Capital i square root of −1</td></tr>
<tr><td><code>||</code></td><td>Concatenation, a||3 becomes a3</td></tr>
<tr><td><code>infinity</code></td><td>Used instead of the symbol ∞</td></tr>
<tr><td>capitals</td><td>Most packages now start with a capital letter, as do <code>Array</code>, <code>Vector</code> <code>Matrix</code>. A capital letter used in <code>Diff</code>, <code>Int</code>, <code>Sum</code>, etc. prevents evaluation so that the data can be seen and checked.</td></tr>
</table>

The books by Heck (1996), Garvan (2002), Richards (2002), Aratyn & Rasinariu (2006), and Shingareva & Lizarraga-Celaya (2007) describe the many different approaches to using Maple.

A.3 General syntax

Maple remembers all the values and parameters used in a calculation so if you inadvertently use the same name in a second calculation its current values will be used and you may not realize that this has happened. It is, therefore, a good habit to start a Maple session by typing

```
> restart:
```

as this removes all memory of previous calculations. This is usually needed when you start a new calculation in the same worksheet.

Brackets play a vital part in clarifying what expression you intend to calculate. In particular after a division all the denominator must be between brackets if there is more than one term. For example,

`> (2+x)/ x^2;` means $\dfrac{2+x}{x^2}$

`> (2+x)/(x^2+x);` means $\dfrac{2+x}{x^2+x}$

`> (2+x)/x^2+x;` means $\dfrac{2+x}{x^2}+x$

When manipulating algebraic expressions, the syntax always has the form

```
> Your_name_for_an_expression:= expression ;
```

The := operator *assigns* the expression to have the name you choose; the semicolon ends the line. When you push the *return* or *enter* button the expression is evaluated and the result printed. For example,

```
> f01:= 2*x;
```

The reason for *naming* an expression is that it can be used in a later calculation. If an *instruction* or mathematical *function* is to be used then this is named first and everything else needed for its evaluation is enclosed between brackets, i.e.

Operation_name(expression, options1, option 2, option3, etc) ;

For example,

```
> simplify( (x^2)^(1/2), symbolic) ;
> diff( x^2, x) ;
> sin( x /(1+x) ) ;
> exp( -x^2 ) ;
```

A.4 Packages

You can use Maple without loading a package but sometimes these are needed. A package is a set of specialized operations, for example, to manipulate matrices and calculate determinants, eigenvalues, and eigenvectors, a linear algebra package is used. This package is used by typing

```
> with(LinearAlgebra) :
```

and notice the use of capitals; this is important. There are older defunct packages named with lower case letters. More common is the

```
> with(plots):
```

package that is needed for many plots other than the simplest.

A.5 Converting units

As an example of using in-built functions the `convert` and `units` packages are illustrated; `convert` is quite general, it will convert from trig to exponential forms of an expression and so forth, but with the `units` option it will convert from one set of units to another. It will do this numerically or dimensionally. For instance to convert 30 mph into km/s the instruction is

```
> convert(30.0,units, mph, km/s);
```

$$0.0134112000$$

```
> convert(1,units, kJ, electronvolt);
```

$$6.241506363\ 10^{21}$$

To check on unit dimensions use the `base=true` option to give dimensions explicitly.

```
> metres_per_sec:= convert(m/s, dimensions, base=true);
  metres_per_sec:= convert(m/s, dimensions);
```

$$metres\_per\_sec := \frac{length}{time}$$

$$metres\_per\_sec := speed$$

```
> Newton_metres:= convert(N*m, dimensions, base=true);
  Newton_metres:= convert(N*m, dimensions);
```

$$Newton\_metres := \frac{mass\ length^2}{time^2}$$

$$Newton\_metres := energy$$

When converting energy units the `energy=true` option has to be used. For example to convert 20 wavenumbers (cm^{-1}) to MHz,

```
> convert(20.0, units, cm^(-1), megahertz, energy = true);
```

$$5.995849160\ 10^5$$

A.6 Defining your own function

Defining a function, although it is simple to do, always seems to give students new to Maple a disproportionate amount of angst.

The syntax for *definition* and *usage* is different.

The function is defined by giving it a name followed by `:=` then its variables, placed between brackets if there are more than one, and then followed by `->`. The function is used by naming it immediately followed by brackets containing its variables. For example,

```
> g:= x-> sin(x)^3/ln(x)^2;                # define a function g
```

$$g := x \rightarrow \frac{\sin(x)^3}{\ln(x)^2}$$

```
> g(3);                    # call the function with the value 3
```

$$\frac{\sin(3)^3}{\ln(3)^2}$$

```
> g(y^3);                  # call with algebraic expression
```

$$\frac{\sin(y^3)^3}{\ln(y^3)^2}$$

```
> f:=(x, y)-> sin(x)*exp(-y);              # two variables
```

$$f := (x, y) \rightarrow \sin(x)\, e^{-y}$$

```
> f(2,3);
```

$$\sin(2)\, e^{-3}$$

A.7 Two examples of plotting

There are several different routines for plotting, but the simplest is shown here. There are also many options to customize a plot and these are listed under the plot(options) on the help pages. The Maple Plotting Guide gives an illustration example of each type of plot.

The next Maple line illustrates the syntax to plot from $x = -3$ to 3 and y from -5 to 5. The x and y values, if present, are always placed first after the function being plotted. It is not necessary to specify the y-axis values but if you want to see just a part of the plot then this is a useful to limit the calculation just to values you specify. Alternatively the view option can be used. This has the form `view = [ -6..6,-8..8]` and is placed after `x= ..` etc. and separated by a comma from any other options.

```
> plot(3+x-x^4,x=-3..3,-5..5);                # left-hand plot A.1
  plot(3+x-x^4,x=-3..3,view=[-1..2,-3..4]);
                                              # right-hand plot A.1
```

To plot two or more functions on the same graph, place them between square brackets, for example,

```
>plot([ x^2, x^3, x^4 ],x=-3..2, color=[red, blue, green]);
```

and using square brackets ensures that the functions will be plotted in the same order as the colours, so they can be identified. As well as colours different line styles can be used, for example, by adding `, linestyle=[dash,solid,dot]`.

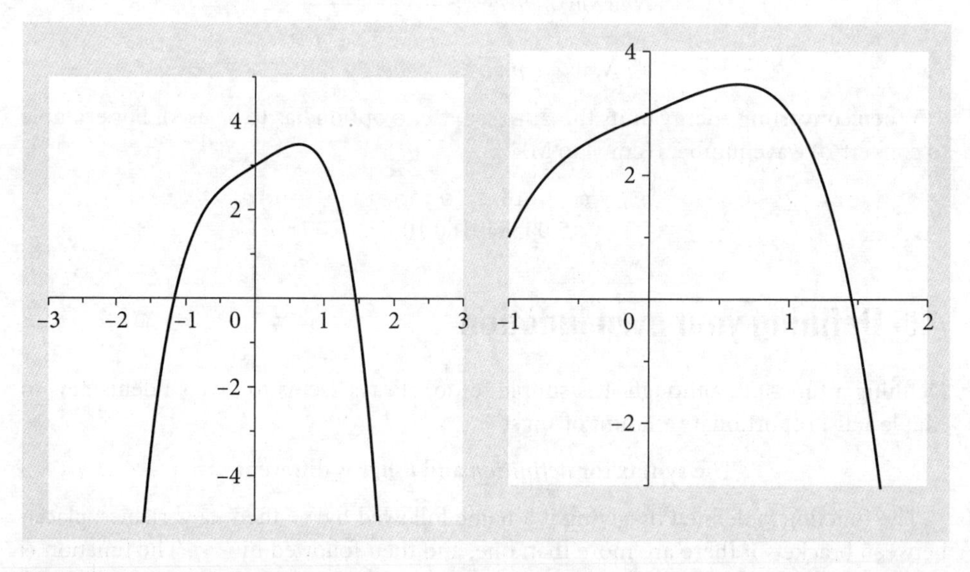

Fig. A.1 Plotting simple functions see A.7

A.8 Examples using expand and factor

Assign an expression $(x + y)^5$ to a name such as `a1`, expand it, then factor it back again.

```
> restart:                  # use restart to reset all parameters
> a1:= (x+y)^5;
```

$$a1 := (x + y)^5$$

```
> a2:= expand(a1);
```

$$a2 := x^5 + 5x^4y + 10x^3y^2 + 10x^2y^3 + 5xy^4 + y^5$$

```
> factor(a2);                                  # factor a2
```

$$(x + y)^5$$

```
> expand( ( 41*x^2+x+1)^2*(2*x-1) );
```

$$3362x^5 - 1517x^4 + 84x^3 - 79x^2 - 1$$

```
> a2:=(1 + 1/( x+2) )^2; # note use of brackets after / sign
```

$$a2 := \left(1 + \frac{1}{x+2}\right)^2$$

```
> expand(a2);
```

$$1 + \frac{2}{x+2} + \frac{1}{(x+2)^2}$$

A.9 simplify, normal, rationalize, collect, combine, % = value (%)

Maple can apply identities to simplify many lengthy mathematical expressions, such as trigonometric ones. Notice that the syntax for powers of trig functions is not the same as strict mathematical notation.

```
> f01:=cos(x)^5 + sin(x)^4 + 2*cos(x)^2 - 2*sin(x)^2 -
  cos(2*x);
```

$$f01 := \cos(x)^5 + \sin(x)^4 + 2\cos(x)^2 - 2\sin(x)^2 - \cos(2x)$$

```
> expand(f01); simplify(f01);
```

$$\cos(x)^5 + \sin(x)^4 - 2\cos(x)^2 + 1$$
$$\cos(x)^4(\cos(x) + 1)$$

The expression `% = value(%)` is a general way of making a neat output. Use it only in the same line as the expression.

```
> f01: % = value( simplify(%) );
```

$$\cos(x)^5 + \sin(x)^4 + 2\cos(x)^2 - 2\sin(x)^2 - \cos(2x) = \cos(x)^4(\cos(x) + 1)$$

```
> 2*(x+y)/(4*x+4*y);
```

$$\frac{2(x+y)}{4x+4y}$$

```
> normal(2*(x+y)/(4*x+4*y)); # normal clears common factors
```

$$\frac{1}{2}$$

```
> (x^3-y^3)/(x^2+x-y-y^2);
```

$$\frac{x^3 - y^3}{x^2 + x - y - y^2}$$

```
> normal( (x^3-y^3)/(x^2+x-y-y^2) );
```

$$\frac{x^2 + xy + y^2}{x + 1 + y}$$

```
> expr:= (x^2)^(1/2); # clearly square root of x^2 is x
```

$$expr := \sqrt{x^2}$$

```
> simplify(expr);
```

$\operatorname{csgn}(x)\, x$

but the result returned is complex sign of x times x. The expected answer is obtained with

```
> simplify(expr, symbolic); # symbolic keyword forces answer
```

$$x$$

The instruction `rationalize` is good at removing fractions

```
> 2/( 2 - sqrt(2) ): % = value( rationalize( %) );
```

$$\frac{2}{2-\sqrt{2}} = 2 + \sqrt{2}$$

You may have to try different approaches to get the answer in the form you want. All the answers below are equivalent

```
> fun:= (x/2+1/2*sqrt(5*x))^3/(x-1);
  expan:= expand(fun);
  ratnlize:= rationalize(fun);
  smpl:= simplify(fun);
  expan_ratnl:= rationalize(expand(fun));
```

$$fun := \frac{\left(\frac{1}{2}x + \frac{1}{2}\sqrt{5}\sqrt{x}\right)^3}{x-1}$$

$$expan := \frac{1}{8}\frac{x^3}{x-1} + \frac{3\sqrt{5}x^{5/2}}{8\ x-1} + \frac{15}{8}\frac{x^2}{x-1} + \frac{5}{8}\frac{\sqrt{5}x^{3/2}}{x-1}$$

$$ratnlize := \frac{(x+\sqrt{5}\sqrt{x})^3}{8x-8}$$

$$smpl := \frac{1}{8}\frac{(x+\sqrt{5}\sqrt{x})^3}{x-1}$$

$$expan\_ratnl := \frac{x^3 + 3\sqrt{5}x^{5/2} + 15x^2 + 5\sqrt{5}x^{3/2}}{8x-8}$$

The instruction `collect` can be used to simplify expressions; in this example we obtain a polynomial in x with coefficients in y and then the opposite, you can collect on lots of things, x, x^2, etc. as well as *sin, cos, trig, exp, ln*, etc. See the help pages for details.

```
> expand((x+2*y+1)^3);
  poly_in_x:= collect((x+2*y+1)^3,x);
  poly_in_y:= collect((x+2*y+1)^3,y);
```

$$x^3 + 6x^2y + 3x^2 + 12xy^2 + 12xy + 3x + 8y^3 + 12y^2 + 6y + 1$$
$$poly\_in\_x := x^3 + (6y+3)x^2 + ((2y+1)(4y+2) + (2y+1)^2)x + (2y+1)^3$$
$$poly\_in\_y := 8y^3 + (12x+12)y^2 + ((x+1)(4x+4) + 2(x+1)^2)y + (x+1)^3$$

`combine` is also useful in manipulating expressions, the second term, in this case, exp, gives the type of answer wanted

```
> exp(x)^3*exp(-y)*exp(ln(x)+1):% = value( combine(%,exp));
```

$$(e^x)^3 e^{-y} e^{\ln(x)+1} = x\, e^{3x-y+1}$$

A.10 Substitutions and evaluations: subs and algsubs, evalf

To substitute x for a number using expression `f01`, the parameter x is made equal to 1;

```
> f01:=cos(x)^5 + sin(x)^4 + 2*cos(x)^2 - 2*sin(x)^2 - cos(2*x):
> subs(x = 1.0, f01 );
```

$$\cos(1.0)^5 + \sin(1.0)^4 + 2\cos(1.0)^2 - 2\sin(1.0)^2 - \cos(2.0)$$

To evaluate `f01` numerically with $x = 1$ and to five decimal places

```
> evalf( subs( x = 1,f01) ,5);
```

$$0.13125$$

Next to substitute $\exp(-x)$ for x use `subs` as

```
> f02:=subs( x = exp(-x),f01);
```

$$f02 := \cos(e^{-x})^5 + \sin(e^{-x})^4 + 2\cos(e^{-x})^2 - 2\sin(e^{-x})^2 - \cos(2\,e^{-x})$$

Several operations can be done together, for example, starting by converting `f02` to exponential form, expanding it, and then simplifying.

```
> simplify( expand( convert( f02, exp ) ) );
```

$$\frac{1}{8}\cos(4\,e^{-x}) + \frac{5}{8}\cos(e^{-x}) + \frac{5}{16}\cos(3\,e^{-x}) + \frac{1}{2}\cos(2\,e^{-x})\ \frac{1}{16}\cos(5\,e^{-x}) + \frac{3}{8}$$

In the next example subs does not work

```
> subs( x*y^2=s, x^3*y^4 );
```

$$x^3y^4$$

instead try `algsubs` which performs algebraic substitution not just pattern matching,

```
> algsubs(x*y^2=s, x^3*y^4);
```

$$s^2x$$

Use `algsubs` to limit to terms less than x^2 in a summation

```
> sumxk:=sum( x^k, k=-5..5 );
```

$$sumxk := 1 + \frac{1}{x^5} + \frac{1}{x^4} + \frac{1}{x^3} + \frac{1}{x^2} + \frac{1}{x} + x + x^2 + x^3 + x^4 + x^5$$

```
> limited_series:= algsubs(x^2=0,sumxk);
```

$$limited\_series := 1 + \frac{1}{x^5} + \frac{1}{x^4} + \frac{1}{x^3} + \frac{1}{x^2} + \frac{1}{x} + x$$

A.11 convert and evala

```
> convert( cos(x)*sin(x), exp); simplify(%);
```

$$-\frac{1}{2}I\left(\frac{1}{2}e^{Ix} + \frac{1}{2}e^{-Ix}\right)(e^{Ix} - e^{-Ix})$$

$$\cos(x)\sin(x)$$

If you don't get the answer you want, try 'expanding' before 'simplifying',

```
> expand( convert( cos(x)*sin(x), exp) ); simplify(%);
```

$$-\frac{1}{4}I(e^{Ix})^2 + \frac{\frac{1}{4}I}{(e^{Ix})^2}$$

$$\frac{1}{2}\sin(2x)$$

Next, convert a couple of real numbers to exact fractions

```
> convert([1.5,0.5255],rational);
```

$$\left[\frac{3}{2}, \frac{1051}{2000}\right]$$

The result of series expansion cannot be used directly in another calculation, so use `convert` to `polynom` to help to get it into a useful form.

```
> f01:= series(exp(-a*x),x); convert( f01 ,polynom);
```

$$f01 := 1 - ax + \frac{1}{2}a^2x^2 - \frac{1}{6}a^3x^3 + \frac{1}{24}a^4x^4 - \frac{1}{120}a^5x^5 + O(x^6)$$

$$1 - ax + \frac{1}{2}a^2x^2 - \frac{1}{6}a^3x^3 + \frac{1}{24}a^4x^4 - \frac{1}{120}a^5x^5$$

Convert to partial fraction in x is shown in two ways

```
> f:= (1+1/(x-b))^2; convert(f, parfrac, x);
```

$$f := \left(1 + \frac{1}{x-b}\right)^2$$

$$1 + \frac{2}{x-b} + \frac{1}{(x-b)^2}$$

```
> (1+1/(x-b))^2: % = value( convert(%, parfrac,x) );
```

$$\left(1+\frac{1}{x-b}\right)^2 = 1+\frac{2}{x-b}+\frac{1}{(x-b)^2}$$

Evaluate f algebraically

```
> f; evala(f);
```

$$\left(1+\frac{1}{x-b}\right)^2$$

$$\frac{(x-b+1)^2}{(x-b)^2}$$

A.12 map

The command map enables one to operate on a whole set of data with a rule such as 'make every value real'. For example, if there is an array of numbers called `eigval` then

```
> map(x-> Re(x), eigval);
```

means apply the operation 'make every value of eigval real'. Similarly the instruction

```
> map(x-> 1/x^3, eigval);
```

will take the reciprocal of the cube of every number in `eigval`, provided a number is not zero; if it is an error message is produced.

The mapping can also be symbolic, as in

```
> map( x-> tan(x^2), x*ln(y) + cos(y) );
```

$$\tan(x^2 \ln(y)^2) + \tan(\cos(y)^2)$$

This may not be the result you anticipated therefore map has to be used carefully. Each term such as $x\ln(y)$ or $\cos(y)$ is changed using the generic rule $x \to \tan(x^2)$.

A.13 Numerical calculations

Integer expressions return integers, any real number forces (mostly) a real number. Answers are obtained to unlimited precision unlike your hand calculator.

```
> 32^50 * 12^13;                # cannot be done on a calculator
```

193577606016250065595970579104537070862177624928632764870822832599435438264767773409 28

```
> 32^70 / 12^10;               # answer remains exact as a fraction
```

21872507247830119243725022271176213653531694308932124364257706064099529991993759232235131770 23053824 / 59049

If one of the numbers is a decimal then a real number is returned

```
> 32.0^70 / 12^10;
```

$$3.704128309\ 10^{94}$$

```
> fract1:=( 2^3/3^(20) )*sqrt(3);
```

$$fract1 := \frac{8}{3486784401}\sqrt{3}$$

```
> evalf( fract1,8 );              # answer to 7 places
```

$$3.973978600\ 10^{-9}$$

```
> evalf(Pi,20); # evaluated to 19 decimal places
```

3.1415926535897932385

If evaluation gives only a few decimal places look in 'options' in the 'tools' menu and alter the default values.

A.14 Sum (algebraic summation), add (numerical summation), and product

Define a summation call it s, use a capital S in Sum to prevent initial evaluation

```
> Sum( ( 1+i) /(1+i^3) ,i = 0..infinity ): %= value(%);
```

$$\sum_{i=0}^{\infty} \frac{1+i}{1+i^3} = 1 + \frac{1}{3}\sqrt{3}\pi \tanh\left(\frac{1}{2}\pi\sqrt{3}\right)$$

The Maple instruction add calculates numerical value of a sum, it does not try a algebraic evaluation. Add (with capital A) does not exist

```
> a:= add( ( 1+i) /(1+i^4) ,i = 0..10 );
```

$$a := \frac{921347056664139330022 1}{406266489388192000884 9}$$

Define an expression f then Sum.

```
> f:=( 1+i) /(1+i^4);
```

$$f := \frac{1+i}{1+i^4}$$

```
> Sum(f, i = 0 ..10 ): % = value(%);
```

$$\sum_{i=0}^{10} \frac{1+i}{1+i^4} = \frac{9213470566641393300221}{4062664893881920008849}$$

or evaluate to a real number using evalf

```
> Sum(f, i = 0 ..10 ): % = evalf(%);
```

$$\sum_{i=0}^{10} \frac{1+i}{1+i^4} = 2.267839166$$

Make a sum to infinity,

```
> d:= sum(1/k^2, k = 1.. infinity);
```

$$d := \frac{1}{6}\pi^2$$

Define and evaluate a product

```
> f01:= Product( (i^2+3*i-11) /( i+3), i = 0..10 );
```

$$f01 := \prod_{i=0}^{10} \left(\frac{i^2 + 3\,i - 11}{i+3}\right)$$

```
> f01:= product( (i^2+3*i-11) /( i+3) ,i = 0..10);
```

$$f01 := -\frac{7781706512657}{40435200}$$

```
> numerical_value:= evalf(f01);
```

$$numerical\_value := -1.924488197\ 10^5$$

A.15 Differentiation and integration

Define a function and differentiate by x, call the answer f_prime. Using a capital D just shows the operation without doing anything.

```
> f:=x * sin( a*x ) + b * x^2;
```

$$f := x\sin(a\,x) + b\,x^2$$

```
> Diff(f,x); f_prime:= diff(f,x);
```

$$\frac{\partial}{\partial x}(x\sin(a\,x) + b\,x^2)$$

$$f\_prime := \sin(a\,x) + x\cos(a\,x)\,a + 2\,b\,x$$

Use `%= value(%)` to make a nice equation.

```
> Diff(f,x) : % = value( %);
```

$$\frac{\partial}{\partial x}(x\sin(a\,x) + b\,x^2) = \sin(a\,x) + x\cos(a\,x)\,a + 2\,b\,x$$

```
> diff(f,a);                # differentiate wrt a, just for a change !
```

$$x^2\cos(a\,x)$$

Make d^2y/dx^2. Notice the `x, x` term to indicate differentiation twice

```
> Diff(f, x, x);
```

$$\frac{\partial^2}{\partial x^2}(x\sin(a\,x) + b\,x^2)$$

```
> diff(f,x,x);
```

$$2\cos(a\,x)\,a - x\sin(a\,x)\,a^2 + 2\,b$$

To differentiate `f`, n times use the `$` symbol, in this case a formal result is returned

```
> Diff(f,x$n): % = value( %);
```

$$\frac{\partial^n}{\partial x^n}(x\sin(a\,x) + b\,x^2) = \frac{1}{2}\frac{1}{a}(e^{1/2\,I\,n\,\pi}(-x\sin(a\,x)\,(-a)^{n+1} + \sin(a\,x)$$
$$+ I((-1)^n - 1)\cos(a\,x))\,x\,a^{n+1} - n\,((-I\sin(a\,x) + \cos(a\,x))\,(-a)^n$$
$$+ (\cos(a\,x) + I\sin(a\,x))a^n))) + b\,\mathrm{pochhammer}(3-n, n)x^{2-n}$$

```
> Diff(f,x$4): % = value( %);        # same as diff(f,x,x,x,x) but
                                       neater
```

$$\frac{\partial^4}{\partial x^4}(x\sin(a\,x) + b\,x^2) = -4\,a^3\cos(a\,x) + a^4\,x\sin(a\,x)$$

A user-defined function is differentiated as shown next

```
> g:= x -> sin(x)^3/ln(x)^2;              # define a function g
> Diff( g(x),x ): % = value( %);          # differentiate g wrt x
```

$$\frac{d}{dx}\left(\frac{\sin(x)^3}{\ln(x)^2}\right) = \frac{3\sin(x)^2\cos(x)}{\ln(x)^2} - \frac{2\sin(x)^3}{\ln(x)^3\,x}$$

A.16 The D operator

Differentiation can also be done with the `D` operator but is not as simple as using `diff`. The function name is `g`.

```
> g:= x-> sin(x)^3/ln(x)^2;               # define a function g
> first_deriv:= D(g)(x);
  second_deriv:= (D@@2)(g)(x);
```

$$first\_deriv := \frac{3\sin(x)^2\cos(x)}{\ln(x)^2} - \frac{2\sin(x)^3}{\ln(x)^3\,x}$$

$$second\_deriv := \frac{6\sin(x)\cos(x)^2}{\ln(x)^2} - \frac{12\sin(x)^2\cos(x)}{\ln(x)^3\,x} - \frac{3\sin(x)^3}{\ln(x)^2} + \frac{6\sin(x)^3}{\ln(x)^4\,x^2} + \frac{2\sin(x)^3}{\ln(x)^3\,x^2}$$

A.17 **Integration**

Integration syntax follows the standard format of the function followed by the variable with which to integrate.

```
> f:= x * sin( a*x ) + b * x^2:
> Int(f, x): % = value(%);
```

$$\int (x\sin(a\,x) + b\,x^2)\,dx = \frac{\sin(a\,x) - x\cos(a\,x)a}{a^2} + \frac{1}{3}b\,x^3$$

Integrating with limits puts the limits after the variable. If integers are used, integer results are returned.

```
> Int(sin(a*x), x= -1/2..2): %= value(%);
```

$$\int_{-1/2}^{2} \sin(a\,x)\,dx = -\frac{-\cos\left(\frac{1}{2}a\right) + \cos(2\,a)}{a}$$

```
> Int(sin(a*x), x= -0.5..2): %= value(%);
```

$$\int_{-.500000}^{2} \sin(a\,x)\,dx = -\frac{1.000000(-1.000000\cos(0.500000a) + \cos(2.000000a))}{a}$$

```
> Int( r*exp( -r^2 ), r = 0..infinity):% = value( % );
```

$$\int_{0}^{\infty} r\,e^{-r^2}\,dr = \frac{1}{2}$$

If an integral is unknown, using `evalf` can sometimes force a numerical answer

```
> f:= sqrt(1 + x^7);
  integral_f:= int(f,x =1..7);
  Int(f, x= 1..7): %= evalf(%);
```

$$f := \sqrt{1 + x^7}$$

$$integral\_f := -\frac{2}{9}\,\mathrm{hypergeom}\left(\left[-\frac{9}{14}, -\frac{1}{2}\right], \left[\frac{5}{14}\right], -1\right) + \frac{4802}{9}\sqrt{7}\,\mathrm{hypergeom}\left(\left[-\frac{9}{14}, -\frac{1}{2}\right], \left[\frac{5}{14}\right], -\frac{1}{823543}\right)$$

$$\int_{1}^{7} \sqrt{1 + x^7}\,dx = 1411.621078000$$

Try a double integral of `x*sin(x+y)`, first by x, then the result by y. This can be done in two steps or all at once.

```
> by_x:= int( x*sin(y+x), x);
```

$$by\_x := \sin(y + x) - (y + x)\cos(y + x) + y\cos(y + x)$$

```
> by_y:= int( by_x, y);
```

$$by\_y := \cos(y + x) + x\sin(y + x)$$

Note the syntax for double integration done all at once

```
> Int( Int(x*sin(y+x),x) ,y ): % = value(%);
```

$$\iint x\sin(y + x)\,dx\,dy = -\cos(y + x) - x\sin(y + x)$$

A.18 solve, fsolve, and unapply

Solving equations algebraically and making functions of the answers.

```
> ans:= solve(a*x^2+b*x+c=0,x);
```

$$ans := -\frac{1}{2}\frac{b-\sqrt{b^2-4\,a\,c}}{a}, -\frac{1}{2}\frac{b+\sqrt{b^2-4\,a\,c}}{a}$$

```
> ans[1]; ans[2]; # the result is a list of 2 elements
```

$$-\frac{1}{2}\frac{b-\sqrt{b^2-4\,a\,c}}{a}$$

$$-\frac{1}{2}\frac{b+\sqrt{b^2-4\,a\,c}}{a}$$

Strangely the `unapply` operation makes a function of the answers. `func1` is made into a function of *a*, *b* and *c*.

```
> funcA:= unapply( ans[1],a);
  func1:= unapply( ans[1],a, b, c);
```

$$funcA := a \to -\frac{1}{2}\frac{b-\sqrt{b^2-4\,a\,c}}{a}$$

$$func1 := (a, b, c) \to -\frac{1}{2}\frac{b-\sqrt{b^2-4\,a\,c}}{a}$$

Now put some values into the functions

```
> funcA(2); func1(1,b,c); simplify( func1(2, 4, 3) );
```

$$-\frac{1}{4}\,b+\frac{1}{4}\sqrt{b^2-8\,c}$$

$$-\frac{1}{2}\,b+\frac{1}{2}\sqrt{b^2-4\,c}$$

$$-1+\frac{1}{2}\mathrm{I}\sqrt{2}$$

The equation below, `3+x-x^4 = 0`, has complex roots, `evalf` finds them all, notice the syntax to do this, we make a list of values [value1, value2, etc.] by placing results in square brackets separated by commas or we could do each separately which is often easier to read. The function could be plotted to find the real roots to convince ourselves as to the values and is shown in Fig. A.1.

```
> ans:= solve(3 + x - x^4 = 0,x):
  ans[1]; ans[2]; ans[3]; ans[4];
```

$$\mathrm{RootOf}(-3-\_Z+\_Z^4, index=1)$$
$$\mathrm{RootOf}(-3-\_Z+\_Z^4, index=2)$$
$$\mathrm{RootOf}(-3-\_Z+\_Z^4, index=3)$$
$$\mathrm{RootOf}(-3-\_Z+\_Z^4, index=4)$$

```
> evalf( [ ans[1], ans[2], ans[3],ans[4] ]); # all together
```

$$[1.452626879, -0.1442958693+1.324149775\mathrm{I}, -1.164035140,$$
$$-0.1442958693-1.324149775\mathrm{I}]$$

```
> ans1:= evalf( ans[1]);
  ans2:= evalf( ans[2]);
```

$$ans1 := 1.452627$$
$$ans2 := -0.144296+1.324150\,\mathrm{I}$$

The function `fsolve` finds numerical solutions. Notice only the real roots are found this way.

```
> fsolve(-3-x+x^4=0,x);
```

$$-1.164035, 1.452627$$

An example of a pair of simultaneous equations; note syntax using curly brackets which denoted a set `{ }`. Square brackets `[ ]` could also be used.

```
> eqns:= 3*y^2 + 2*y*x = 3, 3*x + 4*y = 0 ;
```

$$eqns := \{3\,y^2 + 2\,y\,x = 3, 3\,x + 4\,y = 0\}$$

```
> ans:=solve( { eqns },{ x, y} ); # solve for x and y
```

$$ans := \{y = 3, x = -4\},\{y = -3, x = 4\}$$

```
> ans[1]; ans[2];             # answers are not of use in this form
```

$$\{y = 3, x = -4\}$$
$$\{y = -3, x = 4\}$$

Note how to extract x and y values, `rhs` means get right-hand side of the result, `x_is` and `y_is` can now be used in other calculations.

```
> ans[2][1];
  ans[2][2];
  x_is:= rhs( ans[2][1]);
  y_is:= rhs( ans[2][2]);
```

$$y = -3$$
$$x = 4$$
$$x\_is := -3$$
$$y\_is := 4$$

A.19 Solving differential equations, dsolve

To solve the first-order rate equations for $A \rightarrow B$, let `Ca` be the concentration of A and `Cb` that of B. `Ca` must be defined as a function of time `Ca(t)` for Maple to know what to integrate with respect to.

```
> fa:= diff(Ca(t),t) = - k*Ca(t);                    # define equation
```

$$fa := \frac{d}{dt} Ca(t) = -k\, Ca(t)$$

```
> dsolve(fa); # no initial condition, so constant _C1 is produced
```

$$Ca(t) = \_C1\, e^{-k\,t}$$

Now put in initial conditions `Ca(0) = C0`, which means that the concentration at zero time is `C0`. The initial conditions are placed in brackets next to the equation.

```
> inits:= Ca(0) = C0 ;
> dsolve( { fa, inits});
```

$$Ca(t) = C0\, e^{-k\,t}$$

Now look at the appearance of species B forming from A with rate constant k.

```
> fb:= diff( Cb(t),t ) = + k * Ca(t); # define equation for B
```

$$fb := \frac{d}{dt} Cb(t) = k\, Ca(t)$$

When solving the equations as a pair, also add the initial conditions, use `ans[1]`, `ans[2]` to get at the answers

```
> inits:= Ca(0)= C0,Cb(0)= 0 ;
> ans:= dsolve( { fa, fb, inits});
```

$$ans := \{Ca(t) = C0\, e^{-k\,t}, Cb(t) = -C0\, e^{-k\,t} + C0\}$$

```
> ans[1]; ans[2];        #this is an easier way to look at answers
```

$$Ca(t) = C0\, e^{-k\,t}$$
$$Cb(t) = -C0\, e^{-k\,t} + C0$$

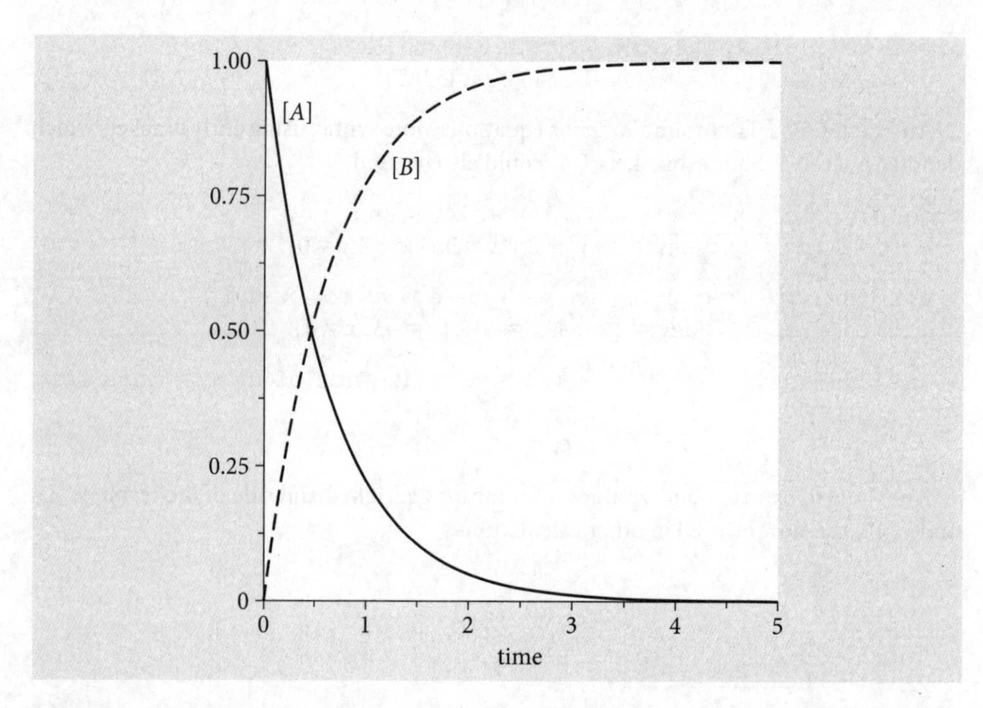

Fig. A.2 Exponential growth and decay

Make a function of the answer for use later on.

```
> conc_A:= unapply( rhs(ans[2]),t);
```

$$conc\_A := t \rightarrow -C0\, e^{-k\,t} + C0$$

Plot results (Fig. A.2), assume some numbers for `k` and `C0`, and use `rhs` to get at the right-hand side of the equations

```
> k:= 1.5; C0:= 1.0;
  plot( [ rhs(ans[1]), conc_A(t)], t= 0..5,
                                   linestyle = [dash, solid]);
```

A.20 Plotting functions

Very often it is useful to make your own function particularly if you want to plot or calculate with different values for its parameters.

Defining functions using the `->` operator, call the function `func` then plot it (Fig. A.3)

```
> func:= x-> exp(-x^2) + sin(x^2) + 2;
```

$$func := x \rightarrow e^{-x^2} + \sin(x^2) + 2$$

```
> plot(func(s),s=0..3*Pi,
       colour=[blue],view=[0..10, 0..4],numpoints =1000,
       axis=[tickmarks=[5, subticks=1]] );
```

In the function x is a dummy variable and we can substitute some other expression such as $x + x^2$ then and plot again. This time some more options are included (Fig. A.4).

```
> func(x+x^2);
```

$$e^{-(x+x^2)^2} + \sin((x + x^2)^2) + 2$$

```
> plot( [func(x+x^2),func(x)],x=0..3,0..4,
     color=[red, blue], axes= boxed, thickness= [1, 2],
     labels=["x ","y"], linestyle=[solid, dash],
     title = "func(x) dashed line and func(x+x^2) solid line",
     titlefont=[TIMES,ITALIC,12],
     axis=[ tickmarks=[5, subticks =1] ]     );
```

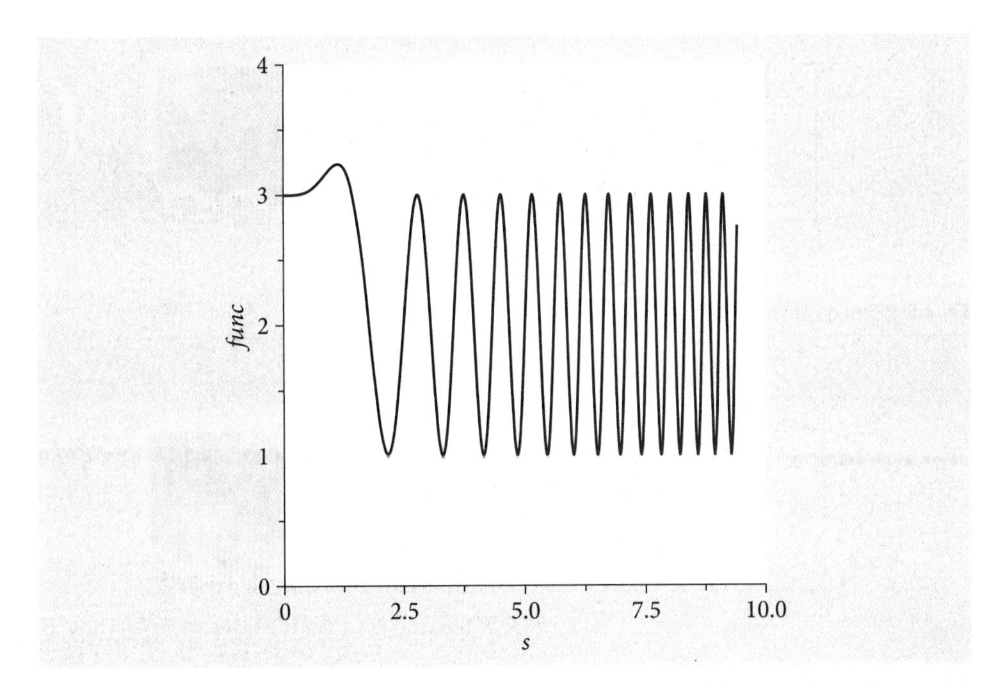

Fig. A.3 Plotting options with increasing frequency sine wave

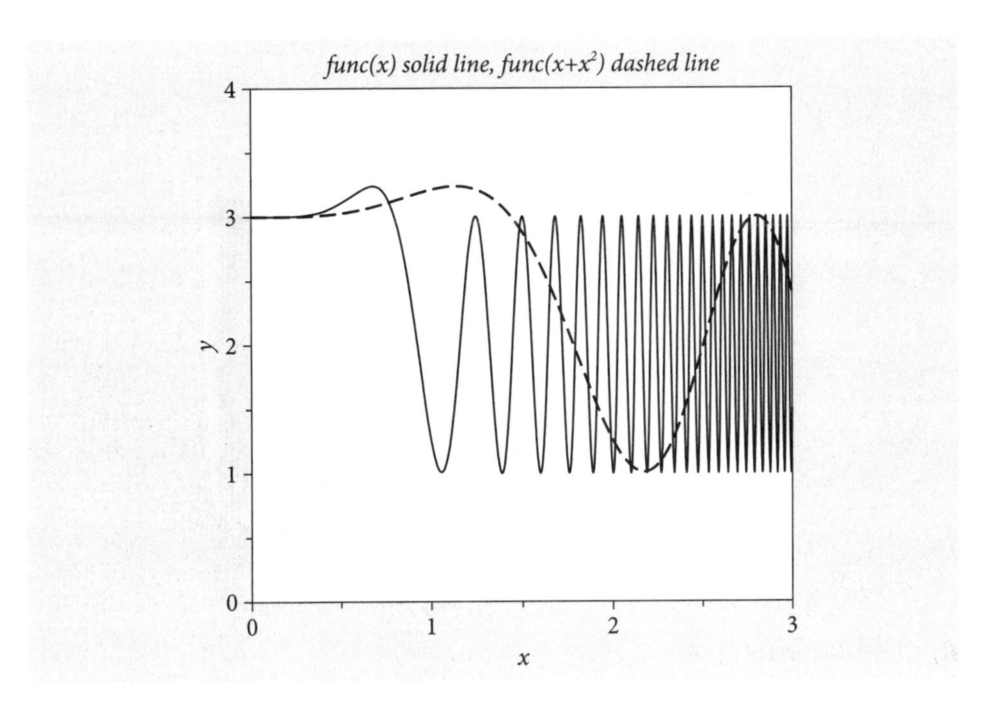

Fig. A.4 Plotting options

If a package is missing the instructions are returned without comment!

```
> logplot(func(x),x=0..5); # no plot: must define with(plots)
```

$$logplot(e^{-x^2} + \sin(x^2) + 2, x = 0..5)$$

If log plots are needed the instruction `with(plots)` must be added.

If not enough data points are defined in a plot then errors appear as in the top graph, Fig. A.5.

```
> with(plots):
> logplot(func(x),x=0..20);                          #Fig. A.5
> logplot(func(x),x=0..20,numpoints=1000);           #Fig. A.6
> semilogplot(func(x),x= 0.05..5, view=[0.5..4, 0.9..4] );
                                                     #Fig. A.7
```

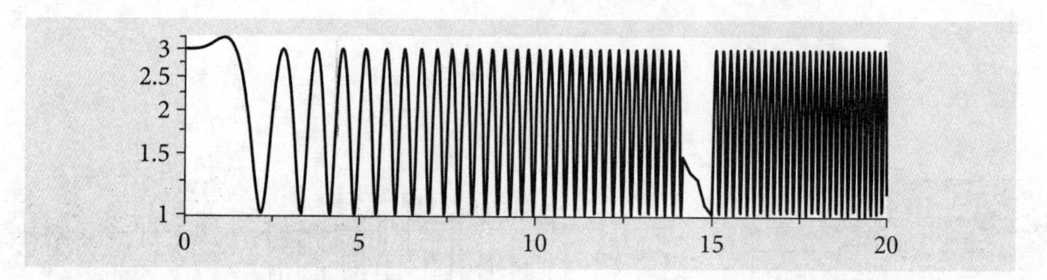

Fig. A.5 Plotting with an insufficient number of points

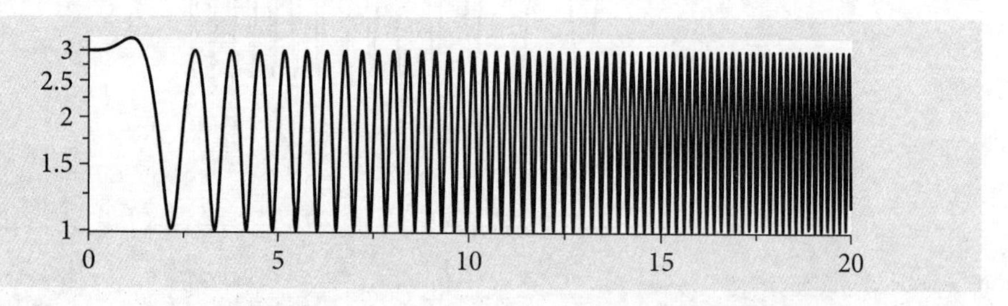

Fig. A.6 Plotting with numpoints = 1000

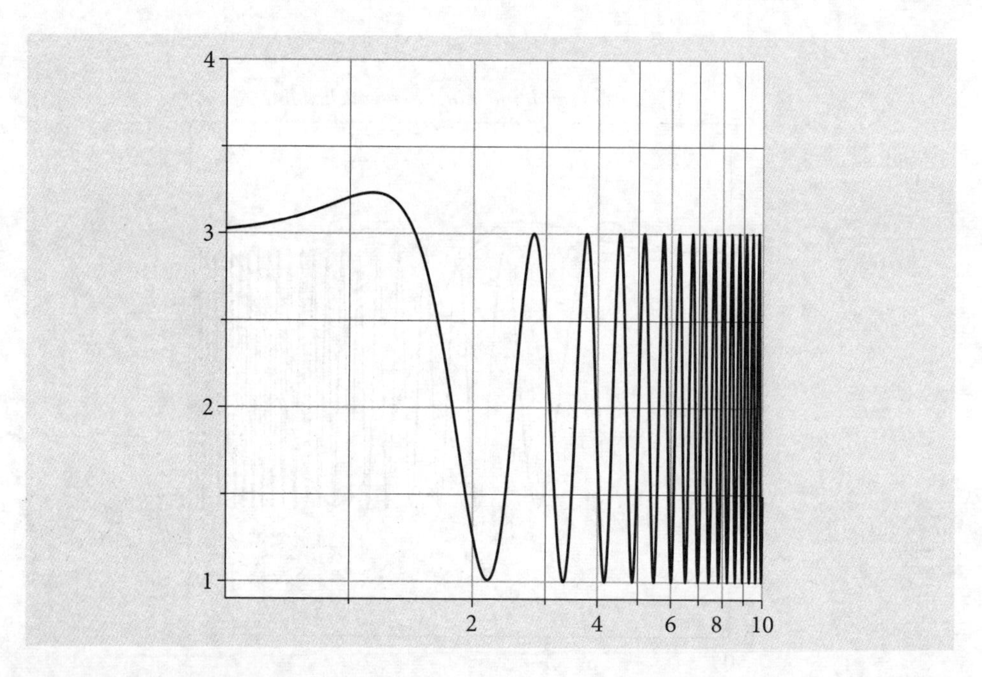

Fig. A.7 Semi-log plot of Fig. A.6

A.21 **Plotting data**

Pairs of data points are made into a list, separated from one another as shown below. These are plotted as points so that the `style` and `symbol` options are used and also `symbolsize` is defined.

```
> pnts:=[ [0,0],[1,2.5],[2,3.4],[3,2.0],[4,1.5],[5,0.25] ];
```

$$pnts := [[0, 0], [1, 2.5], [2, 3.4], [3, 2.0], [4, 1.5], [5, 0.25]]$$

```
> plot(pnts, style= point, symbol= circle, symbolsize= 20,
                                     view=[0..6,-1..4]); # Fig. A.8
```

Data can be input from other programs, see the Maple user manual, chapter 10, for examples of how to do this. This manual can be found by typing *input* into the help menu and looking down the list of topics produced.

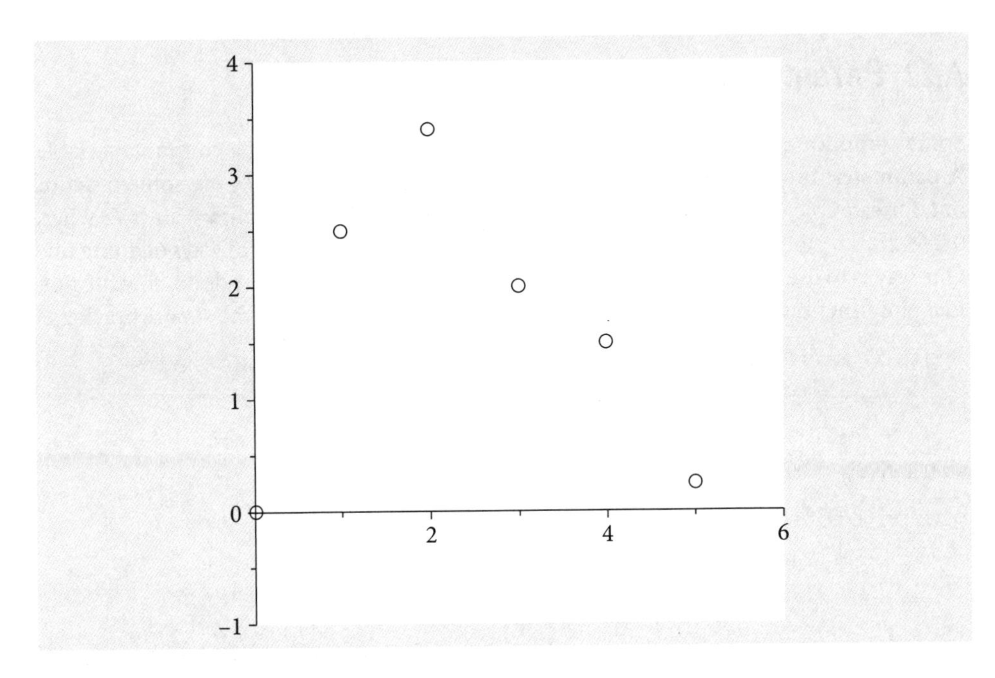

Fig. A.8 Plotting pairs of points

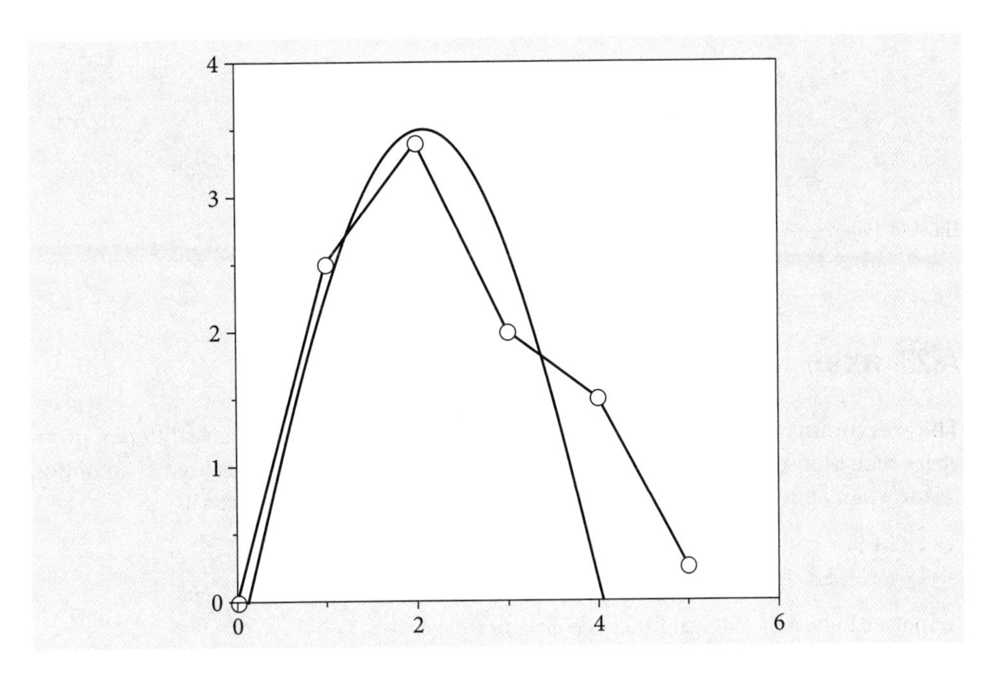

Fig. A.9 Three superimposed plots using `display`.

Use `display` to plot multiple plots of different types, the `with (plots)`, package must be loaded. This is useful when different types of graphs are needed, data points and a function for example. Notice that we give the plots a name, `p1` and `p2`, and suppress printing with the colon : then use `display` to get the graph, also the view size is limited and boxed axes used. The `filledcircle` instruction and `color= white` over-plots with white so that the circle appears solid over the line.

```
> p1:= plot(pnts, style=point, symbol=circle,
                                symbolsize=30 ):
  p11:= plot(pnts, style=point, symbol= solidcircle,
                                 symbolsize=30,color=white ):
  p2:= plot( pnts ):
  p3:= plot( 3.5*sin(x*4/5-0.1),x = 0..6):
  display(p2,p3,p11,p1, view=[0..6,0..4], axes= boxed );
                                                # Fig. A.9
```

A.22 Parametric plots

Some complicated functions are defined with two functions having a common variable. A parametric function is defined as $x = f(t)$ and $y = g(t)$ where f and g are some function of t, for example, the ellipse $x^2 + 9y^2 = 1$ can be written as $x = \cos(t)$ and $y = \sin(t)/3$ where $0 \le t \le 2\pi$. To plot the function, a parametric plot is used. Maple has two ways of doing this. One way is to use `implicitplot`, the other just uses `plot` but depends on how the normal plot function is defined; notice that the `t = ..` is placed *inside* the square brackets.

```
> plot([cos(t),sin(t)/3,t=0..2*Pi], scaling= constrained,
        axis=[ tickmarks=[5, subticks =1] ],
        view=[-1.25..1.25,-0.45..0.45]);
```

alternatively `implicitplot` can be used

```
> with(plots):
> implicitplot(x^2 + 9*y^2 = 1, x=-1..1, y=-1..1,
               scaling = constrained, tickmarks=[5,5],
               view=[-1.25..1.25,-0.45..0.45]);
```

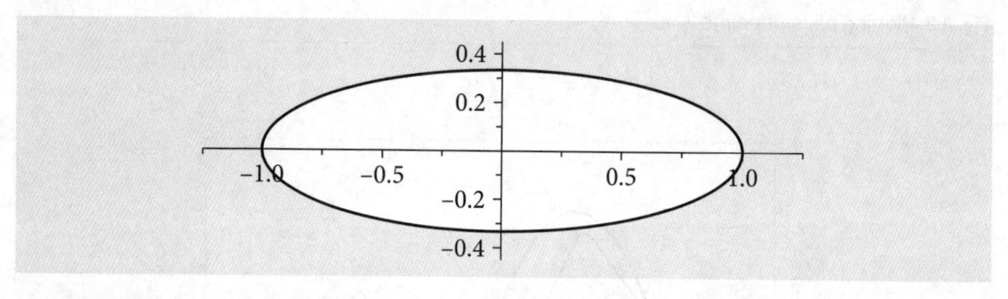

Fig. A.10 Parametric plot

A.23 axes

The axes command allows you to customize a plot by allowing axes to have different properties such as number of tick-marks, colours, or to be linear or log. It is used as an option inside a plot command. The command can be used on an individual axis as

```
axis[1]=[gridlines=[10,thickness=2, subticks=false,
colour=blue],mode=log]
```

to make a log x axis with gridlines. To alter both axes use

`axes=[tickmarks=[5, subticks=1]]` or `axes= boxed` etc.

A.24 Sequences

The following expression a sequence and is mostly used to make names or strings, values are separated by commas

```
> seq1:= 1, 2, 3, 4;
```

$$seq1 := 1, 2, 3, 4$$

```
> seq2:= a, b, c, d;
```

$$seq2 := a, b, c, d$$

To make a list to hold x, y, z coordinates, say, the `seq` command is very useful;

```
> new_xyz:=[ seq([0,0,0], i = 1..5) ];
```

$$new\_xyz := [[0, 0, 0], [0, 0, 0], [0, 0, 0], [0, 0, 0], [0, 0, 0]]$$

```
> new_xyz[3];                              # choose any value
```

$$[0, 0, 0]$$

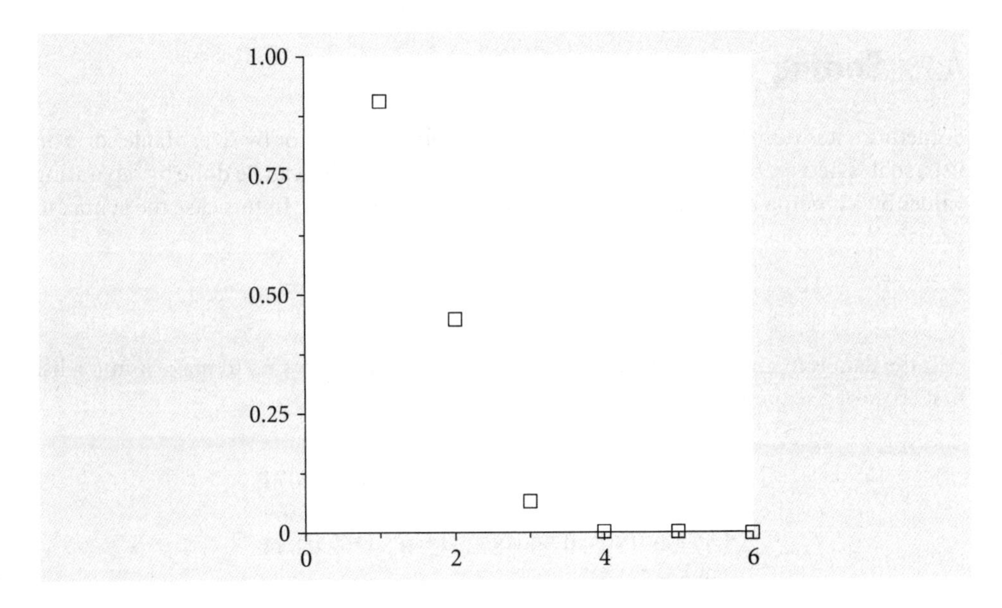

Fig. A.11 Plotting a sequence of points

The new values which must be a vector or part of a matrix can then be entered into each element.

```
> v1:= < 3|4 |5 >;
```

$$v1 := [3 \quad 4 \quad 5]$$

```
> new_xyz[4]:= v1;
```

$$new\_xyz_4 := [3 \quad 4 \quad 5]$$

The `seq` command is also very useful in plotting or in loops. `AA` is a sequence of pairs of values that is then plotted and shown in Fig. A.11.

```
> AA:= seq( [i, exp(-i^3/10)], i=1..9 );
  plot([AA], style= point, symbol= box,
        symbolsize= 20, view=[0..6,0..1]);
```

$$AA := [1, e^{-\frac{1}{10}}], [2, e^{-\frac{4}{5}}], [3, e^{-\frac{27}{10}}], [4, e^{-\frac{32}{5}}], [5, e^{-\frac{25}{2}}], [6, e^{-\frac{108}{5}}], [7, e^{-\frac{343}{10}}], [8, e^{-\frac{256}{5}}], [9, e^{-\frac{729}{10}}]$$

A.25 Arrays, vectors, and matrices

These data structures are defined with a range of values and called by addressing them with their name followed by square brackets. The names `Array`, `Vector`, and `Matrix` must have a capital letter; Maple has older versions of these, which are called with lower case first letter, but these are no longer recommended.

```
> V:= Vector(4);
```

$$V := \begin{bmatrix} 0 \\ 0 \\ 0 \\ 0 \end{bmatrix}$$

```
> M:= Matrix(3,4);
```

$$M := \begin{bmatrix} 0 & 0 & 0 & 0 \\ 0 & 0 & 0 & 0 \\ 0 & 0 & 0 & 0 \end{bmatrix}$$

```
> A:= Array(1..3);
```

$$A := [0 \quad 0 \quad 0 \quad 0]$$

The index runs from 1 to n for each dimension so that `V[1]` is the first element and `V[4]` the last. The matrix has two indices, `M[1,1]` up to `M[4,4]` in this instance.

A.26 Sorting

Sometimes it is necessary to sort results, for example by energy or by size. Maple sorts on lists, so it is necessary to have the data in this form. Usually this can be done by separating values by a comma and enclosing the whole in square brackets. In this case the syntax to use is

```
> sort( [ 10,1,30,15] );
```

$$[1, 10, 15, 30]$$

If the data is in an array then it may be necessary to use `convert` to make it into a list first. To sort a sequence of random numbers

```
> sort( [seq( rand()/1e12,i=1..10)]);
```

$$[0.0224241705, 0.1931398164, 0.3864083074, 0.3957188605, 0.4122862858, 0.4275520569, 0.6946071893, 0.8001874845, 0.8426226844, 0.9964172142]$$

Sorting can be in ascending or descending order either using options `'>'` or `'<'`, or the keywords, `ascending` or `descending`.

To sort two sets of data, the second set based on the ordering of the first, an index is made and sorted and the resulting indexes used to sort the second set of data. If the first array of numbers is called `X` and the second set of data a matrix `Q` then the code below can be used. The results are put into `XX` and `QQ`. It is assumed that array `X` and matrix `Q` have already been defined and contain numbers. The number of data points is `num` as is the size of `Q` and `QQ`.

```
# sort keep as pairs by using index, called jndx
  indx := [ seq( i, i=1..num ) ]:
  jndx:= sort( indx, (i,j)-> evalb( X[i] < X[j] ));
# use jndx to make new array with data in sorted order.
  XX := Array([seq(X[i],i= jndx)]):  # sorted data XX
# now sort second set of data with index from first
  QQ:= Matrix(1..num,1..num):           # use to store sorted data
  for i from 1 to num do                # sort using index
    for j from 1 to num do
        QQ[ i,j ]:= Q[i,jndx[j] ];
    end do
  end do;
```

A.27 Verify

Verify can be used to compare two matrices, `m1` and `m2`. If they are the same, a value of 'true' is returned, which means that verify can be used in an `if.. then do something.. end if` statement to check if matrices are the same or not. Verify can be used to check other objects if necessary; see Maple's help pages.

```
> verify( m1, m2, 'Matrix' );
```

or, for example,

```
> if verify( m1, m2, 'Matrix' ) then
    print("identical") else print("different") end if;
```

A.28 Programming: `for` and `while .. do` loops, and `if..then` statements

When a calculation has to be repeated then `for .. do` loops and `while.. do` loops are used. To check on certain conditions then `if .. then .. else .. end if` statements are needed. The statements always have the same general outline, viz.

```
for .. do
 statement 1;
 statement 2;
 etc;
end do;
```

If a semicolon is put after the `end do;` then the statements in the loop are printed; if a colon, nothing is printed and a specific print statement has to be included in the loop. For example, to check if $j^3 > 20$ as j is increased from 1 and to print the result a loop is needed, such as the one below. The value of `j` is automatically incremented by 1 each time round the loop.

```
> for j from 1 to 3 do
    if j^3 > 20 then print(j^3) else print(-j^3) end if;
  end do:
```

$$-1$$
$$-8$$
$$27$$

Unlike the `for` loop, in a `while` loop the index `j` has to be updated. The `printf` statement uses formats to output a table of values. The printing codes `"%s etc"` are explained on the `printf` help pages.

```
> j:= 180:
  while j >= 0 do
    printf(" %s %3.3a %s %3.4g \n",
                  "sin of ", j , "deg is", sin(j*Pi/180));
    j:= j - 45 ;
  end do:
     sin of 180 deg is 0
     sin of 135 deg is 0.7071
     sin of 90 deg is 1
     sin of 45 deg is 0.7071
     sin of 0 deg is 0
```

The `for` and `while` instructions can be combined as in

```
> q:= 10:
  for i from 1 to 5 while q > 8 do
    print(i^2); q:= q - 1
  end do:
```

which prints 1 and 4 as output.

To fill a matrix with values use

```
> M:= Matrix(6,5):
  for j from 1 to 6 do
    for k from 1 to 5 do
      M[j, k]:= j + k - 1
    end do;
  end do;
> M;
```

$$\begin{bmatrix} 1 & 2 & 3 & 4 & 5 \\ 2 & 3 & 4 & 5 & 6 \\ 3 & 4 & 5 & 6 & 7 \\ 4 & 5 & 6 & 7 & 8 \\ 5 & 6 & 7 & 8 & 9 \\ 6 & 7 & 8 & 9 & 10 \end{bmatrix}$$

A.29 Procedures

Sometimes a set of instructions needs to be grouped together so that they can be reused. In this case a procedure is needed. The general form is

```
name:= proc()
  several lines of instructions etc;
end proc;
```

An example is shown below.

```
> Q:= proc(a, b)              # note that no semicolon is used here
  local c;                    # has value only inside proc
  c:= Pi:
  if a > b then exp(-a*b*c) else print('error') end if;
  end proc:
> Q(20,3);                    # call procedure with a=20, b=3
```

$$e^{-60\pi}$$

```
> Q(3,20);
```

error

A.30 Concatenation

Concatenation can be used as a very neat way of making a polynomial. It joins two things together and the result is then interpreted as if you had typed it. Concatenation is illustrated, then the result is made into a function and finally an example is given to make a general polynomial. The `||` is a concatenation operator which joins `a` to the number produced by `k = 1..5`. (The instruction `cat(a,k)` could also be used instead of `a||k`).

```
> func:= a0 + add( a||k * x^k, k = 1..5);
```

$$func := a0 + a1\ x + a2\ x^2 + a3\ x^3 + a4\ x^4 + a5\ x^5$$

```
> func:= x-> a0 + add( a||k*x^k, k = 1..5);
```

$$func := x \rightarrow a0 + add(a \,||\, k\ x^k, k = 1\,..5)$$

```
> fx:= func(x);
```

$$fx := a0 + a1\ x + a2\ x^2 + a3\ x^3 + a4\ x^4 + a5\ x^5$$

```
> fsinx:= func(sin(x));
```

$$fsinx := a0 + a1\ \sin(x) + a2\ \sin(x)^2 + a3\ \sin(x)^3 + a4\ \sin(x)^4 + a5\ \sin(x)^5$$

A.31 Sets and lists

A set is enclosed in `{ , , }` and a list in `[ , , ]`. Maple may re-order members of a set but not a list or sequence. Any element may be recalled by placing square brackets after the name.

```
> a_list:=[1,3,55,2];
```

$$a\_list := [1, 3, 55, 2]$$

To extract the third member or sub-part of a whole list

```
> a_list[3]; a_list[2..4];
```

$$55$$

$$[3, 55, 2]$$

```
> b_list:=[x,x^2,x^3,x^4,x^5];
```

$$b\_list := [x, x^2, x^3, x^4, x^5]$$

```
> b_list[3];
```

$$x^3$$

Algebra can be performed on the whole list

```
> expand(b_list*x);
```

$$[x^2, x^3, x^4, x^5, x^6]$$

To make a nice equation, name and extract one element and then make it a function use

```
> b_list/sin(x): % = value(expand( %) ); new_list:= rhs(%);
  third_element:= new_list[3];
  func:= unapply( new_list[3], x );
```

$$\frac{[x, x^2, x^3, x^4, x^5]}{\sin(x)} = \left[\frac{x}{\sin(x)}, \frac{x^2}{\sin(x)}, \frac{x^3}{\sin(x)}, \frac{x^4}{\sin(x)}, \frac{x^5}{\sin(x)}\right]$$

$$new\_list := \left[\frac{x}{\sin(x)}, \frac{x^2}{\sin(x)}, \frac{x^3}{\sin(x)}, \frac{x^4}{\sin(x)}, \frac{x^5}{\sin(x)}\right]$$

$$third\_element := \frac{x^3}{\sin(x)}$$

$$func := x \to \frac{x^3}{\sin(x)}$$

Join lists using the `op()` command

```
> joined_list:= [op(a_list),op(b_list)];
```

$$joined\_list := [1, 3, 55, 2, x, x^2, x^3, x^4, x^5]$$

A.32 Example of integration and more complex plotting

If there are two electrons and they occupy distinct orbitals ψ_1 and ψ_2, in the simple orbital approximation the wavefunction is $\psi_1\psi_2$. The joint probability density would be $\psi_1^*\psi_2\psi_1\psi_2^*$ if it were not for the Pauli exclusion principle, which makes the states symmetric or anti-symmetric to exchanging electrons.

A triplet state of an atom or molecule has unpaired spins while in a singlet they are paired. The triplet is lower in energy than the corresponding singlet. One reason for this is that the electrons try to 'avoid' one another in the triplet; see Salem (1982) for a detailed discussion. This means that the *spatial* part of the triplet wavefunction is asymmetric to exchange of the electrons;

$$\varphi_T = \psi_1(1)\psi_2(2) - \psi_1(2)\psi_2(1)$$

and the singlet spatial part is symmetric

$$\varphi_S = \psi_1(1)\psi_2(2) + \psi_1(2)\psi_2(1).$$

The numbers in brackets refer to the electrons' coordinate. As the total wavefunction is anti-symmetric to exchange of electrons the corresponding spin parts have the opposite symmetry to the spatial parts. The probability density is $|\varphi^*\varphi|$ or just the square φ^2 if the wavefunctions are real.

Plot the spatial parts of the probability density of the singlet and triplet state formed from levels with quantum numbers 2 and 3 to confirm their symmetry. Use the wavefunction for a well of side L which is $\psi = N\sin(n\pi x/L)$, where N is the normalization.

Strategy: The wavefunctions will have to be normalized first. As they are real, this is the integral $N^2\int_0^L \sin^2(\pi nx/L)dx = 1$ and this will depend on the quantum number, n. The exchange of electrons is made by exchanging the coordinates; electron 1 has coordinates x_1 and electron 2 has x_2.

Solution: The normalization can be done easily by converting to the exponential form before integrating. The calculation and plots are shown below. The list `bits` contains plotting parameters, you can change these to customize the plot. The `spacecurve` plots a three-dimensional curve, in this case a straight line $x_1 = x_2$. `display` plots different graphs together and these are saved as objects `p1`, `p2` etc. If you click on the plots then drag the mouse you can see the three-dimensional shape of the surface plotted. The functions $1 - \varphi^2$ are plotted to better emphasize the shading.

```
> restart: with(plots):                # particle in well
> assume (n, integer)
> Int(sin(n*Pi*x/L)^2,x=0..L): %=value(%):
                              N:= unapply(1/sqrt(rhs(%)), n );
  psi:=(n,x)-> N(n)*sin(n*Pi*x/L);
```

$$N := n \rightarrow \sqrt{\frac{2}{L}}$$

$$\psi := (n, x) \rightarrow N(n)\sin\left(\frac{n\,\pi\,x}{L}\right)$$

```
> # bits contains plotting parameters
  bits:= axes=framed, orientation =[0,180], shading = zhue,
     style = surfacecontour, font =[courier,12], grid =[50,50],
     tickmarks = [5, 5, 5], view =[0..L,0..L,-15..0]:
  L:= 1.0;                                     # length & Q nums
  n:= 2; m:= 3;
  triplet:= plot3d(
          1- ( psi(n,x1)*psi(m,x2) - psi(n,x2)*psi(m,x1))^2,
                                        x1 = 0..L, x2 = 0..L, bits):
  singlet:= plot3d(
          1- ( psi(n,x1)*psi(m,x2) + psi(n,x2)*psi(m,x1) )^2,
                                        x1 = 0..L, x2 = 0..L, bits):
  p2:=spacecurve([t,t,-15,t=0..L],linestyle=dash,
                                                    color=black):
  display(triplet,p2);
  display(singlet,p2);
```

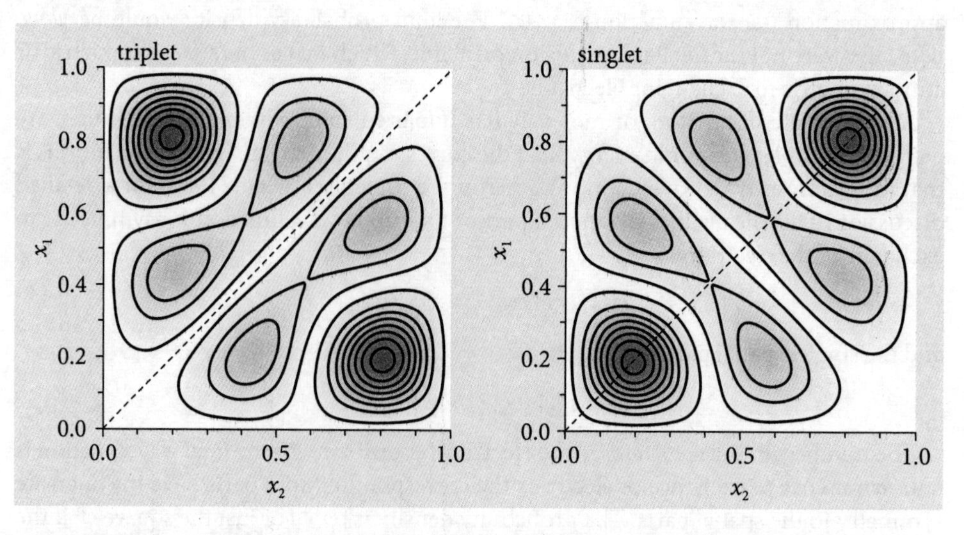

Fig. A.12 Plot showing the probability of finding one electron at the position x_1 if the other electron is at x_2 depending upon whether the electrons are in a triplet (left) or singlet state. Because the electrons crowd together in a singlet, this has higher energy.

Maple and Mathematica Syntax Conversion

The table below gives a translation of some frequently used Maple expressions into equivalent or nearly equivalent Mathematica syntax. It is assumed that you are more familiar with Mathematica than Maple, consequently Mathematica options are not added and there may also be alternative syntax and instructions that could be used.

The two programs have many similar features: simplify, integrate, etc. and their syntax is related but with important differences. Maple uses round brackets except for indexing Arrays, Vectors or Matrices, for example `sin(x)`, `vec[2]`, whereas Mathematica uses square brackets everywhere and starts all instructions with capital letters, e.g., `Sin[x]`. Maple uses small letters for standard mathematical functions but some capitals for packages; see below. Both languages are case sensitive.

All Maple assignments use `:=` but Mathematica uses = and Maple lines end with `;` if a result is to be printed on the screen or `:` to calculate but not to print, whereas Mathematica has no line terminator. Multiplication is always shown in Maple, e.g., `3*x` but the `*` is optional in Mathematica.

Maple packages such as `plots` or `LinearAlgebra` are usually added at the start of a calculation and have the syntax `with(...)`, for example, `with(LinearAlgebra):` or `with(orthopoly);`. In Mathematica, packages are loaded for example, as `<<LinearAlgebra'MatrixManipulation'`

In Maple it is advisable to start each new set of calculations with **`restart:`** because all values are remembered. All values are remembered in Mathematica also so you quit and restart the kernel from the evaluation menu or use Quit[ ]. To calculate in Mathematica you use ShiftReturn and each section is divided into cells so you can recalculate just a specific cell. In Maple a maths line starts on the left of the screen as [> and calculation is continued to the end of the [bracket, which might be one line only. Several maths lines can be contained within one [bracket and to produce a second and more lines *shift return* or *shift enter* is used. You can also add text cells like title, subtitle, etc. in both languages.

Shingareva & Lizarraga-Celaya (2007) give examples of the same calculation in Maple and Mathematica.

	Maple	Mathematica
assign	x:= 3;	x= 3
equality	f:= x^2-3 = 2;	f= x^2-3 == 2
standard functions, sin, cos, tan, log etc.	sin(x^2+2*x)	Sin[x^2+2*x] Sin[x^2+2 x]
range as used in plots, sequences etc.	x = -Pi..3 j = 1..10	{x, -Pi, 3} {j, 1, 10}
i, π, ∞, $\lvert x\rvert$, real, imag	I Pi infinity abs(x) Re(z) Im(z)	I Pi Infinity Abs[x] Re[z] Im[z]
plot two functions	plot([sin(2*x), cos(2*x)], x=-Pi..Pi);	Plot[{ Sin[2*x], Cos[2*x] }, {x,-Pi,Pi}]

plot different graphs together	with(plots):
	# s is a list
	s:= seq([i, sin(i)], i=0..10);
	p1:= plot([s],style=point,symbol=solidcircle,
	symbolsize=20, color=blue);
	p2:= plot(sin(x),x=0..10,colour=black):
	display(p1,p2);
		
		s=Table[{i, Sin[i]}, {i,0,10}]
		p1=Plot[[s],DisplayFunction->Identity];
		
		p2=Plot[Sin[x],{x,0,10},DisplayFunction->Identity];
		Show[p1,p2,DisplayFunction->$DisplayFunction]

user defined functions	f:=(x,y)-> sin(x*y);
	f1:=x-> if x > 1 then sin(x) end if;
	plot('f1(x)',x=-2..2*Pi); # to delay evaluation
	use single quotes. Does not plot x <=1
		f[x_,y_]:= Sin[x*y]
		f1[_x]:= If[x > 1 , Sin[x],0]
		Plot[f1[x],{x,-2,2*Pi}]

numerical value	evalf(f(2,3));	N[f[2,3]]
simplify, factor, expand, combine etc	simplify(f(x, y)); simplify(f(x,y),symbolic);	Simplify[f[x, y]] f[x, y]//Simplify
loops:	s:= 0; for i from 1 to 10 do s:= s + 1 end do;	s = 0 For[i=1, i <=10, i = i+1, s=s + 1]
while loop	s:= 2; while s <=4000 do s:= s^2 end do;	s= 2 While[s <=4000, s= s^2]
if	if x > 0 and y < 1 then s:=1 else s:=0 end if;	If[x > 0 && y < 0, s= 1, s= 0]
solve algebraically	solve(sin(x)^2-sin(x)-3=0);	Solve[sin[x]^2-sin[x]-3==0,x]
solve numerically	evalf(solve(sin(x)^2-sin(x)-3=0)); fsolve(x^2+x-1); # real solutions only	NSolve[sin[x]^2-sin[x]-3==0,x]

<table>
<tr><td>differentiate</td><td>diff(x*sin(x), x);
Diff(x*sin(x),x): %= value(%); # produces question and answer

f:= x-> x*sin(x);
diff(f(x), x);
diff(f(x), x$4);
D(f)(x);</td><td>Derivative[1][x*sin[x]]

f[x_]:= x*sin(x)
Derivative[1][f] f'[x]
Derivative[4][f] f''''[x]
D[f[x],x]</td></tr>
<tr><td>integrate</td><td>int(sin(x),x);
int(sin(x),x = a..b);
Int(f(x),x = a..b): % = value(%);</td><td>Integrate[sin[x],x]
Integrate[sin[x],{x,a,b}]
Integrate[f[x],{x,a,b}]</td></tr>
<tr><td>numerical value</td><td>evalf(int(f(x),x= a..b));</td><td>NIntegrate[f[x],{x,a,b}]
Integrate[f[x],{x,a,b}]//N</td></tr>
<tr><td>vectors & matrices defined</td><td>vec:= Vector(1..3); # elements are zero
f:= j->x^(j-1); vec:= Vector(1..4,f);
vec:= <1|x|x^2|x^3>; #row
vec:= <1,x,x^2,x^3>; #col
mat1:= < <1|2|x>, <3|x|1>, <x|2|3> >;
f:= (i,j)-> i + j:
mat2:= Matrix(1..4,1..3,f); # 4 rows 3 cols</td><td>vec= Table[{i,1,3}]
vec= Table[x^(j-1),{j,1,4}]
vec= {1,x,x^2,x^3}

mat1={ {1,2,x}, {3,x,1}, {x,2,3} }

mat2=Table[i+j,{i,1,3},{j,1,4}]//MatrixForm</td></tr>
<tr><td>matrix & vector multiply</td><td>mat1 . mat2;
vec . vec; # dot product
with(LinearAlgebra):
mat2 . Transpose(mat1);</td><td>mat1 . mat2

Transpose[{{a, b, c}, {x, y, z}}]
mat2 .Transpose[mat1]</td></tr>
<tr><td>select element</td><td>row2_col3:= mat2[2,3];</td><td>row2_col3= mat2[[2,3]]</td></tr>
<tr><td>invert raise to power determinant</td><td>mat2^(-1);
mat2^3
Determinant(mat1);</td><td>Inverse[mat2]
MatrixPower[mat2,3]
Det[mat1]</td></tr>
<tr><td>arrays</td><td>ary:= Array(1..3,1..4); # 3 row 4 cols</td><td>ary= Table[{i,1,4}, {j,1,3}]</td></tr>
<tr><td>differential equations with initial conditions</td><td>dsolve(diff(y(x),x)+y(x) = 2);
eq:= diff(y(x),x,x) + diff(y(x),x) + y(x) = 3;
inits:= D(y)(0)=0, y(0)=1;
dsolve([eq, inits],[y(x)]);</td><td>DSolve[y'[x]+y[x]==2,y[x], x]
eq= y''[x]+y'[x]+y[x]==3
inits= y'[0]==0, y[0]==1
Dsolve[{ eq, inits }, y[x], x]</td></tr>
</table>

<table>
<tr><th></th><th>Maple</th><th>Mathematica</th></tr>
<tr><td>series</td><td># expand to power x^{10} about x=0
series(sin(x^2),x,10);
# expand about x = a to x^{10}.
series(sin(x^2,x = a, 10);
Poly:=convert(series(sin(x^2,x=a,10),polynom);</td><td>
Series[sin[x^2],{x,0,10}]

Series[sin[x^2],{x,a,10}]
Poly:=Normal[Series[sin[x^2],{x,a,10}]]</td></tr>
<tr><td>random numbers</td><td>seed:= randomize();</td><td></td></tr>
<tr><td>uniform distribution,</td><td>rand()/1e12; # real number 0 to 1
r:= rand(a..b): r(); # integer a & b</td><td>RandomReal[{0,1}]
RandomReal[]
RandomInteger[{a,b}]</td></tr>
<tr><td>Normal and other distributions</td><td>with(RandomTools):
r:= Generate(distribution(Normal(mu,sig)),
makeproc=true):
v:= Vector(10,r); # 10 values</td><td>

v= RandomReal[NormalDistribution[mu,sig],10]</td></tr>
<tr><td>eigenvalues & eigenvectors</td><td>with(LinearAlgebra):
mat1:=< <1|2|3>,<3|-2|1>,<6|4|2> >;
eigvals:= Eigenvalues(mat);
(eigvals,eigvecs):= Eigenvectors(mat);</td><td>
mat1={{1,2,3},{3,-2,1},{6,4,2}}
eigvals= Eigenvalues[mat1]
eigsys= Eigensystem[mat1]
eigsys[[1]] # eigenvalues
eigsys[[2]] # eigen vectors</td></tr>
<tr><td>map</td><td>f:= a + x;
map(y^2, f); # produces y(a)^2+y(x)^2
map(x-> x^2,f); # produces a^2+x^2
map(x-> Re(x), eigvals); # get real parts of vector or matrix
map(x-> Im(x), eigvals); # get imag parts</td><td>f= a + x
Map[y^2, f] # produces (y^2)[a]+(y^2)[x]</td></tr>
</table>

References

Abramowicz, M. & Stegun, I. (1965), *Handbook of Mathematical Functions*, Dover.
Acheson, D. (1997), *From Calculus to Chaos*, Oxford University Press.
Adam, J. A. (2003), *Mathematics in Nature*, Princeton University press.
Alexander, R. M. (1996), *Optima for Animals*, Princeton University Press.
Allen, J. & Eberly, J. H. (1987), *Optical Resonance and Two Level Atoms*, Dover.
Allen, M. P. & Tilldesley, D. J. (1987), *Computer Simulation of Liquids*, Oxford University Press.
Altringham, J. (1996), *Bats: Biology and Behaviour*, Oxford University Press.
Aratyn, H. & Rasinariu, C. (2006), *A Short Course in Mathematical Methods with Maple*, World Scientific.
Arkfen, G. (1970), *Mathematical Methods for Physicists*, 2nd edition, Academic.
Atkins, P. (2001), *Elements of Physical Chemistry*, 3rd edition, Oxford University Press.
Atkins, P. & de Paula, J. (2006), *Physical Chemistry*, Oxford University Press.
Atkins, P. & Friedmann, R. (1997), *Molecular Quantum Mechanics*, 3rd edition, Oxford University Press.
Barlow, R. J. (1989), *Statistics: A Guide to the Use of Statistical Methods in the Physical Sciences*, Wiley.
Barnes, B. & Fulford, G. (2002), *Mathematical Modelling with Case Studies*, Taylor & Francis.
Barrow, G. M. (1979), *Physical Chemistry*, 4th edition, McGraw-Hill.
Bell, G. (1978), 'Models for the Specific Adhesion of Cells to Cells', *Science* 200: 618.
Bennett, R. G. & Kellogg, R. E. (1967), *Progress in Reaction Kinetics*, vol. 4, edited by G. Porter, Pergamon.
Bernal, J. D. (1973), *The Extension of Man*, Paladin.
Bernstein, R. B. (1982), *Chemical Dynamics via Molecular Beam and Laser Techniques*, Oxford Science, Clarendon Press.
Bertinotti, F. & Giacomello, G. (1956), 'The Structure of Heterocyclic Compounds Containing Nitrogen. I. Crystal and Molecular Structure of s-Tetrazine', *Acta Cryst.* 9: 510.
Bevington, P. R. (1969), *Data Analysis and Error Reduction for the Physical Sciences*, McGraw-Hill.
Bevington, P. R. & Robinson, D. K. (2003), *Data Analysis and Error Reduction for the Physical Sciences*, 3rd edition, McGraw-Hill.
Birks, J. B. (1970), *Photophysics of Aromatics Molecules*, Wiley.
Bishop, D. M. (1993), *Group Theory and Chemistry*, Dover.
Blatner, D. (1999), *The Joy of Pi*, Walker/Bloomsbury.
Blundell, S. (2001), *Magnetism in Condensed Matter*, Oxford University Press.
Boas, M. L. (1983), *Mathematical Methods in the Physical Sciences*, 2nd edition, Wiley.
Bracewell, R. (1986), *The Fourier Transform and its Applications*, McGraw-Hill International.
Britton, N. (2003), *Essential Mathematical Biology*, Springer.
Brockwell, D., Paci, E., Zinober, R., Beddard, G., Olmsted, P., Smith, D., & Radford, S. (2003), 'Pulling Geometry Defines the Mechanical Resistance of a *B*-Sheet Protein', *Nature Structural Biology*, 10: 731.
Bronson, R. (1994), *Schaum's Outline of Theory and Problems, Differential Equations*, McGraw Hill.
Brown, R. (1828), 'A brief account of microscopical observations made in the months of June, July and August, 1827, on the particles contained in the pollen of plants; and on the general existence of active molecules in organic and inorganic bodies', *Philosophical Magazine* 4: 161.
Butkov, E. (1968), *Mathematical Physics*, Addison-Wesley.
Carlsaw, H. S. & Jaeger, J. C. (1959), *Conduction of Heat in Solids*, Clarendon Press.
Carrington, A. & McLachlan, A. (1969), *Introduction to Magnetic Resonance*, Harper.
Carter, R. (1997), *Molecular Symmetry and Group Theory*, Wiley.
Cherepanov, D. A. & Junge, W. (2001), 'Viscoelastic Dynamics of Actin Filaments Coupled to Rotary F-ATPase: Curvature as an Indicator of the Torque', *Biophys. J.* 81: 1234.
Christie, M., Norrish, R., & Porter, G. (1952), 'The Recombination of Atoms. I. Iodine Atoms in the Rare Gases', *Proceedings of the Royal Society London* 216A: 152.
Cohen, E. R., Cvitas, T., Frey, J. G., & Kuchitsu, K. (2007), *Quantities, Units and Symbols in Physical Chemistry*, 3rd edition, RSC Press. [Mills et al. 1993 is earlier edition.]
Cohen-Tannoudji, C., Diu, B., & Laloë, F. (1977), *Quantum Mechanics*, Wiley-Interscience.
Cooley, J. W. & Tukey, J. W. (1965), 'An Algorithm for Machine Calculation of Complex Fourier Series', *Mathematics of Computation* 19: 297.
Cotton, F. A. (1990), *Chemical Applications of Group Theory*, 3rd edition, Wiley.
Crank, J. (1979), *The Mathematics of Diffusion*, Oxford University Press.
Daune, M. (1999), *Molecular Biophysics*, Oxford University Press.
Davidson, N. (1962), *Statistical Mechanics*, McGraw Hill.
Dence, J. (1975), *Mathematical Techniques in Chemistry*, Wiley.
Evans, E. & Ritchie, K. (1997), 'Dynamic Strength of Molecular Adhesion Bonds', *Biophysics J.* 72: 1541.
Eyring, H., Lin, S. H., & Lin, S. M. (1980), *Basic Chemical Kinetics*, Wiley.

Finkelstein, A. & Ptitsyn, O. (2002), *Protein Physics*, Academic Press.
Flugge, S. (1999), *Practical Quantum Mechanics*, Springer.
Flygare, W. (1978), *Molecular Structure and Dynamics*, Prentice-Hall.
Foote, C. (2005), *Atomic Physics*, Oxford University Press.
Förster, T. (1959), 'Transfer Mechanisms of Electronic Excitation', *Discussion Faraday Soc.* 27: 7.
Frenkel. D. & Smit, B. (1996), *Understanding Molecular Simulation*, Academic Press.
Garvan, F. (2002), *The Maple Book*, Chapman & Hall/CRC.
Gerrard, A. & Burch, J. M. (1975), *Introduction to Matrix Methods in Optics*, Wiley.
Giacovazzo, C., Monaco, H. L., Viterbo, D., Scordari, F., Gilli, G., Zanotti, G., & Catti, M. (1992), *IUCR Texts on Crystallography: vol. 2 Fundamentals of Crystallography*, edited by C. Giacovazzo, Oxford University Press.
Gibbons, C., Montgomery, M. G., Leslie, A. G. W., & Walker, J. E. (2000), 'The Structure of the Central Stalk in Bovine F1-ATPase at 2.4 Å Resolution', *Nature Structural Biology* 7: 1055.
Gibbs, J. W. (1899), 'Fourier's Series', *Nature* 59: 606.
Gillespie, D. J. (1977), 'Exact Stochastic Simulation of Coupled Chemical Reactions', *Phys. Chem.* 81: 2340.
Gillespie, D. J. (2007), 'Stochastic Simulation of Chemical Kinetics', *Annual Reviews of Physical Chemistry* 58: 35.
Goldberg, S. (1986), *Probability, An Introduction*, Dover.
Goldstein, H. (1980), *Classical Mechanics*, Addison-Wesley.
Gordy, W., Smith, W., & Trambarulo, R. (1953), *Microwave Spectroscopy*, Dover.
Gorry, P. A. (1990), 'General Least-Squares Smoothing and Differentiation by the Convolution (Savitzky-Golay) Method', *Analytical Chemistry* 63: 570.
Gould, H., Tobochnik, J., & Christian, W. (2007), *An Introduction to Computer Simulation Methods*, 3rd edition, Pearson.
Gow, G. M. (1964), *A Course in Pure Mathematics*, English Universities Press.
Günther, H. (1992), *NMR Spectroscopy*, Wiley.
Haken, H. (1978), *Synergetics*, Springer Verlag.
Harwit, M. & Sloane, N. J. A. (1979), *Hadamard Transform Optics*, Academic Press.
Heck, A. (1996), *Introduction to Maple*, Springer. [Note that some of the Maple syntax has changed since this book was published, in particular, ditto, ", is replaced by %]
Herzberg, G. (1950), *Spectra of Diatomic Molecules*, 2nd edition, Van Nostrand Reinhold.
Herzberg, G. (1964), *Infrared and Raman Spectra of Polyatomic Molecules*, 11th printing, Van Nostrand Reinhold.
Hines, W. & Montgomery, D. C. (1990), *Probability and Statistics in Engineering and Management Science*, 3rd edition, Wiley.
Hirst, D. M. (1990), *A Computational Approach to Chemistry*, Blackwell.
Jackson, M. B. (2006), *Molecular & Cellular Biophysics*, Cambridge University Press.
James, J. F. (1995), *A Student's Guide to Fourier Transforms*, Cambridge University Press.
Jeffery, A. (1990), *Linear Algebra and Ordinary Differential Equations*, Blackwell Scientific.
Jolliffe, I. T. (2002), *Principal Component Analysis*, Springer.
Kermack, W. O. & McKendrick, A. G. (1927), 'A Contribution to the Mathematical Theory of Epidemics', *Proceedings of the Royal Society London A* 115: 700.
Knobel, P. (2000), *An Introduction to the Mathematical Theory of Waves*, American Mathematical Society.
Koryta, J., Dvorak, J., & Kavan, L. (1993), *Principles of Electrochemistry*, Wiley.
Kosloff, R. J. (1988), 'Time-Dependent Quantum-Mechanical Methods for Molecular Dynamics', *Phys. Chem.* 92: 2087.
Krauth, W. (2006), *Algorithms & Computations*, Oxford University Press.
Kuhn, H. & Fursterling, H.-D. (1999), *Physical Chemistry: Understanding Molecules, Molecular Assemblies, Supramolecular Machines*, Wiley.
Lacowicz, J. (2004), *Principles of Fluorescence Spectroscopy*, Springer.
Lamb, H. (1947), *Dynamics*, Cambridge University Press.
Lengyel, I., Rabai, G., & Epstein, I. J. (1990), 'Experimental and Modeling Study of Oscillations in the Chlorine Dioxide-Iodine-Malonic Acid Reaction', *Amer. Chem. Soc.* 112: 9104.
Levine, I. N. (2001), *Quantum Chemistry*, 5th edition, Prentice Hall.
Levitt, M. (2001), *Spin Dynamics: Basics of Nuclear Magnetic Resonance*, Wiley.
Linthorne, N. (2006), 'A New Angle on Throwing', *Physics World* 19(June): 29.
Lotka, A. (1925), *Elements of physical biology*, Williams and Wilkins (Reprinted 1956 by Dover as *Elements of mathematical biology*).
Margenau, H. & Murphy, G. (1943), *The Mathematics of Physics & Chemistry*, Van Nostrand.
Marshall, A. G. (1978), *Biophysical Chemistry, Principles, Techniques and Applications*, Wiley.
Marshall, A. G. (1982), *Fourier, Hadamard and Hilbert Transforms in Chemistry*, Plenum.
Martin, R. & Davidson, E. (1988), 'Electronic Structure of the Sodium Trimer', *Molecular Physics* 35: 1713.
May, R. (1974), 'Biological Populations with Non-overlapping Generations: Stable Points, Stable Cycles, and Chaos', *Science* 186: 645.

May, R. (1987), 'Chaos and the Dynamics of Biological Populations', *Proceedings of the Royal Society, London A* 413: 27.

May, R. & Oster, G. (1976), 'Bifurcations and Dynamic Complexity in Simple Ecological Models', *American Naturalist* 110: 573.

Maynard Smith, J. (1995), *The Theory of Evolution*, Cambridge University Press.

McKie, D. & McKie, C. (1992), *Essentials of Crystallography*, Blackwell.

McQuarrie, D. A. & Simon, J. D. (1997), *Physical Chemistry, A molecular Approach*, University Science Books.

Metropolis, N., Rosenbluth, A. M., Teller, A., & Teller, E. J. (1953), 'Equation of State Calculations by Fast Computing Machines', *Chem. Phys.* 21: 1087.

Miller, J. & Miller, J. C. (2005), *Statistics and Chemometrics for Analytical Chemistry*, Prentice Hall.

Mills, I., Cvitas, T., Homann, K., Kallay, N., & Kuchitsu, K. (1993), *Quantities, Units and Symbols in Physical Chemistry*, 2nd edition, Blackwell.

Molloy, K. (2004), *Group Theory for Chemists*, Horwood.

Morse, P. M. (1929), 'Diatomic Molecules According to the Wave Mechanics. II. Vibrational Levels', *Physical Review* 34: 57.

Murray, J. D. (2002), *Mathematical Biology 1 An Introduction*, Springer.

Murrell, J. N. & Jenkins, A. D. (1994), *Properties of Liquids and Solutions*, 2nd edition, Wiley.

Newman, E. & Barkema, G. (1999), *Monte Carlo Methods in Statistical Physics*, Clarendon Press.

O'Connor, D. V. & Phillips, D. (1984), *Time-Correlated Single Photon Counting*, Academic Press.

Onsager, L. (1944), 'Crystal Statistics, I. A Two-Dimensional Model with an Order-Disorder Transition', *Phys. Rev.* 65: 117.

Pain, H. (1993), *The Physics of Vibrations and Waves*, Wiley.

Pänke, O., Cherepanov, D. A., Gumbiowski, K., Engelbrecht, S., & Junge, W. (2001), 'Viscoelastic Dynamics of Actin Filaments Coupled to Rotary F-ATPase: Angular Torque Profile of the Enzyme', *Biophysics J.* 81: 1220.

Parratt, L. G. (1971), *Probability and Experimental Errors in Science*, Dover.

Penrose, R. (1990), *The Emperor's New Mind*, Oxford University Press.

Perkus, J. K. (2001), *Mathematics of Genome Analysis*, Cambridge University Press.

Perrin, C. L. (1970), *Mathematics for Chemists*, Wiley-Interscience.

Phillips, D. & Barber, J. (2006), *The Life and Scientific Legacy of George Porter*, Imperial College Press.

Pilling, M. J. & Seakins, P. (2005), *Reaction Kinetics*, Oxford University Press.

Polanyi, J. & Schreiber, J. (1977), 'The Reaction F + H_2*HF + H: A Case Study in Reaction Dynamics', *Faraday Disc. Chem. Soc.* 62: 267.

Porter, G. & Smith, J. (1961), 'The Recombination of Atoms. III. Temperature Coefficients of Iodine Atom Recombination', *Proceedings of the Royal Society London, Series A* 261: 28.

Porter, G. (1967), *Nobel Symposium 5. Fast Reactions and Primary. Processes in Reaction Kinetics*, edited by Claesson, Interscience.

Praly, J. & Lemieux, R. (1987), 'Influence of Solvent on the Magnitude of the Anomeric Effect', *Canadian. J. Chem.* 65: 213.

Prest, W., Flannery, B., Teutolsky, S., & Vetterling, W. (1986), *Numerical Recipes*, Cambridge University Press.

RasMol http://rasmol.org/, http://www.bernstein-plus-sons.com/software/rasmol/, or via 'software tools' link at http://www.rcsb.org/pdb/home/home.do

Ratner, M. & Schatz, G. (2001), *Introduction to Quantum Mechanics in Chemistry*, Prentice Hall.

Reif, M., Gautel, M., Oesterhelt, F., Fernandez, J. M., & Gaub, H. E. (1977), 'Reversible Unfolding of Individual Titin Immunoglobulin Domains by AFM', *Science* 276: 1109.

Richards, D. (2002), *Advanced Mathematical Method with Maple*, Cambridge University Press.

Rigby, M., Smith, E. B., Wakenham, W. A., & Maitland, G. C. (1986), *The Forces between Molecules*, Oxford Science Clarendon Press.

Sage, G. & Klemperer, W. J. (1963), 'Far-Infrared Spectrum and Barrier to Internal Rotation of Ethyl Fluoride', *Chem. Phys.* 39: 371.

Sahimi, M. (1994), *Applications of Percolation Theory*, Taylor & Francis.

Salem, L. (1982), *Electrons in Chemical Reactions: First Principles*, Wiley.

Sanders, J. K. M. & Hunter, B. K. (1987), *Modern NMR Spectroscopy, A Guide for Chemists*, Oxford University Press.

Scott, S. K. (1995), *Oscillations, Waves and Chaos in Chemical Kinetics*, Oxford University Press.

Senent, M. L., Smeyers, Y. G., Dominquez-Gómes, R., & Villa, M. J. (2000), 'Ab Initio Determination of the Far Infrared Spectra of Some Isotopic Varieties of Ethanol', *Chem. Phys.* 112: 5809.

Shey, H. (1993), *Div, Grad, Curl and All That*, W. Norton.

Shingareva, I. & Lizarraga-Celaya, C. (2007), *Maple and Mathematica: A Problem Solving Approach for Mathematics*, Springer-Verlag.

Siegman, A. (1986), *Lasers*, University Science Books.

Simmons, J. (1971), 'Echolocation in Bats: Signal Processing of Echoes for Target Range', *Science* 171: 925.

Squires, G. L. (1995), *Problems in Quantum Mechanics*, Cambridge University Press.

Steinfeld, J. (1981), *Molecules and Radiation*, MIT Press.

Steinfeld, J., Fransisco, J., & Hase, W. (1999), *Chemical Kinetics & Dynamics*, Prentice Hall.
Stephenson, G. (1996), *Partial Differential Equations for Scientists and Engineers*, Imperial College Press.
Steward, E. G. (1987), *Fourier Optics: An Introduction*, 2nd edition, Wiley.
Stewart, I. (1998), *The Magical Maze*, Phœnix.
Stone, A. J. (1996), *The Theory of Intermolecular Forces*, Clarendon Press.
Strogatz, S. (1994), *Nonlinear Dynamics and Chaos*, Perseus Books.
Svelto, O. (1982), *Principles of Lasers*, Plenum Press.
Swalen, J. & Ibers, J. J. (1962), 'Potential Function for the Inversion of Ammonia', *Chem. Phys.* 36: 1914.
Szabo, A. & Ostlund, N. (1982), *Modern Quantum Chemistry*, Dover.
Tronrud, D. E., Schmid, M. F., & Matthews, B. W. (1986), *J. Mol. Biol.* 188: 443.
Turro, N. (1978), *Molecular Photochemistry*, Benjamin-Cummings.
Turton, D. A., Reid, G. D. & Beddard, G. S. (2003), 'Accurate Analysis of Fluorescence Decays from Single Molecules in Photon Counting Experiments', *Analytical Chemistry* 75: 4182.
Verlet, L. (1967), 'Computer "Experiments" on Classical Fluids. I. Thermodynamical Properties of Lennard-Jones Molecules', *Physical Review* 159: 98.
Verlet, L. (1967), 'Computer "Experiments" on Classical Fluids. II. Equilibrium Correlation Functions', *Physical Review* 165: 201.
Vincent, A. (2001), *Molecular Symmetry and Group Theory*, Wiley.
Volterra, V. (1926), 'Fluctuations in the abundance of a species considered mathematically', *Nature* 118: 558.
Vrakking, M. J., Villeneuve, D. M., & Stolow, A. (1996), 'Observation of Fractional Revivals of a Molecular Wave Packet', *Physical Review* 54: R37.
Waller, A. & Liddington, R. C. (1990), 'Refinement of a Partially Oxygenated T State Human Haemoglobin at 1.5 Å Resolution', *Acta Crystallogr. B* 46: 409.
Weast, R. C. (ed.) *CRC Handbook of Chemistry and Physics*, Chemical Rubber Co. [Any edition will do. This book is colloquially known as the 'Rubber Book'.]
Weisstein, E. W. 'Box-Muller Transformation', www.mathworld.wolfram.com
Whittaker, J. B. (ed.) (2007), *Imaging in Molecular Dynamics: Technology and Applications*, Cambridge University Press.
Wyckoff, R. (1969), *Crystal Structures*, vol. 6 (1), 2nd edition, Wiley.
Yariv, A. (1975), *Quantum Electronics*, 2nd edition, Wiley.
Zewail, A. (1994), *Femtochemistry: Ultrafast Dynamics of the Chemical Bond*, Vols 1 & 2, World Scientific.
Zimm. B. H. & Bragg, J. K. J. (1959), 'Theory of the Phase Transition between Helix and Random Coil in Polypeptide Chains', *Chem. Phys.* 31: 526.

Index

W

X

Z